Handbook of Experimental Pharmacology

Volume 95/I

W0043897

Peptide Growth Factors and Their Receptors I

Contributors

K.-I. Arai, N. Arai, A. Baird, N. Beru, P. Böhlen,
D.F. Bowen-Pope, R.A. Bradshaw, A.W. Burgess, G. Carpenter,
K.P. Cavanaugh, K.B. Clairmont, S. Corvera, M.P. Czech,
E. Goldwasser, M. Hatakeyama, T. Hirano, T. Honjo, J.N. Ihle,
T. Kishimoto, S. Nagata, S.P. Nissley, E.W. Raines, M.M. Rechler,
A.B. Roberts, R. Ross, J.A. Schmidt, C.J. Sherr, D. Smith,
M.B. Sporn, E.R. Stanley, K. Takatsu, T. Taniguchi, M.J. Tocci,
M.I. Wahl, T. Yokota, K.A. Yagaloff, A. Zlotnik

Editors

Michael B. Sporn and Anita B. Roberts

Springer-Verlag
Berlin Heidelberg New York
London Paris Tokyo
Hong Kong

Michael B. Sporn, M. D.

Anita B. Roberts, Ph. D.

Laboratory of Chemoprevention
National Cancer Institute
Bethesda, MD 20892
USA

With 101 Figures

ISBN 978-3-642-49297-6   ISBN 978-3-642-49295-2 (eBook)
DOI 10.1007/978-3-642-49295-2

Library of Congress Cataloging-in-Publication Data. Peptide growth factors and their recep-
tors/contributors. K.-I. Arai ... [et al.]; editors, Michael B. Sporn and Anita B. Roberts. p.cm. – (Handbook of experimental pharmacology: v. 95) Contributors for v. 2, J.F. Battey and others. Includes
bibliographical references. ISBN 978-3-642-49297-6   1.
Growth factors. 2. Growth factors – Receptors. I. Arai, Ken-Ichi. II. Sporn, Michael B. III. Roberts,
Anita B. IV. Battey, James F. V. Series. [DNLM: 1. Cell Communication. 2. Growth Substances. 3. Peptides. 4. Receptors, Endogenous Substances. W1 HA51L v. 95/QU 68 P42404] QP905.H3 vol. 95
[QP552.G76] 615'. 1 s-dc20 [574.87'6] DNLM/DLC for Library of Congress.

2127/3130-543210 – Printed on acid-free paper

List of Contributors

K.-I. ARAI, Department of Molecular Biology, DNAX Research Institute of Molecular and Cellular Biology, 901 California Avenue, Palo Alto, CA 94304-1104, USA

N. ARAI, Department of Molecular Biology, DNAX Research Institute of Molecular and Cellular Biology, 901 California Avenue, Palo Alto, CA 94304-1104, USA

A. BAIRD, Department of Molecular and Cellular Growth Biology, Whittier Institute for Diabetes and Endocrinology, 9894 Genesee Ave., La Jolla, CA 92037, USA

N. BERU, Department of Medicine, Section of Hematology/Oncology, University of Chicago 5841 S. Maryland Ave. Chicago, IL 60637, USA

P. BÖHLEN, Medical Research Division, American Cyanamid, Lederle Laboratories, N. Middletown Road, Pearl River, NY 10965, USA

D. F. BOWEN-POPE, Department of Pathology, University of Washington School of Medicine, Health Sciences Building, SM-30, Seattle, WA 98195, USA

R. A. BRADSHAW, Department of Biological Chemistry, California College of Medicine, University of California, Irvine, Irvine, CA 92717, USA

A. W. BURGESS, Melbourne Tumour Biology Branch, Ludwig Institute for Cancer Research, Post Office Royal Melbourne Hospital, Victoria 3050, Australia

G. CARPENTER, Department of Biochemistry, School of Medicine, Vanderbilt University, Nashville, TN 37232-0146, USA

K. P. CAVANAUGH, Department of Biological Chemistry, California College of Medicine, University of California, Irvine, Irvine, CA 92717, USA

K. B. CLAIRMONT, Department of Biochemistry, University of Massachusetts Medical Center, 55 Lake Avenue North, Worcester, MA 01655, USA

S. CORVERA, Department of Pathology and Laboratory Medicine, University of Pennsylvania Medical School, Philadelphia, PA 19104, USA

M. P. Czech, Department of Biochemistry, University of Massachusetts Medical Center, 55 Lake Avenue North, Worcester, MA 01655, USA

E. Goldwasser, Department of Biochemistry and Molecular Biology, The University of Chicago, 920 E. 58th Street, Chicago, IL 60637, USA

M. Hatakeyama, Institute for Molecular and Cellular Biology, Osaka University, 1–3 Yamada-oka, Suita-shi, Osaka 565, Japan

T. Hirano, Division of Molecular Oncology, Biomedical Research Center, Osaka University Medical School, 4-3-57, Nakanoshima, Kita-ku, Osaka 530, Japan

T. Honjo, Department of Medical Chemistry, Faculty of Medicine, Kyoto University, Yoshida, Sakyo-ku, Kyoto 606, Japan

J. N. Ihle, Department of Biochemistry, St. Jude Children's Research Hospital, 332 N. Lauderdale, P.O. Box 318, Memphis, TN 38101, USA

T. Kishimoto, Institute for Molecular and Cellular Biology, Osaka University, Division of Immunology, Yamadaoka-1–3, Suita-shi, Osaka 565, Japan

S. Nagata, Department of Molecular Biology, Osaka Bioscience Institute, 6-2-4 Furuedai, Suita, Osaka 565, Japan

S. P. Nissley, Endocrinology Section, Metabolism Branch, National Cancer Institute, NIH, Bldg. 10, Room 4N115, Bethesda, MD 20892, USA

E. W. Raines, Department of Pathology, School of Medicine, University of Washington, Health Sciences Building, SM-30, Seattle, WA 98195, USA

M. M. Rechler, Growth and Development Section, Molecular, Cellular, and Nutritional Endocrinology Branch, National Institute of Diabetes and Digestive and Kidney Disease, NIH, Building 10, Room 8D14, Bethesda, MD 20892, USA

A. B. Roberts, Laboratory of Chemoprevention, National Cancer Institute, Bldg. 41, Room C–629, Bethesda, MD 20892, USA

R. Ross, Department of Pathology, School of Medicine, University of Washington, Health Sciences Building, SM-30, Seattle, WA 98195, USA

J. A. Schmidt, Biochemical and Molecular Pathology, Merck Sharp and Dohme Research Laboratories, P.O. Box 2000, Rahway, NJ 07065-0900, USA

C. J. Sherr, Howard Hughes Medical Institute and Department of Tumor Cell Biology, St. Jude Children's Research Hospital, 332 N. Lauderdale, Memphis, TN 38105-0318, USA

D. Smith, Department of Medicine, Section of Hematology/Oncology, University of Chicago, 5841 S. Maryland Ave., Chicago, IL 60637, USA

M. B. Sporn, Laboratory of Chemoprevention, National Cancer Institute, Bldg. 41, Room C–629, Bethesda, MD 20892, USA

E. R. Stanley, Department of Developmental Biology and Cancer, Albert Einstein College of Medicine, Bronx, NY 10461, USA

K. Takatsu, Department of Biology, Institute for Medical Immunology, Kumamoto University Medical School, 2-2-1 Honjo, Kumamoto 860, Japan

T. Taniguchi, Institute for Molecular and Cellular Biology, Osaka University, 1–3 Yamada-oka, Suita-shi, Osaka 565, Japan

M. J. Tocci, Biochemical and Molecular Pathology, Merck Sharp and Dohme Research Laboratories, P.O. Box 2000, Rahway, NJ 07065-0900, USA

M. I. Wahl, Department of Biochemistry, School of Medicine, Vanderbilt University, Nashville, TN 37232-0146, USA

T. Yokota, Department of Molecular Biology, DNAX Research Institute of Molecular and Cellular Biology, 901 California Avenue, Palo Alto, CA 94304-1104, USA

K. A. Yagaloff, Department of Biochemistry, University of Massachusetts Medical Center, 55 Lake Avenue North, Worcester, MA 01655, USA

A. Zlotnik, Department of Immunology, DNAX Research Institute of Molecular and Cellular Biology, 901 California Avenue, Palo Alto, CA 94304-1104, USA

Preface

This two-volume treatise, the collected effort of more than 50 authors, represents the first comprehensive survey of the chemistry and biology of the set of molecules known as peptide growth factors. Although there have been many symposia on this topic, and numerous publications of reviews dealing with selected subsets of growth factors, the entire field has never been covered in a single treatise. It is essential to do this at the present time, as the number of journal articles on peptide growth factors now makes it almost impossible for any one person to stay informed on this subject by reading the primary literature. At the same time it is becoming increasingly apparent that these substances are of universal importance in biology and medicine and that the original classification of these molecules, based on the laboratory setting of their discovery, as "growth factors," "lymphokines," "cytokines," or "colony-stimulating factors," was quite artifactual; they are in fact the basis of a common language for intercellular communication. As a set they affect essentially every cell in the body, and in this regard they provide the basis to develop a unified science of cell biology, germane to all of biomedical research.

This treatise is divided into four main sections. After three introductory chapters, its principal focus is the detailed description of each of the major peptide growth factors in 26 individual chapters. These chapters provide essential information on the primary structure, gene structure, gene regulation, cell surface receptors, biological activity, and potential therapeutic applications of each growth factor, to the extent that these are known. The last two sections of these volumes deal with the coordinate actions of sets of growth factors, since it is clear that to understand their physiology, one must consider the interactions between these peptides. There are six chapters on specific cells and tissues, including bone marrow, brain, bone and cartilage, lymphocytes, macrophages, and connective tissue. The final six chapters consider the role of growth factors in controlling fundamental processes that pertain to many different cells and tissues, such as proteolysis, inflammation and repair, angiogenesis, and embryogenesis.

The editors accept full responsibility for the table of contents, and apologize for any significant omissions. We have deliberately excluded classic peptide hormones whose actions are principally endocrine. We have not included many peptides whose biological activities may have been described, but which have not yet been purified to homogeneity and sequenced. Essentially all of the molecules included in this treatise act at specific cell surface receptors, although we have included one peptide (glia-derived neurotrophic factor) which is a protease inhibitor. Because of the importance of growth-inhibitory actions of peptides, we have included chapters on two new sets of inhibitors (mammary-derived growth inhibitor and pentapeptide growth inhibitors) for which there is only very preliminary knowledge of their structure, function, and mechanism of action.

The present era in research on peptide growth factors is indeed an exciting one. The biological activities resulting from the actions of these substances have been known for a very long time, as described in John Hunter's detailed descriptions of the healing of severed tendons and gunshot wounds, made over 200 years ago. Modern cell and tissue culture owes its inception to the actions of growth factors present in the embryo extracts used almost 100 years ago by the pioneers of this technique. However, it has been the recent introduction of new methods for purification of peptides, and the cloning and expression of genes with recombinant DNA technology, which have truly revolutionized this field. These methods provide the scientific basis for the present treatise and the current excitement in this field. They also have provided the practical methodology for the creation of a whole new biotechnology industry, which has made major commitments to develop peptide growth factors as clinical therapeutic agents. Applications being developed include wound healing and other aspects of soft-tissue repair, repair of bone and cartilage, immunosuppression, enhancement of immune cell function, enhancement of bone marrow function in many disease states, and prevention and treatment of many proliferative diseases, including atherosclerosis and cancer. Since many of the peptides described in these pages function as growth factors in the embryo, they raise hopes that they may be used some day to arrest or reverse the ravages of aging and degenerative disease.

There is no question that the peptides described here are of fundamental importance for understanding the behavior of all cells, and that they will be of major importance in the practice of clinical medicine in the years to come. These volumes, then, are a celebration not only of new basic knowledge, but also of the new therapeutic potential of this entire family of molecules, which have such intense potency to make cells move, grow, divide, and differentiate. We hope that this treatise will be of value to both scientists and clinicians in their pursuit and application of new knowledge in this promising area.

We wish to express our appreciation to many individuals who have participated in this venture from its inception. We are greatly indebted to Pedro Cuatrecasas, who initially suggested the writing of this treatise and has been a constant and enthusiastic supporter. We thank all of the authors for their devoted efforts to assimilate the huge literature in their respective fields and to condense this information into readable single chapters. Our secretary, Karen Moran, has been of invaluable help with numerous aspects of the organization and publication of these volumes. Finally, we would like to express our gratitude to Doris Walker and the staff at Springer-Verlag for all of their efforts in bringing this treatise to publication.

MICHAEL B. SPORN
ANITA B. ROBERTS

Contents

CHAPTER 5

Platelet-Derived Growth Factor
E. W. RAINES, D. F. BOWEN-POPE, and R. ROSS. With 5 Figures 173

CHAPTER 6

Insulin-Like Growth Factors

CHAPTER 7

Fibroblast Growth Factors

CHAPTER 8

The Transforming Growth Factor-βs

CHAPTER 9

CHAPTER 10

Interleukin-2

CHAPTER 11

Interleukin-3

CHAPTER 12

Interleukin-4

CHAPTER 15

CHAPTER 16

Granulocyte Colony-Stimulating Factor
S. NAGATA. With 11 Figures 699

CHAPTER 17

CHAPTER 18

Contents of Companion Volume 95, Part II

Section A: Introduction

CHAPTER 1

The Multifunctional Nature of Peptide Growth Factors

M. B. Sporn and A. B. Roberts

A. Introduction

It might seem paradoxical, perhaps even capricious and quixotic, to introduce this first comprehensive treatise on peptide growth factors with a statement that in reality there is no such thing as a peptide growth factor. Indeed, this treatise does concern itself with an extremely potent set of regulators of cell growth, all of which are peptides, many newly discovered. However, at the beginning of these two volumes, it is important to emphasize that all the peptides considered here actually are elements of a complex biological signaling language, providing the basis for intercellular communication in multicellular organisms. Thus, the "peptide growth factors" that are the topic of this treatise in reality are peptide signaling molecules. They often promote cell growth, but they also can inhibit it; moreover, they regulate many other critical cellular functions, such as control of differentiation and many other processes that have little to do with growth itself. Like the symbols or alphabet of a language or code, the meaning of the action of these peptide signaling molecules can only be understood in context with other symbols. Thus, all peptide growth factors act in sets, and to understand their actions, one must always consider the biological context in which they act.

Historically, because of interest in a particular problem, investigators have attempted to purify and characterize specific peptides associated with specific biological activities. However, the original context of the discovery of any signaling peptide has seldom indicated the extent of its diverse range of action. Thus, interest in defining a peptide in human urine which suppressed gastric acid secretion ("urogastrone") antedated by many years (for a review, see GREGORY and WILLSHIRE 1975) the identification of this substance as the mitogen presently called "epidermal growth factor" (EGF). By now it is also known that this same molecule has potent antimitogenic activity on hair follicle cells, which has led to it use as a defleecing agent for sheep (PANARETTO et al. 1984). Furthermore, although EGF was originally isolated and purified by its ability to cause premature eruption of incisor teeth in mice (COHEN 1962), it has recently been shown that the teeth of newborn mice treated with EGF are abnormally small (RHODES et al. 1987). Most recently, it has been shown that EGF has direct and immediate contractile effects on isolated arterial tissue, which interestingly may be either agonistic or antagonistic, depending on the context of the other effectors acting on the artery (GAN et al. 1987).

Is EGF a growth factor? Clearly it is a very important one in certain situations, while in others it is not. In all situations, however, EGF acts as a peptide signaling molecule by binding to its own receptor. Thus, the highly specific fit of this peptide ligand with its glycoprotein cell membrane receptor has provided a modular regulatory element that the evolutionary process has repeatedly used in many cell types for diverse purposes. One cannot consider control of epidermal growth without considering EGF, but it is also apparent that the biological actions of this peptide extend far beyond its name.

After the initial identification of any peptide growth factor, better methods for purification, and the availability of recombinant material in particular, have always expedited the investigation of its biological activity. The results have been quite startling: for any single amino acid sequence, new activities, including new target cells, have been repeatedly found. In this regard, almost all peptide growth factors considered in these two volumes are "misnamed", in that their biological activities are now known to extend far beyond the original context of their discovery. Thus, as examples, "platelet-derived" growth factor (PDGF) is made by many cells other than platelets, including smooth muscle cells, endothelial cells, and normal and malignant glial cells; "fibroblast" growth factor (FGF) is made by numerous and diverse cell types and has an exceptionally wide range of target cells; "transforming" growth factor-β (TGF-β) has many actions that bear no relationship to its ability to cause phenotypic transformation of rat kidney fibroblasts; many of the "interleukins" (originally defined as signaling molecules controlling activities of cells within the immune system) have profound effects on nonimmune cells as diverse as keratinocytes, chondrocytes, fibroblasts, mesangial cells of the kidney, neuroblasts, and glial cells; "tumor necrosis" factor-α (TNF-α) is a mediator of many phenomena, such as angiogenesis, having nothing to do with tumor necrosis; and inhibins and activins have now been shown to have many actions which do not pertain to control of secretion of pituitary hormones (see the respective chapters on these peptides).

Even the parent molecule for the development of the entire field of peptide growth factors, namely "nerve" growth factor (NGF) has recently been shown to be multifunctional. Modern studies with peptide growth factors began when it was shown that extracts of tumors and salivary glands promoted neurite outgrowth from cultured nerve ganglia (LEVI-MONTALCINI and HAMBURGER 1953; COHEN 1960). NGF was named for this activity, purified to homogeneity, and then sequenced (ANGELETTI and BRADSHAW 1971). Assay of neurite outgrowth has been the classical method for measuring NGF activity. However, the concept that NGF functions solely as a neurotrophic agent is no longer tenable. New studies have shown that NGF can promote human hematopoietic colony growth and differentiation (MATSUDA et al. 1988), and that mRNA for its receptor or immunohistochemically detectable receptor is present on fibroblasts, lymphocytes, and many other cells of mesenchymal origin, unrelated to neural lineage (ERNFORS et al. 1988; BOTHWELL et al. 1989). The recent finding of NGF mRNA and protein in the testis and epididymis of the mouse and rat has led to the suggestion that it may be involved in the maturation, survival, or mobility of spermatozoa (AYER-LELIEVRE et al. 1988).

In no way are we suggesting that the set of peptide growth factors be renamed. Historically, cell biologists have named members of this set of molecules as "growth factors"; immunologists have called them "interleukins," "lymphokines," or "cytokines"; while hematologists have used the term "colony stimulating factors." The present nomenclature is widely used throughout the world's scientific literature, and is even consistent in that in almost every case it reflects the context of the original discovery or isolation of any peptide. Since essentially all of these molecules are multifunctional, it is difficult to conceive of unique new names for them that would be entirely satisfactory; almost all of them are "panregulins." They should be regarded as the alphabet or symbols of a biological regulatory language (SPORN and ROBERTS 1986, 1988). Elucidation of the grammar and syntax of this language is one of the most fundamental problems of modern biology and is the essential subject matter of this treatise. Since any individual peptide growth factor may have been "rediscovered" (and then renamed) for yet another one of its numerous different biological activities, we are including an Appendix at the end of these volumes to reflect these various names and actions. It has only been after the amino acid sequence responsible for a given biological activity has been determined that one can attribute that particular activity to a specific peptide. Fortunately, since one peptide can have so many actions, the number of known peptide growth factors does not appear to be increasing to a level that defies comprehension; on the contrary, the finding that some 20–30 differently named biological activities are the result of the action of either acidic or basic FGF (see Chap. 7 and the Appendix) has provided a great simplification. A similar simplification is now being found for many of the other peptide growth factors.

How can an individual peptide molecule have so many different actions, and how can the nature of its action vary so much depending on its biological context? In the remainder of this and many of the following chapters, these questions will be considered, although ultimately they cannot be answered yet. However, it is apparent that peptide growth factors provide an essential means for a cell to communicate with its immediate environment and to ensure that there is proper local homeostatic balance between the numerous cells that comprise a tissue. Since a cell must adjust its behavior to changes in its environment, it needs mechanisms to provide this adaptation. Thus, cells use sets of peptide growth factors as signaling molecules to communicate with each other and to alter their behavior to respond appropriately to their biological context. Since peptide growth factors act by binding to functional receptors which transduce their signal, the peptides themselves may be viewed as bifunctional molecules: they have both an afferent function (conveying information to the receptor, providing it with information from the outside) as well as an efferent function (inception of the latent biochemical activity of the receptor).

B. Autocrine, Paracrine, and Endocrine Mechanisms of Action

Autocrine and paracrine mechanisms of action (Fig. 1) provide a unifying theme to understand regulatory activities of peptide growth factors (Sporn and Todaro 1980; Heldin and Westermark 1984; Sporn and Roberts 1986). The old paradigm of humoral regulation of cellular function by circulating hormones, produced by individual organs comprising an "endocrine system," no longer can account for many phenomena. It is increasingly recognized that local actions of regulatory peptides, mediated by autocrine or paracrine mechanisms, are of great importance in maintaining cellular homeostasis. Thus many peptide growth factors that have significant actions on cells are produced by those cells themselves, or by immediately adjacent cells, particularly those of the immune system. Indeed, the immune system may be regarded as a mobile endocrine system (Weigent and Blalock 1987; Harrison and Campbell 1988) whose actions can be targeted to specific local sites as required. Peptide growth factors, acting in an autocrine and paracrine fashion, provide the essential mechanism whereby cells communicate with each other in this local milieu.

In this regard, peptide growth factors may serve to mediate or modify the local actions of classical hormones, such as growth hormone, adrenocorticotropic hormone, glucocorticoids, estradiol, thyroxine, and 1,25-dihydroxy vitamin D_3. The global actions of these molecules require local cellular mediators. Thus, a multifunctional peptide such as TGF-β is synthesized in response to treatment with steroids such as estradiol (Komm et al. 1988) or glucocorticoids (F. W. Ruscetti, personal communication); furthermore it can regulate the cellular synthesis of these agents in producer organs such as the adrenal cortex and ovary (see Chap. 8). The identification of the steroid

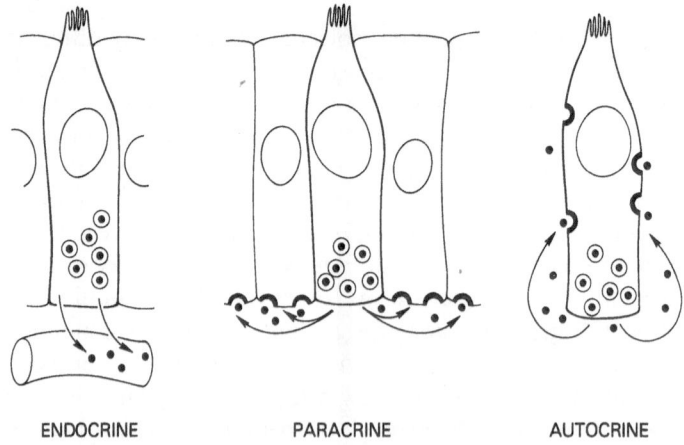

ENDOCRINE   PARACRINE   AUTOCRINE

Fig. 1. Diagrammatic representation of *autocrine*, *paracrine*, and *endocrine* secretion. Peptide growth factors are shown in latent form within the cell. The thickened *semicircular* regions of the cell membrane represent receptor sites. (From Sporn and Todaro 1980)

hormone receptor superfamily (GREEN and CHAMBON 1986; PETKOVICH et al. 1987; EVANS 1988; MINGHETTI and NORMAN 1988) as a group of nuclear transcription factors, interacting with enhancer elements of specific genes, has provided an important conceptual unification in this area. It now remains to be determined how these molecules interact with the entire group of peptide growth factors, to achieve appropriate regulation at the local cellular level. The mechanism of local cellular regulation by classical endocrine molecules involves their interface with autocrine and paracrine mechanisms of action of peptide growth factors, and elucidation of these details is a major current problem for the fields of cellular and molecular biology.

C. Range of Target Cells for Peptide Growth Factors

As we have noted above, not only may the type of action exerted by an individual peptide growth factor on a given cell type be much broader than the context of its original discovery, but the range of cells which respond to the peptide may also be much greater than was found originally. This is particularly true for the set of peptides known as "interleukins," which were first isolated from cells of the immune system and defined as signaling molecules controlling activities within the immune system. As each of the interleukins described in these volumes (IL-1 to IL-6) is studied in detail, it is becoming increasingly apparent that they regulate many processes beyond the classical immune cell functions. In particular, IL-1 and IL-6 have an unusually broad spectrum of target cells: IL-1 acts on cells as diverse as keratinocytes, fibroblasts, astrocytes, and chondrocytes (see Chap. 9), while IL-6 regulates protein synthesis in hepatocytes and neuron-like chromaffin cells, in addition to controlling growth and differentiation of B lymphocytes (see Chap. 14).

The counterpart of the "interleukin" concept is also important for understanding immune phenomena, in that peptides originally isolated from nonimmune cells and defined by their actions on nonimmune cells, such as TGF-β, have subsequently been found to have potent immunoregulatory actions. Thus, TGF-β is at least 10,000-fold more potent (on a molar basis) than cyclosporin as a suppressor of T-cell function (KEHRL et al. 1986b); furthermore it is an extremely potent inhibitor of immunoglobulin synthesis by B lymphocytes (KEHRL et al. 1986a). Yet another example is IL-6, which is produced by normal human fibroblasts and which regulates the synthesis and secretion of immunoglobulins by B lymphocytes (SEHGAL et al. 1987; KISHIMOTO and HIRANO 1988). Since both TGF-β and IL-6 are also synthesized by immune cells, these molecules provide a means of communication between the immune system and its immediate neighbors.

D. Contextuality of Action

Even within a given cell type, the type of action exerted by a specific growth factor may be contextual, and depend on the set of other effectors and receptors that are present. One of the first examples of this principle was found with TGF-β: in a particular fibroblastic cell line, TGF-β was found to inhibit cell proliferation in the presence of EGF, while it stimulated proliferation in the presence of PDGF; all other conditions were held constant in these experiments (ROBERTS et al. 1985). Many other examples of this phenomenon have been found recently, particularly with immune and hematopoietic cells. Thus, IL-2, a potent mitogen for T lymphocytes, can act as an antimitogen on these cells if the IL-2 receptor is overexpressed (HATAKEYAMA et al. 1985). In hematopoietic progenitor cells, IL-4 may augment the mitogenic effects of macrophage (M-CSF) and granulocyte colony-stimulating factors (G-CSF) and erythropoietin, but antagonize the mitogenic action of IL-3 (RENNICK et al. 1987). There are numerous examples of the action of tumor necrosis factor which are dependent on the presence or absence of other peptides, particularly IL-1 and TGF-β (see Chap. 20, also LE and VILCEK 1987; ESPEVIK et al. 1988).

Whether a given peptide stimulates or inhibits growth or synthesis of specific proteins often depends on the state of development or differentiation of its target cells. Thus, the growth of fibroblasts from very early human embryos is stimulated by TGF-β, while the growth of the same cells from older embryos is inhibited by this peptide (HILL et al. 1986). In a similar fashion, TGF-β stimulates synthesis of type II collagen in cells of the very early chondrocytic lineage, but inhibits type II collagen synthesis in these same cells at a later stage of development (SEYEDIN et al. 1987; ROSEN et al. 1988).

Although this contextual multifunctional action of peptide growth factors has created a certain amount of controversy, it really is not surprising in a broader biological perspective. Indeed, the notion that a cell or organism must be coupled in a meaningful, contextual manner to its environment is fundamental to all biological systems. It follows then, that the nature of the action of any regulatory molecule (or more complex regulatory unit) may depend on its particular biological context. Thus, at the molecular and cellular level, thyroid hormone can either upregulate or downregulate the genetic expression of the same myosin heavy chain gene, depending on the nature of the specific muscle in which this action occurs (IZUMO et al. 1986): the fast IIA myosin heavy chain gene is induced by thyroid hormone in the slow-twitch soleus muscle, but deinduced in fast-twitch muscles such as the masseter. Even at the organismic level, similar phenomena may occur. For example, in two closely located South African islands, a rock lobster and a snail have reversed their predator-prey relationships: on one island the lobster preys on the snail, while on the nearby island it is the snail that preys on the lobster (BARKAI and McQUAID 1988). Considering that peptide growth factors are signaling elements of an intercellular language, it is hardly surprising that their actions, too, are contextual. Peptide growth factors should not be considered as equivalents of metallic parts of a machine, with a unique action. Rather, they serve as a means to convey information from one cell to another, and their action in this regard is obviously contextual.

E. Nuclear Transcription Factors are also Multifunctional

One of the principal actions of many of the peptide growth factors described in this treatise is to control gene transcription; this concept is documented at length in the ensuing chapters. Although the sequential signaling pathways between any cell surface receptor and the nucleus are poorly understood, there has been great progress in elucidating the role of a group of acidic proteins that control transcription of specific genes in the nucleus itself. It is now clear that these nuclear transcription factors themselves are multifunctional proteins, acting in sets in a contextual manner to control DNA transcription by RNA polymerase II. In a formal way the specific interactions of these transcription factors with specific nucleotide regulatory sequences of genes may be considered analogous to the interactions of peptide growth factors with their receptors (CURRAN and FRANZA 1988).

A new conceptual synthesis, linking the actions of peptide growth factors and transcription factors, has come from study of the action of oncogenes, an approach which has proven to be consistently useful in previous growth factor research. Just as the original studies establishing the identity of the *sis* gene product with the B chain of PDGF and of the *erb* B gene product with the EGF receptor led to a major conceptual unification several years ago, current studies of the *fos* and *jun* oncogene products, and their role as nuclear transcription factors, now provide a unifying theme to explain mechanisms whereby growth factors control gene activity. The *fos* gene is a prototype of the immediate early genes activated by many peptide growth factors, and it has been suggested that its product may act as a "third messenger" to couple cytoplasmic signals to nuclear transcriptional events (VERMA and SASSONE-CORSI 1987). Indeed, the *fos* protein interacts with the *jun* protein, p39, to stimulate DNA transcription at AP-1 binding sites (CHIU et al. 1988; RAU-SCHER et al. 1988); stimulation of the cell with growth signals results in increased synthesis of several proteins related to both *fos* and *jun*. Although each of these proteins may bind independently to DNA, it appears that the association of *fos* and *jun* proteins as dimers or multimers (CHIU et al. 1988; SASSONE-CORSI et al. 1988a) leads to particularly effective transcription at a palindromic AP-1 site; the motif known as a "leucine zipper" and found on both *fos* and *jun* (as well as *myc*) proteins has been implicated in this activation of transcription (LANDSCHULZ et al. 1988; KOUZARIDES and ZIFF 1988).

As more nuclear transcription factors are isolated and sequenced and their specific sites of interaction with regulatory elements on DNA are determined, it becomes increasingly apparent that they are the nuclear analogs of peptide growth factors that act at the cell surface. Thus, there are many similarities between the nuclear "ligand-receptor" system and the one found on the cell membrane. Both systems are multifunctional; a single transcription factor, like a growth factor, may serve either to enhance or repress transcriptional activity (SHORT 1988). In both systems an individual ligand may be used to stimulate cell growth, to inhibit cell growth, or to regulate yet other processes that have little to do with growth in any way (SANTORO et al. 1988); this variety of responses always depends on the context of other effectors that are present. Both systems use dimeric or oligomeric interaction of ligands with receptors to effect a response (SCHLESSINGER 1988; KUMAR and CHAMBON 1988; SASSONE-CORSI et al. 1988 a). Both systems are autoregulatory: it has been shown that both the *fos* and *jun* proteins, as well as several peptide growth factors, regulate their own gene expression (SASSONE-CORSI et al. 1988 b; ANGEL et al. 1988; COFFEY et al. 1987; PAULSSON et al. 1987; VAN OBBERGHEN-SCHILLING et al. 1988).

To continue the analogy, we would suggest that the set of nuclear transcription factors should also be considered as the symbols or the alphabet of a language or a code. Thus, peptide growth factors may be considered as components of a language mediating communication between cells, while the transcription factors are used for communication and integration within the cell itself. A principal concern of current cell biology is to determine how the two systems, one at the cell membrane, the other in the nucleus, communicate with each other. How are the signals from the cell surface integrated in the nucleus, and in turn, how does the nucleus alter its program of gene transcription to respond appropriately to the environment?

F. Role of Extracellular Matrix in Mediating Interactions Between Cells

Its extracellular matrix provides a critical interface between a cell and its environment, and it is not surprising that the study of interactions between peptide growth factors and the extracellular matrix is an important area for answering some of the questions that were just posed. There is a "dynamic reciprocity" (BISSELL et al. 1982) between the extracellular matrix and the intracellular cytoskeleton and nuclear matrix, with each controlling the other, as discussed in Chap. 34. Extracellular matrix molecules themselves can act as growth signals. Thus, fibronectin and laminin can induce DNA synthesis, while collagen may provide a signal to terminate proliferation. All of these matrix molecules interact with the cell via specific receptors known as integrins (RUOSLAHTI and PIERSCHBACHER 1987; HYNES 1987; BUCK and HORWITZ 1987), which act as transducers of their growth regulatory functions. Proteolytic enzymes, such as plasminogen activator and collagenase, which control turnover of the extracellular matrix, are also critical regulators of cell proliferation (see Chap. 34 and 35).

Many of the growth factors that are reviewed in the following chapters are regulators of the formation and destruction of extracellular matrix; peptide growth factors may control expression of either the genes for matrix molecules themselves, or for the α- and β-chains of integrins, as well as expression of genes for key proteolytic enzymes (or their inhibitors, in turn) which regulate the degradation of matrix molecules. Thus, the chapters on EGF, PDGF, FGF, TGF-β, and IL-1 all deal with mechanisms, particularly with regulation of gene expression, for control of synthesis and degradation of extracellular matrix. Many aspects of the multifunctional action of growth factors can only be understood in reference to the interaction of cells with their matrix; this is particularly significant in bone marrow (Chap. 29) and bone and cartilage (Chap. 31).

Although a great deal is known about the interactions between peptide growth factors and the matrix molecules involved in "heterophilic" adhesive intercellular interactions (those involving a matrix molecule and its separate receptor, such as fibronectin and the fibronectin receptor), much less is known about the relationships between peptide growth factors and another important set of cellular adhesion molecules, generally known as CAMs or cadherins, which mediate "homophilic" interactions between cells. In the case of the CAMs and cadherins, a single type of molecule acts as a bridge between two cells, and they have been shown to have an important role in controlling morphogenesis in animal tissues (TAKEICHI 1988; EDELMAN 1988). Considering that CAMs/cadherins and peptide growth factors (MERCOLA and STILES 1988; see also Chap. 39) are both intimately involved in embryogenesis, there must be a mechanistic interface between them, but at present there is little knowledge in this regard. The regulation of the differential expression of multiple cadherins during embryonic development must be under some type of genetic control (TAKEICHI 1988; EDELMAN 1988), and it is reasonable to expect that peptide growth factors might participate in some way in this process.

G. Therapeutic Implications

The multifunctionality of peptide growth factors is strikingly illustrated in studies of the molecular events in inflammation, angiogenesis, tissue repair, and cancer (SPORN and ROBERTS 1986; FOLKMAN and KLAGSBRUN 1987). Thus, biochemical mechanisms by which cells and tissues respond to injury and initiate the repair process or form a growing tumor mass involve the processes of cell migration, suppression of immune function, and activation of angiogenesis and fibrosis. The action of peptide growth factors provides a common mechanistic link between all of these processes, especially since it is now clear that the same growth factors that play a key role in carcinogenesis are expressed physiologically by the cells that mediate inflammation and repair, namely platelets, macrophages, and lymphocytes. These peptides include TGF-α, TGF-β, PDGF, FGF, IL-1, and others, all of which are involved in control of collagen breakdown, the recruitment and formation of new fibroblasts, the formation of new collagen and other matrix substances, and the formation of new blood vessels.

It is clear that peptide growth factors are critical determinants of almost every aspect of the response of all tissues to injury, and of the ensuing process of repair. As such, they will have important therapeutic applications. The peptides whose structures and functions are described in the following chapters undoubtedly will have a major impact on the practice of clinical medicine in coming years. Almost every one is being made by recombinant DNA techniques (some in very large quantities), and collectively they form the basis of a substantial portion of the new biotechnology industry. Their practical applications are just beginning to be explored. In vivo uses of these peptides are discussed at length in many of the chapters; they include numerous indications for repair of soft and hard tissues, immunosuppression, enhancement of immune cell function, enhancement of bone marrow function in numerous disease states, and treatment of many proliferative diseases, including cancer. As new clinical studies for one particular indication proceed, they yield unexpected new data on the multifunctionality of growth factors; a striking example is the recent report that GM-CSF (being used for treatment of aplastic anemia) can also markedly lower serum cholesterol (Nimer et al. 1988).

Many new areas still require further exploration. Although the use of growth factors for regeneration of skeletal and cardiac muscle, as well as neuronal tissue, has not yet been achieved in a practical manner, the striking advances that have already been made in so many other areas have created an environment in which these possibilities, too, are now being seriously investigated. It is too early to tell whether the achievement of these goals will depend on the discovery of new peptides, or perhaps even the use of known molecules in new combinations or sets. The multifunctional nature of so many of the peptides described in these two volumes provides a definite note of optimism that further practical uses will be found for these agents in the future.

References

Angel P, Hattori K, Smeal T, Karin M (1988) The *jun* proto-oncogene is positively autoregulated by its product, *jun*/AP-1. Cell 55:875–885

Angeletti RH, Bradshaw RA (1971) Nerve growth factor from mouse submaxillary glands: amino acid sequence. Proc Natl Acad Sci USA 68:2417–240

Ayer-LeLievre CS, Olson L, Ebendal T, Hallböök F, Persson H (1988) Nerve growth factor mRNA and protein in the testis and epididymis of mouse and rat. Proc Natl Acad Sci USA 85:2628–2632

Barkai A, McQuaid C (1988) Predator-prey role reversal in a marine benthic ecosystem. Science 242:62–64

Bissell MJ, Hall G, Parry G (1982) How does the extracellular matrix direct gene expression? J Theor Biol 99:31–68

Bothwell M, Patterson SL, Schatteman GC, Thompson S, Underwood R, Claude P, Holbrook K, Byers M, Gown AM (1989) Regulated expression of nerve growth factor receptors in mesenchymal cells. Submitted

Buck CA, Horwitz AF (1987) Cell surface receptors for extracellular matrix molecules. Annu Rev Cell Biol 3:179–205

Chiu R, Boyle WJ, Meek J, Smeal T, Hunter T, Karin M (1988) The c-*fos* protein interacts with c-*jun*/AP-1 to stimulate transcription of AP-1 responsive genes. Cell 54:541–552

Coffey RJ Jr, Derynck R, Wilcox JN, Bringman TS, Goustin AS, Moses HL, Pittelkow MR (1987) Production and auto-induction of transforming growth factor-α in human keratinocytes. Nature 328:817–820

Cohen S (1960) Purification of a nerve-growth promoting protein from the mouse salivary gland and its neurotoxic antiserum. Proc Natl Acad Sci USA 46:302–311

Cohen S (1962) Isolation of a mouse submaxillary gland protein accelerating incisor eruption and eyelid opening in the newborn animal. J Biol Chem 237:1555–1562

Curran T, Franza BR (1988) *Fos* and *jun*: the AP-1 connection. Cell 55:395–397

Edelmann GM (1988) Topobiology: an introduction to molecular embryology. Basic Books, New York

Ernfors P, Hallböök F, Ebendal T, Shooter EM, Radeke MJ, Misko TP, Persson H (1988) Developmental and regional expression of β-nerve growth factor receptor mRNA in the chick and rat. Neuron 1:983–996

Espevik T, Figari IS, Ranges GE, Palladino MA (1988) Transforming growth factor-β and recombinant human tumor necrosis factor-α reciprocally regulate the generation of lymphokine-activated killer cell activity. J Immunol 140:2312–2316

Evans RM (1988) The steroid and thyroid hormone receptor superfamily. Science 240:889–895

Folkman J, Klagsbrun M (1987) Angiogenesis factors. Science 235:442–447

Gan BS, Hollenberg MD, MacCannell KL, Lederis K, Winkler ME, Derynck R (1987) Distinct vascular actions of epidermal growth factor-urogastrone and transforming growth factor-alpha. J Pharmacol Exp Ther 242:331–337

Green S, Chambon P (1986) A superfamily of potentially oncogenic hormone receptors. Nature 324:615–617

Gregory H, Willshire IR (1975) The isolation of the urogastrones – inhibitors of gastric acid secretion – from human urine. Hoppe-Seylers Z Physiol Chem 356:1765–1774

Harrison LC, Campbell IL (1988) Cytokines: an expanding network of immuno-inflammatory hormones. Mol Endocrinol 2:1151–1156

Hatakeyama M, Minamoto S, Uchiyama T, Hardy RR, Yamada G, Taniguchi T (1985) Reconstitution of functional receptor for human interleukin-2 in mouse cells. Nature 318:467–470

Heldin C-H, Westermark B (1984) Growth factors: mechanism of action and relation to oncogenes. Cell 37:9–20

Hill DL, Strain AJ, Elstow SF, Swenne I, Milner RDG (1986) Bi-functional action of transforming growth factor-β on DNA synthesis in early passage human fetal fibroblasts. J Cell Physiol 128:322–328

Hynes RO (1987) Integrins: a family of cell surface receptors. Cell 48:549–554

Izumo S, Nadal-Ginard B, Mahdavi V (1986) All members of the MHC multigene family respond to thyroid hormone in a highly tissue-specific manner. Science 231:597–600

Kehrl JH, Roberts AB, Wakefield LM, Jakowlew S, Sporn MB, Fauci AS (1986a) Transforming growth factor-beta is an important immunomodulatory protein for human B lymphocytes. J Immunol 137:3855–3860

Kehrl JH, Wakefield LM, Roberts AB, Jakowlew S, Alvarez-Mon M, Derynck R, Sporn MB, Fauci AS (1986b) Production of transforming growth factor-beta by human T lymphocytes and its potential role in the regulation of T cell growth. J Exp Med 163:1037–1050

Kishimoto T, Hirano T (1988) Molecular regulation of B lymphocyte response. Annu Rev Immunol 6:485–512

Komm BS, Terpening CM, Ben DJ, Graeme KA, Gallegos A, Korc M, Greene GL, O'Malley BW, Haussler MR (1988) Estrogen binding, receptor mRNA, and biologic response in osteoblast-like osteosarcoma cells. Science 241:81–86

Kouzarides T, Ziff E (1988) The role of the leucine zipper in the *fos-jun* interaction. Nature 336:646–651

Kumar V, Chambon P (1988) The estrogen receptor binds tightly to its responsive element as a ligand-induced homodimer. Cell 55:145–156

Landschulz WH, Johnson PF, McKnight SL (1988) The leucine zipper: a hypothetical structure common to a new class of DNA binding proteins. Science 240:1759–1764

Le J, Vilcek J (1987) Biology of disease: tumor necrosis factor and interleukin 1: Cytokines with multiple overlapping biological activities. Lab Invest 56:234–248

Levi-Montalcini R, Hamburger V (1953) A diffusable agent of mouse sarcoma producing hyperplasia of sympathetic ganglia and hyperneurotization of viscera in the chick embryo. J Exp Zool 123:233–288

Matsuda H, Coughlin MD, Bienenstock J, Denburg JA (1988) Nerve growth factor promotes human hemopoietic colony growth and differentiation. Proc Natl Acad Sci USA 85:6508–6512

Mercola M, Stiles CD (1988) Growth factor superfamilies and mammalian embryogenesis. Development 102:451–460

Minghetti PP, Norman AW (1988) 1,25 $(OH)_2$-Vitamin D_3 receptors: gene regulation and genetic circuitry. FASEB J 2:3043–3053

Nimer SD, Champlin RE, Golde DW (1988) Serum cholesterol-lowering activity of granulocyte-macrophage colony-stimulating factor. JAMA 260:3297–3300

Panaretto BA, Leish Z, Moore GPM, Robertson DM (1984) Inhibition of DNA synthesis in dermal tissue of Merino sheep treated with depilatory doses of mouse epidermal growth factor. J Endocrinol 100:25–31

Paulsson Y, Hammacher A, Heldin C-H, Westermark B (1987) Possible positive autocrine feedback in the prereplicative phase of human fibroblasts. Nature 328:715–717

Petkovich M, Brand NJ, Krust A, Chambon P (1987) A human retinoic acid receptor which belongs to the family of nuclear receptors. Nature 330:444-450

Rauscher FJ III, Cohen DR, Curran T, Bos TJ, Vogt PK, Bohmann D, Tjian R, Franza BR Jr (1988) *Fos*-associated protein p39 is the product of the *jun* proto-oncogene. Science 240:1010–1016

Rennick D, Yang G, Muller-Sieburg C, Smith C, Arai N, Takabe Y, Gemmel L (1987) Interleukin 4 (B-cell stimulatory factor 1) can enhance or antagonize the factor-dependent growth of hemopoietic progenitor cells. Proc Natl Acad Sci USA 86:6889–6893

Rhodes JA, Fitzgibbon DH, Macchiarulo PA, Murphy RA (1987) Epidermal growth factor-induced precocious incisor eruption is associated with decreased tooth size. Dev Biol 121:247–252

Roberts AB, Anzano MA, Wakefield LM, Roche NS, Stern DF, Sporn MB (1985) Type β transforming growth factor: a bifunctional regulator of cellular growth. Proc Natl Acad Sci USA 82:119–123

Rosen DM, Stempien SA, Thompson AY, Seyedin SM (1988) Transforming growth factor-β modulates the expression of osteoblast and chondroblast phenotypes in vitro. J Cell Physiol 134:337–346

Ruoslahti E, Pierschbacher MD (1987) New perspectives in cell adhesion. Science 238:491–497

Santoro C, Mermod N, Andrews PC, Tjian R (1988) A family of human CCAAT-box binding proteins active in transcription and DNA replication: cloning and expression of multiple cDNAs. Nature 334:218–224

Sassone-Corsi P, Ransone LJ, Lamph WW, Verma IM (1988a) Direct interaction between *fos* and *jun* nuclear oncoproteins: role of the leucine zipper domain. Nature 336:692–694

Sassone-Corsi P, Sisson JC, Verma IM (1988b) Transcriptional autoregulation of the proto-oncogene *fos*. Nature 334:314–319

Schlessinger J (1988) Signal transduction by allosteric receptor oligomerization. Trends Biochem Sci 13:443–447

Sehgal PB, May LT, Tamm I, Vilcek J (1987) Human β_2 interferon and B-cell differentiation factor BSF-2 are identical. Science 235:731–732

Seyedin SM, Segarini PR, Rosen DM, Thompson AY, Bentz H, Graycar J (1987) Cartilage-inducing factor-B is a unique protein structurally and functionally related to transforming growth factor-β. J Biol Chem 262:1946–1949

Short NJ (1988) Regulation of transcription: flexible interpretation. Nature 334:192–193

Sporn MB, Roberts AB (1986) Peptide growth factors and inflammation, tissue repair and cancer. J Clin Invest 78:329–332

Sporn MB, Roberts AB (1988) Peptide growth factors are multifunctional. Nature 332:217–219

Sporn MB, Todaro GJ (1980) Autocrine secretion and malignant transformation of cells. N Engl J Med 303:878–880

Takeichi M (1988) The cadherins: cell-cell adhesion molecules controlling animal morphogenesis. Development 102:639–655

Van Obberghen-Schilling E, Roche NS, Flanders KC, Sporn MB, Roberts AB (1988) Transforming growth factor $\beta1$ positively regulates its own expression in normal and transformed cells. J Biol Chem 263:7741–7746

Verma I, Sassone-Corsi P (1987) Proto-oncogene *fos*: complex but versatile regulation. Cell 51:513–514

Weigent DA, Blalock JE (1987) Interactions between the neuroendocrine and immune systems: common hormones and receptors. Immunol Rev 100:79–108

CHAPTER 2

Isolation and Characterization of Growth Factors

R. A. BRADSHAW and K. P. CAVANAUGH

A. Introduction

Polypeptide growth factors represent a diverse group of hormone–like agents that affect a variety of cellular processes including metabolic regulation, cell division, extension of processes (and other morphological changes), and the maintenance of viability (JAMES and BRADSHAW 1984). They differ from their more classical counterparts in that they are often synthesized in a variety of cells and usually enjoy a spectrum of target tissues. They also often use non-systemic transport mechanisms. However, they are highly similar to (even in-distinguishable from) other classic polypeptide hormones in the manner in which they interact with target tissues. These similarities include the obligatory requirement for cell surface receptors, the nature of the transmembrane signals generated and their effects on gene expression including the genes affected. Therefore, while it is convenient to consider this group of substances in a separate category (hence the retention of the term "polypeptide growth factor"), it is clear that they are an important part of the endocrine system.

The isolation and characterization of polypeptide growth factors parallels the major developments in protein chemistry itself. The earliest factors to be identified were only scantily characterized with functional information far outstripping molecular characterization. When structural information be-came available, it was first obtained with substances that could be isolated in relatively large amounts. Only as the methods of protein chemistry advanced and more sensitive techniques became available (beginning of the early 1970s) were smaller and smaller quantities required to produce detailed structural in-formation. With the advent of cloning techniques it became possible to obtain this information even with vanishingly small samples of the hormone. In turn these methodologies, through the creation of expression systems, have al-lowed the production of large amounts of any identified factor. Thus the isola-tion and structural characterization of polypeptide growth factor now only depends upon the ingenuity of the investigator to first craft an appropriate as-say to identify the activity (and hence the causative agent) to be ultimately as-sured of obtaining complete characterization, usually within a relatively short time span.

In this chapter we briefly review the isolation and characterization of some of the major polypeptide growth factors, noting in each case how this in-formation extended our knowledge of them at the time the experiments were reported.

B. Insulin

The identification of insulin by BANTING and BEST (1922) as the principal agent for regulating blood sugar in higher vertebrates and its use in ameliorating most forms of diabetes mellitus is one of the great success stories of medical research. It is somewhat less widely appreciated that insulin is also an important agent in regulating cell growth both in vivo and in vitro and can certainly be considered to be a polypeptide growth factor as well as a classical hormone of the endocrine system. As such, it is really the first such factor to be identified and obtained in a homogeneous preparation. Not inappropriately it was also the first to be characterized with respect to structure, serving as the pioneering model for sequence analysis by SANGER and colleagues (RYLE et al. 1955 and references cited therein) and as the first hormonal polypeptide structure to be solved by single crystal X-ray crystallographic techniques by HODGKIN and her colleagues (BLUNDELL et al. 1971). Thus it has been a landmark substance.

Insulin is relatively easily obtained in crystalline form from pancreatic extracts (ABEL 1926). The considerable amounts of the hormone normally present in this tissue are due to its physiology and the necessity of the organism to respond to rapid increases in blood sugar by the bolus release of the hormone. Its physiological actions as a growth factor are less well understood although there is substantial literature on the use of insulin in tissue culture systems (PAUL 1970). In fact it is a nearly indispensable agent in defined media for growing any cell type in culture. In may also have as yet unappreciated activities as judged, for example, by its unexplained presence in the central nervous system (LEROITH et al. 1988).

C. Nerve Growth Factor

Although identified about 25 years after insulin, nerve growth factor (NGF) also occupies an important place in the chronology of growth factor research. Following its identification as a humoral substance produced by certain tumor cells (BUEKER 1948; LEVI-MONTALCINI and HAMBURGER 1951, 1953), it was designated as a nerve growth-stimulating factor which was eventually shortened to its present name. Thus it became the progenitor of the term "growth factor." It has been suggested on occasion by those who have the narrower view that the term growth factor should be reserved for those substances with mitogenic activity that this was an unfortunate accident of history. (NGF has been described as a mitogen but only under very special circumstances.) This arbitrary distinction seems unwarranted, particularly in the light of more recent findings that a variety of factors, such as fibroblast growth factor (FGF) and epidermal growth factor (EGF), can stimulate both hypertrophic and hyperplastic events in different cell types (MORRISON et al. 1986, 1987).

The study of NGF also led to the all important identification of EGF *(vide infra)*. Both occur in relatively high concentrations in the adult male mouse submaxillary glands and in the prostate of several higher vertebrates (HARPER et al. 1979), and for many years were the only known "growth factors" (with the exception of insulin as noted above). Because of the overall importance of these factors to the development of the field their isolation and characterization will be described in some detail.

The ability to stimulate neurite proliferation from selected peripheral nerves, an important activity of NGF, was the basis of the observations that led to its initial discovery. In a series of seminal experiments beginning with the transplantation of sarcomas 37 and 180 into host animals, RITA LEVI-MONTALCINI and VICTOR HAMBURGER ascertained that these cells produced a soluble factor whose activity could be demonstrated in an in vitro assay (LEVI-MONTALCINI and HAMBURGER 1951, 1953). LEVI-MONTALCINI and her colleagues went on to provide a detailed description of the action of NGF on peripheral nerves (LEVI-MONTALCINI and ANGELETTI 1968) and to make the important discovery that antibodies against crude preparations of NGF were capable of ablating the sympathetic nervous system in newborn animals (LEVI-MONTALCINI and BOOKER 1960).

The first attempts to purify NGF from the sarcoma tissues were performed by STANLEY COHEN. These experiments, although themselves unsuccessful, led to the identification of snake venoms and the adult male mouse submaxillary gland as far richer sources (COHEN 1959, 1960). However, the first demonstrably homogeneous preparations of NGF were not obtained until several years later and were the subject of two independent efforts (VARON et al. 1967, BOCCHINI and ANGELETTI 1969). The report of VARON et al. (1967) described a high molecular weight form of NGF (which they designated as $7S$ NGF after its sedimentation coefficient) eventually characterized as a noncovalent aggregate containing three types of polypeptide chains (α, β, and γ). Although the stoichiometry of the complex has been the subject of some debate, it is generally considered to contain two α- and two γ-subunits associated with the β-subunit, which is itself a noncovalent dimer of identical polypeptides (ANGELETTI et al. 1971). The complex also contains one to two atoms of zinc (PATTISON and DUNN 1975). This preparation turned out to be an accurate description of the form of NGF that occurs in the granulated tubule cells of the submandibular gland of the mouse and of the form that is secreted into the saliva of that animal (GREENE and SHOOTER 1980).

The other preparation, described by BOCCHINI and ANGELETTI (1969), was much smaller with a reported molecular mass of approximately 30 000 Da, as opposed to the 130 000 Da suggested for $7S$ NGF, and a sedimentation coefficient of $2.5S$.[1] As with the higher molecular weight form, the preparation began with a gel filtration step of the gland extract, which partially purifies the $7S$ complex, but is followed by a dialysis at pH 5 and chromatography in carboxymethyl cellulose for the final purification of the dissociated subunits. On-

---

[1] This preparation was often referred to as $2.5S$ NGF in the early literature to distinguish it from β-NGF, isolated from dissociated, homogeneous $7S$ NGF. This distinction has now been largely dropped.

ly the β-subunit is obtained in essentially homogeneous form; further purifications are required to achieve homogeneous preparations of the α- and γ-subunits (JENG et al. 1979). The effectiveness of the ion exchange chromatography reflects the highly basic character of the hormonally active β-subunit (BOCCHINI and ANGELETTI 1969). Modifications of both procedures have been subsequently introduced (JENG and BRADSHAW 1978).

The high concentrations of NGF in the mouse submaxillary gland have rendered it the overwhelming choice for most studies with this growth factor. However, lesser amounts of NGF have been obtained from various snake venoms (HOGUE-ANGELETTI and BRADSHAW 1979), prostate (HARPER et al. 1979), and a variety of other sources, where much smaller amounts have been at least identified by highly sensitive immunological or nucleic-acid based techniques, including some that have been genetically engineered. These last sources may well eventually become the most important for both research and therapeutic purposes.

D. Epidermal Growth Factor

The discovery and ultimately the first preparations of EGF arose out of studies on NGF from the mouse submaxillary gland by COHEN (1962). He observed that injections of submaxillary gland extracts into neonatal rats resulted in premature eyelid opening and incisor eruption as compared to control animals. The causative agent was identified as a small polypeptide (53 amino acids) that could be obtained in homogeneous form from extracts of that tissue using ion exchange chromatography and gel filtration (TAYLOR et al. 1970). As with NGF, the hormonally active EGF could be isolated from extracts of this tissue as a high molecular weight form (HMW-EGF) that contains two copies of an esteropeptidase that was catalytically similar to the γ-subunit of NGF (TAYLOR et al. 1970). The hormone could be readily dissociated from the HMW-EGF complex.

EGF has now been isolated in and/or purified from a number of other tissues including prostate and kidney (RUBIN and BRADSHAW 1984; RALL et al. 1985). It is homologous to a number of other peptides, including transforming growth factor-α described below (MARQUARDT et al. 1983, 1984). However, as with NGF, the mouse protein is still the most readily obtainable and is the most common form used for experimentation with this growth factor.

E. Insulin-Like Growth Factors

The insulin-like growth factors (IGF-I and -II) were first identified and then isolated from human plasma. They were initially recognized in the late 1950s when it was demonstrated that antisera against insulin could not suppress all of the insulin-like activity present in human blood (FROESCH et al. 1967). The agent(s) responsible were designated nonsuppressible insulin-like activity (NSILA), a name that persisted for some 20 years. Other workers, studying the mediators of growth hormone in the periphery, identified substances they designated as "sulfation factors" in recognition of their ability to induce the incorporation of sulfate into proteoglycans (DAUGHADAY and KIPNIS 1966). Subsequently these agents were renamed somatomedins (DAUGHADAY et al. 1972), and three (A, B, and C) were found to be under growth hormone control; somatomedin B, which was later identified as a protease inhibitor, was subsequently dropped from this list.

In yet a third study, PIERSON and TEMIN (1972) identified a substance produced by cultured rat cells that they designated multiplication stimulating activity (MSA). Unlike NSILA and the somatomedins, MSA was clearly appreciated to be a growth factor directly.

In 1976, RINDERKNECHT and HUMBEL (1976a) purified the NSILA to homogeneity and demonstrated that there were two related molecules in the preparation. Sequence analysis of both (RINDERKNECHT and HUMBEL 1976b, 1978) indicated a striking structural similarity to insulin (as well as to each other) and they were renamed insulin-like growth factors I and II. Eventually it was shown that these molecules (particularly IGF-I) was responsible for somatomedin activity and that MSA was equivalent to rat IGF-II (RUBIN et al. 1982). Thus the various activities, NSILA, somatomedin and MSA, were all attributable to two substances, IGF-I and -II.

These molecules were relatively difficult to obtain in homogeneous form as witnessed by the protracted period of time required to obtain homogeneous preparations. The protocols finally devised to produce these molecules involved gel filtration, ion exchange chromatography, and preparative polyacrylamide gel electrophoresis. Their dissociation from carrier protein complexes, with which they occur under normal conditions in the circulation, were the physical basis for the dialysis and ultrafiltration steps that characterized early protocols. Cloned nucleic acid sequences for the IGFs from human as well as other species have now been reported, allowing the production of IGF from genetically engineered expression vectors (RALL et al. 1987).

F. Platelet-Derived Growth Factor

Platelet-derived growth factor (PDGF) is a glycosylated dimer composed of two disulfide-linked, homologous polypeptide chains designated A and B (JOHNSSON et al. 1982; BETSHOLTZ et al. 1986). The molecular weight, which varies slightly according to the degree of glycosylation, is approximately 30000. PDGF stimulates the proliferation of various cells, including fibroblasts, smooth muscle cells, and glial cells. PDGF is synthesized and secreted by platelets and other normal cells as well as by transformed cells (ROSS 1987). Although most purification studies have utilized human platelets as the source of material, PDGF has also been isolated from porcine platelets, human placenta, and cultured transformed and normal cells (MAREZ et al. 1987; STROOBANT and WATERFIELD 1984; NISTER et al. 1988; HELDIN et al. 1986; COLLINS et al. 1985; WESTERMARK et al. 1986).

The first attempts to purify human PDGF were hindered by low tissue levels, the limited availability of platelets, and the apparant heterogeneity of purified PDGF preparations due to variable glycosylation. Four groups, working independently, utilized outdated human platelets in devising the first purification schemes for isolating PDGF (Antoniades et al. 1979; Heldin et al. 1979; Ross et al. 1979; Deuel et al. 1981).

Three physical properties of PDGF, namely its cationic pI, its heat stability, and its hydrophobic nature, were exploited in the first purification schemes. The high isoelectric point (approximately 10) of PDGF rendered ion exchange chromatography a very efficient purification step and was utilized in each of the original protocols. The degree of purification resulting from charge separation was quite high, yielding a 500-fold increase in specific activity in one study (Raines and Ross 1982). The resistance to heat denaturation was utilized by Antoniades et al. (1979) and Deuel et al. (1981), who eliminated several contaminating mitogenic activities from their platelet extracts by heat treatment. Most platelet mitogens, but not PDGF, are denatured by such treatment. Hydrophobic chromatography, which has played an important role in many growth factor purification protocols, was utilized in the form of phenyl Sepharose (Raines and Ross 1982) and blue Sepharose (Heldin et al. 1979; Deuel et al. 1981) in purifying PDGF. These techniques, in conjunction with more conventional separation techniques such as gel filtration, isoelectric focusing, and preparative sodium dodecylsulfate polyacrylamide gel electrophoresis (SDS-PAGE), formed the basis for the original PDGF purification protocols.

The amounts of PDGF purified in most of the early schemes were quite low due to recoveries that were generally less than 10%. Raines and Ross (1982) reported a purification procedure that included chromatography on heparin-Sepharose and recovered 21% of the biological activity, a significant improvement over previous yields. Several groups subsequently developed large-scale purification protocols for PDGF that basically utilized the same techniques described in the original reports (Heldin et al. 1981; Deuel et al. 1981).

G. Fibroblast Growth Factors

The FGF family represents one of the most important groups in the endocrine scheme by virtue of their wide distribution, broad spectrum of biological activities (including both hypertrophic and hyperplastic responses), and the extraordinarily low levels required for activity. They are also curious for the fact that the two parent members of the family, designated acidic and basic FGF, have the properties of intracellular proteins and their genes are devoid of apparent leader sequences, raising the interesting question of how they function, particularly as extracellular agents (Thomas 1988).

The designation FGF was introduced by Gospodarowicz and colleagues (for a description, see Gospodarowicz et al. 1987) for a substance identified as an apparently homogeneous substance from bovine pituitary glands. The initial protocol consisted of a gel filtration column preceded and followed by two ion exchange chromatographies on carboxymethyl cellulose (GOSPODAROWICZ 1975). Initially it was suggested to be of approximately 13000 Da with a basic isoelectric point. A similar substance was found in bovine brain, isolated with basically the same protocol (GOSPODAROWICZ et al. 1978). Interestingly, both preparations were found to have half-maximal activity in ranges similar to that reported for other mitogens such as EGF and PDGF.

In 1978, WESTALL et al. reported that the principal mitogenically active components of the bovine brain FGF preparations were three fragments derived from myelin basic protein, a structural protein of the myelin sheath. This contention was disproven by THOMAS et al. (1980) who demonstrated that the myelin basic protein fragments could be quantitatively separated from the mitogenic activity by immunoaffinity chromatography and/or isoelectric focusing and that the active principal was associated in their preparations with an acidic component. The same workers subsequently showed (LEMMON et al. 1982) that the brain preparations actually contained two components, the aforementioned acidic molecule and a basic component. The latter was highly similar in all properties to that found in the pituitary preparations. Structural studies by several groups (GIMENEZ-GALLEGO et al. 1985, 1986; ESCH et al. 1985) eventually demonstrated that the acidic and basic FGFs of bovine brain were indeed homologous (aproximately 50% identities) and that both molecules were also found in the pituitary (although not in the original pituitary preparations). The acidic component was demonstrated to be the same as another mitogen, earlier identified from this source by ARMELIN (1973).

Interestingly and importantly, the FGFs were independently identified by several groups using a variety of different cultured cells and different bioassays. Among the more important were eye-derived growth factor, retinal growth factor, brain-derived growth factor, and endothelial cell growth factor (THOMAS and GIMENEZ-GALLEGO 1986). In most cases, two forms were identified which ultimately turned out to be the same as the acidic and basic forms of FGF. These various independent isolations underscore the broad spectrum of specificity of a- and b-FGF, and support the wide range of activities demonstrated with various cultured cells by GOSPODAROWICZ and his colleagues (1987). These studies were also responsible for identifying the high affinity of FGFs for heparin, which provided the basis for a quick and convenient method for purification that for some researchers replaced the preparative gel electrophoresis method introduced by LEMMON and BRADSHAW (1983). The heparin method enables substantial additional purification above that provided by the protocol of GOSPODAROWICZ (1975). In some instances the heparin chromatography has been further assisted by high-performance liquid chromatography (HPLC; MORRISON et al. 1986) although this technique generally leads to poor recoveries.

As with other growth factors it is now clear that a- and b-FGF are part of a larger family (THOMAS 1988), some members of which have been identified by their oncogenic capacities. Characterization of these molecules, which as normal growth factors remains to be accomplished, may be facilitated by the identification of new and better sources of these substances for purification. This is not unimportant in view of the fact that bovine brain and pituitary, still the most common source of these agents, are not particularly rich sources.

H. Transforming Growth Factors

The transforming growth factors (TGF-α and TGF-β) are two structurally unrelated proteins that were originally copurified in what was believed to be a homogeneous preparation. Subsequently, the biological activity was subfractionated using HPLC into two pools, which turned out to be as distinct functionally as they were structurally. It was found that although purified TGF-α was able to transform cells, as determined by the induction of small colonies of normal rat kidney (NRK) cells in soft agar, TGF-β alone could not produce such a response. However, the simultaneous use of either TGF-α (or EGF) with TGF-β produced a greater degree of transformation than that produced by TGF-α alone (ANZANO et al. 1983). The capacity of EGF to replace TGF-α in this system is apparently due to the limited structural relatedness (about 30% identities) between TGF-α and EGF and the resulting ability of the two factors to interact with the same receptors (MARQUARDT et al. 1983, 1984). This crossreaction is typically used to detect the presence of TGF-α during purification schemes.

I. TGF-α

TGF-α is synthesized by a variety of transformed and normal cells, including retrovirus-transformed fibroblasts (TODARO et al. 1985), cultured cells derived from carcinomas or sarcomas (DERYNCK et al. 1987), normal skin keratinocytes (COFFEY et al. 1987), and normal brain cells (WILCOX and DERYNCK 1988). TGF-α is also expressed during fetal development in several tissues.

A variety of molecular weight forms of TGF-α, ranging from 5000 to 20000, have been found in conditioned medium. All are believed to result from differential proteolytic cleavage and glycosylation of a transmembrane precursor. A low molecular weight form of TGF-α was first purified to homogeneity by MARQUARDT and TODARO (1982) from medium conditioned by the human melanoma tumor line A2058. They separated the high and low molecular weight forms of TGF-α using gel filtration. Further purification of the low molecular weight form was achieved using two successive applications of reverse-phase HPLC. Following the hydrophobic separation, they recovered 73% of the activity present in the original low molecular weight pool. Nearly identical techniques were used to purify TGF-α from transformed rodent cells and from human melanoma cell lines (MARQUARDT et al. 1983; MASSAGUÉ 1983).

The larger 18 000- to 20 000-Da forms also present in medium conditioned by various cells including human melanoma cells and human bronchogenic carcinoma cells and in human urine are sometimes the predominant species of TGF-α present (MARQUARDT and TODARO 1982; TODARO et al. 1980; SHERWIN et al. 1983; DE LARCO et al. 1985; LINSLEY et al. 1985; LUETTEKE and MICHALOPOULOS 1985; MASSAGUÉ 1983; RICHMOND et al. 1985; TWARDZIK et al. 1982, 1985; TWARDZIK 1985). DART et al. (1985) have isolated intracellular high molecular weight forms of TGF-α (about 18 000 Da in mass) from a human rhabdomyosarcoma cell line and purified one to near homogeneity utilizing gel filtration, cation exchange chromatography, and successive applications of reverse-phase columns.

II. TGF-β

TGF-β was first purified to apparent homogeneity following acid/ethanol extraction of human platelets and placenta and bovine kidney (ASSOIAN et al. 1983; FROLIK et al. 1983; ROBERTS et al. 1983). ASSOIAN et al. (1983) took advantage of the high specific activity of TGF-β in platelet extracts (about 100-fold greater than that of other nonneoplastic tissues) and the aberrant elution of TGF-β during gel filtration to purify the protein. This protocol was composed of two successive gel filtration steps. During the first separation, performed in 1 M acetic acid, TGF-β eluted with proteins about half its molecular weight and was thus separated from proteins with masses similar to its own. Subsequently, the biologically active fractions were rechromatographed on the same molecular sizing column in the presence of urea, which causes TGF-β to elute according to its true mass. The protein is therefore separated from the lower molecular weight contaminants present. FROLIK et al. (1983) and ROBERTS et al. (1983) utilized ion exchange and hydrophobic chromatography in addition to gel filtration in designing their purification schemes. MASSAGUÉ (1984) used these techniques along with gradient gel electrophoresis to first purify TGF-β from an abnormal tissue source, i.e., medium conditioned by a transformed cell line.

TGF-β has also been purified under different names. For example, SEYEDIN et al. (1985) utilized gel filtration, cation exchange and hydrophobic chromatography to purify two factors they named cartilage-inducing factors A and B (CIF) from bovine bone. Both of these have subsequently been shown to be different forms of TGF-β (SEYEDIN et al. 1986), now known as types 1 and 2. CHEIFETZ et al. (1987) subsequently isolated TGF-β2 from porcine platelets. They utilized the procedure developed by ASSOIAN et al. (1983) based on gel filtration followed by hydrophobic chromatography. Although the bioactive protein purified in this manner appeared homogeneous when analyzed on SDS-PAGE, it could be further resolved into three components by another hydrophobic chromatographic separation. Based on N-terminal amino acid sequence analyses, the form eluting first was shown to be identical to that of human, mouse, and bovine TGF-β, now referred to as TGF-β1. The form eluting third contained an N-terminal sequence with partial identity to

TGF-β1. This form, TGF-β2, is 71% homologous to TGF-β1 (Marquardt et al. 1987; De Martin et al. 1987). Another form of porcine TGF-β, eluting between TGF-β1 and TGF-β2, is a heterodimer composed of one TGF-β1 chain and one TGF-β2 chain and is known as TGF-β1.2.

I. Interleukins

The interleukins (IL-1–IL-7) are glycoproteins that regulate several biological activities, and are particularly important in enhancing the immune response. Several of the interleukins are capable of eliciting multiple biological effects from different target cells. For example, some of the actions of IL-1 include stimulating secretion of IL-2, B-cell differentiation, prostaglandin release, and osteoclast activation (Smith et al. 1980; Wood 1979; Dewhirst et al. 1985). IL-2 stimulates T-lymphocyte proliferation, natural killer cell cytotoxicity and secretion of interferon and B-cell growth factor (Robb 1985; Henney et al. 1981; Farrar et al. 1981; Leibson et al. 1981). IL-3 (multi-colony-stimulating factor, CSF) is a hemopoietin that promotes the growth of various cells including the progenitors of granulocytes, macrophages, and erythrocytes (Hapel and Young 1988). Thus it is related closely to the CSFs. IL-4, IL-5, and IL-6 collectively promote the growth and differentiation of several tissues including B lymphocytes, T cells, hepatocytes, and plasmacytomas (Ohara et al. 1987; Harada et al. 1985; Sanderson et al. 1986; Nordan et al. 1987; Van Snick et al. 1986).

The ability of a single molecule to elicit such different effects has complicated purification efforts: multiple laboratories, working independently, were in several instances simultaneously purifying the identical protein under a different name. Two additional problems hindering purification efforts included: (1) the very low rate of secretion of interleukins, resulting in a lack of sufficient material to purify; and (2) the microheterogeneity of the interleukins particularly with regard to glycosylation, resulting in a loss of material when conventional chromatographic procedures were utilized. The first problem was partially resolved when it was discovered that the secretion of interleukins was increased dramatically when cells were treated with phorbol esters. The development of immunoaffinity chromatography helped in some instances to resolve the problem of microheterogeneity, resulting in an efficient purification with high recovery of material. Some of the purification procedures developed for the interleukins are summarized in Table 1.

J. Hemopoietic Growth Factors

Five well-known hemopoietic growth factors have been purified to homogeneity in the last decade. These include erythropoietin, which promotes the maturation of erythroblasts, and three colony-stimulating factors, (CSFs), which control formation of granulocyte and macrophage colonies from bone marrow progenitor cells. CSF-1 (M-CSF) and CSF-2 (G-CSF) induce the

Table 1. Purification techniques utilized for the interleukins

Factor	Source of purified factor	Purification techniques utilized[a] (yield %)	References
IL-1	Murine macrophage tumor cell line	ASP; HC; SE; IE (2)	MIZEL and MIZEL (1981)
	Murine macrophage tumor cell line	SE; IE; GGE (2)	MIZEL et al. (1983)
	Murine macrophage tumor cell line	SE; IAC (20)	MIZEL et al. (1983)
	Murine macrophage tumor cell line	SE; IE; GGE (2)	DUKOVICH and MIZEL (1985)
	Human mononuclear blood cell (MBC)	HC; IE; SE (1.3)	LACHMAN et al. (1985)
	Human leukemic monocytes	SA; IE; SE (NR)	VAN DAMME et al. (1985)
	MBC	HC; IE; DC (31)	KRONHEIM et al. (1985)
	MBC	IE; SE (30)	GERY and SCHMIDT (1985)
IL-2	Human lymphocytes	ASP; IE; SE; DC (19)	WELTE et al. (1982)
	Murine T-lymphoma cell line	SE; HC; IE (27)	RIENDEAU et al. (1983)
	Human transformed T cell (JURKAT)	IAC (63)	ROBB et al. (1983)
	JURKAT	SE; HC (80)	BOHLEN et al. (1983)
	Human lymphocytes	SE; HC; DC (55)	KNIEP et al. (1984)
	JURKAT	HC; SE; IAC; CF (87)	ROBB et al. (1984)
	Murine T-lymphoma cell line	SE; HC; IE (27)	PAETKAU et al. (1985)
IL-3	Myelomonocytic leukemia cell line	ASP; IE; HAP; SE; HC (8)	IHLE et al. (1982)
	Myelomonocytic leukemia cell line	ASP; HC; SE; IE; PBC; NA (4)	CLARK-LEWIS et al. (1984)
	Murine spleen cells	ASP; HC; SE; IE (4)	CUTLER et al. (1985)
	Myelomonocytic leukemia cell line	IAC (97)	ZILTENER et al. (1987)
IL-4	Thymoma cell line	TMS; IE; HC (8)	GRABSTEIN et al. (1986)
	Thymoma cell line	IAC; HC (89)	OHARA et al. (1987)
IL-5	Murine T cell hybridoma	ASP; IE; DC; HAP; SE; GGE (3.8)	TAKATSU et al. (1985)
	Murine T cell hybridoma	ASP; IE; DC; HAP; SE (16)	HARADA et al. (1985)
	Murine T cell clone	SE; LLC; HC (NR)	SANDERSON et al. (1986)
IL-6	Human leukemic cell line	SE; CF; HC (25)	HIRANO et al. (1985)
	Murine helper T-cell clone	SA; SE; IE; HC (10)	VAN SNICK et al. (1986)
	Murine macrophage cell line	TMS-CpG; SE; HC (14)	NORDAN et al. (1987)
	Human leukocytes	SA; IAC; SE; IE; HC (NR)	VAN DAMME et al. (1988)

[a] Abbreviations: *ASP*, ammonium sulfate precipitation; *CF*, chromatofocusing; *DC*, dye adsorbent chromatography; *GGE*, gradient gel electrophoresis; *HAP*, hydroxylapatite chromatography; *HC*, hydrophobic chromatography; *IAC*, immunoaffinity chromatography; *IE*, ion exchange chromatography; *LLC*, lentil lectin chromatography; *NA*, neuraminidase treatment; *PBC*, phenylboronate chromatography; *SA*, silicic acid adsorption; *SE*, size exclusion chromatography; *TMS*, trimethylsilyl-silica adsorption; *TMS-CpG*, trimethylsilyl-glass bead adsorption; *NR*, not reported.

Table 2. Purification methods applied to the hemopoietic growth factors

Factor	Source of purified factor	Purification techniques utilized [a] (yield %)	References
Erythropoietin	Sheep plasma	CP; MA; ASP; EP; IE (NR)	Goldwasser and Kung (1971)
	Human urine	IE; SE; EP; HAP (21)	Miyake et al. (1977)
	Human urine	SE; IAC (63)	Sasaki et al. (1987)
CSF-1 (M-CSF)	Murine L cells	IE; SE; CAS; GGE (25)	Stanley and Heard (1977)
	Murine L cells	IE; SE; CAS; GGE; CP (22–45)	Stanley and Guilbert (1981)
	Human pancreatic carcinoma cells	IE; SE (11)	Wu et al. (1979)
	Murine L cells	IE; CAS; CP (20)	Stanley (1985)
	Human urine	IE; HC; HAP; CF (<1)	Hatake et al. (1985)
	Human T lymphoblastoid cell line	IE; SE; CAS; HC (<1)	Takahashi et al. (1988)
	Human pancreatic carcinoma cells	IAC; LLC; HC; CP (5)	Csejtey and Boosman (1986)
	Human urine	IE; SA; HC; GGE (27)	Tao et al. (1987)
CSF-2 (G-CSF)	Cultured murine lung cells	SE; HC; ASP (34)	Nicola et al. (1983)
	Human squamous cell line CHU-2	SE; HC (NR)	Nomura et al. (1986)
CSF-3 (GM-CSF)	Cultured murine lung	IE; CAS; CP; GGE (8)	Burgess et al. (1977)
	Cultured murine lung	SE; CAS; ASP; HC; IE (2)	Burgess and Nice (1985)
			Burgess et al. (1986)
	Human T cell line	IE; SE; ASP; HC (NR)	Wong et al. (1985)
	Human Hodgkin's tumor cell line	SE; CAS; CP; HC (8)	Byrne et al. (1986)
	Human fibrous histiocytoma cell line	IE; SE; HC (3)	Erickson-Miller et al. (1988)

[a] Abbreviations: *ASP*, ammonium sulfate precipitation; *CAS*, concanavalin A chromatography; *CF*, chromatofocusing; *CP*, calcium phosphate adsorption; *EP*, ethanol precipitation; *GGE*, gradient gel electrophoresis; *HAP*, hydroxylapatite chromatography; *HC*, hydrophobic chromatography; *IAC*, immunoaffinity chromatography; *IE*, ion exchange chromatography; *LLC*, lentil lectin chromatography; *MA*, methylated albumin Kieselgur chromatography; *SA*, silicic acid adsorption; *SE*, size exclusion chromatography; *NR*, not reported.

formation of macrophage and granulocyte colonies, respectively, while CSF-3 (GM-CSF) induces both types of colonies from progenitor cells (BURGESS et al. 1977; STANLEY and HEARD 1977; NICOLA et al. 1983). A variety of cell types produce these factors, which are glycoproteins ranging in mass from about 18 000 to 90 000 Da. The degree of glycosylation of these factors is often variable, a situation which has complicated purification efforts. Thus, a sample may show a single sequence but appear heterogeneous when analyzed by SDS-PAGE, isoelectric focusing, or concanavalin A affinity chromatography. Purification of these factors has been achieved using conventional methods including salting out, gel filtration, and affinity, ion exchange and hydrophobic chromatography, as summarized in Table 2.

K. Concluding Remarks

This chapter has summarized the isolation procedures for most of the better known polypeptide growth factors. In the main, these purifications have used conventional technologies that have been applied equally effectively to other protein isolation schemes. With the exception of heparin Sepharose, affinity-based procedures have not been employed extensively. Nonetheless, future protocols may well take advantage of such approaches, particularly ones based on immunoadsorbants.

It remains an interesting question whether there are large numbers of additional polypeptide growth factors yet to be discovered. There are a substantial number of reports of activities that have been identified that have not yet been characterized in any detail. Experience suggests that many, perhaps most, are related or identical to molecules already obtained and that, like the classical hormones, the number of polypeptide growth factors operating in higher eukaryotes will be quite a proscribed family. At the same time it is safe to predict that at least a few more polypeptide growth factors await identification.

References

Abel JJ (1926) Crystalline insulin. Proc Natl Acad Sci USA 12:132–136

Angeletti RH, Bradshaw RA, Wade RD (1971) Subunit structure and amino acid composition of mouse submaxillary gland nerve growth factor. Biochemistry 10:463–469

Antoniades HN, Scher CD, Stiles CD (1979) Purification of human platelet-derived growth factor. Proc Natl Acad Sci USA 76:1809–1813

Anzano MA, Roberts AB, Smith JM, Sporn MB, De Larco JE (1983) Sarcoma growth factor from conditioned medium of virally transformed cells is composed of both type α and type β transforming growth factors. Proc Natl Acad Sci USA 80:6264–6268

Armelin HA (1973) Pituitary extracts and steroid hormones in the control of 3T3 cell growth. Proc Natl Acad Sci USA 70:2702–2706

Assoian RK, Komoriya A, Meyers CA, Miller DM, Sporn MB (1983) Transforming growth factor-β in human platelets: identification of a major storage site, purification, and characterization. J Biol Chem 11:7155–7160

Banting FG, Best CH (1922) The internal secretion of the pancreas. J Lab Clin Med 7:251–266

Betsholtz C, Johnsson A, Heldin C-H, Westermark B, Lind P, Urdea MS, Eddy R, Shows TB, Philpott K, Mellor AL, Knott TJ, Scott J (1986) cDNA sequence and chromosomal localization of human platelet-derived growth factor A-chain and its expression in tumour cell lines. Nature 320:695–699

Blundell TL, Cutfield JF, Cutfield SM, Dodson EJ, Dodson GG, Hodgkin DC, Mercola DA, Vijayan M (1971) Atomic positions in rhombohedral 2-zinc insulin crystals. Nature 231:506–511

Bocchini V, Angeletti PU (1969) The nerve growth factor: purification as a 30,000-molecular-weight protein. Proc Natl Acad Sci USA 64:787–794

Bohlen P, Esch F, Wegemer D, Salk P, Dennert G (1983) Isolation and partial characterization of human T-cell growth factor. Biochem Biophys Res Commun 117:623–630

Bueker ED (1948) Implantation of tumors in the hind limb field of the embryonic chick and developmental responses of the lumbosacral nervous system. Anat Rec 102:369–390

Burgess AW, Nice EC (1985) Murine granulocyte-macrophage colony-stimulating factor. Methods Enzymol 116:588–600

Burgess AW, Camakaris J, Metcalf D (1977) Purification and properties of colony-stimulating factor from mouse lung-conditioned medium. J Biol Chem 252:1998–2003

Burgess AW, Metcalf D, Sparrow LG, Simpson RJ, Nice EC (1986) Granulocyte/macrophage colony-stimulating factor from mouse lung conditioned medium. Biochem J 235:805–814

Byrne PV, Heit WF, March CJ (1986) Human granulocyte-macrophage colony-stimulating factor purified from a Hodgkin's tumor cell line. Biochim Biophys Acta 874:266–273

Cheifetz S, Weatherbee JA, Tsang ML-S, Anderson JK, Mole JE, Lucas R, Massagué J (1987) The transforming growth factor-β system, a complex pattern of cross-reactive ligands and receptors. Cell 48:409–415

Clark-Lewis I, Kent SBH, Schrader JW (1984) Purification to apparent homogeneity of a factor stimulating the growth of multiple lineages of hemopoietic cells. J Biol Chem 259:7488–7494

Coffey RJ, Derynck R, Wilcox JN, Bringman TS, Goustin AS, Moses HL, Pittelkow MR (1987) Production and autoinduction of transforming growth factor-α in human keratinocytes. Nature 328:817–820

Cohen S (1959) Purification and metabolic effects of a nerve growth-promoting protein from snake venom. J Biol Chem 234:1129–1137

Cohen S (1960) Purification of a nerve-growth promoting protein from the mouse salivary gland and its neurocytotoxic antiserum. Proc Natl Acad Sci USA 46:302–311

Cohen S (1962) Isolation of a mouse submaxillary gland protein accelerating incisor eruption and eyelid opening in the newborn animal. J Biol Chem 237:1555–1562

Collins T, Ginsburg D, Boss JM, Orkin SH, Pober JS (1985) Cultured human endothelial cells express platelet-derived growth factor B chain: cDNA cloning and structural analysis. Nature 316:748–750

Csejtey J, Boosman A (1986) Purification of human macrophage colony stimulating factor (CSF-1) from medium conditioned by pancreatic carcinoma cells. Biochem Biophys Res Commun 138:238–245

Cutler RL, Metcalf D, Nicola NA, Johnson GR (1985) Purification of a multipotential colony-stimulating factor from pokeweed mitogen-stimulated mouse spleen cell conditioned medium. J Biol Chem 260:6579–6587

Dart LL, Smith DM, Meyers CA, Sporn MB, Frolik CA (1985) Factors from a human tumor cell: characterization of transforming growth factor β and identification of high molecular weight transforming growth factor α. Biochemistry 24:5925–5931

Daughaday WH, Kipnis D (1966) The growth-promoting and anti-insulin actions of somatotropin. Recent Prog Horm Res 22:49–99

Daughaday WH, Hall K, Raben M, Salmon WD Jr, Van den Brande JL, Van Wyk JJ (1972) Somatomedin: proposed designation for sulphation factor. Nature 235:107

De Larco J, Pigott DA, Lazarus JA (1985) Ectopic peptides released by a human melanoma cell line that modulate the transformed phenotype. Proc Natl Acad Sci USA 82:5015–5019

De Martin R, Haendler B, Hofer-Warbinek R, Gaugitsch H, Wrann M, Schlusener H, Seifert JM, Bodmer S, Fontana A, Hofer E (1987) Complementary DNA for human glioblastoma-derived T cell suppressor factor, a novel member of the transforming growth factor-β gene family. EMBO J 6:3673–3677

Derynck R, Goeddel DV, Ullrich A, Gutterman JV, Williams RD, Bringman TS, Berger WH (1987) Synthesis of mRNAs for transforming growth factors α and β and the epidermal growth factor receptor by human tumors. Cancer Res 47:707–712

Deuel TF, Huang JS, Proffitt RT, Baenziger JU, Chang D, Kennedy BB (1981) Human platelet-derived growth factor: purification and resolution into two active protein fractions. J Biol Chem 256:8896–8899

Dewhirst FE, Stashenko PP, Mole JE, Tsurumachi T (1985) Purification and partial sequence of human osteoclast-activating factor: identity with interleukin 1β. J Immunol 135:2562–2568

Dukovich M, Mizel SB (1985) Murine interleukin 1. Methods Enzymol 116:480–492

Erickson-Miller EL, Abboud CN, Brennan JK (1988) Purification of GCT cell-derived human colony-stimulating factors. Exp Hematol 16:184–189

Esch F, Baird A, Ling N, Ueno N, Hill F, Denoroy L, Klepper R, Gospodarowicz D, Bohlen P, Guillemin R (1985) Primary structure of bovine pituitary basic fibroblast growth factor (FGF) and comparison with the amino-terminal sequence of bovine brain acidic FGF. Proc Natl Acad Sci USA 82:6507–6511

Farrar WL, Johnson HM, Farrar JJ (1981) Regulation of the production of immune interferon and cytotoxic T lymphocytes by interleukin 2. J Immunol 126:1120–1125

Froesch ER, Burgi H, Muller WA, Humbel RE, Jakob A, Labhart A (1967) Non-suppressible insulin-like activity of human serum: purification, physicochemical and biological properties and its relation to total serum ILA. Recent Prog Horm Res 23:565–616

Frolik CA, Dart LL, Meyers CA, Smith DM, Sporn MB (1983) Purification and initial characterization of a type β transforming growth factor from human placenta. Proc Natl Acad Sci USA 80:3676–3680

Gery I, Schmidt JA (1985) Human interleukin-1. Methods Enzymol 116:456–467

Gimenez-Gallego G, Rodney C, Bennett C, Rios-Candelore M, DiSalvo J, Thomas K (1985) Brain-derived acidic fibroblast growth factor: complete amino acid sequence and homologies. Science 230:1385–1388

Gimenez-Gallego G, Conn G, Hatcher VB, Thomas KA (1986) The complete amino acid sequence of human brain-derived acidic fibroblast growth factor. Biochem Biophys Res Commun 138:611–617

Goldwasser E, Kung K-H (1971) Purification of erythropoietin. Proc Natl Acad Sci USA 68:697–698

Gospodarowicz D (1975) Purification of a fibroblast growth factor from bovine pituitary. J Biol Chem 250:2515–2520

Gospodarowicz D, Bialecki H, Greenburg G (1978) Purification of the fibroblast growth factor activity from bovine brain. J Biol Chem 253:3736–3743

Gospodarowicz D, Ferrara N, Schweigerer L, Neufeld G (1987) Structural characterization and biological functions of fibroblast growth factor. Endocr Rev 8:95–114

Grabstein K, Eisenman J, Mochizuki D, Shanebeck K, Conlon P, Hopp T, March C, Gillis S (1986) Purification to homogeneity of B cell stimulating factor. J Exp Med 163:1405–1414

Greene LA, Shooter EM (1980) The nerve growth factor: biochemistry, synthesis and mechanism of action. Annu Rev Neurosci 3:353–402

Hapel AJ, Young IG (1988) Molecular biology of interleukin 3: a multilineage hemopoietic growth regulator. In: Schrader JW (ed) Lymphokines, vol 15. Academic, San Diego

Harada N, Kikuchi Y, Tominaga A, Takaki S, Takatsu K (1985) BCGF II activity on activated B cells of a purified murine T cell-replacing factor (TRF) from a T cell hybridoma (B151K12). J Immunol 134:3944–3951

Harper GP, Barde YA, Burnstock G, Carstairs JR, Dennison ME, Suda K, Vernon CA (1979) Guinea pig prostate is a rich source of nerve growth factor. Nature 279:160–162

Hatake K, Motoyoshi K, Ishizaka Y, Sato M (1985) Purification of human urinary colony-stimulating factor by high-performance liquid chromatography. J Chromatogr 344:339–344

Heldin C-H, Westermark B, Wasteson A (1979) Platelet-derived growth factor: purification and partial characterization. Proc Natl Acad Sci USA 76:3722–3726

Heldin C-H, Westermark B, Wasteson A (1981) Platelet-derived growth factor: isolation by a large-scale procedure and analysis of subunit composition. Biochem J 193:907–913

Heldin C-H, Johnsson A, Wennergren S, Wernstedt C, Betsholtz C, Westermark B (1986) A human osteosarcoma cell line secretes a growth factor structurally related to a homodimer of PDGF A-chains. Nature 319:511–514

Henney CS, Kuribaysahi K, Kern DE, Gillis S (1981) Interleukin-2 augments natural killer cell activity. Nature 291:335–336

Hirano T, Taga T, Nakano N, Yasukawa K, Kashiwamura S, Shimizu K, Nakajima K, Pyun KH, Kishimoto T (1985) Purification to homogeneity and characterization of human B-cell differentiation factor (BCDF or BSF-2). Proc Natl Acad Sci USA 82:5490–5494

Hogue-Angeletti RA, Bradshaw RA (1979) Nerve growth factor in snake venoms. In: Lee CY (ed) Snake venoms. Springer, Berlin Heidelberg New York, pp 276–294 (Handbook of experimental pharmacology, vol 52)

Ihle JN, Keller J, Henderson L, Klein F, Palaszynski E (1982) Procedures for the purification of interleukin 3 to homogeneity. J Immunol 129:2431–2436

James R, Bradshaw RA (1984) Polypeptide growth factors. Annu Rev Biochem 53:259–292

Jeng I, Bradshaw RA (1978) The preparation of nerve growth factor. In: Marks N, Rodnight R (eds) Research methods in neurochemistry, vol 4. Plenum, New York, pp 265–288

Jeng I, Andres RY, Bradshaw RA (1979) Mouse nerve growth factor: a rapid isolation procedure for the a and γ subunits. Anal Biochem 92:482–488

Johnsson A, Heldin C-H, Westermark B, Wasteson A (1982) Platelet-derived growth factor: identification of constituent polypeptide chains. Biochem Biophys Res Commun 104:66–74

Kniep EM, Kniep B, Grote W, Conradt HS, Monner DA, Muhlradt PF (1984) Purification of the T lymphocyte growth factor interleukin-2 from culture media of human peripheral blood leukocytes (buffy coats). Eur J Biochem 143:199–203

Kronheim SR, March CJ, Erb SK, Conlon PJ, Mochizuki DY, Hopp TP (1985) Human interleukin 1: purification to homogeneity. J Exp Med 161:490–502

Lachman LB, Shih L-CN, Brown DC (1985) Interleukin 1 from human leukemic monocytes. Methods Enzymol 116:467–479

Leibson HJ, Marrack P, Kappler JW (1981) B cell helper factors. I. Requirement for both interleukin 2 and another 40,000 mol wt factor. J Exp Med 154:1681–1693

Lemmon SK, Bradshaw RA (1983) Purification and partial characterization of bovine pituitary fibroblast growth factor. J Cell Biochem 21:195–208

Lemmon SK, Riley MC, Thomas KA, Hoover GA, Maciag T, Bradshaw RA (1982) Bovine fibroblast growth factor: comparison of brain and pituitary preparations. J Cell Biol 95:162–169

LeRoith D, Adamo M, Shemer J, Waldbillig R, Lesniak MA, dePablo F, Hart C, Roth J (1988) Insulin-related materials in the nervous system of vertebrates and non-vertebrates: possible extrapancreatic production. Horm Metab Res 20:411–420

Levi-Montalcini R, Angeletti PU (1968) Nerve growth factor. Physiol Rev 48:534–569

Levi-Montalcini R, Booker B (1960) Destruction of the sympathetic ganglia in mammals by an anti-serum to a nerve growth protein. Proc Natl Acad Sci USA 46:384–391

Levi-Montalcini R, Hamburger V (1951) Selective growth stimulating effects of mouse sarcoma on the sensory and sympathetic nervous system of the chick embryo. J Exp Zool 116:321–363

Levi-Montalcini R, Hamburger V (1953) A diffusible agent of mouse sarcoma producing hyperplasia of sympathetic ganglia and hyperneurotization of viscera in the chick embryo. J Exp Zool 123:233–288

Linsley PS, Hargreaves WR, Twardzik DR, Todaro GJ (1985) Detection of larger polypeptides structurally and functionally related to type I transforming growth factor. Proc Natl Acad Sci USA 82:356–360

Luetteke NC, Michalopoulos GK (1985) Partial purification and characterization of a hepatocyte growth factor produced by rat hepatocellular carcinoma cells. Cancer Res 45:6331–6337

Marez A, N'guyen T, Chevallier B, Clement G, Dauchel MC, Barritault D (1987) Platelet derived growth factor is present in human placenta: purification from an industrially processed fraction. Biochimie 69:125–129

Marquardt H, Todaro GJ (1982) Human transforming growth factor: production by a melanoma cell line, purification, and initial characterization. J Biol Chem 257:5220–5225

Marquardt H, Hunkapiller MW, Hood LE, Twardzik DR, De Larco JE, Stephenson JR, Todaro GJ (1983) Transforming growth factors produced by retrovirustransformed rodent fibroblasts and human melanoma cells: amino acid sequence homology with epidermal growth factor. Proc Natl Acad Sci USA 80:4684–4688

Marquardt H, Hunkapiller MW, Hood LE, Todaro GJ (1984) Rat transforming growth factor type 1: Structure and relation to epidermal growth factor. Science 223:1079–1082

Marquardt H, Lioubin MN, Ikeda T (1987) Complete amino acid sequence of human transforming growth factor type $\beta2$. J Biol Chem 262:12127–12131

Massagué J (1983) Epidermal growth factor-like transforming growth factor: isolation, chemical characterization, and potentiation by other transforming factors from feline sarcoma virus-transformed rat cells. J Biol Chem 258:13606–13613

Massagué J (1984) Type β transforming growth factor from feline sarcoma virus-transformed rat cells. isolation and biological properties. J Biol Chem 259:9756–9761

Miyake T, Kung CK-H, Goldwasser E (1977) Purification of human erythropoietin. J Biol Chem 252:5558–5564

Mizel SB, Mizel D (1981) Purification to apparent homogeneity of murine interleukin 1. J Immunol 126:834–837

Mizel SB, Dukovich M, Rothstein J (1983) Preparation of goat antibodies against interleukin 1: use of an immunoadsorbent to purify interleukin 1. J Immunol 131:1834–1837

Morrison RS, Sharma A, de Vellis J, Bradshaw RA (1986) Basic fibroblast growth factor supports the survival of cerebral cortical neurons in primary culture. Proc Natl Acad Sci USA 83:7537–7541

Morrison RS, Kornblum HI, Leslie FM, Bradshaw RA (1987) Trophic stimulation of cultured neurons from neonatal rat brain by epidermal growth factor. Science 238:72–75

Nicola NA, Metcalf D, Matsumoto M, Johnson GR (1983) Purification of a factor inducing differentiation in murine myelomonocytic leukemia cells. J Biol Chem 258:9017–9023

Nister M, Hammacher A, Mellstrom K, Siegbahn A, Ronnstrand L, Westermark B, Heldin C-H (1988) A glioma-derived PDGF A chain homodimer has different functional activities from a PDGF AB heterodimer purified from human platelets. Cell 52:791–799

Nomura H, Imazeki I, Oheda M, Kubota N, Tamura M, Ono M, Ueyama Y, Asano S (1986) Purification and characterization of human granulocyte colony-stimulating factor (G-CSF). EMBO J 5:871–876

Nordan RP, Pumphrey JG, Rudikoff S (1987) Purification and NH2-terminal sequence of a plasmacytoma growth factor derived from the murine macrophage cell line P3 88D1. J Immunol 139:813–817

Ohara J, Coligan JE, Zoon K, Maloy WL, Paul WE (1987) High-efficiency purification and chemical characterization of B cell stimulatory factor-1/interleukin 4. J Immunol 139:1127–1134

Paetkau V, Riendeau D, Bleackley RC (1985) Murine interleukin 2. Methods Enzymol 116:526–539

Pattison SE, Dunn MF (1975) On the relationship of zinc ion to the structure and function of the 7S NGF protein. Biochemistry 14:2733–2739

Paul J (1970) Cell and tissue culture, 4th edn. Churchill Livingston, Edinburgh

Pierson RW Jr, Temin HM (1972) The partial purification from calf serum of a fraction with multiplication-stimulating activity for chicken fibroblasts in cell culture and with non-suppressible insulin-like activity. J Cell Physiol 79:319–329

Raines EW, Ross R (1982) Platelet-derived growth factor: high yield purification and evidence for multiple forms. J Biol Chem 257:5154–5160

Rall LB, Scott J, Bell GI, Crawford RJ, Penschow JD, Niall HD, Coghlan JP (1985) Mouse prepro-epidermal growth factor synthesis by the kidney and other tissues. Nature 313:228–231

Rall LB, Scott J, Bell GI (1987) Human insulin-like growth factor I and II messenger RNA: isolation of complementary DNA and analysis of expression. Methods Enzymo 146:239–248

Richmond A, Lawson DH, Nixon DW, Chawla RK (1985) Characterization of autostimulatory and transforming growth factors from human melanoma cells. Cancer Res 45:6390–6394

Riendeau D, Harnish DG, Bleackley RC, Paetkau V (1983) Purification of mouse interleukin 2 to apparent homogeneity. J Biol Chem 258:12114–12117

Rinderknecht E, Humbel RE (1976a) Polypeptides with nonsuppresible insulin-like and cell-growth promoting activitites in human serum: isolation, chemical characterization, and some biological properties of forms I and II. Proc Natl Acad Sci USA 73:2365–2369

Rinderknecht E, Humbel RE (1976b) Amino-terminal sequences of two polypeptides from human serum with nonsuppressible insulin-like and cell-growth-promoting activities: evidence for structural homology with insulin B chain. Proc Natl Acad Sci USA 73:4379–4381

Rinderknecht E, Humbel RE (1978) Primary structure of human insulin-like growth factor II. FEBS Lett 89:283–286

Robb RJ (1985) Human interleukin 2. Methods Enzymol 116:493–525

Robb RJ, Kutny RM, Chowdhry V (1983) Purification and partial sequence analysis of human T-cell growth factor. Proc Natl Acad Sci USA 80:5990–5994

Robb RJ, Kutny RM, Panico M, Morris HR, Chowdhry V (1984) Amino acid sequence and post-translational modification of human interleukin 2. Proc Natl Acad Sci USA 81:6486–6490

Roberts AB, Anzano MA, Meyers CA, Wideman J, Blacher R, Pan Y-C, Stein S, Lehrman SR, Smith JM, Lamb LC, Sporn MB (1983) Purification and properties of a type β transforming growth factor from bovine kidney. Biochemistry 22:5692–5698

Ross R (1987) Platelet-derived growth factor. Annu Rev Med 38:71–79

Ross R, Vogel A, Davies P, Raines E, Kariya B, Rivest M, Gustafson C, Glomset J (1979) The platelet-derived growth factor and plasma control cell proliferation. In: Sato GH, Ross R (eds) Hormones and cell culture. Cold Spring Harbor, New York, pp 5–16

Rubin JS, Bradshaw RA (1984) Preparation of guinea pig prostate epidermal growth factor. In: Barnes DW, Sirbasku DA, Sato GH (eds) Methods for preparation of media, supplements and substrata for serum free animal cell culture, vol 1. Liss, New York, pp 139–145

Rubin JS, Mariz I, Jacobs JW, Daughaday WH, Bradshaw RA (1982) Isolation and partial sequence analysis of rat basic somatomedin. Endocrinology 110:734–740

Ryle AP, Sanger F, Smith LF, Kitai R (1955) The disulphide bonds of insulin. Biochem J 60:541–556

Sanderson CJ, O'Garra A, Warren DJ, Klaus GGB (1986) Eosinophil differentiation factor also has B-cell growth factor activity: proposed name interleukin 4. Proc Natl Acad Sci USA 83:437–440

Sasaki R, Yanagawa S-I, Chiba H (1987) Isolation of human erythropoietin with monoclonal antibodies. Methods Enzymol 147:328–340

Seyedin SM, Thomas TC, Thompson AY, Rosen DM, Piez KA (1985) Purification and characterization of two cartilage-inducing factors from bovine demineralized bone. Proc Natl Acad Sci USA 82:2267–2271

Seyedin SM, Thompson AY, Bentz H, Rosen DM, McPherson JM, Conti A, Siegel NR, Galluppi GR, Piez KA (1986) Cartilage-inducing factor-A: apparent identity to transforming growth factor-β. J Biol Chem 261:5693–5695

Sherwin SA, Twardzik DR, Bohn WH, Cockley KD, Todaro GJ (1983) High molecular weight transforming growth factor activity in the urine of patients with disseminated cancer. Cancer Res 43:403–407

Smith KA, Lachman LB, Oppenheim JJ, Favata MF (1980) The functional relationship of the interleukins. J Exp Med 151:1551–1556

Stanley ER (1985) The macrophage colony-stimulating factor, CSF-1. Methods Enzymol 116:564–587

Stanley ER, Guilbert LJ (1981) Methods for the purification, assay, characterization and target cell binding of a colony stimulating factor (CSF-1). J Immunol Methods 42:253–284

Stanley ER, Heard PM (1977) Factors regulating macrophage production and growth. J Biol Chem 252:4305–4312

Stroobant P, Waterfield MD (1984) Purification and properties of porcine platelet-derived growth factor. EMBO J 12:2963–2967

Takahashi M, Hong Y-M, Yasuda S, Takano M, Kawai K, Nakai S, Hirai Y (1988) Macrophage colony-stimulating factor is produced by human T lymphoblastoid cell line, CEM-ON: identification by amino-terminal amino acid sequence analysis. Biochem Biophys Res Commun 152:1401–1409

Takatsu K, Harada N, Hara Y, Takahama Y, Yamada G, Dobashi K, Hamaoka T (1985) Purification and physicochemical characterization of murine T cell-replacing factor (TRF). J Immunol 134:382–389

Tao X, Gao G, Zhang H-Z, Zhu D-X, Boersma A, Lamblin G, Han K-K (1987) Isolation and characterization of human urinary colony-stimulating factor. Biol Chem Hoppe Seyler 368:187–194

Taylor JM, Cohen S, Mitchell WM (1970) Epidermal growth factor: high and low molecular weight forms. Proc Natl Acad Sci USA 67:164–171

Thomas KA (1988) Transforming potential of fibroblast growth factor genes. Trends Biochem Sci 13:327–328

Thomas KA, Gimenez-Gallego G (1986) Fibroblast growth factors: broad spectrum mitogens with potent angiogenic activity. Trends Biochem Sci 11:81–84

Thomas KA, Riley MC, Lemmon SK, Baglan NC, Bradshaw RA (1980) Brain fibroblast growth factor: non-identity with myelin basic protein fragments. J Biol Chem 255:5517–5520

Todaro GJ, Fryling C, De Larco JE (1980) Transforming growth factors produced by certain human tumor cells: polypeptides that interact with epidermal growth factor receptors. Proc Natl Acad Sci USA 77:5258–5262

Todaro GJ, Lee DC, Webb NR, Rose TM, Brown JP (1985) Rat type-a transforming growth factor: structure and possible function as a membrane receptor. In: Feramisco J, Ozanne B, Stiles C (eds) Cancer cells, growth factors and transformation. Cold Spring Harbor Laboratory, Cold Spring Harbor NY

Twardzik DR (1985) Differential expression of transforming growth factor-α during prenatal development of the mouse. Cancer Res 45:5413–5416

Twardzik DR, Ranchalis JE, Todaro GJ (1982) Mouse embryonic transforming growth factors related to those isolated from tumor cells. Cancer Res 42:590–593

Twardzik DR, Kimball ES, Sherwin SA, Ranchalis JE, Todaro GJ (1985) Comparison of growth factors functionally related to epidermal growth factor in the urine of normal and human tumor-bearing athymic mice. Cancer Res 45:1934–1939

Van Damme J, De Ley M, Opdenakker G, Billiau A, De Somer P (1985) Homogeneous interferon-inducing 22K factor is related to endogenous pyrogen and interleukin-1. Nature 314:266–268

Van Damme J, Van Beumen J, Decock B, Van Snick J, De Ley M, Billiau A (1988) Separation and comparison of two monokines with lymphocyte-activating factor activity: IL-1b and hybridoma growth factor (HGF): identification of leukocyte-derived HGF as IL-6. J Immunol 140:1534–1541

Van Snick J, Cayphas S, Vink A, Uyttenhove C, Coulie PG, Rubira MR, Simpson RJ (1986) Purification and NH2-terminal amino acid sequence of a T-cell derived lymphokine with growth factor activity for B-cell hybridomas. Proc Natl Acad Sci USA 83:9679–9683

Varon S, Nomura J, Shooter EM (1967) The isolation of the mouse nerve growth factor protein in a high molecular weight form. Biochemistry 6:2202–2209

Welte K, Wang CY, Mertelsmann R, Venuta S, Feldman SP, Moore MS (1982) Purification of human interleukin 2 to apparent homogeneity and its molecular heterogeneity. J Exp Med 156:454–464

Westall FC, Lennon VA, Gospodarowicz D (1978) Brain-derived fibroblast growth factor: identity with a fragment of the basic protein of myelin. Proc Natl Acad Sci USA 75:4675–4678

Westermark B, Johnsson A, Paulsson Y, Betsholtz C, Heldin C-H, Herlyn'M, Rodeck U, Koprowski H (1986) Human melanoma cell lines of primary and metastatic origin express the genes encoding the chains of platelet-derived growth factor (PDGF) and produce a PDGF-like growth factor. Proc Natl Acad Sci USA 83:7197–7200

Wilcox JN, Derynck R (1988) Developmental expression of transforming growth factors alpha and beta in mouse fetus. Mol Cell Biol 8:3415–3422

Wong GG, Witek JS, Temple PA, Wilkens KM, Leary AC, Luxenberg DP, Jones SS, Brown EL, Kay RM, Orr EC, Shoemaker C, Golde DW, Kaufman RJ, Hewick RM, Wang EA, Clark SC (1985) Human GM-CSF: molecular cloning of the complementary DNA and purification of the natural and recombinant proteins. Science 228:810–815

Wood DD (1979) Mechanism of action of human B cell-activating factor I. Comparison of the plaque-stimulating activity with thymocyte-stimulating activity. J Immunol 123:2400–2407

Wu M-C, Cini JK, Yunis AA (1979) Purification of a colony-stimulating factor from cultured pancreatic carcinoma cells. J Biol Chem 254:6226–6228

Ziltener HJ, Clark-Lewis I, Hood LE, Kent SBH, Schrader JW (1987) Antipeptide antibodies of predetermined specificity recognize and neutralize the bioactivity of the pan-specific hemopoietin interleukin 3. J Immunol 138:1099–1104

CHAPTER 3

Properties and Regulation of Receptors for Growth Factors

M. P. Czech, K. B. Clairmont, K. A. Yagaloff, and S. Corvera

A. Introduction

It has long been known that the biological actions of peptide growth factors are mediated by cell surface receptors, but only within the last few years has intimate knowledge of these membrane glycoprotein structures become available. This new insight has been derived from isolated cDNA clones encoding numerous growth factor receptors (Morgan et al. 1987; Ullrich et al. 1984, 1985, 1986; Coussens et al. 1985; Bargmann et al. 1986a, b; Yarden et al. 1986, 1987; Sherr et al. 1985; Nikaido et al. 1984; MacDonald et al. 1988), in combination with biochemical and immunological approaches toward receptor characterization. Amino acid sequences of growth factor receptors deduced from cDNA isolates have also provided the framework for the development of many important new reagents for their study, as exemplified by antireceptor peptide antibodies. Futhermore, the capability of expressing native and mutant receptor structures at high levels in cultured cells by transfection with plasmids harboring receptor cDNA inserts has led to the generation of excellent new cellular model systems for future studies. Recent success with the production of milligram quantities of receptor proteins that can be isolated from released baculovirus particles has also been achieved (Herrera et al. 1988). Thus, it is now reasonable to expect that major efforts designed to determine three-dimensional structures of the growth factor receptors are in progress. Such detailed information will most probably be available within the next decade.

The recent explosive increase in structural information about the growth factor receptors has led to at least three general insights about this group of glycoproteins. First, the basic architectural design of the growth factor receptors is remarkably simple and constant. Each growth factor receptor polypeptide that has so far been elucidated contains: (1) a single extracellular region that includes one or more ligand binding domains, (2) a single linear hydrophobic peptide region that makes one pass through the membrane bilayer, and (3) a single linear peptide sequence that resides in the cytoplasmic domain of the cell. This general unitary structure contrasts with a number of hormone receptor systems such as the catecholamine receptors (Kobilka et al. 1987), which contain multiple polypeptide extracellular, transmembrane, and cytoplasmic segments. Multiple interrupted linear sequences in these three cellular regions also characterize membrane transporter proteins (Muekler et al.

1985). The simple design of the growth factor receptor structures must function to catalyze signal transduction across the membrane. How this receptor mechanism is served by the simple architecture of the growth factor receptors is not well understood at the molecular level at present.

A second general insight into the growth factor receptors obtained in the last few years is the concept that these receptors can be subdivided into several groups of structurally or functionally related proteins. Three types of relationships have been identified among the growth factor receptors: (1) Receptor isoforms with very similar primary structures and presumed functions, exemplified by two insulin receptor structures which vary significantly only in a short stretch of amino acids (ULLRICH et al. 1985; EBINA et al. 1985). (2) Receptor families consisting of members with extensive sequence similarity but measurably distinct functional capabilities, exemplified by the epidermal growth factor (EGF) receptor (ULLRICH et al. 1984) and the HER-2/*neu* glycoprotein (COUSSENS et al. 1985; BARGMANN et al. 1986b). These receptors share about 50% sequence identity, contain similar intrinsic tyrosine kinase domains, but exhibit different ligand binding capabilities (EGF does not bind the HER-2/*neu* glycoprotein). (3) Receptor groups consisting of members with divergent protein structures that share the ability to bind similar peptide growth factors, exemplified by the type I and type II insulin-like growth factor (IGF) receptors. These receptors exhibit no significant amino acid sequence similarity, but both bind IGF-II with high affinity. These types of isoform, family, or group relationships are of course not unique to the growth factor receptors, but are also characteristic of many other cellular proteins.

A third concept of growth factors receptor structure relates to evidence that independent receptor polypeptide domains retain their specific functions even when separated from the intact native glycoprotein. For example, serum forms of several receptors have been observed which exhibit high-affinity binding of ligands (BARAN et al. 1988; BEGUIN et al. 1988; GAVIN et al. 1972; WEBER et al. 1984; KIESS et al. 1987a; CAUSIN et al. 1988; MACDONALD et al. 1989; MCGUFFIN et al. 1976; HERINGTON 1985; BAUMANN et al. 1986). These serum species appear to be truncated extracellular domains of cell membrane receptors which retain their ligand binding functions. Genetically engineered truncated forms of the cytoplasmic domains of the insulin receptor autophosphorylate and exhibit tyrosine kinase activity with kinetics similar to the native receptor (HERRERA et al. 1988). Furthermore, linkage of the extracellular domain of the insulin receptor with the cytoplasmic domain of the EGF receptor yields a receptor structure capable of insulin-stimulated tyrosine kinase activity (RIEDEL et al. 1986). A similar chimera between the extracellular domain of the EGF receptor and the cytoplasmic portion of the *neu*/EGF receptor 2 protein exhibits kinase activity (LEE et al. 1989). This remarkable interchangeability of receptor domains with retention of function may be very useful in probing mechanisms of transmembrane signaling.

It is now well established that the growth factor receptors are the targets of numerous cellular control mechanisms designed to modify their activities. The segregation of specific biologic functions with discrete independent receptor segments or domains provides a rationale for investigating the biochemical basis of this regulation. These regulated activities include not only peptide ligand binding and signal transduction, but also ligand uptake into cells, receptor sorting and recycling among membranes, and cleavage of extracellular domains which release receptor fragments into the medium or serum. The aim of this chapter is to review in some detail the various molecular motifs that constitute each domain of the growth factor receptors – extracellular, transmembrane, and cytoplasmic – as a framework for understanding regulatory mechanisms that modulate receptor function. Although at present there remain large gaps in our knowledge of these mechanisms at the molecular level, substantial new information is available. Emphasis will be placed on the molecular elements in each receptor domain that might serve as targets or mediators of regulatory control mechanisms.

B. Extracellular Domain Structures of Growth Factor Receptors

Analysis of the deduced amino acid sequences and the locations of structural elements within the extracellular domains of the growth factor receptors illustrates the existence of receptor families with members encoded by related genes. Figure 1 shows the relative sizes and some important structural features of a number of such growth factor receptor gene products characterized in detail. Three families of related receptor proteins with similar primary structures are identified. These are the insulin/IGF-I, EGF/*neu*, and platelet-derived growth factor/colony-stimulating factor 1 (PDGF/CSF-1) receptor families, each consisting of member receptors exhibiting significant size and sequence similarities.

The insulin receptor family as we know it at present includes two nearly identical insulin receptors (ULLRICH et al. 1985; EBINA et al. 1987) and one IGF-I receptor (ULLRICH et al. 1986). These receptors are characterized by a heterotetrameric structure consisting of two α- and two β-subunits (MASSAGUÉ and CZECH 1980, 1982) derived from precursor polypeptides (shown in Fig. 1) containing both subunits (RONNET et al. 1984). The degree of overall sequence identity between the extracellular domains of the insulin and IGF-I receptors is about 50%. This receptor class is characterized by a single similarly located cysteine-rich region (residues 155–333 and 148–323 in the insulin and IGF I receptors, respectively). Outside this cysteine-rich region the two receptors structures are also very similar in respect to the positions of the cysteines, glycosylation sites, and α-β subunit cleavage site (Fig. 1). This general structure for the insulin/IGF-I receptors contrasts with that of the EGF receptor family which consists of the EGF receptor (ULLRICH et al. 1985) and the HER-2/*neu* protein (BARGMANN et al. 1986b). This latter receptor family is characterized by the presence of two separate cysteine-rich regions in the extracellular domain (residues 187–337 and 499–636 in the EGF receptor). Sig-

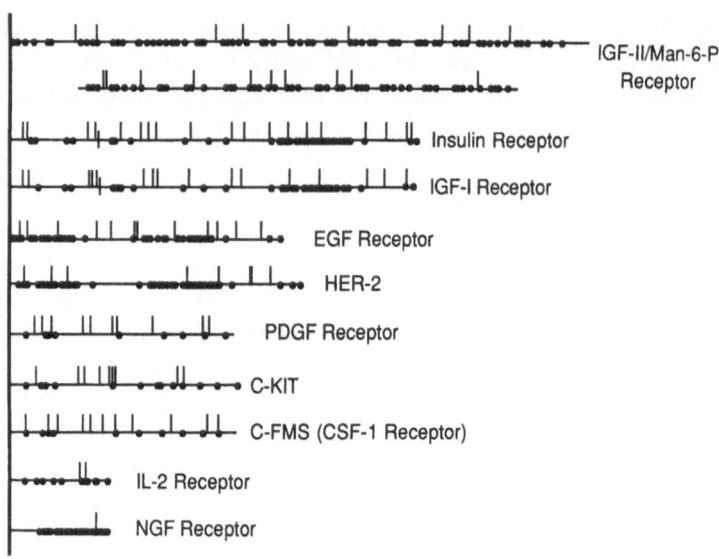

Fig. 1. Schematic representation of the structure of growth factor extracellular domains showing relative mass and the location of cysteine residues (—•—), glycosylation sites (—⊥—), and α–β subunit cleavage sites (—+—). The extracellular domains of the IGF-II/Man-6-P receptor are shown on two lines. The *bold line* to the *left* of the figure represents the putative location of the cellular membrane. The primary structure data used to generate the figure are as follows: human IGF-II/Man-6-P receptor (MORGAN et al.(1987), human insulin receptor (ULLRICH et al. 1985), human IGF-I receptor (ULLRICH et al. 1986), human EGF receptor (ULLRICH et al. 1984), rat HER-2/*neu* (COUSSENS et al. 1985; BARGMANN et al. 1986b), mouse PDGF receptor (YARDEN et al. 1986), human c-*kit* (YARDEN et al. 1987), human CSF-1 receptor (c-*fms*) (SHERR et al. 1985; LEONARD et al. 1984), human IL-2 receptor (NIKAIDO et al. 1984), and the human NGF receptor (JOHNSON et al. 1986)

nificant structural similarities in the locations of the remaining cysteine residues and most of the putative N-linked glycosylation sites are also apparent in comparing the two receptors in this family.

A third receptor family differs in general structure from the first two in that it lacks cysteine-rich regions. Instead, it is characterized by a conserved arrangement of dispersed cysteines. Included in this family are the PDGF receptor (YARDEN et al. 1986), the CSF-1 receptor, also referred to as the c-*fms* gene product (SHERR et al. 1985), and the c-*kit* product (YARDEN et al. 1987). Of these, only c-*kit* lacks an identified ligand. The locations of several putative N-linked glycosylation sites are generally conserved in members of this family.

Three growth factor receptors shown in Fig.1 fail to fit into the above known growth factor receptor families: they are the IGF-II/mannose-6-phosphate (Man-6-P) receptor (MORGAN et al. 1987; MACDONALD et al. 1988), the interleukin (IL) 2 receptor (LEONARD et al. 1984; NIKAIDO et al. 1984), and the nerve growth factor receptor (JOHNSON et al. 1986; RADEKE et al. 1987). Based on the common ability to bind IGFs, the IGF-II receptor might be expected to be structurally similar to the insulin and IGF-I receptors.

However, in contrast to the heterotetrameric structures of the insulin and IGF-I receptors, the type II IGF receptor appears to exist as a single, large glycoprotein with almost all of its mass in the extracellular domain. The rat (MacDonald et al. 1988) and human (Morgan et al. 1987) IGF-II receptors were shown to have high sequence identity to the bovine Man-6-P receptor (Lobel et al. 1988). That the same receptor protein binds both ligands was demonstrated by affinity purification on pentamannosyl-6-phosphate-(PMP)-Sepharose, followed by affinity labeling with ^{125}I-IGF-II (MacDonald et al. 1988). This receptor may differ from other growth factor receptors illustrated in Fig. 1 with respect to signaling function. We found that IGF-II elicits a biological response in H-35 cells even after blockade of its receptor with antireceptor immunoglobulin (Mottola and Czech 1984), a result confirmed by others in L6 cells (Kiess et al. 1987b). Although some reports have suggested selective signaling by this receptor in certain cell types (Nishomoro et al. 1987; Hari et al. 1987), the weight of the published data seems to indicate that metabolic and growth effects of IGF-II are mediated through IGF-I or insulin receptors, not IGF-II/Man-6-P receptors. The only known receptor with functional and structural properties similar to the IGF-II/Man-6-P receptor is a much smaller, cation-dependent Man-6-P receptor (Pohlmann et al. 1987). The extracellular domain of the latter protein is similar in general structure to each of 15 repeating domains of the IGF-II/Man-6-P receptor which contain multiple, similarly spaced cysteines. However, the cation-dependent Man-6-P receptor does not bind IGF-II with detectable affinity.

The extracellular domain of the nerve growth factor receptor contains a single cysteine-rich region and one glycosylation site. It is clearly distinct from the PDGF receptor family, but might be related to the EGF and insulin receptor families based on the similar size of their cysteine-rich domains. The IL-2 receptor lacks the cysteine-rich regions of the EGF, insulin, and NGF receptors, but has a greater cysteine composition than the PDGF receptor. The IL-2 receptor depicted here is an 55-kDa subunit denoted as the Tac antigen, which binds IL-2 with low affinity. There is also a recently discovered 75-kDa subunit which binds ligand with intermediate affinity (Teshigawara et al. 1987). The heterodimer of an α- and β-subunit may compose the high-affinity IL-2 receptor. Thus the IL-2 receptor appears distinct and unique in structure from the other receptors so far characterized.

C. Ligand Binding Regions of Extracellular Domains

The extracellular domain of a growth factor receptor must carry out at least two functions. The first is to bind the appropriate growth factor with high affinity, and the second is to communicate across the transmembrane domain the signal that ligand has been bound. Despite the cloning and sequencing of numerous receptors during the last decade, little direct evidence has become available addressing these two issues at the molecular level. In the past year, a number of different approaches have provided the first insight into the ligand binding regions of growth factor receptor extracellular domains. Site-directed mutagenesis is proving to be a valuable technique in such studies. For example, the ligand affinity of expressed native and mutant IL-2 receptor constructs has been evaluated (Rusk et al. 1988; Robb et al. 1988). Using this approach Rusk et al. (1988) found that all cysteine residues except those at amino acids 192 and 225 were important for binding capability. These workers also identified the disulfide bonds in this receptor as linking cysteines 3 and 147, 131 and 163. Cysteines 28 and 30 are linked to 59 and 61, but exact pairing was not determined. Thus disulfide bonding appears critical for ligand binding in this receptor. Furthermore, a deletion mutant ending beyond cysteine 163 bound ligand normally. Further mutants revealed that regions containing residues 1–6 and 35–43 are essential for detectable ligand binding, while residues 158–160 are required only for high-affinity binding (Robb et al. 1988). Possibly, the regions consisting of amino acids 1–6 and 35–43 constitute the binding site while 158–160 are important in binding of the α- and β-subunits.

Recent studies on the EGF receptor indicate the ligand binding domain is not located within the cysteine-rich segments of the receptor structure. Schlessinger and coworkers (Lax et al. 1988) obtained compelling evidence that binding occurs at a site between the cysteine-rich domains by analyzing tryptic peptides derived from EGF receptor covalently crosslinked to ^{125}I-EGF. These results are reinforced by experiments from the same laboratory in which receptor domains from the chicken versus human EGF receptor were switched by recombinant DNA techniques (Lax et al. 1989). The chicken EGF receptor binds EGF with very low affinity but binds transforming growth factor-α (TGF-α) with high affinity. These characteristics could be conferred to the human EGF receptor structure by substituting the chicken EGF receptor segment between the two cysteine-rich domains. Whether the cysteine-rich domains are required to flank the binding region or whether the latter peptide region could bind ligand independently is an interesting question for further investigation.

Another system where a receptor binds a ligand with high affinity in one species but not in another is the IGF-II/Man-6-P receptor. While in mammals it has been demonstrated that a single 205-kDA monomeric receptor is capable of binding both IGF-II and Man-6-P-containing ligands with high affinity, in chickens the predominant receptor for IGF-II is a heterotetrameric IGF-I-type receptor. We have recently investigated the presence of Man-6-P receptors in chickens and frogs and compared their ligand binding properties to the IGF-II/Man-6-P receptor found in rats. Receptors were affinity purified on PMP-Sepharose by incubating detergent extracts from rat placenta, chicken liver, and frog liver overnight at 4° C with PMP-Sepharose. Fractions were then eluted from the resin with buffer alone, buffer with 5 mM Man-1-P, buffer with 5 mM glucose-6-phosphate, and finally buffer with 5 mM Man-6-P. One portion of each fraction was electrophoresed by sodium dodecylsulface polyacrylamide get electrophoresis (SDS-PAGE), then silver stained. A second fraction was incubated with 2 nM ^{125}I-IGF-II,

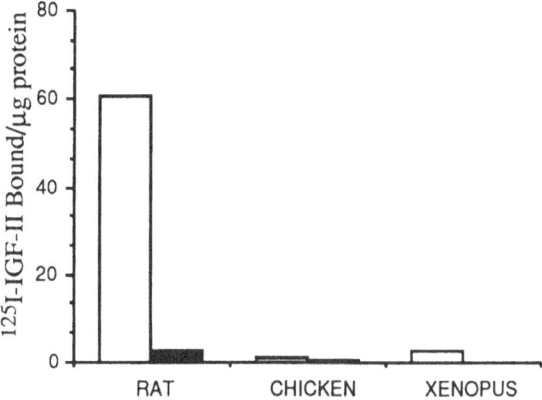

Fig. 2. Rat[125]I- IGF-II was added to purified Man-6-P receptors isolated from rat placenta, chicken liver, and frog liver membranes. Binding of the ligand was measured in the presence (*open bars*) or absence of 500 nM unlabeled IGF-II (*shaded bars*) using the polyethylene glycol precipitation method. Similar results were obtained when the labeled ligand was chicken [125]I- IGF-II, which was a gift from Dr. F.J. Ballard (CSIRO)

crosslinked with disuccinimidyl suberate, and run on SDS-PAGE. In the silver stain, a 250-kDa band was present in the Man-6-P-eluted fraction from each species, but only the receptor purified from rat could be affinity labeled by ^{125}I-IGF-II (K.B. CLAIRMONT and M.P. CZECH, submitted for publication).

To further investigate the interaction of IGF-II with non-mammalian Man-6-P receptors, direct binding studies were conducted on the purified receptor preparations using the polyethylene glycol precipitation method. As shown in Fig. 2, only the rat receptor showed detectable affinity for rat or chicken ^{125}I-IGF-II. Therefore, the Man-6-P receptor has apparently gained the capacity to bind IGF-II with high affinity relatively recently in evolution, following the divergence of mammals and birds. This implies further that in chickens, a species known to possess IGF-II, this peptide mediates biological responses through another receptor, presumably the type I IGF receptor structure. Experiments designed to switch extracellular IGF-II/Man-6-P receptor domain segments between species should be useful in determining the IGF-II binding domain of the mammalian IGF-II/Man-6-P receptor.

D. Extracellular Receptor Domains as Serum Receptors

The presence of growth factor receptors in serum is a phenomenon that has been observed by many groups, and may be a general occurrence. Soluble receptor forms in the circulation have been observed for the IL-2 (HARRINGTON et al. 1988; BARAN et al. 1988), transferrin (BEGUIN et al. 1988), EGF (WEBER et al. 1984), and IGF-II/Man-6-P (KIESS et al. 1987a; CAUSIN et al. 1988; MACDONALD et al. 1989) receptors. In addition, the insulin (GAVIN et

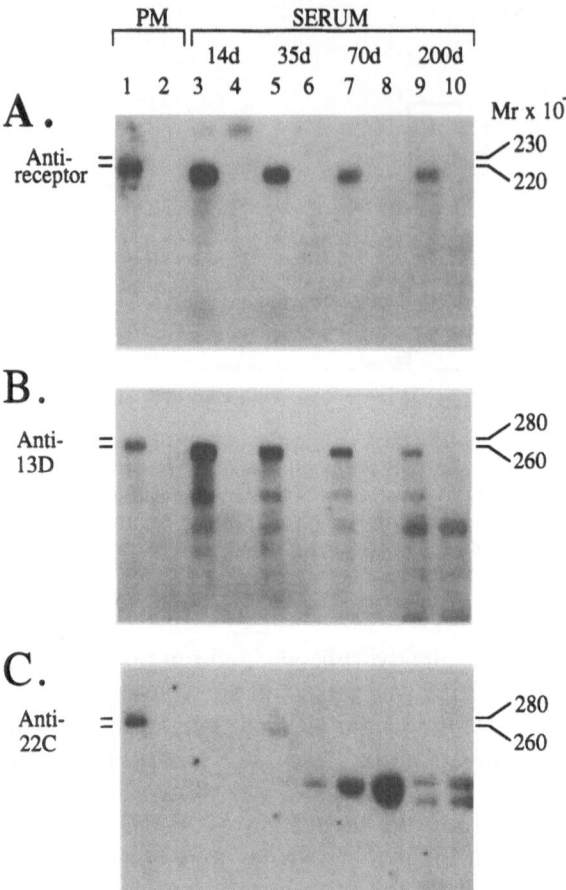

Fig. 3 A–C. Characterization of cellular and serum forms of the IGF-II/Man-6-P receptor through the use of antipeptide antibodies specific for the extracellular (anti-13D) or cytoplasmic (anti-22C) domains of the receptor. A Triton X-100 extract of rat placental plasma membranes (day 25 of gestation, 0.4 mg, lanes *1* and *2*) or 1 ml serum isolated from rats aged 14 days (lanes *3* and *4*), 35 days (lanes *5* and *6*), 70 days (lanes *7* and *8*) or 200 days (lanes *9* and *10*) were immunoadsorbed with antireceptor immunoglobulin Affigel (*odd numbered lanes*), or nonimmune immunoglobulin Affigel (*even numbered lanes*). Prior to electrophoresis, washed resin pellets were heated at 100° C in the absence of dithiothreitol (DTT) (**A**) or reduced with DTT and alkylated with iodoacetic acid (**B, C**). The immunoblots were then probed with polyclonal antireceptor immunoglobulin (**A**), antiextracellular domain peptide (**B**), or anticytoplasmic domain peptide (**C**). From the immunoblots shown, several conclusions can be reached. First, the serum forms of the receptor do not contain the epitope recognized by the anticytoplasmic peptide antibody. Second, the concentration of receptor in serum is developmentally regulated such that it decreases with age beyond 14 days post partum. Third, while the receptor is present as only one species when not reduced, a number of lower molecular weight (*Mr*) species are produced upon reductive alkylation. Several of these lower molecular weight species are capable of binding ligand

al. 1972) and growth hormone (McGUFFIN et al. 1976; HERINGTON 1985; BAUMANN et al. 1986) receptors have been observed to be secreted from cells in culture. It is possible that soluble receptor forms act as serum carrier or binding proteins for their respective ligands, although the serum receptor levels may not be high enough to contribute significantly to such activity under most conditions. The levels of serum IL-2 receptor have been shown to be elevated in patients with active malignant lymphomas (HARRINGTON et al. 1988). What biological purpose receptors in serum serve, if any, is not known. In the case of the IGF-II/Man-6-P receptor, an interesting hypothesis has been suggested. This receptor participates in a membrane recycling pathway that involves Golgi, endosomal, and plasma membrane compartments. Since the IGF-II/Man-6-P receptor may not enter the lysosome normally, it has been suggested that it might be cleaved from the cell surface by a proteolytic enzyme, released into serum, and then taken up by another receptor for degradation (CAUSIN et al. 1988; MACDONALD et al. 1989). The presence of receptor fragments in urine is consistent with the concept that the receptor is taken into the kidneys for degradation (CAUSIN et al. 1988).

The serum receptors are known to bind ligands with high affinity. Studies in our laboratory were conducted to test the hypothesis that the soluble IGF-II/Man-6-P receptor exists as the extracellular domain of truncated cell membrane receptors. Antipeptide antibodies were prepared in rabbits against deduced peptide sequences in the extracellular (anti-13D) and cytoplasmic (anti-22C) domains of the rat IGF-II/Man-6-P receptor. Serum IGF-II/Man-6-P receptors retained the ability to be recognized by antibodies to the extracellular domain peptide, but were not recognized by antibody to the cytoplasmic domain peptide (Fig. 3). These data are consistent with the notion that this serum receptor is identical to the extracellular portion of the surface receptor, but devoid of the cytoplasmic domain. It seems likely that the serum IGF-II/Man-6-P receptor also lacks the transmembrane segment, but this is not proven. The results suggest that the serum IGF-II/Man-6-P receptor is a cleavage product of a cell membrane receptor precursor, but altered mRNA splicing cannot be eliminated as a possible mechanism of truncated receptor production.

E. Multiple Ligand Binding Capabilities of Growth Factor Receptors

An already large and growing number of growth factor receptor structures are related to each other based on overlapping binding affinities for the same growth factor peptide ligands. As outlined in Table 1, three insulin/IGF receptors, two PDGF receptors, and three TGF-β receptors are members of such groups. Four receptors among those listed in Table 1 bind two or more peptide ligands with equal or near equal affinity: IGF-I receptor (IGF-I and IGF-II), EGF receptor (EGF and TGF-α), B-type PDGF receptor (PDGF-AB and -BB), and type 3 TGF-β receptor (TGF-β1 and TGF-β2). In each of

Table 1. Multiple ligand binding capabilities of growth factor receptors

Group	Type	Common designation	Tyrosine kinase	Relative affinity for ligands
Insulin-like growth factor receptors	I	Insulin receptor	Yes	Insulin $\gg$ IGF-II $>$ IGF-I
	I	IGF-I receptor	Yes	IGF-I $\geq$ IGF-II $\gg$ insulin
	II	IGF-II/Man-6-P receptor	No	IGF-II $\gg$ IGF-I
EGF receptor	1	EGF receptor	Yes	EGF = TGF-α
	2	*neu*/EGF receptor	Yes	?
PDGF receptors	A	PDGF receptor	Yes	PDGF-AB $>$ PDGF-AA $>$ PDGF-BB
	B	PDGF receptor	Yes	PDGF-AA = PDGF-BB
TGF-β receptors	1	TGF-β receptor	?	TGF-β1 $>$ TGF-β1.2 $>$ TGF-β2
	2	TGF-β receptor	?	TGF-β1 $>$ TGF-β1.2 $>$ TGF-β2
	3	TGF-β receptor	?	TGF-β1 = TGF-β1.2 = TGF-β2
	4	TGF-β receptor	?	TGF-β1 $>$ TGF-β2

these cases, biological signaling occurs in response to all of the high-affinity ligands at low concentrations. Peptides that bind the growth factor receptors with low affinities also trigger signal transduction at higher concentrations where receptor binding occurs. Thus, signaling occurs at similar fractional occupancies of the receptors by ligands.

It is also apparent from Table 1 that several growth factors bind multiple receptor structures. For example, the heterodimeric form of PDGF, PDGF-AB, binds to two distinct PDGF receptor glycoproteins (Escobedo et al. 1988; Heldin et al. 1988). Similarly, TGF-β1 binds specifically to four distinct receptor proteins that can be identified by affinity crosslinking (Cheifetz et al. 1988a, b). It is also common for a peptide growth factor to exhibit markedly divergent affinities for the various receptors they bind. For example, insulin binds to the insulin receptor with high affinity $k_d = 10^{-9}$ M), while its affinity for the type I IGF-I receptor is two orders of magnitude lower. Insulin has no detectable affinity for the IGF-II/Man-6-P receptor. Although the detailed implications of this phenomenon may not be fully appreciated, one role it may play is the segregation of similar biological response pathways among different cell types which respond to different ligands. This is illustrated by insulin and IGF-I which bind with high affinity to their own type I receptors, but with low affinity to the heterologous type I receptors. A number of cell types express only one of these two receptors, or high concentrations of one with very low concentrations of the other (MASSAGUÉ and CZECH 1982). The type I receptors exhibit very similar tyrosine kinase domains and initiate identical or nearly identical biological signaling pathways (YU et al. 1986), leading to modulation of the same target transport and enzyme systems. Thus, certain cells, e.g., Sertoli cells and fibroblasts, are programmed to respond to physiological concentrations of IGF-I while others, e.g., rat liver and fat cells, are programmed to respond to low levels of insulin.

Among the systems described in detail to date, none of the peptide growth factors or their known synthetic analogs act as antagonists of biological signaling upon binding to their receptors. It may be possible that such antagonists can be prepared synthetically by altering amino acid sequences within the native peptide ligands. Such antagonists might be expected to have antimitogenic effects by virtue of their ability to inhibit the signaling of one or more growth factor receptors. It may also be possible to synthesize peptide agonists or antagonists that display altered selectivity for one or more receptor structures within a group.

F. Regulation of Cell Surface Ligand Binding

Regulation of binding sites on growth factor receptor occurs by numerous mechanisms which change the number or apparent affinity of receptors on the cell surface. These include increased or decreased synthesis of receptors, changes in the relative distribution of receptors on the cell surface versus internal membranes, and possible receptor conformation changes that alter ligand affinity. Such changes are often associated with the binding of ligand itself to a given receptor. This phenomenon, known as downregulation, usually leads to decreased cellular receptor numbers. A mechanism most often associated with ligand-mediated downregulation appears to be increased rate of receptor degradation (GREEN and OLEFSKY 1982). Insulin binding to its receptor also appears to exhibit negative cooperativity (DeMEYTS et al. 1976). The cellular basis of downregulation of the individual growth factor receptors by their homologous ligand will be treated within several of the individual chapters that follow.

Regulation of growth factor receptor binding domains is also mediated by heterologous ligands acting through their own distinct receptor systems. A major mode by which ligand binding sites on growth factor receptors are regulated by such receptor crosstalk or transmodulation involves changes in distribution between the cell surface membrane and intracellular compartments. Acute modulation of this type appears to involve changes in rates of receptor recycling between these membrane systems. An example is the modulation of IGF-II/Man-6-P receptors mediated by the insulin receptor. Work in our laboratory (OPPENHEIMER et al. 1983; OKA et al. 1984; OKA and CZECH 1986) and others (Wardzala et al. 1984) has revealed a marked increase in the number of adipocyte and H-35 cell plasma membrane IGF-II/Man-6-P receptors at the expense of receptors in intracellular pools in response to insulin action.

This action of insulin to cause redistribution of IGF-II/Man-6-P receptors to the cell surface membrane reflects a general paradigm in cell biology. Experiments in our laboratory have shown at least three other receptor proteins undergo similar membrane redistribution reactions in response to insulin. These include the transferrin receptor (Davis and Czech 1986; Davis et al. 1986a), the low-density lipoprotein receptor, and the α_2-macroglobulin receptor. In each case, an increased number of these receptor proteins can be directly monitored on the surface of intact target cells upon addition of the hormone. Furthermore, other growth factor hormones that activate specific receptor kinases also appear to modulate the membrane distribution of these proteins. For example, we found that addition of PDGF, EGF, IGF-I, or insulin to A431 cells in culture caused a rapid expression of transferrin receptors on the cell surface as monitored by specific antireceptor antibody (Davis and Czech 1986). The effect of these growth factors on the number of transferrin receptors expressed on the cell surface exhibits a similar rapid time course to that of insulin action on the IGF-II/Man-6-P receptor. All of the membrane receptors modulated by insulin in this manner are known to recycle constitutively between the plasma membrane and endosomal membranes. Thus it would appear that insulin and other growth factors modulate one or more steps in the recycling process. Importantly, we have also been able to demonstrate that membrane redistribution of IGF-II/Man-6-P and transferrin receptors is associated with increased uptake of their respective ligands. Thus, the receptors for IGF-II and transferrin appear to be regulated by insulin and the other growth factors for the purpose of enhancing cellular uptake of these ligands.

We have studied the regulation of transferrin receptor and the type II IGF receptor systems in order to gain more insight into the steps of the recycling pathway affected by insulin. The transferrin receptor system offers an advantage in evaluating the recycling process in that transferrin remains bound to its receptor throughout the entire process of internalization and recycling back to plasma membrane. Thus, reliable estimates of receptor internalization (endocytosis) and exocytosis rates can be obtained by measuring labeled transferrin uptake and release under certain carefully controlled experimental conditions. We have shown that IGF-I-stimulated membrane redistribution of transferrin receptors results from both an increased receptor exocytosis rate and a decreased receptor endocytosis rate when normalized per cell surface number of receptors (Davis et al. 1987). Thus, while the absolute number of transferrin receptors internalized per unit time increased somewhat in response to IGF-I, the rate of receptor internalization normalized to the increased number of available cell surface receptors is lower after hormone treatment (Davis et al. 1987). Taken together, these experiments indicate that control by insulin on transferrin receptor recycling occurs such that both the endocytotic and exocytotic pathways are affected in opposite directions.

G. Transmembrane Domains

A common characteristic of all growth factor receptors is the presence of a unique sequence of 23–25 hydrophobic amino acids which potentially constitutes a membrane spanning region. Analysis of the amino acid composition of the transmembrane domains of 15 growth factor receptors indicates a prevalence of five hydrophobic amino acids: leucine (31%), valine (19%), isoleucine (11%), alanine (10%), and glycine (9%). Locations of sequence identities among these receptor transmembrane segments are boxed in Fig. 4. The transmembrane domains of growth factor receptors may have several possible functions. Two obvious functions are to link the extracellular and intracellular domains of receptors, and to restrict their location to cellular membranes. This region may also play an active role in the transmission of the ligand binding signal to the intracellular domain. Such a role could involve structural features which facilitate movements of receptors within the membrane or conformational changes in the receptor proteins. Alternatively, the transmembrane domain may play a passive role in receptor signal transduction, merely anchoring the receptor to the membrane.

If the transmembrane domains of growth factor receptors were to play an active role in signal transduction, they might be expected to possess a precise

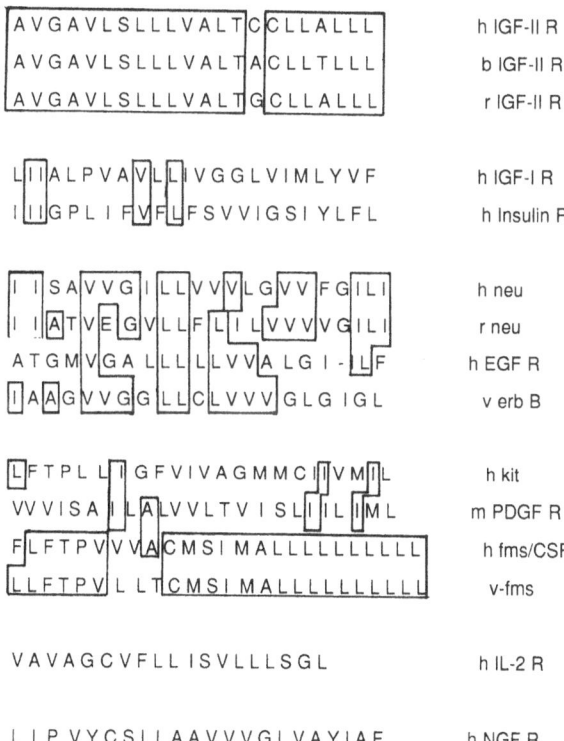

Fig. 4. Amino acid sequences of putative transmembrane domains of 15 growth factor receptor structures. The *boxes* denote sequence identities among receptor groups

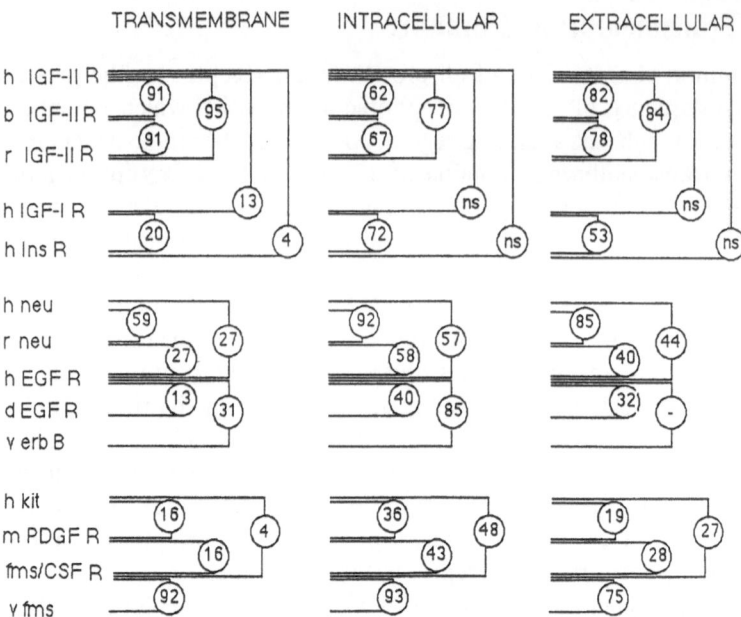

Fig. 5. Relative sequence identities among growth factor receptor extracellular, transmembrane, and intracellular domains. *Numbers* within the circles are the percentage
sequence identities between two receptors domains compared. The receptors compared
are denoted by *brackets*

secondary structure determined by a specific sequence of amino acids to fulfill
this function. The analysis of sequence identities among intracellular, transmembrane, and extracellular domains of species variants of the same receptor,
or among receptors belonging to the same family, is shown in Fig. 5. It can be
seen that among receptors belonging to the insulin/IGF-I, EGF/*neu*/*erb* B,
and *kit*/PDGF/*fms* families, the percentage of identical residues in the transmembrane domains is in general significantly lower than the intracellular or
extracellular domains. For example, the comparison between the human and
rat variants of the *neu* protooncogene reveals a 92% and 85% sequence
identity for the intracellular and extracellular domains respectively, but only a
59% sequency identity in the transmembrane region. An exception is found in
the comparison between the human *fms*/CSF-1 receptor and the feline *fms* sequence, where the percentage identity is lowest in the extracellular domain.

The lack of as much sequence identity in the transmembrane domain as in
the extracellular and intracellular regions of most growth factor receptors suggests that a precise sequence in this region is not required for receptor function. However, an active role for the transmembrane domain of at least some
growth factor receptors was suggested by the finding that a single point mutation in the transmembrane domain was sufficient to confer transforming
potential to the *neu* oncogene (BARGMANN et al. 1986a). It was hypothesized
that this mutation might stimulate the tyrosine kinase activity of the cytoplas-

mic domain, producing continuous receptor activation in the absence of ligand binding. However, similar mutations in the EGF receptor did not give rise to transforming capability nor did they modulate the ability of EGF to activate the receptor tyrosine kinase (KASHLER et al. 1988). Thus, these data indicate that in the case of the EGF receptor, the transmembrane domain may play a minor role in the transmission of information between the intracellular and extracellular domains of the receptor.

Interestingly, the analysis of species variants of the IGF-II/Man-6-P receptor reveals a much higher degree of conservation in the sequence of the transmembrane domain than in the sequences of both the intracellular and extracellular domains (Fig. 5). This analysis suggests the transmembrane segment may be an important element of this receptor's properties. As mentioned previously, this receptor may not be capable of a transmembrane signaling function. Its primary role appears to be internalization and intracellular transport of proteins containing the Man-6-P recognition marker. Whether or not the transmembrane domain of recycling receptors plays on active role in their function is not known.

H. Cytoplasmic Tyrosine Kinase Domain Structures

The receptor cytoplasmic domains play an essential role in propagating the ligand binding event to cellular response pathways. While the mechanism whereby signal propagation occurs is poorly understood, the activation of intrinsic receptor tyrosine kinases may represent the first cellular response to cell surface binding of these growth factors (YARDEN and ULLRICH 1988). The cytoplasmic and tyrosine kinase domains of nine growth factor receptors are illustrated in Fig. 6. Generally, the tyrosine kinase region is contained within a single linear sequence. However, a subclass of growth factor receptors, which includes the PDGF receptor, the c-*fms*/CSF-1 receptor, and the protooncogene c-*kit*, contains a stretch of approximately 70–100 amino acids bisecting the conserved tyrosine kinase domain (COUSSENS et al. 1985; YARDEN et al. 1986, 1987). This conformation may play a role in determining the substrate specificity of these receptor tyrosine kinases, or in the regulation of their catalytic activity (YARDEN and ULLRICH 1988).

Extensive amino acid sequence similarities are apparent among the cytoplasmic domains of the growth factor receptors in the regions which contain the tyrosine kinase activity (thick lines in Fig. 6). These regions begin at a putative ATP binding site which contains a similarly positioned conserved lysine residue (denoted by asterisks in Fig. 6), localized 40–60 amino acids from the transmembrane domain. The sequence similarities continue for approximately 225 amino acids toward the carboxy terminus.

The mechanism whereby the tyrosine kinase activity of the growth factor receptors operates to elicit specific cellular responses is not known. One possibility is that the enzyme catalyzes the phosphorylation of exogenous substrates, and thereby modulates their activity. An alternative possibility is that

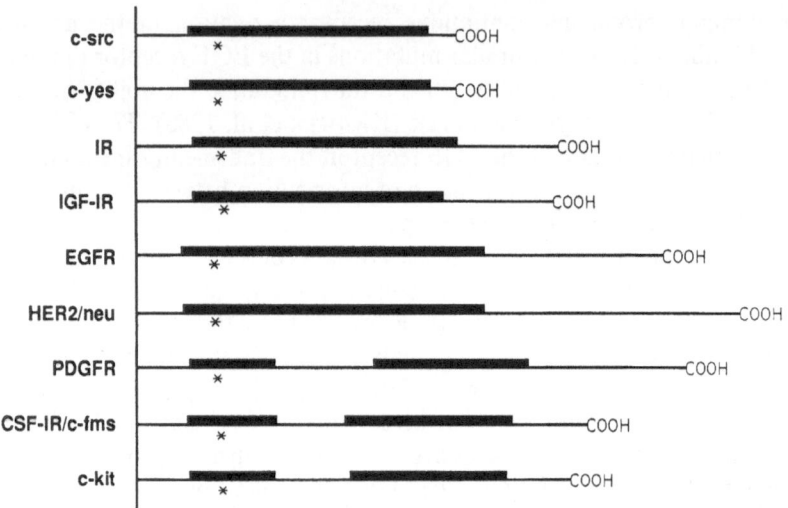

Fig. 6. Comparison of the cytoplasmic domain of tyrosine protein kinases. Relative lengths of the cytoplasmic domains of various tyrosine kinases are shown. The highly related catalytic domain are shown by *thick lines*, while asterisks identify the position of the conserved lysine residue in the putative ATP binding sites. Human sequences were used for this comparison, except for the PDGF receptor sequence which is from mouse

the primary role of the enzyme is to catalyze autophosphorylation of the receptor, thereby inducing a conformation in which the receptor can interact with effector molecules. In either case, the tyrosine kinase activity would be crucial for signal transduction. That this is the case has been supported by site-directed mutagenesis of the ATP binding sites of several receptor tyrosine kinases. Mutation of lysine 1030 to alanine in the insulin receptor resulted in the inhibition of receptor tyrosine kinase activity and the inability of trans-fected cells to respond to insulin (EBINA et al. 1987; CHOU et al. 1987). Another approach has been the microinjection of site specific monoclonal antibodies. Microinjection of antiphosphotyrosine antibodies also decreased insulin responsiveness in *Xenopus* oocytes (MORGAN and ROTH 1987). Similarly, mutation of lysine 721 in the EGF receptor ATP binding site resulted in its failure to mediate EGF-stimulated phosphatidyl inositol turnover, Ca^{2+} in-flux, Na^+/H^+ exchange, and DNA synthesis in NIH 3T3 cells (MOOLENAAR et al. 1988). In addition, this lysine mutant failed to exhibit EGF-stimulated tyrosine kinase activity and receptor autophosphorylation. These experiments have provided strong evidence for the importance of the tyrosine kinase ac-tivity in receptor function.

I. Regulation of Receptor Tyrosine Kinases

The molecular mechanism whereby the interaction of ligand with a receptor extracellular domain results in changes in the tyrosine kinase activity of the cytoplasmic domain are not completely understood. In the case of the EGF receptor, this activation process may involve receptor oligomerization and transmodulation. EGF receptor dimerization occurs in response to EGF binding and correlates with the activation of its tyrosine kinase (YARDEN and ULLRICH 1988; YARDEN and SCHLESSINGER 1985; HONEGGER et al. 1989). This dimerization is accompanied by a shift from low-affinity $(10^{-8}M)$ to high-affinity $(10^{-9}M)$ EGF binding. It has been suggested that this affinity change reflects the stabilization of EGF receptor dimers and the subsequent activation of the receptor tyrosine kinase (YARDEN and SCHLESSINGER 1985; SCHLESSINGER 1988). The importance of oligomerization in signal transduction through the EGF receptor is supported by the finding that receptor immobilization on a resin prevents oligomerization as well as kinase activation (YARDEN and SCHLESSINGER 1985). In addition, it has been observed that treatment of cells with phorbol esters or PDGF results in the inhibition of high-affinity EGF binding, and in the inhibition of EGF-stimulated receptor tyrosine kinase activity (SCHLESSINGER 1988; COUNTAWAY et al. 1989).

The activity of the receptor tyrosine kinases can be modulated by the phosphorylation of specific sites in the cytoplasmic domain. For example, in the case of the insulin and IGF-I receptors, tyrosine phosphorylation leads to marked stimulation of the cytoplasmic tyrosine kinase activity. Tyrosine phosphorylation of these receptors, and consequently their tyrosine kinase activity, may be mediated by an intermolecular reaction or by intramolecular phosphorylation. An example of the later is the phosphorylation and activation of the insulin receptor by the *src* kinase (YU et al. 1985). The importance of tyrosine autophosphorylation in the regulation of receptor kinase activity and function is demonstrated by experiments in which mutagenesis of insulin receptor tyrosines 1150 and 1151 to phenylalanines resulted in inhibition of both tyrosine kinase activity and insulin-stimulated glucose uptake (ULLRICH et al. 1985; ELLIS et al. 1986).

The correlation between tyrosine autophosphorylation and increased receptor kinase activity does not appear to be universal. In the case of the EGF receptor, the effect of autophosphorylation on receptor kinase activity has been controversial. DOWNWARD et al. (1985) found no effect, while BERTICS and GILL (1985) reported an apparent activation of receptor kinase activity by autophosphorylation. Recently, HONEGGER et al. (1988 a, b), using site-directed mutagenesis of EGF receptor tyrosines 1068, 1148, and 1173, observed only minor effects of these mutations on the catalytic activity of the receptor. These data suggest a lack of involvement of autophosphorylation in EGF receptor kinase regulation, and point out differences between the mechanism of activation of EGF and insulin receptors. While hormone binding to the extracellular domain activates the cytoplasmic tyrosine kinase of both receptors, EGF receptor kinase enhancement may solely be an indirect effect of ligand-induced receptor dimerization. The insulin receptor is already dimeric with respect to the tyrosine kinase domain and appears to require this dimeric structure for kinase activation (BOENI-SCHNETZLER et al. 1988). However, tyrosine phosphorylation also plays a major role in insulin receptor kinase activation. An important aspect of insulin and IGF-I receptor activation through tyrosine autophosphorylation is that kinase activation is retained even after dissociation of ligand from the receptor. Thus, a critical step in the deactivation of some growth factor signals is the dephosphorylation of the cytoplasmic domain of their receptors.

Receptor tyrosine kinase activity can also be modulated by serine/ threonine phosphorylation. In contrast to tyrosine phosphorylation, serine/threonine phosphorylation appears to inhibit receptor tyrosine kinase activity. For example, treatment of intact cells with phorbol esters leads to the phosphorylation of multiple serine and threonine sites on the EGF receptor. This phosphorylation correlates with the inhibition of ligand-stimulated receptor tyrosine phosphorylation (DAVIS and CZECH 1985; IWASHITA and FOX 1984; HUNTER et al. 1984; WHITELY and GLASER 1986; COCHET et al. 1984), and with the conversion of high-affinity EGF receptors to a low-affinity state (SCHLESSINGER 1988; DAVIS 1988). The major protein kinase C substrate site of the EGF receptor is threonine 654, which is located 9 amino acids from the cytoplasmic side of the transmembrane domain. Other sites phosphorylated in response to phorbol esters appear to be substrates of other kinases. Similar serine/threonine phosphorylations of the EGF receptor occur in response to treatment of ^{32}P-labeled cells with PDGF, which also causes decreases in EGF receptor affinity and tyrosine kinase activity (SCHLESSINGER 1988; COUNTAWAY et al. 1989).

An important question has been whether the changes in the affinity of the EGF receptor in response to phorbol esters or PDGF are a direct consequence of the phosphorylation of threonine 654 by protein kinase C. To answer this question, DAVIS (1988) use site-directed mutagenesis to replace EGF receptor threonine 654 with alanine 654 and tested the regulation of ligand affinity and tyrosine kinase activation of the mutant receptor in CHO cells which lack endogenous EGF receptors. Cells transfected with either wild-type human EGF receptor threonine 654 or the alanine 654 receptor cDNA expressed high- and low-affinity receptors for EGF on the cell surface. In addition, high-affinity binding of EGF to wild-type or mutant receptors was inhibited similarly by phorbol ester treatment. However, tyrosine kinase inhibition by phorbol ester treatment of cells was observed only for the wild type receptor. Similar results were obtained by LIVNEH et al. (1988) who mutated threonine 654 to tyrosine 654. Thus, these data suggest that the regulation of the affinity of the EGF receptors by phorbol esters is not secondary to the phosphorylation of threonine 654. These results contrast with those of LIN et al. (1986) who reported that an alanine 654 mutant in Rat 1 fibroblasts resulted in an inhibition of the effect of phorbol esters to decrease EGF cell surface binding. HEISERMANN and GILL (1988) and NORTHWOOD and DAVIS (1989) have identified an additional phorbol ester-stimulated phosphorylation site on the EGF receptor, threonine 669, as well as several phosphorylated serine residues (671, 1046/1047). Threonine 669 does not appear to be a substrate for protein kinase C, suggesting activation of other kinases by phorbol esters. The role of these latter serine/threonine phosphorylation reactions in regulating EGF receptor function has not been determined.

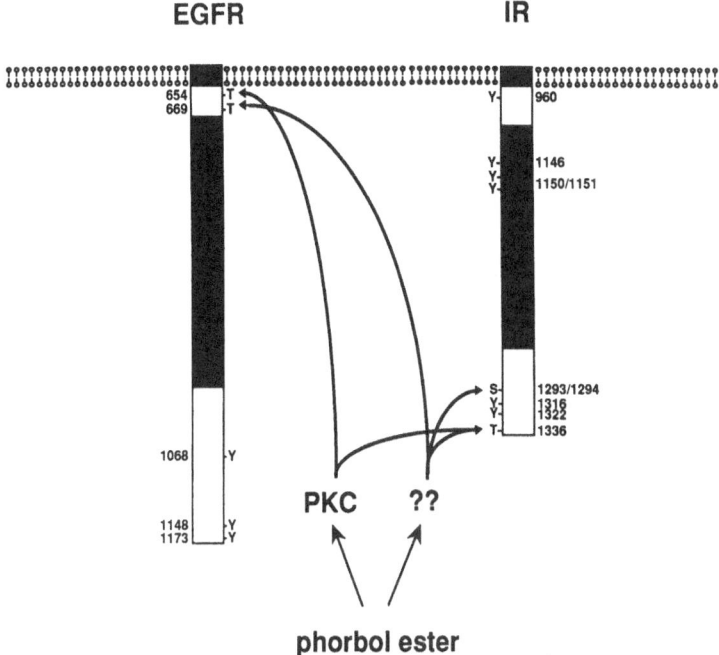

Fig. 7. Identification of major serine and threonine phosphorylation sites in the cytoplasmic domains of EGF (*EGFR*) and insulin receptor (*IR*) tyrosine kinases. Two major phosphorylation sites on the EGF receptor stimulated by phorbol esters in intact cells are located near the plasma membrane (threonine 654 and threonine 669), while phorbol ester-stimulated phosphorylation sites on the insulin receptor are located near the carboxy terminus (threonine 1336 and serine 1293/1294). The phosphorylation sites may be mediated by a serine kinase(s) distinct from protein kinase C because this enzyme in vitro phosphorylates only threonine 654 (EGF receptor) and threonine 1336 (insulin receptor) when incubated with purified receptor preparations. Major tyrosine kinase autophosphorylation sites are also identified in the figure

Serine phosphorylation also appears to play an important role in regulating the insulin receptor tyrosine kinase. The β-subunit of the insulin receptor undergoes an increase in serine/threonine phosphorylation after treatment of intact cells with either phorbol esters or cyclic AMP analogs (JACOBS et al. 1983; TAKAYAMA et al. 1984; STADTMAUER and ROSEN 1986), suggesting potential roles for protein kinase C and cyclic AMP-dependent protein kinase in insulin receptor phosphorylation. The cyclic AMP- and phorbol ester-stimulated receptor phosphorylation inhibits insulin receptor tyrosine kinase activity. Indeed, direct phosphorylation of purified insulin receptor by protein kinase C in vitro results in significant inhibition of the insulin-stimulated tyrosine kinase activity (BOLLAG et al. 1986). In contrast, the purified insulin receptor is poorly phosphorylated by and does not contain a direct phosphorylation site for cyclic AMP-dependent protein kinase. These data suggest that other kinases or phosphatases are involved in mediating the effects of cyclic AMP on insulin receptor phosphorylation.

Recently, LEWIS et al. (1989 a, b) and SMITH et al. (SMITH et al. 1988; SMITH and SALE 1988) have obtained affinity purified insulin receptor preparations with an associated insulin-stimulated serine kinase activity. Interestingly, the serine kinase activities in their in vitro preparations do not appear to be similar to protein kinase C- or cyclic AMP-dependent protein kinase as assessed by their ability to phosphorylate exogenously added peptide substrates. LEWIS et al. (1989 a, b) have determined the location of two of the major serine/threonine sites phosphorylated by the insulin receptor-associated serine kinase: threonine 1336, which is also phosphorylated by protein kinase C in vitro, and serine 1293/1294. These residues are also major serine/threonine phosphorylation sites in insulin receptors in intact cells treated with insulin or phorbol esters. It is proposed that phosphorylation at either or both of these sites may be responsible for the inhibition of receptor tyrosine kinase by phorbol ester. A plausible hypothesis is that insulin-mediated serine/threonine phosphorylation of the insulin receptor leads to desensitization, although this has not been rigorously tested as yet. This same hypothesis apply to other growth factor receptors.

Figure 7 shows the present state of information on the phosphorylation sites on the EGF and insulin receptors. Interestingly, serine/threonine phosphorylation of the EGF receptor is localized to amino acids near the transmembrane domain, while similar phosphorylations on the insulin receptor occur on amino acids close to the carboxy terminus. The ability of phorbol esters to enhance phosphorylation of serine and threonine sites results from either the direct action of of protein kinase C (EGF receptor threonine 654 and insulin receptor threonine 1336), or the actions of other unknown serine/threonine kinases. These may be part of a cascade initiated by the activation of protein kinase C. Alternatively, they may represent independent phorbol ester-sensitive enzymes.

J. Cytoplasmic Domains Lacking Tyrosine Kinase Sequences

For receptors lacking a tyrosine kinase domain (IGF-II/Man-6-P, IL-2, and NGF receptors in Fig. 8), there is no information about which structural regions within the cytoplasmic domains confer signaling or other functions. It is possible that the short cytoplasmic tails of these receptors associate with other membrane proteins which catalyze such functions as signal transduction and membrane sorting. The IL-2 receptor is an example of this concept (TESHIGAWARA et al. 1987). Such associated proteins might exhibit intrinsic tyrosine kinase or other enzyme activities. Two of the three receptors in this group, the IGF-II/Man-6-P and IL-2 receptors, share the common feature of a high density of consensus sequences for specific protein kinase-mediated phosphorylation sites. As illustrated in Fig. 8, both the IL-2 and IGF-II/Man-6-P receptor cytoplasmic domains reveal multiple serine or threonine residues flanked by basic amino acids. Such serine or threonine residues are predicted to be substrates for the cyclic AMP-dependent protein kinase or protein

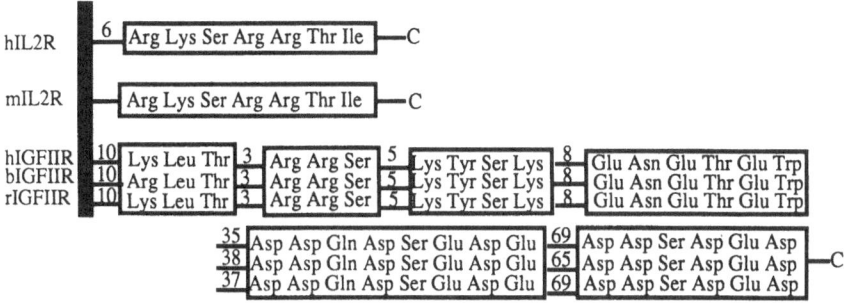

Fig. 8. Conserved consensus sequences of potential serine and threonine phosphorylation sites within the cytoplasmic domains of the IL-2 and IGF-II/Main-6-P receptors. Potential phosphorylation sites (serine or theonine residues) are in *boxes*. Intervening amino acids sequences are depicted as *horizontal bars* and *numbered* to designate number of residues. *Vertical bar* designates putative location of membrane

kinase C. In addition, the IGF-II/Man-6-P receptor exhibits two excellent consensus sequences for casein kinase II substrates, i.e., serines flanked by acidic asparate or glutamate residues.

The functional role of the multiple phosphorylation sites in the IL-2, IGF-II/Man-6-P, and other receptors (e.g., transferrin receptor) lacking tyrosine kinase domains, if any, are unknown. However, they may relate to mechanisms whereby their membrane dynamics or localization are regulated. For example, serine phosphorylation of transferrin receptors is increased by addition of EGF or phorbol esters to intact cells. It has been found that the major phosphorylation site in this receptor is serine 24 (DAVIS et al. 1986b). However, substitution of alanine at this site did not prevent upregulation of the mutant transferrin receptor by phorbol esters (DAVIS and MEISNER 1987).

Probably the IGF-II/Man-6-P receptor has been studied in most detail in respect to its serine/threonine phosphorylation. Experiments carried out with ^{32}P-labeled cells followed by subcellular fractionation and immunoprecipitation of IGF-II/Man-6-P receptors have revealed several interesting aspects of receptor phosphorylation (CORVERA and CZECH 1985; CORVERA et al. 1988 a, b). It has been found that IGF-II/Man-6-P receptors present in control adipocyte plasma membranes have much higher ^{32}P content than receptors in intracellular membranes. Tryptic peptide mapping of ^{32}P-labeled IGF-II receptors derived from H-35 hepatoma cells revealed three major phosphorylated peptides. Phosphoamino acid analysis or IGF-II/Man-6-P receptors isolated from ^{32}P-labeled adipocytes or H-35 hepatoma cells indicated that the phosphorylated residues are serine and threonine.

In order to investigate which kinases might be responsible for the phosphorylation of IGF-II/Man-6-P receptors in intact cells, experiments were performed with purified receptors and preparations of various purified kinases (cyclic AMP-dependent protein kinase, casein kinase II, protein kinase C, and phosphorylase kinase). Significant phosphorylation of purified receptors was catalyzed by casein kinase II but not the other kinases tested

CONTROL   INSULIN

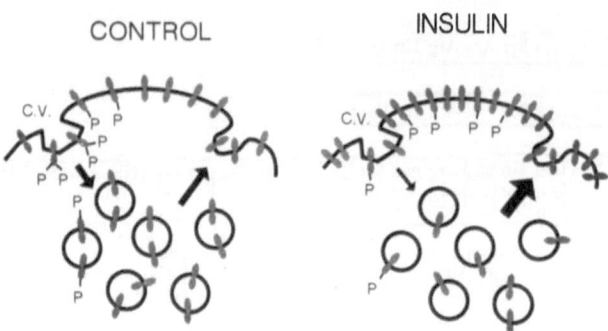

Fig. 9. Hypothetical relationship between insulin-stimulated cell surface IGF-II/Man-6-P receptor and receptor phosphorylation state. In control adipocytes, most IGF-II/Man-6-P receptors (90% or more) reside in intracellular membrane fractions, whereas addition of insulin causes a marked redistribution of receptors to the plasma membrane. Associated with this upregulation of IGF-II/Man-6-P receptors, the phosphorylation state of a subpopulation of these receptors in the plasma membrane fraction decreases substantially. This subpopulation of receptors appears to be associated with the clathrin-enriched, detergent-insoluble aspect of isolated plasma membranes (see text for further details). A direct causal relationship between the regulation of IGF-II/Man-6-P receptor phosphorylation and its membrane dynamics has not yet been demonstrated

(Corvera et al. 1988a). Both threonine and serine residues on the receptor were phosphorylated in vitro by purified casein kinase II. Furthermore, bidimensional peptide mapping revealed that the casein kinase II-catalyzed phosphorylation of the IGF-II/Man-6-P receptor involved a tryptic phosphopeptide found in receptors obtained from cells labeled in vivo with [^{32}P]phosphate. These findings are consistent with the presence of the two conserved consensus sequences for casein kinase II in the cytoplasmic domain of the receptor (Fig. 8). More recent studies on the phosphorylation of the IGF-II/Man-6-P receptor have revealed a marked heterogeneity in the phosphorylation state of receptor populations within isolated adipocytes. A highly phosphorylated form, containing 4–5 phosphates/receptor, was found to reside in a Triton X-100 insoluble plasma membrane fraction. A less phosphorylated form was found to be soluble in this detergent (Corvera et al. 1988b).

It has been hypothesized that the phosphorylation of the IGF-II/Man-6-P receptor is involved in regulating the movement and subcellular distribution of this molecule. This hypothesis is based on the finding that addition of insulin to adipocytes or H-35 hepatoma cells caused a rapid redistribution of receptors among intracellular and plasma membranes, and simultaneously decreases the estimated specific activity of [^{32}P]phosphate in plasma membrane IGF-II/Man-6-P receptors. Importantly, IGF-II/Man-6-P receptors isolated from insulin-treated H-35 cells or adipocytes were phosphorylated in vitro by casein kinase II to a greater extent than receptors isolated from control cells. Thus, the insulin-regulated phosphorylation sites on the IGF-II receptor appear to serve as substrates in vivo for casein kinase II or an enzyme with similar substrate specificity (CORVERA et al. 1988 a). A relationship between IGF-II/Man-6-P receptor phosphorylation and its movement among cellular membranes is also consistent with the finding that the highly phosphorylated form of the receptor copurifies with the heavy chain of clathrin, suggesting that receptor phosphorylation may be involved in the concentration of receptors in clathrin coated structures (CORVERA et al. 1988 b).

A model which exemplifies a current hypothesis on the role of phosphorylation in IGF-II/Man-6-P receptor transit and on the effects of insulin on the phosphorylation and distribution of the receptor is shown in Fig. 9. According to this model, the distribution of IGF-II/Man-6-P receptors between endosomal and plasma membranes is modulated by an increased rate of exocytosis and a decreased rate of endocytosis in response to insulin. The latter effect is hypothesized to be related to the insulin-mediated dephosphorylation of plasma membrane IGF-II/Man-6-P receptors. The identification of the precise sites that are phosphorylated in this receptor in vivo, and their modification by site-directed mutagenesis will be required to directly test this hypothesis.

K. Conclusions

The last several years have established a number of new and important concepts about the structure and function of growth factor receptors. Three particularly important achievements are:

1. The primary structures of a large number of growth factor receptors have been deduced from cDNAs encoding these proteins. This has been accomplished in parallel with the development of numerous powerful reagents and methods for probing the structure-function relationships among these proteins.
2. Tyrosine kinase activities are intrinsic to many growth factor receptor cytoplasmic domains, and appear to be necessary for signaling biological responses. The receptors lacking such domains may interact with other cellular proteins that may contain such activity, or other activities competent to initiate signal propagation.
3. Regulation of both cell surface receptor ligand binding and tyrosine kinase activity is associated with multiple tyrosine and serine phosphorylation sites on the cytoplasmic domain of several growth factor receptors.

Among the many unanswered questions about the growth factor receptors left to solve, two deserve particular emphasis. The first relates to the molecular pathways utilized for signal transduction, and ultimately to the modulation of cellular processes. It seems clear that several signaling modes mediate growth factor receptor action, but the molecular mechanisms involved in linking receptor activation to subsequent steps are completely unknown. The second question relates to the molecular basis whereby phosphorylation of the cytoplasmic domain modifies receptor structure and modulates both receptor tyrosine kinase activity and the distribution of receptors among membrane compartments. Gleaning answers to these questions will challenge current available technologies, but will be critically important in our attempts to understand the complexities of cell function and proliferation.

References

Baran D, Korner M, Theze J (1988) Characterization of the soluble murine IL-2 R and estimation of its affinity for IL-2. J Immunol 141:539–546

Bargmann CI, Hung MC, Weinberg RA (1986a) Multiple independent activations of the neu oncogene by a point mutation altering the transmembrane domain of p 185, Cell 45:649–657

Bargmann CI, Hung MC, Weinberg RA (1986b) The Neu oncogene encodes an epidermal growth factor receptor-related protein. Nature 319:226–234

Baumann G, Stolar MW, Amburn K, Barsano CP, DoVries BC (1986) A specific growth hormone-binding protein in human plasma: initial characterization J Clin Endocrinol Metab 62:134–141

Beguin Y, Heubers HA, Joxphson B, Finch CA (1988) Transferrin receptors in rat plasma. Proc Natl Acad Sci USA 85:637–640

Bertics PJ, Gill GN (1985) Self-phosphorylation enhances the protein-tyrosine kinase activity of the epidermal growth factor receptor. J Biol Chem 260:14642–14647

Bollag GE, Roth RA, Beaudoin J, Mochly-Rosen D, Koshland DE (1986) Protein kinase C directly phosphorylates the insulin receptor in vitro and reduces its protein-tyrosine kinase activity. Proc Natl Acad Sci USA 83:5822–5824

Böni-Schnetzler M, Kaligian A, DelVecchio R, Pilch P (1988) Ligand dependent inter-subunit association within the insulin receptor complex activates its intrinsic kinase activity. J Biol Chem 263:6822–6828

Causin C, Waheed A, Braulka T, Junghans U, Maly P, Humbel RE, VonFigura K (1988) Mannose-6-phosphate/insulin-like growth factor II-binding proteins in human serum and urine. Biochem J 252:795–799

Cheifetz S, Bassols A, Stanley K, Ohta M, Greenberger J, Massagué J (1988a) Heterodimeric TGF-β: biological properties and interaction with three types of cell surface receptors. J Biol Chem 263:10783–10789

Cheifetz S, Ling N, Guillemin R, Massagué J (1988b) A surface component on GH3 pituitary cells that recognize TGF-β activin, and inhibin. J Biol Chem 263:17225–17228

Chou CK, Dull TJ, Russell DS, Gherzi R, Lebwohl D, Ullrich A, Rosen OM (1987) Human insulin receptors mutated at the ATP-binding site lack protein tyrosine .kinase activity and fail to mediate post receptor effects of insulin. J Biol Chem 262:1842–1847

Cochet C, Gill GN, Meisenhelder J, Cooper JA, Hunter T (1984) C-kinase phosphory-lates the epidermal growth factor receptor and reduces its epidermal growth factor stimulated tyrosine protein kinase activity. J Biol Chem 259:2553–2558

Corvera S, Czech MP (1985) Mechanism of insulin action on membrane protein recy-cling: a selective decrease in the phosphorylation state of insulin-like growth fac-tor II receptors in the cell surface membrane. Proc Natl Acad Sci USA 82:7314–7318

Corvera S, Roach PJ, DePaoli-Roach AA, Czech MP (1988a) Insulin action inhibits insulin-like growth factor-II receptor phosphorylation in H-35 hepatoma cells. IGF-II receptors isolated from insulin-treated cells exhibit enhanced in vitro phosphorylation by casein kinase II. J Biol Chem 263:3116–3122

Corvera S, Folander K, Clairmont KB, Czech MP (1988b) A highly phosphorylated subpopulation of insulin-like growth factor II/mannose-6-phosphate receptors is concentrated in a clathrin-enriched plasma membrane fraction. Proc Natl Acad Sci USA 85:7567–7571

Countaway JL, Girones N, Davis RJ (1989) Reconstitution of epidermal growth factor receptor transmodulation by platelet-derived growth factor in Chinese hamster ovary cells. (submitted)

Coussens L, Yang-Feng TL, Liao YC, Chen E, Gray A, McGrath J, Seeburg PH, Liberman TA, Schlessinger J, Francke U, Levinson A, Ullrich A (1985) Tyrosine kinase receptor with extensive homology to EGF receptor shares chromosomal location with Neu oncogene. Science 230:1132–1139

Davis RJ (1988) Independent mechanisms account for the regulation by protein kinase C of the epidermal growth factor receptor affinity and tyrosine-protein kinase activity. J Biol Chem 263:9462–9469

Davis RJ, Czech MP (1985) Tumor-promoting phorbol esters cause the phosphorylation of epidermal growth factor receptors in normal human fibroblasts at threonine 654. Proc Natl Acad Sci USA 82:1974–1978

Davis RJ, Czech MP (1986) Regulation of transferrin receptor expression at the cell surface by insulin-like growth factors, epidermal growth factor and platelet derived growth factor. EMBO J 5:653–658

Davis RJ, Meisner H (1987) Regulation of transferrin receptor cycling by protein kinase C is independent of receptor phosphorylation at serine 24 in Swiss 3T3 fibroblasts. J Biol Chem 262:16041–16047

Davis RJ, Corvera S, Czech MP (1986a) Insulin stimulates cellular iron uptake and causes the redistribution of intracellular transferrin receptors to the plasma membrane. J. Biol Chem 261:8708–8711

Davis RJ, Johnson GL, Kelleher DJ, Anderson JA, Mole JE, Czech MP (1986b) Identification of serine 24 as the unique site on the transferrin receptor phosphorylated by protein kinase C. J Biol Chem 261:9034–9041

Davis RJ, Faucher M, Racaniello LK, Carruthers A, Czech MP (1987) Insulin-like growth factor I and epidermal growth factor regulate the expression of transferrin receptors at the cell surface by distinct mechanisms. J Biol Chem 262:13126–13134

DeMeyts P, Bianco AR, Roth J (1976) Site-site interactions among insulin receptors. J. Biol Chem 251:1877–1888

Downward J, Waterfield MD, Parker PJ (1985) Autophosphorylation and protein kinase C phosphorylation of the epidermal growth factor receptor. J Biol Chem 260:14538–14546

Ebina Y, Ellis L, Jarnagin K, Edery M, Graf L, Clauser E, Ou J-H, Masiarz F, Kan YW, Goldfine ID, Roth RA, Rutter WJ (1985) The human insulin receptor cDNA: the structural basis for hormone-activated transmembrane signalling. Cell 40:747

Ebina Y, Araki E, Taira M, Shimada F, Mori M, Craik CS, Siddle K, Pierce SB, Roth RA, Rutter WJ (1987) Replacement of lysine residue 1030 in the putative ATP-binding region of the insulin receptor abolishes the insulin- and antibody-stimulated glucose uptake and receptor kinase activity. Proc Natl Acad Sci USA 84:704–708

Ellis L, Clauser E, Morgan DO, Edery M, Roth RA, Rutter WJ (1986) Replacement of insulin receptor tyrosine residues 1162 and 1163 compromises insulin-stimulated kinase activity and uptake of 2-deoxyglucose. Cell 45:721–732

Escobedo JA, Navankasatussas S, Cousens LS, Coughlin SR, Bell GI, Williams LT (1988) A common PDGF receptor is activated by homodimeric A and B forms of PDGF. Science 240:1532–1534

Gavin JR, Buell DN, Roth J (1972) Water-soluble insulin receptors from human lymphocytes. Science 178:168–169

Green A, Olefsky JM (1982) Evidence for insulin-induced internalization and degradation of insulin receptors in rat adipocytes. Proc Natl Acad Sci USA 79:427–431

Hari J, Pierce SB, Morgan DO, Sara V, Smith MC, Roth RA (1987) The receptor for insulin-like growth factor II mediates an insulin-like response. EMBO J 6:3371

Harrington DS, Patil K, Lai PK, Yasuda NN, Armitag JO, Ip SH, Weisenburger DD, Linder J, Purillo DT (1988) Soluble interleukin 2 receptors in patients with malignant lymphoma. Arch Pathol Lab Med 112:597–601

Heisermann GJ, Gill GN (1988) Epidermal growth factor receptor threonine and serine residues phosphorylated in vivo. J Biol Chem 263:13152–13158

Heldin CH, Backstrom G, Ostman A, Hammacher A, Ronnstrand L, Rubin K, Nister M, Westermark B (1988) Binding of different dimeric forms of PDGF to human fibroblasts: evidence for two separate receptor types. EMBO J 7:1387–1393

Herington AC (1985) Evidence for the specific binding of growth hormone to a receptor-like protein in rabbit serum. Mol Cell Endocrinol 41:153–161

Herrera R, Lebowhl D, Garcia de Herreros A, Kallen RG, Rosen OM (1988) Synthesis purification and characterization of the cytoplasmic domain of the human insulin receptor using a baculovirus expression system. J Biol Chem 263:5560–5568

Honegger A, Dull TJ, Bellot F, Van Obberghen E, Szapary D, Schmidt A, Ullrich A, Schlesinger J (1988a) Biological activities of EGF-receptor mutants with individually altered autophosphorylation sites. EMBO J 7:3045–3052

Honegger A, Dull TJ, Szapary D, Komoriya A, Kris R, Ullrich A, Schlessinger J (1988b) Kinetic parameters of the protein tyrosine kinase activity of EGF-receptor mutants with individually altered autophosphorylation sites. EMBO J 7:3053–3060

Honegger AM, Kris RM, Ullrich A, Schlessinger J (1989) Evidence that autophosphorylation of solubilized EGF-receptors is mediated by intermolecular cross phosphorylation. Proc Natl Acad Sci USa 86:925–929

Hunter T, Ling N, Cooper JA (1984) Protein kinase C phosphorylates the EGF receptor at a threonine residue close to the cytoplasmic face of the plasma membrane. Nature 311:480–483

Iwashita S, Fox CF (1984) Epidermal growth factor and potent phorbol tumor promoters induce epidermal growth factor receptor phosphorylation in a similar but distinctively different manner in human epidermoid carcinoma A431 cells. J Biol Chem 259:2559–2567

Jacobs S, Sahyoun NE, Saltiel AR, Cuatrecasas P (1983) Phorbol esters stimulate the phosphorylation of receptors for insulin and somatomedin C. Proc Natl Acad Sci USA 80:6211–6213

Johnson D, Lanahan A, Buck CR, Sehgal A, Morgan C, Marzer E, Bothwell M, Chao M (1986) Expression and structure of the human NGF receptor. Cell 47:545–554

Kashler'O, Szapary D, Bellot F, Ullrich A, Schlessinger J, Schmidt A (1988) Ligand-induced stimulation of epidermal growth factor receptor mutants with altered transmembrane regions. Proc Natl Acad Sci USA 85:9567–9571

Kiess W, Greenstein LA, White RM, Lee L, Rechler MM, Nissley SP (1987a) Type II insulin-like growth factor is present in rat serum. Proc Natl Acad Sci USA 84:7720–7724

Kiess W, Haskell JF, Lee L, Greenstein LA, Miller BE, Aarons AL, Rechler MM, Nissley SP (1987b) An antibody that blocks insulin-like growth factor binding to the type II IGF receptor is neither an agonist nor an inhibitor of IGF-stimulated biologic resposes in L6 myoblasts. J Biol Chem 262:12745–12751

Kobilka BK, Frielle T, Collins S, Yano-Feno T, Kobilka TS, Fancke U, Lefkowitz RJ, Caron MG (1987) An intronless gene encoding a potential member of the family of receptors coupled to guanine nucleotide regulatory proteins. Nature 329:75

Lax I, Burgess WH, Bellot F, Ullrich A, Schlessinger J, Givol D (1988) Localization of a major receptor-binding domain for epidermal growth factor by affinity labelling. Mol Cell Biol 8:1831–1834

Lax I, Howk R, Bellot F, Ullrich A, Givol D, Schlessinger J (1989) Functional analysis of the ligand binding site of EGF receptor utilizing chimeric chicken/human receptor molecules. EMBO J 8:421–428

Lee J, Dull TJ, Lax I, Schlessinger J, Ullrich A (1989) HER2 cytoplasmic domain generates normal mitogenic and transforming signals in chimeric receptor. EMBO J 8:167–173

Leonard WJ, Depper JM, Crabtree GR, Rudikoff S, Pumphrey J, Robb RJ, Kronke M, Svetlik PB, Peffer NJ, Waldman TA, Greene WC (1984) Molecular cloning expression of cDNA for the human interleukin-2 receptor. Nature 311:626–631

Lewis RE, Cao L, Perregaux D, Czech MP (1989a) Threonine 1326 of the human insulin receptor is a major target for phosphorylation by protein kinase C. (submitted)

Lewis RE, Wu GP, MacDonald RG, Czech MP (1989b) An insulin-sensitive serine/threonine kinase activity phosphorylates and copurifies with the insulin receptor. (submitted)

Lin CR, Chen WS, Lazar CS, Carpenter CD, Gill GN, Evans RM, Rosenfeld MG (1986) Protein kinase C phosphorylation at Thr 654 of the unoccupied EGF receptor and EGF binding regulate functional receptor loss by independent mechanisms. Cell 44:839–848

Livneh E, Dull TJ, Berent E, Prywes R, Ullrich A, Schlessinger J (1988) Release of a phorbol ester-induced mitogenic block by mutation at Thr-654 of the epidermal growth factor receptor. Mol Cell Bio 8:2302–2308

Lobel P, Dahms NM, Kornfeld S (1988) Cloning and sequence analysis of the cation independent mannose-6-phosphate receptor. J Biol Chem 263:12705–12713

MacDonald RG, Pfeffer SR, Coussens W, Tepper MA, Brocklebank CM, Mole JE, Anderson JK, Chen E, Czech MP, Ullrich A (1988) A single receptor binds both insulin-like growth factor II and mannose-6-phosphate. Science 239:1134–1137

MacDonald RG, Tepper MA, Clairmont KB, Perregaux SB, Czech MP (1989) Serum form of the rat insulin-like growth factor II/mannose-6-phosphate receptor is truncated in the carboxyl terminal domain. J Biol Chem 264:3256–3261

Massagué J, Czech MP (1982) The subunit structure of two distinct receptors for insulin-like growth factors I and II and their relationships to the insulin receptor. J Biol Chem 257:5038–5045

Massagué J, Pilch PF, Czech MP (1980) Electrophoretic resolution of three major insulin receptor structures with unique subunit stoichiometries. Proc Natl Acad Sci USA 77:7137–7141

McGuffin WL, Gavin JR, Lesniak MA, Gordon P, Roth J (1976) Water soluble specific growth hormone binding sites from cultured human lymphocytes: preparation and partial characterization. Endocrinology 98:1401–1407

Moolenaar WH, Bierman AJ, Tilly BC, Verlaan I, Defize LHK, Honegger AM, Ullrich A, Schlessinger JA (1988) Point mutation at the ATP-binding site of the EGF-receptor abolishes signal transduction. EMBOJ 7:707–710

Morgan DO, Roth RA (1987) Acute insulin action requires insulin receptor kinase activity: introduction of an inhibitory monoclonal antibody into mammalian cells blocks the rapid effects of insulin. Proc Natl Acad Sci USA 84:41–45

Morgan DO, Edman JC, Standring DN, Freid VA, Smith MC, Roth RA, Rutter WJ (1987) Insulin-like growth factor II receptor as a multifunctional binding protein. Nature 329:301–307

Mottola C, Czech MP (1984) The type II insulin-like growth factor receptor does not mediate increased DNA synthesis in H-35 hepatoma cells. J Biol Chem 259:12705–12713

Muekler M, Caruso C, Baldwin SA, Panico M, Blench I, Morris HR, Allard WJ, Lienhard GE, Lodish HF (1985) Sequence and structure of a human glucose transporter. Science 229:941

Nikaido T, Shimizu A, Ishida N, Sabe H, Teshigawara K, Maeda M, Uchiyama T, Yodoi J, Honjo T (1984) Molecular cloning of cDNA encoding human interleukin-2 receptor. Nature 311:631–635

Nishomoro I, Hara Y, Ogara E, Kojima I (1987) Insulin-like growth factor II stimulates calcium influx in competent Balb 3T3 cells primed with the epidermal growth factor. J Biol Chem 262:12120–12126

Northwood IC, Davis RJ (1989) Protein kinase C inhibition of the epidermal growth factor receptor tyrosine kinase activity is independent of the oligomeric state of the receptor. J Biol Chem 264:5746–5750

Oka Y, Czech MP (1986) The type-II insulin-like growth factor receptor is internalized and recycles in the absence of ligand. J Biol Chem 261:9090–9093

Oka Y, Mottola C, Oppenheimer CL, Czech MP (1984) Insulin activates the appearance of insulin-like growth factor II receptors on the adipocyte cell surface. Proc Natl Acad Sci USA 81:4028–4032

Oppenheimer CL, Pessin JE, Massague J, Gitomer W, Czech MP (1983) Insulin action rapidly modulates the apparent affinity of the insulin-like growth factor II receptor. J Biol Chem 258:4824–4830

Pohlmann R, Nagel G, Schmidt B, Stein M, Lorkowski G, Krentler G, Cully J, Meyer HE, Grzeschik KH, Mersmann G, Hasilik A, von Figura K (1987) Cloning of a cDNA encoding the human cation-dependent mannose 6-phosphate receptor. Proc Natl Acad Sci USA 84:5575–5579

Radeke MJ, Misko TP, Hsu C, Herzenberg LA, Shooter EM (1987) Gene transfer and molecular cloning of the rat nerve growth factor receptor. Nature 325:593–597

Riedel H, Dull TJ, Ullrich A (1986) A chimeric receptor allows insulin to stimulate tyrosine kinase activity of epidermal growth factor receptor. Nature 324:68–70

Robb RJ, Rusk CM, Neeper MP (1988) Structure-function relationships for the interleukin 2 receptor: location of ligand and antibody binding sites on the Tac receptor chain by mutational analysis. Proc Natl Acad Sci USA 85:5654–5658

Ronnet GV, Knutson VP, Kohanski RA, Simpson TL, Lane MD (1984) Role of glycosylation in the processing of newly translated insulin receptor in 3T3-L1 adipocytes. J Biol Chem 259:4566–4575

Rusk CM, Neeper MP, Kuo LM, Robb RJ (1988) Structure function relationship for the IL-2 receptor system structure-activity analysis of modified and truncated forms of Tac receptor protein: site specific mutagenisis of cysteine residues. J Immunol 140:2249–2259

Schlessinger J (1988) The epidermal growth factor receptor as a multifunctional allosteric protein. Biochemistry 27:3119–3123

Sherr CJ, Rettenmier CW, Sacca R, Roussel MF, Look AT, Stanley ER (1985) The c–fms proto-oncogene product is related to the receptor for the mononuclear phagocyte growth factor, CSF-1. Cell 41:665–676

Smith DM, Sale GL (1988) Evidence that a novel serine kinase catalyses phosphorylation of the insulin receptor in an insulin dependent and tyrosine kinase dependent manner. Biochem J 256:903–909

Smith DM, King MJ, Sale GL (1988) Two systems in vitro that show insulin stimulated serine kinase towards the insulin receptor. Biochem J 250:509–519

Stadtmauer L, Rosen OM (1986) Increasing cAMP content of IM-9 cells alters the phosphorylation state and protein kinase activity of the insulin receptor. J Biol Chem 261:3402–3407

Takayama S, White MF, Lauris V, Kahn CR (1984) Phorbol esters modulate insulin receptor phosphorylation and insulin action in cultured hepatoma cells. Proc Natl Acad Sci USA 81:7797–7801

Teshigawara K, Wang HM, Kato K, Smith KA (1987) Interleukin 2 high-affinity receptor expression requires two distinct binding proteins. J Exp Med 165:223–238

Ullrich A, Coussens L, Hayflick JS, Dull TJ, Gray A, Tam AW, Lee J, Yarden Y, Libermann TA, Schlessinger J, Downward J, Mayes ELV, Whittle N, Waterfield MD, Seeburg PH (1984) Human epidermal growth factor receptor cDNA sequence and aberrant expression of the amplified gene in A431 epidermoid carcinoma cells. Nature 309:418–425

Ullrich A, Bell JR, Chen EY, Herrera R, Petruzzelli LM, Dull TJ, Gray A, Coussens L, Liao YC, Tsubokawa M, Mason A, Seeburg PH, Grunfield C, Rosen OM, Ramachandran J (1985) Human insulin receptor and its relationship to the tyrosine kinase family of oncogenes. Nature 313:756–761

Ullrich A, Gray A, Tam AW, Yang-Feng T, Tsubokawa M, Collins C, Henzel W, LeBon T, Kathuria S, Chen E, Jacobs S, Franckes U, Ramachandran J, Fujita-Yamaguchi Y (1986) Insulin-like growth factor I receptor primary structure: comparison with insulin receptor suggests structural determinants that define functional specifics. EMBO J 5:2503–2512

Wardzala LJ, Simpson IA, Rechler MM, Cushman SW (1984) Potential mechanism of the stimulatory action of insulin on insulin-like growth factor II binding to the isolated rat adipose cell. Apparent redistribution of receptors cycling between a large intracellular pool and the plasma membrane. J Biol Chem 259:8378–8383

Weber W, Gill GN, Spiess J (1984) Production of an epidermal growth factor receptor-related protein. Science 224:294–297

Whitely B, Glaser J (1986) Epidermal growth factor (EGF) promotes phosphorylation at Thr654 of the EGF-receptor: possible role of protein kinase C in homologous regulation of the EGF-receptor. J Cell Biol 103:1355–1362

Yarden Y, Schlessinger J (1985) The EGF receptor kinase: evidence for allosteric activation and intramolecular self-phosphorylation. Ciba Found Symp 116:23–45

Yarden Y, Ullrich A (1988) Growth factor receptor tyrosine kinases. Annu Rev Biochem 57:443–478

Yarden Y, Escobedo JA, Kuang WJ, Yang-Feng TL, Daniel TO, Tremble PM, Chen EY, Ando ME, Harkins RN, Francke U, Fried VA, Ullrich A, Williams LT (1986) Structure of the receptor for platelet-derived growth factor helps define a family of closely related growth factor receptors. Nature 323:226–232

Yarden Y, Kuang WS, Yang-Feng T, Coussens L, Munemits S, Dull TS, Chen E, Schlessinger J, Francke U, Ullrich A (1987) Human proto-oncogene cKit: a new cell surface receptor tyrosine kinase from an unidentified ligand. EMBO J 6:3341–3351

Yu K-T, Werth DK, Pastan IH, Czech MP (1985) Src kinase catalyzes the phosphorylation and activation of the insulin receptor kinase. J Biol Chem 260:5838–5846

Yu K-T, Peters MA, Czech MP (1986) Similar control mechanisms regulate the insulin and type I insulin-like growth factor receptor kinases. J Biol Chem 261: 11341–11349

Section B: Individual Growth Factors and Their Receptors

CHAPTER 4

The Epidermal Growth Factor Family

G. Carpenter and M. I. Wahl

A. Introduction

During the purification of nerve growth factor from mouse submaxillary glands by carboxymethyl cellulose chromatography, distinct biological activities were noticed when a fraction that did not contain nerve growth promoting activity was injected into newborn mice (Cohen 1960; Levi-Montalcini and Cohen 1960). These new responses included precocious opening of the eyelids, eruption of incisors, an inhibition of hair growth, and "stunting" of growth. Two years later Cohen (1962) reported the isolation of a heat-stable protein from mouse submaxillary glands, termed "tooth-lid factor", which reproduced the developmental effects on the eyelids and incisors. Histological studies indicated that the precocious eyelid opening in "factor-treated" animals was the consequence of a generalized increase in epidermal thickening and keratinization (Cohen and Elliott 1963). When studies performed in vitro with explants of either skin or epidermis demonstrated a direct effect on epidermal growth, the tooth-lid factor was termed epidermal growth factor or EGF (Cohen 1964).

Since these seminal studies, and particularly within the last decade, knowledge concerning EGF and its mechanism of action at the cell and molecular level has increased tremendously, providing, at times, particular insights into basic properties and functions of cells. However, a more definitive understanding of the natural role of endogenous EGF in animal physiology has not been forthcoming. Transforming growth factor-α (TGF-α) was discovered as an EGF-like biological activity (Todaro and DeLarco 1976; DeLarco and Todaro 1978) and originally termed sarcoma growth factor. Subsequently, TGF-α was isolated from the sarcoma preparations, based on its EGF-like activity (Anzano et al. 1983), and was shown to have a distinct, but EGF-like sequence (Marquardt et al. 1983, 1984). In the early 1980s, pox virus genes encoding an EGF-like molecule were identified (Blomquist et al. 1984; J. P. Brown et al. 1985; Reisener 1985) and in 1989 the structure and activity of amphiregulin, isolated from the media of phorbol ester-treated tumor cells, were shown to be EGF-like (Shoyab et al. 1989). Whether this family continues to expand remains to be seen, but it is not unlikely.

A comprehensive review of this growth factor family has not been published since 1980, but numerous reviews on focused subtopics have appeared in the last few years. Therefore the choice has been made to concentrate this review on areas that have not been reviewed recently. As nearly 900 papers per year are now published on EGF and its relatives, the authors recognize the hopelessness of presenting either a comprehensive review or an up to date portrait of a rapidly evolving subject.

B. Structural Properties of the Growth Factors

During the past decade two major areas of information regarding the structure of EGF have been developed. One of these concerns the elucidation of other gene products that have a primary sequence similar to EGF. The second area of significant information comes from high resolution nuclear magnetic resonance (NMR) studies of the tertiary structure of EGF. Information from each of these will be covered in this section, but data comparing the biological properties of the different EGF-like molecules will be covered later (Sects. C, D, F, G).

I. Amino Acid Sequences

Two characteristics define members of the EGF family of mitogens. One property is the capacity to mimic biological activities of EGF and to compete with ^{125}I-EGF in radioreceptor assays. The second feature is amino acid sequence similarity to EGF. No activity measurements have been performed with the Shope or myxoma growth factors. Their inclusion in the EGF family is based on nucleic acid sequence data and biological properties of the viruses. Amino acid sequence data for each member of this family are shown in Fig. 1. The 11 residues which are invariant in the family are shown enclosed in this figure. Also shown are the cysteine pairs which form three disulfide bonds, as determined for mouse EGF (C.R SAVAGE et al. 1973), human TGF-α (WINKLER et al. 1986), and the Shope fibroma molecule (Y. Z. LIN et al. 1988). Of the 11 conserved residues, six are cysteinyls and it is these residues and their relative spacing that defines the primary structure of the EGF family of mitogens. Overall, the level of sequence identity within all members of this growth factor family is approximately 20%. However, weighting of the conserved cysteine residues (a reasonable practice) would increase the level of sequence identity. Sequence similarity calculations based on conservative changes would be considerably higher.

If sequence conservation is scored for each subgroup in this family, the level of identity of the various EGFs is approximately 50%, the TGF-αs approximately 90%, and the three pox virus growth factors about 28%. (In the case of the last subgroup, it should be noted that the Shope and myxoma growth factor sequences are 73% identical.) The high level of sequence conservation in the TGF-α subgroup is notable, particularly as human and rodent species are compared. The level of sequence conservation for TGF-α and EGF within any one species is substantially lower: human 42%, rat 36%. Changes in the TGF-α structural sequences, at least after the appearance of rodents,

```
                    1         10        20        30        40        50
EGF
 human       NSDSE CPLSHDGYCL HDG VCMYIEALDKYA - - - CNCVVGYIGERCQYRDLKWWELR
 mouse       NSYPGCPSSYDGYCLNGGVCMHIESLDSYT - - - CNCVIGYSGDRCQTRDLRWWELR
 rat         NSNTGCPPSYDGYCLNGGVCMYVESVDRYV - - - CNCVIGYIGERCQHRDLR*
 guinea pig  QDAPGCPPSHDGYCLHGGVCMHIESLNTYA - - - CNCVIGYVGERCEHQDLDLWE
TGFα
 human       VVSHFNDCPDSHTQFCFH - GTCRFLVQEDKPA - - CVCHSGYVGARCEHADLLA
 rat         VVSHFHKCPDSHTQYCFH - GTCRFLVQEEKPA - - CVCHSGYVGVRCEHADLLA
Amphiregulin
 human  S1...N39 RKKKNPCNAEFQNFCIH - GECKYIEHLEAVT - - CKCQQEYFGERCGEK
Pox Virus GF
 vaccinia D1...D19 IPAIRLCGPEGDGYCLH - GDCIHARDIDGMY - - CRCSHGYTGIRCQHVVLVDYQRSENPNT...
 Shope   M1...I26 VKHVKVCNHDYENYCLNNGTCFTI - ALDNVSITPFCVCRINYEGSRCQFINLVTY
 myxoma  M1...I30 IKRIKLCNDDYKNYCLNNGTCFTV - ALNNVSLNPFCACHINYVGSRCQFINLITIK
```

Fig. 1. Amino acid sequences of the EGF family of mitogens. The residue numbering of EGF is shown at the top and the positioning of disulfide bonds (C. R. SAVAGE et al. 1973; WINKLER et al. 1986; Y. Z. LIN et al. 1988) is indicated at the bottom by solid lines connecting the appropriate cysteine residues. Identical amino acid residues for all members of the mitogen family are shown within boxes. The direct amino acid sequence of human EGF is from GREGORY (1975), of mouse EGF from C. R. SAVAGE et al. (1972), and of rat EGF from SIMPSON et al. (1985). The asterisk at the end of the rat EGF sequence shows that the cDNA for rat EGF (DOROW and SIMPSON 1988) indicates a sequence of WWNWR following the C terminus of the determined protein sequence. The sequence of guinea pig EGF, with a few residues filled in by matching, was determined by R. BRADSHAW and colleagues (personal communication) and is cited in SIMPSON et al. (1985). The direct, but partial, sequence of human TGF-α is from MARQUARDT et al. (1983) and is completed and confirmed by the deduced sequence reported by DERYNCK et al. (1984). The direct rat TGF-α sequence is from MARQUARDT et al. (1984). The direct sequence of amphiregulin is from SHOYAB et al. (1989). The sequence of the vaccinia virus growth factor is from partial N-terminal sequencing of the protein by STROBBANT et al. (1985) and the deduced sequence reported by VENKATESAN et al. (1982). N-terminal residues 2–18 are known, but are not shown for this growth factor. Also, the exact C terminus has not been determined, but is thought to be threonine as shown. The deduced amino acid sequences of the Shope (CHANG et al. 1987) and myxoma (UPTON et al. 1987) growth factors are shown. Since the peptides corresponding to these gene sequences have not been identified or sequenced, the location of the N terminus is unclear

would seem to have been much more restricted than changes in the structural sequences for EGF. Whether this implies differences in function for EGF and TGF-α or is related to some property of their respective precursor molecules (Sect. C) is not clear.

It should be noted, in regard to the sequences shown in Fig. 1, that the C-terminal residues frequently are arginine or lysine, preferred sites for trypsin-like proteases. The relative "shortness" of the C termini of some of these protein sequences, particularly amphiregulin which lacks the highly conserved leucine residue at position 47, may be due to the formation of truncated proteins by protease activity during isolation procedures. As nucleic acid sequences for these growth factors become available, this issue may be resolved. For example, a partial cDNA sequence for rat EGF has been reported recently (DOROW and SIMPSON 1988). This sequence shows that the C-terminal residue in the protein sequence (Fig. 1) is followed by Trp-Trp-Asn-Trp-Arg-His-(X)n. Cleavage between the Arg-His residues of this sequence would generate a 53-residue polypeptide similar in size and C-terminal sequence to the human and mouse EGFs. In regard to the guinea pig EGF sequence with a glutamic acid (E) at the C terminus, a protease which cleaves the last two amino acids from the C terminus of mouse EGF (leaving a C-terminal glutamic acid) exists in tissue homogenates and is active unless a low pH is maintained during the initial purification steps (C. R. SAVAGE et al. 1972). A similar activity might explain the seemingly premature C terminus of guinea pig EGF.

There are two practical issues to note from these sequence data. One point is that EGFs and TGF-αs are distinct in amino acid compositions. Therefore, amino acid analysis is a very sensitive method only not to ascertain the purity of these growth factors, but also to differentiate TGF-α and EGF isolated from the same species. For example, mouse EGF lacks phenylalanine, alanine, and lysine, while mouse TGF-α lacks isoleucine, methionine, and tryptophan.

The second practical issue concerns sites available for radiolabeling with [125]I. Most all radiolabeling studies utilize procedures which depend on the availability of tyrosyl side chains. While it is recognized that the covalent modification of hormones may drastically affect the biological activity of these molecules, only iodinated mouse EGF has been well tested in this regard and shown to be as active as the native molecule (CARPENTER et al. 1975). While the EGFs have three (guinea pig) or five (mouse, rat, human) tyrosine residues as potential labeling sites, the TGF-αs have considerably fewer tyrosyls – one in the case of human TGF-α and two within the rat sequence. The single tyrosine in human TGF-α and one of the two in rat TGF-α occur at a highly conserved position – tyrosine 37 (Fig.1). One must be concerned, therefore, with the potential effects that iodinating this residue might have on biological activity, particularly in the case of human TGF-α. The alternate tyrosyl available in the rodent TGF-α at position 13 also is reasonably well conserved. The only other amino acid at this position within the EGF family is phenylalanine in human TGF-α (Fig. 1), a conservative substitution. (See

Sect. B. IV. 1, 2 and B. V. 1 for additional information on the roles of tyrosine residues 13 and 37 in the structure and function of EGF and TGF-α.)

II. Related Sequences

The cysteinyl-bounded core sequence of the EGF family of mitogens has a consensus sequence of $CX_7CX_{4-5}CX_{10-13}CXCX_8C$ and is 36–40 residues in length. Based on this general arrangement of cysteine residues and a unit length of about 40 amino acids, EGF-like motifs have been identified in a number of proteins that do not function as mitogens. These EGF-like motifs have the general arrangement of $CX_{4-14}CX_{3-8}CX_{4-14}CXCX_{8-14}C$ and are found, freqeuntly as repeating structures, in secreted or transmembrane molecules. A list of these proteins, their known or suggested function, and the number of EGF-like repeat units are shown in Table 1. The list includes different groups of molecules: proteins that are either proteases or protease cofactors, proteins involved in interactions between cells or between cells and the extracellular matrix, molecules having roles in development, and several proteins with distinct or unknown functions.

The organization of genes for several of these proteins has been reported. In most cases discrete exons provide the structural information for each of these EGF-like sequence motifs: factor X (LEYTUS et al. 1986), factor IX (ANSON et al. 1984; YOSHITAKE et al. 1985), protein C (D. C. FOSTER et al. 1985; PLUTZKY et al. 1986), tissue-type (NY et al. 1984) and urokinase-type (NAGAMINE et al. 1984) plasminogen activators, the EGF precursor (BELL et al. 1986), and the low-density lipoprotein (LDL) receptor (SÜDHOF et al. 1985a, b). The uromodulin gene encodes the two EGF-like motifs in a single exon (PENNICA et al. 1987), while the human thrombomodulin gene is without introns to isolate any of the structural information, including the six EGF-like motifs (JACKMAN et al. 1987). The general arrangement of a single EGF-like repeat encoded by one exon is unlike the gene for EGF in which two exons encode the growth factor (Sect. D. II. 1). Rather it resembles the EGF-like repeats in the EGF gene which are all encoded by a single exon. In the genes of lower eukaryotes encoding proteins that contain many EGF-like repeats, the multiple EGF-like sequences are divided into a few exons, with one exon containing a majority of the repeat units. Thirty EGF-like repeats are contained in one exon of the notch gene (KIDD et al. 1986) and eight repeats in one exon of the *lin-12* gene (YOCHEM et al. 1988).

To date no EGF-like motifs have been reported for single cell organisms. This, together with the presence of these cysteine-rich sequences in molecules that are involved in macromolecular interactions, has suggested a general protein-protein recognition/interaction function for these sequences. The evolutionary distribution of these motifs may, particularly in the lower eukaryotes, suggest that the presence of a diffusable growth factor in higher organisms has, as its evolutionary base, a transmembrane protein involved in cell-cell interactions. Based on NMR studies of EGF, APPELLA et al. (1988) have suggested a general structure for these EGF-like regions in various proteins.

Table 1. Occurrence of EGF-related sequences in nonmitogenic proteins

Protein	Function	EGF-like units	References
Factor XII	Serine protease	2	McMULLEN and FUJIKAWA (1985)
Factor X	Serine protease	2	LEYTUS et al. (1984), FUNG et al. (1985)
Factor IX	Serine protease	2	KURACHI and DAVIE (1982)
Factor VII	Serine protease	2	HAGEN et al. (1986), TAKEYA et al. (1988)
Protein C	Serine protease	2	FOSTER and DAVIE (1984), LONG et al. (1984)
Protein S	Protein C cofactor	4	DAHLBÄCK et al. (1986), HOSKINS et al. (1987)
Protein Z	Unknown	1	HØJRUP et al. (1985)
Complement C1r	Serine protease	1	ARLAUD et al. (1987)
Urokinase-type plasminogen activator	Protease	1	GÜNZLER et al. (1982)
Tissue-type plasminogen activator	Protease	1	PENNICA et al. (1983)
Thrombomodulin	Thrombin cofactor	6	JACKMAN et al. (1987), WEN et al. (1987), SUZUKI et al. (1987)
Thrombospondin	Extracellular interactions	3	LAWLER and HYNES (1986)
Uromodulin (Tamm-Horsfall urinary glycoprotein)	Immunosuppression	2	PENNICA et al. (1987)
Chondroitin SO_4 proteoglycan	Extracellular interactions	2	KRUSIUS et al. (1987)
Cytotactin	Extracellular interactions	4	JONES et al. (1988)
Entactin (nidogen)	Extracellular interactions	6	DURKIN et al. (1988), MANN et al. (1989)
EGF precursor	proEGF	9	GRAY et al. (1983), SCOTT et al. (1983), DOOLITTLE et al. (1984)
LDL receptor	Cholesterol transport	3	RUSSELL et al. (1984), SÜDHOFF et al. (1985a, b)
LDL receptor-related protein	Cholesterol transport	6	HERZ et al. (1988)-
Laminin A chain	Cell adhesion	15	SASAKI et al. (1988)
Laminin B2 chain	Cell adhesion	10	SASAKI and YAMADA (1987), PIKKARAINEN et al. (1988)
Laminin B1 chain (mammalian)	Cell adhesion	13	SASAKI et al. (1987), PIKKARAINEN et al. (1987)
Laminin B1 chain (*Drosophila*)	Cell adhesion	13	MONTELL and GOODMAN (1988)
Lymph node homing receptor	Cell adhesion	1	SIEGELMAN et al. (1989), LASKY et al. (1989)
Endothelial leukocyte adhesion molecule-1	Cell adhesion	1	BEVILACQUA et al. (1989)

Table 1 (continued)

Protein	Function	EGF like units	References
GMP-140	Cell adhesion	1	JOHNSON et al. (1989)
Malarial antigen	Sexual stage surface protein	4	KASLOW et al. (1988)
Sea urchin gene	Development	9	HURSH et al. (1987)
Notch gene (*Drosophila*)	Neurogenesis	36	WHARTON et al. (1986)
Lin-12 gene (*C. elegans*)	Binary decisions in development	13	GREENWALD (1985), YOCHEM et al. (1988)
Delta gene (*Drosophila*)	Neurogenesis	9	VÄSSIN et al. (1987)
95F gene (*Drosophila*)	Neurogenesis	5	KNUST et al. (1987)

Many of the mammalian proteins listed in Table 1 contain a β-hydroxyaspartic acid or β-hydroxyasparagine residue within the EGF-like motifs: factor X (MCMULLEN et al. 1983 b), factor IX (MCMULLEN et al. 1983 a), protein S (STENFLO et al. 1987), protein C (DRAKENBERG et al. 1983), factor VII (MCMULLEN et al. 1983 a), LDL receptor (STENFLO et al. 1988), thrombomodulin (STENFLO et al. 1988), complement factor C1r (PRZYSIECKI et al. 1987), and protein Z (HØJRUP et al. 1985). These posttranslational modifications occur at a conserved aspartic acid or asparagine residue that is usually the second residue following the third cysteinyl in the general consensus sequence for the EGF-like motifs (for a recent summary of these sequences see REES et al. 1988). Interestingly, this conserved hydroxylation site is not present in the sequences for mature EGF or related growth factors, but is present within some of the EGF-like motifs of other nonmitogenic proteins listed in Table 1. However, whether hydroxylation occurs at these site has not been determined. This includes some of the repeat units in the EGF precursor (Sect. C. I). Several studies (SUGO et al. 1984; MORITA et al. 1984; ÖHLIN and STENFLO 1987; ÖHLIN et al. 1988) indicate that these hydroxylated residues are involved in the formation of a Ca^{2+} binding site ($K_d \approx 0.1$ mM). The physiological significance of one of these hydroxylation sites is indicated by site-directed mutagenesis of the aspartic acid residue in factor IX (REES et al. 1988). This abolished the clotting activity of this protein. Also, DETON et al. (1988) indicate that point mutations of a conserved glycine residue within the EGF-like domain of factor IX are strongly correlated with and may be the cause of mild hemophilia B.

III. Physical Properties

Physical properties of EGF were described by TAYLOR et al. (1972) and HOLLADAY et al. (1976) and have been discussed in a previous review (CARPENTER 1981). There has not been published research in this area since that time. All the EGF and EGF-like molecules are heat and acid stable and have acidic isoelectric points. HOLLADAY et al. (1976) examined the thermodynamics of unfolding of mouse EGF in solutions of guanidinium hydrochloride and concluded that EGF is one of the most stable proteins described, $\Delta G(25°) = 16 \pm 7$ kcal/mol. However, a subsequent study by DeMARCO et al. (1983), using a different technique to measure unfolding, suggested that the stability, $\Delta G(25°) = 8.2$ kcal/mol, actually is at the lower range of the previous measurements.

IV. High Resolution Structure

A major area of recent progress has been the application of high resolution NMR spectroscopy to determine the three-dimensional solution structure of EGF. Although numerous attempts have been made to crystallize EGF for X-ray diffraction studies, only one recent report (HIGUCHI et al. 1988) indicates success in the preparation of human EGF crystals; a 4-Å diffraction pattern was published.

1. Polypeptide Backbone

Circular dichroism studies of HOLLADAY et al. (1976) indicated that mouse EGF had no α-helical content, but did contain approximately 22% β-sheet structure, suggesting that most of the polypeptide backbone was in an unordered conformation. Consistent with this, early NMR studies provided evidence for an antiparallel β-sheet in the polypeptide backbone of mouse EGF (MAYO 1984, 1985). Subsequent NMR studies of human EGF identified residues 18–23 and 28–34 as the major sites of antiparallel β-sheet conformation (CARVER et al. 1986; MAKINO et al. 1987). MONTELIONE et al. (1986) extended this analysis of β-sheet structure in mouse EGF to include residues 2–4, 37–38, and 44–45 in addition to the residues 19–23 and 28–32 previously described for human EGF. These studies, together with the localization of disulfide bonds (C. R. SAVAGE et al. 1973; WINKLER et al. 1986), define what is presently known about the secondary and tertiary structure of EGF.

The initial NMR studies were followed by the publication of most all ^{1}H NMR assignments (KOHDA and INAGAKI 1988; MONTELIONE et al. 1988; MAYO et al. 1988) and the application of distance geometry, energy minimization, and molecular dynamics computations to produce models of the three-dimensional structure of human EGF (COOKE et al. 1987), mouse EGF (MONTELIONE et al. 1987; KOHDA et al. 1988), and rat EGF (MAYO et al. 1989). Before describing the results of these studies it should be noted that all NMR measurements have been performed at pH 3 due to the insolubility of concentrated EGF solutions. With only small discrepancies, these studies have produced a three-dimensional representation of the EGF polypeptide backbone. This structure for mouse EGF is shown in Fig. 2A (MONTELIONE et al. 1987). Additional aspects of the spatial relationships of individual residues in mouse EGF are shown in Fig. 2B (MAYO et al. 1987).

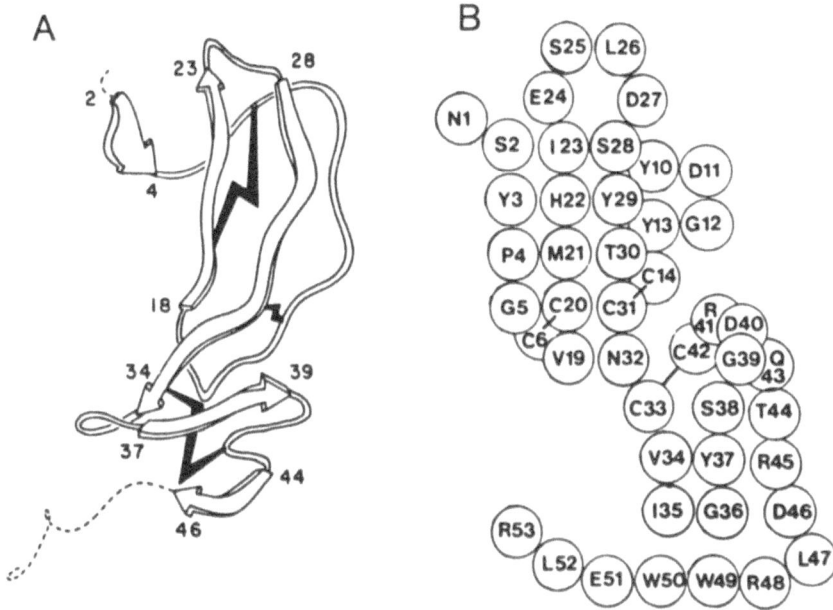

Fig. 2 A, B. Polypeptide backbone structure of mouse EGF. **A** Schematic drawing of the polypeptide chain fold (adapted from MONTELIONE et al. 1987). *Arrowed ribbons* indicate the position and direction of the β-sheets, *dotted lines* designate terminal segments for which data was not reliable, and disulfide bonds are indicated by darkened "*lightning bolts.*" **B** Suggested proximity relationships of residues within the folded polypeptide backbone (from MAYO et al. 1987). Disulfide bonds are indicated by *straight lines* connecting the cysteine residues

At the N terminus there is a short segment of β-structure (residues 2–4), followed by a substantial segment of irregular, multibend structure (residues 5–15), some of which may be in a right-handed helix-like conformation. The next two residues form a turn structure. This description would indicate that most of the first disulfide loop of EGF (residues Cys5→Cys20) contains little well-ordered structure. Beginning at residue 18, a β-sheet structure begins and extends to residue 23. This segment of β-sheet structure is followed by a type I bend (residues 25–27) and then another segment of β-sheet structure encompassing residues 28–33. These two β-sheet structures are in close proximity, with hydrogen bonding between the segments, and are in an antiparallel orientation. The second disulfide loop, therefore, is characterized by a significant level of ordered β-type structure.

The final segment of EGF, residues 34–53, is described by NMR data as forming a double hairpin structure. Within this segment of the EGF molecule there is a type II β-bend at residues 34–36 and a short region of β-sheet structure, residues 37–38. This is followed by a turn region, residues 41–43, and another short segment of β-sheet structure (residues 45–46) which is hydrogen bonded and in an antiparallel orientation to the preceding β-sheet. The small, third disulfide loop is characterized by substantial β-sheet conformation. The final C-terminal residues (48–53) are thought to exist in an unordered state.

Two groups (MONTELIONE et al. 1987; COOKE et al. 1987) treated the EGF molecule as composed of somewhat separated (i.e., noninteracting) domains of residues 1–33 and residues 34–53. This distinction was based on the absence of significant nuclear Overhauser effect (NOE) signals between residues in the two segments. Interestingly, mature EGF and TGF-α are each encoded by two exons that interrupt both sequences at residue 32 (see Sect. D. II). KOHDA et al. (1988), however, have reported a number of interdomain NOE signals and their model is somewhat more compact and globular, particularly in the close positioning of residues 13 and 41 (where interdomain NOE signals were detected). In their EGF model, the loop at residues 39–43 is drawn closer to the N-terminal domain.

The central features of EGF structure are the three intrachain disulfide bonds and the areas of antiparallel β-sheet structure in the C-terminal and N-terminal domains. The orientation and proximity of the β-sheet regions has been aided by the assignment of slowly exchanging amide protons, suggestive of hydrogen bond participation.

By comparison with NOE signals defining the arrangement of the peptide backbone in the central portion of the EGF molecule, relatively few NOE crosspeaks are available to define positioning of the N- and C-terminal "tails" relative to the rest of the molecule. For this reason, most all studies have considered that these portions of the EGF molecule are relatively unrestrained. Although a short β-sheet region in the N terminus was detected in human EGF (COOKE et al. 1987), mouse EGF (MONTELIONE et al. 1987), and rat EGF (MAYO et al. 1989), MAKINO et al. (1987) did not detect signals for this structure in human EGF, nor did KOHDA et al. (1988) for mouse EGF, nor MONTELIONE et al. (1989) for human TGF-α. Signals for this N-terminal strand are then either weak or not detectable, suggesting that this region exists in an equilibrium of multiple backbone conformations and only transiently interacts with the central portion of the molecule. NOE signals for the C terminus, i.e., residues 48–53, have not been detected and most models consider this segment of the polypeptide backbone to be unconstrained in conformation.

2. Aromatic Clusters

Photochemical (photo-CIDNP) NMR studies (DEMARCO et al. 1983; MAYO et al. 1986) to determine the accessibility of aromatic side chains to a "reporter" dye, as well as the conventional NMR studies cited above, have concluded that nearly all aromatic side chains are present on the surface of EGF and are quite solvent accessible. Given the small size of EGF this is not unexpected. The data also indicate that many of these aromatic side chains are grouped in clusters forming hydrophobic pockets on the EGF molecule. Such aromatic clusters were reported to be present on the surface of the insulin molecule and to perhaps have an important role in hormone-receptor interactions (BLUNDELL et al. 1971). In the case of mouse EGF, MAYO (1984, 1985) and KOHDA et al. (1988) suggest that the side chains of Tyr10, Tyr13, His22, and Tyr29 form a hydrophobic pocket in the N-terminal domain of the growth factor (Fig. 2B). To a lesser extent Tyr37, Trp49, and Trp50, together with Val34 and Ile35, may form a hydrophobic pocket in the C-terminal domain. The N-terminal hydrophobic cluster has been substantiated by others, but the C-terminal cluster has not. The proximity of aromatic side chains at residues 10 and 13 (within the irregular multibend loop, residues 6–17) to the aromatic side chains at residues 22 and 29 (within the major segments of antiparallel β-sheet structure, residues 18–34) helps to position these two large regions of the N-terminal domain.

3. TGF-α and EGF Comparisons

MONTELIONE et al. (1989) have reported initial results of NMR studies of the structure of human TGF-α. The general features of TGF-α seem to be very similar to the structural models presented for EGF. Perhaps the most significant difference noticed between TGF-α and EGF in this study relates to the structural dynamics of the two molecules. In EGF, many of the amide proton/deuteron exchange rates, including those for residues involved in hydrogen bonding within the β-sheets, are very low (CARVER et al. 1986; MONTELIONE et al. 1988; MAYO et al. 1989). By comparison the amide proton exchange rates for many of the residues involved in hydrogen bonding of the β-sheets of TGF-α are reported to be much higher. This may suggest that, at least under these experimental conditions (pH 3), the TGF-α structure is more flexible than the EGF structure.

4. Hybrid Molecules

As mentioned in the previous section NMR studies have suggested that EGF consists of independently folded N-terminal (residues 1–33) and C-terminal (residues 34–53) domains. PURCHIO et al. (1987) have utilized recombinant DNA techniques to provide support for this by creating a hybrid gene encoding the C-terminal segment of human TGF-α and the N-terminal segment of the vaccinia virus growth factor (VGF). In the number system shown in Fig. 1, the hybrid gene encodes Val − 2 to Val32 from the human TGF-α sequence and Cys33 to Arg52 of the VGF sequence. The hybrid TGF-α/VGF growth factor produced from this gene was as biologically active in mitogenesis assays as either TGF-α or VGF. The success of this hybrid molecule is reasonably predicted by the organization of two exons coding for mature TGF-α or EGF which interrupt the sequence at residue 32 (Sect. D. II).

V. Structure-Function Relationships

Two categories of experiments have been applied to elucidate the manner in which structural alterations of the EGF molecule may affect functional properties of the growth factor. The most recently developed of these is the application of site-directed mutagenesis. The second is the analysis of fragments of the EGF molecule produced by several means – proteolysis, chemical cleavage, or peptide synthesis. Since site-directed mutagenesis involves the smallest change and usually occurs within the context of the intact molecule, those studies will be covered first. The fragment studies are more complex in interpretation for several reasons and will be considered subsequently.

1. Site-Directed Mutagenesis

The results of site-directed mutagenesis studies of EGF and TGF-α are summarized in Table 2. These analyses have centered on either the highly conserved residues (Fig. 1) or residues indicated by NMR data to be involved in β-sheet structures. In addition to the conserved cysteine residues, which are known to be required for activity, interest has centered on the highly conserved leucine at position 47. All substitutions at this position, including conservative replacements with valine or isoleucine, drastically reduce growth factor activity. Tyr37 also is highly conserved and is thought to participate in hydrogen bonding between β-sheets. Interestingly, replacement of this residue with another aromatic amino acid (phenylalanine) is without effect on activity, but reduction of the side chain (alanine, glycine) markedly decreases activity.

The aromatic side chains of residues 10, 13, 22, and 29 are thought to be in close proximity and to form a hydrophobic pocket on the surface of EGF (Sect. B. IV. 2). Single mutations have been characterized for each of these residues (in either EGF or TGF-α), demonstrating that nonconservative replacements are not tolerated in regard to biological activity. Unexpectedly, insertion of glycine for the charged residues at positions 24 and 27 (glutamic and aspartic acid, respectively), which occur at a turn region between two β-sheet strands, are reasonably well tolerated. In contrast, the conservative replacement of isoleucine at position 23 with threonine results in a large decrease in biological activity. Nonconservative changes for the aspartic acid at position 46, which is a conserved residue in all the EGFs and TGF-αs, are reasonably tolerated though there are some contradictions in the results. DEFEO-JONES et al. (1988) inserted an alanine at this position and lost nearly all activity, but LAZAR et al. (1988) found the opposite result for the same change.

2. Growth Factor Fragments

Other approaches have been used to assess structure-function relationships in the EGF-like family of molecules. These include analyses of chemically synthesized fragments of EGF or TGF-α or modification, usually by proteolysis, of the intact growth factors. Fragments are found in the course of growth factor isolation or result from deliberate chemical or enzymatic treatments of the intact molecules.

Table 2. Site-directed mutagenesis of EGF and TGF-α

Growth factor	Mutagenesis	Activity[a] (% wild type)	References
Mouse EGF	Leu47 → Val	27	RAY et al.
	Leu47 → Ser	12	(1988)
Mouse EGF	Met21 → Lys[c]	100[b]	BURGESS et al. (1988)
Human EGF	Met21 → Leu	100	SUMI et al. (1985)
Human EGF	Met21 → Norleucine	100[b]	KOIDE et al. (1988)
Human EGF	Pro7 → Thr	55	ENGLER et
	Glu24 → Gly	86	al. (1988)
	Asp27 → Gly	48	
	Tyr29 → Gly	17	
	Leu47 → His	7	
Human EGF	Met21 → Thr	36	CAMPION et
	Tyr22 → Asp	8	al. (1989)
	Ile23 → Thr	3	
	Tyr29 → Gly	17	
Human TGF-α[d]	Asp46 → Ala	125	LAZAR et
	Asp46 → Asn	83[e]	al. (1988)
	Asp46 → Glu	20	
	Asp46 → Ser	49	
	Leu47 → Ala	<1	
	Leu47 → Ile	<1	
	Leu47 → Met	<1	
Human GF-α[d]	His10 → Ala	26	DEFEO-
	Phe13 → Ala	1	JONES et
	Asp46 → Ala	4	al. (1988)
	Tyr37 → Ala	4	
	Tyr37 → Phe	250	
	Tyr37 → Phe, Lys28 → Arg	111	
	Cys6, 14, 20, 31, 33, 42 → Ala	0	
	Cys6, 14, 33, 42 → Ala	0	

[a] Competitive binding assay data were used to assess activity relative to control molecules (intact EGF or TGF-α).
[b] Mitogenesis assays only reported.
[c] Expressed in truncated EGF, residues 4–48.
[d] The numbering system of EGF shown in Table 1 is followed.
[e] Average of two points – 126% and 40%.

a) Synthetic Peptides

Several groups have successfully synthesized biologically active molecules corresponding to a complete sequence of one of the EGF-like growth factors: mouse EGF (AKAJI et al. 1985; HEATH and MERRIFIELD 1986), human TGF-α (TAM et al. 1986), rat TGF-α (TAM et al. 1984), and the Shope virus growth factor (Y. Z. LIN et al. 1988). This demonstrates that proper folding of the polypeptide chain and formation of the correct disulfides can occur in vitro. A similar conclusion was reached previously by C. R. SAVAGE et al. (1973) who reduced native EGF and, following reoxidation, quantitatively recovered biologically active material.

There is understandable interest in determining whether defined fragments of the EGF structure might be capable of acting as agonists or antagonists of EGF. One study (KOMORIYA et al. 1984) reported limited success with peptides corresponding to mouse EGF sequences Cys20→Cys31 (no disulfide) and Tyr14→Cys31 (one disulfide). Also, Leu15→Arg53, containing two disulfides corresponding to the second and third disulfide loops, was reported to have EGF-like activity (HEATH and MERRIFIELD 1986) as was the rat TGF-α sequence Cys33→Cys→42 (no disulfide) (NESTOR et al. 1985). However, the potency displayed by any of these synthetic peptides was very low, about 0.01% of the potency of the intact molecule. Subsequent studies have suggested even lower or nondetectable activities for these and other synthetic fragments corresponding to the N terminus, C terminus, and each disulfide loop, and for one peptide corresponding to the first and second disulfide loops (KOMORIYA et al. 1984; NESTOR et al. 1985; HEATH and MERRIFIELD 1986; DEFEO-JONES et al. 1988; DARLAK et al. 1988). The results of these studies suggest that it will probably be necessary to include all three disulfide loops in any synthetic derivatives of this growth factor that retain significant biological activity.

b) Chemical Modification

Aside from the iodination of unidentified tyrosine residues, the only chemical modification studies have involved the single methionine at position 21 in the EGF sequence. Initial reports (SCHECHTER et al. 1979; SCHREIBER et al. 1981; YARDEN et al. 1982) indicated that cyanogen bromide-treated EGF, which should be cleaved in the polypeptide backbone between Met20 and the next residue, was capable of binding to the EGF receptor with high affinity and inducing protein phosphorylation and other early responses provoked by intact EGF. The cyanogen-bromide treated growth factor did not, however, stimulate DNA synthesis, a late response. A more recent study (BURGESS et al. 1988) has failed to confirm these results. They found that cyanogen bromide-treated EGF exhibited only very low affinity binding to the EGF receptor, approximately 100-fold less than that of native EGF. The discrepancy in these results seems attributable to the fact that a significant amount of EGF is not cleaved in the cyanogen bromide treatment and that only the data of BURGESS et al. (1988) are based on careful separation of the modified and unmodified reaction products. The earlier studies may have been compromised by the presence of intact EGF in the cyanogen bromide-treated EGF preparations.

Intact EGF, when stored in vitro, undergoes oxidation of methionine to produce methionine sulfoxide (HEATH and MERRIFIELD 1986; GEORGE-NASCIMENTO et al. 1988; RIEMAN et al. 1987). This modification does not significantly alter biological activity.

c) N-Terminal Modifications

The effects of different types of modifications at the N terminus of EGF and TGF-α are available to assess the functional role of this portion of the molecule. Three types of alterations have been described – additions, truncations, and replacements. Each will be described in the following paragraphs.

Several covalent additions at the N terminus have been made to produce conjugates for other types of studies. Fluorescein (HAIGLER et al. 1978; CHATELIER et al. 1986) and ferritin (HAIGLER et al. 1979) conjugates of mouse EGF were constructed for morphologic studies, while ricin and diphtheria A chain conjugates (CAWLEY et al. 1980) were produced for toxicity studies. Neither the fluorescein nor the toxin conjugations (which would add 20 000–30 000 Da to the N terminus) decreased ligand binding to the EGF receptor. Even the addition of ferritin (approximately 450 000 Da) to the N terminus did not significantly perturb growth factor-receptor interactions and only slightly (25%) reduced the mitogenic activity of the conjugated EGF. In a study of recombinant TGF-α molecules, DEFEO-JONES et al. (1988) expressed the growth factor as a fusion protein containing seven additional residues (from β-galactosidase) at the N terminus. This fusion protein was equivalent to native TGF-α in both receptor binding and mitogenesis assays. Obviously (Fig. 1), the vaccinia virus growth factor is extended by an additional 18 amino acids, compared to the TGF-αs, at the N-terminal portion of the molecule and this does not affect its capacity to bind to the EGF receptor or to act as a mitogen. The N-terminal half of amphiregulin contains non-growth factor-related sequences, but this protein is also truncated at the C terminus and does display reduced receptor interaction compared to EGF (SHOYAB et al. 1989).

Deletions of residues at the N terminus seem to occur either in vivo or during growth factor isolation and involve the loss of up to three amino acids. During reverse-phase high-performance liquid chromatography (HPLC) analysis of purified mouse EGF, two groups (MATRISIAN et al. 1982; BURGESS et al. 1982) noticed two distinct species termed α and β. Further anion-exchange HPLC analysis of the α-EGF species indicated the presence of other molecular species (BURGESS et al. 1983). Subsequent chemical analyses showed that α-EGF is the intact 53-residue polypeptide and that the structure of β-EGF corresponded to residues 2–53 (PETRIDES et al. 1984; DIAUGUSTINE et al. 1985; HYVER et al. 1985). MAYO and BURKE (1987) and RIEMEN et al. (1987) subsequently described a second N-terminal fragment, termed γ-EGF, that contains residues 3–53 of murine EGF. SIMPSON et al. (1985) have described similar species of rat EGF plus a new fragment lacking the first three amino acids. Lastly, an engineered fragment of human TGF-α lacking the first seven amino acids has been reported by DEFEO-JONES et al. (1989). This group also created a hybrid protein by replacing the N-terminal seven residues of human TGF-α with a seven-residue sequence of different amino acids from β-galactosidase. Though one early report (MATRISIAN et al. 1982) suggested that the biological activity of α-EGF was significantly greater than that of β-EGF, most all subsequent studies agree that all the N-terminal deletions of EGF and TGF-α and the one N-terminal replacement of TGF-α are approximately equipotent in regard to biological activity.

Finally, MAYO and BURKE (1987) have used NMR spectrosocpy to investigate the structures of α-, β-, and γ-mouse EGF. They suggest that the structure of the polypeptide backbone in these forms of EGF is not significantly different. They do, however, suggest that the β- and γ-species, based on more rapid backbone amide proton/deuteron exchange rates, have more open and less stable structures than α-EGF.

In summary, there is no evidence that any N-terminal structures or sequence prior to the first cysteine residue exerts a significant influence on the function of EGF or TGF-α.

d) C-Terminal Fragments

Inspection of the amino acid sequences for the different members of the EGF family indicates considerable heterogeneity, in terms of sequence similarity and length, at the C terminus. Amphiregulin, which has the shortest N terminus compared to other members of the EGF family and lacks the conserved and biologically important leucine at position 47, has the lowest binding affinity for the EGF receptor (SHOYAB et al. 1989). Rat EGF, however, does interact with the EGF receptor as well as mouse EGF (SIMPSON et al. 1985; SCHAUDIES and SAVAGE 1986), suggesting that the C-terminal Trp-Trp-Glu-Leu-Arg sequence is not a significant factor in determining the biological activity of mouse and human EGF. (As mentioned earlier in Sect. B. I, however, the cDNA sequence of rat EGF indicates that a Trp-Trp-Asn-Trp-Arg-His sequence follows the reported C-terminal arginine of the protein sequence for rat EGF; DOROW and SIMPSON 1988). The conclusion suggested by the studies with rat EGF is supported by studies of EGF fragments lacking the final one, two, or three residues at the C terminus (SAVAGE and COHEN 1972; COHEN et al. 1975; HOLLENBERG and GREGORY 1980; SAVAGE and HARPER 1981; MOUNT et al. 1985; GREGORY et al. 1988; GEORGE-NASCIMENTO et al. 1988; BURGESS et al. 1988). All studies agree that removal up to three residues from the C terminus of EGF does not affect receptor binding or mitogenic activity.

The lysine or arginine residue at position 48 provides a convenient site for trypsin proteolysis to produce an EGF fragment corresponding to residues 1–48. Two studies (COHEN et al. 1975; GREGORY et al. 1988) showed that the 1–48 fragment has significantly reduced biological potency (5- to 10-fold) in in vitro receptor binding and mitogenesis assays. At higher concentrations, however, this EGF fragment was as active as the intact molecule. When tested in the newborn mouse eye-opening assay, the 1–48 fragment and the intact growth factor had equivalent activity (SAVAGE et al. 1972). One study (BURGESS et al. 1988) indicates the opposite – that the 1–48 EGF fragment does not have reduced potency when assayed in vitro. These authors suggest that, unless carefully monitored by HPLC analysis, trypsinization of intact

EGF can also cleave at Arg45, producing an additional fragment corresponding to residues 1–45 which they describe as inactive in binding and mitogenesis assays. GREGORY et al. (1988) have shown that a protease fragment corresponding to residues 1–42 of human EGF is inactive. It is possible that the prior studies utilized a mixture of fragments 1–48 (active) and 1–45 (inactive).

Three reports (HOLLENBERG and GREGORY 1980; BURGESS et al. 1988; GREGORY et al. 1988) added to analysis of the C terminus. Tryptic fragments 1–48 of mouse or human EGF have been further digested with carboxypeptidase to produce fragments corresponding to residues 1–47 and residues 1–46. The 1–47 fragment has reduced potency (approximately 10% compared to the native growth factor), but the 1–46 fragment displayed very little activity in any assay, its potency being less than 1% of the intact molecule. These studies indicate that removal of residues C-terminal to the conserved Leu47 does not have a drastic effect on growth factor activity. However, Leu47 is necessary for the biological function of the EGF family of growth factors. This point is consistent with site-directed mutagenesis studies of Leu47 in both EGF and TGF-α (Table 2). Proton NMR studies demonstrate that the conformation of the 1–48 EGF fragment does not differ significantly from that of the intact molecule, but the 1–46 fragment does show a significant perturbation in overall structure (DeMARCO et al. 1986).

These results do not necessarily mean that the C-terminal sequences beyond Leu47 are without function. MAYO et al. (1987) have used proton NMR spectroscopy to monitor the interaction of EGF with detergent micelles, used as a model of biological membranes. The experiments suggest that a cluster of aromatic residues (Trp49, Trp50, and Tyr37) are involved in interactions between mouse EGF and the micelles. Similar studies with the 1–48 fragment of mouse or rat EGF, both of which do not have the tryptophan residues, indicate a cluster near Tyr10 (known from other NMR studies to be within an aromatic cluster; Sect. B. IV. 3) may predominate in these fragments to mediate interactions with micelles. These are results of interest as it is possible that some hormones may interact with the lipid bilayer either prior to or during interaction with their receptors.

C. Precursor Molecules

cDNA cloning has led to elucidation of the sequence of the primary translation products encoding several of the growth factors shown in Table 1. The complex and unanticipated structure of these precursor molecules has produced what is almost a separate field, the biology of these precursor molecules. The different precursor structures and their biological properties will be discussed in the following sections.

I. EGF

Two groups (SCOTT et al. 1983; GRAY et al. 1983) reported the cDNA sequence and deduced protein sequence of the mouse EGF precursor. Sub-

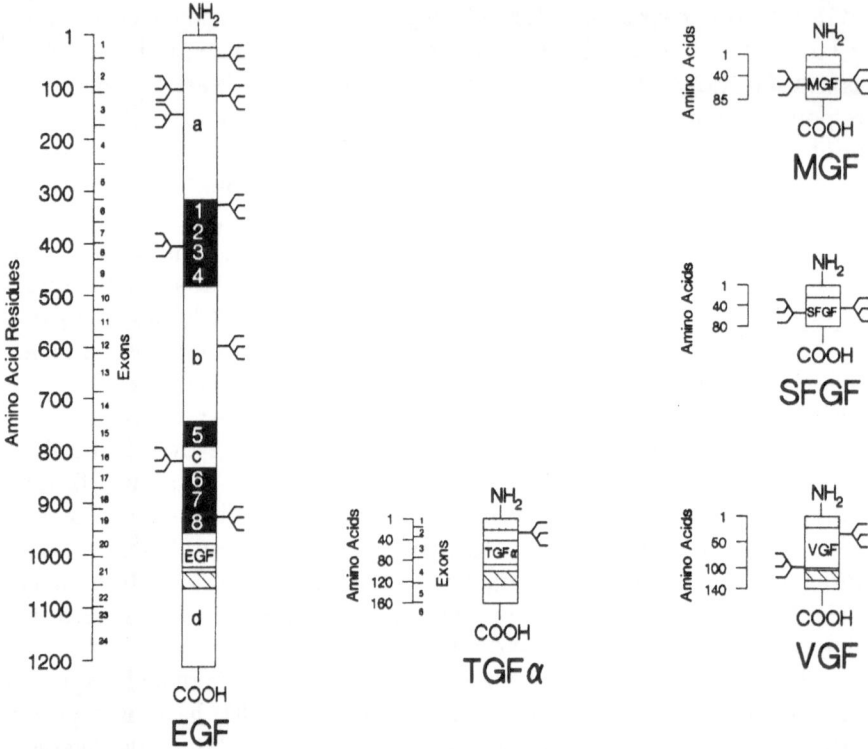

Fig. 3. Structural organization of precursor proteins for EGF and related growth factors. The precursor structures shown are for human EGF (BELL et al. 1986), human TGF-α (DERYNCK et al. 1984, 1985), vaccinia growth factor (VENKATESAN et al. 1982; STROOBANT et al. 1985), Shope fibroma growth factor (CHANG et al. 1987), and myxoma growth factor (UPTON et al. 1987). The latter two molecules are deduced from DNA sequences – the proteins have not been detected. *Crosshatched areas* indicate a C-terminal hydrophobic sequence that functions as a membrane anchor. *Dotted,* N-terminal areas indicate hydrophobic signal sequences that are likely removed from the precursor. Potential sites for N-linked glycosylation are indicated by *branched structures.* Within the EGF precursor, EGF-like repeats are designated by numbers *1–8*

sequently, BELL et al. (1986) described the precursor for human EGF. Features of the precursor molecule for human EGF are shown in Fig. 3. The size and structural organization of the precursor for mouse EGF are extremely similar. The most surprising aspect of these precursors is their size – 1200 amino acid residues (130000 Da) encoding a mature growth factor of 53 residues (6000 Da). The mouse and human EGF precursors are approximately 66% identical in sequence. In comparison, the identity between the low molecular weight mature mouse and human EGF sequences is 70%, suggesting that the entire precursor molecule is as conserved as the biologically active growth factor sequence.

The sequence for mature EGF is located near the C terminus of the precursor (residues 971–1023). In the mouse and human precursor molecules, the C and N termini of the mature EGF sequences are bound by arginine/aspara gine and arginine/histidine residues, respectively. Thus the mature 53-residue EGF molecule could be released from the precursor by an arginine-specific endopeptidase. However, the exact processing event has not been demonstrated.

Hydropathy plots identify two stretches of hydrophobic sequences in the EGF precursors. One is at the N terminus of the precursor and is presumed to be a signal sequence related to biosynthesis that is removed shortly after translation. The second hydrophobic sequence occurs near the C terminus of the precursor following the sequence for mature EGF. This hydrophobic sequence is followed by a short sequence that contains a high number of basic amino acids typical of a "stop transfer" sequence found in transmembrane proteins. It has been suggested, based on the sequence data alone, that the EGF precursor is a transmembrane protein anchored in the membrane by this single transmembrane sequence. Experimental evidence exists for the membrane association of the EGF precursor molecule in transfected cells (BELL et al. 1986; MROCZKOWSKI et al. 1988, 1989), plus kidney and uterine tissue (DiAUGUSTINE et al. 1988). Consistent with the transmembrane structure of the EGF precursor are the presence of canonical sites for N-linked glycosylation. There are nine of these sites in the putative extracellular domain of the human EGF precursor and six in the mouse precursor. Two of these potential glycosylation sites are conserved in the mouse and human precursor molecules, corresponding to asparagine residues 104 and 404 in the human precursor. MROCZKOWSKI et al. (1989) have provided biochemical evidence that the precursor is, in fact, a glycoprotein.

The putative cytoplasmic domain of the EGF precursor has no known function or sequence similarity to other proteins. Also, this domain has the lowest sequence identity (47%) of any major regions of the mouse and human EGF precursor molecules. In contrast, the external domain sequences are 70% identical. In the initial cDNA sequence reports (SCOTT et al. 1983; GRAY et al. 1983), computer analysis of the mouse EGF precursor showed a number of internal repeats of the EGF-like sequence. This computer analysis was extended by DOOLITTLE et al. (1984) and the subsequent cDNA sequence of the human precursor (BELL et al. 1986) has also added to these data. There are eight EGF-like sequences in addition to the mature EGF sequence in both the mouse and human precursors. Many of these repeat units occur in tandem and seem to be organized into two large groups separated by approximately 260 amino acid residues (region b in Fig. 3). Although the precursor structure suggests that low molecular weight EGF-like molecules might be generated from one or more of the repeat units, there is no evidence for this. Unlike the mature EGF sequence, none of the repeat units are bounded by basic amino acid residues that might function as typical protease cleavage sites.

Interestingly, three of the repeat units (numbers 2, 7, 8 in Fig. 3) resemble some EGF-like sequences in non-mitogenic proteins in regard to the presence of a conserved aspartic acid/asparagine residue within the EGF consensus sequence. This residue is the second amino acid following the third cysteine residue in the consensus sequence. In many of the nonmitogenic proteins having EGF-like units, this aspartic acid/asparagine residue is hydroxylated and may function as a calcium binding site (see Sect. B. II). It is not known whether this posttranslation modification actually occurs in the EGF precursor.

Several proteins have been identified by sequence similarity with portions of the EGF precursor molecule. The most significant of these is the low density lipoprotein (LDL) receptor. A large segment of EGF precursor, including repeats 3 and 4, region b, and repeat 5, is approximately 33% identical to an extracellular segment of the LDL receptor (RUSSELL et al. 1984; SÜDHOF et al. 1985 a, b). Interestingly, this central portion of the human EGF precursor, residues 340→756 (including region b), contains the highest level of sequence identity (82%) with the mouse EGF precursor. A lesser, but statistically significant, sequence relationship was reported between the N-terminal region a of the EGF precursor and a portion of the atrial natriuretic precursor (HAYASHIDA and MIYATA 1985). Two groups (SCHNEIDER et al. 1984; MCCLELLAND et al. 1984) have reported a low level of sequence similarity between the transferrin receptor and the EGF precursor. This similarity is within the same region of the EGF precursor as the LDL receptor sequence similarity. Lastly, BALDWIN (1985) has reported that portions of v-*mos*, the oncogene of the Moloney murine sarcoma virus, are related to the EGF precursor. The greatest similarity was between a C-terminal segment of *mos* and a portion of the cytoplasmic domain of the EGF precursor.

While functions for the EGF precursor molecule, aside from the generation of mature EGF, remain speculative, two points deserve note. MROCZKOWSKI et al. (1989) indicate that the high molecular weight ($M_r = 130000$) precursor is biologically active, both in supporting cell growth and in radioreceptor competition assays with ^{125}I-EGF. Also, the intact precursor molecule is reported (RALL et al. 1985) to accumulate in the convoluted portion of the distal tubules in the kidney where it may account for nearly 1% of the cell protein (BELL et al. 1986). Since this portion of the kidney is involved in regulating urine composition, particularly Cl^-, Na^+, and other ions, it has been speculated that the precursor may be involved in these transport processes, acting perhaps as a receptor (PFEFFER and ULLRICH 1985).

Since the high molecular weight EGF precursor is capable of binding to the EGF receptor, an intriguing idea is that perhaps this is a primitive system used for cell-cell recognition and/or adhesion in lower organisms. This would, of course, require that the membrane-anchored EGF precursor be capable of receptor recognition, but the notion is consistent with the known functions of membrane-anchored proteins in lower eukaryotes that contain EGF-like sequence repeats (Sect. B. III).

II. TGF-α

Sequences of the precursor molecules for human (DERYNCK et al. 1984) and rat (LEE et al. 1985b) TGF-α have been derived from cDNA cloning and are extremely similar. The structural organization of the precursor for human

TGF-α is shown in Fig. 3. The TGF-α precursors are much smaller than the EGF precursor molecules, containing no growth factor-like repeat units. Similarities do exist in the presence of two hydrophobic sequences – a signal peptide sequence at the N terminus and a potential transmembrane sequence following the mature TGF-α sequence. This second hydrophobic sequence is followed by several basic residues suggestive of a "stop transfer" sequence. While all mature EGF molecules are bounded by Arg-X sequences, TGF-α sequences are bounded by Ala-Val residues, indicating that different proteases are necessary to produce each of the mature growth factors.

The sequences for the precursor molecules in rat and human are very identical – 92%. The rat and human sequences for mature TGF-α, allowing one gap for matching, are also 92% identical. The N-terminal sequences in the rat and human precursors are 82% identical, while the C-terminal sequences that follow the mature TGF-α sequences are 99% identical. By comparison, the C-terminal sequences in the human and mouse EGF precursor are only 47% identical this is; the least conserved region of the EGF precursor. The putative cytoplasmic domain of the transmembrane TGF-α precursor is unlike the corresponding segment of the EGF precursor not only in amino acid sequence, but also in the presence of a high number of cysteine residues. Interestingly, the second hydrophobic (transmembrane) sequence in the TGF-α precursor is bounded by pairs of basic amino acids, typical proteolytic cleavage sites in hormone precursors. Protease activity at these sites could generate the extracellular TGF-α-containing domain and the cysteine-rich, 32-residue cytoplasmic domain.

Sequence data indicate the presence of one canonical site for N-linked glycosylation, N-terminal to the mature TGF-α sequence. Biochemical data support the prediction that the TGF-α precursor is membrane-localized (GENTRY et al. 1987; TEIXIDO et al. 1987; BRINGMAN et al. 1987) and that it is modified by both N-linked and O-linked carbohydrate (BRINGMAN et al. 1987; TEIXIDO et al. 1987; TEIXIDO and MASSAGUÉ 1988). The studies of BRINGMAN et al. (1987) indicate an additional modification of the precursor molecule – palmitoylation. As the fatty acyl group is sensitive to hydroxylamine, indicative of a thiolester bond, it is likely that acylation occurs at one or more of the cysteine residues in the C-terminal segment of the precursor. This modification occurs in certain other membrane-associated proteins and may have a role in anchoring the molecule to the membrane lipid bilayer.

While a number of researchers have reported the presence of high molecular weight forms of TGF-α-related protein in the conditioned media of transformed cells or in the urine of cancer patients, the molecular nature of these proteins has not been characterized. More recent studies have shown that a soluble glycoprotein of 17–19 kDa that reacts with antibodies to mature TGF-α, but not with antibodies to the cytoplasmic domain of the TGF-α precursor, is released into the media of transformed cells by endogenous proteases (IGNOTZ et al. 1986; BRINGMAN et al. 1987; GENTRY et al. 1987; TEIXIDO and MASSAGUÉ 1988; LUETTEKE et al. 1988). Thus, transformed cells produce the membrane-bound TGF-α precursor (≈ 25 kDa), a soluble fragment corresponding to the extracellular domain (18–21 kDa), and the fully processed mature form of TGF-α (6 kDa).

The 18 to 21-kDa fragment is biologically active (IGNOTZ et al. 1986; TEIXIDO and MASSAGUÉ 1988) and could be generated by protease activity at at least two possible sites between the C terminus of the mature TGF-α sequence and the transmembrane domain. One site is at the C terminus of mature TGF-α, a Ala-Val sequence, while the other site is a dibasic Lys-Lys sequence just before the transmembrane sequence. Either or both of these sites, or an unknown site, could be used to generate the soluble precursor. Addition of elastase, a protease that cleaves between Ala-Val sequences, has been shown to generate the 6-kDa mature TGF-α molecule from the soluble precursor in vitro (IGNOTZ et al. 1986). There are elastase-like Ala-Val sequences bounding both the N and C termini of the mature TGF-α sequence. It is not known, however, whether elastase could generate mature TGF-α from the membrane-bound, intact TGF-α precursor. Nor is it known which protease(s) actually carries out processing of precursor(s) to the mature growth factor molecule in vivo.

III. Pox Virus Growth Factors

The structural organization of the precursor molecule of the vaccinia virus growth factor (VGF) is shown in Fig. 3. This is based on the gene sequence reported by VENKATESAN et al. (1982) and the protein sequence reported by STROOBANT et al. (1985). CHANG et al. (1988) have described some of the events involved in the biosynthesis of the VGF precursor molecule.

The predicted VGF precursor is similar to the TGF-α precursor with the exceptions of a cysteine-rich cytoplasmic domain and elastase-like or dibasic protease sites bounding the mature growth factor sequences. Also, there are two N-glycosylation canonical sequences within the mature growth factor sequence. Biochemical data indicate that at least one glycosylation site is utilized. Therefore, the mature VGF is unlike EGF and TGF-α in that it is a glycoprotein.

There are few data describing processing of the VGF precursor, but infected cells contain a large soluble molecule which corresponds by amino acid composition data to the extracellular domain (minus the signal peptide sequence, which is cleaved during biosynthesis) of the precursor molecule (STROOBANT et al. 1985). This fragment, which contains approximately 77 amino acid residues, is not known to be further processed to a size comparable to that of mature EGF or TGF-α. If this occurred it would require removal of about 18 amino acids from the N terminus and 9 residues from the C terminus of the 77-residue molecule. Such processing would remove both glycosylation sites. However, the 77-residue glycoprotein is functionally active in various growth factor assays (STROOBANT et al. 1985).

Gene sequences of the Shope fibroma (CHANG et al. 1987) and myxoma (UPTON et al. 1987) viruses predict EGF-like translation products of 80 and 85 amino acids, respectively. Both of the putative products have N-terminal signal sequences, but neither has a hydrophobic sequence near the C terminus that might function as a membrane anchor. These proteins, therefore, may be secreted directly into the media of infected cells. At this time neither of the predicted proteins of these genes has been identified in virus-infected cells. The sequence information of these EGF-like genes contains two canonical sites for N-linked glycosylation. Unlike the vaccinia glycosylation sites, however, these sites for potential glycosylation in the Shope and myxoma proteins are within the 50-residue, conserved EGF-like sequence.

D. Growth Factor Genes

I. Chromosomal Localization

In humans the gene encoding EGF is located on chromosome 4 in the q25→q27 region (BRISSENDEN et al. 1984; ZABEL et al. 1985; MORTON et al. 1986), while in mice the EGF gene is present on chromosome 3 (ZABEL et al. 1985). The gene for TGF-α is located on the short arm of human chromosome 2 in the p11→p13 region (BRISSENDEN et al. 1985; TRICOLI et al. 1986).

The genes for the pox virus growth factors are located within the viral DNA genome. The VGF gene is located within an inverted terminal repeat (VENKATESAN et al. 1982), while the Shope fibroma (CHANG et al. 1987) and myoxma (UPTON et al. 1987) growth factor genes are present in a central portion of the genome. Though not sequenced, a gene coding for a potential EGF-like growth factor has been detected within an inverted terminal repeat sequence of the *Molluscum contagiosum* pox virus (PORTER and ARCHARD 1987).

II. Gene Organization

The only extensive study of the nonviral EGF family of genes has been reported by BELL et al. (1986) for the human EGF gene. A more preliminary report has described the exonic organization of the human TGF-α gene (DERYNCK et al. 1985).

1. Human EGF

The human EGF gene is estimated to span approximately 120 kilobases (kb) and to include 24 exons and 23 introns, all of which interrupt protein coding sequences. As indicated in Fig. 3, the numerous exons are similar in size, with the exception of the first and last which contain approximately 500 and 1000 base pairs (bp), respectively. The internal exons contain between 79 and 228 bp with most containing approximately 150 bp. In this gene each of the eight EGF-like repeats is encoded by a distinct exon (numbers 6–9, 15, and 17–19). Mature EGF, in contrast, is composed of two exons (20 and 21) which interrupt the mature EGF sequence at Asn32 (see Table 1). Interestingly, this site corresponds to that in which NMR studies have suggested that EGF is composed of two somewhat independently folded domains (residues 1–33 and 34–53, see Sect. B. IV. 1). Exon 21 encodes the C-terminal half of EGF plus the hydrophobic membrane-anchor sequence.

Exons 8–15 of the human EGF gene correspond to an equivalent number of exons within the gene for the LDL receptor (SÜDHOF et al. 1985a, b). In these two genes these exons are 33% identical in sequence and are interrupted at the same amino acids in both genes by five introns. Four other introns occur at dissimilar points in this related segment of the two genes. The notion that the EGF precursor also might function as a receptor (PFEFFER and ULLRICH 1985) is based in part on this similarity to the LDL receptor.

Dot matrix analysis of EGF precursor sequences indicates internal homology between residues 43–479 (exons 2–9) and residues 480–952 (exons 10–19) (BELL et al. 1986). Previously, DOOLITTLE et al. (1984) reported a 30% identity between residues 121–310 and 561–746 which occur in the non-EGF-like regions (Fig. 3, segments a and b) of the EGF precursor. It has been proposed that these two large regions of internal homology result from the duplication of a block of eight or nine exons (BELL et al. 1986). HAYASHIDA and MIYATA (1985) had also noticed these internal regions of homology in the mouse EGF precursor sequence.

2. Human TGF-α

The human TGF-α gene is reported to contain six exons (DERYNCK et al. 1985). Interstingly, the coding sequence for mature TGF-α is composed of two exons, numbers 3 and 4 (Fig. 3). These exons interrupt the TGF-α sequence at Val32 (see Table 1). This is the exact point at which the coding sequence for mature EGF is interrupted by two exons. Similar to the second EGF-coding exon, the exon encoding the C-terminal half of TGF-α also encodes the hydrophobic membrane anchor sequence.

III. Gene Expression

1. EGF

The primary transcript of the EGF gene is predicted to be approximately 110 kb, while the mature mRNA detected on Northern gels is 4.7–4.9 kb (BELL et al. 1986; GUBITS et al. 1986; POPLIKER et al. 1987; DIAUGUSTINE et al. 1988). Factors regulating expression of the EGF gene remain unknown, though transcription occurs preferentially in certain tissues. The human EGF gene contains a CAAT site and TATA box close to the 5′ end of the cDNA sequence (BELL et al. 1986), suggesting transcription of this gene involves mechanisms known in other systems for RNA polymerase II.

Nearly all studies of EGF mRNA have been performed with mouse tissue and show that gene expression is highest in the submaxillary gland of the male mouse. The mRNA abundance in the submaxillary gland of the female mouse is approximately 10-fold lower. mRNA levels in the kidney are also quite high, perhaps 50% of that in the male mouse submaxillary gland, but without evidence of sexual dimorphism (RALL et al. 1985). In situ hybridization studies show that in the submaxillary glands, EGF mRNA is localized to cells that comprise the granular convoluted tubules and that mRNA levels can be enhanced by androgens or triiodothyronine (GRESLIK et al. 1985). Also, mRNA levels are age dependent, increasing from nondetectable levels within the first 10 days after birth to moderate levels by 20 days for both submaxillary glands and kidneys (GRESLIK et al. 1985; POPLIKER et al. 1987). EGF transcripts have not been detected in any fetal tissue and the gene is not considered to be active until after birth.

Within the kidney, autoradiographic analysis localized EGF precursor mRNA to cells that constitute the straight and beginning of the convoluted portion of the distal tubules (RALL et al. 1985). EGF mRNA levels were not detectable in the proximal tubules or collecting ducts. It is in this area of the kidney that the protein precursor probably accumulates due to the high level of mRNA and apparent lack of rapid processing.

Lower levels of mRNA encoding preproEGF were also reported to be present in the oral cavity, mammary gland, pancreas, duodenum, pituitary, lung, spleen, brain, ovary (RALL et al. 1985), and uterus (DIAUGUSTINE et al. 1988) where levels were increased by estrogen treatment. With the possible exception of the oral cavity, levels of preproEGF mRNA in these tissues are less than 0,01% of the content of this mRNA in the male submaxillary gland. Other than the kidney (BELL et al. 1986), sites of EGF mRNA formation in human tissues have not been reported. Cell lines that produce EGF mRNA or mature EGF from endogenous genes have not been identified. However, SOVOVÁ et al. (1988) have reported two renal carcinoma cell lines which express the 130-kDa EGF precursor.

2. TGF-α

Not unlike EGF transcription, the primary transcript of the TGF-α gene is estimated at about 70–100 kb (DERYNCK 1988) and the mature mRNA, as detected by Northern blot analysis, is approximately 4.8 kb. (DERYNCK et al. 1984; LEE et al. 1985a, b). Analysis of sequences 5' to the transcription start site, however, indicates the promoter region of the human TGF-α gene is somewhat different from the EGF promoter in that there are no discernible CAAT or TATA sequences (JAKOBOVITS et al. 1988). These sequences are thought to be necessary for positioning of RNA polymerase II at a unique transcription start site. Despite the absence of these elements, RNA transcription from the human TGF-α gene does seem to start at a unique site approximately 62 bases before the start codon for preproTGF-α. Thus, transcription of the TGF-α gene may involve alternate unrecognized sequences for the efficient positioning of RNA polymerase. Upstream from the transcription start site are several Sp1 binding sites and further upstream are several AP-2 binding sites, indicating that transcription may be regulated, at least in part, by known transcription factors.

TGF-α mRNA has been detected in a wide variety of transformed cell lines (DERYNCK et al. 1987; J.J. SMITH et al. 1987). The frequency of TGF-α mRNA expression is high in carcinomas, particularly squamous cell and renal carcinomas, but it is not detectable in hematopoietic tumor cell lines. In the MCF-7 breast carcinoma cell line, TGF-α mRNA levels are increased 2- to 3-fold by estrogen administration in vitro or in nude mice (BATES et al. 1988). Few nontransformed cultured cells have been found to express TGF-α mRNA – only primary cultures of keratinocytes (COFFEY et al. 1987) and macrophages (MADTES et al. 1988; RAPPOLEE et al. 1988 b). In the latter case, TGF-α transcripts were detected following activation of the macrophage population in vitro with lipopolysaccharide. In the former case, messenger RNA levels were increased by the addition of either TGF-α or EGF to the keratinocyte culture media, suggesting an unusual autoinduction phenomenon that could be of significance for amplification of growth factor production in a cell population. In a subsequent study (PITTELKOW et al. 1989), it was shown that protein kinase C activators also increased levels of TGF-α mRNA in cultured keratinocytes.

The expression of TGF-α mRNA in cultured cells seems to reflect its expression in tissues: tumors (DERYNCK et al. 1987; BATES et al. 1988), normal epidermis (COFFEY et al. 1987), psoriatic epidermis (5-fold higher than uninvolved skin; ELDER et al. 1989), wound macrophages (RAPPOLEE et al. 1988 b), gastric mucosa (BEAUCHAMP et al. 1989), and brain (LEE et al. 1985a), where transcripts were localized to the caudate nucleus, dentate gyrus, anterior olfactory nuclei, olfactory bulb, and tuberculum olfactorium (WILCOX and DERYNCK 1988a).

Unlike EGF, mRNA for TGF-α is expressed prior to birth though not continuously. RAPPOLEE et al. (1988a) examined TGF-α mRNA in the preimplantation mouse embryo. Their results indicate that unfertilized oocytes contain maternal transcripts for TGF-α which disappear following fertilization. Embryo transcripts appear later in the blastocyst stage. In other studies LEE et al. (1985a) and WILCOX and DERYNCK (1988b) reported that following implantation and embryo development in the rat and mouse there is a peak of TGF-α mRNA levels at about 9 days. A more detailed examination of individual mouse embryo tissues by in situ hybridization showed that cells transcribing TGF-α were found in placenta, otic vesicle, oral cavity, pharyngeal pouch, bronchial arches, and kidneys (WILCOX and DERYNCK 1988b). Other reports, using similar techniques, are not in agreement, however. HAN et al. (1987, 1988) report that transcripts for TGF-α are not found in the embryo at all, but rather are present in the maternal decidua. They suggest that embryo tissue may become contaminated with maternal transcripts during experimental handling or, perhaps, in vivo due to autolysis of the decidua.

3. Pox Virus Growth Factors

All members of this growth factor group are expressed as early genes, being produced in the first few hours of viral infection (VENKATESAN et al. 1982; CHANG et al. 1987). Analysis of mRNA encoding the vaccinia growth factor (VGF) indicates that no splicing occurs and that the mRNA contains short 5' and 3' untranslated regions (YUEN and MOSS 1986). The 5' promoter region of the VGF gene contains AT-rich clusters and possibly equivalents of CAAT and TATA sequences (VENKATESAN et al. 1982). BROYLES et al. (1988) have purified to apparent homogeneity two polypeptides which are thought to function as a dimer in the initiation of transcription of vaccinia early genes, including the VGF gene. Also, a factor directing correct chain termination of this gene has been purified (SHUMAN et al. 1987).

E. The EGF Receptor

Most authors agree that the biological activities of all members of the EGF family of mitogens are mediated by a common receptor molecule, known as the EGF receptor. However, given the fact that there are preliminary reports of differences in the activities of EGF and TGF-α and that receptor subtypes are known or suspected in other hormone systems, there is an "open window" on this question for the EGF receptor. There is no experimental evidence, at present, for this view and if EGF receptor subtypes occur, they are likely to be due to subtle, but perhaps important, differences in posttranslational modifications of the receptor, as opposed to distinct gene products.

Since a number of current reviews are available on the topic of the EGF receptor, this section is abbreviated and reader referred to those articles for additional data (CARPENTER 1987; GILL et al. 1987; YARDEN and ULLRICH 1988; SCHLESSINGER 1988a, b; STAROS et al. 1989). Specifically, these reviews should be sought for analysis of kinase regulation mechanisms. We have tried to concentrate on manuscripts published since 1987.

I. Receptor Structure

The EGF receptor was isolated early in the 1980s from a human carcinoma cell line (A–431) that overexpresses the receptor approximately 20- to 50-fold (COHEN et al. 1980, 1982a). A receptor molecule having similar characteristics was later purified from normal tissue, mouse liver (COHEN et al. 1982b). During this time biochemical evidence (CARPENTER et al. 1978, 1979) suggested that the EGF receptor was closely associated with a ligand-activated protein kinase activity specific for tyrosine residues (USHIRO and COHEN 1980). The purified receptor was demonstrated to contain an ATP binding site by covalent affinity labeling (BUHROW et al. 1982). These biochemical studies suggested that the EGF receptor was an intrinsic membrane glycoprotein of 170000 Da, that it contained a high-affinity binding site specific for EGF, and that the binding site portion of the receptor was physically associated, perhaps in the same polypeptide chain, with an EGF-activatable tyrosine kinase. These data also showed that the 170000-Da receptor was a substrate for the EGF tyrosine kinase activity.

The primary structure of the EGF receptor, determined by cDNA cloning and sequencing of receptor mRNA from A–431 cells (ULLRICH et al. 1984), localized tyrosine kinase sequences in the cytoplasmic portion of the EGF receptor polypeptide chain. The EGF receptor has now been cloned and sequenced from avian cells (LAX et al. 1988a) and a putative EGF receptor gene, based on sequence homology only, has been reported for *Drosophila* (LIVNEH et al. 1985; WADSWORTH et al. 1985). The human and avian EGF receptor are 78% identical. The structural organization of the EGF receptor, based on cDNA sequences and posttranslational modifications, is shown in Fig. 4 and is subdivided into linear domains for the purpose of discussion.

The extracellular ligand-binding domain of the EGF receptor is characterized by: (1) a high content of cysteine (about 10%) that could give rise to 25 disulfide bonds (the number and pairings are unknown); and (2) a high content of N-linked oligosaccharide, accounting for 30% of the mass of this domain. Based on crosslinking experiments with ^{125}I-EGF, LAX et al. (1988b) have suggested that region III, located between the two cysteine-rich regions, is a major determinant of the ligand binding site. Preliminary modeling studies to predict higher order polypeptide structure in the external domain have been reported (FISHLEIGH et al. 1987; BAJAJ et al. 1987). Curiously, the chicken EGF receptor is reported to bind TGF-α much more effectively (100-fold) than EGF (LAX et al. 1988a).

Glycosylation studies of the receptor indicate the absence of O-linked sugar and the presence of 10–11 N-linked oligosaccharide chains of which approximately three or four are the high-mannose type and the rest complex type (MAYES and WATERFIELD 1984; SODERQUIST and CARPENTER 1984; CUMMINGS et al. 1985). TODDERUD and CARPENTER (1988) have reported the presence of mannose phosphate on the EGF receptor. This carbohydrate chain modification has not been described for other membrane proteins, but is known as a key modification for the sorting of lysosomal enzymes to their correct intracellular compartment. Following EGF binding, the ligand-EGF receptor complex is rapidly internalized and both ligand and receptor are sorted to lysosomes where they are degraded (see Sect. E. II. 2). Based on the lysosomal enzyme systems, the mannose phosphate on the EGF receptor should be located on the high-mannose type oligosaccharides.

Hydropathy analysis indicates the presence of one internal hydrophobic sequence (domain V) in both the human (ULLRICH et al. 1984) and avian (LAX et al. 1988a) EGF receptors. This sequence, considered to be the only membrane-spanning region in the mature receptor, is followed by a "stop transfer" sequence enriched in basic amino acid residues.

The C-terminal half of the EGF receptor encodes tyrosine kinase sequences (domain VII) shared with other proteins having this enzymatic activity. The homology is extremely high (>95%) between the human EGF receptor and the avian v-*erb* B oncogene derived from the chicken EGF receptor

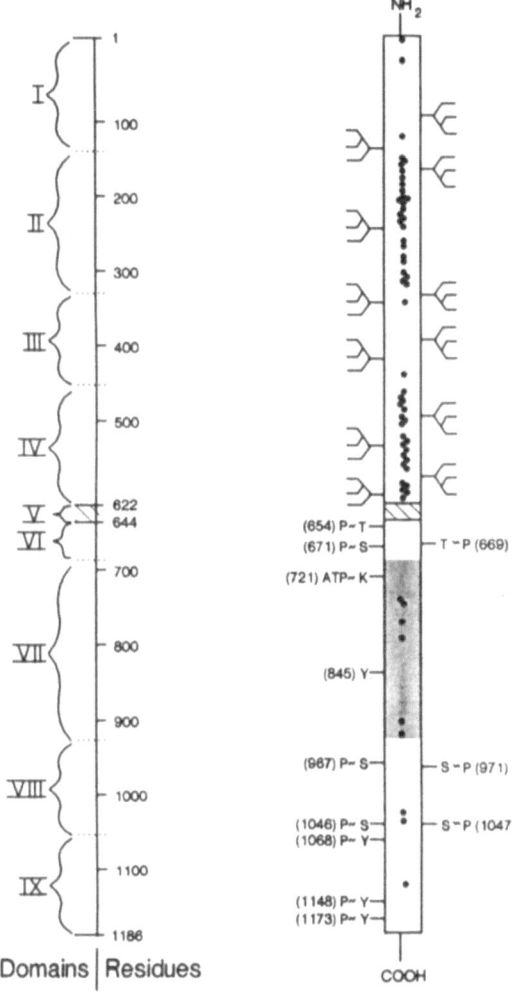

Fig. 4. Structural organization of the mature human EGF receptor. The *crosshatched area* indicates the hydrophobic membrane anchor sequence, *tree-like structures* indicate potential sites for N-linked glycosylation in the extracellular domain, *filled dots* indicate the positioning of cysteine residues, and the *dotted area* indicates the *src*-like tyrosine kinase domain (from ULLRICH et al. 1984). $P \sim Y$ designates demonstrated autophosphorylation sites (DOWNWARD et al. 1984), $P \sim T$ (654) designates a C-kinase phosphorylation site (HUNTER et al. 1984; DAVIS and CZECH 1985), other serine/threonine phosphorylation sites are from the report of HEISERMAN and GILL (1988). The lysine residue (721) involved in ATP binding within the kinase domain is from RUSSO et al. (1985). A short, N-terminal signal sequence which is removed during translation is not shown, but is indicated in the cDNA sequence

(ULLRICH et al. 1984; YAMAMOTO et al. 1983). In addition to encoding a tyrosine kinase activity, regulated by EGF binding to the external domain, the cytoplasmic region of the receptor can be phosphorylated at multiple sites. The indicated tyrosine residues in domain IX are sites of receptor autophosphorylation (DOWNWARD et al. 1984). Domains VI and VII on either side of the tyrosine kinase domain contain multiple sites of serine/threonine phosphorylation (HEISERMANN and GILL 1988). Phosphorylation of most all these serine/threonine sites is catalyzed by protein kinases that have yet to be identified. The exception is Thr654 which is phosphorylated by protein kinase C (HUNTER et al. 1984; DAVIS and CZECH 1985). It is likely that protein kinase C phosphorylates other residues as well. The capacity of protein kinase C phosphorylation to modulate functions of the EGF receptor is discussed below (Sect. E. III).

II. Receptor Gene

1. Chromosomal Localization

In humans, the EGF receptor gene is located on chromosome 7 in the p14→p12 region near a heritable fragile site 7p11 (MERLINO et al. 1985). In the mouse, the EGF receptor gene is on chromosome 11 (ZABEL et al. 1984), where it is closely linked to the α-globin locus (SILVER et al. 1985). In the chicken, the EGF receptor gene location has not been clearly identified. The *Drosophila* EGF receptor homolog maps to position 57F on chromosome 2.

2. Gene Organization

a) Structural Information

A preliminary report describing the human EGF receptor gene has been published by HALEY et al. (1987). The gene is estimated to cover approximately 110 kb of DNA and to include 26 exons encoding the receptor sequence. The first exon contains the 5' untranslated leader sequence and nucleotides coding for the signal sequence and first five amino acids of the mature receptor molecule (reported also by ISHII et al. 1985). Exon 15 encodes the single transmembrane sequence plus a small number of residues on each side of this sequence. Most all of the extracellular domain of the EGF receptor, therefore, is encoded by exons 2–14 with the two cysteine-rich regions coded by exons 5–8 and 12–15. ULLRICH et al. (1984) suggest a repeat unit of approximately 170 amino acid residues within these two cysteine-rich units. TOH et al. (1985) have noted a larger (≈ 300 residue) region of internal homology within the extracellular domain, residues 1–300 and 320–620 approximately. The data have suggested to these authors that the extracellular domain has arisen by duplication of a single cysteine-rich region. The tyrosine kinase domain is encoded by exons 15–22 and the C-terminal autophosphorylation sites are encoded by exons 22–26. The 26 exons are separated by 25 introns which range from 270 bp to 20 000 bp.

b) Regulatory Information

The promoter region of the human EGF receptor gene was identified and sequenced by ISHII et al. (1985) and later by HALEY et al. (1987). Deletion analysis of the promoter region was reported by HALEY et al. (1987) and A. C. JOHNSON et al. (1988a). These studies and the work of KAGEYAMA et al. (1988b) and HUDSON et al. (1989) indicated that a short region (-151 to -20 is sufficient for promotion and regulation of transcription. A larger region (-481 to -16) as well as untranslated sequences within the first exon may contribute to receptor gene expression. The promoter region is $G+C$ rich (88%), there are several transcription start sites, and there are no characteristic CAAT or TATA sequences. Within the larger regulatory region, which contains elements capable of both positive and negative control of receptor transcription, there are four binding sites for the transcription factor Sp1. Sp1 is required for maximal transcriptional activity. A second and novel transcription factor for the EGF receptor gene has been identified (KAGEYAMA et al. 1988a) and purified (KAGEYAMA et al. 1988b). The 5' regulatory region has been demonstrated to contain sites sensitive to S1 nuclease and DNAse, suggestive of active transcription (A. C. JOHNSON et al. 1988b).

3. Gene Expression

The EGF receptor is expressed in nearly all adult tissues with the singular exception of cells in the hematopoietic system. During fetal life, the EGF receptor is expressed, but these studies have not, as yet, utilized nucleic acid probes to carefully measure expression in different tissues. The primary RNA transcript for the EGF receptor would seem to be approximately 100 kb, while Northern blots of RNA from cell lines or tissues show mature mRNA species of 10.5 kb and 5.8 kb (ULLRICH et al. 1984). In the A–431 cell line a 2.8-kb mRNA is also detected and seems to have arisen from a truncated, translocated EGF receptor gene. In these cells this abbreviated mRNA produces a secreted protein corresponding to the external domain of the EGF receptor. The two large mRNAs differ in 3' untranslated sequences as different transcription polyadenylation sites are used.

Hormonal factors which produce modulations of EGF receptor gene transcription are discussed in Sect. E. IV. 1. Overexpression of the receptor gene frequently accompanies the transformed phenotype, particularly in the case of carcinoma (see Sect. G. 11). Based on ligand binding studies, the EGF receptor has been detected in fish (S. Cohen, personal communication) and *Xenopus* hepatocytes (WOLFFE et al. 1985), suggesting a broad evolutionary function.

III. Receptor Life Cycle

1. Biosynthesis

A diagrammatic outline of steps in the biosynthesis, ligand binding, and subsequent endocytic processing of the EGF receptor is shown in Fig. 5. As discussed elsewhere (Sect. E. II. 3), there is little information available to define

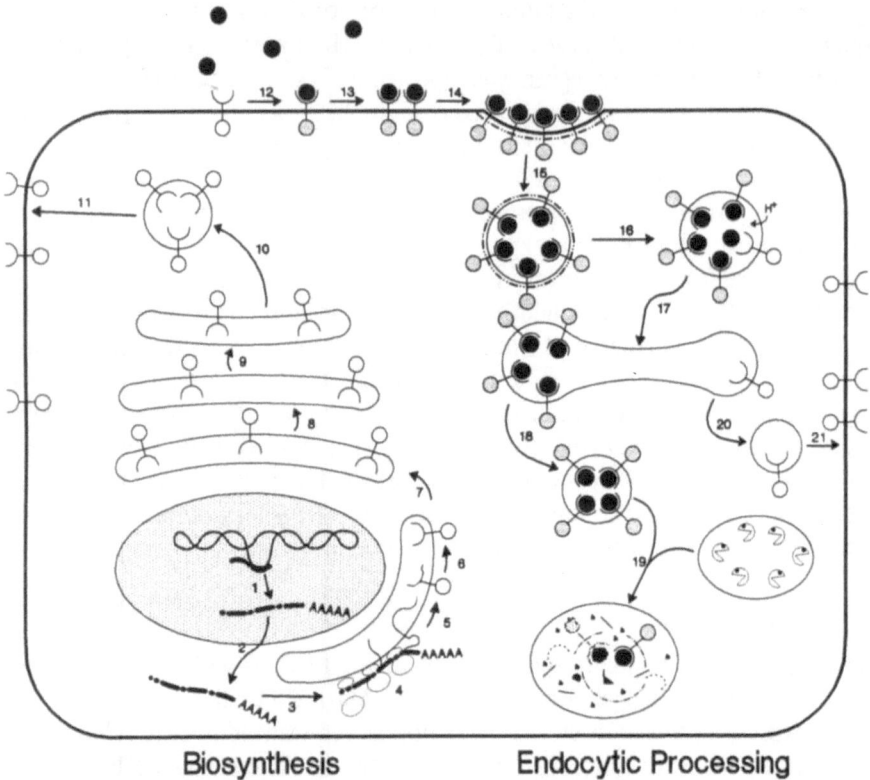

Biosynthesis   **Endocytic Processing**

Fig. 5. Life cycle of the EGF receptor

regulatory events for transcription of EGF receptor mRNA (step 1) or its splicing and maturation (step 2). Three groups have reported that EGF increases the level of mRNA for its receptor (CLARK et al. 1985; EARP et al. 1986; KUDLOW et al. 1986). Two groups, BJORGE and KUDLOW (1987) and EARP et al. (1988), suggest that this EGF effect is mediated, at least in part, by activation of protein kinase C, which would be a response to EGF activation of phospholipase C (Sect. E. III. 4). CLARK et al. (1985) indicate that the EGF mechanism is a posttranscriptional effect on receptor mRNA, and JINNO et al. (1988) show that EGF increases the EGF receptor mRNA half-life from 1.6 h to more than 6 h. In the latter experiments EGF also increased the half-life of mRNA for tubulin and actin. Other data indicate that transcription of the receptor gene may be modulated by EGF, 12-*O*-tetradecanoylphorbol-13-acetate (TPA), cyclic AMP, and retinoic acid (HUDSON et al. 1988) or TPA, glucocorticoids, retinoic acid, and serum in a cell-type dependent manner (HALEY et al. 1987). These stimulatory effects are also observed at the level of translation (steps 4–5) for EGF (CLARK et al. 1985; KUDLOW et al. 1986; EARP et al. 1986; DEPALO and DAS 1988) and retinoic acid (OBERG et al. 1988). In contrast, OBERG and CARPENTER (1989) report that dexamethasone decreases EGF receptor protein synthesis. A. C. JOHNSON et al. (1988c) have observed that EGF receptor mRNA levels decrease during the first 12 h after partial hepatectomy, but then increase substantially after 24 h.

Translation of the EGF receptor mRNA occurs coincident with N-type glycosylation within the lumen of the endoplasmic reticulum (SODERQUIST and CARPENTER 1984). If glycosylation is blocked with tunicamycin, the non-glycosylated receptor is not processed from the endoplasmic reticulum and is biologically inactive in terms of ^{125}I-EGF binding capacity (SODERQUIST and CARPENTER 1984). Interestingly, the newly synthesized EGF receptor (following step 5) is also inactive in ligand binding assays even though glycosylation is allowed to occur (SLIEKER and LANE 1985). The newly synthesized but inactive receptor acquires ligand binding capactiy (step 6) approximately 30 min later, while in the late endoplasmic reticulum (SLIEKER et al. 1986). These authors suggest that rearrangement of disulfide bonds may be the process driving this posttranslational activation of the receptor.

In the Golgi, the N-linked oligosaccharide chains are processed (steps 8 and 9) so that approximately seven become complex chains and three remain high-mannose type chains (CUMMINGS et al. 1985). Determinants of these processing events are unclear but do not seem to require the membrane-spanning or cytoplasmic domain of the receptor (SODERQUIST et al. 1988). Formation of complex-type oligosaccharide chains can be blocked with swainsonine, but this does not affect receptor transport to the cell surface (steps 10 and 11) or receptor activity (SODERQUIST and CARPENTER 1984). At the cell surface EGF receptors are randomly distributed in the plasma membrane and not localized to specific areas (HAIGLER et al. 1979).

2. Endocytosis and Degradation

Ligand binding to the EGF receptor (step 12) is rapidly followed by clustering (steps 13 and 14) of EGF-receptor complexes at coated pit areas on the plasma membrane (HAIGLER et al. 1978; SCHLESSINGER et al. 1978; HAIGLER et al. 1979). The ligand-receptor complexes are then internalized (step 15) and after sorting through several intracellular compartments (steps 16–19) eventually reach the lysosome where both the ligand (CARPENTER and COHEN 1976) and receptor (STOSCHECK and CARPENTER 1984; DECKER 1984; BEGUINOT et al. 1984) are degraded. EGF induces a 10-fold increase in receptor degradation which is inconsistent with a major level of recycling of internalized receptors (steps 20 and 21), though this has been proposed to occur at an undetermined but probably low level in certain cells (DUNN et al. 1986; MURTHY et al. 1986; TESLENKO et al. 1987; GLADHAUG and CHRISTOFFERSEN 1988). The mechanism by which internalized receptors are sorted to separate routes for degradation or recycling is not known. Clearly, however, the internalized EGF receptor is sorted differently than the internalized LDL or transferrin receptors which recycle at a high level. Recently, TODDERUD and CARPENTER (1988) have shown the presence of mannose phosphate on the EGF receptor. This could suggest that the known mechanism for delivery of lysosomal enzymes, either from the Golgi or from the external media, to the lysosome could function to route the internalized EGF receptor.

After uncoating of the initial endocytic vesicles (step 16), acidification of the vesicles occur. Though low pH is known to dissociate ^{125}I-EGF bound to the receptor at the cell surface (HAIGLER et al. 1980), it has not been demonstrated that growth factor dissociation from the receptor occurs in endocytic vesicles and one group (SORKING et al. 1988) has argued that dissociation does not occur. Two studies (COHEN and FAVA 1985; KAY et al. 1986) showed that EGF receptor derived from internalized vesicles remains active in terms of tyrosine kinase activity in vitro. Another report (CARPENTIER et al. 1987) showed by morphologic methods that the internalized receptor remains phosphorylated at tyrosine residues for up to 60 min after endocytosis. If EGF were dissociated from the receptor, one would expect rapid dephosphorylation.

The influence of mutations of the EGF receptor on receptor internalization and/or degradation is discussed below (Sect. E. IV. 2).

IV. Receptor Function

Most of the biochemical and molecular information concerning EGF receptor function has been obtained by investigating effects of EGF in a relatively small number of cell types. The most studied cell line has been the A–431 cell, derived from a human epidermoid carcinoma (GIARD et al. 1973), which expresses a high number ($\approx 2 \times 10^6$) of EGF receptors per cell (FABRICANT et al. 1977; HAIGLER et al. 1978). The overexpression of the EGF receptor in this cell line has greatly facilitated a molecular understanding of EGF's interactions with its receptor, as biochemical responses to the formation of growth factor-receptor complexes are amplified. This includes signal transduction mechanisms. Many of the "early" mitogenic responses observed with EGF-treated nontransformed cells are reproduced in A–431 cells. However, the long-term response of A–431 cells to EGF in vitro is growth inhibition, although the cells are stimulated to grow by EGF in vivo or in soft agar. The biology of the growth response of this cell line to EGF is complex and is discussed in more detail in a later section (Sect. G. II. A).

In addition to the A–431 cell line, various nontransformed cell lines have been used as model systems for EGF regulation of cell proliferation. While these cell types are stimulated to proliferate by EGF under most cell culture conditions, they express a lower complement of EGF receptors, usually in the range of $0.5–10 \times 10^4$ receptors per cell. Since the EGF receptor is in relatively low abundance on these cell types, biological and biochemical sequelae of EGF treatment, particularly receptor–proximal signaling mechanisms, are more difficult to observe experimentally. As a starting point, however, the receptor-overexpressing cell lines have been extremely helpful and may provide information directly relevant to the high percentage of carcinomas that overexpress the EGF receptor in vivo (Sect. G. II. A). As mentioned earlier, the reader is referred to recent reviews (CARPENTER 1987; GILL et al. 1987;

SCHLESSINGER 1988a,b; STAROS et al. 1989) for background data concerning EGF receptor function and biochemistry.

1. Heterologous Receptor Expression

Since human EGF receptor cDNA has become available, a number of studies have utilized this reagent to analyze receptor structure and function. Receptor cDNA has been transfected into and expressed in recipient cells (usually rodent) that have no endogenous EGF receptor or levels that are very low (less than 10^4). Therefore, responses to EGF are expected to be mediated by the exogenous EGF receptor. There has, however, been some confusion about the functional contribution of endogenous EGF receptors to the EGF-induced biological responses in some of the recipient cell lines.

An important result of these experiments is that the transfected receptors are responsive to EGF in these heterologous systems. This includes two studies of hematopoietic cells transfected with the EGF receptor (VON RÜDEN and WAGNER 1988; PIERCE et al. 1988). Since there is no evidence in vivo or in vitro that these cell types have endogenous EGF receptors or respond to EGF, the experimental results indicate that information for the specificity of EGF responsiveness is present in the EGF-receptor complex.

EGF receptors expressed by transfected cell lines have been examined for the mediation of a variety of biochemical properties (DiFIORE et al. 1987; PIERCE et al. 1988; GREENFIELD et al. 1988; VELU et al. 1987; PANDIELLA et al. 1988; VON RÜDEN and WAGNER 1988). Ligand binding studies show that transfected EGF receptors are expressed at levels of $3-150 \times 10^4$ receptors per cell. Scatchard analysis of ^{125}I-EGF binding data results in a curvilinear plot interpreted to represent two discrete receptor populations with high and low affinities for the ligand, similar to the result obtained with cells expressing endogenous EGF receptors. The biochemical basis for two (or more) receptor affinity states remains unknown. Several studies have examined the relationship between high-level expression of the EGF receptor in heterologous cell lines and transformation. Those results are summarized in Sect. G. II. B.

2. Mutants

The functional significance of structural characteristics of the EGF receptor has been studied by using mutant receptor constructs in transfected cells. A variety of mutant EGF receptors have been made to address specific aspects of receptor biology, such as autophosphorylation mechanisms, membrane mobility, dimerization, intracellular signaling mechanisms, endocytosis, transmodulation by protein kinase C, and growth stimulation. Furthermore, truncated receptors and chimeric molecules have been constructed to examine the importance of receptor domains in the regulation of receptor function, cell growth, and transformation. A description of receptor mutants appears in Table 3.

Table 3. EGF receptor mutants

Mutation types and sites	References
Site-directed mutants	
Single	
Thr654 → Ala	C. R. Lin et al. (1986), Glenney et al. (1988), Davis (1988)
Thr654 → Tyr	Livneh et al. (1988)
Lys721 → Ala	Honegger et al. (1987a, b), Moolenaar et al. (1988)
Tyr1173 → Phe	Honegger et al. (1988a, b)
Tyr1173 → Ser	Honegger et al. (1988a, b)
Tyr1148 → Phe	Honegger et al. (1988a, b)
Tyr1068 → Phe	Honegger et al. (1988a, b)
Double	
Thr654 → Ala & Lys721 → Met	Chen et al. (1987), Glenney et al. (1988)
Thr654 → Ala & Tyr1173 → Phe	Bertics et al. (1988)
Insertion mutants	
Ile708, (Gly-Arg-Pro-Ile), Pro709	Prywes et al. (1986), Livneh et al. (1987)
Ser888, (Val-Asp-Arg-Ser), Lys889	Prywes et al. (1986)
Deletion mutants	
Single	
ΔLys1155 → Ala1186	Khazaie et al. (1988)
ΔAla1124 → Ala1186	Livneh et al. (1986a, b), Honegger et al. (1988a, b)
ΔThr1061 → Ala1186	Khazaie et al. (1988)
ΔVal1023 → Ala1186	Glenney et al. (1988)
ΔAla743 → Ala1186	Livneh et al. (1986a, b)
ΔXXX726 → Ala1186[a]	Prywes et al. (1986)
ΔThr654 → Ala1186	Livneh et al. (1986a, b)
ΔThr624 → Ala1186	Livneh et al. (1986a)
ΔLeu1 → Cys555	Khazaie et al. (1988)
Δ20 kDa C terminus[b]	S. Clark et al. (1988)
Double	
ΔLeu1 → Cys555 & Lys1155 → Ala1186	Khazaie et al. (1988)
ΔLeu1 → Cys555 & Leu1063 → Ala1186	Khazaie et al. (1988)
Chimeric molecules	

Chimeric molecules

Domains

Ligand binding	Trans-membrane	Tyrosine kinase	
hEGF-R	hEGF-R	v-*abl*	Prywes et al. (1986)
hEGF-R	hEGF-R	v-*erb* B	Riedel et al. (1987)
hIns-R	hEGF-R	hEGF-R	Riedel et al. (1986)
hIL-2-R	hEGF-R	hEGF-R	Bernard et al. (1987)

[a] Residues 661 → 724 translated out of frame.
[b] Exact truncation site not specified.

a) Site-Directed Mutants

α) *Lysine 721*

Because of the parallel with many oncogene products as well as other growth factor receptors, the most intriguing aspect of the EGF receptor is the biological role of the tyrosine kinase activity. Based on comparison with other kinase sequences (ULLRICH et al. 1984) and affinity labeling data (RUSSO et al. 1985) of the EGF receptor, the lysine residue at position 721 in the EGF receptor was determined to be a conserved residue involved in ATP binding. Mutation of this residue, therefore, would be expected to have significant biological consequences, if the kinase activity is fundamental to EGF signal transduction.

EGF receptor cDNA having a Lys721→Ala mutation was transfected into cells and compared with wild-type receptor transfectants (HONEGGER et al. 1987a, b). Analysis of receptor kinase activity demonstrated that, although the Ala721 receptor bound EGF, it was incapable of EGF-dependent autophosphorylation or exogenous substrate phosphorylation; whereas, the wild-type receptor demonstrated both activities. Both wild-type and mutant EGF receptors were rapidly phosphorylated following stimulation of protein kinase C activity with the phorbol ester TPA.

The Ala721 mutant and wild-type receptors displayed similar ligand binding characteristics, including rapid internalization and degradation of bound ligand. However, a distinct difference was noted in receptor biology. In most cell types, EGF binding produces a gradual downregulation of the EGF receptor due to enhanced receptor degradation. In contrast, cell surface binding capacity and receptor degradation did not decrease following EGF addition to cells expressing the mutant Ala721 receptor. Recycling of the mutant receptor was proposed to account for receptor internalization and ligand degradation in the absence of receptor degradation. This indicated that tyrosine kinase activity was not essential for receptor internalization, but was essential for processing of the internalized receptor to the lysosome.

Another receptor mutant with two alterations, Lys721→Met plus Thr654→Ala, also displayed a lack of tyrosine kinase activity (CHEN et al. 1987; GLENNEY et al. 1988). A prior study (C. R. LIN et al. 1986) had shown that the single Thr654→Ala mutation did not affect tyrosine kinase activity. Ligand binding characteristics were similar for wild-type, the Ala654 mutant, and the double mutant (Met721 plus Ala654) receptors. Furthermore, ligand-dependent dimerization was noted for wild-type and mutant receptor constructs. Again, ligand stimulation of EGF receptor downregulation was abrogated in the kinase-deficient mutant. However, internalization of the EGF receptor following EGF binding was significantly impaired in the double mutant but was unaffected in the control single mutant, Ala654. The investigators also showed that microinjection of an antibody to phosphotyrosine significantly inhibited internalization of wild-type receptors and Ala654 receptors, as judged by anti-EGF receptor antibodies and fluorescence spectroscopy. These authors concluded that Lys721 and tyrosine kinase activity were necessary for the internalization process.

These contrasting results regarding ligand-stimulated receptor internalization in kinase-deficient receptors are difficult to reconcile. However, the two mutants are somewhat different (one is a double mutant), the techniques employed to measure internalization were not the same, and the host cell lines were different.

Receptor–distal responses to EGF binding were examined in cells expressing kinase-deficient mutants (CHEN et al. 1987; MOOLENAAR et al. 1988; HONEGGER et al. 1987 b). These responses to EGF included: stimulation of inositol phosphate formation, increased intracellular free Ca^{2+}, enhanced expression of c-*myc* and c-*fos* mRNA, cytosolic alkalinization, and DNA synthesis. Cells expressing kinase-dificient receptors failed to generate any of these responses after EGF treatment, whereas positive responses were recorded for cells expressing wild-type receptors. These results show that EGF receptor tyrosine kinase activity is an absolute requirement for the generation of biochemical responses to EGF, including mitogenesis. The data do not, however, permit the conclusion that exogenous substrate substrate tyrosine phosphorylation is necessary for those responses. Autophosphorylation is blocked in these mutants and that might be a prerequisite for any signal transduction systems, including those not dependent on exogenous substrate phosphorylation.

β) Threonine 654

A common observation in the studies of EGF receptor function has been the capacity of protein kinase C agonists to decrease ^{125}I-EGF binding capacity, particularly the high-affinity EGF receptor binding sites. Transmodulation of EGF receptor binding capacity has been documented after treatment of cells with direct activators of protein kinase C, such as phorbol esters or analogs of diacylglycerol, or with other growth factors, such as PDGF or bombesin, which are thought to activate protein kinase C indirectly.

In addition to effects on ^{125}I-EGF binding, protein kinase C modulation of the EGF receptor leads to a decrease in receptor tyrosine kinase activity and an inhibition of postreceptor signaling mechanisms stimulated by EGF. This suggests that protein kinase C is an important physiological regulator of EGF receptor function (for reviews on this topic, see CARPENTER 1987; GILL et al. 1987).

The mechanism of EGF receptor transmodulation by protein kinase C is thought to be direct phosphorylation of the receptor on serine and threonine residues. A major protein kinase C phosphorylation site in the EGF receptor is Thr654, (HUNTER et al. 1984; DAVIS and CZECH 1985), a residue within the cytoplasmic domain of the receptor, but close to the plasma membrane. To investigate the role of this phosphorylation site in receptor transmodulation, site-directed Thr654 mutant receptors were expressed in heterologous cells.

EGF receptor constructs containing either Thr654→Ala (C. R. LIN et al. 1986) or Thr654→Tyr (LIVNEH et al. 1988) were expressed in recipient cells and were capable of binding ^{125}I-EGF with characteristics similar to cells transfected with the wild-type receptor. Addition of EGF to cells bearing these mutant receptors led to receptor dimerization, ligand internalization, receptor downregulation, receptor autophosphorylation, and DNA synthesis. Thus, EGF receptors altered at Thr654 are capable of mediating a mitogenic response to EGF.

Transmodulation of the mutant receptors by protein kinase C was measured by ^{125}I-EGF binding studies following treatment of intact cells with phorbol ester. C. R. LIN et al. (1986) and LIVNEH et al. (1988) reported that phorbol ester treatment affected neither high- nor low-affinity ^{125}I-EGF binding in cells expressing the mutant receptors, whereas a large decrease was detected in high-affinity binding to cells having the wild-type receptor. In contrast, DAVIS (1988), using different recipient cells, found that phorbol ester treatment provoked a substantial loss of high-affinity binding capacity in cells expressing the Ala654 mutant receptor, similar to the effect observed for the wild-type receptor.

Regulation of EGF receptor tyrosine kinase activity by protein kinase C was examined in intact cells and in vitro using the Ala654 and Tyr654 mutants. The wild-type and both mutant receptors were phosphorylated at serine/threonine residues after TPA stimulation of protein kinase C activity in intact cells (DAVIS 1988; LIVNEH et al. 1988) or in vitro (DAVIS 1988). Analysis of tryptic phosphopeptides of the EGF receptors demonstrated that TPA stimulated phosphorylation at multiple sites on both wild-type and mutant receptors. However, no phosphopeptide corresponding to the altered site (Ala654, Tyr654) of the mutant receptors was detected. LIVNEH et al. (1988) reported that both wild-type and Tyr654 receptors had diminished autophosphorylation activity in an imunorecipitate tyrosine kinase assay. DAVIS (1988) measured receptor autophosphorylation in cells treated with TPA prior to EGF. No protein kinase C inhibition of EGF receptor autophosphorylation was detected in the Ala654 mutant, whereas autophosphorylation of the wild-type receptor was inhibited. Furthermore, receptor phosphorylation in vitro by protein kinase C resulted in decreased levels of EGF-dependent receptor autophosphorylation and exogenous substrate phosphorylation in the wild-type receptor only.

Clearly, additional investigation of the Thr654 receptor mutant is required to resolve the contradictory data regarding the role of phosphorylation of this residue in ligand binding and tyrosine kinase activities of the receptor. DAVIS (1988) suggests that there are other protein kinase C sites on the EGF receptor that have a significant role(s) in transmodulation of the receptor.

γ) Autophosphorylation Sites

EGF-dependent sites of receptor autophosphorylation are known to be tyrosine residues 1173, 1148, and 1068, located at the C terminus of the receptor (DOWNWARD et al. 1984). Since receptor autophosphorylation occurs rapidly after EGF binding and insulin is known to increase tyrosine kinase activity of the insulin receptor toward exogenous substrates, this has been postulated on the basis of kinetic evidence to also occur in the EGF receptor

(BERTICS and GILL 1985). The importance of autophosphorylation in control of the mitogenic signaling cascade is an important issue.

EGF receptor constructs (HONEGGER et al. 1988 a, b; BERTICS et al. 1988) with altered single phosphorylation sites (Table 3) have been expressed in receptor-deficient recipient cells. Also, as shown in Table 3, a variety of C-terminal receptor deletion mutants have been constructed and expressed in recipient cells. Certain of these deletion mutants, lacking short portions of the C terminus ($\Delta 1155 \rightarrow 1186$, $\Delta 1124 \rightarrow 1186$, $\Delta 20$ kDa C terminus), have been used to study autophosphorylation.

Cells bearing receptors altered for single autophosphorylation sites or having short C-terminal deletions appear to be similar to the wild-type receptor in regard to posttranslational processing, ligand binding, autophosphorylation, internalization, and downregulation. Also, EGF-dependent cell responses such as increased serine phosphorylation of ribosomal protein S6 and DNA synthesis are preserved in cells expressing receptor autophosphorylation site mutants or receptor deletions which lack these known autophosphorylation sites.

Wild-type and mutant receptor-expressing cells have been analyzed for sites of EGF-stimulated autophosphorylation. Receptor autophosphorylation was detected in each of the site-directed mutants lacking a single autophosphorylation site (HONEGGER et al. 1988 a, b; DAVIS 1988; BERTICS et al. 1988). Analysis of tryptic phosphopeptides from autophosphorylated receptors demonstrated the loss of a single phosphopeptide corresponding to each of the tyrosine site-directed mutants (HONEGGER et al. 1988 b; DAVIS 1988; BERTICS et al. 1988). The absence of one autophosphorylation site should not preclude phosphorylation at other sites unless the autophosphorylation mechanism is ordered. Autophosphorylation also occurred in the 20-kDa C-terminus deletion mutant that lacks all three major autophosphorylation sites (S. CLARK et al. 1988). This result is not unexpected. A 150-kDa EGF receptor fragment, produced in cell homogenates (COHEN et al. 1982a) or in vitro (GATES and KING 1985, 1988) by a Ca^{2+}-activated protease and lacking the three major tyrosine phosphorylation sites of the native receptor, was known to autophosphorylate at other, unmapped, tyrosine residues in response to EGF. A different C-terminal proteolytic fragment of the EGF receptor (≈ 140 kDa), was reported to bind EGF but not to undergo autophosphorylation (SEGER et al. 1988). If this is accurate then the C-terminal region (≈ 10 kDa) retained in the 150-kDa receptor fragment should contain structural information essential for autophosphorylation.

Autophosphorylation had been proposed from kinetic studies to be a mechanism controlling access of exogenous substrates to the tyrosine kinase active site (BERTICS and GILL 1985; GILL et al. 1987). This is based on the known substrate and pseudosubstrate regulation of other protein kinases (for a synopsis see HARDIE 1988). An important issue, therefore, is whether exogenous substrate phosphorylation is affected by mutations at the autophosphorylation sites. BERTICS et al. (1988) measured a three-fold higher K_m for tyrosine phosphorylation of an exogenous peptide substrate in the double mutant receptor (Thr654→Ala plus Tyr1173→Phe) compared to the single mutant (Thr654→Ala) receptor. This indicated that the Phe1173 mutation reduced the affinity of the EGF receptor for exogenous substrates. A small decrease (20%) in V_{max} for a peptide substrate was also reported for the Phe1173 receptor. The effect of the Tyr1173→Phe mutation was interpreted by the authors as consistent with the idea that unphosphorylated (i.e., Tyr or Phe) sites bind effectively to the receptor kinase active site and thereby limit access of exogenous substrate to the active site. Autophosphorylation of Tyr1173, but not Phe1173, would decrease binding of the phosphorylated C-terminal receptor domain to the active site and thereby increase access of exogenous substrates to the active site.

HONEGGER et al. (1988 b) examined the phosphorylation of peptide substrates with single autophosphorylation site mutants (Tyr1173→Phe, Tyr1173→Ser, Tyr1148→Phe, Tyr1068→Phe) and the deletion mutant Δ1124→1186, which lacks all three of these tyrosyls. They detected a slight decrease (20%) in K_m for exogenous peptide substrates in the mutant receptors compared to the wild-type receptor. No difference was observed in the V_{max} of mutant receptors for the substrate. Though differing in interpretation with the prior study in assuming that the Phe1173 site does not bind to the kinase catalytic site, and the unphosphorylated Tyr1173 does bind to the kinase site, a similar conclusion was reached. The authors reasoned that the unphosphorylated tyrosine autophosphorylation sites act as competitive substrates for the receptor kinase and that phosphorylation of these sites makes the C terminus a less competitive substrate.

b) Deletion Mutants

The EGF receptor is composed of three topological domains that mediate different receptor activities. Analyses of the influence of alterations in receptor domain structure have involved two approaches: creating N- or C-terminal deletions of the receptor and constructing chimeric receptors. The former will be reviewed in this section and the latter will be discussed in the next section.

Deletion analyses of the EGF receptor, as with single residue mutations, have focused on the intracellular domain. Progressive deletions from the C terminus of EGF receptor cDNA were made and the truncated proteins were expressed in transfected cells (Table 3).

Receptors having deletions of up to 533 residues from the C terminus (Ala1186) were processed, delivered to the cell surface, and capable of binding ^{125}I-EGF (KHAZAI et al. 1988; LIVNEH et al. 1986a,b; HONEGGER et al. 1988a,b; GLENNEY et al. 1988; PRYWES et al. 1986; CLARK et al. 1988). Scatchard analysis of growth factor binding suggested that the truncated mutants Δ1155→1186, Δ1124→1186, and Δ20 kDa C terminus had high and low ^{125}I-EGF binding affinities, but that the mutant receptors Δ1061→1186, Δ743→1186, Δ726→1186, and Δ654→1186 displayed only a low binding affinity binding site. Deletion of the entire intracellular and transmembrane domains, Δ624→1186, resulted in a secreted form of the receptor.

The deletion mutant $\Delta654 \rightarrow 1186$ was capable of mediating ligand-induced endocytosis, albeit at a slower rate than the wild type. As noted previously, deletion of the C-terminal region containing the receptor autophosphorylation sites did not abrogate the ligand-stimulated mitogenic response. Larger deletions removing part or all of the kinase region ($\Delta743 \rightarrow 1186$, $\Delta726 \rightarrow 1186$, $\Delta654 \rightarrow 1186$) yielded receptors incapable of stimulating cellular responses.

A naturally occurring form of the EGF receptor which resembles a double deletion mutant is found in cells transformed by the avian erythroblastosis virus (AEV). The v-*erb* B transforming gene of this virus encodes what seems to be an EGF receptor having a deletion of nearly all of the extracellular ligand binding domain and a second deletion of the last two autophosphorylation sites at the C terminus (YAMAMOTO et al. 1983). Constructs of the EGF receptor that mimic one or both of these truncations have been produced and expressed in heterologous cells (KHAZAIR et al. 1988) to test the hypothesis that truncations of the native EGF receptor will yield an oncogenic protein.

The capacity of various N- and C-terminal truncated EGF receptor molecules to induce a transformed phenotype was assayed in fibroblasts and erythroblasts, (KHAZAIE et al. 1988), cell types sensitive to activated c-*erb* B- or v-*erb* B-mediated transformation (MAIHLE and KUNG 1989). Non-transformed fibroblasts were transfected with single deletion mutants having C-terminal truncations ($\Delta1155 \rightarrow 1186$; $\Delta1061 \rightarrow 1186$) or an N-terminal truncation ($\Delta1 \rightarrow 555$). Also, double deletion mutants having both N- and C-terminal truncations ($\Delta1 \rightarrow 555$ plus $\Delta1155 \rightarrow 1186$; $\Delta1 \rightarrow 555$ plus $\Delta1061 \rightarrow 1186$), or the wild-type receptor were transfected. All transfectants were then screened for the transformed phenotype. The double deletion mutant $\Delta1$-555 plus $\Delta1155 \rightarrow 1186$ would be comparable to v-*erb* B in structure, while the single deletion mutant $\Delta1 \rightarrow 555$ would resemble the activated c-*erb* B oncogene. The results of the experiments in which transformation parameters were measured in transfected fibroblasts or erythroblasts were suggestive of an increase toward the transformed phenotype, particularly in those cells having an EGF receptor with a short (32-residue) C-terminal deletion. (This is a complicated interpretation as similar C-terminal alterations in activated *erb* B oncogene of the avian leukosis virus have differential effects on fibroblast and erythroblast transformation; for review see MAIHLE and KUNG 1989.) Also, in many of the assays the intact EGF receptor scored positive in the presence of EGF. None of the truncated receptors mimicked v-*erb* B in a quantitative manner. Therefore, the notion that N- and/or C-terminal deletions of the EGF receptor can produce an oncogenic protein remains unproven.

c) Chimeric Molecules

Domains of the EGF receptor were joined to domains of oncogenic proteins or other receptors to yield chimeric molecules. This approach has the potential to reveal structural requirements for necessary interactions between tyrosine kinase, transmembrane, and ligand-binding domains for a receptor to transduce a mitogenic signal or for an oncogenic protein to act as a transforming agent. As shown in Table 3, two classes of hybrid molecules were constructed. In one instance the tyrosine kinase domain from either the v-*erb* B (RIEDEL et al. 1987) or v-*abl* (PRYWES et al. 1986) oncogene was joined to the transmembrane and extracellular domains of the EGF receptor. In the other group, the external domain of either the insulin (RIEDEL et al. 1986) or the interleukin-2 (IL-2; BERNARD et al. 1987) receptor was added to the transmembrane and tyrosine kinase domains of the EGF receptor.

In the latter cases, involving the external domains of other growth factors, the chimeric molecules were tested for different properties. The hybrid receptor containing the insulin binding domain was reported to activate the EGF tyrosine kinase domain in an insulin-dependent manner. This would suggest a common kinase activation mechanism for the insulin and EGF receptors. The chimeric receptor containing the ligand binding domain of the IL-2 receptor was not assayed for IL-2-enhanced tyrosine kinase activity. It was reported, however, that this hybrid molecule bound radiolabeled IL-2 with an affinity intermediate between the high and low affinities of the IL-2 receptor. Also, the authors reported a preliminary study indicating that this chimeric receptor was oncogenic. The basis for this observation is not clear.

The other pair of chimeric molecules involved replacing the kinase cytoplasmic domain of the EGF receptor with an oncogene (v-*erb* B or v-*abl*) tyrosine kinase domain. In the case of the v-*erb* B substitution, the results were complex. EGF induced autophosphorylation of the hybrid molecule, but cells expressing the hybrid molecules were growth inhibited by EGF in monolayer cultures. In contrast, these same cells were stimulated by EGF in focus formation and soft agar colony formation assays. Actually, this seeming contradiction is not unlike the biological response of EGF receptor-overexpressing cells such as A431 cells. A431 cells are growth inhibited by EGF in monolayer culture, but are growth stimulated by EGF in soft agar or in vivo (see Sect. G. II. A). The recipient cells in these experiments contained endogenous EGF receptors and this complicated several of the assays, such as soft agar colony formation. Hybrid receptors containing the v-*abl* tyrosine kinase domain were not activated by EGF, though they did bind radiolabeled EGF.

3. Receptor Activation

EGF binding to its receptor induces rapid changes in the enzymatic characteristics of the receptor's tyrosine kinase activity toward exogenous substrates, decreasing the K_m for phosphorylation substrates and increasing the V_{max} of the reaction (reviewed in CARPENTER 1987; GILL et al. 1987). As discussed above (Sect. E. IV. 2. a.), receptor autophosphorylation is rapidly stimulated upon ligand binding and may have functional consequences in the stimulation of kinase activity towards exogenous substrates. The mechanism by which ligand binding at the extracellular domain of the receptor affects the cytoplasmic domain kinase activity is a fundamental question in the study of receptor activation. This is a special topological problem to understand conceptually, since the only connection between the two domains is a single transmembrane α-helix. For a different view, based on more complex algorithms for transmembrane sequences see STAROS et al. (1989).

An early observation in the analysis of ligand-regulated receptor dynamics was that EGF induced a change in distribution of EGF receptors on the cell surface, from a population of dispersed receptors to an oligomeric or "clustered" state. Since receptor clusters were observed in coated pits on the cell surface and within internalized vesicles, oligomerization was believed to be part of the process of receptor internalization. Recently, the involvement of receptor-receptor interactions in ligand stimulation of receptor function has focused on the capacity of EGF to induce dimerization of receptors and the relationship of dimerization to the mechanism of ligand induction of tyrosine kinase activity have been discussed (see SCHLESSINGER 1988 a, b). The model of these studies is, perhaps, the tetrameric insulin receptor which can be thought of as a dimer of two monomers, each having an α- and a β-chain.

Ligand-induced dimerization has been investigated either by chemical crosslinking of closely associated receptor molecules or by separation of monomeric and oligomeric receptors with nondenaturing physical methods. Dimerization may influence many of the observed receptor biochemical characteristics, including interconversion between low- and high-affinity states, receptor autophosphorylation, and exogenous substrate phosphorylation. Crosslinking experiments using cells, membranes, or purified receptor preparations have demonstrated that EGF induces a rapid and reversible dimerization of the EGF receptor (BÖNI-SCHNETZLER and PILCH 1987; YARDEN and SCHLESSINGER 1987a, b; COCHET et al. 1988; NORTHWOOD and DAVIS 1988; FANGER et al. 1989). Dimerization of purified EGF receptor was dependent upon ligand concentration, temperature, and type of detergent employed (YARDEN and SCHLESSINGER 1987b). Analyses of ^{125}I-EGF binding indicated a higher binding affinity of the dimeric receptors than of the monomeric receptors (BÖNI-SCHNETZLER and PILCH 1987; YARDEN and SCHLESSINGER 1987a). Furthermore, BÖNI-SCHNETZLER and PILCH (1987) reported that EGF receptor autophosphorylation activity in the absence of EGF was 4-fold higher in the dimeric receptors than in the monomeric species. However, addition of EGF stimulated the monomeric receptor activity approximately 5-fold but yielded only a small stimulation of dimeric receptor activity. YARDEN and SCHLESSINGER (1987a, b) observed that both receptor dimerization and autophosphorylation activity increased linearly with the square of the receptor concentration, and that the dimeric receptor exhibited a higher phosphorylation activity than the monomer. Curiously, only a small fraction of the total receptor population is ever observed in the dimer form.

These results suggest that the EGF receptor exists in interconvertible aggregation states including monomeric, dimeric, and possibly oligomeric forms. Receptor dimerization in the presence of EGF suggests a mechanism for the generation of high-affinity receptors in the intact cell. Furthermore, since EGF shifts the equilibrium from the monomeric to the dimeric state and the dimeric receptors possess a higher level of kinase activity, dimerization is thought to represent an "allosteric" mechanism for regulation of kinase activity (SCHLESSINGER 1988 a, b). There are alternative interpretations. For example dimers may by a mechanism to stabilize activated receptor kinases (for this and other interpretations see STAROS et al. 1989). The monomer-dimer studies do not, however, resolve the basic question of how this allosteric mechanism is transmitted by a single transmembrane helix, unless activation of the kinase activity simply occurs upon contact of two cytoplasmic domains of EGF receptor molecules. This mechanism is not helped by the differences of opinion about the intra- or intermolecular nature of the autophosphorylation process, though these are not mutually exclusive. Also, there is no information at present to elucidate the mechanism that drives dimer formation or conversion of dimers to monomers or to the larger aggregates observed in the process of endocytosis.

4. Receptor Substrates

Since tyrosine kinase activity is necessary for the biological activity of the EGF receptor, a critical area for understanding the biochemical mechanisms of growth control is the identification of physiological substrates for this and other receptor tyrosine kinases. Although the EGF receptor is known to phosphorylate a variety of exogenous substrates in vitro, cellular substrates for the EGF receptor have been difficult to identify, except as radioactive spots following electrophoresis of cell extracts. Recent evidence suggests that several known proteins are rapidly phosphorylated in cells directly or indirectly after EGF treatment. Whether biological activity of these proteins is altered by phosphorylation and mediates cellular responses to EGF is not known.

A variety of proteins have been reported to undergo phosphorylation after the addition of EGF to cells. These proteins include: polyoma middle T antigen (SEGAWA and ITO 1983), the *ras* oncogene protein (KAMATA and FERAMISCO 1984), an 81-kDa homolog of the microvillus core protein ezrin (GOULD et al. 1986), type 1 protein phosphatase (CHAN et al. 1988), the human glucocorticoid receptor (RAO and FOX 1987), and the activated *erb* B2 (*neu*) gene product, which is closely related to the EGF receptor (AKIYAMA et al. 1988; STERN and KAMPS 1988, KING et al. 1988; KOKAI et al. 1988). Furthermore, a number of reports detail activation of serine-specific kinases after EGF treatment, suggesting that a cascade of kinases may occur during growth stimulation. These serine kinases, which appear to be activated indirectly by the EGF receptor kinase (none are known to contain phosphotyrosine) include: ribosomal protein S6 kinase (C.J. SMITH et al. 1979, 1980; NOVAK-HOFER and THOMAS 1984, 1985; JENOE et al. 1988; MUTOH et al. 1988); the protooncogene gene product *raf*-1 (MORRISON et al. 1988); a unique serine kinase from A431 cells (GIUGNI et al. 1988); Ca^{2+}/calmodulin-dependent kinase II (OHTA et al. 1988); casein kinase II (SOMMERCORN et al. 1987); and microtubule-associated protein 2 kinase (HOSHI et al. 1988; SATO et al. 1988).

The initial direct substrate identified for the EGF receptor was a 35-kDa protein (SAWYER and COHEN 1985) now known as lipocortin I or calpactin II.

This protein is a member of a family of proteins distinctive for their capacity to bind phospholipids in the presence of calcium (reviewed in BRUGGE 1986; KRETSINGER and CREUTZ 1986; CROMPTON et al. 1988). Individual members of this family have been independently identified in diverse systems.

Several roles related to phospholipid binding have been postulated for these proteins. One interesting proposed function of lipocortin I is as a glucocorticoid-induced protein that inhibits phospholipase A_2 activity (FLOWER and BLACKWELL 1979; BLACKWELL et al. 1980; F. HIRATA et al. 1980). Phospholipase A_2-mediated hydrolysis of phospholipids yields arachidonate, a precursor for prostaglandins and leukotrienes, which are mediators of inflammation (as well as many other responses). Inhibition of phospholipase A_2 activity by lipocortin I may yield an antiinflammatory response. However, since the mechanism of lipocortin I inhibition of phospholipase A_2 activity is through substrate sequestration (F. F. DAVIDSON et al. 1987; HAIGLER et al. 1987; SCHLAEPFER and HAIGLER 1987), the specificity of this inhibition is uncertain. Lipocortin I was identified as a major substrate for the EGF receptor in A–431 cell membranes, and purified based on the capacity to bind to A–431 cell membranes in the presence of Ca^{2+} and to undergo EGF-dependent phosphorylation (FAVA and COHEN 1984). Independently, PEPINSKY et al. (1986) purified and sequenced glucocorticoid-induced lipocortin I from rat peritoneal exudates as an inhibitor of phospholipase A_2 activity. Cloning and sequencing of human lipocortin I cDNA and porcine lung lipocortin I protein revealed a single canonical tyrosine phosphorylation site located near the N terminus of the molecule (WALLNER et al. 1986; DE et al. 1986). However, the relationship between tyrosine phosphorylation and functional activity of lipocortin I remains uncertain. Increased prostaglandin formation is a frequent response of cells and tissues to EGF (see Table 5).

Another substrate for the EGF receptor kinase may be involved in the hydrolysis of phospholipids. EGF rapidly stimulates the formation of two important intracellular second messengers, inositol-1,4,5-trisphosphate (IP_3) and 1,2-diacylglycerol (DAG) in a variety of cell types. Initially, this observation was made only for A–431 and other transformed cell lines expressing high levels of the EGF receptor (SAWYER and COHEN 1981; K. B. SMITH et al. 1983; PIKE and EAKES 1987; WAHL et al. 1987; HEPLER et al. 1987; PANDIELLA et al. 1987; WAHL and CARPENTER 1988), but it has been confirmed in both nontransformed cells, e.g., hepatocytes and keratinocytes, and EGF receptor-transfected cells (R. M. JOHNSON and GARRISON 1987; MOSCAT et al. 1988; TAKASU et al. 1988; HESKETH et al. 1988; R. OLSEN et al. 1988; PANDIELLA et al. 1988).

Phospholipase C is the key enzyme in this hormone-sensitive pathway. This enzyme hydrolyzes phosphatidylinositol-4,5-bisphosphate to produce IP_3 and DAG, which stimulate increases in intracellular free Ca^{2+} and protein kinase C activity, respectively. In order to determine whether phospholipase C is a substrate for the EGF receptor, cell proteins phosphorylated on tyrosine residues were separated from total cell proteins using a phosphotyrosine antibody matrix (WAHL et al. 1988). Since increased amounts of phospholipase C activity were immunoisolated from EGF-treated cells, it was likely that EGF stimulated tyrosine phosphorylation of phospholipase C. This result is consistent with studies by CHEN et al. (1987) and MOOLENAAR et al. (1988) which examined EGF-induced inositol phosphate metabolism in cells transfected with wild-type and tyrosine kinase-deficient EGF receptors. Both studies reported that tyrosine kinase activity was necessary for EGF to increase inositol phosphate metabolism. The relationship between tyrosine kinase activity and phospholipase C has been clarified by immunoprecipitation of phospholipase C isozymes with specific monoclonal antibodies (WAHL et al. 1989; MEISENHELDER et al. 1989; MARGOLIS et al. 1989). A 145-kDa phospholipase C isozyme, phospholipase C-II, was found to be phosphorylated on tyrosine residues in an EGF-dependent manner. This phospholipase C isozyme is also an efficient phosphorylation substrate of the EGF receptor kinase in vitro (NISHIBE et al. 1989). However, the functional role of tyrosine phosphorylation in mediating phospholipase C activity is unknown.

5. Activation of Gene Expression

The major, but not exclusive, site for reception of intracellular signals generated by the formation of growth factor-receptor complexes is, obviously, the nucleus where gene transcription and DNA replication are stimulated. Since little is known about EGF signals for DNA replication, we have focused on transcriptional control by EGF. The types of genes subject to modulated expression by growth factors fall into two broad categories: those involved in the mitogenic response and those involved in more specialized and non-mitogenic responses.

a) Non-Growth Regulation Genes

In this latter category are two clear examples of genes whose increased expression by EGF is unrelated to mitogenic events. The most studied of these is the EGF induction of prolactin mRNA in pituitary tumor cell lines. Early studies (SCHONBRUNN et al. 1980; C. C. JOHNSON et al. 1980) demonstrated that EGF stimulated prolactin production in these cells, but had an inhibitory effect on cell growth. MURDOCH et al. (1982) showed that EGF increased prolactin mRNA levels and that a 10-fold increase in transcription of this gene occurred 1 h after addition of EGF. Increased prolactin mRNA could be detected within a few minutes of growth factor addition. This remains one of the most rapid transcriptional activations by EGF that is known. An EGF-responsive 5' promoter region of the prolactin gene was identified (SUPOWIT et al. 1984) and localized to a short 5' sequence of approximately 45–50 bp (ELSHOLTZ et al. 1986) approximately 300 bp upstream from the transcription initiation site. The position and orientation independence of this *cis*-active sequence suggested an enhancer function in mediating transcription of the prolactin gene by EGF. Interestingly, this 5' element also responded to the protein kinase C

activator TPA (ELSHOLTZ et al. 1986), and TPA and the calcium ionophore
A23187 produced a synergistic increase in prolactin mRNA transcription
equivalent to the level obtained with EGF (MURDOCH et al. 1985). Other
studies have shown that EGF stimulation of prolactin mRNA is dependent on
the presence of extracellular calcium (WHITE and BANCROFT 1983) and calmodulin activity (WHITE 1985). Since EGF induction of prolactin mRNA
does not require protein synthesis, the growth factor signaling and receiving
systems for this gene depend on molecules that preexist within the cell.
Whether EGF action in this system is actually mediated by growth factor activation of protein kinase C (potentially brought about by increased
phospholipase C activity) and growth factor stimulation of intracellular calcium levels remains to be proven.

In pheochromocytoma cell lines, EGF promotes increased synthesis of
tyrosine hydroxylase (GOODMAN et al. 1980), the rate-limiting enzyme in the
biosynthesis of catecholamines. A 300-bp region 5' to the tyrosine hydroxylase
coding sequence $(+27/-272)$ was identified as an EGF-responsive element
(LEWIS and CHIKARAISHI 1987). Nucleotide sequences in this region resemble
sequences identified in the prolactin promoter as mediating EGF responsiveness.

b) Growth Regulation Genes

α) Types of Genes

These EGF-sensitive genes fall into two categories – genes which control cell
proliferation or the induction of cell proliferation and genes which are regulated by growth status. Genes in the latter category are usually activated
rather slowly (several hours) by EGF and this induction requires protein
synthesis. Examples would include the following EGF-enhanced mRNAs:
glycolytic enzymes (MATRISIAN et al. 1985), the secreted metalloprotease
transin (MATRISIAN et al. 1986a; MACHIDA et al. 1988; KERR et al. 1988), plasminogen activator (STOPPELLI et al. 1986), ornithine decarboxylase (JASKULSKI
et al. 1988), an ADP/ATP carrier (BATTINI et al. 1987), and fibronectin
(BLATTI et al. 1988). Some of these may have important functions in growth
control in vivo. For example, transin expression is higher in malignant than
benign tumors (MATRISIAN et al. 1986b). Both transin and plasminogen activator may be of significance in the invasive and metastatic behavior of carcinomas.

A second set of EGF-activated genes may be of greater relevance to the
process by which cell proliferation is initiated and regulated. EGF-increased
transcription of these genes occurs rapidly ($\sim$30–60 min) and is independent
of new protein synthesis. These genes include: the glucose transporter gene
(HIRAKI et al. 1988), the knox-24 gene encoding a "zinc finger" putative transcription factor (LEMAIRE et al. 1988), the gene for the nuclear antigen cyclin
(JASKULSKI et al. 1988), the calcyclin gene which encodes an S100-like putative
calcium-binding protein (CALABRETTA et al. 1986; FERRARI et al. 1987;
GHEZZO et al. 1988), the β- and γ-actin genes (ELDER et al. 1984; JASKULSKI

et al. 1988), the retrovirus-like VL30 gene(s) (D.N. FOSTER et al. 1982; HODGSON et al. 1983; RODLAND et al. 1986, 1988), the c-*jun* gene (QUANTIN and BREATHNACH 1988), which encodes an AP-1-like transcription factor thought to form a complex with c-*fos* (for a review see CURRAN and FRANZA 1988), and the c-*fos* and c-*myc* genes, which encode nuclear proteins suspected of having functions in transcription (MÜLLER et al. 1984, 1985; BRAVO et al. 1985; GREENBERG et al. 1985; COFFEY Jr et al. 1989; HELDIN and WESTERMARK 1988). In many instances the induction of these genes by EGF is potentiated by cycloheximide, suggesting a posttranscriptional mechanism of action. However, evidence also exists that EGF treatment increases transcription per se. The overall pattern of the rapidly induced genes seems clear. The growth factor initially activates genes that in many instances code for transcription factors that may be required to accelerate general mRNA synthesis.

β) Mechanism of c-fos Activation

Various authors have indicated that the EGF pathway for c-*fos* and/or c-*myc* induction involves protein kinase C activation (GREENBERG et al. 1985; BRAVO et al. 1985, 1987), increased intracellular calcium levels (BRAVO et al. 1985, 1987), protein kinase S6 activation (BLACKSHEAR et al. 1987), or increased cyclic AMP levels (GREENBERG et al. 1985; RAN et al. 1986). Most authors have agreed that there are multiple EGF-dependent signals arriving in the nucleus that participate, perhaps cooperatively, in the induction of c-*fos* and c-*myc* transcription. While 5′ mapping studies have separated a calcium-sensitive element from an EGF-responsive element, the protein kinase C sensitive element does not separate from the EGF sensitive site (FISCH et al. 1987; SHENG et al. 1988). The EGF response element (also sensitive to TPA, nerve growth factor, fibroblast growth factor) is approximately 300 bp upstream from the c-*fos* transcription start site and functions as an enhancer. This site is termed the dyad symmetry element (DSE) or serum response element (SRE) (for a review see VERMA and SASSONE-CORSI 1987). Basal transcription of the c-*fos* gene is controlled by a short sequence about 60 bp from the transcription start site. Interestingly, this site, which is not responsive to growth factors, resembles cyclic AMP-regulated promoters in other systems. As mentioned above, two studies have associated increased cyclic AMP levels with EGF induction of c-*fos*. Several groups have begun the process of purifying proteins that bind to the DSE/SRE and which may mediate growth factor stimulation of c-*fos*. One of these reports (PRYWES and ROEDER 1986) has detected a factor present in EGF-treated A–431 cells (but not in TPA- or A23187-treated cells) which binds to the c-*fos* DSE. The role this factor plays in EGF enhancement of c-*fos* transcription remains to be determined.

There is evidence that protein kinase activity has a role in the activation of gene expression. PRYWES et al. (1988) have provided evidence that a DSE/SRE factor is phosphorylated when A–431 cells are treated with EGF for 15 min. The phosphorylation sites on this protein are serine residues and there appear to be multiple sites of phosphorylation. The phosphoprotein ($M_r \approx 64$ kDa) binds to the DSE and binding is decreased by pretreatment of the factor with alkaline phosphatase. While phosphatase pretreatment prevented binding of the factor to DNA, phosphatase treatment after binding of the factor to DNA was without effect. The authors propose that phosphorylation is necessary for this factor to bind to the DSE, but phosphorylation does not control transcriptional activation per se. ZINN et al. (1988) investigated the influence of 2-aminopurine, a reported protein kinase inhibitor, on gene expression. The inhibitor reduced the induction of c-*fos* and c-*myc* transcription by serum.

BARBAR and VERMA (1987) have reported that the c-*fos* protein is also subject to phosphorylation when cells were treated with TPA or serum (EGF was not assayed). The phosphorylation sites are serine residues and up to five phosphorylated residues are present on each c-*fos* molecule. The functional consequence of these phosphorylations remains speculative, however.

Finally, MAHADEVAN et al. (1988) have detected the rapid appearance of 33-kDa and 15-kDa phosphoproteins in the nucleus of cells treated with various mitogens, including EGF.

An area of continuing controversy is the issue of whether or not EGF and/or the EGF receptor reach the nucleus, following ligand-induced internalization. Two reports reach such a conclusion (RAKOWICZ-SZULCZYNSKI et al. 1986; RAPER et al. 1987), but the techniques employed were not convincing on what is a difficult technical point – fine point intracellular localization. Morphologic studies have been negative.

F. Physiology of the EGF Family

The attempt in this section is to concentrate on physiological aspects of these growth factors with special attention to issues and facts that have not been reviewed before. Thus, components of the general mitogenic response are omitted and instead we present the numerous nonmitogenic responses that have been observed with these growth factors. Also, this section concentrates on studies conducted with intact animals.

I. EGF

1. Distribution in Fluids/Secretions

Sensitive radioimmunoassays for human EGF have been available for some time and a body of information has developed describing the distribution of this growth factor in various body fluids and tissues. The amounts in tissues are low and it is difficult to know whether the observed data are physiologically relevant or whether they simply represent contamination from fluids which seem comparatively rich in EGF. In regard to the presence of EGF in tissues, it might be more reliable to examine EGF mRNA levels to know whether a certain tissue actually produces the growth factor (see Sect. D. III. 1).

The data in Table 4 are a compilation of studies documenting the concentrations of EGF in various body fluids or secretions. Data from nonhuman species are not shown unless no human results are available. It is clear from these studies that EGF is widely distributed in nearly all body fluids. With the

Table 4. Distribution of EGF in human body fluids/secretions

Fluid/secretion	Amount (ng/ml)		References
	Mean	Range	
Amniotic fluid	0.25	0.06–1.1	BARKA et al. (1978)
	NA[a]	0.10–0.96	D'SOUZA et al. (1985)
	0.53	0.2–1.02	WEAVER et al. (1988)
Aqueous humor[b]			
males	10.3	0.7–60	SHINODA et al. (1988)
female	4.9	0.9–18	
Blood components			
plasma (platelet-rich)	0.27	0.14–0.39	OKA and ORTH (1983)
plasma (platelet-poor)	<0.02	<0.02	
serum (PRP-derived)	0.27	0.11–0.46	
serum (PPP-derived)	<0.02	<0.02	
plasma (platelet-rich)	0.3	NA	A. P. SAVAGE et al. (1986)
plasma (platelet-poor)	0.02	NA	
serum (sep. at 30 min)	0.2	NA	
serum (sep. at 270 min)	0.7	NA	
plasma	0.35	0.2–0.7	KUROBE et al. (1986)
serum	1.14	0.55–1.5	
plasma	NA	0.2–0.8	HAYASHI and SAKAMOTO (1988)
Brunner's glands fluid	1.6	NA	KIRKEGAARD et al. (1984)
Cerebrospinal fluid	0.17	NA	Y. HIRATA et al. (1982c)
Mammary fluids			
breast cyst fluid	241	5–945	JASPAR and FRANCHIMONT (1985)
breast cyst fluid	548	5–1950	COLLETTE et al. (1986)
breast cyst fluid	111	0.25–433	BOCCARDO et al. (1988)
breast fluid[c] (EGF)	205	62–654	CONNOLLY and ROSE (1988)
breast fluid[c] (TGF-α)	5.1	0–50	
Mammary secretions			
milk	80	35–140	MORAN et al. (1983)
precolostrum	NA	130–800	BEARDMORE et al. (1983)
colostrum	NA	35–438	
milk	NA	20–110	
milk	107	NA	JASPAR and FRANCHIMONT (1985)
colostrum	197	NA	
colostrum	NA	25–38	JANSSON et al. (1985)
milk	NA	5.2–11.5	
milk	65	37–107	PESONEN et al. (1987)
milk (EGF)	140	32–600	CONNOLLY and ROSE (1988)
milk (TGF-α)	0.8	0–8.4	
Ovarian follicle fluid[d]	7.0	NA	HSU et al. (1987)
Pancreatic fluid	2.3	2.2–2.5	Y. HIRATA et al. (1982b)
Peritoneal fluid	37	NA	DELEON et al. (1986)

[a] Data not available (NA).
[b] Data from mice.
[c] Nipple aspirates, nonpregnant.
[d] Data from porcine species.

Table 4 (continued)

Fluid/secretion	Amount (ng/ml)		References
	Mean	Range	
Prostatic fluid	272	NA	GREGORY et al. (1986)
Saliva	9.2	5.6–16.8	STARKEY and ORTH (1977)
	NA	0.3–5.0	DAGOGO-JACK et al. (1985)
	3.2	NA	OHMURA et al. (1987)
	3.0	1.5–5.0	PESONEN et al. (1987)
	NA	0.9–3.0	HAYASHI and SAKAMOTO (1988)
Seminal fluid	36.4	5–150	ELSON et al. (1984)
	50	22–115	PESONEN et al. (1987)
	48	NA	HIRATA et al. (1987)
	101	4–331	RICHARDS et al. (1988)
Sweat (armpit)	25	1.2–125	PESONEN et al. (1987)
(breast)	1.0	0.7–2.5	
Tears	1.5	0.7–2.2	PESONEN et al. (1987)
	NA	9.5–27	HAYASHI and SAKAMOTO (1988)
Urine	88	29–272	STARKEY and ORTH (1977)
	NA	6–150	GREGORY et al. (1977)
	80	10–100	PESONEN et al. (1987)
	NA	11–100	HAYASHI and SAKAMOTO (1988)
(24 h)	[11.8	38][e]	MATTILA et al. (1985)
(24 h–14 days)	[9.8	72]	
(15–30 days)	[9.8	115]	
(1 month–1 year)	[25.6	177]	
(1–2 years)	[67.4	235]	
(2–5 years)	[69.9	168]	
(6–10 years)	[70.2	101]	
(10–16 years)	[82.6	70]	
(20–29 years)	[32.4	19.5][e]	UCHIHASHI et al. (1982)
(30–39 years)	[12.2	11.4]	
(40–49 years)	[14.8	16.6]	
(50–59 years)	[9.6	18.1]	
(60–69 years)	[6.4	10.2]	
(70–79 years)	[5.4	6.6]	
Uterine luminal fluid[b]	1.5	1.2–1.7	DiAUGUSTINE et al. (1988)

[e] Mean values normalized for urinary creatinine content [ng EGF/mg creatinine].

exception of the age-dependent changes in urinary EGF, there are no data to indicate that these fluid concentrations are significantly altered in various physiological circumstances, including pathologies. Many of the fluids (for example saliva, mammary fluids and secretions, prostatic and seminal fluids, urine) contain rather high levels of the hormone. The data on EGF concentrations in blood components, in contrast, suggest very low levels. Apparently, EGF is not a circulating hormone, but is present within platelet granules. One study (MACNEIL et al. 1988) has reported that thrombin causes the release of platelet EGF. Given the wide distribution and in some cases high levels of EGF in body fluids, it is likely that these fluids are a significant source of EGF for many cells in vivo. In viewing the data in Table 4, it is necessary to remember that many of the radioimmunoassay reagents might crossreact with TGF-α. Studies with reagents specific for TGF-α have not been reported.

2. Biological Responses

Due to the absence of well defined sites of synthesis that could be removed surgically, it is not possible to perform classical hormone ablation experiments with EGF or any other growth factor. Therefore, the physiological role of endogenous EGF in any animal remains unknown. More complex approaches, such as the use of antisense probes or transgenic animals, jet have to be worked out. Perhaps more than any other growth factor, EGF has been available for a number of years in quantities sufficient for administration to large and small animals. A body of data exists describing the results of exogenous EGF administration. One purpose of Table 5 is to present, in outline

Table 5. Physiological responses to exogenous EGF

Tissue	Response(s)	References
Adipocytes	↑ lipogenesis	HAYSTEAD and HARDIE (1986)
	↑ acetyl-CoA carboxylase	HAYSTEAD and HARDIE (1986)
	↓ differentiation	SERRERO (1987)
	↓ induction of aromatase	EVANS et al. (1987)
Adrenal gland	↑ cortisol synthesis	SINGH-ASA and WATERS (1983), SINGH-ASA et al. (1985)
	↑ epinephrine, norepinephrine	KENNEDY et al. (1986)
Amnion	↑ PG H_2 synthase	CASEY et al. (1988)
	↑ PGE_2	MITCHELL (1987), ZAKAR and OLSON (1988), CASEY et al. (1987)
Bone	↑ cell proliferation	CENTRELLA et al. (1987)
	↓ nodule formation	ANTOSZ et al. (1987)
	↓ collagen synthesis	HATA et al. (1984)
	↑ PGE_2	TASHJIAN and LEVINE (1978), YOKOTA et al. (1986)
	↓ PTH-stimulated adenylate cyclase	GUTIERREZ et al. (1987)
	↑ bone resorption	TASHJIAN and LEVINE (1978), RAISZ et al. (1980)
	↑ collagenase	CHIKUMA et al. (1984)
Crop-sac (pigeon)	↑ mucosal epithelial growth	ANDERSON et al. (1987)

PG, prostaglandin; PTH, parathyroid hormone; CRH, corticotropin-releasing hormone; FSH, follicle-stimulating hormone; LH, luteinizing hormone; ACTH, adrenocorticotropic hormone; hCG, human chorionic gonadotropin; T_3, 3,5,3'-triiodothyronine; T_4, thyroxine; rT_3, 3,3',5'-triiodothyronine; T_2, 3,3'-diiodothyronine; TSH, thyroid-stimulating hormone.

Table 5 (continued)

Tissue	Response(s)	References
Endo-metrium	↓ epithelial cell proliferation	GERSCHENSON et al. (1979), HOFMANN et al. (1988), KORC et al. (1986)
	↑ smooth muscle proliferation	GOSPODAROWICZ et al. (1981), CLEMMONS (1984), REILLY et al. (1987)
Ear	↓ ear development	HOATH (1986), HOATH and PICKENS (1987)
Eye	↑ neonate eyelid opening	COHEN (1962)
	↑ migration endothelial cells	K. WATANABE et al. (1987)
	↑ proliferation endothelial cells	GOSPODAROWICZ et al. (1977b), RAYMOND et al. (1986)
	↑ proliferation lens cells	REDDAN and WILSON-DZIEDZIC (1983)
	↑ corneal wound repair	C. R. SAVAGE and COHEN (1973), HO et al. (1974), DANIELE et al. (1979), GOSPODAROWICZ and GREENBURG (1979), FABRICANT et al. (1982), WOOST et al. (1985), BRIGHTWELL et al. (1985), SINGH and FOSTER (1987), REIM et al. (1988)
Heart	↑ beating rate	RABKIN et al. (1987)
	↑ proliferation mesenchymal cells	BALK et al. (1982)
Hypothal-amus	↑ CRH release	LUGER et al. (1988)
Gastroin-testinal tract	↓ acid secretion	BOWER et al. (1975), FINKE et al. (1985), DEMBINSKI et al. (1986), SHAW et al. (1987)
	↓ food consumption	PANARETTO et al. (1982)
	↑ PGE$_2$	CHIBA et al. (1982)
	↑ ulcer healing	KONTUREK et al. (1988a, b)
	↑ ulcer prevention	KONTUREK et al. (1981a, b), P. S. OLSEN et al. (1984), SAKAMOTO et al. (1985)
	↑ neonatal maturation/growth	BEAULIEU and CALVERT (1981), MALO and MÉNARD (1982), CALVERT et al. (1982, OKA et al. (1983), O'LOUGHLIN et al. (1985), DEMBINSKI and JOHNSON (1985), ARSENAULT and MÉNARD (1987), POLLACK et al. (1987), FALCONER (1987), PUCCIO and LEHY (1988), MÉNARD et al. (1988)
	↑ cell proliferation (adult)	SCHEVING et al. (1979, 1980), L. R. JOHNSON and GUTHRIE (1980), DEMBINSKI et al. (1982), AL-NAFUSSI and WRIGHT (1982), CHABOT et al. (1983), ULSHEN et al. (1986), FINNEY et al. (1987), JACOBS et al. (1988)

Table 5 (continued)

Tissue	Response(s)	References
Kidney	↓ kidney development	HOATH (1986)
	↑ proliferation proximal tubular cells	NORMAN et al. (1987)
	↑ PGE_2, glomerular mesangial cells	MARGOLIS et al. (1988)
	↓ glomerular filtration rate	HARRIS et al. (1988)
	↓ renal blood flow	HARRIS et al. (1988)
	↓ vasopressin-induced water flow	BREYER et al. (1988)
Liver	↑ fat accumulation	HEIMBERG et al. (1965), YOSHIMOTO et al. (1983), VAARTIES et al. (1985), HOLLAND and HARDIE (1985)
	↑ cell proliferation	BUCHER et al. (1978)
	↑ regeneration	R. OLSEN et al. (1988)
	↑↓ neonatal maturation	HOATH (1986), OPLETA et al. (1987)
	↑ gluconeogenesis	SOLEY and HOLLENBERG (1987)
	↑ glycogen synthesis	FREEMARK (1986)
	↑ cell migration	BADE and FEINDLER (1988)
	↑ hepatocyte proliferation	RICHMAN et al. (1976), McGOWAN et al. (1981), SAND et al. (1985), CRUISE and MICHALOPOULOS (1985), TOMOMURA et al. (1987)
Mammary	↑ collagen	SALOMON et al. (1981)
	↑ ductal morphogenesis	COLEMAN et al. (1988)
	↑ lobuloalveolar development	TONELLI and SOROF (1980, 1981), VONDERHAAR (1987)
	↓↑ casein	TAKETANI and OKA (1983a), ARAKAWA et al. (1985), VONDERHAAR and NAKHASI (1986)
	↓↑ α-lactalbumin	TAKETANI and OKA (1983a, b), SANKARAN and TOPPER (1983, 1984), KOMURA et al. (1986), NICHOLAS et al. (1988)
	↑ cell proliferation	TURKINGTON (1969), PASCO et al. (1982), IMAGAWA et al. (1985), McGRATH et al. (1985)
Meso-thelium	↑ cell proliferation	CONNELL and RHEINWALD (1983), GABRIELSON et al. (1988)
	↓ differentiation	CONNELL and RHEINWALD (1983), KIM et al. (1987)
Immune system	↑ macrophage chemotaxis	LASKIN et al. (1981)
	↑ macrophage phagocytosis	LASKIN et al. (1980)
	↑ interferon-γ T cells	H. M. JOHNSON and TORRES (1985)
	↑ proliferation T cells	ACRES et al. (1985)
	↓ immune suppressor activity T cells	AUNE (1985), DEAN et al. (1987)
Nerve	↑ proliferation glia	BRUNCK et al. (1976)
	↑ proliferation astrocytes	LEUTZ and SCHACHNER (1981), GUENTERT-LAUBER and HONEGGER (1985)

Table 5 (continued)

Tissue	Response(s)	References
Nerve	↑ proliferation astroglia	AVOLA et al. (1988)
	↑ myelin basic protein	ALMAZAN et al. (1985)
	↑ neuron survival and process outgrowth	R. S. MORRISON et al. (1987)
	↑ tyrosine hydroxylase	GOODMAN et al. (1980)
Oral cavity	↑ mucosal proliferation	STEIDLER and READE (1980, 1981)
Ovary	↓ action of Mullerian inhibition substance	UENO et al. (1988)
	↓ inhibin production	FRANCHIMONT et al. (1986), ZHIWEN et al. (1987)
	↑↓ steroid formation	C. C. JOHNSON et al. (1980), HSUEH et al. (1981), SCHOMBERG et al. (1983), KNECHT and CATT (1983), SHAW et al. (1985), PULLEY and MARRONE (1986), FRANCHIMONT et al. (1986), TAPANAINEN et al. (1987), TRZECIAK et al. (1987)
	↑ gonadotropin binding	MONDSCHEIN and SCHOMBERG (1981), MAY et al. (1987)
	↑ oocyte maturation	DEKEL and SHERIZLY (1985), DOWNS et al. (1988)
	↓ differentiation theca interstitial cells	ERICKSON and CASE (1983)
	↑ follicular atresia	RADFORD et al. (1987a)
	↓ follicle formation	LINTERN-MOORE et al. (1981), CAHILL et al. (1982)
	↑ granulosa cell proliferation	GOSPODAROWICZ et al. (1977a), GOSPODAROWICZ and BIALECKI (1978), GOSPODAROWICZ et al. (1979), BERTONCELLO et al. (1981, 1982), TAPANAINEN et al. (1987)
Palate	↓ palatal shelf fusion (increased proliferation epithelium and mesenchyme)	HASSELL (1975), HASSELL and PRATT (1977), BEDRICK and LADDA (1978), TYLER and PRATT (1980), YONEDA and PRATT (1981), GROVE and PRATT (1983, 1984)
	↑ glycosaminoglycans	TURLEY et al. (1985), PISANO and GREENE (1987)
	↑ collagen	SILVER et al. (1984)
Pancreas	↑ amylase secretion	LOGSDON and WILLIAMS (1983a, b)
	↑ insulin synthesis	CHATTERJEE et al. (1986)
	↑ chemical carcinogenesis	CHESTER et al. (1986), MALT et al. (1987)
Pituitary	↑ FSH secretion	RADFORD et al. (1987a)
	↓↑ LH secretion	MIYAKE et al. (1985), RADFORD et al. (1987b), PRZYLIPIAK et al. (1988)
	↓ GH_4C_1 cell proliferation	SCHONBRUNN et al. (1980), C. C. JOHNSON et al. (1980)

Table 5 (continued)

Tissue	Response(s)	References
Pituitary	↑ prolactin	SCHONBRUNN et al. (1980), C. C. JOHNSON et al. (1980), MURDOCH et al. (1982)
	↑↓ growth hormone	SCHONBRUNN et al. (1980), IKEDA et al. (1984), MOORE et al. (1984)
	↑ ACTH	POLK et al. (1987)
Placenta	↑ placental lactogen	MOORE et al. (1984), MORRISH et al. (1987), MARUO et al. (1987)
	↑ hCG	BENEVNISTE et al. (1978), Y. HIRATA et al. (1982a), MORRISH et al. (1987), MARUO et al. (1987)
	↑ progesterone	BAHN et al. (1980), RITVOS (1988)
Prostate	↑ cell proliferation	MCKEEHAN et al. (1984)
Rectum	↑ chemical carcinogenesis	KINGSNORTH et al. (1985)
Respiratory tract	↑ fetal lung development	CATTERTON et al. (1979), SUNDELL et al. (1980)
	↑ lung epithelial proliferation	SUNDELL et al. (1980), GOLDIN et al. (1980)
	↓ lung liquid secretion	KENNEDY et al. (1986)
	↑ tracheal epithelia proliferation	GOLDIN et al. (1980)
Salivary glands	↑ cell proliferation	INOUE et al. (1986), REDMAN et al. (1988)
Skeletal muscle	↑ protein synthesis	HARPER et al. (1987)
Skin	↑ keratinization	COHEN and ELLIOTT (1963)
	↑ epidermal disulfide content	FRATI et al. (1972), ROBERTSON and BLECHER (1986)
	↑ epidermal proliferation	COHEN and ELLIOTT (1963), COHEN (1965), REINWALD and GREEN (1977), STEIDLER and READE (1980), MOORE et al. (1985)
	↑ chemical carcinogenesis	REYNOLDS et al. (1965), ROSE et al. (1976)
	↓ hair development	COHEN (1962), REYNOLDS et al. (1965), MOORE et al. (1981a, b, 1982, 1983)
	↑ sebaceous gland cell proliferation	MCDONALD et al. (1983), MATIAS et al. (1983)
	↑ hair follicle regression (decreased bulb cell mitosis)	MOORE et al. (1985), HOLLIS and CHAPMAN (1987)
	↓↑ dermal proliferation	COHEN and ELLIOTT (1963), PANARETTO et al. (1984), LAATO et al. (1986a, b), BUCKLEY et al. (1985, 1987)
	↑ wound healing	J. D. FRANKLIN and LYNCH (1979), GREEN et al. (1979), NIALL et al. (1982), BUCKLEY et al. (1985), LAATO et al. (1986a, b), G. C. BROWN et al. (1986), T. J. FRANKLIN et al. (1986)

Table 5 (continued)

Tissue	Response(s)	References
Somatic growth	↓ neonate body weight and size	Cohen (1962), Hoath (1986), Aulerich et al. (1988), Hoath et al. (1988)
Testes	↑ inhibin production	Morris et al. (1988)
	↓↑ androgen synthesis	Hsueh et al. (1981), Welsh et al. (1982), Verhoeven and Cailleau (1986), Ascoli et al. (1987), Mallea et al. (1987)
	↑ ornithine decarboxylase	Stastny and Cohen (1972)
	↑ sperm production	Tsutsumi et al. (1986)
	↓ hCG receptor level	Freeman and Ascoli (1982)
Thymus	↑ epithelial proliferation	Nieburgs et al. (1987)
Thyroid	↓ thyroxine (T_3, T_4)	Moore et al. (1984), Corcoran et al. (1986), Ahren (1987)
	↑ thyroxine metabolites (rT_3, T_2)	Corcoran et al. (1986)
	↑ TSH	Corcoran et al. (1986)
	↑ cell proliferation	Roger and Dumont (1982), Wester-mark et al. (1983), Eggo et al. (1984), Schatz et al. (1986), Zerek-Melen et al. (1987), Takasu et al. (1987)
	↓ differentiated functions	Roger and Dumont (1982), Wester-mark et al. (1983), Eggo et al. (1984), Takasu et al. (1987)
Tooth	↑ neonate incisor eruption	Cohen (1962)
	↑↓ cell proliferation	Steidler and Reade (1981), Partanen et al. (1985), Topham et al. (1987), Rihtniemi and Thesleff (1987)
Uterus	↑ contractions	Gardner et al. (1987)
	↑ epithelial proliferation	Tomooka et al. (1986)
Vagina	↑ neonate vaginal opening	Imada et al. (1987)
Vascular system	↑ endothelial proliferation	Knauer and Cunningham (1983), McAusland et al. (1985)
	↑ endothelial migration	McAusland et al. (1985)
	↑ endothelial PGI_2	Ristimäki et al. (1988)
	↑ smooth muscle proliferation	Bhargava et al. (1979), Tomita et al. (1986), Grosenbaugh et al. (1988)
	↑ smooth muscle PGI_2	Bailey et al. (1985)
	↑ angiogenesis	Gospodarowicz et al. (1979), Schreiber et al. (1986)
	↑ blood flow	Gan et al. (1987a, b), Carter et al. 1988)
	↓ $PGF_{2\alpha}$, norepinephrine-induced smooth muscle contraction	Gan et al. (1987a, b)
	↑ cardiac output	Carter et al. (1988)
	↓ total peripheral resistance	Carter et al. (1988)
	↓ mean arterial pressure	Carter et al. (1988)
	↑ vasocontraction	Berk et al. (1985), Muramatsu et al. (1985, 1986)

form, studies which suggest a number of biological activities for EGF that seem not related to mitogenesis. Similar results, i.e., nonmitogenic responses, have been obtained in cell and organ culture systems and these are also included. Also listed in Table 5 are references to various clinical studies, in the area of wound repair, that have been performed with EGF. There is not space to describe in more detail the results of these interesting studies.

Given the wide distribution of EGF in the body, the sensitivity of epithelial populations to growth stimulation by EGF, and the need of many of these epithelia to undergo renewal at a high rate (estimated at several hundred thousand cell divisions per second in vivo), a reasonable hypothesis is that EGF serves to ensure this renewal process in the intact animal.

II. TGF-α

In a wide variety of cell lines, TGF-α acts as a mitogenic agent for EGF-responsive cells, stimulating tyrosine phosphorylation of the EGF receptor and other events associated with the mitogenic response. Like EGF, TGF-α stimulates a nonmitogenic response, bone resorption in vitro (IBBOTSON et al. 1985, 1986; TASHJIAN et al. 1985, P. H. STERN et al. 1985). Administration of TGF-α to animals also produces EGF-like responses including: precocious eyelid opening (J. M. SMITH et al. 1985), growth retardation and slow hair growth (TAM 1985), induction of testicular ornithine decarboxylase (NAKHLA and TAM 1985) in the newborn mouse, inhibition of gastric acid secretion (RHODES et al. 1986), and wound healing (SCHULTZ et al. 1987).

Although most comparative studies have not found a significant difference in potency or activity between EGF and TGF-α, there are several reports indicating differences in potency or activity do exist. SCHREIBER et al. (1986) reported that TGF-α is more potent than EGF in promoting angiogenesis in a hamster cheek pouch assay. P.H. STERN et al. (1985) and IBBOTSON et al. (1986) indicate that TGF-α is more potent than EGF in bone resorption assays in vitro. The largest quantitative difference between EGF and TGF-α is reported for their capacities to bind to the chicken EGF receptor (LAX et al. 1988 a). The chicken EGF receptor bound TGF-α with an approximately 100-fold stronger affinity than EGF. Differences in potency of these agonists may be more apparent than real unless a number of quantitative issues are clearly defined. More intriguing are reported differences in the activities of EGF and TGF-α. One such case is the study by BARRANDON and GREEN (1987) in which TGF-α promoted a greater increase in keratinocyte colony expansion than EGF. In these experiments, cells grown in the presence of EGF never reached the colony size achieved by cells grown in the presence of TGF-α. This effect was independent of growth factor concentration in that no concentration of EGF was able to mimic the degree of colony expansion achieved in the presence of TGF-α. In a very different system GAN et al. (1987a) have described another seemingly qualitative difference between TGF-α and EGF. Measuring blood flow in the femoral arterial bed of dogs, the authors reported that both EGF and TGF-α increased blood flow, but did not change systemic blood pressure or control venous pressure. However, differences between the two growth factors were noted in terms of cross–desensitization. While administration of EGF produced desensitization to a subsequent bolus of EGF, it did not desensitize to subsequent treatment with TGF-α. Also, initial treatment with TGF-α did not produce desensitization to the subsequent administration of TGF-α. Whether these two interesting reports lead to clear differences in the actions of TGF-α and EGF will be interesting to observe.

III. Amphiregulin

Amphiregulin is a glycoprotein produced in response to treatment of the human breast carcinoma cell line MCF-7 with phorbol-12-myristate-13-acetate, a potent tumor promoter and protein kinase C activator (SHOYAB et al. 1988). It is reported that amphiregulin inhibits the growth of some cell lines and stimulates the growth of others (SHOYAB et al. 1988; 1989).

IV. Pox Virus Growth Factors

Pox viruses are a family of large DNA viruses having approximately 100 genes and replicating in the cytoplasm of infected cells. Frequently, these viruses induce replication of cells (either infected or nearby cells), but can be cytolytic depending on the particular virus. The vaccinia virus is cytolytic, but is also associated with hyperplasia of cells near the lesion. In late 1984 (BLOMQUIST et al. 1984) and early 1985 (J. P. BROWN et al. 1985; REISNER 1985) three groups reported a sequence similarity between EGF and a portion of the 19-kDa product of one of the vaccinia early genes. The biochemical existence of a virally encoded growth factor was demonstrated (STROOBANT et al. 1985; TWARDZIK et al. 1985) using the media from cells infected by the vaccinia virus. Also, the purified VGF was demonstrated to compete with ^{125}I-EGF in radioreceptor assays, to induce autophosphorylation of the EGF receptor, and to stimulate DNA synthesis in EGF-sensitive cell lines (STROOBANT et al. 1985; TWARDZIK et al. 1985; KING et al. 1986). These data provided intriguing evidence that perhaps the capacity of the vaccinia virus to induce proliferation of neighboring uninfected cells is due to the production and secretion of the virally encoded growth factor. Also, it was suggested that this capacity to stimulate uninfected cells might render them more susceptible to viral infection and/or reproduction and thus contribute to the spread of the virus in a local cell population. Later studies utilizing a deletion mutation in the growth factor gene of the virus provided evidence for a role of this gene product in both viral virulence (BULLER et al. 1988a) and the capacity of the virus to induce proliferation in neighboring cells (BULLER et al. 1988b).

   At least three other members of the pox virus family have, on the basis of gene sequencing and/or hybridization, the potential to encode similar EGF-like growth factors and each of these viruses produces a profound effect on growth control in the host. The Shope virus (CHANG et al. 1987) induces benign fibromas in rabbits, the *Molluscum contagiosum* virus (PORTER and

ARCHARD 1987) induces benign tumor-like epidermal lesions in humans, and the myxoma virus (UPTON et al. 1987) the causative agent of myxomatosis produces fibromatous dermal lesions in its natural host, rabbits. Obviously, the potential is high for a significant biological role of the EGF-like growth factors encoded by these viruses.

Like EGF and TGF-α, the topical application of purified VGF has been demonstrated to be capable of stimulating the reepithelization of an experimentally induced skin lesion, i.e., second degree burns, in pigs (SCHULTZ et al. 1987).

G. Role in Transformation

Because of the structural and functional similarities that exist between known oncogenes and growth factors, growth factor receptors, and other potential key components of the normal growth control pathway, a current view is that any regulatory molecule in this mechanism can potentially function as a transforming protein; through overexpression, expression at inappropriate times, or mutational alterations of function. For the mitogen pathway described in this article, evidence to support this hypothesis exists at several levels: the ligand, the receptor, a substrate (phospholipase C) of the receptor kinase, and gene products induced by the growth factor. The first two of these areas are described in the sections that follow. Also to be kept in mind is that tumor registries show that approximately 90% of all tumors arise from epithelial cells. Other than EGF there are few known mitogens that have receptors on epithelial cells and are capable of stimulating epithelial cell growth. The fibroblast growth factors and bombesin are the other known epithelial mitogens, but are less characterized than EGF, especially in vivo.

I. Growth Factor Studies

1. Correlative Information

As has been discussed previously (Sect. D. III. 2), TGF-α is expressed by a wide variety of cells transformed in vitro or derived from tumors, as well as tumor tissue (see DERYNCK et al. 1987; DERYNCK 1988). The intriguing notion is that these tumor cells secrete TGF-α and bear EGF receptors forming an autocrine loop that enhances cell proliferation and escape from an exogenous growth factor requirement. Experimental evidence to support this perhaps simplified idea has not been produced, however. Though EGF acts as a tumor promoter in vivo (Table 5: skin, rectum, pancreas) and promotes transformation in vitro, the continous presence of EGF or TGF-α does not produce cellular transformation. Though many transformed cells do produce TGF-α, no one has successfully altered the transformed phenotype of a tumor cell by the addition of antibodies to TGF-α (see Sect. G. I. B). This experiment also does not work if blocking antibodies to the EGF receptor are employed. There may be technical problems in the execution of these "immunosympathectomy" experiments, such as low-affinity antibodies, or there may be more theoretical reasons why the experiments do not work. A plausible idea is that growth factor expression occurs at an early stage in transformation, allowing more time for the normal or partially transformed cell to acquire a critical transforming event. EGF has been shown to extend the lifespan of cells in vitro (RHEINWALD and GREEN 1977; GOSPODAROWICZ and BIALECKI 1978). Also, it is possible that the TGF-α secreted by tumor cells serves some other function in vivo such as angiogenesis (SCHREIBER et al. 1986). To date antisense experiments to block TGF-α synthesis in transformed cells have not been reported.

2. Transfection Studies

Several groups have transfected the EGF or TGF-α gene into various non-transformed cell lines and then asked whether the transfected and growth factor-expressing cells display a transformed phenotype. One study (FINZI et al. 1987), using recipient NIH 3T3 cells for the TGF-α gene, showed no effect in any transformation assay. Another study (S. WATANABE et al. 1987) showed that transfected normal rat kidney (NRK) cells produced TGF-α, but did not show that these same cells were transformed by any criteria. A third study (STERN et al. 1987) with FR3T3 fibroblasts produced several clones expressing EGF. In culture assays, these transfectants exhibited high saturation densities and elevated levels of glucose uptake, but failed to form colonies well in soft agar. When assayed for tumor formation in nude mice, the EGF-producing FR3T3 cells gave rise to tumors, but weakly in comparison to ras-transformed FR3T3 cells. The EGF transfectants had a much longer (3-fold) latency than the ras-transfected cells. ROSENTHAL et al. (1986) had previously achieved a similar result by introducing the TGF-α gene into Rat-1 fibroblasts. TGF-α-producing cells yielded significantly more colonies than the control cells in soft agar assays, but far fewer than ras-transformed cells. Tumor formation assays in nude mice gave similar results – long latency and small size compared to ras-transformed cells. Lastly, FINZI et al. (1988) have transfected the TGF-α gene into epithelial cells and achieved somewhat better results in terms of larger, but benign tumor development.

While these experiments are encouraging, it is not clear whether endogenous growth factor expression per se alters the phenotype toward the transformed state or permits selection of cells that are more transformed for other reasons from a large population. None of the studies have utilized an inducible promoter to test this question.

II. Receptor Studies

1. Correlative Information

There are two pieces of information suggesting that the EGF receptor and transformation are related. The first and strongest is the high level of homology observed between the primary sequences of the human EGF receptor and the avian v-*erb* B oncogene product (ULLRICH et al. 1984). Within large regions of the cytoplasmic domain of the two proteins, sequence identity is 90%. Compared to the EGF receptor, the v-*erb* B protein has a large N-terminal deletion representing nearly all of the EGF receptor external domain and a smaller truncation at the C terminus representing the primary autophosphorylation site of the EGF receptor (YAMAMOTO et al. 1983). Cloning and sequencing of the chicken EGF receptor almost assures that the EGF receptor gene is the protooncogene from which v-*erb* B sequences were derived. In another avian tumor virus system, avian leukosis virus, the virus causes transformation by insertion of viral transcription sequences within the host EGF receptor gene. This produces a truncated protein (activated c-*erb* B) similar to the v-*erb* B protein (for review see MAIHLE and KUNG 1989). In the leukosis virus system, however, the activated c-*erb* B molecule has an N-terminal truncation, similar to the v-*erb* B protein, but does not have a C-terminal truncation.

A second point of relationship between the EGF receptor and transformation is the frequent overexpression of the EGF receptor in transformed cells derived from tumors, particularly, but not limited to carcinomas. Many of the cells which overexpress the EGF receptor are growth inhibited by EGF in cell culture (GILL and LAZAR 1981; N. KAMATA et al. 1986). However, when these cells are grown in soft agar (K. LEE et al. 1987) or in athymic mice (GINSBURG and VONDERHAAR 1985; OZAWA et al. 1987b), EGF stimulates proliferation. The study of OZAWA et al. (1987b) is of note. Tumor formation from EGF receptor-overexpressing human carcinoma lines, including A431 cells, was increased 2.6- to 6.3-fold when the animals received a steady infusion of EGF. Thus, the EGF growth inhibition of these cells lines in culture may be quite misleading in terms of their biology in vivo.

Papers by XU et al. (1984) and DERYNCK et al. (1987) contain lists of tumor cell lines and tumor tissues in which EGF receptor overexpression is found. There are several major tissues in which altered EGF receptor expression in tumors has been noted. NEAL et al. (1985) have reported that for bladder carcinoma tissue about 58% of the specimens have high EGF receptor levels. The expression was highest in invasive or poorly differentiated carcinomas (85%) and lower in superficial (30%) or moderately differentiated (37%) carcinomas. BERGER et al. (1987a), MESSING et al. (1987), and OZANNE et al. (1986) obtained similar results. In esophageal carcinomas increased EGF receptor levels have been noted for 71% (OZAWA et al. 1987a), 50% (HUNTS et al. 1985), 14% (LU et al. 1988), and 8% (HOLLSTEIN et al. 1988) of the specimens.

Cell lines derived from head and neck tumors seem to have perhaps the highest EGF receptor number (COWLEY et al. 1986; YAMAMOTO et al. 1986). Tissue studies confirm this, as OZANNE et al. (1986) reported that 100% (12/12) of tumor specimens studied overexpressed the EGF receptor. GUSTERSON et al. (1985) reported that about 58% of histiocytoma samples overexpressed the EGF receptor.

In the lung, carcinomas often overexpress the EGF receptor: 80% (HUNTS et al. 1985), 65% (CERNY et al. 1986), 58% (VEAL et al. 1987), and 42%

(BERGER et al. 1987b). HAEDER et al. (1988) found that 40% of cell lines derived from lung carcinoma overexpressed the EGF receptor. The percentage was much higher if only squamous cell lung carcinomas were considered – 68%–100%. Significant levels of EGF receptor expression were also noted in all these studies for adenocarcinomas and large cell carcinomas. However, EGF receptor expression in small cell lung carcinoma was not detectable (0%).

Mammary tumors display a high percentage of increased EGF receptor levels: 60% (SPITZER et al. 1988), 47% (FITZPATRICK et al. 1984), 20%–42% (SAINSBURY et al. 1985a, b, c), 22% (DELARUE et al. 1988), 32% (CAPPELLETTI et al. 1988), 28% (NICHOLSON et al. 1988), and 14% (MACIAS et al. 1987). As pointed out by SAINSBURY et al. (1985a, b, c), there is an inverse correlation between estrogen receptor levels and EGF receptor levels. In all reported studies the incidence of EGF receptor expression is 2- to 10-fold higher in estrogen receptor negative cells than in estrogen receptor positive cells. Also, frequency of increased EGF receptor expression increases with tumor grade (grade I, 17%; II, 30%; III 53%) and tumor size (SAINSBURY et al. 1985c). N. E. DAVIDSON et al. (1987) report similar relationships between EGF and estrogen receptor appearance in breast cancer cell lines.

Brain tumors and particularly glioblastoma multiforme, as first noted by LIBERMANN et al. (1984, 1985), have high levels of EGF receptor expression. In this tumor, the incidence of high EGF receptor is reported at 49% by WONG et al. (1987), but 71%, 80%, and 89% by HUMPHREY et al. (1988), LIBERMANN et al. (1984), and MALDEN et al. (1988), respectively.

REAL et al. (1986) have reported an extensive study of EGF receptor expression in a number of different tumor cell lines. Their results are of particular interest in regard to melanoma cell lines. In cells classified as "early" nonpigmented melanomas, high expression of the EGF receptor was noted in 71% of the cell lines (total 21). In contrast, in 14 "late" pigmented melanoma lines only one was found to display EGF receptors.

2. Transfection Studies

Analogous to transfection experiments with the growth factor genes, an attempt has been made to transfect the EGF receptor into nontransformed cells to determine if overexpression of the gene may lead to a transformed phenotype. In all these studies (VELU et al. 1987; DiFIORE et al. 1987; RIEDEL et al. 1988) the recipient cells were NIH 3T3. In each study the selected recipient cells displayed high levels of EGF receptors (≈ 200000 receptors per cell or greater) and in the presence of EGF had significant growth stimulation compared to untransfected cells. This included colony formation in soft agar, but in only one study (VELU et al. 1987) was tumorigenesis in nude mice determined. In that experiment, which involved a limited number of animals, the EGF receptor-transfected cells formed tumors and EGF injection of the animals decreased latency by a factor of two. In any case, the latency period was long – about 30 days. Unfortunately, none of the studies included, for comparison, a known transformed NIH 3T3 cell line, such as the *ras* transformants used in the growth factor transfection assays. Therefore, it is not possible to judge the real significance of these experiments in quantitative terms. As a beginning step, however, the EGF dependence of the transfectants in the soft agar and nude mouse experiments is important, particularly in light of the capacity of EGF to stimulate human tumor cell lines overexpressing the EGF receptor in the same type of assays.

In related studies, von RÜDEN and WAGNER (1988) and PIERCE et al. (1988) transfected the EGF receptor into hematopoietic cell lines dependent on interleukin (IL-3) for growth. Interestingly, hematopoietic cell lines transfected with the EGF receptor responded, in terms of growth stimulation, to exogenous EGF and this abrogated their IL-3 requirement. Transfection of the EGF receptor into primary bone marrow cells, however, did not relieve the IL-3 dependence even though they could be stimulated by EGF. In one of these studies (PIERCE et al. 1988), the transfected myeloid cells produced high numbers of EGF receptors, equivalent to approximately 60% of ^{125}I-EGF binding sites present on A431 cells. However, these myeloid cells bearing high levels of EGF receptor did not display evidence of the transformed phenotype.

Acknowledgments. The authors are grateful to Kerby Oberg for assistance with the figures, and to Susan Heaver, Sue Carpenter, and David Sullins for processing the manuscript. Also, we appreciate advance copies of manuscripts from several investigators. The authors are supported by research grants from the National Cancer Institute (CA43720, CA24071) and the National Institutes of Health (HL14214), plus training support from NIH grants DKO7563 and GMO7347.

References

Acres RB, Lamb JR, Feldman M (1985) Effects of platelet-derived growth factor and epidermal growth factor on antigen-induced proliferation of human T-cell lines. Immunology 54:9–16

Ahrén B (1987) Epidermal growth factor (EGF) inhibits stimulated thyroid hormone secretion in the mouse. Peptides 8:743–745

Akaji K, Fuki N, Yamjima H, Kyozo H, Hayashi K, Mizuta K, Aono M, Moriga M (1985) Studies on peptides. CXXVII. Synthesis of a tripentacontapeptide with epidermal growth factor activity. Chem Pharm Bull 33:184–201

Akiyama T, Saito T, Ogawara H, Toyoshima K, Yamamoto T (1988) Tumor promoter and epidermal growth factor stimulate phosphorylation of the *c-erbB-2* gene product in MKN-7 human adenocarcinoma cells. Mol Cell Biol 8:1019–1026

Almazan G, Honegger P, Matthieu J-M, Guentert-Lauber B (1985) Epidermal growth factor and bovine growth hormone stimulate differentiation and myelination of brain cell aggregates in culture. Dev Brain Res 21:257–264

Al-Nafussi AI, Wright NA (1982) The effect of epidermal growth factor (EGF) on cell proliferation of the gastrointestinal mucosa in rodents. Virchows Arch [B] 40:63–69

Anderson TR, Mayer GL, Hebert N, Nicoll CS (1987) Interactions among prolactin, epidermal growth factor, and proinsulin on the growth and morphology of the pigeon corp-sac mucosal epithelium in vivo. Endocrinology 120:1258–1264

Anson DS, Choo KH, Rees DJG, Giannelli F, Gould K, Huddleston JA, Bronlee GF (1984) The gene structure of human anti-haemophilic factor IX. EMBO J 3:1053–1060

Antosz ME, Bellows CG, Aubin JE (1987) Biphasic effects of epidermal growth factor on bone nodule formation by isolated rat calvaria cells in vitro. J Bone Mineral Res 2:385–393

Anzano MA, Roberts AB, Smith MJ, Sporn MB, De Larco JE (1983) Sarcoma growth factor from conditioned medium of virally transformed cells is composed of both type α and type β transforming growth factors. Proc Natl Acad Sci USA 80:6264–6268

Appella E, Weber IT, Blasi F (1988) Structure and function of epidermal growth factor-like regions in proteins. FEBS Lett 231:1–4

Arakawa M, Perry JW, Cossu MF, Oka T (1985) Further characterization of the inhibition of casein production in a primary mouse mammary epithelial cell culture by epidermal growth factor. Antagonism by cyclic AMP. Exp Cell Res 158:111–118

Arlawd GJ, Willia AC, Gagnon J (1987) Complete amino acid sequence of the A chain of human complement-classical-pathway enzyme C1r. Biochem J 241:711–720

Arsenault P, Ménard D (1987) Stimulatory effects of epidermal growth factor on deoxyribonucleic acid synthesis in the gastrointestinal tract of the suckling mouse. Comp Biochem Physiol [B] 86:123–127

Ascoli M, Euffa J, Segaloff DL (1987) Epidermal growth factor activates steroid biosynthesis in cultured Leydig tumor cells without affecting the levels of cAMP and potentiates the activation of steroid biosynthesis by choriogonadotropin and cAMP. J Biol Chem 262:9196–9203

Aulerich RJ, Bursian SJ, Napolitano AC (1988) Biological effects of epidermal growth factor and 2,3,7,8-tetrachlorodibenzo-p-dioxin on developmental parameters of neonatal mink. Arch Environ Contam Toxicol 17:27–31

Aune TM (1985) Inhibition of soluble immune response suppressor activity by growth factors. Proc Natl Acad Sci USA 82:6260–6264

Avola R, Condorelli DR, Surrentino S, Turpeenoja L, Costa A, Giuffrida-Stella AM (1988) Effect of epidermal growth factor and insulin on DNA, RNA, and cytoskeletal protein labeling in primary rat astroglial cell cultures. J Neurosci Res 19:230–238

Bade EG, Feindler S (1988) Liver epithelial cell migration induced by epidermal growth factor or transforming growth factor alpha is associated with changes in the gene expression of secreted proteins. In Vitro Cell Dev Biol 24:149–154

Bahn RS, Speeg KV Jr, Ascoli M, Rabin D (1980) Epidermal growth factor stimulates production of progesterone in cultured human choriocarcinoma cells. Endocrinology 107:2121–2123

Bailey JM, Muza B, Hla T, Salata K (1985) Restoration of prostacyclin synthese in vascular smooth muscle cells after aspirin treatment: regulation by epidermal growth factor. J Lipid Res 26:54–61

Bajaj M, Waterfield MD, Schlessinger J, Taylor WR, Blundell T (1987) On the tertiary structure of the extracellular domains of the epidermal growth factor and insulin receptors. Biochim Biophys Acta 916:220–226

Baldwin GS (1985) Epidermal growth factor precursor is related to the translation product of the Moloney sarcoma virus oncogene *mos*. Proc Natl Acad Sci USA 82:1921–1925

Balk SD, Shiu RPC, LaFleur MM, Young LL (1982) Epidermal growth factor and insulin cause normal chicken heart mesenchymal cells to proliferate like their Rous sarcoma virus-infected counterparts. Proc Natl Acad Sci USA 79:1154–1157

Barbar JR, Verma IM (1987) Modification of *fos* proteins: phosphorylation of *c-fos*, but not *v-fos*, is stimulated by 12-tetradecanoyl-phorbol-13-acetate and serum. Mol Cell Biol 7:2201–2211

Barka T, van der Noen H, Gresik EW, Kerenyi T (1978) Immunoreactive epidermal growth factor in human amniotic fluid. Mt Sinai J Med (NY) 45:679–684

Barrandon Y, Green H (1987) Cell migration is essential for the sustained growth of keratinocyte colonies: the roles of transforming growth factor-α and epidermal growth factor. Cell 50:1131–1137

Bates SB, Davidson NE, Valverius EM, Freter CE, Dickson RB, Tam JP, Kudlow JE, Lippman ME, Salomon DS (1988) Expression of transforming growth factor α and its messenger RNA in human breast cancer: its regulation by estrogen and its possible functional significance. Mol Endocrinol 2:543–555

Battini R, Ferrari S, Kaczmarek L, Calabretta B, Chen S-t, Baserga R (1987) Molecular cloning of a cDNA for a human ADP/ATP carrier which is growth-regulated. J Biol Chem 262:4355–4359

Beardmore JM, Lewis-Jones DI, Richards RC (1983) Urogastrone and lactose concentrations in precolostrum, colostrum, and milk. Pediatr Res 17:825–828

Beauchamp RD, Barnard JA, McCutchen CM, Cherner JA, Coffey RJ (1989) Localization of TGFα and its receptor in gastric mucosal cells: implications for a regulatory role in acid secretion and mucosal renewal. J Clin Invest 84:1017–1023

Beaulieu J-F, Calvert R (1981) The effect of epidermal growth factor (EGF) on the differentiation of the rough endoplasmic reticulum in fetal mouse small intestine in organ culture. J Histochem Cytochem 29:765–770

Bedrick AD, Ladda RL (1978) Epidermal growth factor potentiates cortisone-induced cleft palate in the mouse. Teratology 17:13–18

Beguinot L, Lyall RM, Willingham ME, Pastan I (1984) Down-regulation of the epidermal growth factor receptor in KB cells is due to receptor internalization and subsequent degradation in lysosomes. Proc Natl Acad Sci USA 81:2384–2388

Bell GI, Fong NM, Stempie NM, Wormsted MA, Caput D, Ku L, Urdea MS, Rall LB, Sanchez-Pescador R (1986) Human epidermal growth factor precursor: cDNA sequence, expression in vitro and gene organization. Nucleic Acids Res 14:8427–8446

Benveniste R, Speeg KV Jr, Carpenter G, Cohen S, Linder T, Rabinowitz D (1978) Epidermal growth factor stimulates secretion of human chorionic gonadotropin by cultured human choriocarcinoma cells. J Clin Endocrinol Metab 46:169–172

Berger MS, Greenfield C, Gullick WJ, Haley J, Downward J, Neal DE, Harris AL, Waterfield MD (1987a) Evaluation of epidermal growth factor receptors in bladder tumours. Br J Cancer 56:533–537

Berger MS, Gullick WJ, Greenfield C, Evans S, Addis BJ, Waterfield MD (1987b) Epidermal growth factor receptors in lung tumours. J Pathol 152:297–307

Berk BC, Brock TA, Webb RC, Taubman MB, Atkinson WJ, Gimbrone MA Jr, Alexander RW (1985) Epidermal growth factor, a vascular smooth muscle mitogen, induces rat aortic contraction. J Clin Invest 75:1083–1086

Bernard O, De St. Groth BF, Ullrich A, Greca W, Schlessinger J (1987) High-affinity interleukin 2 binding by an oncogenic hybrid interleukin 2-epidermal growth factor receptor molecule. Proc Natl Acad Sci USA 84:2125–2129

Bertics PJ, Gill GN (1985) Self-phosphorylation enhances the protein-tyrosine kinase activity of the epidermal growth factor receptor. J Biol Chem 260:14642–14647

Bertics PJ, Chen WS, Hubler L, Lazar CS, Rosenfeld MG, Gill GN (1988) Alteration of epidermal growth factor receptor activity by mutation of its primary carboxyl-terminal site of tyrosine phosphorylation. J Biol Chem 263:3610–3617

Bertoncello I, Bradley TR, Hodgson GS (1981) Clonal agar culture of normal primary explanted bovine granulosa cells. Cell Biol Int Rep 5:169–178

Bertoncello I, Bradley TR, Chamley WA, Hodgson GS (1982) The characteristics of an anchorage-independent clonal agar assay for primary explanted bovine granulosa cells. J Cell Physiol 113:224–230

Bevilacqua MP, Stengelin S, Gimbrone MA Jr, Seed B (1989) Endothelial leukocyte adhesion molecule 1: an inducible receptor for neutrophils related to complement regulatory proteins and lectins. Science 243:1160–1165

Bhargava G, Rifas L, Makman MH (1979) Presence of epidermal growth factor receptors and influence of epidermal growth factor on proliferation and aging in cultured smooth muscle cells. J Cell Physiol 100:365–374

Bjorge JD, Kudlow JE (1987) Epidermal growth factor receptor synthesis is stimulated by phorbol ester and epidermal growth factor. Evidence for a common mechanism. J Biol Chem 262:6615–6622

Blackshear PJ, Stumpo DJ, Huang J-K, Nemenoff RA, Spach DH (1987) Protein kinase C-dependent and -independent pathways of proto-oncogene induction in human astrocytoma cells. J Biol Chem 262:7774–7781

Blackwell GI, Carnuccio R, Di Rosa M, Flower RJ, Parente L, Persico P (1980) Macrocortin: a polypeptide causing the anti-phospholipase effect of glucocorticoids. Nature 287:147–149

Blatti SP, Foster DN, Ranganathan G, Moses HL (1988) Induction of fibronectin gene transcription and mRNA is a primary response to growth-factor stimulation of AKR-2B cells. Proc Natl Acad Sci USA 85:1119–1123

Blomquist MC, Hunt LT, Barker WC (1984) Vaccinia virus 19-kilodalton protein: relationship to several mammalian proteins, including two growth factors. Proc Natl Acad Sci USA 81:7363–7367

Blundell TL, Cutfield JF, Cutfield SM, Dodson EJ, Dodson GG, Hodgkin DC, Mercola DA, Vijayan M (1971) Atomic positions in rhombohedral 2-zinc insulin crystals. Nature 231:506–511

Boccardo F, Valenti G, Zanardi S, Cerruti G, Fassio T, Bruzzi P, De Francesc V, Barreca A, Del Monte P, Minuto F (1988) Epidermal growth factor in breast cyst fluid: relationship with intracystic cation and androgen conjugate content. Cancer Res 48:5860–5863

Böni-Schnetzler M, Pilch PF (1987) Mechanism of epidermal growth factor receptor autophosphorylation and high-affinity binding. Proc Natl Acad Sci USA 84:7832–7836

Bower JM, Camble R, Gregory H, Gerring EL, Willshire IR (1975) The inhibition of gastric acid secretion by epidermal growth factor. Experientia 31:825–826

Bravo R, Burckhardt J, Curran T, Müller R (1985) Stimulation and inhibition of growth by EGF in different A431 cell clones is accompanied by the rapid induction of c-fos and c-myc proto-oncogenes. EMBO J 4:1193–1197

Bravo R, MacDonald-Bravo H, Müller R, Hübsch D, Almendral JM (1987) Bombesin induces c-fos and c-myc expression in quiescent Swiss 3T3 cells. Comparative study with other mitogens. Exp Cell Res 170:103–115

Breyer MD, Jacobson HR, Breyer JA (1988) Epidermal growth factor inhibits the hydrosomatic effect of vasopressin in the isolated perfused rabbit cortical collecting tubule. J Clin Invest 82:1313–1320

Brightwell JR, Riddle SL, Eiferman RA, Valenzuela P, Barr PJ, Merryweather JP, Schultz GS (1985) Biosynthetic human EGF accelerates healing of neodecadron-treated primate corneas. Invest Opthalmol Vis Sci 26:105–110

Bringman TS, Lindquist PB, Derynck R (1987) Different transforming growth factor-α species are derived from a glycosylated and palmitoylated transmembrane precursor. Cell 48:429–440

Brissenden JE, Ullrich A, Francke U (1984) Human chromosomal mapping of genes for insulin-like growth factors I and II and epidermal growth factor. Nature 310:781–784

Brissenden JE, Derynck R, Francke U (1985) Mapping of transforming growth factor α gene on human chromosome 2 close to the breakpoint of the Burkitt's lymphoma t (2;8) variant. Cancer Res 45:5593–5597

Brown GL, Curtsinger L III, Brightwell JR, Ackerman DM, Tobin GR, Polk HC Jr, George-Nascimento C, Valenzuela P, Schultz GS (1986) Enhancement of epidermal regeneration by biosynthetic epidermal growth factor. J Exp Med 163:1319–1324

Brown JP, Twardzik DR, Marquardt H, Todaro GJ (1985) Vaccinia virus encodes a polypeptide homologous to epidermal growth factor and transforming growth factor. Nature 313:491–492

Broyles SS, Yuen L, Shuman S, Moss B (1988) Purification of a factor required for transcription of vaccinia virus early genes. J Biol Chem 263:10754–10760

Brugge JS (1986) The p35/p36 substrates of protein-tyrosine kinases as inhibitors of phospholipase A_2. Cell 46:149–150

Brunk U, Schellens J, Westermark B (1976) Influence of epidermal growth factor (EGF) on ruffling activity, pinocytosis and proliferation of cultivated human glia cells. Exp Cell Res 103:295–302

Bucher NLR, Patel U, Cohen S (1978) Hormonal factors and liver growth. Adv Enzymol Regul 16:205–213

Buckley A, Davidson JM, Kamerath CD, Wolt TB, Woodward SC (1985) Sustained release of epidermal growth factor accelerates wound repair. Proc Natl Acad Sci USA 82:7340–7344

Buckley A, Davidson JM, Kamerath CD, Woodward SC (1987) Epidermal growth factor increases granulation tissue formation dose dependently. J Surg Res 43:322–328

Buhrow SA, Cohen S, Staros JV (1982) Affinity labeling of the protein kinase associated with the epidermal growth factor receptor in membrane vesicles from A431 cells. J Biol Chem 257:4019–4022

Buller RNL, Chakrabarti S, Cooper JA, Twardzik DR, Moss B (1988a) Deletion of the vaccinia virus growth factor gene reduces virus virulence. J Virol 62:866–874

Buller RNL, Chakrabarti S, Moss B, Fredrickson T (1988b) Cell proliferative response to vaccinia virus is mediated by VGF. Virology 164:182–192

Burgess AW, Knesel J, Sparrow LG, Nicola NA, Nice EC (1982) Two forms of murine epidermal growth factor: rapid separation by using reverse-phase HPLC. Proc Natl Acad Sci USA 79:5753–5757

Burgess AW, Lloyd CJ, Nice EC (1983) Murine epidermal growth factor: heterogeneity on high resolution ion-exchange chromatography. EMBO J 2:2065–2069

Burgess AW, Lloyd CJ, Smith S, Stanley E, Walker F, Fabri L, Simpson RJ, Nice EC (1988) Murine epidermal growth factor: structure and function. Biochemistry 27:4977–4985

Cahill L, Dolling M, Young R, Chamley W, Thorburn G (1982) The effect of epidermal growth factor (EGF) on follicular growth in the ovine fetus. Proc 14th Annu Conf Australian Soc. Reprod Biol, p 78

Calabretta B, Battini R, Kaczmarek L, de Riel JK, Baserga R (1986) Molecular cloning of the cDNA for a growth factor-inducible gene with strong homology to S-100, a calcium-binding protein. J Biol Chem 261:12628–12632

Calvert R, Beaulieu J-F, Ménard (1982) Epidermal growth factor (EGF) accelerates the maturation of fetal mouse intestinal mucosa in utero. Experientia 38:1096–1097

Campion SR, Matsunami RK, Engler DA, Stevens A, Niyogi SK (1989) Site-directed mutagenesis of human epidermal growth factor: biochemical properties of β-loop mutants. (submitted)

Cappelletti V, Brivia M, Miodini P, Granata G, Coradini D, Di Fronzo G (1988) Simultaneous estimation of epidermal growth factor receptors and steroid receptors in a series of 136 resectable primary breast tumors. Tumour Biol 9:200–211

Carpenter G (1981) Epidermal growth factor. In: Baserga R (ed) Tissue growth factors. Springer, Berlin Heidelberg New York, pp 89–132 (Handbook of experimental pharmacology, vol 57)

Carpenter G (1987) Receptors for epidermal growth factor and other polypeptide mitogens. Annu Rev Biochem 56:881–894

Carpenter G, Cohen S (1976) ^{125}I-Labeled human epidermal growth factor (hEGF): binding, internalization, and degradation in human fibroblasts. J Cell Biol 71:159–171

Carpenter G, Lembach KJ, Morrison MM, Cohen S (1975) Characterization of the binding of ^{125}I-labeled epidermal growth factor to human fibroblasts. J Biol Chem 250:4297–4304

Carpenter G, King L Jr, Cohen S (1978) Epidermal growth factor stimulates phosphorylation in membrane preparations in vitro. Nature 276:409–410

Carpenter G, King L Jr, Cohen S (1979) Rapid enhancement of protein phosphorylation in A-431 cell membrane preparations by epidermal growth factor. J Biol Chem 254:4884–4891

Carpentier J-L, White MF, Orci L, Kahn RC (1987) Direct visualization of the phosphorylated epidermal growth factor receptor during its internalization in A-431 cells. J Cell Biol 105:2751–2762

Carter NB, Fawcett AA, Hales JRS, Moore GPM, Panaretto BA (1988) Circulatory effects of a depilatory dose of mouse epidermal growth factor in sheep. J Physiol (Lond) 403:27–39

Carver JA, Cooke RM, Esposito G, Campbell ID, Gregory H, Sheard B (1986) A high resolution ^{1}H NMR study of the solution structure of human epidermal growth factor. FEBS Lett 205:77–81

Casey ML, Mitchell MD, MacDonald PC (1987) Epidermal growth factor-stimulated prostaglandin E_2 production in human amnion cells: specificity and nonesterified arachidonic acid dependency. Mol Cell Endocrinol 53:169–176

Casey ML, Korte K, MacDonald PC (1988) Epidermal growth factor stimulation of prostaglandin E_2 biosynthesis in amnion cells. Induction of prostaglandin H_2 synthase. J Biol Chem 263:7846–7854

Catterton WZ, Escobedo MB, Sexson WR, Gray ME, Sundell HW, Stahlman MT (1979) Effect of epidermal growth factor on lung maturation in fetal rabbits. Pediatr Res 13:104–108

Cawley DB, Herschman HR, Gilliland DG, Collier RJ (1980) Epidermal growth factor-toxin A chain conjugates: EGF-ricin A is a potent toxin while EGF-diphtheria fragment A is nontoxic. Cell 22:563–570

Centrella M, McCarthy TL, Canalis E (1987) Mitogenesis in fetal rat bone cells simultaneously exposed to type β transforming growth factor and other growth regulators. FASEB J 1:312–317

Cerny T, Barnes DM, Hasleton P, Barber PV, Healy K, Gullick W, Thatcher N (1986) Expression of epidermal growth factor receptor (EGF-R) in human lung tumours. Br J Cancer 54:265–269

Chabot JG, Payet N, Hugon JS (1983) Effects of epidermal growth factor (EGF) on adult mouse small intestine in vivo and in organ culture. Comp Biochem Physiol [A] 74:247–252

Chan CP, McNall SJ, Krebs EG, Fischer EH (1988) Stimulation of protein phosphatase activity by insulin and growth factors in 3T3 cells. Proc Natl Acad Sci USA 85:6257–6261

Chang W, Upton C, Hu S-L, Purchio AF, McFadden G (1987) The genome of the Shope fibroma virus, a tumorigenic poxvirus, contains a growth factor gene with sequence similarity to those encoding epidermal growth factor and transforming growth factor alpha. Mol Cell Biol 7:535–540

Chang W, Lim JG, Hellström I, Gentry LE (1988) Characterization of Vaccinia virus growth factor biosynthetic pathway with an antipeptide antiserum. J Virol 62:1080–1083

Chatelier RC, Ashcroft RG, Lloyd CJ, Nice EC, Whitehead RH, Sawyer WH, Burgess AW (1986) Binding of fluoresceinated epidermal growth factor to A431 cell subpopulations studies using a model-independent analysis of flow cytometric fluorescence data. EMBO J 5:1181–1186

Chatterjee AK, Sieradzki J, Schatz H (1986) Epidermal growth factor stimulates (Pro-) insulin biosynthesis and ^{3}H-thymidine incorporation in isolated pancreatic rat islets. Horm Metab Res 18:873–874

Chen WS, Lazar CS, Poenie M, Tsien RY, Gill GN, Rosenfeld MG (1987) Requirement for intrinsic protein tyrosine kinase in the immediate and late actions of the EGF receptor. Nature 328:820–823

Chester JR, Gaissert HA, Ross JS, Malt RA (1986) Pancreatic cancer in the Syrian hamster produced by N-Nitrosobis(2-Oxoropyl)-amine: cocarcinogenic effect of epidermal growth factor. Cancer Res 46:2954–2957

Chiba T, Hirata Y, Taminato T, Kadowaki S, Matsukura S, Fujita T (1982) Epidermal growth factor stimulates prostaglandin E release from isolated perfused rat stomach. Biochem Biophys Res Commun 105:370–374

Chikuma T, Kato T, Hiramatsu M, Kanayama S, Kumegawa M (1984) Effect of epidermal growth factor on dipeptidyl-aminopeptidase and collagenase-like peptidase activities in cloned osteoblastic cells. J Biochem 95:283–286

Clark AJ, Ishii S, Richert N, Merlino GT, Pastan I (1985) Epidermal growth factor regulates the expression of its own receptor. Proc Natl Acad Sci USA 82:8374–8378

Clark S, Cheng DJ, Hsuan JJ, Haley JD, Waterfield MD (1988) Loss of three major autophosphorylation sites in the EGF receptor does not block the mitogenic action of EGF. J Cell Physiol 134:421–428

Clemmons DR (1984) Interaction of circulating cell-derived and plasma growth factors in stimulating cultured smooth muscle cell replication. J Cell Physiol 121:425–430

Cochet C, Kashles O, Chambaz EM, Borrello I, King CR, Schlessinger J (1988) Demonstration of epidermal growth factor-induced receptor dimerization in living cells using a chemical covalent cross-linking agent. J Biol Chem 263:3290–3295

Coffey RJ Jr, Derynck R, Wilcox JN, Bringman TS, Goustin AS, Moses HL, Pittelkow MR (1987) Production and auto-induction of transforming growth factor-α in human keratinocytes. Nature 328:817–820

Coffey RJ Jr, Bascom CC, Sipes NJ, Graves-Deal R, Weissman BE, Moses HL (1989) Selective inhibition of growth-related gene expression in murine keratinocytes by transforming growth factor β. Mol Cell Biol 8:3088–3093

Cohen S (1960) Purification of a nerve-growth promoting protein from the mouse salivary gland and its neuro-cytotoxic antiserum. Proc Natl Acad Sci USA 46:302–311

Cohen S (1962) Isolation of a mouse submaxillary gland protein accelerating incisor eruption and eyelid opening in the new-born animal. J Biol Chem 237:1555–1562

Cohen S (1964) Isolation and biological effects of an epidermal growth-stimulating protein. In: Rutter WJ (ed) Metabolic control mechanisms in animal cells. National Cancer Institute Monograph 13, pp 13–27

Cohen S (1965) The stimulation of epidermal proliferation by a specific protein (EGF). Dev Biol 12:394–407

Cohen S, Elliott G (1963) The stimulation of epidermal keratinization by a protein isolated from the submaxillary gland of the mouse. J Invest Dermatol 40:1–5

Cohen S, Fava RA (1985) Internalization of functional epidermal growth factor: receptor/kinase complexes in A-431 cells. J Biol Chem 260:12351–12358

Cohen S, Carpenter G, Lembach KJ (1975) Interaction of epidermal growth factor (EGF) with cultured fibroblasts. Adv Metab Disord 8:265–284

Cohen S, Carpenter G, King L Jr (1980) Epidermal growth factor-receptor-kinase interactions. Co-purification of receptor and epidermal growth factor-enhanced phosphorylation activity. J Biol Chem 255:4834–4842

Cohen S, Ushiro H, Stoscheck C, Chinkers M (1982a) A native 170000 epidermal growth factor receptor-kinase complex from shed plasma membrane vesicles. J Biol Chem 257:1523–1531

Cohen S, Fava R, Sawyer ST (1982b) Purification and characterization of epidermal growth factor receptor protein kinase from normal mouse liver. Proc Natl Acad Sci USA 79:6237–6241

Coleman S, Silberstein GB, Daniel CW (1988) Ductal morphogenesis in the mouse mammary gland: evidence supporting a role of epidermal growth factor. Dev Biol 127:304–315

Collette J, Hendrick J-C, Jaspar J-M, Franchimont P (1986) Presence of α-lactalbumin, epidermal growth factor, epithelial membrane antigen, and gross cystic disease fluid protein (15000 daltons) in breast cyst fluid. Cancer Res 46:3728–3733

Connell ND, Rheinwald JG (1983) Regulation of the cytoskeleton in mesothelial cells: reversible loss of keratin and increase in vimentin during rapid growth in culture. Cell 34:245–254

Connolly JM, Rose DP (1988) Epidermal growth factor-like proteins in breast fluid and human milk. Life Sci 42:1751–1756

Cooke RM, Wilkinson AJ, Baron M, Pastore A, Tappin MJ, Campbell ID, Gregory H, Sheard B (1987) The solution structure of human epidermal growth factor. Nature 327:339–341

Corcoran JM, Waters MJ, Eastman CJ, Jorgensen G (1986) Epidermal growth factor: effect on circulating thyroid hormone levels in sheep. Endocrinology 119:214–217

Cowley GP, Smith JA, Gusterson BA (1986) Increased EGF receptors on human squamous carcinoma cell lines. Br J Cancer 53:223–229

Crompton MR, Moss SE, Crumpton MJ (1988) Diversity in the lipocortin/calpactin family. Cell 55:1–3

Cruise JL, Michalopoulos G (1985) Norepinephrine and epidermal growth factor: dynamics of their interaction in the stimulation of hepatocyte DNA synthesis. J Cell Physiol 125:45–50

Cummings RD, Soderquist AM, Carpenter G (1985) The oliosaccharide moieties of the epidermal growth factor receptor in A-431 cells. Presence of complex-type N-linked chains that contain terminal N-acetylgalactosamine. J Biol Chem 260:11944–11952

Curran T, Franza BR Jr (1988) Fos and jun: the AP-1 connection. Cell 55:395–397

Dagogo-Jack S, Atkinson S, Kendall-Taylor P (1985) Homologous radioimmunoassay for epidermal growth factor in human saliva. J Immunoassay 6:125–136

Dahlbäck B, Lundevall Å, Stenfle J (1986) Primary structure of bovine vitamin K-dependent protein S. Proc Natl Acad Sci USA 83:4199–4203

Daniele S, Frati L, Fiore C, Santoni G (1979) The effect of the epidermal growth factor (EGF) on the corneal epithelium in humans. Exp Ophthalmol 210:159–165

Darlak K, Franklin G, Woost P, Sonnenfeld E, Twardzik D, Spatola A, Schultz G (1988) Assessment of biological activity of synthetic fragments of transforming growth factor-alpha. J Cell Biochem 36:341–352

Davidson FF, Dennis EW, Powell M, Glenney JR Jr (1987) Inhibition of phospholipase A_2 by "lipocortins" and calpactins. An effect of binding to substrate phospholipids. J Biol Chem 262:1698–1705

Davidson NE, Gelmann EP, Lippman ME, Dickson RB (1987) Epidermal growth factor receptor gene expression in estrogen receptor-positive and negative human breast cancer cell lines. Mol Endocrinol 1:216–223

Davis RJ (1988) Independent mechanisms account for the regulation by protein kinase C of the epidermal growth factor receptor affinity and tyrosine-protein kinase activity. J Biol Chem 263:9462–9469

Davis RJ, Czech MP (1985) Tumor promoting phorbol diesters cause the phosphorylation of the epidermal growth factor receptor in normal human fibroblasts. Proc Natl Acad Sci USA 82:1974–1978

De BK, Misono KS, Lukas TJ, Mroczkowski B, Cohen S (1986) A calcium-dependent 35-kilodalton substrate for epidermal growth factor receptor/kinase isolated from normal tissue. J Biol Chem 261:13784–13792

Dean DH, Hiramoto RN, Ghanta VK (1987) Modulation of immune response. A possible role for murine salivary epidermal and nerve growth factors. J Periodontol 58:498–500

Decker SJ (1984) Effects of epidermal growth factor and 12-O-tetradecanoylphorbol-13-acetate on metabolism of the epidermal growth factor receptor in normal human fibroblasts. Mol Cell Biol 4:1718–1724

Defeo-Jones D, Tai JY, Wegrzyn RJ, Vuocolo GA, Baker AE, Payne LS, Garsky VM, Oliff A, Riemen MW (1988) Structure-function analysis of synthetic and recombinant derivatives of transforming growth factor alpha. Mol Cell Biol 8:2999–3007

Dekel N, Sherizly I (1985) Epidermal growth factor induces maturation of rat follicle-enclosed oocytes. Endocrinology 116:406–409

De Larco JE, Todaro GJ (1978) Growth factors from murine sarcoma virus-transformed cells. Proc Natl Acad Sci USA 75:4001–4005

Delarue JC, Friedman S, Mouriesse H, May-Levin F, Sancho-Garnier H, Contesso G (1988) Epidermal growth factor receptor in human breast cancers: correlation with estrogen and progesterone receptors. Breast Cancer Res Treat 11:173–178

De Leon FD, Vijayakumar R, Brown M, Rao CV, Yussman MA, Schultz G (1986) Peritoneal fluid volume, estrogen, progesterone, prostaglandin, and epidermal growth factor concentrations in patients with and without endometriosis. Obstet Gynecol 68:189–194

DeMarco A, Menegatti E, Guarneri M (1983) Epidermal growth factor: exposure and dynamics of the aromatic side chains as investigated by photo-CIDNP and variable temperature ^{1}H-NMR. FEBS Lett 159:201–206

DeMarco A, Mayo KH, Bartolotti F, Scalia S, Menegatti E, Kaptein R (1986) Proton NMR and photochemically induced dynamic nuclear polarization studies of peptide fragments obtained by controlled proteolysis of mouse epidermal growth factor. J Biol Chem 261:13501–13516

Dembinski AB, Johnson LR (1985) Effect of epidermal growth factor on the development of rat gastric mucosa. Endocrinology 116:90–94

Dembinski A, Gregory H, Konturek SJ, Polanski (1982) Trophic action of epidermal growth factor on the pancreas and gastroduodenal mucosa in rats. J Physiol [Lond] 325:35–42

Dembinski A, Drozdowicz D, Gregory H, Konturek SJ, Warzecha Z (1986) Inhibition of acid formation by epidermal growth factor in the isolated rabbit gastric glands. J Physiol [Lond] 378:347–357

Denton PH, Fowlkes DM, Lord ST, Reisner HM (1988) Hemophilia B Durham: a mutation in the first EGF-like domain of factor IX that is characterized by polymerase chain reaction. Blood 72:1407–1411

DePalo L, Das M (1988) Epidermal growth factor-induced stimulation of epidermal growth factor-receptor synthesis in human cytotrophoblasts and A431 carcinoma cells. Cancer Res 48:1105–1109

Derynck R (1988) Transforming growth factor α. Cell 54:593–595

Derynck R, Roberts AB, Winkler ME, Chen EY, Goeddel DV (1984) Human transforming growth factor-α: precursor structure and expression in E. coli. Cell 38:287–297

Derynck R, Roberts AB, Eaton DH, Winkler ME, Goeddel DV (1985) Human transforming growth factor-α: precursor sequence, gene structure, and heterologous expression. Cancer Cells 3:79–86

Derynck R, Goeddel DV, Ullrich A, Gutterman JV, Williams RD, Bringman TS, Berger WH (1987) Synthesis of messenger RNAs for transforming growth factors α and β and the epidermal growth factor receptor by human tumors. Cancer Res 47:707–712

DiAugistine RP, Walker MP, Klappner DG, Grove RI, Willis WD, Harvan DJ, Hernandez O (1985) β-Epidermal growth factor is the des-asparaginyl form of the polypeptide. J Biol Chem 260:2807–2811

DiAugistine RP, Petrusz P, Bell GI, Brown CF, Korach KS, McLachlan JA, Teng CT (1988) Influence of estrogens on mouse uterine epidermal growth factor precursor protein and messenger RNA. Endocrinology 122:2355–2363

DiFiore PP, Pierce JH, Fleming TP, Hazan R, Ullrich A, Kling CR, Schlessinger J, Aaronson SA (1987) Overexpression of the human EGF receptor confers an EGF-dependent transformed phenotype to NIH 3T3 cells. Cell 51:1063–1070

Doolittle RF, Feng DF, Johnson MS (1984) Computer-based characterization of epidermal growth factor precursor. Nature 307:558–560

Dorow DS, Simpson RJ (1988) Cloning and sequence analysis of a cDNA for rat epidermal growth factor. Nucleic Acids Res 16:9338

Downs SM, Daniel SAJ, Eppig JJ (1988) Induction of maturation in cumulus cell-enclosed mouse oocytes by follicle-stimulating hormone and epidermal growth factor: evidence for a positive stimulus of somatic cell origin. J Exp Zool 245:86–96

Downward J, Parker P, Waterfield MD (1984) Autophosphorylation sites on the epidermal growth factor receptor. Nature 311:483–485

Drakenberg T, Fernlung P, Roepstorff P, Stenflo J (1983) β-Hydroxyaspartic acid in vitamin K-dependent protein C. Proc Natl Acad Sci USA 80:1802–1806

D'Souza SW, Haigh R, Micklewright L, Donnai P, Keys A (1985) Amniotic fluid epidermal growth factor and placental weight at term. Lancet II (8449):272–273

Dunn WA, Connolly TP, Hubbard AL (1986) Receptor-mediated endocytosis of epidermal growth factor by rat hepatocytes: receptor pathway. J Cell Biol 102:24–36

Durkin ME, Chakravarti S, Bartos BB, Liu S-M, Friedman RL, Chung AE (1988) Amino acid sequence and domain structure of entactin. Homology with epidermal growth factor precursor and low density lipoprotein receptor. J Cell Biol 107:2749–2756

Earp HS, Austin KS, Blaisdell J, Rubin RA, Nelson KG, Lee LW, Grisham JW (1986) Epidermal growth factor (EGF) stimulates EGF receptor synthesis. J Biol Chem 261:4777–4780

Earp HS, Hepler JR, Petch LA, Miller A, Berry AR, Harris J, Raymond VW, McCune BK, Lee LW, Grisham JW, Harden TK (1988) Epidermal growth factor (EGF) and hormones stimulate phosphoinositide hydrolysis and increase EGF receptor protein synthesis and mRNA levels in rat liver epithelial cells. Evidence for protein kinase C-dependent and -independent pathways. J Biol Chem 263:13868–13874

Eggo MC, Bachrach LK, Fayet G, Errick J, Kudlow JE, Cohen MF, Burrow GN (1984) The effects of growth factors and serum on DNA synthesis and differentiation in thyroid cells in culture. Mol Cell Endocrinol 38:141–150

Elder JT, Fisher GJ, Lindquist PB, Bennett GL, Derynck R, Pittelkow MR, Coffey RJ, Ellingsworth L, Voorhees JJ (1989) Overexpression of transforming growth factor-α in psoriatic epidermis. Science 243:811–814

Elder PK, Schmidt LJ, Ono T, Getz MJ (1984) Specific stimulation of actin gene transcription by epidermal growth factor and cycloheximide. Proc Natl Acad Sci USA 81:7476–7480

Elsholtz HP, Mangalam HJ, Potter E, Albert VR, Supowit S, Evans RM, Rosenfeld MG (1986) Two different cis-active elements transfer the transcriptional effects of both EGF and phorbol esters. Science 234:1552–1557

Elson SD, Browne CA, Thorburn GD (1984) Idendification of epidermal growth factor-like activity in human male reproductive tissues and fluids. J Clin Endocrinol Metab 58:589–594

Engler DA, Matsunami RK, Campion SR, Stringer CD, Stevens A, Niyogi SK (1988) Cloning of authentic human epidermal growth factor as a bacterial secretory protein and its initial structure-function analysis by site-directed mutagenesis. J Biol Chem 263:12384–12390

Erickson GF, Case E (1983) Epidermal growth factor antagonizes ovarian theca-interstitial cytodifferentiation. Mol Cell Endocrinol 31:71–76

Evans CT, Corbin CJ, Saunders CT, Merrill JC, Simpson ER, Mendelson CR (1987) Regulation of estrogen biosynthesis in human adipose stromal cells. Effects of dibutyryl cyclic AMP, epidermal growth factor, and phorbol esters on the synthesis of aromatase cytochrome P-450. J Biol Chem 262:6914–6920

Fabricant RN, De Larco JE, Todaro GJ (1977) Nerve growth factor receptors on human melanoma cells in culture. Proc Natl Acad Sci USA 74:565–569

Fabricant R, Salisbury JD, Berkowitz RA, Kaufman HE (1982) Regenerative effects of epidermal growth factor after penetrating keratoplasty in primates. Arch Ophthalmol 100:994–995

Falconer J (1987) Oral epidermal growth factor is trophic for the stomach in the neonatal rat. Biol Neonate 52:347–350

Fanger BO, Stephens JE, Staros JV (1989) High-yield trapping of EGF-induced receptor dimers by chemical cross-linking. FASEB J 3:71–75

Fava RA, Cohen S (1984) Isolation of a calcium-dependent 35 kilodalton substrate for the epidermal growth factor receptor/kinase from A-431 cells. J Biol Chem 259:2636–2645

Ferrari S, Calabretta B, de Riel JK, Battini R, Ghezzo F, Lauret E, Griffin C, Emanuel BS, Gurrieri F, Baserga R (1987) Structural and functional analysis of a growth-regulated gene, the human calcyclin. J Biol Chem 262:8325–8332

Finke U, Rutten M, Murphy RA, Silen W (1985) Effects of epidermal growth factor on acid secretion from guinea pig gastric mucosa: in vitro analysis. Gastroenterology 88:1175–1182

Finney KJ, Ince P, Appleton DR, Sunter JP, Watson AJ (1987) A trophic effect of epidermal growth factor (EGF) on rat colonic mucosa in organ culture. Cell Tissue Kinet 20:43–56

Finzi E, Fleming T, Segatto O, Pennington DY, Bringman TS, Derynck R, Aaronson SA (1987) The human transforming growth factor type α coding sequence is not a direct-acting oncogene when overexpressed in NIH 3T3 cells. Proc Natl Acad Sci USA 84:3733–3737

Finzi E, Kilkenny A, Strickland JE, Balaschak M, Bringman T, Derynck R, Aaronson S, Yuspa SH (1988) TGFα stimulates growth of skin papillomas by autocrine and paracrine mechanisms but does not cause neoplastic progression. Mol Carcinogen 1:7–12

Fisch TM, Prywes R, Roeder RG (1987) c-fos Sequences necessary for basal expression and induction by epidermal growth factor, 12-O-tetradecanoyl phorbol-13-acetate, and the calcium ionophore. Mol Cell Biol 7:3490–3502

Fishleigh RV, Robson B, Garnier J, Finn PW (1987) Studies on rationales for an expert system approach to the interpretation of protein sequence data. Preliminary analysis of the human epidermal growth factor receptor. FEBS Lett 214:219–225

Fitzpatrick SL, Brightwell J, Wittliff JL, Barrows GH, Schultz GS (1984) Epidermal growth factor binding by breast tumor biopsies and relationship to estrogen receptor and progestin receptor levels. Cancer Res 44:3448–3453

Flower RJ, Blackwell GJ (1979) Anti-inflammatory steroids induce biosynthesis of a phospholipase A_2 inhibitor which prevents prostaglandin generation. Nature 278:456–459

Foster D, Davie EW (1984) Characterization of a cDNA coding for human protein C. Proc Natl Acad Sci USA 81:4766–4770

Foster DC, Yoshitake S, Davie EW (1985) The nucleotide sequence of the gene for human protein C. Proc Natl Acad Sci USA 82:4673–4677

Foster DN, Schmidt LJ, Hodgson CP, Moses HL, Geta MJ (1982) Polyadenylylated RNA complementary to a mouse retrovirus-like multigene family is rapidly and specifically induced by epidermal growth factor stimulation of quiescent cells. Proc Natl Acad Sci USA 79:7317–7321

Franchimont P, Hazee-Hagelstein MT, Charlet-Renard C, Jaspar JM (1986) Effect of mouse epidermal growth factor on DNA and protein synthesis, progesterone and inhibin production by bovine granulosa cells in culture. Acta Endocrinol 111:122–127

Franklin JD, Lynch JB (1979) Effects of topical applications of epidermal growth factor on wound healing. Experimental study on rabbit ears. Plast Reconstr Surg 64:766–770

Franklin TJ, Gregory H, Morris WP (1986) Acceleration of wound healing by recombinant human urogastrone (epidermal growth factor). J Lab Clin Med 108:103–108

Frati C, Covelli I, Mozzi R, Frati L (1972) Mechanism of action of the epidermal growth factor. Effect on the sulfhydryl and disulphide groups content of the mouse epidermis during keratinization. Cell Differ 1:239–244

Freeman DA, Ascoli M (1982) Desensitization of steroidogenesis in cultured Leydig tumor cells: role of cholesterol. Proc Natl Acad Sci USA 79:7796–7800

Freemark M (1986) Epidermal growth factor stimulates glycogen synthesis in fetal rat hepatocytes: comparison with the glycogenic effects of insulin-like growth factor I and insulin. Endocrinology 119:522–526

Fung MR, Hay CW, MacGillivary RTA (1985) Characterization of an almost full-length cDNA coding for human blood coagulation factor X. Proc Natl Acad Sci USA 82:3591–3595

Gabrielson EW, Gerwin BI, Harris CC, Roberts AB, Sporn MB, Lechner JR (1988) Stimulation of DNA synthesis in cultured primary human mesothelial cells by specific growth factors. FASEB J 2:2717–2721

Gan BS, Hollenberg MD, MacCannell KL, Lederis K, Winkler ME, Derynck R (1987a) Distinct vascular actions of epidermal growth factor-urogastrone and transforming growth factor-α. J Pharmacol Exp Ther 242:331–337

Gan BS, MacCannell KL, Hollenberg MD (1987b) Epidermal growth factor-urogastrone causes vasodilation in the anesthetized dog. J Clin Invest 80:199–206

Gardner RM, Lingham RB, Stancel GM (1987) Contractions of the isolated uterus stimulated by epidermal growth factor. FASEB J 1:224–228

Gates RE, King LE Jr (1985) Different forms of the epidermal growth factor receptor kinase have different autophosphorylation sites. Biochemistry 24:5209–5215

Gates RE, King LE Jr (1988) Alkaline hydrolysis and multiple site autophosphorylation differ for two forms of the epidermal growth factor receptor. Biochem Biophys Res Commun 153:183–190

Gentry LE, Twardzik DR, Lim GJ, Ranchalis JE, Lee DC (1987) Expression and characterization of transforming growth factor α precursor protein in transfected mammalian cells. Mol Cell Biol 7:1585–1591

George-Nascimento C, Gyenes A, Halloran SM, Merryweather J, Valenzuela P, Steimer KS, Masiarg FR, Randolph A (1988) Characterization of recombinant human epidermal growth factor produced in yeast. Biochemistry 27:797–802

Gerschenson LE, Conner EA, Yang J, Andersson M (1979) Hormonal regulation of proliferation in two populations of rabbit endometrial cells in culture. Life Sci 24:1337–1344

Ghezzo F, Lauret E, Ferrari S, Baserga R (1988) Growth factor regulation of the promoter for calcylin, a growth-regulated gene. J Biol Chem 263:4758–4763

Giard DJ, Aaronson SA, Todaro GJ, Arnstein P, Kersey JH, Dosik H, Parks WP (1973) In vitro cultivation of human tumors: establishment of cell lines derived from a series of solid tumors. JNCI 51:1417–1423

Gill GN, Lazar CS (1981) Increased phosphotyrosine content and inhibition of proliferation in EGF-treated A-431 cells. Nature 293:305–307

Gill GN, Bertics PJ, Stanton JB (1987) Epidermal growth factor and its receptor. Mol Cell Endorinol 51:169–186

Ginsburg E, Vonderhaar BK (1985) Epidermal growth factor stimulates the growth of A431 tumors in athymic mice. Cancer Lett 28:143–150

Giugni TD, Chen K, Cohen C (1988) Activation of a cytosolic serine protein kinase by epidermal growth factor. J Biol Chem 263:18988–18995

Gladhaug IP, Christoffersen T (1988) Rapid constitutive internalization and externalization of epidermal growth factor receptors in isolated rat hepatocytes. Monensin inhibits receptor externalization and reduces the capacity for continued endocytosis of epidermal growth factor. J Biol Chem 263:12199–12203

Glenney JR Jr, Chen WS, Lazar CS, Walton GM, Zokas LM, Rosenfeld MG, Gill GN (1988) Ligand-induced endocytosis to the EGF receptor is blocked by mutational inactivation and by microinjection of anti-phosphotyrosine antibodies. Cell 52:675–684

Goldin GV, Opperman LA (1980) Induction of supernumerary tracheal buds and the stimulation of DNA synthesis in the embryonic chick lung and trachea by epidermal growth factor. J Embryo Exp Morphol 60:235–243

Goodman R, Slater E, Herschman HR (1980) Epidermal growth factor induces tyrosine hydroxylase in a clonal pheochromocytoma cell line, PC-G2. J Cell Biol 84:495–500

Gospodarowicz D, Bialecki (1978) The effects of the epidermal and fibroblast growth factors on the replicative lifespan of cultured bovine granulosa cells. Endocrinology 103:854–865

Gospodarowicz D, Bialecki H (1979) Fibroblast and epidermal growth factors are mitogenic agents for cultured granulosa cells of rodent, porcine, and human origin. Endocrinology 104:757–764

Gospodarowicz D, Greenburg G (1979) The effects of epidermal and fibroblast growth factors on the repair of corneal endothelial wounds in bovine corneas maintained in organ culture. Exp Eye Res 28:147–157

Gospodarowicz D, Ill CR, Birdwell CR (1977a) Effects of fibroblast and epidermal growth factors on ovarian cell proliferation in vitro. I. Characterization of the response of granulosa cells to FGF and EGF. Endocrinology 100:1108–1120

Gospodarowicz D, Mescher AL, Birdwell CR (1977b) Stimulation of corneal endothelial cell proliferation in vitro by fibroblast and epidermal growth factors. Exp Eye Res 25:75–89

Gospodarowicz D, Bialecki H, Thakral TK (1979) The angiogenic activity of the fibroblast and epidermal growth factor. Exp Eye Res 28:501–514

Gospodarowicz D, Hirabayashi K, Giguère L, Tauber J-P (1981) Factors controlling the proliferative rate, final cell density, and life span of bovine vascular smooth muscle cells in culture. J Cell Biol 89:569–578

Gould KL, Cooper JA, Bretscher A, Hunter T (1986) The protein-tyrosine kinase substrate, p81, is homologous to a chicken microvillar core protein. J Cell Biol 102:660–669

Gray A, Dull TJ, Ullrich A (1983) Nucleotide sequence of epidermal growth factor cDNA predicts a 128000-molecular weight protein precursor. Nature 303:722–725

Green H, Kehinde O, Thomas J (1979) Growth of cultured human epidermal cells into multiple epithelia suitable for grafting. Proc Natl Acad Sci USA 76:5665–5668

Greenberg ME, Greene LA, Ziff EG (1985) Nerve growth factor and epidermal growth factor induce rapid transient changes in proto-oncogene transcription in PC12 cells. J Biol Chem 260:14101-14110

Greenfield C, Patel G, Clark S, Jones N, Waterfield MD (1988) Expression of the human EGF receptor with ligand-stimulatable kinase activity in insect cells using a baculovirus vector. EMBO J 7:139–146

Greenwald I (1985) lin-12, A nematode homeotic gene, is homologous to a set of mammalian proteins that includes epidermal growth factor. Cell 43:583–590

Gregory H (1975) Isolation and structure of urogastrone and its relationship to epidermal growth factor. Nature 257:325–327

Gregory H, Holmes JE, Willshire IR (1977) Urogastrone levels in the urine of normal adult humans. J Clin Endocrinol Metab 45:668–672

Gregory H, Willshire IR, Kavanagh JP, Blacklook NJ, Chowdury S, Richards RC (1986) Urogastrone-epidermal growth factor concentrations in prostatic fluid of normal individuals and patients with benign prostatic hypertrophy. Clin Sci 70:359–363

Gregory H, Thomas CE, Young JA, Willshire IR, Garner A (1988) The contribution of the C-terminal undecapeptide sequence of urogastrone-epidermal growth factor to its biological action. Regu Pept 22:217–226

Greslik EW, Gubits RM, Barka T (1985) In situ localization of mRNA for epidermal growth factor in the submandibular gland of the mouse. J Histochem Cytochem 33:1235–1240

Grosenbaugh DA, Amoss MS, Hood DM, Morgan SJ, Williams JD (1988) Epidermal growth factor-mediated effects on equine vascular smooth muscle cells. Am J Physiol 255:C447–C451

Grove RI, Pratt RM (1983) Growth and differentiation of embryonic mouse palatal epithelial cells in primary culture. Exp Cell Res 148:195–205

Grove RI, Pratt R (1984) Influence of epidermal growth factor and cyclic AMP on growth and differentiation of palatal epithelial cells in culture. Dev Biol 106:427–437

Gubits RM, Shaw PA, Gresik EW, Onetti-Muda A, Barka T (1986) Epidermal growth factor gene expression is regulated differently in the mouse kidney and submaxillary gland. Endocrinology 119:1382–1387

Guentert-Lauber B, Honegger P (1985) Responsiveness of astrocytes in serum–free aggregate cultures to epidermal growth factor: dependence on the cell cycle and the epidermal growth factor concentration. Dev Neurosci 7:286–295

Günzler WA, Steffens GJ, Otting F, Klim S-MA, Frankus E, Flohé L (1982) The complete primary structure of high molecular mass urokinase from human urine. The complete amino acid sequence of the A chain. Hoppe-Seyler's Z Physiol Chem 363:1155–1165

Gusterson B, Cowley G, McIlhinney J, Ozanne B, Fisher C, Reeves B (1985) Evidence for increased epidermal growth factor receptors in human sarcomas. Int J Cancer 36:689–693

Gutierrez GE, Mundy GR, Derynck R, Hewlett EL, Katz MS (1987) Inhibition of parathyroid hormone-responsive adenylate cyclase in clonal osteoblast-like cells by transforming growth factor α and epidermal growth factor. J Biol Chem 262:15845–15850

Haeder M, Rotsch M, Bepler G, Hennig C, Havemann K, Heimann B, Moelling K (1988) Epidermal growth factor receptor expression in human lung cancer cell lines. Cancer Res 48:1132–1136

Hagen FS, Gray CL, O'Hara P, Grant FJ, Saari GC, Woodbury RG, Hart CE, Insley M, Kistel W, Kurachi K, Davie EW (1986) Characterization of a cDNA coding for human factor VII. Proc Natl Acad Sci USA 83:2412–2416

Haigler H, Ash JR, Singer SJ, Cohen S (1978) Visualization by fluorescence of the binding and internalization of epidermal growth factor in human carcinoma cells A-431. Proc Natl Acad Sci USA 75:3317–3321

Haigler HT, McKanna JA, Cohen S (1979) Direct visualization of the binding and internalization of a ferritin conjugate of epidermal growth factor in human carcinoma cells A-431. J Cell Biol 81:382–395

Haigler HT, Maxfield FR, Willingham FR, Pastan I (1980) Dansylcadavarine inhibits internalization of ^{125}I-epidermal growth factor in Balb 3T3 cells. J Biol Chem 255:1239–1241

Haigler HT, Schlaepfer DD, Burgess WH (1987) Characterization of lipocortin I and an immunologically unrelated 33-kDa protein as epidermal growth factor receptor/kinase substrates and phospholipase A_2 inhibitors. J Biol Chem 262:6921–6930

Haley J, Whittle N, Bennett P, Kinchington D, Ullrich A, Waterfield MD (1987) The human EGF receptor gene: structure of the 110kb locus and identification of sequences regulating its transcription. Oncogene Res 1:375–396

Han VKM, Hunter ES, Pratt RM, Zendegui JG, Lee DC (1987) Expression of rat transforming growth factor alpha mRNA during development occurs predominantly in the maternal decidua. Mol Cell Biol 7:2335–2343

Han VKM, Di'Ercole AJ, Lee DC (1988) Expression of transforming growth factor alpha during development. Can J Physiol Pharmacol 66:1113–1121

Hardie G (1988) Pseudosubstrates turn off protein kinases. Nature 335:592–593

Harper JMM, Soar JB, Buttery PJ (1987) Changes in protein metabolism of bovine primary muscle cultures on treatment with growth hormone, insulin, insulin-like growth factor I or epidermal growth factor. J Endocrinol 112:87–96

Harris RC, Hoover RL, Jacobson HR, Badr KF (1988) Evidence for glomerular actions of epidermal growth factor in the rat. J Clin Invest 82:1028–1039

Hassell JR (1975) The development of rat palatal shelves in vitro an ultrastructural analysis of the inhibition of epithelial cell death and palate fusion by the epidermal growth factor. Dev Biol 45:90–102

Hassell JR, Pratt RM (1977) Elevated levels of cAMP alters the effect of epidermal growth factor in vitro on programmed cell death in the secondary palatal epithelium. Exp Cell Res 106:55–62

Hata R-I, Hori H, Nagai Y, Tanaka S, Kondo M, Hiramatsu M, Utsumi N, Kumegawa M (1984) Selective inhibition of type I collagen synthesis in osteoblastic cells by epidermal growth factor. Endocrinology 115:867–876

Hayashi T, Sakamoto S (1988) Radioimmunoassay of human epidermal growth factor-hEGF levels in human body fluids. J Pharmacobiodyn 11:146–151

Hayashida H, Miyata T (1985) Sequence similarity between epidermal growth factor precursor and atrial natriuretic factor precursor. FEBS Lett 185:125–128

Haystead AJ, Hardie DG (1986) Both insulin and epidermal growth factor stimulate lipogenesis and acetyl-CoA carboxylase activity in isolated adipocytes. Importance of homogenization procedure in avoiding artifacts in acetyl-CoA carboxylase assay. Biochem J 234:279–284

Heath WF, Merrifield RB (1986) A synthetic approach to structure-function relationships in the murine epidermal growth factor molecule. Proc Natl Acad Sci USA 83:6367–6371

Heimberg M, Weinstein I, LeQuire VS, Cohen S (1965) The induction of fatty liver in neonatal animals by a purified protein (EGF) from mouse submaxillary gland. Life Sci 4:1625–1633

Heiserman GJ, Gill GN (1988) Epidermal growth factor receptor threonine and serine residues phosphorylated in vivo. J Biol Chem 263:13152–13158

Heldin N-E, Westermark B (1988) Epidermal growth factor, but not thyrotropin, stimulates the expression of c-fos and c-myc messenger ribonucleic acid in porcine thyroid follicle cells in primary culture. Endocrinology 122:1042–1046

Helper JR, Nakahata N, Lovenberg TW, DiGuiseppi J, Herman B, Harden TK (1987) Epidermal growth factor stimulates the rapid accumulation of inositol (1,4,5)-trisphosphate and a rise in cytosolic calcium mobilized from intracellular stores in A431 cells. J Biol Chem 262:2951–2956

Herz J, Hamann U, Rogne S, Myklebost O, Gausepohl H, Stanley KK (1988) Surface location and high affinity for calcium of a 500-kd liver membrane protein closely related to the LDL receptor suggest a physiolgical role as lipoprotein receptor. EMBO J 7:4119–4127

Hesketh TR, Morris JDH, Moore JP, Metcalfe JC (1988) Ca^{2+} and pH responses to sequential additions of mitogens in single 3T3 fibroblasts: correlations with DNA synthesis. J Biol Chem 263:11879–11886

Higuchi Y, Morimoto Y, Horinaka A, Yasuoka N (1988) Crystallization and preliminary X-ray studies on human epidermal growth factor. J Biochem 103:905–906

Hiraki Y, Rosen OM, Birbaum MJ (1988) Growth factors rapidly induce expression of the glucose transporter gene. J Biol Chem 263:13655–13662

Hirata F, Schiffmann E, Venkatasubramaniam K, Salomon D, Axelrod J (1980) A phospholipase A_2 inhibitory protein in rabbit neutrophils induced by glucocorticoids. Proc Natl Acad Sci USA 77:2533–2536

Hirata Y, Moore GW, Bertagna C, Orth DN (1980) Plasma concentrations of immunoreactive human epidermal growth factor (urogastrone) in man. J Clin Endocrinol Metab 50:440–444

Hirata Y, Sueoka S, Uchihashi M, Yoshimoto Y, Fuijita T, Matsukura S, Motoyama T (1982a) Specific binding sites for epidermal growth factor and its effect on human chorionic gonadotrophin secretion by cultured tumour cell lines: comparison between trophoblastic and non-trophoblastic cells. Acta Endocrinol 101:281–286

Hirata Y, Uchihashi M, Nakajima M, Fujita T, Matsukura S (1982b) Immunoreactive human epidermal growth factor in human pancreatic juice. J Clin Endocrinol Metab 54:1242–1245

Hirata Y, Uchihashi M, Nakajima H, Fujita T, Matsukura S (1982c) Presence of human epidermal growth factor in human cerebrospinal fluid. J Clin Endocrinol Metab 55:1174–1177

Hirata Y, Uchihashi M, Hazama M, Fujita T (1987) Epidermal growth factor in human seminal plasma. Horm Metab Res 19:35–37

Ho PC, Davis WH, Elliott JH, Cohen S (1974) Kinetics of corneal epithelial regeneration and epidermal growth factor. Invest Ophthalmol 13:804–809

Hoath SB (1986) Treatment of the neonatal rat with epidermal growth factor: differences in time and organ response. Pediatr Res 20:468–472

Hoath SB, Pickens WL (1987) Effect of thyroid hormone and epidermal growth factor on tactile hair development and craniofacial morphogenesis in the postnatal rat. J Craniofac Genet Dev Biol 7:161–167

Hoath SB, Pickens WL, Donnelly MM (1988) Epidermal growth factor-induced growth retardation in the newborn rat: quantitation and relation to changes in skin temperature and viscoelasticity. Growth Dev Aging 52:77–83

Hodgson CP, Elder PK, Ono T, Foster DN, Getz MJ (1983) Structure and expression of mouse VL30 genes. Mol Cell Biol 3:2221–2231

Hofmann J, Kunzmann R, Drescher A, Hackenberg R, Hölzel F, Schulz K-D (1988) Growth regulation of human endometrial carcinoma cells in vitro by steroid hormones and growth factors. Prog Cancer Res Ther 35:452–455

Højrup P, Jensen MS, Peterson TE (1985) Amino acid sequence of bovine protein Z: a vitamin K-dependent serine protease homolog. FEBS Lett 184:333–337

Holladay LA, Savage CR Jr, Cohen S, Puett D (1976) Conformation and unfolding thermodynamics of epidermal growth factor and derivatives. Biochemistry 15:2624–2633

Holland R, Hardie DG (1985) Both insulin and epidermal growth factor stimulate fatty acid synthesis and increase phosphorylation of acetyl-CoA carboxylase and ATP-citrate lyase in isolated hepatocytes. FEBS Lett 181:308–312

Hollenberg M, Gregory H (1980) Epidermal growth factor-urogastrone: biological activity and receptor binding derivatives. Mol Pharmacol 17:314–320

Hollis DE, Chapmann RE (1987) Apoptosis in wool follicles during mouse epidermal growth-factor (mEGF)-induced catagen regression. J Invest Dermatol 88:455–458

Hollstein MC, Smits AM, Galiana C, Yamasaki H, Bos JL, Mandard A, Partensky C, Montesano R (1988) Amplification of epidermal growth factor receptor gene but no evidence of ras mutations in primary human esophageal cancers. Cancer Res 48:5119–5123

Honegger AM, Dull TJ, Felder S, Van Obberghen E, Bellot F, Szapary D, Schmidt A, Ullrich A, Schlessinger J (1987a) Point mutation at the ATP binding site of EGF receptor abolishes protein-tyrosine kinase activity and alters cellular routing. Cell 51:199–209

Honegger AM, Szapary D, Schmidt A, Lyall R, Van Obberghen E, Dull TJ, Ullrich A, Schlessinger J (1987b) A mutant epidermal growth factor receptor with defective protein tyrosine kinase is unable to stimulate proto-oncogene expression and DNA synthesis. Mol Cell Biol 7:4568–4571

Honegger A, Dull TJ, Bellott F, Van Obberghen E, Szapary D, Schmidt A, Ullrich A, Schlessinger J (1988a) Biological activities of EGF-receptor: mutant with individually altered autophosphorylation sites. EMBO J 7:3045–3052

Honegger A, Dull TJ, Szapary D, Komoriya A, Kris R, Ullrich A, Schlessinger J (1988b) Kinetic parameters of the protein tyrosine kinase activity of EGF-receptor mutants with individually altered autophosphorylation sites. EMBO J 7:3053–3063

Hoshi M, Nishida E, Sakai H (1988) Activation of a Ca^{2+}-inhibitable protein kinase that phosphorylates microtubule-associated protein 2 in vitro by growth factors, phorbol esters, and serum in quiescent cultured human fibroblasts. J Biol Chem 263:5396–5401

Hoskins J, Norman DK, Bechman RJ, Long GL (1987) Cloning and characterization of human liver cDNA encoding a protein S precursor. Proc Natl Acad Sci USA 84:349–353

Hsu C-J, Holmes SD, Hammone JM (1987) Ovarian epidermal growth factor-like activity. Concentrations in porcine follicular fluid during follicular enlargement. Biochem Biophys Res Communl 47:242–247

Hsueh AJW, Welsh TH, Jones PBC (1981) Inhibition of ovarian and testicular steroidogenesis by epidermal growth factor. Endocrinology 108:2002–2004

Hudson LG, Santon JB, Gill GN (1989) Regulation of epidermal growth factor receptor gene expression. Mol Endocrinol 3:400–408

Humphrey PA, Wong AJ, Vogelstein B, Friedman HS, Werner MH, Bigner DD, Bigner SH (1988) Amplification and expression of the epidermal growth factor receptor gene in human glioma xenografts. Cancer Res 48:2231–2238

Hunter T, Ling N, Cooper JA (1984) Protein kinase C phosphorylation of the EGF receptor at a threonine residue close to the cytoplasmic face of the plasma membrane. Nature 311:480–483

Hunts J, Ueda M, Ozawa S, Abe O, Pastan I, Shimizu N (1985) Hyperproduction and gene amplification of the epidermal growth factor receptor in squamous cell carcinomas. Jpn J Cancer Res 76:663–666

Hursh DA, Andrews ME, Raff RA (1987) A sea urchin gene encodes a polypeptide homologous to epidermal growth factor. Science 237:1487–1490

Hyver KJ, Campana JE, Cotter RJ, Fenselau C (1985) Mass spectral analysis of murine epidermal growth factor. Biochem Biophys Res Commun 130:1287–1293

Ibbotson KJ, Twardzik DR, D'Souza SM, Hargreaves WR, Todaro GJ, Mundy GR (1985) Stimulation of bone resorption in vitro by synthetic transforming growth factor-alpha. Science 228:1007–1009

Ibbotson KJ, Harrod J, Gowen M, D'Souza S, Smith DD, Winkler ME, Derynck R, Mundy GR (1986) Human recombinant transforming growth factor α stimulates bone resorption and inhibits formation in vitro. Proc Natl Acad Sci USA 83:2228–2232

Ignotz RA, Kelly B, Davis RJ, Massagué J (1986) Biologically active precursor for transforming growth factor type α, released by retrovirally transformed cells. Proc Natl Acad Sci USA 83:6307–6311

Ikeda H, Mitsuhashi T, Kubota K, Kuzuya N, Uchimura H (1984) Epidermal growth factor stimulates growth hormone secretion from superfused rat adenohypophyseal fragments. Endocrinology 115:556–558

Imada O, Hayashi N, Masamoto K, Kasuga S, Fuwa T, Nakagawa S (1987) Long-latency growth-promoting activity of EGF when administered to mice at the neonatal stage. Am J Physiol 253:E251–E254

Imagawa W, Tomooka Y, Hamamoto S, Nandi S (1985) Stimulation of mammary epithelial cell growth in vitro: interaction of epidermal growth factor and mammogenic hormones. Endocrinology 116:1514–1524

Inoue H, Kikuchi K, Nishino M (1986) Effects of epidermal growth factor on the synthesis of DNA and polyamine in isoproterenol-stimulated murine parotid gland. J Biochem 100:605–613

Ishii S, Xu Y-H, Stratton RH, Roc BA, Merlino GT, Pastan I (1985) Characterization and sequence of the promoter region of the human epidermal growth factor receptor gene. Proc Natl Acad Sci USA 82:4920–4924

Jackman RW, Beeler DL, Futze L, Soff G, Rosenberg RD (1987) Human thrombomodulin gene is intron depleted: nucleic acid sequences of the cDNA and gene predict protein structure and suggest sites of regulatory control. Proc Natl Acad Sci USA 84:6425–6429

Jacobs DO, Evans DA, Mealy K, O'Dwyer ST, Smith RJ, Wilmore MD (1988) Combined effects of glutamine and epidermal growth factor on the rat intestine. Surgery 104:358–364

Jakobovits EB, Schlokat V, Vaunice JL, Derynck R, Levinson AD (1988) The human transforming growth factor alpha promoter directs transcription initiation from a single site in the absence of a TATA sequence. Mol Cell Biol 8:5549–5554

Jansson L, Karlson FA, Westermark B (1985) Mitogenic activity and epidermal growth factor content in human milk. Acta Paediatr Scand 74:250–253

Jaskulski D, Gatti C, Travali S, Calabretta S, Baserga R (1988) Regulation of the proliferating cell nuclear antigen cyclin and thymidine kinase mRNA levels by growth factors. J Biol Chem 263:10175–10179

Jaspar JM, Franchimont P (1985) Radioimmunoassay of human epidermal growth factor in human breast cyst fluid. Eur J Cancer Clin Oncol 21:1343–1348

Jenö P, Ballou LM, Novak-Hofer I, Thomas G (1988) Identification and characterization of a mitogen-activated S6 kinase. Proc Natl Acad Sci USA 85:406–410

Jinno Y, Merlino GT, Pastan I (1988) A novel effect of EGF on mRNA stability. Nucleic Acids Res 16:4957–4966

Johnson AC, Ishii S, Jinno Y, Pastan I, Merlino GT (1988a) Epidermal growth factor receptor gene promoter. Deletion analysis and identification of nuclear protein binding sites. J Biol Chem 263:5693–5699

Johnson AC, Jinno Y, Merlino GT (1988b) Modulation of epidermal growth factor receptor proto-oncogene transcription by a promoter site sensitive to S1 nuclease. Mol Cell Biol 8:4174–4184

Johnson AC, Garfield SH, Merlino GT, Pastan I (1988c) Expression of epidermal growth factor receptor proto-oncogene mRNA in regenerating rat liver. Biochem Biophys Res Commun 150:412–418

Johnson CC, Dawson WE, Turner JT, Wyche JH (1980) Regulation of rat ovarian cell growth and steroid secretion. J Cell Biol 86:483–489

Johnson GI, Cook RG, McEver RP (1989) Cloning of GMP-140 a granule membrane protein of platelets and endothelium: sequence similarity to proteins involved in cell adhesion and inflammation Cell 56:1033–1044

Johnson HM, Torres BA (1985) Peptide growth factors PDGF, EGF, and FGF regulate interferon-y production. J Immunol 134:2824–2826

Johnson LR, Guthrie PD (1980) Stimulation of rat oxyntic gland mucosal growth by epidermal growth factor. Am J Physiol 238:G45–G49

Johnson RM, Garrison JC (1987) Epidermal growth factor and angiotensin II stimulate formation of inositol 1,4,5- and inositol 1,3,4-trisphosphate in hepatocytes. Differential inhibition by pertussis toxin and phorbol 12-myristate 13-acetate. J Biol Chem 262:17285–17293

Jones FS, Burgoonn MP, Horrman S, Crossin KL, Cunningham BA, Edelman GM (1988) A cDNA clone for cytotactin contains sequences similar to epidermal growth factor-like repeats and segments of fibronectin and fibrinogen. Proc Natl Acad Sci USA 85:1286–2190

Kageyama R, Merlino GT, Pastan I (1988a) A transcription factor active on the epidermal growth factor receptor gene. Proc Natl Acad Sci USA 85:5016–5020

Kageyama R, Merlino GT, Pastan I (1988b) Epidermal growth factor (EGF) receptor gene transcription, Requirement for Sp1 and an EGF receptor specific factor. J Biol Chem 263:6329–6336

Kamata N, Chida K, Rikimaru K, Horikoshi M, Enomoto S, Kuroki T (1986) Growth-inibitory effects of epidermal growth factor and overexpression of its receptors on human squamous cell carcinomas in culture. Cancer Res 46:1648–1653

Kamata T, Feramisco JR (1984) Epidermal growth factor stimulates guanine nucleotide binding activity and phosphorylation of ras oncogene proteins. Nature 310:147–150

Kaslow DC, Quakyi IA, Lyin C, Raum MG, Keister DB, Coligan JE, McCutchan TF, Miller LH (1988) A vaccine candidate from the sexual stage of human malaria that contains EGF-like domains. Nature 333:74–76

Kay DG, Lai WH, Uchihashi M, Khan MN, Posner BI, Bergeron JM (1986) Epidermal growth factor receptor kinase translocation and activation in vivo. J Biol Chem 261:8473–8480

Kennedy KA, Wilton P, Mellander M, Rojas J, Sundell H (1986) Effect of epidermal growth factor on lung liquid secretion in fetal sheep. J Dev Physiol 8:421–433

Kerr LD, Olashaw NE, Matrisian LM (1988) Transforming growth factor β1 and cAMP inhibit transcription of epidermal growth factor- and oncogene induced transin RNA. J Biol Chem 263:16999–17005

Khazaie K, Dull TJ, Graf T, Schlessinger J, Ullrich A, Beug H, Vennstrom B (1988) Truncation of the human EGF receptor leads to differential transforming potentials in primary avian fibroblasts and erythroblasts. EMBO J 7:3061–3071

Kidd S, Kelley MR, Young MW (1986) Sequence of the Notch locus of Drosophila melanogaster: relationship of the encoded protein to mammalian clotting and growth factors. Mol Cell Biol 6:3094–3108

Kim KH, Stelmach V, Javors J, Fuchs E (1987) Regulation of human mesothelial cell differentiation: opposing roles of retinoids and epidermal growth factor in the expression of intermediate filament proteins. J Cell Biol 105:3039–3051

King CR, Borrello I, Bellot F, Comoglio P, Schlessinger J (1988) EGF binding to its receptor triggers a rapid tyrosine phosphorylation of the erbB-2 protein in the mammary tumor cell line SK-BR-3. EMBO J 7:1647–1651

King CS, Cooper JA, Moos B, Twardzik DR (1986) Vaccinia virus growth factor stimulates tyrosine kinase activity of A431 cell epidermal growth factor receptors. Mol Cell Biol 6:332–336

Kingsnorth AN, Abu-Khalaf M, Ross JS, Malt RA (1985) Potentiation of 1,2-dimethylhydrazine-induced anal carcinoma by epidermal growth factor in mice. Surgery 97:696–700

Kirkegaard P, Olsen PS, Nexo E, Holst JJ, Poulsen SS (1984) Effect of vasoactive intestinal polypeptide and somatostatin on secretion of epidermal growth factor and bicarbonate from Brunner's glands. Gut 24:1225–1229

Knauer DJ, Cunningham DD (1983) A reevaluation of the response of human umbilical vein endothelial cells to certain growth factors. J Cell Physiol 117:397–406

Knecht M, Catt KJ (1983) Modulation of cAMP-mediated differentiation in ovarian granulosa cells by epidermal growth factor and platelet-derived growth factor. J Biol Chem 158:2789–2794

Knust E, Dietrich U, Tepass U, Bremer KA, Weigel D, Vässin H, Campos-Ortega JA (1987) EGF homologous sequences encoded in the genome of *Drosophila melanogaster,* and their relation to neurogenic genes. EMBO J 6:761–766

Kohda D, Inagaki F (1988) Complete sequence-specific ^{1}H nuclear magnetic resonance assignments for mouse epidermal growth factors. J Biochem 103:554–571

Kohda D, Go N, Hayashi K, Inagaki F (1988) Tertiary structure of mouse epidermal growth factor determined by two-dimensional ^{1}H NMR. J Biochem 103:741–743

Koide H, Yokoyama S, Kawai G, Ha J-M, Oka T, Kawai S, Miyake T, Fuma T, Miyazawa T (1988) Biosynthesis of a protein containing a nonprotein amino acid by *Escherichia coli:* L-2-aminohexanoic acid at position 21 in human epidermal growth factor. Proc Natl Acad Sci USA 85:6237–6241

Kokai Y, Dobashi K, Weiner DB, Myers JN, Nowell PC, Greene MI (1988) Phosphorylation process induced by epidermal growth factor alters the oncogenic and cellular *neu* (NGL) gene products. Proc Natl Acad Sci USA 85:5389–5393

Komoriya A, Hortsch M, Meyers C, Smith M, Kanety H, Schlessinger J (1984) Biologically active synthetic fragments of epidermal growth factor: localization of a major receptor-binding region. Proc Natl Acad Sci USA 81:1351–1355

Komura H, Wakimoto H, Chen C-F, Terakawa N, Aono T, Tanizawa O, Matsumoto K (1986) Retinoic acid enhances cell responses to epidermal growth factor in mouse mammary gland in culture. Endocrinology 118:1530–1536

Konturek SJ, Brzozowski T, Piastucki I, Dembinski A, Radecki T, Dembinska-Kiec A, Zmuda A, Gregory H (1981a) Role of mucosal prostaglandins and DNA synthesis in gastric cytoprotection by luminal epidermal growth factor. Gut 22:927–932

Konturek SJ, Radecki T, Brzozowski T, Piastucki I, Dembinski A, Dembinska-Kiec A, Zmuda A, Gryglewski R, Gregory H (1981b) Gastric cytoprotection by epidermal growth factor. Role of endogenous prostaglandins and DNA synthesis. Gastroenterology 81:438–443

Konturek SJ, Brzozowski T, Dembinski A, Warzecha A, Drozdowica D (1988a) Comparison of solcoseryl and epidermal growth factors (EGF) in healing of chronic gastroduodenal ulcerations and mucosal growth in rats. Hepatogastroenterology 35:25–29

Konturek SJ, Dembinski A, Warzecha Z, Brzozowski T, Gregory H (1988b) Role of epidermal growth factor in healing of chronic gastroduodenal ulcers in rats. Gastroenterology 94:1300–1307

Korc M, Padilla J, Grosso D (1986) Epidermal growth factor inhibits the proliferation of a human endometrial carcinoma cell line. J Clin Endocrinol Metab 62:874–880

Kretsinger RH, Creutz CE (1986) Consensus in exocytosis. Nature 320:573

Krusius T, Gehlsen KR, Ruoslahti E (1987) A fibroblast chondroitin sulfate proteoglycan core protein contains lectin-like and growth factor-like sequences. J Biol Chem 262:13120–13125

Kudlow JE, Cheung C-YM, Bjorge JD (1986) Epidermal growth factor stimulates the synthesis of its own receptor in a human breast cancer cell line. J Biol Chem 261:4134–4138

Kurachi K, Davie EW (1982) Isolation and characterization of a cDNA coding for factor IX. Proc Natl Acad Sci USA 79:6461–6464
Kurobe M, Tokida N, Furukawa S, Ishikawa E, Hayashi K (1986) Development of a
  sensitive enzyme immunoassay for human epidermal growth factor (Urogastrone).
  Clin Chim Acta 156:51–60
Laato M, Niinikoski J, Gerdin B, Label L (1986a) Stimulation of wound healing by
  epidermal growth factor. A dose-dependent effect. Ann Surg 203:379–381
Laato M, Niinikoski J, Lundberg C, Arfors K-E (1986b) Effect of epidermal growth
  factor (EGF) on experimental granulation tissue. J Surg Res 41:252–255
Laskin DL, Laskin JD, Weinstein B, Carchman RA (1980) Modulation of
  phagocytosis by tumor promoters and epidermal growth factor in normal and
  transformed macrophages. Cancer Res 40:1028–1035
Laskin DL, Laskin JD, Weinstein IB, Carchman RA (1981) Induction of chemotaxis in
  mouse peritoneal macrophages by phorbol ester tumor promoters. Cancer Res
  41:1923–1928
Lasky LA, Singer MS, Yechock TA, Dawbenko D, Fennie C, Rodriguez H, Nguyen T,
  Stachel S, Rosen SD (1989) Cloning of a lymphocyte homing receptor reveals a lectin domain. Cell 56:1045–1055
Lawler J, Hynes RO (1986) The structure of human thrombospondin, a adhesive
  glycoprotein with multiple calcium-binding sites and homologies with several different proteins. J Cell Biol 103:1635–1648
Lax I, Johnson A, Hawk R, Snap J, Bellot F, Winkler M, Ullrich A, Vennstrom B,
  Schlessinger J, Givol D (1988a) Chicken epidermal growth factor (EGF) receptor:
  cDNA cloning, expression in mouse cells, and differential binding of EGF and
  transforming growth factor alpha. Mol Cell Biol 8:1970–1978
Lax I, Burgess WH, Bellot F, Ullrich A, Schlessinger J, Givol D (1988b) Localization
  of a major receptor-binding domain for epidermal growth factor by affinity labeling. Mol Cell Biol 8:1831–1834
Lazar E, Watanabe S, Dalton S, Sporn MB (1988) Transforming growth factor α:
  mutation of aspartic acid 47 and leucine 48 results in different biological activities.
  Mol Cell Biol 8:1247–1252
Lee DC, Rockford R, Todaro GJ, Villarreal LP (1985a) Developmental expression of
  rat transforming growth factor-α mRNA. Mol Cell Biol 5:3644–3646
Lee DC, Rose TM, Webb NR, Todaro GJ (1985b) Cloning and sequence analysis of a
  cDNA for rat transforming growth factor-α. Nature 313:489–491
Lee K, Tanaka M, Hatanaka M, Kuze F (1987) Reciprocal effects of epidermal growth
  factor and transforming growth factor β on the anchorage-dependent and independent growth of A431 epidermoid carcinoma cells. Exp Cell Res 173:156–162
Lemaire P, Revelant O, Bravo R, Charnay P (1988) Two mouse genes encoding potential transcription factors with identical DNA-binding domains are activated by
  growth factors in cultured cells. Proc Natl Acad Sci USA 85:4691–4695
Leutz A, Schachner M (1981) Epidermal growth factor stimulates DNA-synthesis of
  astrocytes in primary cerebellar cultures. Cell Tissue Res 220:393–404
Levi-Montalcini R, Cohen S (1960) Effects of the extract of the mouse submaxillary
  salivary glands on the sympathetic system of mammals. Ann NY Acad Sci
  85:324–341
Lewis EJ, Chikaraishi DM (1987) Regulated expression of the tyrosine hydroxylase
  gene by epidermal growth factor. Mol Cell Biol 7:3332–3336
Leytus SP, Chung DW, Kisiel W, Kurachi K, Davie EW (1984) Characterization of a
  cDNA coding for human factor X. Proc Natl Acad Sci USA 81:3699–3702
Leytus SP, Foster DC, Kurachi K, Davie EW (1986) Gene for human factor X: a blood
  coagulation factor whose gene organization is essentially identical with that of factor IX and C. Biochemistry 25:5098–5102
Libermann TA, Razon N, Bartal AD, Yarden Y, Schlessinger J, Soreq H (1984) Expression of epidermal growth factor receptors in human brain tumors. Cancer Res
  44:753–760

Libermann TA, Nusbaum HR, Razon N, Kris R, Lax I, Soreq H, Whittle N, Waterfield MD, Ullrich A, Schlessinger J (1985) Amplification, enhanced expression and possible rearrangement of EGF receptor gene in primary human brain tumours of glial origin. Nature 313:144–147

Lin CR, Chen WS, Lazar CS, Carpenter CD, Gill GN, Evans RM, Rosenfeld MC (1986) Protein kinase C phosphorylation at Thr654 of the unoccupied EGF receptor and EGF binding regulate functional receptor and EGF binding regulated functional receptor and EGF binding regulate functional receptor loss by independent mechanisms. Cell 44:839–848

Lin Y-Z, Caporaso G, Chang P-Y, Ke X-H, Tam JP (1988) Synthesis of a biologically active tumor growth factor from the predicted DNA sequence of Shope fibroma virus. Biochemistry 27:5640–5645

Lintern-Moore S, Moore GPM, Panaretto BA, Robertson D (1981) Follicular development in the neonatal mouse ovary; effect of epidermal growth factor. Acta Endocrinol (Copenh) 96:123–126

Livneh E, Glazer L, Segal D, Schlessinger J, Shilo B-Z (1985) The Drosophila EGF receptor gene homolog: conservation of both hormone binding and kinase domains. Cell 40:599–607

Livneh E, Prywes R, Kashles O, Reiss N, Mory Y, Ullrich A, Schlessinger J (1986a) Reconstitution of human epidermal growth factor receptor and its deletion mutants in cultured hamster cells. J Biol Chem 261:12490–12497

Livneh E, Beneviste M, Prywes R, Felder S, Kam Z, Schlessinger J (1986b) Large deletions in the cytoplasmic kinase domain of the epidermal growth factor receptor do not affect its lateral mobility. J Cell Biol 103:327–331

Livneh E, Reiss N, Berent E, Ullrich A, Schlessinger J (1987) An insertional mutant of the epidermal growth factor receptor allows dissection of diverse receptor functions. EMBO J 6:2669–2676

Livneh E, Dull TJ, Bereut E, Prywes R, Ullrich A, Schlessinger J (1988) Release of a phorbol ester-induced mitogenic block by mutation at Thr654 of the epidermal growth factor receptor. Mol Cell Biol 8:2302–2308

Logsdon CD, Williams JA (1983a) Pancreatic acini in short-term culture: regulation by EGF, carbachol, insulin, and corticosterone. Am J Physiol 244:G675–G682

Logsdon CD, Williams JA (1983b) Epidermal growth factor binding and biologic effects on mouse pancreatic acini. Gastroenterology 85:339–345

Long GL, Belagaje RM, MacGillivary RTA (1984) Cloning and sequencing of liver cDNA coding for bovine protein C. Proc Natl Acad Sci USA 81:5653–5656

Lu A-H, Hsieh L-L, Luo F-C, Weinstein IB (1988) Amplification of the EGF receptor and c-myc genes in human esophageal cancers. Int J Cancer 42:502–505

Luetteke NC, Michalopoulos GK, Teixido J, Gilmore R, Massagué J, Lee DC (1988) Characterization of high molecular weight transforming growth factor α produced by rat hepatocellular carcinoma cells. Biochemistry 27:6487–6494

Luger A, Calogero AE, Kalogeras K, Gallucci WT, Gold PW, Loriaux DL, Chrousos GP (1988) Interaction of epidermal growth factor with the hypothalamic-pituitary-adrenal axis: potential physiologic relevance. J Clin Endocrinol Metab 66:334–337

Machida CM, Muldoon LL, Rodland KD, Magun BE (1988) Transcriptional modulating of transin gene expression by epidermal growth factor and transforming growth factor β. Mol Cell Biol 8:2479–2483

Macias A, Azavedo E, Hägerström T, Klintenbert C, Pérez R, Skook L (1987) Prognostic significance of the receptor for epidermal growth factor in human mammary carcinomas. Anticancer Res 7:459–464

MacNeil S, Dawson RA, Crocker G, Barton CH, Hanford L, Metcalfe R, McGurk M, Munro DS (1988) Extracellular calmodulin and its association with epidermal growth factor in normal body fluids. J Endocrinol 118:501–509

Madtes DK, Raines EW, Sakariassen KS, Assoian RK, Sporn MB, Bell GD, Ross R (1988) Induction of transforming growth factor-α in activated human alveolar macrophages. Cell 53:285–293

Mahadevan LC, Heath JK, Leichtfried FE, Cumming DVE, Hirst EMA, Foulkes JG (1988) Rapid appearance of novel phosphoproteins in the nuclei of mitogen-stimulated fibroblasts. Oncogene 2:249–257

Maihle NJ, Kung H-J (1989) c-*erb B* and the EGF receptor: a molecule with dual identity. Biochim Biophys Acta 948:287–304

Makino K, Morimoto M, Nishi M, Sakamoto S, Tamura A, Inooka H, Akasaka K (1987) Proton nuclear magnetic resonance study on the solution conformation of human epidermal growth factor. Proc Natl Acad Sci USA 84:7841–7845

Malden LT, Novak U, Kaye AH, Burgess AW (1988) Selective amplification of the cytoplasmic domain of the epidermal growth factor receptor gene in glioblastoma multiforme. Cancer Res 48:2711–2714

Mallea LE, Machado AJ, Navaroli F, Rommerts FFG (1987) Modulation of stimulatory action of follicle stimulating hormone (FSH) and inhibitory action of epidermal growth factor (EGF) on aromatase activity in Sertoli cells by calcium. FEBS Lett 218:143–147

Malo C, Ménard D (1982) Influence of epidermal growth factor on the development of suckling mouse intestinal mucosa. Gastroenterology 83:28–35

Malt RA, Chester JR, Gaissert HA, Ross JS (1987) Augmentation of chemically induced pancreatic and bronchial cancers by epidermal growth factor. Gut 28:249–251

Mann K, Deutzmann R, Aumailley M, Timpl R, Raimondi L, Jamada Y, Pan T-C, Conway D, Chu M-L (1989) Amino acid sequence of mouse nidogen, a multi-domain basement membrane protein with binding activity for laminin, collagen IV, and cells. EMBO J 8:65–75

Margolis BL, Bonventre JV, Kremer SG, Kudlow JE, Skorecki KL (1988) Epidermal growth factor is synergistic with phorbol esters and vasopressin in stimulating arachidonate release and prostaglandin production in renal glomerular mesangial cells. Biochem J 249:587–592

Margolis B, Rhee SG, Felder S, Mervic M, Lyall R, Levitzki A, Ullrich A, Zilberstein A, Schlessinger J (1989) EGF induces tyrosine phosphorylation of phospholipase c-II: A potential mechanism for EGF receptor signaling. Cell 57:1101–1107

Marquardt H, Hunkapiller MW, Hood LE, Twardzik DR, De Larco JE, Stephenson JR, Todaro GJ (1983) Transforming growth factors produced by retrovirus-transformed rodent fibroblasts and human melanoma cells: Amino acid sequence homology with epidermal growth factor. Proc Natl Acad Sci USA 80:4684–4688

Marquardt H, Hunkapiller MW, Hood LE, Todaro GJ (1984) Rat transforming growth factor type 1: structure and relation to epidermal growth factor. Science 223:1079–1082

Maruo T, Matsuo H, Oishi T, Hayashi M, Nishino R, Mochizuki M (1987) Induction of differentiated trophoblast function by epidermal growth factor: relation of immunohistochemically detected cellular epidermal growth factor receptor levels. J Clin Endocrinol Metab 64:744–750

Matias JR, Orentreich N (1983) Stimulation of hamster sebaceous glands by epidermal growth factor. J Invest Dermatol 80:516–519

Matrisian LM, Larsen BR, Finch JS, Magun BE (1982) Further purification of epidermal growth factor by high-performance liquid chromatography. Anal Biochem 125:339–351

Matrisian LM, Rautmann G, Magun BE, Breathnach R (1985) Epidermal growth factor or serum stimulation of rat fibroblasts induces an elevation in mRNA levels for lactate dehydrogenase and other glycolytic enzymes. Nucleic Acids Res 13:711–726

Matrisian LM, Leroy P, Ruhlmann C, Gesnel M-C, Breathnach R (1986a) Isolation of the oncogene and epidermal growth factor-induced transin gene: complex control in rat fibroblasts. Mol Cell Biol 6:1679–1686

Matrisian LM, Bowden GT, Krieg P, Fürstenberger G, Briand J-P, Leroy P, Breathnach R (1986b) The mRNA coding for the secreted protease transin is expressed more abundantly in malignant than in benign tumors. Proc Natl Acad Sci USA 83:9412–9417

Mattila A-L, Perheentupa J, Pesonen K, Viinikka L (1985) Epidermal growth factor in human urine from birth to puberty. J Clin Endocrinol Metab 61:997–1000

May JV, Buck PA, Schomberg DW (1987) Epidermal growth factor enhances [^{125}I]iodo-follicle-stimulating hormone binding by cultured porcine granulosa cells. Endocrinology 120:2413–2420

Mayes ELV, Waterfield MD (1984) Biosynthesis of the growth factor receptor in A431 cells. EMBO J 3:531–537

Mayo K (1984) Epidermal growth factor from the mouse. Structural characterization by proton nuclear magnetic resonance and nuclear Overhauser experiments at 500 MHz. Biochemistry 23:3960–3973

Mayo KH (1985) Epidermal growth factor from the mouse. Physical evidence for a tiered β-sheet domain: two-dimensional NMR correlated spectroscopy and nuclear Overhauser experiments on backbone amide protons. Biochemistry 24:3783–3794

Mayo KH, Burke C (1987) Structural and dynamical comparison of α, β, and γ forms of murine epidermal growth factor. Eur J Biochem 169:201–207

Mayo KH, Schandies P, Savage CR, De Marco A, Kaptein R (1986) Structural characterization and exposure of aromatic residues in epidermal growth factor from rat. Biochem J 239:13–18

Mayo KH, De Marco A, Menegatti E, Kaptein R (1987) Interaction of epidermal growth factor with micelles monitored by photochemically induced dynamic nuclear polarization-^{1}H NMR spectroscopy. J Biol Chem 262:14899–14904

Mayo KH, Cavalli RC, Peters AR, Boelens R, Kaptein R (1989) Sequence-specific ^{1}H-n.m.r. assignments and peptide backbone conformation in rat epidermal growth factor. Biochem J 257:197–205

McAuslan BR, Bener V, Reilly W, Moss BA (1985) New functions of epidermal growth factor: stimulation of capillary endothelial cell migration and matrix dependent proliferation. Cell Biol Int Rep 9:175–182

McClelland A, Kuhn LC, Ruddle FH (1984) The human transferrin receptor gene: genomic organization, and the complete primary structure of the receptor deduced from a cDNA sequence. Cell 39:267–274

McDonald BJ, Waters MJ, Richards MD, Thorburn GD, Hopkins PS (1983) Effect of epidermal growth factor on wool fibre morphology and skin histology. Res Vet Sci 35:91–99

McGowan JA, Strain AJ, Bucher NLR (1981) DNA synthesis in primary cultures of adult rat hepatocytes in a defined medium: effects of epidermal growth factor, insulin, glucagon, and cyclic-AMP. J Cell Physiol 108:353–363

McGrath M, Palmer S, Nandi S (1985) Differential response of normal rat mammary epithelial cells to mammogenic hormones and EGF. J Cell Physiol 125:182–191

McKeehan WL, Adams PS, Rosser MP (1984) Direct mitogenic effects of insulin, epidermal growth factor, glucocorticoid, cholera toxin, unknown pituitary factors and possibly prolactin, but not androgen, on normal rat prostate epithelial cells in serum-free, primary cell culture. Cancer Res 44:1998–2010

McMullen BA, Fujikawa K (1985) Amino acid sequence of the heavy chain of human α-factor XIIa (activated Hageman factor). J Biol Chem 260:5328–5341

McMullen BA, Fujikawa K, Kisiel W (1983a) The occurrence of β-hydroxyaspartic acid in the vitamin K-dependent blood coagulation zymogens. Biochem Biophys Res Commun 115:8–14

McMullen BA, Fujikawa K, Kisiel W, Sasagawa T, Howald WN, Kwa EY, Weinstein B (1983b) Complete amino acid sequence of the light chain of human blood coagulation factor X: evidence for identification of residue 63 as β-hydroxyaspartic acid. Biochemistry 22:2875–2884

Meisenhelder J, Suh P-G, Rhee SG, Hunter T (1989) Phospholipase C-γ is a substrate for the PDGF and EGF receptor protein-tyrosine kinases in vivo and in vitro. Cell 57:1109–1122

Ménard D, Arsenault P, Pothier P (1988) Biologic effects of epidermal growth factor in human fetal jejunum. Gastroenterology 94:656–663

Merlino GT, Ishii S, Whang-Peng J, Knutsen T, Young-Hua X, Clark AJL, Stratton RH, Wilson RK, Ma DP, Roe BA, Hunts JH, Shimizu N, Pastan I (1985) Structure and localization of genes encoding aberrant and normal epidermal growth factor receptor RNAs from A431 human carcinoma cells. Mol Cell Biol 5:1722–1734

Messing EM, Hanson P, Ulrich P, Erturk E (1987) Epidermal growth factor-interactions with normal and malignant urothelium: in vivo and in situ studies. J Urol 138:1329–1335

Mitchell MD (1987) Epidermal growth factor actions on arachidonic acid metabolism in human amnion cells. Biochim Biophys Acta 928:240–242

Miyake A, Tasaka K, Otsuka S, Kohmura H, Wakimoto H, Aono T (1985) Epidermal growth factor stimulates secretion of rat pituitary luteinizing hormone in vitro. Acta Endocinol 108:175–178

Mondschein JS, Schomberg DW (1981) Growth factors modulate gonadotropin receptor induction in granulosa cell cultures. Science 211:1179–1180

Montelione GT, Wüthrich K, Nice EC, Burgess AW, Scheraga HA (1986) Identification of two anti-parallel β-sheet conformations in the solution structure of murine epidermal growth factor by proton magnetic resonance. Proc Natl Acad Sci USA 83:8594–8598

Montelione GT, Wüthrich K, Nice EC, Burgess AW, Scheraga HA (1987) Solution structure of murine epidermal growth factor: determination of the polypeptide backbone chain-fold by nuclear magnetic resonance and distance geometry. Proc Natl Acad Sci USA 84:5226–5230

Montelione GT, Wüthrich K, Scheraga HA (1988) Sequence-specific ^{1}H NMR assignments and identification of slowly exchanging amide protons in murine epidermal growth factor. Biochemistry 27:2235–2243

Montelione GT, Winkler ME, Burton LE, Rinderknecht E, Sporn MB, Wagner G (1989) Sequence-specific ^{1}H-NMR assignments and identification of two small antiparallel β-sheets in the solution structure of recombinant human transforming growth factor α. Proc Natl Acad Sci USA 86:1519–1523

Montell DM, Goodman CS (1988) Drosophila substrate adhesion molecule: sequence of laminin B1 chain reveals domains of homology with mouse. Cell 53:463–473

Moolenaar WH, Bierman AJ, Tilly BC, Verlaan I, Defize LHK, Honegger AM, Ullrich A, Schlessinger J (1988) A point mutation at the ATP-binding site of the EGF receptor abolishes signal transduction. EMBO J 8:707–710

Moore GPM, Panaretto BA, Robertson D (1981 a) Effects of epidermal growth factor on hair growth in the mouse. J Endocrinol 88:293–299

Moore GPM, Panaretto BA, Robertson D (1981 b) Epidermal growth factor causes shedding of the fleece of merino sheep. Search 12:128–129

Moore GPM, Panaretto BA, Robertson D (1982) Inhibition of wool growth in merino sheep following administration of mouse epidermal growth factor and a derivative. Aust J Biol Sci 35:163–172

Moore GPM, Panaretto BA, Robertson D (1983) Epidermal growth factor delays the development of the epidermis and hair follicles of mice during growth of the first coat. Anat Rec 205:47–55

Moore GPM, Panaretto BA, Wallace ALC (1984) Treatment of ewes at different stages of pregnancy with epidermal growth factor: effects on wool growth and plasma concentrations of growth hormone, prolactin, placental lactogen and thyroxine and on foetal development. Acta Endocrinol (Copenh) 105:558–566

Moore GPM, Panaretto BA, Carter NB (1985) Epidermal hyperplasia and wool follicle regression in sheep infused with epidermal growth factor. J Invest Dermatol 84:172–175

Moran JR, Courtney ME, Orth DN, Vaughan R, Coy S, Mount CD, Sherrell BJ, Greene HL (1983) Epidermal growth factor in human milk: daily production and diurnal variation during early lactation in mothers delivering at term and at premature gestation. J Pediatr 103:402–405

Morita T, Issacs BS, Esmon CT, Johnson AE (1984) Derivatives of blood coagulation factor IX contain a high affinity Ca^{2+}-binding site that lacks T-carboxyglutamic acid. J Biol Chem 259:5698–5704

Morris PL, Vale WW, Cappel S, Bardin CW (1988) Inhibin production by primary Sertoli cell-enriched cultures: regulation by follicle-stimulating hormone, androgens, and epidermal growth factor. Endocrinology 122:717–725

Morrish DW, Bhardwaj D, Dabbagh LK, Marusyk H, Siy O (1987) Epidermal growth factor induces differentiation and secretion of human chorionic gonadotropin and placental lactogen in normal human placenta. J Clin Endocrinol Metab 65:1282–1290

Morrison DK, Kaplan DR, Rapp U, Roberts TM (1988) Signal transduction from membrane to cytoplasm: Growth factors and membrane-bound oncogene products increase Raf-1 phosphorylation and associated protein kinase activity. Proc Natl Acad Sci USA 85:8855–8859

Morrison RS, Kornblum HI, Leslie FM, Bradshaw RA (1987) Trophic stimulation of cultured neurons from neonatal rat brain by epidermal growth factor. Science 238:72–75

Morton CC, Byers MG, Nakai H, Bell GI, Shows TB (1986) Human genes for insulin-like growth factors I and II and epidermal growth factor are located on 12q22q24.1, 11q15, and 4q25-q27, respectively. Cytogenet Cell Genet 41:245–249

Moscat J, Molley DJ, Fleming TP, Aaronson SA (1988) Epidermal growth factor activates phosphoinositide turnover and protein kinase C in BALB/MK keratinocytes. Mol Endocrinol 2:799–805

Mount CD, Lukas TJ, Orth DN (1985) Purification and characterization of epidermal growth factor (β-urogastrone) and epidermal growth factor fragments from large volumes of human urine. Arch Biochem Biophys 240:33–42

Mroczkowski B, Reich M, Whittaker J, Bell GI, Cohen S (1988) Expression of human epidermal growth factor precursor cDNA in transfected mouse NIH 3T3 cells. Proc Natl Acad Sci USA 85:126–130

Mroczkowski B, Reich M, Chen K, Bell GI, Cohen S (1989) Recombinant human EGF precursor is a glycosylated membrane protein with biological activity. Mol Cell Biol 9:2771–2778

Müller R, Bravo R, Burckhardt J, Curran T (1984) Induction of c-*fos* gene and protein by growth factors precedes activation of c-*myc*. Nature 312:716–720

Müller R, Curran T, Burckhardt J, Rüther U, Wagner EF, Bravo R (1985) Evidence for a role of the c-*fos* proto-oncogene in both differentiation and growth control. Cancer Cells 3:289–300

Muramatsu I, Hollenberg MD, Lederis K (1985) Vascular actions of epidermal growth factor-urogastrone: possible relationship to prostaglandin production. Can J Physiol Pharmacol 63:994–999

Muramatsu I, Hollenberg MD, Lederis K (1986) Modulation by epidermal growth factor – urogastrone of contraction in isolated canine helical mesenteric arterial strips. Can J Physiol Pharmacol 64:1561–1565

Murdoch GH, Potter E, Nicolaisen AK, Evans RM, Rosenfeld MG (1982) Epidermal growth factor rapidly stimulates prolactin gene transcription. Nature 300:192–194

Murdoch GH, Waterman M, Evans RM, Resenfeld MG (1985) Molecular mechanisms of phorbol ester, thyrotropin-releasing hormone, and growth factor stimulation of prolactin gene transcription. J Biol Chem 260:11852–11858

Murthy U, Basu M, Sen-Majumadr A, Das M (1986) Perinuclear location and recycling of epidermal growth factor receptor kinase: immunofluorescent visualization using antibodies directed to kinase and extracellular domains. J Cell Biol 103:333–343

Mutoh T, Rudkin RR, Koizumi S, Guroff G (1988) Nerve growth factor, a differentiating agent, and epidermal growth factor, a mitogen, increase the activities of different S6 kinases in PC12 cells. J Biol Chem 263:15853–15856

Nagamine Y, Pearson D, Alters MS, Reich E (1984) cDNA and gene nucleotide sequence of porcine plasminogen activator. Nucleic Acids Res 12:9525–9541

Nakhla AM, Tam JP (1985) Transforming growth factor is a potent stimulator of testicular ornithine decarboxylase in immature mouse. Biochem Biophys Res Commun 132:1180–1186

Neal DE, Bennett MK, Hall RR, Marsh C, Abel PD, Sainsbury JRC, Harris AL (1985) Epidermal growth factor receptors in human bladder cancer: comparison of invasive and superficial tumours. Lancet i (8425):366–368

Nestor JJ, Newman SR, De Lustro B, Todaro GJ, Schreiber AB (1985) A synthetic fragment of rat transforming growth factor α with receptor binding and antigenic properties. Biochem Biophys Res Commun 129:226–232

Niall M, Ryan GB, O'Brien B McC (1982) The effect of epidermal growth factor on wound healing in mice. J Surg Res 33:164–169

Nicholas KR, Sankaran L, Kulski JK, Chomczynski P, Qasba P (1988) Comparison of some biological effects of epidermal growth factor and commercial serum albumin on the induction of α-lactalbumin in rat and rabbit mammary explants. J Endocrinol 119:133–139

Nicholson S, Sainsbury JRC, Needham GK, Chambers P, Farndon JR, Harris AL (1988) Quantitative assays of epidermal growth factor receptor in human breast cancer: cut-off points of clinical relevance. Int J Cancer 42:36–41

Nieburgs AC, Korn JH, Picciano PT, Cohen S (1987) Thymic epithelium in vitro. Regulation of growth and mediator production by epidermal growth factor. Cell Immunol 108:396–404

Nishibe S, Wahl MI, Rhee SG, Carpenter C (1989) Tyrosine phosphorylation of phospholipase C-II in vitro by the epidermal growth factor receptors. J Biol Chem 264:10335–10338

Norman J, Badie-Dezfooly B, Nord EP, Kurtz I, Schlosser J, Chaudhari A, Fine LG (1987) EGF-induced mitogenesis in proximal tubular cells: potentiation by angiotensin II. Am J Physiol 253:F299–309

Northwood IC, Davis RJ (1988) Activation of the epidermal growth factor receptor tyrosine protein kinase in the absence of receptor oligomerization. J Biol Chem 263:7450–7453

Novak-Hofer I, Thomas G (1984) An activated S6 kinase in extracts from serum- and epidermal growth factor-stimulated Swiss 3T3 cells. J Biol Chem 259:5995–6000

Novak-Hofer I, Thomas G (1985) Epidermal growth factor-mediated activation of an S6 kinase in Swiss mouse 3T3 cells. J Biol Chem 260:10314–10319

Ny T, Elgh F, Lund B (1984) The structure of the human tissue type plasminogen activator gene: correlation of intron and exon structures to functional and structural domains. Proc Natl Acad Sci USA 81:5355–5359

Oberg KC, Carpener G (1989) Dexamethasone acts as a negative regulator of epidermal growth factor receptor synthesis in fegal rat lung cells. Mol Endocrinol 3:915–922

Oberg KC, Soderquist AM, Carpenter G (1988) Accumulation of epidermal growth factor receptors in retinoic acid-treated fetal rat lung cells is due to enhanced receptor synthesis. Mol Endocrinol 2:959–965

Öhlin A-K, Stenflo J (1987) Calcium-dependent interaction between the epidermal growth factor precursor-like region of human protein C and a monoclonal antibody. J Biol Chem 262:13798–13804

Öhlin A-K, Linse S, Stenflo J (1988) Calcium binding to the epidermal growth factor homology region of bovine protein C. J Biol Chem 263:7411–7417

Ohmura E, Emoto N, Tsushima T, Watanabe S, Takeuchi T, Kawamura M, Shigemoto M, Shizume K (1987) Salivary immunoreactive human epidermal growth factor (IR-hEGF) in patients with peptic ulcer disease. Hepatogastroenterology 34:160–163

Ohta Y, Ohba T, Fukunaga K, Miyamoto E (1988) Serum and growth factors rapidly elicit phosphorylation of the Ca^{2+}/calmodulin-dependent protein kinase II in intact quiescent rat 3Y1 cells. J Biol Chem 263:11540–11547

Oka Y, Orth DN (1983) Human plasma epidermal growth factor/β-urogastrone is associated with blood platelets. J Clin Invest 72:249–259

Oka Y, Ghishan FK, Greene HL, Orth DN (1983) Effect of mouse epidermal growth factor/urogastrone on the functional maturation of rat intestine. Endocrinology 112:940–944

O'Loughlin EV, Chung M, Hollenberg M, Hayden J, Zahavi I, Gall DG (1985) Effect of epidermal growth factor on ontogeny of the gastrointestinal tract. Am J Physiol 249:G674–678

Olsen PS, Poulsen SS, Kirkegaard P, Nexo (1984) Role of submandibular saliva and epidermal growth factor in gastric cytoprotection. Gastroenterology 87:103–108

Olsen PS, Boesby S, Kirkegaard P, Therkelsen K, Almdal T, Poulsen SS, Nexo E (1988) Influence of epidermal growth factor on liver regeneration after partial hepatectomy in rats. Hepatology 8:992–996

Olsen R, Santone K, Melder D, Oakes SG, Abraham R, Powis G (1988) An increase in intracellular free Ca^{2+} associated with serum-free growth stimulation of Swiss 3T3 fibroblasts by epidermal growth factor in the presence of bradykinin. J Biol Chem 263:18030–18035

Opleta K, O'Loughlin EV, Shaffer EA, Hayden J, Hollenberg M, Gall DB (1987) Effect of epidermal growth factor on growth and postnatal development of the rabbit liver. Am J Physiol 253:G622–G626

Ozanne B, Richards S, Hendler F, Burns D, Gusterson B (1986) Overexpression of the EGF receptor is a hallmark of squamous cell carcinomas. J Pathol 149:9–14

Ozawa S, Ueda M, Ando N, Abe O, Shimizu N (1987a) High incidence of EGF receptor hyperproduction in esophageal squamous-cell carcinomas. Int J Cancer 39:333–337

Ozawa S, Ueda M, Ando N, Abe O, Hirai M, Shimizu N (1987b) Stimulation by EGF of the growth of EGF receptor-hyperproducing tumor cells in athymic mice. Int J Cancer 40:706–710

Panaretto BA, Moore GPM, Robertson DM (1982) Plasma concentrations and urinary excretion of mouse epidermal growth factor associated with the inhibition of food consumption and of wool growth in Merino wethers. J Endocrinol 94:191–202

Panaretto BA, Leish Z, Moore GPM, Robertson DM (1984) Inhibition of DNA synthesis in dermal tissue of merino sheep treated with depilatory doses of mouse epidermal growth factor. J Endocrinol 100:25–31

Pandiella A, Malgaroli A, Meldolesi J, Vicentini LM (1987) EGF raises cytosolic Ca^{2+} in A431 and Swiss 3T3 cells by a dual mechanism. Redistribution from intracellular stores and stimulated influx. Exp Cell Res 170:175–185

Pandiella A, Beguinot L, Velu TJ, Meldolesi J (1988) Transmembrane signalling at epidermal growth factor receptors overexpressed in NIH 3T3 cells. Biochem J 254:223–228

Partanen AM, Ekblom P, Thesless I (1985) Epidermal growth factor inhibits morphogenesis and cell differentiation in cultured mouse embryonic teeth. Dev Biol 111:84–94

Pasco D, Quan A, Smith S, Nandi S (1982) Effect of hormones and EGF on proliferation of rat mammary epithelium enriched for alveoli. An in vitro study. Exp Cell Res 141:313–324

Pennica D, Holmes WE, Kohr WJ, Harkins RN, Vehar GA, Ward CA, Bennett WF, Yelverton E, Seeburg PH, Heyneker HL, Goeddel DV (1983) Cloning and expression of human tissue-type plasminogen activator cDNA in *E. coli*. Nature 301:214–219

Pennica D, Kohr WJ, Kuang W-J, Glaister D, Aggarwal BB, Chen EY, Goeddel DV (1987) Identification of human uromodulin as the Tamm-Horsfall urinary glycoprotein. Science 236:83–88

Pepinsky RB, Sinclair LK, Browning JL, Mattaliano RJ, Smart JE, Chow EP, Falbel T, Ribolini A, Garwin JL, Wallner BP (1986) Purification and partial sequence analysis of a 37-kDa protein that inhibits phospholipase A_2 activity from rat peritoneal exudates. J Biol Chem 261:4239–4246

Pesonen K, Viinikka L, Koskimies A, Banks AR, Nicolson M, Perheentupa J (1987) Size heterogeneity of epidermal growth factor in human body fluids. Life Sci 40:2489–2494

Petrides PE, Böhlen P, Shively JE (1984) Chemical characterization of the two forms of epidermal growth factor in murine saliva. Biochem Biophys Res Commun 125:218–228

Pfeffer S, Ullrich A (1985) Epidermal growth factor. Is the precursor receptor a? Nature 313:184

Pierce JH, Ruggiero M, Fleming TP, DiFiore PP, Greenberger JS, Varticovski L, Schlessinger J, Rovera G, Aaronson SA (1988) Signal transduction through the EGF receptor transfected in IL-3-dependent hematopoietic cells. Science 239:628–631

Pike LJ, Eakes AT (1987) Epidermal growth factor stimulates the production of phosphatidylinositol monophosphate and the breakdown of polyphosphoinositides in A431 cells. J Biol Chem 262:1644–1651

Pikkarainen T, Eddy R, Fukushima Y, Byers M, Shows T, Pihlahaniae T, Saraste M, Tryggvasoro K (1987) Human laminin B1 chain. A multidomain protein with gene (Lamb1) locus in the q22 region of chromosome 7. J Biol Chem 262:10454–10462

Pikkarainen T, Kallunki T, Tryggvason K (1988) Human laminin B2 chain. Comparison of the complete amino acid sequence with the B1 chain reveals variability in sequence homology between different structural domains. J Biol Chem 263:6751–6758

Pisano MM, Greene RM (1987) Epidermal growth factor potentiates the induction of ornithine decarboxylase activity by prostaglandins in embryonic palate mesenchymal cells: effects on cell proliferation and glycosaminoglycan synthesis. Dev Biol 122:419–431

Pittelkow MR, Lindquist PB, Derynck R, Abraham RT, Graves-Deal R, Coffey RJ (1989) Induction of transforming growth factor-α expression in human keratinocytes by phorbol esters. J Biol Chem 264:5164–5171

Plutzky J, Hoskins JA, Long GL, Crabtree GR (1986) Evolution and organization of the human protein C gene. Proc Natl Acad Sci USA 83:546–550

Polk DH, Ervin MG, Padbury JF, Lam RW, Reviczky AL, Fisher DA (1987) Epidermal growth factor acts as a corticotropin-releasing factor in chronically catheterized fetal lambs. J Clin Invest 79:984–988

Pollack PF, Goda T, Colony PC, Edmond J, Thornburg W, Korc M, Koldovsky O (1987) Effects of enterally fed epidermal growth factor on the small and large intestine of the suckling rat. Regul Pept 17:121–132

Popliker M, Shatz A, Avaine A, Ullrich A, Schlessinger J, Webb CG (1987) Onset of endogenous synthesis of epidermal growth factor in neonatal mice. Dev Biol 119:38–44

Porter CD, Archard LC (1987) Characterization and physical mapping of *Molluscum contagiosum* virus DNA and location of a sequence capable of encoding a conserved domain of epidermal growth factor. J Gen Virol 68:673–682

Prywes R, Roeder RG (1986) Inducible binding of a factor to the c-*fos* enhancer. Cell 47:777-784

Prywes R, Livneh E, Ullrich A, Schlessinger J (1986) Mutations in the cytoplasmic domain of EGF receptor affect EGF binding and receptor internalization. EMBO J 5:2179–2190

Prywes R, Dutta A, Cromlish JA, Roeder RG (1988) Phosphorylation of serum response factor, a factor that binds to the serum response element of the c-*fos* enhancer. Proc Natl Acad Sci USA 85:7206–7210

Przylipiak A, Kiesel L, Rabe T, Helm K, Przylipiak M, Runnebaum B (1988) Epidermal growth factor stimulates luteinizing hormone and arachidonic acid release in rat pituitary cells. Mol Cell Endocrinol 57:157–162

Przysiecki CT, Staggers JE, Ramjit HG, Musson DG, Stern DG, Bennett CD, Friedman PA (1987) Occurence of β-hydroxylated asparagine residues in non-vitamin K-dependent proteins containing epidermal growth factor-like domains. Proc Natl Acad Sci USA 84:7856–7860

Puccio F, Lehy T (1988) Oral administration of epidermal growth factor in suckling rats stimulates cell DNA synthesis in fundic and antral gastric mucosae as well as in intestinal mucosa and pancreas. Regul Pept 20:53–64

Pulley DD, Marrone BL (1986) Inhibitory action of epidermal growth factor on progesterone biosynthesis in hen granulosa cells during short term culture: two sites of action. Endocrinology 118:2284–2291

Purchio AF, Twardzik DR, Bruce AG, Wizental L, Ranchalis JE, Hu S-L, Todaro G (1987) Synthesis of an active hybrid growth factor (GF) in bacteria: transforming GF-α/vaccinia GF fusion protein. Gene 60:175–182

Quantin B, Breathnach R (1988) Epidermal growth factor stimulates transcription of the c-*jun* proto-oncogene in rat fibroblasts. Nature 334:538–539

Rabkin SW, Sunga P, Myrdal S (1987) The effect of epidermal growth factor on chronotropic response in cardiac cells in culture. Biochem Biophys Res Commun 146:889–897

Radford HM, Avenell JA, Panaretto BA (1987a) Some effects of epidermal growth factor on reproductive function in Merino sheep. J Reprod Fertil 80:113–118

Radford HM, Panaretto BA, Avenell JA, Turnbull KE (1987b) Effect of mouse epidermal growth factor on plasma concentrations of FSH, LH and progesterone and on oestrus, ovulation and ovulation rate in Merino ewes. J Reprod Fertil 80:383–393

Raisz LG, Simmons HA, Sandberg AL, Canalis E (1980) Direct stimulation of bone resorption by epidermal growth factor. Endocrinology 107:270–273

Rakowicz-Szulczynski EM, Rodeck U, Herlyn M, Koprowski H (1986) Chromatin binding of epidermal growth factor, nerve growth factor, and platelet-derived growth factor in cells bearing the appropriate surface receptors. Proc Natl Acad Sci USA 83:3728–3732

Rall LB, Scott J, Bell GI, Crawford RJ, Penschow JD, Niall HD, Coghlan JP (1985) Mouse prepro-epidermal growth factor synthesis by the kidney and other tissues. Nature 313:228–231

Ran W, Dean M, Levine RA, Henkle C, Campisi J (1986) Induction of c-fos and c-myc mRNA by epidermal growth factor or calcium ionophore is cAMP dependent. Proc Natl Acad Sci USA 83:8216–8220

Rao KVS, Fox CF (1987) Epidermal growth factor stimulates tyrosine phosphorylation of human glucocorticoid receptor in cultured cells. Biochem Biophys Res Commun 144:512–519

Raper SE, Burwen SJ, Barker ME, Jones AL (1987) Translocation of epidermal growth factor to the hepatocyte nucleus during rat liver regeneration. Gastroenterology 92:1243–1250

Rappolee DA, Brenner CA, Schultz R, Mark D, Werb Z (1988a) Developmental expression of PDGF, TGFα, and TGF-β genes in preimplantation mouse embryos. Science 241:1823–1825

Rappolee DA, Mark D, Banda MJ, Werb Z (1988b) Wound macrophages express TGF-α and other growth factors in vivo: analysis by mRNA phenotyping. Science 241:708–712

Ray P, Moy FJ, Montelione GT, Liu J-F, Narang SA, Scheraga HA, Wu R (1988) Structure-function studies of murine epidermal growth factor: expression and site-directed mutagenesis of epidermal growth factor gene. Biochemistry 27:7289–7295

Raymond GM, Jumblatt MM, Bartels SP, Neufeld AH (1986) Rabbit corneal endothelial cells in vitro: effects of EGF. Invest Ophthalmol Vis Sci 27:474–479

Real FX, Rettig WJ, Chesa PG, Melamed MR, Old LJ, Mendelsohn J (1986) Expression of epidermal growth factor receptor in human cultured cells and tissues: relationship to cell lineage and stage of differentiation. Cancer Res 46:4726–4731

Reddan JR, Wilson-Dziedzic D (1983) Insulin growth factor and epidermal growth factor trigger mitosis in lenses cultured in a serum-free medium. Invest Ophthalmol Vis Sci 24:409–416

Redman RS, Quissel DO, Barzen KA (1988) Effects of dexamethasone, epidermal growth factor, and retinoic acid on rat submandibular acinar–intercalated duct complexes in primary culture. In Vitro Cell Div Biol 24:734–742

Rees DJG, Jones IM, Handford PA, Walter SJ, Esnouf MP, Smith KJ, Brownlee GG (1988) The role of β-hydroxyaspartate and adjacent carboxylate and adjacent carboxylate residues in the first EGF domain of human factor IX. EMBO J 7:2053–2061

Reilly CF, Fritze LMS, Rosenberg RD (1987) Antiproliferative effects of heparin on vascular smooth muscle cells are reversed by epidermal growth factor. J Cell Physiol 131:149–157

Reim M, Busse S, Leber M, Schulz C (1988) Effect of epidermal growth factor in severe experimental alkali burns. Ophthalmic Res 20:327–331

Reisner AH (1985) Similarity between the vaccinia virus 19K early protein and epidermal growth factor. Nature 313:801–803

Reynolds VH, Boehm FH, Cohen S (1965) Enhancement of chemical carcinogenesis by an epidermal growth factor. Surg Forum 16:108–109

Rheinwald JG, Green H (1977) Epidermal growth factor and the multiplication of cultured human epidermal keratinocytes. Nature 265:421–424

Rhodes JA, Tam JP, Finke V, Saunders M, Bernanke J, Silen W, Murphy RA (1986) Transforming growth factor α inhibits secretion of gastric acid. Proc Natl Acad Sci USA 83:3844–3846

Richards RC, Lewis-Jones DI, Walker JM, Desmond AD (1988) Epidermal growth factor (urogastrone) in human seminal plasma from fertile and infertile males. Fertil Steril 50:640–643

Richman RA, Claus TH, Pilkis SJ, Friedman DL (1976) Hormonal stimulation of DNA synthesis in primary cultures of adult rat hepatocytes. Proc Natl Acad Sci USA 73:3589–3593

Riedel H, Dull TJ, Schlessinger J, Ullrich A (1986) A chimeric receptor allows insulin to stimulate tyrosine kinase activity of epidermal growth factor receptor. Nature 324:68–70

Riedel H, Schlessinger J, Ullrich A (1987) A chimeric, ligand-binding v-erbB/EGF receptor retains transforming potential. Science 236:197–200

Riedel H, Massoglia S, Schlessinger J, Ullrich A (1988) Ligand activation of overexpressed epidermal growth factor receptors transforms NIH 3T3 mouse fibroblasts. Proc Natl Acad Sci USA 85:1477–1481

Riemen MW, Wegrzyn RJ, Baker AE, Hurni WM, Bennett CD, Oliff A, Stein RB (1987) Isolation of multiple biologically and chemically diverse species of epidermal growth factor. Peptides 8:877–885

Rihtniemi L, Thesleff I (1987) An autoradiographic study on the effect of epidermal growth factor on cell proliferation in erupting mouse incisors. Arch Oral Biol 32:859–863

Ristimäki A, Ylikorkala O, Perheentupa J, Viinikka L (1988) Epidermal growth factor stimulates prostacyclin production by cultured human vascular endothelial cells. Thromb Haemost 59:248–250

Ritvos O (1988) Modulation of steroidogenesis in choriocarcinoma cells by cholera toxin, phorbol ester, epidermal growth factor and insulin-like growth factor I. Mol Cell Endocrinol 59:125–133

Robertson NW, Blecher ST (1986) Epidermal growth factor (EGF) affects sulphydryl and disulphide levels in cultured mouse skin: possible relationship between effects of EGF and the tabby gene on thiols. Biochem Cell Biol 65:658–667

Rodland KD, Jue SF, Magun BE (1986) Regulation of VL30 gene expression by activators of protein kinase C. J Biol Chem 261:5029–5031

Rodland KD, Muldoon LL, Dinh T-H, Magun BE (1988) Independent transcriptional regulation of a single VL30 element by epidermal growth factor and activators of protein kinase C. Mol Cell Biol 8:2247–2250

Roger PP, Dumont JE (1982) Epidermal growth factor controls the proliferation and the expression of differentiation in canine thyroid cells in primary culture. FEBS Lett 144:209–212

Rose SP, Stahn R, Passovoy DS, Herschman H (1976) Epidermal growth factor enhancement of skin tumor induction in mice. Experientia 32:913–915

Rosenthal A, Lindquist PB, Bringman TS, Goeddel DV, Derynck R (1986) Expression in rat fibroblasts of a human transforming growth factor-α cDNA results in transformation. Cell 46:301–309

Russell DW, Schneider WJ, Yamamoto T, Luskey KL, Brown MS, Goldstein JL (1984) Domain map of the LDL receptor: sequence homology with the epidermal growth factor precursor. Cell 37:577–585

Russo MW, Lukas TJ, Cohen S, Staros JV (1985) Identification of residues in the nucleotide binding site of the epidermal growth factor receptor/kinase. J Biol Chem 260:5205–5208

Sainsbury JRC, Farndon JR, Sherbet GV, Harris AL (1985a) Epidermal growth factor receptors and oestrogen receptors in human breast cancer. Lancet I (8425):364–366

Sainsbury JRC, Farndon JR, Harris AL, Sherbet GV (1985b) Epidermal growth factor receptors on human breast cancers. Br J Surg 72:186–188

Sainsbury JRC, Malcolm AJ, Appleton DR, Farndon JR, Harris AL (1985c) Presence of epidermal growth factor receptor as an indicator of poor prognosis in patients with breast cancer. J Clin Pathol 38:1225–1228

Sakamoto T, Swierczek JS, Ogden WD, Thompson JC (1985) Cytoprotective effect of pentagastrin and epidermal growth factor on stress ulcer formation. Possible role of somatostatin. Ann Surg 201:290–295

Salomon DS, Liotta LA, Kidwell WR (1981) Differential response to growth factor by rat mammary epithelium plated on different collagen substrata in serum-free medium. Proc Natl Acad Sci USA 78:382–386

Sand T-E, Bronstad G, Digernes V, Killi A, Amara W, Refsnes M, Christoffersen T (1985) Quantitative aspects of the effects of insulin, epidermal growth factor and dexamethasone on DNA synthesis in cultured adult rat hepatocytes. Acta Endocrinol (Copenh) 109:369–377

Sankaran L, Topper YJ (1983) Selective enhancement of the induction of α-lactalbumin activity in rat mammary explants by epidermal growth factor. Biochem Biophys Res Commun 117:524–529

Sankaran L, Topper YJ (1984) Prolactin-induced α-lactalbumin activity in mammary explants from pregnant rabbits. A role for epidermal growth factor and glucocorticoids. Biochem J 217:833–837

Sasaki M, Yamada Y (1987) The laminin B2 chain has a multidomain structure homologous to the B1 chain. J Biol Chem 262:17111-17117

Sasaki M, Kato S, Kohno K, Martin GR, Yamada Y (1987) Sequence of the cDNA encoding the laminin B1 chain reveals a multidomain protein containing cysteine-rich repeats. Proc Natl Acad Sci USA 84:935–939

Sasaki M, Klienman HK, Huber H, Deutzmann R, Yamada Y (1988) Laminin, a multidomain protein. The A chain has a unique globular domain and homology with the basement membrane proteoglycan and the laminin B chains. J Biol Chem 263:16536–16544

Sato C, Nishizawa K, Nakayama T, Ohtsuka K, Nakamura H, Kobayashi T, Inagaki M (1988) Rapid phosphorylation of MAP-2-related cytoplasmic and nuclear M 300000 protein by serine kinases after growth stimulation in quiescent cells. Exp Cell Res 175:136–147

Savage AP, Chatterjee VK, Gregory H, Bloom SR (1986) Epidermal growth factor in blood. Regu Pept 16:199–206

Savage CR Jr, Cohen S (1972) Epidermal growth factor and a new derivative. Rapid isolation procedures and biological and chemical characterization. J Biol Chem 247:7609–7611

Savage CR Jr, Cohen S (1973) Proliferation of corneal epithelium induced by epidermal growth factor. Exp Eye Res 15:361–366

Savage CR Jr, Harper R (1981) Human epidermal growth factor urogastrone: rapid purification procedure and partial characterization. Anal Biochem 111:195–202

Savage CR Jr, Inagami T, Cohen S (1972) The primary structure of epidermal growth factor. J Biol Chem 247:7612–7672

Savage CR Jr, Hash JH, Cohen S (1973) Epidermal growth factor: location of disfulfide bonds. J Biol Chem 248:7669–7672

Sawyer ST, Cohen S (1981) Enhancement of calcium uptake and phosphatidylinositol turnover by epidermal growth factor in A-431 cells. Biochemistry 20:6280–6286

Sawyer ST, Cohen S (1985) Epidermal growth factor stimulates the phosphorylation of the calcium-dependent 35000 dalton substrate in intact A-431 cells. J Biol Chem 260:8233–8236

Schaudies RP, Savage CR Jr (1986) Isolation of rat epidermal growth factor (r-EGF): chemical, biological and immunological comparisons with mouse and human EGF. Comp Biochem Physiol [B] 846:497–505

Schatz H, Pschierer-Berg K, Nickel J-A, Bär R, Müller F, Bretzel RG, Müller H, Stracke H (1986) Assay for thyroid growth stimulating immunoglobulins: stimulation of [^{3}H]thymidine incorporation into isolated thyroid follicles by TSH, EGF, and immunoglobulins from goitrous patients in an iodine-deficient region. Acta Endocrinol (Copenh) 112:523–530

Schechter Y, Hernaez L, Schlessinger J, Cuatrecasas P (1979) Local aggregation of hormone-receptor complexes is required for activation by epidermal growth factor. Nature 278:835–838

Scheving LA, Yeh YC, Tsai TH, Scheving LE (1979) Circadian phase-dependent stimulatory effects of epidermal growth factor on deoxyribonucleic acid synthesis in the tongue, esophagus, and stomach of the adult male mouse. Endocrinology 105:1475–1480

Scheving LA, Yeh YC, Tsai TH, Scheving LE (1980) Circadian phase-dependent stimulatory effects of epidermal growth factor on deoxyribonucleic acid synthesis in the duodenum, jejunum, illeum, caecum, colon, and rectum of the adult male mouse. Endocrinology 106:1498–1503

Schlaepfer DD, Haigler HT (1987) Characterization of Ca^{2+}-dependent phospholipid binding and phosphorylation of lipocortin I. J Biol Chem 262:6931–6937

Schlessinger J (1988a) The epidermal growth factor receptor as a multifunctional allosteric protein. Biochemistry 27:3119–3123

Schlessinger J (1988b) Signal transduction by allosteric receptor oligomerization. Trends Biol Sci 13:443–447

Schlessinger J, Shechter Y, Willingham MC, Pastan I (1978) Direct visualization of binding, aggregation, and internalization of insulin and epidermal growth factor on living fibroblastic cells. Proc Natl Acad Sci USA 75:2659–2663

Schneider C, Owen MJ, Banville D, Williams JG (1984) Primary structure of human transferrin receptor deduced from the mRNA sequence. Nature 311:675–678

Schomberg DW, May JV, Mondschein JS (1983) Interations between hormones and growth factors in the regulation of granulosa cell differentiation in vitro. J Steroid Biochem 19:291–295

Schonbrunn A, Krasnoff M, Westendorf JM, Tashjian AH Jr (1980) Epidermal growth factor and thytrotropin-releasing hormone act similarly on a clonal pituitary cell strain. Modulation of hormone production and inhibition of cell proliferation. J Cell Biol 85:786–797

Schreiber AB, Yarden Y, Schlessinger J (1981) A non-mitogenic analogue of epidermal growth factor enhances the phosphorylation of endogenous membrane proteins. Biochem Biophys Res Commun 101:517–523

Schreiber AB, Winkler ME, Derynck R (1986) Transforming growth factor-α: a more potent angiogenic mediator than epidermal growth factor. Science 232:1250–1253

Schultz GS, White M, Mitchell R, Brown G, Lynch J, Twardzik DR, Todaro GJ (1987) Epithelial wound healing enhanced by transforming growth factor-α and vaccinia growth factor. Science 235:350–352

Scott J, Urdea M, Quiroga M, Sanchez-Pescador R, Fong N, Selby M, Rutter WJ, Bell GI (1983) Structure of a mouse submaxillary messenger RNA encoding epidermal growth factor and seven related proteins. Science 221:236–240

Segawa K, Ito Y (1983) Enhancement of polyoma virus middle T antigen tyrosine phosphorylation by epidermal growth factor. Nature 304:742–744

Seger R, Yarden Y, Kashles O, Goldblatt D, Schlessinger J, Shaltiel S (1988) The epidermal growth factor receptor as a substrate for a kinase-splitting membranal proteinase. J Biol Chem 263:3496–3500

Serrero G (1987) EGF inhibits the differentiation of adipocyte precursors in primary cultures. Biochem Biophys Res Commun 146:194–202

Shaw G, Jorgensen GI, Tweedale R, Tennison M, Waters MJ (1985) Effect of epidermal growth factor on reproductive function of ewes. J Endocrinol 107:429–436

Shaw GP, Hatt JF, Anderson NG, Hanson PJ (1987) Action of epidermal growth factor on acid secretion by rat isolated parietal cells. Biochem J 244:699–704

Sheng M, Dougan ST, McFadden G, Greenberg ME (1988) Calcium and growth factor pathways of c-*fos* transcriptional activation require distinct upstream regulatory sequences. Mol Cell Biol 8:2787–2796

Shinoda I, Tokida N, Kurobe M, Furukawa S, Hayashi K (1988) Demonstration of a considerable amount of mouse epidermal growth factor in aqueous humor. Biochem Int 17:243–248

Shoyab M, McDonald VL, Bradley JG, Todaro GJ (1988) Amphiregulin: a bifunctional growth-modulating glycoprotein produced by the phorbol 12-myristate 13-acetate-treated human breast adenocarcinoma cell line MCF-7. Proc Natl Acad Sci USA 85:6528–6532

Shoyab M, Plowman GD, McDonald VL, Bradley JG, Todaro GJ (1989) Structure and function of human amphiregulin: a member of the epidermal growth factor family. Science 243:1074–1076

Shuman S, Broyles SS, Moss G (1987) Purification and characterization of a transcription termination factor from vaccinia virus. J Biol Chem 262:12372–12380

Siegelman MH, van de Rijn M, Weissman IL (1989) Mouse lymph node honing receptor cDNA clone encodes aglycoprotein revealing tandem interaction domains. Science 243:1165–1172

Silver J, Whitney JB, Kozak C, Hollis G, Kirsch I (1985) Erb-B is linked to the alpha-globin locus on mouse chromosome 11. Mol Cell Biol 5:1784–1786

Silver MH, Murray JC, Pratt RM (1984) Epidermal growth factor stimulates type-V collagen synthesis in cultured murine palatal shelves. Differentiation 27:205–208

Simpson RJ, Smith JA, Moritz RL, O'Hara MR, Rudland PS, Morrison JR, Lloyd CJ, Grego B, Burgess AW, Nice EC (1985) Rat epidermal growth factor: complete amino acid sequence. Homology with the corresponding murine and human proteins; isolation of a form truncated at both ends with full in vivo biological activity. Eur J Biochem 153:629–637

Singh G, Foster CS (1987) Epidermal growth factor in alkali-burned corneal epithelial wound healing. J Ophthalmol 103:802–807

Singh-asa P, Waters MJ (1983) Stimulation of adrenal cortisol biosynthesis by EGF. Mol Cell Endocrinol 30:189–199

Singh-asa P, Waters MJ, Wilce PA (1985) A mechanism for the in vitro stimulation of adrenal cortisol biosynthesis by epidermal growth factor. Int J Biochem 17:857–862

Slieker LJ, Lane MD (1985) Post-translational processing of the epidermal growth factor receptor. Glycosylation-dependent acquisition of ligand-binding capacity. J Biol Chem 260:687–690

Slieker LJ, Martensen TM, Lane MD (1986) Synthesis of epidermal growth factor receptor in human A431 cells. J Biol Chem 261:15233–15241

Smith CJ, Wejksnora PJ, Warner JR, Rubin CS, Rosen OM (1979) Insulin-stimulated protein phosphorylation in 3T3-L1 preadipocytes. Proc Natl Acad Sci USA 76:2725–2729

Smith CJ, Rubin CS, Rosen OM (1980) Insulin-treated 3T3-L1 adipocytes and cell-free extracts derived from them incorporate ^{32}P into ribosomal protein S6. Proc Natl Acad Sci USA 77:2641–2645

Smith JJ, Derynck R, Korc M (1987) Production of transforming growth factor α in human pancreatic cancer cells: evidence for a superagonist autocrine cycle. Proc Natl Acad Sci USA 84:7567–7570

Smith JM, Sporn MB, Roberts AB, Derynck R, Winkler M, Gregory H (1985) Human transforming growth factor-α causes precocious eyelid opening in newborn mice. Nature 315:515–516

Smith KB, Losonczy I, Sahai A, Pannerselvam M, Fehmel P, Salomon DS (1983) Effect of 12-O-tretradecanoylphorbol-13-acetate (TPA) on the growth inhibitory and increased phosphatidylinositol (PI) responses induced by epidermal growth factor (EGF) in A431 cells. J Cell Physiol 117:91–100

Soderquist AM, Carpenter G (1984) Glycosylation of the epidermal growth factor receptor in A-431 cells. The contribution of carbohydrate to receptor function. J Biol Chem 259:12586–12594

Soderquist AM, Stoscheck C, Carpenter G (1988) Similarities in glycosylation and transport between the secreted and plasma membrane forms of the epidermal growth factor receptor in A-431 cells. J Cell Physiol 136:447–454

Soley M, Hollenberg MD (1987) Epidermal growth factor (urogastrone)-stimulated glyconeogenesis in isolated mouse hepatocytes. Arch Biochem Biophys 255:136–146

Sommercorn J, Mulligan JA, Lozeman FJ, Krebs EG (1987) Activation of casein kinase II in response to insulin and to epidermal growth factor. Proc Natl Acad Sci USA 84:8834–8838

Sorkin AD, Teslanko LV, Nikolsky NN (1988) The endocytosis of epidermal growth factor in A431 cells: a pH of microenvironment and the dynamics of receptor complex dissociation. Exp Cell Res 175:192–205

Sovová V, Vydra J, Cerna H, Sloncova E, Damkova M, Jakoubkova J, Hlozanek I (1988) Analysis of epidermal growth factor and epidermal growth factor receptor expression in human renal carcinoma cell cultures. Folia Biol (Praha) 34:233–239

Spitzer E, Koepke K, Kunde D, Grosse R (1988) EGF binding is quantitatively related to growth in node-positive breast cancer. Breast Cancer Res Treat 12:45–49

Stahlman MT, Gray ME, Chytil F, Sundell H (1988) Effect of retinol on fetal lamb tracheal epithelium, with and without epidermal growth factor. A model for the effect of retinol on the healing lung of human premature infants. Lab Invest 59:25–35

Starkey RH, Orth DN (1977) Radioimmunoassay of human epidermal growth factor (Urogastrone). J Clin Endocrinol Metab 45:1144–1153

Staros JV, Fanger BO, Faulkner LA, Palaszewski PP, Russo MW (1989) Mechanism of transmembrane signaling by the epidermal growth factor receptor/kinase. In: Moudgil VK (ed) Receptor phosphorylation. CRC, Boca Raton FL, pp 227–242

Stastny M, Cohen S (1972) The stimulation of ornithine decarboxylase activity in testes of the neonatal mouse. Biochim Biophys Acta 261:177–180

Steidler NE, Reade PC (1980) Histomorphological effects of epidermal growth factor on skin and oral mucosa in neonatal mice. Arch Oral Biol 25:37–43

Steidler NE, Reade PC (1981) Epidermal growth factor and proliferation of odontogenic cells in culture. J Dent Res 60:1977–1982

Stenflo J, Lundwall A, Dahlbäck B (1987) β-hydroxy-asparagine in domains homologous to the epidermal growth factor precursor in vitamin K-dependent protein S, Proc Natl Acad Sci USA 84:368–372

Stenflo J, Ohlin A-K, Owen WB, Schneider JR (1988) β-hydroxy-aspartic acid or β-hydroxy-asparagine in bovine low density lipoprotein receptor and in bovine thrombomodulin. J Biol Chem 263:21–24

Stern DF, Hare DL, Cecchini MA, Weinberg RA (1987) Construction of a novel oncogene based on synthetic sequences encoding epidermal growth factor. Science 235:321–324

Stern DF, Kamps MP (1988) EGF-stimulated tyrosine phosphorylation of p185[neu]: a potential model for receptor interactions. EMBO J 7:995–1001

Stern PH, Krieger NS, Niessenson RA, Williams RD, Winkler ME, Derynck R, Strewler GJ (1975) Human transforming growth factor-alpha stimulates bone resorption in vitro. J Clin Invest 76:2016–2019

Stoppelli MP, Verde P, Grimaldi G, Locatelli EK, Blasi F (1986) Increase in urokinase plasminogen activator mRNA synthesis in human carcinoma cells is a primary effect of the potent tumor promoter, phorbol myristate acetate. J Cell Biol 102:1235–1241

Stoscheck CM, Carpenter G (1984) "Down regulation" of epidermal growth factors receptors: direct demonstration of receptor degradation in human fibroblasts. J Cell Biol 98:1048–1053

Stroobant P, Rice AP, Bullick WJ, Cheng DJ, Kerr IM, Waterfield MD (1980) Purification and characterization of vaccinia virus growth factor. Cell 42:383–393

Südhof TC, Goldstein JL, Brown MS, Russell DW (1985a) The LDL receptor gene: a mosaic of exons shared with different proteins. Science 228:815–822

Südhof TC, Russell DW, Goldstein JL, Brown MS, Sanchez-Pescador R, Bell GI (1985b) Cassette of eight exons shared by genes for LDL receptor and EGF precursor. Science 228:893–895

Sugo T, Bjärk I, Holmgren A, Stenflo J (1984) Calcium-binding properties of bovine factor X lacking the γ-carboxyglutamic acid-containing region. J Biol Chem 259:5705–5710

Sumi S, Akira H, Shintaro Y, Kenichi M, Atsushi K, Shizutoshi N, Masanori S (1985) Overproduction of human epidermal growth factor/urogastrone in Escherichia coli and demonstration of its full biological activities. J Biotechnol 2:59–74

Sundell HW, Gray ME, Serenius FS, Escobedo MB, Stahlman MT (1980) Effects of epidermal growth factor on lung maturation in fetal lambs. Am J Pathol 110:706–726

Supowit SC, Potter E, Evans RM, Rosenfeld MG (1984) Polypeptide hormone regulation of gene transcription: specific 5′ genomic sequences are required for epidermal growth factor and phorbol ester regulation of prolactin gene expression. Proc Natl Acad Sci USA 81:2975–2979

Suzuki K, Kusumoto H, Deyashiki Y, Nishioka J, Maruyama I, Zushi M, Kawahara S, Honda G, Yammoto S, Horiguchi S (1987) Structure and expression of human thrombomodulin, a thrombin receptor on endothelium acting as a cofactor for protein C activation. EMBO J 6:1891–1897

Takasu N, Shimizu Y, Yamada T (1987) Tumour promoter 12-O-tetradecanoyl-phorbol 13-acetate and epidermal growth factor stimulate proliferation and inhibit differentiation of porcine thyroid cells in primary culture. J Endocrinol 113:485–487

Takasu N, Takasu M, Yamada T, Shimizu Y (1988) Epidermal growth factor (EGF) produces inositol phosphates and increases cytoplasmic free calcium in cultured porcine thyroid cells. Biochem Biophys Res Commun 151:530–534

Taketani Y, Oka T (1983a) Epidermal growth factor stimulates cell proliferation and inhibits functional differentiation of mouse mammary epithelial cells in culture. Endocrinology 113:871–877

Taketani Y, Oka T (1983b) Possible physiological role of epidermal growth factor in the development of the mouse mammary gland during prenancy. FEBS Lett 152:256–260

Takeya H, Kawabata S, Nakagawa K, Yamamichi Y, Miyake T, Iwanaga S, Takao T, Shimonishi Y (1988) Bovine factor VII. Its purification and complete amino acid sequence. J Biol Chem 263:14868–14877

Tam JP (1985) Physiological effects of transforming growth factor in the newborn mouse. Science 229:673–675

Tam JP, Marquardt H, Rosberger DF, Wong TW, Todaro GJ (1984) Synthesis of biologically active rat transforming growth factor I. Nature 309:376–378

Tam JP, Sheikh MA, Salomon DS, Ossowski L (1986) Efficient synthesis of human type α transforming growth factor: its physical and biological characterization. Proc Natl Acad Sci USA 83:8082–8086

Tapanainen J, Leinonen PJ, Tapanainen P (1987) Regulation of human granulosa-luteal cell progesterone production and proliferation by gonadotropins and growth factors. Ferti Steril 48:578–580

Tashjian AH Jr, Levine L (1978) Epidermal growth factor stimulates prostaglandin production and bone resorption in cultured mouse calvaria. Biochem Biophys Res Commun 85:966–975

Tashjian AH Jr, Voelkel EF, Lazzaro M, Singer FR, Roberts AB, Derynck R, Winkler ME, Levine L (1985) α and β human transforming growth factors stimulate prostaglandin production and bone resorption in cultured mouse calavaria. Proc Natl Acad Sci USA 82:4535–4538

Taylor JM, Mitchell WM, Cohen S (1972) Epidermal growth factor. Physical and chemical properties. J Biol Chem 247:5928–5934

Teixido J, Massagué J (1988) Structural properties of a soluble bioactive precursor for transforming growth factor-α. J Biol Chem 263:3924–3929

Teixido J, Gilmore R, Lee DC, Massagué J (1987) Integral membrane glycoprotein properties of the prohormone pro-transforming growth factor-α. Nature 326:883–885

Teslenko LV, Kornilova ES, Sorkin AD, Nikolsky NN (1987) Recycling of epidermal growth factor in A431 cells. FEBS Lett 221:105–109

Todaro GJ, De Larco JE (1976) Transformation by murine and feline sarcoma viruses specifically blocks binding of epidermal growth factor to cells. Nature 264:26–31

Todderud G, Carpenter G (1988) Presence of mannose phosphate on the epidermal growth factor receptor in A-431 cells. J Biol Chem 263:17893–17896

Toh H, Hayashida H, Kikuno R, Yasunaga T, Miyata T (1985) Sequence similarity between the EGF receptor and α_1-acid glycoprotein. Nature 314:199

Tomita M, Hirata Y, Uchihashi M, Fuhita T (1986) Characterization of epidermal growth factor receptors in cultured vascular smooth muscle cells of rat aorta. Endocrinol Jpn 33:177–184

Tomomura A, Sawada N, Sattler GL, Kleinman HK, Pitot H (1987) The control of DNA synthesis in primary cultures of hepatocytes from adult and young rats: inter-actions of extracellular matrix components, epidermal growth factor, and the cell cycle. J Cell Physiol 130:221–227

Tomooka Y, DiAugustine RP, McLachlan JA (1986) Proliferation of mouse uterine epithelial cells in vitro. Endocrinology 118:1011–1018

Tonelli QJ, Sorof S (1980) Epidermal growth factor requirement for development of cultured mammary gland. Nature 285:250–252

Tonelli QJ, Sorof S (1981) Expression of a phenotype of normal differentiation in cul-tured mammary glands is promoted by epidermal growth factor and blocked by cyclic adenine nucleotide and prostaglandins. Differentiation 20:253–259

Topham RT, Chiego DJ, Gattone VH II, Hinton DA, Klein RM (1987) The effect of epidermal growth factor on neonatal incisor differentiation in the mouse. Dev Biol 124:532–543

Tricoli JV, Nakai H, Byers MG, Bell GI, Shows TB (1986) The gene for human trans-forming growth factor α is on the short arm of chromosome 2. Cytogenet Cell Genet 42:94–98

Trzeciak WH, Duda T, Waterman MR, Simpson ER (1987) Effects of epidermal growth factor on the synthesis of the cholesterol side-chain cleavage enzyme com-plex in rat ovarian granulosa cells in primary culture. Mol Cell Endocrinol 52:43–50

Tsutsumi O, Kurachi H, Oka T (1986) A physiological role of epidermal growth factor in male reproductive function. Science 233:975–977

Turkington RW (1969) The role of epithelial growth factor in mammary gland development in vitro. Exp Cell Res 57:79–85

Turley EA, Hollenberg MD, Pratt RM (1985) Effect of epidermal growth factor-urogastrone on glycosaminoglycan synthesis and accumulation in vitro in the developing mouse palate. Differentiation 28:279–285

Twardzik DR, Brown JP, Ranchalis JE, Todaro GJ, Moss B (1985) Vaccinia virus-infected cells release a novel polypeptide functionally related to transforming and epidermal growth factors. Proc Natl Acad Sci USA 82:5300–5304

Tyler MS, Pratt RM (1980) Effect of epidermal growth factor on secondary palatal epithelium in vitro: tissue isolation and recombination studies. J Embryo Exp Morpho 58:93–106

Uchihashi M, Hirata Y, Fujita T, Matsukura S (1982) Age-related decrease of urinary excretion of human epidermal growth factor (hEGF). Life Sci 31:679–683

Ueno S, Manganaro TF, Donahoe PK (1988) Human recombinant Mullerian inhibiting substance inhibition of rat oocyte meiosis is reversed by epidermal growth factor in vitro. Endocrinology 123:1652–1658

Ullrich A, Coussens L, Hayflick JS, Dull TJ, Gray A, Tam AW, Lee J, Yarden Y, Libermann TA, Schlessinger J, Downward J, Mayes ELV, Whittle N, Waterfield MD, Seegurg PH (1984) Human epidermal growth factor receptor cDNA sequence and aberrant expression of the amplified gene in A431 epidermoid carcinoma cells. Nature 309:418–424

Ulshen MH, Lyn-Coo LE, Raasch RH (1986) Effects of intraluminal epidermal growth factor on mucosal proliferation in the small intestine of adult rats. Gastroenterology 91:1134–1140

Upton C, Macen JL, McFadden G (1987) Mapping and sequencing of a gene from myxoma virus that is related to those encoding epidermal growth factor and transforming growth factor alpha. J Virol 61:1271–1275

Ushiro H, Cohen S (1980) Identification of phosphotyrosine as a product of epidermal growth factor-activated protein kinase in A-431 cell membranes. J Biol Chem 255:8363–8365

Vaartjes WJ, de Haas CGM, van den Bergh SG (1985) Differential short-term effects of growth factors on fatty acid synthesis in isolated rat-liver cells. Biochem Biophys Res Commun 131:449–455

Vässin H, Bremer KA, Knust E, Campos-Ortega JA (1987) The neurogenic gene delta of *Drosophila melanogaster* is expressed in neurogenic territories and encodes a putative transmembrane protein with EGF-like repeats. EMBO J 6:3431–3440

Veal D, Ashcroft T, Marsh C, Gibson GJ, Harris AL (1987) Epidermal growth factor receptors in non-small cell lung cancer. Br J Cancer 55:513–516

Velu TJ, Beguinot L, Vass WC, Willingham MC, Merline GT, Pastan I, Lowy DR (1987) Epidermal growth factor-dependent transformation by a human EGF receptor proto-oncogene. Science 238:1408–1410

Venkatesan S, Gershowitz A, Moss B (1982) Complete nucleotide sequences of two adjacent early vaccinia virus genes located within the inverted terminal repetition. J Virol 44:637–646

Verhoeven G, Cailleau J (1986) Stimulatory effects of epidermal growth factor on steroidogenesis in Leydig cells. Mol Cell Endocrinol 47:99–106

Verma IM, Sassone-Corsi P (1987) Proto-oncogene *fos:* complex but versatile regulation. Cell 51:513–514

Vonderhaar BK (1987) Local effects of EGF, α-TGF, and EGF-like growth factors on lobuloalveolar development of the mouse mammary gland in vivo. J Cell Physiol 132:581–584

Vonderhaar BK, Nakhasi HL (1986) Bifunctional activity of epidermal growth factor on α- and K-casein gene expression in rodent mammary glands in vitro. Endocrinology 119:1178–1184

von Rüden T, Wagner EF (1988) Expression of functional human EGF receptor on murine bone marrow cells. EMBO J 7:2749–2756

Wadsworth SC, Vincent WS, Bilodeau-Wentworth D (1985) A Drosophila genomic sequence with homology to human epidermal growth factor receptor. Nature 314:178–180

Wahl M, Carpenter G (1988) Regulation of epidermal growth factor-stimulated formation of inositol phosphates in A-431 cells by calcium and protein kinase C. J Biol Chem 263:7581–7590

Wahl MI, Sweatt D, Carpenter G (1987) Epidermal growth factor (EGF) stimulates inositol trisphosphate formation in cells which overexpress the EGF receptor. Biochem Biophys Res Commun 142:688–695

Wahl MI, Daniel TO, Carpenter G (1988) Antiphosphotyrosine recovery of phospholipase C activity after EGF treatment of A-431 cells. Science 241:968–970

Wahl MI, Nishibe S, Pann-Ghill S, Rhee SG, Carpenter G (1989) Epidermal growth factor stimulates tyrosine phosphorylation of phospholipase C-II indenpendently of receptor internalization and extracellular calcium. Proc Natl Acad Sci USA 86:1568–1572

Wallner BP, Mattaliano RJ, Hession C, Cate RL, Tizard R, Sinclair LK, Foeller C, Chow EP, Browning JL, Ramachandran KL, Pepinsky RB (1986) Cloning and expression of human lipocortin, a phosphlipase A_2 inhibitor with potential anti-inflammatory activity. Nature 320:77–81

Watanabe K, Nakagawa S, Nishida T (1987) Stimulatory effects of fibronectin and EGF on migration of corneal epithelial cells. Invest Ophthalmol Vis Sci 28:205–211

Watanabe S, Lazar E, Sporn MB (1987) Transformation of normal rat kidney (NRK) cells by an infections retrovirus carrying a synthetic rat type α transforming growth factor gene. Proc Natl Acad Sci USA 84:1258–1262

Weaver LT, Freiberg E, Israel EJ, Walker WA (1988) Epidermal growth factor in human amniotic fluid. Gastroenterology 95:1436

Welsh TH Jr, Hsueh AJW (1982) Mechanism of the inhibitory action of epidermal growth factor on testicular androgen biosynthesis in vitro. Endocrinology 110:1498–1506

Wen D, Dittman WA, Ye RD, Deaven LL, Majerus PW, Sadler JE (1987) Human thrombomodulin: complete cDNA sequence and chromosome localization of the gene. Biochemistry 26:4350–4357

Westermark K, Karlsson FA, Westermark B (1983) Epidermal growth factor modulates thyroid growth and function in culture. Endocrinology 112:1680–1686

Wharton KA, Johansen KM, Xu T, Artavanis-Tsakonas S (1986) Nucleotide sequence from the neurogenic locus notch implies a gene product that shares homology with proteins containing EGF-like repeats. Cell 43:567–581

White BA (1985) Evidence for a role of calmodulin in the regulation of prolactin gene expression. J Biol Chem 260:1213–1217

White BA, Bancroft FC (1983) Epidermal growth factor and thyrotropin-releasing hormone interact synergistically with calcium to regulate prolactin mRNA levels. J Biol Chem 258:4618–4622

Wilcox JN, Derynck R (1988a) Localization of cells synthesizing transforming growth factor-alpha mRNA in the mouse brain. J Neurosci 8:1901–1904

Wilcox JN, Derynck R (1988b) Developmental expression of transforming growth factor alpha and beta in mouse fetus. Mol Cell Biol 8:3415–3422

Winkler ME, Bringman T, Marks BJ (1986) The purification of fully active recombinant transforming growth factor α produced in Escherichia coli. J Biol Chem 261:13838–13843

Wolffe AP, Bersimbaev RI, Tata JR (1985) Inhibition by estradiol of binding and mitogenic effect of epidermal growth factor in primary cultures of Xenopus hepatocytes. Mol Cell Endocrinol 40:167–173

Wong AJ, Bigner SH, Bigner DD, Kinzler KW, Hamilton SR, Vogelstein B (1987) Increased expression of the epidermal growth factor receptor gene in malignant gliomas is invariably associated with gene amplification. Proc Natl Acad Sci USA 84:6899–6903

Woost PG, Brightwell J, Eiferman RA, Schultz GS (1985) Effect of growth factors with dexamethasone on healing of rabbit corneal stromal incisions. Exp Eye Res 40:47–60

Xu Y-H, Richert N, Ito S, Merlino GT, Pastan I (1984) Characterization of epidermal growth factor receptor gene expression in malignant and normal human cell lines. Proc Natl Acad Sci USA 81:7308–7312

Yamamoto T, Nishida T, Miyajima N, Kawai S, Ooi T, Toyoshima K (1983) The erb B gene of avian erythroblastosis virus is a member of the *src* gene family. Cell 35:71–78

Yamamoto T, Kamata N, Kawano H, Shimizu S, Kuroki T, Toyoshima K, Rikimaru K, Nomura N, Ishizaki R, Pastan I, Gamou S, Shimizu N (1986) High incidence of amplification of the epidermal growth factor receptor gene in human squamous carcinoma cell lines. Cancer Res 46:414–416

Yarden Y, Schlessinger J (1987a) Self-phosphorylation of epidermal growth factor receptor: evidence for a model of intermolecular allosteric activaion. Biochemistry 26:1434–1442

Yarden Y, Schlessinger J (1987b) Epidermal growth factor induces rapid, reversible aggregation of the purified epidermal growth factor receptor. Biochemistry 26:1443–1451

Yarden Y, Ullrich A (1988) Growth factor receptor tyrosine kinases. Annu Rev Biochem 57:443–478

Yarden Y, Schreiber AB, Schlessinger J (1982) A non-mitogenic analogue of epidermal growth factor induces early responses mediated by epidermal growth factor. J Cell Biol 92:687–693

Yochem J, Weston K, Greenwald I (1988) The *Caenorhabditis elegans lin*-12 gene encodes a trans-membrane protein with overall similarity to *Drosophila notch*. Nature 335:547–550

Yokota K, Kusaka M, Ohshima T, Yamamoto S, Kurihara N, Yoshino T, Kumegawa M (1986) Stimulation of prostaglandin E_2 synthesis in cloned osteoblastic cells of mouse (MC373–E1) by epidermal growth factor. J Biol Chem 261:15410–15415

Yoneda T, Pratt RM (1981) Mesenchymal cells from the human embryonic palate are highly responsive to epidermal growth factor. Science 213:563–565

Yoshimoto K, Nakamura T, Ishihara A (1983) Reciprocal effects of epidermal growth factor on key lipogenic enzymes in primary cultures of adult rat hepatocytes. Induction of glucose-6-phosphate dehydrogenase and suppression of malic enzyme and lipogenesis. J Biol Chem 258:12355–12360

Yoshitake S, Schach BG, Foster DC, Davie EW, Kurachi K (1985) Nucleotide sequence of the gene for human factor IX (antihemophilic factor B). Biochemistry 24:3736–3750

Yuen L, Moss B (1986) Multiple 3' ends of mRNA encoding vaccinia virus growth factor occur within a series of repeated sequences downstream of T clusters. J Virol 60:320–323

Zabel BU, Fournier REK, Lalley PA, Naylor SL, Sakaguchi AY (1984) Cellular homologs of the avian erythroblastosis virus erb-A and erb-B genes are asyntenic and mouse but asyntenic in man. Proc Natl Acad Sci USA 81:4874–4878

Zabel BU, Eddy RL, Lalley PA, Scott J, Bell GI, Shows TB (1985) Chromosomal locations of the human and mouse genes for precursors of epidermal growth factor and the β subunit of nerve growth factor. Proc Natl Acad Sci USA 82:469–473

Zakar T, Olson DM (1988) The action of epidermal growth factor on human amnion prostaglandin E_2 output. Can J Physiol Pharmacol 66:769–775

Zerek-Melen G, Lewinski A, Pawlikowski M, Sewerynek E, Kunert-Radek J (1987) Influence of somatostatin and epidermal growth factor (EGF) on the proliferation of follicular cells in the organ-cultured rat thyroid. Res Exp Med (Berl) 187:415–421

Zhiwen A, Herington AC, Carson RS, Findley JK, Burger HG (1987) Direct inhibition of rat granulosa cell inhibin production by epidermal growth factor in vitro. Mol Cell Endocrinol 54:213–220

Zinn K, Keller A, Whittemore L-A, Maniatis T (1988) 2-Aminopurine selectively inhibits the induction of β-interferon, c-*fos*, and c-*myc* gene expression. Science 240:210–213

CHAPTER 5

Platelet-Derived Growth Factor

E. W. RAINES, D. F. BOWEN-POPE, and R. ROSS

A. Introduction

Platelet-derived growth factor (PDGF) is a ubiquitous mitogen that was
originally discovered because it was the principal source of growth factor activity for mesenchymal cells in culture that was present in whole blood serum
and missing in cell-free, plasma–derived serum (KOHLER and LIPTON 1974;
ROSS et al. 1974). The possibility that platelets might serve as a source for this
growth factor activity had been postulated by BALK (1971). However, not until it was demonstrated that platelets could restore all of the growth factor activity missing from plasma did it become clear that these cells were the source
of not only PDGF, but of other growth factors also present in whole blood
serum. These include an epidermal growth factor-like molecule (OKA and
ORTH 1983), an angiogenic molecule (MIYAZONO et al. 1987), and transforming growth factor-β (TGF-β; ASSOIAN et al. 1983; CHILDS et al. 1982), all of
which appear to be sequestered within α-granules (WHITE 1974; WITTE et al.
1978; KAPLAN et al. 1979 a, b; GERRARD et al. 1980).

PDGF, which constitutes the major mitogenic activity in platelets, is a
growth factor principally for connective-tissue-forming cells such as dermal
fibroblasts (RUTHERFORD and ROSS 1976), glial cells (BUSCH et al. 1976;
WESTERMARK and WASTESON 1976), and arterial smooth muscle cells (ROSS et
al. 1974). As a major mitogen for connective tissue cells, PDGF may play
numerous roles in both normal biology and in disease. It may be important in
various phases of embryogenesis and development, wound repair, neoplasia,
and fibrotic responses associated with inflammatory diseases such as pulmonary fibrosis, myelofibrosis, rheumatoid arthritis, and atherosclerosis, the
disease responsible for approximately 50% of all deaths in Western society.
The term platelet-derived growth factor is now known to be inaccurate, since
it has been determined that PDGF can be formed by a large number of transformed or neoplastic cells (HELDIN et al. 1986; ROSS et al. 1986; DEUEL 1987),
as well as by many different diploid cells. The latter include activated arterial
endothelial cells (DICORLETO and BOWEN-POPE 1983), activated monocyte/
macrophages (SHIMOKADO et al. 1985; MARTINET et al. 1986), newborn rat
arterial smooth muscle cells (SEIFERT et al. 1984; NILSSON et al. 1985), stimulated fibroblasts (PAULSSON et al. 1987; RAINES et al. 1989), and
cytotrophoblasts (GOUSTIN et al. 1985).

This chapter will describe our present knowledge concerning the nature of
the molecules of PDGF; their specific cell-surface, high-affinity receptors;

binding proteins; the biochemical and cellular mechanisms of action of these molecules; and their biological activities in vitro and in vivo. It is important to note, as will be discussed in this chapter, that PDGF can be assembled in at least three isoforms: PDGF-AA, PDGF-BB, and PDGF-AB. Each of these isoforms binds to a different spectrum of receptor phenotypes, whose distribution vary with cell type. Consequently it will be important in future studies to determine which PDGF isoform is being evaluated, since emerging data (discussed below) demonstrate that their biological activities, although similar in some circumstances, may be quite different in others. Finally, potential clinical applications of this molecule will be examined in terms of the roles it may play to correct situations such as inadequate wound repair, versus the possibility of controlling its activity in neoplastic cells and in atherosclerosis.

B. PDGF Molecules

I. Multiple Forms and Amino Acid Sequence

1. Two Distinct but Homologous Chains Comprise PDGF

PDGF, as isolated from human platelets (ANTONIADES 1981; HELDIN et al. 1981 a) and platelet-rich plasma (DEUEL et al. 1981 a; RAINES and ROSS 1982), exhibits multiple molecular weight forms ranging in size from 28000 to 35000 Da, as determined by sodium dodecylsulfate–polyacrylamide gel electrophoresis (SDS-PAGE). Each PDGF molecule contains 16 cysteine residues, and reduction of PDGF produces inactive, lower molecular weight polypeptides ranging from 12000 to 18000 Da. Amino acid sequence analysis of both active PDGF and the major inactive peptides following reduction revealed two primary peptide sequences and suggested that active PDGF is composed of two distinct but homologous polypeptide chains, denoted A and B, linked by disulfide bonds (ANTONIADES and HUNKAPILLER 1983; WATER-FIELD et al. 1983; JOHNSSON et al. 1984). Analysis of the two major forms of active human PDGF demonstrated approximately equal proportions of PDGF A and B chains, but it was unclear whether PDGF is a heterodimer or a mixture of homodimers (ANTONIADES and HUNKAPILLER 1983; JOHNSSON et al. 1984). Recently it has been demonstrated that the major part of PDGF purified from human platelets occurs as a heterodimer of the A and B chains based on a combination of immunochemical, biochemical, and chromatographic behavior. Using immunoprecipitation with chain-specific antisera, susceptibility to mild acid hydrolysis and two different chromatographic systems, reversed-phase, high-performance chromatography, and immobilized metal ion affinity chromatography, HAMMACHER and coworkers (1988 a) have estimated that approximately 70% of PDGF purified from lysed human platelet pellets is PDGF-AB heterodimer, with the remainder being PDGF-BB homodimer. This is in reasonable agreement with our sequence analysis of PDGF, isolated from human platelet-rich plasma using chain-specific antibodies, which demonstrates only 5%–25% PDGF-BB homodimer with the major form of PDGF being PDGF-AB (HART et al. 1989 b). Purification and analysis of porcine PDGF demonstrated a single N-terminal sequence homologous with the B chain of human PDGF (STROOBANT and WATERFIELD 1984). PDGF-like factors have also been purified from the conditioned media of human tumor cell lines, and in several cases were found to be homodimers of chains similar to the A chain of PDGF (HELDIN et al. 1986; WESTERMARK et al. 1986; HAMMACHER et al. 1988 b). PDGF-AA has also been identified in human platelets (HART et al. 1989 b). Thus all possible dimeric combinations of PDGF chains (PDGF-AB, PDGF-BB, and PDGF-AA) have been found.

The amino acid sequences shown in Fig. 1 have been deduced from the nucleotide sequence of cloned cDNA for the A and B chains of PDGF. All eight cysteines are conserved within the mature chains, implying a similar tertiary structure. Maintenance of this tertiary structure by intrachain and/or interchain disulfide bonds is necessary for maintenance of the biological activity of PDGF (ANTONIADES 1984; HELDIN et al. 1979; RAINES and ROSS 1982). There is also a significant degree of homology between the two chains, 51% over the predicted 109 amino acids of the mature B chain (see Fig. 1). The B chain of PDGF is 93% homologous with p28^{sis}, the transforming protein of simian sarcoma virus (SSV) (DOOLITTLE et al. 1983; WATERFIELD et al. 1983; JOHNSSON et al. 1984). p28^{sis} is dimerized after synthesis and then cleaved both in the N and C termini, yielding a biologically active molecule structurally similar to B chain homodimer (ROBBINS et al. 1983). The exact processing site for the C terminus of the mature B chain is not known; however, amino acid sequencing of human PDGF suggests that the B chain ends with threonine residue 190 in the precursor sequence, resulting in a 109-amino-acid mature B chain (Fig. 1; JOHNSSON et al. 1984). The mature A and B chains are hydrophilic proteins and contain a relatively large number of basic residues, findings that are consistent with the pI (9.8-10.0) of human PDGF (HELDIN et al. 1977; ROSS et al. 1979; ANTONIADES et al. 1979; DEUEL et al. 1981 a).

The A and B chains of PDGF are both synthesized as precursors. The N-terminal region of the propeptides shares 50% homology over the first half (exon 2) but only 15% homology over the second half (exon 3), and neither contains any cysteine residues. The resultant N-terminal "pro" region of the B chain is acidic, while the A chain has approximately equivalent numbers of acidic and basic residues. In addition to a classical hydrophobic signal sequence preceding the N-terminus of the propeptides, both the A and B chains have a 12-amino-acid hydrophobic region beginning 28 and 34 residues from the N terminus of the mature A and B chain, respectively, with only one difference (Ile/Val) within this sequence (Fig. 1 B). Only the B chain also has "pro" sequences C-terminal to the mature protein which share only 10% homology with the C terminus of the long form of the A chain.

The configuration of the A chain is more complex. Analyses of multiple clones of the A chain of PDGF isolated from cDNA libraries from a human clonal glioma cell line and normal human umbilical vein endothelial cells have identified clones containing a 69 base pair (bp) deletion with the predicted C

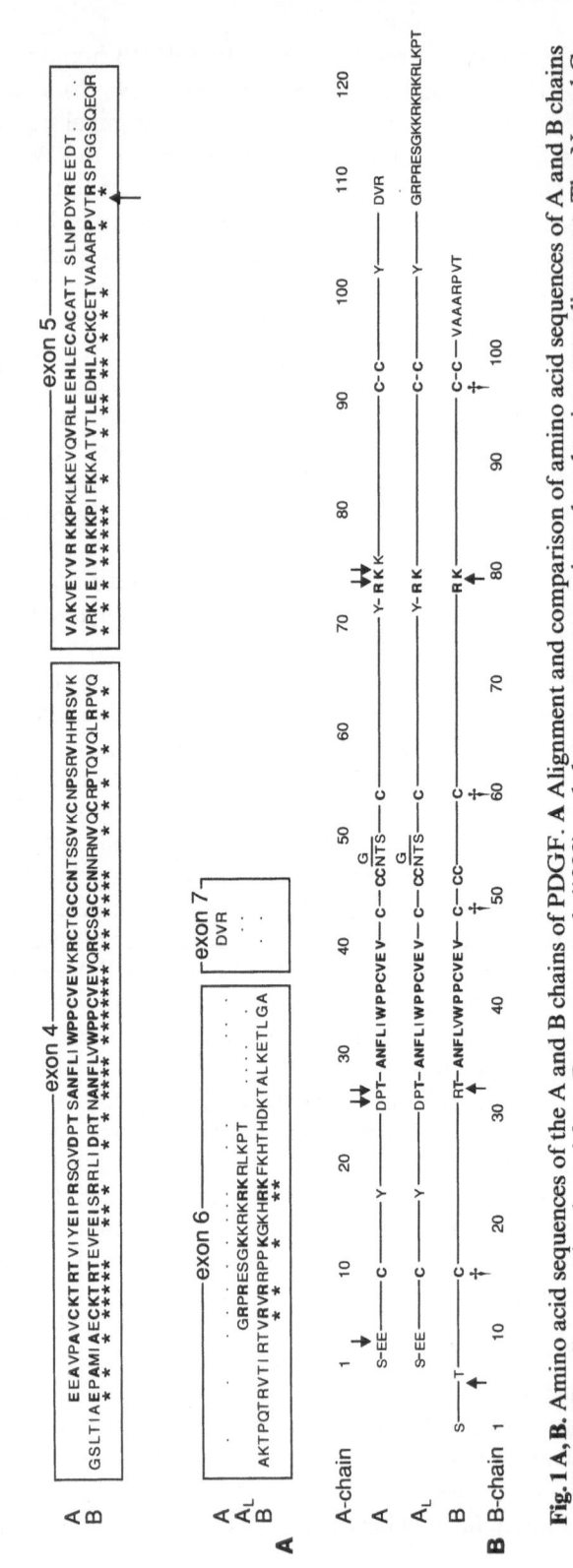

Fig. 1 A, B. Amino acid sequences of the A and B chains of PDGF. **A** Alignment and comparison of amino acid sequences of A and B chains of PDGF. The alignment is adapted from Bonthron et al. (1988) and *dots* represent gaps introduced to improve alignment. The N- and C-terminal propeptide cleavage sites are shown by *arrows*. *Boxes* indicate exons and each exon is *numbered above*. *Asterisks* indicate matching residues. For the A chain of PDGF, *A* indicates the short form of the A chain isolated from human endothelial cells (Collins et al. 1987a; Tong et al. 1987), and A_L indicates the long form of PDGF isolated from a human clonal glioma cell line (Betsholtz et al. 1986b). **B** Structural features of the mature A and B chains of PDGF. The mature A and B chains of PDGF have been aligned to highlight common and unique structural features as indicated in the text. Cleavage sites detected from N-terminal sequencing are shown by *arrows*. Potential sites of asparagine-linked glycosylation are marked by a *horizontal line* with a *G above* the asparagine residue. Cysteine residues with a *dagger below* them indicate those cysteines shown to be required for maintenance of transforming activity in *v-sis* (Giese et al. 1987; Sauer and Donoghue 1988)

terminus of the protein (BETSHOLTZ et al. 1986 b; COLLINS et al. 1987 a; TONG et al. 1987; RORSMAN et al. 1988). These clones are believed to be the result of differential splicing, and the predicted proteins differ significantly in size and charge of the C-terminal portion of the A chain. Clone D1, isolated from a human clonal glioma cell line, encodes a 211-amino-acid PDGF A-chain precursor (denoted as A_L in Fig. 1) with an extremely basic C terminus (BETSHOLTZ et al. 1986 b) whereas clones 13.1 and 15.1, also isolated from the human glioma cell line (RORSMAN et al. 1988), and the cDNA clones isolated from human umbilical vein endothelial cells (COLLINS et al. 1987 a; TONG et al. 1987; BONTHRON et al. 1988) encode an A-chain precursor which is 15 residues smaller (196 residues) and lacks the basic C terminus (see Fig. 1). Using the two types of clones in expression systems, it has been confirmed that both clones encode biologically active PDGF A-chain proteins (COLLINS et al. 1987 a; BECKMANN et al. 1988; BYWATER et al. 1988). It would be predicted that the presence of the basic region encoded by clone D1 would have profound effects on the physicochemical properties of the molecule and potentially its susceptibility to proteolytic cleavage. However, at present the only direct evidence that the 211-amino-acid form of the A chain (A_L) may be expressed remains the isolation of the original glioma clone D1 (BETSHOLTZ et al. 1986 b) and a homologous clone from a *Xenopus* oocyte library (MERCOLA et al. 1988). Comparable clones were not found in human endothelial cells (COLLINS et al. 1987 a; TONG et al. 1987). Amino acid sequence data for the A chain of PDGF from human platelets do not contain the C-terminal region in question (JOHNSSON et al. 1984). It remains possible, therefore, that in addition to PDGF-AB, PDGF-AA, and PDGF-BB (Fig. 1), homodimers and heterodimers containing PDGF-A_L may also exist.

2. Structural Heterogeneity of PDGF

Analysis of both the major active forms of PDGF and the inactive peptides derived by reduction revealed multiple N-terminal sequences in all of them (ANTONIADES and HUNKAPILLER 1983; WATERFIELD et al. 1983; JOHNSSON et al. 1984). In addition to the major sequences of the A and B chains, shortened forms of both chains were identified. Quantitative analysis of the abundance of these shortened forms derived from the major active forms of PDGF demonstrated that 25%–56% of the N-terminal sequences were the result of cleavage of N-terminal sequences (ANTONIADES and HUNKAPILLER 1983; JOHNSSON et al. 1984). With residue 1 assigned to the N-terminal residue of the mature chains (Fig. 1 B), major species resulting from N-terminal cleavage would include mature A-chain molecules beginning with Glu at residue 4, Thr at residue 27, and Lys at residue 74. For the B chain, major species would contain N termini of Thr at residue 6, Thr at residue 33, and Lys at residue 70. These cleavages significantly contribute to the size heterogeneity detected in purified PDGF. Cleavage of the B chain at Thr residue 33 and Lys residue 80 is also associated with a loss of mitogenic activity (RAINES et al., submitted).

Both the A and B chains of PDGF are derived from precursor proteins (Fig. 1 and Sect. B. V). However, no peptides corresponding to sequences from the nucleotide sequences located N-terminal (A or B chain) or C-terminal (B chain) to the mature protein have been found in PDGF purified from platelets (ANTONIADES and HUNKAPILLER 1983; WATERFIELD et al. 1983, JOHNSSON et al. 1984). Therefore, the presence of precursor chains does not appear to be a significant contributor to the observed heterogeneity of purified PDGF.

Carbohydrate analysis of purified PDGF has demonstrated that 4%–7% of PDGF is neutral or amino sugars (DEUEL et al. 1981 a). Sugar residues detected include mannose, galactose, glucosamine, galactosamine, fucose, and possibly sialic acid. A possible glycosylation site, Asn-Thr-Ser, exists at residue 48 of the mature A chain (Fig. 1 B). The discrepancy between the predicted molecular weight of the A chain (12 000) and the molecular weight estimated from SDS-PAGE (16 000–18 000) may be due to glycosylation. No sites for N-linked glycosylation exist on the mature B chain. The predicted molecular weight of the B chain (12 000) also differs from the molecular weight estimated from SDS-PAGE (14 000–16 000) and consequently must be due to O-linked glycosylation or some other posttranslational modification, or to anomalous behavior on SDS-PAGE.

3. Structural Features of the PDGF Molecules

The PDGF A chain has several unique characteristics. As mentioned above, only the mature A chain has a site for N-linked glycosylation at residue 48 (JOHNSSON et al. 1984). The A chain also has a peptide bond (Asp-Pro) at positions 25 and 26 from the N terminus of the mature protein that is susceptible to mild acid hydrolysis (HELDIN et al. 1986; HAMMACHER et al. 1988 a, b). This characteristic has been utilized to help distinguish homodimers and heterodimers (HAMMACHER et al. 1988 a). Another interesting feature of the A chain is that the Glu-Glu sequence at positions 3 and 4 of the mature A chain are also present in platelet factor 4 (DEUEL et al. 1977) and β-thromboglobulin (BEGG et al. 1978), two peptides which, like PDGF, are stored in the α-granule of the platelet. One of the shortened forms of the A chain begins with Glu at residue 4. A significant difference between the PDGF A and B chains is that the mature B chain contains no tyrosine (Fig. 1). Consequently only the A chain of PDGF is labeled in PDGF-AB with ^{125}I unless the Bolton-Hunter reagent is used (BOLTON and HUNTER 1973).

The roles of individual cysteine residues have been investigated in p28^{sis}, the transforming gene product of v-sis which is 97% homologous to the B chain of PDGF. These investigations have demonstrated that the cysteines at positions 16, 49, 60, and 97 in the mature protein are required for maintenance of transforming activity and, presumably, interaction with the PDGF receptor (GIESE et al. 1987; SAUER and DONOGHUE 1988). However, cysteines 43, 52, 53, and 99 can be mutated to serine with maintenance of transforming activity. Alteration of any of the cysteines decreases dimer formation. In addition, mutant forms of v-*sis* in which residues 1–15 or 98–109 in the mature protein have been deleted are capable of inducing transformation (SAUER et al. 1986; SAUER and DONOGHUE 1988). Removal of the first 15 residues of the mature B chain had no effect on transforming efficiency (SAUER et al. 1986), while removal of the C-terminal 11 residues resulted in a mutant with a transforming efficiency approximately 250-fold lower than that of the wild type, but still significantly above background (SAUER and DONOGHUE 1988). Removal of six C-terminal residues had no detectable effect on biological activity (HANNINK et al. 1986). Although the C-terminal 11 residues appear to be dispensable, they clearly contribute to either the stability or the receptor-binding properties of the biologically active protein.

4. Structural Conservation of the PDGF Molecule

Very little information exists on PDGF molecules for species other than humans. However, several lines of evidence suggest that the PDGF molecules are highly conserved. Direct evidence for conservation comes from DNA blotting of *Xenopus laevis* genomic DNA and N-terminal sequence analysis of porcine PDGF. Distinct A- and B-chain genes were detected in *Xenopus laevis* genomic DNA using the human A-chain cDNA probe and the v-*sis* cDNA probe (MERCOLA et al. 1988). The coding region of A-chain clones isolated from oocyte and gastrula stage cDNA libraries from *Xenopus* shows an overall sequence identity of 73% over the coding region, and 53% and 65% sequence identity in the 5′ and 3′ untranslated regions, respectively. The predicted protein products of the longest oocyte clone and the human cDNA differ significantly only by an 11-amino-acid insertion in the C-terminal region, which introduces a 9-amino-acid hydrophobic domain into the normally hydrophilic C terminus, most likely in exon 6. Interestingly, both the long and short forms of the A chain (Fig. 1) were also detected in the cDNA from *Xenopus*.

The N-terminal sequence for 11 and 15 residues of porcine PDGF is identical to that of the human B chain with three conservative substitutions (STROOBANT and WATERFIELD 1984). Interestingly, no A-chain-related sequence was detected, although the possibility of the presence of a blocked N terminus has not been eliminated. Porcine PDGF is clearly different in some respects from human PDGF. Purified porcine PDGF has an estimated molecular weight of 38 000, and it migrates in SDS-PAGE with a higher apparent molecular weight than human PDGF (STROOBANT and WATERFIELD 1984). Porcine PDGF also has much higher levels of proline, glycine, histidine, and arginine by amino acid analysis. The predominance of PDGF-BB in whole blood serum is not limited to pigs. Using the specificity of binding of PDGF isoforms (HART et al. 1988), binding assays on whole blood serum from different species demonstrated that PDGF-BB is the predominant isoform in sera from all species tested (mouse, rat, pig, cow, sheep, dog, chicken) except primates (BOWEN-POPE et al., 1989).

A phylogenetic analysis of PDGF by radioreceptor assay determined that all tested specimens of clotted whole blood from phylum Chordata contain a

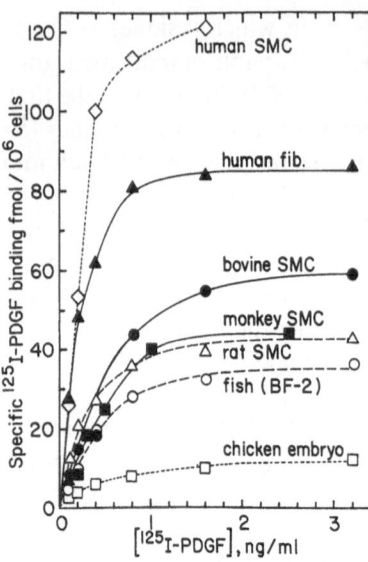

Fig. 2. Specific binding of ^{125}I-PDGF to cultured cells from different species. Monospecific cultures of each cell type were prepared in 24-well culture trays. When the cultures were confluent, the medium was replaced with culture medium containing 2% plasma-derived serum or 2% calf whole blood serum from which the PDGF had been removed by cation exchange chromatography. After 2 days, the cultures were rinsed once with ice-cold saline, then incubated for 4 h at 4° C with gentle oscillation in 1 ml of medium containing 0.25% bovine serum albumin and the concentration of ^{125}I-PDGF shown on the abscissa, with ^{125}I-PDGF plus a 100-fold excess of unlabeled PDGF to determine nonspecific binding (always less than 15% of total binding), or with medium alone to determine cell number by electronic particle counting. The cell types examined were: adult human aortic smooth muscle (*open diamond*); adult human foreskin fibroblasts (*closed diamond*): bovine aortic smooth muscle (*closed circle*); adult monkey (*Macaca nemestrina*) aortic smooth muscle (*closed square*); adult rat aortic smooth muscle (*open triangle*); bluegill fry fish cell line BF-2 (American Type Culture Collection) (*open circle*); secondary cultures of chicken embryo cells (*open square*). (From Bowen-Pope et al. 1985)

homolog capable of competing with labeled human PDGF for specific receptor binding to mouse cells (Singh et al. 1982). Sera from tunicates to lower species in the chordate line of evolution, and sera from all tested animals of the arthropod line of development were negative. The appearance of PDGF-like activity in whole blood serum does coincide with the formation of a pressurized vascular system in vertebrates (Singh et al. 1982). Labeled human PDGF is also able to bind to cells from human, mouse, rat, chicken, and fish with the same apparent affinity (see Fig. 2). These observations support significant structural homology, at least in the receptor binding site for chordates. In addition, monospecific antisera made to human PDGF recognize a PDGF analog in a wide variety of species, including rabbit, rat, mouse, chicken, and horse (Raines and Ross 1985).

A functional homolog of PDGF and its receptor appears to be present in even more primitive animals. PDGF stimulates chemotaxis and nucleic acid synthesis in the protozoan *Tetrahymena* (Andersen et al. 1984) and depresses

synthesis in the protozoan *Tetrahymena* (ANDERSEN et al. 1984) and depresses the feeding response of the small freshwater coelenterate *Hydra* elicited by *S*-methylglutathione (HANAI et al. 1987). The *Hydra* response to PDGF is eliminated by the addition of anti-PDGF immunoglobulin (IgG) or chemical reduction of PDGF, both of which prevent PDGF from binding to its cell-surface receptor on responsive cells. *Hydra* has strong regenerating potential when it is excised at its body column (WEBSTER 1971). Potent depressing activities are released from excised animals, and this activity is inhibited by anti-PDGF (K. HANAI, E. W. RAINES, and R. ROSS, unpublished observations). Though further investigations are required, PDGF may be involved in the wound healing response even in the primitive coelenterate *Hydra*.

II. Gene Structure of the A and B Chains of PDGF

1. Characteristics of the A-Chain Gene

The A-chain gene contains at least seven exons spanning 22-24 kilobases (kb) of genomic DNA (Fig. 3). However, since full-length cDNA clones for the 5' and 3' sequences have not been isolated, the possibility of additional exons cannot be excluded. The unusually long 5' untranslated region and the signal peptide constitute the first exon; exons 2 and 3 encode the N-terminal propeptide; exons 4 and 5 encode most of the mature protein; exon 6 contains 69 bp, which encode the C-terminal amino acids of the long form of the A chain (A_L); and exon 7 encodes the three C-terminal residues of the A chain deduced from endothelial cell cDNA and the 3' untranslated sequences (Fig. 1; BETSHOLTZ et al. 1986b; COLLINS et al. 1987a; TONG et al. 1987; RORSMAN et al. 1988; BONTHRON et al. 1988).

On the basis of somatic cell hybrid chromosome segregation, the gene was assigned to the proximal long arm of chromosome 7 (BETSHOLTZ et al. 1986b). By in situ hybridization, the A chain has been sublocalized to 7p21-p22 using the 1.75-kb endothelial-cell-derived cDNA clone dT1.1, which lacks 69 bp (corresponding to exon 6) in the 3' protein coding sequence (BONTHRON et al. 1988). However, using two different cDNA probes, the 1.3-kb clone D1 (which contains exon 6) and clone 13.1 (which lacks exon 6) isolated from the clonal human glioma cell line U-343 MGa, STENMAN and colleagues (1988) have sublocalized the A chain to 7q11.23. Using clone 13.1 as a probe, they noted an additional minor peak of grains at the terminal end of 7p. At present, the meaning of these differences in sublocalization is unclear. Analysis of the cDNA clones on genomic Southern blots demonstrates that a single PDGF A chain gene is present in the human genome (RORSMAN et al. 1988).

Three PDGF A chain transcripts of 1.9, 2.3, and 2.8 kb have been detected in human tumor cell lines (BETSHOLTZ et al. 1986b, 1987) and in normal human endothelial cells in culture (COLLINS et al. 1987a, b; TONG et al. 1987; BONTHRON et al. 1988). Exon 6, which contains the 69-bp region encoding the extremely basic C-terminal portion of the A chain (A_L, Fig. 1), was found in oly one of four cDNA clones isolated from the human glioma cell line

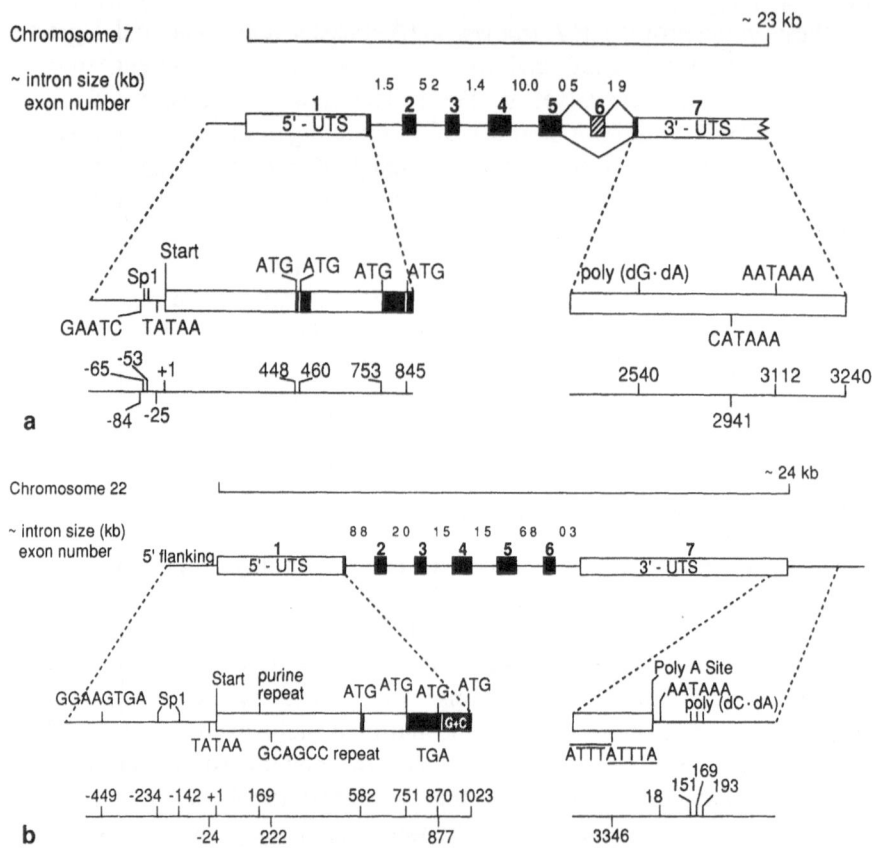

Fig. 3 a, b. The gene structure of the A and B chains of PDGF. The diagram shows the intron–exon structures of the PDGF A and B chains and indicates possible regulatory regions in the 5′ and 3′ untranslated regions, as discussed in the text. Translated sequences are indicated by *solid boxes,* except for exon 6 of the A chain (*hatched box*), which codes for the C terminus of the long form of the A chain (A_L) and is found in a minority of the PDGF A-chain transcripts. **a** PDGF A-chain gene. The numbering sequence used in this figure assigns $+1$ to the mRNA start site determined by RORSMAN et al. (1988). **b** PDGF B-chain gene. The numbering sequence used in this figure assigns $+1$ to the mRNA start site determined by RAO et al. (1988). The *shaded area* preceding the coding sequence for the propeptide indicates the $G+C$ region shown by RAO and colleagues (1988) to inhibit translation

(BETSHOLTZ et al. 1986b; RORSMAN et al. 1988) and in none of the cDNA clones isolates from normal human endothelial cells (COLLINS et al. 1987a; TONG et al. 1987). The absence of exon 6 in other isolated cDNA clones suggests alternative splicing and usage of this exon in a minority of the PDGF A-chain transcripts. This is supported by Northern blot analysis of RNA from the glioma cell line U-343 MGa clone 2:6 and other tumor cell lines and human endothelial cells. Using probes specific for exon 6, the same three transcripts were detected principally in the glioma cell line, to a lesser extent in other tumor cell lines that hybridized with similar intensities when probed with the cDNA probe (RORSMAN et al. 1987). Thus it is possible that the alternative PDGF A-chain (A_L) mRNA occurs only in certain tumor cells in humans, although it has also been found in *Xenopus* oocytes (MERCOLA et al. 1988).

An unusual feature of the A-chain mRNA is the presence of a long 5′ untranslated region containing three AUG triplets upstream of the authentic initiator codon (Fig. 3; BONTHRON et al. 1988; RORSMAN et al. 1988). The first two AUG are in close proximity in the same reading frame and are closely followed by a stop codon. The third AUG could encode a 31-amino-acid peptide which would terminate two nucleotides before the AUG encoding the precursor A chain. If the third coding region were translated, backward movement of the ribosome by one nucleotide would be required to reinitiate translation of the precursor A chain (KOZAK 1986). It would be predicted that the complicated leader sequence might impair translation. The 5′ untranslated regions would also be expected to have substantial secondary structure due to their high $G + C$ content, and therefore negatively affect translation (HUNT 1985). The 5′ end of the A chain also fits the criteria for islands of nonmethylated, CpG-rich DNA, which have been described in the 5′ end of a number of genes (BIRD et al. 1985). It has been proposed that these areas may be protected from methylation and located near areas of regulatory significance. A putative promoter region is found within this area including a TATAA sequence 870 bp upstream from the initiation site for translation of the A chain (BETSHOLTZ et al. 1986b; RORSMAN et al. 1988; BONTHRON et al. 1988). Primer extension analyses place the transcriptional start site approximately 25-36 bp downstream of the TATAA box (RORSMAN et al. 1988; BONTHRON et al. 1988). The 5′ untranslated region of the A chain also contains more than one copy of the consensus binding sequence for transcription factor Sp1 (BRIGGS et al. 1986) upstream of the cap site. The inability to detect additional extrension products in the 370 bp downstream of this site suggests that multiple transcriptional start sites downstream from the TATAA are unlikely to account for the heterogeneity in the A-chain transcript size (BONTHRON et al. 1988).

A more likely source of the A-chain mRNA heterogeneity is alternate use of polyadenylation sites or alternative splicing of exons other than exon 6. An authentic polyadenylation signal (AATAAA) is situated 182 bp and 234 bp downstream of the end of cDNA clones isolated from glioma and endothelial cells, respectively (RORSMAN et al. 1988; BONTHRON et al. 1988). Two AACAAA sequences were found in the cDNA clones immediately following a $(dGA)_{12}$ repeat. Whether any of these are used as polyadenylation signals remains to be determined.

2. Characteristics of the B-Chain Gene

The complete extent of the human B-chain transcript has been defined by a combination of cDNA cloning, nuclease S1 mapping, and primer extension analysis (CHIU et al. 1984; JOSEPHS et al. 1984; COLLINS et al. 1985; RAO et al. 1986). The PDGF B chain is encoded by the c-*sis* protooncogene, the cellular homolog of the oncogene transduced by both the SSV and Parodi-Irgenes feline sarcoma virus (GELMANN et al. 1981; BESMER et al. 1983; DEVARE et al. 1983; DOOLITTLE et al. 1983; WATERFIELD et al. 1983; CHIU et al. 1984; JOHNSSON et al. 1984; JOSEPHS et al. 1984). The B chain also contains seven exons spanning 24 kb of genomic DNA and has a number of similarities with the A-chain gene (Fig. 3 B). The first exon contains the long 5′ untranslated region (1022 bp) and the signal sequence; the propeptide and the first two amino acids of the mature protein constitute exons 2 and 3; exons 4 and 5 encode the mature chain plus nine amino acids of the C-terminal propeptide (assuming the C terminus of the mature protein is threonine 109 after carboxypeptidase removal of arginine 110; JOHNSSON et al. 1984); exon 6 encodes the remainder of the C-terminal propeptide; and exon 7 encodes the entire 3′ untranslated region, which is also unusually long (1625 bp).

Analyses of somatic cell hybrids have allowed assignment of the B chain of PDGF to human chromosome 22 (DALLA FAVERA et al. 1982; SWAN et al. 1982). Hybrids between thymidine kinase-deficient mouse cells and human fibroblasts, carrying a translocation of the region q11-qter of chromosome 22 to chromosome 17, demonstrated further localization on region 22q11-qter (DALLA FAVERA et al. 1982). Southern blot analysis using a cDNA probe encompassing exon 4 detected a single band, establishing that the PDGF B chain exists as a single locus within the human genome (CHIU et al. 1984).

The PDGF B chain is transcriptionally active in several normal cell types as well as a number of transformed cell lines. Using a v-*sis* cDNA probe, a 4.2-kb transcript was shown to be expressed in some glioblastomas and fibrosarcomas, but not in the normal counterparts of such tumors (EVA et al. 1982), and in the neoplastic T-cell line HUT 102 (WESTIN et al. 1982). Northern analysis of cytoplasmic RNA from cultured human umbilical vein endothelial cells, HUT 102 cells, and a human osteosarcoma cell line using the v-*sis* or c-*sis* probe demonstrated a prominent band of 3.5-3.7 kb and minor bands of 2.6-2.7 kb and 1.3 kb, but no hybridizing bands were detected in RNA from human fibroblasts or a B-lymphoblastoid cell line (BARRETT et al. 1984; COLLINS et al. 1985). The apparent discrepancy in transcript size is likely to be due to differences in method, reference markers, and gel systems.

Within the long 5′ untranslated region of the B chain, specific sequences have been defined which may be important in the regulation of transcription and translation (RATNER et al. 1987; RAO et al. 1988). Two features of the 5′ untranslated region are unusual: it is highly rich in $G+C$ (70% overall, 90% for positions 331–450, and 85% for positions 840–1022, the area immediately upstream of the PDGF coding region), and translation begins at the fourth most 5′ proximal ATG codon (Fig. 3 B). The three upstream ATG codons initiate open reading frames of 5, 42, and 10 codons, respectively. Interestingly, both the v-*sis* (the transforming retrovirus) and the HUT 102 cDNA (capable of experimental transformation) delete the other upstream ATG codons and the $G+C$-rich region immediately preceding the coding region (DEVARE et al. 1983; HANNINK and DONOGHUE 1984; CLARKE et al. 1984). The sequences surrounding all four of the ATG codons poorly match the consensus signal for translational initiation (KOZAK 1986).

Other features of the 5' untranslated region (Fig. 3 B) include a cluster of specific short repeat sequences (between positions 222 and 266) and a polypurine tract of 23 nucleotides (between positions 162 and 185). A TATAA box was identified 24 bp upstream of the proposed transcriptional start site using mRNA isolated from human placental and EJ tumor cells (RAO et al. 1986). Other studies have assigned the cap site within close proximity based on analysis of feline and human transcripts (VAN DEN OUWELAND et al. 1987) and a clone from human mononuclear cells (RATNER et al. 1985, 1987). In all cases, the reported start site is unique. Transfection experiments, using constructs containing 4 kb of genomic DNA upstream of the proposed cap site fused to the bacterial CAT coding sequence, demonstrated that promoter activity was present in this DNA fragment. Deletion from position -220 to -7, containing the TATAA box, induced negligible levels of CAT activity (RAO et al. 1988). In agreement with these findings, significant transcriptional promoter activity was found when positions -379 to 23 were cloned into a CAT construct (RATNER et al. 1987). A minimal promoter region, including sequences extending only 42 bp upstream of the TATA signal, has been identified in uninduced K562 cells (a hematopoietic cell line; PECH et al. 1989). Analyses of deletion mutants in the promoter region have identified both positive and negative regulatory elements in phorbol ester-induced expression of the PDGF B-chain of these cells. The endogenous activity of the B-chain promoter in their studies of cells transfected with the minimal promoter sequence suggests it is constitutively active in uninduced K562 cells and normal fibroblasts. However, the lack of detectable PDGF B-chain transcripts in both of these cells suggests that other levels of regulation must exist. At -234 and -142 (using the start site of RAO et al. 1986), a decanucleotide sequence which is identical to the Sp1 binding consensus sequence was identified. Located at position -449 is a purine-rich region containing the sequence GGAAGTGA, which is identical to the sequence in the adenovirus E1a enhancer and is also found in human interferon-α and -β genes (HEARING and SHENK 1983; GOODBOURN et al. 1985).

More detailed studies of the role of particular sequences of the 5' untranslated region have demonstrated a potent inhibitory effect on translation (RATNER et al. 1987; RAO et al. 1988). RATNER and collaborators (1987) used a wheat germ translation system to analyze the effect of various sequences on translation efficiency in vitro. Deletion of sequences from positions -14 to 886 or 136 to 970 (using the start site of RAO et al. 1986) resulted in the highest level of translation as determined by evaluation of PDGF-immunoprecipitable protein by SDS-PAGE. Deletion of -14 to 769 generated approximately 10-fold less PDGF product than the larger deletions. Both RNA expression and protein translation were evaluated for various deletions in the 5' untranslated region by RAO and coworkers (1988). Utilizing both the PDGF promoter region and the SV40 early promoter with a CAT reporter system, they demonstrated that the 5' untranslated region has no effect on RNA expression but could inhibit translation by up to 30-fold. Addition of the SV40 enhancer was able to partially overcome the inhibition. The 140-bp,

G + C-rich sequence immediately upstream of the PDGF open reading frame was nearly as effective as the entire 5' untranslated region in inhibiting translation. However, none of the individual deletions was as effective as deletion of the entire 5' untranslated region. The role of the upstream AUG sites was also evaluated with deletion constructs. The presence of three, two, or one of the upstream AUG sites has no observable effect, either negative or positive, on translation. Together, these studies suggest that the primary cause of inhibition is the strong secondary structure in the 5' untranslated region and particularly localized to the 140-bp, G + C-rich region immediately preceding the PDGF open reading frame (Fig. 3 B). There is growing evidence that expression of genes may be inhibited at the translational level by stable secondary structures in mRNAs in the vicinity of the initiator ATG codon (Kozak 1986).

Sequence analysis of the 3' noncoding region of the genomic DNA revealed a consensus polyadenylation signal, AATAAA, 18 bp downstream of the site of polyadenylation observed in the cDNA clones (Rao et al. 1988). Other characteristics of the 3' untranslated region include the putative enhancer-like poly(dC·dA), or "TG elements" (Hamada et al. 1984a, b) and the repeat ATTTA which are present in a 3' untranslated region of the mRNA for certain lymphokines, cytokines, and protooncogenes, and appear to mediate selective mRNA degradation (Shaw and Kamen 1986).

3. Comparison of the A- and B-Chain Genes

If exon 1 encodes all of the 5' untranslated region of the A chain and exon 7 encodes all of the 3' untranslated region, then the A- and B-chain genes are of similar size and share similar exon structures, with the principle differences being the use of exon 6 and the size of the introns (Figs. 1, 3). The close structural relationship suggests that they stem from a common ancestral gene.

III. Expression and Secretion of PDGF by Normal Cells

1. PDGF is Expressed at Low or Undetectable Levels in Normal Cells

As shown in Table 1, PDGF A and/or B chain is expressed in cells from epithelial tissue, connective tissue, muscle, nervous tissue, and cells of hematopoietic origin. However, the expression of PDGF in normal cells appears to be tightly regulated. Endothelial cells scraped from fresh bovine aorta or human umbilical vein contain only 10% and 1.3%, respectively, as much B-chain-hybridizing mRNA as do the same cells grown in culture (Barrrett et al. 1984). Further, under culture conditions in which the cells organize into tube-like structures more closely approximating their state in vivo, expression of both the A and B chain of PDGF decreases (Jaye et al. 1985; Tong et al. 1987). Similarly, freshly isolated monocytes or alveolar macrophages from normal nonsmokers express low or undetectable levels of PDGF B chain as compared with activated monocytes or alveolar macrophages (Mornex et al. 1986). Analysis of whole tissues, including normal carotid arteries and

Table 1. Tissues and normal cultured cells: expression of PDGF molecules and their receptors

Cell type	PDGF molecules			PDGF receptors				Citations
	PDGF	mRNA A	mRNA B	Binding of AB	AA	BB	mRNA β	
Endothelial/epithelial cells								
Endothelial cells								
Aortic, bovine	+	−	++	−	−			DiCorleto and Bowen-Pope (1983), Barrett et al. (1984), Collins et al. (1987b), Bowen-Pope and Ross (1982), Kazlauskas and DiCorleto (1985)
Aortic, monkey	+	+	+	−	−			Our unpublished data
Hyperplastic, human in glioblastoma multiforme							+	Hermansson et al. (1988)
Iliac, human	++	++	++					Sitaras et al. (1987)
Renal microvascular, human	++	++	++					Daniel et al. (1986), Starksen et al. (1987), Barrett et al. (1984)
Umbilical vein, human	+	+	+	−				Collins et al. (1987a, b), Heldin et al. (1981b), Our unpublished data
Venous, human	+	+	+				−	Limanni et al. (1988)
Epithelial cells								
Kidney, African green monkey (BSC-1)	+	−	+				−	Kartha et al. (1988)
Kidney, canine (MDCK)				−				Our unpublished data
Melanocytes, human	−							Westermark et al. (1986)
Retinal pigment, rat	+						−	Campochiaro et al. (1989)

Table 1 (continued)

Cell type	PDGF molecules			PDGF receptors				Citations
	PDGF	mRNA A	mRNA B	Binding of AB	AA	BB	mRNA β	
Mesothelial cells								
Pleural, human		+	−					Gerwin et al. (1987)
Trophoblast cells								
Cytotrophoblasts, human (first-trimester placenta)				+				Goustin et al. (1985)
Connective tissue cells								
Chondrocytes								
Articular, porcine				+				Our unpublished data
Fibroblasts								
Bone marrow, human				+				Bowen-Pope et al. (1984b), Rosenfeld et al. (1985)
Dermal, human				+				Heldin et al. (1981b), Rosenfeld et al. (1984)
				+	+	+		Hart et al. (1988)
				+	+	+		Seifert et al. (1989)
							+	Gronwald et al. (1988)
+ interleukin 1	+	+						Raines et al. (1989)
Embryo, chicken			−	+		+		Bowen-Pope et al. (1985)
Foreskin, human				+	+			Heldin et al. (1981b), Bowen-Pope and Ross (1982)
Foreskin, human + EGF or PDGF	+	+	−	+				Heldin et al. (1988), Hart et al. (1988)
							+	Claesson-Welsh et al. (1988)
								Paulsson et al. (1987)
Lung, human				+				Bowen-Pope et al. (1984b)

Cell type						Reference
Fibroblastoid cell lines						
BF-2, fish				+		Bowen-Pope et al. (1985)
BHK, hamster				−	−	Gronwald et al. (1988)
C3H10T 1/2, mouse				+	−	Bowen-Pope et al. (1984b)
CHO, hamster				−	−	Claesson-Welsh et al. (1988)
NR6, mouse				+		Bowen-Pope and Ross (1982)
NRK, rat				+		Bowen-Pope et al. (1984b)
3T3, mouse				+		Heldin et al. (1981b), DiCorleto and Bowen-Pope (1983)
BALB/c 3T3, mouse			+	+		Bowen-Pope et al. (1984b)
NIH 3T3, mouse			+	+		Bowen-Pope et al. (1984b)
Swiss 3T3, mouse			+	+	+	Seifert et al. (1989)
			+	+	+	Bowen-Pope and Ross (1982), Huang et al. (1982)
				+	+	Seifert et al. (1989)
Muscle cells						
Myoblasts						
Skeletal, rat (L6J1)	+	+				Sejersen et al. (1986)
Primary, rat		+				Sejersen et al. (1986)
Vascular smooth muscle						
Aortic, baboon				+		Valente et al. (1988)
Aortic, bovine	−	+	+			Williams et al. (1982), DiCorleto and Bowen-Pope (1983)
Aortic, human	+	+		+	+	Libby et al. (1988)
				+	+	Bowen-Pope and Ross (1982)
Aortic, monkey				+	+	Our unpublished data
Aortic, porcine				+		Bowen-Pope and Ross (1982)
						Heldin et al. (1981b), DiCorleto and Bowen-Pope (1983)
Aortic, adult rat	±	+		+		Seifert et al. (1984)
	±	+				Majesky et al. (1988)
	+					Sjolund et al. (1988)

Table 1 (continued)

Cell type	PDGF molecules			PDGF receptors				Citations
	PDGF	mRNA A	mRNA B	Binding of AB	AA	BB	mRNA β	
Aortic, adult rat	+	+	−	+	+	+	+	SEJERSEN et al. (1986), R. A. SEIFERT et al. (unpublished data)
"neointimal"	++			+	+		+	WALKER et al. (1986)
Aortic, newborn rat	++	+	+	−	−	+	+	SEIFERT et al. (1984), MAJESKY et al. (1988), R. A. SEIFERT et al. (unpublished data)
Carotid, human	++	++	−					LIBBY et al. (1988)
Femoral, human	++	++	−					LIBBY et al. (1988)
Other smooth muscle								
Mesangial cells, rat	++	+		+				ABBOUD et al. (1987)
Mesangial cells, human	++	+	+					SHULTZ et al. (1988)
Nervous tissue cells								
Astrocytes								
Type I, rat	+	+	+					NOBLE et al. (1988), RICHARDSON et al. (1988)
Glial cells								
Human				+				HELDIN et al. (1981b)
Meningeal cells								
Primary, rat	−	+	−					RICHARDSON et al. (1988)
Hematopoietic cells								
Erythrocytes								
Peripheral blood, baboon				−				BOWEN-POPE et al. (1984a)

Cell / tissue				References
Lymphocytes				
Peripheral blood, human			—	Heldin et al. (1981b)
Macrophages				
Alveolar, human + LPS	+		+	Shimokado et al. (1985), Mornex et al. (1986)
Peritoneal, human	+		+	Shimokado et al. (1985)
Monocytes				
Peripheral blood, freshly isolated, human	—	—	—	Claesson-Welsh et al. (1988); Martinet et al. (1986); —
Cultured + LPS	—	—	+	Our unpublished data; Martinet et al. (1986)
Cultured + PMA	+	+	+	Sariban and Kufe (1988)
Tissues				
Aorta				
Human	+	+	+	Barrett and Benditt (1988)
Monkey	+	+	+	Barrett and Benditt (1987)
Newborn rat media	+	+	+	Our unpublished data
Adult rat media	±	±	±	Majesky et al. (1988); Majesky et al. (1988); Sjolund et al. (1988)
Carotid				
Intima (endarterectomy), human	+	+	+	Barrett and Benditt (1987)
Normal media, rat	±	±		Barrett and Benditt (1987)
Neointimal cells, rat (after balloon injury)	±	±		Barrett and Benditt (1987)
Embryo				
Unfertilized oocyte, mouse	+	+	+	Rappollee et al. (1988)
Blastocyst, mouse	+	+		Rappollee et al. (1988)
Xenopus	—			Mercola et al. (1988)

Table 1 (continued)

Cell type	PDGF molecules			PDGF receptors					Citations
	PDGF	mRNA A	mRNA B	Binding of			mRNA β		
				AB	AA	BB			
Endothelium									
Bovine			±						BARRETT et al. (1984)
Human umbilical vein		+	±						BARRETT et al. (1984)
Placenta									
Second trimester, human	+	+	+				+		GOUSTIN et al. (1985)
									TAYLOR and Williams (1988)

LPS, endotoxin; PMA, phorbol myristic acetate.

thoracic aorta (from human, rhesus monkey, and rat), have demonstrated mRNA levels 1%–5% of those seen with activated cultured cells from these tissues (BARRETT and BENDITT 1987; MAJESKY et al. 1988). Low levels of B chain have also been reported in human heart and kidney (BARRETT and BENDITT 1987). A-chain mRNA is also expressed in freshly isolated rat and human aortic medial smooth muscle cells at levels of 5%–30% of those observed in the same cells in culture (BARRETT and BENDITT 1988; MAJESKY et al. 1988; SJOLUND et al. 1988). For both genes, a gradient of expression was observed across the normal vessel wall. PDGF A-chain transcripts were primarily associated with the smooth-muscle-rich media, and the B-chain transcripts were principally expressed in the adventitia and associated with cell-type-specific markers characteristic of endothelial cells and macrophages of the microvasculature of the adventitia and vasa vasorum (BARRETT and BENDITT 1988).

2. Inducible Expression and Secretion of PDGF

The increased expression of PDGF in cultured normal cells, as compared with freshly isolated cells, and the decreased expression in endothelial cell cultures associated with the formation of tube-like structures suggest that the PDGF genes are expressed in response to perturbations associated with culturing. Differences in the matrix components to which the cells are attached and metabolic alterations in response to the different culture media may be involved in PDGF gene induction. The resulting expression of PDGF and secretion of PDGF activity into the media is constitutive. In the case of endothelial cells, this constitutive secretion of PDGF is almost exclusively into the basal compartment (ZERWES and RISAU 1987). The level of constitutive expression and secretion of PDGF does appear to be affected by the state of the cells when isolated in vivo. Medial smooth muscle cells isolated from 13- to 18-day-old rats (pups) but not 3-month-old animals (adults) (SEIFERT et al. 1984; MAJESKY et al. 1988) and intimal smooth muscle cells isolated from rat carotid arteries 2 weeks after injury with a balloon catheter (WALKER et al. 1986) secreted higher levels of PDGF into the media. The pup rat smooth muscle cells were shown to accumulate higher levels of B-chain mRNA in culture than adult cells with apparently equivalent levels of A-chain transcripts (MAJESKY et al. 1988). In the pup smooth muscle cells, the increased secretion of PDGF in culture was stable over up to 20 passages in vitro (SEIFERT et al. 1984).

The expression and secretion of PDGF by cultured normal cells can also be modulated by a number of factors. For rat smooth muscle cells, the constitutive induction of PDGF A chain can be further modulated by alteration of the substrate that the cells are grown on and cell density (SJOLUND et al. 1988). In cultured endothelial cells, constitutive secretion of PDGF activity is decreased by certain modified low-density lipoproteins and fish oils, both being dependent on free radical oxidative processes for inhibition (FOX and DICORLETO 1986, 1988; FOX et al. 1987), and by forskolin, a direct activator of adenylate cyclase (DANIEL et al. 1987). All of these examples appear to result in "stable" alteration of the constitutive level of expression and/or secretion of PDGF.

A number of examples also exist for transient induction of PDGF in response to physiologic mediators that would be present at sites of injury, such as coagulation factors, transforming growth factor-β (TGF-β, released from activated platelets and monocyte/macrophages), and tumor necrosis factor-α (TNF-α, released by activated macrophages). Physiologic concentrations of thrombin and activated factor X, two products of the coagulation cascade, increase the rate of production of PDGF activity by up to 9-fold (HARLAN et al. 1986; GAJDUSEK et al. 1986). Release of a significant amount of PDGF activity is observed by 1.5 h in response to thrombin (HARLAN et al. 1986) and precedes induction of PDGF A- and B-chain transcripts, which is maximal at 4 h (DANIEL et al. 1986; STARKSEN et al. 1987). The release of PDGF observed by 1.5 h after thrombin addition is not diminished by inhibition of protein synthesis (HARLAN et al. 1986). These data suggest that thrombin stimulates either the release of PDGF from an intracellular storage pool, or stimulates the conversion of preformed precursor into active, secreted PDGF. Thrombin also increases the PDGF A- and B-chain transcript levels with maximal stimulation 4 h after addition of thrombin (DANIEL et al. 1986; STARKSEN et al. 1987). By nuclear run-on analysis, thrombin stimulated B-chain transcription rate and had little or no detectable effect on A-chain transcription (KAVANAUGH et al. 1988). Thus thrombin has two transient effects on the release of active PDGF: an initial release from an intracellular storage pool, and a transient increase in transcript levels which is followed by secretion of PDGF.

The effect of thrombin is mimicked by exposure of microvascular endothelial cells to biologically active phorbol esters (PMA) (DANIEL et al. 1986). However, the relative efficacies are different: PMA is the most effective in inducing B-chain transcripts, while thrombin more effectively increases A-chain transcripts (STARKSEN et al. 1987). TGF-β increases the transcriptional rate of both the A and B chains of PDGF (KAVANAUGH et al. 1988). As compared with thrombin and PMA, TGF-β was most effective in increasing the level of A-chain transcripts in microvascular endothelial cells, increasing levels 25-fold above control levels. However, TGF-β is less effective than either PMA or thrombin in increasing B-chain transcripts (STARKSEN et al. 1987). The enhanced expression of A- and B-chain transcripts by TGF-β is also prolonged (peak 4 h, duration 48 h) as compared with thrombin and PMA (peak 4 h, duration 8 h) (STARKSEN et al. 1987). These differences suggest that the forms of PDGF produced by endothelial cells vary with the inducing stimulus. This possibility has not been confirmed at the protein level. TNF-α addition to human umbilical vein endothelial cell cultures also induces a transient increase in A- and B-chain transcription followed by increased secretion of PDGF activity (HAJJAR et al. 1987; our unpublished observations).

Basal constitutive expression, as well as transient induction, of the PDGF genes is attenuated by agents that increase cellular cyclic AMP levels (DANIEL et al. 1987; STARKSEN et al. 1987). Forskolin, a direct activator of the catalytic subunit of adenylate cyclase, blocks basal transcription of PDGF B chain in microvascular endothelial cells, but has no demonstrable effect on low basal levels of PDGF A-chain mRNA (STARKSEN et al. 1987). Pretreatment of cultures with forskolin prior to addition of thrombin and TGF-β blocked the increases in B-chain transcriptional rate, but did not inhibit increased A-chain transcription induced by TGF-β (KAVANAUGH et al. 1988).

A- and B-chain transcript stability also appears to vary with the state of the cells. PDGF A- and B-chain transcipts are not detectable in freshly isolated monocytes (SARIBAN and KUFE 1988). However, treatment of the monocytes with cycloheximide resulted in appearance of PDGF B-chain mRNA only. Both PDGF A chain and B chain are induced in monocytes by PMA, and the combination of PMA and cycloheximide resulted in further increases in both PDGF A- and B-chain transcripts.

PDGF itself can induce expression of its mRNA in cultured fibroblasts (PAULSSON et al. 1987). Neither the A nor the B chain of PDGF is normally expressed by cultured fibroblasts. However, following the addition of PDGF or epidermal growth factor (EGF), PDGF A-chain mRNA is transiently induced and PDGF activity is released into the media. Interleukin-1 (IL-1), a potent lymphokine released by activated macrophages, also induces transient expression of PDGF A-chain mRNA with subsequent release of PDGF-AA (RAINES et al. 1989). Addition of anti-PDGF antibodies completely inhibited the mitogenic activity previously ascribed to IL-1.

IV. Expression and Secretion of PDGF in Transformed Cells

Secretion of PDGF by transformed cells is very common and includes cell types responsive to PDGF as well as PDGF-unresponsive cell types (Table 2, and reviewed in Ross et al. 1986). In vitro transformation of cells by a number of transforming agents also results in secretion of PDGF activity into the media (BOWEN-POPE et al. 1984b).

Analysis of clonal variation in the production of PDGF and expression of PDGF receptors in human malignant glioma demonstrated considerable variation (NISTER et al. 1986). Long-term culturing and cloning resulted in selection of sublines with a higher level of growth factor release, implying that the production of PDGF is of selective growth advantage. Analysis of a large number of human malignant glioma cell lines for PDGF mRNA, PDGF secretion, PDGF-AB binding, and expression of the PDGF β-subunit demonstrated considerable variability in expression in different cell lines (Table 1, and NISTER et al. 1988b). No amplification or structural rearrangements of the genes, as determined by Southern blot hybridization, could explain the varying expression of PDGF A- and B-chain transcripts. These data suggest a more complex situation in which both autocrine and paracrine growth stimulation would be important in the growth of tumor cells.

Expression and secretion of PDGF in transformed cells not responsive to PDGF is also extremely variable (Table 2). If PDGF has any effect on these cells, it must be an indirect one. In vitro evidence exists for an indirect effect of PDGF production on the clonal growth of the erythroleukemic cell lines

Table 2. Tumor cells: expression of PDGF molecules and their receptors

Cell type	PDGF molecules			PDGF receptors				Citations
	PDGF	mRNA A	mRNA B	Binding of AB	AA	BB	mRNA β	
Tumors of epithelial cells (carcinomas)								
Bladder, human								
T-24	+		+	−				BOWEN-POPE et al. (1984b), Our unpublished data
EJ	+	+	+					IGARASHI et al. (1987)
Colon, human								
COLO-201, COLO-205	+	+	+				−	SARIBAN et al. (1988b)
Epidermoid, human								
A431	+			−			−	HELDIN et al. (1981b), HUANG et al. (1982), BOWEN-POPE and ROSS (1982), YARDEN et al. (1986), GRONWALD et al. (1988)
Gastric, human								
KATO III	+	+	+				−	SARIBAN et al. (1988b)
Liver, human								
Hep G2	+ +	+	+	−				BOWEN-POPE et al. (1984b), Our unpublished data
Lung, human								
CALU-1	+	+	+				−	SARIBAN et al. (1988b)
Squamous, U-1754	+							HELDIN et al. (1981b)
U-1810	+	+	+	−				BETSHOLTZ et al. (1987)

Cell line					Reference
Mammary, human					
BT-20, MCF-7, ZR-75-1	+	+	+	–	SARIBAN et al. (1988b)
CAMA-1	+	+	(+)		PERES et al. (1987)
MCF-7, MDA-MB-231	+	+	+		PERES et al. (1987), BRONZERT et al. (1987)
MDA-MB-157, T47D, ZR-75-1, HBL-100, BT-20	+	–	+		PERES et al. (1987), SARIBAN et al. (1988b)
MDA-MB-134	++	–	–		PERES et al. (1987)
MDA-MB-468	+	–	+		PERES et al. (1987)
Melanoma, human					
SW-691, WM-9	–	–	–		WESTERMARK et al. (1986)
WM-115	++	+	+		WESTERMARK et al. (1986)
WM-239A, WM-266-4	+	+	–		WESTERMARK et al. (1986)
Mesothelioma, human					
DND, JMN, MT-1, MT-3, VAMT-1	+	+	+		GERWIN et al. (1987)
HUT-28	+	++	+		GERWIN et al. (1987)
HUT-290	++	–	–		GERWIN et al. (1987)
Ovarian, human					
ARM, DUN, MAC, SAM	+	+	+	–	SARIBAN et al. (1988b)
Thyroid, human					
SW136				–	HELDIN et al. (1981b)
Tumors of connective tissue					
Fibrosarcoma, human					
HT-1080	+				PANTAZIS et al. (1985)
Osteosarcoma, human					
MG-63	+	+	+		BISHAYEE et al. (1986); WOMER et al. (1987); SEIFERT et al. (1989)
4-393OS	+				HELDIN et al. (1981b)

Table 2 (continued)

Cell type	PDGF molecules			PDGF receptors				Citations
	PDGF	mRNA A	mRNA B	Binding of AB	AA	BB	mRNA β	
U-2OS	+ +							BETSHOLTZ et al. (1983)
								HELDIN et al. (1980b, 1986)
U-1810	+	+	+	+			+	NISTER et al. (1988b)
								COLLINS et al. (1987a)
Transformed connective tissue cells								
SV40-transformed BHK, WI-38, 3T3	+		+					BETSHOLTZ et al. (1985)
	+							STROOBANT et al. (1985)
	+							BOWEN-POPE et al. (1984b)
	+							BLEIBERG et al. (1985)
SSV-transformed 3T3, NRK	+			−				DEUEL et al. (1983)
	+							OWEN et al. (1984)
	+							ROBBINS et al. (1983)
	+							BOWEN-POPE et al. (1984b)
Adenovirus-transformed rat embryo cells	+							BOWEN-POPE et al. (1984b)
Maloney murine sarcoma virus-transformed BALB/3T3	+			−				BOWEN-POPE et al. (1984b)
Methylcholanthrene-transformed mouse line (C3H/MCA C115)	−			+				BOWEN-POPE et al. (1984b)
Tumors of mixed cells								
Nephroblastoma								
Wilms' tumor, human	+	+	−					FRAIZER et al. (1987)

Cell / tissue						References
Teratocarcinoma						
Mouse endoderm-like PC13, F9, PSA5E, PYS2	+					GUDAS et al. (1983), RIZZINO and BOWEN-POPE (1985)
After differentiation	±			+		GUDAS et al. (1983), RIZZINO and BOWEN-POPE (1985)
Tumors of muscle cells						
Rhabdomyosarcoma, human						
RD	+					BETSHOLTZ et al. (1983)
Tumors of nervous tissue						
Glioblastoma, human						
4-251 MG	+	+	+	+		HELDIN et al. (1981b)
Gliomas						
157	+	++	+.	+	+	RICHARDSON et al. (1988)
U-105 MG, U-118 MG, U-343 MG, U-343 MGa, U-373 MG	−	++	−	+	−	NISTER et al. (1988b)
U-138 MG, U-489 MG	++	++	+	+	++	NISTER et al. (1988b)
U-178 MG, U-1242 MG	++	++	+	+	+	NISTER et al. (1988b)
U-251 AgCl, U-563 MG	−	(+)	+	+	−	NISTER et al. (1988b)
U-251 MGO	+	+	+	+	−	NISTER et al. (1988b)
U-251 MGsp	−	+	+	±	−	NISTER et al. (1988b)
U-343 MGaCl2	+	+	+	±	−	BETSHOLTZ et al. (1983), HAMMACHER et al. (1988b), NISTER et al. (1988b)
U-372 MG, U-399 MG, U-410 MG, U-1796 MG	−	+	+	+	+	NISTER et al. (1988b)
Neuroblastomas						
Human SH	+		−			HELDIN et al. (1981b)
Mouse 2A	+	+				VAN ZOELEN et al. (1985)

Table 2 (continued)

Cell type	PDGF molecules			PDGF receptors				Citations
	PDGF	mRNA A	mRNA B	Binding of			mRNA β	
				AB	AA	BB		
Tumors of hematopoietic cells								
Erythroleukemic cells								
HEL, human + PMA	+	±	+	—				Papayannopoulou et al. (1987), Weich et al. (1987)
K562, human + PMA	+	+	+	—				Papayannopoulou et al. (1987), Alitalo (1987)
+ butyrate		—	+					Alitalo et al. (1987)
Promyelocytic								
HL-60, human + PMA		++	++					Alitalo et al. (1987), Pantazis et al. (1986), Sariban et al. (1988a)
		++	++					
Promonocytic								
U-937 + PMA		+						Alitalo et al. (1987)

OCIM2 and HEL (PAPAYANNOPOULOU et al. 1987). Both of these cell lines secrete PDGF, and their growth is enhanced in the presence of fibroblasts, presumably due to secretion of a mediator by the fibroblasts in response to stimulation by PDGF.

V. Processing and Cellular Localization of PDGF Isoforms

1. Simian Sarcoma Virus-Transformed Cells

The near identity of p28^{sis} [the transforming product of simian sarcoma virus (SSV)] with the B chain of PDGF, and the high level of expression of v-*sis* in SSV-transformed cells have provided a system to evaluate the biosynthetic pathway of a homodimer highly related to PDGF-BB and to examine the role of particular portions of the molecule in this process. The predicted size of the initial translation product of v-*sis* is 30–33 kDa, which includes a sequence of 18 hydrophobic amino acids derived from the simian sarcoma-associated virus (SSAV) envelope gene. In place of the envelope-derived hydrophobic sequence, c-*sis* has a comparable 15-amino-acid hydrophobic sequence. Metabolic labeling studies in SSV-transformed marmoset cells (HF/SSV) suggest that the initial translation product is processed by clipping of the putative signal peptide, N-linked glycosylation, and removal of N- and C-terminal sequences to give an 11-kDa polypeptide, which is disulfide-bonded to form a 24-kDa dimer (ROBBINS et al. 1983, 1985). Antibodies against the N- and C-terminal sequences that are removed during processing do neutralize the mitogenic activity in lysates of SSV-transformed human fibroblasts, suggesting that proteolytic processing is not necessary for biological activity (LEAL et al. 1985). Mutational analysis has suggested that the N- and C-terminal portions of v-*sis* are not necessary for transformation of 3T3 cells (KING et al. 1985). However, these regions are greater than 90% conserved between new world primates and humans, and may play a role in the storage of PDGF in different cells.

Data regarding the secretion of *sis*-related peptides in SSV-transformed cells support the possibility that these peptides remain membrane associated. Secreted PDGF-like molecules have been detected in media conditioned by SSV-transformed 3T3 cells (BOWEN-POPE et al. 1984b; HUANG et al. 1984b), normal rat kidney (NRK) cells (BOWEN-POPE et al. 1984b; OWEN et al. 1984), and marmoset HF cells (JOHNSSON et al. 1985a). However, in experiments with SSV-transformed human fibroblasts, ROBBINS et al. (1985) found that less than 1% of the *sis*-related peptides radiolabeled during a 4-h period were found in the culture medium. Eighty per cent of the *sis*-related peptides remained associated with membrane fractions, 10% of this in association with the outer surface of the plasma membrane. Using the same cells, JOHNSSON et al. (1985a) reported that during a 2-day collection period, 5% of the PDGF activity was present in the cell lysate and 95% in the conditioned medium. However, JOHNSSON et al. (1985a) used cell lysates that had been centrifuged after lysis and thus probably depleted of membrane-associated species.

At least some cells transformed by SSV release a proteoglycan-like molecule, gp200^{sis}, which appears to be highly specific for transformation by this virus (Thiel et al. 1981). This molecule is specifically recognized by anti-PDGF antibodies (Thiel and Hafenrichter 1984), and the deglycosylated molecule consists of two dimers with molecular weights of 26000 and 28000 (Klein and Thiel 1988). Mitogenic activity was also demonstrated to be associated with this highly glycosylated molecule. It is unclear, however, what percentage of the v-*sis* proteins are represented by this form. Most other studies would not have detected this species, as it does not enter SDS-PAGE gels used in prior analyses.

The predicted secondary structure and hydropathy index of p28^{sis} are consistent with a water-soluble globular protein, about one-third α-helical, containing no membrane-spanning segments (Doolittle et al. 1983). Analysis of p28^{sis} mutants has provided more information on processing and structure. Several gene fusions that encode transmembrane forms of the v-*sis* gene product were constructed by fusing a membrane anchor domain from the *G* gene of vesicular stomatitis virus to the v-*sis* gene (Hannink and Donoghue 1986). These membrane-anchored forms are properly folded as indicated by dimerization, glycosylation, and N-terminal proteolytic processing. Removal of the N-linked glycosylation site from the v-*sis* product did not prevent cell-surface transport, and a correlation between cell-surface expression of membrane-anchored v-*sis* gene products and transformation was demonstrated. On the other hand, if transport across the endoplasmic reticulum is blocked by the introduction of a charged amino acid residue within the signal sequence, the protein does not dimerize, is not secreted, and no longer has the ability to transform NIH 3T3 cells (Lee et al. 1987). This mutant protein localizes to the nucleus as demonstrated by cell fractionation and indirect immunofluorescence. An amino acid sequence within the C-terminal "pro" sequence (residues 237-255) outside the region required for transformation was shown to contain a nuclear transport signal. Using a heat-inducible promoter, it was demonstrated that within 30 min of induction, v-*sis* protein accumulates in the nucleus.

2. c-*sis*-Transformed Cells

Analysis of the processing of PDGF B chain in cells transfected with c-*sis* demonstrated processing similar to that of v-*sis* (Igarashi et al. 1987). Like the v-*sis* gene product, the B-chain precursor undergoes N-linked glycosylation and removal of the N-terminal "pro" sequence. The primary translation product of c-*sis* is a 26-kDa protein, as compared with p28^{v-sis}, which possesses an additional 17 amino acids of *env* gene origin. These amino acid residues would account for the larger size of p28^{v-sis} as compared with p26^{c-sis}. In these studies of c-*sis*-transfected COS-1 cells, the 35-kDa dimer of the B chain containing the C-terminal propeptide but lacking the N-terminal "pro" sequence was the major species, and cleavage of the precursor to the size of the fully processed PDGF B-chain dimer was not detected.

Analysis of clones of normal human fibroblasts transfected with c-*sis* cDNA demonstrated variability in the protein products (STEVENS et al. 1988). One clone (3-6-3) contained predominantly an apparent PDGF dimer of 21-kDa. Another clone (3-3-2) contained only an apparent PDGF monomer of 12-kDa, which was shown to account for all of the mitogenic activity present in the cells. The dimer and monomer were associated with the pelletable cellular membrane fraction of a postnuclear supernatant of sonicated cells. These observations again suggest the possibility that a monomer can either interact with the receptor, or is able to form noncovalent dimers capable of interacting with the receptor.

Other studies using Rat 1 cell cultures also demonstrated more abundant cell-associated B-chain protein (BYWATER et al. 1988). A major 30-kDa doublet and a less abundant doublet of 24-kDa were identified. Pulse chase experiments demonstrated that after a 15-min labeling period, all of the 30-kDa protein was cell associated. After a 3-h chase, the 30-kDa protein was detected in the medium and the cell-associated form was predominantly 24-kDa. All of these forms were dimers.

3. PDGF A-Chain Transfectants

Under the influence of the same promoter, the PDGF A chain is 20-fold less efficient in inducing transformation of NIH 3T3 cells than the PDGF B chain (BECKMANN et al. 1988). There are also marked differences in the secretory and mitogenic properties of the two chains, which may account for the observed differences in transformation potential. During a 3-h labeling period, PDGF B chain was not detected in tissue culture fluids, while 70%–80% of the total PDGF A chain synthesized during this period was released into the supernatant. No difference in compartmentalization of the longer and shorter PDGF A-chain products (A and A_L, Fig. 1) was detected. The major products of A_L were doublets of 22-23-kDa and 17-18-kDa (under reducing conditions), representing the propeptide and the cleaved mature protein. The shorter PDGF A-chain product was expressed as two major products of approximately 20 and 16-kDa. No significant differences in the specific mitogenic activities of the two PDGF A-chain variants were detected. Low levels of expression of the shorter PDGF A-chain cDNA precluded direct comparison of their transforming activities. Comparison of the mitogenic activities of the A-chain and B-chain products suggested that for the test cells, NIH 3T3, PDGF B-chain homodimers were more mitogenic than PDGF A-chain homodimers. PDGF-AA, however, is not intrinsically less mitogenic than PDGF-BB, as the mitogenic potency of different isoforms varies with cell type and parallels the ability of the cells to bind the different isoforms (SEIFERT et al. 1989). However, this difference in binding of the different isoforms may contribute to the observed variations in transformation potential.

Transfection of Rat 1 cell cultures demonstrated precursor molecules of 36 and 44-kDa and the processed PDGF-AA homodimer of 31-kDa in cell lysates and medium. As previously described for NIH 3T3 cells, PDGF-AA was predominantly secreted while PDGF-BB was principally cell associated. Both mRNA and protein levels in the A- and B-chain transfectant were comparable, and the differences in cellular localization appeared to reflect variation in processing and not different rates of synthesis.

It was also reported that expression of A and A_L in monkey COS cells resulted in secretion of PDGF activity only when A_L was used (Collins et al. 1987a). Expression of the short form of the A chain was also low in NIH 3T3 cells. However, where detectable, it had similar properties to A_L (Beckmann et al. 1988). In subsequent studies in Rat I cells, transfectants of the short form of the A chain synthesized and secreted 10 times more A-chain protein than A_L cDNA transfectants, in spite of similar mRNA levels, possibly indicating lower translation rate or increased turnover (Bywater et al. 1988). The reasons for the discrepancies are most likely differences in the expression systems. More significantly, the original suggestion that only the PDGF-A_L variant is secreted does not hold for other systems.

The predicted A-chain protein encoded by A_L contains a sequence near its C terminus similar to the nuclear targeting signal (NTS) previously identified in the B chain in exon 6 (Lee et al. 1987). This sequence is capable of targeting a nonsecreted form of the A chain to the nucleus as well as other cytoplasmic proteins to which the NTS was fused to the 3' end of the coding region (Maher et al. 1989). The possibility that growth factors may have a direct role in the nucleus has been a controversial topic (Burwen and Jones 1987). PDGF and B-chain proteins have been described in the nuclei of cells bearing PDGF receptors (Rakowicz-Szulezynska et al. 1986) and SSV-transformed cells (Yeh et al. 1987). The alternative splicing of the NTS in exon 6 of the PDGF A chain may represent a level of control of nuclear localization and possibly activity within the nucleus.

4. Cells Expressing Recombinant PDGF-AB Heterodimers

Following transfection of CHO cells with a plasmid encoding the precursors of both PDGF A and B chains, all three dimeric combinations of PDGF chains were found in the cell medium (Ostman et al. 1988). In the cell lysate, the PDGF antisera recognized components of 33, 30, and 24-kDa; after reduction, 23-, 17-, 16-, 14-, and 12-kDa species were detected. Analysis with chain-specific antisera demonstrated that the 23-, 17,- and 14-kDa species were A-chain specific, and the 16- and 12-kDa species were related to the B chain. Pulse chase experiments demonstrated that dimerization occurs rapidly after synthesis and precedes proteolytic processing. As demonstrated in several other systems, the 24-kDa PDGF-BB homodimer was found predominantly in cell lysates and not in the cell media.

5. Transformed Cells

Analysis of the synthesis and secretion of PDGF-related proteins in many transformed cells preceded our knowledge of the complexity of the multiple forms of PDGF. A recent analysis of a human malignant glioma cell line, U-343 MGa clone 2:6, previously used for the cloning of the A chain of PDGF (BETSHOLTZ et al. 1986b), demonstrated that the cells secreted all three isoforms of PDGF (HAMMACHER et al. 1988b). Although particular PDGF isoforms have been purified from specific cell lines, as in PDGF-AA isolated for U-2OS cells (HELDIN et al. 1986), these are most likely not the only form made by these cells. B-chain species have been observed in the media of U-2OS cells (our unpublished observations).

Analysis of immunoprecipitates of proteins recognized by antisera against purified PDGF demonstrated dimerized species consistent with precursor and mature PDGF homodimers and/or heterodimers in fibrosarcoma and glioblastoma cells (PANTAZIS et al. 1985), human osteosarcoma cells (GRAVES et al. 1984), and a number of malignant epithelial cell lines (SARIBAN et al. 1988b). In addition, other high molecular weight species, larger than the predicted size of precursor proteins, were also observed. High molecular weight species have also been detected in extracts of a number of transformed and nontransformed cells using anti-peptide antisera to sequences in both the A and B chains of PDGF (NIMAN et al. 1984). Possible explanations for these high molecular weight species include glycosylation, as has been demonstrated for gp200^{sis} in SSV-transformed cells (KLEIN and THIEL 1988), or complexes of PDGF with binding proteins.

6. Normal Cells

Very few data exist on the processing and cellular localization of PDGF in normal cells. However, in human alveolar macrophages, a major species detected in metabolic labeling studies appears to be a monomer of 12-13-kDa (SHIMOKADO et al. 1985). Since activated alveolar macrophages can express both the A and B chains of PDGF (our unpublished observations), it is unclear which PDGF isoform this represents.

Another example of a normal cell whose secreted products differ from the major species detected in transfected and transformed cells is the primate aortic smooth muscle cell (VALENTE et al. 1988). These cells transcribe both PDGF A- and B-chain genes in culture, but do not secrete detectable mitogenic activity. Analysis of medium following metabolic labeling demonstrated 12- and 100-kDa species which were specifically immunoprecipitated with anti-PDGF. Their mobility did not change after reduction, and no dimeric 30-kDa PDGF-like protein was detected.

C. PDGF Receptors

Even before PDGF was purified to homogeneity and receptor binding studies became possible, there was little doubt that as a large polypeptide factor, it would prove to have a cell-surface receptor through which its actions were mediated. Several studies during the early 1980s quickly established many points of apparent similarity between the PDGF receptor and the EGF receptor, including ligand-stimulated tyrosine kinase activity, similar size, and ability to stimulate many intracellular events, including intracellular phosphorylation and entry into the cell cycle. During the last year, however, we have had to alter substantially many of our early concepts of PDGF receptor structure. The old and new models of receptor structure are contrasted below.

I. General Models of Receptor Structure and Properties

1. Old and New Models of the Structure of the PDGF Receptor

a) Initial Model of Receptor Structure: A Single Cell-Surface Monomer

Prior to 1988, studies of PDGF binding had always yielded results which were interpreted in terms of a single form of PDGF receptor. This receptor was described as a 170000- to 180000-Da membrane glycoprotein with an extracellular ligand-binding domain and an intracellular tyrosine kinase domain.

b) Postulation of Multiple PDGF Receptors

The true complexity of the PDGF receptor began to unfold only as new and more specific reagents became available, including recombinant PDGF-AA and PDGF-BB, monoclonal antibodies against the PDGF receptor, and cDNA clones for a PDGF receptor. The patterns of binding (and binding competition) of different isoforms of PDGF and anti-receptor monoclonal antibodies initially indicated that human fibroblasts expressed at least two classes of PDGF receptor: an abundant form which binds only PDGF-BB ("B receptor") and a much less abundant form which binds all three isoforms ("A/B receptor") (HART et al. 1988; also called the "A receptor" by HELDIN et al. 1988).

c) Current Model of Receptor Structure: A Dimer of Two Possible Subunits

The studies that led to the two-receptor model described above were performed entirely using diploid human fibroblasts. SEIFERT et al. (1989) surveyed patterns of receptor expression on additional cell types, and found cell types (e.g., MG-63 human osteosarcoma cells) that express a much larger ratio of ^{125}I-PDGF-AA to ^{125}I-PDGF-BB binding. These cells thus allow better evaluation of the properties of the hypothesized A/B receptor. Patterns of crosscompetition, saturation binding, and antibody binding obtained using MG-63 cells are not explicable by the original two-receptor model. We now propose (SEIFERT et al. 1989) that the high-affinity PDGF receptor is a dimer recruited from separate pools of α-subunits (which can bind A or B chains) or β-subunits (which can bind only B chains). In the absence of PDGF, the subunits exist as separate monomers or as unstable dimers. Two receptor subunits

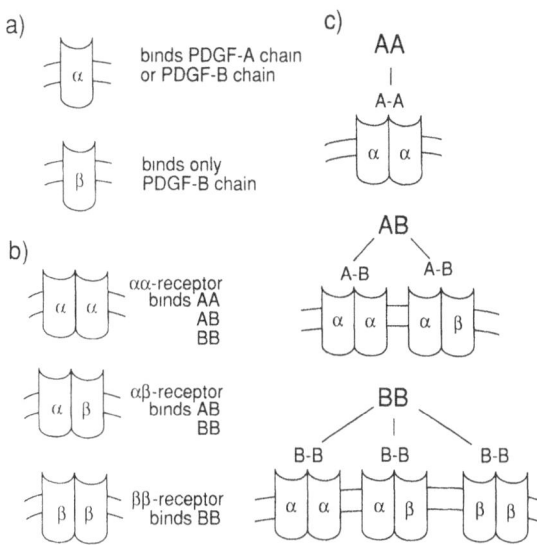

Fig. 4a–c. The PDGF receptor subunit model. **a** Receptor subunits. Two types of PDGF receptor subunit exist: α-subunits, which can bind PDGF A and B chains, and β-subnits, which can bind only PDGF B chains. Before ligand binding, PDGF receptor subunits are present on the cell surface as independent monomers or weakly associated dimers. **b** Receptor types. The two PDGF receptor subunits can form three types of high-affinity PDGF receptors: αα-receptors, αβ-receptors, and ββ-receptors. **c** Receptor binding. PDGF-AA can only bind to PDGF αα-receptors, while PDGF-AB can bind to either PDGF αα-receptors or αβ-receptors. PDGF-BB can bind to all three PDGF receptor classes. (Adapted from SEIFERT et al. 1989)

together can form a high-affinity complex with PDGF, presumably with one receptor subunit binding each of the chains of the dimeric ligand. As indicated in Fig. 4, three possible dimeric receptors can form, depending on which PDGF isoform is added. Information about the sequence and/or physical properties of the two proposed subunits is presented in Sects. C. III and C. V.

Two pieces of binding data argue against the monomeric two-receptor model proposed above, which predicts that PDGF-AA and PDGF-AB should behave identically since they both were proposed to bind to the "A/B receptor." Studies with MG-63 cells (SEIFERT et al. 1989) clearly show that this is not the case, as indicated by the following: (1) PDGF-AA and PDGF-AB compete completely for each other's binding, but PDGF-AB competes for more ^{125}I-PDGF-BB binding than does PDGF-AA. (2) PDGF-AB can downregulate a significant portion of monoclonal ^{125}I-PR7212 binding, but PDGF-AA has no effect.

Direct evidence for ligand-induced subunit dimerization or stabilization is obtained in co-immunoprecipitation experiments. Mouse receptors were metabolically labeled with [35S]methionine, solubilized, mixed with unlabeled solubilized extracts of human fibroblasts, then incubated with different isoforms of PDGF to permit formation of high-affinity dimeric ligand-receptor complexes. The complexes were immunoprecipitated with monoclonal PR7212 (which recognizes human but not mouse β-subunits) and visualized by autoradiography after SDS-PAGE. Labeled mouse subunits were immunoprecipitated only if PDGF-AB or PDGF-BB had been included during the coincubation of the soluble extracts to induce or stabilize dimerization with the human β-subunits. Incubation of the soluble extracts with PDGF-AA, which does not bind to β-subunits, did not result in immunoprecipitation of labeled mouse subunits by PR7212.

Why was the proposed multimeric structure of the PDGF receptor not observed in earlier studies? Noncovalent association of subunits, which we now propose, would not have been detected by the published sizing studies since the presence of SDS during analysis by SDS-PAGE would have dissociated the subunits. Gel filtration of solubilized ^{125}I-PDGF-receptor complexes and/or nondissociating conditions have indicated a receptor size of 200 000 Da (HELDIN et al. 1983). Determination of size by this technique is not very accurate, and although the result was interpreted to indicate that the receptor was monomeric, we suspect that the receptor was in fact a dimer in these experiments. We are currently attempting to use other nondenaturing techniques to evaluate receptor dimerization.

II. Binding Properties of the PDGF Receptor

1. Practical or Technical Concerns

The many factors which affect the measurement of ^{125}I-PDGF-AB binding are discussed in detail in BOWEN-POPE and ROSS (1985). Some of the ways these factors can influence the apparent affinity or other properties of binding are discussed below.

2. Receptor Specificity

No other growth factor or substance tested to date has been clearly demonstrated to bind to the PDGF receptor with high affinity. Two types of antagonists of PDGF binding have been reported: basic proteins and suramin. Very basic proteins such as histone, polylysine, and protamine inhibit PDGF binding, but only at relatively high concentrations, probably by binding with low affinity to the PDGF receptor (HUANG et al. 1982). The most potent of these is protamine, with an $I_{0.5}$ of 2µg/ml (HUANG et al. 1982, 1984c). Although HUANG et al. (1982) reported that protamine dissociates prebound ^{125}I-PDGF, we have found that it does this only very incompletely.

Suramin is a small acidic molecule which inhibits many protein-protein interactions. Its utility for investigating PDGF binding is that it can efficiently strip off prebound PDGF (GARRETT et al. 1984; WILLIAMS et al. 1984; HUANG and HUANG 1988). Its disadvantage is that it is not specific for PDGF (HAWKING 1978) and, when used at 37° C for several hours, becomes toxic at concentrations very close to those necessary for efficient dissociation of PDGF. Since suramin is very negatively charged, it was initially assumed that its effects on PDGF binding would be restricted to the cell surface. Evidence that this may not be true is central to arguments concerning the nature and location of PDGF binding in SSV-transformed cells and will be discussed in Sect. C. X. 3. a below.

3. Receptor Affinity and Number

a) Saturation Binding Determination

Saturation binding experiments of the type shown in Fig. 2 clearly demonstrate the presence of saturable high-affinity binding sites for ^{125}I-PDGF-AB on cells from diverse species. There is no obvious difference in the apparent affinity of binding of human ^{125}I-PDGF-AB to cells from human, fish, rodent, or avian species, indicating that the interaction between PDGF and its receptor has been highly conserved during evolution.

Although it is clear that PDGF binds with high affinity to its receptor, there is no general agreement as to the exact value of the dissociation constant (K_d). Estimates have ranged from less than or equal to 10^{-11} M (BOWEN-POPE and ROSS 1982) to 10^{-10} M (WILLIAMS et al. 1982) to 10^{-9} M (HELDIN et al. 1981 b; HUANG et al. 1982). Saturation binding analyses of this sort give good estimates of receptor number (subject to the accuracy with which the true specific activity of the labeled PDGF can be known), but the true affinity of the binding is difficult to determine. Since most of the artifacts which can affect determination of binding affinity (see below) result in overestimation of K_d, we believe that the best estimate is 10^{-11} M or lower. Two aspects of the measurements are particularly problematic: (1) It is difficult to achieve equilibrium binding of low concentrations of ^{125}I-PDGF to intact cells at 4° C. This problem is discussed in detail in BOWEN-POPE and ROSS (1985). (2) The high affinity of PDGF binding and the large numbers of PDGF receptors per cell can result in serious depletion of ligand concentrations during the binding incubation. Similar experiments with ^{125}I-PDGF-AA and ^{125}I-PDGF-BB demonstrate that these isoforms also bind with high affinity. The affinity of ^{125}I-PDGF-BB binding seems to be somewhat lower than the affinity of ^{125}I-PDGF-AB binding.

b) Kinetic Constants

In theory, affinity and dissociation constants can be calculated from measured rates of association and dissociation. This has not been done for PDGF. As is discussed below, the rate of dissociation is too slow to measure and the rate of association is limited by the physical state of the membrane.

α) *Association*

Binding of subsaturating concentrations is relatively slow at 4° C, and can take more than 4 h to approach steady state. Gentle rocking or swirling increases the rate of binding, consistent with a partial limitation of binding by local depletion of ligand. Binding is considerably faster at 37° C (WILLIAMS et al. 1984; BOWEN-POPE and ROSS 1985), reaching maximal values by 45 min.

Thereafter, degradation of bound ^{125}I-PDGF and concomitant degradation of receptors begin to decrease cell-associated ^{125}I-PDGF. The large temperature dependence of PDGF binding might reflect the influence of membrane fluidity on the ability of receptor subunits to dimerize rapidly enough to stabilize the initial low-affinity interaction between one ligand chain and one receptor subunit.

β) Dissociation

When ^{125}I-PDGF is bound at 4° C for 2–4 h, then rinsed and incubated at 4° C either in medium alone or in medium containing saturating concentrations of unlabeled PDGF, there is no observable dissociation of bound ^{125}I-PDGF over a period of 6 h. This is true both for ^{125}I-PDGF-AB (BOWEN-POPE and ROSS 1985) and for ^{125}I-PDGF-BB and ^{125}I-PDGF-AA (D. F. BOWEN-POPE et al., unpublished observations). The bound ^{125}I-PDGF is still at the cell surface since it can be dissociated with acidified saline (BOWEN-POPE et al. 1983). It is not clear why WILLIAMS et al. (1984) observed that binding becomes nondissociable only after warming to 37° C.

III. Cloning and Expression of Receptor cDNAs

1. The Receptor β-Subunit cDNA

a) Sequence of the Receptor β-Subunit cDNA

The mouse PDGF receptor cDNA was obtained from cDNA libraries from mouse placenta and cultured 3T3 cells. The library was screened with oligonucleotide probes prepared against amino acid sequences obtained from the N terminus, and tryptic fragments of receptors purified from cultured 3T3 cells via affinity chromatography on wheat germ agglutinin, anti-phosphotyrosine, and elution from preparative SDS-PAGE (YARDEN et al. 1986). The human sequence was obtained by probing human fibroblast cDNA libraries with oligonucleotides from the mouse sequence (CLAESSON-WELSH et al. 1988; GRONWALD et al. 1988). The 5449-nucleotide sequence reported by CLAESSON-WELSH et al. includes a poly-A addition signal at the 5′ end, and terminates with a stretch of As. The 5570-nucleotide sequence reported by GRONWALD et al. extends 170 nucleotides further 5′ (Fig. 5). The composite sequence probably does not include the very 5′ end of the mRNA since the known sequence is smaller than the mRNA species of 5.7 kb (abundant) and 4.8 kb (much less abundant) detected in human fibroblasts (GRONWALD et al. 1988).

cDNA sequences from both species contain relatively long 5′ and 3′ non-translated regions. Since the very 5′ end of the mRNA has not been cloned, the length of the 5′ untranslated regions can only be estimated as at least 105 bases in the mouse (YARDEN et al. 1986) and at least 356 bases in the human (GRONWALD et al. 1988). Actual lengths may be 200 bases longer. The length of the 3′ untranslated region is 1678 nucleotides in the mouse (YARDEN et al. 1986) and 1945 nucleotides in the human (CLAESSON-WELSH et al. 1988).

Although the sequence of the 3' region is relatively poorly conserved overall, there is a stretch of 150 nucleotides that is 85% identical between mouse and human, and this stretch contains an ATTTA sequence that characterizes many unstable mRNA species (SHAW and KAMEN 1986).

b) Amino Acid Sequence of the β-Subunit Deduced from the cDNA: Conservation and Homologies

The predicted amino acid sequences of the mouse and human PDGF receptors are very similar (87% overall amino acid sequence similarity) and show striking homology to two other cDNA sequences: that of c-*fms* (the colony-stimulating factor CSF-1 receptor) and that of c-*kit*, a receptor-like sequence whose ligand has not yet been identified (YARDEN et al. 1987). YARDEN et al. (1986) proposed that these three genes define a distinct family of growth factor receptors. The members of this family share the following characteristics: (a) Members show conservation of cysteine residues in the extracellular domain. This spacing of cysteines suggests that the extracellular portion of the receptor consists of five immunoglobulin-like domains, i.e., that the PDGF receptor is a member of the immunoglobulin superfamily (CLAESSON-WELSH et al. 1988). (b) Members have a "split kinase" region in which an unrelated sequence of approximately 100 amino acids interrupts the cytoplasmic sequence homologous to the conserved region of other tyrosine kinases. The amino acid sequence of the kinase insert is only poorly conserved between members of the split kinase family.

c) Expression of Transfected β-Subunit cDNA

ESCOBEDO et al. (1988b) have expressed the mouse PDGF receptor cDNA in CHO cells and demonstrated that the transfected gene reconstitutes a normal responsiveness to PDGF by all criteria examined, including stimulation of intracellular phosphorylation, phosphatidylinositol turnover, and [³H]thymidine incorporation.

Human β-subunit cDNA has been expressed in two cell lines by three groups. We expressed the cDNA in a subclone of BHK-570 cells, which do not bind any isoform of PDGF (GRONWALD et al. 1988). Three subclones expressing differing levels of the human receptor were analyzed. All bound ^{125}I-PDGF-BB and monoclonal ^{125}I-PR7212 with affinities comparable to that of binding to human dermal fibroblasts. ^{125}I-PDGF-AA did not bind at all, and binding of ^{125}I-PDGF-AB was of much lower affinity than to fibroblasts. The nature of the low-affinity ^{125}I-PDGF-AB binding is not clear. Since no α-subunit is available to bind the A chain of the ligand, the low-affinity binding may represent binding of the PDGF to a single β-subunit of the receptor via its single B chain. We have also transfected a mouse myoblast cell line (MM-14) and obtained a comparable binding pattern, although the level of expression of the transfected receptor was much lower (D. F. BOWEN-POPE et al., unpublished observations). These results are essentially in agreement with the

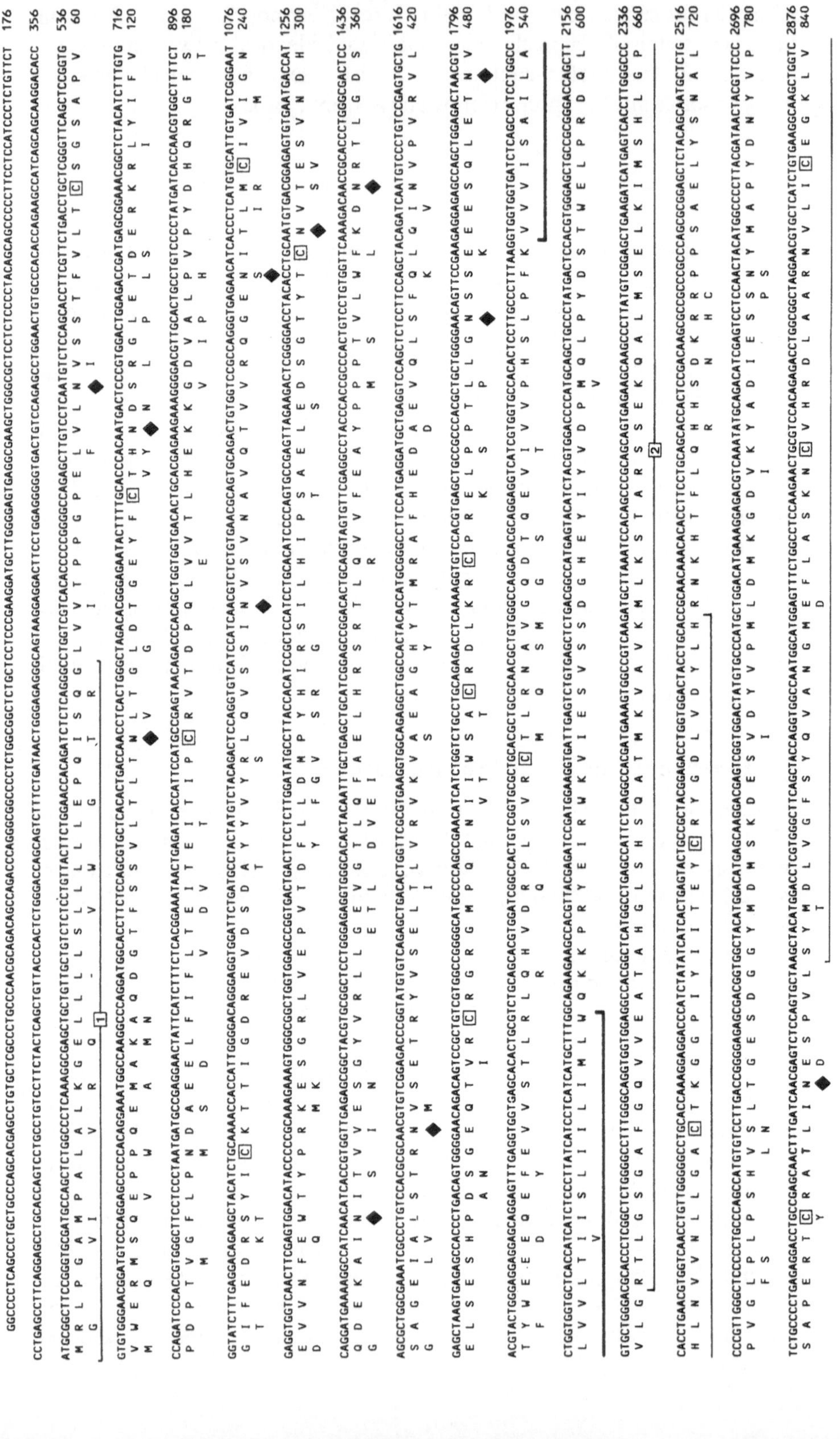

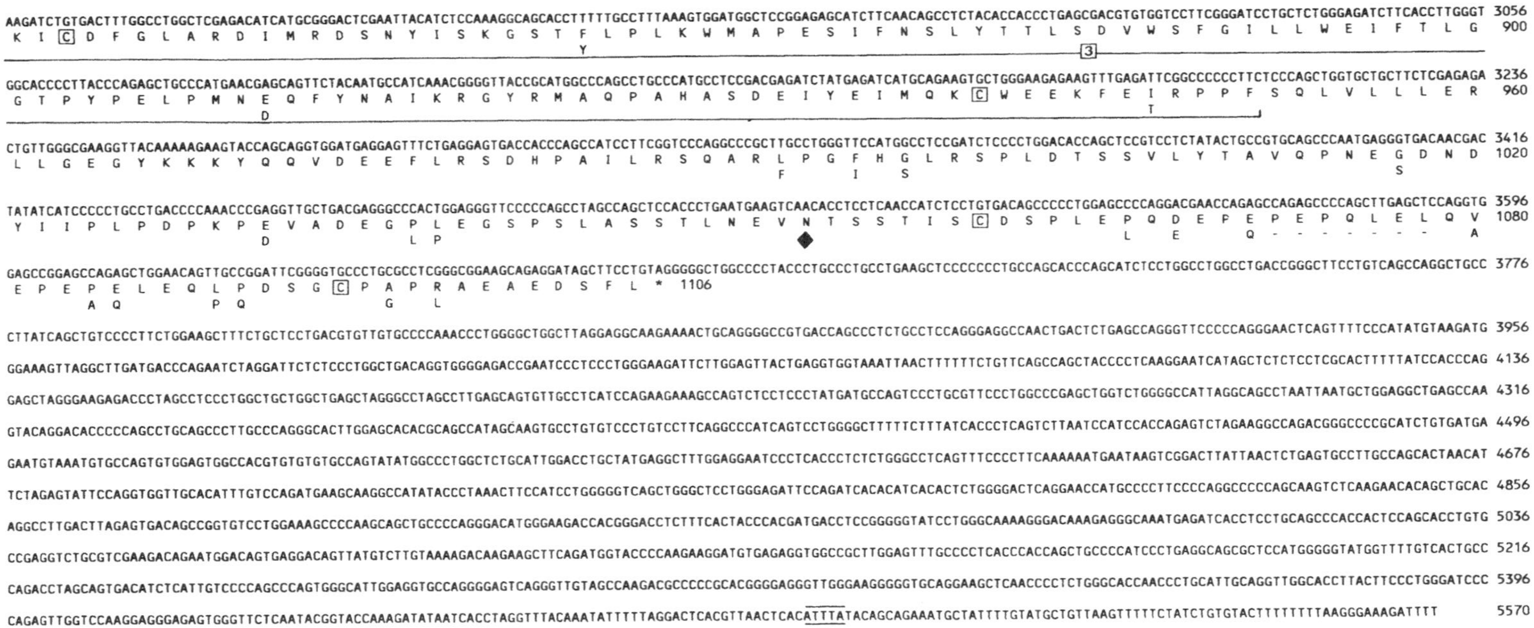

Fig 5. Nucleotide sequence of the human PDGF receptor and comparison of the deduced amino acid sequence with the mouse PDGF receptor sequence (YARDEN et al. 1986). For maximal alignment of the mouse and human amino acid sequences, two gaps were inserted into the mouse sequence (positions 18 and 1072–1078). Nucleotides and amino acids of the human PDGF receptor are numbered at the right. Amino acid numbering starts at the N-terminal methionine. Residues of the mouse PDGF receptor that differ from the human sequence are depicted below the human amino acid sequence. Potential sites of asparagine-linked glycosylation are marked by solid diamonds below the asparagine residue. Cysteine residues are boxed. The putative transmembrane domain is underlined with a heavy line. Other domains are underlined and labeled with boxed numbers: 1, signal peptide domain; 2, 3, the split tyrosine kinase domain. An AUUUA sequence motif is demarcated at the 3' end of the cDNA. The stop codon is indicated by an asterisk. (From GRONWALD et al. 1988)

results obtained by Claesson-Welsh et al. (1988) using CHO cells transfected with the human β-subunit cDNA. The results differ from those of Escobedo et al. (1988c), who reported that CHO cells transfected with a cDNA for the human receptor can bind PDGF-AA with high affinity. The explanation for these differences is not clear. Possibly the high-affinity binding in the latter studies reflects the expression of a finite level of *endogenous* α-subunits by the CHO host cells. Some clones and subclones of CHO cells do express receptors (e.g., see Bowen-Pope et al. 1985). Although the parental cell population did not express detectable PDGF receptors, it is possible that the clones chosen for analysis (selected for high PDGF binding) did express the endogenous gene.

2. The Receptor α-Subunit cDNA

The α-subunit of the receptor seems to be the product of a separate gene. Evidence for this includes the following: (a) The binding phenotype of CHO, BHK, and mouse myoblast cells transfected with the cloned (β-subunit) cDNA suggests that, in these cells, this gene is not sufficient to generate the α-subunit phenotype. This does not necessarily mean that the α-subunit is the product of a different gene, since the host cells into which the cDNA was transfected may have lacked accessory components or processing systems necessary to convert the transfected gene product into an α-subunit. (b) To determine the role of the cloned (β-subunit) gene in creating α-subunits, we used an antisense construct to specifically eliminate products of this gene. When transfected into cells expressing both α- and β-subunits, this construct eliminated expression of β-subunits without decreasing expression of α-subunits (J. D. Kelly and D. F. Bowen-Pope, unpublished observations). (c) Using low-stringency hybridization with the β-subunit cDNA to screen a cDNA library prepared from MG-63 cells (rich in α-subunits), we have obtained a cDNA whose sequence is homologous, but not identical, to that of the β-subunit cDNA (J. D. Kelly et al., unpublished observations). The cDNA detects 6.4-kb mRNA only in cell types which express α-subunits. Expression of the transfected cDNA generates the α-subunit phenotype. The α-subunit cDNA has been independently cloned by Matsui et al. (1989).

IV. The Receptor Subunit Genes

The gene for the β-subunit has been mapped to the long arm of human chromosome 5 (5q31-32) by in situ hybridization with cDNA and by Southern blot analysis of hamster/human chimeras expressing limited numbers of human chromosomes (Yarden et al. 1986). The gene for the CSF-1 receptor (c-*fms*) was mapped to an adjacent band of chromosome 5 (5q33.2-34). It has been shown by genomic cloning that less than 500 nucleotides separate the 3′ end of the PDGF receptor gene from the 5′ end of the CSF-receptor gene (Roberts et al. 1988), suggesting that the two genes arose through duplication. Both genes have a long 3′ exon encoding the end of the translated sequence and the 3′ untranslated sequence (ROBERTS et al. 1988). The intron/exon borders of this exon, as well as the two upstream exons sequenced, correspond to intron/exon borders in the CSF receptor gene. The α-subunit gene has been mapped to the long arm of chromosome 4 (4q11-12; MATSUI et al. 1989; R. G. K. GRONWALD et al., submitted).

V. Physical Characteristics of PDGF Receptor Proteins

1. Size of the PDGF Receptor Based on Affinity Crosslinking Studies

The size and structure of the PDGF receptor was first studied using autophosphorylation (see below) and affinity crosslinking. Crosslinking studies identified labeled ^{125}I-PDGF-receptor complexes of about 190000-Da in several cell types (GLENN et al. 1982; HELDIN et al. 1983; WILLIAMS et al. 1984). Subtraction of the molecular weight of the ^{125}I-PDGF (30000) yielded an estimated molecular weight of about 160000. Reduction did not significantly affect the apparent size of the receptor, so it was concluded that the receptor did not consist of disulfide-bonded subunits. The crosslinking studies were done with preparations of PDGF in which the major labeled species would have been ^{125}I-PDGF-AB with the ^{125}I on the A chain. Since an A chain can bind only to an α-subunit, it is probable that the subunit visualized in these studies was the α-subunit, although it is possible that the labeled A chain would be close enough to a β-subunit to be crosslinked. This technique would probably not have the size resolution to distinguish between the two subunits (which differ by only about 4000-Da) if both were present on the same lane (see below). Crosslinking studies using radioiodinated pure PDGF isoforms gave crosslinked bands that seemed to indicate a relatively large difference in size between the "A receptor" (α-subunit), which was calculated to be 165000, and the "B receptor" (β-subunit), estimated to be 160000. It is likely that the small size estimated for the α-subunit in these studies results from partial degradation of the α-subunit or from effects of crosslinking per se.

2. Size of the Mature Receptor Subunits Based on Metabolic Labeling Data

We now believe that the 170- to 190-kDa receptor protein identified above represents a subunit of a receptor which, in its high-affinity state, exists as a noncovalently linked dimer. The α- and β-subunits appear to be very similar in size; both fall within the 170- to 190-kDa range, which encompasses lab-to-lab (and protocol-to-protocol) differences in size estimation. In our own experience, the two subunits cannot be satisfactorily distinguished as separate bands after SDS-PAGE. At least three factors contribute to the difficulty: (1) the size difference is small (about 5000-Da in human fibroblasts); (2) the bands are somewhat broad (probably reflecting glycosylation differences); and (3) the relative abundance of the two subunits is very different (in human fibroblasts, the β-subunit is at least 10-fold more abundant), making it difficult to distinguish the less abundant subunit as a separate band.

At present, our best estimate of the relative sizes of the subunits is derived from an experiment in which human cells were metabolically labeled with [^{35}S]cysteine and [^{35}S]methionine, and immunoprecipitated with the subunit-specific monoclonal antibodies PR7212 (which recognizes only primate β-subunits; SEIFERT et al., 1989) and PR292 (which recognizes only primate α-subunits; BLACKWOOD et al., in preparation). When analyzed by SDS-PAGE on adjacent lanes, the apparent sizes of the subunits were 180000 (α-subunit) and 185000 (β-subunit) (BLACKWOOD et al., in preparation). Note that we have not used the frequent convention of designating the larger subunit as the "α-subunit." The subunits were named before we knew about the very small difference in size and reflect the mnemonic device α (can bind A chain) and β (can bind only B chain).

3. Isoelectric Point of Receptor Subunits

Both PDGF receptor subunits are acidic, with isoelectric points estimated as 4.2 (FRACKELTON et al. 1984), 4.5 (CLAESSON-WELSH et al. 1987), and 5.3 (PIKE et al. 1983; KAZLAUSKAS et al. 1988).

VI. Biosynthesis and Turnover of Receptor Subunits

1. Predictions about β-Subunit Biosynthesis Based on cDNA Sequence

The cDNA sequences of mouse (YARDEN et al. 1986) and human (GRONWALD et al. 1988; CLAESSON-WELSH et al. 1988) β-subunits have been published. The cDNA sequences predict initial translation products of 1098 amino acids (mouse) and 1106 amino acids (human). Cleavage of consensus signal peptides of 31 (mouse) and 32 (human) amino acids would yield predicted protein backbones of 1067 (mouse) and 1074 (human) amino acids, i.e., about 120000-Da. When analyzed by SDS-PAGE, the apparent molecular weight of the human core protein obtained by in vitro translation is 135000 rather than 120000 (CLAESSON-WELSH et al. 1988). Since all of the experimental size data on receptor biosynthesis are obtained by SDS-PAGE and since the PDGF receptor seems to be one of those proteins which do not migrate according to their true molecular weight, it is clear that the apparent sizes of the receptor precursors or deglycosylated forms must be considered to be only approximate.

There are 11 potential sites (Asn-X-Ser/Thr) of N-linked glycosylation in the extracellular domain, of which all except one are in identical positions in mouse and human sequences. Studies with chemical and enzymatic deglycosylation of the receptor indicate that the mature 180000-Da form of the receptor includes about 30000–40000-Da N-linked carbohydrate and about 10000-Da of O-linked carbohydrate (CLAESSON-WELSH et al. 1987, 1988; KEATING and WILLIAMS 1987). Sequence data from the purified mouse receptor indicated the presence of covalently linked ubiquitin (8000-Da) (YARDEN et al. 1986). The location, time of addition, or significance of this modification is not clear.

2. Experimentally Determined Rates of Biosynthesis and Turnover of the β-Subunit

When cultured cells are metabolically pulse-labeled with [^{35}S]cysteine, the first product that can be immunoprecipitated by antiphosphotyrosine (DANIEL et al. 1987), rabbit anti-β-subunit peptide antisera (KEATING and WILLIAMS 1987; HUANG and HUANG 1988), monoclonal anti-β-subunit antibodies (HART et al. 1987), or rabbit antireceptor antiserum (CLAESSON-WELSH et al. 1987) has a molecular weight of approximately 160000. This form is probably a precursor with the high-mannose form of N-linked oligosaccharide. Evidence for this includes the following: (a) The 160000-Da form is intracellular, as judged by inaccessibility to external antibodies (HART et al. 1987). (b) The 160000-Da form is sensitive to cleavage to 135000-Da by endoglycosidase H (which is active only on hybrid CHO), and a 168000-Da form accumulates in the presence of swainsonine (which prevents CHO maturation by inhibiting mannosidase II) (KEATING and WILLIAMS 1987; CLAESSON-WELSH et al. 1987). In solubilized cell extracts, both the 160000-Da precursor and the 180000-Da mature forms show PDGF-stimulated phosphorylation (HART et al. 1987; KEATING and WILLIAMS 1987), indicating that the 160-kDa "precursor" form of the receptor is able to bind and respond to PDGF even though, in intact cells, the membrane barrier prevents its exposure to the ligand.

Metabolically labeled 160000-Da precursor can be detected after a 15-min pulse labeling with [^{35}S]cysteine. This is converted to a cell-surface 180-kDa form with a half-time of about 30–60 min (HART et al. 1987; KEATING and WILLIAMS 1987). In the absence of PDGF, the half-life of a metabolically labeled cell-surface PDGF receptor is about 3–4 h (HART et al. 1987; HUANG and HUANG 1988). This measured half-life is consistent with the observed rate of recovery of cell-surface receptors on fibroblasts after transient exposure to PDGF (HELDIN et al. 1982). The half-life of an unstimulated PDGF receptor is thus much shorter than the 10 h half-life of insulin or EGF receptors (KRUPP et al. 1982; RONNETT et al. 1983; STOSCHECK and CARPENTER 1984).

When PDGF is added, the half-life of the 180000-Da form decreases to about 30 min (KEATING and WILLIAMS 1987; HUANG and HUANG 1988; HART et al. 1989a). This is consistent with studies of the rates of degradation of radiolabeled PDGF (HELDIN et al. 1982; BOWEN-POPE and ROSS 1982; NILSSON et al. 1983; ROSENFELD et al. 1984) and indicates that the receptor is degraded after internalization rather than being recycled.

3. Biosynthesis of the α-Subunit

Since the existence of this subunit was unsuspected until very recently and specific reagents (monoclonal antibodies and cDNA probes) have only recently been obtained, much less is known about the biosynthesis of the α-subunit than about the β-subunit. We have directly compared the biosynthesis of the α- and β-subunits by using the α-subunit-specific monoclonal antibody PR292 and β-subunit-specific monoclonal PR7212 to immunoprecipitate metabolically labeled subunits from the MG-63 cell line, which expresses approximately equal numbers of the two subunits (BLACKWOOD et al., in preparation). The general pattern of glycosylation and processing is very similar to that of the β-subunit but with smaller corresponding molecules. In these direct comparisons, the sizes were estimated to be as follows: mature, 180000 for α-subunit versus 185000 for β-subunit; and precursor, 160000 for α-subunit versus 170000 for β-subunit. There is some evidence that receptor processing and turnover may be altered in cells that synthesize both a PDGF-like molecule and PDGF receptors (see below).

VII. Purification of Receptor Proteins

PDGF receptor proteins have been purified from several sources using procedures involving affinity chromatography on wheat germ agglutinin (WGA) and antiphosphotyrosine Sepharose. DANIEL et al. (1985) added PDGF to intact mouse BALB/c 3T3 cells to stimulate tyrosine phosphorylation of the receptor. Membranes were prepared and solubilized in octylglucoside and receptors purified by affinity chromatography on antiphosphotyrosine Sepharose and WGA Sepharose. To obtain PDGF receptor for amino acid sequencing, a preparative SDS-PAGE step was added (YARDEN et al. 1986). BISHAYEE et al. (1986) stimulated receptor phosphorylation in vitro using membranes prepared from human MG-63 cells, solubilized with NP-40 and purified over antiphosphotyrosine Sepharose and diethylaminoethanolcellulose. RONSSTRAND et al. (1987) used pig uterus as a starting material, and purified Triton-solubilized receptors by sequential steps of WGA-Sepharose, fast protein liquid chromatography (FPLC)-Mono-Q, and antiphosphotyrosine-Sepharose. For the last step, receptor tyrosine phosphorylation was allowed to occur overnight at 4° C in the presence of ATP but the absence of PDGF. Apparently the receptors were autophosphorylated on tyrosine via basal tyrosine kinase activity.

VIII. Activities of PDGF Receptors

1. Ligand-Induced Internalization

All three isoforms of PDGF are internalized and degraded as determined by lysosome-dependent degradation of iodinated PDGF. Early studies with ^{125}I-PDGF-AB indicated that surface-bound ^{125}I-PDGF is internalized with a half-time of 15 min (ROSENFELD et al. 1985), and trichloroacetic acid soluble radioactivity appears in the culture medium with a half-time of 60–90 min (HUANG et al. 1983; NILSSON et al. 1983; ROSENFELD et al. 1985). ^{125}I-PDGF-AA (HART et al. 1989a) and ^{125}I-PDGF-BB (D. F. BOWEN-POPE et al., unpublished observations) are internalized and degraded with similar kinetics. Studies with metabolically labeled receptors indicate that the occupied receptors themselves are degraded rather than recycled. After ligand binding, the metabolically labeled receptor rapidly ceases to be recognized by antisera, presumably as a result of degradation (CLAESSON-WELSH et al. 1987; KEATING and WILLIAMS 1987; HUANG and HUANG 1988; HART et al. 1989a).

2. Ligand-Induced Receptor Autophosphorylation

Addition of PDGF to membrane preparations in the presence of [^{32}P]ATP stimulates the phosphorylation of a protein of 170000–185000-Da (EK and HELDIN 1982; NISHIMURA et al. 1982; HELDIN et al. 1983; PIKE et al. 1983; BISHAYEE et al. 1986). This phosphorylation is easily detected on autoradiographs without enrichment of receptors, and provided the first presumptive evidence for the size of the PDGF receptor. Purified receptors retain PDGF-stimulated tyrosine kinase activity (BISHAYEE et al. 1986; RONNSTRAND et al. 1987; NISTER et al. 1988 b).

When purified isoforms of PDGF are used, the extent of phosphorylation observed is proportional to the number of receptor subunits able to bind the isoform. Thus, receptor phosphorylation in human fibroblasts (which have many β- but few α-subunits) is well stimulated by PDGF-BB but not by PDGF-AA (HART et al. 1988; NISTER et al. 1988a). When intact cells are prelabeled with [^{32}P]orthophosphate, the receptors incorporate ^{32}P, but only on serine and threonine, with most of the label on serine rather than on threonine as for the EGF receptor (KAZLAUSKAS et al. 1988). Addition of any of the three isoforms of PDGF stimulated the incorporation of ^{32}P onto tyrosine but not serine or threonine residues (FRACKELTON et al. 1984; KAZ-LAUSKAS et al. 1988). The extent of receptor phosphorylation is proportional to the number of binding sites for each isoform. Use of subunit-specific monoclonal antibodies to specifically immunoprecipitate α- or β-subunits supports the conclusion that both are autophosphorylated in response to ligand binding (BLACKWOOD et al., in preparation).

Mutations in three regions of the receptor cDNA (the ATP-binding site, the transmembrane region, and the C terminus) have been reported to result in loss of receptor kinase activity (ESCOBEDO et al. 1988a). The mutant receptors did not mediate any detectable biological response to PDGF except downregulation after ligand binding. As described in Sect. C. III. 1. b above, the kinase region of the PDGF receptor is interrupted by a 104-amino-acid sequence with no kinase homology. ESCOBEDO and WILLIAMS (1988) have reported that deletion of this region does not decrease the kinase activity of the receptor but does diminish its ability to stimulate a mitogenic response. The authors speculated that the insert region may be involved in determining the substrate specificity of the receptor kinase.

IX. Pattern of Expression of PDGF Receptors

1. Numbers of Receptors on Different Cell Types

The great majority of information about the binding and biological effects of PDGF has been obtained using PDGF purified from human platelets. In light of our current understanding of the PDGF receptor, these data have two general limitations. First, the PDGF was not always functionally pure and often contained significant amounts of TGF-β and possibly other growth factors. As will be described below, TGF-β can affect PDGF receptor expression. Second, even truly "pure" human platelet PDGF is a mixture of PDGF isoforms, containing predominantly PDGF-AB, but also significant PDGF-BB (HAMMACHER et al. 1988a; BOWEN-POPE et al. 1989; HART et al. 1989b). The commonly used radioiodination methods, which react with tyrosine residues, will iodinate the PDGF-AB component (via one of three tyrosines on the A chain) but not the PDGF-BB component (the B chain contains no tyrosine residues). The most significant consequence of this for binding studies is the potential for greatly underestimating, or completely missing, PDGF receptors composed of β-subunits. This can be illustrated by two examples: (1) Human dermal fibroblasts express 10- to 20-fold more β-subunits than α-subunits. Since binding of ^{125}I-PDGF-AB, the classic reagent, requires formation of $\alpha\alpha$-receptors or $\alpha\beta$-receptors, binding of radioactivity will be limited by the availability of α-subunits, and the great excess of β-subunits will go undetected. Use of ^{125}I-PDGF-BB (radiolabeled as described below) detects at least 7-fold more PDGF receptors on fibroblasts (HART et al. 1988; SEIFERT et al. 1989). (2) Cell types which express β-subunits but no α-subunits will not bind ^{125}I-PDGF-AB with high affinity, and would be incorrectly concluded to express no PDGF receptors. This is true for the BHK cells transfected with the human β-subunit cDNA (GRONWALD et al. 1988) and for neonatal rat smooth muscle cells (R. A. SEIFERT and D. F. BOWEN-POPE, unpublished observations). It is not yet clear how many other putatively receptor-negative cell types will be found to express β-subunits and bind PDGF-BB.

Tables 1 and 2 list many (but not all) reports of PDGF receptor expression by different cell types from different species. With only a few exceptions, the binding data were obtained using ^{125}I-PDGF-AB. It should be remembered that by this criterion, cells expressing only β-subunits would be incorrectly classed as receptor negative. It is not clear if there are a significant number of cell types with this phenotype. Tables 1 and 2 indicate that, in general, expression of PDGF receptors is characteristic of "connective tissue" cells and not of epithelial, endothelial, or mature hematopoietic cells. There may be some exceptions to this general rule. For example, capillary endothelial cells within malignant human glioblastoma multiforme tumors have been reported to express PDGF receptor β-subunits based on immunocytochemical and in situ studies (HERMANSSON et al. 1988).

2. Relative Levels of Expression of the Two Receptor Subunits in Different Cell Types

The numbers of each of the two receptor subunits expressed per cell vary widely from cell type to cell type, both in absolute numbers and in proportions. At one extreme, pup rat smooth muscle cells seem to express only β-subunits. Most of the cell types examined, including all of the diploid fibroblasts, have been found to express at least 5 times as many β-subunits as α-subunits. Absolute numbers of receptors vary from a low of about 20 000 β-subunits in U-2OS human osteosarcoma to about 270 000 β-subunits in human dermal fibroblasts. A third "group" of cells express comparable levels of α-subunits and β-subunits. This group includes mouse 3T3 cells and human MG-63 osteosarcomas. This group of cells is particularly useful for comparing the properties of α- and β-subunits since α-subunits constitute an easily detectable proportion of total receptor subunits. To date we have not found a naturally occurring cell type that expresses only α-subunits.

The number of receptors detected using ^{125}I-PDGF-AB was already substantial compared with the number of receptors for other growth factors. For example 3T3 cells bind about 50 000 molecules of ^{125}I-EGF per cell compared to 100 000–200 000 molecules of ^{125}I-PDGF-AB. When we include the number of β-subunits, and consider that it takes two subunits to constitute one PDGF binding site, the number of PDGF receptor subunits (α plus β) is very large (300 000 on human fibroblasts and 530 000 on Swiss 3T3 cells).

X. Regulation of Expression or Properties of PDGF Receptors

1. Regulation of the Affinity of the PDGF Receptor

In contrast to many circumstances in which the affinity of the EGF receptor can be altered, e.g., in response to PDGF binding (BOWEN-POPE et al., 1983; COLLINS et al. 1983), there are very few instances in which the affinity of the PDGF receptor seems to be altered. Thus, protein kinase C is activated in response to several growth factors and has been implicated in the regulation of the affinity of the EGF receptor. Addition of tumor promoters (exogenous activators of kinase C) does not significantly affect PDGF binding, but does synergize to some extent with PDGF in stimulating mitogenesis (EIDE et al. 1986).

Considerable circumstantial evidence indicates that the glycolipid gangliosides affect, or are affected by, cell proliferation. Addition of specific gangliosides (e.g., G_{m1}) to the culture medium inhibits the mitogenic response of 3T3 cells to PDGF without affecting response to EGF (BREMER et al. 1984). This inhibition is correlated with an increase in the affinity of PDGF binding to treated cells and with a decrease in the ability of PDGF to stimulate receptor autophosphorylation. It is possible that these changes reflect the influence of the gangliosides on the membrane environment of the receptor, possibly by altering the way in which receptor subunits come together to form active dimers.

2. Acute Regulation of PDGF Receptor Expression

RIZZINO et al. (1988) have reported that binding of TGF-β, EGF, fibroblast growth factors (FGF), and PDGF all decrease as cell density increases in culture. For PDGF, the effect was about 35%. Insulin-like growth factor I (IGF-I) has been reported to cause a detectable increase in the numbers of PDGF receptors on rat aortic smooth muscle cells during the first 3 h after treatment (Pfeifle et al. 1987). Expression of the two PDGF receptor subunits on 3T3 cells can be rapidly and separately regulated by several growth factors. After exposure to TGF-β, EGF, FGF, or serum, binding of ^{125}I-PDGF-AA remains unchanged for about 1.5 h, then decreases sharply, becoming almost undetectable by 24 h (Gronwald et al. 1989). Expression of the β-subunit is regulated by TGF-β in the opposite direction. Receptor mRNA detected by the β-subunit cDNA probe increased 2-fold during this period, and expression of β-subunits (measured as PDGF-AA-noncompetable ^{125}I-PDGF-BB binding) increased by about 50%. One possible mechanism through which the α-subunits might be lost would be the selective induction of PDGF-AA secretion by TGF-β resulting in selective autocrine downregulation of α-subunits. Indeed, TGF-β has been proposed to stimulate AKR-2B cells by inducing a PDGF autocrine system (Leof et al. 1986). As is detailed in Gronwald et al. (1989), the evidence indicates that such a mechanism does not account for the decrease in ^{125}I-PDGF-AA binding. The ability of TGF-β to selectively regulate the expression of the two receptor subunits could be of physiological significance in regulating responsiveness to PDGF by selectively altering responsiveness to different isoforms. TGF-β treatment, which is itself slightly stimulatory, has no effect on mitogenic responsiveness to PDGF-BB, but greatly reduces stimulation by PDGF-AA (Gronwald et al., 1989).

3. Regulation of PDGF Receptor Expression in Potential PDGF Autocrine Systems

a) Receptor Expression in Cells Transformed by SSV

Since the oncogene of simian sarcoma virus (SSV) is derived from the B chain of PDGF, it seemed likely that transformation by SSV was mediated through autocrine stimulation of PDGF receptors by the retrovirally encoded PDGF. This is supported by the observations that: (a) SSV-transformed cells produce a protein able to bind to PDGF receptors (Bowen-Pope et al. 1984b; Huang et al. 1984b; Owen et al. 1984; Johnsson et al. 1985a); (b) high-level expression of the normal form of the PDGF B chain can also transform cells with PDGF receptors (Gazit et al. 1984; Clarke et al. 1984); (c) exogenously added PDGF can elicit the same phenotype as expression of v-*sis* (Johnsson et al. 1986); and (d) SSV can infect, but not transform, cells which do not express PDGF receptors (Leal et al. 1985).

Although it is clear that transformation by SSV involves PDGF binding to the PDGF receptor, the location of the interaction is a subject of current controversy. Initially it was assumed that the v-*sis* was secreted, then bound to the cell surface. This was supported by the presence of v-*sis* in the culture medium (Bowen-Pope et al. 1984b; Garrett et al. 1984; Huang et al. 1984b; Johnsson et al. 1985b) and by the ability to interfere with the autocrine phenotype using agents which were presumed to act at the cell surface or extracellular compartment: (a) Addition to the culture medium of antibodies against PDGF partially inhibits the proliferation of some SSV-transformed cells (HUANG et al. 1984b; JOHNSSON et al. 1985b). (b) Suramin largely prevents the decrease in PDGF receptors seen in SSV-transformed cells (GARRETT et al. 1984) and restores normal growth properties (BETSHOLTZ et al. 1986a).

The arguments in favor of an *intracellular* interaction between PDGF and its receptor include the following:

1. The receptor, as a membrane protein, and the v-*sis* product, as a secreted protein, are both synthesized on the rough endoplasmic reticulum and processed through the Golgi. Since the immature intracellular form of the receptor is capable of binding PDGF when tested in vitro (see above), the receptor-ligand interaction would seem likely to occur within the cell unless prevented by adverse conditions of pH, membrane composition, etc.

2. Phosphorylated receptor precursors have been detected in SSV-transformed cells (HUANG and HUANG 1988; KEATING and WILLIAMS 1988). If the phosphorylation reflects receptor activation, this would argue strongly for intracellular activation. It should be noted that receptor activation in this compartment may not be capable of initiating a mitogenic response. HANNINK and DONOGHUE (1988) have demonstrated, using v-*sis* expression controlled by the heat shock promoter, that when monensin is used to inhibit maturation of intracellular receptor precursors, the precursor is phosphorylated but expression of c-*fos* is not induced, indicating that this activation of the intracellular precursor autophosphorylation does not seem to be competent to initiate all of the components of a mitogenic response.

3. Metabolic labeling studies have been interpreted to indicate that in SSV-transformed cells, the receptor precursor is degraded by lysosomal hydrolases before its carbohydrate has been fully processed, presumably as a consequence of intracellular occupation by v-*sis* (KEATING and WILLIAMS 1988).

4. The efficacy of suramin in interfering with the autocrine system might not mandate an extracellular occupation since there is some evidence that the effects of suramin may be mediated within the cell rather than at the cell surface, as previously assumed, possibly as a consequence of endocytosis of the suramin (HUANG and HUANG 1988).

Given the evidence for both intracellular and extracellular sites of autocrine receptor-ligand interaction, it seems most likely that the PDGF receptor can be occupied by PDGF both inside or outside the cell, and that the relative biological significance of occupation in either of these two locations may depend on the cell type and/or specific culture conditions.

b) Oncogenic Transformation by Agents Other Than SSV

When fibroblasts and cell lines expressing PDGF receptors are transformed by a variety of agents, they often secrete PDGF into their culture medium and express reduced levels of PDGF receptors (BOWEN-POPE et al. 1984b, and citations therein). It has been proposed that the PDGF produced by these transformed cells may partially account, via an autocrine mechanism, for both the loss of surface receptors and for the independence of these cells from exogenous mitogens. Although it has been shown that the transformed cells secrete sufficient PDGF to account for the loss of observable surface receptors (e.g., Bowen-Pope et al. 1984b), this does not prove that autocrine downregulation is the cause of the decrease. Several lines of evidence now indicate that in some of these cases, decreased receptor synthesis, rather than autocrine downregulation, is in fact the explanation for the loss of receptors (Seifert and Bowen-Pope, unpublished observations).

4. Regulation of PDGF Receptor Expression During Embryogenesis, Development, and Wound Healing

It has been proposed that PDGF-driven autocrine proliferation functions, in a temporally and spatially regulated fashion, to drive some of the growth which occurs during normal development and tissue repair. "Undifferentiated" inner-cell-mass-like embryonal carcinoma cells secrete PDGF and do not bind significant PDGF. Upon induction of "differentiation" by retinoic acid, the cells assume characteristics of primitive endoderm, cease to secrete PDGF, and express PDGF receptors (Gudas et al. 1983; Rizzino and Bowen-Pope 1985). Within the first-trimester human placenta cytotrophoblasts express mRNA for the B chain of PDGF. Since cultured cytotrophoblasts express PDGF receptors, it was proposed that these cells are stimulated by autocrine production of PDGF.

Developmentally regulated switches in PDGF phenotype seem to occur during the final stages of growth of the aorta. Smooth muscle cells cultured from the aorta of newborn ("pup") rats secrete PDGF, proliferate rapidly even in the absence of added PDGF, and express reduced levels of cell-surface PDGF receptors (Seifert et al. 1984). We proposed that the growth of the aorta during this period might be driven, in part, by autocrine secretion of, and response to, PDGF. The proposed PDGF autocrine system seems to be reactivated in smooth muscle cells from adult arteries in response to injury (Walker et al. 1986).

5. Receptor Expression In Vivo

Very little is known to date about expression of PDGF receptors in vivo. Accumulating evidence suggests that receptor expression may be low in normal tissues and increased in response to various forms of disturbance, including the adaptation of the cells to culture. Human skin fibroblasts, one of the classic receptor-positive cells in culture, were found to express few receptor β-subunits in vivo and to show progressive increases in receptor expression as the cells were maintained in culture over several days (Terracio et al. 1988). In the pig uterus, receptor expression was highest in the endometrium, particularly in the stromal cells around the myometrial glands, and in small arteries. Myometrial cells in culture progressively increased levels of receptor expression (TERRACIO et al. 1988). Smooth muscle cells in normal large arteries have been reported to express low levels of receptors, but smooth muscle cells in atherosclerotic lesions express higher levels (RUBIN et al. 1988 b; WILCOX et al. 1988). In the latter study, most of the receptor β-subunit mRNA in advanced lesions was detected in scattered cells in the intima, which differed in several ways from well-differentiated smooth muscle cells and were described as "mesenchymal-appearing intimal cells" (WILCOX et al. 1988). An early fibrous lesion showed higher levels of receptor expression. Receptor expression by endothelial cells was not detected in any of the above studies, but has been reported in capillary endothelial cells in malignant human glioblastoma multiforme (HERMANSSON et al. 1988).

D. In Vivo Clearance of PDGF and PDGF-Binding Proteins

I. Rapid In Vivo Clearance of PDGF-AB

PDGF was discovered based on the greatly reduced mitogenic potency of serum prepared from cell-free plasma-derived serum as compared with the potency of serum prepared from whole blood (BALK et al. 1973; KOHLER and LIPTON 1974; ROSS et al. 1974). It is now known that PDGF constitutes only approximately 50% of the mitogenic activity released by activated human platelets (HELDIN et al. 1981 c). However, no PDGF is detectable in plasma using a sensitive radioreceptor assay for PDGF, and it therefore appears to be sequestered within the platelet (BOWEN-POPE et al. 1984 a). Intravenous injection of ^{125}I-PDGF-AB into normal baboons resulted in rapid clearance of PDGF from the plasma (half-time 2 min) (BOWEN-POPE et al. 1984 a). Unlabeled PDGF ($\approx 95\%$ PDGF-AB and 5% PDGF-BB) was cleared with comparable kinetics, suggesting that the rapid clearance was not due to iodination damage. In addition, PDGF present in freeze-thaw lysates of platelets ($\approx 85\%$ PDGF-AB and 15% PDGF-BB) was cleared equally rapidly and therefore the rapid clearance does not appear to be due to structural alteration of the molecule during purification.

II. PDGF-Binding Proteins

One possible mediator of this rapid clearance, which would be consistent with the concept that PDGF acts as a local mitogen, is the presence in plasma of specific binding proteins for PDGF which inhibit the binding of PDGF to its cell-surface receptor (BOWEN-POPE et al. 1984 a; RAINES et al. 1984; RAINES and ROSS 1988). It has been shown that under conditions of low concentrations of PDGF and high concentrations of plasma (i.e., conditions existing in vivo at sites peripheral to sites of platelet release), 80%–85% of the PDGF would be bound to, and inhibited by, the plasma components. PDGF binding proteins were first identified because they interfered with the measurement of PDGF levels in plasma or serum by radioreceptor assay (BOWEN-POPE et al.

1984a; RAINES et al. 1984) and by radioimmunoassay (HUANG et al. 1983). These plasma fractions do not act by blocking or permanently altering the properties of the PDGF cell surface receptor, but rather appear to block the site on PDGF responsible for receptor binding or alter the PDGF molecule so that it is no longer able to bind to its receptor.

Further characterization has demonstrated specific association of ^{125}I-PDGF with plasma fractions by gel filtration and SDS-PAGE (HUANG et al. 1983, 1984a; RAINES et al. 1984). One of the PDGF binding proteins has been identified as α_2-macroglobulin (HUANG et al. 1984a; RAINES et al. 1984). Antisera to α_2-macroglobulin removed only 50% of the PDGF-binding protein activity from plasma, and it therefore appears to account for only part of the plasma binding activity for PDGF. Other species distinct from α_2-macroglobulin with molecular weights of 40000 and 150000 have also been identified. These studies were performed with ^{125}I-PDGF-AB, but binding proteins in plasma also block the binding of PDGF-AA and PDGF-BB to cells (our unpublished observations).

By preventing binding of PDGF to its receptor on responsive cells, binding proteins may alter its activity. It is conceivable that the levels of binding proteins may regulate the amount of unbound and active PDGF. In addition to plasma, cells such as monocyte/macrophages synthesize and secrete binding proteins for PDGF, and their production of binding proteins may be regulated (SHIMOKADO et al. 1985; RAINES and ROSS 1988). By altering the molecular size of PDGF molecules by complex formation, the accessibility of PDGF to target cells may be affected and ultimately clearance from the circulation may be modulated. It is also possible that complex formation between PDGF and its binding protein may protect it against degradative enzymes or other mechanisms of inactivation and thus prolong its potential availability.

E. Biochemical and Cellular Mechanism of Action

The interaction of PDGF with its cell-surface receptors is rapidly translated into activation of intracellular pathways, initiating a sequence of events that eventually results in a specific cellular response. It is unclear which of these rapidly triggered cellular events are necessary to induce the cellular responses that eventually lead to directed movement, smooth muscle contraction, and ultimately cell cycle progression. Many of the early events may simply be a consequence of PDGF-receptor interaction and not essential for the signaling process. Currently, the biochemical mechanism by which these pleiotropic responses are induced is unknown. PDGF-induced responses which have been the focus of studies of possible signaling mechanisms are discussed below.

I. Coordinate Control of Cell Proliferation by PDGF and Plasma Components

The capacity of PDGF to induce cell doubling may be modified by other biological components in the extracellular environment, particularly plasma components, that act in a coordinate fashion to promote the optimum mitogenic response (PLEDGER et al. 1977; VOGEL et al. 1978). STILES and colleagues (STILES et al. 1979; SMITH and STILES 1981) have proposed the concept that mitogenesis is induced in two stages: induction of "competence" by growth factors like PDGF, followed by "progression" through the cell cycle that is dependent on plasma components. In the absence of plasma factors, PDGF-treated 3T3 cells remain growth arrested after treatment with PDGF alone. The active components of plasma in this system can be replaced by EGF and IGFs (STILES et al. 1979; LEOF et al. 1982). The absolute requirement for "progression" factors to release PDGF-treated cells from growth arrest and to induce several rounds of cell division appears to be dependent on the cell system evaluated and the growth media. Glial cells in a defined medium containing PDGF undergo several rounds of division without the addition of progression factors (HELDIN et al. 1980). However, in the 3T3 system, it has been demonstrated that when cells are briefly exposed to PDGF and then fused with untreated cells, the resulting heterokaryons become competent to replicate their DNA (SMITH and STILES 1981). Cytoplasts are able to transfer this growth response, but protein synthesis is not required. However, when RNA synthesis is blocked during PDGF treatment, the capacity to transfer the PDGF growth signal to untreated cells is lost. The evaluation of the PDGF-induced "competence" state has led to experiments to further define the PDGF-stimulated early genes (see below).

II. Phospholipase Activation and Prostaglandin Metabolism

One of the earliest effects of such PDGF ligand-receptor interactions is the rapid induction of increased turnover of phosphatidylinositol, release of free arachidonic acid, and formation of new eicosanoids including prostaglandins I_2 and E_2 (HABENICHT et al. 1985). This early release of free arachidonic acid is probably due to activation of both phospholipase A_2 (SHIER 1980; SHIER and DURKIN 1982) and phospholipase C (HABENICHT et al. 1981, 1985, 1986). Phospholipase activation not only leads to release of free arachidonic acid, but also to formation of diacylglycerol (COUGHLIN et al. 1980; HABENICHT et al. 1981; SAWYER and COHEN 1981; ROZENGURT et al. 1983). This activation of a PDGF-sensitive phospholipase C/diglyceride lipase pathway results in the complete breakdown of phosphatidylinositol and the subsequent breakdown of phosphatidylinositol-4,5-bisphosphate (PIP_2) (CHU et al. 1985; HASEGAWA-SASAKI 1985; MACDONALD et al. 1987; MATUOKA et al. 1988). The turnover of PIP_2 is felt to represent a crucial step in stimulating cells to enter S and subsequently divide. MATUOKA et al. (1988) have demonstrated that a monoclonal antibody to PIP_2, when microinjected into the cytoplasm of NIH

3T3 cells before or after exposure to PDGF, could completely abolish [³H]thymidine incorporation normally induced by PDGF exposure, suggesting that PIP₂ breakdown is a crucial component for induction of cell proliferation by mitogens such as PDGF. Observations such as these also prompted investigators to determine whether G protein activation within the plasma membrane, or in the subplasmalemmal component of the cells was associated with PDGF receptor-ligand interaction and phospholipase C-mediated mitogenesis. Studies by HASEGAWA-SASAKI et al. (1988) and by LETTERIO et al. (1986), using inhibitors of G protein synthesis or comparison with vasopressin- or bombesin-elicited hydroloysis of polyphosphoinositides, suggest that G proteins are not involved in PDGF-induced mitogenesis.

Phosphatidylinositol-induced breakdown is correlated not only with an increase in free arachidonic acid, but also with cell-cycle-dependent changes in stearic acid and in glycerol metabolism. Both arachidonic and stearic acids are incorporated into newly formed phosphatidic acid and polyphosphatidylinositol within 60 min after PDGF stimulation of 3T3 cells (HABENICHT et al. 1985). Thus PDGF will also induce de novo phosphatidylinositol synthesis within an hour after PDGF binding to its receptor, and ultimately PDGF stimulates incorporation of glycerol into all of the major phospholipids and into triacylglycerol during S phase in the cell cycle. PDGF therefore has an early effect on phosphatidylinositol synthesis, later stimulation of arachidonic acid incorporation into triacylglycerol, and later still induction of de novo phosphatidylcholine, phosphatidylethanolamine, and triacylglycerol synthesis. These PDGF effects are undoubtedly important in the generation of new membranes and in increased fatty acid turnover in rapidly growing cells.

III. Modulation of Ion Flux

Another important component of the PDGF-induced early changes that are closely related to the formation of new diacylglycerol is the increase in intracellular Ca^{2+}. When PDGF binds to its receptor on cells such as 3T3 cells, there is a decrease in the intracellular content of Ca^{2+} with an apparent increased rate of efflux of preloaded Ca^{2+} and a decrease in residual intracellular Ca^{2+} that remains after this efflux. These changes are consistent with a rapid release of Ca^{2+} from an intracellular Ca^{2+} pool, and it has been postulated by ROZENGURT et al. (1984) and by FRANTZ (1985) that Ca^{2+} is an important component of the induction of the mitogenic signal in these cells (LOPEZ-RIVAS et al. 1987). Such an intracellular shift of Ca^{2+} was observed by BERK et al. (1986) upon exposure of vascular smooth muscle to PDGF.

In a similar vein, it has been suggested by PARIS and POUYSSEGUR (1984) that changes in the Na^+/H^+ antiport in cells is a critical component in inducing mitogenesis. Changes in intracellular pH which influence the exchange of Na^+ and H^+ may be critical in growth-factor-induced signaling of mitogenesis. Such a Na^+/H^+ exchange under physiologic conditions would be responsible for a sustained alkalinization of the cytoplasm of cells. These two investigators have suggested that conformational changes of the Na^+/H^+

antiporter may be critical in inducing mitogenesis. Several other laboratories have examined the role of cytoplasmic pH and exchange of Ca^{2+} and Na^{+} ions, and their studies also support the notion that cytoplasmic pH changes associated with these are important in mitogenesis (MOOLENAAR et al. 1984; HESKETH et al. 1988; LUCAS et al. 1988; MARTINEZ et al. 1988). The inter-relationship between the Ca^{2+} exchange noted above and the role of Na^{+}/H^{+} exchange, and the resultant increase in pH within the cells, is somewhat con-troversial. However, there is general agreement that a common signal result-ing from the binding of growth factors such as PDGF to their receptors that follows the breakdown of inositol phospholipids is the increased cellular alkalinity that results from these ionic shifts. The role of these various ionic in-termediates (and in particular calcium oscillations) within the cell in inducing cellular mitogenesis have been reviewed by BERRIDGE et al. (1988).

IV. Tyrosine Phosphorylation

PDGF also induces a host of other changes, some of which are directly as-sociated with the receptor for PDGF. When PDGF binds to its receptor, it in-duces phosphorylation of tyrosine moieties on the receptor due to activation of a tyrosine kinase (EK et al. 1982; NISHIMURA et al. 1982; PIKE et al. 1983; KAZLAUSKAS and COOPER 1988; PASQUALE et al. 1988). It has been suggested that such autophosphorylation of the receptor on tyrosine moieties is critical to mitogenesis. As discussed in previous sections, in receptor mutants altering these tyrosine kinase regions, mitogenesis is ablated. Nevertheless, in other mutated forms of the β-subunit of the PDGF receptor where the tyrosine kinase region, which is split in the PDGF receptor, is joined by removing the region between these two sites, the ability of the receptor to induce mitogenesis is decreased with no change in kinase activity (ESCOBEDO and WILLIAMS 1988). Thus, although tyrosine phosphorylation on the receptor ap-pears to be a necessary event in inducing mitogenesis, it does not appear to be sufficient in itself.

PDGF also induces phosphorylation of a number of other proteins in the cells, not only on tyrosine moieties but on threonine (DAVIS and CZECH 1985) as well as a host of other intracellular proteins, including $pp60^{c\text{-}sarc}$ (GOULD and HUNTER 1988) and others (COOPER et al. 1982). The phosphorylation of some of these intracellular proteins is mediated by protein kinase C activation (KAZLAUSKAS et al. 1988; KAZLAUSKAS and COOPER 1988), in which proteins of varying molecular weights (in the case of C kinase, a protein denoted as p42) are phosphorylated as a result of PDGF-stimulated protein kinase C ac-tivation.

V. PDGF Induction of "Early Genes"

As discussed in Sect. E. I, transfer of PDGF-induced "competence" in 3T3 cells requires transcription of unique genes (SMITH and STILES 1981), but the mitogenic response to PDGF and serum growth factors is accompanied by little change in the overall rate of transcription or translation during cell growth (reviewed in ROLLINS and STILES 1988). This paradox has been partially resolved by the discovery that PDGF and other growth factors regulate the expression of discrete low-abundance and labile gene products (STILES 1983; ROLLINS and STILES 1988). These PDGF-inducible gene sequences account for 0.1%–0.3% of the total genes transcribed in 3T3 cells (COCHRAN et al. 1983; LINZER and NATHANS 1983). Inhibition of protein synthesis blocks progression of the cells into S but does not block induction of these genes, demonstrating that their induction by PDGF is not a consequence of induction of cell proliferation. However, data indicate that the PDGF-inducible genes are not coordinately regulated by a common intracellular signal (HALL and STILES 1987; ROLLINS et al. 1987).

Two of these genes are the protooncogenes c-*myc* and c-*fos* (KELLY et al. 1983; COCHRAN et al. 1984; GREENBERG and ZIFF 1984), whose products seem to act as intracellular mediators of the mitogenic response to PDGF. Expression of c-*myc* and c-*fos* within PDGF-stimulated cells is not sufficient, or even necessary in particular instances, for growth to occur (COUGHLIN et al. 1985; ZULLO et al. 1985). However, constitutive expression of c-*myc* reduces the requirement for PDGF (ARMELIN et al. 1984). The biochemical functions of c-*myc* and c-*fos* proteins within the nucleus are not known. A report that c-*myc* protein facilitates DNA replication in an in vitro system (STUDZINSKI et al. 1986) is interesting but not yet demonstrated within the cell. c-*fos* protein has been shown to *bind* to specific DNA sequences and may also be involved in the regulation of other gene expression (DISTEL et al. 1987).

PDGF directly regulates the expression of 10–30 independent "early" genes in 3T3 cells (COCHRAN et al. 1983, 1984; KELLY et al. 1983). The function of the protein products of these genes has not yet been determined. However, one of these, *JE*, has been shown to code for a protein with cytokine-like properties (ROLLINS et al. 1988) and another, *KC*, codes for an α-granule-related protein corresponding to a factor described as "melanoma growth-stimulating activity" (OQUENDO et al. 1989). Unlike c-*myc* and c-*fos*, whose products act on the nucleus, *JE* and *KC* may have quite different functions. Their target cells and biological activities remain to be determined. Another gene induced with kinetics similar to c-*fos* is the gene encoding the facilitated glucose transporter protein (HIRAKI et al. 1988).

PDGF has also been shown to induce, although more slowly, β-fibroblast interferon and (2′,5′)-oligoadenylate synthetase (ZULLO et al. 1985). The interferon gene responds to PDGF only in confluent monolayer cultures, whereas the *myc* and *JE* genes are inducible in either subconfluent or confluent cultures. The translation of these genes acts to inhibit cell growth and may function as feedback inhibitors of the mitogenic response to PDGF.

F. Biological Activity of PDGF in Vitro and In Vivo

I. Direct and Indirect Effects of PDGF on Cell Growth In Vitro

1. Direct Mitogenic Response to PDGF

The mitogenic potency of the different isoforms of PDGF depends on the cell type, and parallels the ability of the cells to bind the different isoforms (SEIFERT et al. 1989). 3T3 cells, which express high levels of both subunits and thus bind high levels of all three isoforms, are equally stimulated by all three isoforms. By contrast, human dermal fibroblasts express few α-subunits, bind relatively little PDGF-AA, and are very poorly stimulated by PDGF-AA. Therefore it seems, at least for mitogenic stimulation, that PDGF-AA is not intrinsically less mitogenic than PDGF-BB, as was implied by NISTER et al. (1988a), and that the poor stimulation may reflect quantitative differences in numbers of binding sites rather than qualitative differences in the activities of the corresponding receptors.

The detection of high-affinity receptors for PDGF has been limited to connective tissue cells such as dermal fibroblasts, vascular smooth muscle, glial cells, and chondrocytes (see Tables 1 and 2). This restriction in cell type coincides with the ability of these cells to respond mitogenically to PDGF. Recently, however, epithelial and endothelial cells have been shown to respond to PDGF. PDGF stimulates DNA synthesis in both human mesothelial cells (GABRIELSON et al. 1988) and rat islets of Langerhans (SWENNE et al. 1988). The rat lens, an epithelial tissue, required PDGF in culture (with pulsatile delivery of PDGF every 5–6 h) for growth and transparency of the lens (BREWITT and CLARK 1988). It therefore appears that both the PDGF target cell populations and the PDGF receptor distribution need to be reevaluated, particularly in relation to the different PDGF isoforms and PDGF receptor subunit distribution.

2. Indirect Effects of PDGF on Cell Growth

In addition to directly stimulating the proliferation of a number of cells, PDGF appears to enhance the proliferative response of other cells by inducing the expression of other genes whose products then amplify the mitogenic response. An example of this is the stimulation of production and secretion of somatomedin-C-like substances by cultured fibroblasts and porcine aortic smooth muscle cells in response to PDGF (CLEMMONS et al. 1981; CLEMMONS and VAN WYK 1985). With aortic smooth muscle cells, which require somatomedin C for optimal growth, a monoclonal antibody to somatomedin C inhibits 85% of the PDGF-induced mitogenic stimulation (CLEMMONS and VAN WYK 1985). Another similar example is the induction of PDGF A-chain expression and secretion in fibroblasts in response to PDGF stimulation (PAULSSON et al. 1987). In both of these examples, the gene products stimulated by PDGF presumably enhance the proliferation of the same cell population responding to PDGF.

Other examples exist in which it appears that the cell population directly responding to PDGF secretes gene products which then enhance the proliferation of other cell types. In such a manner, PDGF enhances in vitro erythropoiesis (DELWICHE et al. 1985). Erythroid colony growth in cultures of whole marrow is enhanced in a dose-dependent manner by increasing concentrations of PDGF, but no enhancement is observed if only nonadherent marrow cells are cultured. The response of the erythroid cells to PDGF can be restored if the nonadherent marrow cells are cultured in the presence of fibroblasts or smooth muscle cells, cells responsive to PDGF. Presumably, PDGF is stimulating the synthesis and/or release of colony stimulating factors from the fibroblasts or smooth muscle cells that then stimulate erythroid colony growth. Another example of indirect enhancement is the ability of PDGF to partially restore the antigen-specific T-cell proliferative response, which has been limited by the reduction of serum (ACRES et al. 1985). In the absence of antigen, PDGF is not mitogenic for T cells, is unable to act synergistically with T-cell growth factor, and does not modulate T-cell surface molecules. The effect can be produced by preincubating the antigen-presenting cells (E-rosette-negative cells, principally B cells and monocytes) with PDGF, which increases major histocompatibility complex (MHC) class II antigens. It is unclear in this example which cell is mediating the effect and whether the enhanced growth is due solely to the increased expression of MHC class II antigens.

II. Other In Vitro Activities of PDGF

1. Directed Cell Migration and Cell Activation in Response to PDGF

The chemotactic response of mesenchymal cells such as fibroblasts, smooth muscle cells, and some cell lines (3T3 and NRK) to PDGF also coincides with their ability to respond mitogenically to PDGF (GROTENDORST et al. 1981, 1982; SEPPA et al. 1982; GROTENDORST 1984). The dose of PDGF required for maximal chemotactic activity coincides with that required for maximal mitogenic stimulation (GROTENDORST et al. 1982; SEPPA et al. 1982). The ability to induce directed migration appears to be retained by PDGF molecules made by a number of different cells, including monocyte/macrophages (MARTINET et al. 1986), SV40-transformed 3T3 cells (BLEIBERG et al. 1985), kidney epithelial cells (KARTHA et al. 1988), and recombinant PDGF-AA and PDGF-BB (our unpublished observations). However, PDGF-AA isolated from the clonal human malignant glioma cell line U-343 MGa clone 2:6 (HAMMACHER et al. 1986b) has been reported to inhibit PDGF-AB/BB chemotaxis and to be unable to induce a chemotactic response in human foreskin fibroblasts (AG1523) (NISTER et al. 1988a). It is unclear which form(s) of the A chain are made by these cells and how specific the observed inhibition is to the cell type examined. It will be important to understand the molecular basis for the observed inhibition, as this represents a major difference in the biological activities reported for the different PDGF isoforms.

PDGF is also reported to be chemotactic for mononuclear cells and neutrophils (DEUEL et al. 1982; WILLIAMS et al. 1983), cells that do not respond mitogenically to PDGF. The induction of monocyte chemotaxis is controversial (GRAVES et al. 1989) and may reflect different subpopulations of cells and/or differences in the PDGF receptor responsible for the observed directed migration of monocytes. Retinal pigment epithelial cells have also been reported to respond chemotactically to PDGF (CAMPOCHIARO and GLASER 1985). The capacity of PDGF to act as a chemoattractant appears to have been extensively conserved, since PDGF is also a chemoattractant for the free-living ciliated protozoan *Tetrahymena* (ANDERSEN et al. 1984).

PDGF also stimulates a dose-dependent granule release by neutrophils and monocytes (WILLIAMS et al. 1983; TZENG et al. 1984, 1985). In neutrophils, this leads to release of superoxide, increased aggregation, and release of lysozyme and vitamin B_{12} binding protein (TZENG et al. 1984). The levels of activation are comparable to those induced by N-formylmethionyl leucylphenylalanine methyl ester (FMLP), a soluble activator of neutrophils. PDGF has also been reported to stimulate neutrophil phagocytosis and block agonist-induced activation of the neutrophil oxidative burst (WILSON et al. 1987). Addition of PDGF to peripheral blood mononuclear cells inhibits human natural killer cell activity in a dose-dependent manner (GERSUK et al. 1986). Thus, in addition to attracting leukocytes to sites where PDGF is released, PDGF is also capable of modulating the activity of the cells and the release of their specific granule constituents.

2. Modification of Cellular Matrix Constituents

In a number of disease processes, connective tissue destruction and deposition have significant clinical implications, as does cellular proliferation. In response to PDGF, protein synthesis as well as collagen synthesis is increased (CANALIS 1981; OWEN et al. 1982). PDGF specifically stimulates formation of type V collagen, and regulates the amount of type III versus type IV collagen synthesized in gingival fibroblasts (NARAYANAN and PAGE 1983). The expression and secretion of thrombospondin, another cell-associated matrix protein thought to be important in cell-matrix interactions (LAWLER 1986; SILVERSTEIN et al. 1986), is also modulated by PDGF (MAJACK et al. 1985, 1987). In human dermal fibroblasts, PDGF also stimulates a 3-fold increase in collagenase activity (CHUA et al. 1985), and stimulation of collagenase secretion is detected 8–10 h after exposure to PDGF (BAUER et al. 1985). The ability of PDGF to stimulate both secretion of matrix components as well as enzymes capable of breaking down the matrix may be important in altering the cellular milieu in preparation for cell movement and mitosis in normal processes such as wound repair and development, as well as in pathologic conditions such as arthritis, liver cirrhosis, and atherosclerosis.

3. Vasoconstriction

PDGF has been shown to induce concentration-dependent contraction of
strips of rat aorta in vitro and, on a molar basis, is more active than
angiotensin II (BERK et al. 1986). In vitro contractions induced by PDGF are
correlated with increased intracellular calcium in vascular smooth muscle
cells. This activity does appear to be dependent upon the vascular bed examined, as rat penetrating intracerebral arterioles isolated from brain parenchyma did not respond to PDGF-AB, PDGF-BB, or total platelet PDGF
(BASSETT et al. 1988). The ability of locally released PDGF to induce
vasoconstriction could be important in vivo at sites in the arterial tree that
may be partially occluded.

III. PDGF In Vivo and Clinical Applications

To date there are relatively few data concerning the actual roles played by
PDGF in vivo. Some data are beginning to accumulate in relation to its role in
several diseases, including atherosclerosis, systemic sclerosis (scleroderma),
and neoplasia. Several other roles have also been suggested and will be discussed below.

As described earlier in this chapter and detailed in Tables 1 and 2,
numerous diploid cells as well as neoplastically transformed cells have the
capacity to express the gene for both chains of PDGF and to synthesize and
secrete one or more of the different isoforms of PDGF. Many of the cells that
can secrete PDGF are also capable of secreting other growth factors such as
EGF, TGF-α, TGF-β, and IGF, and of synthesizing FGF and IL-1 (ROSS et
al. 1986).

As discussed in Sects. F. I. and F. II., PDGF induces directed cell migration (or chemotaxis) and direct and indirect cell proliferation, which are often
associated with reformation of connective tissue. Thus PDGF can be considered to be important in a variety of responses that include wound repair,
normal growth and development, myelofibrosis, pulmonary fibrosis, systemic
sclerosis, atherosclerosis, and neoplasia. Each of these will be discussed below
in relation to data presently available.

1. Wound Repair

Normal wound repair consists of a series of chronologic events that follow
formation of wounds with the associated blood coagulation and thrombosis
that occur during wounding. The earliest cells that appear in wounds are
leukocytes, principally neutrophilic leukocytes, together with monocyte-
derived macrophages. The monocyte-derived macrophage becomes a
dominant cell within a day or two following wounding, and the appearance of
these cells is followed by the ingress and proliferation of capillaries,
fibroblasts, and smooth muscle cells (ROSS 1968).

The entire process begins with the deposition of numerous activated
platelets during the process of wounding. The platelets are activated by

thrombin and other factors present in the blood coagulum. Platelet release of PDGF would readily occur with other factors released from the platelets (such as TGF-β, platelet factor 4, and β-thromboglobulin), which can induce chemotaxis of cells into the wound (SENIOR et al. 1983; DEUEL et al. 1981; WAHL et al. 1987). The subsequent appearance of monocyte/macrophages, capillaries, fibroblasts, and smooth muscle cells also represents possible sources of PDGF (see Sect. B. III). This would permit formation and secretion of PDGF into the wounds at different times during the process of repair.

A limited number of studies to date support the above suggestions. For example, GROTENDORST et al. (1985) demonstrated that when PDGF was added to Hunt-Schilling wound chambers containing collagen gels and placed subcutaneously in rats, the chambers induced an early influx of connective tissue cells, increase in DNA synthesis, and increase in collagen deposition. Similar studies in diabetic animals, which normally exhibit a decreased rate of wound repair, showed that healing was restored to normal by addition of PDGF to these wound chambers. Addition of TGF-β and EGF in combination with PDGF in such chambers demonstrated a clear synergism with a combination of growth factors (LAWRENCE et al. 1986). This study and that of LYNCH et al. (1987) demonstrate the importance of the synergistic, and thus interactive, effects of PDGF when it is combined with other growth factors such as EGF, TGF-β, or IGF-I. SPRUGEL et al. (1987) examined the effect of implantation of PDGF, TGF-β, basic FGF, and EGF in a different kind of wound chamber made of polytetrafluoroethylene that contained collagen or collagen-heparin mixtures and the various combinations of growth factors. Their studies demonstrated that PDGF, basic FGF, and TGF-β each could induce granulation tissue development in normal animals 10 days after implantation of the wound chamber, suggesting that these growth factors may have, at least in part, induced cells such as macrophages, lymphocytes, and others to enter into the wound chambers at an earlier time. It is also possible that these infiltrating cells may have participated in the proliferative response through growth factor formation.

KNIGHTON et al. (1986) examined the effect of platelet extracts on nonhealing cutaneous ulcers in a series of diabetic patients. The ulcers in these patients were treated with substances released from autologous platelets, and repeated applications produced healing of the ulcers, which could have otherwise led to leg amputations.

The normal and injured chick nervous tissue has been used to study alterations in the expression of PDGF receptors following injury using ^{125}I-PDGF-AB binding as a probe (RAIVICH and KREUTZBERG 1987). Injury to the chick sciatic nerve led to a rapid and massive induction of PDGF receptors on the fibroblast-like cells of the injured endoneurium. Comparison of the patterns of PDGF and β-nerve growth factor (β-NGF) binding following injury to the peripheral nerve suggests that different cell-type-selective mechanisms are responsible for specific activation of receptors on endoneurial fibroblasts (PDGF) and Schwann cells (β-NGF). A close correspondence in spatial and temporal pattern of blood-nerve barrier deficiency, endoneural interstitial edema, and induction of massive PDGF binding suggests that there may be a cause and effect relationship. Together these observations suggest that PDGF may be involved in proliferation of injured endoneurial connective tissue.

2. Embryogenesis and Development

Embryogenesis involves cell migration and proliferation, followed at particular stages by terminal differentiation to develop particular organs and tissues. PDGF is capable of inducing both cell migration and proliferation as well as altering cellular gene expression. Previous sections have described data showing in vitro modulation of expression of PDGF and its receptor in an embryonal carcinoma model of embryonic endoderm development, by cytotrophoblasts of the placenta, by smooth muscle cells of the growing aorta of young rats, and by vascular endothelial cells undergoing reversible tube formation (Tables 1, 2). PDGF has also been shown to inhibit differentiation by promoting proliferation of the bipotential glial progenitor cells (O-2A) that produce oligodendrocytes and type 2 astrocytes (RICHARDSON et al. 1988; NOBLE et al. 1988). These investigators have also shown that type 1 astrocytes, the major cell population in the developing optic nerve in addition to O-2A progenitors, are capable of secreting PDGF in vitro. The time course of appearance of PDGF A-chain mRNA in the brain is consistent with the numbers of astrocytes during brain development (RICHARDSON et al. 1988). All of these observations are consistent with, but do not prove, the hypothesis that developmentally regulated production of PDGF and its receptors contributes to autocrine stimulation of proliferation in embryogenesis.

Further support for a role for PDGF in development comes from analysis of tissue expression (Table 1). GOUSTIN et al. (1985) demonstrated B-chain expression in first-trimester human placentas localized to the highly proliferative and invasive cytotrophoblastic shell of the placenta, which also expresses elevated levels of c-*myc* mRNA, while the adjacent syncytiotrophoblasts do not. They also demonstrated that levels of expression of the B chain decrease approximately 10-fold between early first trimester and term. TAYLOR and WILLIAMS (1988) have extended these observations, and demonstrate placental expression of transcripts for both chains of PDGF and the β-subunit of the PDGF receptor throughout human pregnancy that peak during the second trimester. An identical pattern of expression of the β-subunit protein was confirmed by immunoblotting. Placental c-*fos* mRNA, a gene rapidly induced after the interaction of PDGF with its receptor (COCHRAN et al. 1984), also followed a similar temporal pattern. At all gestational ages, the apparent levels of PDGF A-chain mRNA exceeded those of B-chain mRNA by a mean of 6.4-fold.

Evidence for even earlier expression of PDGF has been obtained in *Xenopus* embryos and in preimplantation mouse embryos. In *Xenopus* embryos, where development prior to midblastula transition (6–7 h after fertilization) is directed by maternal mRNAs, two different forms of the PDGF A chain are encoded by maternal mRNAs (MERCOLA et al. 1988). Of approximately 600000 clones screened from an oocyte and gastrula stage cDNA library, five clones hybridized to the A-chain probe and no clones hybridized to the B-chain probe, in spite of detection of both genes in genomic DNA. No evidence for expression of B chain was found with Northern analysis of various stage embryos. Thus, the B-chain mRNA must be present at a much lower level than the A-chain mRNA within the early embryo, if at all. Analysis of A-chain transcripts demonstrated a decrease in abundance during development from oocyte to cleavage stage embryos, with reappearance at late gastrula stage after the onset of embryonic transcription.

Preimplantation mouse embryos, which grow and differentiate in the absence of exogenous factors, and unfertilized ovulated oocytes both contain transcripts for the PDGF A chain (RAPPOLEE et al. 1988). PDGF antigens were also detected in blastocysts by immunocytochemistry. Transcripts for TGF-α were also present in both maternal transcripts in the oocyte and in whole blastocysts, whereas EGF, β-NGF, basic FGF, and granulocyte-colony stimulating factor genes were not transcribed. In contrast, TGF-β1 transcripts and antigen appeared only after fertilization.

The regulated expression of PDGF and its receptor in in vitro models of embryogenesis and in oocytes, developing embryos, and placenta suggests that PDGF is involved at a number of stages of development. Further investigation will be required to understand the actions of PDGF and other growth factors produced during embryonic development (e.g., cell proliferation versus differentiation) and the direction of their action (maternal tissue to support embryonic growth versus intraembryonic targets).

3. Atherosclerosis

The most prevalent causes of death in the United States and western Europe, myocardial infarction and stroke, result from the advanced lesion of atherosclerosis, the fibrous plaque. This intimal proliferative lesion of smooth muscle cells also contains macrophages, T cells, new connective tissue matrix, and large amounts of lipid in hyperlipidemic individuals. The hallmark of the lesion, however, is the marked increase in smooth muscle cells within the intima that are responsible for the large amounts of newly formed connective tissue. It has been postulated that various forms of insult or "injury" result in dysfunction of the endothelial cells of the affected site in the arterial tree, and culminate in this intimal smooth muscle proliferative response (Ross and GLOMSET 1973, 1976; Ross 1986).

As described in previous sections, endothelial injury could result in expression and secretion of PDGF as well as other growth factors. In addition, it has been observed that hypercholesterolemia and/or hypertension both result in increased monocyte attachment to the endothelium and chemotaxis of the monocytes into the intimal space, with their subsequent conversion to macrophages (GERRITY 1981 a, b; FAGGIOTTO et al. 1984; FAGGIOTTO and Ross 1984; ROSENFELD et al. 1987 a, b). Activated macrophages have been clearly defined as a source of growth factors, and particularly PDGF in culture (SHIMOKADO et al. 1985; MORNEX et al. 1986). These emigrating macrophages and altered endothelium may secrete growth factors, including both chains of PDGF, that can induce chemotaxis of smooth muscle cells from the media into the intima followed by their subsequent proliferation, culminating in the intimal proliferative lesions of atherosclerosis. Smooth muscle cells from these lesions have also been found to express the gene for the A chain of PDGF (LIBBY et al. 1988; WILCOX et al. 1988).

In some instances, such as in homocystinemia, endothelial injury is sufficiently great that there may in fact be endothelial desquamation (HARKER et al. 1974, 1976; HARKER and ROSS 1978). Under these circumstances, the sites of endothelial loss will lead to areas of connective tissue exposure and platelet adherence. Increased platelet turnover has been measured under these conditions and may serve as an important source of growth factors in development of lesions in homocystinemic patients. This possibility is supported by in vivo studies in which the endothelial surface is altered by chronic homocystinemia (HARKER et al. 1976), an indwelling catheter (MOORE et al. 1976; FREIDMAN et al. 1977), or hypercholesterolemia (FUSTER et al. 1978). In these experimental models, endothelial denudation is sufficient to cause platelet adherence and aggregation and lesion formation. Inhibition of platelet function in each of these studies prevented lesion formation. Lesions of atherosclerosis that arise at the perianastomotic sites of coronary bypass grafts and those that recur after percutaneous transluminal coronary angioplasty (PTCA) may also be the result of traumatic injury to the artery wall sufficient to induce platelet interactions. Interruption of acute platelet-dependent thrombosis in a baboon model of arterial graft thrombosis supports this possibility. Administration of a small molecular weight inhibitor of thrombin into vascular-graft-bearing baboons abolished ^{111}In-platelet deposition and vascular graft occlusion (HANSON and HARKER 1988).

Direct in vivo data for a specific role for PDGF are more circumstantial. Analysis of human carotid artery lesions demonstrated levels of PDGF B-chain mRNA 5-fold greater than the low level of constitutive expression detected in normal artery (BARRETT and BENDITT 1987). In an attempt to determine the cell types responsible for PDGF expression in lesions, BARRETT and BENDITT (1988) assayed fractions of atherosclerotic plaques for each of the PDGF genes and then serially rehybridized the blots with markers for endothelial cells (von Willebrand factor, vWF), macrophages (c-*fms*), and smooth muscle cells (smooth muscle α-actin). In plaques, PDGF B-chain expression correlated strongly with c-*fms*. There was also a positive correlation between PDGF B-chain and vWF, indicating that some of the PDGF B-chain in lesions could be associated with endothelial cells (approximately 10% by multiple regression analysis). PDGF A-chain expression correlated with smooth muscle actin. Two major limitations of these studies are that the statistical relationships are the basis for cell identification and they are limited to transcript levels, indicating nothing about translation of active protein.

WILCOX et al. (1988) have addressed the issue of more direct cell type identification using in situ hybridization. The predominant cell types found to express PDGF A- and B-chain mRNA in carotid endarterectomy specimens were mesenchymal-appearing intimal cells and endothelial cells in interplaque capillaries, respectively. Expression of the β-subunit of the PDGF receptor was found to be primarily localized in the plaque intimal mesenchymal cells and not medial smooth muscle cells. RUBIN et al. (1988b) also observed pronounced expression of the β-subunit of the PDGF receptor in atherosclerotic plaques as compared with vessels of normal tissues. Areas containing large numbers of foam cells (principally macrophages) and areas of necrosis did not show evidence for PDGF transcription (WILCOX et al. 1988). These data are not necessarily in conflict with those of BARRETT and BENDITT (1988), who demonstrated a high correlation in the same lesions between PDGF B chain and the macrophage marker (c-*fms*). Endothelial cell PDGF B-chain expression may be elevated in the lesions rich in macrophages.

A major drawback of both studies of PDGF mRNA in lesions is that the lesions examined are advanced lesions where proliferation may no longer be a major component. Studies of developing lesions will be required to determine whether the pattern of expression is similar at earlier stages.

4. Neoplasia and Transformed Cells

The simian sarcoma virus (SSV) was one of the first transforming retroviruses to be isolated from nonhuman primates and was originally derived from a woolly monkey that had multiple subcutaneous fibrosarcomas (THIELEN et al. 1971; DEINHARDT et al. 1972; DEINHARDT 1980). The firm nodular masses over the mesentery were composed of "bundles and whorls of spindle-shaped cells," which grew slowly and appeared benign histopathologically. All of the cell types affected in vivo are connective tissue cells which have been shown to express PDGF receptors in culture (see Table 1). Transformation by SSV in vitro has also been shown to be limited to connective tissue cells that express PDGF receptors (LEAL et al. 1985). Bovine endothelial cells and the mink epithelioid cell line CCL64, which do not express PDGF receptors, are not transformed by SSV, even though they support viral replication. As discussed in previous sections, the transformed phenotype was totally or partially blocked by interference with either the transport of the PDGF-like v-*sis* protein to the outside of the cell (LEE et al. 1987) or with the accessibility of PDGF-like v-*sis* to the PDGF receptor with PDGF antibodies or suramin (GARRETT et al. 1984; HUANG et al. 1984b; JOHNSSON et al. 1985b; BETSHOLTZ et al. 1986a). All of these observations are consistent with the hypothesis that the transformation is mediated through the production of a PDGF-like molecule. In fact, proliferation in response to exogenously added PDGF elicits many (or all) of the characteristics of primary transformants such as growth in soft agar (ASSOIAN et al. 1984) and morphological changes comparable to those observed with SSV-transformed cells (JOHNSSON et al. 1985b).

As shown in Table 2 and discussed in previous sections, many cells obtained from naturally occurring human tumors and cells transformed in culture by agents other than SSV synthesize PDGF-like molecules and express the genes for PDGF. NIMAN et al. (1985) have also reported that the urine of many cancer patients contains elevated levels of PDGF-like proteins by immunoblotting. It is unclear, however, whether activation of PDGF expression and secretion is direct and closely associated with the initial transformation event or occurs subsequent to the initiating event. It is also unclear how significant the production of PDGF is to the transformation process and tumor growth and progression. Many of the cells listed in Table 2, such as carcinomas and tumors of hematopoietic cells, express and secrete PDGF but do not appear to have receptors or respond mitogenically to PDGF. If PDGF has any effect on these cells, it must be an indirect one. However, many of these same tumors are associated with a phenomenon termed desmoplasia, an increase in connective tissue formation in the immediate vicinity of lesions (SEEMAYER et al. 1979; KAO et al. 1984).

Many of the cells listed in Table 2, isolated from naturally occurring tumors or cells transformed in vitro, do express PDGF receptors and respond mitogenically to PDGF, as in osteosarcomas and gliomas. Detailed analyses of two sublines of a human glioma line U-343 MG, derived from a glioblastoma multiforme of a 64-year-old man, suggest a common clonal origin for the two sublines, which have markedly different phenotypes (NISTER et al. 1986). PDGF production but lack of detectable PDGF receptors was associated with "immature"-appearing, tightly growing cells. Clones that had large star-shaped cells, resembling normal glial cells, were found to have low or no production of PDGF and a high binding capacity for PDGF. High passage cultures were found to give rise to a higher number of high-PDGF-producing clones and growth rate correlated with PDGF production. However, true clonal variation was suggested by the presence of clones apparently devoid of both PDGF production and PDGF binding. Two interpretations of the clonal variation have been suggested (WESTERMARK et al. 1985). One possibility is that production of PDGF gives the cell a certain selective advantage and occurs as a relatively late event in the pathogenesis of glioma. An alternate possibility is that the primary transforming events hit an immature glial cell that expresses PDGF as part of its genetic program. As the tumor expands, more differentiated clones (expressing PDGF receptors but not PDGF protein) segregate. The data do not allow differentiation between the proposed models, but suggest an equally important role for paracrine growth mechanisms in gliomas.

5. Inflammatory Joint Disease

The proliferation of synovial cells that occurs in arthritis, particularly rheumatoid arthritis, results in the formation of a pannus that can lead to joint fibrosis and many of the crippling aspects of the disease. RUBIN et al. (1988 a) have demonstrated that in chronic synovitis there is an increase in the β-subunit of the PDGF receptor in this inflamed tissue. They observed the presence of PDGF receptors in association with increased staining of the major histocompatibility complex-coded human leukocyte antigen DR and infiltration of macrophages and T lymphocytes in chronic synovial inflammation (RUBIN et al. 1988 b). HAMERMAN et al. (1987) have observed PDGF in the synovial fluid of patients with osteoarthritis. Similarly, we (E.W. RAINES et al., unpublished observations) have also observed that synovial fluids obtained from patients with rheumatoid arthritis contain significant levels of PDGF. The possibility of both autocrine and paracrine stimulation of cell proliferation under these circumstances may be important in this form of fibroproliferative disease associated with chronically inflamed joints.

6. Fibrosis

Fibrotic responses occur in a number of pathologic states in both the lung and bone marrow. A role for PDGF in these processes is primarily based on the presence of cells capable of secreting PDGF clearly associated with fibrotic regions. Interstitial pulmonary fibrosis is a heterogeneous group of disorders involving the lower respiratory tract and characterized by alveolar inflammation and subsequent fibrosis (BRODY and CRAIGHEAD 1976). Although each of these disease processes demonstrates distinguishing histologic features, the fibroproliferative response observed in lung tissue is characterized by an alveolitis consisting of inflammatory cell infiltration, proliferation of mesenchymal and type II epithelial cells, and eventually the accumulation of collagen. The pulmonary macrophage is the predominant inflammatory cell present following injury, and PDGF constitutes the major mitogen for fibroblasts secreted by activated alveolar macrophages (SHIMOKADO et al. 1985). In addition, alveolar macrophages obtained from patients with idiopathic pulmonary fibrosis, sarcoidosis, and histiocytosis X transcribe the gene encoding the B chain of PDGF (MORNEX et al. 1986). Cultured alveolar macrophages isolated from patients with idiopathic pulmonary fibrosis also release increased amounts of PDGF into the culture medium compared to macrophages obtained from normal individuals (MARTINET et al. 1986). However, there is no direct demonstration of expression of PDGF in vivo associated with fibrotic areas.

Fibrosis of the bone marrow is also not associated with a single clinical or pathologic syndrome. It arises from chronic infections, following exposure to a number of toxins, and in association with a variety of hematologic and non-hematologic malignancies (GILBERT 1984). The marrow fibroblasts responsible for the connective tissue deposition are polyclonally derived by chromosomal analysis and therefore not a part of abnormal proliferation (MANIATIS et al. 1969; VAN SLYCK et al. 1970). Marrow fibrosis is therefore considered a secondary phenomenon, and this idea is supported in vivo by the reversibility of marrow fibrosis in acute myelofibrosis after bone marrow transplantation (SMITH et al. 1981). Common clinical features, in particular an excess of megakaryocytes and monocyte/macrophages, and activation of these cells are consistent features shared by all forms of reactive myelofibrosis (GILBERT 1984). Both of these cells are abundant sources of PDGF. Examples of possible abnormal extrusion of the megakaryocytic cellular contents include ineffective megakaryocytopoiesis resulting in a high proportion of developing megakaryocytes that die within the marrow (CASTRO-MALASPINA 1984), defects in granule packaging such as the gray platelet syndrome (BRETON-GORIUS et al. 1981), and megakaryoblastic leukemias (DEN OTTOLANDER et al. 1979). All of these conditions are associated with myelofibrosis, and two groups (BAGLIN et al. 1988; KATOH et al. 1988) have shown decreased levels of platelet mitogenic activity in circulating platelets of patients with myeloproliferative disorders. However, simply an expanded bone marrow pool of megakaryocytes, as seen in immune thrombocytopenia, benign secondary thrombocytoses, and myeloproliferative disorders lacking myelofibrosis, do not lead to bone marrow fibrosis (GROOPMAN 1980).

7. Other Possible Disease Associations

PDGF may also be associated with other disease entities. PDGF activity has been described in platelet lysates derived from platelets of untreated diabetic patients, and the activity disappears after insulin therapy (HAMET et al. 1985). In another study, a series of children with atypical sporadic hemolytic uremia syndrome had detectable PDGF activity in their plasma, in contrast to patients with typical hemolytic uremic syndrome (LEVIN et al. 1986). In patients with a different form of renal disease, for example renal graft rejection, there is an induction of PDGF receptor expression on interstitial cells associated with the rejection process (FELLSTROM et al. 1987). Patients with glomerulonephritis contained demonstrable PDGF (by immunofluorescence with anti-PDGF antisera) in the glomerular capillary walls (FRAMPTON et al. 1988).

A series of studies of patients with scleroderma, a vascular disease in which there is a proliferation of connective tissue surrounding the blood vessels, have suggested a role for PDGF in that disease. TAKEHARA et al. (1987) found that sera from scleroderma patients have increased mitogenic activity for fibroblasts as compared with that of normal controls, and that this mitogenic activity could be decreased if the patients were treated with pharmacologic agents that decreased platelet activity and platelet release. They demonstrated that approximately 70% of the activity present in the serum was due to PDGF, since it could be removed by anti-PDGF IgG. PANDOLFI et al. (1989) measured the PDGF-related mitogenic activity in whole blood serum and platelet-poor serum. They confirmed the increased activity in serum and further demonstrated increased levels in plasma-derived serum in the absence of other signs of platelet activation.

References

Abboud HE, Poptic E, DiCorleto P (1987) Production of platelet-derived growth factor–like protein by rat mesangial cells in culture. J Clin Invest 80:675–683

Acres RB, Lamb JR, Feldmann M (1985) Effects of platelet-derived growth factor and epidermal growth factor on antigen-induced proliferation of human T-cell lines. Immunology 54:9–16

Alitalo R, Andersson LC, Betsholtz C, Nilsson K, Westermark B, Heldin C-H, Alitalo K (1987) Induction of platelet-derived growth factor gene expression during megakaryoblastic and monocytic differentiation of human leukemia cell lines. EMBO J 6:1213–1218

Andersen HA, Flodgaard H, Klenow H, Leick V (1984) Platelet-derived growth factor stimulates chemotaxis and nucleic acid synthesis in the protozoan tetrahymena. Biochim Biophys Acta 782:437–440

Antoniades HN (1981) Human platelet-derived growth factor (PDGF): purification of PDGF-I and PDGF-II and separation of their reduced subunits. Proc Natl Acad Sci USA 78:7314–7317

Antoniades HN, Hunkapiller MW (1983) Human platelet-derived growth factor (PDGF): amino-terminal amino acid sequence. Science 220:963–965

Antoniades HN, Scher CD, Stiles CD (1979) Purification of human platelet-derived growth factor. Proc Natl Acad Sci USA 76:1809–1813

Armelin HA, Armelin MCS, Kelly K, Stewart T, Leder P, Cochran BH, Stiles CD (1984) Functional role for c-myc in mitogenic response to platelet-derived growth factor. Nature 310:655–660

Assoian RK, Komoriya A, Meyers CA, Miller DM, Sporn MB (1983) Transforming growth factor-β in human platelets. Identification of a major storage site, purification, and characterization. J Biol Chem 258:7155–7160

Assoian RK, Grotendorst GR, Miller DM, Sporn MB (1984) Cellular transformation by coordinated action of three peptide growth factors from human platelets. Nature 309:804–806

Baglin TP, Price SM, Boughton BJ (1988) A reversible defect of platelet PDGF content in myeloproliferative disorders. Br J Haematol 69:483–486

Balk SD (1971) Calcium as a regulator of the proliferation of normal but not of transformed chicken fibroblasts in a plasma-containing medium. Proc Natl Acad Sci USA 68:271–275

Balk SD, Whitfield JF, Youdale T, Braun AC (1973) Roles of calcium, serum, plasma, and folic acid in the control of proliferation of normal and Rous sarcoma virus-infected chicken fibroblasts. Proc Natl Acad Sci USA 70:675–679

Barrett TB, Benditt EP (1987) sis (platelet-derived growth factor B chain) gene transcript levels are elevated in human atherosclerotic lesions compared to normal artery. Proc Natl Acad Sci USA 84:1099–1103

Barrett TB, Benditt EP (1988) Platelet-derived growth factor gene expression in human atherosclerotic plaques and normal artery wall. Proc Natl Acad Sci USA 85:2810–2814

Barrett TB, Gajdusek CM, Schwartz SM, McDougall JK, Benditt EP (1984) Expression of the sis gene by endothelial cells in culture and in vivo. Proc Natl Acad Sci USA 81:6772–6774

Bassett JE, Bowen-Pope DF, Takayasu M, Dacey RG Jr (1988) Platelet-derived growth factor does not constrict rat intracerebral arterioles in vitro. Microvasc Res 35:368–373

Bauer EA, Cooper TW, Huang JS, Altman J, Deuel TF (1985) Stimulation of in vitro human skin collagenase expression by platelet-derived growth factor. Proc Natl Acad Sci USA 82:4132–4136

Beckmann MP, Betsholtz C, Heldin C-H, Westermark B, DiMarco E, DiFiore PP, Robbins KC, Aaronson SA (1988) Comparison of biological properties and transforming potential of human PDGF-A and PDGF-B chains. Science 241:1346–1349

Begg GS, Pepper DS, Chesterman CN, Morgan FJ (1978) Complete covalent structure of human β-thromboglobulin. Biochemistry 17:1739–1744

Berk BC, Alexander RW, Brock TA, Gimbrone MA Jr, Webb RC (1986) Vasoconstriction: a new activity for platelet-derived growth factor. Science 232:87–90

Berridge MJ, Cobbold PH, Cuthbertson KSR (1988) Spatial and temporal aspects of cell signalling. Phil Trans R Soc Lond 320:325–343

Besmer P, Snyder HW, Murphy JE, Hardy WD Jr, Parodi A (1983) The Parodi-Irgens feline sarcoma virus and simian sarcoma virus have homologous oncogenes, but in different contexts of the viral genomes. J Virol 46:606–613

Betsholtz C, Heldin C-H, Nister M, Ek B, Wasteson A, Westermark B (1983) Synthesis of a PDGF-like growth factor in human glioma and sarcoma cells suggests the expression of the cellular homologue to the transforming protein of simian sarcoma virus. Biochem Biophys Res Commun 117:176–182

Betsholtz C, Bywater M, Westermark B, Burk RR, Heldin C-H (1985) Expression of the c-sis gene and secretion of a platelet-derived growth factor-like protein by simian virus 40-transformed BHK cells. Biochem Biophys Res Commun 130:753–760

Betsholtz C, Johnsson A, Heldin C-H, Westermark B (1986a) Efficient reversion of simian sarcoma virus-transformation and inhibition of growth factor-induced mitogenesis by suramin. Proc Natl Acad Sci USA 83:6440–6444

Betsholtz C, Johnsson A, Heldin C-H, Westermark B, Lind P, Urdea MS, Eddy R, Shows TB, Philpott K, Mellor AL, Knott TJ, Scott J (1986b) The human platelet-derived growth factor A-chain: complementary DNA sequence, chromosomal localization and expression in tumour cell lines. Nature 323:226–232

Betsholtz C, Bergh J, Bywater M, Pettersson M, Johnsson A, Heldin C-H, Ohlsson R, Knott TJ, Scott J, Bell GI, Westermark B (1987) Expression of multiple growth factors in a human lung cancer cell line. Int J Cancer 39:502–507

Bird A, Taggart M, Frommer M, Miller OJ, Macleod D (1985) A fraction of the mouse genome that is derived from islands of nonmethylated, CpG-rich DNA. Cell 40:91–99

Bishayee S, Ross AH, Womer R, Scher CD (1986) Purified human platelet-derived growth factor receptor has ligand-stimulated tyrosine kinase activity. Proc Natl Acad Sci USA 83:6756–6760

Bleiberg I, Harvey AK, Smale G, Grotendorst GR (1985) Identification of a PDGF-like mitoattractant produced by NIH/3T3 cells after transformation with SV-40. J Cell Physiol 123:161–166

Bolton AE, Hunter WM (1973) The labeling of proteins to high specific radioactivities by conjugation to a ^{125}I-containing acylating agent. Biochem J 133:529–539

Bonthron DT, Morton CC, Orkin SH, Collins T (1988) Platelet-derived growth factor A chain: gene structure, chromosomal location, and basis for alternative mRNA splicing. Proc Natl Acad Sci USA 85:1492–1496

Bowen-Pope DF, Ross R (1982) Platelet-derived growth factor. II. Specific binding to cultured cells. J Biol Chem 257:5161–5171

Bowen-Pope DF, Ross R (1985) Methods for studying the platelet-derived growth factor receptor. Methods Enzymol 109:69–100

Bowen-Pope DF, DiCorleto PE, Ross R (1983) Interactions between the receptors for platelet-derived growth factor and epidermal growth factor. J Cell Biol 96:679–683

Bowen-Pope DF, Malpass TW, Foster DM, Ross R (1984a) Platelet-derived growth factor in vivo: levels, activity, and rate of clearance. Blood 64:458–469

Bowen-Pope DF, Vogel A, Ross R (1984b) Production of platelet-derived growth factor-like molecules and reduced expression of platelet-derived growth factor receptors accompany transformation by a wide spectrum of agents. Proc Natl Acad Sci USA 81:2396–2400

Bowen-Pope DF, Seifert RA, Ross R (1985) The platelet-derived growth factor receptor. In: Boynton AL, Leffert HL (eds) Control of animal cell proliferation. Academic Press, New York, p 281

Bowen-Pope DF, Hart CE, Seifert RA (1989) Sera and conditioned media contain different isoforms of PDGF which bind to different classes of PDGF receptor. J Biol Chem 264:2502–2508

Bremer EG, Hakomori S-I, Bowen-Pope DF, Raines E, Ross R (1984) Ganglioside-mediated modulation of cell growth, growth factor binding, and receptor phosphorylation. J Biol Chem 259:6818–6825

Breton-Gorius J, Vainchenker W, Nurden A, Levy-Toledano S, Caen J (1981) Defective α-granule production in megakaryocytes from gray platelet syndrome. Am J Pathol 102:10–19

Brewitt B, Clark JI (1988) Growth and transparency in the lens, an epithelial tissue, stimulated by pulses of PDGF. Science 242:777–779

Briggs MR, Kadonaga JT, Bell SP, Tjian R (1986) Purification and biochemical characterization of the promoter-specific transcription factor, Spl. Science 234:47–52

Brody AR, Craighead JE (1976) Interstitial associations of cells lining air spaces in human pulmonary fibrosis. Virchows Arch [A] 372:39–49

Bronzert DA, Pantazis P, Antoniades HN, Kasid A, Davidson N, Dickson RB, Lippman ME (1987) Synthesis and secretion of platelet-derived growth factor by human breast cancer cell lines. Proc Natl Acad Sci USA 84:5763–5767

Burwen SJ, Jones AL (1987) The association of polypeptide hormones and growth factors with the nuclei of target cells. Trends Biochem Sci 12:159–162

Busch C, Wasteson A, Westermark B (1976) Release of a cell growth promoting factor from human platelets. Thromb Res 8:493–500

Bywater M, Rorsman F, Bongcam-Rudloff E, Mark G, Hammacher A, Heldin C-H, Westermark B, Betsholtz C (1988) Expression of recombinant platelet-derived growth factor A- and B-chain homodimers in Rat-1 cells and human fibroblasts reveals differences in protein processing and autocrine effects. Mol Cell Biol 8:2753–2762

Campochiaro PA, Glaser BM (1985) Platelet-derived growth factor is chemotactic for human retinal pigment epithelial cells. Arch Ophthalmol 103:576–579

Campochiaro PA, Sugg R, Grotendorst G, Hjelmeland L (1989) Retinal pigment epithelial cells produce PDGF-like proteins and secrete them into their media. Exp Eye Res 49:217–227

Canalis E (1981) Effect of platelet-derived growth factor on DNA and protein synthesis in cultured rat calvaria. Metabolism 30:970–975

Castro-Malaspina H (1984) Pathogenesis of myelofibrosis: role of ineffective megakaryopoiesis and megakaryocyte components. In: Berk PD, Castro-Malaspina H, Wasserman LR (eds) Myelofibrosis and the biology of connective tissue. Liss, New York, p 427

Childs CB, Proper JA, Tucker RF, Moses HL (1982) Serum contains a platelet-derived transforming growth factor. Proc Natl Acad Sci USA 79:5312–5316

Chiu I-M, Reddy EP, Givol D, Robbins KC, Tronick SR, Aaronson SA (1984) Nucleotide sequence analysis identifies the human c-sis proto-oncogene as a structural gene for platelet-derived growth factor. Cell 37:123–129

Chu S-HW, Hoban CJ, Owen AJ, Geyer RP (1985) Platelet-derived growth factor stimulates rapid polyphosphoinositide breakdown in fetal human fibroblasts. J Cell Physiol 124:391–396

Chua CC, Geiman DE, Keller GH, Ladda RL (1985) Induction of collagenase secretion in human fibroblast cultures by growth promoting factors. J Biol Chem 260:5213–5216

Claesson-Welsh L, Ronnstrand L, Heldin C-H (1987) Biosynthesis and intracellular transport of the receptor for platelet-derived growth factor. Proc Natl Acad Sci USA 84:8796–8800

Claesson-Welsh L, Eriksson A, Moren A, Severinsson L, Ek B, Ostman A, Betsholtz C, Heldin C-H (1988) cDNA cloning and expression of a human platelet-derived growth factor (PDGF) receptor specific for a B-chain-containing PDGF molecule. Mol Cell Biol 8:3476–3486

Clarke MF, Westin E, Schmidt D, Josephs SF, Ratner L, Wong-Staal F, Gallo RC, Reitz MS Jr (1984) Transformation of NIH 3T3 cells by a human c-sis cDNA clone. Nature 308:464–467

Clemmons DR, Van Wyk JJ (1985) Evidence for a functional role of endogenously produced somatomedin-like peptides in the regulation of DNA synthesis in cultured human fibroblasts and porcine smooth muscle cells. J Clin Invest 75:1914–1918

Clemmons DR, Underwood LE, Van Wyk JJ (1981) Hormonal control of immunoreactive somatomedin production by cultured human fibroblasts. J Clin Invest 67:10–19

Cochran BH, Reffel AC, Stiles CD (1983) Molecular cloning of gene sequences regulated by platelet-derived growth factor. Cell 33:939–947

Cochran BH, Zullo J, Verma IM, Stiles CD (1984) Expression of the c-fos gene and of a fos-related gene is stimulated by platelet-derived growth factor. Science 226:1080–1082

Collins MKL, Sinnett-Smith JW, Rozengurt E (1983) Platelet-derived growth factor treatment decreases the affinity of the epidermal growth factor receptors of Swiss 3T3 cells. J Biol Chem 258:11689–11693

Collins T, Ginsburg D, Boss JM, Orkin SH, Pober JS (1985) Cultured human endothelial cells express platelet-derived growth factor B chain: cDNA cloning and structural analysis. Nature 316:748–750

Collins T, Bonthron DT, Orkin SH (1987a) Alternative RNA splicing affects function of encoded platelet-derived growth factor A chain. Nature 328:621–624

Collins T, Pober JS, Gimbrone MA Jr, Hammacher A, Betsholtz C, Westermark B, Heldin C-H (1987b) Cultured human endothelial cells express platelet-derived growth factor A chain. Am J Pathol 127:7–12

Cooper JA, Bowen-Pope DF, Raines E, Ross R, Hunter T (1982) Similar effects of platelet-derived growth factor and epidermal growth factor on the phosphorylation of tyrosine in cellular proteins. Cell 31:263–273

Coughlin SR, Moskowitz A, Zetter BR, Antoniades HN, Levine L (1980) Platelet-dependent stimulation of prostacyclin synthesis by platelet-derived growth factor. Nature 288:600–602

Coughlin SR, Lee WMF, Williams PW, Giels GM, Williams LT (1985) c-myc gene expression is stimulated by agents that activate protein kinase C and does not account for the mitogenic effect of PDGF. Cell 43:243–251

Dalla Favera R, Gallo RC, Giallongo A, Croce CM (1982) Chromosomal localization of the human homolog (c-sis of the simian sarcoma virus onc gene). Science 218:686–688

Daniel TO, Tremble PM, Frackelton AR Jr, Williams LT (1985) Purification of the platelet-derived growth factor receptor by using an anti-phosphotyrosine antibody. Proc Natl Acad Sci USA 82:2684–2687

Daniel TO, Gibbs VC, Milfay DF, Garovoy MR, Williams LT (1986) Thrombin stimulates c-sis gene expression in microvascular endothelial cells. J Biol Chem 261:9579–9582

Daniel TO, Gibbs VC, Milfay DF, Williams LT (1987) Agents that increase cAMP accumulation block endothelial c-sis induction by thrombin and transforming growth factor-β. J Biol Chem 262:11893–11896

Davis RJ, Czech MP (1985) Platelet-derived growth factor mimics phorbol diester action on epidermal growth factor receptor phosphorylation at threonine-654. Proc Natl Acad Sci USA 82:4080–4084

Deinhardt F (1980) Biology of primate retroviruses. In: Klein G (ed) Viral oncology. Raven, New York, p 357

Deinhardt F, Wolfe L, Northrop R, Marczynska B, Ogden J, McDonald R, Falk L, Shramek G, Smith R, Deinhardt J (1972) Induction of neoplasms by viruses in Marmoset monkeys. J Med Primatol 7:29–50

Delwiche F, Raines E, Powell J, Ross R, Adamson J (1985) Platelet-derived growth factor enhances in vitro erythropoiesis via stimulation of mesenchymal cells. J Clin Invest 76:137–142

Den Ottolander GT, te Velde J, Brederoo P, Geraedts JPM, Slee PHT, Willemze R, Zwaan FE, Haak HL, Muller HP, Bieger R (1979) Megakaryoblastic leukaemic (acute myelofibrosis): a report of three cases. Br J Haematol 42:9–20

Deuel TF (1987) Polypeptide growth factors: roles in normal and abnormal cell growth. Annu Rev Cell Biol 3:443–492

Deuel TF, Keim PS, Farmer M, Heinrikson RL (1977) Amino acid sequence of human platelet factor 4. Proc Natl Acad Sci USA 74:2256–2258

Deuel TF, Senior RM, Huang JS, Griffin GL (1982) Chemotaxis of monocytes and neutrophils to platelet-derived growth factor. J Clin Invest 69:1046–1049

Deuel TF, Huang JS, Huang SS, Stroobant P, Waterfield MD (1983) Expression of a platelet-derived growth factor-like protein in Simian Sarcoma virus transformed cells. Science 221:1348–1350

Deuel TF, Huang JS, Proffitt RT, Baenziger JU, Chang D, Kennedy BB (1981a) Human platelet-derived growth factor – purification and resolution into two active protein fractions. J Biol Chem 256:8896–8899

Deuel TF, Senior RM, Chang D, Griffin GL, Heinrikson RL, Kaiser ET (1981b) Platelet factor 4 is chemotactic for neutrophils and monocytes. Proc Natl Acad Sci USA 78:4584–4587

Devare SG, Reddy EP, Law DJ, Robbins KC, Aaronson SA (1983) Nucleotide sequence of the simian sarcoma virus genome: demonstration that its acquired cellular sequences encode the transforming gene product p28^{sis}. Proc Natl Acad Sci USA 80:731–735

DiCorleto PE, Bowen-Pope DF (1983) Cultured endothelial cells produce a platelet-derived growth factor-like protein. Proc Natl Acad Sci USA 80:1919–1923

Distel RJ, Ro H-S, Rosen BS, Groves DL, Spiegelman BM (1987) Nucleoprotein complexes that regulate gene expression in adipocyte differentiation: direct participation of c-fos. Cell 49:835–844

Doolittle RF, Hunkapiller MW, Hood LE, Devare SG, Robbins KC, Aaronson SA, Antoniades HN (1983) Simian sarcoma virus onc gene, v-sis, is derived from the gene (or genes) encoding a platelet-derived growth factor. Science 221:275–277

Eide BL, Krebs EG, Ross R, Pike LJ, Bowen-Pope DF (1986) Tumor promoter enhances mitogenesis by PDGF with little effect on PDGF binding. J Cell Physiol 126:254–258

Ek B, Heldin C-H (1982) Characterization of a tyrosine-specific kinase activity in human fibroblast membranes stimulated by platelet-derived growth factor. J Biol Chem 257:10486–10492

Ek B, Westermark B, Wasteson A, Heldin C-H (1982) Stimulation of tyrosine-specific phosphorylation by platelet-derived growth factor. Nature 295:419–420

Escobedo JA, Williams LT (1988) A PDGF receptor domain essential for mitogenesis but not for many other responses to PDGF. Nature 335:85–87

Escobedo JA, Barr PJ, Williams LT (1988a) Role of tyrosine kinase and membrane-spanning domains in signal transduction by the platelet-derived growth factor receptor. Mol Cell Biol 8:5126–5131

Escobedo JA, Keating MT, Ives HE, Williams LT (1988b) Platelet-derived growth factor receptors expressed by cDNA transfection couple to a diverse group of cellular responses associated with cell proliferation. J Biol Chem 263:1482–1487

Escobedo JA, Navankasatussas S, Cousens LS, Coughlin SR, Bell GI, Williams LT (1988c) A common PDGF receptor is activated by homodimeric A and B forms of PDGF. Science 240:1532–1534

Eva A, Robbins KC, Andersen PR, Srinivasan A, Tronick SR, Reddy EP, Ellmore NW, Galen AT, Lautenberger JA, Papas TS, Westin EH, Wong-Staal F, Gallo RC, Aaronson SA (1982) Cellular genes analogous to retroviral onc genes are transcribed in human tumor cells. Nature 295:116–119

Faggiotto A, Ross R (1984) Studies of hypercholesterolemia in the nonhuman primate. II. Fatty streak conversion to fibrous plaque. Arteriosclerosis 4:341–356

Faggiotto A, Ross R, Harker L (1984) Studies of hypercholesterolemia in the nonhuman primate. I. Changes that lead to fatty streak formation. Arteriosclerosis 4:323–340

Fellstrom B, Klareskog L, Larsson E, Tufveson G, Wahlberg J, Ronnstrand L, Heldin C-H, Terracio L, Rubin K (1987) Tissue distribution of macrophages, class II transplantation antigens, and receptors for platelet-derived growth factor in normal and rejected human kidneys. Transplan Proc 19:3625–3627

Fox PL, DiCorleto PE (1986) Modified low density lipoproteins suppress production of a platelet-derived growth factor-like protein by cultured endothelial cells. Proc Natl Acad Sci USA 83:4774–4778

Fox PL, DiCorleto PE (1988) Fish oils inhibit endothelial cell production of platelet-derived growth factor-like protein. Science 241:453–456

Fox PL, Chisolm GM, DiCorleto PE (1987) Lipoprotein-mediated inhibition of endothelial cell production of platelet-derived growth factor-like protein depends on free radical lipid peroxidation. J Biol Chem 262:6046–6054

Frackelton AR, Tremble PM, Williams LT (1984) Evidence for the platelet-derived growth factor-stimulated tyrosine phosphorylation of the platelet-derived growth factor receptor in vivo: immunopurification using a monoclonal antibody to phosphotyrosine. J Biol Chem 259:7909–7915

Fraizer GE, Bowen-Pope DF, Vogel AM (1987) Production of platelet-derived growth factor by cultured Wilms' tumor cells and fetal kidney cells. J Cell Physiol 133:169–174

Frampton G, Hildreth G, Hartley B, Cameron JS, Heldin C-H, Wasteson A (1988) Could platelet-derived growth factor have a role in the pathogenesis of lupus nephritis? Lancet 6:343

Frantz CN (1985) Effects of platelet-derived growth factor on Ca^{2+} in 3T3 cells. Exp Cell Res 158:287–300

Friedman RJ, Stemerman MB, Wenz B, Moore S, Gauldie J, Gent M, Tiell ML, Spaet TH (1977) The effect of thrombocytopenia on experimental atherosclerotic lesion formation in rabbits. Smooth muscle cell proliferation and re-endothelialization. J Clin Invest 60:1191–1201

Fuster V, Bowie EJW, Lewis JC, Fass DN, Owen CA Jr, Brown AL (1978) Resistance to arteriosclerosis in pigs with von Willebrand's disease. J Clin Invest 61:722–730

Gabrielson EW, Gerwin BI, Harris CC, Roberts AB, Sporn MB, Lechner JF (1988) Stimulation of DNA synthesis in cultured primary human mesothelial cells by specific growth factors. FASEB J 2:2717–2721

Gajdusek C, Carbon S, Ross R, Nawroth P, Stern D (1986) Activation of coagulation releases endothelial cell mitogens. J Cell Biol 103:419–428

Garrett JS, Coughlin SR, Niman HL, Tremble PM, Giels GM, Williams LT (1984) Blockade of autocrine stimulation in simian sarcoma virus-transformed cells reverses down-regulation of platelet-derived growth factor receptors. Cell Biol 81:7466–7470

Gazit A, Igarashi H, Chiu I, Srinivasan A, Yaniv A, Tronick SR, Robbins KC, Aaronson SA (1984) Expression of the normal human sis/PDGF-2 coding sequence induces cellular transformation. Cell 39:89–97

Gelmann EP, Wong-Staal F, Kramer RA, Gallo RC (1981) Molecular cloning and comparative analyses of the genomes of simian sarcoma virus and its associated helper virus. Proc Natl Acad Sci USA 78:3373–3377

Gerrard JM, Philipps DR, Rao GHR, Plow EF, Walz DA, Ross R, Harker LA, White JG (1980) Biochemical studies of two patients with the gray platelet syndrome. J Clin Invest 66:102–109

Gerrity RG (1981 a) The role of the monocyte in atherogenesis. I. Transition of blood-borne monocytes into foam cells in fatty lesions. Am J Pathol 103:181–190

Gerrity RG (1981 b) The role of the monocyte in atherogenesis. II. Migration of foam cells from atherosclerotic lesions. Am J Pathol 103:191–200

Gersuk GM, Holloway JM, Chang W-C, Pattengale PK (1986) Inhibition of human natural killer cell activity by platelet-derived growth factor. Nat Immun Cell Growth Regul 5:283–293

Gerwin BI, Lechner JF, Reddel RR, Roberts AB, Robbins KC, Gabrielson EW, Harris CC (1987) Comparison of production of transforming growth factor-β and platelet-derived growth factor by normal human mesothelial cells and mesothelioma. Cancer Res 47:6180–6184

Giese NA, Robbins KC, Aaronson SA (1987) The role of individual cysteine residues in the structure and function of the v-sis gene product. Science 236:1315–1318

Gilbert HS (1984) Myelofibrosis revisited: characterization and classification of myelofibrosis in the setting of myeloproliferative disease. In: Berk PD, Castro-Malaspina H, Wasserman LR (eds) Myelofibrosis and the biology of connective tissue. Liss, New York, p 3

Glenn K, Bowen-Pope DF, Ross R (1982) Platelet-derived growth factor. III. Idenficiation of a platelet-derived growth factor receptor by affinity labeling. J Biol Chem 257:5172–5176

Goodbourn S, Zinn K, Maniatis T (1985) Human β-interferon gene expression is regulated by an inducible enhancer element. Cell 41:509–520

Gould KL, Hunter T (1988) Platelet-derived growth factor induces multisite phosphorylation of pp60$^{c\text{-}src}$ and increases its protein-tyrosine kinase activity. Mol Cell Biol 8:3345–3356

Goustin AS, Betsholtz C, Pfeiffer-Ohlsson S, Persson H, Rydnert J, Bywater M, Holmgren G, Heldin C-H, Westermark B, Ohlsson R (1985) Coexpression of the sis and myc proto-oncogenes in developing human placenta suggests autocrine control of trophoblast growth. Cell 41:301–312

Graves DT, Owen AJ, Barth RK, Tempst P, Winoto A, Fors L, Hood LE, Antoniades HN (1984) Detection of c-sis transcripts and synthesis of PDGF-like proteins by human osteosarcoma cells. Science 226:972–974

Graves DT, Grotendorst GR, Antoniades HN, Schwartz CJ, Valante AJ (1989) Platelet derived growth factor is not chemotactic for peripheral blood monocytes Exp Cell Res 180:497–503

Greenberg ME, Ziff EB (1984) Stimulation of 3T3 cells induces transcription of the c-fos proto-oncogene. Nature 311:433–438

Gronwald RGK, Grant FJ, Haldeman BA, Hart CE, O'Hara PJ, Hagen FS, Ross R, Bowen-Pope DF, Murray MJ (1988) Cloning and expression of a cDNA coding for the human platelet-derived growth factor receptor: evidence for more than one receptor class. Proc Natl Acad Sci USA 85:3435–3439

Gronwald RGK, Seifert RA, Bowen-Pope DF (1989) Differential regulation of expression of two PDGF receptor subunits by transforming growth factor β. J Biol Chem 264:8120–8125

Groopman JE (1980) The pathogenesis of myelofibrosis in myeloproliferative disorders. Ann Intern Med 92:857–858

Grotendorst G (1984) Alteration of the chemotactic response of NIH/3T3 cells to PDGF by growth factors, transformation, and tumor promoters. Cell 36:279–285

Grotendorst G, Seppa HEJ, Kleinman HK, Martin G (1981) Attachment of smooth muscle cells to collagen and their migration toward platelet-derived growth factor. Proc Natl Acad Sci USA 78:3669–3672

Grotendorst GR, Chang T, Seppa HEJ, Kleinman HK, Martin GR (1982) Platelet-derived growth factor is a chemoattractant for vascular smooth muscle cells. J Cell Physiol 113:261–266

Grotendorst GR, Martin GR, Pencev D, Sodek J, Harvey AK (1985) Stimulation of granulation tissue formation by platelet-derived growth factor in normal and diabetic rats. J Clin Invest 76:2323–2329

Gudas LJ, Singh JP, Stiles CD (1983) Secretion of growth regulatory molecules by teratocarcinoma stem cells. In: Silver LM, Martin GR, Strickland S (eds) Teratocarcinoma stem cells. Cold Spring Harbor conferences of cell proliferation, vol 10. Cold Spring Harbor New York, p 229

Habenicht AJR, Glomset JA, King WC, Nist C, Mitchell CD, Ross R (1981) Early changes in phosphatidylinositol and arachidonic acid metabolism in quiescent Swiss 3T3 cells stimulated to divide by platelet-derived growth factor. J Biol Chem 256:12329–12335

Habenicht AJR, Goerig M, Grulich J, Rothe D, Gronwald R, Loth U, Schettler G, Kommerell B, Ross R (1985) Human platelet-derived growth factor stimulates prostaglandin synthesis by activation and by rapid de novo synthesis of cyclooxygenase. J Clin Invest 75:1381–1387

Habenicht AJR, Dresel HA, Goerig M, Weber J-A, Stoehr M, Glomset JA, Ross R (1986) Low-density lipoprotein receptor-dependent prostaglandin synthesis in Swiss 3T3 cells stimulated by platelet-derived growth factor. Proc Natl Acad Sci USA 83:1344–1348

Hajjar KA, Hajjar DP, Silverstein RL, Nachman RL (1987) Tumor necrosis factor mediated release of platelet-derived growth factor from cultured endothelial cells. J Exp Med 166:235–245

Hall DJ, Stiles CD (1987) Platelet-derived growth factor-inducible genes respond differentially to at least two distinct intracellular second messengers. J Biol Chem 262:15302–15308

Hamada H, Petrino MG, Kakunaga T, Seidman M, Stollar BD (1984a) Characterization of genomic poly(dT-dG)-poly(dC-dA) structure, organization, and conformation. Mol Cell Biol 4:2610–2621

Hamada H, Seidman M, Howard BH, Gorman CM (1984b) Enhanced gene expression by the poly(dT-dG)-poly(dC-dA) sequence. Mol Cell Biol 4:2622–2630

Hamet P, Sugimoto H, Umeda F, Lecavalier L, Franks DJ, Orth DN, Chiasson J-L (1985) Abnormalities of platelet-derived growth factors in insulin-dependent diabetes. Metabolism 34:25–31

Hamerman D, Taylor S, Kirschenbaum I, Klagsbrun M, Raines EW, Ross R, Thomas KA (1987) Growth factors with heparin binding affinity in human synovial fluid. Proc Soc Exp Biol Med 186:384–389

Hammacher A, Hellman U, Johnsson A, Ostman A, Gunnarsson K, Westermark B, Wasteson A, Heldin C-H (1988a) A major part of platelet-derived growth factor purified from human platelets is a heterodimer of one A and one B chain. J Biol Chem 263:16493–16498

Hammacher A, Nister M, Westermark B, Heldin C-H (1988b) A human glioma cell line secretes three structurally and functionally different dimeric forms of platelet-derived growth factor. Eur J Biochem 176:179–186

Hanai K, Kato H, Matsuhashi S, Morita A, Raines EW, Ross R (1987) Platelet proteins, including platelet-derived growth factor, specifically depress a subset of the multiple components of the response elicited by glutathione in Hydra. J Cell Biol 104:1675–1681

Hannink M, Donoghue DJ (1984) Requirement for a signal sequence in biological expression of the v-sis oncogene. Science 236:1197–1199

Hannink M, Donoghue DJ (1986) Cell surface expression of membrane-anchored v-sis gene products: glycosylation is not required for cell surface transport. J Cell Biol 103:2311–2322

Hannink M, Donoghue DJ (1988) Autocrine stimulation by the v-sis gene product requires a ligand-receptor interaction at the cell surface. J Cell Biol 107:287–298

Hannink M, Sauer MK, Donoghue DJ (1986) Deletions in the C-terminal coding region of the v-sis gene: dimerization is required for transformation. Mol Cell Biol 6:1304–1314

Hanson SR, Harker LA (1988) Interruption of acute platelet-dependent thrombosis by the synthetic antithrombin D-phenylalanyl-L-prolyl-L-arginyl chloromethyl ketone. Proc Natl Acad Sci USA 85:3184–3188

Harker LA, Ross R (1978) Sulfhydryl-mediated vascular disease. Eur J Clin Invest 8:199

Harker LA, Slichter SJ, Scott CR, Ross R (1974) Homocystinemia: vascular injury and arterial thrombosis. N Engl J Med 291:537–543

Harker LA, Ross R, Slichter SJ, Scott CR (1976) Homocystine-induced arteriosclerosis: the role of endothelial cell injury and platelet response in its genesis. J Clin Invest 58:731–741

Harlan JM, Thompson PJ, Ross R, Bowen-Pope DF (1986) α-Thrombin induces release of PDGF-like molecule(s) by cultured human endothelial cells. J Cell Biol 103:1129–1133

Hart CE, Seifert RA, Ross R, Bowen-Pope DF (1987) Synthesis, phosphorylation, and degradation of multiple forms of the platelet-derived growth factor receptor studied using a monoclonal antibody. J Biol Chem 262:10780–10785

Hart CE, Forstrom JW, Kelly JD, Seifert RA, Smith RA, Ross R, Murray MJ, Bowen-Pope DF (1988) Two classes of PDGF receptor recognize different isoforms of PDGF. Science 240:1529–1531

Hart CE, Forstrom JW, Kelly JD, Smith RA, Ross R, Murray MJ, Bowen-Pope DF (1989a) Biochemical evidence for multiple classes of PDGF receptor. In: Growth factors and their receptors: genetic control and rational applications. Liss, New York. V. 102, pp 297–305

Hart CE, Bailey M, Curtis DA, Osborn S, Raines EW, Ross R, Forstrom JW (1989b) Purification of PDGF-AB and PDGF-BB from human platelet extracts and the identification of all three PDGF dimers in human platelets. Biochemistry (in press)

Hasegawa-Sasaki H (1985) Early changes in inositol lipids and their metabolities induced by platelet-derived growth factor in quiescent Swiss 3T3 cells. Biochem J 232:99–109

Hasegawa-Sasaki H, Lutz F, Sasaki T (1988) Pathway of phospholipase C activation initiated with platelet-derived growth factor is different from that initiated with vasopressin and bombesin. J Biol Chem 263:12970–12976

Hawking F (1978) Suramin: with special reference to onchocerceasis. Adv Pharmacol Chemother 15:289–323

Hearing P, Shenk T (1983) The adenovirus type 5 E1A transcriptional control region contains a duplicated enhancer element. Cell 33:695–703

Heldin C-H, Wasteson A, Westermark B (1977) Partial purification and characterization of platelet factors stimulating the multiplication of normal human glial cells. Exp Cell Res 109:429–437

Heldin C-H, Westermark B, Wasteson A (1979) Platelet-derived growth factor: purification and partial characterization. Proc Natl Acad Sci USA 76:3722–3726

Heldin C-H, Wasteson A, Westermark B (1980a) Growth of normal human glial cells in a defined medium containing platelet-derived growth factor. Proc Natl Acad Sci USA 77:6611–6615

Heldin C-H, Westermark H, Wasteson A (1980b) Chemical and biological properties of a growth factor from human-cultured osteosarcoma cells: resemblance with platelet-derived growth factor. J Cell Physiol 105:235–246

Heldin C-H, Westermark B, Wasteson A (1981a) Platelet-derived growth factor: isolation by a large-scale procedure and analysis of subunit composition. Biochem J 193:907–913

Heldin C-H, Westermark B, Wasteson A (1981b) Specific receptors for platelet-derived growth factor on cells derived from connective tissue and glia. Proc Natl Acad Sci USA 78:3664–3668

Heldin C-H, Westermark B, Wasteson A (1981c) Demonstration of an antibody against platelet-derived growth factor. Exp Cell Res 136:255–261

Heldin C-H, Wasteson A, Westermark B (1982) Interaction of platelet-derived growth factor with its fibroblast receptor. Demonstration of ligand degradation and receptor modulation. J Biol Chem 257:4216–4221

Heldin C-H, Ek B, Ronnstrand L (1983) Characterization of the receptor for platelet-derived growth factor on human fibroblasts: demonstration of an intimate relationship with a 185000-Dalton substrate for the platelet-derived growth factor-stimulated kinase. J Biol Chem 258:10054–10061

Heldin C-H, Johnsson A, Wennergren S, Wernstedt C, Betsholtz C, Westermark B (1986) A human osteosarcoma cell line secretes a growth factor structurally related to a homodimer of PDGF A-chains. Nature 319:511–514

Heldin C-H, Backstrom G, Ostman A, Hammacher A, Ronnstrand L, Rubin K, Nister M, Westermark B (1988) Binding of different dimeric forms of PDGF to human fibroblasts: evidence for two separate receptor types. EMBO J 7:1387–1393

Hermansson M, Nister M, Betsholtz C, Heldin C-H, Westermark B, Funa K (1988) Endothelial cell hyperplasia in human glioblastoma: co-expression of mRNA for platelet-derived growth factor (PDGF) B chain and PDGF receptor suggests autocrine growth stimulation. Proc Natl Acad Sci USA 85:7748–7752

Hesketh TR, Morris JDH, Moore JP, Metcalfe JC (1988) Ca^{2+} and pH responses to sequential additions of mitogens in single 3T3 fibroblasts: correlations with DNA synthesis. J Biol Chem 263:11879–11886

Hiraki Y, Rosen OM, Birnbaum MJ (1988) Growth factors rapidly induce expression of the glucose transporter gene. J Biol Chem 263:13655–13662

Huang JS, Huang JS, Kennedy B, Deuel TF (1982) Platelet-derived growth factor: specific binding to target cells. J Biol Chem 257:8130–8136

Huang SS, Huang JS (1988) Rapid turnover of the platelet-derived growth factor receptor in sis-transformed cells and reversal by suramin. J Biol Chem 263:12608–12618

Huang JS, Huang SS, Deuel TF (1983) Human platelet-derived growth factor: radioimmunoassay and discovery of a specific plasma-binding protein. J Cell Biol 97:383–388

Huang JS, Huang SS, Deuel TF (1984a) Specific covalent binding of platelet-derived growth factor to human plasma alpha-2-macroglobulin. Proc Natl Acad Sci USA 81:342–346

Huang JS, Huang SS, Deuel TF (1984b) Transforming protein of simian sarcoma virus stimulates autocrine growth of SSV-transformed cells through PDGF cell-surface receptors. Cell 39:79–87

Huang JS, Nishimura J, Huang SS, Deuel TF (1984c) Protamine inhibits platelet derived growth factor receptor activity but not epidermal growth factor activity. J Cell Biochem 26:205–220

Hunt T (1985) False starts in translational control of gene expression. Nature 316:580–581

Igarashi H, Rao CD, Siroff M, Leal F, Robbins KC, Aaronson SA (1987) Detection of PDGF-2 homodimers in human tumor cells. Oncogene 1:79–85

Jaye M, McConathy E, Drohan W, Tong B, Deuel T, Maciag T (1985) Modulation of the sis gene transcript during endothelial cell differentiation in vitro. Science 228:882–885

Johnsson A, Betsholtz C, Heldin C-H, Westermark B (1985b) Antibodies against platelet-derived growth factor inhibit acute transformation by simian sarcoma virus. Nature 317:438–440

Johnsson A, Betsholtz C, Heldin C-H, Westermark B (1986) The phenotypic characteristics of simian sarcoma virus-transformed human fibroblasts suggest that the v-sis gene product acts solely as a PDGF receptor agonist in cell transformation. EMBO J 5:1535–1541

Johnsson A, Betsholtz C, von der Helm K, Heldin C-H, Westermark B (1985a) Platelet-derived growth factor agonist activity of a secreted form of the v-sis oncogene product. Proc Natl Acad Sci USA 82:1721–1725

Johnsson A, Heldin C-H, Wasteson A, Westermark B, Deuel TF, Huang JS, Seeburg PH, Gray A, Ullrich A, Scrace G, Stroobant P, Waterfield MD (1984) The c-sis gene encodes a precursor of the B chain of platelet-derived growth factor. EMBO J 3:921–928

Josephs SF, Guo C, Ratner L, Wong-Staal F (1984) Human-proto-oncogene nucleotide sequences corresponding to the transforming region of simian sarcoma virus. Science 223:487–491

Kao RT, Hall J, Engel L, Stern R (1984) The matrix of human breast tumor cells is mitogenic for fibroblasts. Am J Pathol 115:109–116

Kaplan DR, Chao FC, Stiles CD, Antoniades HN, Scher CD (1979a) Platelet alpha-granules contain a growth factor for fibroblasts. Blood 53:1043–1052

Kaplan KL, Broekman MJ, Chernoff A, Lesznik GR, Drillings M (1979b) Platelet alpha-granule proteins: studies on release and subcellular localization. Blood 53:604–618

Kartha S, Bradham DM, Grotendorst GR, Toback FG (1988) Kidney epithelial cells express c-sis protooncogene and secret PDGF-like protein. Am J Physiol 255:F800–F806

Katoh O, Kimura A, Kuramoto A (1988) Platelet-derived growth factor is decreased in patients with myeloproliferative disorders. Am J Hematol 27:276–280

Kavanaugh WM, Harsh GR, Starksen NF, Rocco CM, Williams LT (1988) Transcriptional regulation of the A and B chain genes of platelet-derived growth factor in microvascular endothelial cells. J Biol Chem 263:8470–8472

Kazlauskas A, Bowen-Pope D, Seifert R, Hart CE, Cooper JA (1988) Different effects of homo- and heterodimers of platelet-derived growth factor A and B chains on human and mouse fibroblasts. EMBO J 7:3727–3735

Kazlauskas A, Cooper JA (1988) Protein kinase C mediates platelet-derived growth factor-induced tyrosine phosphorylation of p42. J Cell Biol 106:1395–1402

Kazlauskas A, DiCorleto PE (1985) Cultured endothelial cells do not respond to a platelet-derived growth-factor-like protein in an autocrine manner. Biochim Biophys Acta 846:405–412

Keating MT, Williams LT (1987) Processing of the platelet-derived growth factor receptor: biosynthetic and degradation studies using anti-receptor antibodies. J Biol Chem 262:7932–7937

Keating MT, Williams LT (1988) Autocrine stimulation of intracellular PDGF receptors in v-sis-transformed cells. Science 239:914–916

Kelly K, Cochran BH, Stiles CD, Leder P (1983) Cell-specific regulation of c-myc gene by lymphocyte mitogens and platelet-derived growth factor. Cell 35:603–610

King CR, Giese NA, Robbins KC, Aaronson SA (1985) In vitro mutagenesis of the v-sis transforming gene defines functional domains of its growth factor-related product. Proc Natl Acad Sci USA 82:5295–5299

Klein R, Thiel H-J (1988) Highly glycosylated PDGF-like molecule secreted by simian sarcoma virus-transformed cells. Virology 164:403–410

Knighton DR, Fiegel VD, Austin LL, Ciresi KF, Butler EL (1986) Classification and treatment of chronic nonhealing wounds. Ann Surg 204:322–330

Kohler N, Lipton A (1974) Platelets as a source of fibroblast growth-promoting activity. Exp Cell Res 87:297–301

Kozak M (1986) Bifunctional messenger RNAs in eukaryotes. Cell 47:481–483

Krupp MN, Connolly DT, Lane MD (1982) Synthesis, turnover, and down-regulation of epidermal growth factor receptors in human A431 epidermoid carcinoma cells and skin fibroblasts. J Biol Chem 257:11489–11496

Lawler J (1986) The structural and functional properties of thrombospondin. Blood 67:1197–1209

Lawrence WT, Sporn MB, Gorschboth C, Norton JA, Grotendorst GR (1986) The reversal of an Adriamycin induced healing impairment with chemoattractants and growth factors. Ann Surg 203:142–147

Leal F, Williams LT, Robbins KC, Aaronson SA (1985) Evidence that the v-sis gene product transforms by interaction with the receptor for platelet-derived growth factor. Science 230:327–330

Lee BA, Maher DW, Hannink M, Donoghue DJ (1987) Identification of a signal for nuclear targeting in platelet-derived growth-factor-related molecules. Mol Cell Biol 7:3527–3537

Leof EB, Wharton W, Van Wyk JJ, Pledger WJ (1982) Epidermal growth factor (EGF) and somatomedin C regulate G1 progression in competent Balb/c-3T3 cells. Exp Cell Res 141:107–115

Leof EB, Proper JA, Goustin AS, Shipley GD, DiCorleto PE, Moses HL (1986) Induction of c-sis mRNA and activity similar to pletelet-derived growth factor by transforming growth factor β: a proposed model for indirect mitogenesis involving autocrine activity. Proc Natl Acad Sci USA 83:2453–2457

Letterio JJ, Coughlin SR, Williams LT (1986) Pertussis toxin-sensitive pathway in the stimulation of c-myc expression and DNA synthesis by bombesin. Science 234:1117–1119

Levin M, Walters MDS, Waterfield MD, Stroobant P, Cheng D, Barratt TM (1986) Platelet-derived growth factor as possible mediators of vascular proliferation in the sporadic haemolytic uraemic syndrome. Lancet Oct 11.:830–833

Libby P, Warner SJC, Salomon RN, Birinyi LK (1988) Production of platelet-derived growth factor-like mitogen by smooth-muscle cells from human atheroma. N Engl J Med 318:1493–1498

Limanni A, Fleming T, Molina R, Hufnagel H, Cunningham RE, Cruess DF, Sharefkin JB (1988) Expression of genes for platelet-derived growth factor in adult human venous endothelium. A possible non-platelet-dependent cause of intimal hyperplasia in vein grafts and perianastomotic areas of vascular prostheses. J Vasc Surg 7:10–20

Linzer DIH, Nathans D (1983) Growth-related changes in specific mRNAs of cultured mouse cells. Proc Natl Acad Sci USA 80:4271–4275

Lopez-Rivas A, Mendoza SA, Nanberg E, Sinnett-Smith J, Rozengurt E (1987) Ca^{2+}-mobilizing actions of platelet-derived growth factor differ from those of bombesin and vasopressin in Swiss 3T3 mouse cells. Proc Natl Acad Sci USA 84:5768–5772

Lucas CA, Gillies RJ, Olson JE, Giuliano KA, Martinez R, Sneider JM (1988) Intracellular acidification inhibits the proliferative response in BALB/c-3T3 cells. J Cell Physiol 136:161–167

Lynch SE, Nixon JC, Colvin RB, Antoniades HN (1987) Role of platelet-derived growth factor in wound healing: synergistic effects with other growth factors. Proc Natl Acad Sci USA 84:7696–7700

MacDonald ML, Mack KF, Glomset JA (1987) Regulation of phosphoinositide phosphorylation in Swiss 3T3 cells stimulated by platelet-derived growth factor. J Biol Chem 262:1105–1110

Maher DW, Lee BA, Donoghue DJ (1989) The alternatively spliced exon of the platelet derived growth factor. A chain encodes a nuclear targeting signal. Mol Cell Biol 9:2251–2253

Majack RA, Cook SC, Bornstein P (1985) Platelet-derived growth factor and heparin-like glycosaminoglycans regulate thrombospondin synthesis and deposition in the matrix by smooth muscle cells. J Cell Biol 101:1059–1070

Majack RA, Mildbrandt J, Dixit VM (1987) Induction of thrombospondin messenger RNA levels occurs as an immediate primary response to platelet-derived growth factor. J Biol Chem 262:8821–8825

Majesky MW, Benditt EP, Schwartz SM (1988) Expression and developmental control of platelet-derived growth factor A-chain and B-chain/sis genes in rat aortic smooth muscle cells. Proc Natl Acad Sci USA 85:1524–1528

Maniatis AK, Amsel S, Mitus WJ, Coleman N (1969) Chromosome pattern of bone marrow fibroblasts in patients with chronic granulocytic leukaemia. Nature 222:1278–1279

Martinet Y, Bitterman PB, Mornex J-F, Grotendorst GR, Martin GR, Crystal RG (1986) Activated human monocytes express the c-sis proto-oncogene and release a mediator showing PDGF-like activity. Nature 319:158–160

Martinez R, Gillies RJ, Giuliano KA (1988) Effect of serum on intracellular pH of BALB/c-3T3 cells: serum deprivation causes changes in sensitivity of cells to serum. J Cell Physiol 136:154–160

Matsui T, Heidaran M, Miki T, Popescu N, La Rochelle W, Kraus M, Pierce J, Aaronson S (1989) Isolation of a novel receptor cDNA establishes the existence of two PDGF receptor genes. Science 243:800–804

Matuoka K, Fukami K, Nakanishi O, Kawai S, Takenawa T (1988) Mitogenesis in response to PDGF and bombesin abolished by microinjection of antibody to PIP_2. Science 239:640–644

Mercola M, Melton DA, Stiles CD (1988) Platelet-derived growth factor A chain is maternally encoded in Xenopus embryos. Science 241:1223–1225

Miyazono K, Okabe T, Urabe A, Takaku F, Heldin C-H (1987) Purification and properties of an endothelial cell growth factor from human platelets. J Biol Chem 262:4098–4103

Moolenar WH, Tertoolen LGJ, de Laat SW (1984) Phorbol ester and diacylglycerol mimic growth factors in raising cytoplasmic pH. Nature 312:371–374

Moore S, Friedman RJ, Singal DP, Gauldie MA, Blajchman MA, Roberts RS (1976) Inhibition of injury induced thromboatherosclerotic lesions by anti-platelet serum in rabbits. Thromb Haemost 35:70–81

Mornex J-F, Martinet Y, Yamauchi K, Bitterman PB, Grotendorst GR, Chytil-Weir A, Martin GR, Crystal RG (1986) Spontaneous expression of the c-sis gene and release of platelet-derived growth factor like molecule by human alveolar macrophages. J Clin Invest 78:61–66

Narayanan AS, Page RC (1983) Biosynthesis and regulation of type V collagen in diploid human fibroblasts. J Biol Chem 258:11694–11699

Nilsson J, Thyberg J, Heldin C-H, Westermark B, Wasteson A (1983) Surface binding and internalization of platelet-derived growth factor in human fibroblasts. Proc Natl Acad Sci USA 80:5592–5596

Nilsson J, Sjolund M, Palmberg L, Thyberg J, Heldin C-H (1985) Arterial smooth muscle cells in primary culture produce a platelet-derived growth factor-like protein. Proc Natl Acad Sci USA 82:4418–4422

Niman HL, Houghten RA, Bowen-Pope DF (1984) Detection of high molecular weight forms of platelet-derived growth factor by sequence-specific antisera. Science 226:701–703

Niman HL, Thompson AMH, Yu A, Markman M, Willems JJ, Herwig KR, Habib NA, Wood CB, Houghten RA, Lerner RA (1985) Anti-peptide antibodies detect oncogene-related proteins in urine. Proc Natl Acad Sci USA 82:7924–7928

Nishimura J, Huang JS, Deuel TF (1982) Platelet-derived growth factor stimulates tyrosine-specific protein kinase activity in Swiss mouse 3T3 cell membranes. Proc Natl Acad Sci USA 79:4303–4307

Nister M, Hammacher A, Mellstrom K, Siegbahn A, Ronnstrand L, Westermark B, Heldin C-H (1988a) A glioma-derived PDGF A chain homodimer has different functional activities than a PDGF AB heterodimer purified from human platelets. Cell 52:791–799

Nister M, Heldin C-H, Westermark B (1986) Clonal variation in the production of a platelet-derived growth factor-like growth factor and expression of corresponding receptors in a human malignant glioma. Cancer Res 46:332–337

Nister M, Libermann TA, Betsholtz C, Pettersson M, Claesson-Welsh L, Heldin C-H, Schlessinger J, Westermark B (1988b) Expression of messenger RNAs for platelet-derived growth factor and transforming growth factor-α and their receptors in human malignant glioma cell lines. Cancer Res 48:3910–3918

Noble M, Murray K, Stroobant P, Waterfield MD, Riddle P (1988) Platelet-derived growth factor promotes division and motility and inhibits premature differentiation of the oligodendrocyte/type-2 astrocyte progenitor cell. Nature 333:560–565

Oka Y, Orth DN (1983) Human plasma epidermal growth factor/beta-urogastrone is associated with blood platelets. J Clin Invest 72:249–259

Oquendo P, Alberta J, Wen D, Graycar JL, Derynck R, Stiles CD (1989) The platelet-derived growth factor-inducible KC gene encodes a secretory protein related to platelet α-granule proteins. J Biol Chem 264:4133–4137

Ostman A, Rall L, Hammacher A, Wormstead MA, Coit D, Valenzuela P, Betsholtz C, Westermark B, Heldin C-H (1988) Synthesis and assembly of a functionally active recombinant platelet-derived growth factor AB heterodimer. J Biol Chem 263:16202–16208

Owen AJ, Geyer RP, Antoniades HN (1982) Human platelet-derived growth factor stimulates amino acid transport and protein synthesis by human diploid fibroblasts in plasma-free media. Proc Natl Acad Sci USA 79:3203–3207

Owen AJ, Pantazis P, Antoniades HN (1984) Simian sarcoma virus-transformed cells secrete a mitogen identical to platelet-derived growth factor. Science 225:54–56

Pandolfi A, Florita A, Altomare G, Pigatto P, Donati MB, Poggi A (1989) Increased plasma levels of platelet-derived growth factor (PDGF) activity in patients with progressive systemic sclerosis (PSS). Proc Soc Exp Biol Med 191:1–4

Pantazis P, Peliccii PG, Dalla-Favera R, Antoniades HN (1985) Synthesis and secretion of proteins resembling platelet-derived growth factor by human glioblastoma and fibrosarcoma cells in culture. Proc Natl Acad Sci USA 82:2404–2408

Pantazis P, Sariban E, Kufe D, Antoniades HN (1986) Induction of c-sis gene expression and synthesis of platelet-derived growth factor in human myeloid leukemia cells during monocytic differentiation. Proc Natl Acad Sci USA 83:6455–6459

Papayannopoulou T, Raines E, Collins S, Nakamoto B, Tweeddale M, Ross R (1987) Constitutive and inducible secretion of platelet-derived growth factor analogs by human leukemic cell lines coexpressing erythroid and megakaryocytic markers. J Clin Invest 79:859–866

Paris S, Pouyssegur J (1984) Growth factors activate the Na^+/H^+ antiporter in quiescent fibroblasts by increasing its affinity for intracellular H^+. J Biol Chem 259:10989–10994

Pasquale EB, Maher PA, Singer SJ (1988) Comparative study of tyrosine phosphorylation of proteins in Swiss 3T3 fibroblasts stimulated by a variety of mitogenic agents. J Cell Physiol 137:146–156

Paulsson Y, Hammacher A, Heldin C-H, Westermark B (1987) Possible positive autocrine feedback in the prereplicative phase of human fibroblasts. Nature 328:715–717

Pech M, Rao CD, Rubbins KC, Aaronson SA (1989) Functional identification of regulatory elements within the promoter region of platelet-derived growth factor 2. Mol Cell Biol 9:396–405

Peres R, Betsholtz C, Westermark B, Heldin C-H (1987) Frequent expression of growth factors for mesenchymal cells in human mammary carcinoma cells lines. Cancer Res 47:3425–3429

Pfeifle B, Boeder H, Ditschuneit H (1987) Interaction of receptors for insulin-like growth factor I, platelet-derived growth factor, and fibroblast growth factor in rat aortic cells. Endocrinol 120:2251–2258

Pike LJ, Bowen-Pope D, Ross R, Krebs EG (1983) Characterization of platelet-derived growth factor-stimulated phosphorylation in cell membranes. J Biol Chem 258:9383–9390

Pledger WJ, Stiles CD, Antoniades HN, Scher CD (1977) Induction of DNA synthesis in BALB/c3T3 cells by serum components: reevaluation of the commitment process. Proc Natl Acad Sci USA 74:4481–4485

Raines EW, Ross R (1982) Platelet-derived growth factor. I. High yield purification and evidence for multiple forms. J Biol Chem 257:5154–5160

Raines EW, Ross R (1985) Purification of human platelet-derived growth factor. Methods Enzymol 109:749–773

Raines EW, Ross R (1988) Identification and assay of platelet-derived growth factor-binding proteins. Methods Enzymol 147:48–64

Raines EW, Bowen-Pope DF, Ross R (1984) Plasma binding proteins for platelet-derived growth factor that inhibit its binding to cell-surface receptors. Proc Natl Acad Sci USA 81:3424–3428

Raines EW, Dower SK, Ross R (1989) IL-1 mitogenic activity for fibroblasts and smooth muscle cells is due to PDGF-AA. Science 243:393–396

Raivich G, Kreutzberg GW (1987) Expression of growth factor receptors in injured nervous tissue. II. Induction of specific platelet-derived growth factor binding in the injured PNS is associated with a breakdown in the blood-nerve barrier and endoneurial interstitial oedema. J Neurocytol 16:701–711

Rakowicz-Szulazynska EM, Rodeck U, Herlyn M, Koprowski H (1986) Chromatin binding of epidermal growth factor, nerve growth factor, and platelet-derived growth factor in cells bearing the appropriate surface receptors. Proc Natl Acad Sci USA 83:3728–3732

Rao CD, Igarashi H, Chiu I-M, Robbins KC, Aaronson SA (1986) Structure and sequence of the human c-sis/platelet-derived growth factor 2 (SIS/PDGF2) transcriptional unit. Proc Natl Acad Sci USA 83:2392–2396

Rao CD, Pech M, Robbins KC, Aaronson SA (1988) The 5' untranslated sequence of the c-sis/platelet-derived growth factor 2 transcript is a potent translational inhibitor. Mol Cell Biol 8:284–292

Rappolee DA, Brenner CA, Schultz R, Mark D, Werb Z (1988) Development expression of PDGF, TGF-α, and TGF-β genes in preimplantation mouse embryos. Science 241:1823–1825

Ratner L, Josephs SF, Jarrett R, Reitz MS, Wong-Staal F (1985) Nucleotide sequence of transforming human c-sis cDNA clones with homology to platelet-derived growth factor. Nucleic Acids Res 13:5007–5018

Ratner L, Thielan B, Collins T (1987) Sequences of the 5' portion of the human c-sis gene: characterization of the transcriptional promoter and regulation of expression of the protein product by 5' untranslated mRNA sequences. Nucleic Acids Res 15:6017–6036

Richardson WD, Pringle N, Mosley MJ, Westermark B, Dubois-Dalcq M (1988) A role for platelet-derived growth factor in normal gliogenesis in the central nervous system. Cell 53:309–319

Rizzino A, Bowen-Pope DF (1985) Production of PDGF-like factors by embryonal carcinoma cells and response to PDGF by endoderm-like cells. Dev Biol 110:15–22

Rizzino A, Kazakoff P, Ruff E, Kuszynski C, Nebelsick J (1988) Regulatory effects of cell density on the binding of transforming growth factor β, epidermal growth factor, platelet-derived growth factor, and fibroblast growth factor. Cancer Res 48:4266–4271

Robbins KC, Antoniades HN, Devare SG, Hunkapiller MW, Aaronson SA (1983) Structural and immunological similarities between simian sarcoma virus gene product(s) and human platelet-derived growth factor. Nature 305:605–608

Robbins KC, Leal F, Pierce JA, Aaronson SA (1985) The v-sis/PDGF-2 transforming gene product localizes to cell membranes but is not a secretory protein. EMBO J 4:1783–1792

Roberts WM, Look AT, Roussel MF, Sherr CJ (1988) Tandem linkage of human CSF-1 receptor (c-fms) and PDGF receptor genes. Cell 55:655–661

Rollins BJ, Stiles CD (1988) Regulation of c-myc and c-fos proto-oncogene expression by animal cell growth factors. In Vitro 24:81–84

Rollins BJ, Morrison ED, Stiles CD (1987) A cell-cycle constraint on the regulation of gene expression by platelet-derived growth factor. Science 238:1269–1271

Rollins BJ, Morrison ED, Stiles CD (1988) Cloning and expression of JE, a gene inducible by platelet-derived growth factor and whose product has cytokine-like properties. Proc Natl Acad Sci USA 85:3738–3742

Ronnett GV, Tennekoon G, Knutson VP, Lane MD (1983) Kinetics of insulin receptor transit to and removal from the plasma membrane: effect of insulin-induced downregulation in 3T3-L1 adipocytes. J Biol Chem 258:283–290

Ronnstrand L, Beckmann MP, Faulders B, Ostman A, Ek B, Heldin C-H (1987) Purification of the receptor for platelet-derived growth factor from porcine uterus. J Biol Chem 262:2929–2932

Ronnstrand L, Terracio L, Claesson-Welsh L, Heldin C-H, Rubin K (1988) Characterization of two monoclonal antibodies reactive with the external domain of the platelet-derived growth factor receptor. J Biol Chem 263:10429–10435

Rorsman F, Bywater M, Knott TJ, Scott J, Betsholtz C (1988) Structural characterization of the human platelet-derived growth factor A-chain cDNA and gene: alternative exon usage predicts two different precursor proteins. Mol Cell Biol 8:571–577

Rosenfeld ME, Bowen-Pope DF, Ross R (1984) Platelet-derived growth factor: morphologic and biochemical studies of binding, internalization, and degradation. J Cell Physiol 121:263–274

Rosenfeld M, Keating A, Bowen-Pope DF, Singer JW, Ross R (1985) Responsiveness of the in vitro hematopoietic microenvironment to platelet-derived growth factor. Leuk Res 9:427–434

Rosenfeld ME, Tsukada T, Chait A, Bierman EL, Gown AM, Ross R (1987a) Fatty streak expansion and maturation in Watanabe heritable hyperlipemic and comparably hypercholesterolemic fat-fed rabbits. Arteriosclerosis 7:24–34

Rosenfeld ME, Tsukada T, Gown AM, Ross R (1987b) Fatty streak initiation in Watanabe heritable hyperlipemic and comparably hypercholesterolemic fat-fed rabbits. Arteriosclerosis 7:9–23

Ross R (1968) The fibroblast and wound repair. Biol Rev 43:51–96

Ross R (1981) Atherosclerosis – a problem of the biology of arterial wall cells and their interactions with blood components. Arteriosclerosis 1:293–311

Ross R (1986) The pathogenesis of atherosclerosis: an update. N Engl J Med 314:488–500

Ross R, Glomset JA (1973) Arteriosclerosis and the arterial smooth muscle cell. Science 180:1332–1339

Ross R, Glomset JA (1976) The pathogenesis of atherosclerosis. N Engl J Med 295:369–377, 420–425

Ross R, Glomset JA, Kariya B, Harker L (1974) A platelet-dependent serum factor that stimulates the proliferation of arterial smooth muscle cells in vitro. Proc Natl Acad Sci USA 71:1207–1210

Ross R, Vogel A, Davies P, Raines E, Kariya B, Rivest MJ, Gustafson C, Glomset J (1979) The platelet-derived growth factor and plasma control cell proliferation. In: Hormones in cell culture. Cold Spring Harbor conferences on cell proliferation, vol 6. Cold Spring Harbor, New York, p 27

Ross R, Raines EW, Bowen-Pope DF (1986) The biology of platelet-derived growth factor. Cell 46:155–169

Rozengurt E, Stroobant P, Waterfield MD, Deuel TF, Keechan M (1983) Platelet-derived growth factor elicits cyclic AMP accumulation in Swiss 3T3 cells: role of prostaglandin production. Cell 34:265–272

Rozengurt E, Rodriquez-Pena A, Coombs M, Sinnett-Smith J (1984) Diacylglycerol stimulates DNA synthesis and cell division in mouse 3T3 cells: role of Ca^{2+}-sensitive phospholipid-dependent protein kinase. Proc Natl Acad Sci USA 81:5748–5752

Rubin K, Terracio L, Ronnstrand L, Heldin C-H, Klareskog L (1988a) Expression of plateled-derived growth factor receptors is induced on connective tissue cells during chronic synovial inflammation. Scand J Immunol 27:285–294

Rubin K, Hansson GK, Ronnstrand L, Claesson-Welsh L, Fellstrom B, Tingstrom A, Larsson E, Klareskog L, Heldin C-H, Terracio L (1988b) Induction of B-type receptors for platelet-derived growth factor in vascular inflammation: possible implications for development of vascular proliferative lesions. Lancet June 18:1353–1356

Rutherford RB, Ross R (1976) Platelet factors stimulate fibroblasts and smooth muscle cells quiescent in plasma serum to proliferate. J Cell Biol 69:196–203

Sariban E, Kufe D (1988) Expression of the platelet-derived growth factor 1 and 2 genes in human myeloid cell lines and monocytes. Cancer Res 48:4498–4502

Sariban E, Mitchell T, Rambaldi A, Kufe DW (1988a) C-sis but not c-fos gene expression is lineage specific in human myeloid cells. Blood 71:488–493

Sariban E, Sitaras NM, Antoniades HN, Kufe DW, Pantazis P (1988b) Expression of platelet-derived growth factor (PDGF)-related transcripts and synthesis of biologically active PDGF-like proteins by haman malignant epithelial cell lines. J Clin Invest 82:1157–1164

Sauer MK, Donoghue DJ (1988) Identification of nonessential disulfide bonds and altered conformations in the v-sis protein, a homolog of the B chain of platelet-derived growth factor. Mol Cell Biol 8:1011–1018

Sauer MK, Hannink M, Donoghue DJ (1986) Deletions in the N-terminal coding region of the v-sis gene: determination of the minimal transforming region. J Virol 59:292–300

Sawyer ST, Cohen S (1981) Enhancement of calcium uptake and phosphatidylinositol turnover by epidermal growth factor in A-431 cells. Biochemistry 20:6280–6286

Seemayer TA, Lagace R, Schurch W, Tremblay G (1979) Myofibroblasts in the stroma of invasive and metastatic carcinoma. Am J Surg Pathol 3:525–533

Seifert RA, Schwartz SM, Bowen-Pope DF (1984) Developmentally regulated production of platelet-derived growth factor-like molecules. Nature 311:669–671

Seifert RA, Hart CE, Phillips PE, Forstrom JW, Ross R, Murray MJ, Bowen-Pope DF (1989) Two different subunits associate to create isoform-specific platelet-derived growth factor receptors. J Biol Chem 264:8771–8778

Sejersen T, Betsholtz C, Sjolund M, Heldin C-H, Westermark B, Thyberg J (1986) Rat skeletal myoblasts and arterial smooth muscle cells express the gene for the A chain but not the gene for the B chain (c-sis) of platelet-derived growth factor (PDGF) and produce a PDGF-like protein. Proc Natl Acad Sci USA 83:6844–6848

Senior RM, Griffin GL, Huang JS, Walz DA, Deuel TF (1983) Chemotactic activity of platelet alpha granule proteins for fibroblasts. J Cell Biol 96:382–385

Seppa H, Grotendorst G, Seppa S, Schiffmann E, Martin GR (1982) Platelet-derived growth factor is chemotactic for fibroblasts. J Cell Biol 92:584–588

Shaw G, Kamen R (1986) A conserved AU sequence from the 3'untranslated region of GM-CSF mRNA mediates selective mRNA degradation. Cell 46:659–667

Shier WT (1980) Serum stimulation of phospholipase A_2 and prostaglandin release in 3T3 cells is associated with platelet-derived growth-promoting activity. Proc Natl Acad Sci USA 77:137–141

Shier WT, Durkin JP (1982) Role of stimulation of arachidonic acid release in the proliferative response of 3T3 mouse fibroblasts to platelet-derived growth factor. J Cell Physiol 112:171–181

Shimokado K, Raines EW, Madtes DK, Barrett TB, Benditt EP, Ross R (1985) A significant part of macrophage-derived growth factor consists of at least two forms of PDGF. Cell 43:277–286

Shultz PJ, DiCorleto PE, Silver BJ, Abboud HE (1988) Mesangial cells express PDGF mRNAs and proliferate in response to PDGF. Am J Physiol 255:F674–F684

Silverstein RL, Leung LLK, Nachman RL (1986) Thrombospondin: a versatile multifunctional glycoprotein. Arteriosclerosis 6:245–253

Singh JP, Chaikin MA, Stiles CD (1982) Phylogenetic analysis of platelet-derived growth factor by radio-receptor assay. J Cell Biol 95:667–671

Sitaris NM, Sariban E, Pantazis P, Zetter B, Antoniades HN (1987) Human iliac artery endothelial cells express both genes encoding the chains of platelet-derived growth factor (PDGF) and synthesize PDGF-like mitogen. J Cell Physiol 132:376–380

Sjolund M, Hedin U, Sejersen T, Heldin C-H, Thyberg J (1988) Arterial smooth muscle cells express platelet-derived growth factor (PDGF) A chain mRNA, secrete a PDGF-like mitogen, and bind exogenous PDGF in a phenotype- and growth state-dependent manner. J Cell Biol 106:403–413

Smith JC, Stiles CD (1981) Cytoplasmic transfer of the mitogenic response to platelet-derived growth factor. Proc Natl Acad Sci USA 78:4363–4367

Smith JW, Shulman HM, Thomas ED, Fefer A, Buckner CD (1981) Bone marrow transplantation for acute myelosclerosis. Cancer 48:2198–2203

Sprugel KH, McPherson JM, Clowes AW, Ross R (1987) Effects of growth factors in vivo. I. Cell ingrowth into porous subcutaneous chambers. Am J Pathol 129:601–613

Starksen NF, Harsh GR, Gibbs VC, Williams LT (1987) Regulated expression of the platelet-derived growth factor A chain gene in microvascular endothelial cells. J Biol Chem 262:14381–14384

Stenman G, Rorsman F, Betsholtz C (1988) Sublocalization of the human PDGF A-chain gene to chromosome 7, band q11.23, by in situ hybridization. Exp Cell Res 178:180–184

Stevens CW, Brondyk WH, Burgess JA, Manoharan TH, Hane BG, Fahl WE (1988) Partially transformed, anchorage-independent human diploid fibroblasts result from overexpression of the c-sis oncogene: mitogenic activity of an apparent monomeric platelet-derived growth factor 2 species. Mol Cell Biol 8:2089–2096

Stiles CD (1983) The molecular biology of platelet-derived growth factor. Cell 33:653–655

Stiles CD, Capone GT, Scher CD, Antoniades HN, Van Wyk JJ, Pledger WJ (1979) Dual control of cell growth by somatomedins and platelet-derived growth factor. Proc Natl Acad Sci USA 76:1279–1283

Stoscheck CM, Carpenter G (1984) Down regulation of epidermal growth factor receptors: direct demonstration of receptor degradation in human fibroblasts. J Cell Biol 98:1048–1053

Stroobant P, Waterfield MD (1984) Purification and properties of porcine platelet-derived growth factor. EMBO J 3:2963–2967

Stroobant P, Gullick WJ, Waterfield MD, Rozengurt E (1985) Highly purified fibroblast-derived growth factor, an SV-40-transformed fibroblast-secreted mitogen, is closely related to platelet-derived growth factor. EMBO J 4:1945–1949

Studzinski GP, Brelvi ZS, Feldman SC, Watt RA (1986) Participation of c-myc protein in DNA synthesis of human cells. Science 234:467–470

Swan DC, McBride OW, Robbins KC, Keithley DA, Reddy EP, Aaronson SA (1982) Chromosomal mapping of the simian sarcoma virus onc gene analogue in human cells. Proc Natl Acad Sci USA 79:4691–4695

Swenne I, Heldin C-H, Hill DJ, Hellerstrom C (1988) Effects of platelet-derived growth factor and somatomedin-C/insulin-like growth factor I on the deoxyribonucleic acid replication of fetal rat Islets of Langerhans in tissue culture. Endocrinology 122:214–218

Takehara K, Grotendorst GR, Silver R, LeRoy EC (1987) Dipyridamole decreases platelet-derived growth factor levels in human serum. Arteriosclerosis 7:152–158

Taylor RN, Williams LT (1988) Development expression of platelet-derived growth factor and its receptor in the human placenta. Mol Endocrinol 2:627–632

Terracio L, Ronnstrand L, Tingstrom A, Rubin K, Claesson-Welsh L, Funa K, Heldin C-H (1988) Induction of platelet-derived growth factor receptor expression in smooth muscle cells and fibroblasts upon tissue culturing. J Cell Biol 107:1947–1957

Thiel H-J, Hafenrichter R (1984) Simian sarcoma virus transformation-specific glycopeptide: immunological relationship to human platelet-derived growth factor. Virology 136:414–424

Thiel H-J, Matthews TJ, Broughton EM, Butchko AW, Bolognesi DP (1981) Detection of a transformation-specific glycopeptide in SSV-infected cells. Virology 112:642–650

Thielen GH, Gould D, Fowler M, Dungworth DL (1971) C-type virus in tumor tissue of a wooly monkey (Lagothrix spp.) with fibrosarcoma. J Natl Cancer Inst 47:881–889

Tong BD, Auer DE, Jaye M, Kaplow JM, Ricca G, McConathy E, Drohan W, Deuel TF (1987) cDNA clones reveal differences between human glial and endothelial cell platelet-derived growth factor A-chains. Nature 328:619–621

Tzeng DY, Deuel TF, Huang JS, Baehner RL (1985) Platelet-derived growth factor promotes human peripheral monocyte activation. Blood 66:179–183

Tzeng DY, Deuel TF, Huang JS, Senior RM, Boxer LA, Baehner RL (1984) Platelet-derived growth factor promotes polymorphonuclear leukocyte activation. Blood 64:1123–1128

Valente AJ, Delgado R, Metter JD, Cho C, Sprague EA, Schwartz CJ, Graves DT (1988) Cultured primate aortic smooth muscle cells express both the PDGF-A and PDGF-B genes but do not secrete mitogenic activity or dimeric platelet-derived growth factor protein. J Cell Physiol 136:479–485

Van den Eijnden-van Raaij AJM, van Maurik P, Boonstra J, van Zoelen EJJ, de Laat SW (1988) Ultrastructural localization of platelet-derived growth factor and related factors in normal and transformed cells. Exp Cell Res 178:479–492

Van den Ouweland AMW, van Groningen JJM, Schalken JA, van Neck HW, Bloemers HPJ, van de Ven WJM (1987) Genetic organization of the c-sis transcription unit. Nucleic Acids Res 15:959–970

Van Slyck EJ, Weiss L, Dully M (1970) Chromosomal evidence for the secondary role of fibroblastic proliferation in acute myelofibrosis. Blood 36:729–735

van Zoelen EJJ, van de Ven WJM, Franssen HJ, van Oostwaard TMJ, van der Saag PT, Heldin C-H, de Laat SW (1985) Neuroblastoma cells express c-sis and produce a transforming growth factor antigenically related to the platelet-derived growth factor. Mol Cell Biol 5:2289–2297

Vogel A, Raines E, Kariya B, Rivest M-J, Ross R (1978) Coordinate control of 3T3 cell proliferation by platelet-derived growth factor and plasma components. Proc Natl Acad Sci USA 75:2810–2814

Wahl SM, Hunt DA, Wakefield LM, McCartney-Francis N, Wahl LM, Roberts AB, Sporn MB (1987) Transforming growth factor type β induces monocyte chemotaxis and growth factor production. Proc Natl Acad Sci USA 84:5788–5792

Walker LN, Bowen-Pope DF, Ross R, Reidy MA (1986) Production of platelet-derived growth factor-like molecules by cultured arterial smooth muscle cells accompanies proliferation after arterial injury. Proc Natl Acad Sci USA 83:7311–7315

Waterfield MD, Scrace GT, Whittle N, Stroobant P, Johnsson A, Wasteson A, Westermark B, Heldin C-H, Huang JS, Deuel TF (1983) Platelet-derived growth factor is structurally related to the putative transforming protein p^{28sis} of simian sarcoma virus. Nature 304:35–39

Webster G (1971) Morphogenesis and pattern formation in hydroids. Biol Rev 46:1–46

Weich HA, Herbst D, Schairer HU, Hoppe J (1987) Platelet-derived growth factor. Phorbol ester induces the expression of the B-chain but not of the A-chain in HEL cells. FEBS Lett 213:89–94

Westermark B, Wasteson A (1976) A platelet factor stimulating human normal glial cells. Exp Cell Res 98:170–174

Westermark B, Nister M, Heldin C-H (1985) Growth factors and oncogenes in human malignant glioma. Neurol Clin 3:785–799

Westermark B, Johnsson A, Paulsson Y, Betsholtz C, Heldin C-H, Herlyn M, Rodeck U, Koprowski H (1986) Human melanoma cell lines of primary and metastatic origin express the genes encoding the chains of platelet-derived growth factor (PDGF) and produce a PDGF-like growth factor. Proc Natl Acad Sci USA 83:7197–7200

Westin EH, Wong-Staal F, Gelmann EP, Dalla Favera R, Papas TS, Lautenberger JA, Eva A, Reddy EP, Tronick SR, Aaronson SA, Gallo RC (1982) Expression of cellular homologues of retroviral onc genes in human hematopoietic cells. Proc Natl Acad Sci USA 79:2490–2494

White JG (1974) Physiochemical dissection of platelet structure physiology. In: Baldini MG, Ebbe S (eds) Platelets: production, function, transfusion, and storage. Grune and Stratton, New York, p 235

Wilcox JN, Smith KM, Williams LT, Schwartz SM, Gordon D (1988) Platelet-derived growth factor mRNA detection in human atherosclerotic plaques by in situ hybridization. J Clin Invest 82:1134–1143

Williams LT, Tremble P, Antoniades HN (1982) Platelet-derived growth factor binds specifically to receptors on vascular smooth muscle cells and the binding becomes nondissociable. Proc Natl Acad Sci USA 79:5867–5870

Williams LT, Antoniades HN, Goetzl EJ (1983) Platelet-derived growth factor stimulates mouse 3T3 cell mitogenesis and leukocyte chemotaxis through different structural determinants. J Clin Invest 72:1759–1763

Williams LT, Tremble PM, Lavin MF, Sunday ME (1984) Platelet-derived growth factor receptors form a high affinity state in membrane preparations. Kinetics and affinity cross-linking studies. J Biol Chem 259:5287–5294

Wilson E, Laster SM, Gooding LR, Lambeth JD (1987) Platelet-derived growth factor stimulates phagocytosis and blocks agonist-induced activation of the neutrophil oxidative burst: a possible cellular mechanism to protect against oxygen radical damage. Proc Natl Acad Sci USA 84:2213–2217

Witte LD, Kaplan KL, Nossel HL, Lages BA, Weiss HJ, Goodman DS (1978) Studies of the release from human platelets of the growth factor for cultured human arterial smooth muscle cells. Circ Res 42:402–409

Womer RB, Frick K, Mitchell CD, Ross AH, Bishayee S, Scher CD (1987) PDGF induces c-myc expression in MG-63 human osteosarcoma cells but does not stimulate cell replication. J Cell Physiol 132:65–72

Yarden Y, Escobedo JA, Kuang W-J, Yang-Feng TL, Daniel TO, Tremble PM, Chen EY, Ando ME, Harkins RN, Francke U, Fried VA, Ullrich A, Williams LT (1986) Structure of the receptor for platelet-derived growth factor helps define a family of closely related growth factor receptors. Nature 323:226–232

Yarden Y. Kuang W-J, Yang-Feng T, Coussens L, Munemitsu S, Dull TJ, Chen E, Schlessinger J, Francke U, Ullrich A (1987) Human proto-oncogene c-kit: a new cell surface receptor tyrosine kinase for an unidentified ligand. EMBO J 6:3341–3351

Yeh HJ, Pierce GF, Deuel TF (1984) Ultrastructural localization of a platelet-derived growth factor/v-sis-related protein(s) in cytoplasm and nucleus of simian sarcoma virus-transformed cells. Proc Natl Acad Sci 84:2317–2321

Zerwes H-G, Risau W (1987) Polarized secretion of a platelet-derived growth factor-like chemotactic factor by endothelial cells in vitro. J Cell Biol 105:2037–2041

Zullo JN, Cochran BH, Huang AS, Stiles CD (1985) Platelet-derived growth factor and double-stranded ribonucleic acids stimulate expression of the same genes in 3T3 cells. Cell 43:793–800

CHAPTER 6

Insulin-Like Growth Factors

M. M. RECHLER and S. P. NISSLEY

A. Overview

Until recently, insulin-like growth factors (or, as they were originally known, somatomedins) were thought to be produced in the liver in response to growth hormone, to circulate in the blood, and to mediate the effects of growth hormone (GH) on skeletal cartilage to promote bone elongation during childhood (SALMON and DAUGHADAY 1957). Recent insights have necessitated revision of the somatomedin hypothesis in several important respects:

1. The insulin-like growth factor (IGF) family consists of two related polypeptides, IGF-I and IGF-II. [Peptides identical to IGF-I were initially designated somatomedin A, somatomedin C and basic somatomedin; rat IGF-II was initially designated MSA or multiplication stimulating activity (DAUGHADAY et al. 1987a)]. Despite similar chemical structure and in vitro activity, IGF-I and IGF-II have different in vivo activities. IGF-I is GH dependent, and fulfills many of the criteria of a somatomedin or mediator of GH action. IGF-II is minimally GH dependent, and has little effect on skeletal cartilage. It may play a role in fetal development and in the central nervous system.

2. The IGFs are synthesized in many tissues in fetal and adult animals. Their actions may be local (autocrine or paracrine) or endocrine (circulate to a distant target organ). Thus, the physiologically relevant IGF levels may be tissue levels rather than plasma levels.

3. The biological actions of IGFs are not limited to mitogenesis. They also may induce differentiation or promote the expression of differentiated functions. The precise biological response is determined by the developmental state of the cell, as well as the presence of other hormones and growth factors.

4. The biological actions of IGFs are potentially mediated by either of two specific receptors. The IGF-I receptor (or type I IGF receptor) contains a cytoplasmic tyrosine kinase domain, and binds both IGF-I and IGF-II. Studies with blocking antibodies to the receptor suggest that it mediates many, if not all, of the known effects of IGF-I and IGF-II. The IGF-II/Mannose-6-phosphate (Man-6-P) receptor (or type II IGF receptor) has a completely different structure. It lacks a tyrosine kinase domain, and has a small cytoplasmic domain similar to other cycling receptors. This receptor binds lysosomal proteins that contain a mannose-6-phosphate recognition marker in addition to binding IGF-II. Although it has been claimed that the

IGF-II receptor mediates classic IGF actions, the structural dissimilarity from the type I receptor requires that it use a different set of signalling mechanisms.

5. The IGFs in blood, other extracellular fluids, and cell culture media occur complexed to specific binding proteins. At least three IGF binding proteins have been purified and cloned. The total number of binding proteins is not known. Most of the IGFs in human and rat plasma are carried in a 150-kDa GH-dependent binding protein complex. It consists of a glycosylated ligand-binding subunit (~ 50-kDa) and a second acid-labile nonbinding subunit (~ 100-kDa). Lower molecular weight binding proteins occur in fetal plasma and extracellular fluid. These are GH independent and are not glycosylated. Two distinct classes of GH-independent binding proteins have been cloned. Although the role of the IGF binding proteins is not fully understood, they have the capacity to modulate IGF action. In most cases, the IGF-binding protein complex does not bind to cell receptors and is inactive. In some cases, however, binding proteins act synergistically with IGF-I to stimulate mitogenesis. Whether these different biological activities are properties of different binding proteins or of binding proteins that have undergone different posttranslational modifications remains to be determined.

This article will attempt to summarize current insights into the three components of the IGF system – the proteins, their receptors, and binding proteins – as a basis for understanding the participation of IGFs in physiology and disease. We will focus on selected topics of intense current interest and activity. The reader is referred to other reviews for discussion of the earlier literature and for areas not discussed in the present article: Humbel (1984); Nissley and Rechler (1984a); Van Wyk (1984); Rechler and Nissley (1985b); Froesch et al. (1985); Baxter (1986).

B. Genes and Proteins

I. IGF-I and IGF-II Proteins

1. IGF-I in Different Species

As reviewed by Humbel (1984), IGF-I and IGF-II are single-chain polypeptides containing three disulfide bonds that are closely related to each other and to human insulin. Mature IGF-I (70 amino acids) and IGF-II (67 residues) are highly conserved in different species (Figs. 1, 2). Human, bovine and porcine IGF-I are identical. Rat and mouse IGF-I differ from human IGF-I at three and four positions, respectively. Chicken IGF-I is identical to human IGF-I at 30/31 amino-terminal residues. There is no evidence for variants within the coding sequence of the mature IGF-I molecule; however, alternate splicing of human and rodent IGF-I mRNAs generates precursor proteins with different propeptides (see below).

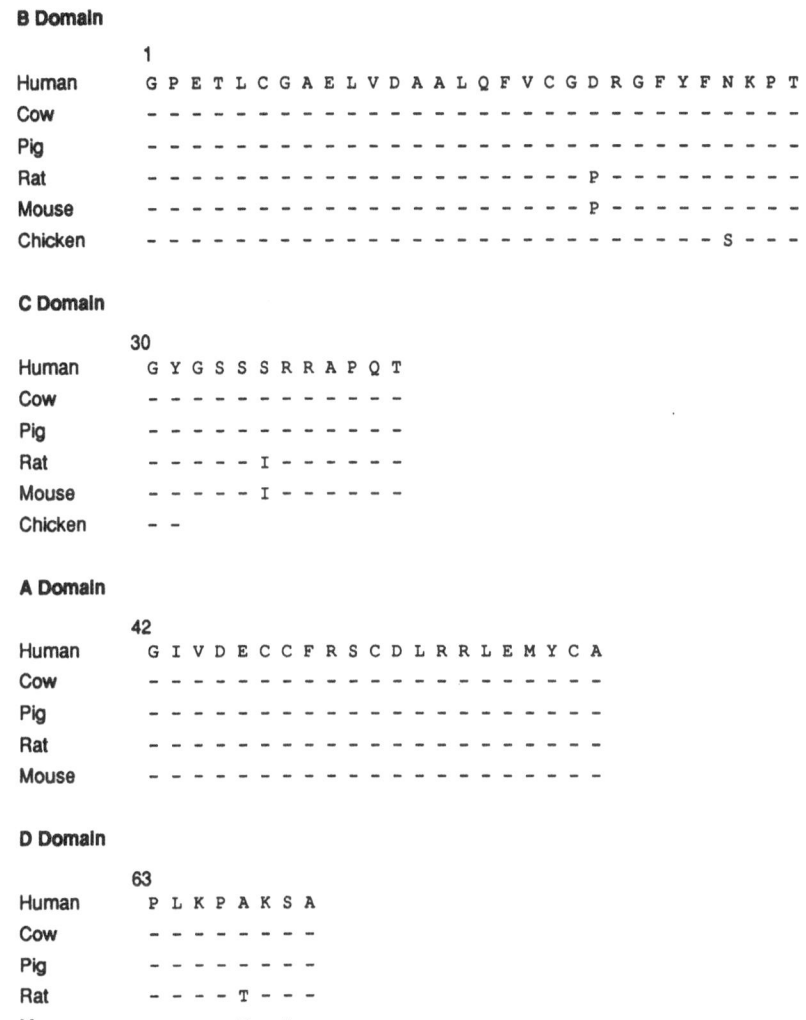

Fig. 1. Amino acid sequences of IGF-I from different mammalian and avian species. The sequence of human IGF-I is indicated by the single letter amino acid code grouped by domain. The B domain begins at residue 1, the C domain at residue 30, the A domain at residue 42, and the D domain at residue 63. Identical residues in other species are indicated by a *dash*. Different residues are designated by the appropriate letter. Sequences are based on protein sequences for IGF-I from human, cow, and chicken and are derived from nucleotide sequences for pig, mouse, and rat IGF-I. Sequences are from the following sources: human, HUMBEL (1984); cow serum, HONEGGER and HUMBEL (1986); cow colostrum, FRANCIS et al. (1988); pig, TAVAKKOL et al. (1988); mouse, BELL et al. (1986); rat, MURPHY et al. (1987b), CASELLA et al. (1987), ROBERTS et al. (1987b); chicken, DAWE et al. (1988)

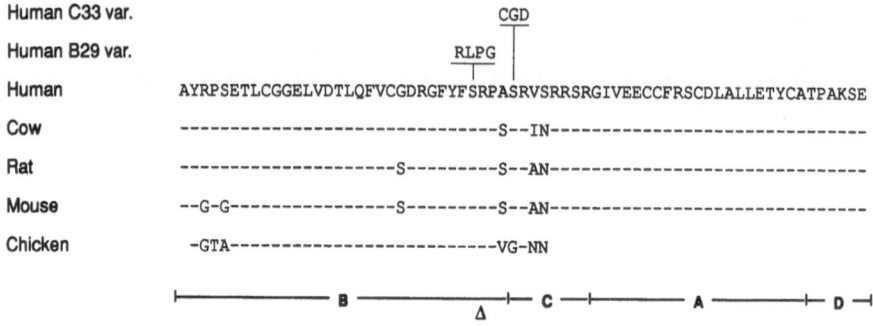

Fig. 2. Amino acid sequences of IGF-II from different species and human IGF-II variant proteins. The sequence of IGF-II from human serum is indicated by the single letter amino acid code (Humbel 1984). The domain boundaries are indicated *below*. The *open triangle* indicates a possible splice junction. Identical residues are indicated by *dashes*; residues different from human IGF-II are designated. Only the variant tetrapeptide and tripeptide are shown for the human B29 and C33 variant proteins. The sequence of the human C33 variant protein is taken from the protein sequence of Zum-stein et al. (1985). The sequence of the human B29 variant is derived from the nucleotide sequence of Jansen et al. (1985) and Le Bouc et al. (1987). The corresponding protein has been identified in serum (Zumstein et al. 1985). The bovine IGF-II sequence is from protein sequencing of IGF-II in serum (Honegger and Humbel 1986) and colostrum (Francis et al. 1988). [Isoleucine C35 is based on the results of Francis et al. (1988) and is compatible with the amino acid composition of Honegger and Humbel (1986)]. Rat IGF-II was determined from protein sequence (Marquardt et al. 1981) and nucleotide sequence (Dull et al. 1984; Whitfield et al. 1984; Soares et al. 1985; Frunzio et al. 1986). [Residue C33 was determined as serine by Soares et al. (1985), instead of glycine as reported by Marquardt et al. (1981).] The mouse IGF-II sequence is based on the nucleotide sequence (Stempien et al. 1986). The chicken IGF-II sequence is from the protein sequence (Dawe et al. 1988)

2. IGF-II Variant Proteins

IGF-II also is highly conserved, but shows somewhat greater variability within the coding sequence of the mature peptide (Fig. 2). Differences in variant human IGF-II molecules and in IGF-II from different species tend to cluster at the juncture of the B and C domains, close to an RNA splice site. Human IGF-II variants occur at residues B29 [in which an Arg-Leu-Pro-Gly (RLPG) tetrapeptide replaces serine (Jansen et al. 1985), presumably through the use of an alternate splice acceptor] and at C33 [in which the tripeptide Cys-Gly-Asp (CGD) replaces serine (Zumstein et al. 1985)]. The C33 variant is thought to represent allelic variation rather than alternate splicing, since no suitable splice site is present.

Bovine, rat and mouse IGF-II differ from human IGF-II at positions B32, C35, and C36. In addition, rat and mouse IGF-II have a glycine to serine substitution at position 22, and mouse IGF-II also substitutes arginine and serine for glycine at positions 3 and 5, respectively. B22 glycine is found in insulins from all species except hystricomorphs in IGF-I, and in human and bovine IGF-II (Humbel 1984).

Two amino-terminal sequences were identified for chicken IGF-II. The major sequence differs from human IGF-II in the first 5 positions, is identical from residues 6 to 31, but then differs at 4 of the next 5 residues. The minor chicken IGF-II sequence is identical to human IGF-II at residues 1–7.

3. Structure-Function Analysis of IGF-I

BAYNE et al. (1987, 1988) have designed vectors suitable for cassette mutagenesis of IGF-I and capable of high-efficiency expression and processing in yeast. By analysis of mutant IGF-I molecules, they have identified sites that are involved in the recognition of IGF-I by IGF-I receptors, IGF-II/Man-6-P receptors, and IGF binding proteins. The three proteins that bind IGF-I recognize different domains on the IGF-I molecule.

a) Binding Protein Domain

Substitution of residues Glu3, Thr4, Gln15, and Phe16 of the B domain of IGF-I with Gln3, Ala4, Tyr15, and Leu16 reduced the affinity for IGF binding proteins 600-fold (BAYNE et al. 1988). This is consistent with results using chemically synthesized hybrid molecules of IGF-I and insulin that showed that the B domain of IGF-I contained determinants recognized by IGF binding proteins (DE VROEDE et al. 1985), and the observation that truncated IGF-I molecules lacking the first three residues of the B domain had greatly reduced affinity for binding proteins (SZABO et al. 1988).

b) IGF-I Receptor Domain

A cluster of aromatic residues in the B domain (Phe23, Tyr24, Phe25), especially at position 24, are important for binding to the IGF-I receptor. Substitution of Tyr24 by Leu24 or Ser24 decreases the affinity for the IGF-I receptor 32- and 16-fold, respectively, and decreases the mitogenic potency in mouse L cells 29-fold (CASCIERI et al. 1988 a).

The C domain of IGF-I assumes a conformation such that Tyr31 is in proximity to the aromatic cluster at residues B23–25. Truncation of the C domain (including deletion of Tyr31) results in a 30-fold decrease in affinity for the IGF-I receptor (BAYNE et al. 1989). This is consistent with the results of TSENG et al. (1987) showing that two-chain molecules containing the A and B domains of IGF-I were only 10%–15% as potent as single-chain IGF-I in binding to the IGF-I receptor and in stimulating DNA synthesis through this receptor. MALY and LUTHI (1988) demonstrated that Tyr24 and Tyr31 became inaccessible to iodination by chloramine T when IGF-I was complexed to purified IGF-I receptor, suggesting that these residues formed part of the binding site.

c) IGF-II/Man-6-P Receptor Domain

The IGF-II/Man-6-P receptor appears to recognize several regions of the A domain of IGF-I (CASCIERI et al. 1989). IGF-I mutants in which Thr49, Ser50, Ile51 are substituted for Phe49, Arg50, Ser51 have 100-fold lower affinity for the IGF-II/Man-6-P receptor. [Tyr 55, Gln56]IGF-I has 7-fold higher affinity than native [Arg55, Arg56]IGF-I for the receptor. Since the uncharged residues Ala, Leu occur in the corresponding positions in IGF-II, the presence of charged residues in IGF-I may be responsible for its lower affinity for the IGF-II/Man-6-P receptor.

II. Biosynthetic Precursors

1. IGF-II

The existence and structure of biosynthetic precursors for rat IGF-II were demonstrated in cell-free translation experiments (ACQUAVIVA et al. 1982) and by biosynthetic labeling of intact BRL-3A cells (YANG et al. 1985a, b; RECHLER et al. 1985). In a reticulocyte lysate cell-free translation system, poly A-enriched BRL-3A RNA directed the translation of a protein with an apparent molecular weight of 22-kDa that is specifically immunoprecipitated by antibodies to mature rat IGF-II. Inclusion of microsomal membranes containing signal peptidase decreased the size of the precursor to 20-kDa by cotranslational removal of the signal peptide. When intact cells were labeled with [^{35}S]cysteine, a 20-kDa protein (pro-rat IGF-II) was labeled rapidly intracellularly, decreased with time, and did not appear in the medium. Intermediate (15- and 8-kDa) and mature (7-kDa) forms increased in the media with time. Amino-terminal sequence analysis of biosynthetically labeled pro-rat IGF-II established directly that the mature 67-residue peptide is located at the amino terminus of the precursor (YANG et al. 1985a).

The structure of the rat IGF-II precursor was confirmed by the sequence of cDNA clones for human IGF-II (BELL et al. 1984; DULL et al. 1984) and rat IGF-II (WHITFIELD et al. 1984; SOARES et al. 1985). The human and rat IGF-II carboxy-terminal E-domain propeptides are identical at 68/89 residues (Fig. 3).

Stable higher molecular weight forms of IGF-II (8.7 and 16-kDa) had been identified in BRL-3A conditioned media by MOSES et al. (1980a, b) and shown to bind to IGF receptors, antibodies, and binding proteins, and to stimulate thymidine incorporation (Table 1). GOWAN et al. (1987) purified the 8.7-kDa and 7.5-kDa forms by high-performance liquid chromatography (HPLC), and showed that they bound with equal potency to IGF binding proteins.

Stable partially processed precursor forms of human IGF-II also have been identified and characterized. ZUMSTEIN et al. (1985) purified a 10-kDa molecule from human serum that contained a 21-residue E-domain extension (and an unrelated alleleic substitution at C33). Approximately 15% of the IGF-II in human serum is 10-kDa. The 10-kDa molecule is immunoreactive, and has ~35% the activity of authentic IGF-II in the fat pad bioassay for

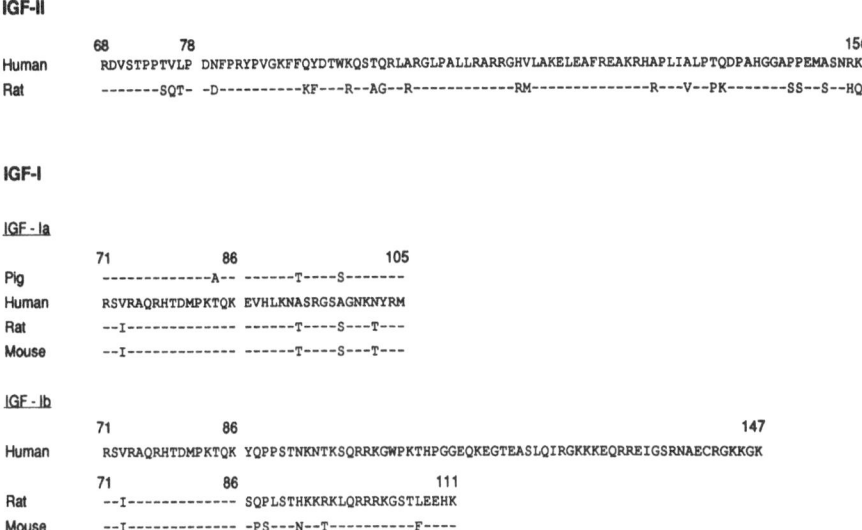

Fig. 3. Amino acid sequence of the E-domain propeptides of IGF-II (*top*) and IGF-I (*bottom*) in different species. *Top,* the sequence of the human IGF-II E-domain (residues 68–156) is derived from the nucleotide sequences of BELL et al. (1984) and JANSEN et al. (1985). The sequence of the rat IGF-II E-domain is based on the nucleotide sequences of DULL et al. (1984); WHITFIELD et al. (1984); SOARES et al. (1985). *Dashes* indicate identical residues; only different residues are indicated. The space between residues 78 and 79 represents an exon-intron boundary. *Bottom,* IGF-I mRNAs encoding different E-domain propeptides were described by JANSEN et al. (1983) and ROTWEIN (1986), and have been designated IGF-Ia and IGF-Ib, respectively. These variants are generated by alternative splicing between residues 86 and 87. Counterparts of human IGF-Ia have been identified in pig (TAVAKKOL et al. 1988), rat (ROBERTS et al. 1987b), and mouse (BELL et al. 1986). Counterparts of human IGF-Ib have been identified in rat (ROBERTS et al. 1987b) and mouse (BELL et al. 1986), although they arise by different mechanisms than human IGF-Ib (see text)

insulin-like activity. GOWAN et al. (1987) purified a 15-kDa molecule from human serum that binds normally to antibodies, IGF binding proteins and the IGF-II/Man-6-P receptor. The 15-kDa protein also binds to IGF-I receptors and stimulates DNA synthesis in 12-week human fetal fibroblasts with equal or greater potency to 7.5-kDa IGF-II or IGF-I (J. PERDUE, personal communication).

Little is known of the enzymatic steps involved in processing the carboxy-terminal propeptide. HYLKA et al. (1985) isolated a fragment of pro-rat IGF-II corresponding to the 40 carboxy-terminal residues (117–156). The sequence preceding this peptide, Arg-Arg-Gly-Arg (RRGR), is susceptible to tryptic cleavage. Chemically synthesized fragment 117–156 has mitogenic activity for NIL-8 hamster fibroblasts at high concentrations ($>10\ \mu g/ml$). Antibodies raised to the synthetic peptide do not recognize the 16- and 8.7-kDa species, as expected if cleavage of the carboxy-terminal fragment (4432 Da) is an early step in processing of the precursor (HYLKA et al. 1987).

Table 1. Protein variants of IGF-I and IGF-II and their biological activities

	cDNA	Protein	Activity
Allelic variation			
Human IGF-II C33 (CGD)	–	A[a, b]	Radioimmunoassay; insulin-like activity[c]
Alternate splicing			
Human IGF-II B29 (RLPG)	B	A[b]	–
Human IGF-Ia	C	D, E	Immunoreactivity (D, E)[d]
Human IGF-Ib	F	G[e]	Radioimmunoassay; IGF-I receptor assay; binding protein assay; thymidine incorporation (G)
Rat, mouse IGF-Ia and IGF-Ib	H, I	–	–
Partial processing of E domain			
Rat IGF-II, 8.7 kDa	–	J, K[f]	Binding protein assay, receptor assay, thymidine incorporation (J); binding protein assay (K)[g]
Human IGF-II, 15 kDa	–	K[b]	IGF-II/Man-6-P receptor assay, binding protein assay, thymidine incorporation (K); IGF-I receptor assay (L)
Human IGF-II, 10 kDa	–	A[b]	Radioimmunoassay; insulin-like activity[c]

[a] *References:* A, ZUMSTEIN et al. (1985); B, JANSEN et al. (1985), LE BOUC et al. (1987), HOPPENER et al. (1988); C, JANSEN et al. (1983); D, MILLS et al. (1986); E, POWELL et al. (1987); F, ROTWEIN (1986); G, CLEMMONS and SHAW (1986); H, BELL et al. (1986); I, ROBERTS et al. (1987b); J, MOSES et al. (1980a, b); K, GOWAN et al. (1987); L, J. PERDUE (personal communication).
[b] Human Serum.
[c] Protein purified from human serum contains both the C33 substitution and a 10-kDa carboxy-terminal extension (A). It gave nonparallel displacement in the IGF-II radioimmunoassay, and had 35% the insulin-like activity of IGF-II in a rat epididymal fat bad bioassay (A).
[d] MILLS et al. (1986) specifically immunoprecipitated a 14-kDa protein using a monoclonal antiserum (Sm 1.2) that does not distinguish IGF-II, IGF-Ia, and IGF-Ib. POWELL et al. (1987) observed a 19-kDa protein using antibodies to the E peptide of IGF-Ia. Thus, the proteins identified in both studies can only be provisionally identified as 11.7-kDa pro-IGF-Ia.
[e] Purified from human fibroblast conditioned media (G). Relationship of amino acid sequence to that of IGF-Ib was cited in CASELLA et al. (1986) and confirmed by D. Clemmons (personal communication).
[f] BRL-3A conditioned media.
[g] The 8.7-kDa rat IGF-II was equipotent with 7.5-kDa rat IGF-II (K).

2. IGF-I

Nucleotide sequence analysis indicates that IGF-I, like IGF-II, is synthesized as a biosynthetic precursor with the mature peptide at the amino terminus and the E-domain propeptide at the carboxy terminus. In contrast to IGF-II, precursors with two distinct propeptides occur in man, rat, and mouse, and arise by alternate RNA splicing. In each species, one precursor corresponds to the human IGF-I cDNA clone first described by JANSEN et al. (1983) and designated IGF-Ia. The second form arises by different mechanisms in human (ROTWEIN 1986) and rodent genes. Rodent IGF-Ib propeptides are more closely related to each other than to the human IGF-Ib propeptide. The sequences of the different E peptides are presented in Fig. 3. The splice variations that generate these sequences will be discussed below.

Suggestive evidence has been presented indicating that these potential precursors are translated into protein:

1. CLEMMONS and SHAW (1986) purified a somatomedin from human fibroblast conditioned media that reacts with antibodies to IGF-I, is mitogenic for BALB/c3T3 and human fibroblasts, and binds to placental membrane IGF-I receptors and amniotic fluid IGF binding proteins. On sodium dodecylsulfate (SDS)-gel electrophoresis, a major protein of 21-kDa was identified, as well as a minor 11-kDa component. The amino acid composition of the 21-kDa protein is more consistent with pro-IGF-Ib than pro-IGF-Ia (ROTWEIN 1986). This has been confirmed by amino acid sequencing (cited in CASELLA et al. 1987; D. R. CLEMMONS, personal communication).

2. MILLS et al. (1986) demonstrated a 14-kDa protein in reticulocyte lysate translations directed by RNA from 10- to 13-week human placenta that was immunoprecipitated by a monoclonal antibody to IGF-I. The size of this protein is consistent with the expected size of prepro-IGF-Ia, 11.7-kDa, but the antibody used exhibits 60% crossreactivity with IGF-II and does not distinguish between IGF-Ia and IGF-Ib.

3. POWELL et al. (1987) used antibodies raised to the E peptide of IGF-Ia to demonstrate immunoreactive material in uremic serum having an apparent molecular weight of 13-kDa on neutral gel filtration, and 19-kDa by immunoblotting. Although this may represent pro-IGF-Ia, the precursor size estimated by immunoblotting is considerably larger than the formula weight of pro-IGF-Ia.

III. IGF-II Gene Expression

1. Multiple IGF-II RNA Transcripts from One Gene

a) RNAs

cDNA clones have been obtained for rat (DULL et al. 1984; WHITFIELD et al. 1984; SOARES et al. 1985), human (BELL et al. 1984; JANSEN et al. 1985; DE PAGTER-HOLTHUIZEN et al. 1987), and mouse (STEMPIEN et al. 1986) IGF-II. Inserts from these clones were used as hybridization probes for Northern blots of RNA from different tissues and cells. Both rat and human IGF-II genes are expressed as multiple RNA transcripts ranging in size from 1 to 4.6 kilobases (kb) for rat IGF-II and 2.2- to 6.0-kb for human IGF-II. In rat cells, the most abundant IGF-II RNA is 3.5-kb; distinct 4.6, 2.2, and 1.0-kb forms, as well as other less distinct species, have been observed. All of these RNAs exhibit similar tissue-specific and development-specific regulation (see below).

Human fetal liver expresses a major 6.0-kb IGF-II mRNA, and less abundant 4.8- and 2.2-kb RNAs (De Pagter-Holthuizen et al. 1987, 1988). Adult liver expresses small amounts of a 5.3-kb IGF-II RNA (Bell et al. 1985; De Pagter-Holthuizen et al. 1987). The 6.0-, and 2.2-kb RNAs are observed in other fetal and adult tissues (J. Scott et al. 1985; Gray et al. 1987). By contrast, the 5.3-kb RNA only is seen in adult liver (J. Scott et al. 1985; De Pagter-Holthuizen et al. 1987; Gray et al. 1987; Irminger et al. 1987), and appears to arise from a tissue-specific and development-specific promoter.

b) Genes

The IGF-II gene is located on human chromosome 11 (Brissenden et al. 1984; Tricoli et al. 1984) and rat chromosome 1 (Soares et al. 1986). It is contiguous to the human insulin gene (Bell et al. 1985; Kittur et al. 1985) and the rat insulin II gene (Soares et al. 1986) and has the same polarity. The IGF-II gene is separated from the poly A site of the insulin gene by < 15-kb (see below).

The rat IGF-II gene is about 15-kb. Its organization is illustrated schematically in Fig. 4. It contains three coding exons: exon 1, containing the prepeptide and the B domain; exon 2, containing the end of the B domain, the complete C, A, and D domains, and the start of the E domain; and exon 3, containing the remainder of the E domain and the 3′ untranslated region.

The human IGF-II gene spans about 28-kb. The organization of the coding region of the human IGF-II gene (Fig. 4) is similar to that of the rat IGF-II gene, and the rat and human IGF-I genes (Dull et al. 1984; Frunzio et al. 1986).

c) Multiple Promoters and 5′ Untranslated Regions

α) Rat

Frunzio et al. (1986) and Soares et al. (1986) demonstrated that transcription of the rat IGF-II gene originates at two noncontiguous promoters (P2 and P1) and generates mRNAs with alternate 5′ noncoding exons, designated exon −2 (1121 nucleotides, nt) and exon −1 (85 nt), respectively (Fig. 4). Hybridization with specific oligonucleotide probes established that the most abundant 3.5-kb RNA originates at P1, the larger 4.6-kb RNA at P2 (Frunzio et al. 1986). The 3.5-kb RNA is about 10-fold more abundant than the 4.6-kb RNA; chloramphenicol acetyltransferase assays (Evans et al. 1988) and nuclear run-on experiments (R. Frunzio, unpublished results) suggest that this reflects the greater efficiency of promoter P1 than P2. Both the exon −2 and exon −1 promoters have a TATA box and Sp1 binding sites (Frunzio et al. 1986; Soares et al. 1986). Exon −2 (Frunzio et al. 1986; Soares et al. 1986) also has a CAAT sequence. Evans et al. (1988) have performed a detailed analysis of the exon −1 promoter and demonstrated that it contains four GC regions between ATA (nucleotide −30) and nucleotide

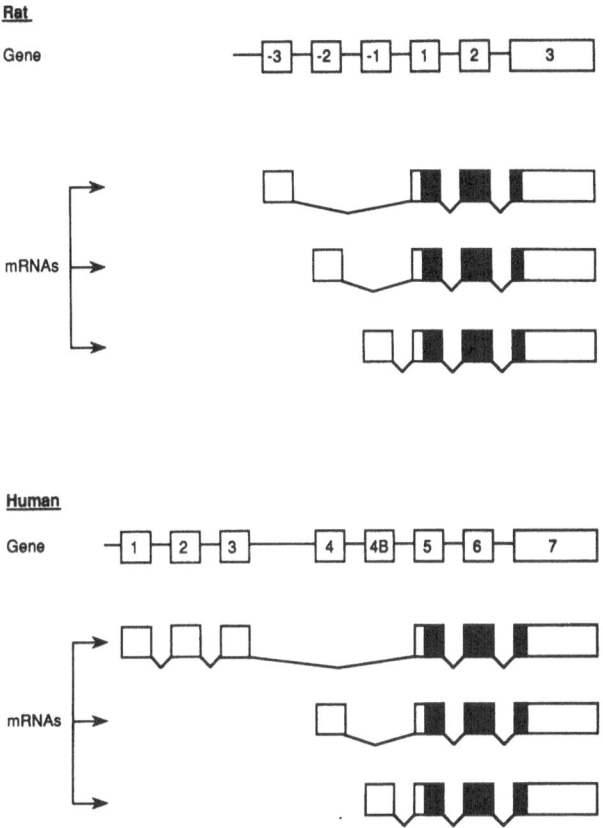

Fig. 4. Schematic representation of the organization of the rat (*top*) and human (*bottom*) IGF-II genes. Exons are indicated by *open boxes,* and numbered as in the text. Introns are indicated by a *thin line.* Transcripts arising at each of three promoters in rat and human and terminating at the 3′ distal end of exons 3 (rat) or 7 (human) are shown. Coding regions are *solid,* untranslated region are *open.* Exons and introns are not drawn to scale. The organization of the rat IGF-II genome is based on the results of FRUNZIO et al. (1986), SOARES et al. (1986), and UENO et al. (1987). The organization of the human IGF-II genome is based on the results of DE PAGTER-HOLTHUIZEN et al. (1987, 1988)

−128. Transcription factor Sp1 binds to the four GC boxes; its binding is abolished by methylation of Gs.

Recently, UENO et al. (1987, 1988a) isolated a cDNA clone from a rat hepatoma AH60C library that contained a third 5′ untranslated region for rat IGF-II. This sequence, which we designate exon −3 (Fig. 4), is located 1.4-kb upstream from the exon −2 promoter. S1 protection experiments suggest that the 5′ end of exon −3 is heterogeneous (260–371 nt). Primer extension and Northern anaylsis suggest that this represents a complete 5′ untranslated region. cDNA clones having exon −3 as the 5′ untranslated region were isolated with a frequency similar to that of exon −2 clones, that is about 5% as frequently as clones containing exon −1. The exon −3 promoter region lacks TATA and GC sequences. [It does, however, have dyad symmetry (which may mimic the TATA box) and ATTGGG repeats that are homologous to CAAT in the dihydrofolate reductase gene.] Exon −3 probes hybridize to a 3.8-kb RNA in neonatal rat tissues (UENO et al. 1988a). Thus, the rat IGF-II gene consists of six exons, three noncoding exons (each with their own promoter), and three coding exons. In general, no significant differences in tissue- or development-specific expression of the three promoters have been observed. UENO et al. (1988a, b) have observed differential promoter expression in different rat hepatoma ascites tumors.

β) Human

The human IGF-II gene also has a complex structure, consisting of eight exons and three promoters. The three coding exons (exons 5, 6, and 7) are directly analogous to those of the rat IGF-II gene (Fig. 4). Messenger RNAs containing three 5′ untranslated sequences have been observed: one, corresponding to exons 1, 2, and 3; a second corresponding to exon 4; and a third corresponding to exon 4B. The original cDNA clones isolated by BELL et al. (1984) and JANSEN et al. (1985) from adult liver cDNA libraries had a 5′ untranslated region corresponding to exon 3. The genomic clone of DULL et al. (1984) had a different 5′ untranslated region (exon 4, corresponding to rat exon −2).

In 1986, DE PAGTER-HOLTHUIZEN et al. identified cDNA clones containing exons 2 and exon 3. In 1987, DE PAGTER-HOLTHUIZEN et al., isolated cDNA clones arising from two distinct promoters: exon 4 (from a HepG2 library, corresponding to the 5′ untranslated region of DULL et al. 1984) and exons 1, 2, and 3 (from an adult liver library). Exon 1 is located only 1.4-kb from the 3′ end of the insulin gene. cDNA clones containing exon 4 as the 5′ untranslated region also were isolated from placenta by LE BOUC et al. (1987) and SHEN et al. (1988), and from adult hypothalamus by IRMINGER et al. (1987).

Exon 4B (homologous to rat exon −1) was identified and localized in genomic clones (SUSSENBACH et al. 1988; DE PAGTER-HOLTHUIZEN et al. 1988), and in a placental cDNA clone (LE BOUC et al. 1987).

The major IGF-II RNAs expressed in fetal liver and most other tissues, 6.0 and 4.8-kb, correspond to 5′ untranslated exons 4 (1164 nt) and 4B (∼100 nt), respectively. The 2.2-kb RNA also arises at the exon 4 promoter, but is truncated at its 3′ end (DE PAGTER-HOLTHUIZEN et al. 1988). The region preceding the exon 4 promoter contains TATA, CAAT, and Sp1 sequences (DE PAGTER-HOLTHUIZEN et al. 1987). The region preceding the exon 4B promoter contains a potential TATA box and two Sp1 sites (SUSSENBACH et al. 1988).

The 5.3-kb RNA expressed in adult liver appears to arise from a promoter upstream from exon 1. Exon 1 is spliced to noncoding exons 2 and 3 before being spliced to the common coding exons (exons 5, 6, and 7). Exon 1 lacks TATA and CAAT boxes, but has a potential Sp1 binding site (DE PAGTER-HOLTHUIZEN et al. 1987).

γ) *Mouse*

ROTWEIN and POLLOCK (1987) reported that the mouse IGF-II gene has three coding exons and at least four 5' noncoding exons. Exons −1 and −2 are directly homologous to the corresponding rat exons. Exon −3 has not been characterized. Exon −4 is homologous to human exon 2. A mouse exon −4 oligonucleotide probe hybridizes to rat genomic DNA, but not to BRL-3A cell RNA (A. BROWN and L. CHIARIOTTI, unpublished results). A probe corresponding to human exon 1 did not hybridize either to rat genomic DNA or to BRL-3A RNA. This suggests that the rat genome cannot express mouse exon −4/human exon 2 because it lacks human exon 1 and the corresponding promoter region.

d) Alternate 3' Polyadenylation Sites

Size heterogeneity of the rat (CHIARIOTTI et al. 1988) and human (DE PAGTER-HOLTHUIZEN et al. 1988) IGF-II genes arises primarily from the use of different polyadenylation signals within a long 3' untranslated region. Both rat exon 3 (3147 nt) and human exon 7 (4037 nt) contain long 3' untranslated regions: 2910 and 3800 nt, respectively. The distal polyA site in both species is the canonical sequence AATAAA. Its use was demonstrated by isolation of cDNA clones containing a polyA tract. Use of this polyA site accounts for the longer rat IGF-II mRNAs (4.6- and 3.5-kb, respectively) and the longer human IGF-II mRNAs (6.0-, 5.3-, and 4.8-kb). The 3' untranslated regions of the human and rat IGF-II genes contain repetitive sequences of alternating purine and pyrimidine bases. This stretch is longer in human (∼ 700 nt) than in the rat (∼ 80 nt); it also is present in the mouse IGF-II gene (STEMPIEN et al. 1986). The rat IGF-II 3' untranslated sequence also contains a palindrome, and inverted repeats within the 3' untranslated region and between the 3' untranslated region and coding exon 2.

There is suggestive but not conclusive evidence that at least one proximal polyA site is used in the rat and human IGF-II genes. In rat exon 3, a cluster of potential nonconsensus polyA sites occurs 286–406 nt beyond the TGA translation stop codon that could generate mRNAs of about 1-kb in conjunction with the exon −1 promoter. One cDNA clone encoding an mRNA of this size contained a polyA tail, but this may reflect an A-rich sequence in the genome. In human exon 7, a consensus polyA site is present that could generate a 2.2-kb RNA arising at the exon 4 promoter.

e) 3' Distal Variant RNAs

Recently, variant RNAs have been recognized in human (DE PAGTER-HOLTHUIZEN et al. 1988) and rat (CHIARIOTTI et al. 1988) that contain sequences corresponding to the distal approximately 1.8-kb of the 3' untranslated region of the IGF-II gene. These RNAs do not contain IGF-II coding sequences. The precise 5' end of the human RNA has been mapped by nuclease protection (DE PAGTER-HOLTHUIZEN et al. 1988), and a similar terminus has been identified in the rat IGF-II gene by primer extension (Matsuguchi et al. 1989; L. Chiariotti et al., unpublished results). The sequences downstream from the putative initiation site are homologous in rat and human for about 50 base pairs (bp). In the rat, but not the human IGF-II gene, a consensus splice acceptor site is present 40 nucleotides upstream from the proposed 5' end of the mRNA (Chiariotti et al. 1988), but the primer extension results suggest that splicing does not occur at this site. The variant mRNA may arise by transcription initiation or by precise endonucleolytic cleavage at this site.

Expression of rat and human variant IGF-II RNAs in different tissues at different ages closely corresponds to that of the major IGF-II mRNAs (see below). If the variant RNAs arise by transcription initiation, the basis for this coordinate regulation will be of considerable interest. The human IGF-II variant RNA has an open reading frame encoding a potential 113-residue protein. [The size of this potential protein is based on a revision of the sequence of De Pagter-Holthuizen et al. (1988) which contains a 75-bp deletion at its 3' end (Sussenbach 1989).] The rat IGF-II variant RNA also possesses several large open reading frames, but none of them correspond to the potential protein encoded by the human IGF-II variant RNA. Thus, the significance of the variant RNAs and whether they encode an additional protein remains to be determined.

f) Translatability of IGF-II mRNAs

Size-fractionated rat BRL-3A cell RNA, about 1.2-kb, directs the synthesis of prepro-IGF-II in a cell-free translation system (Whitfield et al. 1984; Rechler et al. 1985). By contrast, size-fractionated BRL-3A cell RNA enriched in 3.5- to 4.6-kb IGF-II RNA stimulated the translation of other proteins, but did not stimulate translation of prepro-IGF-II (Graham et al. 1986). Since the 3.5-kb and 1.2-kb IGF-II RNAs arise at the exon -1 promoter, they differ only in the length of their 3' untranslated regions. This raised the intriguing possibility that the most abundant IGF-II RNA species may require a separate activation step in order to be translated.

Attempts to elucidate the basis for the differences in translatability of these mRNAs have thus far not been successful. Both 1.2- and 3.5–4.6-kb IGF-II RNAs are present on polysomes, indicating that translation is initiated and the nascent peptide chain elongated (R. Frunzio, unpublished results). Analysis is complicated by the fact that one or more intermediate size IGF-II RNAs (e.g., 2-kb) appear to be translated. Further study of IGF-II gene expression in intact cells will be required to resolve this question.

2. Tissue and Developmental Expression

a) Considerations in IGF Protein Determinations

For purposes of this discussion, the presence of IGF-I or IGF-II mRNA is taken as the principal index of gene expression. It still needs to be established that these RNAs are translated, and the resulting protein precursors processed to biologically active forms. Nonetheless, mRNA expression provides stronger evidence of synthesis in a given tissue than does extraction of proteins or immunostaining, either of which might represent accumulation rather than synthesis.

To obtain meaningful determinations of IGF protein content, it is essential that IGFs be completely resolved from their binding proteins prior to radioligand assays by acid dissociation combined with size fractionation by gel filtration or HPLC. Unless it is demonstrated that a putative IGF pool does not contain binding protein activity, one cannot be certain that apparent IGF activity represents authentic IGF rather than a false-positive result arising from the presence of binding protein in the sample. Shortcut approaches such as acid ethanol extraction may be satisfactory in some but not all cases (DAUGHADAY et al. 1987 b). The burden of proof remains with the investigator to establish the validity of the methods employed.

b) Rat

α) Fetal

The rat IGF-II gene is expressed in many fetal (14–21 days' gestation) and neonatal (2–11 days postnatal) tissues (Table 2). Abundant IGF-II RNA was observed in total RNA from fetal liver, muscle, intestine, skin, lung, and thymus; lower levels were present in heart, kidney, and brain (BROWN et al. 1986). The 3.5- and 4.6-kb RNAs were present in all tissues, with the 3.5-kb RNA more abundant, indicating that both the exon -1 and exon -2 promoters were expressed with similar relative efficiences in all tissues tested. RNAs arising from the exon -3 promoter (UENO et al. 1988 a) and the 1.8-kb variant RNA (CHIARIOTTI et al. 1988) also appear to be coordinately regulated in different tissues. IGF-II RNAs also are expressed at high levels in rat placenta beginning on day 12 of gestation (GRAY et al. 1987; ROMANUS et al. 1988).

RNAs extracted from fetal rat tissues containing IGF-II RNA by hybridization analysis (i.e., liver, muscle, intestine, lung, and stomach, as well as rat placenta) direct the synthesis of 22-kDa prepro-IGF-II in cell-free translation (ROMANUS et al. 1988). Immunoreactive 7.5-kDa IGF-II was extracted from fetal liver, limb, lung, intestine, and brain, and at lower levels, from heart and kidney. The IGF-II protein levels are in good general agreement with the relative hybridization signal in different tissues (ROMANUS et al. 1988). Although it is not known which of the IGF-II RNA species are translatable, since all species are regulated coordinately, differential translatability would not affect the relative levels of IGF-II protein expression.

β) Adult

In all nonneural tissues, hybridizable IGF-II RNA is greatly decreased by the third week after birth (Table 2). Consistent with this developmental pattern,

RNA extracted from adult rat liver, muscle, and intestine failed to direct the synthesis of prepro-IGF-II (ROMANUS et al. 1988). BECK et al. (1988) reported that the negative developmental regulation of IGF-II mRNA in liver may be related to glucocorticoids. Intraperitoneal injection of high doses of cortisone to 9-day-old rats accelerated the loss of IGF-II mRNA from the liver at days 11 and 13, determined by both Northern blot and in situ hybridization. Cortisone treatment also was associated with a significant (25%) weight loss.

In contrast to nonneural tissues, IGF-II RNA persists in adult brain and spinal cord (Table 2). When MURPHY et al. (1987a) examined 10 µg of polyA$^+$ RNA from adult rat tissues by Northern blot hybridization and quantitative densitometry, IGF-II RNA was more abundant in brain than in any peripheral tissue. Under these sensitive assay conditions, IGF-II RNA was detected at low levels in heart ($\sim$5-fold lower than brain), uterus, kidney, skeletal muscle, and liver, but was not detected in lung, ovary, testes, and mammary gland.

Table 2. Tissue expression of rat IGF-II mRNAs

Tissue	Fetal/neonatal (positive)	Adult (negative)	Adult (positive)
Liver	A, B, C, D, E, F, G[a]	A, B, C, D, E, F	
Muscle	A, B, G	A, B, D[b]	
Skin	B	B	
Lung	A, B, C, D, G	B, C, D, E, H	
Intestine	B, C, G	B, C	
Thymus	B	B	
Heart	A, B, D, G	B, D	
Kidney	B, D, G	B, D, E	
Ovary		H	
Testes		H	
Mammary gland		H	
Brain	A, B, C, D, G, I		B, C, D, E, G, H, I, J, K
Spinal cord			D, E

[a] *References:* A, SOARES et al. (1985) – neonatal; B, BROWN et al. (1986) – 16 and 21 days' gestation, 2 and 11 days postnatal; C, LUND et al. (1986) – 14–17 days' gestation; D, SOARES et al. (1986) – 18 and 20 days' gestation, 2 and 4 days postnatal; E, GRAY et al. (1987) – 16–21 days' gestation; F, GRAHAM et al. (1986) – neonatal (2, 12, 15 days); G, UENO et al. (1988a) – neonatal; H, MURPHY et al. (1987b) – adult; 10 µg polyA + RNA; low amounts were observed in heart, uterus, skeletal muscle, kidney, and liver; I, ROTWEIN et al. (1988); J, HYNES et al. (1987); K, HYNES et al. (1988).
[b] Muscle. Muscle was positive in reference (E) using 4 µg polyA + RNA and a mixed probe (coding region and exon -2).

γ) *Embryonic*

IGF-II RNA and protein are expressed early in mouse embryonic development. This was demonstrated in: (a) undifferentiated stem cells, equivalent to 6-day embryos; (b) differentiated embryonal cells with features of parietal or visceral endoderm; (c) extraembryonic membranes containing mesoderm (NAGARAJAN et al. 1985; HEATH and SHI 1986; SOARES et al. 1985; RECHLER et al. 1988).

STYLIANOPOULOU et al. (1988a) examined rat embryos from the early somite stage (day 10) to day 16 by in situ hybridization using an IGF-II-specific probe. Most tissues derived from mesoderm were positive, including, sclerotome (axial skeleton), myotome (muscles of the trunk), and dermatome (dermis). Whereas muscle was positive at all stages of differentiation, cartilage precursors were strongly positive but decreased as they differentiated prior to ossification. The urogenital system (derived from intermediate mesoderm) was weakly positive. By contrast, few ectoderm- and endoderm-derived tissues were positive. These included rudimentary anterior and posterior pituitary (surface ectoderm), choroid plexus (neural tube), and head mesenchyme (derived from neural crest, and contributing to bone and muscle of the face and mouth). Other neurectoderm-derived central and peripheral nerves were negative. Liver and bronchiolar epithelium were the only endoderm-derived positive tissues. Whether the positive hybridization signal in the liver resulted from hepatocytes or erythroid or perisinusoidal cells was not determined. BECK et al. (1987) independently made similar observations. They also noted a strong positive signal with 18-day yolk sac, especially its mesodermal component. IGF-I oligonucleotides gave only faint hybridization to the same fetal tissues.

δ) *Central Nervous System*

Early studies (reviewed in NISSLEY and RECHLER 1984a) identified immunoreactive IGF-II in human cerebrospinal fluid (HASELBACHER and HUMBEL 1982) and in extracts of human brain (HASELBACHER et al. 1985), and demonstrated IGF synthesis by rat brain explants (BINOUX et al. 1981). This suggested that IGFs and IGF-II in particular were synthesized in the brain and might have special functions in the nervous system. Formal demonstration that the immunoreactivity corresponded to authentic IGF molecules came from the purification of IGF-II and a truncated IGF-I from fetal and adult human brain (SARA et al. 1986; CARLSSON-SKWIRUT et al. 1986). Proof that these IGF molecules were synthesized in brain came from the demonstration of mRNAs for IGF-II and IGF-I (see citations in Tables 2, 3, and 6).

Recent in situ hybridization studies have localized IGF-II mRNA in adult rat brain to the choroid plexus and leptomeninges (STYLIANOPOULOU et al. 1988b; HYNES et al. 1988). Strong hybridization was observed throughout the choroid plexus, but was not observed in other brain regions. STYLIANOPOULOU et al. (1988b) estimated that choroid plexus contained at least 200 times more IGF-II mRNA than the surrounding brain tissue. This is consistent with studies of rat embryogenesis, in which no signal was observed in brain regions other than the choroid plexus (Beck et al. 1987; Stylianopoulou et al. 1988 a). These results suggest that IGF-II is synthesized in the choroid plexus, secreted to the cerebrospinal fluid, and taken up by neuronal and glial cells throughout the brain.

c) Mouse

Mouse IGF-II is highly homologous to rat IGF-II (Stempien et al. 1986). Mouse IGF-II RNAs in placenta (Stempien et al. 1986) and neonatal liver (Soares et al. 1985; Romanus et al. 1988) comigrate with rat IGF-II species on formaldehyde-agarose gels. The major RNA is 3.5-kb, and the minor species is 4.6-kb.

d) Human

α) Fetal

IGF-II RNAs arising at the exon 4 and exon 4B promoters (i.e., 6.0, 4.8, and 2.2-kb) were observed in many fetal tissues from 7 to 20 weeks of gestation (Table 3). High levels were observed in liver, adrenal, and skeletal muscle; intermediate levels were seen in kidney, skin, and pancreas; and lower levels were seen in intestine, heart, lung, stomach, and spleen. Hybridization was detected in fetal brain (20 weeks) by Gray et al. (1987) and in fetal brain stem (17–20 weeks) by Han et al. (1988 b). However, hybridization was not observed in 7- to 10-week fetal brain (J. Scott et al. 1985), fetal brain of unspecified age (De Pagter-Holthuizen et al. 1988), or 17- to 20-week cerebral cortex and hypothalamus (Han et al. 1988 b).

IGF-II mRNA was 100–600 times more abundant than IGF-I RNA in liver, intestine, adrenal, skin, kidney, and pancreas (Han et al. 1988 b). Similarly, Hossenlopp et al. (1987) observed that explants of human fetal liver in culture synthesize 7 to 47 times as much IGF-II as IGF-I.

IGF-II RNAs, 6.0- and 4.8-kb, are expressed at a high level in placenta (J. Scott et al. 1985; Reeve et al. 1985; Shen et al. 1986; Gray et al. 1987; Irminger et al. 1987; Han et al. 1988 b). Umbilical cord, amnion, trophoblast, and yolk sac also express IGF-II RNA (J. Scott et al. 1985).

β) Adult

In contrast to the rat, IGF-II RNAs also are expressed in many tissues in human adults. Levels of IGF-II RNA in adult liver (J. Scott et al. 1985; De Pagter-Holthuizen et al. 1987, 1988) and kidney (J. Scott et al. 1985), however, are considerably lower than in the corresponding fetal tissues.

Liver uniquely expresses a 5.3-kb RNA (arising from the exon 1 promoter) (J. Scott et al. 1985; De Pagter-Holthuizen et al. 1987; Gray et al. 1987; Irminger et al. 1987), but not 6.0- and 4.8-kb RNAs. The 5.3-kb RNA is not seen in any other tissue.

Table 3. Tissue expression of human IGF-II mRNAs

Fetal	Adult
High Liver (A, B, C, D, E, F)[a] Adrenal (A, E, F, I) Skeletal muscle (A, E, F) Intermediate Kidney (A, B, C, E, F) Skin (F) Pancreas (F) Low Intestine (A, F) Lung (A, B, F) Heart (A, F) Stomach (F) Spleen (F) Brain[b]	Liver (A, B, D, E, G, H) Kidney (A, B, G, H) Skin (B) Peripheral nerve (B) Muscle (B) Colon (B) Uterus (B) Stomach (B) Hypothalamus (G) Adrenal (G) Granulosa cells (I)

[a] *References:* A, J. SCOTT et al. (1985) – 7–10 weeks fetal; B, GRAY et al. (1987) – 20 weeks; C, SOARES et al. (1986) – 18 weeks; D, DE PAGTER-HOLTHUIZEN et al. (1987) – fetal age not specified; E, DE PAGTER-HOLTHUIZEN et al. (1988) – fetal age not specified; F, HAN et al. (1988b) – 16–20 weeks fetal; adrenal, muscle, and liver are based on Northern blots; other results are from dot blots; G, IRMINGER et al. (1987); H, BELL et al. (1985); I, VOUTILAINEN and MILLER (1987) – 17–20 weeks fetal adrenal cultured in vitro; ovarian granulosa cells were cultured from hyperstimulated infertility patients.

[b] Fetal brain positive: 4 μg polyA[+] RNA (GRAY et al. 1987); brain stem (HAN et al. 1988b). However, fetal brain was negative in other reports: J. SCOTT et al. (1985); DE PAGTER-HOLTHUIZEN et al. (1988); hypothalamus and cerebral cortex negative (HAN et al. 1988b).

The 6.0- and 4.8-kb RNAs were observed in adult skin ≫ peripheral nerve and muscle > kidney, colon, uterus, and stomach (GRAY et al. 1987). Adrenal (IRMINGER et al. 1987), hypothalamus (IRMINGER et al. 1987), and cultured ovarian granulosa cells (VOUTILAINEN and MILLER 1987) also were positive.

γ) Hybridization Histochemistry

HAN et al. (1987a) examined the cellular localization of IGF-II mRNA in different tissues from 16- to 20-week human fetuses using hybridization with synthetic oligonucleotides. Positive hybridization was observed in all tissues except cerebral cortex. A stronger hybridization signal was seen with IGF-II than IGF-I oligonucleotides, consistent with the results of Northern analysis that suggested that IGF-II is more abundant than IGF-I in human fetal tissues. The hybridization signal was localized to connective tissue, capsules, tissue sheaths, and septae. These results demonstrate that IGF-II RNA is present in cells of mesenchymal origin but, given the limitations of in situ analysis, they do not exclude the presence of IGF-II RNA at lower abundance in parenchymal cells. Immunocytochemical analysis of similar sections of human fetal tissues using antibodies that do not discriminate between IGF-I and IGF-II showed a different distribution, with positive staining in both epithelial and mesenchymal cells (HAN et al. 1987b).

3. Hormonal Regulation

HYNES et al. (1987) reported that brain IGF-II RNA was decreased 4- to 5-fold by hypophysectomy; ubiquitin RNA was not affected. IGF-II RNA levels were not restored by intraperitoneal injection of human GH. However, a 25%–50% increase in IGF-II RNA was observed 8 h after the injection of human GH into the lateral ventricle. TURNER et al. (1988) demonstrated that adult rats inoculated with a GH- and prolactin-secreting tumor (GH_3) showed a 4- to 6-fold increase in IGF-II mRNA in skeletal and cardiac muscle, associated with increased weight of the organs. Hepatic IGF-II RNA was not increased in these animals.

BECK et al. (1988) injected 9-day-old rats intraperitoneally with high doses of cortisone. IGF-II mRNA (assessed by Northern blot and in situ hybridization) was markedly decreased in liver at 11 and 13 days, compared to control animals. A smaller decrease was observed in muscle, but no significant decrease was seen in choroid plexus.

VOUTILAINEN and MILLER (1987) reported that IGF-II mRNA levels were regulated by follicle-stimulating hormone (FSH) and dibutyryl cyclic AMP in cultured human granulosa cells (obtained from women treated for infertility with clomiphene and gonadotropins), and by adrenocorticotropin (ACTH) and dibutyryl cyclic AMP in cultured human fetal adrenals (17–20 weeks' gestation). IGF-II RNA decreased in untreated granulosa cells between 6 and 18 days in culture; FSH, chorionic gonadotropin and dibutyryl cyclic AMP prevented this decrease. The maximum increase (day 12) relative to maximal IGF-II levels in control cells (day 6) was less than 2-fold. It is unclear whether this reflects regulation of IGF-II RNA or preservation of cell viability. Similarly, addition of ACTH or dibutyryl cyclic AMP to fetal adrenal cells after 10 days of cultivation increased IGF-II RNA levels. However, since the effect of prolonged cultivation of adrenal cells on IGF-II levels was not described, it is not known whether this represents a specific effect on IGF-II RNA or nonspecific effects on cell viability.

IGF-II RNA is not increased during liver regeneration following partial hepatectomy (UENO et al. 1988b; NORSTEDT et al. 1988).

4. Tumors

a) Human

GRAY et al. (1987) surveyed 58 tumors and tumor cell lines and noted that 87% expressed high levels of IGF-II RNA. High levels of IGF-II RNA have been reported in Wilms' tumor, rhabdomyosarcoma, hepatoblastoma,

Table 4. Expression of IGF-II and IGF-I RNAs in human tumors

IGF-II

 Wilms' tumor: 12/12 (A)[a], 4/4 (B), 1/1 (C), 2/2 (D)[b]
 Rhabdomyosarcoma: 2/2 (A), 1/1 (C)
 Hepatoblastoma: 1/1 (C)[c]
 Liposarcoma: 10/11 (E)
 Colon: 8/20 (E)
 Pheochromocytoma: 8/8 (F); 3/3 (D)[b, d]
 Neuroblastoma: 1/2 (C), 2/8 (F), 0/1 (B), 0/1 (A)
 Leiomyoma: 7/7 (G)
 Leiomyosarcoma: 9/10 (G), 1/1 (H)[b, e]
 Hepatocellular carcinoma: 6/11 (I)[f], 9/40 (J)
 Breast: 24/26 (K)
 Miscellaneous[g]
 Negative[h]

IGF-I

 Leiomyoma: 6/7 (G)[b]
 Leiomyosarcoma: 6/10 (G)[b]
 Liposarcoma: (E)[i]
 Colon: (E)[j]

---

[a] *References:* A, J. SCOTT et al. (1985); B, REEVE et al. (1985); C, GRAY et al. (1987); D, HASELBACHER et al. (1987); E, TRICOLI et al. (1986); F, EL-BADRY et al. (1989); G, HOPPENER et al. (1988); H, DAUGHADAY et al. (1988); I, SHAPIRO et al. (1987); J, CARIANI et al. (1988); K, YEE et al. (1988); L, ROTH et al. (1987a).
[b] Assayed by protein in addition to RNA.
[c] Hepatoblastoma cell line HepG2 also is positive (A, C).
[d] In extracts of tumors, 65% of IGF-II exists as a 10-kDa protein.
[e] 75% of IGF-II in serum and tumor extracts was 12 kDa. Symptoms of hypoglycemia.
[f] 5/5 in patients with hypoglycemia, 1/6 in patients without hypoglycemia.
[g] 4/4 teratocarcinoma, 2/2 Ewing's sarcoma (low levels) (A); bladder, mammary, neurofibroma, VIPoma, squamous cell (C); hemangiopericytoma (L).
[h] 0/1 retinoblastoma (A); 0/3 malignant fibrous histiocytoma (G); fibrosarcoma, leiomyosarcoma, melanoma (E).
[i] 55% have modest increase in IGF-I. All of these have increased IGF-II.
[j] 20% have modest increase in IGF-I. All of these have increased IGF-II.

liposarcoma, colon carcinoma, pheochromocytoma, some neuroblastomas, leiomyomas, and leiomyosarcomas (Table 4). They arise from the promoters associated with exons 4 and 4B that are expressed in fetal liver, and not from the adult liver promoter associated with exon 1. Expression of immunoreactive IGF-II has been confirmed for pheochromocytomas (HASELBACHER et al. 1987) and one leiomyosarcoma (DAUGHADAY et al. 1988). For both classes of tumors, 50%–70% of the IGF-II immunoreactivity was partially processed IGF-II precursor, 10–12-kDa. For some tumors, overproduction of IGF-II mRNA is associated with clinical hypoglycemia. These include hemangiopericytomas (KAHN 1980; GORDEN et al. 1981; ROTH et al. 1987a), one leiomyosarcoma (reclassified as a fibrosarcoma) (DAUGHADAY et al. 1988), and some hepatocellular carcinomas (SHAPIRO et al. 1988).

b) Rat

Increased rat IGF-II RNA was observed in five of seven rat medullary thyroid carcinomas (HOPPENER et al. 1987). UENO et al. (1988b) reported increased IGF-II RNA expression in six of six chemically induced liver carcinomas and in four of six transplantable rat ascites tumors. NORSTEDT et al. (1988) reported similar results after feeding 2-acetylaminofluorene: IGF-II RNA was not detected in preneoplastic nodules, but was present in six of nine hepatic carcinomas (associated with markedly reduced levels of IGF-I RNA).

c) Woodchuck

Woodchucks have a high endemic rate of persistent infection with woodchuck hepatitis virus (WHV), resulting in a nearly 100% incidence of multifocal hepatocellular carcinoma by 3–5 years of age. FU et al. (1988) noted that half of 30 tumors from 13 woodchucks expressed high levels of IGF-II RNA. These tumors did not express WHV genes, whereas the other 50% of tumors expressed WHV genes but not the IGF-II gene. The two groups of tumors also differed morphologically. The authors propose that WHV-positive tumors may arise from differentiated hepatocytes, and IGF-II-positive cells from hepatocyte precursors (oval cells).

IV. IGF-I Gene Expression

1. Multiple IGF-I Transcripts from a Single Gene

a) Overview

The IGF-I gene has many features in common with the IGF-II gene. These include a common arrangement of coding exons: one exon, exon 2, contains the prepeptide and B domain; a second, exon 3, contains the remainder of the B domain, the complete C, A, and D domains, and part of the E domain; and a third exon contains the carboxy terminus of the E domain propeptide and the 3′ untranslated region. [Assuming that translation is initiated at Met −22, exon 1 and the first 15 bp of exon 2 would represent the 5′ untranslated region.] Unlike the IGF-II gene, a large intron separates residue B25 of exon 2 from residue B26 of exon 3: >20-kb in human IGF-I (BELL et al. 1985; ROTWEIN et al. 1986), 50-kb in rat IGF-I (SHIMATSU and ROTWEIN 1987a).

   Alternate RNA splicing generates IGF-I precursors with different E-domain propeptides (Fig. 5). One E-domain sequence, designated IGF-Ia, is observed in human, rat, and mouse IGF-I. The other sequence, designated

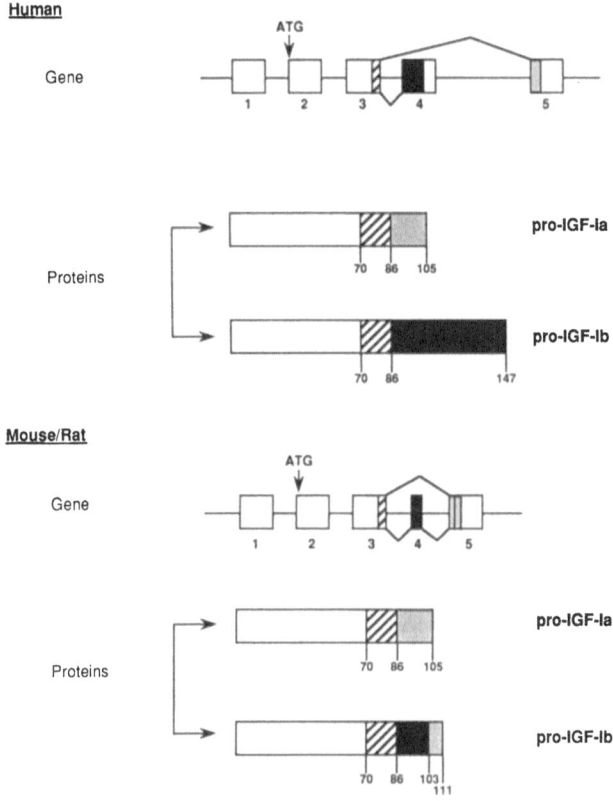

Fig. 5. Alternative splicing of IGF-I mRNAs generates alternative IGF-I propeptides. The human (*top*) and rodent (*bottom*) IGF-I genes contain five exons. For simplification, we assume that translation is initiated at the ATG codon at Met −22, 15 bp from the 5′ end of exon 3. [Whether Met −22 represents the translation initiation site in vivo is not known. IGF-I mRNAs in human, rat and mouse also contain other methionine residues in the same reading frame. ROTWEIN et al. (1987a) demonstrated the translation can initiate at Met −48 of human IGF-I in a wheat germ translation system, but this need not correspond to the site used in vivo.] Exon 2 contains the prepeptide and residues 1–25 of the B domain. Exon 3 contains the remainder of the B domain, the complete C, A, and D domains, and residues 70–86 of the E domain (*hatched*). Exons 4 and 5 contain alternate sequences for the distal E domain and the 3′ untranslated region. The IGF-Ia precursor (105 amino acids) arises in humans and rodents by splicing of exon 3 to exon 5 (the protein-coding region of exon 5 is *stippled*). The human IGF-Ib precursor (147 amino acids) arises by splicing exon 3 to exon 4 (the protein-coding region of exon 4 is *solid*). The mouse and rat IGF-Ib precursor (111 amino acids) arises by splicing exon 3 to the 52-bp exon 4 (*solid*) which is spliced to exon 5. Pro-IGF-Ib terminates earlier in exon 5 than does pro-IGF-Ia (at the *vertical lines* within the stippled area). This schematic drawing is based on the results of JANSEN et al. (1983), ROTWEIN (1986), ROTWEIN et al. (1986), BELL et al. (1986), ROBERTS et al. (1987b), and SHIMATSU and ROTWEIN (1987a)

IGF-Ib, arises by different splicing mechanisms in the human and rodent genes. Although three distinct 5' untranslated regions have been identified in rat IGF-I, none are complete and only one has been mapped in the rat genome. Like IGF-II, multiple size IGF-I RNAs have been observed in human (BELL et al. 1985; ROTWEIN 1986), mouse (BELL et al. 1986), and rat (ROBERTS et al. 1986; LUND et al. 1986). They range in size from 0.8–1.2-kb to 7.0–7.5-kb. Most of the size heterogeneity occurs at the 3' end of the mRNA (P. K. LUND, personal communication).

b) Gene Location

The IGF-I gene appears to be a single-copy gene in human (ROTWEIN et al. 1986) and rat (SHIMATSU and ROTWEIN 1987a). It is located on human chromosome 12. Chromosome 12 is related to chromosome 11, the site of the IGF-II gene, and may have arisen by gene duplication. The two chromosomes contain related loci: c-Ha-*ras* I and *LDH-A* (chromosome 11), c-Ki-*ras* and *LDH-B* (chromosome 12) (BRISSENDEN et al. 1984; TRICOLI et al. 1984).

c) 3' Heterogeneity

α) Human

ROTWEIN (ROTWEIN 1986; ROTWEIN et al. 1986) first appreciated that human IGF-I exons 1, 2, and 3 (containing the first 16 amino acid residues of the E domain) were alternately spliced to exon 4 (containing 61 residues of E domain plus the 3' untranslated region) or exon 5 (containing 19 residues of E domain plus the 3' untranslated region). Splicing of exon 3 to exon 5 generates the 35-residue E domain of IGF-Ia (JANSEN et al. 1983), whereas splicing of exon 3 to exon 4 generates the 77-residue E domain of IGF-Ib (ROTWEIN 1986).

Hybridization of E-domain-specific probes to Northern blots of human liver suggested that both IGF-Ia and IGF-Ib sequences are present in all size classes of IGF-I RNAs (ROTWEIN et al. 1986). Recent results of HOPPENER et al. (1988) suggest that this may not always be the case. In leiomyoma and adult liver, exon 5 probes hybridize to 7.6-kb but not to 1.4-kb IGF-I RNA, whereas an exon 4 probe hybridizes to the 1.4-kb but not to the 7.6-kb RNA. A prominent 1.1-kb RNA in placenta hybridizes to an exon 5 probe but not to an exon 4 probe.

β) Rodent

Mouse and rat cDNA clones corresponding to human IGF-Ia were identified by BELL et al. (1986) and ROBERTS et al. (1987b), and the corresponding exon 5 was identified in rat IGF-I genomic clones by SHIMATSU and ROTWEIN (1987a). In addition, BELL et al. (1986) and ROBERTS et al. (1987b) identified a second set of clones, designated IGF-Ib, that contained a 52-base insertion after residue 16 of the E domain, that is, at the site of the exon-intron junction in IGF-Ia. The 52-base insertion changes the reading frame and consequently alters the amino acid sequence of the carboxy terminus of the E domain of IGF-Ib. Following the insertion, the nucleotide sequence resumes with the sequence of exon 5. SHIMATSU and ROTWEIN (1987a) confirmed that the 52-base insertion in rat IGF-Ib represents a short exon 4, which is spliced to exon 5.

In the mouse, BELL et al. (1986) isolated twice as many cDNA clones containing IGF-Ia as IGF-Ib. In the rat, LOWE et al. (1988) observed in RNAse protection assays that IGF-Ib mRNA is much less frequent than IGF-Ia RNA in liver (13%) and in other tissues (2%–5%).

d) 5′ Heterogeneity

cDNA clones with three distinct 5′ untranslated regions have been identified in the rat, and have been designated classes A, B, and C (Fig.6, Table 5)

Table 5. Rat IGF-I transcripts with alternate 5′ untranslated region

	Class C	Class B	Class A
References[a]	A, B, C, D	E, F	A
Divergence from class A[b]	15 bp from Met − 22	72 bp from Met − 22	NA[c]
Properties	–	–	40-bp inverted repeat with 3′ end (A)
Homology with other species	Human[d]	Mouse[e]	Mouse[f]
Abundance in adult tissues (G)	Most abundant in all tissues[g]	Only liver; low levels	Some tissues; variable amounts[h]
GH[i]	2–3 × all tissues	6 ×	Liver 7 ×; kidney, lung 1 ×

[a] *References:* A, ROBERTS et al. (1987a); B, CASELLA et al. (1987); C, SHIMATSU and ROTWEIN (1987a); D, R. FRUNZIO (unpublished results); E, ROBERTS et al. (1987b); F, BELL et al. (1986) mouse; G, LOWE et al. (1987).
[b] Ribonuclease protection was performed using a riboprobe fragment containing 98 bp of class A 5′ untranslated region and 224 bp of coding region (322-bp fragment). The class B mRNA protects a fragment 26 nucleotides shorter (72 bp untranslated + 224 bp coding = 296 nucleotides). The class C mRNA protects a fragment that is 57 nucleotides shorter (15 nucleotides 5′ untranslated region + 224 nucleotides coding region = 239 nucleotides). See Fig. 6.
[c] NA, not applicable.
[d] 95% homology to human IGF-I from Met − 48 (C).
[e] Homologous for 143 bp from Met − 22 (F).
[f] Homologous for 72 bp from Met − 22 (F).
[g] Heart, lung, brain, testes, liver, muscle, stomach, kidney.
[h] Moderate (liver, 26% of total); low (lung, testes, stomach, kidney; 4–13% of total); negative (brain, muscle, heart).
[i] Fold increase after treatment of hypophysectomized rats with GH.

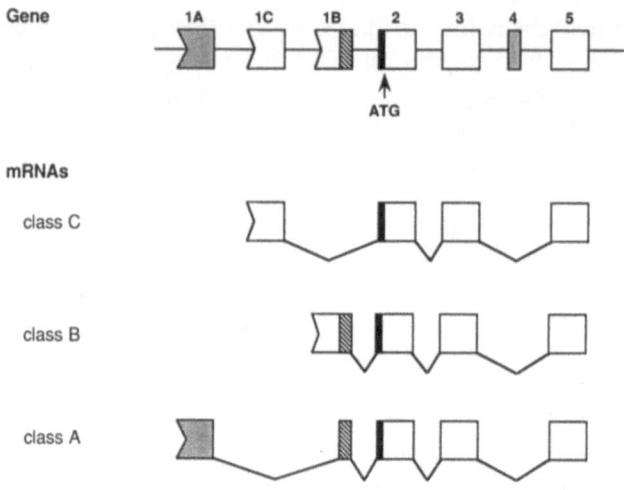

Fig. 6. Proposed origin of rat IGF-I RNA transcripts with different 5′ sequences. The rat IGF-I gene is illustrated at the *top*. Exons 2, 3, 4, and 5 are as shown in Fig. 5. Exon 2 contains 15 nucleotides of presumed 5′ untranslated region common to all IGF-I mRNAs (*solid*). The exons are not drawn to scale. Three alternative 5′ untranslated exons are shown: exons 1A, 1B, and 1C. They are *notched* to indicate that they are incomplete at the 5′ end. Exon 1C (SHIMATSU and ROTWEIN 1987a) and exon 1B (R. FRUNZIO et al., unpublished results) have been identified in the rat IGF-I gene. Exon 1A has not yet been demonstrated in the genome. Three IGF-1 mRNA transcripts are shown that have different 5′ exons. For simplicity, only RNAs in which exon 3 is spliced to exon 5 (IGF-Ia) are shown. Class C mRNA (CASELLA et al. 1987; ROBERTS et al. 1987a; SHIMATSU and ROTWEIN 1987a; contains 5′ untranslated exon 1C. Class B mRNA (ROBERTS et al. 1987b) contains 5′ untranslated exon 1B. Class A mRNA (ROBERTS et al. 1987a) contains 57 nucleotides in common with class B mRNA (*hatched*), arising from exon 1B, and an additional unique approximately 30 nucleotides (*stippled*) which presumably arise from exon 1A. [Although the genomic nucleotide sequence confirms the sequence of the 5′ end of the class B cDNA, it does not contain a consensus splice acceptor where the class B and class A sequences diverge, so that the origin of the class A cDNA is presently unclear (C. B. BRUNI, personal communication.)] The structural relationships between the three 5′ mRNAs are based on solution-hybridization ribonuclease protection experiments of LOWE et al. (1987) using a probe containing the class A 5′ untranslated region

(LOWE et al. 1987). Using solution-hybridization and ribonuclease protection assays with a class A riboprobe fragment, they established that the class C sequence diverges from that of class A and class B 15-bp upstream from Met −22 at a site established as an intron-exon boundary in the human IGF-I gene. cDNA clones corresponding to class C have been identified by CASELLA et al. (1987), ROBERTS et al. (1987a), SHIMATSU and ROTWEIN (1987a) and R. FRUNZIO (unpublished results). The first 370-bp of the rat class C mRNA are 95% homologous to human IGF-I cDNA clones (ROTWEIN et al. 1986). SHIMATSU and ROTWEIN (1987b) also identified a 186-bp deletion in this region that probably results from alternate splicing. The class C 5′ untranslated region has been located in the rat genome, about 5-kb upstream from exon 2 (SHIMATSU and ROTWEIN 1987a).

cDNA clones corresponding to class B (793-bp 5' untranslated region; ROBERTS et al. 1987b) and class A (102-bp 5' untranslated region; ROBERTS et al. 1987a) share a common sequence for 57 bp (following the 15-bp in exon 2 that both share with class C), and then diverge. The mouse 5' untranslated region is highly homologous to the rat class B cDNA sequence (BELL et al. 1986).

The 40-nucleotide sequence that is unique to the class A 5' untranslated region is complementary to a 41-nucleotide sequence in the 3' untranslated region common to all IGF-I RNAs (ROBERTS et al. 1987a). Formation of this stable duplex ($\Delta G = -75$ kcal/mol) might affect the translation of the class A mRNAs.

Hybridization with specific probes for each of the 5' untranslated regions has demonstrated that class B and class C 5' sequences are represented in RNAs of all size categories. Oligonucleotides corresponding to the unique region of the class A mRNA failed to hybridize to Northern blots (LOWE et al. 1987), possibly because of the inverted repeat.

e) Polyadenylation Sites

BELL et al. (1986) obtained six of seven mouse IGF-I cDNA clones with polyA tracts added at four different sites within 200-bp of the translation termination site. SHIMATSU and ROTWEIN (1987a) provided indirect evidence from ribonuclease protection experiments for the use of four polyA sites 300–1100 bp beyond the translation termination codon of the rat IGF-I gene. LE BOUC et al. (1986) identified an IGF-I cDNA clone that contained a polyA tract 380 nt beyond the translation termination signal; a corresponding consensus polyA addition signal was not identified. The IGF-Ia clone of JANSEN et al. (1983) originates at a dA-rich sequence in the genome.

f) Transcript Size

Transcripts of 0.8–1.2, 1.7–2.0, and 7.0–7.8-kb have been observed in rat, human, and mouse tissues. Variable amounts of intermediate forms, e.g., 2.6- and 3.9-kb, also have been seen.

The major size heterogeneity of IGF-I RNAs occurs at the 3' end. RNAs containing the rat class B and C 5' untranslated regions (LOWE et al. 1987) and human (ROTWEIN 1986) and rat (LOWE et al. 1988) RNAs containing the IGF-Ia and IGF-Ib 3' untranslated regions are present in all size categories. LOWE et al. (1988) proposed that selection of 5' and 3' exons is independent, generating at least six possible RNA combinations.

2. Tissue and Developmental Expression

a) Tissue Distribution

IGF-I RNA is expressed in all tissues tested (Table 6). The richest source is adult rat liver, which contains approximately 30 times more IGF-I RNA than the highest level in other tissues (MURPHY et al. 1987a). The greater abundance of IGF-I RNA in adult liver is consistent with the suggestion that the liver is the major source of circulating IGF-I (SCHWANDER et al. 1983; SCOTT et al. 1985a). By contrast, levels of IGF-I peptide extracted from dif-

Table 6. Expression of IGF-I mRNA in adult rat tissues

Tissue	References[a]
Liver	A, B, C, D, E, F, G, H[c], I[c], J[b], K[b], N[b], O
Uterus	G
Lung	B, D, F, G, H[c], O
Ovary	G, O
Kidney	D, E, F, G, H[c], J[b], O
Heart	D, F, G, H[c]
Testes	C, E, F, G, H[c], O
Pancreas	E, H[c]
Stomach/intes- tine	B, F, O
Skeletal muscle	F, G
Mammary gland	G, O
Brain	B, E[d], F, G, H[c]
Spleen	E, H[c]
Placenta[b]	L, N
Cartilage	M
Pituitary	P

[a] *References:* A, ROBERTS et al. (1986); B, LUND et al. (1986); C, CASELLA et al. (1987); D, ROBERTS et al. (1987b); E, SHIMATSU and ROTWEIN (1987a)[e]; F, LOWE et al. (1987)[e]; G, MURPHY et al. (1987a); H, MATHEWS et al. (1986)[c, e]; I, BELL et al. (1986)[c]; J, BELL et al. (1985)[b]; K, ROTWEIN (1986)[b]; L, WANG et al. (1988a)[b]; M, ISGAARD et al. (1988b)[e]; N, HOPPENER et al. (1988)[b]; O, HOYT et al. (1988); P, FAGIN et al. (1988).
[b] Human.
[c] Mouse.
[d] Faint signal.
[e] Based on nuclease protection experiments.

ferent tissues were similar in liver, lung, kidney, and testes (D'ERCOLE et al. 1984).

MURPHY et al. (1987a) observed that uterus is the second richest source of IGF-I RNA. Other positive tissues include testes, kidney, heart, pancreas, and skeletal muscle in the rat, and human placental syncytiotrophoblast.

LOWE et al. (1987) characterized the 5' and 3' heterogeneity of IGF-I mRNAs expressed in different rat tissues using solution hybridization and RNAse protection. The class C 5' untranslated region was expressed in all tissues examined (heart, lung, brain, testes, liver, muscle, stomach, and kidney), and is more abundant than class B or A transcripts. The class B 5' untranslated region is found only in liver, and in low abundance. Class A transcripts are present in moderate abundance in liver (27% of total), in smaller amounts in lung, testes, stomach, and kidney (4%–13%), and are not present in brain, skeletal muscle, and heart.

IGF-Ia is the major 3' variant in all tissues (LOWE et al. 1988). IGF-Ib accounts for only 13% of the total IGF-I mRNAs in liver, and 2%–5% of the total IGF-I RNA in the eight other tissues tested.

b) Development

IGF-I RNA is lower in fetal (LUND et al. 1986; HOYT et al. 1988) and neonatal (ROBERTS et al. 1986; NORSTEDT et al. 1987; HOYT et al. 1988) rat liver than in adult liver, suggesting a developmental regulation opposite to that seen for rat IGF-II RNA and similar to that seen for immunoreactive IGF-I in rat serum (RECHLER et al. 1985).

IGF-I mRNA was detected in several 15- to 19-days' gestation fetal rat tissues (LUND et al. 1986; HOYT et al. 1988), including intestine/stomach, liver, lung, and brain. Higher levels were present in fetal lung than in the adult lung (LUND et al. 1986; HOYT et al. 1988). IGF-I mRNA levels in intestine/stomach and brain did not change appreciably with developmental age.

Using solution hybridization techniques, ROTWEIN et al. (1987b) demonstrated the presence of IGF-I mRNA in 11-day whole rat embryos. Levels increased 8.6-fold between day 11 and day 13, but reached only 12% of the level in adult rat liver.

3. Regulation of IGF-I mRNA

a) Growth Hormone

α) Liver

All sizes of IGF-I RNA in rat liver are regulated by growth hormone (GH). They are decreased by hypophysectomy, and restored by GH treatment (ROBERTS et al. 1986; MURPHY et al. 1987b; HYNES et al. 1987). GH also increased IGF-I RNA in the *lit/lit* GH-deficient mouse (MATHEWS et al. 1986).

Transcription run-off experiments demonstrated that the increase in steady-state mRNA after GH resulted from an increase in transcription rate (MATHEWS et al. 1986). Unlike the rat, GH appeared to induce the 0.8-kb but not the 7.5-kb IGF-I RNA in the *lit/lit* mouse. NORSTEDT and MOLLER (1987) showed that GH caused a 3-fold increase in IGF-I mRNA in hepatocytes from normal and hypophysectomized rats, suggesting that GH acts directly on liver cells. The increase in IGF-I mRNA was slower in hepatocyte cultures (24–32 h) than in hypophysectomized rats (8 h) (NORSTEDT et al. 1987). Levels of IGF-I mRNA in cultured normal liver cells were only 15% of those in fresh isolates.

β) Extrahepatic

GH treatment of hypophysectomized rats also increased IGF-I RNA in muscle (TURNER et al. 1988), heart, lung, and kidney (ROBERTS et al. 1987b; MURPHY et al. 1987b); pancreas (HYNES et al. 1987); uterus (MURPHY and FRIESEN 1988); and rat costal cartilage (ISGAARD et al. 1988b) by as much as 9-fold. Pulsatile GH administration, simulating the secretory pattern in males, gave greater stimulation of IGF-I mRNA in cartilage and muscle, but not in liver (ISGAARD et al. 1988a). Pituitary IGF-I mRNA is increased in rats bearing GH-secreting (GH$_3$) tumors (FAGIN et al. 1988). These results agree with the observation of D'ERCOLE et al. (1984) that GH increased extractable immunoreactive IGF-I in kidney, lung, heart, and testes. BORTZ et al. (1988) noted that GH increased both IGF-I mRNA and immunoreactive IGF-I in the principal cells of the kidney collecting ducts. By contrast, in the GH-deficient *lit/lit* mouse, GH gave only a small increase in IGF-I mRNA in brain and heart, and no increase in pancreas, spleen, lung, testes, and kidney.

GH regulation of IGF-I in the brain is less clearcut. D'ERCOLE et al. (1984) observed no decrease in immunoreactive IGF-I following hypophysectomy, and a minimal increase after treatment with ovine GH. HYNES et al. (1987) reported a 4-fold decrease in IGF-I mRNA after hypophysectomy, which was normalized 4 h (but not 8 h) after intraventricular administration of human GH.

γ) Selectivity for 5' and 3' Transcripts

LOWE et al. (1987) analyzed the effect of GH on the different 5' and 3' IGF-I RNA transcripts. The largest induction was observed for the class B and class A 5' transcripts in liver (6- to 7-fold). By contrast, class C transcripts only were induced 2- to 3-fold in all tissues, including liver, and class A transcripts in kidney and lung were not induced by GH. The greatest induction of 3' transcripts was an 8-fold induction of transcript Ib in liver (LOWE et al. 1988). Liver transcript Ia, and transcripts Ia and Ib in heart, lung, and kidney only were induced 1.3- to 2.4-fold by GH. Thus, GH exerts selective effects on different IGF-I mRNA transcripts in different tissues.

b) Hormones Other than GH

Other hormonal regulators of IGF-I mRNA have been described in addition to GH. These include: (a) estrogens in the uterus; (b) ovine prolactin in the liver; and (c) other growth factors in cultured human fibroblasts.

α) Estrogens

MURPHY et al. (1987c) reported that a single injection of 17β-estradiol to ovariectomized rats (with intact pituitaries or after hypophysectomy) increased uterine IGF-I mRNA by 14- to 20-fold within 6 h without affecting IGF-I mRNA in the liver or kidney. Immunoreactive IGF-I was increased 2-fold in the uterus, with no change in serum IGF-I. The authors propose that estrogens promote uterine proliferation by stimulating the production of local synthesis of IGF-I.

β) Ovine Prolactin

Acute injection of ovine prolactin (not contaminated by ovine GH) caused a 15-fold increase in liver IGF-I mRNA and a modest increase in serum IGF-I (MURPHY et al. 1988). Ovine prolactin is about 50% as potent as oGH. Chronic treatment with ovine prolactin gave similar but smaller effects. The physiological relevance of these small effects is unclear. In contrast to these results, ISGAARD et al. (1988b) reported that rat prolactin did not increase hepatic IGF-I mRNA.

c) Nonhormonal

α) Nutrition

Plasma IGF-I is decreased in fasting (CLEMMONS et al. 1981a) and increased by refeeding (CLEMMONS et al. 1985). EMLER and SCHALCH (1987) demonstrated similar changes in the major IGF-I mRNA transcript (8-kb) in liver: 60% decrease after a 30-h fast; 18-fold increase over fasting level after 30-h refeeding.

β) Diabetes

SCOTT and BAXTER (1986) demonstrated that streptozotocin-induced diabetes decreased plasma IGF-I and IGF-I synthesis by cultured hepatocytes. GOLDSTEIN et al. (1988) compared the effects of different doses of streptozotocin. At the two highest doses tested (144 and 288 mg/kg), doses that induced severe ketonemia and weight loss, a profound and parallel decrease occurred in serum IGF-I, liver IGF-I, and liver IGF-I mRNA.

γ) Injury

Compensatory Renal Hypertrophy. Following unilateral nephrectomy, the remaining kidney is enlarged and contains increased extractable IGF-I (STILES et al. 1985). FAGIN and MELMED (1987) observed a 4–6 fold increase in IGF-I mRNA, but this was not confirmed by LAJARA et al. (1989). Weight gain was normal, and serum IGF-I and liver IGF-I mRNA were not affected. These results suggest that local or circulating renotropic factors may be involved in the stimulation of IGF-I mRNA.

Muscle. JENNISCHE et al. (1987) reported an increase in immunoreactive IGF-I in skeletal muscle following ischemic injury. NORSTEDT et al. (1987) demonstrated a major increase in IGF-I mRNA on the second day following injury. Levels promptly returned to normal.

Liver Regeneration. No change in IGF-I mRNA was observed following partial hepatectomy (NORSTEDT et al. 1988).

4. Tumors

In contrast to IGF-II, which is expressed at high levels in many tumor types, only a few examples have been reported of IGF-I RNA expression (Table 4). These include moderate expression in some colon carcinomas and liposarcomas (TRICOLI et al. 1986), and in leiomyomas and leiomyosarcomas (HOPPENER et al. 1988). IGF-I RNA is expressed in small cell lung carcinoma cell lines (NAKANISHI et al. 1988), and in rat medullary thyroid carcinoma (HOPPENER et al. 1987).

C. Receptors

I. Introduction

Competitive binding experiments showed that IGF receptors are distinct from insulin receptors, and that there are two types of IGF receptors based on their relative affinities for IGF-I versus IGF-II and whether they recognize insulin. Later, a physical basis for these two IGF receptor subtypes was provided by affinity crosslinking experiments which showed that the receptor which bound IGF-I with higher affinity than IGF-II and recognized insulin had a binding subunit of 130-kDa, whereas the receptor that preferred IGF-II was a single 250-kDa protein without subunits. The receptor which prefers IGF-I over IGF-II and weakly binds insulin has been called the type I IGF receptor or the IGF-I receptor. The receptor which binds IGF-II 100-fold better than IGF-I and does not recognize insulin has been designated the type II receptor, the IGF-II receptor or, more recently, the IGF-II/mannose-6-phosphate (Man-6-P) receptor. Biosynthetic labeling experiments demonstrated a 95-kDa β-subunit of the IGF-I receptor in addition to the 130-kDa α-subunit, suggesting an α_2, β_2 heterodimeric structure for the complete receptor. The IGF-I receptor was, therefore, similar to the insulin receptor in structure. Immunologic experiments also indicated that the two receptors were related. Competitive binding and affinity crosslinking experiments demonstrated that the two IGF receptors were widely distributed in different tissues and on the surface of many cells in culture. Analysis of purified receptor structure by SDS gel electrophoresis supported the results of the affinity crosslinking experiments. Finally, the IGF-I receptor was shown to undergo ligand-induced autophosphorylation on tyrosine residues of the β-subunit.

These results have been discussed in earlier reviews (RECHLER and NISSLEY 1985a, b; NISSLEY and RECHLER 1984a, b). In this section, we will focus on more recent developments in understanding the structure and function of IGF receptors.

II. IGF-I Receptor

1. Structure

Close similarity of the insulin receptor and the IGF-I receptor was indicated by results from affinity crosslinking studies, biosynthetic labeling, receptor purification, ligand-induced activation of tyrosine kinase activity of the β-subunit, and immunologic crossreactivity (reviewed in RECHLER and NISSLEY 1985b). It was not surprising, therefore, when molecular cloning and sequencing of the human IGF-I receptor cDNA revealed extensive homologies between the two receptors (ULLRICH et al. 1986) (Fig. 7). Excluding the signal

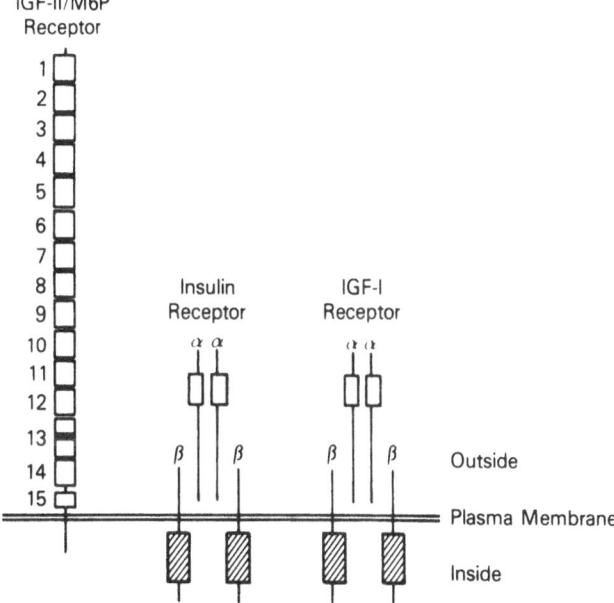

Fig. 7. Receptor structure of the IGF-II/mannose-6-phosphate (*M6P*) receptor, insulin receptor, and IGF-I receptor. The structures are based on molecular cloning data of MORGAN et al. (1987), EBINA et al. (1985), and ULLRICH et al. (1986), respectively, and the figure is adapted from MORGAN et al. (1987). Repeat sequences in the extracellular domain of the IGF-II/Man-6-P receptor are indicated by *open boxes.* The *dark bar* in repeat 13 corresponds to a 43 amino acid segment which is homologous to the type II region of fibronectin. The *open boxes* in the extracellular domains of the insulin and IGF-I receptors correspond to cysteine-rich regions; the *hatched boxes* in the cytoplasmic portion of the β-subunits represent tyrosine kinase domains. The α- and β-subunits of the insulin receptor and the IGF-I receptor are connected by disulfide bonds (not shown)

peptide, the IGF-I receptor precursor protein consists of 1337 amino acids compared to 1343 for the insulin receptor. The $\alpha\beta$ proreceptor polypeptide contains the α-subunit at the amino terminus separated from the carboxy-terminal β-subunit by an Arg-Lys-Arg-Arg cleavage sequence after residue 706. The α-subunit contained a single cysteine-rich region and 11 potential N-glycosylation sites. Overall, 15 out of 16 potential N-linked glycosylation sites in the α-subunit and extracellular domain of the β-subunit were located in nearly identical positions in the insulin receptor and IGF-I receptor. A single 24-amino-acid hydrophobic sequence (906–929) in the β-subunit probably represents the membrane spanning domain. The tyrosine kinase domain located in the cytoplasmic portion of the β-subunit of the IGF-I receptor was 84% homologous with the tyrosine kinase domain of the insulin receptor. The oncogene v-*ros* shows equal homology ($\sim 50\%$) with the tyrosine kinase domains of the insulin receptor and the IGF-I receptor. This homology is too distant for the insulin receptor or the IGF-I receptor to be the protooncogene for v-*ros*. IGF-I receptor and insulin receptor mRNAs displayed different size patterns in the same tissue. The IGF-I receptor gene is located on chromosome 15, whereas the insulin receptor is on chromosome 19.

The IGF-I receptor cDNA from human placenta was expressed in Chinese hamster ovary (CHO) cells and displayed high-affinity IGF-I binding ($K_d = 1.5 \times 10^{-9}$) (STEELE-PERKINS et al. 1988). The expressed receptor bound IGF-II with 3-fold less affinity than IGF-I and weakly recognized insulin. IGF-I stimulated autophosphorylation of the β-subunit of the expressed receptor as well as phosphorylation of cellular substrates. IGF-I stimulated glucose uptake, glycogen synthesis, and DNA synthesis via the transfected IGF-I receptor. Thus, the cDNA clone encodes a functional IGF-I receptor.

2. Receptor Heterogeneity

Variability in binding affinity for IGF-I and IGF-II and in reactivity with receptor antibodies suggested that IGF-I receptors are heterogeneous. Thus, the chick embryo fibroblast IGF-I receptor has equal binding for IGF-I and IGF-II rather than the more typical preference for IGF-I (RECHLER et al. 1980). In IM9 lymphocytes and human placenta, IGF-II bound with high affinity to an IGF-I receptor (HINTZ et al. 1984; MISRA et al. 1986). The monoclonal antibody αIR-3 which blocked binding of IGF-I tracer to the IM9 and human placenta IGF-I receptor did not block binding of IGF-II tracer to this receptor (CASELLA et al. 1986; MISRA et al. 1986). CASELLA et al. (1986) provided evidence that in human placenta there were two binding sites on the same population of type I receptors, one which bound IGF-I better than IGF-II and was inhibitable by αIR-3, and a second site that preferred IGF-II over IGF-I and was not inhibited by αIR-3.

IGF-I receptors also have been distinguished on the basis of reactivity with receptor antibodies. Thus, a monoclonal antibody (5D9) raised against the human placental insulin receptor completely blocked the binding of radiolabeled IGF-I to IM9 lymphocytes and solubilized IGF-I receptors from human placenta but did not inhibit the binding to human hepatoma cells (HepG2), human fibroblasts, or human muscle cells even at 100-fold higher concentration (MORGAN and ROTH 1986). A polyclonal antibody against the human insulin receptor (B2) inhibited 90% of ^{125}I-IGF-I binding to IM9 lymphocytes but only partially inhibited its binding to human placenta (KASUGA et al. 1983). Using this same antiserum, JONAS and HARRISON (1985) immunoprecipitated IGF-I receptors from solubilized human placental membranes which showed relatively lower binding affinity for IGF-I than the residual IGF-I receptor which was not precipitated by the antiserum. They provided evidence that low-affinity form of the receptor could be converted to the high-affinity form by limited reduction of disulfide bonds; reactivity to antiserum B2 also was lost (JONAS and HARRISON 1986). Size heterogeneity has also been noted for the β-subunit of the IGF-I receptor in the same cell line (KULL et al. 1983; JACOBS et al. 1983a; MORGAN et al. 1986).

It is not clear whether these heterogeneous forms of the type I receptor represent translation products from different type I receptor mRNAs or post-translation modifications including glycosylation and extent of disulfide bond formation.

3. Biosynthesis

IGF-I receptor biosynthesis was studied by biosynthetically labeling cells with [^{35}S]methionine, solubilizing the cells with detergent, and immunoprecipitating the receptor with a specific monoclonal antibody (αIR-3). When HepG2 human hepatoma cells were radiolabeled for 30 min followed by chase with unlabeled medium, a 190-kDa species was identified initially, followed by the appearance of a more faintly labeled species of slightly higher molecular weight (DURONIO et al. 1988). The 190-kDa band represented the αβ precursor molecule. The 190-kDa precursor species contained N-linked high mannose oligosaccharides (endo-H sensitive) whereas the larger, more faintly labeled band contained primarily processed, complex type oligosaccharide (endo-H insensitive). As labeling in these two precursor species diminished, the intensity of the α- and β-subunit bands increased, indicating cleavage of the αβ precursor.

A series of inhibitors of oligosaccharide-processing enzymes (N-methyldeoxynojirimycin or MedJN, manno-1-deoxynojirimycin or MandJN, swainsonine) and an inhibitor of translocation of newly synthesized protein through the Golgi (monensin) have been utilized to define the functional role of receptor oligosaccharide in the IGF-I receptor (JACOBS et al. 1983a; DURONIO et al. 1986, 1988). In the process of N-linked glycosylation in the endoplasmic reticulum, the oligosaccharide that is transferred to asparagine residues from a dolichol precursor is $Glc_3Man_9GlcNAc_2$. Next, glucose residues are removed by glucosidases (inhibited by MedJN) and then mannose residues are removed by mannosidase I (inhibited by MandJN) yielding $Man_5GlcNAc_2$. A GlcNAc residue is then added before additional mannose residues are removed by mannosidase II (inhibited by swainsonine). Inhibition by MedJN and MandJN results in glycoproteins containing oligosaccharides of the high mannose type, differing only in the presence or absence of glucose residues. Inhibition by swainsonine results in hybrid-type structures containing both mannose-rich oligosaccharide and terminally glycosylated oligosaccharides (complex type). Inhibition of movement of newly synthesized proteins through the Golgi by monensin is accompanied by inhibition of terminal glycosylation and proteolytic cleavage.

In IM9 lymphocytes, addition of monensin caused the accumulation of the 180- to 190-kDa $\alpha\beta$ precursor. This precursor existed as a disulfide linked dimer $(\alpha\beta)_2$ and some of these molecules reached the cell surface (JACOBS et al. 1983a). Although there were differences in the effects of MedJN, MandJN, and swainsonine in HepG2 cells, the resulting receptors were processed by proteolytic cleavage, translocated to the cell surface, and exhibited ligand-dependent autophosphorylation of the β-subunit of the receptor (DURONIO et al. 1986, 1988). Therefore, the presence of all the sugars normally added to form complex type oligosaccharides is not essential for receptor processing, translocation to the cell surface, or tyrosine kinase activity.

4. Tyrosine Kinase Activity

JACOBS et al. (1983b) first demonstrated IGF-I-dependent phosphorylation of the β-subunit of the IGF-I receptor by immunoprecipitating the receptor with a specific monoclonal antibody (αIR-3) following labeling of intact IM9 lymphocytes with $H_3{}^{32}PO_4$. It was also possible to demonstrate IGF-I-dependent phosphorylation of the β-subunit of a partially purified IGF-I receptor preparation using $[\gamma\text{-}{}^{32}P]ATP$. In the latter experiment, phosphoamino acid analysis indicated that phosphorylation occurred on tyrosine residues. In intact cells, IGF-I addition resulted in increased phosphorylation on serine and threonine residues in addition to tyrosine (SHEMER et al. 1987a). Recently BEGUINOT et al. (1988) have provided evidence that in L6 skeletal muscle cells, insulin caused phosphorylation on tyrosine residues of the IGF-I receptor by acting through the insulin receptor. Insulin caused 50% of maximal stimulation of IGF-I receptor phosphorylation at concentrations of insulin too low to bind to the IGF-I receptor.

The IGF-I-dependent phosphorylation of the β-subunit of the IGF-I receptor as well as IGF-I-dependent phosphorylation of an artificial substrate, $(Glu^4, Tyr^1)_n$, was examined in detail using a partially purified IGF-I receptor preparation from a rat liver cell line (BRL-3A2) with few insulin receptors (SASAKI et al. 1985). When $[\gamma\text{-}{}^{32}P]ATP$ was added to the receptor preparation following incubation with IGF-I, there was a lag in the phosphorylation of $(Glu^4, Tyr^1)_n$. This lag was eliminated by preincubating with ATP before addition of $[\gamma\text{-}{}^{32}P]ATP$, consistent with activation of the receptor tyrosine kinase by autophosphorylation. The activation of the receptor tyrosine kinase by autophosphorylation was shown to be an intramolecular reaction. The divalent cation requirement for the autophosphorylation reaction was best satisfied by Mn^{2+}, while Mg^{2+} was preferred for substrate phosphorylation by the receptor tyrosine kinase. Later, other investigators demonstrated IGF-I-dependent phosphorylation of the β-subunit on tyrosine residues of the IGF-I receptor as well as receptor-mediated phosphorylation of exogenous substrates, using more highly purified receptor preparations (YU et al. 1986; MORGAN et al. 1986; TOLLEFSEN et al. 1987; LE BON et al. 1986). By brief treatment of human placental membranes with dithiothreitol at pH 8.5, FELTZ et al. (1988) produced α,β heterodimers of the IGF-I receptor that bound IGF-I with high affinity, but IGF-I did not stimulate autophosphorylation of the β-subunit of these heterodimers. The α_2,β_2 heterotetrameric structure was required for IGF-I induced β-subunit phosphorylation.

Many feature of the IGF-I receptor tyrosine kinase are very similar to the insulin receptor tyrosine kinase. However, SAHAL et al. (1988) demonstrated differences between the tyrosine kinases of highly purified insulin receptors and IGF-I receptors from human placenta. While $(Glu^4, Tyr^1)_n$ and $(Glu^6, Tyr^1)_n$ were good substrates for both receptor kinases, the ratio of phosphate incorporated into $(Glu^4, Tyr^1)_n$ compared to $(Glu^6, Ala^3, Tyr^1)_n$ was high (~ 4) for the IGF-I receptor kinase and low (~ 1) for the insulin receptor kinase. In addition, the insulin receptor kinase was at least 10-fold more sensitive than the IGF-I receptor kinase to inhibition by $(Tyr, Ala, Glu)_n$, and $(Tyr-Ala-Glu)_n$ of β-subunit autophosphorylation and phosphorylation of exogenous substrate.

5. Mediation of Biological Responses to IGF-I and IGF-II

Comparison of dose-response curves of IGF-I and IGF-II for stimulating various biological responses in cells in culture has revealed in most cases that higher concentrations of IGF-II than IGF-I are required to produce the same response. These data suggest that most biological responses to the IGFs are mediated through the IGF-I receptor rather than the IGF-II receptor. More definitive conclusions are drawn from experiments which utilize receptor antibodies that block binding of IGFs to the IGF-I receptor. KULL et al. (1983) developed a monoclonal antibody (αIR-3) against the IGF-I receptor using as antigen a population of receptors that had been purified from human placenta by insulin affinity chromatography. This monoclonal antibody was initially used in biosynthetic labeling studies to immunoprecipitate the IGF-I receptor (KULL et al. 1983), but was later recognized to also block ^{125}I-IGF-I binding to the IGF-I receptor (FLIER and MOSES 1985; JACOBS et al. 1986). The antibody is not directed precisely against the IGF-I binding site since IGF-I competes poorly for radiolabeled αIR-3 and lower concentrations of αIR-3 were required to compete for binding of radiolabeled αIR-3 than for binding of ^{125}I-IGF-I (FLIER and MOSES 1985; JACOBS et al. 1986). In human fibroblasts in culture, αIR-3 has been shown to block IGF-I-stimulated [^{3}H]thymidine incorporation into DNA (FLIER et al. 1986; VAN WYK et al. 1985) and α-aminoisobutyric acid (AIB) transport (CHAIKEN et al. 1986). The monoclonal antibody also blocked stimulation of [³H]thymidine incorporation into DNA by IGF-II in human fibroblasts (Conover et al. 1986; Furlanetto et al. 1987), showing that both IGF-I and IGF-II act through the IGF-I receptor to stimulate DNA synthesis.

The ability of αIR-3 to inhibit biological responses to insulin in human fibroblasts differed depending upon the source of the fibroblasts. In embryonic lung fibroblasts, αIR-3 blocked stimulation of DNA synthesis by insulin, but in dermal fibroblast strains, αIR-3 did not block stimulation of DNA synthesis and AIB transport by low concentrations of insulin, but only inhibited high concentrations of insulin that are required for binding to the IGF-I receptor. These results suggest that in some fibroblasts, insulin stimulates DNA synthesis and AIB transport by acting through both the insulin receptor and the IGF-I receptor. Indeed, insulin had been reported earlier to stimulate growth responses in certain cell lines by acting through the insulin receptor (Koontz and Iwahashi 1981; Nagarajan and Anderson 1982).

6. Signaling

a) Polyphosphoinositide Hydrolysis, 1,2-Diacylglycerol Generation, Coupling Through a G Protein

The immediate steps involved in coupling the ligand-occupied IGF-I receptor to postreceptor events (nutrient transport, ion flux, synthesis of macromolecules, etc.) remain largely undefined. In Chinese hamster lung fibroblasts, the mitogenic effect of α-thrombin and EGF are potentiated by high concentrations of insulin, presumably acting through the IGF-I receptor. Whereas thrombin stimulated polyphosphoinositide hydrolysis, EGF or EGF plus insulin did not (L'Allemain and Pouyssegur 1986). Similarly, the stimulation of phosphoinositide hydrolysis by thrombin was not increased by addition of insulin. Pertussis toxin blocked thrombin-induced DNA synthesis but not the potentiation by insulin (Chambard et al. 1987). Thus, in Chinese hamster lung fibroblasts, the thrombin receptor appears to be coupled to hydrolysis of polyphosphoinositides by phospholipase C through a G protein as an early step in stimulation of DNA synthesis, whereas insulin acting through the IGF-I receptor potentiates the mitogenic effect of thrombin by acting through another pathway.

Brenner-Gati et al. (1988) recently reported that IGF-I caused a 540% increase in 1,2-diacyglycerol content in a rat thyroid cell line (FRTL-5), and thyroid-stimulating hormone (TSH) plus IGF-I synergistically increased 1,2-diacylglycerol content to 1890% of control. 1,2-Diacylglycerol is an endogenous activator of protein kinase C. TSH could be replaced with forskolin or 8-bromocyclic AMP, suggesting an interaction between lipid and adenylate cyclase signaling systems in the stimulation of DNA synthesis in these cells. However, the stimulation of 1,2-diacylglycerol content by IGF-I or IGF-I plus TSH was not rapid (no increase in 1,2-diacylglycerol content during 6 h exposure to TSH plus insulin), suggesting either a very delayed stimulation of phospholipase C hydrolysis of lipid or mediation at the level of transcription to induce enzymes of lipid synthesis.

NISHIMOTO et al. (1987b) reported that in BALB/c 3T3 cells, pertussis toxin inhibited the stimulation by IGF-I of [³H]thymidine incorporation into DNA and Ca^{2+} influx, suggesting involvement of a G protein in the response to IGF-I.

b) Tyrosine Phosphorylation

Intrinsic tyrosine kinase activity is a common property of many growth factor receptors as well as at the transforming proteins of retroviruses, leading to the attractive proposal that protein phosphorylation is somehow involved in the signaling pathway for growth stimulation caused by growth factors and the products of oncogenes. MORGAN and ROTH (1987) introduced a monoclonal antibody against the tyrosine kinase domain of the insulin receptor and IGF-I receptor into TA1 mouse adipocytes and blocked the stimulation of glucose uptake by IGF-I, IGF-II, and insulin. Since the rank order of potency in stimulating glucose uptake in these cells was IGF-I > IGF-II > insulin, this response presumably was mediated by the IGF-I receptor. The monoclonal antibody results suggest that the tyrosine kinase domain of the IGF-I receptor is important. In contrast, STEELE-PERKINS et al. (1988) observed that αIR-3, a monoclonal antibody to the IGF-I receptor, stimulated several biological responses (glucose uptake, glycogen synthesis, DNA synthesis) in CHO cells that had been transfected with the cDNA for the IGF-I receptor, and that this stimulation occurred without demonstrable autophosphorylation of the β-subunit of the IGF-I receptor and without phosphorylation of cellular substrates.

There are two mechanisms whereby the IGF-I induced receptor autophosphorylation might lead to signaling of the biological response. IGF-I-induced autophosphorylation might induce a conformational change in the receptor which would lead to coupling to other membrane components or to second messengers in the cytoplasm. According to this conformational model, signaling would occur without phosphorylation of these cellular components. Alternatively, activation of the receptor tyrosine kinase by IGF-I might result in phosphorylation on tyrosine residues of membrane components or second messengers, leading to a cascade of phosphorylations which produce the various biological responses to IGF-I. With this latter model in mind, investigators have utilized phosphotyrosine antibodies to look for cellular components that are rapidly phosphorylated on tyrosine residues in response to IGF-I. A 185-kDa protein has been shown to be rapidly phosphorylated in response to IGF-I acting through the IGF-I receptor in several cell lines (IZUMI et al. 1987; SHEMER et al. 1987a). Interestingly, the phosphorylation of the 185-kDa protein also has been observed in response to insulin, and peptide maps of these phosphoproteins are consistent with them being identical. Other proteins which were rapidly phosphorylated in response to IGF-I are pp42, pp80, pp105, pp160, and pp240 (Madoff et al. 1988; Kadowaki et al. 1987; Shemer et al. 1987 b; Hunter et al. 1985).

III. IGF-II/Man-6-P Receptor

1. Identification of the IGF-II Receptor as the Cation-Independent Man-6-P Receptor

a) Molecular Cloning

Morgan et al. (1987) reported the primary structure of the human IGF-II receptor based on cloning and sequencing of the receptor cDNA (Fig. 7). The open reading frame encodes a protein with a predicted M_r of 274 353 which included a 40-residue segment having the characteristics of a cleavable signal sequence. There is one major hydrophobic segment of 23 residues which presumably represents the transmembrane domain of the receptor. Ninety-two per cent of the receptor is extracellular. There are 19 N-linked glycosylation sites in the extracellular domain. The extracellular domain consists of 15 conserved repeats of about 150 residues which are only 20% identical but share a highly conserved pattern of eight cysteine residues. There is an insertion of 43 amino acids in repeat 13 that is homologous to the type II region of fibronectin. The cytoplasmic domain is hydrophilic and includes several potential phosphorylation sites on tyrosine, threonine, and serine residues. It has no homology with known protein kinases.

A startling discovery was made when the overall amino acid sequence of the IGF-II receptor was compared with sequences in the data bank; the human IGF-II receptor was found to be 80% homologous with the bovine cation-independent Man-6-P receptor which recognizes mannose-6-phosphate residues on acid hydrolases and targets these enzymes to lysosomes (Lobel et al. 1988). Later, the human Man-6-P receptor was reported to be 99.4% homologous to the human IGF-II receptor (Oshima et al. 1988) and the rat IGF-II receptor was found to be 79% homologous to the bovine Man-6-P receptor (MacDonald et al. 1988). These reports strongly suggest that the same receptor binds two different classes of ligands, proteins bearing the mannose-6-phosphate recognition marker such as lysosomal enzymes and IGF-II.

b) Biochemical and Immunochemical Characterization

Further evidence that the IGF-II receptor and the Man-6-P receptor are identical has been provided by biochemical experiments. Man-6-P receptor purified by affinity chromatography on lysosomal enzyme or phosphomannan affinity columns bound IGF-II with high affinity and a binding stoichiometry close to 1.0 (Tong et al. 1988; Kiess et al. 1988 a; Braulke et al. 1988). Conversely, IGF-II receptor purified on an IGF-II affinity column bound quantitatively to lysosomal enzyme or phosphomannan affinity columns and was eluted with mannose-6-phosphate (ROTH et al. 1987b; MAC-DONALD et al. 1988; KIESS et al. 1988a).

In addition, mannose-6-phosphate itself and a lysosomal enzyme (β-galactosidase) were found to influence the binding of IGF-II. Thus mannose-6-phosphate increased the binding of ^{125}I-IGF-II to receptors that had been purified on an IGF-II affinity column (ROTH et al. 1987b; MACDONALD et al. 1988), and β-galactosidase inhibited the binding of ^{125}I-IGF-II to a bovine Man-6-P receptor (KIESS et al. 1988a). Although the paradox of mannose-6-phosphate and β-galactosidase having opposite effects on IGF-II binding remains unexplained, these observations strongly suggest that IGF-II and ligands that bind to the mannose-6-phosphate recognition site are binding to the same receptor molecule. The binding sites for the two classes of ligands are distinct, since a monoclonal antibody against the Man-6-P receptor that blocked the binding of ^{125}I-IGF-II by 88% had no effect on the binding of ^{125}I-pentamannosyl phosphate-substituted bovine serum albumin (PMP BSA) (BRAULKE et al. 1988). Also, the neoglycoprotein ^{125}I-PMP-lys-aprotinin and ^{125}I-IGF-II were crosslinked to different tryptic fragments of the Man-6-P receptor (WAHEED et al. 1988).

Antibodies that had been originally raised against either the Man-6-P receptor or the IGF-II receptor were shown to recognize the other receptor (MORGAN et al. 1987; ROTH et al. 1987b; KIESS et al. 1988a). Receptors prepared by IGF-II affinity chromatography or β-galactosidase affinity chromatography from the same source behaved identically in immunoprecipitation assays which utilized a panel of five antisera and monoclonal antibodies that had been raised against either the IGF-II receptor or the Man-6-P receptor (KIESS et al. 1988a). Thus the results of both molecular cloning and experiments which have compared the biochemical and immunologic properties of the two receptors lead to the conclusion that the IGF-II receptor also binds proteins containing the mannose 6-phosphate recognition marker.

c) Role of IGF-II/Man-6-P Receptor in Targeting of Lysosomal Enzymes

The role of the IGF-II/Man-6-P receptor in targeting acid hydrolases to lysosomes has been recently reviewed (VON FIGURA and HASILIK 1986; KORN-FELD 1987; PFEFFER 1988). In the endoplasmic reticulum, high mannose oligosaccharides are attached to some asparagine residues in acid hydrolases. In the *cis* Golgi, lysosomal hydrolases are specifically recognized by *N*-acetylglucosamine phosphotransferase which catalyses the attachment of GlcNAc-1-P to the 6-hydroxyl group of mannose residues. In a second reaction, GlcNAc is removed, leaving mannose-6-phosphate residues in the lysosomal enzyme molecules. The mannose-6-phosphate residues serve as a recognition marker for the Man-6-P receptor. In the *trans* Golgi network, secreted proteins and plasma membrane proteins are segregated from acid hydrolases and membrane proteins destined for lysosomes. The binding of acid hydrolases to the Man-6-P receptor enables the hydrolases to move into an acidic, reticular-vesicular structure adjacent to the Golgi complex (GRIFFITHS et al. 1988). Most of the cellular Man-6-P receptors are found in this structure. In the acid environment of the prelysosomal compartment, acid hydrolases dissociate from the Man-6-P receptor to form the contents of mature lysosomes while the Man-6-P receptors recycle back to the Golgi complex to bind more lysosomal enzymes. Man-6-P receptors are not found in mature lysosomes. During the segregation process in the *trans* Golgi network, 10%–15% of lysosomal enzymes are secreted rather than being targeted to lysosomes and a minor population of Man-6-P receptors travels to the plasma membrane. These extracellular hydrolases bind to cell surface Man-6-P receptors and travel to lysosomes by way of endosomes and the acidic prelysosomal compartment. Thus, cell surface Man-6-P receptors are in equilibrium with the bulk of intracellular Man-6-P receptors and cycle continuously between the plasma membrane, the prelysosomal compartment, and the *trans* Golgi network at a rate that is independent of occupancy by IGF-II or lysosomal enzymes.

A second Man-6-P receptor has been identified which requires divalent cations for the binding of lysosomal hydrolases via the Man-6-P recognition marker (cation-dependent Man-6-P receptor) (reviewed by VON FIGURA and HASILIK 1986; KORNFELD 1987; PFEFFER 1988). This 46-kDa receptor exists in the membrane as a dimer. The extracellular domain of the 46-kDa receptor possesses a single copy of the 150 amino acid cysteine-rich domain that is repeated 15 times in the cation-independent Man-6-P receptor (DAHMS et al. 1987; POHLMANN et al. 1987). The cation-dependent Man-6-P receptor does not bind IGF-II (TONG et al. 1988; KIESS et al. 1988 a).

A fundamental question is whether there are important interactions between the two classes of ligands that bind to the IGF-II/Man-6-P receptor. The observations that mannose-6-phosphate increased the binding of ^{125}I-IGF-II to some preparations of the IGF-II/Man-6-P receptor (ROTH et al. 1987 b; MACDONALD et al. 1988) and that the lysosomal enzyme β-galactosidase inhibited binding of ^{125}I-IGF-II (KIESS et al. 1988 a) raise the possibility that mannose-6-phosphate and lysosomal enzymes could influence signaling by IGF-II through the IGF-II/Man-6-P receptor. Conversely, IGF-II partially inhibited the binding of ^{125}I-β-galactosidase to pure IGF-II/Man-6-P receptor and inhibited the uptake of ^{125}I-β-galactosidase by cells in culture (KIESS et al. 1988 b). Thus, in the remodeling of tissues such as bone where lysosomal enzymes are secreted by the cell into a localized extracellular compartment, IGF-II might influence this process by regulating the cellular uptake of lysosomal enzymes. Alternatively, in a cell that is producing IGF-II, endogenous IGF-II might affect the routing of newly synthesized lysosomal enzymes by inhibiting the binding to the intracellular receptor.

2. Receptor Size

The cation-independent Man-6-P receptor was referred to as the 215-kDa receptor whereas the IGF-II receptor was usually reported to be 250-260-kDa by SDS gel electrophoresis. The calculated molecular mass of the protein portion of the receptor based on sequencing of the cDNA is 270-kDa (MORGAN et al. 1987; OSHIMA et al. 1988; LOBEL et al. 1988). Addition of carbohydrate may increase the size of the receptor glycoprotein to as much as 300-kDa (LOBEL et al. 1988). LOBEL et al. (1988) reported that an antibody raised against a carboxy-terminal peptide that was synthesized on the basis of the molecular cloning data recognized the purified receptor. Thus, the deduced carboxy terminus is present in the mature receptor. Also, MORGAN et al. (1987) reported that the amino-terminal amino acid sequence of the purified receptor commences within several residues of the 40-residue signal peptide. Thus, it is very likely that the molecular mass of the protein portion of the mature receptor is 270-kDa.

3. Binding Affinity for IGF-I

There has been considerable variability in the reported affinity of IGF-I for the IGF-II/Man-6-P receptor, ranging from 0% to 20% as potent as IGF-II (RECHLER et al. 1980; DAUGHADAY et al. 1981; PILISTINE et al. 1984; ROSENFELD et al. 1987; SCOTT and BAXTER 1987; EWTON et al. 1987; BARENTON et al. 1987; TALLY et al. 1987a; BALLARD et al. 1988). Part of this variability in IGF-I affinity toward the IGF-II/Man-6-P receptor is explained by low-level contamination of some plasma-derived IGF-I preparations with IGF-II. Thus a preparation of IGF-I reported by ROSENFELD et al. (1987) to show 10% the potency of IGF-II for binding to the IGF-II/Man-6-P receptor on BRL-3A rat liver cell membranes was later noted by TALLY et al. (1987a) to have been contaminated with a small amount of IGF-II which could be removed by HPLC ion exchange chromatography. Similarly, other investigators have reported that natural IGF-I preparations, highly purified from human Cohn fraction IV (BLUM et al. 1986; SCOTT and BAXTER 1987) or bovine colostrum (BALLARD et al. 1988), showed 1% or less the potency of IGF-II. SCOTT and BAXTER (1987) reported that recombinant IGF-I was equipotent to plasma-derived IGF-I which showed 1% the potency of IGF-II in binding to highly purified rat liver IGF-II/Man-6-P receptor. We have confirmed this relative potency of IGF-I and IGF-II using highly purified IGF-II receptor from rat placenta and recombinant IGF-I (unpublished experiments). Thus, it seems unlikely that the greater potencies of plasma-derived IGF-I compared to recombinant or synthetic IGF-I (ROSENFELD et al. 1987; EWTON et al. 1987) can be ascribed to differences in the intrinsic properties of these molecules. Rather, the greater reactivity of the natural IGF-I preparations toward the IGF-II receptor is probably due to low-level contamination with IGF-II.

It is more difficult to explain reports of zero binding of IGF-I to the IGF-II receptor. BALLARD et al. (1988) have pointed out that if an IGF-II binding protein is present in the medium which binds IGF-I with higher affinity than the cell surface IGF-II/Man-6-P receptor, then addition of unlabeled IGF-I to monolayer cultures in which ^{125}I-IGF-II binding is being measured may actually cause increased binding of the tracer to the cell surface IGF-II/Man-6-P receptor rather than competing for binding. The net result will be to make it appear as though IGF-I does not compete for binding to the IGF-II/Man-6-P receptor. Of course, it is also possible that in different tissues or species IGF-II/Man-6-P receptors possess different intrinsic binding characteristics for IGF-I as suggested by TALLY et al. (1987a). More experiments need to be conducted with highly purified preparations of IGF-II/Man-6-P receptor from different tissues and species before definite conclusions can be drawn.

4. Biosynthesis

SAHAGIAN and NEUFELD (1983) studied the biosynthesis of the Man-6-P receptor in Chinese hamster ovary (CHO) cells by labeling with radiolabeled amino acids, sugars, and [^{32}P]phosphate, followed by immunoprecipitation of the radiolabeled receptor with a specific antiserum. After 30 min of labeling, the receptor was sensitive to digestion with endo-H and the endo-H sensitivity was lost between 30 and 180 min of chase with unlabeled medium, consistent with conversion of asparagine-linked oligosaccharides from a high mannose type to a complex type. [^{32}P]phosphate was incorporated into serine residues of the receptor. Only the endo-H-resistant form of the receptor was phosphorylated and the turnover of the receptor associated [^{32}P]phosphate was faster than the turnover of the receptor (half-life, $t_{1/2}$ 2.3 h versus 16 h).

MACDONALD and CZECH (1985) examined the biosynthesis of the IGF-II receptor in H35 rat hepatoma cells using similar methods. After a 30-min pulse with radiolabeled methionine, a 245-kDa species was observed which became maximally labeled after 10–20 min of chase. The 245-kDa species disappeared as a 250-kDa species appeared and became maximally labeled after 4 h of chase, establishing a precursor:product relationship for the 245- and 250-kDa molecules. The 245-kDa receptor precursor was converted to 232-kDa upon treatment with endo-H, and a 232-kDa species was also observed when biosynthetic labeling was performed in the presence of tunicamycin which inhibits the cotranslational addition of N-linked oligosaccharides. The 250-kDa mature receptor was resistant to endo-H but was converted to a 245-kDa species after treatment with neuraminidase, indicating that the 250-kDa form contains complex oligosaccharides with terminal sialic acid residues. MACDONALD and CZECH (1985) were able to affinity crosslink ^{125}I-IGF-II to the 245-kDa species produced by neuraminidase treatment of the mature receptor but were unable to crosslink ^{125}I-IGF-II to the 232-kDa species observed in the presence of tunicamycin. They concluded that terminal sialic acid residues were not essential for IGF-II binding function of the receptor, but that glycosylation was required for formation of a functional binding site. By contrast, in C6 glial cells, KIESS et al. (1987c) found that ^{125}I-IGF-II did bind to the receptor precursor which accumulated in the presence of tunicamycin. In addition, IGF-II was found to bind with equal affinity to the mature receptor and to a receptor that had been stripped of N-linked glycoprotein by treatment with N-glycanase. Since KIESS et al. (1987c) were not able to assess the binding affinity of IGF-II for the receptor precursor formed in the presence of tunicamycin, it is still possible that N-linked glycosylation is required for proper folding of the protein portion of the receptor and that once this is accomplished, the carbohydrate can be removed without adversely affecting the IGF-II binding function.

5. Regulation

Insulin stimulates a rapid increase in IGF-II binding in adipocytes, H35 hepatoma cells, and pancreatic acini (SCHOENLE et al. 1976; KING et al. 1982; OPPENHEIMER et al. 1983; POTAU et al. 1984). This effect is mediated by the insulin receptor since low concentrations of insulin (0.1 ng/ml) were effective and an antibody directed against the insulin receptor mimicked the effect of insulin (KING et al. 1982). Although initially it was thought that the increase in ^{125}I-IGF-II binding resulted from an increase in binding affinity for ^{125}I-IGF-II (KING et al. 1982; OPPENHEIMER et al. 1983), later experiments conclusively demonstrated that insulin actually caused an increase in cell surface IGF-II receptor number by causing a shift in receptors from a large intracellular pool (Golgi and low density microsomal fraction) to the cell surface. This was determined by performing Scatchard analysis of IGF-II binding to intact adipocytes in the presence of KCN (WARDZALA et al. 1984) and by immunoblotting experiments which employed an antiserum specific for the IGF-II receptor (OKA et al. 1984). KADOTA et al. (1986, 1987a, b) have shown that vanadate plus H_2O_2 (vanadate peroxides) mimicked the effect of insulin in producing a shift of IGF-II/Man-6-P receptors to the cell surface. The addition of vanadate and H_2O_2 to intact cells stimulated tyrosine kinase activity of insulin receptors purified from these cells by wheat germ agglutinin affinity chromatography, consistent with insulin receptor kinase activity being important for IGF-II/Man-6-P receptor redistribution.

CORVERA and CZECH (1985) have shown that the insulin-induced redistribution of the IGF-II/Man-6-P receptor is correlated with the level of phosphorylation of the cell surface receptor. In rat adipocytes, cell surface IGF-II/Man-6-P receptors were found to be more extensively phosphorylated than receptors in the low density microsomal fraction, and insulin addition brought about a decrease in the level of phosphorylation of the cell surface receptor. It is not clear whether the state of IGF-II/Man-6-P receptor phosphorylation determines receptor distribution or is a secondary event. Receptor phosphorylation in intact cells was on serine and threonine residues with no evidence for phosphorylation on tyrosine residues (CORVERA et al. 1988a). Among several well characterized protein kinases tested for ability to phosphorylate purified IGF-II/Man-6-P receptor, only casein kinase II was active, suggesting that this enzyme or a related enzyme may be accounting for phosphorylation in intact cells (CORVERA et al. 1988a). Phosphorylation on tyrosine residues of the IGF-II/Man-6-P receptor has been observed in detergent-treated plasma membrane preparations (CORVERA et al. 1986), and membranes prepared from cells treated with insulin exhibited decreased tyrosine phosphorylation (CORVERA et al. 1988 b). It is difficult to relate these *in vitro* observations of tyrosine phosphorylation to insulin regulation of IGF-II/Man-6-P receptor phosphorylation in intact cells, which is exclusively on serine and threonine residues (CORVERA et al. 1988 a).

The stimulation by insulin of IGF-II/Man-6-P receptor redistribution to the cell surface is qualitatively similar to the insulin-induced shift in the glucose transporter. However, there are qualitative and quantitative differences in the two processes suggesting different pathways. Chloroquine or NH_4Cl, agents known to block recycling of receptors to the cell surface by inhibiting acidification of intracellular vesicles, reversed the insulin-stimulated increase in adipocyte cell surface IGF-II/Man-6-P receptor but did not reverse the insulin-stimulated redistribution of glucose transporters (OKA et al. 1987). There were also quantitative differences between the insulin-induced redistribution of the two membrane components in sensitivity and response time to insulin as well as the magnitude of the redistribution (APPELL et al. 1988). In addition, fractionation of a rat adipocyte low density microsomal preparation by sucrose gradient centrifugation and agarose gel electrophoresis revealed that the glucose transporter and the IGF-II receptor did not comigrate perfectly on the agarose gel, suggesting different vesicle fractions for the two membrane components (JAMES et al. 1987).

In cultures of normal rat hepatocytes, IGF-II/Man-6-P receptor number per cell was inversely related to cell density; sparse cultures exhibited 6-fold higher receptor density than dense cultures (SCOTT et al. 1988). In contrast, three rat hepatoma cell lines showed high numbers of receptors that were independent of cell density. Since proliferative functions are prominent in sparse cultures of normal hepatocytes and in hepatoma cell lines whereas confluent cultures of normal hepatocytes exhibit differentiated functions, IGF-II/Man-6-P receptor density correlates with the state of cell proliferation in this model.

The IGF-II/Man-6-P receptor cycles continuously between the cell surface and intracellular compartments, like the low density lipoprotein, asialoglycoprotein, and transferrin receptors. The rate of internalization is not affected by binding of either lysosomal enzymes or IGF-II (OKA et al. 1985; CZECH 1986; BRAULKE et al. 1987). Consequently, the IGF-II/Man-6-P receptor is not downregulated by ligand, unlike the IGF-I receptor.

Growth hormone (GH) treatment of hypophysectomized rats caused an increase in IGF-II/Man-6-P receptor number on rat liver membranes (BRYSON and BAXTER 1987). However, there was no effect of GH treatment in normal rats and the level of the receptor was not lower in hypophysectomized rats than in normals.

6. Signaling by the Receptor

For many cells in culture, stimulation of growth responses by IGF-II appears to proceed via the IGF-I receptor. Suggestive evidence is provided by biological dose-response curves characteristic of ligand binding to the IGF-I receptor

(IGF-I > IGF-II ≫ insulin). More compelling evidence for IGF-I receptor involvement is provided by experiments which utilized blocking antibodies directed against the IGF-I-receptor (CONOVER et al. 1986; FURLANETTO et al. 1987) or the IGF-II/Man-6-P receptor (KIESS et al. 1987b). However, there are scattered reports suggesting signaling by IGF-II via the IGF-II/Man-6-P receptor. KOJIMA and his colleagues have provided evidence that low concentrations (100 pM) of IGF-II stimulated Ca^{2+} influx in 3T3 mouse fibroblasts that were pretreated sequentially with platelet-derived growth factor (PDGF) and epidermal growth factor (EGF) (primed, competent cells) (NISHIMOTO et al. 1987a; KOJIMA et al. 1988; MATSUNAGA et al. 1988). This has been demonstrated by several techniques: aequorin loading, measurement of ^{45}Ca influx rate, and tight seal patch technique. IGF-I was reported to be less effective than IGF-II in stimulating Ca^{2+} influx, although the relative potency of IGF-I and IGF-II was not clearly defined in these reports (NISHIMOTO et al. 1987a, b). Signaling through a G protein was implicated because pertussis toxin blocked IGF-II-stimulated Ca^{2+} influx and GTP-γS inhibited IGF-II binding to 3T3 cell membranes (NISHIMOTO et al. 1987a). An IGF-II receptor antibody (R-II PAB1) also stimulated Ca^{2+} influx (KOJIMA et al. 1988), providing further evidence for involvement of the IGF-II/Man-6-P receptor. KOJIMA and colleagues have proposed that the stimulation of Ca^{2+} influx by IGF-II is a signal leading to DNA synthesis. Low concentrations of IGF-II (1 nM) stimulated [^{3}H]thymidine incorporation into DNA in primed, competent 3T3 cells. This stimulation was blocked by pertussis toxin (NISHIMOTO et al. 1987a), and the IGF-II receptor antibody (R-II PAB1) mimicked the effect of IGF-II (KOJIMA et al. 1988). This proposal that IGF-II stimulates DNA synthesis in these cells by acting through the IGF-II receptor is not supported by our data showing that in primed, competent 3T3 cells, IGF-II was much less potent than IGF-I in stimulating DNA synthesis, suggesting signaling through the IGF-I receptor (B. LINDER and S. P. NISSLEY, unpublished results).

HAMMERMAN and his colleagues have reported stimulation of biological responses by very low concentrations of IGF-II (< 1 nM), utilizing canine proximal tubular segments and brush border and basolateral membranes (HAMMERMAN and GAVIN 1984; MELLAS et al. 1986; ROGERS and HAMMERMAN 1988). IGF-II, but not IGF-I at 10 nM, stimulated Na^+/H^+ exchange across the brush border of proximal tubule segments, increased phosphorylation of proteins in basolateral membranes, and stimulated phospholipase C activity (inositol trisphosphate and diacylglycerol generation) in basolateral membranes. HAMMERMAN proposed that stimulation of protein kinase C by diacylglycerol accounts for the phosphorylation of membrane proteins and activation of the Na^+/H^+ antiporter.

A human hepatoma cell line (HepG2) displayed receptors for IGFs and insulin and all three polypeptides stimulated glycogen synthesis in these cells (VERSPOHL et al. 1984). HARI et al. (1987) reported that blocking antibodies for the insulin receptor (MC51) and the IGF-I receptor (αIR-3) only partially blocked the stimulation of glycogen synthesis by IGF-II. An antibody directed against the human IGF-II receptor stimulated glycogen synthesis by 2.5-fold. Moreover, the Fab fragment of this antibody, which was 20% less potent than the intact immunoglobulin G (IgG) in inhibiting the binding of ^{125}I-IGF-II, was also 20% less potent in stimulating glycogen synthesis.

HALL and her colleagues have studied an erythroleukemia cell line (K562) which displayed IGF-II/Man-6-P and insulin receptors but not IGF-I receptors and a subclone (K562/cl 1) which showed greater binding of ^{125}I-IGF-II and less binding of ^{125}I-insulin compared to the parental line (TALLY et al. 1987b). Using a clonal growth assay in semisolid agar, they reported that, in the parental cells, IGF-II and insulin were equipotent and IGF-I was 100-fold less potent. In the subclone, insulin was less active than IGF-II and IGF-I at 100 ng/ml did not stimulate. The relative potency of IGF-II versus IGF-I together with an apparent absence of the IGF-I receptor suggested that stimulation of clonal growth by IGF-II was via the IGF-II/Man-6-P receptor. A limitation of this study is that, because of low cloning efficiency, a minor population of cells was responding to IGF-II.

The K562 erythroleukemia cell line carried in different laboratories may not display uniform properties. BLANCHARD et al. (1988) reported the IGF-II was only slightly more potent than IGF-I in stimulating [^{3}H]thymidine incorporation into DNA. (Recombinant IGF-I was actually more potent than plasma-derived IGF-II.) Mediation via the IGF-II/Man-6-P receptor was favored by these authors in part because of failure to demonstrate the IGF-I receptor. HIZUKA et al. (1987a) were able to demonstrate the presence of IGF-I receptors on K562 cells by affinity crosslinking and showed that IGF-I stimulated [^{3}H]thymidine incorporation into DNA and an increase in cell number.

Taken together, these reports demonstrate stimulation of biological responses by low concentrations of IGF-II and in some cases IGF-II was shown to be much more potent than IGF-I. In addition, in two of these reports, IGF-II/Man-6-P receptor antibodies have been used to strengthen the argument for signaling via the IGF-II/Man-6-P receptor. More experiments which utilize receptor antibodies need to be performed and the signaling roles of the high-affinity binding site for IGF-II on the IGF-I receptor (CASELLA et al. 1986; HINTZ et al. 1984) should be explored before we can definitely conclude that the IGF-II-Man-6-P receptor has functions in addition to targeting of lysosomal enzymes to lysosomes and internalization of IGF-II.

7. The IGF-II/Man-6-P Receptor is Present in the Circulation

WHITE et al. (1982) identified in fetal rat serum a component capable of specifically binding ^{125}I-IGF-II that was considerably larger than either the 150-kDa or 40-kDa IGF binding proteins. This binding component had binding characteristics of the IGF-II receptor; IGF-I was much less potent than IGF-II in competing for binding of ^{125}I-IGF-II and insulin was inactive (KIESS et al. 1987a). Affinity crosslinking of ^{125}I-IGF-II to the serum binding component demonstrated a band on SDS gel electrophoresis only slightly smaller than the membrane-derived IGF-II receptor. An antiserum specific for the IGF-II receptor inhibited the binding of ^{125}I-IGF-II to the large serum binding component and on immunoblotting the antiserum identified a band that was approximately 10-kDa smaller than the membrane-derived receptor. Immunoblotting of sera from rats of different ages demonstrated that the levels of the circulating receptor were high in fetal sera and in the sera of young pups (1-5 µg/ml) but declined dramatically between age 20 and 40 days. The serum IGF-II receptor has also been conclusively demonstrated in humans (CAUSIN et al. 1988) and monkeys (GELATO et al. 1988), and it seems likely that a large IGF-II binding component described in fetal sheep serum is the IGF-II receptor (HEY et al. 1987; M.C. GELATO, personal communication).

The slightly smaller size of the circulating receptor compared to the membrane-derived receptor is consistent with the serum receptor representing the extracellular domain of the IGF-II receptor. Thus, the extracellular domain of the receptor has been shown to be approximately the size of the circulating receptor both by molecular cloning of the IGF-II/Man-6-P receptor cDNA (MORGAN et al. 1987; OSHIMA et al. 1988) and by earlier experiments which measured the size of the cytoplasmic domain of the receptor by protease treatment of coated vesicles bearing the receptor (SAHAGIAN and STEER 1985; VON FIGURA et al. 1985). Therefore, the circulating receptor could arise by enzymatic cleavage from the cell surface.

The function of the circulating receptor is unknown. In the rat, the serum receptor is not a significant carrier of IGF-II (WHITE et al. 1982); this function is served by the IGF serum binding proteins which are present in higher concentration than the receptor. However, in fetal monkey (GELATO et al. 1988) and sheep serum (HEY et al. 1987), the serum receptor carries as much as 20%–40% of circulating IGF-II. The role of the serum receptor in binding serum lysosomal enzymes has not been assessed. If present in extracellular fluid, the receptor could modulate the binding of IGF-II and lysosomal enzymes to cell surface IGF-II/Man-6-P receptors.

8. Developmental Expression of Tissue Receptor

The level of total IGF-II/Man-6-P receptor in various tissues of the developing rat has been measured by immunoblotting using highly purified receptor as standard (SKLAR et al. 1989). In fetal tissues, the receptor concentration ranged from 0.16% to 1.7% of extracted protein in brain and heart, respectively. There was a striking developmental pattern observed with a sharp decline which began before birth in most tissues but after 5 days postnatal in others. There was very little change with development seen in brain and spleen. It is interesting that this developmental pattern seen in most tissues is very similar to the developmental pattern of serum IGF-II (MOSES et al. 1980c) and tissue IGF-II mRNA expression (BROWN et al. 1986), suggesting an important role for IGF-II and the IGF-II/Man-6-P receptor in fetal growth and development.

D. Binding Proteins

I. Introduction

Early in the development of the somatomedin field, investigators recognized that the IGFs in human and rat serum existed as part of larger protein complexes, rather than as 7.5-kDa free IGFs (reviewed in NISSLEY and RECHLER 1984a). Gel filtration of adult serum at neutral pH established that the IGFs occurred predominantly as a 150-kDa complex. Incubation of serum at acid pH dissociated IGF from the binding protein, and irreversibly denatured the binding protein complex into an approximately 40-kDa binding component. In contrast to the 150-kDa complex seen in adult serum, native binding protein complexes of about 40-kDa were observed: (a) in GH-deficient rats and humans (MOSES et al. 1976; WHITE et al. 1981); (b) in fetal rats (WHITE et al. 1982), humans (D'ERCOLE et al. 1980), mice (D'ERCOLE and UNDERWOOD 1980), and sheep (BUTLER and GLUCKMAN 1986); (c) in extravascular fluids such as amniotic fluid and cerebrospinal fluid; and (d) in cell culture media (NISSLEY and RECHLER 1984a). These preliminary observations raised several fundamental questions. Were the approximately 40-kDa native binding proteins and the approximately 40-kDa binding subunit of the 150-kDa complex the same? What was the structure of the 150-kDa binding complex? What was the basis of its regulation by GH?

In the past few years, considerable progress has been made in defining the structural components of the IGF-binding protein complexes. This involved the development of better assays to study the heterogeneity of IGF binding proteins, the purification of several binding proteins, the generation of specific antisera, and most recently the identification of cDNA clones for several of the binding proteins. Current evidence points to the existence of at least three IGF binding proteins: two low molecular weight GH-independent binding proteins, exemplified by the binding proteins from human amniotic fluid (BP-25) and BRL-3A cells (BP-3A), and the GH-dependent binding subunit (BP-53) of the 150-kDa binding protein [1]. A second acid-labile subunit of the 150-kDa binding protein complex that does not bind IGFs also has been identified and partially purified.

The IGF binding proteins play a potentially pivotal role in modulating the actions of IGF-I and IGF-II with their receptors. Although in most cases the IGF-binding protein complex does not bind to receptors and is biologically inactive, certain IGF binding proteins enhance the mitogenic activity of IGF-I. The basis for this difference is unknown. One possibility is that distinct binding proteins exist with different functional capabilities. These might arise from as yet undiscovered genes, by alternate processing of known binding-protein mRNAs, or by posttranslational modification of the protein. This section will describe recent efforts to define the IGF binding proteins in order to understand their functional potential.

---

[1] At a workshop on IGF-binding proteins held in Vancouver, B.C. in June, 1989, it was proposed to use the designations IGFBP-1 for BP-25, IGFBP-2 for BP-3A, and IGFBP-3 for BP-53.

II. Assays for IGF Binding Proteins

1. Binding of ^{125}I-Labeled IGFs

Initially, IGF binding proteins were identified by the ability of ^{125}I-labeled IGFs to form high molecular weight complexes with serum or media proteins (reviewed in NISSLEY and RECHLER 1984a). Most simply, this was accomplished by incubating IGF tracer with serum, and fractionating the mixture by gel filtration (e.g., Sephadex G-200) at neutral pH. Since only those binding proteins having available binding sites for IGFs could form complexes, these methods only detected unsaturated native binding proteins or saturated binding proteins after dissociation of endogenous IGFs at acid pH.

An alternative approach used activated charcoal to adsorb free IGF tracer and separate unbound tracer from binding protein complexes (reviewed in NISSLEY and RECHLER 1984a), making it possible to process larger numbers of samples and to estimate binding capacities and affinities. In some cases, however, separation of complex and free IGF is imperfect, so that the results are not quantitative (BAXTER et al. 1987). Despite these reservations, the charcoal assay is a useful qualitative approach to establishing the presence of binding proteins.

A third alternative also has been described (MARTIN and BAXTER 1986; BAXTER et al. 1987). At the end of the binding incubation, samples are incubated with antibody to the binding protein (see below), and antigen-antibody complexes precipitated with polyethylene glycol. This has the advantage of examining the binding properties of individual binding proteins in more complex mixtures. Complexes of IGF and BP-53 also have been precipitated by addition of concanavalin A followed by the addition of polyethylene glycol (MARTIN and BAXTER 1986).

2. Affinity Crosslinking and SDS Gel Electrophoresis

This method was initially described by HASELBACHER et al. (1980) and extensively used by WILKINS and D'ERCOLE (1985) to characterize IGF binding proteins in human serum. Like the previous methods, it only recognizes unoccupied IGF binding sites. It has the advantage that individual binding proteins may be identified according to size. Disadvantages of the affinity crosslinking technique include the generation of complex profiles by multiple crosslinking events and inefficiency of the crosslinking reaction.

3. Radioimmunoassay

Antisera have been developed to several purified IGF binding proteins (see below). Many of the antisera recognize both unoccupied binding proteins and IGF-binding protein complexes (ROMANUS et al. 1986; POVOA et al. 1987). Specific radioimmunoassays have been developed (RUTANEN et al. 1982; MARTIN and BAXTER 1985; ROMANUS et al. 1986; POVOA et al. 1987) using either ^{125}I-labeled binding protein or ^{125}I-IGF covalently coupled to binding protein as radioligand.

4. SDS Gel Electrophoresis and Electroblotting

HOSSENLOPP et al. (1986 b) devised a powerful method, ligand blotting, for characterizing IGF binding proteins following electrophoresis on SDS gel electrophoresis and blotting to nitrocellulose. In contrast to immunoblotting, in which only immunoreactive proteins are identified by binding specific antibodies, ligand blotting identifies all binding proteins by their ability to bind ^{125}I-IGFs. For ligand blots, gels must be run under nonreducing conditions, since even low concentrations of reducing agents interfere with IGF binding.

Ligand blotting has several potential advantages: (a) endogenous IGFs are dissociated by SDS, so that saturated and unsaturated binding proteins are examined; (b) it eliminates crosslinking to other proteins or formation of multimers; (c) the molecular weight reflects that of the binding component alone, rather than a crosslinked complex; (d) parallel blots may be compared by ligand binding and antibody binding; (e) individual proteins may be isolated and their binding specificity examined.

Using this technique, HARDOUIN et al. (1987) has identified five binding components in human serum: 41.5- and 38.5-kDa major components, and 34-, 30-, and 24-kDa minor components. This is illustrated schematically in Fig. 8. The 41.5/38.5-kDa binding proteins are GH dependent and present in the 150-kDa complex. The human 34-kDa and 30-kDa proteins arise from the 40-kDa region of neutral gel filtration columns, and may correspond to the proteins of similar size found in human cerebrospinal fluid and amniotic fluid, respectively. Whether the 24-kDa protein represents another binding protein or a fragment of one of the known proteins is presently unresolved.

The profile of IGF binding proteins determined by ligand blotting in serum or other biological fluids identifies all of the binding proteins that are present and their relative abundance. Individual proteins may be further characterized by enzymatic deglycosylation to identify glycosylated species, incubation with specific antisera, and determination of binding specificity. If binding subunits of the 150-kDa binding protein are present, neutral gel filtration still is required to determine whether they are combined with an acid-labile subunit to form a 150-kDa binding protein complex.

5. Hybridization

The recent availability of cDNA clones for several IGF binding proteins (see below) makes possible the study of the expression of specific IGF binding protein mRNAs by Northern blot and in situ hybridization.

6. Nomenclature

At present, three distinct IGF binding proteins have been purified and cloned. We have designated these as BP-53, BP-25, and BP-3A. *BP-53* (IGFBP-3) re-

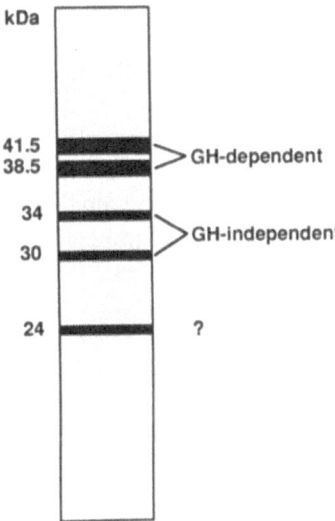

Fig. 8. Schematic diagram of IGF binding components in human serum demonstrated by SDS gel electrophoresis (nonreducing conditions) and electroblotting. Proteins were identified by specific binding of ^{125}I-IGF-I. The 41.5- and 38.5-kDa components occur in the 150-kDa complex. The are GH-dependent glycoproteins that presumably differ in their N-linked oligosaccharides. The 34- and 30-kDa species are predominantly found in the 40-kDa region of a neutral gel filtration column. The 34-kDa protein may be identical to the major binding protein in cerebrospinal fluid, and the 30-kDa protein may be identical to the major IGF binding protein in amniotic fluid. The nature of the 24-kDa complex is presently unclear. The same five binding protein species are present in all cells and biological fluids examined, although in varying proportions (adapted from HARDOUIN et al. 1987)

presents the glycosylated binding component of the GH-dependent binding protein complex in human serum. Rat and porcine counterparts are referred to as rat and porcine BP-53. *BP-25* (IGFBP-1) represents the nonglycosylated binding protein synthesized by human secretory endometrium and decidua, and purified from human amniotic fluid, human placenta and placental membranes (designated placental protein 12 or PP12), and HepG2 cells. Irrespective of its source, this protein will be designated BP-25. *BP-3A* (IGFBP-2) is the nonglycosylated binding protein isolated from the BRL-3A rat liver cell line. A related protein has been identified in a bovine kidney cell line. BP-53, BP-25, and BP-3A have been identified in man and rat (RECHLER et al. 1989).

Biological fluids and cell culture media, however, may contain multiple binding proteins differing in size, antigenic properties, and binding specificity. Complexity may arise from differently glycosylated forms of BP-53, and from proteolytic fragmentation. Analysis by ligand blotting and affinity crosslinking gives different apparent molecular weights, since ligand blotting identifies the binding protein alone but is performed under nonreducing conditions, whereas affinity crosslinking may be performed under reducing or nonreducing conditions but includes the ligand in the complex. (The crosslinked binding protein-IGF complex may not be completely unfolded by SDS, so that its apparent molecular weight may not necessarily be the sum of the two component proteins.) In discussing particular studies, we will refer to the apparent molecular weights from whichever analytic method was used, and try, where possible, to relate these to the purified components.

7. Differentiation from IGF-II/Man-6-P Receptors

IGF binding proteins larger than the 150-kDa GH-dependent binding protein complex have been observed in serum from fetal rats (WHITE et al. 1982) and sheep (HEY et al. 1987). These proteins have been identified by immunoblotting with antireceptor antibodies as the IGF-II/Man-6-P receptor in fetal rat (KIESS et al. 1987a), monkey (GELATO et al. 1988), human (CAUSIN et al. 1988), and sheep (M. C. GELATO, unpublished results) sera. The large binding protein in sheep serum is unusual in that it appears to be stable at acid pH.

III. Purification of Binding Protein Components and Development of Radioimmunoassays

1. GH-Dependent Binding Proteins

a) Binding Subunit

α) Human

The acid-stable binding subunit of the 150-kDa binding protein complex of human plasma was purified from Cohn fraction IV paste by MARTIN and BAXTER (1986). After extraction with acetic acid, the dissociated endogenous IGF-I was adsorbed to SP-Sephadex at pH 3. The supernate was neutralized, and the binding protein purified by affinity chromatography on IGF-II agarose and reversed phase HPLC.

The purified binding protein preparation contained a major protein of 53-kDa and a minor protein of 47-kDa (SDS gel electrophoresis, nonreducing conditions), which decreased in size to 43- and 40-kDa following disulfide reduction. Both proteins bound ^{125}I-IGF-I and were recognized by antibodies to the purified binding subunit after SDS gel electrophoresis and electroblotting (BAXTER et al. 1986). Only a single amino-terminal amino acid sequence was obtained for the first 15 residues (BAXTER et al. 1986; BAXTER and MARTIN 1987), suggesting that the two bands are variants of the same protein which we will designate as BP-53. BP-53 is a glycoprotein (see below). It binds IGF-I and IGF-II stoichiometrically (0.9 residues per 53-kDa) and with similar high affinity $(2-3 \times 10^{10} \ M^{-1})$.

β) Rat

BAXTER and MARTIN (1987) and ZAPF et al. (1988) purified the GH-dependent IGF binding protein subunit from rat serum using a similar scheme. Doublet bands (50-kDa major, 56-kDa minor, nonreduced; 44- and 48 kDa, reduced) were observed on SDS gel electrophoresis (BAXTER and MARTIN 1987). The purified preparation, which we will designate as rat BP-53, bound IGF-I and IGF-II with high affinity ($7\text{--}9 \times 10^{10}M^{-1}$) and a stoichiometry of 0.5 mol/50-kDa binding subunit.

The sequence of the 31 amino-terminal amino acids of rat BP-53 is highly homologous to that of human BP-53 (Fig. 9). ZAPF et al. (1988) also purified a 30- to 32-kDa binding protein from rat serum that had the same amino-terminal sequence, suggesting that it is an amino-terminal fragment of rat BP-53. Residues 8–28 of the rat and human binding proteins are identical, as are residues 35–41 (rat) and 33–39 (human) (WOOD et al. 1988). In addition, residues 12–30 of rat BP-53 are 60% identical to residues 2–21 of the BRL-3A

GH-dependent Binding Proteins:

Amino Terminus

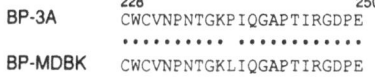

```
          1                                        39
h BP-53   GASSGGLGPVVRCEPCDARALAQCAPPPAV--CAELVREPG
          ..    .  ...................   .  .......
r BP-53   GAGAVGAGPVVRCEPCDARALAQCAPPPXAPACTELVREPG
          . |||.|.....
p BP-53   GKGAVGAGPVVR
```

GH-independent Binding Proteins:

Amino Terminus

```
          1                                                  44
BP-3A     EVLFRCPPCTPERLAACGPPPDA---------PCAELVREPGCGCCSVCARQE
          ...........  ....  ...  .      ..  ..............  .. .
BP-MDBK   EVLFRCPPCTPESLAACKPPPGAAAGPAGDARVPC-ELVREPGCGCCSVFARLE
```

Carboxy Terminus

```
          228                 250
BP-3A     CWCVNPNTGKPIQGAPTIRGDPE
          ..........  ...........
BP-MDBK   CWCVNPNTGKLIQGAPTIRGDPE
```

Fig. 9. *Top,* Comparison of the amino-terminal amino acid sequences of mammalian GH-dependent binding proteins. The sequence for human BP-53 is based on the nucleotide sequence of WOOD et al. (1988). The sequence for rat BP-53 is taken from the amino acid sequences of BAXTER and MARTIN (1987) (residues 1–15) and ZAPF et al. (1988) (residues 1–31). The sequence of porcine BP-53 is from the amino acid sequence of WALTON et al. (1988). Residues that are identical in rat, human, and pig BP-53 are indicated by *dots.* Residues that are identical in rat and porcine BP-53 but differ in human are indicated by *vertical lines.* The *X* in the rat BP-53 sequence represents an unidentified residue *Bottom,* Comparison of the amino-terminal and carboxy-terminal amino acid sequence of BP-3A [deduced from the nucleotide sequence of BROWN et al. 1989)] and the 40-kDa binding protein from the MDBK bovine kidney cell line (based on protein sequence data from SZABO et al. 1988). Identical residues are indicated by *dots.* Gaps are indicated by *dashes. Numbers (above)* refer to the sequence of the mature BP-3A protein

binding protein (BP-3A) and 50% identical to residues 4–22 of the human amniotic fluid carrier protein (BP-25).

γ) Pig

WALTON et al. (1989) used an analogous purification scheme to purify the acid-stable binding subunit of the 150-kDa IGF binding protein in pig serum. Major and minor proteins of 50- and 45-kDa were obtained, as well as faint 23-, 25-, and 29-kDa forms. Affinity crosslinking demonstrated the expected 50- and 57-kDa complexes, a major 27-kDa complex, and a 75-kDa complex whose significance is unclear. The amino-terminal amino acid sequence is identical to that of rat BP-53 at 11/12 positions, and to human BP-53 at 8/12 positions (Fig. 9). ^{125}I-IGF-I binds to a site on pig BP-53 that has equal affinity for IGF-I and IGF-II. Surprisingly, ^{125}I-IGF-II appears to bind to a second site that has a 25-fold lower affinity for IGF-I than for IGF-II.

b) Acid-Labile Subunit

Using ion exchange chromatography (DEAE Sephadex) instead of acid to dissociate the 150-kDa complex of human serum, FURLANETTO (1980) obtained the first preliminary evidence for the existence of an acid-labile nonbinding subunit. The IGF binding subunit eluted with 0.15 M NaCl (peak II), whereas the 0.6 M NaCl eluate (peak III) contained a component that did not itself bind IGF-I, but which combined with the binding subunit (DEAE peak II or acid-dissociated) to reconstitute a 150-kDa IGF-binding protein complex. The presence of binding protein complex was inferred from the size of IGF-I immunoreactivity.

BAXTER (1988) confirmed Furlanetto's observation using improved methodologies. The acid-labile subunit was purified by DEAE Sephadex chromatography, followed by affinity chromatography on IGF-II agarose loaded with purified BP-53. Binding protein complexes were assayed using a radioimmunoassay for BP-53 that measures both the 50-kDa subunit and the 150-kDa complex. HPLC gel permeation columns (Superose 12) were used to rapidly identify a shift in size of the BP-53 subunit from 50- to 150-kDa when recombined with the acid-labile subunit.

The size of the acid-labile subunit was estimated as 100–110-kDa, suggesting that the native 150-kDa complex consists of one molecule each of the 110-kDa subunit, BP-53, and IGF-I or IGF-II. The acid-labile subunit binds to complexes of IGF and BP-53, but does not bind to free IGF, free BP-53, free BP-25, or BP-25 complexed to IGF. The 110-kDa subunit is present in excess in adult human plasma, accounting for the fact that all BP-53 occurs in a 150-kDa complex.

Formation of the 150-kDa complex is obligately dependent on GH, which cannot be replaced by IGF-I. Infusion of GH or IGF-I in hypophysectomized rats restored serum levels of BP-53; however, formation of the 150-kDa complex was seen after GH but not IGF-I infusion (ZAPF et al. 1985, 1989). This suggests that the 110-kDa protein is regulated by GH, although a specific radioimmunoassay for the 110-kDa protein will be required to determine this directly.

c) GH-Dependent Binding Proteins are Glycoproteins

FURLANETTO (1980) reported that the 150-kDa complex from human serum bound to concanavalin A Sepharose and was eluted by α-methylmannoside. He suggested that binding was through the acid-labile subunit and not the binding subunit. WILKINS and D'ERCOLE (1985), analyzing crosslinked ^{125}I-IGF complexes by SDS gel electrophoresis and autoradiography, and GRANT et al. (1987b), analyzing the concanavalin A eluate by Sephadex G-200 chromatography and IGF binding, presented more direct evidence that the 150-kDa binding protein complex, but not the 40-kDa complex, in human plasma binds specifically to concanavalin A.

MARTIN and BAXTER (1986) definitively demonstrated that human BP-53 was a glycoprotein. (a) The purified protein stained with periodic acid Schiff. (b) ^{125}I-IGF-binding protein complexes were precipitated by concanavalin A-polyethylene glycol. (c) BP-53 bound to concanavalin A ($\sim 40\%$ elution with $0.5\ M$ α-methylmannoside) and wheat germ ($\sim 50\%$ recovery with N-acetylglucosamine) but not to $H.\ pomatia$ lectin. These results indicate that the oligosaccharide chains contain mannose and/or glucose (concanavalin A) and N-acetylglucosamine (wheat germ), but not N-acetylgalactosamine ($H.\ pomatia$ lectin). (Binding to wheat germ cannot be attributed to sialic acid because this sugar is destroyed by the acid conditions used during the purification of BP-53.) Absence of detectable N-acetylgalactosamine makes it unlikely that O-linked sugars are present.

The nucleotide sequence of human BP-53 identifies three potential N-glycosylation sites (WOOD et al. 1988). Failure to recover asparagine from two peptides that contain these sites was interpreted as suggesting that these sites are in fact glycosylated.

Rat BP-53 also is a glycoprotein containing N-linked oligosaccharides. (a) Binding activity can be assayed using concanavalin A-polyethylene glycol precipitation (BAXTER and MARTIN 1987). (b) The rat serum binding subunit binds to concanavalin A (YANG et al. 1989). (c) N-glycanase treatment of the purified rat BP-53 and its fragment reduces their apparent molecular weights from 42/45-kDa to 37-kDa and from 32-kDa to 26-kDa, respectively (ZAPF et al. 1989; YANG et al. 1989). Enzymatic deglycosylation of BP-53 with N-glycanase does not destroy its ability to bind IGF-I (YANG et al. 1989).

2. Non-GH-Dependent Binding Proteins

a) Human

A 35-kDa binding protein was purified from term human amniotic fluid by POVOA et al. (1984) using salt fractionation, hydrophobic chromatography, and anion exchange chromatography, and by Baxter et al. (1987) using IGF-I agarose affinity chromatography followed by reverse phase HPLC. The binding protein also was purified from 16–22 weeks' gestation amniotic fluid by Drop et al. (1984a). The purified amniotic fluid binding protein had an apparent molecular mass of 34 kDa (reduced) or 28 kDa (nonreduced) (Baxter et al. 1987) and an isoelectric point of 4.3 (Povoa et al. 1984) or 5.3 (Busby et al. 1988a). The amino acid composition and sequence of the first ten amino acids were reported by Povoa et al. (1984); subsequently, about 40% of the amino acid sequence was determined (Brewer et al. 1988a). Binding activity and immunoreactivity were stable to heating: boiling 10 min or incubating at 60° C for 1 h at pH 2.2–8.5 (Drop et al. 1984a; Busby et al. 1988a). The purified binding protein had an affinity of $0.6-2 \times 10^{10} \ M^{-1}$ for IGF-I (Baxter et al. 1987; Busby et al. 1988a). IGF-II was 60% as potent as IGF-I (Baxter et al. 1987). [D'Ercole et al. (1985) had estimated from affinity crosslinking experiments that the amniotic fluid binding protein had a considerably greater affinity for IGF-I than for IGF-II. This difference may reflect the fact that crosslinking only looks at a small subset of the binding molecules.]

Busby et al. (1988a) modified the purification procedure of Povoa et al. (1984) and obtained two physicochemically similar binding proteins with different biological properties. The phenyl-Sepharose eluate was applied to DEAE-cellulose: binding activity eluted at 100 mM NaCl (peak B) and at 250 mM NaCl (peak C). Both binding proteins were purified to homogeneity on HPLC. They had the same size, the same amino acid composition, the same affinity for IGF-I, and the same 28 amino-terminal amino acids. However, the peak B binding protein adheres to cells and stimulates the mitogenic activity of IGF-I, whereas the peak C protein does not adhere to cells and inhibits IGF-I mitogenic activity (see below). The difference between peak B and C binding proteins is unknown, but may represent posttranslational modification. Peak B, but not peak C, forms disulfide linked multimers.

Povoa et al. (1985) also purified a similar if not identical binding protein from the conditioned media of the human hepatocarcinoma cell line HepG2 using a two-step purification: immunoaffinity chromatography (with immunoglobulin prepared from antiserum to the amniotic fluid binding protein) and size fractionation. The sequence of the ten amino-terminal amino acids is identical to that of the amniotic fluid binding protein. The identity of these proteins has been confirmed by cDNA cloning (see below).

It also was recognized that two proteins purified from human placenta and placental membranes (placental protein 12, PP12; Koistinen et al. 1986) and human endometrium (pregnancy-associated α_1-globulin; Bell and Keyte 1988) had the same amino terminus as the amniotic fluid binding protein, and bound ^{125}I-IGF-I. [Despite its name, PP12 is not a placental protein, but rather the product of secretory endometrium (Rutanen et al. 1986) and decidua (Rutanen et al. 1985).] Although it was reported initially that residues 11 and 12 (Asp, Glu) of PP12 (Koistinen et al. 1986) differed from those of the human amniotic fluid binding protein (Baxter et al. 1986; Busby et al. 1988a) and the pregnancy-associated α_1-globulin (Bell and Keyte

1988), this difference has not been confirmed by cDNA cloning (see below). Thus, the human amniotic fluid binding protein, HepG2 binding protein, PP12, and the pregnancy-associated α_1-globulin appear to be identical. We will designate this protein, irrespective of its source, as BP-25.

b) Rat

An IGF binding protein was purified from the serum-free conditioned media of BRL-3A rat liver cells by acid gel filtration followed by IGF-II affinity chromatography (LYONS and SMITH 1986; MOTTOLA et al. 1986) or HPLC (WANG et al. 1988b). It had an apparent molecular mass of 34–36-kDa (reduced), 31.5–33-kDa (nonreduced). The amino-terminal amino acid sequence of residues 1–10 (LYONS and SMITH 1986) and of residues 3–34 of a truncated protein (MOTTOLA et al. 1986) were reported. Recently, about 70% of the complete amino acid sequence was determined (WANG et al. 1988b; BROWN et al. 1989). The purified BRL-3A binding protein, which we shall refer to as BP-3A, binds IGF-I and IGF-II with similar high affinity ($K_a \sim 1.5 \times 10^9 \, M^{-1}$) using either IGF-I or IGF-II as tracers (WANG et al. 1988b; BROWN et al. 1989).

IV. Cloning and Expression of IGF Binding Proteins

1. Human BP-25

Several groups recently cloned BP-25 from cDNA libraries of human decidua (BREWER et al. 1988a, b; JULKUNEN et al. 1988), placenta (BRINKMAN et al. 1988a), liver (BRINKMAN et al. 1988a), and HepG2 hepatocarcinoma cells (LEE et al. 1988). Each clone encodes an identical 259 amino acid binding protein precursor, consisting of a 25-residue prepeptide and a 234-residue mature protein. The deduced amino acid sequence agrees with that determined by protein sequencing (BREWER et al. 1988a). The cDNA clones have been expressed in *E. coli* (BREWER et al. 1988a) and *cos* cells (BRINKMAN et al. 1988) and translated in vitro (JULKUNEN et al. 1988), yielding proteins that are immunoreactive (BREWER et al. 1988a; BRINKMAN et al. 1988a; JULKUNEN et al. 1988) and bind ^{125}I-IGF-I (BRINKMAN et al. 1988a).

BP-25 mRNA is about 1.5-kb, including 56–164 bp of 5' untranslated region and 564–612 bp of 3' untranslated region culminating in a consensus polyA signal (AATAAA) and a polyA tail. BP-25 cDNA probes hybridize to a single RNA species about 1.5-kb on Northern blots, suggesting that the reported cDNA clones are approximately full length.

The deduced sequence for BP-25 contains 18 cysteine residues, clustered at the amino-terminal and carboxy-terminal ends. The hydrophobic cysteine-rich amino-terminal domain has been suggested as a possible IGF binding domain (LEE et al. 1988; JULKUNEN et al. 1988). BP-25 lacks potential sites of *N*-glycosylation. It contains an Arg-Gly-Asp (RGD) sequence at the carboxy

terminus (BREWER et al. 1988a; JULKUNEN et al. 1988), a potential cell attachment sequence similar to that found in fibronectin and other matrix proteins
and recognized by integrin receptors (RUOSLAHTI and PIERSCHBACHER 1987).
The RGD sequence may be involved in the adhesion of BP-25 to cells (see
below). JULKUNEN et al. (1988) have called attention to two features of BP-25
RNA that are associated with rapidly turning over proteins: (1) a PEST sequence (Pro-Glu-Ser-Thr), associated with proteins that have a short intracellular half-life, and (2) four ATTTA sequences in the 3' untranslated
region, associated with mRNAs having a short half-life. They note that this is
consistent with the short half-life of BP-25 protein in the circulation following
insulin administration (see below).

The BP-25 gene is present as a single copy (BREWER et al. 1988a;
BRINKMAN et al. 1988a; JULKUNEN et al. 1988) and is located on human chromosome 7 (BRINKMAN et al. 1988a). It spans 5–6 kb, and contains four exons
including a 5' untranslated region of 165 nt (BRINKMAN et al. 1988b; CUBBAGE
et al. 1989).

BP-25 RNA is expressed in decidua (BREWER et al. 1988a; JULKUNEN et al.
1988) and placental membranes (BRINKMAN et al. 1988a), but not in placenta
itself (BRINKMAN et al. 1988a; JULKUNEN et al. 1988). Levels are higher in term
decidua than in early pregnancy decidua, which are higher than in nonpregnant secretory endometrium (JULKUNEN et al. 1988). These results are
consistent with earlier studies of the immunoprecipitation of biosynthetically
labeled BP-25 proteins from explants of secretory and proliferative
endometrium, placenta, and decidua (RUTANEN et al. 1985, 1986).

BP-25 mRNA also is present at low concentrations in fetal (BRINKMAN et
al. 1988a) and adult (JULKUNEN et al. 1988) liver; RNA levels are 5–10 times
lower in adult liver (BRINKMAN et al. 1988a). It is not detected in other 14- to
16-week fetal tissues (BRINKMAN et al. 1988a) or in adult proliferative
endometrium, kidney, or adrenal (JULKUNEN et al. 1988). It is present in
HepG2 cells (LEE et al. 1988; BRINKMAN et al. 1988a). In the rat, BP-25 mRNA
is present in fetal liver and decreases after birth (G. T. OOI, unpublished
results).

2. Rat BP-3A

BROWN et al. (1989) obtained a clone encoding BP-3A from a BRL-3A cDNA
library. It encodes a precursor protein of 304 amino acids, comprising a 34-
residue prepeptide and a 270-residue mature protein (29.6-kDa). The mature
protein contains 18 cysteines and no N-glycosylation sites. It contains an Arg-
Gly-Asp (RGD) sequence near the carboxy terminus. An RNA transcript of
the cDNA synthesized in vitro and injected into *Xenopus oocytes* directs the
synthesis of a full-size binding protein that is immunoprecipitated by
antibodies to purified BP-3A, and binds IGF-I and IGF-II with equal high affinity, similar to the binding specificity of purified BP-3A.

BP-3A cDNA probes hybridize to a 2-kb mRNA in BRL-3A cells and in
multiple fetal rat tissues (BROWN et al. 1989). Levels of this mRNA are greatly

reduced in the corresponding nonneural adult tissues, but persist in adult brain stem, hypothalamus, and cerebral cortex (A. BROWN, unpublished results). In situ hybridization localizes the BP-3A mRNA to the choroid plexus of adult brain (L. TSENG, unpublished results). The developmental regulation of BP-3A mRNA in nonneural tissues and its selective expression in the choroid plexus of adult brain parallel similar results obtained for rat IGF-II (BROWN et al. 1986; SOARES et al. 1986; STYLIANOPOULOU et al. 1988b; HYNES et al. 1988).

The partial amino acid sequence of an IGF binding protein isolated from the MDBK bovine kidney cell line (SZABO et al. 1988) is virtually identical to that of BP-3A (Fig. 9), strongly suggesting that this protein is the bovine homolog of BP-3A.

3. Human BP-53

WOOD et al. (1988) isolated cDNA clones for BP-53 from a human liver cDNA library. The 2.5-kb nucleotide sequence appears to represent a full-length mRNA, since the cDNA probe hybridizes to a 2.5-kb RNA on a Northern blot of adult human liver. The mRNA contains 109 nucleotides of 5′ untranslated region, an open reading frame of at least 876 bases, and 1.5-kb of 3′ untranslated region followed by a polyA tract. The open reading frame encodes a protein with a presumed signal peptide of 27 amino acids, and a mature protein of 264 amino acids (28.7 kDa). This is considerably smaller than the apparent molecular mass of 43 kDa observed on SDS gel electrophoresis under reducing conditions (MARTIN and BAXTER 1986), reflecting extensive glycosylation. The nucleotide sequence encodes three possible N-glycosylation sites (with indirect evidence that at least two are used) and possible O-glycosylation sites (which probably are not used). In contrast to BP-25 and BP-3A, BP-53 does not contain an RGD sequence. Transfection of mammalian kidney cells with a plasmid containing the BP-53 coding region yields a protein of appropriate size that binds ^{125}I-IGF with high affinity and is immunoprecipitated by antibodies to BP-53.

4. Relationship Among the Three Cloned IGF Binding Proteins

The amino acid sequences determined for the three cloned IGF binding proteins establish that they are distinct members of a family of related proteins (Fig. 10, Table 7). Each mature binding protein has 18 cysteines, 17 of which are precisely aligned. Based on amino acid sequence homologies, the binding proteins may be divided into three regions. The amino-terminal portion is most highly conserved: about 50% amino acid identity, and 12/12 cysteines aligned for the three proteins. The sequence of BP-53 is somewhat closer to that of rat BP-3A than to human BP-25, including a stretch of 13 identical amino acids. In the carboxy-terminal region, BP-3A and BP-25 are more homologous to each other ($\sim 50\%$) than either is with BP-53. However,

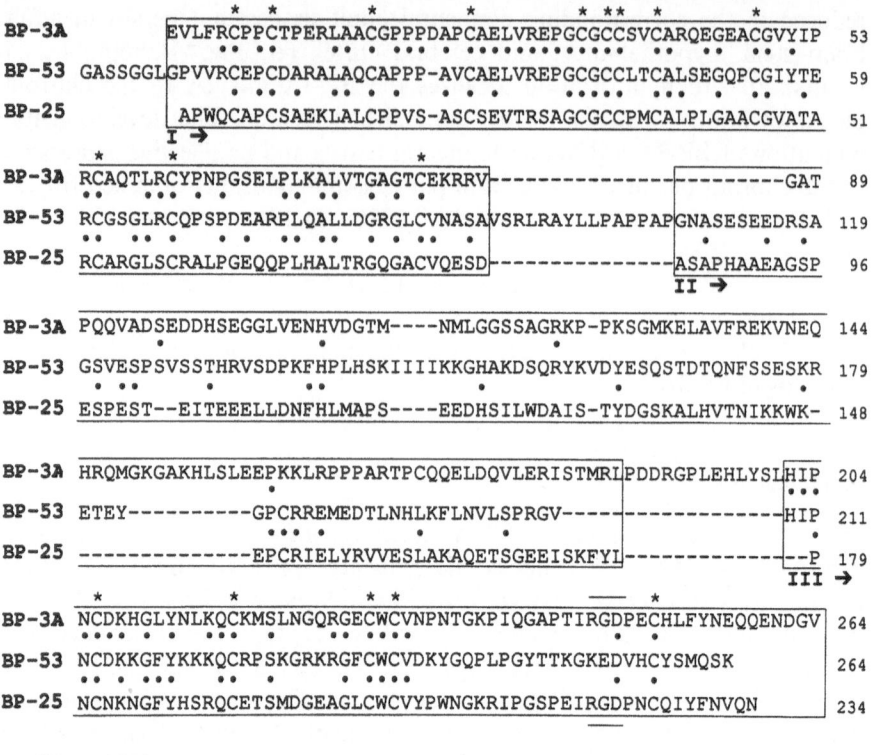

Fig. 10. Sequence homologies of cloned IGF binding proteins. Amino acid sequences of mature BP-3A, BP-53, and BP-25 were determined from the respective nucleotide sequence (see text). Sequences have been aligned to maximize cysteine pairing and homologies, and to minimize the introduction of gaps. *Dots* indicate residues that are identical in BP-3A and BP-53 (*upper*), and in BP-53 and BP-25 (*lower*). The 17 aligned cysteines are indicated by *asterisks*. The *boxes* identify three regions (I, II, III) of similar homology that are demarcated by an insertion in one sequence. The RGD sequences in BP-3A and BP-25 are *underlined/overlined*. The indicated domains correspond well with the genomic organization of BP-25 (BRINKMAN et al. 1988b; CUBBAGE et al. 1989). Exon 2 begins within residue 92 (Glu, E); exon 3 with residue 149 (Glu, E); and exon 4 with residue 192 (Cyc, C)

5/5 cysteine residues in the carboxy-terminal region can be aligned in the three sequences. An RGD sequence is present near the carboxy terminus of the two GH-independent binding proteins but is not present in BP-53. By contrast, the three binding proteins have negligible homology in the middle region (which corresponds approximately to exons 2 and 3). It contains one cysteine, which can be aligned between the two GH-independent binding proteins, but not between them and BP-53. The middle segment of BP-53 contains the three potential *N*-glycosylation sites.

Table 7. Comparison of amino acid sequence identities

	Amino terminus		Middle		Carboxy terminus	
	(%)	(n)	(%)	(n)	(%)	(n)
BP-53/BP-25	48	40/84	19	18/94	32	17/53
BP-53/BP-3A	52	45/86	4	4/92	38	21/56
BP-25/BP-3A	45	38/84	16	14/85	50	28/56

Comparisons are based on the sequences shown in Fig. 10. The amino-terminal, middle, and carboxy-terminal regions correspond to the boxed regions I, II, and II in the figure.

V. Regulation of Individual IGF Binding Proteins in Serum

1. Human

a) BP-25

Radioimmunoassays for BP-25 were developed by RUTANEN et al. (1982), DROP et al. (1984b), POVOA et al. (1987), BAXTER et al. (1987), and BUSBY et al. (1988b) using ^{125}I-labeled BP-25. Purified BP-53 did not crossreact (BAXTER et al. 1987). Occupied and unoccupied binding proteins were equally reactive (POVOA et al. 1987; BUSBY et al. 1988b). Orangutan serum was reactive, but not serum from rat, mouse, cow, pig, sheep, rabbit, elephant, or chicken (DROP et al. 1984b; BUSBY et al. 1988b).

Midterm amniotic fluid (20–22 weeks) contained high levels of BP-25, about 50 µg/ml (RUTANEN et al. 1982; POVOA et al. 1987), approximately two times higher than at term (DROP et al. 1984a). Levels in cord serum are 100–200 times lower than in term amniotic fluid (RUTANEN et al. 1982; POVOA et al. 1987). Levels in fetal serum (18–22 weeks gestation) are 10 times higher than those in term cord plasma (DROP et al. 1984b), which are 5–10 times higher than those in adult plasma (POVOA et al. 1987; DROP et al. 1984b). Serum BP-25 falls continuously from birth to puberty, an age dependence opposite to that of IGF-I and BP-53 (HALL et al. 1988). BP-25 levels are increased 5-fold in pregnancy (DROP et al. 1984b; HALL et al. 1986); levels are maximal at 22–23 weeks, and decrease before term (RUTANEN et al. 1982). BP-25 levels are increased 5-fold in renal failure (POVOA et al. 1987; DROP et al. 1984b) and are decreased in Cushing's disease (HALL et al. 1987).

BP-25 levels in adult serum also are regulated by metabolic factors (for example, fasting, insulin, anorexia nervosa, and diabetes mellitus) and by growth hormone.

1. *Metabolic*. BAXTER and COWELL (1987) reported a 13-fold increase in BP-25 levels between midnight and 8 A.M. in the serum of 12 normal and three GH-deficient children. This is not a true circadian rhythm, but instead is related to feeding time (COTTERILL et al. 1988). The increase is blunted by feeding, and sustained by prolonged fasting. HALL et al. (1988) and BUSBY et al. (1988 b) have obtained similar results. The rise in BP-25 is inversely related to plasma insulin, and is not related to cortisol levels (COTTERILL et al. 1988).

BP-25 levels are increased 2-fold in anorexia nervosa, despite an increase in GH levels (HALL et al. 1986). SUIKKARI et al. (1988) observed a 2.5- to 4-fold increase in immunoreactive BP-25 in insulin-dependent diabetes mellitus, and a 63% decrease in insulinomas that was reversed by successful surgical removal of the tumor. A euglycemic insulin clamp decreased BP-25 levels by 40%–70% in patients with insulin-dependent and non-insulin-dependent diabetes mellitus, and in controls ($t_{1/2} = 1.5$–2h). HALL et al. (1987) also observed that the half-life of BP-25 in patients with insulin-dependent diabetes mellitus after an insulin clamp was considerably shorter (1–1.5 h) than the 24-h half-life observed after parturition. The rapid disappearance of BP-25 from the circulation after insulin administration is thought to reflect increased degradation, clearance, or redistribution, rather than decreased synthesis. These authors suggest that BP-25 might have a shuttle/transport role, contrasted with a storage function for BP-53.

2. *GH*. Early studies (HINTZ et al. 1981) suggested that the affinity of low molecular weight binding proteins for IGF-I was increased in GH deficiency, although it was not known whether this represented an increase in the mass of higher affinity binding proteins or simply lower occupancy by IGFs. POVOA et al. (1987) and HALL et al. (1988) observed an inverse relationship between immunoreactive BP-25 and GH: BP-25 levels were increased 2-fold in GH deficiency and decreased 2-fold in acromegaly. BUSBY et al. (1988 b) observed a similar increase in hypopituitarism, but no significant change in acromegaly. They did observe a 2-fold decrease in BP-25 after administration of pharmacologic doses of GH to obese subjects. HARDOUIN et al. (1987) observed increases in 30-kDa (BP-25) and 34-kDa components in hypopituitary plasma by ligand blotting. BAXTER and COWELL (1987), however, did not observe a correlation between nocturnal GH pulses and BP-25 levels, or an increase in basal or peak BP-25 levels in GH deficiency. Whether the changes in BP-25 observed with GH are primary effects, or whether they are secondary to changes in glucose or insulin is not known.

3. *Tumors*. BP-25 levels were increased in 15/16 patients with primary liver cancer, 5/10 cirrhotics, 10/68 with other gastrointestinal malignancies (RUTANEN et al. 1984), and in 64% of ovarian cancers, with postoperative normalization in 77% (IINO et al. 1986). BP-25 levels were occasionally increased in trophoblastic disease and tumors, but did not correlate with chorionic gonadotropin levels (RUTANEN et al. 1982).

b) BP-53

MARTIN and BAXTER (1985; BAXTER and MARTIN 1986) developed a radioimmunoassay for BP-53 using polyclonal antibodies raised in rabbits and a covalent complex of ^{125}I-IGF-I:BP-53 as radioligand. Immunoreactive BP-53 was identified in the 150-kDa peak, but not in the 40-kDa peak, after neutral gel filtration of human serum. Occupancy with IGF-I or IGF-II or acidification did not affect the immunoreactivity of BP-53. Purified BP-25 showed <0.1% crossreactivity (BAXTER et al. 1987). The antiserum is species specific, reacting with serum from higher primates but not with serum from lower primates or nonprimate mammals.

Immunoreactive BP-53 is low in childhood (2.4 µg/ml at age 3), increases with puberty (8.2 µg/ml at age 14–15), and gradually decreases thereafter (BAXTER and MARTIN 1986). Levels vary directly with GH levels: a 2.2-fold increase in acromegaly and a 50%–80% decrease in GH deficiency. BP-53 showed no diurnal variation (BAXTER and COWELL 1987). In contrast to BP-25, BP-53 levels are similar in plasma and amniotic fluid (BAXTER et al. 1987). BP-53 increases in the third trimester of pregnancy. Levels are decreased in poorly controlled diabetes mellitus (as is IGF-I), normal in hypothyroidism, and slightly elevated in renal failure.

2. Rat BP-3A

Polyclonal antibodies were raised in rabbits to purified BP-3A (LYONS and SMITH 1986; MOTTOLA et al. 1986; ROMANUS et al. 1986, 1987). Antibodies to BP-3A recognize binding proteins in fetal and neonatal rat serum by radioimmunoassay (ROMANUS et al. 1986), by immunoblotting (LYONS and SMITH 1986; YANG et al. 1989), and by immunoprecipitation (YANG et al. 1989), but do not react with adult rat serum.

Two polyclonal antisera to BP-3A block the binding of IGF tracer to IGF binding proteins from the H35 rat hepatoma cell line, but react poorly with the H35 binding protein on immunoblot (>6-fold less than BP-3A) or enzyme-linked immunosorbent assay (>20-fold less potent than BP-3A) (MOTTOLA et al. 1986). This suggests than the H35 binding protein is antigenically distinct from BP-3A, and that the antibodies that inhibit IGF tracer binding to it represent a minor component of the antisera. [Recent results indicate that the predominant H35 binding protein is rat BP-25 (Y.W.-H. YANG, unpublished results)].

3. Pig BP-53

WALTON and ETHERTON (1988) developed a radioimmunoassay to the pig GH-dependent IGF binding subunit. The antiserum is species specific. It recognizes the 150-kDa binding protein complex in normal pig serum, which is absent in hypophysectomized pig serum and is restored after GH treatment. Immunoreactivity also was detected in porcine amniotic fluid, allantoic fluid, colostrum, milk, and follicular fluid.

VI. Fragments of IGF Binding Proteins that Bind IGFs

1. BP-53

In studies of IGF binding proteins in human serum using affinity crosslinking, WILKINS and D'ERCOLE (1985) observed 23- and 28-kDa complexes that appeared to arise from the 150-kDa GH-dependent complex. They proposed that these complexes represented one IGF binding protein subunit ($\sim$16-kDa) coupled to one IGF molecule. However, purification and nucleotide sequencing have established that the binding subunit, BP-53, has a formula weight (nonglycosylated) of 28.7-kDa, and an apparent molecular mass of 43-kDa (glycosylated).

OOI and HERINGTON (1986) purified a 16-kDa IGF binding protein (21.5-kDa crosslinked complex) from human serum that inhibited the stimulation of lipogenesis in rat epididymal adipocytes by IGF-I via the insulin receptor. Several properties suggest than the purified inhibitor is part of BP-53: (a) it is a glycoprotein; (b) it is recognized by antibodies to BP-53 (BAXTER and MARTIN (1986); (c) it contains several tryptic peptides that are present in BP-53 but not in BP-25 (G. T. OOI, personal communication); (d) BAXTER and MARTIN (1986) noted that prolonged storage of purified BP-53 at acid pH at 2° C generates 26 (major) and 22-kDa (minor) species on nonreduced SDS gels.

ZAPF et al. (1989) purified glycoproteins of 42/45-kDa and 30–32-kDa from adult rat serum. The 31 amino-terminal amino acid residues of the two forms are identical, suggesting that the 30- to 32-kDa protein is a fragment of the 42/45-kDa binding subunit. The binding protein fragment remains capable of binding IGF-I. Enzymatic deglycosylation of rat BP53 also does not affect its ability to bind IGF-I (ZAPF et al. 1989; YANG et al. 1989).

2. BP-25

During the purification of PP12 from human placenta, lower molecular weight fractions with BP-25 immunoreactivity were noted. A major fraction of 25-kDa and minor fractions of 22 and 19-kDa were purified by HPLC (HUHTALA et al. 1986). All three fractions reacted with antibodies to BP-25 and had the same amino-terminal amino acid sequence as intact BP-25. The 25-kDa fragment bound ^{125}I-IGF-I specifically in an enzyme-linked immunosorbent assay. These results suggest that the amino-terminal region of BP-25 contains the IGF binding domain.

3. BP-3A

WANG et al. (1988 b) purified an IGF binding protein fragment from BRL-3A conditioned media with an apparent molecular mass of 14-kDa. By amino acid sequence, it corresponds to residues 148–270 of BP-3A. Although the 14-kDa fragment does not react with antibodies to BP-3A, it binds IGF-I and IGF-II with similar specificity but lower affinity than the intact binding protein. HPLC and affinity crosslinking identify a binding species smaller than the native binding protein, thereby excluding trace contamination with intact BP-3A. Assuming that the amino acid sequence was determined for the same fragment whose binding properties were examined, these results suggest that the binding domain is located in the carboxy-terminal half of the molecule.

4. Identification of a Possible Binding Domain

The previous results suggest that glycosylation is not essential to the binding of IGF-I by BP-53, and that the amino-terminal fragments of BP-53 and BP-25 and the carboxy-terminal fragment of BP-3A appear to posses binding activity. These results might be explained if the binding domain were in the middle of the respective binding proteins. This is unlikely because the amino acid sequence of the middle region of the three binding proteins is least highly conserved. It seems more likely that the binding domain is at the cysteine-rich amino terminus, which is most highly conserved. For example, residues 31–43 of human BP-53 are identical in rat BP-3A. Expression of clones encoding truncated binding proteins should resolve this apparent contradiction.

VII. Binding Specificity of IGF Binding Proteins

All IGF binding proteins specifically bind IGF-I and IGF-II, but do not bind insulin (NISSLEY and RECHLER 1984a). The affinities of the purified binding proteins range from $7-9 \times 10^{10} \ M^{-1}$ for rat BP-53 (BAXTER and MARTIN 1987) to $1.5 \times 10^9 \ M^{-1}$ (WANG et al. 1988b). Based on competitive binding experiments using both IGF-I and IGF-II tracers, BINOUX et al. (1982) suggested that binding proteins from different sources (and even from the same source using different tracers) had different affinities for IGF-I and IGF-II. For example, using ^{125}I-IGF-I, rat liver and human fetal liver had preferential affinity for IGF-I, whereas using IGF-II tracer, adult human liver, human serum, and human cerebrospinal fluid had preferential affinity for IGF-II. (Using IGF-I tracer, the latter fluids had equal or greater affinity for IGF-I).

General support for this concept came from studies using binding proteins from fluids containing one predominant binding protein. For example, using IGF-II tracer, human cerebrospinal fluid (containing predominantly a 34-kDa binding protein) had a 40-fold higher affinity for IGF-II than for recombinant IGF-I (BINOUX et al. 1986); conditioned media from HepG2 hepatoma cells (containing predominantly BP-25) had equal affinity for IGF-I, and for human and rat IGF-II (MOSES et al. 1983). Adult rat serum contains unsaturated binding proteins that have preferential affinity for IGF-II (demonstrated using IGF-II tracer), whereas acid-stripped rat serum binds IGF-I and IGF-II with equal affinity (using either IGF-I or IGF-II as tracer) (YANG et al. 1989).

Attempts to address this question in biological fluids containing multiple binding proteins have been more difficult to interpret. Competitive binding experiments using binding proteins fractionated on SDS gel electrophoresis and electroblotted to nitrocellulose show apparent differences in binding specificity for the 41.5- and 38.5-kDa binding subunits of human serum (HARDOUIN et al. 1987). However, the 34-kDa serum protein shows a greater preferential affinity for IGF-II in competitive binding experiments after chromatofocusing than after SDS polyacrylamide gel electrophoresis and electroblotting, suggesting that exposure to SDS may alter the binding affinity. In fluids such as serum that contain multiple binding proteins with potentially different specificities, the observed binding specificity is a composite of that of the individual components.

The most reliable results have been obtained with purified binding proteins or binding protein subunits and are summarized in Table 8. Two general patterns have been observed: (1) binding proteins that have similar affinities for IGF-I and IGF-II when IGF-I or IGF-II tracers are used; and (2)

Table 8. Comparison of ligand binding specificities of IGF binding proteins[a]

	^{125}I-IGF-I	^{125}I-IGF-II
I. BP-53		
Human (A)[b]	I = II	I = II
Rat (B)	I = II	I = II
Pig (C)	I = II	II ≫ I
II. BP-25		
Human amniotic fluid (D)	I = II	I = II
HepG2 (E)[c]	NT[d]	I = II
III. BP-3A		
Rat BP-3A (F, G)	I = II	I = II
MDBK (H)	I = II	II ≫ I
IV. Unpurified		
Human CSF	I = II (I)	II ≫ I (J, K)
Adult rat serum:[e]		
Unsaturated	NT	II ≫ I
Acid-stripped	I = II	I = II

[a] Ligand binding specificities were determined in competitive binding studies using purified proteins unless otherwise noted. "I = II" indicates that the potencies of IGF-I and IGF-II differ by no more than 2-fold. "II ≫ I" designates binding proteins having > 10-fold higher affinity for IGF-II.
[b] *References:* A, MARTIN and BAXTER (1986); B, BAXTER and MARTIN (1987); C, WALTON et al. (1988); D, BAXTER et al. (1987); E, MOSES et al. (1983); F, WANG et al. (1988b); G, BROWN et al. (1989); H, SZABO et al. (1988); I, BINOUX et al. (1982); J, HOSENLOPP et al. (1986a); K, BINOUX et al. (1986).
[c] Conditioned media; IGF-II (purified from human plasma by R. HUMBEL) and MSA III-2 are equipotent.
[d] NT, not tested.
[e] YANG et al. (1989).

binding proteins that have preferential affinity for IGF-II when IGF-II tracer is used. In some cases, the two IGF tracers give different results with the same purified binding protein (e.g., porcine BP-53 and the bovine MDBK cell binding protein). This may represent an unresolved mixture of two proteins with different binding specificities, or overlapping (and mutually exclusive) binding sites for IGF-I and IGF-II tracer on a single binding protein. Proof of the latter hypothesis must await expression of a suitable cloned binding protein.

VIII. Occurrence and Distribution of IGF Binding Proteins

1. Biological Fluids

a) Human Serum

Using affinity crosslinking and SDS gel electrophoresis, WILKINS and D'ERCOLE (1985) identified multiple IGF binding protein complexes in human serum. A 35- to 43-kDa triplet was identified that was increased in hypopituitary patients, was associated with the 40-kDa peak on Sephadex G-200, and did not bind to concanavalin A. In addition, multiple GH-dependent binding proteins were identified that bound to concanavalin A. These included a 24- to 28-kDa doublet, and faint bands of 50, 80, 110, 135, and 160-kDa. The 24- to 28-kDa complexes most likely arise by proteolysis of BP-53. The higher molecular weight forms may arise by multiple crosslinking events (of IGF to BP-53 to the acid-labile subunit), with partial degradation of one or more components.

    HARDOUIN et al. (1987) used ligand blotting to study the different IGF binding proteins in human serum (Fig. 8). Major forms of 41.5- and 38.5-kDa arise from the 150-kDa complex; minor 34-, 30-, and 24-kDa proteins arise from the 40-kDa region. The 41.5- and 38.5-kDa proteins are increased in acromegaly and decreased in hypopituitarism; the 34- and 30-kDa proteins are increased in hypopituitarism. (Although the latter result was interpreted as inverse GH regulation, this has not been established since GH replacement has not been studied.) By analogy with results in rat serum (ZAPF et al. 1989; YANG et al. 1989), the 41.5- and 38.5-kDa proteins most likely differ only in their glycosylation. The 30-kDa and 34-kDa serum binding proteins may be identical to proteins of the same size in amniotic fluid and cerebrospinal fluid, respectively. The nature of the 24-kDa protein is not presently known.

    The different binding proteins in human serum have different binding specificities. After SDS gel electrophoresis and electroblotting, or chromatofocusing and acid gel filtration, the 38.5-kDa binding protein has 6–7 times higher affinity for IGF-II than IGF-I, whereas the 41.5-kDa protein has a slightly higher affinity for IGF-I. This implies that differences in glycosylation may alter the binding specificity. The 34-kDa binding protein also has a higher affinity for IGF-II than IGF-I.

Binding proteins similar to those seen in serum have been identified in culture media and other biological fluids, although in different proportions (see below). The same five proteins are present in media conditioned by adult and fetal human liver explants, in fetal human serum, in amniotic fluid (where the 30-kDa form predominates), and in cerebrospinal fluid (where the 34-kDa forms predominates). The fact that most fluids contain a mixture of binding components (including constituents of the 150-kDa complex) but do not form a 150-kDa complex suggests that synthesis of the acid-labile subunit is limited.

b) Rat Serum

Similar proteins have been identified in rat serum (Hossenlopp et al. 1987): a triplet of GH-dependent binding proteins (43-, 41-, and 39-kDa), and 32-, 29-, and 23-kDa forms. The 43/41/39-kDa proteins are present in the 150-kDa complex and are N-glycosylated (Zapf et al. 1989; Yang et al. 1989). The 32-kDa complex in rat serum may represent a glycosylated fragment of the 43/41/39-kDa binding proteins, with which it shares a common amino-terminal sequence. The 24-kDa and 29-kDa complexes are not glycosylated. Their relationship to the larger purified binding proteins, if any, is not known.

c) Amniotic Fluid

BP-25 is present in amniotic fluid at 100–500 times higher levels than in maternal or fetal serum (Rutanen et al. 1982; Drop et al. 1984a; Povoa et al. 1987). Peak values in amniotic fluid occur in midgestation (weeks 18–22) (Drop et al. 1984a). BP-25 also is present in menstrual fluid, suggesting that it is secreted to the uterine lining and amniotic fluid rather than the blood stream. Amniotic fluid also contains BP-53 in a 1:15 molar ratio relative to BP-25 (Baxter et al. 1987). By contrast, in plasma, BP-53 is 10–100 times more abundant than BP-25. Both BP-25 and BP-53 levels decrease with increasing fetal maturity. In contrast to plasma, in which BP-53 is saturated, neither BP-53 (100 nM) nor BP-25 (1500 nM) is saturated by IGF-I and IGF-II (10–15 nM) in amniotic fluid (Baxter et al. 1987; Merimee et al. 1984).

d) Cerebrospinal Fluid

Using the ligand blot technique, Hossenlopp et al. (1986a) demonstrated that a 34-kDa binding protein is the predominant species in human cerebrospinal fluid, although other IGF binding proteins are evident when tested a higher concentration. It chromatofocuses at pH 5.0, and has a 40-fold higher affinity for IGF-II than for recombinant IGF-I (Binoux et al. 1986; Hossenlopp et al. 1986a)[2]. A 34-kDa binding protein that focuses at pH 5.0 and binds IGF-II

---

[2] Cerebrospinal fluid appears to contain two 34-kDA binding proteins: one that cross-reacts with antibodies to rat BP-3A (J. A. Romanus, unpublished results), and one whose amino acid sequence is unrelated to BP-3A (M. Binoux, personal communication).

with high affinity also is present in the small complex of human serum (HARDOUIN et al. 1987), but its immunologic relationship, if any, to the protein in cerebrospinal fluid has not been established.

e) Lymph

Lymph contains about 10% as much binding protein as serum, all of it 40 kDa by neutral gel filtration (BINOUX and HOSSENLOPP 1988). The same five binding components present in human serum were identified by ligand blotting, although lymph contained relatively less of the 41.5- and 38.5-kDa forms. After acid gel filtration, IGF-II tracer bound to a site with an 8-fold higher affinity for IGF-II than IGF-I, similar to the specificity of the small complex of human serum. The authors suggest that a barrier may prevent the exit of the 150-kDa species from the vascular compartment, and propose that the 150-kDa complex is a circulating reservoir of IGFs.

f) Milk

BAXTER et al. (1984) provided indirect evidence for the presence of 150-kDa IGF-binding protein complexes in human milk. Neutral gel filtration of fresh 1-day postpartum milk contained similar amounts of IGF-I immunoreactivity (after acid ethanol extraction) in the 150-kDa peak as in the 40-kDa peak. SIMMEN et al. (1988) observed immunoreactive IGF-I in 150-kDa complexes in porcine milk and colostrum. Thus, milk appears to be the only biological fluid other than serum in which the 150-kDa complex has been observed.

g) Follicular Fluid

Follicular fluid from women hyperstimulated for in vitro fertilization was reactive in the radioimmunoassay for BP-25 (SEPPALA et al. 1984). WALTON and ETHERTON (1989) detected BP-53 immunoreactivity in pig follicular fluid.

h) Free IGF in Saliva and Other Fluids

COSTIGAN et al. (1988) reported IGF-I at about 2 ng/ml in human saliva, without any evidence for the existence of binding protein. They note that corticosteroids also exist in free form in saliva, and that this is the basis for a clinically useful assay. Significant components of free IGF have been reported in urine (30%; HIZUKA et al. 1987b), milk ($\sim$20%; BAXTER et al. 1984), and human serum (19%; GULER et al. 1987).

2. Tissues

a) Liver

SCHWANDER et al. (1983) demonstrated that perfused rat liver synthesized 35-kDa IGF binding proteins. Levels of binding protein, estimated after acid gel filtration and affinity crosslinking, were not affected by the presence of GH in the perfusate.

SCOTT et al. (1985a, b) examined IGF binding protein production by primary rat hepatocytes. Hepatocyte binding proteins differed subtly in isoelectric point and binding specificity from rat serum binding proteins (SCOTT et al. 1985a). Binding proteins were decreased by hypophysectomy, and restored by in vivo GH replacement, but not by in vitro GH or insulin (SCOTT et al. 1985b). In vitro GH modestly increased binding protein synthesis in hepatocytes from normal rats. Which binding protein species were affected was not determined.

HOSSENLOPP et al. (1987) characterized IGF binding proteins in explants from fetal and adult human and rat liver. Ligand blotting identified the same protein species observed in serum: 41.5-, 38.5-, 34, 30, and 23-kDa (human); 43, 41, 39, 32, 29, 24-kDa (rat). The higher molecular weight forms (41.5 and 38.5-kDa in human liver; 43, 41, and 39-kDa in rat liver) were relatively decreased compared with serum. Despite the presence of all binding components, no 150-kDa complexes were observed.

b) Secretory Endometrium and Decidua

Placental protein 12 (PP12), originally purified from human placenta and adjacent membranes and thought to be pregnancy and placenta specific, is identical with BP-25. Biosynthetic labeling of endometrial explants demonstrated that PP12 is synthesized by maternal endometrium (decidua) but not by trophoblast (RUTANEN et al. 1985). PP12 also is synthesized at higher levels by explants of secretory endometrium than of proliferative endometrium (RUTANEN et al. 1986). Incubation with progesterone in vitro increases BP-25 synthesis (RUTANEN et al. 1986).

c) Pituitary Gland and Brain Explants

Explants of anterior pituitary, neurointermediate lobe, hypothalamus, cerebral cortex, and cerebellum synthesize and secrete IGF binding proteins (BINOUX et al. 1981). After incubation with IGF tracer and neutral gel filtration, tracer bound to the 40-kDa region, although the profile was more heterogeneous than observed with media from liver explants. Which binding components are present is unknown.

3. Cells

a) Human Fibroblasts

CLEMMONS et al. (1981b) observed IGF-I immunoreactivity in a 150-kDa fraction in human fibroblast conditioned media, suggesting that these cells synthesized high molecular weight IGF binding proteins as well as IGF-I. ADAMS et al. (1984) directly demonstrated IGF binding activity in human fibroblast conditioned medium. Binding activity was present in the void volume of a neutral G-150 column, and a 67-kDa crosslinked complex was observed after SDS gel electrophoresis (reducing conditions). The binding proteins were synthesized by fibroblasts rather than stored, since cycloheximide decreased binding activity in the medium. CLEMMONS et al. (1986) demonstrated 43-, 36-, and 25-kDa IGF-binding protein complexes in fibroblast conditioned media by affinity crosslinking and autoradiography; the 43-kDa complex also was observed on the cell surface. Antibody to purified BP-25 (CLEMMONS et al. 1987) immunoprecipitated 50% of the complexes formed by incubation of ^{125}I-IGF with human fibroblast conditioned media. Although the size of the immunoprecipitated complex was not examined, it seems likely that it was 43-kDa. BUSBY et al. (1988a) identified a 43-kDa complex on porcine smooth muscle cells after they had been incubated with one form of purified BP-25 (DEAE peak B).

MARTIN and BAXTER (1988) recently reevaluated this question using antibodies to BP-53. Conditioned medium from neonatal skin fibroblasts formed crosslinked complexes of 60-, 42-, and 37-kDa (reducing conditions). The 60- and 42-kDa complexes were immunoprecipitated by antibody to BP-53 but not by antibody to BP-25. This identification of the 42-kDa complex as a fragment of BP-53 differs from the results of CLEMMONS et al. (1987) and BUSBY et al. (1988a) which suggested that it is a complex of BP-25. The basis for this discrepancy is unclear. Finally, the 37-kDa complex was not precipitated by antibody to BP-53 or BP-25, suggesting that it might represent a distinct species (MARTIN and BAXTER 1988).

Using the radioimmunoassay for BP-53, MARTIN and BAXTER (1988) examined the regulation of BP-53 by serum and EGF. Incubation with calf serum increased immunoreactive material 8-fold relative to serum-free media. Interestingly, this was 150-kDa material, suggesting that serum also induced the acid-labile subunit. EGF alone gave a 2- to 3-fold increase in BP-53 but did not generate 150-kDa material.

b) Endothelial Cells

BAR et al. (1987) characterized IGF binding proteins synthesized by cultured endothelial cells from different species. Human umbilical vein endothelial cells (primary cultures) synthesize proteins in serum-containing or serum-free medium that crossreact with antibodies to BP-53. The binding proteins are smaller than those in serum (major peak 32–40-kDa, minor peak of 22–28-kDa). Neither serum, insulin, nor growth hormone regulates binding protein production.

Similar sized binding proteins are produced by bovine endothelial cell lines (microvessel and macrovessel) and by rat epididymal fat, although it was not possible to establish an immunologic relationship to BP-53 because of the species specificity of the antiserum. Bovine endothelial cell binding proteins had slightly higher affinity for IGF-I than IGF-II (especially with IGF-II tracer). They had a moderate affinity for heparin, a lower affinity for heparan sulfate, and no affinity for other glycosaminoglycans.

c) Cells Preferentially Synthesizing one Binding Protein Species

α) BRL-3A

The predominant binding protein in BRL-3A cells, BP-3A, has been purified and cloned. It is a 30-kDa, nonglycosylated binding protein that binds IGF-I and IGF-II with equal affinity. It is the major binding protein in fetal and neonatal rat serum.

β) MDBK Cells

The MDBK bovine kidney cell line (Szabo et al. 1988) synthesizes a 40-kDa binding protein (reducing conditions) that appears to be the bovine homolog of rat BP-3A. Partial amino acid sequence analysis indicates that 38/44 residues at the amino terminus and 21/22 residues at the carboxy terminus are identical to BP-3A (Fig. 9). IGF-I and IGF-II tracers appear to bind to different sites on the MDBK binding protein: IGF-I tracer binds to a site with equal affinity for IGF-I and IGF-II, whereas IGF-II tracer binds to a site that has preferential affinity of IGF-II. Whether these represent overlapping sites on the same binding protein molecule or discrete sites on separate molecules is not known.

γ) HepG2 cells

HepG2 human hepatocarcinoma cells synthesize BP-25 that is identical to the human amniotic fluid and endometrial binding proteins in amino-terminal amino acid sequence (Povoa et al. 1985) and in nucleotide sequence (Lee et al. 1988). IGF-I and IGF-II inhibit IGF-II tracer binding with equal affinity (Moses et al. 1983), as reported by Baxter et al. (1987) for BP-25 purified from amniotic fluid.

4. Clinical States

a) Hypopituitarism

Serum from GH-deficient humans and rats lacks 150-kDa IGF-binding protein complexes (Moses et al. 1976; White et al. 1981). Hintz et al. (1981) reported a 156% increase in binding of IGF-I tracer to the unsaturated 40-kDa serum binding proteins of 21 GH-deficient children. The binding increase appeared to result from a 3-fold increase in binding affinity for IGF-I. Similarly, Wilkins and D'Ercole (1985) observed an increase in unsaturated 35- to 40-kDa binding protein complexes by affinity crosslinking to serum from hypopituitary patients. Hardouin et al. (1987), using ligand blot techniques, demonstrated that the 41.5- and 38.5-kDa GH-dependent binding components were decreased in serum from hypopituitary patients, whereas the 34- and 30-kDa GH-independent components were increased. Povoa et al. (1987) reported a 2-fold increase in immunoreactive BP-25 in GH-deficient patients. Despite this increase, BP-25 constitutes only about 10% of the total IGF binding activity in GH-deficient patients (Povoa et al. 1987; Hardouin et al. 1987). Baxter and Cowell (1987), however, did not observe a change in BP-25 levels with GH deficiency.

b) Diabetes Mellitus

Immunoreactive BP-53 is decreased 40% in serum from patients with poorly controlled non-insulin-dependent diabetes mellitus (BAXTER and MARTIN 1986). Immunoreactive BP-25 is increased 2.5- to 4-fold in insulin-dependent diabetes mellitus (SUIKKARI et al. 1988).

In streptozotocin-diabetic rats, presumptive BP-53 proteins are decreased and rat BP-25 1S increased (UNTERMAN et al. 1988 and unpublished results). Synthesis of total IGF binding proteins was decreased in hepatocytes cultured from streptozotocin-diabetic rats (SCOTT and BAXTER 1986).

IX. Role of IGF Binding Proteins

1. In Vivo

Virtually all of the IGFs in blood and in tissue fluids (with the possible exception of saliva) exist complexed to IGF binding proteins. Although the IGFs are present in blood at high levels, and prolonged infusion of recombinant IGF-I partially replaces GH in hypophysectomized rats (GULER et al. 1988), the physiological significance of circulating IGFs has not been established. Conceivably, the major actions of IGFs may be local (paracrine and autocrine) rather than endocrine.

The main carrier of IGFs in the vascular compartment is the 150-kDa binding protein complex. In the absence of the 150-kDa complex (for example, in GH deficiency), the half-life of IGF in the blood is greatly shortened (NISSLEY and RECHLER 1984a; ZAPF et al. 1986). It has been suggested that the 150-kDa complex represents a storage form of IGFs that cannot leave the bloodstream. If circulating IGFs are biologically functional, a mechanism is required to allow the IGFs or the IGF-binding protein complex to leave the vascular space. Endothelial cells possess receptors for IGFs, but it is doubtful that the IGF-binding protein complex would bind to the receptors (see below). Since there is little evidence for the existence of the 150-kDa complex outside the blood, possibly IGFs or IGF-BP-53 complexes are released from the 150-kDa complex by an unknown mechanism.

Low molecular weight IGF binding proteins are present in plasma and are the predominant IGF binding proteins in extracellular fluid. Their smaller size allows them to exit from the vascular compartment. The short half-life of BP-25 after insulin administration (HALL et al. 1987; SUIKKARI et al. 1988) has suggested that BP-25 may serve to transport IGFs from the 150-kDa complex to cells.

Most available in vitro evidence suggests that IGF-binding protein complexes are not recognized by IGF receptors and hence are biologically inert (see below). If this is the case, then a mechanism would be required for target tissues to extract active IGFs from the complex. For example, proteolysis (CHATELAIN et al. 1983) or proteolysis plus heparin (CLEMMONS et al. 1983) increase the accessibility of IGFs in a binding protein complex to antibodies without dissociating the complex. The weak affinity of the endothelial cell binding protein (presumably a fragment of BP-53) for heparin is particularly interesting in this regard (BAR et al. 1987). Alternatively, as discussed below, IGF-binding protein complexes may interact with the cell surface, possibly through an RGD adhesion sequence that is recognized by extracellular matrix receptors, or through some as yet unidentified recognition feature (e.g., fatty acid adduct). BUSBY et al. (1988a) have provided evidence for high affinity ($ED_{50} \sim 1$ nM), specific and saturable binding of a purified binding protein (BP-25 DEAE peak B) to the cell surface.

In the tissues, the low molecular weight IGF binding proteins are pivotally placed to be important modulators of IGF action. The GH-independent binding proteins and the binding component of the 150-kDa complex (BP-53) that is not complexed to the acid-labile subunit may serve this function. These binding proteins are widely produced, typically in excess relative to the IGFs. They may exert stimulatory or inhibitory action (see below).

2. In Vitro

a) Inhibition of Receptor-Binding and Biological Activity

In the screening of acid gel-filtered conditioned media for IGF production by radioimmunoassay or radioreceptor assay (e.g., RECHLER et al. 1979; ADAMS et al. 1983), inhibition of binding of radiolabeled IGFs frequently was observed at the position of IGF binding proteins as well as that of the free IGFs. This suggested that ^{125}I-IGF-binding protein complexes did not bind to IGF receptors and antibodies.

A second observation indicating that IGFs complexed to binding proteins did not interact with insulin receptors was the fact that, although human plasma contained high levels of IGF-I and IGF-II, people were not hypoglycemic. When the binding capacity of the IGF binding proteins is exceeded by a rapid intravenous infusion of IGFs, a transient increase in free IGF occurs that results in hypoglycemia (ZAPF et al. 1986; GULER et al. 1987).

Inhibition of IGF biological activity was demonstrated in in vitro studies using partially purified binding protein preparations. Acid-stripped binding proteins from human serum inhibited IGF-stimulated but not insulin-stimulated lipogenesis and 3-O-methylglucose transport in isolated rat adipocytes, two effects that are mediated via the insulin receptor (ZAPF et al. 1979). (IGF binding proteins bind IGFs but do not bind insulin.) Partially purified amniotic fluid binding protein inhibited sulfate incorporation into proteoglycans in rabbit articular cartilage, an effect mediated by an IGF receptor (DROP et al. 1979). KNAUER and SMITH (1980) then showed that purified binding protein from BRL-3A cells (BP-3A) inhibited the stimulation of DNA synthesis in chick embryo fibroblasts through the IGF-I receptor, as well as the binding of radiolabeled IGF to this receptor. By contrast, the mitogenic effects of insulin, which also acts through the IGF-I receptor, were unaffected by the presence

Table 9. Inhibition of IGF binding and action by purified IGF binding proteins

Binding protein	Assay	References[a]
BP-3A	Thymidine incorporation (chick embryo fibroblasts)[b]	A
	Binding to IGF-I receptor (chick embryo fibroblasts)	A
	Binding to IGF-II receptor (H35 cells, mouse liver membranes)	B
	Binding to IGF-I receptor (rat astroglia cells)	C
	Thymidine incorporation (rat astroglia cells)	C
H35	Binding to IGF-II receptor (H35 cells, mouse liver membranes)	B
	Anchorage independent proliferation (NRK cells)[c]	D
BP-25	Binding to IGF-I receptor (human secretory endometrium)	E
	Binding to IGF-I receptor (JEG3 choriocarcinoma)[d]	F
	AIB uptake (JEG3 choriocarcinoma)[e]	F
	IGF-I stimulated thymidine incorporation (multiple cells)[f]	G, H
BP-53	Thymidine incorporation (human fibroblasts)[g]	I
	Lipogenesis (rat epididymal adipocytes)[h]	J

[a] *References:* A, KNAUER and SMITH (1980); B, MOTTOLA et al. (1986); C, HAN et al. (1988a); D, MASSAGUE et al. (1985); E, RUTANEN et al. (1988); F, RITVOS et al. (1988); G, ELGIN et al. (1987); H, BUSBY et al. (1988a); I, DE MELLOW and BAXTER (1988); J, OOI and HERINGTON (1986).

[b] Stimulation by insulin is not affected.

[c] Assay was performed in the presence of transforming growth factor-β, EGF, and 10% calf serum. Binding protein presumably inhibits stimulation by IGF in calf serum. Inhibition was overcome by IGF-II.

[d] Binding protein complex is in the supernate.

[e] AIB, α-aminoisobutyric acid. Basal and stimulated uptake are inhibited.

[f] DEAE peak C of BP-25 purified from amniotic fluid. Coincubation with IGF-I.

[g] Inhibition requires coincubation with IGF-I. By contrast, stimulation is observed if BP-53 is preincubated with fibroblasts for 8–24 h before the addition of IGF-I.

[h] 16-kDa fragment of BP-53 purified from human serum. Mediated by insulin receptors.

of binding proteins. The authors concluded that the IGF binding protein complex was biologically inactive because it did not bind to receptor.

Subsequently, inhibitory effects of purified IGF binding proteins on different biological actions of IGFs have been described (Tabel 9). Purified BP-3A, BP-25, BP-53, a fragment of BP-53, and the H35 binding protein have been used. The activities inhibited include stimulation of DNA synthesis, anchorage-independent proliferation, α-aminoisobutyric acid uptake, and lipogenesis. The effects inhibited include those mediated by IGF receptors as well as insulin receptors. Binding to IGF-I receptors and IGF-II/Man-6-P receptors was inhibited. In two cases, subtle differences in binding protein (ELGIN et al. 1987; BUSBY et al. 1988a) or assay conditions (DE MELLOW and BAXTER 1988) changed the observed biological effect from inhibition to stimulation (see below).

b) Anomalous Effects of IGF Binding Proteins
on IGF-I Binding to Receptors

Anomalous binding of IGF-I tracer to monolayer cultures of human
fibroblasts was associated with the formation of IGF-I-binding protein complexes in the binding incubation medium (DE VROEDE et al. 1986) or on the cell
surface (CLEMMONS et al. 1986). Unlike IGF-I tracer binding to the IGF-I
receptor of suspended fibroblasts, when binding was performed to cell
monolayers, a paradoxical increase of IGF-I tracer binding was observed at
low concentrations of added IGF-I; inhibition required higher concentrations
of IGF-I, and no inhibition was seen with insulin or the antireceptor antibody
αIR-3. Using affinity crosslinking techniques, CLEMMONS et al. (1986)
identified a 43-kDa IGF-I-binding protein complex on the cell surface after
IGF-I binding to cell monolayers but not to suspended cells.

The 43-kDa binding protein complex and the altered IGF-I binding
properties could be transferred from human fibroblast conditioned media to
secondary cultures of porcine smooth muscle cells (PSMC), cells which exhibit
normal IGF-I tracer binding to the IGF-I receptors on cell monolayers
although they synthesize a 35-kDa binding protein. Incubation with PSMC
media did not affect IGF-I binding. Although human fibroblast conditioned
media contain 36- and 25-kDa complexes as well as the 43-kDa complex, only
the 43-kDa complex was transferred. Why binding proteins in human
fibroblast conditioned media but not from PSMC conditioned media adhere
to the PSMC surface is unclear.

Cycloheximide treatment inhibited synthesis of the 35-kDa binding
protein and depleted the crosslinked 43-kDa complex from the cell surface
(CLEMMONS et al. 1987). IGF-I binding to cycloheximide-treated cells was
normal. Following removal of the cycloheximide, paradoxical binding slowly
returns, provided serum was present in the incubation medium.

Thus, anomalous binding of IGF-I tracer to fibroblast monolayers seems
associated with the presence of a 43-kDa binding protein complex, suggesting
that binding proteins are capable of modulating IGF-I binding to receptors.
There is conflicting evidence as to whether the 43-kDa complex is a fragment
of BP-53 (DE MELLOW and BAXTER 1988), since it reacts with antibodies to
BP-53, or whether it is a form of BP-25 (BUSBY et al. 1988a), since incubation
of PSMC with DEAE peak B of BP-25 increased IGF-I tracer binding and
generated a 43-kDa complex on the cell surface. In addition, radiolabeled BP-
25 (peak B) binds to saturable sites on PSMC. A physicochemically indistinguishable second DEAE peak (peak C) did not increase IGF-I tracer
binding or adhere to PSMC (BUSBY et al. 1988a).

c) Stimulation of Mitogenesis

DEAE peak B BP-25 from amniotic fluid acted synergistically with IGF-I to
stimulate thymidine incorporation into DNA in porcine smooth muscle, chick
embryo fibroblasts, mouse embryo fibroblasts, and human skin fibroblasts
(ELGIN et al. 1987; BUSBY et al. 1988a). Cells were grown in serum-containing

medium, switched to serum-free medium (or not switched), and then incubated with purified binding protein and peptides in the presence of 1% platelet-poor plasma for 36 h. Coincubation with binding protein and IGF-I increased DNA synthesis approximately 4-fold over IGF-I alone. Stimulation was detectable at 2 ng/ml of binding protein, and maximal at 100 ng/ml. DEAE peak B had no effect on basal or insulin-stimulated DNA synthesis.

By contrast, a closely related second component of BP-25, DEAE peak C (see above), inhibited basal thymidine incorporation and the stimulation of DNA synthesis by IGF-I plus DEAE peak B. The authors suggest that the synergistic effects of peak B result from its adherence to the cell surface and its ability to increase cell-associated IGF-I, although how it enhances IGF-I stimulated DNA synthesis is not known. The molecular basis for the adherence of the peak B binding protein is unknown. It has been suggested that it might involve an RGD attachment sequence (BREWER et al. 1988a) or a fatty acid adduct (BUSBY et al. 1988b). Finally, the difference between the peak B and peak C proteins resulting in opposite cellular adhesion and biological properties is unknown.

A second example of possible binding protein synergism with IGF-I recently has been reported by DE MELLOW and BAXTER (1988). Purified BP-53, preincubated with human fibroblasts for 8–24 h, caused a 2-fold stimulation of DNA synthesis. Basal DNA synthesis was not affected, although the specificity of the effect for IGF-I stimulated mitogenesis was not established. Interestingly, coincubation of BP-53 and IGF-I inhibited IGF-I-stimulated DNA synthesis in the same cells. The ED_{50} for inhibition and stimulation of mitogenesis were similar. Since BP-53 does not contain an RGD sequence, adherence to a cell integrin receptor is unlikely to be involved in its stimulatory action. Thus, BP-25 and BP-53 may exert their synergistic effects with IGF-I on DNA synthesis via different mechanisms.

X. Sites on IGF-I Recognized by IGF Binding Proteins

1. Two-Chain Insulin-IGF-I Hybrid Molecules

DE VROEDE et al. (1985) noted that a chemically synthesized two-chain hybrid insulin molecule in which the B domain of IGF-I was substituted for the B chain of insulin and combined with the A chain of insulin was recognized by IGF binding proteins, in contrast to insulin itself which was not. This suggested that at least some determinants recognized by the IGF binding proteins are located within the B-domain of IGF-I. The B_{IGF-I} hybrid molecule had 30 times lower potency than IGF-I, suggesting that either the single-chain structure or other molecular determinants contributed importantly to recognition by the binding proteins.

2. Truncated Destri-IGF-I

Truncated IGF-I molecules lacking the amino-terminal three amino acids (destri-IGF-I) were isolated from bovine colostrum (FRANCIS et al. 1988) and from fetal and adult human brain (SARA et al. 1986; CARLSSON-SKWIRUT et al. 1986). Destri-IGF-I purified from bovine colostrum is more potent than IGF-I in stimulating DNA and protein synthesis and in inhibiting protein breakdown in L6 myoblasts (BALLARD et al. 1987). The increased potency was confirmed using chemically synthesized destri-IGF-I and the protein synthesis assay. The increased potency did not result from increased binding of synthetic destri-IGF-I to either type I or type II IGF receptors in L6 myoblasts (BALLARD et al. 1987). [These results differ from the report of SARA et al. (1986) that destri-IGF-I isolated from human fetal brain was 5 times more potent than IGF-I in the human fetal brain radioreceptor assay.]

An alternative mechanism to account for the increased biological potency of destri-IGF-I was suggested by studies of the interaction of destri-IGF-I with binding proteins purified from the bovine MDBK cell line (SZABO et al. 1988). ^{125}I-labeled destri-IGF-I bound poorly, and unlabeled destri-IGF-I at high concentrations failed to inhibit the binding of IGF-I or IGF-II tracer to the MDBK binding protein. If L6 cell binding proteins also do not bind destri-IGF-I (and presumably form an inactive complex), destri-IGF-I might have increased bioavailability and greater biological activity than native IGF-I.

3. Site-Directed Mutagenesis

BAYNE et al. (1988) generated single-chain variant IGF-I molecules that had greatly decreased recognition by IGF binding proteins from human serum by site-directed mutagenesis. An IGF-I mutant in which residues B1–17 of insulin replace residues B1–16 of IGF-I no longer binds to acid-stripped human serum. Substitution of IGF-I residues B3, B4, B15, and B16 with the corresponding residues of insulin reduces the binding affinity 600-fold. The B1–16 mutant and the B3, 4, 15, 16 mutant had a similar low potency with native rat serum (CASCIERI et al. 1988b). By contrast, substitution of residues B3, 4 or B15, 16 caused only a 4-fold decrease in binding. The B1–16 mutant and the B3, 4, 15, 16 mutant bind normally to the IGF-I receptor and give normal stimulation of DNA synthesis in A10 rat smooth muscle cells.

Mutants B1–16 and B3, 4, 15, 16 are cleared from the rat circulation 4 times faster than IGF-I (CASCIERI et al. 1988b). Despite this, following intravenous injection, they are 4-fold and 2-fold more potent than IGF-I in stimulating glucose incorporation into glycogen in rat diaphragm via the IGF-I receptor (CASCIERI et al. 1988b). These results are consistent with the interpretation that IGF binding proteins normally inhibit the ability of IGF-I to exert its biological actions via the IGF-I receptor. Although the mutant molecules are cleared more rapidly from the circulation, they reach the tissues in a noncomplexed (and hence more active) form, so that the net effect is stimulatory.

E. Biological Role and Clinical Implications

I. Diverse Biological Roles

Like other polypeptide growth factors, the IGFs appear to have diverse biological roles. As mitogenic agents, they participate in physiological growth in the developing child, the fetus, and the embryo. They may participate in the physiological hypertrophy of specific organs (for example, estrogen stimulation of uterine IGF-I gene expression may contribute to estrogen-stimulated uterine proliferation). The IGFs may participate in wound repair: immunoreactive IGF-I is increased in injured nerve (HANSSON et al. 1986), muscle (JENNISCHE et al. 1987; NORSTEDT et al. 1987), and endothelium (HANSSON et al. 1987). Local infusion of antibodies to IGF-I inhibited regeneration of the sciatic nerve following crush injury, and administration of IGF-I accelerated nerve regeneration (KANJE et al. 1989). The IGFs are increased in, and may contribute to, compensatory hypertrophy of some organs, for example following unilateral nephrectomy (STILES et al. 1985; FAGIN and MELMED 1987), although they do not appear to be involved in liver regeneration following partial hepatectomy (NORSTEDT et al. 1988; UENO et al. 1988b).

The IGFs also may contribute to neoplastic cell proliferation (Table 4). Elevated levels of IGF-II and less commonly IGF-I have been observed in different tumors. IGF overexpression does not appear to act as a transforming oncogene, that is, to trigger formation of a tumor. Nonetheless, it may contribute to proliferation of a tumor. Blockade of IGF-I receptors (which mediate the mitogenic effects of both IGF-I and IGF-II) with antireceptor antibodies inhibits autonomous cell growth in a small cell lung carcinoma cell line that produces IGF-I (NAKANISHI et al. 1988) and a neuroblastoma cell line that produces IGF-II (EL-BADRY et al. 1989).

The biological actions of IGFs are not limited to growth. The IGFs have been shown to promote the chemotaxis of endothelial cells (GRANT et al. 1987a) and melanoma cells (STRACKE et al. 1988). They promote the differentiation of, and the expression of differentiated functions in, an increasing number of cell types. Examples include: chondroblasts, myoblasts, and osteoblasts (reviewed in FROESCH et al. 1985), neuroblasts (RECIO-PINTO and ISHII 1984; RECIO-PINTO et al. 1986), the lens, hematopoietic cells, and the gonads. In the lens, lentropin, a protein factor in vitreous humor that promotes the elongation of lens fibers, has been shown to be immunologically related to IGFs (BEEBE et al. 1987). IGFs act synergistically with erythropoietin to stimulate erythropoiesis (KURTZ et al. 1982; DAINIAK and KRECZKO 1985; Claustres et al. 1987), and with granulocyte-monocyte colony stimulating factor to stimulate granulopoiesis and granulocyte maturation (MERCHAV et al. 1988). In the gonads, the IGFs act synergistically with the appropriate gonadotropic hormones (follicle-stimulating hormone, luteinizing hormone) to promote steroidogenesis in granulosa cells (ADASHI et al. 1985), Leydig cells (MORERA et al. 1987), and ovarian theca-interstitial cells (HERNANDEZ et al. 1988). The effects of IGFs on differentiation have been shown to occur in the absence of any growth-promoting effects in myoblasts (TURO and FLORINI 1982) and rat granulosa cells (ADASHI et al. 1985).

II. In Vivo Actions: IGF-I

Despite their chemical similarities and similar in vitro activities, IGF-I and IGF-II have distinct physiological regulation and in vivo activities. As originally proposed by the somatomedin hypothesis (SALMON and DAUGHADAY 1957), IGF-I appears, at least in part, to mediate the action of GH on skeletal cartilage. According to the somatomedin hypothesis, the effects of GH on cartilage are mediated by substances that became known as somatomedins or insulin-like growth factors (DAUGHADAY et al. 1987a) which circulate in the blood and act on chondrocytes at the epiphysial growth plates of long bones. Support for this hypothesis came from the demonstration by SCHOENLE et al. (1982) that infusion of purified IGF-I into hypophysec- tomized rats for 6 days using an osmotic minipump induced weight gain and increased tibial epiphyseal width, classic effects of GH. More quantitative analysis of these results (ZAPF et al. 1985) and results using recombinant IGF- I preparations (SKOTTNER et al. 1987; GULER et al. 1988), however, demon- strated that IGF-I was considerably less effective than GH when compared at doses giving comparable concentrations of plasma IGF-I. IGF-I and GH also had different effects on the weight of different organs (GULER et al. 1988). Thus, IGF-I only partially substituted for GH.

Resolution of this paradox came from the important experiments of ZEZULAK and GREEN (1986) demonstrating that GH exerted dual effects in an adipogenic fibroblast cell line (3T3-442A): (a) GH acted directly to promote the differentiation of stem cells to preadipocytes which were more responsive to IGF-I, and (b) GH stimulated the synthesis of IGF-I by these cells, result- ing in a net clonal expansion. ISAKSSON et al. (1987) has provided considerable evidence that this model applies to the effects of GH on cartilage: (a) Local in- fusion of GH (into the epiphyseal cartilage or femoral artery) caused widening of the tibial epiphysial cartilage (and new bone deposition, measured by tetracyline staining) on the same but not the opposite side, suggesting that the infused GH acted locally. (b) GH stimulated the synthesis of immunoreactive IGF-I in the proliferative zone of the tibial growth plate (NILSSON et al. 1986). (c) Early proliferative chondrocytes possess GH receptors (BARNARD et al. 1988) and are stimulated to form large colonies by GH (LINDAHL et al. 1987). Thus, GH may promote longitudinal bone growth by acting directly to stimu- late the differentiation of stem cells, and indirectly to stimulate the synthesis of IGF-I. The differentiation/clonal expansion hypothesis of GH action is direct- ly analogous to that proposed for hematopoietic and immune cell lineages.

The biological actions of IGF-I are not limited to promoting skeletal growth. As stated above, IGF-I may participate in the response to tissue in- jury in nerve, muscle, and endothelial cells. It acts synergistically with erythropoietin and granulocyte-monocyte colony stimulating factor to stimu- late erythropoiesis and granulopoiesis. And, IGF-I interacts with other hormones such as the gonadotropins.

III. In Vivo Actions: IGF-II

The in vivo biological role of IGF-II is less well understood. IGF-II levels are less regulated by GH, and IGF-II is considerably less potent than IGF-I in promoting skeletal growth (Schoenle et al. 1982). [The intriguing observation of Stylianopoulou et al. (1988a) that IGF-II mRNA is expressed at high levels in undifferentiated chondroblasts but not in differentiated chondrocytes raises the possibility that IGF-II may play a role in the expansion and differentiation of chondroblasts.] Leading possibilities include a role in fetal/embryonic development, in the central nervous system, and in bone. These possibilities will be briefly considered.

The suggestion that IGF-II has a role in the fetus and embryo is primarily based on the preferential expression of IGF-II mRNA early in development. IGF receptors are present in fetal and embryonic cells, and these cells respond to IGFs in vitro. To date, however, no specific functions in early development have been linked to IGF-II.

Likewise, the suggestion of a central nervous system role for IGF-II was suggested by the predominance of IGF-II in brain and cerebrospinal fluid (Haselbacher and Humbel 1982; Haselbacher et al. 1985), and of IGF-II mRNA in choroid plexus (Stylianopoulou et al. 1988b; Hynes et al. 1988). IGF-I receptors (the presumed mediator of IGF-II action) are widely distributed in brain (Bohannon et al. 1988; Lesniak et al. 1988). High receptor concentrations are found in choroid plexus (like insulin receptors), the median eminence (unlike insulin receptors), cerebral cortex, thalamus, and the limbic system.

Several in vivo observations suggest that brain IGF-II may function both as a mitogen and in regulating gastrointestinal function. Schoenle et al. (1986) reported an infant with marked (+ 10 standard deviation) macrocephaly, principally in the cerebrum, associated with a 5-fold increase in extractable immunoreactive IGF-II from the frontal brain cortex, suggesting that IGF-II may function as a brain growth factor. Tannenbaum et al. (1983) reported that central administration of a partially purified IGF preparation (enriched in IGF-II) reduced food intake by two-thirds and decreased body weight by 8% in 24 h. Lauterio et al. (1987) confirmed this observation using purified IGF-II, although the magnitude of the effects were somewhat smaller (a one-third reduction in food intake, and a 4% decrease in weight). Finally, Mulholland and Debas (1988) reported that central administration of IGF-II rapidly and reversibly decreased pentagastrin-stimulated gastric acid secretion. This effect was blocked by vagotomy, indicating that it is neurally mediated.

A third possible role for IGF-II has been suggested by the demonstration that skeletal growth factor purified from adult human bone matrix is identical to IGF-II (Mohan et al. 1988; Frolik et al. 1988). IGF-II is 10 times as abundant in human bone matrix than IGF-I. Skeletal growth factor/IGF-II is synthesized by osteoblasts, and stimulates the proliferation of preosteoblasts (reviewed in Mohan et al. 1988). It is released during bone resorption, and requires demineralization to be released from bone during purification. Skeletal growth factor has been proposed to couple new bone formation with bone resorption, maintaining bone volume during remodeling.

IV. Conclusion

The past few years have witnessed an explosion of interest in and understanding of the IGFs. Sophisticated molecular tools are now available to study the components of the IGF system. This promises to unlock the remaining mysteries of how the IGF system interacts with other hormones and growth factors, and to more completely delineate its role in physiology and disease.

Acknowledgments. We wish to thank Drs. R. Baxter, W. Wood, T. Etherton, and A. Skottner for making results available prior to publication; Dr. C. C. Orlowski, A. L. Brown, D. E. Graham, Y. W-H. Yang, and R. Thotakura for helpful discussions; and G. Dickstein for secretarial assistance.

References

Acquaviva AM, Bruni CB, Nissley SP, Rechler MM (1982) Cell-free synthesis of rat insulin-like growth factor II. Diabetes 31:656–658

Adams SO, Nissley SP, Greenstein LA, Yang YW-H, Rechler MM (1983) Synthesis of multiplication-stimulating activity (rat insulin-like growth factor II) by rat embryo fibroblasts. Endocrinology 112:979–987

Adams SO, Kapadia M, Mills B, Daughaday WH (1984) Release of insulin-like growth factors and binding protein activity into serum-free medium of cultured human fibroblasts. Endocrinology 115:520–526

Adashi EY, Resnick CE, D'Ercole AJ, Svoboda ME, Van Wyk JJ (1985) Insulin-like growth factors as intraovarian regulators of granulosa cell growth and function. Endocr Rev 6:400–420

Appell KC, Simpson IA, Cushman SW (1988) Characterization of the stimulatory action of insulin on insulin-like growth factor II binding to rat adipose cells. Differences in the mechanism of insulin action on insulin-like growth factor II receptors and glucose transporters. J Biol Chem 263:10824–10829

Ballard FJ, Francis GL, Ross M, Bagley CJ, May B, Wallace JC (1987) Natural and synthetic forms of insulin-like growth factor-1 (IGF-1) and the potent derivative, destripeptide IGF-1: biological activities and receptor binding. Biochem Biophys Res Comm 149:398–404

Ballard FJ, Ross M, Upton FM, Francis GL (1988) Specific binding of insulin-like growth factors 1 and 2 to the type 1 and type 2 receptors respectively. Biochem J 249:721–726

Bar RS, Harrison LC, Baxter RC, Boes M, Dake BL, Booth B, Cox A (1987) Production of IGF-binding proteins by vascular endothelial cells. Biochem Biophys Res Commun 148:734–739

Barenton B, Guyda HJ, Goodyer CG, Polychronakos C, Posner BI (1987) Specificity of insulin-like growth factor binding to the type-II IGF receptors in rabbit mammary gland and hypophysectomized rat liver. Biochem Biophys Res Commun 149:555–561

Barnard R, Haynes KM, Werther GA, Waters MJ (1988) The ontogeny of growth hormone receptors in the rabbit tibia. Endocrinology 122:2562–2569

Baxter RC (1986) The somatomedins: insulin-like growth factors. Adv Clin Chem 25:49–115

Baxter RC (1988) Characterization of the acid-labile subunit of the growth hormone-dependent insulin-like growth factor binding protein complex. J Clin Endocrinol Metab 67:265–272

Baxter RC, Cowell CT (1987) Diurnal rhythm of growth hormone-independent binding protein for insulin-like growth factors in human plasma. J Clin Endocrinol Metab 65:432–440

Baxter RC, Martin JL (1986) Radioimmunoassay of growth hormone-dependent insulinlike growth factor binding protein in human plasma. J Clin Invest 78:1504–1512

Baxter RC, Martin JL (1987) Binding proteins for insulin-like growth factors in adult rat serum. Comparison with other human and rat binding proteins. Biochem Biophys Res Commun 147:408–415

Baxter RC, Zaltsman Z, Turtle JR (1984) Immunoreactive somatomedin-C/insulin-like growth factor I and its binding protein in human milk. J Clin Endocrinol Metab 58:955–959

Baxter RC, Martin JL, Tyler MI, Howden ME (1986) Growth hormone-dependent insulin-like growth factor (IGF) binding protein from human plasma differs from other human IGF binding proteins. Biochem Biophys Res Commun 139:1256–1261

Baxter RC, Martin JL, Wood MH (1987) Two immunoreactive binding proteins for insulin-like growth factors in human amniotic fluid: relationship to fetal maturity. J Clin Endocrinol Metab 65:423–431

Bayne ML, Cascieri MA, Kelder B, Applebaum J, Chicchi G, Shapiro JA, Pasleau F, Kopchick JJ (1987) Expression of a synthetic gene encoding human insulin-like growth factor I in cultured mouse fibroblasts. Proc Natl Acad Sci USA 84:2638–2642

Bayne ML, Applebaum J, Chicchi GG, Hayes NS, Green BG, Cascieri MA (1988) Structural analogs of human insulin-like growth factor I with reduced affinity for serum binding proteins and the type 2 insulin-like growth factor receptor. J Biol Chem 263:6233–6239

Bayne ML, Applebaum J, Underwood D, Chicchi GG, Green BG, Hayes NS, Cascieri MA (1989) The C region of human insulin-like growth factor (IGF) I is required for high affinity binding to the type I IGF receptor. J Biol Chem 264:11004–11008

Beck F, Samani NJ, Penschow JD, Thorley B, Tregear GW, Coghlan JP (1987) Histochemical localization of IGF-I and -II mRNA in the developing rat embryo. Development 101:175–184

Beck F, Samani NJ, Senior P, Byrne S, Morgan K, Gebhard R, Brammar WJ (1988) Control of IGF-II mRNA levels by glucocorticoids in the neonatal rat. J Mol Endocrinol 1:R5–R8

Beebe DC, Silver MH, Belcher KS, Van Wyk JJ, Svoboda ME, Zelenka PS (1987) Lentropin, a protein that controls lens fiber formation, is related functionally and immunologically to the insulin-like growth factors. Proc Natl Acad Sci USA 84:2327–2330

Beguinot F, Smith RJ, Kahn CR, Maron R, Moses AC, White MF (1988) Phosphorylation of insulin-like growth factor I receptor by insulin receptor tyrosine kinase in intact cultured skeletal muscle cells. Biochemistry 27:3222–3228

Bell GI, Merryweather JP, Sanchez-Pescador R, Stempien MM, Priestley L, Scott J, Rall LB (1984) Sequence of a cDNA clone encoding human preproinsulin-like growth factor II. Nature 310:775–777

Bell GI, Gerhard DS, Fong NM, Sanchez-Pescador R, Rall LB (1985) Isolation of the human insulin-like growth factor genes: insulin-like growth factor II and insulin genes are contiguous. Proc Natl Acad Sci USA 82:6450–6454

Bell GI, Stempien MM, Fong NM, Ball LB (1986) Sequences of liver cDNAs encoding two different mouse insulin-like growth factor I precursors. Nucleic Acids Res 14:7873–7882

Bell SC, Keyte JW (1988) N-terminal amino acid sequence of human pregnancy-associated endometrial alpha-1-globulin, an endometrial insulin-like growth factor (IGF) binding protein-evidence for two small molecular weight IGF binding proteins. Endocrinology 123:1202–1204

Binoux M, Hossenlopp P (1988) Insulin-like growth factor (IGF) and IGF-binding proteins: comparison of human serum and lymph. J Clin Endocrinol Metab 67:509–514

Binoux M, Hossenlopp P, Lassare C, Hardouin N (1981) Production of insulin-like growth factors and their carrier by rat pituitary gland and brain explants in culture. FEBS Lett 124:178–184

Binoux M, Hardouin S, Lassare C, Hossenlopp P (1982) Evidence for production by the liver of two IGF binding proteins with similar molecular weights but different affinities for IGF I and IGF II. Their relations with serum and cerebrospinal fluid IGF binding proteins. J Clin Endocrinol Metab 55:600–602

Binoux M, Lassarre C, Gourmelen M (1986) Specific assay for insulin-like growth factor (IGF-II) using the IGF binding proteins extracted from human cerebrospinal fluid. J Clin Endocrinol Metab 63:1151–1155

Blanchard MM, Barenton B, Sullivan A, Foster B, Guyda HJ, Posner BI (1988) Characterization of the insulin-like growth factor (IGF) receptor in K562 erythroleukemia cells; evidence for a biological function for the type II IGF receptor. Mol Cell Endocrinol 56:235–244

Blum WF, Ranke MB, Bierich JR (1986) Isolation and partial characterization of six somatomedin-like peptides from human plasma Cohn fraction IV. Acta Endocrinol (Copenh) 111:271–284

Bohannon NJ, Corp ES, Wilcox BJ, Figlewicz DP, Dorsa DM, Baskin DG (1988) Localization of bindings sites for insulin-like growth factor-I (IGF-I) in the rat brain by quantitative autoradiography. Brain Res 444:205–213

Bortz JD, Rotwein P, De Vol D, Bechtel PJ, Hansen VA, Hammermann MR (1988) Focal expression of insulin-like growth factor I in rat kidney collecting duct. J Cell Biol 107:811–819

Braulke T, Gartung C, Hasilik A, von Figura K (1987) Is movement of mannose 6-phosphate-specific receptor triggered by binding of lysosomal enzymes? J Cell Biol 104:1735–1742

Braulke T, Causin C, Waheed A, Junghans U, Hasilik A, Maly P, Humbel RE, von Figura K (1988) Mannose 6-phosphate/insulin-like growth factor II receptor: distinct binding sites for mannose 6-phosphate and insulin-like growth factor II. Biochem Biophys Res Commun 150:1287–1293

Brenner-Gati L, Berg KA, Gershengorn MA (1988) Thyroid-stimulating hormone and insulin-like growth factor-1 synergize to elevate 1,2-diacylglycerol in rat thyroid cells. J Clin Invest 82:1144–1148

Brewer MT, Stetler GL, Squires CH, Thompson RC, Busby WH, Clemmons DR (1988a) Cloning, characterization, and expression of a human insulin-like growth factor binding protein. Biochem Biophys Res Commun 152:1289–1297

Brewer MT, Stetler GL, Squires CH, Thompson RC, Busby WH, Clemmons DR (1988b) Cloning, characterization, and expression of a human insulin-like growth factor binding protein (erratum). Biochem Biophys Res Commun 155:1485–1485

Brinkman A, Groffen C, Kortleve DJ, van Kessel AG, Drop SLS (1988a) Isolation and characterization of a cDNA encoding the low molecular weight insulin-like growth factor binding protein (IBP-1). EMBO J 7:2417–2423

Brinkman A, Groffen CAH, Kortleve DJ, Drop SLS (1988b) Organization of the gene encoding the insulin-like growth factor binding protein IBP-1. Biochem Biophys Res Commun 157:898–907

Brissenden JE, Ullrich A, Francke U (1984) Human chromosomal mapping of genes for insulin-like growth factors I and II and epidermal growth factor. Nature 310:781–784

Brown AL, Graham DE, Nissley SP, Hill DJ, Strain AJ, Rechler MM (1986) Developmental regulation of insulin-like growth factor II mRNA in different rat tissues. J Biol Chem 261:13144–13150

Brown AL, Chiariotti L, Orlowski C, Mehlman T, Burgess WH, Ackerman EJ, Bruni CB, Rechler MM (1989) Nucleotide sequence and expression of a cDNA clone encoding a fetal rat binding protein for insulin-like growth factors. J Biol Chem 264:5148–5154

Bryson JM, Baxter RC (1987) High-affinity receptor for insulin-like growth factor II in rat liver: properties and regulation in vivo. J Endocrinol 113:27–35

Busby WH, Klapper DG, Clemmons DR (1988a) Purification of a 31 000-dalton insulin-like growth factor binding protein from human amniotic fluid. Isolation of two forms with different biologic actions. J Biol Chem 263:14203–14210

Busby WH, Snyder DK, Clemmons DR (1988b) Radioimmunoassay of a 26 000-dalton plasma insulin-like growth factor-binding protein: control by nutritional variables. J Clin Endocrinol Metab 67:1225–1230

Butler JH, Gluckman PD (1986) Circulating insulin-like growth factor-binding proteins in fetal, neonatal and adult sheep. J Endocrinol 109:333–338

Cariani E, Lasserre C, Seurin D, Hamelin B, Kemeny F, Franco D, Czech MP, Ullrich A, Brechot C (1988) Differential expression of insulin-like growth factor II mRNA in human primary liver cancers, benign liver tumors, and liver cirrhosis. Cancer Res 48:6844–6849

Carlsson-Skwirut C, Jornvall H, Homgren A, Andersson C, Bergman T, Lundquist G, Sjogren B, Sara VR (1986) Isolation and characterization of variant IGF-1 as well as IGF-2 from adult human brain. FEBS Lett 201:46–50

Cascieri MA, Chicchi GG, Applebaum J, Hayes NS, Green BG, Bayne ML (1988a) Mutants of human insulin-like growth factor I with reduced affinity of the type I insulin-like growth factor receptor. Biochemistry 27:3229–3299

Cascieri MA, Saperstein R, Hayes NS, Green BG, Chicchi GG, Applebaum J, Bayne ML (1988b) Serum half-life and bilogical activity of mutants of human insulin-like growth factor I which do not bind to serum binding proteins. Endocrinology 123:373–381

Cascieri MA, Chicchi GG, Applebaum J, Green BG, Hayes NS, Bayne ML (1989) Structural analogs of human insulin-like growth factor (IGF) I with altered affinity for type 2 IGF receptors. J Biol Chem 264:2199–2202

Casella SJ, Han VK, D'Ercole AJ, Svoboda ME, Van Wyk JJ (1986) Insulin-like growth factor II binding to the type I somatomedin receptor. J Biol Chem 261: 9268–9273

Casella SJ, Smith EP, Van Wyk JJ, Joseph DR, Hynes MA, Hoyt EC, Lund PK (1987) Isolation of rat testis cDNAs encoding an insulin-like growth factor I precursor. DNA 6:325–330

Causin C, Waheed A, Braulke T, Junghans U, Maly P, Humbel RE, von Figura K (1988) Mannose 6-phosphate/insulin-like growth factor II-binding proteins in human serum and urine. Biochem J 252:795–799

Chaiken RL, Moses AC, Usher P, Flier JS (1986) Insulin stimulation of aminoisobutyric acid transport in human skin fibroblasts is mediated through both insulin and type I insulin-like growth factor receptors. J Clin Endocrinol Metab 63:1181–1184

Chambard JC, Paris S, L'Allemain G, Pouyssegur J (1987) Two growth factor signalling pathways in fibroblasts distinguished by pertussis toxin. Nature 326:800–803

Chatelain PG, Van Wyk JJ, Copeland KC, Blethen SL, Underwood LE (1983) Effect of in vitro action of serum proteases or exposure to acid on measurable immunoreactive somatomedin-C in serum. J Clin Endocrinol Metab 56:376–383

Chiariotti L, Brown AL, Frunzio R, Clemmons DR, Rechler MM, Bruni CB (1988) Structure of the rat insulin-like growth factor II transcriptional unit: heterogeneous transcripts are generated from two promoters by use of multiple polyadenylation sites and differential ribonucleic acid splicing. Mol Endocrinol 2:1115–1126

Claustres M, Chatelain P, Sultan C (1987) Insulin-like growth factor I stimulates human erythroid colony formation in vitro. J Clin Endocrinol Metab 65:78–82

Clemmons DR, Shaw DS (1986) Purification and biologic properties of fibroblast somatomedin. J Biol Chem 261:10293–10298

Clemmons DR, Klibanski A, Underwood LE, McArthur JW, Ridgway EC, Beitins IZ, Van Wyk JJ (1981a) Reduction of plasma immunoreactive somatomedin C during fasting in humans. J Clin Endocrinol Metab 53:1247–1250

Clemmons DR, Underwood LE, Van Wyk JJ (1981b) Hormonal control of immunoreactive somatomedin production by cultured human fibroblasts. J Clin Invest 67:10–19

Clemmons DR, Underwood LE, Chatelain PG, Van Wyk JJ (1983) Liberation of immunoreactive somatomedin-C from its binding proteins by proteolytic enzymes and heparin. J Clin Endocrinol Metab 56:384–389

Clemmons DR, Underwood LE, Dickerson RN, Brown RO, Hak LJ, MacPhee RD, Heizer WD (1985) Use of plasma somatomedin-C/insulin-like growth factor I measurements to monitor the response to nutritional repletion in malnourished patients. Am J Clin Nur 41:191–198

Clemmons DR, Elgin RG, Han VK, Casella SJ, D'Ercole AJ, Van Wyk JJ (1986) Cultured fibroblast monolayers secrete a protein that alters the cellular binding of somatomedin-C/insulinlike growth factor I. J Clin Invest 77:1548–1556

Clemmons DR, Han VKM, Elgin RG, D'Ercole AJ (1987) Alterations in the synthesis of a fibroblast surface associated 35K protein modulates the binding of somatomedin-C/insulin-like growth factor I. Mol Endocrinol 1:339–347

Conover CA, Misra P, Hintz RL, Rosenfeld RG (1986) Effect of an anti-insulin-like growth factor I receptor antibody on insulin-like growth factor II stimulation of DNA synthesis in human fibroblasts. Biochem Biophys Res Commun 139:501–508

Corvera S, Czech MP (1985) Mechanism of insulin action on membrane protein recycling: a selective decrease in the phosphorylation state of insulin-like growth factor II receptors in the cell surface membrane. Proc Natl Acad Sci USA 82:7314–7318

Corvera S, Whitehead RE, Mottola C, Czech MP (1986) The insulin-like growth factor II receptor is phosphorylated by a tyrosine kinase in adipocyte plasma membranes. J Biol Chem 261:7675–7679

Corvera S, Roach PJ, DePaoli-Roach AA, Czech MP (1988a) Insulin action inhibits insulin-like growth factor-II (IGF-II) receptor phosphorylation in H-35 hepatoma cells. J Biol Chem 263:3116–3122

Corvera S, Yagaloff KA, Whitehead RE, Czech MP (1988b) Tyrosine phosphorylation of the receptor for insulin-like growth factor II is inhibited in plasma membranes from insulin-treated rat adipocytes. Biochem J 250:47–52

Costigan DC, Guyda HJ, Posner BI (1988) Free insulin-like growth factor I (IGF-I) and IGF-II in human saliva. J Clin Endocrinol Metab 66:1014–1018

Cotterill AM Cowell CT, Baxter RC, McNeil D, Silinik M (1988) Regulation of the growth hormone-independent growth factor-binding protein in children. J Clin Endocrinol Metab 67:882–887

Cubbage ML, Suwanichkul A, Powell DR (1989) Structure of the human chromosomal gene for the 25 kilodalton insulin-like growth factor binding protein. Mol Endocrinol 3:846–851

Dahms NM, Lobel P, Breitmeyer J, Chirgwin JM, Kornfeld S (1987) 46 kd Mannose 6-phosphate receptor: cloning, expression, and homology to the 215 kd mannose 6-phosphate receptor. Cell 50:181–192

Dainiak N, Kreczko S (1985) Interactions of insulin, insulinlike growth factor II, and platelet-derived growth factor in erythropoietic culture. J Clin Invest 76:1237–1242

Daughaday WH, Mariz IK, Trivedi B (1981) A preferential binding site for insulin-like growth factor II in human and rat placental membranes. J Clin Endocrinol Metab 53:282–288

Daughaday WH, Hall K, Salmon WD, Van den Brande JL, Van Wyk JJ (1987a) On the nomenclature of the somatomedins and insulin-like growth factors. Endocrinology 121:1911–1912

Daughaday WH, Kapadia M, Mariz I (1987b) Serum somatomedin binding proteins: physiologic significance and interference in radioligand assay. J Lab Clin Med 109:355–363

Daughaday WH, Emanuele MA, Brooks MH, Barbato AL, Kapadia M, Rotwein P (1988) Synthesis and secretion of insulin-like growth factor II by a leiomyosarcoma with associated hypoglycemia. N Engl J Med 319:1434–1440

Dawe SR, Francis GL, McNamara PJ, Wallace JC, Ballard FJ (1988) Purification, partial sequences and properties of chicken insulin-like growth factors. J Endocrinol 117:173–181

D'Ercole AJ, Underwood LE (1980) Ontogeny of somatomedin during development in the mouse. Serum concentrations, molecular forms, binding proteins, and tissue receptors. Dev Biol 79:33–45

D'Ercole AJ, Willson DF, Underwood LE (1980) Changes in the circulating form of serum somatomedin-C during fetal life. J Clin Endocrinol Metab 51:674–676

D'Ercole AJ, Stiles AD, Underwood LE (1984) Tissue concentrations of somatomedin C: further evidence for multiple sites of synthesis and paracrine or autocrine mechanisms of action. Proc Natl Acad Sci USA 81:935–939

D'Ercole AJ, Drop SLS, Kortleve DJ (1985) Somatomedin-C/insulin-like growth factor I-binding proteins in human amniotic fluid and in fetal and postnatal blood: evidence of immunological homology. J Clin Endocrinol Metab 61:612–617

De Mellow JSM, Baxter RC (1988) Growth hormone-dependent insulin-like growth factor (IGF) binding protein both inhibits and potentiates IGF-I-stimulated DNA synthesis in human skin fibroblasts. Biochem Biophys Res Commun 156:199–204

De Pagter-Holthuizen P, van Schaik FMA, Verduijn GM, van Ommen GJB, Bouma BN, Jansen M, Sussenbach JS (1986) Organization of the human genes for like growth factors I and II. FEBS Lett 195:179–184

De Pagter-Holthuizen P, Jansen M, van Schaik FMA, van der Kammen R, Oosterwijk C, Van den Brande JL, Sussenbach JS (1987) The human insulin-like growth factor II gene contains two development-specific promoters. FEBS Lett 214:259–264

De Pagter-Holthuizen P, Jansen M, van der Kammen RA, van Schaik FMA, Sussenbach JS (1988) Differential expression of the human insulin-like growth factor II gene. Characterization of the IGF-II mRNAs and an mRNA encoding a putative IGF-II associated protein. Biochim Biophys Acta 950:282–295

De Vroede MA, Rechler MM, Nissley SP, Joshi S, Burke GT, Katsoyannis PG (1985) hybrid molecules containing the B-domain of insulin-like growth factor I (IGF-I) are recognized by IGF carrier proteins. Proc Natl Acad Sci USA 82:3010–3014

De Vroede MA, Tseng LY, Katsoyannis PG, Nissley SP, Rechler MM (1986) Modulation of insulinlike growth factor I binding to human fibroblast monolayer cultures by insulinlike growth factor carrier proteins released to the incubation media. J Clin Invest 77:602–613

Drop SLS, Valiquette G, Guyda HJ, Corvol MT, Posner BI (1979) Partial purification and characterization of a binding protein for insulin-like activity (ILAs) in human amniotic fluid: a possible inhibitor of insulin-like activity. Acta Endocrinol (Copenh) 90:505–518

Drop SLS, Kortleve DJ, Guyda HJ (1984a) Isolation of a somatomedin-binding protein from preterm amniotic fluid. Development of a radioimmunoassay. J Clin Endocrinol Metab 59:899–907

Drop SLS, Kortleve DJ, Guyda HJ, Posner BI (1984b) Immunoassay of a somatomedin-binding protein from human amniotic fluid: levels in fetal, neonatal, and adult sera. J Clin Endocrinol Metab 59:908–915

Dull TJ, Gray A, Hayflick JS, Ullrich A (1984) Insulin-like growth factor II precursor gene organization in relation to insulin gene family. Nature 310:777–781

Duronio V, Jacobs S, Cuatrecasas P (1986) Complete glycosylation of the insulin and insulin-like growth factor I receptors is not necessary for their biosynthesis and function. J Biol Chem 261:970–975

Duronio V, Jacobs S, Romero PA, Herscovics A (1988) Effects of inhibitors of N-linked oligosaccharide processing on the biosynthesis and function of insulin and insulin-like growth factor-I receptors. J Biol Chem 263:5436–5445

Ebina Y, Ellis L, Jarnagin K, Edery M, Graf L, Clauser E, Ou J-H, Masiarz F, Kan YW, Goldfine ID et al. (1985) The human insulin receptor cDNA: the structural basis for hormone-activated transmembrane signalling. Cell 40:747–758

El-Badry OM, Romanus JA, Helman LJ, Cooper MJ, Rechler MM, Israel MA (1989) Autonomous growth of a human neuroblastoma cell line is mediated by insulin-like growth factor II. J Clin Invest 84:829–839

Elgin GR, Busby WH, Clemmons DR (1987) An insulin-like growth factor (IGF) binding protein enhances the biologic response to IGF-I. Proc Natl Acad Sci USA 84:3254–3258

Emler CA, Schalch DS (1987) Nutritionally-induced changes in hepatic insulin-like growth factor I (IGF-I) gene expression in rats. Endocrinology 120:832–834

Evans T, DeChiara T, Efstratiadis A (1988) A promoter of the rat insulin-like growth factor II gene consists of minimal control elements. J Mol Biol 199:61–81

Ewton DZ, Falen SL, Florini JR (1987) The type II insulin-like growth factor (IGF) receptor has low affinity for IGF-I analogs: pleiotypic actions of IGFs on myoblasts are apparently mediated by the type I receptor. Endocrinology 120:115–123

Fagin JA, Melmed S (1987) Relative increase in insulin-like growth factor I messenger ribonucleic acid levels in compensatory renal hypertrophy. Endocrinology 120:718–724

Fagin JA, Brown A, Melmed S (1988) Regulation of pituitary insulin-like growth factor-I messenger ribonucleic acid levels in rats harboring somatomammotropic tumors: implications for growth hormone autoregulation. Endocrinology 122:2204–2210

Feltz SM, Swanson ML, Wemmie JA, Pessin JE (1988) Functional properties of an isolated alphabeta heterodimeric human placenta insulin-like growth factor 1 receptor complex. Biochemistry 27:3234–3242

Flier JS, Moses AC (1985) Characterization of monoclonal antibodies to the IGF-I receptor that inhibit IGF-I binding to human cells. Biochem Biophys Res Commun 127:929–936

Flier JS, Usher P, Moses AC (1986) Monoclonal antibody to the type I insulin-like growth factor (IGF-I) receptor blocks IGF-I receptor-mediated DNA synthesis: clarification of the mitogenic mechanisms of IGF-I and insulin in human skin fibroblasts. Proc Natl Acad Sci USA 83:664–668

Francis GL, Upton FM, Ballard FJ, McNeil KA, Wallace JC (1988) Insulin-like growth factors 1 and 2 in bovine colostrum. Biochem J 251:95–103

Froesch ER, Schmid C, Schwander J, Zapf J (1985) Actions of insulin-like growth factors. Ann Rev Physiol 47:443–467

Frolik CA, Ellis LF, Williams DC (1988) Isolation and characterization of insulin-like growth factor-II from human bone. Biochem Biophys Res Commun 151:1011–1018

Frunzio R, Chiariotti L, Brown AL, Graham DE, Rechler MM, Bruni CB (1986) Structure and expression of the rat insulin-like growth factor II (rIGF-II) gene. rIGF-II RNAs are transcribed from two promoters. J Biol Chem 261:17138–17149

Fu X-X, Su CY, Lee Y, Hintz R, Biempica L, Snyder R, Rogler CE (1988) Insulinlike growth factor II expression and oval cell proliferation associated with hepatocarcinogenesis in woodchuck hepatitis virus carriers. J Virol 62:3422–3430

Furlanetto RW (1980) The somatomedin C binding protein: evidence for a heterologous subunit structure. J Clin Endocrinol Metab 51:12

Furlanetto RW, DiCarlo JN, Wisehart C (1987) The type II insulin-like growth factor receptor does not mediate deoxyribonucleic acid synthesis in human fibroblasts. J Clin Endocrinol Metab 64:1142–1149

Gelato MC, Kiess W, Lee L, Malozowski S, Rechler MM, Nissley SP (1988) The insulin-like growth factor-II/mannose 6-phosphate receptor is present in monkey serum. J Clin Endocrinol Metab 67:669–675

Goldstein S, Sertich GJ, Levan KR, Phillips LS (1988) Nutrition and somatomedin, XIX. Molecular regulation of insulin-like growth factor-1 in streptozotocin-diabetic rats. Mol Endocrinol 2:1093–1100

Gorden P, Hendricks CM, Kahn CR, Megyesi K, Roth J (1981) Hypoglycemia associated with non-islet-cell tumor and insulin-like growth factors: a study of the tumor type. N Engl J Med 305:1452–1455

Gowan LK, Hampton B, Hill DJ, Schlueter RJ, Perdue JF (1987) Purification and characterization of a unique high molecular weight form of insulin-like growth factor II. Endocrinology 121:449–458

Graham DE, Rechler MM, Brown AL, Frunzio R, Romanus JA, Bruni CB, Whitfield HJ, Nissley SP, Seelig S, Berry S (1986) Coordinate development regulation of high and low molecular weight mRNAs for rat insulin-like growth factor II. Proc Natl Acad Sci USA 83:4519–4523

Grant M, Jerdan J, Merimee TJ (1987a) Insulin-like growth factor-I modulates endothelial cell chemotaxis. J Clin Endocrinol Metab 65:370–371

Grant MB, Russell B, Harwood HJ, Merimee TJ (1987b) Separation of the insulin-like growth factor-binding proteins in plasma and purification of the larger molecular weight species. J Clin Endocrinol Metab 64:1060–1065

Gray A, Tam AW, Dull TJ, Hayflick J, Pintar J, Cavenee WK, Koufos A, Ullrich A (1987) Tissue-specific and developmentally regulated transcirption of the insulin-like growth factor 2 gene. DNA 6:283–295

Griffiths G, Hoflack B, Simons K, Mellman I, Kornfeld S (1988) The mannose 6-phosphate receptor and the biogenesis of lysosomes. Cell 52:329–341

Guler H-P, Zapf J, Froesch ER (1987) Short-term metabolic effects of recombinant human insulin-like growth factor I in healthy adults. N Engl J Med 317:137–140

Guler H-P, Zapf J, Scheiwiller E, Froesch ER (1988) Recombinant human insulin-like growth factor I stimulates growth and has distinct effects on organ size in hypophysectomized rats. Proc Natl Acad Sci USA 85:4889–4893

Hall K, Hansson U, Lundin G, Luthman M, Persson B, Povoa G, Stangenberg M, Of-verholm U (1986) Serum levels of somatomedins and somatomedin-binding protein in pregnant women with type I or gestational diabetes and their infants. J Clin Endocrinol Metab 63:1300

Hall K, Brismar K, Povoa G (1987) The role of somatomedin binding protein. In: Isaksson O et al. (ed) Growth hormone – basic and clinical aspects. Elsevier Science, Stockholm, p 415

Hall K, Lundin G, Povoa G (1988) Serum levels of the low molecular weight form of insulin-like growth factor binding protein in healthy subjects and patients with growth hormone deficiency, acromegaly and anorexia nervosa. Acta Endocrinol (Copenh) 118:321–326

Hammerman MR, Gavin JR III (1984) Binding of insulin-like growth factor II and multiplication-stimulating activity-stimulated phosphorylation in basolateral membranes from dog kidney. J Biol Chem 259:13511–13517

Han VKM, D'Ercole AJ, Lund PK (1987a) Cellular localization of somatomedin (insulin-like growth factor) messenger RNA in the human fetus. Science 236:193–197

Han VKM, Hill DJ, Strain AJ, Towle AC, Lauder JM, Underwood LE, D'Ercole AJ (1987b) Identification of somatomedin/insulin-like growth factor immunoreactive cells in the human fetus. Pediatr Res 22:245–249

Han VKM, Lauder JM, D'Ercole AJ (1988a) Rat astroglial somatomedin/insulin-like growth factor binding proteins: characterization and evidence of biologic function. J Neurosci 8:3135–3143

Han VKM, Lund PK, Lee DC, D'Ercole AJ (1988b) Expression of somatomedin/insulin-like growth factor messenger ribonucleic acids in the human fetus: identification, characterization, and tissue distribution. J Clin Endocrinol Metab 66:422–429

Hansson HA, Dahlin LB, Danielsen N, Fryklund L, Nachemson AK, Polleryd P, Rozell B, Skottner A, Stemme S, Lundborg G (1986) Evidence indicating trophic importance of IGF-I in regenerating peripheral nerves. Acta Physiol Scand 126:609–614

Hansson H-A, Jennische E, Skottner A (1987) Regenerating endothelial cells express insulin-like growth factor-I immunoreactivity after arterial injury. Cell Tissue Res 250:499–505

Hardouin S, Hossenlopp P, Segovia B, Seurin D, Portolan G, Lassarre C, Binoux M (1987) Heterogeneity of insulin-like growth factor binding proteins and relationships between structure and affinity. I. Circulating forms in man. Eur J Biochem 170:121–132

Hari J, Pierce SB, Morgan DO, Sara V, Smith MC, Roth RA (1987) The receptor for insulin-like growth factor II mediates an insulin-like response. EMBO J 6:3367–3371

Haselbacher G, Humbel R (1982) Evidence for two species of insulin-like growth factor II (IGF II and "big" IGF II) in human spinal fluid. Endocrinology 110:1822–1824

Haselbacher GK, Andres RY, Humbel RE (1980) Evidence for the synthesis of a somatomedin similar to insulin-like growth factor I by chick embryo liver cells. Eur J Biochem 111:245–250

Haselbacher GK, Schwab ME, Pasi A, Humbel RE (1985) Insulin-like growth factor II (IGF II) in human brain: regional distribution of IGF II and of higher molecular mass forms. Proc Natl Acad Sci USA 82:2153–2157

Haselbacher GK, Irminger JC, Zapf J, Ziegler WH, Humbel RE (1987) Insulin-like growth factor II in human adrenal pheochromocytomas and Wilms tumors: expression at the mRNA and protein level. Proc Natl Acad Sci USA 84:1104–1106

Heath JK, Shi W-K (1986) Developmentally regulated expression of insulin-like growth factors by differentiated murine teratocarcinomas and extraembryonic mesoderm. J Embryol Exp Morphol 95:193–212

Hernandez ER, Resnick CE, Svoboda ME, Van Wyk JJ, Payne DW, Adashi EY (1988) Somatomedin-C/insulin-like growth factor I as an enhancer of androgen biosynthesis by cultured rat ovarian cells. Endocrinology 122:1603–1612

Hey AW, Browne CA, Thorburn GD (1987) Fetal sheep serum contains a high molecular weight insulin-like growth factor (IGF) binding protein that is acid stable and specific for IGF-II. Endocrinology 121:1975–1984

Hintz RL, Liu F, Rosenfeld RG, Kemp SF (1981) Plasma somatomedia-binding proteins in hypopituitarism: changes during growth hormone therapy. J Clin Endocrinol Metab 53:100–104

Hintz RL, Thorsson AV, Enberg G, Hall K (1984) IGF-II binding on human lymphoid cells: demonstration of a common high affinity receptor for insulin like peptides. Biochem Biophys Res Commun 118:774–782

Hizuka N, Sukegawa I, Takano K, Asakawa K, Horikawa R, Tsushima T, Shizume K (1987a) Characterization of insulin-like growth factor I receptors on human erythroleukemia cell line (K-562 cells). Endocrinol Jpn 34:81–88

Hizuka N, Takano K, Tanaka I, Asakawa K, Miyakawa M, Horikawa R, Shizume K (1987b) Demonstration of insulin-like growth factor I in human urine. J Clin Endocrinol Metab 64:1309–1312

Honegger A, Humbel RE (1986) Insulin-like growth factors I and II in fetal and adult bovine serum. J Biol Chem 261:569–575

Hoppener JWM, Steenbergh PH, Slebos RJC, de Pagter-Holthuizen P, Roos BA, Jansen M, Van den Brande JL, Sussenbach JS, Jansz HS, Lips CJM (1987) Expression of insulin-like growth factor-I and -II genes in rat medullary thyroid carcinoma. FEBS Lett 215:122–126

Hoppener JWM, Mosselman S, Roholl PJM, Lambrechts C, Slebos RJC, de Pagter-Holthuizen P, Lips CJM, Jansz HS, Sussenbach JS (1988) Expression of insulin-like growth factor-I and -II genes in human smooth muscle tumours. EMBO J 7:1379–1385

Hossenlopp P, Seurin D, Segovia-Quinson B, Binoux M (1986a) Identification of an insulin-like growth factor-binding protein in human cerebrospinal fluid with a selective affinity for IGF-II. FEBS Lett 208:439–444

Hossenlopp P, Seurin D, Segovia-Quinson B, Hardouin S, Binoux M (1986b) Analysis of serum insulin-like growth factor binding proteins using western blotting: use of the method for tritration of the binding proteins and competitive binding sites. Anal Biochem 154:138–143

Hossenlopp P, Seurin D, Segovia B, Portolan G, Binoux M (1987) Heterogeneity of insulin-like growth factor binding proteins and relationships between structure and
affinity. II. Forms released by human and rat liver in culture. Eur J Biochem
170:133–142

Hoyt EC, Van Wyk JJ, Lund PK (1988) Tissue and development specific regulation of
a complex family of rat insulin-like growth factor I messenger ribonucleic acids.
Mol Endocrinol 2:1077–1086

Huhtala M-L, Koistinen R, Palomaki P, Partanen P, Bohn H, Seppala M (1986)
Biologically active domain in somatomedin-binding protein. Biochem Biophys Res
Commun 141:263–270

Humbel RE (1984) Insulin-like growth factors, somatomedins, and multiplication-
stimulating activity: chemistry. In: Li CH (ed) Hormonal proteins and peptides,
vol 12. Academic, New York, p 57

Hunter T, Alexander CB, Cooper JA (1985) Protein phosphorylation and growth control. Ciba Found Symp 116:188–204

Hylka VW, Teplow DB, Kent SBH, Straus DS (1985) Identification of a peptide fragment from the carboxyl-terminal extension region (E-domain) of rat proinsulin-like
growth factor-II. J Biol Chem 260:14417–14420

Hylka VW, Kent SBH, Straus DS (1987) E-domain peptide of rat proinsulin-like
growth factor II: validation of a radioimmunoassay and measurement in culture
medium and rat serum. Endocrinology 120:2050–2058

Hynes MA, Van Wyk JJ, Brooks PJ, D'Ercole AJ, Jansen J, Lund PK (1987) Growth
hormone dependence of somatomedin-C/insulin-like growth factor-I and in- sulin-like growth factor-II messenger ribonucleic acids. Mol Endocrinol 1:233–242

Hynes MA, Brooks PJ, Van Wyk JJ, Lund PK (1988) Insulin-like growth factor II messenger ribonucleic acids are synthesized in the choroid plexus of the rat brain. Mol
Endocrinol 1:47–54

Iino K, Seppala M, Heinonen PK, Sipponen P, Rutanen EM (1986) Elevated levels of a
somatomedin-binding protein PP12 in patients with ovarian cancer. Cancer
58:2294–2297

Irminger JC, Rosen KM, Humbel RE, Villa-Komaroff L (1987) Tissue-specific expression of insulin-like growth factor II mRNAs with dinstinct 5' untranslated regions.
Proc Natl Acad Sci USA 84:6330–6334

Isaksson OGP, Lindahl A, Nilsson A, Isgaard J (1987) Mechanism of the stimulatory
effect of growth hormone on longitudinal bone growth. Endocr Rev 8:426–438

Isgaard J, Carlsson L, Isaksson OGP, Jansson J-O (1988a) Pulsatile intravenous
growth hormone (GH) infusion to hypophysectomized rats increases insulin-like
growth factor I messenger ribonucleic acid in skeletal tissues more effectively than
continuous GH infusion. Endocrinology 123:2605–2610

Isgaard J, Moller C, Isaksson OGP, Nilsson A, Mathews LS, Norstedt G (1988b)
Regulation of insulin-like growth factor messenger ribonucleic acid in rat growth
plate by growth hormone. Endocrinology 122:1515–1520

Izumi T, White MF, Kadowaki T, Takaku F, Akanuma Y, Kasuga M (1987) Insulin-
like growth factor I rapidly stimulates tyrosine phosphorylation of a Mr 185,000
protein in intact cells. J Biol Chem 262:1282–1287

Jacobs S, Kull FC, Cuatrecasas P (1983a) Monensin blocks the maturation of receptors for insulin and somatomedin C: identification of receptor precursors. Proc Natl
Acad Sci USA 80:1228–1231

Jacobs S, Kull FC, Earp HS, Svoboda ME, Van Wyk JJ, Cuatrecasas P (1983b)
Somatomedin-C stimulates the phosphorylation of the beta-subunit of its own
receptor. J Biol Chem 258:9581–9584

Jacobs S, Cook S, Svoboda M, Van Wyk JJ (1986) Interaction of the monoclonal
antibodies alpha-IR-1 and alpha-IR-3 with insulin and somatomedin-C receptors.
Endocrinology 118:223–226

James DE, Lederman L, Pilch PF (1987) Purification of insulin-dependent exocytic
vesicles containing the glucose transporter. J Biol Chem 262:11817–11824

Jansen M, van Schaik FM, Ricker AT, Bullock B, Woods DE, Gabbay KH, Nussbaum
    AL, Sussenbach JS, Van den Brande JL (1983) Sequence of a cDNA encoding
    human insulin-like growth factor I precursor. Nature 306:609–611
Jansen M, van Schaik FMA, Van Tol H, Van den Brande JL, Sussenbach JS (1985)
    Nucleotide sequences of cDNAs encoding precursors of human insulin-like growth
    factor II (IGF-II) and an IGF-II variant. FEBS Lett 179:243–246
Jennische E, Skottner A, Hansson HA (1987) Satellite cells express the trophic factor
    IGF-I in regenerating skeletal muscle. Acta Physiol Scand 129:9–15
Jonas HA, Harrison LC (1985) The human placenta contains two distinct binding and
    immunoreactive species of insulin-like growth factor-I receptors. J Biol Chem
    260:2288–2294
Jonas HA, Harrison LC (1986) Disulphide reduction alters the immunoreactivity and
    increases the affinity of insulin-like growth-factor-1 in human placenta. Biochem J
    236:417–423
Julkunen M, Koistinen R, Aalto-Setala K, Seppala M, Janne OA, Kontula K (1988)
    Primary structure of human insulin-like growth factor-binding protein/placental
    protein 12 and tissue-specific expression of its mRNA. FEBS Lett 236:295–302
Kadota S, Fantus IG, Hersh B, Posner BI (1986) Vanadate stimulation of IGF binding
    to rat adipocytes. Biochem Biophys Res Commun 138:174–178
Kadota S, Fantus IG, Deragon G, Guyda HJ, Hersh B, Posner BI (1987a) Peroxide(s)
    of vanadium: a novel and potent insulin-mimetic agent which activates the insulin
    receptor kinase. Biochem Biophys Res Commun 147:259–266
Kadota S, Fantus IG, Deragon G, Guyda HJ, Posner BI (1987b) Stimulation of
    insulin-like growth factor II receptor binding and insulin receptor kinase activity in
    rat adipocytes. Effects of vanadate and H_2O_2. J Biol Chem 262:8252–8256
Kadowaki T, Koyasu S, Nishida E, Tobe K, Izumi T, Takaku F, Sakai H, Yahara I,
    Kasuga M (1987) Tyrosine phosphorylation of common and specific sets of cellular
    proteins rapidly induced by insulin, insulin-like growth factor I, and epidermal
    growth factor in an intact cell. J Biol Chem 262:7342–7350
Kahn CR (1980) The riddle of tumour hypoglycemia revisited. Clin Endocrinol Metab
    9:335–360
Kanje M, Skottner A, Sjoberg J, Lundborg G (1989) Insulin-like growth factor I (IGF-
    I) stimulates regeneration of the rat sciatic nerve. Brain Res 486:396–398
Kasuga M, Sasaki N, Kahn CR, Nissley SP, Rechler MM (1983) Antireceptor antibo-
    dies as probes of insulinlike growth factor receptor structure. J Clin Invest
    72:1459–1469
Kiess W, Greenstein LA, White RM, Lee L, Rechler MM, Nissley SP (1987a) Type II
    insulin-like growth factor receptor is present in rat serum. Proc Natl Acad Sci USA
    84:7720–7724
Kiess W, Haskell JF, Lee L, Greenstein LA, Miller BE, Aarons AL, Rechler MM,
    Nissley SP (1987b) An antibody that blocks insulin-like growth factor (IGF) bind-
    ing to the type II IGF receptor is neither an agonist nor an inhibitor of IGF-
    stimulated biologic responses in L6 myoblasts. J Biol Chem 262:12745–12751
Kiess W, Lee L, Greenstein L, Rechler MM, Nissley SP (1987c) Synthesis and post-
    translational processing of the type II insulin-like growth factor (IGF) receptor in
    C6 glial cells. J Cell Biol 105:111a
Kiess W, Blickenstaff GD, Sklar MM, Thomas CL, Nissley SP, Sahagian GG (1988a)
    Biochemical evidence that the type II insulin-like growth factor receptor is identical
    to the cation-independent mannose 6-phosphate receptor. J Biol Chem
    263:9339–9344
Kiess W, Sklar MM, Thomas CL, Lee L, Nissley SP (1988b) Insulin-like growth fac-
    tor II (IGF-II) inhibits binding of beta-galactosidase to the IGF-II/mannose-6-
    phosphate (M6P) receptor. Program of the Endocrine Society, 70th annual meet-
    ing, p 245 (Abstract 899)
King GL, Rechler MM, Kahn CR (1982) Interactions between the receptors for insulin
    and the insulin-like growth factors on adipocytes. J Biol Chem 257:10001–10006

Kittur SD, Hoppener JWM, Antonarakis SE, Daniels JDJ, Meyers DA, Maestri NE, Jansen M, Korneluk RG, Nelkin BD, Kazazian HH (1985) Linkage map of the short arm of human chromosome 11: location of the genes for catalase, calcitonin, and insulin-like growth factor II. Proc Natl Acad Sci USA 82:5064–5067

Knauer DJ, Smith GL (1980) Inhibition of biological activity of multiplication-stimulating activity of binding to its carrier protein. Proc Natl Acad Sci USA 77:7252–7256

Koistinen R, Kalkkinen N, Huhtala M-L, Seppala M, Bohn H, Rutanen EM (1986) Placental protein 12 is a decidual protein that binds somatomedin and has an identical N-terminal amino acid sequence with somatomedin-binding protein from human amniotic fluid. Endocrinology 118:1375–1378

Kojima I, Nishimoto I, Iiri T, Ogata E, Rosenfeld R (1988) Evidence that type II insulin-like growth factor receptor is coupled to calcium gating system. Biochem Biophys Res Commun 154:9–19

Koontz JW, Iwahashi M (1981) Insulin as a potent, specific growth factor in a rat hepatoma cell line. Science 211:947–949

Kornfeld S (1987) Trafficking of lysosomal enzymes. FASEB J 1:462–468

Kull FC, Jacobs S, Su Y-F, Svoboda ME, Van Wyk JJ, Cuatrecasas P (1983) Monoclonal antibodies to receptors for insulin and somatomedin-C. J Biol Chem 258:6561–6566

Kurtz A, Jelkmann W, Bauer C (1982) A new candidate for the regulation of erythropoiesis. Insulin-like growth factor I. FEBS Lett 149:105–108

L'Allemain G, Pouyssegur J (1986) EGF and insulin action in fibroblasts. Evidence that phosphoinositide hydrolysis is not an essential mitogenic signalling pathway. FEBS Lett 197:344–348

Lajara R, Rotwein P, Bortz JD, Hansen VA, Sadow JL, Betts CR, Rogers SA, Hammerman MR (1989) Dual regulation of insulin-like growth factor I expression during renal hypertrophy. Am J Physiol 257:F252–F261

Lauterio TJ, Marson L, Daughaday WH, Baile CA (1987) Evidence for the role of insulin-like growth factor II (IGF-II) in the control of food intake. Physiol Behav 40:755–758

Le Bon TR, Jacobs S, Cuatrecasas P, Kathuria S, Fujita-Yamaguchi Y (1986) Purification of insulin-like growth factor I receptor from human placental membranes. J Biol Chem 261:7685–7689

Le Bouc Y, Dreyer D, Jaeger F, Binoux M, Sondermeyer P (1986) Complete characterization of the human IGF-I nucleotide sequence isolated from a newly constructed adult liver cDNA library. FEBS Lett 196:108–112

Le Bouc Y, Noguiez P, Sondermeijer P, Dreyer D, Girard F, Binoux M (1987) A new 5'-non-coding region for human placental insulin-like growth factor II mRNA expression. FEBS Lett 222:181–185

Lee Y-L, Hintz RL, James PM, Lee PDK, Shively JE, Powell DR (1988) Insulin-like growth factor (IGF) binding protein complementary deoxyribonucleic acid from human HEP G2 hepatoma cells: predicted protein sequence suggests an IGF binding domain different from those of the IGF-I and IGF-II receptors. Mol Endocrinol 2:401–411

Lesniak MA, Hill JM, Kiess W, Rojeski M, Pert CB, Roth J (1988) Receptors for insulin-like growth factors I and II: autoradiographic localization in rat brain and comparison to receptors for insulin. Endocrinology 123:2089–2099

Lindahl A, Nilsson A, Isaksson OGP (1987) Effects of growth hormone and insulin-like growth factor-I on colony formation of rabbit epiphyseal chondrocytes at different stages of maturation. J Endocrinol 115:263–271

Lobel P, Dahms NM, Kornfeld S (1988) Cloning and sequence analysis of the cation-independent mannose 6-phosphate receptor. J Biol Chem 263:2563–2570

Lowe WL, Roberts CT, Lasky SR, LeRoith D (1987) Differential expression of alternative 5' untranslated regions in mRNAs encoding rat insulin-like growth factor I. Proc Natl Acad Sci USA 84:8946–8950

Lowe WL, Lasky SR, LeRoith D, Roberts CT (1988) Distribution and regulation of rat insulin-like growth factor I messenger ribonucleic acids encoding alternative carboxyterminal E-peptides: evidence for differential processing and regulation in liver. Mol Endocrinol 2:528–535

Lund PK, Moats-Staats BM, Hynes MA, Simmons JG, Jansen M, D'Ercole AJ, Van Wyk JJ (1986) Somatomedin-C/insulin-like growth factor-I and insulin-like growth factor-II mRNAs in rat fetal and adult tissues. J Biol Chem 261:14 539–14 544

Lyons RM, Smith GL (1986) Characterization of multiplication-stimulating activity (MSA) carrier protein. Mol Cell Endocrinol 45:263–270

MacDonald RG, Czech MP (1985) Biosynthesis and processing of the type II insulin-like growth factor receptor in H-35 hepatoma cells. J Biol Chem 260: 11 357–11 365

MacDonald RG, Pfeffer SR, Coussens L, Tepper MA, Brocklebank CM, Mole JE, Anderson JK, Chen E, Czech MP, Ullrich A (1988) A single receptor binds both insulin-like growth factor II and mannose 6-phosphate. Science 239:1134–1137

Madoff DH, Martensen TM, Lane MD (1988) Insulin and insulin-like growth factor 1 stimulate the phosphorylation on tyrosine of a 160-kDa cytosolic protein in 3T3-L1 adipocytes. Biochem J 252:7–15

Maly P, Luthi C (1988) The binding sites of insulin-like growth factor I (IGF I) to type I IGF receptor and to a monoclonal antibody. J Biol Chem 263:7068–7072

Marquardt H, Todaro GJ, Henderson LE, Oroszlan SJ (1981) Purification and primary structure of a polypeptide with multiplication-stimulating activity from rat liver cell cultures. Homology with human insulin-like growth factor II. J Biol Chem 256:6859–6865

Martin JL, Baxter RC (1985) Antibody against acid-stable insulin-like growth factor binding protein detects 150,000 mol. wt. growth hormone-dependent complex in human plasma. J Clin Endocrinol Metab 61:799–801

Martin JL, Baxter RC (1986) Insulin-like growth factor-binding protein from human plasma purification and characterization. J Biol Chem 261:8754–8760

Martin JL, Baxter RC (1988) Insulin-like growth factor-binding proteins (IGF-BPs) produced by human skin fibroblasts: immunological relationship to other human IGF-BPs. Endocrinology 123:1907–1915

Massague J, Kelly B, Mottola C (1985) Stimulation by insulin-like growth factors is required for cellular transformation by type beta transforming growth factor. J Biol Chem 260:4551–4554

Mathews LS, Norstedt G, Palmiter RD (1986) Regulation of insulin-like growth factor I gene expression by growth hormone. Proc Natl Acad Sci USA 83:9343–9347

Matsuguchi T, Takahashi K, Ueno T, Endo H, Yamamoto M (1989) A novel transcription unit within the exon sequence of the rat insulin like growth factor II gene. Biochem Int 18:71–79

Matsunaga H, Nishimoto I, Kojima I, Yamashita N, Kurokawa K, Ogata E (1988) Activation of a calcium permeable cation channel by insulin-like growth factor-II in Balb/c 3T3 cells. Am J Physiol 255:C442–C446

Mellas J, Gavin JR III, Hammerman MR (1986) Multiplication-stimulating activity-induced alkalinization of canine renal proximal tubular cells. J Biol Chem 261:14437–14 4442

Merchav S, Tatarsky I, Hochberg Z (1988) Enhancement of human granulopoiesis in vitro by biosynthetic insulin-like growth factor I/somatomedin C and human growth hormone. J Clin Invest 81:791–797

Merimee TJ, Grant M, Tyson JE (1984) Insulin-like growth factors in amniotic fluid. J Clin Endocrinol Metab 59:753–755

Mills NC, D'Ercole AJ, Underwood LE, Ilan J (1986) Synthesis of somatomedin C/insulin-like growth factor I by human placenta. Mol Biol Rep 11:231–236

Misra P, Hintz RL, Rosenfeld RG (1986) Structural and immunological characterization of insulin-like growth factor II binding to IM-9 cells. J Clin Endocrinol Metab 63:1400–1405

Mohan S, Jennings JC, Linkhart TA, Baylink DJ (1988) Primary structure of human skeletal growth factor: homology with human insulin-like growth factor-II. Biochim Biophys Acta 966:44–55

Morera AM, Chauvin MA, de Peretti E, Binoux M, Benahmed M (1987) Somatomedin C/insulin-like growth factor 1: an intratesticular differentiative factor of Leydig cells? Horm Res 28:50–57

Morgan DO, Roth RA (1986) Identification of a monoclonal antibody which can distinguish between two distinct species of the type I receptor for insulin-like growth factor. Biochem Biophys Res Commun 138:1341–1347

Morgan DO, Roth RA (1987) Acute insulin action requires insulin receptor kinase activity: introduction of an inhibitory monoclonal antibody into mammalian cells blocks the rapid effects of insulin. Proc Natl Acad Sci USA 84:41–45

Morgan DO, Jarnagin K, Roth RA (1986) Purification and characterization of the receptor for insulin-like growth factor I. Biochemistry 25:5560–5564

Morgan DO, Edman JC, Standring DN, Fried VA, Smith MC, Roth RA, Rutter WJ (1987) Insulin-like growth factor II receptor as a multifunctional binding protein. Nature 329:301–307

Moses AC, Nissley SP, Cohen KL, Rechler MM (1976) Specific binding of a somatomedin-like polypeptide in rat serum depends on growth hormone. Nature 263:137–140

Moses AC, Nissley SP, Short PA, Rechler MM (1980a) Immunological cross-reactivity of multiplication stimulating activity polypeptides. Eur J Biochem 103:401–408

Moses AC, Nissley SP, Short PA, Rechler MM, Podskalny JM (1980b) Purification and characterization of multiplication-stimulating activity. Insulin-like growth factors purified from rat-liver-cell-conditioned medium. Eur J Biochem 103:387–400

Moses AC, Nissley SP, Short PA, Rechler MM, White RM, Knight AB, Higa OZ (1980c) Increased levels of multiplication-stimulating activity, an insulin-like growth factor, in fetal rat serum. Proc Natl Acad Sci USA 77:3649–3653

Moses AC, Freinkel AJ, Knowles BB, Aden DP (1983) Demonstration that a human hepatoma cell line produces a specific insulin-like growth factor carrier protein. J Clin Endocrinol Metab 56:1003–1008

Mottola C, MacDonald RG, Brackett JL, Mole JE, Anderson JK, Czech MP (1986) Purification and amino-terminal sequence of an insulin-like growth factor-binding protein secreted by rat liver BRL-3A cells. J Biol Chem 261:11 180–11 188

Mulholland MW, Debas HT (1988) Central nervous system inhibition of pentagastrin-stimulated acid secretion by insulin-like growth factor II. Life Sci 42:2091–2096

Murphy LJ, Friesen HG (1988) Differential effects of estrogen and growth hormone on uterine and hepatic insulin-like growth factor I gene expression in the ovariectomized hypophysectomized rat. Endocrinology 122:325–332

Murphy LJ, Bell GI, Friesen HG (1987a) Tissue distribution of insulin-like growth factor I and II messenger ribonucleic acid in the adult rat. Endocrinology 120:1279–1282

Murphy LJ, Bell GI, Duckworth ML, Friesen HG (1987b) Identification, characterization, and regulatin of a rat complementary deoxyribonucleic acid which encodes insulin-like growth factor-I. Endocrinology 121:684–691

Murphy LJ, Murphy LC, Friesen HG (1987c) Estrogen induces insulin-like growth factor-I expression in the rat uterus. Mol Endocrinol 1:445–450

Murphy LJ, Tachibana K, Friesen HG (1988) Stimulation of hepatic insulin-like growth factor-I gene expression by ovine prolactin: evidence for intrinsic somatogenic activity in the rat. Endocrinology 122:2027–2033

Nagarajan L, Anderson WB (1982) Insulin promotes the growth of F9 embryonal carcinoma cells apparently by acting through its own receptor. Biochem Biophys Res Commun 106:974–980

Nagarajan L, Anderson WB, Nissley SP, Rechler MM, Jetten AM (1985) Production of insulin-like growth factor-II (MSA) by endoderm-like cells derived from embryonal carcinoma cells: possible mediator of embryonic cell growth. J Cell Physiol 124:199–206

Nakanishi Y, Mulshine JL, Kasprzyk PG, Natale RB, Maneckjee R, Avis I, Treston AM, Gazdar AF, Minna JD, Cuttitta F (1988) Insulin-like growth factor-I can mediate autocrine proliferation of human small cell lung cancer cell lines in vitro. J Clin Invest 82:354–359

Nilsson A, Isgaard J, Lindahl A, Dahlstrom A, Skottner A, Isaksson OGP (1986) Regulation by growth hormone of number of chondrocytes containing IGF-I in rat growth plate, Science 233:571–574

Nishimoto I, Hata Y, Ogata E, Kojima I (1987a) Insulin-like growth factor II stimulates calcium influx in competent BALB/c 3T3 cells primed with epidermal growth factor. J Biol Chem 262:12120–12126

Nishimoto I, Ogata E, Kojima I (1987b) Pertussis toxin inhibits the action of insulin-like growth factor-I. Biochem Biophys Res Commun 148:403–411

Nissley SP, Rechler MM (1984a) Insulin-like growth factors: biosynthesis, receptors, and carrier proteins. In: Li CH (ed) Hormonal proteins and peptides, vol 12. Academic, New York, p 127

Nissley SP, Rechler MM (1984b) Somatomedin/insulin-like growth factor tissue receptors. Clin Endocrinol Metab 13:43–67

Norstedt G, Moller C (1987) Growth hormone induction of insulin-like growth factor I messenger RNA in primary cultures of rat liver cells. J Endocrinol 115:135–139

Norstedt G, Andersson G, Edwall D, Eriksson L, Hansson-H-A, Jennische E, Levinovitz A, Moller C (1987) Regulatory aspects of insulin-like growth factor expression in rodents. In: Isaksson O, Binder C, Hall K, Hokfelt B (eds) Growth hormone. Basic and clinical aspects, Elsevier Science, Amsterdam, p 387

Norstedt G, Levinovitz A, Moller C, Eriksson LC, Andersson G (1988) Expression of insulin-like growth factor I (IGF-I) and IGF-II mRNA during hepatic development, proliferation and carcinogenesis in the rat. Carcinogenesis 9:209–213

Oka Y, Czech MP (1986) The type II insulin-like growth factor receptor is internalized and recycles in the absence of ligand. J Biol Chem 261:9090–9093

Oka Y, Mottola C, Oppenheimer CL, Czech MP (1984) Insulin activates the appearance of insulin-like growth factor II receptors on the adipocyte cell surface. Proc Natl Acad Sci USA 81:4028–4032

Oka Y, Rozek LM, Czech MP (1985) Direct demonstration of rapid insulin-like growth factor II receptor internalization and recycling in rat adipocytes. J Biol Chem 260:9435–9442

Oka Y, Kasuga M, Kanazawa Y, Takaku F (1987) Insulin induces chloroquine-sensitive recycling of insulin-like growth factor II receptors but not of glucose transporters in rat adipocytes. J Biol Chem 262:17480–17486

Ooi GT, Herington AC (1986) Covalent cross-linking of insulin-like growth factor-1 to a specific inhibitor from human serum. Biochem Biophys Res Commun 137:411–417

Oppenheimer CL, Pessin JE, Massague J, Gitomer W, Czech MP (1983) Insulin action rapidly modulates the apparent affinity of the insulin-like growth factor II receptor. J Biol Chem 258:4824–4830

Oshima A, Nolan CM, Kyle JW, Grubb JH, Sly WS (1988) The human cation-independent mannose 6-phosphate receptor. Cloning and sequence of the full-length cDNA and expression of functional receptor in cos cells. J Biol Chem 263:2553–2562

Pfeffer SR (1988) Mannose 6-phosphate receptors and their role in targeting proteins to lysosomes. J Membr Biol 103:7–16

Pilistine SJ, Moses AC, Munro HN (1984) Insulin-like growth factor receptor in rat placental membranes. Endocrinology 115:1060–1065

Pohlmann R, Nagel G, Schmidt B, Stein M, Lorkowski G, Krentler C, Cully J, Meyer HE, Grzeschik K-H, Mersmann G et al. (1987) Cloning of a cDNA encoding the human cation-dependent mannose 6-phosphate-specific receptor. Proc Natl Acad Sci USA 84:5575–5579

Potau N, Mossner J, Williams JA, Goldfine ID (1984) Cholecystokinin and insulin regulate insulin-like growth factor II binding to pancreatic receptors; evidence of role for intracellular calcium. Biochem Biophys Res Commun 119:359–364

Povoa G, Enberg G, Jornvall H, Hall K (1984) Isolation and characterization of a so-
    matomedin-binding protein from mid-term human amniotic fluid. Eur J Biochem
    144:199–204
Povoa G, Isaksson M, Jornvall H, Hall K (1985) The somatomedin-binding protein
    isolated from a human hepatoma cell line is identical to the human amniotic fluid
    somatomedin-binding protein. Biochem Biophys Res Commun 128:1071–1078
Povoa G, Roovete A, Hall K (1987) Cross-reaction of serum somatomedin-bin-
    ding protein in a radioimmunoassay developed for somatomedin-binding pro-
    tein isolated from human amniotic fluid. Acta Endocrinol (Copenh) 107:563–570
Powell DR, Lee PDK, Chang D, Liu F, Hintz RL (1987) Antiserum developed for the
    E peptide region of insulin-like growth factor IA prohormone recognizes a serum
    protein by both immunoblot and radioimmunoassay. J Clin Endocrinol Metab
    65:868–875
Rechler MM, Nissley SP (1985a) Receptors for insulin-like growth factors. In: Posner
    BI (ed) Polypeptide hormone receptors. Dekker, New York, p 227
Rechler MM, Nissley SP (1985b) The nature and regulation of the receptors for
    insulin-like growth factors. Annu Rev Physiol 47:425–442
Rechler MM, Eisen HJ, Higa OZ, Nissley SP, Moses AC, Schilling EE, Fennoy I, Bruni
    CB, Phillips LS, Baird KL (1979) Characterization of a somatomedin (insulin-like
    growth factor) synthesized by fetal rat liver organ cultures. J Biol Chem 254:
    7942–7950
Rechler MM, Zapf J, Nissley SP, Froesch ER, Moses AC, Podskalny JM, Schilling EE,
    Humbel RE (1980) Interactions of insulin-like growth factors I and II and
    multiplication-stimulating activity with receptors and serum carrier proteins.
    Endocrinology 107:1451–1459
Rechler MM, Bruni CB, Whitfield HJ, Yang YW-H, Frunzio R, Graham DE, Coligan
    JE, Terrell JE, Acquaviva AM, Nissley SP (1985) Characterization of the biosynthetic
    precursor for rat insulin-like growth factor-II by biosynthetic labeling, radiose-
    quencing, and nucleotide sequence analysis of a cDNA clone. Cancer Cells 3:131–138
Rechler MM, Yang YW-H, Brown AL, Romanus JA, Adams SO, Kiess W, Nissley SP
    (1988) Insulin-like growth factors in fetal growth. In: Bercu BB (ed) Basic and clini-
    cal aspects of growth hormone. Plenum, New York, p 233
Rechler MM, Yang YW-H, Brown AL, Orlowski CC, Ooi GT (1989) Molecular
    characterization of insulin-like growth factor binding proteins in the rat. In: Drop
    SLS, Hintz RL (eds) Insulin-like growth factor binding proteins, excerpta medica
    international congress series, no 881. Elsevier, Amsterdam, p 133
Recio-Pinto E, Ishii DN (1984) Effects of insulin-insulin-like growth factor-II and
    nerve growth factor on neurite outgrowth in cultured human neuroblastoma cells.
    Brain Res 302:323–334
Recio-Pinto E, Rechler MM, Ishii DN (1986) Effects of insulin, insulin-like growth
    factor-II, and nerve growth factor on neurite formation and survival in cultured
    sympathetic and sensory neurons. J Neurosci 6:1211–1219
Reeve AE, Eccles MR, Wilkins RJ, Bell GI, Millow LJ (1985) Expression of insulin-like
    growth factor-II transcripts in Wilms' tumour. Nature 317:258–260
Ritvos O, Ranta T, Jalkanen J, Suikkari A-M, Voutilainen R, Bohn H, Rutanen EM
    (1988) Insulin-like growth factor (IGF) binding protein from human decidua in-
    hibits the binding and biological action of IGF-I in cultured choriocarcinoma cells.
    Endocrinology 122:2150–2157
Roberts CT, Brown AL, Graham DE, Seelig S, Berry S, Gabbay KH, Rechler MM
    (1986) Growth hormone regulates the abundance of insulin-like growth factor I
    RNA in adult rat liver. J Biol Chem 261:10025–10028
Roberts CT, Lasky SR, Lowe WL, LeRoith D (1987a) Rat IGF-I cDNA's contain
    multiple 5'-untranslated regions. Biochem Biophys Res Commun 146:1154–1159
Roberts CT, Lasky SR, Lowe WL, Seaman WT, LeRoith D (1987b) Molecular cloning
    of rat insulin-like growth factor I complementary deoxyribonucleic acids: differen-
    tial messenger ribonucleic acid processing and regulation by growth hormone in ex-
    trahepatic tissues. Mol Endocrinol 1:243–248

Rogers SA, Hammerman MR (1988) Insulin-like growth factor II stimulates production of inositol trisphosphate in proximal tubular basolateral membranes from dog kidney. Proc Natl Acad Sci USA 85:4037–4041

Romanus JA, Terrell JE, Yang YW, Nissley SP, Rechler MM (1986) Insulin-like growth factor carrier proteins in neonatal and adult rat serum are immunologically different: demonstration using a new radioimmunoassay for the carrier protein from BRL-3A rat liver cells. Endocrinology 118:1743–1758

Romanus JA, Yang YW-H, Nissley SP, Rechler MM (1987) Biosynthesis of the low molecular weight carrier protein for insulin-like growth factors in rat liver and fibroblasts. Endocrinology 121:1041–1050

Romanus JA, Yang YW, Adams SO, Sofair AN, Tseng LY, Nissley SP, Rechler MM (1988) Synthesis of insulin-like growth factor II (IGF-II) in fetal rat tissues: translation of IGF-II ribonucleic acid and processing of pre-pro-IGF-II. Endocrinology 122:709–716

Rosenfeld RG, Conover CA, Hodges D, Lee PDK, Misra P, Hintz RL, Li CH (1987) Heterogeneity of insulin-like growth factor-I affinity for the insulin-like growth factor-II receptor: comparison of natural, synthetic and recombinant DNA-derived insulin-like growth factor-I. Biochem Biophys Res Commun 143:199–205

Roth J, Lowe W, Fui ST, Arnold D, Eastmann R, LeRoith D, Keen H (1987a) Messenger RNA for IGF-II is increased in a hemangiopericytoma associated with hypoglycemia. Program of the Endocrine Society, 69th annual meeting, p 187 (Abstract 664)

Roth RA, Stover C, Hari J, Morgan DO, Smith MC, Sara V, Fried VA (1987b) Interactions of the receptor for insulin-like growth factor II with mannose-6-phosphate and antibodies to the mannose-6-phosphate receptor. Biochem Biophys Res Commun 149:600–606

Rotwein P (1986) Two insulin-like growth factor I messenger RNAs are expressed in human liver. Proc Natl Acad Sci USA 83:77–81

Rotwein P, Pollock KM (1987) Structure and expression of the mouse insulin-like growth factor II gene. Program of the Endocrine Society, 69th annual meeting, p 187 (Abstract 666)

Rotwein P, Pollock KM, Didier DK, Krivi GG (1986) Organization and sequence of the human insulin-like growth factor I gene: alternative RNA processing produces two insulin-like growth factor I precursor peptides. J Biol Chem 261: 4828–4832

Rotwein P, Folz RJ, Gordon JI (1987a) Biosynthesis of human insulin-like growth factor I (IGF-I). The primary translation product of IGF-I mRNA contains an unusual 48-amino acid signal peptide. J Biol Chem 262:11807–11812

Rotwein P, Pollock KM, Watson M, Milbrandt JD (1987b) Insulin-like growth factor gene expression during rat embryonic development. Endocrinology 121:2141–2144

Rotwein P, Burgess SK, Milbrandt JD, Krause JE (1988) Differential expression of insulin-like growth factor genes in rat central nervous system. Proc Natl Acad Sci USA 85:265–269

Ruoslahti E, Pierschbacher MD (1987) New perspectives in cell adhesion: RGD and integrins. Science 238:491–497

Rutanen EM, Bohn H, Seppala M (1982) Radioimmunoassay of placental protein 12: levels in amniotic fluid, cord blood, and serum of healthy adults, pregnant women, and patients with trophoblastic disease. Am J Obstet Gynecol 144:460–463

Rutanen EM, Wahlstrom T, Koistinen R, Sipponen P, Jalanko H, Seppala M (1984) Placental protein 12 (PP12) in primary liver cancer and cirrhosis. Tumour Biol 5:95–102

Rutanen EM, Koistinen R, Wahlstrom T, Bohn H, Ranta T, Seppala M (1985) Synthesis of placental protein 12 by human decidua. Endocrinology 116:1304–1309

Rutanen EM, Koistinen R, Sjoberg J, Julkunen M, Wahlstrom T, Bohn H, Seppala M (1986) Synthesis of placental protein 12 by human endometrium. Endocrinology 118:1067–1071

Rutanen EM, Pekonen F, Makinen T (1988) Soluble 34K binding protein inhibits the binding of insulin-like growth factor I to its cell receptors in human secretory phase endometrium: evidence for autocrine/paracrine regulation of growth factor action. J Clin Endocrinol Metab 66:173–180

Sahagian GG, Neufeld EF (1983) Biosynthesis and turnover of the mannose 6-phosphate receptor in cultured Chinese hamster ovary cells. J Biol Chem 258: 7121–7128

Sahagian GG, Steer CJ (1985) Transmembrane orientation of the mannose 6-phosphate receptor in isolated clathrin-coated vesicles. J Biol Chem 260:9838–9842

Sahal D, Ramachandran J, Fujita-Yamaguchi Y (1988) Specificity of tyrosine protein kinases of the structurally related receptors for insulin and insulin-like growth factor I: tyr-containing synthetic polymers as specific inhibitors or substrates. Arch Biochem Biophys 260:416–426

Salmon WD, Daughaday WH (1957) A hormonally controlled serum factor which stimulates sulfate incorporation by cartilage in vitro. J Lab Clin Med 49:825–836

Sara VR, Carlsson-Skwirut C, Andersson C, Hall E, Sjogren B, Holmgren A, Jornvall H (1986) Characterization of somatomedins from human fetal brain: identification of a variant form of insulin-like growth factor I. Proc Natl Acad Sci USA 83:4904–4907

Sasaki N, Rees-Jones RW, Zick Y, Nissley SP, Rechler MM (1985) Characterization of insulin-like growth factor I-stimulated tyrosine kinase activity associated with the beta-subunit of type I insulin-like growth factor receptors of rat liver cells. J Biol Chem 260:9793–9804

Schoenle E, Zapf J, Froesch ER (1986) Binding of non-suppressible insulin-like activity (NSILA) to isolated fat cells: evidence for two separate membrane acceptor sites. FEBS Lett 67:175–179

Schoenle E, Zapf J, Humbel ER, Froesch ER (1982) Insulin-like growth factor I stimulates growth in hypophysectomized rats. Nature 296:252–253

Schoenle EJ, Haselbacher GK, Briner J, Janzer RC, Gammeltoft S, Humbel RE, Prader A (1986) Elevated concentration of IGF II in brain tissue from an infant with macrencephaly. J Pediatr 108:737–740

Schwander JC, Hauri C, Zapf J, Froesch ER (1983) Synthesis and secretion of insulin-like growth factor and its binding protein by the perfused rat liver: dependence on growth hormone status. Endocrinology 113:297–305

Scott CD, Baxter RC (1986) Production of insulin-like growth factor I and its binding protein in rat hepatocytes cultured from diabetic and insulin-treated diabetic rats. Endocrinology 119:2346–2352

Scott CD, Baxter RC (1987) Purification and immunological characterization of the rat liver insulin-like growth factor-II receptor. Endocrinology 120:1–9

Scott CD, Martin JL, Baxter RC (1985a) Production of insulin-like growth factor I and its binding protein by adult rat hepatocytes in primary culture. Endocrinology 116:1094–1101

Scott CD, Martin JL, Baxter RC (1985b) Rat hepatocyte insulin-like growth factor I and binding protein: effect of growth hormone in vitro and in vivo. Endocrinology 116:1102–1107

Scott CD, Taylor JE, Baxter RC (1988) Differential regulation of insulin-like growth factor-II receptors in rat hepatocytes and hepatoma cells. Biochem Biophys Res Comm 151:815–821

Scott J, Cowell J, Robertson ME, Priestly LM, Wadey R, Hopkins B, Pritchard J, Bell GI, Rall LB, Graham CF, et al. (1985) Insulin-like growth factor-II gene expression in Wilms' tumour and embryonic tissues. Nature 317:260–262

Seppala M, Wahlstrom T, Koskimies AI, Tenhunen A, Rutanen EM, Koistinen R, Huhtaniemi I, Bohn H, Stenman U-H (1984) Human preovulatory follicular fluid, luteinized cells of hyperstimulated preovulatory follicles, and corpus luteum contain placental protein 12. J Clin Endocrinol Metab 58:505–510

Shapiro T, Polonsky K, Kew M, Rubinstein A, Bell G (1988) Tumor hypoglycemia is associated with increased expression of the gene for insulin-like growth factor II (Abstract) Clin Res 36:490 A

Shemer J, Adamo M, Wilson GL, Heffez D, Zick Y, LeRoith D (1987a) Insulin and insulin-like growth factor-I stimulate a common endogenous phosphoprotein substrate (pp 185) in intact neuroblastoma cells. J Biol Chem 262:15476–15482

Shemer J, Perrotti N, Roth J, LeRoith D (1987b) Characterization of an endogenous substrate related to insulin and insulin-like growth factor-I receptors in lizard brain. J Biol Chem 262:3436–3439

Shen S-J, Wang C-Y, Nelson KK, Jansen M, Ilan J (1986) Expression of insulin-like growth factor II in human placentas from normal and diabetic pregnancies. Proc Natl Acad Sci USA 83:9179–9182

Shen S-J, Daimon M, Wang C-Y, Jansen M, Ilan J (1988) Isolation of an insulin-like growth factor II cDNA with a unique 5' untranslated region from human placenta. Proc Natl Acad Sci USA 85:1947–1951

Shimatsu A, Rotwein P (1987a) Mosaic evolution of the insulin-like growth factors. J Biol Chem 262:7894–7900

Shimatsu A, Rotwein P (1987b) Seqeunce of two rat insulin-like growth factor I mRNAs differing within the 5' untranslated region. Nucleic Acids Res 15:7196

Simmen FA, Simmen RCM, Reinhart G (1988) Maternal and neonatal somatomedin C/insulin-like growth factor I (IGF-I) during early lactation in the pig. Dev Biol 130:16–27

Sklar MM, Kiess W, Thomas CL, Nissley SP (1989) Developmental expression of the tissue insulin-like growth factor II/mannose 6-phosphate receptor in the rat. Measurement by quantitative immunoblotting. J Biol Chem 264:16733–16738

Skottner A, Clark RG, Robinson ICAF, Fryklund L (1987) Recombinant human insulin-like growth factor: testing the somatomedin hypothesis is hypophysectomized rats. J Endocrinol 112:123–132

Soares MB, Ishii DN, Efstratiadis A (1985) Developmental and tissue-specific expression of a family of transcripts related to rat insulin-like growth factor II mRNA. Nucleic Acids Res 13:1119–1133

Soares MB, Turken A, Ishii D, Mills L, Episkopou V, Cotter S, Zeitlin S, Efstratiadis A (1986) Rat insulin-like growth factor II gene. A single gene with two promoters expressing a multitranscript family. J Mol Biol 192:737–752

Steele-Perkins G, Turner J, Edman JC, Hari J, Pierce SB, Stover C, Rutter WJ, Roth RA (1988) Expression and characterization of a functional human insulin-like growth factor I receptor. J Biol Chem 263:11486–11492

Stempien MM, Fong NM, Rall LB, Bell GI (1986) Sequence of a placental cDNA encoding the mouse insulin-like growth factor II precursor. DNA 5:357–361

Stiles AD, Sosenko IRS, D'Ercole AJ, Smith BT (1985) Relation of kidney tissue somatomedin-C/insulin-like growth factor I to postnephrectomy renal growth in the rat. Endocrinology 117:2397–2401

Stracke ML, Kohn EC, Aznavoorian SA, Wilson LL, Salomon D, Krutzsch HC, Liotta LA, Schiffmann E (1988) Insulin-like growth factors stimulate chemotaxis in human melanoma cells. Biochem Biophys Res Commun 153:1076–1083

Stylianopoulou F, Efstratiadis A, Herbert J, Pintar J (1988a) Pattern of the insulin-like growth factor II gene expression during rat embryogenesis. Development 103:497–506

Stylianopoulou F, Herbert J, Soares MB, Efstratiadis A (1988b) Expression of the insulin-like growth factor II gene in the choroid plexus and the leptomeninges of the adult rat central nervous system. Proc Natl Acad Sci USA 85:141–145

Suikkari AM, Koivisto VA, Rutanen EM, Yki-Jarvinen H, Karonen SL, Seppala M (1988) Insulin regulates the serum levels of low molecular weight insulin-like growth factor-binding protein. J Clin Endocrinol Metab 66:266–272

Sussenbach JS (1989) The gene structure of the insulin-like grwoth factor family. Prog Growth Factor Res 1:33–48

Sussenbach JS, de Pagter-Holthuizen P, Jansen M, Van den Brande JL (1988) Organization and expresion of the genes encoding the human somatomedins. In: Bercu BB (ed) Basic and clinical aspects of growth hormone. Plenum, New York, p 251

Szabo L, Mottershead DG, Ballard FJ, Wallace JC (1988) The bovine insulin-like growth factor (IGF) binding protein purified from conditioned medium requires the N-terminal tripeptide in IGF-1 for binding. Biochem Biophys Res Commun 151:207–214

Tally M, Enberg G, Li CH, Hall K (1987a) The specificity of the human IGF-2 receptor. Biochem Biophys Res Commun 147:1206–1212

Tally M, Li CH, Hall K (1987b) IGF-2 stimulated growth mediated by the somatomedin type 2 receptor. Biochem Biophys Res Commun 148:811–816

Tannenbaum GS, Guyda HJ, Posner BI (1983) Insulin-like growth factors: a role in growth hormone negative feedback and body weight regulation via brain. Science 220:77–79

Tavakkol A, Simmen FA, Simmen RCM (1988) Porcine insulin-like growth factor-I (pIGF-I): complementary deoxyribonucleic acid cloning and uterine expression of messenger ribonucleic acid encoding evolutionarily conserved IGF-I peptides. Mol Endocrinol 2:674–681

Tollefsen SE, Thompson K, Petersen DJ (1987) Separation of the high affinity insulin-like growth factor I receptor from low affinity binding sites by affinity chromatography. J Biol Chem 262:16461–16469

Tong PY, Tollefsen SE, Kornfeld S (1988) The caton-independent mannose 6-phosphate receptor binds insulin-like growth factor II. J Biol Chem 263:2585–2588

Tricoli JV, Rall LB, Scott J, Bell GI, Shows TB (1984) Localization of insulin-like growth factor genes to human chromosomes 11 and 12. Nature 310:784–786

Tricoli JV, Rall LB, Karakousis CP, Herrera L, Petrelli NJ, Bell GI, Shows TB (1986) Enhanced levels of insulin-like growth factor messenger RNA in human colon carcinomas and liposarcomas. Cancer Res 46:6169–6173

Tseng LY, Schwartz GP, Sheikh M, Chen ZZ, Joshi S, Wang J-F, Nissley SP, Burke GT, Katsoyannis PG, Rechler MM (1987) Hybrid molecules containing the A-domain of insulin-like growth factor-I and the B-chain of insulin have increased mitogenic activity relative to insulin. Biochem Biophys Res Commun 149: 672–679

Turner JD, Rotwein P, Novakofski J, Bechtel PJ (1988) Induction of mRNA for IGF-I and -II during growth hormone-stimulated muscle hypertrophy. Am J Physiol 255:E513–E517

Turo KA, Florini JR (1982) Hormonal stimulation of myoblast differentiation in the absence of DNA synthesis. Am J Physiol 12:278–284

Ueno T, Takahashi K, Matsuguchi T, Endo H, Yamamoto M (1987) A new leader exon identified in the rat insulin-like growth factor II gene. Biochem Biophys Res Commun 148:344–349

Ueno T, Takahashi K, Matsuguchi T, Endo H, Yamamoto M (1988a) Transcriptional deviation of the rat insulin-like growth factor II gene initiated at three alternative leader exons between neonatal tissues and ascites hepatomas. Biochim Biophys Acta 950:411–419

Ueno T, Takahashi K, Matsuguchi T, Ikejiri K, Endo H, Yamamoto M (1988b) Reactivation of rat insulin-like growth factor II gene during hepatocarcinogenesis. Carcinogenesis 9:1779–1783

Ullrich A, Gray A, Tam AW, Yang-Feng T, Tsubokawa M, Collins C, Henzel W, Le Bon T, Kathuria S, Chen E et al. (1986) Insulin-like growth factor I receptor primary structure: comparison with insulin receptor suggests structural determinants that define functional specificity. EMBO J 5:2503–2512

Unterman TG, Rajamohan G, Patel K, Uy R (1988) Increased insulin-like growth factor binding protein activity in diabetes mellitus. Program of the Endocrine Society 70th annual meeting, p 151 (Abstract 524)

Van Wyk JJ (1984) The somatomedins: biological actins and physiologic control mechanisms. In: Li CH (ed) Hormonal proteins and peptides, vol 12. Academic, New York, p 82

Van Wyk JJ, Graves DC, Casella SJ, Jacobs S (1985) Evidence from monoclonal antibody studies that insulin stimulates deoxyribonucleic acid synthesis through the type I somatomedin receptor. J Clin Endocrinol Metab 61:639–943

Versphol EJ, Roth RA, Vigneri R, Goldfine ID (1984) Dual regulation of glycogen metabolism by insulin and insulin-like growth factors in human hepatoma cells (HEP-G2). J Clin Invest 74:1436–1443

von Figura K, Hasilik A (1986) Lysosomal enzymes and their receptors. Annu Rev Biochem 55:167–193

von Figura K, Gieselmann V, Hasilik A (1985) Mannose 6-phosphate-specific receptor is a transmembrane protein with a C-terminal extension oriented towards the cytosol. Biochem J 225:543–547

Voutilainen R, Miller WL (1987) Coordinate tropic hormone regulation of mRNAs for insulin-like growth factor II and the cholesterol side-chain-cleavage enzyme, P450ssc, in human steroidogenic tissues. Proc Natl Acad Sci USA 84: 1590–1594

Waheed A, Braulke T, Junghans U, von Figura K (1988) Mannose 6-phosphate/insulin like growth factor II receptor: the two types of ligands bind simultaneously to one receptor at different sites. Biochem Biophys Res commun 152:1248–1254

Walton PE, Etherton TD (1989) Effects of porcine growth hormone and insulin-like growth factor I (IGF-I) on immunoreactive IGF binding protein concentration in pigs. J Endocrinol 120:153–160

Walton PE, Baxter RC, Burleigh BD, Etherton TD (1989) Purification of the serum acidstable insulin-like growth factor binding protein from the pig (sus scrofa). Comp Biochem Physiol [B] 92:561–567

Wang C-Y, Daimon M, Shen S-J, Engelmann GL, Ilan J (1988a) Insulin-like growth factor-I messenger ribonucleic acid in the developing human placenta and in term placenta of diabetics. Mol Endocrinol 2:217–229

Wang J-F, Hampton B, Mehlman T, Burgess WH, Rechler MM (1988b) Isolation of a biologically active fragment from the carboxy-terminus of the fetal rat binding protein for insulin-like growth factors. Biochem Biophys Res Commun 157: 718–726

Wardzala LJ, Simpson IA, Rechler MM, Cushman SW (1984) Potential mechanism of the stimulatory action of insulin on insulin-like growth factor II binding to the isolated rat adipose cell. J Biol Chem 259:8378–8383

White RM, Nissley SP, Moses AC, Rechler MM, Johnsonbaugh RE (1981) The growth hormone dependence of a somatomedin-binding protein in human serum. J Clin Endocrinol Metab 53:49–57

White RM, Nissley SP, Short PA, Rechler MM, Fennoy I (1982) The developmental pattern of a serum binding protein for multiplication-stimulating activity in the rat. J Clin Invest 69:1239–1252

Whitfield HJ, Bruni CB, Frunzio R, Terrell JE, Nissley SP, Rechler MM (1984) Isolation of a cDNA clone encoding rat insulin-like growth factor-II precursor. Nature 312:277–280

Wilkins JR, D'Ercole AJ (1985) Affinity-labeled plasma somatomedin-C/insulinlike growth factor I binding proteins. J Clin Invest 75:1350–1358

Wood WI, Cachianes G, Henzel WJ, Winslow GA, Spencer SA, Hellmiss R, Martin JL, Baxter RC (1988) Cloning and expression of the growth hormone-dependent insulin-like growth factor-binding protein. Mol Endocrinol 2: 1176–1185

Yang YW-H, Rechler MM, Nissley SP, Coligan JE (1985a) Biosynthesis of rat insulin-like growth factor II (rIGF-II) II. Localization of mature rat IGF-II (7,484 daltons) to the amino terminus of the ~20,000 dalton biosynthetic precursor by radiosequence analysis. J Biol Chem 260:2578–2582

Yang YW-H, Romanus JA, Liu T-Y, Nissley SP, Rechler MM (1985b) Biosynthesis of rat insulin-like growth factor II (rIGF-II) I. Immunochemical demonstration of a ~ 20,000 dalton biosynthetic precursor of rat IGF-II in metabolically labeled BRL-3A rat liver cells. J Biol Chem 260:2570–2577

Yang YW-H, Wang J-F, Orlowski CC, Nissley SP, Rechler MM (1989) Structure, specificity and regulation of the insulin-like growth factor binding proteins in adult rat serum. Endocrinology 125:1540–1555

Yee D, Cullen KJ, Paik S, Perdue JF, Hampton B, Schwartz A, Lippman ME, Rosen N (1988) Insulin-like growth factor II mRNA expression in human breast cancer. Cancer Res 48:6691–6696

Yu K-T, Peters MA, Czech MP (1986) Similar control mechanisms regulate the insulin and type I insulin-like growth factor receptor kinases. J Biol Chem 261:11 341–11 349

Zapf J, Schoenle E, Jagars G, Sand I, Grunwald J, Froesch ER (1979) Inhibition of the action of nonsuppressible insulin-like activity on isolated rat fat cells by binding to its carrier protein. J Clin Invest 62:1077–1084

Zapf J, Schoenle E, Froesch ER (1985) In vivo effects of the insulin-like growth factors (IGFs) in the hypophysectomized rat: comparison with human growth hormone and the possible role of the specific IGF carrier proteins. Ciba Found Symp 116:169–187

Zapf J, Hauri C, Waldvogel M, Froesch ER (1986) Acute metabolic effects and half-lives of intravenously administered insulinlike growth factors I and II innormal and hypophysectomized rats. J Clin Invest 77:1768–1775

Zapf J, Born W, Chang J-Y, James P, Froesch ER, Fischer JA (1988) Isolation and NH_2-terminal amino acid sequences of rat serum carrier proteins for insulin-like growth factors. Biochem Biophys Res Commun 156:1187–1194

Zapf J, Hauri C, Waldvogel M, Futo E, Hasler H, Binz K, Guler HP, Schmid C, Froesch ER (1989) Recombinant human insulin-like growth factor I induces its own specific carrier protein in hypophysectomized and diabetic rats. Proc Natl Acad Sci USA 86:3813–3817

Zezulak KM, Green H (1986) The generation of insulin-like growth factor-I-sensitive cells by growth hormone action. Science 233:551–553

Zumstein PP, Luthi C, Humbel RE (1985) Amino acid sequence of a variant pro-form of insulin-like growth factor II. Proc Natl Acad Sci USA 82:3169–3172

CHAPTER 7

Fibroblast Growth Factors

A. BAIRD and P. BÖHLEN

A. Introduction

Growth factors are multifunctional (SPORN and ROBERTS 1988) and the family of fibroblast growth factors (FGFs) is a case in point. Although FGF activities have been known to exist in tissue extracts for almost half a century (TROWELL et al. 1939; HOFFMAN 1940), it has only been in the last 10 years that these activities have been characterized and the identity of many of the factors capable of stimulating fibroblast growth have been established. They include some of the interleukins (ILs), the tumor necrosis factors (TNFs), the platelet-derived growth factors (PDGFs), the transforming growth factors (TGFs), and the FGFs. The pleiotropic nature of each of these growth factors is highlighted by the fact that they share common activities on some cell types but have restricted activities on others. Thus, the capacity of crude or partially purified preparations of tissue and cell extracts (macrophages, as an example) to stimulate fibroblast proliferation is due to the presence of a combination of many distinct families of molecules. Hence the importance of obtaining structural data to distinguish molecules that share FGF activities.

While the biological redundancy of growth factors might seem to preclude the possibility of significant differences between their biological activities, there are many important functional features that distinguish these molecules. The purpose of this chapter is to consolidate the current knowledge regarding molecules initially described as having FGF activity, found later to be potent angiogenic factors, and now collectively known as the fibroblast growth factors (FGFs). Like the families of growth factors listed above, the FGFs have the capacity to stimulate fibroblast proliferation, are structurally related, and have a wide range of biological activities that may prove to have potential clinical importance.

B. Isolation and Characterization of Acidic and Basic FGFs

The existence of substances in brain and pituitary extracts which promote the growth of cultured fibroblasts was established as early as 1940 (TROWELL et al. 1939; HOFFMAN 1940) and effectively rediscovered in the early 1970s (ARMELIN 1973; GOSPODAROWICZ 1975). On the basis of these studies, a mitogen for fibroblasts was partially purified from brain and pituitary extracts and named fibroblast growth factor, FGF (GOSPODAROWICZ 1974, 1975). This molecule was tentatively described as being approximately 12 000 Da in size, acid and temperature sensitive, and with a high (basic) isoelectric point. The existence of a second, potentially distinct mitogen for fibroblasts was detected in these same tissues (THOMAS et al. 1980; GAMBARINI and ARMELIN 1982), and because of its acidic isoelectric point, was tentatively called acidic FGF to distinguish it from its basic counterpart. Over the course of the next 10 years, purification procedures were gradually improved in an effort to establish the identity of these two factors (LEMMON and BRADSHAW 1983). In 1984; the two mitogens, acidic FGF from brain (acidic FGF) (THOMAS et al. 1984), and basic FGF from pituitary (basic FGF) (BÖHLEN et al. 1984), were purified to homogeneity, and it was possible to obtain significant sequence information. It was thus established that the similar biological effects of acidic and basic FGFs were due to the fact that they are structurally related molecules.

Almost simultaneous to the characterization of the FGFs, it was recognized that a significant portion of the FGF-like activities present in tumor extracts bind heparin (SHING et al. 1984). Within a short period of time, it was possible to identify these FGF-like activities as being either acidic or basic FGF (GOSPODAROWICZ et al. 1984; BÖHLEN et al. 1985). Affinity chromatography using immobilized heparin was then applied to the isolation of FGF-like activities found in many tissues (BAIRD et al. 1986b; GOSPODAROWICZ et al. 1987). The method is now widely used as a powerful, yet very simple procedure for the preparation of highly purified native and recombinant FGFs.

I. Amino Acid Sequences of Acidic and Basic FGFs

The complete amino acid sequences of bovine and ovine pituitary basic FGF were established by classical protein sequencing (ESCH et al. 1985; SIMPSON et al. 1987). Basic FGF, as initially isolated from pituitary extracts, was found to be a single-chain protein of 146 amino acids with an apparent molecular mass of 16 500 Da (Fig. 1). The amino-terminal extended forms that were later identified from tissue extracts (UENO et al. 1986a; STORY et al. 1987) and cDNA cloning (ABRAHAM et al. 1986a) are shown in lower case and have identified a 155-amino acid variant which, after processing (removal of amino-terminal methionine) leads to a 154-amino-acid protein. A form extended at the amino terminus to give an even bigger form of basic FGF (157 amino acids) has been recently characterized (SOMMER et al. 1987). These predicted translation products of alternative initiation sites are shown in parentheses to illustrate the significant microheterogeneity of preparations of basic FGF when purified from tissue extracts.

Bovine and human acidic FGF were initially purified from brain extracts and first characterized by classical protein sequencing (GIMENEZ-GALLEGO et al. 1985, 1986a; GAUTSCHI-SOVA et al. 1986). The sequence determination established that acidic FGF is a 140-amino-acid protein with a predicted molecular mass of 15 500 Da. Just as in the case of basic FGF, the protein se-

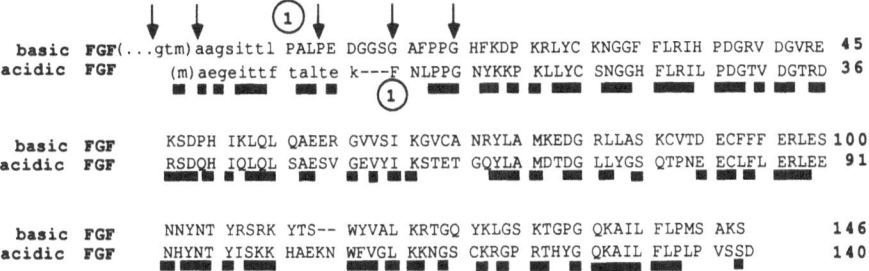

Fig. 1. Primary sequences of human basic and acidic FGFs. The sequences shown in *capital letters* correspond to the proteins that were isolated from tissues according to the procedures outlined by ESCH et al. (1985, 1986) and THOMAS et al. (1984). The sequences in *lower case* correspond to amino-terminal extensions deduced from the cloning of the FGFs (JAYE et al. 1986; ABRAHAM et al. 1986a, b) or from sequence analyses (BURGESS et al. 1986; STORY et al. 1987; UENO et al. 1986a). The sequences in *parentheses* are encoded by the mRNA for FGFs (JAYE et al. 1986; ABRAHAM et al. 1986a, b) and may represent larger molecular weight forms of basic FGF (SOMMER et al. 1987). *Arrows* indicate amino acid cleavage sites and the proteins are *numbered* using the nomenclature of ESCH et al. (1985) as shown. *Black bars* indicate sequence homology

quenced as acidic FGF was in fact a fragment of a larger molecule extended at the amino terminus and known as endothelial cell growth factor, ECGF (JAYE et al. 1986). The sequence of the acidic FGF isolated and characterized from tissue extracts is shown in upper case letters in Fig. 1. The sequences contributing to amino-terminal extensions are shown in lower-case letters. As shown by the black bars in Fig. 1, acidic and basic FGFs are homologous proteins with about 55% sequence identity. Analysis of their structures in a sequence database identified low, but significant homology (19%–25%) with the Il-α and IL-1β, suggesting the existence of a family of structurally related mitogens. This possibility was later confirmed by the discovery of three related factors; the products of *int*-2, *hst/ks*, and *FGF*-5, which are all genes that encode for proteins homologous to acidic and basic FGF (see Sect. B. III). Although the homology between the FGFs and IL-1α and IL-1β is significant, it is relatively low, particularly when compared to all of the members of the FGF family. For this reason, and because the IL-1s share only a few of their biological activities with the FGFs, they are not usually considered to be members of the FGF family.

II. Microheterogeneous Forms of Acidic and Basic FGFs

Purified preparations of basic and acidic FGFs were frequently observed to include several molecular forms of the mitogens. Sequence analysis revealed that some tissue extracts contained, in addition to the FGFs described above, amino-terminally truncated or extended forms of the growth factors. Truncated forms of basic FGF that were lacking the first four (BERTOLINI and HEARN 1987), eleven (UENO et al. 1986b), and fifteen (BAIRD et al. 1985b; GOSPODAROWICZ et al. 1985a, 1986b; UENO et al. 1987) amino acids were found in extracts prepared from a variety of tissues. When tissue extracts were prepared under conditions that inhibit proteolytic activity, however, amino-terminally extended forms of basic FGF containing eight and eleven additional residues were isolated (Sommer et al. 1987; Klagsbrun et al. 1987). These molecular forms are illustrated by the arrows in Fig. 1. The possibility that the microheterogeneity was generated by extraction was further supported by the observation that homologous, amino-terminally truncated forms of acidic FGFs could be purified and characterized from tissue extracts (Gautschi et al. 1986 a, b; Gautschi-Sova et al. 1987; Burgess et al. 1986). Similarly, an extended form of acidic FGF (i.e. ECGF) could be purified from brain when it was extracted under neutral conditions (Burgess et al. 1986). The degradation of the amino-terminal extended forms of acidic and basic FGF yields proteins ranging from 130 to 155 amino acids which are currently regarded as extraction artifacts.

III. An FGF Family of Related Growth Factors

Basic and acidic FGFs are now recognized as belonging to a family of at least five structurally related proteins. Highly significant sequence homology was recently discovered to exist between FGFs and three cellular oncogene products (Fig. 2). The first is a protein encoded by the *hst/ks* oncogene. This DNA was independently discovered from fibroblasts transfected with Kaposi's sarcoma DNA (Yoshida et al. 1987) and a human stomach cancer DNA (Barr et al. 1988). In both instances, the mRNA codes for an identical 206-amino-acid protein whose carboxy-terminal portion has about 40% homology with basic FGF. The second protein is the gene product of the *int*-2 oncogene (Dickson and Peters 1987). This gene codes for a 243-residue protein whose amino terminus is homologous to the FGFs. Although the product of the *int*-2 gene has no known function, it has been implicated in the virally induced formation of mouse mammary tumors and appears to be expressed in a very restricted manner in early mouse embryogenesis. A third protooncogene and fifth member of the FGF family was most recently identified by Zhan et al. (1988) and called *FGF-5*. The protein encoded by this mRNA is 267 amino acids in length and has up to 50% structural homology with acidic and basic FGFs. Even more recently, two novel FGFs called FGF-6 (Marics et al. 1989) and KGF/FGF-7 (Rubin et al. 1989) have been described.

The biological activities of the *int*-2, *hst/ks,* and FGF-5 protooncogenes have not been studied in detail yet and it is unknown to what extent acidic and basic FGFs, FGF-5, and the proteins encoded by *int*-2 and *hst/ks* are biologically related. Similar activities may be expected but it is possible that the carboxy- and amino-terminal extensions convey different activities to the mitogens. It will be of great interest to study the relative roles of the members of this FGF family in normal development and tumor formation. Preliminary studies with the *hst/ks* protein expressed in cos cells has revealed that it can interact with the FGF receptor because peptide antagonists of basic FGF (Baird et al. 1988) inhibit the effects of this protooncogene (Halaban et al.

```
Basic FGF    -----------------------------------------------------------
Acidic FGF   -----------------------------------------------------------
hst/KS3      ----MSGPGTAAVALLPAVLLALLAPWAGRGGAAAPTAPNGTLEAELERRWESLVALS  54
int-2        ----------------------------------------MGLIWLLLLSLLEPSW  16
FGF-5        MSLSFLLLLFFSHLILSAWAHGEKRLAPKGQPGPAATDRNPRGSSSRQSSSSAMSSSS  58
```

```
                                               ▬▬▬
Basic FGF    MAAGSITTLPALPEDGGSGAFPPGHFKDPKRLYCKNG-GFFLRIHPDGRVDGVRE     54
Acidic FGF   MAEGEITTFTALTEK---FNLPPGNYKKPKLLYCSNG-GHFLRILPDGTVDGTRD     51
hst/KS3      LARLPVAAQPKEAAVQSGAGDYLLGIKRLRRLYCNVGIGFHLQALPDGRIGGAHA    109
int-2        PTTGPGTRLRRDAGGRGGVYEHLGGAPRFRKLYCATK--YHLQLHPSGRVNGSLE     69
FGF-5        ASSSPAASLGSQGSGLEQSSFQWSLGARTGSLYCRVGIGFHLQIYPDGKVNGSHE    113
             *    *    *              *  ** ***** * ****** ***** ** *
```

```
                   ▪           ▪ ▪           ▬▬     ▪▪          ▪▪ ▪
Basic FGF    KSDPHIKLQLQAEERGVVSIKGVCANRYLAMKEDGRLLASKCVTDECFFFERLES    109
Acidic FGF   RSDQHIQLQLSAESVGEVYIKSTETGQYLAMDTDGLLYGSQTPNEECLFLERLEE    106
hst/KS3      DTRDSL-LELSPVERGVVSIFGVASRFFVAMSSKGKLYGSPFFTDECTFKEILLP    163
int-2        NSAYSI-LEITAVEGVVAIKGLFSGRYLAMNKRGRLYASDHYNAECEFVERIHE    123
FGF-5        ANMLSV-LEIFAVSQGIVGIRGVFSNKFLAMSKKGKLHASAKFTDDCKFRERFQE    167
             *   ** *** *** ** *** **** * ***** ***** **   **** * *** *
```

```
                 ▪▪ ▪▪  ▪                           ▪      ▪
Basic FGF    NNYNTYRSRKYTS---------------WYVALKRTGQYKLG--SKTGPGQKAI    146
Acidic FGF   NHYNTYISKKHAEKN-------------WFVGLKKNGSCKRG--PRTHYGQKAI    145
hst/KS3      NNYNAYESYKYPG---------------MFIALSKNGKTKKG-NRVSPTMKVT    200
int-2        LGYNTYASRLYRTGSSGPGAQRQPGAQRPWYVSVNGKGRPRRGFKTRRT--QKSS    176
FGF-5        NSYNTYASAIHRTEKTGRE---------WYVALNKRGKAKRGCSPRVKPQHIST    212
             * **** **** *              ***** * ** *** *   * *** *
```

```
                 ▬▬▬
Basic FGF    LFLPMSAKS------------------------------------------------    155
Acidic FGF   LFLPLPVSSD------------------------------------------------    155
hst/KS3      HFLPRL----------------------------------------------------    206
int-2        LFLPRVLGHKDHEMVRLLQSSQPRAPGEGSQPRQRRQKKQSPGDHGKMETLSTRA    229
FGF-5        HFLPRFKQSEQPELSFTVTVPEKKNPPSPIKSKIPLSAPRKNTNSVKYRLKFRFG    267
             *****   *
```

```
Basic FGF    --------------
Acidic FGF   --------------
hst/KS3      --------------
int-2        TPSTQLHTGGLAVA                                             243
FGF-5        --------------
```

Fig. 2. Sequence homologies of the FGF family of growth factors. The primary sequences for human basic FGF (ABRAHAM et al. 1986 b), acidic FGF (JAYE et al. 1986), *hst/ks* (DELLI BOVI et al. 1987; BARR et al. 1988) and FGF-5 (ZHAN et al. 1988) and mouse *int*-2 (DICKSON and PETERS 1987) are presented and aligned to illustrate their structural homology. Sequences of amino acids that have structural identity for all five sequences are indicated by *black bars* and significant homologies are shown by *asterisks*. The numbering system uses the presumed initiation methionines as amino acid 1

1988). Thus, this oncogene product can be expected to possess many of the activities of basic FGF. Indeed, the *hst/ks* protein has the capacity to stimulate fibroblast and endothelial cell proliferation and plasminogen activator release (Delli-Bovi et al. 1988). Like the acidic and basic FGFs, it also binds immobilized heparin, but unlike the FGFs, it is glycosylated (Delli-Bovi et al. 1987, 1988). Similar studies with FGF-5 have established its mitogenic activity and heparin binding activity (Zhan et al. 1988). No information has been reported regarding the biological activities of the *int*-2 protein.

There is one feature that distinguishes some of the members of the FGF family from others. Basic FGF, acidic FGF, and the *int*-2 protein have no obvious signal sequence. As such it seems unlikely that they are secreted in a classical sense (see Sect. G). This is in marked contrast to both the *hst/ks* protein and FGF-5, whose mRNA predicts the existence of a signal peptide. Accordingly, both of these growth factors can be found in conditioned medium when their genes are expressed in cos cells. In the case of the *hst/ks* protein, glycosylation is required for secretion as well (Delli-Bovi et al. 1988).

IV. Molecular Biology of FGFs

cDNAs encoding human acidic FGF (Jaye et al. 1986) and bovine (Abraham et al. 1986a), human (Abraham et al. 1986b,c), and rat (Shimasaki et al. 1988; Kurokawa et al. 1988) basic FGF have been cloned and sequenced and predict the existence of proteins identical to those found by protein sequencing. The nucleotide sequence and genomic organization of the human basic FGF gene (Abraham et al. 1986b) has been described and is similar in organization to the gene for acidic FGF (Abraham et al. 1986c). They are single-copy genes which are located on chromosomes 5 (Jaye et al. 1986) and 4 (Mergia et al. 1986), respectively. The basic FGF gene spans at least 38 kilobases (kb) of genomic DNA with the coding sequence being interrupted by two large introns. This is characteristic of the FGF family of growth factors (Fig. 3). The exon boundaries for exons 1, 2, and 3 of all of the

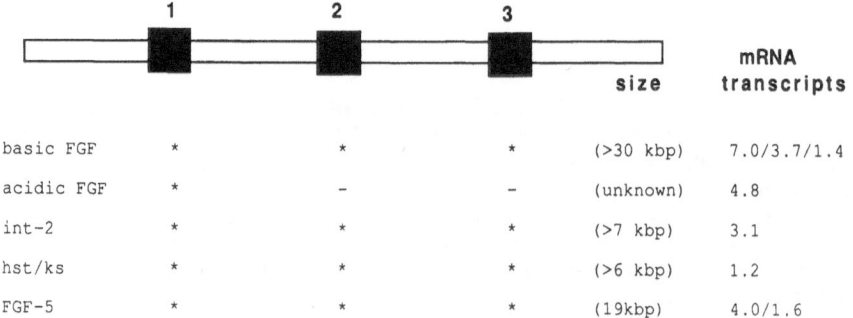

				size	mRNA transcripts
basic FGF	*	*	*	(>30 kbp)	7.0/3.7/1.4
acidic FGF	*	–	–	(unknown)	4.8
int-2	*	*	*	(>7 kbp)	3.1
hst/ks	*	*	*	(>6 kbp)	1.2
FGF-5	*	*	*	(19kbp)	4.0/1.6

Fig. 3. The genes for the FGF family represented diagrammatically to demonstrate the presence of three exons (*black boxes*) of various sizes and illustrate the current known sizes of transcripts found for the mRNA of each gene

known genes of the FGF family (basic FGF, *int*-2, *hst*/*ks,* and FGF-5) align perfectly, with the exception being the exon 1 boundary for basic FGF that is shifted by three nucleotides. The FGF-related genes thus appear to have evolved with virtually no change in the exon structure of their ancestral gene. The DNA sequences which code for basic and acidic FGFs are even more homologous than their protein counterparts, suggesting that the two proteins evolved from a common ancestral gene. Genomic DNAs are transcribed into 7-, 3.7-, and in some instances 1.4-kb mRNAs in the case of basic FGF and a single 4.8-kb transcript for acidic FGF. These mRNAs are readily observed in brain and in some cultured cells, but are virtually undetectable in other tissues.

Basic and acidic FGFs lack a signal peptide consensus sequence that could account for their secretion from cells (JAYE et al. 1986; ABRAHAM et al. 1986c) and it is currently unknown how they get out of the cell. It is thus paradoxical that these mitogens have an extracellular receptor (see Sect. E. II), that they can be detected outside the cell (see Sect. G.I), and that they are associated with the extracellular matrix (see Sect. G. II). Although the significance of this observation remains unclear, not all members of the FGF family of growth factors share this characteristic (see Sect. B III). A second feature of the FGFs is that the polyA$^+$ mRNA levels for both basic and acidic FGFs are low and cDNA clones are found only at very low frequencies even in high quality, large libraries (JAYE et al. 1986; ABRAHAM et al. 1986a, b, c; MERGIA et al. 1986; SHIMASAKI et al. 1988; KUROKAWA et al. 1988). While this suggests that the mRNAs may be unstable, it should be noted that despite the low mRNA levels, the concentrations of both basic and acidic FGFs found in tissues are relatively high. The observation of KIMELMAN et al. (1988) that the mRNAs detected in *Xenopus* are dependent on the size of the cDNA probes raises many unanswered questions pertaining to the processing of basic FGF mRNA and suggests that in some tissues the presence of mRNA may be masked from detection.

The cDNAs that encode the mRNAs for mature basic and acidic FGF possess open reading frames that predict the existence of 155-residue proteins. The open reading frames of the FGF mRNAs correspond exactly to the amino-terminally extended forms of FGFs that have been isolated from tissues (see Fig. 1). It is thought that the methionine codons at the beginning of the FGF sequences represent the translation initiation codons, and that these are removed at the time of protein synthesis. This assignment is not disputed in the case of the mRNA encoding acidic FGF because the initiation codon is immediately preceded by termination codons. In contrast, the situation is by no means clear in the case of the mRNA encoding basic FGF. While structural features of the mRNA for basic FGF and the general homology between the genes for basic and acidic FGFs suggest that the methionine codon corresponds to the initiation codon, there is no nearby stop signal in the known 5′ upstream sequence of its mRNA. Thus, there could potentially exist high molecular weight forms of basic FGF (see Sect. D. II). To date there is no evidence for alternative splicing of FGF mRNA and there is no evidence that any of the microheterogeneous forms have modified biological activity.

V. Homologies Between Species

FGFs have been partially characterized in many vertebrate species including mammals, birds, amphibians, and fish and there is high sequence homology between FGFs of different species. This is particularly evident with the basic FGFs of human (ABRAHAM et al. 1986b), bovine (ESCH et al. 1985), rat (SHIMASAKI et al. 1988), *Xenopus* (KIMELMAN et al. 1988), and ovine (SIMPSON et al. 1987) origin which are exceptionally highly preserved. This high degree of structure conservation strongly implies evolutionary pressure for maintaining important biological functions. Homology between acidic FGFs from various species is also high. Bovine and human acidic FGFs differ in eleven positions, and chicken and human acidic FGFs have identical amino-terminal sequences (RISAU et al. 1988). The proteins encoded by human *hst/ks, FGF-5,* and mouse *int*-2 have not yet been determined in other species.

C. Biosynthesis of FGFs

Many of the cells that respond to FGFs also synthesize the growth factor. Among the cells in this category are fibroblasts, vascular and capillary endothelial cells, smooth muscle cells, granulosa cells, adrenocortical cells, and astrocytes grown in cell culture (Table 1). Such cells, if capable of releas-

Table 1. Distribution of FGF-like mitogens in normal tissues, tumors, and cultured cells

Tissue localization	Pseudonym	Abbreviation	References
Tissues			
Pituitary	Pituitary FGF	FGF	BÖHLEN et al. (1984)
Brain	Brain FGF	FGF	THOMAS et al. (1984)
Brain	Brain-derived growth factor	BNDF	HUANG et al. (1986a, b)
Brain	Heparin-binding growth factor	HBGF	LOBB and FETT (1984)
Hypothalamus	Endothelial cell growth factor	ECGF	LOBB and FETT (1984)
Brain stem	Brain-derived growth factor	BDGF	JAYE et al. (1986)
Retina	Retina-derived growth factor	RDGF	BAIRD et al. (1985c)
Eye	Eye-derived growth factor	EDGF	COURTY et al. (1985)
Ciliary nerve	Ciliary growth factor	CNTF	SCHUBERT et al. (1987)
Kidney	Kidney angiogenic factor	KAF	BAIRD et al. (1985b)
Adrenal	Adrenal growth factor	AGF	GOSPODAROWICZ et al. (1986b)
Corpus luteum	Corpus luteum angiogenic factor	CLAF	GOSPODAROWICZ et al. (1985a)
Ovary	Ovarian growth factor	OGF	MAKRIS et al. (1984)
Placenta	Placental angiogenic factor	PAF	SOMMER et al. (1987)
Liver	Hepatocyte growth factor	HGF	UENO et al. (1986b)
Skeletal muscle	Myogenic growth factor	MGF	KARDAMI et al. (1985)
Myocardium	Heparin-binding growth factors	HBGF	THOMPSON et al. (1986)

Table 1 (continued)

Tissue localization	Pseudonym	Abbre-viation	References
Tissues			
Cartilage	Cartilage-derived growth factor	CDGF	KLAGSBRUN and BECKOFF (1980)
Bone matrix	Bone growth factor	BGF	MATSUO et al. (1987)
Testis	Seminferous growth factor	SGF	BELLVÉ and FEIG (1984)
Prostate	Prostatropin	PGF	STORY et al. (1987)
Thymus	Heparin-binding growth factor	HBGF	GOSPODAROWICZ (1987)
Tumors			
Chondrosarcoma	Tumor-derived growth factor	CDGF	SHING et al. (1984)
Hepatoma	Hepatoma-derived growth factor	HDGF	LOBB et al. (1986a)
Melanoma	Melanoma-derived growth factor	MDGF	LOBB et al. (1986a)
Mammary tumor	Mammary-tumor-derived GF	MTGF	ROWE et al. (1986)
Bladder tumor	Heparin-binding growth factor	HBGF	CHODAK et al. (1986)
Prostate tumor	Prostatic growth factor	PGF	MATSUO et al. (1987)
Others	Tumor angiogenic factor	TAF	FOLKMAN (1972)
Others	Tumor endothelial cell GF	tECGF	
Cells			
Pituitary follicle	Basic fibroblast growth factor	FGF	FERRARA et al. (1987)
Vascular endothelial	Basic fibroblast growth factor	FGF	SCHWEIGERER et al. (1987b)
Capillary endothelial	Basic fibroblast growth factor	FGF	SCHWEIGERER et al. (1987b)
Corneal endothelial	Basic fibroblast growth factor	FGF	KARDAMI et al. (1985)
Smooth muscle	Endothelial cell growth factor	ECGF	WINKLES et al. (1987)
Neurons	Brain-derived growth factor	BDGF	HUANG et al. (1987)
Adrenal cortex	Basic fibroblast growth factor	FGF	SCHWEIGERER et al. (1987a)
Granulosa	Basic fibroblast growth factor	FGF	NEUFELD et al. (1987a)
Fibroblasts	Basic fibroblast growth factor	FGF	MOSCATELLI et al. (1986)
Retinal pigment epithelial	Basic fibroblast growth factor	FGF	SCHWEIGERER et al. (1987b)
Macrophage	Basic fibroblast growth factor	FGF	BAIRD et al. (1985a)
Hepatoma	Basic fibroblast growth factor	FGF	KLAGSBRUN et al. (1986)
Medulloblastoma	Heparin-binding growth factors	HBGF	LOBB et al. (1986a)
Rhabdomyosarcoma	Basic fibroblast growth factor	FGF	SCHWEIGERER et al. (1987a)
Retinoblastoma	Basic fibroblast growth factor	FGF	SCHWEIGERER et al. (1987c)
HeLa	Basic fibroblast growth factor	FGF	MOSCATELLI et al. (1986)
Leukemia	Basic fibroblast growth factor	FGF	MOSCATELLI et al. (1986)
Neuroblastoma	Brain-derived growth factor	BDGF	HUANG et al. (1987)
Glioma	Angiogenic factor	FGF	LIBERMANN et al. (1987)

ing synthesized FGFs, might well stimulate their own proliferation. Whether such an autocrine regulation of growth contributes to physiological cell growth and differentiation remains to be established. This concept has received considerable attention in attempts to explain uncontrolled cellular proliferation in tumors. A large number of tumor cells are known to snythesize and respond to FGFs (glioma, rhabdomyosarcoma, leukemia, hepatoma, melanoma). In this context, it has been suggested that basic FGF-like activity derived from melanomas may contribute to the malignant phenotype of these tumors (HALABAN et al. 1987). The association of acidic FGF, *hst/ks,* and FGF-5 with tumors and tumor-derived cells also supports this hypothesis (BARR et al. 1988; DELLI BOVI et al. 1987; ZHAN et al. 1988; ZEYTIN et al. 1988). Because all FGFs have transforming potential (THOMAS 1988), it will become of paramount importance to understand the processing of their mRNAs to biologically active mitogens and the regulation of their biological activities.

I. Biosynthesis of Acidic FGF

The steps involved in the biosynthesis of acidic FGF can be deduced from its known biochemistry and molecular biology. While the mRNA codes for a 155-residue protein with an amino-terminal methionine, acidic FGF exists in many microheterogeneous forms (see Sect. B. II). Mature acidic FGF, as isolated from tissue under conditions of suppressed proteolytic activity, is a 154-amino-acid protein which possesses an acetylated alanine residue at the amino terminus (BURGESS et al. 1986) and corresponds to the entire mRNA coding sequence. The only processing that occurs is the removal of the amino-terminal methionine and the acetylation of the following alanine residue (see Fig. 1). How and if acidic FGF is secreted is not known, for it lacks a classical consensus signal peptide. Data that suggest the existence of a molecular weight form greater than 154 amino acids have not been confirmed and were obtained on the basis of Western blotting experiments (LOBB et al. 1986b). The discovery of many structurally related FGFs (basic, *hst, int*-2, FGF-5) makes it necessary to eliminate the possibility of antibody crossreactivity. The various truncated forms of acidic FGF in tissue extracts most likely represent a combination of protein processing and suppressible proteolytic degradation by the tissue-derived acid-activated proteinases described for basic FGF (KLAGSBRUN et al. 1987).

II. Biosynthesis of Basic FGF

The biosynthesis of basic FGF is not easily deduced even though properties of the biosynthetic system of basic FGF resemble closely those of acidic FGF. Its mRNA encodes a 155-amino-acid protein that is devoid of a signal peptide sequence. It is then presumably processed into a 154-residue mature non-glycosylated polypeptide with a modified amino terminus (UENO et al. 1986a; STORY et al. 1987). The question whether there is a larger biosynthetic basic FGF precursor protein (possibly containing a signal peptide sequence) cannot be answered in the absence of an unequivocal determination of the location of the translation initiation codon on the basic FGF mRNA.

There are experimental data to support the existence of a form of basic FGF larger than the 154–155 amino acids predicted by the cDNA. SOMMER et al. (1987) sequenced an enzymatic digestion fragment of placental basic FGF whose sequence includes the presumed methionine initiation site, the expected

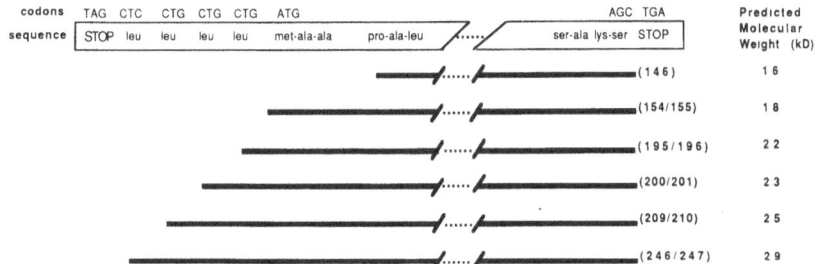

Fig. 4. Microheterogeneity in the molecular form of basic FGF. The 146-amino-acid protein first characterized is derived by proteolytic degradation of several potential translation products. The multiple forms of basic FGF exist as a function of the use of leucine (CTC, CTG) rather than methionine (ATG) initiation codons, resulting in the generation of proteins of 18 000–29 000 Da

amino-terminal sequence of basic FGF, and a two-amino-acid extension of the amino terminus. The isolation of the extended form of FGF is consistent with Western blotting experiments using various anti-basic FGF antibodies that have repeatedly suggested the existence of larger forms of basic FGF (MOSCATELLI et al. 1987a; PRESTA et al. 1988). Although it remains possible that the crossreactivity of antisera with high molecular weight forms of basic FGF reflects their crossreactivity with related protooncogenes like *hst, int*-2 and the FGF-5 gene, FLORKIEWICZ and SOMMER (1989) have recently shown that the translation of mRNA encoding basic FGF may also be initiated by sequences encoding leucine. As shown in Fig. 4, this would predict the existence of the 18 000- to 29 000-Da forms of basic FGF. Interestingly, each of these molecular forms has been described in the literature (GIMENEZ-GALLEGO et al. 1985; MOSCATELLI et al. 1987a; PRESTA et al. 1988). Inasmuch as there is no evidence for a precursor-product relationship between these molecular forms, it appears more likely that independent transcripts encode the 18-kDa mitogen of 154 amino acids (after processing) and the amino-terminal extended variants. The existence and importance of regulatory elements that would regulate transcript length are not known.

D. Distribution of FGFs in Tissues and Cells

Virtually nothing is known about the distribution, biological activity, biochemical characteristics, and protein chemistry of *hst/ks* protein, *int*-2 protein, and FGF-5 than what has been described above (see Sect. B. III). In contrast, considerable progress has been made in our understanding of the biology of acidic and basic FGFs since their characterization in 1985. Mitogens structurally identical or closely similar to one of the several characterized forms of basic and acidic FGFs have been isolated from many normal and tumor tissues (Table 1). Many activities that have previously been assigned various names and acronyms are all FGFs.

Large amounts of basic FGF can be purified from pituitary extracts (approximately 0.5 mg/kg), with most other tissues yielding 10- to 50-fold less.

Basic FGF is also synthesized in many varied normal and tumor cell types when they are grown in cell culture (Table 1). It is not known if these cells also synthesize FGFs in vivo. The distribution of acidic FGF was originally thought to be limited to neural tissues such as brain (LOBB and FETT 1984) and retina (BAIRD et al. 1985a), where the growth factor is present in relatively high concentrations (0.4 mg/kg brain). Only recently has acidic FGF been found in other tissues such as kidney (GAUTSCHI-SOVA et al. 1987), myocardium (THOMPSON et al. 1986; CASSCELLS et al. 1988), and bone (HAUSCHKA et al. 1986). Because acidic FGF is synthesized in various cell types (WINKLES et al. 1987; ZEYTIN et al. 1988), it is reasonable to expect acidic FGF to be detected in additional tissues and cells as well.

I. FGF-Like Growth Factors and Their Identity with FGF

The discovery that the heparin-binding growth factors in tumor cell extracts were in fact indistinguishable from basic FGF left open the possibility that many of the currently undefined activities present in cell and tissue extracts were due to basic or acidic FGF (Table 1). Accordingly, many angiogenic, fibroblast- and endothelial-stimulating, and heparin binding activities were identified within a short period of time. They included the heparin-binding FGF-like growth factors (LOBB et al. 1986c), adrenal angiogenic factor (GOSPODAROWICZ et al. 1986b), corpus luteum angiogenic factor (GOSPODAROWICZ et al. 1985a), cartilage-derived growth factor (LOBB et al. 1986a; KLAGSBRUN and BECKOFF 1980), retina- and eye-derived growth factors (BAIRD et al. 1985a; D'AMORE et al. 1981; D'AMORE and KLAGSBRUN 1984; BARRITAULT et al. 1981; COURTY et al. 1985, 1987), prostatropin (MATSUO et al. 1987; CRABB et al. 1986), kidney-derived angiogenic factor (BAIRD et al. 1985b; RISAU and EKBLOM 1986), endothelial cell growth factor (BURGESS et al. 1986; MACIAG et al. 1982), testes and seminiferous factors (UENO et al. 1987; BELLVÉ and FEIG 1984), brain-derived growth factor (HUANG and HUANG 1986), and many others, all of which were shown to be either basic or acidic FGF. In some extracts, the identity of FGF-like activities has remained undetermined. An example of this is the macrophage-derived growth factor (LEIBOVICH and ROSS 1976) which is probably not one growth factor, but composed of PDGF (MARTINET et al. 1986), basic FGF (BAIRD et al. 1985b), TNF (FRÀTER-SCHROEDER et al. 1987), TGF-β (ASSOIAN et al. 1987), and possibly other currently unidentified angiogenic factors. The proteins that are encoded by *hst/ks* and the FGF-5 gene have not been examined in detail, but can be found in some tumor cell lines and in cells derived from Kaposi's sarcoma (DELLI BOVI et al. 1987) and a human stomach tumor (TAIRA et al. 1987). FGF-5 is expressed in the neonatal brain and at least in some tumors (ZHAN et al. 1988).

II. FGFs in Serum and Physiological Fluids

An important issue that remains unresolved relates to the question whether any of the FGFs are present in serum or in other physiological fluids. Rat serum has been shown to contain an immunoreactive basic FGF-like substance of high molecular weight (MORMÈDE et al. 1985), but those results have been questioned (GAUTHIER et al. 1987). There is independent evidence (PHOTOPOULOS and CONNOLLY 1986; LATHROP et al. 1985b) to support the existence of FGF-like growth factors in serum, because serum treated with immobilized heparin (to which FGFs would presumably bind) permits the differentiation of the muscle cell line BC3H1. Unambiguous data demonstrating the presence of any FGF in serum does not exist. The presently favored view holds that FGFs are not circulating hormones, but rather, locally active paracrine or autocrine tissue factors. Like most growth factors, if indeed FGFs are in serum, it is most likely that they are in a biologically inactive form. This may include being bound to carrier proteins (MORMÈDE et al. 1985), a heparin-like molecule in serum (SNOW et al. 1987) or associated with carrier cells in blood (i.e., macrophages; BAIRD et al. 1985b).

There is little known concerning the presence of FGFs in other body fluids. So far, FGF-like material has been detected in chick embryo vitreous (basic and acidic FGFs; MASCARELLI et al. 1987), in a pathological sample of human ocular fluid (basic FGF; BAIRD et al. 1985b), in synovial fluid (basic and acidic FGFs; HAMERMAN et al. 1987). In follicular fluid (BAIRD et al. 1989a), and in the urine of mice and patients with bladder or kidney cancer (basic FGF; CHODAK et al. 1986, 1988). The evidence that FGFs play any physiological or pathophysiological function in these fluids remains circumstantial.

E. Mechanism of Action of FGFs

I. Target Cells

Many cell types are capable of responding to the FGFs (Table 2). The most remarkable observation is how little is known about the effects of FGFs in vivo. In most cells, FGFs elicit the typical response expected from a growth factor (i.e., increased DNA synthesis and cell division). However, a large number of cell types that respond to FGF with a mitogenic response can also react to the same growth factor in a nonmitogenic manner. Such mitosis-independent activities include the stimulation or suppression of specific cellular protein synthesis, the induction of cellular motility and migration, and changes in differentiated function, cell survival and the onset of senescence (see below). Basic FGF has even been reported as a growth inhibitor for some cultured tumor cells (SCHWEIGERER et al. 1987c). Frequently, the stimulation of cells by FGFs evokes multiple responses within a single target cell population. At this time, very little is known about the factors that direct how the cell will respond to FGFs. The interactions between FGFs with substances like hydrocortisone (GOSPODAROWICZ 1974), estrogens (MORMÈDE and BAIRD 1988), or other growth factors like TNF (FRÀTER-SCHROEDER et al. 1987), TGF-β (BAIRD and DURKIN 1986; FRÀTER-SCHROEDER et al. 1986; SAKSELA et

Table 2. Pleiotropic activities of FGFs on normal cells

Target tissue or cell	In vitro		In vivo[a]	
	Mitogenic activity	Nonmitogenic activity	Mitogenic activity	Nonmitogenic activity
Fibroblasts	GOSPODAROWICZ (1974)	SENIOR et al. (1986)	DAVIDSON et al. (1985)	SPRUGEL et al. (1987)
Vascular endothelial	GOSPODAROWICZ et al. (1976)	GOSPODAROWICZ et al. (1980)	GOSPODAROWICZ et al. (1979)	
Capillary endothelial	GOSPODAROWICZ et al. (1984)	PRESTA et al. (1986)		
Smooth muscle	WINKLES et al. (1987)	MIOH and CHENG (1987)		
Skeletal muscle	CLEGG et al. (1987)	CLEGG et al. (1987)		
Satellite cells	ALLEN et al. (1984)			
Myocytes	LATHROP et al. (1985b)	LATHROP et al. (1985a)		
Chondrocytes	KATO and GOSPODAROWICZ (1984)	KATO and GOSPODAROWICZ (1985)	CUEVAS et al. (1988b)	JENTZSCH et al. (1980)
Bone cells	GLOBUS et al. (1988)	CANALIS et al. (1987)		
Keratinocytes	O'KEEFE et al. (1988)			
Melanocytes	HALABAN et al. (1987)	HALABAN et al. (1987)		
Adrenal cortex	GOSPODAROWICZ et al. (1977)	HORNSBY and GILL (1977)		
Adrenal medulla	None	UNSICKER et al. (1988)		
Granulosa	GOSPODAROWICZ et al. (1985a)	BAIRD and HSUEH (1986)		
Lactotrophs	None	BAIRD et al. (1985)		
Thyrotrophs	None	BAIRD et al. (1985)		
Leydig cells		FAUSER et al. (1988)		
Retinal neurons	None	LIPTON et al. (1988)	None	
CNS neurons	None	WALICKE et al. (1986)	None	SIEVERS et al. (1987)
Peripheral neurons	None	UNSICKER et al. (1987)	None	ANDERSON et al. (1988)
Astrocytes	PETTMANN et al. (1985)	MORRISON et al. (1985)		OTTO et al. (1987)
Oligodendrocytes	SANETO and DE VELLIS (1985)	ROGISTER et al. (1988)		
Schwann cells	PRUSS et al. (1981)	PRUSS et al. (1981)		
Lens epithelial	CUNY et al. (1986)	CHAMBERLAIN and McAVOY (1987)	CUEVAS et al. (1988a)	

Table 2 (continued)

Target tissue or cell	In vitro		In vivo[a]	
	Mitogenic activity	Nonmitogenic activity	Mitogenic activity	Nonmitogenic activity
Corneal epithelial	THOMPSON et al. (1982)	GOSPODAROWICZ et al. (1979)	FREDJ-REYGROBELLET et al. (1987)	
Prostatic epithelial	CRABB et al. (1986)			
Myoblasts	LINKHART et al. (1980)		GOSPODAROWICZ and MESCHER (1981)	
Osteoblasts	TOGARI et al. (1983)			
Neuroblasts	GENSBURGER et al. (1987)		GOSPODAROWICZ and MESCHER (1981)	
Blastema	MESCHER and LOH (1980)	+/–		
Tumor cells[b]	+/–			

[a] The effects of FGFs on cell types in vivo are not necessarily direct and may involve secondary or tertiary responses mediated by other growth factors (i.e., TGF-β, epidermal growth factor) over the time course of the experiments (3–20 days).

[b] FGFs have mitogenic and nonmitogenic effects on a wide range of tumor cell types and have been purified from several tumors. In some instances the mitogens are antimitogenic, emphasizing their pleiotropic activity. Target tumor cells include PC-12 (pheochromocytoma), glioma, cervical carcinoma, mammary tumors, prolactinoma, and transformed fibroblasts. Not all tumor cells produce, have receptors for, or respond to FGFs.

al. 1987), and γ-interferons (BÖHLEN et al. 1987) result in changes in the cellular response to FGFs. In part this is due to the capacity of these factors to change the number of FGF receptors on cells, increase FGF mRNA, and inhibit cell proliferation. The components involved in modulating the cell responsiveness to FGFs are only beginning to be identified.

As shown in Table 2, the pleiotropic activity of FGFs on cells in culture is observed for both its mitogenic and nonmitogenic effects. In these instances, the growth factors can be stimulating cell proliferation, but it usually occurs in conjunction with changes in differentiated function. Thus, adrenocortical cells stimulated with FGF show delayed cell senescence (as measured by their capacity for steroid synthesis), endothelial cells produce plasminogen activator, and chondrocytes produce increased amounts of extracellular matrix (see below). In some instances, the effects of FGFs can occur in the absence of cell proliferation as in the case of prolactin production by normal (BAIRD et al. 1985a) and tumor (MORMÈDE and BAIRD 1988) rat pituitary cells or in the case of the stimulation of neurite outgrowth and survival of neurons in culture (MORRISON et al. 1986; WALICKE et al. 1986; LIPTON et al. 1988).

Significantly less is known about the proliferative effects of FGF on specific cell types in vivo. Many cell types can be shown to proliferate upon administration of exogenous FGFs. It is not clear whether these are direct effects of the growth factor, but from these experiments many analogies can be drawn from the in vitro results. As an example, the administration of exogenous FGF to lesioned articular cartilage stimulates chondrocyte proliferation and extracellular matrix formation (CUEVAS et al. 1988a; DAVIDSON et al. 1985). This effect is also seen in cultured cells (KATO and GOSPODAROWICZ 1984, 1985). Similarly, the administration of basic FGF to a severed sciatic nerve increases Schwann cell number and myelination and its delivery to the wall of the carotid artery increases capillary numbers of the vasa vasorum (see Sect. E. VII).

II. The FGF Receptor

Cells that respond to the FGFs have been shown to possess specific FGF receptor(s). Plasma membrane proteins that bind FGFs with high affinity and high specificity have been identified on several cell types including baby hamster kidney (BHK) cells (MOSCATELLI 1987; NEUFELD and GOSPODAROWICZ 1986), endothelial cells (FRIESEL et al. 1986), smooth muscle cells (WINKLES et al. 1987), fibroblasts (OLWIN and HAUSCHKA 1986; HUANG and HUANG 1986), myoblasts (OLWIN and HAUSCHKA 1986), epithelial lens cells (MOENNER et al. 1986), and PC12 pheochromocytoma (NEUFELD et al. 1987b) as well as a variety of other tumor cells (MOSCATELLI et al. 1987b). The receptor proteins appear to be single chain polypeptides with molecular masses ranging from 110 to 150 kDa, depending on cell type. The receptors can be purified from rat brain (IMAMURA et al. 1988) using a combination of lectin and ligand affinity chromatography and are associated with tyrosine kinase activity (HUANG and

HUANG 1986). The proteins bind FGFs with high affinity ($K_d = 10\text{--}80$ pM), with receptor numbers ranging from 2000 to 80 000 per cell.

In most but not all of the systems studied, the FGF receptor binds basic FGF with higher affinity than acidic FGF. This is consistent with the relative in vitro potencies of the two mitogens when tested in the absence of heparin (Sect. F. II.). On BHK cells, two FGF receptors with estimated molecular weights of 110 and 130 kDa have been identified (NEUFELD and GOSPODAROWICZ 1985; NEUFELD and GOSPODAROWICZ 1986). Both receptor proteins bind basic and acidic FGFs. Additional data suggest that the larger of the two receptors binds preferentially basic FGF while the smaller has somewhat higher affinity for acidic FGF (NEUFELD and GOSPODAROWICZ 1986). Other investigators have been unable to demonstrate the presence of the 110-kDa receptor form on BHK cells (MOSCATELLI 1987). The possibility that these receptors are shared with the proteins encoded by *FGF-5, hst/ks,* and *int*-2 remains speculative until these proteins are available for testing. The capacity of peptide antagonists of basic FGF (BAIRD et al. 1988) to inhibit the effects of *hst/ks* on melanocytes strongly suggests that this protooncogene acts through the FGF receptor (HALABAN et al. 1988).

In a series of elegant studies, MOSCATELLI (1987, 1988) has distinguished between specific high-affinity FGF binding sites and the low-affinity sites. Only the high-affinity binding sites are capable of transmitting the FGF signal to the cell. The low-affinity site is characterized by the fact that it is sensitive to heparinase treatment and that binding is blocked by exogenous heparin. This is not to say that carbohydrates play no function in the binding of FGF to its receptor; from 10 to 25 kDa of the mass of the high-affinity receptor have been attributed to the presence of carbohydrates. These sugars are N-linked, appear necessary for FGF binding, and are not heparin-related (FEIGE and BAIRD 1988 b). Accordingly, the mRNA that encodes the basic FGF receptor (flg) has a predicted molecular weight of only ~ 100 kDa (LEE et al. 1989). It is unclear that this one protein encodes for the receptor for all FGFs (RUTA et al. 1989). BURRUS and OLWIN (1989) have reported the existence of receptors distinct from flg.

III. Signal Transduction

The pathway(s) by which the signal generated by the binding of FGF to its receptor is transmitted to the cell nucleus has been studied in some detail, and although a substantial amount of data is at hand, it is not possible to propose a general signal transducing mechanism for FGFs. The available information is derived from studying varied cell types and it remains to be seen to what extent the findings obtained with one cell type apply to other cells. Different cell types appear to use, at least to some extent, different signaling pathways. FGFs, like other growth factors, have been found to rapidly induce the expression of the *fos* and *myc* protooncogenes which are considered early key events in the cellular response to a growth factor signal (MÜLLER et al. 1984; TSUDA et al. 1986; MAGNALDO et al. 1986; ZEYTIN et al. 1988). Several signal transducing pathways have been implicated in FGF-induced gene activation. They include the induction of (a) adenylate or guanylate cyclases (TSUDA et al. 1986; MIOH and CHENG 1987; RUDLAND et al. 1974); (b) stimulation of phospholipase-induced breakdown of phosphatidyl inositides and the generation of the second messenger diacylglycerol (TSUDA et al. 1985; TAKEYAMA et al. 1986) with subsequent activation of protein kinase C (TSUDA et al. 1985, 1986; TAKEYAMA et al. 1986) and calcium ion influx into the cell (MAGNALDO et al. 1986; TSUDA et al. 1985); and (c) activation of an FGF receptor-associated tyrosine kinase (HUANG and HUANG 1986). Unexpectedly, the generation of diacylglycerol does not seem to be accompanied by the formation of inositol trisphosphate in all of the cell types investigated (MAGNALDO et al. 1986; MIOH and CHENG 1987; CHAMBARD et al. 1987; MOSCAT et al. 1988). In some instances, inositol monophosphate can be formed in response to FGF (MOSCAT et al. 1988), suggesting the possibility of a metabolism of phosphatidyl inositide by phospholipases other than phospholipase C (MAGNALDO et al. 1986; BLANQUET et al. 1988).

Binding of FGF to its receptor is associated with increased protein phosphorylation (MAGNALDO et al. 1986; MIOH and CHENG 1987; PELECH et al. 1986; COUGHLIN et al. 1988). FGFs induce the tyrosine phosphorylation of a unique 90-kDa protein that is not phosphorylated in unstimulated or EGF- and PDGF-stimulated Swiss 3T3 cells (COUGHLIN et al. 1988). Basic FGF can also stimulate the phosphorylation of the S6 ribosomal protein by S6 kinase (MAGNALDO et al. 1986; PELECH et al. 1986), as well as a variety of other cytosolic proteins (BLANQUET et al. 1988; TOGARI et al. 1985), presumably as a result of the activation of a phosphorylating enzyme such as the receptor kinase, protein kinase C, or cyclic nucleotide-dependent protein kinases.

Several additional mechanisms have been proposed to play a part in the transduction of the FGF signal; these include the activation of the Na^+/H^+ antiport (MAGNALDO et al. 1986) and other ion channels (PANET et al. 1986; HALPERIN and LOBB 1987). Because FGFs have been demonstrated to be internalized and slowly degraded in the cell (MOENNER et al. 1987; MOSCATELLI 1988), it is particularly noteworthy that intracellular basic FGF can be translocated to and accumulated in the nucleolus (BOUCHÉ et al. 1987). Whether FGFs have direct effects inside the cell is not known, but remains an intriguing possibility in view of their lack of classical secretion.

IV. Cellular Responses to FGFs

Consistent with the finding that the FGF receptors bind basic and acidic FGFs, most cell types respond to both basic and acidic FGFs in a qualitatively indistinguishable manner. Basic FGF, however, is usually 10- to 100-fold more potent than acidic FGF (GOSPODAROWICZ et al. 1985b, 1986a), likely reflecting the higher affinity of basic FGF for the FGF receptor. Some rare biological systems, however, seem to respond to only one of the FGFs (CUNY et al. 1986; HALABAN et al. 1987), and in others, acidic FGF is equipotent to or even more potent than basic FGF (SIEVERS et al. 1987; FREDJ-REYGROBELLET et al. 1987).

FGFs have been observed to induce a variety of additional responses which may not be readily classified as being part of the mitogenic cascade of events that leads to cell division, but that are integral parts of the target cell response. They include the stimulation of reversible changes of morphology (GOSPODAROWICZ et al. 1986a; KATOH and TAKAYAMA 1984) brought about by cytoskeletal rearrangement (GOSPODAROWICZ et al. 1986a; HERMAN and D'AMORE 1984; DETHLEFSEN et al. 1986), the stimulation of cellular transport systems, polyribosome formation, RNA stabilization and synthesis, general protein synthesis, and the inhibition of protein degradation (GOSPODAROWICZ et al. 1986a; RUDLAND et al. 1974; CANALIS et al. 1987; ROSS and BALLARD 1988). Proteins (e.g., the polyamine-synthesizing enzyme ornithine decarboxylase; TOGARI et al. 1985; LATHROP et al. 1985a) that are required for the cell to proceed through cell division are also induced by basic FGF. The growth factor also induces proteins of unknown function, some of which are major secreted proteins (HAMILTON et al. 1985). Frequently, cells assume a reverse "transformed" phenotype when treated with FGFs with characteristic reduced cell-substratum adhesion, growth in crisscross pattern, increased membrane ruffling, and an ability of cells to grow anchorage-independent in soft agar (GOSPODAROWICZ et al. 1986a; HUANG et al. 1986b; RIZZINO and RUFF 1986; KATO et al. 1987).

F. Structure-Activity Relationships

At present little is known about the structural elements which convey biological activity to the FGFs. Investigators have compared the primary structure of the members of the FGF family to identify common sequences (ESCH et al. 1985; GIMENEZ-GALLEGO et al. 1985; ZHAN et al. 1988; YOSHIDA et al. 1987) and others have used synthetic peptide fragments (BAIRD et al. 1988; SCHUBERT et al. 1987). Point mutations in the primary structure of FGFs are also being used to identify critical amino acid sequences (SENO et al. 1988).

I. Receptor, Heparin, and Copper Binding Domains

Synthetic basic FGF peptide fragments have been used to identify potential receptor and heparin binding sites (BAIRD et al. 1988; SCHUBERT et al. 1987). Peptides corresponding to sequence locations near the amino terminus and the carboxy terminus were found to bind heparin and to displace labeled basic FGF from its high-affinity and low-affinity receptors. Furthermore, at high concentrations those peptides were able to interfere with the biological activity of basic FGF or mimic it in the absence of basic FGF. It was thus possible to identify a critical sequence of amino acids [FGF(105–115)] required for receptor binding. This sequence is outlined by residues 115–124 in the numbering system used in Fig. 2. A similar study with reductive methylation of acidic FGF (HARPER and LOBB 1988) identified Lys118 (Lys133 in Fig. 3) as being involved in the binding of this mitogen to its receptor and heparin.

The role of the cysteine residues in the maintenance of tertiary structure in FGFs is not well understood. Two of the cysteine residues in the FGFs (positions 34, 101) are positionally conserved, suggesting that these cysteines form a disulfide bond which contributes to a stable tertiary structure (THOMAS 1987). However, the presence of a disulfide bond does not seem essential for biological activity of the FGFs. Disulfide reduction in acidic FGF (McKEEHAN and CRABB 1987) and replacement of cysteines in basic FGF by serine residues (SENO et al. 1988) do not drastically affect their biological activity. Whether these cysteines are conserved for other functions remains speculative. One possibility that has not been extensively investigated is that they are involved in the interaction between FGF and copper. Basic FGF has been shown to bind tightly to copper chelate columns, suggesting an important interaction between the metal ion and these mitogens (SHING 1988). This is of particular interest since copper and heparin in combination can elicit an angiogenic response in the rabbit cornea (RAJU et al. 1982). Copper is also known to have a high affinity for free sulfhydryl groups (SULKOWSKI 1985).

II. Interactions with Heparin

One of the hallmarks that distinguishes FGFs from many other families of growth factors is their relatively high affinity for heparin. FGFs are protected from inactivation by enzymes, high temperature, or low pH by the addition of heparin or related highly sulfated glycosaminoglycans (GAGs) (BAIRD et al. 1988; GOSPODAROWICZ and CHENG 1986; ROSENGART et al. 1988a) and cell-derived heparan sulfate (SAKSELA et al. 1988), suggesting two important features of FGF binding to GAGs. First, FGFs can be localized, bound to heparan sulfate on the cell surface, and locally available for the target cell. Second, these interactions can protect the growth factors from the proteolytic degradation that characterizes tissue remodeling, neovascularization, and metastases thus allowing them to function in a metabolically "hostile" environment. Because heparin also potentiates the biological activity of acidic FGF to make it as potent as basic FGF (GIMENEZ-GALLEGO et al. 1986b; UHLRICH et al. 1986; WAGNER and D'AMORE 1986; UNSICKER et al. 1987; RODAN et al. 1987; THOMAS et al. 1985) and appears to stabilize both the *hst/ks* protein (DELLI-BOVI et al. 1988) and FGF-5 (ZHAN et al. 1988), there is a strong line of evidence to indicate that the binding to GAGs underlies the regulation of their activity. Because the potentiating effects of heparin have also been observed with a synthetic heparin pentasaccharide (UHLRICH et al. 1986), it is clear that these effects of GAGs are mediated by the carbohydrate and not a protein contaminant.

III. Glycosylation

Human, but not bovine acidic FGF contains a potential glycosylation site but posttranslational glycosylation of acidic FGF does not seem to occur. Similarly, there is no evidence that basic FGF is glycosylated. A recent study by BLAM et al. (1988) has demonstrated that when basic FGF chimeras are constructed with the signal sequence of human growth hormone, the mitogen will be posttranslationally modified. Unlike chimeras produced by other methods (ROGELJ et al. 1988; JAYE et al. 1988), it is secreted and found in the conditioned media.

IV. Phosphorylation of FGFs

There are consensus sequences for the phosphorylation of basic FGF by the cyclic AMP-dependent protein kinase A and the calcium- and phospholipid-dependent protein kinase C. This led FEIGE et al. (FEIGE and BAIRD 1988a) to investigate the possibility that FGFs are substrates for protein phosphorylation. The results established that basic FGF is a substrate for both protein kinase C and protein kinase A and that acidic FGF is a substrate for protein kinase C alone. The physiological significance of these findings is not clear, even though cells have been shown to synthesize a phosphorylated form of basic FGF. The phosphorylation of basic FGF by protein kinase A is in the receptor binding domain of the growth factor (Thr121 in Fig. 2) and the phosphorylation confers greater receptor affinity to basic FGF. The binding of FGF to heparin masks this amino acid and results in a substitution of targeted amino acids (FEIGE et al. 1989). Under these conditions, basic FGF is no longer phosphorylated in the receptor binding domain and the phosphorylated mitogen is equipotent to the unphosphorylated growth factor in radioreceptor assays.

G. Secretion and Regulation of FGFs

Nothing is known about the secretion and regulation of the *int*-2, *hst/ks*, or FGF-5 proteins at this time, due largely to the fact that they have only been recently discovered. Although very little is known about the secretion of acidic and basic FGFs, recent studies have suggested that their secretion and their activity are highly regulated.

I. Secretion of Basic and Acidic FGFs

The absence of a typical amino-terminal signal peptide consensus sequence required to mediate classical protein secretion agrees with the result that very little basic or acidic FGFs are released into conditioned medium by FGF-producing cells (SCHWEIGERER et al. 1987a). And yet, antibodies to basic FGF can inhibit the autocrine growth of endothelial cells (SAKAGUCHI et al. 1988;

GOSPODAROWICZ et al. 1984; SCHWEIGERER et al. 1987b), suggesting that the growth factor can get out of the cell to bind its receptor. In support of this conclusion, it has recently been established that FGF-synthesizing cells accumulate mitogens in the extracellular matrix which they lay down onto the culture dish (BAIRD and LING 1987; VLODAVSKY et al. 1987a; FOLKMAN et al. 1988). Accordingly, FGFs have been detected in the basement membranes of certain tissues (FOLKMAN et al. 1988). Because FGFs bind tightly to heparin, it is likely that the mitogens bind to the structurally related heparan sulfate proteoglycans present in extracellular matrices. FGFs bound to the extracellular matrix are released into the culture medium when the matrix produced by cultured endothelial cells is treated with heparin, heparan sulfate, or degrading enzymes (BAIRD and LING 1987; VLODAVSKY et al. 1987a). It is intriguing to speculate that a directed secretion of FGF from cells into their basement membranes constitutes a mechanism by which cells could accumulate and store large amounts of mitogen outside the cell. This would reconcile the apparently conflicting observations that there exist relatively high concentrations of FGFs in tissues and yet extremely low levels of intracellular FGF mRNA. How the FGFs are released from the cell and subsequently into the matrix is not known, although both active transport by carrier proteins and passive infusion during development have been proposed.

II. Regulation of FGFs in the Local Environment

FGF-binding heparan sulfate proteoglycans are located in basement membranes and on the surface of many cell types (MOSCATELLI 1987b; ROSENGART et al. 1988a). Thus, FGF bound in the extracellular matrix and to cell-surface GAGs could function as extracellular stored growth factor. Using this knowledge, a mechanism has been proposed to account for the regulation of basic and acidic FGF activity (BAIRD and WALICKE 1989). The model attempts to resolve the apparent discrepancies between the significant quantities of growth factor in tissues and the fact that they are not biologically available. It is based on the premise that both the extracellular matrix and cell associated GAGs are involved in a process to regulate the bioavailability of FGFs. Target cells surrounded by bio-unavailable growth factor (Fig. 5) can "activate" FGFs and make them bioavailable through a combination of several mechanisms. In each instance there is supporting experimental evidence. One such mechanism includes increasing the number of high-affinity receptors in response to a given trophic stimulus (i.e., injury, hypoxia, hormones). This process would remove FGFs from the low-affinity GAG-related binding sites by changing the binding kinetics in favor of the high affinity site ($K_d \sim 10$–50 pM) over the low-affinity sites $K_d \sim 10$–50 nM) (Fig. 5A). Examples of this process include TGF-β and FSH, which can increase the number of FGF receptors on endothelial and granulosa cells, respectively (BAIRD et al. 1989a). The disappearance of FGF receptors as seen during terminal differentiation of skeletal muscle cells in culture (OLWIN and HAUSCHKA 1988) is a similar example since this process ensures that FGFs remain extracellular,

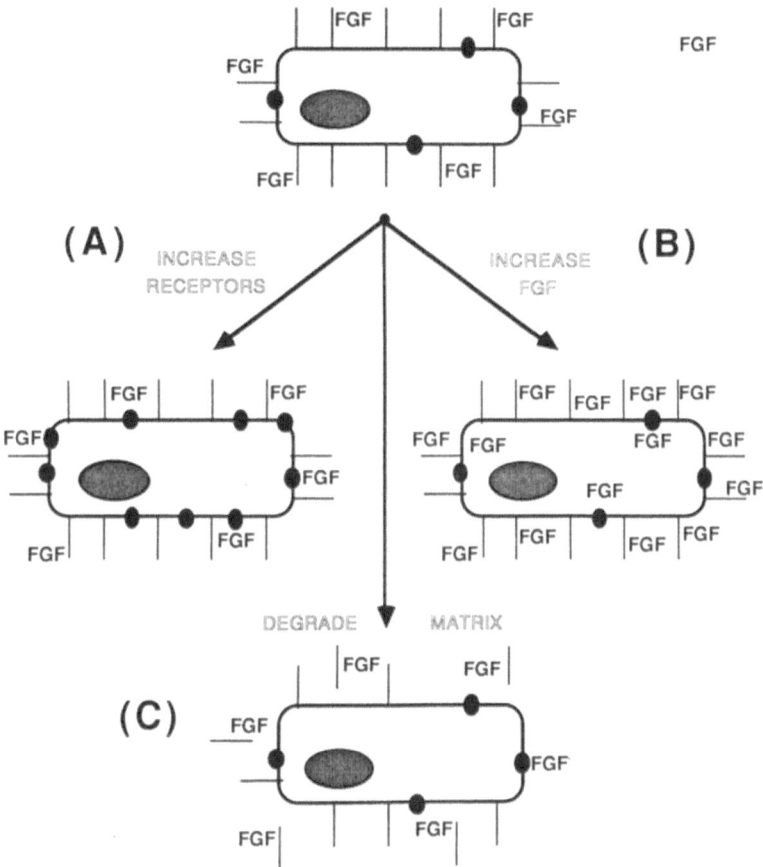

Fig. 5 A–C. Proposed mechanism involved in the regulation of FGF bioavailability. The receptors to FGF are illustrated on the plasma membrane by the *circles*. C The generation of heparin-bound FGFs from the extracellular matrix is illustrated to reflect the possibility that they remain associated with heparin while being delivered from the matrix

matrix associated, and not bioavailable to the target cell. The second mechanism relies on the process of autocrine production of FGFs. Target cells can synthesize their own growth factor (SCHWEIGERER et al. 1987 b; VLODAVSKY et al. 1987 b) or produce a modified form of the protein (FEIGE and BAIRD 1988 a) with potential differences in heparin and/or receptor binding affinities and thus bypass the need to mobilize extracellular FGFs (Fig. 5 B). Finally, the local release of enzymes (collagenase, plasminogen activator, heparinase) capable of degrading the extracellular matrix (MONTESANO et al. 1986; NAKAJIMA et al. 1983; VLODAVSKY et al. 1983) is a third mechanism that could generate a biologically active FGF (Fig. 5 C). It is interesting to note that each of the regulatory steps that have been attributed to the target cell in this model can also be mediated by macrophages. They

synthesize and release TGF-β (ASSOIAN et al. 1987) which in turn will increase
FGF receptors (BAIRD et al. 1988); they synthesize basic FGF (BAIRD et al.
1985b; JOSEPH-SILVERSTEIN et al. 1988) and thus can deliver the growth factor
to the target cell; and, finally, they release extracellular matrix-degrading enzymes (MATZNER et al. 1985; BAR-NER et al. 1985) and thus can make FGFs
available to the local environment. The relative simplicity of this model can
account for most of the experimental observations made with FGFs. Any
event that mobilizes FGFs from the extracellular matrix (i.e., pH, enzymes,
oxygen), will effectively ensure the generation of a biologically active FGF,
and as such a cell growth response.

H. Biological Activities of FGF

One of the more remarkable general insights derived from growth factor
research is the recognition that these trophic factors are multifunctional
(SPORN and ROBERTS 1988). While basic and acidic FGFs were originally discovered on the basis of their mitogenic activity for cultured fibroblasts, it is
now strikingly evident that these molecules act as mitogens on many diverse
cells, mostly of mesenchymal or neuroectodermal origin (Table 2). Equally
important, FGFs possess activities as modifiers of a multitude of nonmitogenic cell functions including chemotactic activity, induction or suppression of cell-specific protein synthesis or secretion, regulation of cellular differentiated functions, and modulation of endocrine and neural functions (see
Sect. E. I). At first glance the multitude of activities of the FGFs may appear
difficult to interpret. However, with the data now at hand many of those activities can be seen in the context of specific physiological functions for local
and highly regulated FGFs. It has thus become of paramount importance to
establish whether the in vitro activities of FGFs correlate with any in vivo
function.

I. FGFs as Angiogenic Factors

FGFs induce new capillary blood vessel growth in various pharmacological
animal models (ESCH et al. 1985; GOSPODAROWICZ et al. 1979, 1985b, 1986b;
THOMAS et al. 1985; LOBB et al. 1985). For example, when implanted together
with a slow release carrier into a pocket of the normally avascular cornea of
the rabbit eye, FGFs cause a massive ingrowth of capillaries from the
periphery of the eye towards the site of the implant. An example of this effect
is shown in Fig. 6. Similar neovascularization is seen when FGFs are applied
to avascular areas of the chick chorioallantoic membrane (CAM) or in the
hamster cheek pouch. FGFs are very potent angiogenic agents, with
picomolar (nanogram) amounts of FGFs required to elicit a response.

The question of whether FGFs are involved in the physiological or
pathological regulation of new blood vessel growth is of great interest.
Although it remains to be established whether physiological neovasculariza-

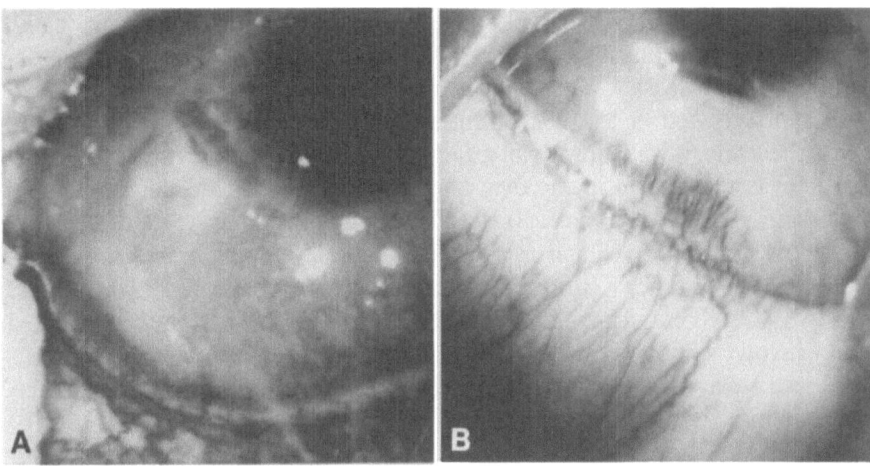

Fig. 6 A, B. Angiogenic effects of FGFs on the rabbit corneal assay. Basic FGF (100 ng) was tested in Elvax pellets (RISAU 1986) and implanted into a female rabbit corneal pouch. Seventeen days later, the effects of control pellets (**A**) were compared to pellets containing FGF (**B**)

tion does indeed depend on FGF, there is considerable experimental evidence to support this notion. First and foremost, exogenously administered FGFs possess all of the biological activities required to elicit neovascularization (see below). Basic FGF, as an example, is also present in relatively high quantities in probably all vascularized tissues, including tumors. Second, acidic FGF and probably basic FGF are expressed at the time when vascularization begins within the developing chick embryo brain (RISAU et al. 1988; RISAU 1986). Likewise, the embryonic mouse kidney elaborates an FGF-like angiogenesis factor which induces vascularization of kidney mesenchyme (RISAU and EKBLOM 1986). Finally, neutralizing anti-basic FGF antibodies administered to developing rat embryos (LIU and NICOLL 1988) or chondrosarcoma-bearing mice (BAIRD et al. 1986 a) were found to inhibit significantly their growth. FGFs also bind to the endothelium in vivo (ROSENGART et al. 1988 b).

1. Angiogenesis and Reproduction

The potential role of FGFs in the regulation of vascularization is also apparent in certain angiogenesis-dependent events in reproduction such as egg implantation, the development and maturation of the corpus luteum, and embryonic development (BAIRD et al. 1989 a; GOSPODAROWICZ et al. 1985 a; SLACK et al. 1987). Angiogenesis plays an important role in these processes, as reflected by the rapid appearance of a dense vascular network around the follicle at the time of ovulation. The initial vascular bed develops to form a new capillary network which invades the previously avascular granulosa cell layer (BASSETT 1943). The corpus luteum, follicles, and follicular fluid also contain angiogenic and endothelial cell mitogenic activities (JAKOB et al. 1977; GOSPODAROWICZ and THAKRAL 1978; KOOS and LEMAIRE 1983; FREDERICK et al. 1984), at least some of which are now identified as basic FGF (GOSPODAROWICZ et al. 1985a; SHIMASAKI et al. 1988; BAIRD et al. 1989a). More recently, it was shown that granulosa cells, which appear to be the target of neovascularization in the ovary when they develop into corpus luteum, synthesize basic FGF (NEUFELD et al. 1987a) and thus may be inducing capillary invasion. The mechanisms by which other developing tissues, e.g., brain (RISAU 1986), kidney (RISAU and EKBLOM 1986), adrenal cortex (TURNER and BAGNARA 1981), and hair follicles (STENN et al. 1988), become vascularized may involve similar events. To date, the concept for an involvement of FGFs in these processes is provocative, but evidence for its participation is still circumstantial.

2. Angiogenesis and the Ischemic Response

One process that has gathered interest is the role of endothelial growth factors in the vessel wall (JOSEPH-SILVERSTEIN and RIFKIN 1987), and the neovascular response that characterizes the formation of collateral vessels in ischemia. FGFs have been implicated in this process for several reasons. First, they can be localized in the heart (CASSCELLS et al. 1988) and brain (LOBB and FETT 1984) and, second, they can stimulate vascularization when injected into the brain (CUEVAS et al. 1988c) or when infused onto the arterial wall (BAIRD et al. 1987). In this latter model, the vasa vasorum of the rat carotid artery was used to demonstrate that, in fact, basic FGF can stimulate vascularization in the cardiovascular system (Fig. 7). Because the administration of FGF to the arterial wall results in a significant neovascular response (BAIRD et al. 1987), there is increasing evidence that local FGFs, activated by processes such as hypoxia, and local changes in pH could elicit neovascularization. In response to ischemic injury, local endogenous FGFs would thus increase the capillary bed that vascularizes the wounded tissue (PATERSON 1938; BRICK 1959).

3. Tumor Angiogenesis

The prominent vascularization of most solid tumors has long attracted the attention of biologists. FOLKMAN (1972), as an example, recognized that unless solid tumors become vascularized, they can remain embedded in tissues for prolonged periods without significant growth. On the basis of this observation, he proposed the existence of a tumor angiogenesis factor (TAF). To date, a specific tumor angiogenesis factor has not been characterized. Instead, several potent angiogenic factors, all of which are found in normal tissues, have been isolated from several tumor tissues or tumor cells. They include transforming growth factors α (TGF-α; SCHREIBER et al. 1986) and β (TGF-β; ROBERTS et al. 1986), TNF (FRÀTER-SCHROEDER et al. 1987), and angiogenin (FETT et al. 1985). Although tumor cells produce FGFs (including *hst/ks* product and FGF-5) and it seems likely that they are participating in tumor-induced neovascularization, FGFs are not alone in supporting tumor progression.

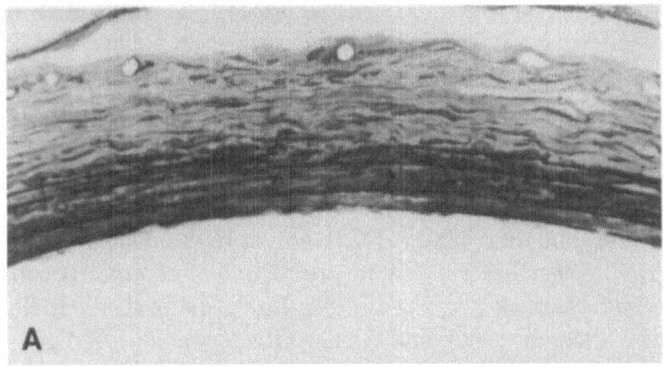

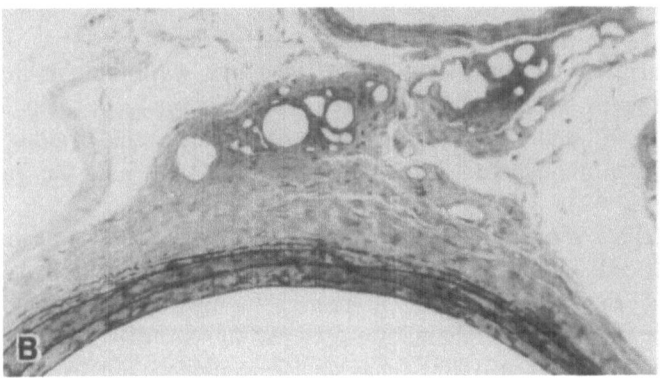

Fig. 7 A, B. FGFs stimulate vascularization of the vasa vasorum. Saline alone (**A**) or containing FGF (**B**) was infused at a rate of 1 ng/μl h for a total of 14 days onto the surface of the rat carotid artery. Tissues were collected, mounted, and photographed. (From CUEVAS et al. in preparation)

4. Cell Biology of Angiogenesis

The mechanism by which FGF-induced angiogenesis occurs has been investigated in considerable detail. Capillary blood vessel formation is a complex process consisting of several critical elements including endothelial cell proliferation, the sprouting of new capillaries, the migration of endothelial cells, and the breakdown of extracellular matrix surrounding existing capillaries. Many investigators currently hold the view that angiogenic factors should be capable of inducing those activities (JOSEPH-SILVERSTEIN and RIFKIN 1987). This is indeed the case for the FGFs, as shown in vivo using wound chambers (SPRUGEL et al. 1987) and by site-directed vascularization (HAYEK et al. 1987; THOMPSON et al. 1988). They can also stimulate the proliferation of vascular endothelial cells from capillary and large blood vessels in vitro (BÖHLEN et al. 1984, 1985; CONN and HATCHER 1984; MACIAG et al. 1982; D'AMORE and KLAGSBRUN 1984; GOSPODAROWICZ et al. 1976) and in vivo (BUNTROCK et al. 1984). FGFs can stimulate the migration of endothelial cells

(PRESTA et al. 1986; HERMAN and D'AMORE 1984; TERRANOVA et al. 1985) and the release of at least two matrix-degrading proteases: collagenase (PRESTA et al. 1986; CHUA et al. 1987) and plasminogen activator (PRESTA et al. 1986; MOSCATELLI 1987; MONTESANO et al. 1986). Moreover, migrating endothelial cells, which produce their own basic FGF, are capable of degrading extracellular matrix proteins (KALEBIC et al. 1983), a result which is consistent with the proposed mechanism of neovascularization. Interestingly, in endothelial cells and fibroblasts, FGF not only stimulates the release of matrix-degrading enzymes but also the production of new matrix components [e.g., collagen (GOSPODAROWICZ et al. 1980; DAVIDSON et al. 1985), fibronectin (GOSPODAROWICZ et al. 1980), and GAGs (GORDON et al. 1985) and of an inhibitor of matrix degradation, plasminogen activator inhibitor (SAKSELA et al. 1987)]. Although the physiological significance of these opposing activities is not clear. FGFs may direct endothelial cells (and fibroblasts) at the site of neovascularization to break down the extracellular scaffolding in an effort to facilitate the formation of new vessels, but eventually participate in the processes to stimulate repair. Clearly, one of the key events is going to be its interaction with the other growth factors participating in the angiogenic response.

The FGFs also promote maintenance of the differentiated state in cultured endothelial cells independent of their effects on proliferation. This includes stimulating the expression of a nonthrombogenic apical cell surface (VLODAVSKY et al. 1979), enhancing a preferential synthesis of certain types of collagen (SCHEFFER et al. 1982), delaying cell senescence (GOSPODAROWICZ et al. 1986a), and increasing cell attachment to the substratum (SCHUBERT et al. 1987). Basic FGF also induces the rearrangement of endothelial cells in vitro into tubular structures resembling blood capillaries (MONTESANO et al. 1986; JAYE et al. 1985) and can stimulate the proliferation of smooth muscle cells (GOSPODAROWICZ et al. 1985b) and of pericytes (BUNTROCK et al. 1984), both of which are associated with the growth of new blood vessels.

II. FGFs in Wound Healing and Tissue Repair

The healing of injured tissue represents another complex and poorly understood process in which many cells and many cellular functions are activated or inhibited in a precisely regulated and timed manner. A considerable body of data suggests that FGFs may also play an endogenous role in the regeneration of tissues. Increased basic FGF has been reported at the site of focal brain wounds (FINKELSTEIN et al. 1988) and it is present in wound fluids (HAMERMAN et al. 1987). There is also an early release of acidic FGF after rat brain injury (NIETO-SAMPEDRO et al. 1988). FGFs stimulate the healing of injured corneal epithelium (FREDJ-REYGROBELLET et al. 1987; PETROUTSOS et al. 1984; MÜLLER et al. 1985) and endothelium (GOSPODAROWICZ and GREENBRUG 1979), and promote the regeneration of cartilage in experimental knee joint wounds (JENTZSCH et al. 1980; CUEVAS et al. 1988a). An example of this latter effect is shown in Fig. 8. Clearly, the effects of FGF on chondrocytes

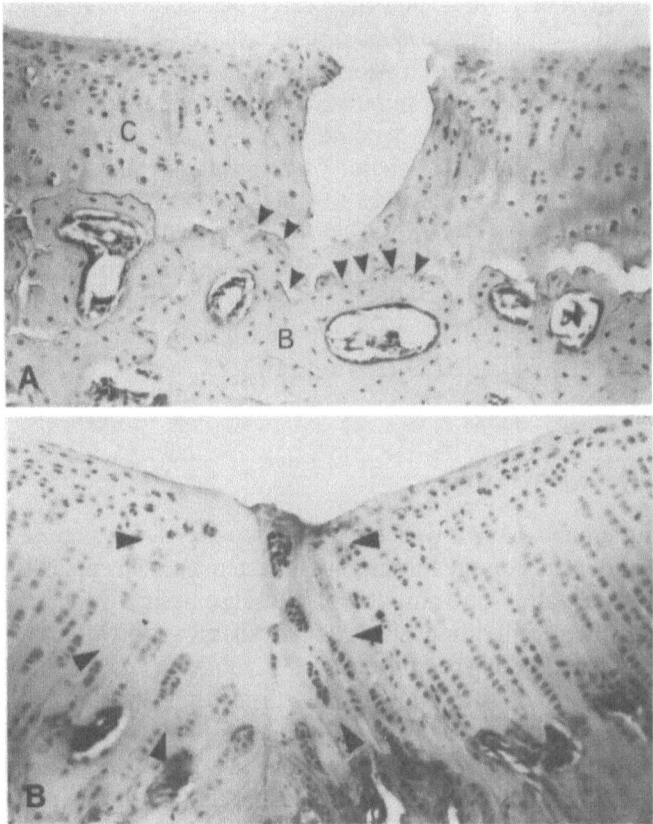

Fig. 8 A, B. FGFs stimulate cartilage repair. Basic FGF was infused onto lesions in rabbit articular cartilage at a concentration of 1 ng/μl h (**B**) and the repair of the tissue compared to with saline (**A**). (From CUEVAS et al. 1988 a)

that are seen in vitro can now be observed in in vivo models. There is a marked increase in the number of chondrocytes as well as in the extracellular matrix. In uninjured knees, FGF has no effect, thus suggesting that the process of injury makes FGF available to the target cell. In a series of similar studies, the capacity of FGFs to stimulate bone cells (RODAN et al. 1987; GLOBUS et al. 1988) has been coupled to their angiogenic effects to enhance angiogenesis in bone graft healing (EPPLEY et al. 1988). These types of studies are now the focus of considerable attention in an effort to establish whether the therapeutic application of angiogenic factors can be used in conjunction with tissue transplantation (HAYEK et al. 1987; THOMPSON et al. 1988) or in gene therapy (ST LOUIS and VERMA 1988).

FGF promotes wound healing processes (BUNTROCK et al. 1984; DAVIDSON et al. 1985) by eliciting a number of cellular responses that are considered beneficial for tissue repair. Among its effects that lead to accelerated healing are the attraction of various important cell types to the wound site (Sprugel et al. 1987), the stimulation of cells to release matrix-degrading enzymes such as collagenase and plasmin activators into the wound site (Chua et al. 1987; Saksela et al. 1987), and the stimulation of fibroblasts to proliferate and to synthesize new extracellular matrix, i.e., collagen and probably other matrix proteins and proteoglycans (Buntrock et al. 1984; Davidson et al. 1985; Sprugel et al. 1987; Buntrock et al. 1982). Wound healing promoted by FGFs is also associated with angiogenesis as reflected by the increased proliferation of endothelial cells in capillary sprouts and of pericytes associated with growing blood vessels (Buntrock et al. 1984). Furthermore, FGFs increase melanocyte and keratinocyte proliferation (Halaban et al. 1987; Fourtanier et al. 1986; O'Keefe et al. 1988).

III. Tissue Regeneration

Certain amphibians have a well developed capacity to regenerate lost limbs. This process depends on the formation of blastema cells from dedifferentiated cells immediately after limb amputation. Because basic FGF has been shown to stimulate the proliferation of blastema cells in vitro (Mescher and Loh 1980) and is a well established mitogen for myoblasts (Olwin and Hauschka 1986; Gospodarowicz 1976) and chondrocytes (Gospodarowicz et al. 1985b; Gospodarowicz 1976), it has been the focus of considerable attention. FGF supports blastema formation in the denervated newt limb (Mescher and Gospodarowicz 1979) and stimulates regrowth of amputated frog limbs (Gospodarowicz and Mescher 1981). The regeneration of the lens from the dorsal iris also depends on the FGF-like stimuli. The effects are mediated by its stimulation of the proliferation of epithelial lens cells (Arruti et al. 1985) and the induction of synthesis of the lens-specific protein crystallin (Chamberlain and McAvoy 1987).

    More recently, the effects of basic FGF in models of regeneration have been examined in further detail in mammalian systems. As an example, several investigators (Cuevas et al. 1988d; Danielsen et al. 1988) have established the ability of FGF to stimulate peripheral nerve regeneration (Fig. 9). The infusion of basic FGF into a gap separating stumps of the severed nerve promotes survival, axonal bridging, and reconnection of the nerve stumps. In this model, the administration of basic FGF greatly enhances recovery. The transport of horseradish peroxidase across the axonal bridge to the spinal cord established that the regenerated tissue had regained at least part of its function. Similar studies have been performed using transected optic nerves (Cuevas et al. 1988b; Sievers et al. 1987) and have established the capacity of basic FGF to support survival of retinal ganglion cells in vivo (see Sect. H. VI.).

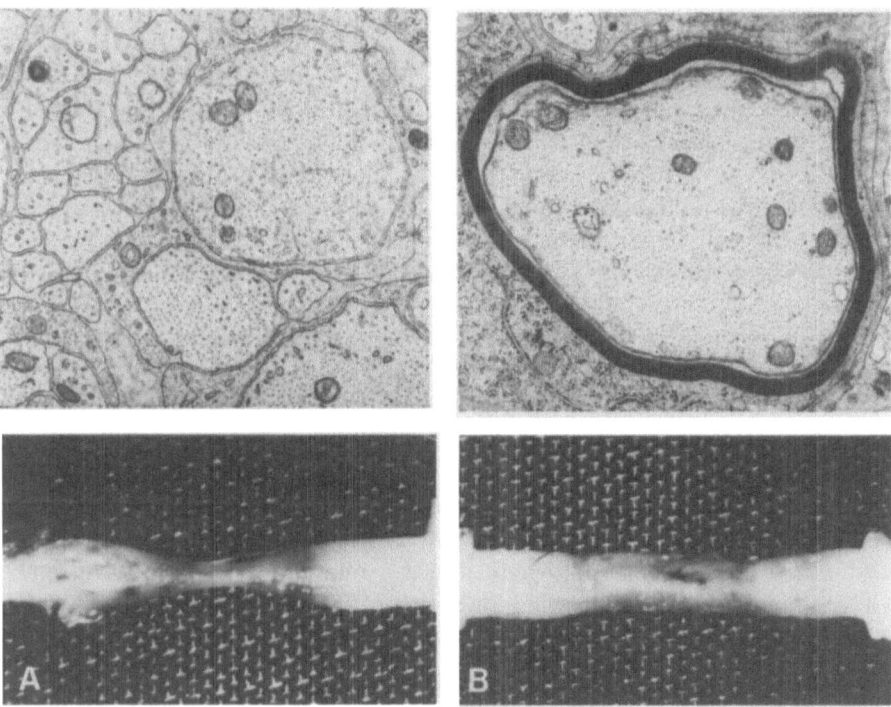

Fig. 9 A, B. Effects of FGFs on peripheral nerve regeneration. Basic FGF was infused into an 8-mm space separating the distal and proximal stumps of the rat sciatic nerve. After 4 weeks the regenerating tissues were collected, photographed and the presence of myelinated and unmyelinated axons examined (**A**) saline; (**B**) FGF, 1 mg/μl h for 28 days). (From CUEVAS et al. 1988 d)

IV. Embryonic Development and Differentiation

The capacity of basic FGFs to stimulate the regeneration of tissues and the repair of wounds led to the hypothesis that FGFs may play a role in the development and differentiation of tissues and there is significant evidence that this is indeed the case. The temporal regulation of basic FGF expression in embryonic chick brain (RISAU et al. 1988; RISAU 1986) and in the kidney mesenchyme of the mouse embryo (RISAU and EKBLOM 1986) is believed to represent the stimulus for the initial vascularization of the developing tissues which then supports their further growth. FGFs could thus conceivably support the proliferation and differentiation of specific tissue cells during embryonic organogenesis. Accordingly, basic FGF mRNA is also expressed in a variety of mouse embryo tissues (ABRAHAM et al. 1986 b). More recent results indicate that endogenous basic FGF may direct cellular differentiation very early in embryogenesis. A growth factor with properties very similar to basic FGF can be detected in yolk and white of unfertilized chicken eggs and in the limb buds and bodies of stage-18 chick embryos (SEED et al. 1988). FGF mRNA can also be detected in the fertilized *Xenopus* egg (KIMELMAN and KIRSCHNER 1987). Because the *Xenopus* egg also transcribes the basic FGF gene and exogenous FGF can mimic the effect of the ventrovegetal signal to induce the differentiation of ectoderm into mesodermal structures in the fertilized *Xenopus* egg (SLACK et al. 1987; GRUNZ et al. 1988), FGF may be a physiological mesoderm-inducing factor in the *Xenopus* embryo. It is interesting to note, however, that the FGF-related protooncogene *int*-2 is also expressed in the developing mouse embryo from early gastrulation to early organogenesis (WILKINSON et al. 1988). Thus, other members of the FGF family of growth factors are most likely participating in the processes that occur during development. The discovery that TGFs also play a role in embryonic growth and tissue differentiation (KIMELMAN and KIRSCHNER 1987) implies a function for many families of factors in these complex developmental processes.

Direct evidence for a role of endogenous FGF in mammalian embryonic differentiation has recently been provided by LIU and NICOLL (1988). When transplanted under the renal capsule of syngeneic hosts, the infusion of antiserum to basic FGF retards the growth and differentiation of endoderm and mesoderm. Thus, it seems likely that molecules like basic FGF are at the same time morphogens, mitogens, and differentiation factors in vivo, and have a function in embryonic development and tissue differentiation.

V. Modulation of Endocrine Function

The endothelium of endocrine tissues plays a critical role in ensuring delivery of hormones to the circulation and is characteristically fenestrated. Endocrine tissues are also highly vascularized, suggesting that, as angiogenic factors, FGFs ensure the maintenance and integrity of the complex endocrine tissue capillary network. FGFs, however, have many effects on the differentiated function of endocrine cells in culture.

Basic FGF can be purified from pituitary extracts in far greater amounts than in other tissues. Although small amounts are found to be releasable from cultured anterior pituitary cells (BAIRD et al. 1985a), there is no evidence for the suggestion that it is a hormone in the classical sense. Instead, there is good evidence from in vitro studies that it is playing a local function in maintaining pituitary function. Basic FGF potentiates thyrotropin releasing factor-stimulated prolactin and thyrotropin secretion from cultured anterior pituitary cells without having an effect on the secretion of other anterior pituitary hormones (BAIRD et al. 1985a). There are no in vivo studies to establish whether the FGFs play a role in modulating endocrine function. This is in spite of the fact that basic FGF is a potent inhibitor of aromatase activity in granulosa cells (ADASHI et al. 1988; CHANNING et al. 1983; BAIRD and HSUEH 1986; FAUSER et al. 1988), Leydig cells (RAESIDE et al. 1988), and fibroblasts (EMOTO and BAIRD 1987), of testoterone synthesis in testes cells (FAUSER et al. 1988), and of LH receptor induction by FSH in vitro. Furthermore, its capacity to stimulate the proliferation and delay cell senescence of adrenocortical cells in vitro (GOSPODAROWICZ et al. 1977; HORNSBY and GILL 1977) has not been examined in vivo. Whether or not FGFs participate in local changes in adrenocortical, ovarian, and hypophysial function remains to be established.

VI. FGFs as Neuronotrophic Factors

FGFs have profound effects on several neuronal cell types. Basic FGF is mitogenic for glial cells such as oligodendrocytes (SANETO and DEVELLIS 1985; ECCLESTON and SILVERBERG 1985), astrocytes (PETTMANN et al. 1985; GOSPODAROWICZ 1976; MORRISON and DEVELLIS 1981), and Schwann cells (PRUSS et al. 1981). It stimulates the nonmitogenic functions of glial cells, such as the migration of astrocytes (SENIOR et al. 1986) and the release of plasminogen activators by astroglial cells (ROGISTER et al. 1988). It also modulates the expression of the intermediate filament protein GFAP (glial fibrillary acidic protein) (MORRISON et al. 1985; WEIBEL et al. 1985), of glutamine synthetase and S100 protein (WEIBEL et al. 1985), all of which represent astrocyte-specific differentiated functions. Basic FGF can also modify the morphological maturation of astrocytes as reflected by rearrangements of intermediate filaments, increased extension of cellular processes typical of mature astrocytes (WEIBEL et al. 1985), and, finally, change astrocyte membrane structure (WOLBURG et al. 1986). FGFs are active on neurons and while there is no evidence that they are mitogenic for these cells, basic FGF can stimulate the proliferation of their precursor cells, the neuroblasts (GENSBURGER et al. 1987). Basic and acidic FGFs can also prolong the survival of various central and peripheral neurons in culture (SCHUBERT et al. 1987; UNSICKER et al. 1987; WALICKE et al. 1986; MORRISON et al. 1986) and stimulate choline acetyltransferase synthesis (UNSICKER et al. 1987) and neurite outgrowth (WALICKE et al. 1986; MORRISON et al. 1986).

Recent studies have sought to determine whether the neuronotrophic effects of FGFs have an in vivo correlate. Rat retinal ganglion neurons survive after transection of the optic nerve when FGFs are either implanted in a gel foam at the site of the proximal stump of the transected optic nerve (FREDJ-REYGROBELLET et al. 1987) or infused (CUEVAS et al. 1988 b) onto the severed optic nerve. In the absence of added FGF, death of retinal ganglial cells is greatly accelerated, thus establishing that basic FGF can support neuronal survival in vivo. This conclusion is supported by the work of ANDERSON et al. (1988), who showed that when basic FGF is administered into the lesioned brain, hippocampal neurons survived that would die in the absence of added FGF. In the peripheral nervous system (Fig. 9), basic FGF applied to the proximal end of a severed sciatic nerve enhances remyelinization of the neuronal sheath and prevents the death of dorsal root ganglion neurons (OTTO et al. 1987; CUEVAS et al. 1988 d; DANIELSEN et al. 1988). These results support the concept that basic FGF is an important neuronotrophic factor, at least when administered exogenously. Although there is no demonstrated neuronotrophic function for local FGFs found in the brain, these results support the experimental observation that both the FGF protein and its mRNA are elevated in CNS tissues.

VII. FGFs and Diseases of Cell Proliferation

Much has been discussed concerning the pleiotropic nature of FGFs, and it will readily appear that, just as they can be implicated in physiological processes and possess considerable therapeutic potential in wound healing, tissue regeneration, and repair, they can also be associated by inference with the etiology of many diseases of cell proliferation. Indeed, for every biological activity of the FGFs, it is possible to implicate these molecules in the pathophysiological progression of a related disease. This includes their potential role as tumor angiogenesis factors (see Sect. H. I) as macrophage-derived angiogenic factors in atherosclerosis (PATERSON 1938; BRICK 1959), and as factors involved in reproduction, embryonic development, and tissue differentiation. The observation that overexpression of FGFs by cells can lead to transformation (ROGELJ et al. 1988; JAYE et al. 1988) is a strong suggestion that these molecules have at least some oncogenic potential (THOMAS 1988) and that they must remain highly regulated. One relatively new course of investigation that is now being examined is the possibility that FGFs contribute to the complications of diabetes. The FGF family of growth factors are candidates for these processes because they are angiogenic, are located in the many tissues affected by the complications of diabetes (eye, kidney, peripheral nerve, etc.), and their regulation is thought to be dependent on interactions with GAGs and the basement membrane (see model in Sect. G. II). In view of the fact that carbohydrates, GAGs, and the extracellular matrix are all modified in diabetes, there is thus considerable circumstantial evidence that at least some of the complications of diabetes might be FGF mediated. If this is the case, then it will be of paramount importance to define the processes that regulate the biological activities of endogenous FGFs and determine the molecular mechanisms that induce their dysfunction in disease.

VIII. Conclusions

The structural characterization of FGFs as distinct molecular entities has opened a chapter in growth factor research. In spite of all of the in vitro studies performed over the last several years, evidence for any physiological functions of endogenous FGFs remains circumstantial and a matter of speculation. Although there is little, if any, evidence that establishes a causative relationship between perturbed FGF regulation and disease, there are many reasons to suspect that excessive FGFs, or the lack of FGFs, may contribute to their pathophysiology. To this end, a variety of potential modulators of FGF activity have been identified which include TGF-β, TNF, γ-interferons, heparin, and related GAGs. There probably also exist a variety of as yet unidentified inhibitory molecules. It should thus be recognized that neither the FGFs, nor in fact any other trophic factor, exert their effects isolated from other hormones and cytokines. Instead, the in vivo response is the result of their combined interactions in a complex homeostatic system. Somewhat akin to a microprocessor, the target cell processes its response to the cumulated parallel and sequential signals that are induced by many trophic stimuli. The time- and concentration-dependent combinations of these stimuli act on the cell to ultimately define the cellular response. It was initially important to identify the components in this homeostatic system. The family of FGFs is just one such component, and their activities and interactions are complex in their own right. The interactions between the FGFs and other growth factors will now need to be addressed before a more unified and realistic appreciation of the role of these trophic factors in growth and differentiation can be realized. Until that time, as the experimental data suggest, the potential applications of FGFs as therapeutic tools in the management of disease will remain only partially uncovered.

References

Abraham JA, Mergia A, Whang JL, Tumolo A, Friedman J, Hjerrield KA, Gospodarowicz D, Fiddes JC (1986a) Nucleotide sequence of a bovine clone encoding the angiogenic protein, basic fibroblast growth factor. Science 233:545–548

Abraham JA, Whang JL, Tumolo A, Mergia A, Friedman J, Gospodarowicz D, Fiddes JC (1986b) Human basic fibroblast growth factor: nucleotide sequence and genomic organization. EMBO J 5:2523–2528

Abraham JA, Whang JL, Tumolo A, Mergia A, Fiddes JC (1986c) Human basic fibroblast growth factor: nucleotide sequence, genomic organization, and expression in mammalian cells. Cold Spring Harbor Symp Quant Biol 51:657–668

Adashi EY, Resnick CE, Croft CS, May JV, Gospodarowicz D (1988) Basic fibroblast growth factor as a regulator of ovarian granulosa cell differentiation: a novel nonmitogenic role. Mol Cell Endocrinol 55:7–14

Allen RE, Dodson MV, Luiten LS (1984) Regulation of skeletal muscle satellite cell proliferation by bovine pituitary fibroblast growth factor. Exp Cell Res 152:154–160

Anderson KJ, Dam D, Lee S, Cotman CW (1988) Basic fibroblast growth factor prevents death of lesioned cholinergic neurons in vivo. Nature 332:360–361

Armelin HA (1973) Pituitary extracts and steroid hormones in the control of 3T3 cell growth. Proc Natl Acad Sci USA 70:2702–2706

Arruti C, Cirillo A, Courtois Y (1985) An eye-derived growth factor regulates epithelial cell proliferation in the cultured lens. Differentiation 28:286–290

Assoian RK, Fleurdelys BE, Stevenson HC, Miller PJ, Madtes DK, Raines EW, Ross R, Sporn M (1987) Expression and secretion of type beta transforming growth factor by activated human macrophages. Proc Natl Acad Sci USA 84:6020–6024

Baird A, Durkin T (1986) Inhibition of endothelial cell proliferation by type β-transforming growth factor: interactions with acidic and basic fibroblast growth factors. Biochem Biophys Res Commun 138:476–482

Baird A, Hsueh AJW (1986) Fibroblast growth factor as an intraovarian hormone: differential regulation of steroidogenesis by an angiogenic factor. Regul Pept 16:243–250

Baird A, Ling N (1987) Fibroblast growth factors are present in the extracellular matrix produced by endothelial cells in vitro: implication for a role of heparinase-like enzymes in the neovascular response. Biochem Biophys Res Commun 142:428–435

Baird A, Walicke PA (1989) Fibroblast growth factors. Br Med Bull 45:438–452

Baird A, Esch F, Gospodarowicz D, Guillemin R (1985a) Retina- and eye-derived endothelial cell growth factors: partial molecular characterization and identity with acidic and basic fibroblast growth factors. Biochemistry 24:7855–7860

Baird A, Esch F, Böhlen P, Ling N, Gospodarowicz D (1985b) Isolation and partial characterization of an endothelial cell growth factor from bovine kidney: homology with basic fibroblast growth factor. Regul Pept 12:201–213

Baird A, Böhlen P, Ling N, Guillemin R (1985c) Radioimmunoassay for fibroblast growth factor (FGF): release by the bovine anterior pituitary in vitro. Regul Pept 10:309–317

Baird A, Culler F, Jones KL, Guillemin R (1985d) Angiogenic factor in human ocular fluid. Lancet 2:563

Baird A, Mormède P, Ying SY, Wehrenberg WB, Ueno N, Ling N, Guillemin R (1985e) A non-mitogenic pituitary function of fibroblast growth factor: regulation of thyrotropin and prolactin secretion. Proc Natl Acad Sci USA 82:5545–5549

Baird A, Mormède P, Böhlen P (1985f) Immunoreactive fibroblast growth factor in cells of peritoneal exudate suggests its identity with macrophage-derived growth factor. Biochem Biophys Res Commun 126:358–364

Baird A, Mormède P, Böhlen P (1986a) Immunoreactive fibroblast growth factor (FGF) in a transplantable chondrosarcoma: inhibition of tumor growth by antibodies to FGF. J Cell Biochem 30:79–85

Baird A, Esch F, Mormède P, Ueno N, Ling N, Böhlen P, Ying SY, Wehrenberg W, Guillemin R (1986b) Molecular characterization of fibroblast growth factor: distribution and biological activities in various tissues. Recent Prog Horm Res 42:143–205

Baird A, Cuevas P, Gonzalez A, Emoto N, Feige JJ, Walicke P (1987) Fibroblast growth factors, the extracellular matrix and the regulation of the neovascular response (Abstract). J Cell Biochem [Suppl] 11A:A192

Baird A, Schubert D, Ling N, Guillemin R (1988) Receptor and heparin-binding domains of basic fibroblast growth factor. Proc Natl Acad Sci USA 85:2324–2328

Baird A, Emoto N, Shimasaki S, Gonzalez AM, Fauser B, Hsueh AJW (1989a) Fibroblast growth factors as local mediators of gonadal function. In: Seventh ovarian workshop, paracrine communication in the ovary: ontogenesis and growth factors, Serono (in press)

Baird A, Feige JJ, Emoto N (1989b) Modulation of FGF mRNA and receptor by transforming growth factor β (in preparation)

Bar-Ner M, Kramer MD, Schirmacher V, Ishai-Michaeli R, Fuks Z, Vlodavsky I (1985) Sequential degradation of heparan sulfate in the subendothelial extracellular matrix by highly metastatic lymphoma cells. In J Cancer 35:483–491

Barr PJ, Cousens LS, Lee-Ng CT, Medina-Selby A, Masiarz FR, Hallewell RA, Chamberlain S, Bradley J, Lee D, Steimer KS, Poulter L, Burlingame AL, Esch F, Baird A (1988) Expression and processing of biologically active fibroblast growth factors in the yeast Saccharomyces cerevisiae. J Biol Chem 263:16471–16478

Barritault D, Arruti C, Courtois Y (1981) Is there a ubiquitous growth factor in the eye? Proliferation induced in different cell types by eye-derived growth factor. Differentiation 18:29–42

Bassett DL (1943) The changes in the vascular pattern of the ovary of the albino rat during the estrous cycle. Am J Anat 73:251–262

Bellvé AP, Feig LA (1984) Cell proliferation in the mammalian testis: biology of the seminiferous growth factor (SGF) Recent Prog Horm Res 40:531–567

Bertolini J, Hearn MTW (1987) Isolation, characterization and tissue localization of an N-terminal-truncated variant of fibroblast growth factor. Mol Cell Endocrinol 51:187–199

Blam SB, Mitchell R, Tischer E, Rubin JS, Silvan M, Silver S, Fiddes JC, Abraham JA, Aaronson SA (1988) Addition of growth hormone secretion signal to basic fibroblast growth factor results in cell transformation and secretion of aberrant forms of the protein. Oncogene 3:129–136

Blanquet PR, Paillard S, Courtois Y (1988) Influence of fibroblast growth factor on phosphorylation and activity of a 34kD lipocortin-like protein in bovine epithelial lens. FEBS Lett 229:183–187

Böhlen P, Baird A, Esch F, Ling N, Gospodarowicz D (1984) Isolation and partial molecular characterization of pituitary fibroblast growth factor. Proc Natl Acad Sci USA 81:5364–5368

Böhlen P, Baird A, Esch F, Gospodarowicz D (1985) Acidic fibroblast growth factor (FGF) from bovine brain: amino-terminal sequence and comparison with basic FGF. EMBO J 4:1951–1956

Böhlen P, Fràter-Schroeder M, Michel T, Jiang ZP (1987) Inhibitors of endothelial cell proliferation. In: Rifkin DB, Klagsbrun M (eds) Angiogenesis. Cold Spring Harbor Lab New York, pp 119–124

Bouché G, Gas N, Prats H, Balsin V, Tauber JP, Teissié J, Amalric F (1987) bFGF enters the nucleolus and stimulates the transcription of ribosomal genes in ABAE cells undergoing GO-G1 transition. Proc Natl Acad Sci USA 84:6770–6774

Brick RC (1959) Electron microscopic observation on capillaries of atherosclerotic aorta. Arch Pathol 67:656–659

Buntrock P, Jentzsch KD, Heder G (1982) Stimulation of wound healing, using brain extract with fibroblast growth factor activity. II. Histological and morphometric examination of cells and capillaries. Exp Pathol 21:62–67

Buntrock P, Buntrock M, Marx I, Kranz D, Jentzsch KD, Heder G (1984) Stimulation of wound healing, using brain extract with fibroblast growth factor activity. III. Electron microscopy, autoradiography, and ultrastructural autoradiography of granulation tissue. Exp Pathol 26:247–254

Burgess WH, Mehlman T, Marshak DR, Fraser BA, Maciag T (1986) Structural evidence that endothelial cell growth factor beta is the precursor of both endothelial cell growth factor alpha and acidic fibroblast growth factor. Proc Natl Acad Sci USA 83:7216–7220

Burrus LW, Olwin BB (1989) Isolation of a receptor for acidic and basic fibroblast growth factor from embryonic chick. J Biol Chem 264:18647–18653

Canalis E, Lorenzo J, Burgess WH, Maciag T (1987) Effects of endothelial cell growth factor on bone remodelling in vitro. J Clin Invest 79:52–58

Casscells W, Speir E, Allen P, Epstein S (1988) Heparin treatment of ischemia reduces infarct-5 and mortality of subsequent coronary ligation. Clin Res 36:266 A

Chambard JC, Paris S, L'Allemain G, Pouyssegur J (1987) Two growth factors signalling pathways in fibroblasts distinguished by pertussis toxin. Nature 326:800–803

Chamberlain CG, McAvoy JW (1987) Evidence that fibroblast growth factor promotes lens fibre differentiation. Curr Eye Res 6:1165–1168

Channing CP, Garrett R, Kroman N, Conn T, Gospodarowicz D (1983) Ability of EGF and FGF to cause a decrease in progesterone secretion by cultured porcine granulosa cells: changes in responsiveness throughout follicular maturation. In: Greenwald GS, Terranova PF (eds) Factors regulating ovarian function. Raven, New York, pp 215–220

Chodak GW, Shing Y, Borge M, Judge SM, Klagsbrun M (1986) Presence of heparin binding growth factor in mouse bladder tumors and urine from mice with bladder cancer. Cancer Res 46:5507–5510

Chodak GW, Hospelhorn V, Judge SM, Mayforth R, Koeppen H, Sasse J (1988) Increased levels of fibroblast growth factor-like activity in urine from patients with bladder or kidney cancer. Cancer Res 48:2083–2088

Chua CC, Barritault D, Geiman DE, Ladda RL (1987) Induction and suppression of type I collagenase in cultured human cells. Coll Relat Res 7:277–284

Clegg CH, Linkhart TA, Olwin BB, Hauschka SD (1987) Growth factor control of skeletal muscle differentiation: commitment to terminal differentiation occurs in G1 phase and is repressed by fibroblast growth factor. J Cell Biol 105:949–956

Conn G, Hatcher VB (1984) The isolation and purification of two anionic endothelial cell growth factors from human brain. Biochem Biophys Res Commun 124:262–268

Coughlin SR, Barr PJ, Cousens LS, Fretto LJ, Williams LT (1988) Acidic and basic fibroblast growth factors stimulate tyrosine kinase activity in vivo. J Biol Chem 263:988–993

Courty J, Chevallier B, Moenner M, Loret C, Lagente O, Böhlen P, Courtois Y, Barritault D (1985) Evidence for FGF-like growth factor in adult bovine retina: analogies with EDGF 1. Biochem Biophys Res Commun 136:102–108

Courty J, Loret C, Chevallier B, Moenner M, Barritault D (1987) Biochemical comparative studies between eye- and brain-derived growth factors. Biochimie 69:511–516

Crabb JW, Armes GL, Johnson CM, McKeehan WL (1986) Characterization of multiple forms of prostatropin (prostate epithelial cell growth factor) from bovine brain. Biochem Biophys Res Commun 136:1155–1161

Cuevas P, Burgos J, Baird A (1988a) Basic fibroblast growth factor (FGF) promotes cartilage repair in vivo. Biochem Biophys Res Commun 156:611–618

Cuevas P, Carceller F, Esteban A, Baird A, Guillemin R (1988b) Basic fibroblast growth factor (bFGF) enhances retinal ganglion cells survival and promotes axonal growth of rat transected optic nerve. Third joint meeting on neurochemical approaches to the unterstanding of cerebral disorders, Copenhagen, June 9–12 (Abstract)

Cuevas P, Baird A, Guillemin R (1988c) Angiogenic response to fibroblast growth factor in the rat brain in vivo. In: Gagliardi R, Benvenuti L (eds) Controversies in EIAB for cerebral ischemia. Monduzzi, Florence, pp 731–737

Cuevas P, Carceller F, Baird A, Guillemin R (1988d) Basic fibroblast growth factor (bFGF) increases peripheral nerve regeneration rate. Seventh general meeting of the European Society of Neurochemistry, Gothenburg, June 12–17 (Abstract)

Cuny R, Jeanny JC, Courtois Y (1986) Lens regeneration from cultured newt irises stimulated by retina-derived growth factors. Differentiation 32:221–229

D'Amore P, Klagsbrun M (1984) Endothelial cell mitogens derived from retina and hypothalamus: biochemical and biological similarities. J Cell Biol 99:1545–1549

D'Amore PA, Glaser BM, Brunson SK, Fenselau AH (1981) Angiogenic activity from bovine retina: partial purification and characterization. Proc Natl Acad Sci USA 78:3068–3072

Danielsen N, Pettmann B, Vahlsing HL, Manthorpe M, Varon S (1988) Fibroblast growth factor effects on peripheral nerve regeneration in a silicone chamber model. J Neurosci Res 20:320–330

Davidson JM, Klagsbrun M, Hill KE, Buckley A, Sullivan R, Brewer PS, Woodward S (1985) Accelerated wound repair, cell proliferation, and collagen accumulation are produced by a cartilage-derived growth factor. J Cell Biol 100:1219–1227

Delli Bovi P, Curatola AM, Kern FG, Greco A, Ittman M, Basilico C (1987) An oncogene isolated by transfection of Kaposi's sarcoma DNA encodes a growth factor that is a member of the FGF family. Cell 50:729–737

Delli-Bovi P, Curatola AM, Newman KM, Sato Y, Moscatelli D, Hewick RM, Rifkin DB, Basilico C (1988) Processing, secretion and biological properties of a novel growth factor of the fibroblast growth factor family with oncogenic potential. Mol Cell Biol 8:2933–2941

Dethlefsen SM, Butterfield C, Ausprunk DH (1986) Structural changes induced in capillary endothelial cells by growth factors: altered distribution of cytoskeletal elements and organelles in cells spreading in the presence of tumor conditioned medium and hypothalamus derived growth factor. Tissue Cell 18:827–837

Dickson C, Peters G (1987) Potential oncogene product related to growth factors. Nature 326:833

Eccleston PA, Silverberg DH (1985) Fibroblast growth factor is a mitogen for oligodendrocytes in vitro. Dev Brain Res 21:315–318

Emoto N, Baird A (1987) The regulation of aromatase activity in cultured human skin fibroblasts (Abstract). Endocrinology 120:321 A

Eppley BL, Doucet M, Connolly DT, Feder J (1988) Enhancement of angiogenesis of bFGF in mandibular bone graft healing in the rabbit. J Oral Maxillofac Surg 46:391–398

Esch F, Baird A, Ling N, Ueno N, Hill F, Denoroy L, Klepper R, Gospodarowicz D, Böhlen P, Guillemin R (1985) Primary structure of bovine pituitary basic fibroblast growth factor (FGF) and comparison with the amino-terminal sequence of bovine acidic FGF. Proc Natl Acad Sci USA 82:6507–6511

Esch F, Ueno N, Baird A, Hill F, Denoroy L, Ling N, Gospodarowicz D, Guillemin R (1986) Primary structure of bovine brain acidic fibroblast growth factor. Biochem Biophys Res Commun 133:554–562

Fauser B, Baird A, Hsueh A (1988) Fibroblast growth factor inhibits luteinizing hormone-stimulated androgen production by cultured rat testicular cells. Endocrinology 123:2935–2941

Feige JJ, Baird A (1988a) Phosphorylation of basic FGF: A new substrate for protein kinase C (Abstract). Endocrinology 122:1237A

Feige JJ, Baird A (1988b) Glycosylation of the basic fibroblast growth factor receptor. The contribution of carbohydrate to receptor function. J Biol Chem 263:14023–14029

Feige JJ, Bradley JD, Fryburg K, Farris J, Cousens LC, Barr PJ, Baird A (1989) Differential effects of heparin, fibronectin, and laminin on the phosphorylation of basic fibroblast growth factor by protein kinase C and the catalytic subunit of protein kinase A. J Cell Biol 109:3105–3114

Ferrara N, Schweigerer L, Neufeld G, Mitchell R, Gospodarowicz D (1987) Pituitary follicular cells produce basic fibroblast growth factor. Proc Natl Acad Sci USA 84:5773–5777

Fett J, Strydom D, Lobb R, Alderman E, Bethune J, Riordan J, Vallee B (1985) Isolation and characterization of angiogenin, an angiogenic protein from human carcinoma cells. Biochemistry 24:5480–5486

Finkelstein SP, Apostolides PJ, Caday CG, Prosser J, Philips MF, Klagsbrun M (1988) Increased basic fibroblast growth factor (bFGF) immunoreactivity at the site of focal brain wounds. Brain Res 460:253–259

Florkiewicz R, Sommer A (1989) The human bFGF gene encodes four polypeptides: three initiate translation from non-ATG codons. Proc Natl Acad Sci USA 86 (in press)

Folkman J (1972) Antiangiogenesis: new concept for therapy. Ann Surg 175:409–416

Folkman J, Klagsbrun M, Sasse J, Wadzinski MG, Ingber D, Vlodavsky I (1988) A heparin-binding angiogenic protein-basic fibroblast growth factor-is stored within basement membrane. Am J Pathol 130:393–400

Fourtanier AY, Courty J, Müller E, Courtois Y, Prunieras M, Barritault D (1986) Eye-derived growth factor isolated from bovine retina and used for epidermal wound healing in vivo. J Invest Dermatol 87:76–80

Fràter-Schroeder M, Risau W, Hallmann R, Gautschi-Sova P, Böhlen P (1987) Tumor necrosis factor-alpha, a potent inhibitor of endothelial cell proliferation in vitro, is angiogenic in vivo. Proc Natl Acad Sci USA 84:5277–5281

Fràter-Schröeder M, Müller G, Birchmeier W, Böhlen P (1986) Transforming growth factor-beta inhibits endothelial cell proliferation. Biochem Biophys Res Commun 137:295–302

Frederick JL, Shimanuki T, DiZerega GS (1984) Initiation of angiogenesis by human follicular fluid. Science 224:389–390

Fredj-Reygrobellet D, Plouet J, Delayre T, Baudoin C, Bourrtet F, Lapalus P (1987) Effects of aFGF and bFGF on wound healing in rabbit corneas. Curr Eye Res 6:1025–1029

Friesel R, Burgess WH, Mehlman T, Maciag T (1986) The characterization of the receptor for endothelial cell growth factor by covalent ligand attachment. J Biol Chem 261:7581–7584

Gambarini AG, Armelin HA (1982) Purification and partial characterization of an acidic fibroblast growth factor from bovine pituitary. J Biol Chem 257:9692–9697

Gauthier T, Maftouh M, Picard C (1987) Rapid enzymatic degradation of (^{125}I) [Tyr^{10}] FGF(1-10) by serum in vitro and involvement in the determination of circulating FGF by RIA. Biochem Biophys Res Commun 145:775–781

Gautschi P, Fràter-Schroeder M, Müller T, Böhlen P (1986a) Chemical and biological characterization of a truncated form of acidic fibroblast growth factor from bovine brain. Eur J Biochem 160:357–361

Gautschi P, Fràter-Schroeder M, Böhlen P (1986b) Partial molecular characterzation of endothelial cell mitogens from human brain: acidic and basic fibroblast growth factors. FEBS Lett 204:203–207

Gautschi-Sova P, Müller T, Böhlen P (1986) Amino acid sequence of human acidic fibroblast growth factor. Biochem Biophys Res Commun 140:874–880

Gautschi-Sova P, Fràter-Schroeder M, Jiang ZP, Böhlen P (1987) Acidic fibroblast growth factor is present in non-neural tissue: isolation and chemical characterization from bovine kidney. Biochemistry 26:5844–5847

Gensburger C, Labourdette G, Sensenbrenner M (1987) Brain basic fibroblast growth factor stimulates proliferation of rat neuronal precursor cells in vitro. FEBS Lett 217:1–5

Gimenez-Gallego G, Rodkey K, Bennett C, Rios-Candelore M, DiSalvo J, Thomas KA (1985) Brain-derived acidic fibroblast growth factor: complete amino acid sequence and homologies. Science 230:1385–1388

Gimenez-Gallego G, Conn G, Hatcher VB, Thomas KA (1986a) The complete amino acid sequence of human brain-derived acidic fibroblast growth factor. Biochem Biophys Res Commun 138:611–617

Gimenez-Gallego G, Conn G, Hatcher VB, Thomas KA (1986b) Human brain-derived acidic and basic fibroblast growth factors: aminoterminal sequences and specific mitogenic activities. Biochem Biophys Res Commun 135:541–548

Globus RK, Patterson-Buckendahl P, Gospodarowicz D (1988) Regulation of bovine bone cell proliferation by fibroblast growth factor and transforming growth factors. Endocrinology 123:98–105

Gordon PB, Conn G, Hatcher VB (1985) Glycosaminoglycan production in cultures of early and late passage human endothelial cells: the influence of an anionic endothelial cell growth factor and the extracellular matrix. J Cell Physiol 125:596–607

Gospodarowicz D (1974) Localization of a fibroblast growth factor and its effect alone and with hydrocortisone on 3T3 cell growth. Nature 249:123–127

Gospodarowicz D (1975) Purification of a fibroblast growth factor from bovine pituitary. J Biol Chem 250:2515–2520

Gospodarowicz D (1976) Humoral control of cell proliferation: the role of fibroblast growth factor in regeneration, angiogenesis, wound healing and neoplastic growth. Prog Clin Biol Res 9:1–19

Gospodarowicz D, Cheng J (1986) Heparin protects basic and acidic FGF from inactivation. J Cell Physiol 128:475–484

Gospodarowicz D, Greenburg G (1979) The effects of epidermal and fibroblast growth factors on the repair of corneal endothelial wounds in bovine corneas maintained in organ culture. Exp Eye Res 28:147–157

Gospodarowicz D, Mescher AL (1981) Fibroblast growth factor and vertebrate regeneration. In: Riccardi VM, Mulvihill JJ (eds) Advances in neurology: neurofibromatosis. Raven, New York, pp 149–171

Gospodarowicz D, Thakral TK (1978) Production of a corpus luteum angiogenic factor responsible for proliferation of capillaries and neovascularization of the corpus luteum. Proc Natl Acad Sci USA 75:847–851

Gospodarowicz D, Moran J, Braun D, Birdwell C (1976) Clonal growth of bovine vascular endothelial cells: fibroblast growth factor as a survival agent. Proc Natl Acad Sci USA 73:4120–4124

Gospodarowicz D, Ill CR, Hornsby PJ, Gill G (1977) Control of bovine adrenal cortical cell proliferation by fibroblast growth factor: lack of effect of epidermal growth
factor. Endocrinology 100:1080–1089

Gospodarowicz D, Bialecki H, Thakral TK (1979) The angiogenic activity of the
fibroblast and epidermal growth factor. Exp Eye Res 28:501–514

Gospodarowicz D, Vlodavsky I, Savion N, Tauber JP (1980) Control of the proliferation and differentiation of vascular endothelial cells by fibroblast growth factor. In:
Bloom F (ed) Peptides: integrators of cell and tissue function. Raven, New York,
pp 1–37

Gospodarowicz D, Cheng J, Lui GM, Baird A, Böhlen P (1984) Isolation of brain
fibroblast growth factor by heparin-Sepharose affinity chromotography: identity
with pituitary fibroblast growth factor. Proc Natl Acad Sci USA 81:6963–6967

Gospodarowicz D, Cheng J, Lui GM, Baird A, Esch F, Böhlen P (1985a) Corpus
luteum angiogenic factor is related to fibroblast growth factor. Endocrinology
117:2383–2391

Gospodarowicz D, Massoglia S, Cheng J, Lui GM, Böhlen P (1985b) Isolation of
(bovine) pituitary fibroblast growth factor purified by fast protein liquid
chromatography (FPLC): partial chemical and biological characterization. J Cell
Physiol 122:323–332

Gospodarowicz D, Neufeld G, Schweigerer L (1986a) Fibroblast growth factor. Mol
Cell Endocrinol 46:187–204

Gospodarowicz D, Baird A, Cheng J, Lui GM, Esch F, Böhlen P (1986b) Isolation of
fibroblast growth factor from bovine adrenal gland: physicochemical and biological
characterization. Endocrinology 118:82–90

Gospodarowicz D, Ferrara N, Schweigerer L, Neufeld G (1987) Structural characterization and biological functions of fibroblast growth factor. Endocrinol Rev
8(2):95–114

Grunz H, McKeehan WL, Knöchel W, Born J, Tiedemann H (1988) Induction of
mesodermal tissues by acidic and basic heparin binding growth factors. Cell Differ
22:183–190

Halaban R, Ghosh S, Baird A (1987) bFGF is the putative natural growth factor for
human melanocytes. In Vitro Cell Dev Biol 23:47–52

Halaban R, Kwon B, Ghosh S, Delli Bovi P, Baird A (1988) bFGF as an autocrine
growth factor in human melanomas. Mol Cell Biol 8:2933–2941

Halperin JA, Lobb RR (1987) Effect of heparin-binding growth factors on monovalent
cation transport in Balb/C 3T3 cells. Biochem Biophys Res Commun 144:115–122

Hamerman D, Taylor S, Kirschenbaum I, Klagsbrun M, Raines EW, Ross R, Thomas
KA (1987) Growth factors with heparin binding affinity in human synovial fluid.
Proc Soc Exp Biol Med 186:384–389

Hamilton RT, Nilsen-Hamilton M, Adams GA (1985) Superinduction by
cycloheximide of mitogen-induced secreted proteins produced by Balb/c 3T3 cells. J
Cell Physiol 123:201–208

Harper JW, Lobb RR (1988) Reductive methylation of lysine residues in acidic
fibroblast growth factor: effect on mitogenic activity and heparin affinity.
Biochemistry 27:671–678

Hauschka PV, Mavrakos AE, Iafrati MD, Doleman S, Klagsbrun M (1986) Growth
factors in bone matrix. Isolation of multiple types by affinity chromatography on
heparin-Sepharose. J Biol Chem 261:12665–12674

Hayek A, Culler FL, Beattie GM, Lopez AD, Cuevas P, Baird A (1987) An in vivo
model for study of the angiogenic effects of basic fibroblast growth factor. Biochem
Biophys Res Commun 147:876–880

Herman IM, D'Amore P (1984) Capillary endothelial cell migration: loss of stress
fibres in response to retina-derived growth factor. J Muscle Res Cell Motil
5:697–709

Hoffman RS (1940) The growth-activating effect of extracts of adult and embryonic tissues of the rat on fibroblast colonies in culture. Growth 4:361–376

Hornsby PJ, Gill GN (1977) Hormonal control of adrenocortical cell proliferation. Desensitization to ACTH and interaction between ACTH and fibroblast growth factor in bovine adrenocortical cell cultures. J Clin Invest 60:342–354

Huang SS, Huang JS (1986) Association of bovine brain-derived growth factor receptor with protein tyrosine kinase activity. J Biol Chem 261:9568–9571

Huang JS, Huang SS, Kuo MD (1986a) Bovine brain-derived growth factor. Purification and characterization of its interaction with responsive cells. J Biol Chem 261:11 600–11 607

Huang SS, Kuo MD, Huang JS (1986b) Transforming growth factor activity of bovine brain-derived growth factor. Biochem Biophys Res Commun 139:619–625

Huang SS, Tsai CC, Adams SP, Huang JS (1987) Neuron localization and neuroblastoma cell expression of brain-derived growth factor. Biochem Biophys Res Commun 144:81–87

Imamura T, Tokita Y, Mitsui Y (1988) Purification of basic FGF receptors from rat brain. Biochem Biophys Res Commun 155:583–590

Jakob W, Jentzsch B, Bauersberger B, Oehme P (1977) Demonstration of angiogenesis activity in corpus luteum of cattle. Exp Pathol 13:231–242

Jaye M, McConathy E, Drohan W, Tong B, Deuel T (1985) Modulation of the sis gene transcript during endothelial cell differentiation in vitro. Science 228:882–885

Jaye M, Howk R, Burgess W, Ricca GA, Chiu IM, Ravera MW, O'Brien SJ, Modi WS, Maciag T, Drohan WN (1986) Human endothelial cell growth factor: cloning, nucleotide sequence, and chromosome localization. Science 233:541–545

Jaye M, Burgess WH, Shaw AB, Drohan WN (1987) Biological equivalence of natural bovine and recombinant human alpha-endothelial cell growth factors. J Biol Chem 252:16612–16617

Jaye M, Lyall RM, Mudd R, Schlessinger J, Sarver N (1988) Expression of acidic fibroblast growth factor cDNA confers growth advantage and tumorigenesis to Swiss 3T3 cells. EMBO J 7:963–969

Jentzsch KD, Wellmitz G, Heder G, Petzold E, Buntrock P, Oehme P (1980) A bovine brain fraction with fibroblast growth factor activity inducing articular cartilage regeneration in vivo. Acta Biol Med Ger 39:967–971

Joseph-Silverstein J, Rifkin DB (1987) Endothelial cell growth factors and the vessel wall. Semin Thromb Hemost 13:504–513

Joseph-Silverstein J, Moscatelli D, Rifkin DB (1988) The development of a quantitative RIA for basic fibroblast growth factor using polyclonal antibodies against the 157 amino acid form of human bFGF. J Immunol Methods 110:183–192

Kalebic T, Garbisa S, Glaser B, Liotta LA (1983) Basement membrane collagen: degradation by migrating endothelial cells. Science 221:281–283

Kardami E, Spector D, Strohman RC (1985) Myogenic growth factor present in skeletal muscle is purified by heparin-affinity chromatography. Proc Natl Acad Sci USA 82:8044–8047

Kato FA, Gospodarowicz DJ (1984) Growth requirements of low-density rabbit costal chondrocyte cultures maintained in serum-free medium. J Cell Physiol 120:354–363

Kato Y, Gospodarowicz D (1985) Sulfated proteoglycan synthesis by confluent cultures of rabbit costal chondrocytes grown in presence of fibroblast growth factor. J Cell Biol 100:477–485

Kato Y, Iwamoto M, Koike T (1987) Fibroblast growth factor stimulates colony formation of differentiated chondrocytes in soft agar. J Cell Physiol 133:491–498

Katoh Y, Takayama S (1984) Characterization of the growth responses of hamster fibroblasts and chondrogenic cells to fibroblast growth factor. Exp Cell Res 150:131–140

Kimelman D, Kirschner M (1987) Synergistic induction of mesoderm by FGF and TGF-beta and the identification of an mRNA coding for FGF in the early Xenopus embryo. Cell 51:869–877

Kimelman D, Abraham JA, Haaparanta T, Palisi TM, Kirschner MW (1988) The presence of fibroblast growth factor in the frog egg: its role as a natural mesoderm inducer. Science 242:1053–1056

Klagsbrun M, Beckoff MC (1980) Purification of cartilage-derived growth factor. J Biol Chem 255:10859–10866

Klagsbrun M, Sasse J, Sullivan R, Smith JA (1986) Human tumor cells synthesize an endothelial growth factor that is structurally related to basic fibroblast growth factor. Proc Natl Acad Sci USA 83:2448–2452

Klagsbrun M, Smith S, Sullivan R, Shing Y, Davidson S, Smith JA, Sasse J (1987) Multiple forms of basic fibroblast growth factor: amino-terminal cleavages by tumor cell- and brain cell-derived acid proteinases. Proc Natl Acad Sci USA 84:1839–1843

Koos RD, Lemaire WJ (1983) Evidence for an angiogenic factor from rat follicles. In: Greenwald GS, Terranova PF (eds) Factors regulating ovarian function. Raven, New York, pp 847–851

Kurokawa T, Seno M, Igarashi K (1988) Nucleotide sequence of rat basic fibroblast growth factor cDNA. Nucleic Acids Res 16(11):5201

Lathrop B, Thomas K, Glaser L (1985a) Control of myogenic differentiation by fibroblast growth factor is mediated by position in the G1 phase of the cell cycle. J Cell Biol 101:2194–2198

Lathrop B, Olson E, Glaser L (1985b) Control by fibroblast growth factor of differentiation in the BC3H1 muscle cell line. J Cell Biol 100:1540–1547

Lee PL, Johnson DE, Cousens LS, Fried VA, Williams LT (1989) Purification and complementary DNA cloning of a receptor for basic fibroblast growth factor. Science 245:57–60

Leibovich SJ, Ross R (1976) A macrophage-dependent factor that stimulates the proliferation of fibroblasts in vitro. Am J Pathol 84:501–514

Lemmon SK, Bradshaw RA (1983) Purification and partial characterization of bovine pituitary fibroblast growth factor. J Cell Biochem 21:195–208

Libermann TA, Friesel R, Jaye M, Lyall RM, Westermark B, Drohan W, Schmidt A, Maciag T, Schlessinger J (1987) An angiogenic growth factor is expressed in human glioma cells. EMBO J 6:1627–1632

Linkhart TA, Clegg CH, Hauschka SD (1980) Control of mouse myoblast commitment to terminal differentiation by mitogens. J Supramol Struct 14:483–498

Lipton SA, Wagner JA, Madison RD, D'Amore PA (1988) Acidic fibroblast growth factor enhances regeneration of processes by postnatal retinal ganglion cells in culture. Proc Natl Acad Sci USA 85:2388–2392

Liu L, Nicoll CS (1988) Evidence for a role of basic fibroblast growth factor in rat embryonic growth and differentiation. Endocrinology 123:2027–2031

Lobb RR, Fett JW (1984) Purification of two distinct growth factors from bovine neural tissue by heparin affinity chromatography. Biochemistry 23:6295–6299

Lobb RR, Alderman ER, Fett JW (1985) Induction of angiogenesis by bovine brain-derived class 1 heparin-binding growth factor. Biochemistry 24:4969–4973

Lobb RR, Sasse J, Sullivan R, Shing Y, D'Amore P, Jacobs J, Klagsbrun M (1986a) Purification and characterization of heparin-binding endothelial cell growth factors. J Biol Chem 261:1924–1928

Lobb RR, Rybak SM, StClair DK, Fett JW (1986b) Lysates of two established human tumor lines contain heparin-binding growth factors related to bovine acidic fibroblast growth factor. Biochem Biophys Res Commun 139:861–867

Lobb RR, Harper JW, Fett JW (1986c) Purification of heparin-binding growth factors. Anal Biochem 154:1–14

Maciag T, Hoover GA, Weinstein R (1982) High and low molecular weight forms of endothelial cell growth factor. J Biol Chem 257:5333–5336

Magnaldo I, L'Allemain G, Chambard JC, Moenner M, Barritault D, Pouyssegur J (1986) The mitogenic signaling pathway of fibroblast growth factor is not mediated through phosphoinositide hydrolysis and protein kinase C activation in hamster fibroblasts. J Biol Chem 261:16916–16922

Marics I, Adelaide J, Raybaud F, Mattei MG, Coulier F, Planche J, De Lapeyriere O, Birnbaum D (1989) Characterization of the HST-related FGF.6 gene, a new member of the fibroblast growth factor gene family. Oncogene 4:335–340

Makris A, Ryan KJ, Yasumizu T, Hill CL, Zetter BR (1984) The nonluteal porcine ovary as a source of angiogenic activity. Endocrinology 115:1672–1677

Martinet Y, Bitterman PB, Mornex JF, Grotendorst GR, Martin GR, Crystal RG (1986) Activated human monocytes express the c-sis proto-oncogene and release a mediator showing PDGF-like activity. Nature 319:158–160

Mascarelli F, Raulais D, Counis MF, Courtois Y (1987) Characterization of acidic and basic fibroblast growth factors in brain, retina and vitreous chick embryo. Biochem Biophys Res Commun 146:478–486

Matsuo Y, Nishi N, Matsui S, Sansberg A, Isaacs JT, Wada F (1987) Heparin binding affinity of rat prostatic growth factor in normal and cancerous prostates: partial purification and characterization of rat prostatic growth factor in Dunning tumor. Cancer Res 47:188–192

Matzner Y, Bar-Ner M, Yahalom J, Ishai-Michaeli R, Ruks Z, Vlodavsky I (1985) Degradation of heparan sulfate in the subendothelial extracellular matrix by a readily released heparanase from human neutrophils. Possible role in invasion through basement membranes. J Clin Invest 76:1306–1313

McKeehan WL, Crabb JW (1987) Isolation and characterization of different molecular and chromatographic forms of heparin-binding growth factor 1 from bovine brain. Anal Biochem 164:563–569

Mergia A, Eddy R, Abraham JA, Fiddes JC, Shows TB (1986) The genes for basic and acidic fibroblast growth factors are on different human chromosomes. Biochem Biophys Res Commun 138:644–651

Mescher AL, Gospodarowicz D (1979) Mitogenic effect of a growth factor derived from myelin of denervated regenerates of newt forelimbs. J Exp Zool 207:497–503

Mescher AL, Loh JJ (1980) Newt forelimb regeneration of blastemas in vitro. Cellular response to explantation and effects of various growth-promoting substances. J Exp Zool 216:235–245

Mioh H, Cheng JK (1987) Acidic heparin binding growth factor transiently activates adenylate cyclase activity in human adult arterial smooth muscle cells. Biochem Biophys Res Commun 146:771–776

Moenner M, Chevallier B, Badet J, Barritault D (1986) Evidence and characterization of the receptor to eye-derived growth factor I, the retinal form of basic fibroblast growth factor, on bovine epithelial lens cells. Proc Natl Acad Sci USA 83:5024–5028

Moenner M, Badet J, Chevallier B, Tardieu M, Courty J, Barritault D (1987) Eye-derived fibroblast growth factors: receptors and early events studies. In: Rifkin DB, Klagsbrun M (eds) Angiogenesis: mechanisms and pathobiology. Current communications in molecular biology. Cold Spring Harbor, New York, pp 52–57

Montesano R, Vasalli JD, Baird A, Guillemin R, Orci L (1986) Basic fibroblast growth factor induces angiogenesis in vitro. Proc Natl Acad Sci USA 83:7297–7301

Mormède P, Baird A (1988) Estrogens, cyclic adenosine 3′, 5′-monophosphate, and phorbol esters modulate the prolactin response of GH3 cells to basic fibroblast growth factor. Endocrinology 122:2265–2271

Mormède P, Baird A, Pigeon P (1985) Immunoreactive fibroblast growth factor (FGF) in rat tissues: molecular weight forms and effects of hypophysectomy. Biochem Biophys Res Commun 128:1108–1113

Morrison RS, DeVellis J (1981) Growth of purified astrocytes in a chemically defined medium. Proc Natl Acad Sci USA 78:7205–7209

Morrison RS, DeVellis J, Lee YL, Bradshaw R, Eng LF (1985) Hormones and growth factors induce the synthesis of glial fibrillary acidic protein in rat brain astrocytes. J Neurosci Res 14:167–176

Morrison RS, Sharma A, DeVellis J, Bradshaw RA (1986) Basic fibroblast growth factor supports the survival of cerebral cortical neurons in primary culture. Proc Natl Acad Sci USA 83:7537–7541

Moscat J, Moreno F, Herrero C, Lopez C, Garcia-Barreno P (1988) Endothelial cell growth factor and ionophore A23187 stimulation of production of inositol phosphates in porcine aorta endothelial cells. Proc Natl Acad Sci USA 85:659–663

Moscatelli D (1987) High and low affinity binding sites for basic fibroblast growth factor on cultured cells: absence of a role for low affinity binding in the stimulation of plasminogen activator production by bovine capillary endothelial cells. J Cell Physiol 131:123–130

Moscatelli D (1988) Metabolism of receptor-bound and matrix-bound basic fibroblast growth factor by bovine capillary endothelial cells. J Cell Biol 107:753–759

Moscatelli D, Presta M, Joseph-Silverstein J, Rifkin DB (1986) Both normal and tumor cells produce basic fibroblast growth factor. J Cell Physiol 129:273–276

Moscatelli D, Joseph-Silverstein J, Manejias R, Rifkin DB (1987a) Mr 25000 heparin-binding protein from guinea pig brain is a high molecular weight form of basic fibroblast growth factor. Proc Natl Acad Sci USA 84:5778–5782

Moscatelli D, Presta M, Joseph-Silverstein J, Rifkin DB (1987b) Presence of basic fibroblast growth factor in a variety of cells and its binding to cells. In: Rifkin DB, Klagsbrun M (eds) Angiogenesis. Current communications in molecular biology. Cold Spring Harbor, New York, pp 47–51

Müller G, Courty J, Courtois Y, Clerc B, Barritault D (1985) Utilisation du facteur de croissance derive de l'oeil dans le traitement des ulceres de cornee. J Fr Ophthalmol 8:187–192

Müller R, Bravo R, Burckhardt J (1984) Induction of c-fos gene and protein by growth factors precedes activation of c-myc. Nature 312:716–720

Nakajima M, Irimura T, Di Ferrante D, Di Ferrante N, Nicolson GL (1983) Heparan sulfate degradation: relation to tumor invasive and metastatic properties of mouse B16 melanoma sublines. Science 220:611–613

Neufeld G, Gospodarowicz D (1985) The identification and partial characterization of the fibroblast growth factor receptor of baby hamster kidney cells. J Biol Chem 260:13860–13868

Neufeld G, Gospodarowicz D (1986) Basic and acidic fibroblast growth factors interact with the same cell surface receptors. J Biol Chem 261:5631–5637

Neufeld G, Ferrara N, Schweigerer L, Mitchell R, Gospodarowicz D (1987a) Bovine granulosa cells produce basic fibroblast growth factor. Endocrinology 121:597–603

Neufeld G, Gospodarowicz D, Dodge L, Fujii D (1987b) Heparin modulation of the neurotrophic effects of acidic and basic fibroblast growth factors and nerve growth factor on PC12 cells. J Cell Physiol 131:131–140

Nieto-Sampedro M, Lim R, Hicklin DJ, Cotman CW (1988) Early release of glia maturation factor and acidic fibroblast growth factor after rat brain injury. Neurosci Lett 86:361–365

O'Keefe EJ, Chiu NL, Payne RE (1988) Stimulation of growth of keratinocytes by bFGF. J Invest Dermatol 90:767–769

Olwin BB, Hauschka SD (1986) Identification of the fibroblast growth factor receptor of Swiss 3T3 cells and mouse skeletal muscle myoblasts. Biochemistry 25:3487–3492

Olwin BB, Hauschka SD (1988) Cell surface fibroblast growth factor and epidermal growth factor receptors are permanently lost during skeletal muscle terminal differentiation in culture. J Cell Biol 107:761–769

Otto D, Unsicker K, Grothe C (1987) Pharmacological effects of nerve growth factor and fibroblast growth factor applied to the transsectioned sciatic nerve on neuron death in adult rat dorsal root ganglia. Neurosci Lett 83:156–160

Panet R, Amir I, Atlan H (1986) Fibroblast growth factor induces a transient net K+ influx carried by the bumetanide-sensitive transporter in quiescent BALB/c 3T3 fibroblasts. Biochem Biophys Acta 859:117–121

Paterson JC (1938) Vascularization and hemorrhage of the intima of atheriosclerotic coronary arteries. Arch Pathol Lab Med 25:313–324

Pelech S, Olwin BB, Krebs EG (1986) Fibroblast growth factor treatment of Swiss 3T3 cells activates a subunit S6 kinase that phosphorylates a synthetic peptide substrate. Proc Natl Acad Sci USA 83:5968–5972

Petroutsos G, Courty J, Guimaraes R, Pouliquen Y, Barritault D, Plouet J, Courtois Y (1984) Comparison of the effects of EGF, pFGF, and EDGF on corneal epithelium wound healing. Curr Eye Res 3:593–598

Pettmann B, Weibel M, Sensenbrenner M, Labourdette G (1985) Purification of two astroglial growth factors from bovine brain. FEBS Lett 189:102–108

Photopoulos G, Connolly JA (1986) Levels of fibroblast growth factor receptors and differentiation in the non-fusing muscle cell line BC3H1. J Cell Biol 103:121 A

Presta M, Moscatelli D, Joseph-Silverstein J, Rifkin DB (1986) Purification from a human hepatoma cell line of a basic fibroblast growth factor-like molecule that stimulates capillary endothelial cell plasminogen activator production, DNA synthesis, and migration. Mol Cell Biol 6:4060–4066

Presta M, Rusnati M, Maier JAM, Ragnotti G (1988) Purification of basic fibroblast growth factor from rat brain: identification of a Mr 22000 immunoreactive form. Biochem Biophys Res Commun 155:1161–1172

Pruss RM, Bartlett PF, Gavrilovic J, Lisak RP, Rattray S (1981) Mitogens for glial cells: a comparison of the response of cultured astrocytes, oligodendrocytes and Schwann cells. Brain Res 254:19–35

Raeside JI, Berthelon M-C, Sanchez P, Saez JM (1988) Stimulation of aromatase activity in immature porcine Leydig cells by fibroblast growth factor (FGF). Biochem Biophys Res Commun 151:163–169

Raju KS, Alessandri G, Ziche M, Gullino PM (1982) Ceruloplasmin, copper ions, and angiogenesis. JNCI 69:1183–1188

Risau W (1986) Developing brain produces an angiogenesis factor. Proc Natl Acad Sci USA 83:3855–3859

Risau W, Ekblom P (1986) Production of a heparin-binding angiogenesis factor by the embryonic kidney. J Cell Biol 103:1101–1107

Risau W, Gautschi-Sova P, Böhlen P (1988) Endothelial cell growth factors in embryonic and adult chick brain are related to human acidic fibroblast growth factor. EMBO J 7:959–962

Rizzino A, Ruff E (1986) Fibroblast growth factor induces the soft agar growth of two non-transformed cell lines. In Vitro Cell Dev Biol 22:749–755

Roberts AB, Sporn MB, Assoian RK, Smith JM, Roche NS, Wakefield LM, Heine UI, Liotta LA, Falanga Y, Kehrl JH, Fauci AS (1986) Transforming growth factor type beta: rapid induction of fibrosis and angiogenesis in vivo and stimulation of collagen formation in vitro. Proc Natl Acad Sci USA 83:4167–4171

Rodan SB, Wesolowski G, Thomas K, Rodan GA (1987) Growth stimulation of rat calvaria osteoblastic cells by acidic fibroblast growth factor. Endocrinology 121:1917–1923

Rogelj S, Weinberg RA, Fanning P, Klagsbrun M (1988) Basic fibroblast growth factor fused to a signal peptide transforms cells. Nature 331:173–175

Rogister B, Leprince P, Pettmann B, Labourdette G, Sensenbrenner M, Moonen G (1988) Brain basic fibroblast growth factor stimulates the release of plasminogen activators by newborn rat cultured astroglial cells. Neurosci Lett 91:321–326

Rosengart TK, Johnson WV, Friesel R, Clark R, Maciag T (1988a) Heparin protects heparin-binding growth factor-I from proteolytic inactivation in vitro. Biochem Biophys Res Commun 152:432–440

Rosengart TK, Kupferschmid JP, Ferrans VJ, Casscells W, Maciag T, Clark RE (1988b) Heparin-binding growth factor-I (endothelial cell growth factor) binds to endothelium in vivo. J Vasc Surg 7:311–317

Ross M, Ballard FJ (1988) Regulation of protein metabolism and DNA synthesis by fibroblast growth factor in BHK-21 cells. Biochem J 249:363–368

Rowe JM, Kasper S, Shiu RP, Friesen HG (1986) Purification and characterization of a human mammary tumor-derived growth factor. Cancer Res 46:1408–1412

Rubin JS, Osada H, Finch PW, Taylor WG, Rudikoff S, Aaronson SA (1989) Purification and characterization of a newly indentified growth factor specific for epithelial cells. Proc Natl Acad Sci USA 86:802–806

Rudland PS, Gospodarowicz D, Seifert W (1974) Activation of guanyl cyclase and intracellular cyclic GMP by fibroblast growth factor. Nature 250:741–743

Ruta M, Burgess W, Givol D, Epstein J, Neiger N, Kaplow J, Crumley G, Dionne C, Jaye M, Schlessinger J (1989) Receptor for acidic fibroblast growth factor is related to tyrosine kinase encoded by the fms-like gene (FLG). Proc Natl Acad Sci USA 86:8722–8726

Sakaguchi M, Kajio T, Kawahara K, Kato K (1988) Antibodies against basic fibroblast growth factor inhibit the autocrine growth of pulmonary artery endothelial cells. FEBS Lett 233:163–166

Saksela O, Moscatelli D, Rifkin DB (1987) The opposing effect of basic fibroblast growth factor and transforming growth factor beta on the regulation of plasminogen activator activity in capillary endothelial cells. J Cell Biol 105:957–963

Saksela O, Moscatelli D, Sommer A, Rifkin DB (1988) Endothelial cell-derived heparan sulfate binds basic fibroblast growth factor and protects it from proteolytic degradation. J Cell Biol 107:743–751

Saneto RP, DeVellis J (1985) Characterization of cultured rat oligodendrocytes proliferating in a serum-free, chemically defined medium. Proc Natl Acad Sci USA 82:3509–3513

Scheffer C, Tseng G, Savion N, Stern R, Gospodarowicz D (1982) Fibroblast growth factor modulates synthesis of collagen in cultured vascular endothelial cells. Eur J Biochem 122:355–360

Schreiber AB, Winkler ME, Derynck R (1986) Transforming growth factor-alpha: a more potent angiogenic mediator than epidermal growth factor. Science 232:1250–1253

Schubert D, Ling N, Baird A (1987) Multiple influences of a heparin-binding growth factor on neuronal development. J Cell Biol 104:635–643

Schweigerer L, Neufeld G, Friedman J, Abraham JA, Fiddes JC, Gospodarowicz D (1987a) Basic fibroblast growth factor: production and growth stimulation in cultured adrenal cortex cells. Endocrinology 120:796–800

Schweigerer L, Neufeld G, Friedman J, Abraham JA, Fiddes JC, Gospodarowicz D (1987b) Capillary endothelial cells express basic fibroblast growth factor, a mitogen that promotes their own growth. Nature 325:257–259

Schweigerer L, Neufeld G, Gospodarowicz D (1987c) Basic fibroblast growth factor as a growth inhibitor for cultured human tumor cells. J Clin Invest 80:1516–1520

Seed J, Olwin BB, Hauschka SD (1988) Fibroblast growth factor levels in the whole embryo and limb bud during chick development. Dev Biol 128:50–57

Senior RM, Huang SS, Griffin GL, Huang JS (1986) Brain-derived growth factor is a chemoattractant for fibroblasts and astroglial cells. Biochem Biophys Res Commun 141:67–72

Seno M, Sasada R, Iwane M, Sudo K, Kurokawa T, Ito K, Igarashi K (1988) Stabilizing basic fibroblast growth factor using protein engineering. Biochem Biophys Res Commun 151:701–708

Shimasaki S, Emoto N, Koba A, Mercado M, Shibata F, Cooksey K, Baird A, Ling N (1988) Complementary DNA cloning and sequencing of rat ovarian basic fibroblast growth factor and tissue distribution study of its mRNA. Biochem Biophys Res Commun 152:717–723

Shing Y (1988) Heparin-copper biaffinity chromatography of fibroblast growth factor. J Biol Chem 263:1296–1299

Shing Y, Folkman J, Sullivan R, Butterfield C, Murray J, Klagsbrun M (1984) Heparin affinity: purification of a tumor-derived capillary endothelial cell growth factor. Science 223:1296–1299

Sievers J, Hausmann B, Unsicker K, Berry M (1987) Fibroblast growth factors promote the survival of adult rat retinal ganglion cells after transection of the optic nerve. Neurosci Lett 76:157–162

Simpson RJ, Moritz RL, Lloyd CJ, Fabri LJ, Nice EC, Rubira MR, Burgess AW (1987) Primary structure of ovine pituitary basic fibroblast growth factor. FEBS Lett 224:128–132

Slack J, Darlington B, Heath H, Godsave S (1987) Mesoderm induction in early Xenopus embryos by heparin-binding growth factors. Nature 326:197–200

Snow AD, Kisilevsky R, Stephens C, Anastassiades T (1987) Electrophoresis of glycosominoglycans isolated from normal human plasma. Direct evidence for the presence of a heparin-like molecule. Biomed Biochim Acta 7:537–546

Sommer A, Brewer MT, Thompson RC, Moscatelli D, Presta M, Rifkin DB (1987) A form of human basic fibroblast growth factor with an extended amino terminus. Biochem Biophys Res Commun 144:543–550

Sporn MB, Roberts AB (1988) Peptide growth factors are multifunctional. Nature 332:217–219

Sprugel KH, McPherson JM, Clowes WA, Ross R (1987) Effects of growth factors in vivo. I. Cell ingrowth into porous subcutaneous chambers. Am J Pathol 129:601–613

StLouis D, Verma IM (1988) An alternative approach to somatic cell gene therapy. Proc Natl Acad Sci USA 85:3150–3154

Stenn KS, Fernandez LA, Tirrell SJ (1988) Initiation of angiogenesis by human follicular fluid. J Invest Dermatol 90:409–411

Story MT, Esch F, Shimasaki S, Sasse J, Jacobs SC, Lawson RK (1987) Amino-terminal sequence of a large form of basic fibroblast growth factor isolated from human benign prostatic hyperplastic tissue. Biochem Biophys Res Commun 142:702–709

Sulkowski E (1985) Purification of proteins by IMAC. Trends in Biotechnoloy 31:1–7

Taira M, Yoshida T, Miyagawa K, Sakamoto H, Terada M, Sugimura T (1987) cDNA sequence of human transforming gene hst and identification of the coding sequence required for transforming activity. Proc Natl Acad Sci USA 84:2980–2984

Takeyama Y, Tanimoto T, Hoshijima M, Kaibuchi K, Ohyanagi H, Saitoh Y, Takai Y (1986) Enhancement of fibroblast growth factor-induced diacylglycerol formation and protein kinase C activation by colon tumor-promoting bile acid in Swiss 3T3 cells. Different modes of action between bile acid and phorbol ester. FEBS Lett 197:339–343

Terranova VP, DiFlorio R, Lyall RM, Hic S, Friesel R, Maciag T (1985) Human endothelial cells are chemotactic to endothelial cell growth factor and heparin. J Cell Biol 101:2330–2334

Thomas KA (1987) Fibroblast growth factors. FASEB J 1:434–440

Thomas KA (1988) Transforming potential of fibroblast growth factor genes. TIBS 13:327–328

Thomas KA, Riley MC, Lemmon SK, Baglan NC, Bradshaw RA (1980) Brain fibroblast growth factor: nonidentity with myelin basic protein fragment. J Bil Chem 255:5517–5520

Thomas KA, Rios-Cadelore M, Fitzpatrick S (1984) Purification and characterization of acidic fibroblast growth factor from bovine brain. Proc Natl Acad Sci USA 81:357–361

Thomas KA, Rios-Candelore M, Gimenez-Gallego G, DiSalvo J, Bennett C, Rodkey K, Fitzpatrick S (1985) Pure brain-derived acidic fibroblast growth factor is a potent angiogenic vascular endothelial cell mitogen with sequence homology to interleukin-1. Proc Natl Acad Sci USA 82:6409–6413

Thompson JA, Anderson KD, DiPietro JM, Zwiebel JA, Zametta A, Anderson WF, Maciag T (1988) Site-directed neovessel formation in vivo. Science 241:1349–1352

Thompson P, Desbordes SM, Girand J, Pauliquen J, Barritault D, Courtois Y (1987) The effect of an eye-derived growth factor (EDGF) on corneal epithelial regeneration. Exp Eye Res 34:191–199

Thompson RW, Wadzinski MG, Sasse J, Klagsbrun M, Folkman J, Shemin RJ, D'Amore P (1986) Isolation of heparin-binding endothelial cell mitogens from normal human myocardium. J Cell Biol 203:300 A

Togari A, Dickens G, Kuzuya H, Guroff G (1985) The effect of fibroblast growth factor on PC12 cells. J Neurosci 5:307–316

Trowell OA, Chir B, Willmer EN (1939) Growth of tissues in vitro. VI. The effects of some tissue extracts on the growth of periosteal fibroblasts. J Exp Biol 16:60–70

Tsuda T, Kaibuchi K, Kawahara Y, Fukuzaki H, Takai Y (1985) Induction of protein kinase C activation and Ca2+ mobilization by fibroblast growth factor in Swiss 3T3 cells. FEBS Lett 191:205–210

Tsuda T, Hamamori Y, Yamashita T, Fukumoto Y, Takai Y (1986) Involvement of three intracellular messenger systems, protein kinase C, calcium ion and cyclic AMP, in the regulation of c-fos gene. FEBS Lett 208:39–42

Turner DC, Bagnara JT (1981) In: General endocrinology. Saunders, Philadelphia, pp 349–386

Ueno N, Baird A, Esch F, Ling N, Guillemin R, (1986 a) Isolation of an amino terminal extended form of basic fibroblast growth factor. Biochem Biophys Res Commun 138:580–588

Ueno N, Baird A, Esch F, Shimasaki S, Ling N, Guillemin R (1986 b) Purification and partial characterization of a mitogenic factor from bovine liver: structural homology with basic fibroblast growth factor. Regul Peptides 16:135–145

Ueno N, Baird A, Esch F, Ling N, Guillemin R (1987) Isolation and partial characterization of basic fibroblast growth factor from bovine testis. Mol Cell Endocrinol 49:189–194

Uhlrich S, Lagente O, Lenfant M, Courtois Y (1986) Effect of heparin on the stimulation of non-vascular cells by human acidic and basic FGF. Biochem Biophys Res Commun 137:1205–1213

Unsicker K, Reichert-Preibsch H, Schmidt R, Pettmann B, Labourdette G, Sensenbrenner M (1987) Astroglial and fibroblast growth factors have neurotrophic functions for cultured peripheral and central nervous system neurons. Proc Natl Acad Sci USA 84:5459–5463

Vlodavsky I, Johnson LK, Greenburg G, Gospodarowicz D (1979) Vascular endothelial cells maintained in the absence of fibroblast growth factor undergo structural and functional alterations that are incompatible with their in vivo differentiated properties. J Cell Biol 83:468–486

Vlodavsky I, Fuks Z, Bar-Ner M, Ariav Y, Schirrmacher V (1983) Lymphoma cell-mediated degradation of sulfated proteoglycans in the subendothelial extracellular matrix: relationship to tumor cell metastasis. Cancer Res 43:2704–2711

Vlodavsky I, Folkman J, Sullivan R, Fridman R, Ishai-Michaeli R, Sasse J, Klagsbrun M (1987a) Endothelial cell-derived basic fibroblast growth factor: synthesis and deposition into subendothelial extracellular matrix. Proc Natl Acad Sci USA 84:2292–2296

Vlodavsky I, Fridman R, Sullivan R, Sasse J, Klagsbrun M (1987b) Aortic endothelial cells synthesize basic fibroblast growth factor which remains cell associated and platelet-derived growth factor-like protein which is secreted. J Cell Physiol 131:402–408

Wagner JA, D'Amore P (1986) Neurite outgrowth induced by an endothelial cell mitogen isolated from retina. J Cell Biol 103:1363–1367

Walicke P, Cowan WM, Ueno N, Baird A, Guillemin R (1986) Fibroblast growth factor promotes survival of dissociated hippocampal neurons and enhances neurite extension. Proc Natl Acad Sci USA 83:3012–3016

Weibel M, Pettmann B, Labourdette G, Miehe M, Bock E, Sensenbrenner M (1985) Morphological and biochemical maturation of rat astroglial growth factor. Int J Dev Neurosci 3:617–630

Wilkinson DG, Peters G, Dickson C, McMahon AP (1988) Expression of the FGF-related proto-oncogene int-2 during gastrulation and neurulation in the mouse. EMBO J 7:691–695

Winkles JA, Friesel R, Burgess WH, Howk R, Mehlman T, Weinstein R, Maciag T (1987) Human vascular smooth muscle cells both express and respond to heparin-binding growth factor I (endothelial cell growth factor). Proc Natl Acad Sci USA 84:7124–7128

Wolburg H, Neuhaus J, Pettmann B, Labourdette G, Sensenbrenner M (1986) Decrease in the density of orthogonal arrays of particles in membranes of cultured rat astroglial cells by the brain fibroblast growth factor. Neurosci Lett 72:25–30

Yoshida T, Miyagawa K, Odagiri H, Sakamoto H, Little PF, Terada M, Sugimura T (1987) Genomic sequence of hst, a transforming gene encoding a protein homologous to fibroblast growth factors and the int-2-encoded protein. Proc Natl Acad Sci USA 84:7305–7309

Zeytin FM, Rusk SF, Baird A, Raymond V, Leff SE, Mandell AJ (1988) Induction of c-fos, calcitonin gene expression, and acidic FGF production in a multipeptide-secreting neuroendocrine cell line. Endocrinology 122:1114–1120

Zhan X, Bates B, Hu X, Goldfarb M (1988) The human FGF-5 oncogene encodes a novel protein related to fibroblast growth factors. Mol Cell Biol 8:3487–3497

CHAPTER 8

The Transforming Growth Factor-βs

A. B. ROBERTS and M. B. SPORN

A. Introduction

Following the initial purification and characterization of transforming growth factor-β (TGF-β) as a homodimeric, 25-kDa peptide (FROLIK et al. 1983; ASSOIAN et al. 1983; ROBERTS et al. 1983a), there has been an exponential increase in knowledge relating to this molecule. The cloning of TGF-β1 and the resultant elucidation of its precursor structure (DERYNCK et al. 1985) have led to the identification of at least four other forms of TGF-β and the definition of a larger gene family comprising many other structurally related, but functionally distinct, regulatory proteins. The original narrow definition of TGF-β, in terms of induction of a transformed phenotype in mesenchymal cells (MOSES et al. 1981; ROBERTS et al. 1981, 1983b), has now been supplanted by the knowledge that this peptide affects many functions in nearly all cells. Immunohistochemical and in situ hybridization studies have identified in vivo sites of TGF-β action; its broad spectrum of cellular targets as well as its multifunctional actions suggest that it has a pivotal control function in many physiological and pathological processes. Elucidation of physiological mechanisms of activation of the latent forms of the TGF-βs and of regulation of expression of the TGF-βs and their receptors are important problems that must still be solved for better understanding of the action of the TGF-βs. In this chapter we summarize current knowledge of the chemistry and complex biology of the growing family of TGF-βs.

B. Chemistry and Molecular Biology of TGF-β

I. Bioassays for TGF-β

TGF-β1 was first purified to homogeneity from human platelets (ASSOIAN et al. 1983), human placenta (FROLIK et al. 1983), and bovine kidney (ROBERTS et al. 1983a). The specific assay used to monitor these purifications, which provided the original name for the peptide, was the ability of TGF-β to induce normal rat kidney (NRK) fibroblasts to grow and form colonies of cells in soft agar in the presence of epidermal growth factor (EGF; ROBERTS et al. 1981). This assay system was originally described by DE LARCO and TODARO (1978) as an activity of "sarcoma growth factor" (SGF); SGF was later found to be a mixture of TGF-β and TGF-α, which belongs to the EGF family and acts through the EGF receptor (ANZANO et al. 1982, 1983). TGF-β stimulates anchorage-independent growth of other fibroblastic cell lines as well, among them AKR-2B cells (MOSES et al. 1981, 1985) and BALB/c 3T3 cells (MASSAGUÉ et al. 1985).

As might be expected from the wide spectrum of biological activities now attributed to TGF-β, purification of putative novel peptides based on many diverse assays unrelated to colony-forming activity has often resulted in reisolation of TGF-βs. Depending on the cell type and the purification scheme employed, either TGF-β1 or its closely related homolog, TGF-β2, have been isolated, as determined by amino acid sequencing. Thus TGF-β1 has been purified by utilizing not only the NRK assay system, but also assays measuring the differentiation of primitive mesenchymal cells into cells expressing a cartilaginous phenotype (SEYEDIN et al. 1985, 1986). TGF-β2 has also been purified based on its induction of a cartilaginous phenotype (SEYEDIN et al. 1985, 1987) as well as by assays measuring the inhibition of growth of Mv1Lu mink lung epithelial cells (CCL-64) (IKEDA et al. 1987) and inhibition of C3H/HeJ mouse thymocyte mitogenesis (WRANN et al. 1987). In addition, peptides related to TGF-β have been partially characterized by assays measuring inhibition of growth of monkey kidney cells (HOLLEY et al. 1980; TUCKER et al. 1984 b) and inhibition of myoblast differentiation (FLORINI et al. 1986). The broad range of biological activities of TGF-β makes it highly likely that yet other activities identified by presumably novel and specific assays will be found to result from TGF-β once their amino acid sequence is determined.

II. Multiple Forms of TGF-β

1. TGF-βs 1 and 2

The original isolation of TGF-β from human platelets resulted in the identification of a single form of the peptide, a 25 000 molecular weight homodimer, now called TGF-β1 (ASSOIAN et al. 1983). Since then, a second form of the peptide, TGF-β2, has been purified from tissues including porcine platelets (CHEIFETZ et al. 1987), bovine bone (SEYEDIN et al. 1985, 1987), human glioblastoma cells (WRANN et al. 1987), and monkey BSC-1 cells (HANKS et al. 1988). Each of these peptides has been cloned, revealing that TGF-β1 is encoded as a 390 amino-acid precursor (DERYNCK et al. 1985) and TFG-β2 as a 412 amino acid precursor (DE MARTIN et al. 1987; MADISEN et al. 1988), each having a signal peptide of 20–23 amino acids at the N terminus. The processed 112 amino acid chains of the two peptides are 72% identical, including conservation of all nine cysteine residues. These two peptides are interchangeable in most biological assays (SEYEDIN et al. 1987; CHEIFETZ et al. 1987; MULÉ et al. 1988), and the use of blocking antibodies specific to either TGF-β1 or 2 suggests that most of the TGF-β-like biological activity secreted by cells in culture is either TGF-β1 or 2, although the presence of other immunologically cross-reactive TGF-βs cannot be ruled out (DANIELPOUR et al.

Table 1. Quantitation of TGF-β1 and TGF-β2 in medium conditioned by cells

Cells	% Reduction in activity[a]		
	Anti-β1	Anti-β2	Anti-β1 + anti-β2
Murine B16 melanoma	41	76	91
Simian BSC-1 kidney	<10	83	93
Canine MDCK kidney	23	75	100
Human PC-3 prostatic carcinoma	<10	95	100
Human WI-38 lung fibroblasts	95	<10	100

[a] Acid-activated media conditioned by the indicated cells were assayed for inhibition of growth of mink lung CCL-64 cells in the presence or absence of blocking antibodies specific for either TGF-β1 or TGF-β2. (Adapted from DANIELPOUR et al. 1989).

1989). In porcine platelets and bovine bone, TGF-β1 is approximately 4 times as abundant as TGF-β2; in contrast, cell lines can be found which secrete a spectrum of TGF-βs ranging from greater than 90% TGF-β1 to greater than 90% TGF-β2 (DANIELPOUR et al. 1989; Table 1). A small amount of TGF-β1.2 heterodimer has been purified from porcine platelets (CHEIFETZ et al. 1987); this form may be secreted by certain cultured cells as well (DANIELPOUR et al. 1989).

While platelets represent the most concentrated natural source of TGF-β1 (20 mg/kg; VAN DEN EIJNDEN-VAN RAAIJ et al. 1988), biologically active recombinant TGF-β1 can now be expressed at high levels (6 mg/l) in Chinese hamster ovary (CHO) cells using gene amplification with dihydrofolate reductase (GENTRY et al. 1987). The expressed gene encodes the entire precursor form of the peptide. The precursor is both glycosylated and phosphorylated (BRUNNER et al. 1988), and the biological activity of the secreted recombinant TGF-β1 is latent, similar to that of the native peptide (see Sect. F). CHO cells process recombinant TGF-β1 appropriately, cleaving the signal peptide between Gly29 and Leu30, while the precursor is cleaved from the mature form at the basic site preceding Ala279 (GENTRY et al. 1988). Use of deletion constructs demonstrates that synthesis of biologically active TGF-β1 can proceed only from the first ATG codon (methionine; see Fig. 1) of the precursor, again implicting the precursor in proper assembly of the disulfide bonds of the mature peptide (WAKEFIELD et al. 1989). Thus far, no attempts to produce biologically active recombinant TGF-β1 in either yeast or bacteria have been successful.

2. Novel TGF-βs Predicted from cDNA Clones

In the past year, three new forms of TGF-β have been identified by screening of cDNA libraries. None of these putative peptides has yet been isolated from natural sources, although Northern blots demonstrate expression of the corresponding mRNAs. TGF-β3 has been identified from cDNA libraries

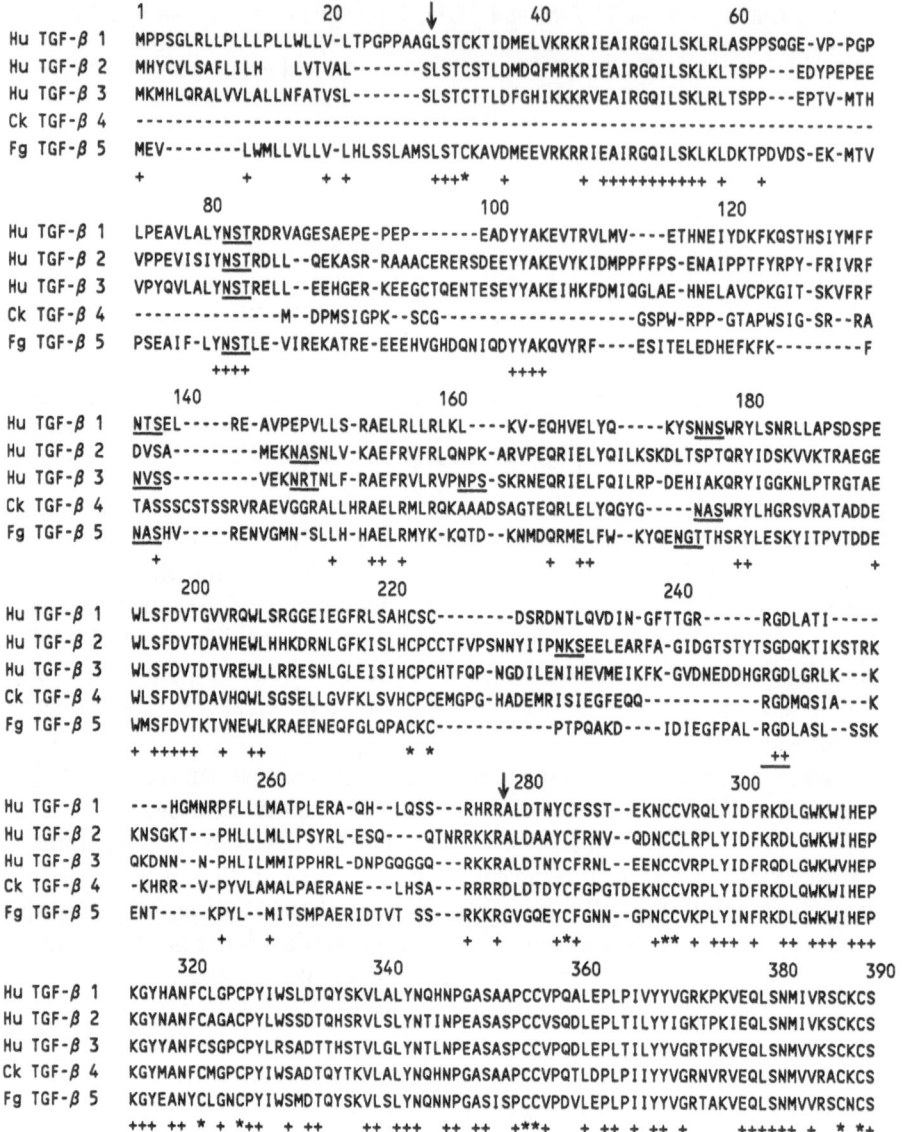

Fig. 1. Amino acid sequences of the TGF-β precursors. *Arrows* indicate the position of proteolytic processing resulting in cleavage of the signal peptide of TGF-β1 and of the mature TGF-βs. +, Amino acids which are conserved in all TGF-βs; * cysteines which are conserved in all TGF-βs; N-linked glycosylation sites are *underlined*, as is the integrin cellular recognition sequence, RGD

derived from human ovary, placenta, umbilical cord, A172 glioblastoma cells, and A673 rhabdomyosarcoma cells (TEN DIKJE et al. 1988a; DERYNCK et al. 1988), from porcine ovary (DERYNCK et al. 1988), and from cultured chick embryo chondrocytes (JAKOWLEW et al. 1988a). It represents the most abundant mRNA expressed in developing chick embryos (JAKOWLEW et al. 1988a) and is also expressed in human umbilical cord, in several human carcinoma cells including A673, A549, and A498 (TEN DIKJE et al. 1988a), as well as in a variety of mesenchymal cells of both human and rodent origin (DERYNCK et al. 1988). TGF-βs 4 and 5 have been cloned from a chicken chondrocyte library (JAKOWLEW et al. 1988b) and from a frog oocyte library (KONDAIAH et al. 1990), respectively. TGF-β4 mRNA is detectable in chick embryo chondrocytes, but is far less abundant than TGF-β3 mRNA in developing embryos or in chick embryo fibroblasts (JAKOWLEW et al. 1988b). TGF-β5 mRNA is expressed in frog embryos beyond the neurula stage and in *Xenopus* tadpole (XTC) cells (KONDAIAH et al. 1990).

The discovery of these new forms of TGF-β has raised many questions concerning a definition of TGF-β. However, several criteria strongly support the designation of these putative proteins as novel TGF-βs, even in the absence of data regarding their biological activity. Like TGF-βs 1 and 2, each of these novel peptides is encoded as a larger precursor. All TGF-β precursors share a region of high homology near the N terminus and show conservation of three cysteine residues in the portion of the precursor that will later be removed by processing (Fig. 1); there is no significant homology between this processed region and the corresponding regions of any of the functionally distinct members of the TGF-β supergene family (see Sect. C). Also noteworthy in the TGF-β precursors are sites for N-linked glycosylation as well as the cel-

Table 2. Comparison of the different types of TGF-β

TGF-β type	No. of amino acids		Processing site[b]	Chromosomal location[c]		mRNA (kb)[d]
	Precursor	Processed		Human	Mouse	
1	390	112	RHRR	19q13	7	2.5
2	412	112	RKKR	1q41	1	4.1, 5.1, 6.5
3	412	112	RKKR	14q24	12	3.0
4	304	114[a]	RRRR			1.7
5	382	112	RKKR			2.5

[a] Based on cleavage at the tetrabasic processing site.
[b] The site of cleavage of the precursor to the processed form of the peptide.
[c] The chromosomal locations of TGF-βs 1, 2, and 3 have been determined by FUJII et al. (1986), BARTON et al. (1988), and TEN DIJKE et al. (1988b), respectively. The locations of TGF-βs 4 and 5 are not known.
[d] The sizes given for TGF-βs 1, 2, and 3 are for human mRNAs, while those for TGF-βs 4 and 5 are for chicken and frog mRNAs, respectively.

lular recognition site for fibronectin/vitronectin, Arg-Gly-Asp (Ruoslahti and Pierschbacher 1986), which is found in all TGF-βs with the exception of TGF-β2 (see Fig. 1). Each of these TGF-βs has a four or five amino acid processing site (Table 2) and each is processed to a 112 amino acid chain, with the exception of TGF-β4 which has an insertion of two amino acids near the N terminus of the mature form (Fig. 1). Two other features which distinguish TGF-βs from the other members of its supergene family are the conservation of all nine cysteine residues in the processed peptide and conservation of the C-terminal sequence, Cys-Lys-Cys-Ser-COOH (Fig. 1), which is found in all TGF-βs with the exception of TGF-β5, in which an asparagine residue replaces the lysine. In contrast, with the exception of the β-chain of inhibin, in other members of the supergene family, only seven of the nine cysteine residues are conserved; and the C-terminal sequence, Cys-X-Cys-Y, never contains lysine in the X position or serine in the Y position, as in the TGF-βs. Finally, the amino acid sequences of the processed TGF-βs are between 60% and 80% identical to each other (Table 3), whereas the homology of the other family members to the TGF-βs is usually less than 40%.

In addition to similarities based on amino acid sequence, new evidence suggests that TGF-βs 3, 4, and 5 have biological activities analogous to TGF-βs 1 and 2. Thus, media conditioned by XTC cells, which express TGF-β5 mRNA at high levels (Rosa et al. 1988; Kondaiah, et al. 1989), or by chick embryo chondrocytes or fibroblasts, which express only TGF-β3 and 4 mRNAs (S. Jakowlew, personal communication), contain TGF-β-like activity which inhibits the growth of CCL-64 cells.

Certain unique features of TGF-β4 deserve mention. It is the only TGF-β to have an insertion of two amino acids in the processed peptide and to have a truncated precursor lacking characteristic signal peptide sequences (Jakowlew et al. 1988b). Thus the TGF-β4 precursor is only 304 amino acids; only one of the three or four potential N-linked glycosylation sites of the other TGF-βs is conserved in TGF-β4 (Fig. 1). These data suggest that TGF-β4 may have a unique biological activity, perhaps functioning *within* the

Table 3. Identities of the different TGF-βs

	TGF-β (% identity)[a]				
	1	2	3	4	5
1	100				
2	71	100			
3	72	76	100		
4	82	64	71	100	
5	76	66	69	72	100

[a] Identities are based on comparisons of human TGF-βs 1, 2, and 3, chicken TGF-β4, and frog TGF-β5.

cell. In addition, the truncated precursor as well as the altered N terminus of the processed peptide may not allow assembly of a latent complex characteristic of the other TGF-βs (see below).

3. Functions of Multiple Forms of TGF-β

Thus far, five different forms of TGF-β have been identified, and most evidence suggests that they share similar biological activities, at least in certain established cell lines. Gene duplications resulting in two or more alternate forms of a peptide growth factor with similar biological activities are rather common. Thus, platelet-derived growth factor (PDGF) exists in three different homodimeric or heterodimeric combinations of the PDGF A chain and B chain (STROOBANT and WATERFIELD 1984; BETSHOLTZ et al. 1986); EGF, vaccinia virus peptide, and TGF-α bind to the same receptor (TODARO et al. 1985), as do acidic and basic fibroblast growth factors (FGF; BAIRD et al. 1986) and the two forms of interleukin-1 (IL-1; CAMERON et al. 1986). An analogous situation is found in the case of the inhibins and activins which have partial homology to TGF-β (see below); inhibins are heterodimers of an α-chain and either of two β-chains (MASON et al. 1985; FORAGE et al. 1986), whereas activins, like PDGF, have been proposed to exist in three different hetero- or homodimeric combinations of the two β-chains of inhibin (LING et al. 1986; VALE et al. 1986).

The evolution of so many different forms of TGF-β with similar biological activities might be the result of differential regulation under control of unique promoters. TGF-βs 1, 2, and 3 are each located on a different chromosome (FUJII et al. 1986; BARTON et al. 1988; TEN DIJKE et al. 1988 b; see Table 2), and different types of TGF-β are expressed in various cells. Thus, whereas the ratio of TGF-β1 to TGF-β2 is about 4:1 in both bovine bone and porcine platelets (SEYEDIN et al. 1985; CHEIFETZ et al. 1987), human platelets contain exclusively TGF-β1 (ASSOIAN et al. 1983), and a human glioblastoma cell line (WRANN et al. 1987) and the human prostatic adenocarcinoma cell line PC-3 (IKEDA et al. 1987) secrete predominantly TGF-β2 (see also Table 1). At the mRNA level, both species- and cell-specific differences in expression have been observed: mouse, chick, and frog embryos express predominantly TGF-βs 1, 3, and 5, respectively (HEINE et al. 1987; JAKOWLEW et al. 1988 b; KONDAIAH et al. 1989), and whereas both TGF-β1 and TGF-β3 are expressed in many mesenchymal cell lines, TGF-β3 mRNA was not detected in several epithelial cell lines which expressed TGF-β1 (DERYNCK et al. 1988).

All of these data suggest that synthesis of the alternate forms of these peptides is regulated by distinct mechanisms. Control of the proteolytic processing and activation of the biological activity of the various TGF-βs may represent other selective control points. The tetrabasic cleavage sites of TGF-βs 1 and 4 are unique, whereas those of TGF-βs 2, 3, and 4 are identical (Table 2). Interestingly, several viruses including the human immunodeficiency viruses HIV-1 and HIV-3 as well as the insulin and insulin-like growth factor-I proreceptors have a similar tetrabasic cleavage site having the sequence Arg-X-Lys/Arg-Arg (YOSHIMASA et al. 1988). Whether all of these proteins might be processed by a similar cellular protease is an intriguing question. Processed TGF-β is still in a biologically inactive, or latent form. Evidence thus far suggests that all latent TGF-β types can be activated by acid treatment in vitro; however, physiological mechanisms of activation of the different latent forms may be specific (see Sect. F). Thus experiments comparing the biological activities of the various types of *processed, activated* TGF-βs may be circumventing physiologically significant activation events. Elucidation of these differential controls represents an exciting area for future research.

4. Conservation of TGF-βs in Different Species

Given the many different forms of TGF-β and the diversity of amino acid substitutions in certain regions of the processed peptide (Fig. 1), it might be expected that there would not be a high degree of conservation of these sequences in different species. To the contrary, the genes for murine (DERYNCK et al. 1986), bovine (VAN OBBERGHEN-SCHILLING et al. 1987), porcine (DERYNCK and RHEE 1987; KONDAIAH et al. 1988), simian (SHARPLES et al. 1987), and chicken TGF-β1 (JAKOWLEW et al. 1988c) have all been cloned and sequenced and found to have an extraordinarily high degree of homology to the human gene; all of the mature, processed peptides are identical, with the exception of the murine peptide which differs only in the substitution of a serine for an alanine at position 75. Processed human (DE MARTIN et al. 1987; MADISEN et al. 1988) and simian (HANKS et al. 1988) TGF-β2s are also identical as are the partial sequences of porcine and bovine TGF-β2s (CHEIFETZ et al. 1987; SEYEDIN et al. 1987). Human (TEN DIJKE et al. 1988a; DERYNCK et al. 1988) and chicken TGF-β3s (JAKOWLEW et al. 1988a) differ only in the substitution of a tyrosine for a pheynlalanine, and human and porcine TGF-β3s differ in only two amino acids (DERYNCK et al. 1988). In addition, comparison of the identities of the different TGF-βs in the regions of the precursor peptide that are later removed by processing demonstrates that within any particular *type* of TGF-β, conservation of this region is greater than 85%, which suggests that the precursor itself might have an important biological function, as will be discussed below with respect to the structure of the latent form of TGF-β.

III. The TGF-β1 Gene

The human TGF-β1 precursor is encoded by seven exons (DERYNCK et al. 1987b; see Fig. 2), and the splice junctions are preserved in the bovine (VAN OBBERGHEN-SCHILLING et al. 1987) and porcine genes (KONDAIAH et al. 1988). The positions of the intron/exon junctions of TGF-β1 are conserved in TGF-βs 2 and 3, with the exception of the first which differs by three nucleotides, suggesting that the various TGF-βs originated from duplication of a common ancestral gene (DERYNCK et al. 1988). Although TGF-β mRNAs

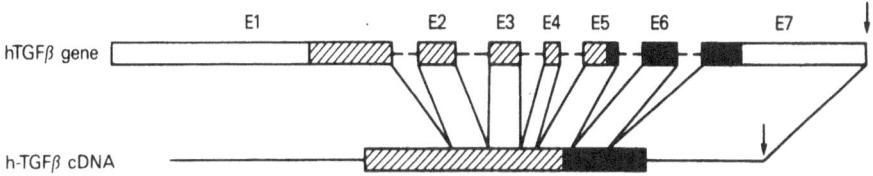

Fig. 2. Intron-exon structure of the human TGF-β1 gene. The *boxed* region of the cDNA represents the coding sequence corresponding to the 390 amino acid human TGF-β1 precursor. It is divided into seven exons as follows: exon 1, amino acids 1-119; exon 2, amino acids 119–172; exon 3, amino acids 173–212; exon 4, amino acids 212–238; exon 5, amino acids 238–287; exon 6, amino acids 287–338; and exon 7, amino acids 339–390. The size of the introns is unknown, but the gene is estimated to be greater than 100 kb. The *arrow* represents the polyadenylation signal. (Compiled from DERYNCK et al. 1987b)

typically contain approximately 1200 nucleotides of coding sequence, the mRNA species range from 1.7 to 6.5 kilobases (kb; Table 2) with extensions at both the 3′ and 5′ ends of the coding sequence. These extensions are quite distinctive for the different TGF-βs; the noncoding regions of TGF-β1 are very G-C rich, whereas those of TGF-β2 are relatively A-T rich, with those of the other TGF-βs falling between these two extremes.

1. Alternate Splicing of the TGF-β1 Gene

In porcine tissues there is evidence for alternate splicing of TGF-β1 with the omission of exons 4 and 5; translation of the resulting mRNA would produce a novel peptide having the TGF-β1 precursor sequences at its N terminus, but having a different sequence in place of the mature TGF-β1 at its C terminus (KONDAIAH et al. 1988). Whether such alternate splicing is important in control of TGF-β transcription or translation is not yet known.

2. Characterization of the Promoter for the Human TGF-β1 Gene

There are two major transcriptional start sites for human TGF-β1 mRNAs, 271 nucleotides apart (KIM et al. 1989a). Two promoter regions have been characterized, one extending 1400 base pairs (bp) upstream of the first transcriptional start site and the second one located between the two start sites (KIM et al. 1989a). Within the upstream promoter, two distinct negative regulatory regions, an enhancer-like region, and a positive regulatory region have been identified (Fig. 3). The negative regulatory regions correspond to the presence of FSE2 negative elements (DISTEL et al. 1987) while the positive regions contain several binding sites for known transcription factors including nuclear factor 1 (NF-1), SP1, and AP-1, which binds to the region of several promoters responsive to phorbol esters (KIM et al. 1989a). Characterization of the promoter regions for the other TGF-β genes as well as the promoter regions of the same TGF-β in different species will be important for understanding the differential expression patterns of the various TGF-βs and of a particular TGF-β in different species.

```
-1362 GGATCCTTAGCAGGGGAGTAACATGGATTTGGAAAGATCACTTTGGCTGCTGTGTGGGGATAGATAAGACGGTGGGAGCCTAGAAAGGAGGCTGGGTTGG

-1262 AAACTCTGGGACAGAAACCCAGAGAGGAAAAGACTGGGCCTGGGGTCTCCAGTGAGTATCAGGGAGTGGGGAATCAGCAGGAGTCTGGTCCCCACCCATC

-1162 CCTCCTTTCCCCTCTCTCTCCTTTCCTGCAGGCTGGCCCCGGCTCCATTTCCAGGTGTGGTCCCAGGACAGCTTTGGCCGCTGCCAGCTTGCAGGCTATG

-1062 GATTTTGCCATGTGCCCAGTAGCCCGGGCACCCACCAGCTGGCCTGCCCCACGTGGCGGCCCCTGGGCAGTTGGCGAGAACAGTTGGCACGGGCTTTCGT

-962 GGGTGGTGGGCCGCAGCTGCTGCATGGGGACACCATCTACAGTGGGGCCGACCGCTATCGCCTGCACACAGCTGCTGGTGGCACCGTGCACCTGGAGATC

-862 GGCCTGCTGCTCCGCAACTTCGACCGCTACGGCGTGGAGTGCTGAGGGACTCTGCCTCCAACGTCACCACCATCCACACCCCGGACACCCAGTGATGGGG

-762 GAGGATGGCACAGTGGTCAAGAGCACAGACTCTAGAGACTGTCAGAGCTGACCCCAGCTAAGGCATGGCACCGCTTCTGTCCTTTCTAGGACCTCGGGGT

-662 CCCTCTGGGCCCAGTTTCCCTATCTGTAAATTGGGGACAGTAAATGTATGGGGTCGCAGGGTGTTGAGTGACAGGAGGCTGCTTAGCCACATGGGAGGTG

-562 CTCAGTAAAGGAGAGCAATTCTTACAGGTGTCTGCCTCCTGACCCTTCCATCCCTCAGGTGTCCTGTTGCCCCCTCCTCCCACTGACACCCTCCGGAGGC

-462 CCCCATGTTGACAGACCCTCCTTCTCCTACCTTGTTTCCCAGCCTGACTCTCCTTCCGTTCTGGGTCCCCCTCCTCTGGTCGGCTCCCCTGTGTCTCATC

-362 CCCCGGATTAAGCCTTCTCCGCCTGGTCCTCTTTCTCTGGTGACCCACACCGCCCGCAAAGCCACAGCGCATCTGGATCACCCGCTTTGGTGGCGCTTGG

-262 CCGCCAGGAGGCAGCACCCTGTTTGGGGGGCGGAGCCGGGGAGCCGCCCCCTTTCCCCCAGGGCTGAAGGGACCCCCCTCGGAGCCCGCCCACGCGAGA

-162 TGAGGACGGTGGCCCAGCCCCCCCATGCCCTCCCCCTGGGGCCGCCCCGCGCTCCCGCCCCGGTGCGCTTCCTGGGTGGGGCCGGGGGCGGCTTCAAAACC

-62 CCCTGCCGACCCAGCCGGTCCCCGCCGCCGCCGCCCCTTCGCGCCCTGGGCCATCTCCCTCCCACCTCCCTCCGCGGAGCAGCCAGACAGCGAGGGCCCCG

39 GCCGGGGGCAGGGGGGACGCCCCGTCCGGGGCACCCCCCCGGCTCTGAGCCGCCCGCGGGGCCGGCCTCGGCCCGGAGCGGAGGAAGGAGTCGCCGAGGA

139 GCAGCCTGAGGCCCCAGAGTCTGAGACGAGCCGCCGCCGCCCCGCCACTGCGGGGAGGAGGGGGAGGAGGAGCGGGAGGAGGGGACGAGCTGGTCGGGAG

239 AAGAGGAAAAAAACTTTTGAGACTTTTCCGTTGCCGCTGGGAGCCGGAGGCGCGGGGACCTCTTGGCGCGACGCTGCCCCGCGAGGAGGCAGGACTTGGG

339 GACCCCAGACCGCCTCCCTTTGCCGCCGGGGACGCTTGCTCCCTCCCTGCCCCCTACACGGCGTCCCTCAGGCGCCCCCATTCCGGACCAGCCCTCGGGA

439 GTCGCCGACCCGGCCTCCCGCAAAGACTTTTCCCCAGACCTCGGGCGCACCCCCTGCACGCCGCCCTTCATCCCCGGCCTGTCTCCTGAGCCCCCGCGCAT

539 CCTAGACCCTTTCTCCTCCAGGAGACGGATCTCTCTCCGACCTGCCACAGATCCCCTATTCAAGACCACCCACCTTCTGGTACCAGATCGCGCCCATCTA

639 GGTTATTTCCGTGGGATACTGAGACACCCCCGGTCCAAGCCTCCCCTCCACCACTGCGCCCTTCTCCCTGAGGAGCCTCAGCTTTCCCTCGAGGCCCTCC

739 TACCTTTTGCCGGGAGACCCCCAGCCCCTGCAGGGGCGGGGCCTCCCCACCACACCAGCCCTGTTCGCGCTCTCGGCAGTGCCGGGGGGGCGCCGCCTCCC
```

Met
839 CCATG

Fig. 3. Nucleotide sequence of the upstream regulatory region of the human TGF-β1 gene. The transcriptional start sites are indicated by *arrows*; the *thick arrows* (at +1, +271) mark the two major transcription initiation points and thin arrows (+470, +525) indicate minor sites. The 5′-most residue of the human TGF-β1 precursor cDNA is designated +1 (Derynck et al. 1985). The repeated CCGCCC sequence and its reverse complement are underlined by a *thick solid line*. *Boxes*, consensus sequences of SP1 binding sites; *filled circles*, the phorbol ester inducible element (AP-1 binding site); *filled triangles* the NF-1 binding site; *inverted triangles*, the FSE2 element. (From Kim et al. 1989a)

C. The TGF-β Supergene Family

There are now many peptides which belong to the TGF-β supergene family by virtue of amino acid homologies, particularly with respect to the conservation of seven of the nine cysteine residues of TGF-β among all known family members. These include the mammalian inhibins (MASON et al. 1985; FORAGE et al. 1986) and activins (LING et al. 1986; VALE et al. 1986), and Mullerian inhibitory substances (MIS; CATE et al. 1986), as well as the predicted products of both a pattern gene in *Drosophila* (the decapentaplegic gene complex, DPP-C; PADGETT et al. 1987) and an amphibian gene expressed in frog oocytes (Vg1; WEEKS and MELTON 1987). Most recently, three new proteins called bone morphogenetic proteins (BMPs) have been added to the family. One subset of these proteins, BMP-2A and -2B, is approximately 75% homologous to DPP-C and may represent the mammalian equivalent of that protein (WOZNEY et al. 1988; WANG et al. 1988). In every case where the information is available, all peptides belonging to this family are encoded as larger precursors; the family resemblance is limited to the C terminus of the precursor corresponding to the processed mature TGF-β (PADGETT et al. 1986). With the exception of MIS, the C-terminal region is cleaved from the precursor at a pair of arginine residues; although the position of this cleavage site varies among the family members, the C terminus of all of the peptides is in the identical position, ending in the sequence Cys-X-Cys-X, but differing in every case from the TGF-β consensus C-terminus of Cys-Lys-Cys-Ser.

A unifying feature of the biology of these peptides is their ability to regulate developmental processes: MIS induces regression of the female rudiments of the developing male reproductive system; the inhibins and activins regulate the activity of the gonadotropin, follicle stimulating hormone (FSH); the BMPs are thought to play a role in the formation of cartilage and bone in vivo; DPP-C directs dorsal-ventral patterning in the developing fly embryo, and Vg1 is postulated to be involved in the process of induction of mesoderm from ectoderm during gastrulation in the amphibian embryo. In amphibians, TGF-β itself (KIMELMAN and KIRSCHNER 1987) has been shown to augment the ability of fibroblast growth factor (FGF) to induce mesoderm (SLACK et al. 1987), and, as will be discussed in greater detail in Sect. G.III, it also plays a pivotal role in morphogenesis and organogenesis in mammalian embryos. It appears that each of these peptides has its own unique receptor. Thus, although TGF-βs 1 and 2 compete for receptor binding (CHEIFETZ et al. 1987; SEGARINI et al. 1987), neither inhibin nor activin can compete for binding of TGF-β1 to a variety of cell types, including pituitary cells (CHEIFETZ et al. 1988a), in which both TGF-β and the activins elicit secretion of follicle-stimulating hormone (FSH), while inhibin antagonizes that activity (YING et al. 1986a, b).

D. TGF-β Receptors

The action of TGF-β is mediated through binding to specific cell membrane receptors. The binding of TGF-β to nearly 150 different cell types and cell

Table 4. Characteristics of TGF-β receptors[a]

Type	Size (kDa)		Modification	Relative binding TGF-β1 : TGF-β2
	Native	Core		
I	65	60	N-linked carbohydrate	16 : 1
II	85–110	70– 95	N-linked carbohydrate	12 : 1
III	250–350	100–140	Heparan sulfate proteoglycan	1 : 1

[a] Data have been compiled from SEGARINI and SEYEDIN (1988) and CHEIFETZ et al. (1988a, b).

lines has been investigated to date and, with only a few exceptions, almost all cells, regardless of their origin, bind TGF-β with affinities in the picomolar concentration range (FROLIK et al. 1984; TUCKER et al. 1984a; MASSAGUÉ and LIKE 1985; WAKEFIELD et al. 1987). Various methods based on use of chloramine-T (FROLIK et al. 1984), the Bolton-Hunter reagent (TUCKER et al. 1984a), or lactoperoxidase: glucose oxidase (MASSAGUÉ and LIKE 1985) have been used to radioiodinate TGF-β; none of these methods have been found to destroy the biological activity of TGF-β1, though lactoperoxidase is the method of choice for TGF-β2 (DANIELPOUR et al. 1989).

Crosslinking of TGF-β to membrane receptors with disuccinimidyl suberate has revealed three distinct classes of integral cell membrane components which bind TGF-β specifically and with high affinity (see Table 4). Class I components are 65-kDa in all species, whereas class II components range from 85-kDa in rodent cells to 95-kDa in monkey and human cells to 110-kDa in chicken cells; the binding of the various forms of TGF-β to both class I and class II receptors is in the order TGF-β1 > TGF-β1.2 > TGF-β2 (CHEIFETZ et al. 1988a, b). In contrast, all three forms of TGF-β bind equivalently to class III receptors which represent the most abundant cross-linked species; this form is generally considered to be dimeric (MASSAGUÉ 1985) and to be composed of preteoglycan subunits of 250–350-kDa (SEGARINI and SEYEDIN 1988; CHEIFETZ et al. 1988a). Another subset of this high molecular weight component has also been described which preferentially binds TGF-β2 (SEGARINI et al. 1987). Most frequently, all three classes of these binding proteins coexist on cells.

Class I and II components, like most growth factor receptors, are glycoproteins: most of the carbohydrate is N-linked and contributes approximately 5-kDa and 15–20-kDa to the mass of components I and II, respectively (CHEIFETZ et al. 1988a). In contrast to all other known polypeptide receptors, the class III protein is a proteoglycan consisting predominantly of heparan sulfate glycosaminoglycan chains with a smaller amount of chondroitin or dermatan sulfate attached to a core protein of approximately 100–140-kDa; the binding site for TGF-β resides in this core protein (SEGARINI and SEYEDIN 1988; CHEIFETZ et al. 1988a). This interesting finding has led to speculation that class III receptors may, in some novel way, both bind to cell matrix components and mediate organization of the cytoskeleton, two functions which are in accordance with the biological activity of TGF-β on many mesenchymal cells.

There is considerable controversey concerning the roles of the various classes of proteins which crosslink to TGF-β. MASSAGUÉ and coworkers have proposed that class III receptors mediate all functions of TGF-β in which TGF-βs 1 and 2 have been shown to be equipotent; this includes regulation of extracellular matrix as well as most effects on growth and differentiation (CHEIFETZ et al. 1987, 1988a, b). Moreover, they suggest that biological activities specific to TGF-β1 are mediated through the class I and II receptors; these would include the reported selective inhibitory activity of TGF-β1 on growth of either B6SUt-A multipotential hematopoietic progenitor cells (CHEIFETZ et al. 1988b) or endothelial cells (JENNINGS et al. 1988). In contrast, SEGARINI and coworkers (1989) have shown that class III binding is not exhibited by primary epithelial, endothelial, and lymphoid cells and may not be necessary for many biological activities of TGF-β: thus, cells such as L-6 myoblasts (SEGARINI et al. 1989) and primary lymphocytes (KEHRL et al. 1989), which respond equally well to TGF-βs 1 and 2, have only class I and II receptors. Confirmatory evidence that type I receptors are the mediators of TGF-β action comes from the work of BOYD and MASSAGUÉ (1989), demonstrating that selection of CCL 64 cell mutants resistant to the action of TGF-β results in the isolation of cell lines which have selectively lost the type I TGF-β binding.

Although their occurrence is rather rare, several neoplastic cells appear to lack these putative TGF-β receptors; these include the PC12 rat pheochromocytoma, human retinoblastoma cells, and several leukemic cell lines (KIMCHI et al. 1988; KELLER et al. 1989). In each of these cases, the lack of receptor proteins correlates with resistance of the cells to the inhibitory effects of TGF-β, and it has been proposed that loss of TGF-β receptors might be a mechanism whereby preneoplastic cells could progress to tumor cells by escaping from negative growth control (SPORN and ROBERTS 1985).

Yet another anomaly of TGF-β binding in tumor cells has been reported. Although the binding of TGF-β to cells has been shown to be specific, a novel 70- to 74-kDa complex has been reported on GH$_3$ rat pituitary tumor cells which binds not only TGF-β1, but also TGF-β2, activin AB, and inhibin with lower affinity (CHEIFETZ et al. 1988c). The biological function of this complex is not known.

E. Antibodies to TGF-β

I. Antibodies that Block Biological Activity

The use of antibodies in the study of TGF-β action has been severely limited by the difficulty of raising high-titer blocking antibodies. The basis for the difficulty is not understood. However, procedures have now been developed for reliable immunization of both rabbits and turkeys with uncoupled TGF-β, resulting in high-titer blocking antisera specific for TGF-β1 or TGF-β2 (DANIELPOUR et al. 1989) as well as antisera that crossreact with both peptides

(KESKI-OJA et al. 1987). The type-specific antibodies have been utilized to quantitate the relative amounts of TGF-βs 1 and 2 in media and other biological fluids (ROSA et al. 1988; DANIELPOUR et al. 1989; CONNOR et al. 1989).

II. Antibodies Used in Immunohistochemical Localization of TGF-β

To circumvent some of the problems associated with the generation of antibodies to the intact native peptide and to acquire a set of reagents for immunohistochemical studies, antibodies have been raised to synthetic peptides representing various regions of the TGF-β molecule and its precursor. Some of these antibodies have been raised against internal peptides synthesized in such a manner that a loop structure could be generated (ELLINGSWORTH et al. 1986; FLANDERS et al. 1988); this was accomplished either by coupling the protein to carrier via both ends (in this case tyrosine was used as the coupling residue) or by joining terminal cysteine residues of the peptides in disulfide linkages.

Antisera raised against peptides representing different regions of TGF-β1 have been used to detect the native peptide in a variety of assays including immunoblots, radioimmunoassays, and receptor binding assays. These studies suggest that the C-terminal region of the molecule, which is the most highly conserved region among all the members of the TGF-β family of proteins (PADGETT et al. 1987), might be essential to receptor binding (FLANDERS et al. 1988).

The most important application of the antibodies raised against the TGF-β peptides has been in immunohistochemical studies. The peptide antisera react better with fixed, denatured TGF-β in paraffin sections of tissues than do antibodies raised against the native protein. The two antisera used most extensively have been raised to two different synthetic preparations of a peptide representing the N-terminal 30 amino acids of TGF-β1 (ELLINGSWORTH et al. 1986; HEINE et al. 1987; FLANDERS et al. 1989; THOMPSON et al. 1989); surprisingly, these two antisera have distinctly different staining patterns, suggesting that they recognize different epitopes and that the secreted form of TGF-β may be conformationally different from the intracellular form (FLANDERS et al. 1989; THOMPSON et al. 1989).

F. Latent Forms of TGF-β

TGF-β is released from degranulating platelets and secreted from nearly all cells in a biologically inactive form which is unable to bind to cellular receptors and is not recognized by antibodies to TGF-β (PIRCHER et al. 1986; MIYAZONO et al. 1988; WAKEFIELD et al. 1988); the peptide can be activated by acidification, alkalinization, or action of chaotropic agents in vitro (LAWRENCE et al. 1985; LYONS et al. 1988). In contrast, TGF-β, as it is typically purified from platelets or bone, has been exposed to conditions that strip it of other proteins and permanently activate it. It is now realized that, since most cells have receptors for TGF-β (WAKEFIELD et al. 1987), physiological control of the activation of this "latent," biologically inert form of TGF-β is important in both paracrine actions of TGF-β (such as might follow release of the peptide from platelets in a wound) and potential autocrine actions of the peptide. An example of this is the A549 human lung carcinoma cell line which has been found to secrete relatively high amounts of latent TGF-β yet which grows well in the presence of the latent peptide. In contrast, active TGF-β is a potent growth inhibitor of these cells, suggesting that their uncontrolled growth may result, in part, from loss of the ability to activate the TGF-β they secrete (WAKEFIELD et al. 1987). Thus, the demonstration that a particular cell has receptors for TGF-β and secretes latent TGF-β is no longer sufficient to suggest autocrine action; in addition, a mechanism for activation of the latent form must be identified (WAKEFIELD et al. 1987). It is equally clear that elucidation of the physiological mechanisms of activation of the various forms of TGF-β is central to understanding potential selective actions of the different types of TGF-β in vivo.

The structure of the latent TGF-β1 complex of platelets has recently been elucidated (MIYAZONO et al. 1988; WAKEFIELD et al. 1988) and found to consist of three components: (1) a 125- to 160-kDa TGF-β modulator protein belonging to a family of proteins with EGF-like repeats; (2) a latency protein consisting of the remainder of the precursor (the precursor minus both the N-terminal 29 amino acid signal peptide sequence and the C-terminal 112 amino acids of mature TGF-β1), and (3) mature, processed dimeric TGF-β1 (Fig. 4). Analysis of the complex under both reducing and nonreducing conditions suggests that the latency protein is glycosylated and, like processed TGF-β1, also forms a disulfide-bonded dimer of approximately 75 kDa (gp75); in addition, it is covalently linked via a disulfide bond to the modulator protein (MIYAZONO et al. 1988). Whether the latency protein might have an independent biological role, once released from the complex, is important to consider. As discussed above (Sect. B. II. 4), this protein is very highly conserved between species: for example, the N-terminal portions of the precursors of TGF-βs 1 and 3 are 91% and 85% conserved between human and chicken (JAKOWLEW et al. 1988 a, c), respectively. Moreover, the presence in the latency protein of the Arg-Gly-Asp integrin recognition sequence (RUOSLAHTI and PIERSCHBACHER 1986) in all TGF-βs with the exception of TGF-β2 suggests that either the latent complex or the free latency protein might bind to cells in a specific manner (Fig. 1); latent TGF-β1 also binds to heparin Sepharose, suggesting that there is a specific heparin binding site on the processed precursor (WAKEFIELD et al. 1989).

The role of the modulator protein in the latent platelet complex is unclear. Recombinant TGF-β expressed in CHO cells or in a human renal carcinoma cell line is also secreted in a latent form; however, this latent form is distinguished from the natural form of the complex by the absence of the modulator protein, indicating that it is not necessary to confer latency on TGF-β (GENTRY et al. 1987; WAKEFIELD et al. 1989; Fig. 4).

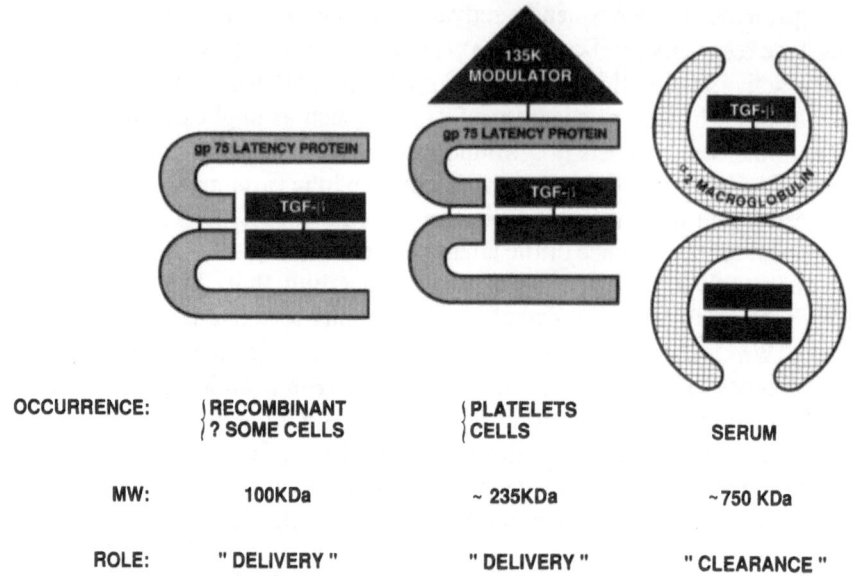

Fig. 4. Latent forms of TGF-β. Latent TGF-β1 secreted by recombinant cells is a non-covalent complex of the 75-kDa dimeric glycoprotein latency protein derived from the TGF-β precursor and processed dimeric 25-kDa TGF-β. As secreted from cells and released from platelets, the complex contains in addition a 135-kDa modulator protein. The complex of α_2-macroglobulin with mature dimeric TGF-β is also latent, and probably functions as a clearance mechanism. See text Sect. F for details. (From Wakefield et al. 1989)

Physiological mechanisms for activation of the latent form are under investigation. Miyazono and Heldin (1989) have demonstrated that the N-linked glycosylation sites of the latency protein of TGF-β1 (see Fig. 1) play a critical role in latency; thus treatment of latent TGF-β1 with endoglycosidase F or sialidase and addition of sialic acid or mannose-6-phosphate result in dose-dependent activation. Lyons et al. (1988) have shown that proteases such as plasmin and cathepsin D can also partially activate latent TGF-β. Alternatively, local acid microenvironments might also be sufficient for activation. Thus, activated macrophages, as might be found in a healing wound, secrete sialidase (Pilatte et al. 1987) and proteases and can lower the pH to 4 (Silver et al. 1988), all of which could contribute to activation of latent TGF-β. At high concentrations, activation of the recombinant latent complex by acidification in vitro is reversible, demonstrating that it does not result in structural modifications of the component subunits, but only a weakening of the noncovalent interactions which stabilize the complex (Wakefield et al. 1989).

In serum, a latent form of TGF-β1 consisting of a complex with α_2-macroglobulin has been described by O'Connor-McCourt and Wakefield (1987). This is distinguished from the latent forms of TGF-β1 secreted from platelets and released from cells in that it contains only processed TGF-β1

(Fig. 4); α_2-macroglobulin will not form a complex with latent platelet TGF-β1. For these reasons, it has been proposed that this form of TGF-β in serum might function as a scavenging mechanism for excess TGF-β, serving to restrict the local action of the activated peptide, as at the site of a wound (HUANG et al. 1988; O'CONNOR-MCCOURT and WAKEFIELD 1987).

G. Biological Activity of the TGF-βs

I. Multifunctional Nature of TGF-β Action

TGF-β is the prototypical multifunctional growth factor (SPORN and ROBERTS 1988 a, b). The nature of its action on a particular target cell is critically dependent on many parameters including the cell type and its state of differentiation, the growth conditions, and on other growth factors present (SPORN et al. 1987). For example: (1) TGF-β1 stimulates the growth of NRK cells in the presence of EGF in soft agar, but both inhibits the growth and antagonizes the mitogenic action of EGF on the same cells in monolayer culture (ROBERTS et al. 1985). (2) In Fischer rat 3T3 cells transfected with a *myc* gene, TGF-β1 can either stimulate or inhibit the growth of the cells in soft agar depending on whether PDGF or EGF, respectively, are also present in the assay (ROBERTS et al. 1985). (3) TGF-β1 stimulates the growth of fibroblasts from very early human fetuses, but inhibits the growth of fibroblasts derived from fetuses of somewhat later gestational age (HILL et al. 1986). (4) TGF-β1 stimulates differentiation of certain cell types such as bronchial epithelial cells (MASUI et al. 1986) and prechondrocytes (SEYEDIN et al. 1985), while it inhibits differentiation of others such as adipocytes (IGNOTZ and MASSAGUÉ 1985) and myoblasts (MASSAGUÉ et al. 1986; OLSON et al. 1986; FLORINI et al. 1986). (5) TGF-βs 1 and 2 stimulate primitive mesenchymal cells to differentiate and express a cartilaginous phenotype, but treatment of chondrocytes with TGF-β leads to suppression of cartilage markers such as synthesis of type II collagen (ROSEN et al. 1988). These are but a few examples of the multifunctionality of TGF-β action. Clearly, the diversity of its effects together with the almost universal ability of cells to respond to TGF-β place this peptide in a unique position with regard to regulation of normal and pathologic physiology. Some of the major effects of TGF-β on specialized cells and tissues will be discussed below.

II. Activity of TGF-β on Specialized Cells and Tissues

1. Mesenchyme: Control of Extracellular Matrix

Immunohistochemical staining of both embryonic (ELLINGSWORTH et al. 1986; HEINE et al. 1987) and adult tissues (FLANDERS et al. 1989; THOMPSON et al. 1989) shows localization of TGF-β in mesenchyme. Recent in vitro investigations suggest that the many different activities of TGF-β on mesenchymal cells are possibly only different aspects of a complex scheme whereby

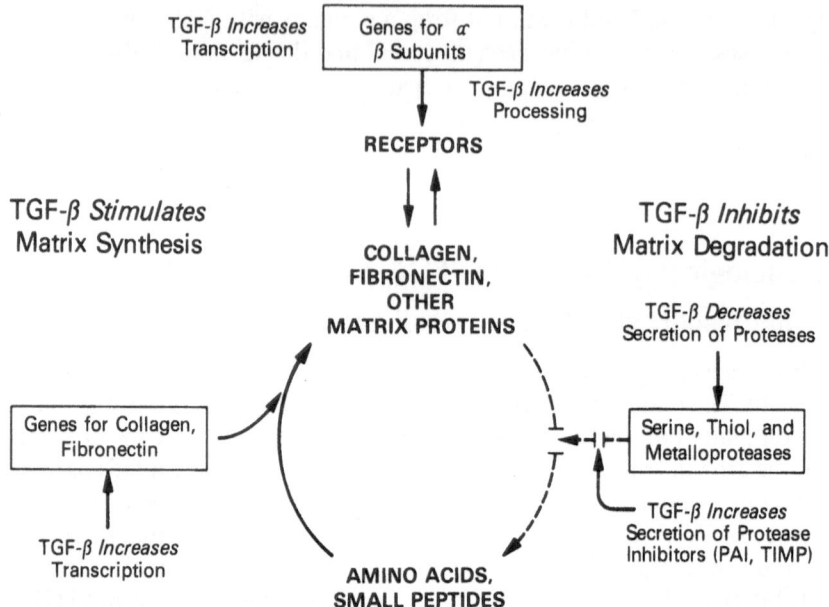

Fig. 5. Mechanisms of TGF-β enhancement of both the accumulation of extracellular matrix proteins and the integrins, including enhancement of synthesis of matrix proteins, inhibition of the degradation of matrix proteins, and enhanced synthesis of the receptors for matrix proteins. Details are provided in the text in Sect. G. II. 1. *PAI*, plasminogen activator inhibitor; *TIMP*, tissue inhibitor of metalloendoproteases

TGF-β serves to increase accumulation of and response of cells to extracellular matrix. Thus, as represented diagrammatically in Fig. 5, TGF-β has been shown to (1) activate gene transcription and increase synthesis and secretion of matrix proteins, (2) decrease synthesis of proteolytic enzymes that degrade matrix proteins and increase synthesis of protease inhibitors that block the activity of these enzymes, and (3) increase both the transcription, translation, and processing of cellular receptors for matrix proteins. The multiple levels at which TGF-β acts suggest that control of matrix interactions of cells likely represents one of the principal mechanisms by which the peptide controls growth, differentiation, and function of mesenchymal cells.

Effects of TGF-β on increased synthesis of matrix proteins have been demonstrated both in vivo and in vitro. Many fibroblastic cell lines of human, mouse, rat, and chicken origin show increased synthesis of type I collagen and fibronectin following treatment with TGF-β (IGNOTZ and MASSAGUÉ 1986; ROBERTS et al. 1986; WRANA et al. 1986). The effect is specific for TGF-β: EGF and PDGF are ineffective. These observations have now been extended to include many other more specialized matrix proteins, among them types III, IV, and V collagen (VARGA et al. 1987; ROSSI et al. 1988; MADRI et al. 1988), thrombospondin (PENTTINEN et al. 1988), osteopontin (NODA et al. 1988a), tenascin (PEARSON et al. 1988), elastin (LIU and DAVIDSON 1988), osteonectin/SPARC (NODA and RODAN 1987; WRANA et al. 1988), and chondroitin/dermatan sulfate proteoglycans (CHEN et al. 1987; SEYEDIN et al. 1985; FALANGA et al. 1987; MORALES and ROBERTS 1988; BASSOLS and MASSAGUÉ 1988; HIRAKI et al. 1988). In every instance where it has been examined, TGF-βs 1 and 2 are equipotent in induction of matrix protein synthesis.

More than one mechanism may contribute to the enhancement of the synthesis of matrix proteins by TGF-β. Several studies have shown that TGF-β increases mRNA levels for types I, III, and V collagen and fibronectin (IGNOTZ et al. 1987; VARGA et al. 1987; DEAN et al. 1988; ROSSI et al. 1988; PENTTINEN et al. 1988). Experiments have shown that TGF-β can directly stimulate the activity of the mouse α2(I) collagen promoter (ROSSI et al. 1987) and the fibronectin promoter (DEAN et al. 1989); in both cases a NF-1 binding site has been implicated in mediating the effects of TGF-β, although it is clear that other regions of the promoters are involved as well. Stimulation of the collagen promoter is specific for TGF-β (types 1 and 2 are interchangeable), and the ED_{50} of 3–5 pM TGF-β for activation of the transfected construct in cells is consistent with the concentration of TGF-β effective in induction of collagen synthesis. In addition to this increased synthesis of matrix protein mRNAs, RAGHOW et al. (1987) and PENTTINEN et al. (1988) have demonstrated that, under certain growth conditions, increased mRNA levels also result, in part, from stabilization of the mRNAs by TGF-β.

TGF-β also regulates expression of another set of genes which control the degradation of the newly synthesized matrix proteins. This control is exerted at two levels: (1) decreased synthesis and secretion of proteases and (2) increased synthesis and secretion of protease inhibitors. Again, there is substantial evidence supporting a direct role of TGF-β in altering mRNA levels of the affected genes. Depending on the cell type, TGF-β treatment has been shown to result in decreased synthesis of a thiol protease, major excreted protein (CHIANG and NILSEN-HAMILTON 1986), of a serine protease, plasminogen activator (LAIHO et al. 1986; LUND et al. 1987), and of the metalloproteinases, collagenase (EDWARDS et al. 1987; OVERALL et al. 1989), elastase (REDINI et al. 1988), and transin/stromelysin, which acts on noncollagenous matrix proteins (MATRISIAN et al. 1986). TGF-β controls the activity of these proteases in yet another way; it stimulates cellular secretion of protease inhibitors. Thus far, there is evidence for greatly increased secretion of plasminogen activator inhibitor (LAIHO et al. 1986; LUND et al. 1987) and of tissue inhibitor of metalloproteinases (TIMP; EDWARDS et al. 1987; OVERALL et al. 1989) when cells are treated with TGF-β.

TGF-β acts in yet a third way to increase interaction of cells with extracellular matrix. IGNOTZ and MASSAGUÉ (1987) and C.J. ROBERTS et al. (1988) have demonstrated that both TGF-β1 and TGF-β2 specifically increase the expression of cell membrane receptors for cell adhesion proteins. These receptors, termed "integrins" (HYNES 1987) represent transmembrane links between the extracellular matrix and the cytoskeletal elements of the cell. They are a family of homologous glycoprotein dimers consisting of a larger α-subunit, which confers ligand specificity, and a smaller β-subunit, which appears to be common to several different cell adhesion receptors (HYNES 1987); TGF-β regulates both the α- and the β-subunits of the fibronectin receptor complex by distinct mechanisms (C. J. ROBERTS et al. 1988). Recent evidence suggests that TGF-β can specifically and selectively regulate expression of the different α-subunits, as demonstrated by analysis of the relative expression of $\alpha_{1-6}\beta_1$ receptors in several different normal and neoplastic cell lines (HEINO et al. 1989). The coordinate regulation by TGF-β of synthesis of both fibronectin and its receptor complex is necessary for assembly of fibronectin into the pericellular matrix (McDONALD et al. 1987). Probably as a result of its ability to increase integrin expression, TGF-β treatment has been shown to promote increased adhesion to fibronectin and collagen substrates in a variety of cell types of both mammalian and avian species (IGNOTZ and MASSAGUÉ 1987).

2. Muscle

TGF-β1 is a potent regulator of myogenesis in in vitro models using cell lines exhibiting properties of both skeletal and smooth muscle, reversibly inhibiting fusion and expression of muscle-specific genes including the muscle isozyme of creatine kinase in L6 (FLORINI et al. 1986), L6E9 (MASSAGUÉ et al. 1986), and C-2 and BC_3H1 myoblasts (OLSON et al. 1986) as well as in embryonic myoblasts and skeletal muscle myoblasts (ALLEN and BOXHORN 1987). With the exception of satellite cells whose growth is inhibited by TGF-β, these effects of TGF-β1 on differentiation occur in the absence of any effects on cellular proliferation. Although the physiological significance of the inhibition by TGF-β of differentiation of muscle cells is unknown, it has been proposed that it could play a role in muscle regeneration, in prevention of precocious fusion of embryonic myoblasts, and in prevention of fusion of satellite cells into the main body of muscle at the time of initial myogenesis.

Immunohistochemical studies show strong intracellular staining of cardiac myocytes (THOMPSON et al. 1989); this staining rapidly disappears following experimental myocardial infarction, but becomes intensified in the border zone of the infarcted tissue (THOMPSON et al. 1988). These data implicate TGF-β in maintenance and repair of these cells. Evidence that TGF-β induces hypertrophy and polyploidy in cultured vascular smooth muscle cells (OWENS et al. 1988) suggests that it might function similarly in terminally differentiated cardiac myocytes.

3. Skeleton

Bone represents the most abundant source of TGF-β in the body; it is present at about 0.3 mg/kg, or about 1/10th that obtainable from platelets (SEYEDIN et al. 1985, 1986, 1987). This is in marked contrast to the approximately 3–5µg TGF-β/kg characteristic of soft tissues such as placenta or kidney (FROLIK et al. 1983; ROBERTS et al. 1983a). Just as platelet TGF-β is important in tissue repair and controls the actions of several different cell types which participate in the process (see Sect. I.I), so it is postulated that TGF-β in bone plays a central role in the formation and continuous remodeling of mineralized tissues by its coordinated actions on mesenchymal precursor cells, chondrocytes, osteoblasts, and osteoclasts (CENTRELLA et al. 1988).

In the mouse embryo, immunohistochemical studies show that TGF-β plays a role in formation of the axial skeleton (HEINE et al. 1987). TGF-β staining can also be seen in older embryos (15–18 days' gestation) in the cytoplasm of osteoblasts in centers of endochondral ossification and in areas of intramembraneous ossification of flat bones, as in the calvarium (HEINE et al. 1987). Localization of TGF-β in both osteoclasts and osteoblasts has been described in developing human long bones and calvarial bones following in situ hybridization of TGF-β1 probes (SANDBERG et al. 1988a, b). Several lines of evidence suggest that the mechanisms of TGF-β action in bone formation in the embryo are reiterated in the processes of bone growth and remodeling and in repair of bone injury: (1) TGF-β is found in adult bone matrix (SEYEDIN et al. 1985, 1986) and appears at the time of endochondral ossification in an in vivo model of bone formation (CARRINGTON et al. 1988); (2) cultured fetal bovine bone osteoblasts as well as rat osteosarcoma cells have high mRNA levels of TGF-β and secrete relatively high concentrations of TGF-β (ROBEY et al. 1987); and (3) levels of TGF-β1 mRNA are elevated in healing fractures at the time of mineralization (G. NEMETH and M. BOLANDER, personal communication).

Mechanistically, it has been shown that TGF-β is mitogenic for osteoblasts in culture (CENTRELLA et al. 1987; ROBEY et al. 1987). This is highly significant when compared with the inhibitory effects of the peptide on the majority of cultured cells. There are conflicting reports on the ability of TGF-β to stimulate collagen synthesis in osteoblasts: CENTRELLA et al. (1987) showed increased collagen synthesis following TGF-β treatment of osteoblasts derived from rat calvaria, whereas ROBEY et al. (1987), using fetal bovine bone osteoblasts, were unable to show augmented synthesis and postulated that the increased collagen accumulation is secondary to its effects on the proliferation of the cells. TGF-β activity is increased in cultures of fetal rat calvaria and in calvarial cells incubated with agents that stimulate bone resorption, such as parathyroid hormone, 1,25-dihydroxyvitamin D_3, and IL-1 (PETKOVICH et al. 1987; PFEILSCHRIFTER and MUNDY 1987); moreover, other data showing that TGF-β inhibits the formation of osteoclasts in bone marrow cultures (CHENU et al. 1988) further suggest that it may act as a local regulator of bone remodeling. These observations, showing that TGF-β has effects on both osteoclasts and osteoblasts, have led to the proposal that it is involved in the strict coupling of the processes of bone resorption and bone formation characteristic of the remodeling process in adult bone (PFEILSCHRIFTER and MUNDY 1987). Based on the existing data, the simplest model for the action of TGF-β in bone remodeling would be that the local acidic, proteolytic environment created by osteoclastic activity would result in activation of matrix-associated latent TGF-β (OREFFO et al. 1988). The activated TGF-β would then inhibit new osteoclast formation and expand the local population of osteoblasts, resulting in deposition of matrix to serve as a center for mineralization.

4. Immune Cells and Hematopoiesis

Both TGF-βs 1 and 2 are potent immunoregulatory agents enhancing monocyte function and suppressing lymphocyte proliferation and function. Their ability to inhibit proliferation of T and B lymphocytes at femtomolar concentrations (KEHRL et al. 1986a, b 1989) make them significantly more active than the T-cell-specific immunosuppressant cyclosporin A (KRONKE et al. 1984). TGF-β also inhibits proliferation of thymocytes (RISTOW 1986), suppresses the activity of natural killer cells (ROOK et al. 1986), and inhibits the in vitro generation, but not the activity, of both lymphokine-activated killer (LAK) cells and allospecific cytotoxic T lymphocytes (CTL; MULÉ et al. 1988; ESPEVIK et al. 1988). It also antagonizes the effects of the interleukins IL-1, IL-2, and IL-3 (WAHL et al. 1988; KEHRL et al. 1986a, b; OHTA et al. 1987); however, its ability to suppress secretion by B cells of immunoglobulin G (IgG) and IgM even in the presence of IL-2 concentrations sufficiently high to overcome its antiproliferative effects demonstrates that the action of TGF-β cannot be explained simply as an "anti-interleukin."

TGF-β not only opposes the actions of several of the interleukins, but also antagonizes the action of other immunoregulatory agents such as tumor necrosis factor (TNF) and the interferons. Thus ROOK et al. (1986) demonstrated that TGF-β blunted the boosting of NKC cytolysis by treatment of large granular lymphocytes with interferon-α; TGF-β also suppresses the effects of interferon-γ on the induction of class II major histocompatibility antigens in both lymphoid and nonlymphoid cells (CZARNIECKI et al. 1988). RANGES et al. (1987) demonstrated both that TGF-β inhibits production of TNF-α in mixed lymphocyte cultures and that its suppressive effects on generation of CTL can be partially reversed by addition of TNF-α. TGF-βs 1 or 2 also inhibit production of interferon-γ, TNF-α, and TNF-β during generation of LAK cells by IL-2; as in generation of CTL, effects of TGF-β on generation of LAK cells can be reversed by TNF-α (ESPEVIK et al. 1988).

Although most of its effects on immune cells are inhibitory, new data suggest that TGF-β may play a critical role in isotype switching in B lymphocytes. In contrast to its inhibition of secretion of IgG and IgM by activated B cells (KEHRL et al. 1986a), COFFMAN and coworkers (1989) have found that TGF-β cooperates with IL-2 and IL-5 to upregulate secretion of IgA by splenic lymphocytes, resulting in an isotype switch from 0.1% IgA to greater than 10% IgA. Whether TGF-β might play a role in generation of IgA-secreting plasma cells such as those localized in respiratory mucosa and in lymph nodes in the gastrointestinal tract (Peyers patches) is currently being investigated. Conversely, treatment of activated IgA-secreting B cells with TGF-β suppresses IgA secretion (COFFMAN et al. 1989); this is yet another example of the bifunctional nature of TGF-β.

While the above effects result from exogenous TGF-β and clearly demonstrate the ability of immune cells to respond to the peptide, other data suggest that TGF-β functions as an autoregulatory lymphokine that limits T-cell clonal expansion (KEHRL et al. 1986b). Thus, activation of T cells with phytohemagglutinin increases cellular expression of TGF-β receptors and increases both TGF-β1 mRNA and secretion of TGF-β1 peptide; the delay in secretion of TGF-β1 suggests that it might function to terminate the activation response.

TGF-β can also suppress immune cell function in vivo. Patients with glioblastoma often have impaired cell-mediated immune responses and their sera can inhibit mitogen- and antigen-induced proliferation of normal T lymphocytes (WRANN et al. 1987). Purification of the immune-suppressive factor from a glioblastoma cell line revealed it to be TGF-β2 (DE MARTIN et al. 1987). In the absence of additional data, it cannot be ruled out that other tumor types or even other glioblastomas might preferentially secrete other forms of TGF-β with similar immune-suppressive effects.

In contrast to its effects on lymphocytes, TGF-β activates several functions of monocytes/macrophages. It is a potent chemoattractant for monocytes at concentrations ranging from 0.04 to 0.4 pM (WAHL et al. 1987); at higher concentrations it stimulates monocytes to express higher mRNA levels for several growth-regulatory peptides including TNF-α, PDGF-B, TGF-α, FGF, IL-1, and TGF-β itself (WAHL et al. 1987; MCCARTNEY-FRANCIS et al. 1989), In contrast to resting lymphocytes in which levels of TGF-β1 mRNA are undetectable, monocytes express TGF-β1 mRNA constitutively, yet only in response to activation do they secrete the peptide (ASSOIAN et al. 1987). Finally, TGF-β has yet another highly selective effect on macrophages resulting in deactivation or suppression of release of H_2O_2 (TSUNAWAKI et al. 1988). This inhibition of the respiratory burst capacity by activated macrophages may reflect yet another coordinated response of TGF-β in wound healing in which the growth factors secreted by the macrophages are allowed to act, while the potentially detrimental killing capacity of the cells is ablated. Taken together, the effects of TGF-β on lymphocytes and monocytes indicate that it suppresses destructive aspects of the inflammatory response while facilitating the anabolic effects of macrophage-derived growth factors on tissue repair.

In addition to its effects on lymphocytes and monocytes, TGF-β also controls hematopoiesis in bone marrow culture systems. It suppresses the growth of less mature hematopietic cell populations which have a high proliferative capacity, but does not affect the growth of more differentiated cells. Specifically, the proliferation of murine bone marrow progenitor cells induced by IL-3 and of human bone marrow cells induced by IL-3 and granulocyte/macrophage-colony stimulating factor (GM-CSF) was inhibited reversibly by both TGF-βs 1 and 2, whereas the proliferation of murine bone marrow cells induced by GM-CSF or of human marrow cells induced with G-CSF was not inhibited (CHENU et al. 1988; KELLER et al. 1988, 1989; HAMPSON et al. 1988). Moreover, TGF-β seems to modulate the effects of GM-CSF on the granulocyte/macrophage precursor cell, shifting the balance towards the granulocytic lineage (CHENU et al. 1988). Both human (MITJAVILA et al. 1988), and murine (ISHIBASHI et al. 1987) megakaryocytopoiesis is also inhibited by TGF-β, suggesting a possible negative autocrine loop involving platelets.

In summary, TGF-β has profound effects on proliferation and differentiation of hematopoietic stem cells; in addition, several of the differentiated cell types secrete TGF-β. In contrast to the selective effects of TGF-β1 on murine hematopoiesis reported by Ohta and coworkers (1987), all other investigators who have tested the action of TGF-βs 1 and 2 have reported equivalent effects not only on hematopoiesis, but also on freshly isolated mononuclear cells.

5. Steroidogenic Cells – Ovary, Testis, Adrenal

There is now a wealth of literature on the effects of TGF-β on steroidogenic cells, among them adrenocortical cells, thecal and granulosa cells of the ovary, and Leydig and Sertoli cells of the testis. The effects of TGF-β on steroidogenesis in these cells are independent of proliferative effects since it has no effect on their growth. Its effects on steroid synthesis are cell dependent: it inhibits steroidogenesis in adrenocortical cells (HOTTA and BAIRD 1986; FEIGE et al. 1987) and Leydig cells (AVALLET et al. 1987; T. LIN et al. 1987; MORERA et al. 1988) but enhances the activity of follicle stimulating hormone (FSH) on aromatase activity in granulosa cells (YING et al. 1986b; HUTCHINSON et al. 1987). Although most investigations have focused on effects of exogenous TGF-β on these cells, it is clear that the peptide also acts endogenously as an autocrine or paracrine mediator of cell function: thecal cells (SKINNER et al. 1987) and Sertoli cells (BENAHMED et al. 1988) both secrete TGF-β and follicular fluid has been reported to contain between 25 and 100 ng TGF-β/ml (RUEGSEGGER VEIT and ASSOIAN 1988).

Cellular interactions between thecal cells and granulosa cells have been implicated in control of oocyte development; interactions between these two major cell types of the follicle are mediated through steroids as well as through growth factors. Ovarian thecal cells have been shown to secrete TGF-β and it has been proposed that it could act in a paracrine fashion to alter granulosa cell differentiation (SKINNER et al. 1987). Specific effects of TGF-β on ovarian granulosa cells include stimulation of the FSH-induced maturation of the cells by amplification of the effects of the gonadotropin on aromatase activity (YING et al. 1986b; ADASHI and RESNICK 1986) and progesterone production (KNECHT et al. 1987; DODSON and SCHOMBERG 1987). TGF-β has a biphasic effect on expression of luteinizing hormone (LH) receptors, increasing the receptor content in the presence of low concentrations of FSH and insulin and decreasing the receptor content and subsequently limiting development of granulosa cells when gonadotropin concentrations are increased (KNECHT et al. 1987). TGF-β also has a direct effect on the oocyte, stimulating its meiotic maturation (FENG et al. 1988); the high concentrations of TGF-β in follicular fluid (RUEGSEGGER VEIT and ASSOIAN 1988) suggest that it may regulate the selection and maturation of oocytes during follicular development in vivo.

With respect to its role in the testis, a TGF-β-like peptide has been characterized in the medium of cultured porcine Sertoli cells (BENHAMED et al. 1988). Secretion of this peptide is regulated by hormones: FSH treatment decreases secretion to undetectable levels while secretion increases following treatment with estradiol, dexamethasone, or thyroxine. Analogous to a possible paracrine role of TGF-β in control of thecal-granulosa cell interactions in the ovary, TGF-β secreted by Sertoli cells has been implicated in paracrine control of Leydig cell steroidogenesis (AVALLET et al. 1987; T. LIN et al. 1987; BENHAMED et al. 1988). This control is exerted at several levels. TGF-β reduces expression of (human) chorionic gonadotropin (hCG) receptors on Leydig cells by almost 70% and reduces the cyclic AMP response to hCG by almost 50%. It also directly affects steroid synthesis, acting at a site prior to pregnenolone formation, since conversion of exogenous pregnenolone to testosterone was increased in cells treated with TGF-β even though the steroidogeneic response to hCG was severely depressed (AVALLET et al. 1987).

In adrenocortical cells, TGF-β inhibits both basal and adrenocorticotropin (ACTH)-stimulated cortisol production by 50%–60% (HOTTA and BAIRD 1986; FEIGE et al. 1987) and decreases angiotensin II-stimulated cortisol synthesis by 70%–90%, in this case by decreasing the expression of angiotensin II receptors on the cells one-half (FEIGE et al. 1987). Again, TGF-β has direct effects on steroidogenesis. By suppressing expression of receptors for low-density lipoprotein, it decreases the availability of exogenous lipoprotein, which is the major source of cholesterol for adrenocorticoid biosynthesis (HOTTA and BAIRD 1986). It also reduces expression of the relatively short-lived 17α-hydroxylase, a key enzyme in the synthesis of 17α-hydroxylated end products such as cortisol (FEIGE et al. 1987). Yet another aspect of control in this system is the ability of ACTH to upregulate (about 2-fold) TGF-β receptors selectively on adrenocortical cells (COCHET et al. 1988).

While TGF-β was added exogenously in all of the above studies of cultured adrenocortical cells, immunohistochemical staining of adult mouse adrenal cortex suggests that it might play a role in homeostatic control of steroidogenesis in this tissue. Thus staining for TGF-β is localized to the inner zones of the cortex, the zona fasiculata, and the zona reticularis (THOMPSON et al. 1989); these two zones of the cortex are under control of ACTH, and cortisol synthesis is restricted to this region.

Although indirect, the effects of TGF-β on pituitary hormone synthesis must also be considered in its overall effects on steroidogenesis. TGF-β stimulates FSH secretion by cultured pituitary cells, in this respect mimicking the effects of activin, but antagonizing the effects of inhibin (YING et al. 1986a). Whether TGF-β also has effects on synthesis of other pituitary hormones or of hypothalamic releasing factors remains to be investigated.

6. Endothelium

Many of the processes, such as embryogenesis, inflammation and repair, and carcinogenesis, in which TGF-β plays a prominent role are dependent on either vasculogenesis or angiogenesis. Indeed, TGF-β1 appears to be angiogenic in a variety of assay systems in vivo, including a localized response at the site of injection (ROBERTS et al. 1985) and the rabbit corneal pocket assay (FIEGEL and KNIGHTON 1988). However, due to the many participating cell types, it cannot be determined whether new blood vessel formation is a direct effect of the peptide on endothelial cells, or whether it results from indirect stimulation of endothelial cells by products of other target cells for TGF-β action such as macrophages (see Sect. B. II. 3).

TGF-β1 has many and varied actions on endothelial cells in vitro, and attempts to extrapolate these results to in vivo systems underscore the artificial nature of culture systems in which single cell populations are grown on plastic. Thus in vitro, TGF-β1 opposes the action of fibroblast growth factor (FGF), one of the major mitogens for endothelial cells, and inhibits both the proliferation of endothelial cells in monolayer culture (BAIRD and DURKIN 1986; FRÀTER-SCHRÖDER et al. 1986) and migration of the cells in monolayer or in a standard motility assay (HEIMARK et al. 1986; MÜLLER et al. 1987). Cells studied include both capillary and large vessel endothelial cells from several different species. TGF-β2 is almost 100 times less active than TGF-β1 in inhibition of growth (JENNINGS et al. 1988); since the two TGF-β homologs have equivalent activity on fibroblasts and immune cells, a switching of TGF-β types, for example during the course of healing, may facilitate angiogenesis. Curiously, TGF-β1 has been found to stimulate the growth of endothelial cells passaged by mechanical means, rather than by trypsinization (BLAKE and FALANGA 1988), suggesting that inhibitory effects of the peptide may be expressed only after damage to cell surface proteins by trypsin.

TGF-β1 is also a strong inhibitor of phorbol ester-induced invasion of endothelial cells into collagen gels (MÜLLER et al. 1987), an assay that has been used as an in vitro model of angiogenesis (MONTESANO and ORCI 1985). However, when endothelial cells are cultured three-dimensionally *in* collagen gels, TGF-β1 no longer inhibits proliferation but instead promotes contraction of the gel and organization of endothelial cells into tube-like structures (MADRI et al. 1988), mimicking, in part, its effects on angiogenesis in vivo.

TGF-β1 also modulates gene expression by endothelial cells resulting in increased synthesis of matrix proteins including fibronectin and types IV and V collagen (MÜLLER et al. 1987; MADRI et al. 1988), increased synthesis of an endothelial type plasminogen activator inhibitor (SAKSELA et al. 1987), and increased synthesis of mRNAs encoding the A and B chains of PDGF (STARKSEN et al. 1987; DANIEL et al. 1987) and FGF (A. BAIRD, personal communication). These effects on extracellular matrix and on growth factor secretion probably are important in interactions between endothelial cells and smooth muscle cells in vivo. TGF-β has density-dependent effects on growth of vascular smooth muscle cells (ASSOIAN and SPORN 1986; MAJACK 1987), and is expressed by those cells following vascular injury (MAJESKY et al. 1989), The observation that coculture of endothelial cells with either smooth muscle cells or pericytes results in activation of latent TGF-β1 secreted by the cells suggests that site-dependent activation of TGF-β in vivo along the growing shaft of a vessel may be an important control point in its effects on angiogenesis (A. ORLEDGE and P. D'AMORE, personal communication).

7 Liver

Hepatocytes, like endothelium, are normally in a quiescent state in the adult animal. The potent growth inhibitory activity of TGF-β on hepatocytes in vitro suggests that it might play in role in maintaining quiescence in vivo (NAKAMURA et al. 1985; CARR et al. 1986; MCMAHON et al. 1986). Not only cultured primary hepatocytes but also hepatocytes removed from partially hepatectomized animals or from discrete preneoplastic nodules and neoplastic hepatocytes are all inhibited by TGF-β (STRAIN et al. 1987; WOLLENBERG et al. 1987).

Even though many hepatocytes do not replicate during the entire adult life of an animal, they retain the capacity to replicate and do so in response to loss of liver tissue, as can be achieved experimentally by partial hepatectomy. There is evidence that TGF-β functions in an inhibitory paracrine fashion during liver regeneration, perhaps to prevent uncontrolled hepatocyte growth (BRAUN et al. 1988). Following partial hepatectomy, mRNA levels for TGF-β1 increase approximately 8-fold by 72 h, well after the major wave of hepatocyte mitosis; the mRNA is localized to nonparenchymal sinusoidal cells, particularly in fractions enriched for endothelial cells (BRAUN et al. 1988). RUSSELL and coworkers (1988) have shown that intravenously injected TGF-βs 1 or 2 (0.5 μg), of which 60% is estimated to be cleared by the liver in the first pass (COFFEY et al. 1987), substantially reduce the fraction of hepatocytes synthesizing DNA in a partially hepatectomized animal; however, daily injections of TGF-β for 5 days did not affect the mass of the liver 8 days after surgery. These data are consistent with an intrinsic role for TGF-β in negative control of liver growth and repair, and suggest that loss of inhibitory growth pathways could be a factor in malignant transformation of this tissue (see Sect. I. II).

8. Other Epithelia

The effects of TGF-β on hepatocytes are but one specific example of its growth-inhibitory effects on epithelial cells. Many epithelial tissues exist in a dynamic state in which highly regulated mechanisms exist to control cell proliferation and differentiation. TGF-β suppresses the growth of most epithelial cells in culture including keratinocytes (MOSES et al. 1985; SHIPLEY et al. 1986; REISS and SARTORELLI 1987; COFFEY et al. 1988 a, b; WILKE et al. 1988), bronchial epithelial cells (MASUI et al. 1986), tracheal epithelial cells (JETTEN et al. 1986), intestinal epithelial cells (KUROKOWA et al. 1987), rat prostate epithelial cells (MCKEEHAN and ADAMS 1988), and renal proximal tubular cells (FINE et al. 1985). However, exceptions have been found: TGF-β stimulates the growth of certain epithelial cells such as human mesothelial cells (GABRIELSON et al. 1988) and prostate cells (E. KAIGHN, personal communication). Since control of growth and differentiation appear to be integrally linked in epithelial cells, inhibition of proliferation by TGF-β is often accompanied by terminal differentiation of the cells. Thus TGF-β induces expression of markers of terminal squamous cell differentiation in tracheal and bronchial epithelial cells (JETTEN et al. 1986; MASUI et al. 1986) and promotes the development of differentiated function in intestinal epithelial cells (KUROKAWA et al. 1987).

Effects of TGF-β on epithelial cells have been most extensively studied in cultured keratinocytes. Inhibition of the growth of these cells by TGF-β is reversible and is different from the growth arrest and subsequent terminal differentiation induced by high calcium concentrations (COFFEY et al. 1988a, b; WILKE et al. 1988). Treatment with 1 mM calcium induces expression of several protein markers of differentiation which are not expressed following growth arrest with TGF-β, demonstrating that TGF-β can uncouple growth arrest and induction of differentiation. Moreover, the observations that keratinocytes express TGF-β1 mRNA (COFFEY et al. 1988b) and that secretion of TGF-β2 is markedly elevated following a shift to high calcium concentrations (A. GLICK and S. YUSPA, personal communication) suggest that the peptide might be involved in autocrine maintenance of a nonproliferative state of epithelia. Similarly, treatment of mouse epidermis with the tumor promoter 12-O-tetradecanoylphorbol-13-acetate results in rapid and transient induction of expression of TGF-β1 mRNA; in situ hybridization shows that this induction is localized to the immediate suprabasal layer of the epidermis, indicating that TGF-β1 expression is initiated after the commitment to terminal differentiation (AKHURST et al. 1988).

Other studies demonstrate that the effects of TGF-β on keratinocytes are quite unique, perhaps representing yet another of its many-faceted effects on wound healing. Thus, in human keratinocytes, TGF-β stimulates expression of specific keratins characteristic of regenerative maturation rather than of terminal differentiation, supporting a physiological role in controlling the balance between cell division, migration, and maturation during epithelial wound healing (MANSBRIDGE and HANAWALT 1988).

Results of in vivo studies of murine tissues strengthen the proposed role for TGF-β in control of differentiation of epithelia. Using an antibody (LC 1-30) to the N-terminal amino acids of TGF-β1, FLANDERS et al. (1989) show staining for TGF-β in epithelia of diverse adult tissues including dermis and colon, and in embryonic intestine and hair follicles (K. FLANDERS, personal communication); in situ hybridization techniques also revealed an abundance of TGF-β1 mRNA in differentiating epithelia undergoing morphogenesis (LEHNERT and AKHURST 1988). Evidence that selective expression of TGF-β by epithelial cells can be correlated with differentiation is provided by several studies: (1) BARNARD et al. (1989) have demonstrated an inverse correlation in isolated intestinal enterocytes between mitotic labeling and expression of TGF-β1 mRNA. (2) SILBERSTEIN and DANIEL (1987), using slow-release implants containing TGF-β, have shown strong local inhibition of mammary growth and morphogenesis, resulting in complete inhibition of end bud formation. (3) Immunohistochemical staining of TGF-β in branching mammary epithelia (G. SILBERSTEIN and C. DANIEL, personal communication) and bronchial epithelia (U. HEINE, personal communication) is most intense at crotches of the branches where proliferation is suppressed and deposition of matrix proteins is most concentrated.

III. The Role of TGF-β in Embryogenesis

Mechanisms operative in the processes of wound healing and carcinogenesis have long been thought to recapitulate mechanisms operative in embryonic development. The central role of TGF-β in wounding and carcinogenesis (see Sect. I), the almost universal distribution of the TGF-β receptor on cells (WAKEFIELD et al. 1987), and the potent effects of the growth factor in control of cell migration, growth, differentiation, and function and regulation of extracellular matrix, strongly implicate it in embryonic development (A. B. ROBERTS et al. 1988). Indeed, with the exception of TGF-β2 for which no data are available, all of the other TGF-β genes are strongly expressed during embryogenesis. Expression of TGF-β1 mRNA first appears after fertilization (RAPPOLEE et al. 1988a) and remains high throughout the remainder of the development of the mouse embryo (HEINE et al. 1987) and on into neonatal and adult life (THOMPSON et al. 1989). Expression of TGF-β3 mRNA is high in human umbilical cord (TEN DIJKE et al. 1988a) and expression of both TGF-βs 3 and 4 is detected in cultured chick embryo chondrocytes and fibroblasts (JAKOWLEW et al. 1988a, b). Finally, TGF-β5 was cloned from a frog oocyte cDNA library and is most strongly expressed from the neurula stage onward (KONDAIAH et al. 1989).

Studies in a variety of species have shown that both undifferentiated embryonic tissue and more differentiated embryonic cell types such as chondrocytes, fibroblasts, and osteoblasts respond to TGF-β1 or 2, demonstrating that they express TGF-β receptors. Embryonal carcinoma cell lines have been used as an experimental model of early embryogenesis; although the stem cells express only very low levels of TGF-β receptors and do not appear to respond to TGF-β, their differentiated derivatives express 16- to 40-fold higher receptor levels and are growth inhibited by TGF-β (RIZZINO 1987). Changes in TGF-β receptors might also be the basis of the observation that the growth of very early human embryo fibroblasts is stimulated by TGF-β1, whereas that of later stage human (HILL et al. 1986) or murine fibroblasts (ANZANO et al. 1986) is inhibited. Fetal bovine osteoblasts (ROBEY et al. 1987) and mesenchymal precursors of chondrocytes (SEYEDIN et al. 1985) are also stimulated to grow by TGF-β1. Amphibian embryonic tissue is also responsive to TGF-β; ectodermal explants from frog blastulae are induced to form mesoderm either by TGF-β2 alone (ROSA et al. 1988) or by the combination of TGF-β1 plus FGF (KIMELMAN and KIRSCHNER 1987).

Many of these responsive embryonic cell types also secrete TGF-βs, possibly implicating the peptides in an autocrine mechanism of regulation of the growth and differentiation of embryonic tissues. Accordingly, fetal bovine osteoblasts secrete rather high levels of TGF-β (ROBEY et al. 1987), and evidence for autocrine action is provided by preliminary studies demonstrating an inhibition of the basal growth rate by blocking antibodies to TGF-β1 (ERNST et al. 1988). Cultured chicken embryo chondrocytes and fibroblasts (S. JAKOWLEW, personal communication) as well as *Xenopus* tadpole cells (ROSA et al. 1988) also secrete TGF-β-like activity into their medium, but its effect on the cells is not known.

Detailed studies of the immunolocalization and in situ hybridization of TGF-β1 in the developing mouse embryo suggest both autocrine and paracrine mechanisms of action. Using in situ hybridization, WILCOX and DERYNCK (1988) have demonstrated prominent expression of TGF-β1 mRNA in hematopoietic cells of early mouse embryos, in agreement with the immunohistochemical localization in fetal bovine liver described by EL-LINGSWORTH et al. (1986). In later mouse embryos, LEHNERT and AKHURST (1988) have shown in situ hybridization of a TGF-β1 probe in fetal bone in both perichondral osteocytes and osteocytes involved in intramembraneous ossification; these same cells stain for TGF-β protein, suggesting autocrine action (ELLINGSWORTH et al. 1986; HEINE et al. 1987). Similar patterns of in situ hybridization have been observed in developing human long bones and calvaria (SANDBERG et al. 1988 a, b). TGF-β1 mRNA and immunohistochemical staining also colocalize in the submucosa of the developing intestine and in the cushion tissue of the developing heart valves. In contrast to these potential examples of autocrine action of TGF-β, in most differentiating tissues which have both epithelial and mesenchymal components, TGF-β1 mRNA is expressed in the epithelial components, whereas the protein is localized to the underlying mesenchymal elements (LEHNERT and AKURST 1988; HEINE et al. 1987); examples of such tissues are the developing hair follicles of the snout, the developing tooth bud, and the submandibular gland. The simplest interpretation of these data is that TGF-β1 is synthesized by the epithelial cells of these tissues, secreted, and localized in the mesenchyme. Since TGF-β1 is secreted in a latent form (Sect. F), it is possible that latent TGF-β1 might bind to mesenchymal cells via a heparin binding site (WAKEFIELD et al. 1989) or via the cellular integrin recognition sequence, Arg-Gly-Asp (Figs. 1, 4) which are present in the latency protein; mechanisms such as these could localize the peptide to its target cells in a stable, inactive state. Once activated in situ, mature TGF-β1 could bind to its receptors and trigger the appropriate response pattern.

The results of HEINE et al. (1987) clearly demonstrate that TGF-β is localized in a unique pattern, not only spatially, but also temporally, in the developing mouse embryo, correlating with specific morphogenetic and histogenetic events. For example, the pattern of TGF-β staining in the developing somites changes as the somites mature, demonstrating that TGF-β contributes to the segmentation of the axial skeleton: staining is uniform throughout the primitive somite, but subsequently localizes in the sclerotome and dermatome as development progresses, and finally in the area defining the centrum of the future definitive vertebrae. The rapidly changing staining pattern for TGF-β which accompanies maturation of the hair follicles also suggests a dynamic role for the peptide in control of epithelial-mesenchymal interactions, intensifying in the mesenchyme surrounding the hair follicles as the follicles form and disappearing as they mature.

Important areas for future investigation are study of the comparative roles of the various forms of TGF-β in embryonic development. Although the major focus thus far has been on TGF-β1, TGF-β3 and TGF-β5 mRNAs are the most prominent species in the chicken embryo (JAKOWLEW et al. 1988 a) and frog embryo, respectively (KONDAIAH et al. 1989). Development of antibodies that can selectively recognize these other TGF-βs is critical to further studies of the role of TGF-β in embryogenesis. For example, recently developed peptide antibodies specific for TGF-β3 and its precursor show intense localization in both neurons and myocytes of the mouse embryo; the staining of neuronal cells has not been observed with antibodies to TGF-β1, suggesting a specific function of TGF-β3 in these cells (K. FLANDERS and S. JAKOWLEW, personal communication).

H. Biochemical Mechanisms of Action

TGF-β has several effects on cellular metabolism that are shared with other growth factors and are therefore unlikely to account for its specific actions. These include stimulation of glucose uptake (INMAN and COLOWICK 1985), increased amino acid transport (BOERNER et al. 1985; RACKER et al. 1985), and stimulation of the synthesis of prostaglandins (TASHJIAN et al. 1985) and of cytoplasmic actin (LEOF et al. 1986 a).

The ability of TGF-β to modulate expression of receptors for other growth factors and hormones appears to be cell specific and is therefore more likely to be related to its mode of action on those cells. As examples, TGF-β increases synthesis of EGF receptors in NRK cells (ASSOIAN et al. 1984) and in cultured rat granulosa cells (FENG et al. 1986); in both cases, the effects correlate well with the effects of TGF-β on the cells. TGF-β also blocks the ability of IL-2 to upregulate its own receptor as well as the transferrin receptor on T lymphocytes (KEHRL et al. 1986 b), consistent with its proposed role in control of clonal expansion of T-cells, and downregulates expression of angiotensin II receptors on adrenalcorticocal cells (FEIGE et al. 1987) and of receptors for hCG on Leydig cells (AVALLET et al. 1987) as one aspect of its inhibition of steroidogenesis in those cells.

I. Receptor Signaling

TGF-β interferes with the action of many different mitogens including PDGF and EGF (ROBERTS et al. 1985; ANZANO et al. 1986), FGF (BAIRD and DURKIN 1986), and insulin-like growth factor I (HILL et al. 1986), suggesting that it might interfere with the signaling from these receptors. The receptors for each of these growth factors are tyrosine kinases that can be stimulated by binding of the appropriate ligand; the TGF-β receptors have no such demonstrable activity (FANGER et al. 1986; LIBBY et al. 1986). Examination of the effects of TGF-β on other intracellular signals further distinguishes its mode of action from that of growth factors linked to tyrosine kinase activity. As examples, induction of S6 kinase activity in MvlLu cells by EGF (LIKE and MASSAGUÉ 1986) and induction of signals such as phospholipid turnover, activation of protein kinase C, Na^+/H^+ antiport activity, or expression of *myc* or *fos* in the nucleus of Chinese hamster lung fibroblasts stimulated by thrombin or FGF (CHAMARD and POUYSSÉGUR 1987) are not blocked by simultaneous treatment of the cells with concentrations of TGF-β that completely block the mitogenic activity of those peptides. Although TGF-β does affect transcription of c-*myc* in select carcinoma cell lines (FERNANDEZ-POL et al. 1987; MULDER et al. 1988), overall, these data implicate TGF-β in novel signaling pathways which converge in the nucleus to block DNA synthesis at some step distal to those already examined.

In mouse embryo-derived AKR-2B fibroblasts, TGF-β results in delayed mitogenesis which is thought to be mediated indirectly by induction of c-*sis* mRNA and secretion of PDGF-like proteins (SHIPLEY et al. 1985; LEOF et al. 1986b); the PDGF-regulated genes c-*myc* and c-*fos* were also induced with delayed kinetics by TGF-β. In certain leukemia cell lines. TGF-β induces selective expression of PDGF A chain (MAKELA et al. 1987), whereas in endothelial cells it induces expression of both the A and B chains of PDGF (STARKSEN et al. 1987; DANIEL et al. 1987).

In AKR-2B cells, several lines of evidence implicate a GTP-binding protein in the mitogenic response to TGF-β (MURTHY et al. 1988): (1) Pertussis toxin, which is known to block receptor-mediated signal transduction by ADP-ribosylation of the α-subunit of certain G proteins (SPIEGEL 1987), inhibits the mitogenic action of TGF-β. (2) TGF-β stimulates membrane GTPase activity. (3) Gpp(NH)p, which is known to dissociate G proteins, reduces the receptor-binding affinity of TGF-β to 17% of control values. The latter provides strong evidence for a functional association between a G protein and the TGF-β receptor and argues against an indirect effect of pertussis toxin on PDGF-like peptides.

Analysis of the inhibitory effects of TGF-β on the cell cycle of a variety of different cell types demonstrates that it results in arrest of the cells in the middle to late G_1 phase, delaying entry into S phase. Thus flow microfluorimetry of regenerating endothelial cells shows that TGF-β delays entry into S phase, but that cells that have passed the G_1/S border or are in G_2 are no longer responsive to inhibition (HEIMARK et al. 1986). Similar studies using flow cytometric and autoradiographic analysis of primary cultures of human prokeratinocytes (SHIPLEY et al. 1986) and rat hepatocytes (NAKAMURA et al. 1985; P. LIN et al. 1987) also showed delayed progression of cells from G_1 to S phase. Others have arrived at similar conclusions by examination of effects of TGF-β on induction of various cell cycle markers. Thus induction of thymidine kinase, a marker of entry into S phase, was repressed by TGF-β in Chinese hamster lung fibroblasts (CHAMBARD and POUYSSÉGUR 1987) and induction of the transferrin receptor, which normally occurs in late G_1 phase of the cell cycle, was repressed in activated human B lymphocytes treated with TGF-β (SMELAND et al. 1987); TGF-β had no effect on expression of any of the well characterized markers normally associated with G_0/G_1 transition in either of these cells.

II. Transcriptional Activation of Genes by TGF-β

The end result of TGF-β action on a target cell is a change in its pattern of gene transcription. The transcription of many genes is affected. Some of these genes, such as those for growth factors or their receptors and those encoding nuclear protooncogenes or enzymes involved in DNA synthesis, are directly involved in cell replication (as discussed above); others are components of the complex scheme by which TGF-β regulates extracellular matrix, as discussed in Sect. G. II. 1. Depending on the gene, transcription can be either increased or decreased by TGF-β; effects of TGF-β on mRNA stability can also contribute to the increased levels of mRNA (RAGHOW et al. 1987; PENTTINEN et al. 1988).

It is now known that transcriptional control is achieved through the interaction of *trans*-acting transcription factors with specific *cis*-acting DNA promoter elements. The availability of chimeric constructs in which the promoter of a gene responsive to regulation by TGF-β is linked to a reporter gene such as the bacterial enzyme chloramphenicol acetyltransferase (CAT) is beginning to shed light on the specific molecular mechanisms of transcriptional control by TGF-β. Data are available for five genes: $\alpha2(I)$ collagen (ROSSI et al. 1988), fibronectin (DEAN et al. 1988, 1989), collagenase and tissue inhibitor of metalloendoproteinase (TIMP; J. HEATH, personal communication), and TGF-$\beta1$ (KIM et al. 1989a, b). Analysis of TGF-β effects on transcription of the type I collagen promoter are the most straightforward, demonstrating that a segment of the promoter overlapping an NF-1 binding site is needed for TGF-β stimulation; a 3-bp substitution mutation abolishing NF-1 binding to this site ablates TGF-β activation (ROSSI et al. 1988). Although one might speculate that, since NF-1 acts as both a transcriptional factor and a DNA replication factor, it could mediate both the changes in gene transcription and the changes in cell proliferation induced by TGF-β, results from the other systems suggest that the story is far more complicated. For example, deletion of the NF-1 site in the fibronectin promoter (DEAN et al. 1989) and the TIMP promoter (J. HEATH, personal communication) reduce only partially the TGF-β-dependent activation of the promoter; in the TGF-$\beta1$ promoter, the presence of the NF-1 site is not sufficient for autoactivation and optimal activation requires the presence of an additional region of 130 bp upstream of the NF-1 site (KIM et al. 1989b). These date suggest that there will be gene-specific mechanisms mediating the response to TGF-β, and it will be important to determine if the effects of TGF-β are mediated through modification of preexisting transcription factors, or through the induction of specific transcription factors.

III. TGF-β as a Potential Mediator of the Action of Retinoids and Steroids

Although still preliminary, new data suggest that the actions on cells of low molecular weight molecules such as the retinoids and phorbol esters and of hormones such as estrogen and dexamethasone might be mediated, at least in part, by TGF-β. Data in support of this are the induction by phorbol esters of the expression of TGF-β1 in mouse epidermis (Akhurst and Ballmain 1988) and of TGF-β2 in cultured keratinocytes (A. Glick and S. Yuspa, personal communication); the induction by retinoic acid of expression of TGF-β2 mRNA in primary cultures of mouse keratinocytes (A. Glick and S. Yuspa, personal communication) and of TGF-β3 mRNA in cultured chicken embryo chondrocytes (S. Jakowlew, personal communication); the induction by tamoxifen of TGF-β2 secretion by the human prostatic carcinoma PC-3 cell line (Ikeda et al. 1987); the induction by estrogen of TGF-β1 mRNA in human HOS TE85 osteosarcoma cells (Komm et al. 1988); and the induction by dexamethasone of the expression of TGF-β1 mRNA in T lymphocytes (F. Ruscetti, personal communication). In several of these examples, the biological effects of TGF-β and the hormone modifier are similar. Thus both tamoxifen and TGF-β inhibit the growth of breast cancer cell lines (Knabbe et al. 1987), both estrogen and TGF-β induce collagen synthesis in bone cells (Komm et al. 1988), and both retinoic acid and TGF-β inhibit proliferatin of various epithelial cells (Roberts and Sporn 1984). With the recent discovery and cloning of the superfamily of regulatory proteins that includes receptors for the steroid hormones including vitamin D_3, for thyroid hormone, and for retinoic acid (Evans 1988), it should be possible to test directly at the molecular level whether these *trans*-acting enhancer factors might lead to transcriptional activation of the various TGF-β promoters. If so, the relationships between the actions of hormones, vitamins, oncogenes, and growth factors on cells will have come full circle, having a firm mechanistic link at the level of control of gene transcription.

J. Clinical Relevance of TGF-β in Healing and Disease

The best examples of the multifunctionality of TGF-β action are found in analyses of its concerted actions on the many cell types involved in inflammation and repair as well as in pathologic states including carcinogenesis (for review see A. B. Roberts et al. 1988). TGF-β regulates many of the processes common to both tissue repair and to disease (Haddow 1972; Dvorak 1986), including angiogenesis, chemotaxis, fibroblast proliferation, and the controlled synthesis and degradation of matrix proteins such as collagen and fibronectin; moreover, in many instances, the responsive cell types participating in these processes are also capable of secreting TGF-β.

I. The Role of TGF-β in Repair Processes

The presence of relatively large amounts of TGF-β in platelets (Assoian et al. 1983; Van den Eijnden-van Raaij et al. 1988) and in bone (Seyedin et al. 1985, 1986) implicates the peptide in both repair of soft tissues and remodeling/repair of hard tissues. Many of the individual cellular response patterns contributing to these processes have already been described earlier in this chapter. For soft tissue healing, these include chemotactic attraction and activation of monocytes/macrophages, suppression of immune cell responses (Sect. G. II. 4), both direct and indirect effects on endothelium (Sect. G. II. 6), both chemotactic attraction (POSTLETHWAITE et al. 1987) and activation of fibroblasts to enhance extracellular matrix (Sect. G. II. 1), and, in certain cases, effects on keratinocytes (Sect. G. II. 8). An additional effect of TGF-β on fibroblasts that also contributes to its effects on wound healing is stimulation of the ability of the cells to contract provisional wound matrix (MONTESANO and ORCI 1988). For remodeling and repair of mineralized tissues, many of these same responses are involved in addition to specific effects on suppression of osteoclast differentiation and enhancement of osteoblast proliferation and function (Sect. G. II. 3). Individually, effects of TGF-β on the various participating cell types could be described as being either inhibitory or stimulatory; yet, collectively, all the various actions of the peptide synergize to augment the overall healing or remodeling process.

Many mechanisms have evolved to extend the action of TGF-β in repair processes. Following its initial release from platelet α-granules (ASSOIAN and SPORN 1986), other participating cell types including activated macrophages (ASSOIAN et al. 1987; RAPPOLEE et al. 1988 b) and lymphocytes (KEHRL et al. 1986 b) as well as fibroblasts (ANZANO et al. 1986), smooth muscle cells (MAJESKY et al. 1989), and osteoblasts (ROBEY et al. 1987) can secrete TGF-β. In addition, the ability of TGF-β to autoinduce its own synthesis in macrophages (MCCARTNEY-FRANCIS et al. 1988), lymphocytes (J. KEHRL, personal communication), and fibroblasts (VAN OBBERGHEN-SCHILLING et al. 1988) could serve to sustain its action. Mechanisms for the activation and deactivation of TGF-β at the site of injury can also be postulated: the proteolytic, acidic environment in the vicinity of activated macrophages would favor activation of latent TGF-β (SILVER et al. 1988), whereas excess TGF-β could be scavenged by α_2-macroglobulin (O'CONNOR-MCCOURT and WAKEFIELD 1987) as previously described in Sect. F. Not only is TGF-β clearly implicated as an *intrinsic, endogenous* component of the healing process (CROMACK et al. 1987), but several experiments in which TGF-β was injected locally in vivo demonstrated that the protein by itself can initiate the cascade of events characteristic of wound healing. Using the wound-healing model of HUNT et al. (1967), TGF-β, injected into wire mesh Schilling-Hunt chambers implanted subcutaneously in the backs of adult rats, accelerated the accumulation of total protein, collagen, and DNA (SPORN et al. 1983; LAWRENCE et al. 1986); similar results have been obtained with TGF-β in implants of Teflon chambers (SPRUGEL et al. 1987). Subsequent experiments in which nanogram qunatities of TGF-β were injected subcutaneously into the nape of the neck of newborn mice demonstrated that TGF-β rapidly induced formation of a localized nodule of granulation tissue characterized by an influx of inflammatory cells, fibroblast proliferation with accompanying elaboration of connective tissue proteins, and formation of new blood vessels (ROBERTS et al. 1986). It is important to note that this process was reversible; the histology of sections removed from animals which had been injected with TGF-β for 3 days to give an optimal response and then left untreated for an additional 5 days was indistinguishable from that of control sections.

In contrast to adult wounds which derive their strength from collagen fibrils, fetal wounds heal without scarring; the acute inflammatory response is absent, few fibroblasts participate, and the matrix is composed of proteoglycan, rather than collagen (DePalma et al. 1987). Interestingly, TGF-β1 implanted in rabbit fetuses elicited an adult-type response pattern with fibroblast proliferation and collagen accumulation, thus documenting the responsiveness of the fetal system to adult repair signals (Krummel et al. 1988). These data suggest that fetal healing may involve independent mechanisms in which TGF-β is not a participant, or, alternatively, in which other types of TGF-β might play a role; this will be an important area for future investigation once the new forms of TGF-β become available in larger amounts and once specific immunohistochemical reagents are developed.

As discussed previously, evidence suggests that TGF-β also plays a role in repair of other soft tissues such as the heart (Sect. G. II. 2) and the liver (Sect. G. II. 7). Thus, following ligation of the left coronary artery, immunohistochemical staining for TGF-β first disappears in myocytes in the infarcted area, but later is intensified in myocytes at the margin of the damaged tissue (Thompson et al. 1988). In the regenerating liver, the elevation of TGF-β1 mRNA after the first round of compensatory cell division suggests that the peptide might serve to limit the proliferative response (Braun et al. 1988).

Fracture healing is characterized by a tissue response pattern which mimics that of the initial stages of wound healing, and by subsequent mineralization processes which mimic embryonic bone formation, suggesting that TGF-β will play an important role in the process. Thus fracture results in hematoma and clot formation followed by invasion by inflammatory cells, revascularization, and the resorption of necrotic tissue to form granulation tissue. Later stages involve proliferation of osteoprogenitor cells within the periosteum and endosteum followed by chondrification and calcification of the callus and replacement with lamellar bone by the process of endochondral ossification. After fracture of the femur in rats, TGF-β1 mRNA levels become elevated both early and again around day 10 when other markers of osteoblast activity are detected (G. Nemeth and M. Bolander, personal communication). Local injection of TGF-β1 at the site of a healing fracture or adjacent to a growing femur in neonatal rats results in proliferation of the cells of the periosteum and endosteum with extensive cartilage formation (M. Joyce and M. Bolander, personal communication).

In summary, it is well established that repair of both soft and hard tissue requires a series of properly orchestrated events including cell replication and differentiation, chemotaxis, and controlled synthesis and degradation of connective tissue proteins. We propose that TGF-β itself, either directly or indirectly, can account for many of these aspects.

II. The Role of TGF-β in Carcinogenesis and Other Proliferative Diseases

Theoretically, carcinogenesis could as likely result from failure of cells to respond to negative growth factors which normally function to control their growth as from increased responsiveness to autocrine-acting mitogens (SPORN and ROBERTS 1985). There is now substantial evidence that TGF-β acts to control the growth of many normal epithelia and lymphoid cells (see Sect. G. II. 4, 7, 8). Conversely, many transformed keratinocytes (SHIPLEY et al. 1986), hepatocytes (McMAHON et al. 1986), bronchial carcinoma cells (JETTEN et al. 1986), squamous cell carcinomas (REISS and SARTORELLI 1987), leukemia cells (KELLER et al. 1988b), and retinoblastoma cells (KIMCHI et al. 1988) are no longer inhibited by TGF-β. Similar loss of responsiveness is found following viral transformation of cells, as in transformation of NIH 3T3 (ANZANO et al. 1985) or Fischer rat 3T3 cells with the ras oncogene (STERN et al. 1986) or of B lymphocytes with Epstein-Barr virus (BLOMHOFF et al. 1987).

In at least one example, loss of responsiveness to TGF-β is based on loss of mechanisms of activation of latent TGF-β. Thus A549 human lung carcinoma cells divide at a high rate even though they secrete substantial amounts of latent TGF-β; growth of these cells is potently inhibited by their own conditioned medium, after the TGF-β in that medium has been activated by treatment with acid (WAKEFIELD et al. 1987). In certain other cases, it has been possible to correlate lack of responsiveness to TGF-β with an absence of TGF-β receptors; this is found in the rat PC-12 pheochromocytoma, human retinoblastoma cells, and several leukemic cell lines (KIMCHI et al. 1988; KELLER et al. 1989). However, many other nonresponsive cell lines exhibit either normal receptor numbers or even higher receptor binding levels (CHINKERS 1987; HAMPSON et al. 1988), suggesting that alterations in TGF-β signaling pathways might have occurred.

Many tumor cells express higher levels of TGF-β than their normal counterparts. As examples, DERYNCK et al. (1987a) found elevated levels of TGF-β1 mRNA in tumor tissue compared to adjacent normal tissue; ras-transformed cells both express higher levels of TGF-β1 mRNA (JAKOWLEW et al. 1988d) and secrete elevated levels of the protein (ANZANO et al. 1985) than the parental cells, and peripheral mononuclear cells from adult T-cell leukemia patients expressed higher levels of both TGF-β1 mRNA and protein compared to controls (NIITSU et al. 1988).

Since many tumor cells secrete TGF-β but have lost the ability to respond to the peptide, it has been postulated that TGF-β might stimulate tumor growth indirectly via paracrine effects on stromal elements of the tumor. Like granulation tissue of healing wounds, tumor stroma is characterized by an influx of inflammatory cells, neovascularization, and elaboration of connective tissue (DVORAK 1986). The tumor, by continuously secreting TGF-β, can utilize the host's natural repair mechanisms to support its growth and to impair immune surveillance.

It is too early to identify specific roles for TGF-β in other proliferative diseases, but we would predict, based its multifaceted ability to elaborate connective tissue by control of both matrix protein synthesis and proteolysis (see Fig. 5), that aberrant expression of TGF-β would be associated with a variety of connective tissue diseases including keloids, cirrhosis, atherosclerosis, pulmonary fibrosis, scleroderma, and rheumatoid arthritis (SPORN and ROBERTS 1986). Initial investigations have found TGF-β levels to be high in synovial fluid of rheumatoid arthritis patients and, although it is not mitogenic for synoviocytes, TGF-β probably plays a role in the reparative aspects of the disease involving fibrosis and scar formation (LAFYATIS et al. 1989; REMMERS et al. 1989). In a fibrotic disease of the eye, proliferative vitreoretinopathy, levels of TGF-β2 in the vitreous of diseased eyes show a strong correlation with the severity of the disease (CONNOR et al. 1989). Retinal pigment epithelial cells, which are thought to play a central role in the disease process, also secrete predominantly TGF-β2 in culture. Based on these promising results, further studies of the roles of TGF-β in a variety of proliferative disorders are currently underway using immunohistochemical techniques to localize the peptide in the diseased tissues.

III. Therapeutic Opportunities

It is clear that TGF-β plays a role in many cell-mediated processes and that these can be manipulated by exogenous peptide. With the advent of recombinant techniques for producing large quantities of both the processed protein and possibly also its latent form, it is hoped that therapeutic approaches to repair of soft and hard tissues and to immune suppression can be developed.

In vivo studies have already demonstrated the possibility of initiating a cellular response analogous to wound healing by local application of TGF-β (SPORN et al. 1983; LAWRENCE et al. 1985). More recently, it has been shown that TGF-β, applied topically in a collagen gel, can accelerate healing and increase tensile strength of incisional wounds in rats (MUSTOE et al. 1987) and that when injected in a hyaluronic acid vehicle into the eyes of rabbits directly onto the site of a retinal tear, it can stimulate a localized granulation response which results in closure of the tear (SMIDDY et al. 1989). Such studies hold great promise that, following the development of an appropriate delivery system, TGF-β might have practical application.

The in vivo use of TGF-β for repair of cartilage and bone is not yet as advanced. New studies have shown a time-dependent appearance of mRNA for TGF-β1 at a fracture site in a rat and have localized the peptide immunohistochemically in the callus and periosteum of the healing fracture (M. BOLANDER, G. NEMETH, M. JOYCE, unpublished data). Injections of TGF-β1 into the periosteal area of the femur of young rats have caused significant formation of new cartilage (M. JOYCE and M. BOLANDER, unpublished data), while similar experiments on parietal bone have stimulated periosteal bone formation, resulting in a thickening of the calvarium (NODA et al. 1988b).

However, it is not yet clear whether exogenous TGF-β itself can be used to accelerate the repair of damaged cartilage or bone in vivo. Such an application may depend on the concomitant use of other peptide growth factors. In this regard it should be noted that two new peptide regulators of bone formation, known as bone morphogenetic proteins (BMP) 2A and 3, have been identified as members of the TGF-β supergene family (WOZNEY et al. 1988). These peptides can induce cartilage and bone formation when implanted in rats. Their combined use with TGF-β remains to be investigated.

Another important clinical application of TGF-β that is being pursued is suppression of the immune response. The potent inhibitory effects of TGF-β on immune cells in vitro, together with the immune suppression of glioblastoma patients assumed to be a function of TGF-β secretion by their own tumor cells in vivo (WRANN et al. 1987), suggest the practicality of such an approach. A particularly important application would be the use of TGF-β to prevent rejection of organ transplants, and there is active investigation of this problem at the present time.

Finally, one can conceive of totally new, and essentially unexplored, therapeutic possibilities, particularly in the areas of regeneration of muscle and nerve. Could TGF-β induce repair or regeneration of skeletal or cardiac muscle? Could TGF-β accelerate regeneration of peripheral nerves, and might it have a role in promoting regenerative phenomena in the central nervous system? These are all problems of the future. The recent immunohistochemical localization of TGF-β1 in normal cardiac myocytes, as well as at sites of repair of cardiac injury (THOMPSON et al. 1988) suggests there is some new, as yet unknown, function for TGF-β1 in the heart. The observed intense immunohistochemical staining of TGF-β3 in embryonic neurons and both skeletal and cardiac myocytes (K. FLANDERS and S. JAKOWLEW, unpublished data) suggest further study of its use to repair these cells when injured in the adult. The concept that tissue repair involves a reprogramming of embryonic processes is a very old one. The importance of TGF-β in controlling many aspects of embryogenesis provides optimism that it may have further new therapeutic applications for promotion of tissue repair.

Acknowledgments. We have been privileged to have had an outstanding group of collaborators to work with us on TGF-β. Those whose contributions have been cited here include Mario Anzano, Richard Assoian, Mark Bolander, Ward Casscells, David Danielpour, Linda Dart, Igor Dawid, Benoit de Crombrugghe, Rik Derynck, Kathleen Flanders, Charles Frolik, Bert Glaser, Adam Glick, Ursula Heine, Sonia Jakowlew, John Kehrl, Seong-Jin Kim, Paturu Kondaiah, Arthur Levinson, Doug Melton, Chester Meyers, Maureen O'Connor-McCourt, Hari Reddi, Pamela Robey, Nanette Roche, Frédéric Rosa, Ellen Van Obberghen-Schilling, Diane Smith, Joseph Smith, John Termine, Nancy Thompson, Lalage Wakefield, and Ronald Wilder. Their contributions have been invaluable to the progress that has been made. Lastly, we wish to thank Cephas Swamidoss for his invaluable help in searching the ever-growing TGF-β literature.

References

Adashi EY, Resnick CE (1986) Antagonistic interactions of transforming growth factors in the regulation of granulosa cell differentiation. Endocrinology 119:1879–1881

Akhurst RJ, Fee F, Balmain A (1988) Localized production of TGF-β mRNA in tumour promoter-stimulated mouse epidermis. Nature 331:363–365

Allen RE, Boxhorn LK (1987) Inhibition of skeletal muscle satellite cell differentiation by transforming growth factor-beta. J Cell Physiol 133:567–572

Anzano MA, Roberts AB, Meyers CA, Komoriya A, Lamb LC, Smith JM, Sporn MB (1982) Synergistic interaction between two classes of transforming growth factors from murine sarcoma cells. Cancer Res 42:4776–4778

Anzano MA, Roberts AB, Smith JM, Sporn MB, De Larco JE (1983) Sarcoma growth factor from conditioned medium is composed of both type alpha and type beta transforming growth factors. Proc Natl Acad Sci USA 80:6264–6268

Anzano MA, Roberts AB, De Larco JE, Wakefield LM, Assoian RK, Roche NS, Smith JM, Lazarus JE, Sporn MB (1985) Increased secretion of type β transforming growth factor accompanies viral transformation of cells. Mol Cell Biol 5:242–247

Anzano MA, Roberts AB, Sporn MB (1986) Anchorage-independent growth of primary rat embryo cells is induced by platelet-derived growth factor and inhibited by type-beta transforming growth factor. J Cell Physiol 126:312–318

Assoian RK, Fleurdelys BE, Stevenson HC, Miller PJ, Madtes DK, Raines EW, Ross R, Sporn MB (1987) Expression and secretion of type beta transforming growth factor by activated human macrophages. Proc Natl Acad Sci USA 84:6020–6024

Assoian RK, Komoriya A, Meyers CA, Miller DM, Sporn MB (1983) Transforming growth factor-beta in human platelets. J Biol Chem 258:7155–7160

Assoian RK, Frolik CA, Roberts AB, Miller DM, Sporn MB (1984) Transforming growth factor beta controls receptor levels for epidermal growth factor in NRK fibroblasts. Cell 36:35–41

Assoian RK, Sporn MB (1986) Type-beta transforming growth factor in human platelets: release during platelet degranulation and action on vascular smooth muscle cells. J Cell Biol 102:1217–1223

Avallet O, Vigier M, Perrard-Sapori MH, Saez JM (1987) Transforming growth factor-β inhibits Leydig cell functions. Biochem Biophys Res Commun 146:575–581

Baird A, Esch F, Mormede P, Ueno N, Ling N, Böhlen P, Ying S-Y, Wehrenberg WB, Guillemin R (1986) Molecular characterization of fibroblast growth factor: distribution and biological activities in various tissues. Recent Prog Horm Res 42:143–205

Baird A, Durkin T (1986) Inhibition of endothelial cell proliferation by type-beta transforming growth factor: interactions with acidic and basic fibroblast growth factors. Biochem Biophys Res Commun 138:476–482

Barnard JA, Beauchamp RD, Coffey RJ, Moses HL (1989) Regulation of intestinal epithelial cell growth by transforming growth factor-beta. Proc Natl Acad Sci USA 86:1578–1582

Barton DE, Foellmer BE, Du J, Tamm J, Derynck R, Francke U (1988) Chromosomal locations of TGF-β's 2 and 3 in man and mouse: dispersion of the TGF-β family. Oncogene Res 3:323–331

Bassols A, Massagué J (1988) Transforming growth factor-β regulates the expression and structure of extracellular matrix chondroitin/dermatan sulfate proteoglycans. J Biol Chem 263:3039–3045

Benahmed M, Cochet C, Keramidas M, Chauvin MA, Morera AM (1988) Evidence for a FSH dependent secretion of a receptor reactive transforming growth factor-β-like material by immature Sertoli cells in primary culture. Biochem Biophys Res Commun 154:1222–1231

Betsholtz C, Johnsson A, Heldin C-H, Westermark B, Lind P, Urdea MS, Eddy RR, Shows TB, Philpott K, Mellor AL, Knott TJ, Scott J (1986) cDNA sequence and chromosomal localization of human platelet-derived growth factor A-chain and its expression in tumor cell lines. Nature 320:695–699

Blake AG, Falanga V (1988) Serum modulates the effect of transforming growth factor beta on endothelial cells. J Invest Dermatol 90:547

Blomhoff HK, Smeland E, Mustafa AS, Godal T, Ohlsson R (1987) Epstein-Barr virus mediates a switch in responsiveness to transforming growth factor, type beta, in cells of the B cell lineage. Eur J Immunol 17:299–301

Boerner P, Resnick RJ, Racker E (1985) Stimulation of glycolysis and amino acid uptake in NRK-49F cells by transforming growth factor beta and epidermal growth factor. Proc Natl Acad Sci USA 82:1350–1353

Boyd FT, Massagué J (1989) Transforming growth factor-β inhibition of epithelial cell proliferation linked to the expression of a 53-kDa membrane receptor. J Biol Chem 264:2272–2278

Braun L, Mead JE, Panzica M, Mikumo R, Bell GT, Fausto N (1988) Transforming growth factor-β mRNA increases during liver regeneration: a possible paracrine mechanism of growth regulation. Proc Natl Acad Sci USA 85:1539–1543

Brunner AM, Gentry LE, Cooper JA, Purchio AF (1988) Recombinant type 1 transforming growth factor-β precursor produced in Chinese hamster ovary cells is glycosylated and phosphorylated. Mol Cell Biol 8:2229–2232

Cameron PM, Limjuco GA, Chin J, Silberstein L, Schmidt JA (1986) Purification to homogeniety and amino acid sequence analysis of two anionic species of human interleukin 1. J Exp Med 164:237–250

Carr BI, Hayashi I, Branum EL, Moses HL (1986) Inhibition of DNA synthesis in rat hepatocytes by platelet-derived type β transforming growth factor. Cancer Res 46:2330–2334

Carrington JL, Roberts AB, Flanders KC, Roche NS, Reddi HA (1988) Accumulation, localization and compartmentation of transforming growth factor-β during endochondral bone development. J Cell Biol 107:1969–1975

Cate RL, Mattaliano RJ, Hession C, Tizard R, Farber NM, Cheung A, Ninfa EG, Frey AZ, Gash DJ, Chow EP, Fisher RA, Bertonis JM, Torres G,Wallner BP, Ramachandran KL, Ragin RC, Manganaro TF, MacLaughlin DT, Donahoe PK (1986) Isolation of the bovine and human genes for Mullerian inhibiting substance and expression of the human gene in animal cells. Cell 45:685–698

Centrella M, McCarthy TL, Canalis E (1987) Transforming growth factor beta is a bifunctional regulator of replication and collagen synthesis in osteoblast-enriched cell cultures from fetal rat bone. J Biol Chem 262:2869–2874

Centrella M, McCarthy TL, Canalis E (1988) Skeletal tissue and transforming growth factor-β. FASEB J 2:3066–3073

Chambard JC, Pouysségur J (1987) TGF-β inhibits growth factor-induced DNA synthesis in hamster fibroblasts without affecting the early mitogenic events. J Cell Physiol 135:101–107

Cheifetz S, Weatherbee JA, Tsang MLS, Anderson JK, Mole JE, Lucas R, Massagué J (1987) The transforming growth factor-beta system, a complex pattern of cross-reactive ligands and receptors. Cell 48:409–415

Cheifetz S, Andres JL, Massagué J (1988a) The transforming growth factor-β receptor type III is a membrane proteoglycan. J Biol Chem 263:16984–16991

Cheifetz S, Bassols A, Stanley K, Ohta M, Greenberger J, Massagué J (1988b) Heterodimeric transforming growth factor-β. Biological properties and interaction with three types of cell surface receptors. J Biol Chem 263:10783–10789

Cheifetz S, Ling N, Guillemin R, Massagué J (1988c) A surface component on GH$_3$ pituitary cells that recognizes TGF-β, activin and inhibin. J Biol Chem 263:16984–16991

Chen J-K, Hoshi H, McKeehan WL (1987) Transforming growth factor-β specifically stimulates synthesis of proteoglycan in human arterial smooth muscle cells. Proc Natl Acad Sci USA 84:5287–5291

Chenu C, Pfeilschrifter J, Mundy GR, Roodman GD (1988) Transforming growth factor-β inhibits formation of osteoclast-like cells in long-term human marrow cultures. Proc Natl Acad Sci USA 85:5683–5687

Chiang CP, Nilsen-Hamilton M (1986) Opposite and selective effects of epidermal growth factor and human platelet transforming growth factor-β on the production of secreted proteins by murine 3T3 cells and human fibroblasts. J Biol Chem 261:10478–10481

Chinkers M (1987) Isolation and characterization of mink lung epithelial cell mutants resistant to transforming grwoth factor-β. J Cell Physiol 130:1–5

Cochet C, Feige J-J, Chambaz EM (1988) Bovine adrenocortical cells exhibit high affinity transforming growth factor-β receptors which are regulated by adrenocorticotropin. J Biol Chem 263:5707–5713

Coffey RJ, Kost LJ, Lyons RM, Moses HL, LaRusso NF (1987) Hepatic processing of transforming growth factor β in the rat. J Clin Invest 80:750–757

Coffey RJ, Bascom CC, Sipes NJ, Graves-Deal R, Weissman BE, Moses HL (1988a) Selective inhibition of growth-related gene expression in murine keratinocytes by transforming growth factor-β. Mol Cell Biol 8:3088–3093

Coffey RJ, Sipes NJ, Bascom CC, Graves-Deal R, Pennington CY, Weissman BE, Moses HL (1988b) Growth modulation of mouse keratinocytes by transforming growth factors. Cancer Res 48:1596–1602

Coffman RL, Lebman DA, Shrader B (1989) Transforming growth factor-β specifically enhances IgA production by lipolysaccharide-stimulated murine B lymphocytes. J Exp Med 170:1039–1044

Connor TB, Roberts AB, Sporn MB, Danielpour D, Dart LL, Michels RG, de Bustros S, Enger C, Glaser BM (1989) Correlation of fibrosis and transforming growth factor-beta type 2 levels in the eye. J Clin Invest 83:1661–1666

Cromack DT, Sporn MB, Roberts AB, Merino MJ, Dart LL, Norton JA (1987) Transforming growth factor-beta levels in rat wound chambers. J Surg Res 42:622–628

Czarniecki CW, Chiu HH, Wong GHW, McCabe SM, Palladino MA (1988) Transforming growth fractor-β1 modulates the expression in class II histocompatability antigens on human cells. J Immunol 140:4217–4223

Daniel TO, Gibbs VC, Milfay DF, Williams LT (1987) Agents that increase cAMP accumulation block endothelial c-*sis* induction by thrombin and transforming growth factor-β. J Biol Chem 262:11893–11896

Danielpour D, Dart LL, Flanders KC, Roberts AB, Sporn MB (1989) Immunodetection and quantitation of the two foms of transforming growth factor-beta (TGF-beta 1 and TGF-beta 2) secreted by cells in culture. J Cell Physiol 138: 79–86

Dean DC, Newby RF, Bourgeois S (1988) Regulation of fibronectin biosynthesis by dexamethasone, transforming growth factor-β, and cAMP in human cell lines. J Cell Biol 106:2159–2170

Dean DC, Newby RF, Bourgeois S (1989) Identification of transforming growth factor-β1 responsive elements in the fibronectin and SV40 early gene promoters (to be published)

de Martin R, Haendler B, Hofer-Warbinek R, Gaugitsch H, Wrann M, Schlusener H, Seifert JM, Bodmer S, Fontana A, Hofer E (1987) Complementary DNA for human glioblastoma-derived T cell suppressor factor, a novel member of the transforming growth factor-β gene family. EMBO J 6:3673–3677

De Larco JE, Todaro GJ (1978) Growth factors from murine sarcoma virus-transformed cells. Proc Natl Acad Sci USA 75:4001–4005

DePalma RL, Krummel TM, Nelson JM (1987) Fetal wound matrix is composed of proteoglycan rather than collagen. Surg Forum 38:626–628

Derynck R, Rhee L (1987) Sequence of the porcine transforming growth factor-beta precursor. Nucleic Acids Res 15:3187

Derynck R, Jarrett JA, Chen EY, Eaton DH, Bell JR, Assoian RK, Roberts AB, Sporn MB, Goeddel DV (1985) Human transforming growth factor-beta cDNA sequence and expression in tumor cell lines. Nature 316:701–705

Derynck R, Jarrett JA, Chen EY, Goeddel DV (1986) The murine transforming growth factor-beta precursor. J Biol Chem 261:4377–4379

Derynck R, Goeddel DV, Ullrich A, Gutterman JU, Williams RD, Bringman TS, Berger WH (1987a) Synthesis of messenger RNAs for transforming growth factors alpha and β and the epidermal growth factor receptor by human tumors. Cancer Res 47:707–712

Derynck R, Rhee L, Chen EY, Van Tilburg A (1987b) Intron-exon structure of human transforming growth factor-β precursor gene. Nucleic Acids Res 15:3188–3189

Derynck R, Lindquist PB, Lee A, Wen D, Tamm J, Graycar JL, Rhee L, Mason AJ, Miller DA, Coffey RJ, Moses HL, Chen EY (1988) A new type of transforming growth factor-β, TGF-β3. EMBO J 7:3737–3743

Distel RJ, Ro HS, Rosen BS, Groves DL, Spiegelman BM (1987) Nucleoprotein complexes that regulate gene expression in adipocyte differentiation: direct participation of c-fos. Cell 49:835–844

Dodson WC, Schomberg DW (1987) The effect of transforming growth factor-β on follicle-stimulating hormone-induced differentiation of cultured rat granulosa cells. Endocrinology 120:512–516

Dvorak HF (1986) Tumors: wounds that do not heal. N Engl J Med 315:1650–1659

Edwards DR, Murphy G, Reynolds JJ, Whitman SE, Docherty AJP, Angel P, Heath JK (1987) Transforming growth factor beta modulates the expression of collagenase and metalloproteinase inhibitor. EMBO J 6:1899–1904

Ellingsworth LR, Brennan JE, Fok K, Rosen DM, Bentz H, Piez KA, Seyedin SM (1986) Antibodies to the N-terminal portion of cartilage-inducing factor A and transforming growth factor beta. J Biol Chem 261:12362–12367

Ernst M, Schmid C, Frandenfelt C, Froesch ER (1988) Estradiol stimulation of osteoblast proliferation in vitro: mediator roles for TGF-β, PE₂, insulin-like growth factor (IGF)? Calcif Tissue Int 42 (Suppl):A30

Espevik T, Figari IS, Ranges GE, Palladino MA (1988) Transforming growth factor-β1 (TGF-β1) and recombinant human tumor necrosis factor-alpha reciprocally regulate the generation of lymphokine-activated killer cell activity. J Immunol 140:2312–2316

Evans RM (1988) The steroid and thyroid hormone receptor superfamily. Science 240:889–895

Falanga V, Tiegs SL, Alstadt SP, Roberts AB, Sporn MB (1987) Transforming growth factor-beta: selective increase in glycosaminoglycan synthesis by cultures of fibroblasts from patients with progressive systemic sclerosis. J Invest Dermatol 89:100–104

Fanger BO, Wakefield LM, Sporn MB (1986) Structure and properties of the cellular receptor for transforming growth factor type beta. Biochemistry 25:3083–3091

Feng P, Catt KJ, Knecht M (1986) Transforming growth factor β regulates the inhibitory actions of epidermal growth factor during granulosa cell differentiation. J Biol Chem 261:14167–14170

Feng P, Catt KJ, Knecht M (1988) Transforming growth factor-β stimulates meiotic maturation of the rat oocyte. Endocrinology 122:181–186

Feige J-J, Cochet C, Rainey WE, Madani C, Chambaz EM (1987) Type β transforming growth factor affects adrenocortical cell-differentiated functions. J Biol Chem 262:13491–13495

Fernandez-Pol JAS, Talkad VD, Klos DJ, Hamilton PD (1987) Suppression of the EGF-dependent induction of c-myc proto-oncogene expression by transforming growth factor-β in a human breast carcinoma cell line. Biochem Biophys Res Comm 144:1197–1205

Fiegel VD, Knighton DR (1988) Transforming growth factor-beta causes indirect angiogenesis by recruiting monocytes (Abstract) FASEB J 2:A1601

Fine LG, Holley RW, Nasri H, Badie-Dezfooly B (1985) BSC-1 growth inhibitor transforms a mitogenic stimulus into a hypertrophic stimulus for renal proximal tubular cells: relationship to Na⁺/H⁺ antiport activity. Proc Natl Acad Sci USA 82:6163–6166

Flanders KC, Roberts AB, Ling N, Fleurdelys BE, Sporn MB (1988) Antibodies to peptide determinants in transforming growth factor-beta and their applications. Biochemistry 27:739–746

Flanders KC, Thompson NL, Cissel DS, Ellingsworth LR, Roberts AB, Sporn MB (1989) Transforming growth factor β1: histochemical localization with antibodies to different epitopes. J Cell Biol 108:653–660

Florini JR, Roberts AB, Ewton DZ, Falen SL, Flanders KC, Sporn MB (1986) Transforming growth factor-beta. A very potent inhibitor of myoblast differentiation, identical to the differentiation inhibitor secreted by Buffalo rat liver cells. J Biol Chem 261:16 509–16 513

Forage RG, Ring JM, Brown RW, McInerney BV, Cobon GS, Gregson RP, Robertson DM, Morgan FJ, Hearn MTW, Findlay JK, Wettenhall REH, Burger HG, de Kretser DM (1986) Cloning and sequence analysis of cDNA species coding for the two subunits of inhibin from bovine follicular fluid. Proc Natl Acad Sci USA 83:3091–3095

Fràter-Schröder M, Müller G, Birchmeier W, Böhlen P (1986) Transforming growth factor-beta inhibits endothelial cell proliferation. Biochem Biophys Res Commun 137:295–302

Frolik CA, Dart LL, Meyers CA, Smith DM, Sporn MB (1983) Purification and initial characterization of a type beta transforming growth factor from human placenta. Proc Natl Acad Sci USA 80:3676–3680

Frolik CA, Wakefield LM, Smith DM, Sporn MB (1984) Characterization of a membrane receptor for transforming growth factor-β in normal rat kidney fibroblasts. J Biol Chem 259:10995–11 000

Fujii D, Brissenden JE, Derynck R, Franke U (1986) Transforming growth factor-β gene maps to human chromosome 19 long arm and to mouse chromosome 7. Somatic Cell Mol Genet 12:281–288

Gabrielson EW, Gerwin BI, Harris CC, Roberts AB, Sporn MB, Lechner JF (1988) Stimulation of DNA synthesis in cultured primary human mesothelial cells by specific growth factors. FASEB J 2:2717–2721

Gentry LE, Webb NR, Lim GJ, Brunner AM, Ranchalis JE, Twardzik DR, Lioubin MN, Marquardt H, Purchio AF (1987) Type 1 transforming growth factor beta: amplified expression and secretion of mature and precursor polypeptides in Chinese hamster ovary cells. Mol Cell Biol 7:3418–3427

Gentry LE, Lioubin MN, Purchio AF, Marquardt H (1988) Molecular events in the processing of recombinant type 1 pre-pro-transforming growth factor beta to the mature polypeptide. Mol Cell Biol 8:4162–4168

Haddow A (1972) Molecular repair, wound healing, and carcinogenesis: tumor production a possible overhealing? Adv Can Res 16:181–234

Hampson J, Ponting ILO, Roberts AB, Dexter TM (1988) The effects of TGF-β on haemopoietic cells. Growth Factors 1:193–202

Hanks SK, Armour R, Baldwin JH, Maldonado F, Spiess J, Holley RW (1988) Amino acid sequence of the BSC-1 cell growth inhibitor (polyergin) deduced from the nucleotide sequence of the cDNA. Proc Natl Acad Sci USA 85: 79–83

Heimark RL, Twardzik DR, Schwartz SM (1986) Inhibition of endothelial cell regeneration by type-beta transforming growth factor from platelets. Science 233:1078–1080

Heine UI, Flanders K, Roberts AB, Munoz EF, Sporn MB (1987) Role of transforming growth factor-β in the development of the mouse embryo. J Cell Biol 105:2861–2876

Heino J, Ignotz RA, Hemler ME, Crouse C, Massagué J (1989) Regulation of cell adhesion receptors by transforming growth factor-β. J Biol Chem 264:380–388

Hill DJ, Strain AJ, Elstow SF, Swenne I, Milner RDG (1986) Bi-functional action of transforming growth factor-β on DNA synthesis in early passage human fetal fibroblasts. J Cell Physiol 128:322–328

Hirake Y, Inoue H, Hirai R, Kato Y, Suzuki F (1988) Effect of transforming growth factor-β on cell proliferation and glycosaminoglycan synthesis by rabbit growth plate chondrocytes in culture. Biochim Biophys Acta 969:91–99

Holley RW, Böhlen P, Fava R, Baldwin JH, Kleeman G, Armour R (1980) Purification of kidney epithelial cell growth inhibitors. Proc Natl Acad Sci USA 77:5989–5992

Hotta M, Baird A (1986) Differential effects of transforming growth factor type β on the growth and function of adrenocortical cells *in vitro*. Proc Natl Acad Sci USA 83:7795–7799

Huang SS, O'Grady P, Huang JS (1988) Human transforming growth factor-β: alpha$_2$-macroglobulin complex is a latent form of transforming growth factor-β. J Biol Chem 263:1535–1541

Hunt TK, Twomey P, Zederfelt B, Dunphy JE (1976) Respiratory gas tensions and pH in healing wounds. Am J Surg 114:302–307

Hutchinson LA, Findlay JK, deVos FL, Robertson DM (1987) Effects of bovine inhibin, transforming growth factor-β and bovine activin-A on granulosa cell differentiation. Biochem Biophys Res Commun 146:1405–1412

Hynes RO (1987) Integrins: a family of cell surface receptors. Cell 48:549–554

Ignotz RA, Massagué J (1985) Type β transforming growth factor controls the adipogenic differentiation of 3T3 fibroblasts. Proc Natl Acad Sci USA 82:8530–8534

Ignotz RA, Massagué J (1986) Transforming growth factor-beta stimulates the expression of fibronectin and collagen and their incorporation into the extracellular matrix. J Biol Chem 261:4337–4345

Ignotz RA, Endo T, Massagué J (1987) Regulation of fibronectin and type I collagen mRNA levels by transforming growth factor-β. J Biol Chem 262:6443–6446

Ikeda T, Lioubin MN, Marquardt H (1987) Human transforming growth factor type β2; production by a prostatic adenocarcinoma cell line, purification, and initial characterization. Biochemistry 26:2406–2410

Inman WH, Colowick SP (1985) Stimulation of glucose uptake by transforming growth factor-beta: evidence for the requirement of epidermal growth factor receptor activation. Proc Natl Acad Sci USA 82:1346–1349

Ishibashi T, Miller SL, Burstein SA (1987) Type β transforming growth factor is a potent inhibitor of murine megakaryocytopoiesis in vitro. Blood 69:1737–1741

Jakowlew SB, Dillard PJ, Kondaiah P, Sporn MB, Roberts AB (1988a) Complementary deoxyribonucleic acid cloning of a novel transforming growth factor-β messenger ribonucleic acid from chick embryo chondrocytes. Mol Endocrinol 2:747–755

Jakowlew SB, Dillard PJ, Sporn MB, Roberts AB (1988b) Complementary deoxyribonucleic acid cloning of an mRNA encoding transforming growth factor-beta 4 from chicken embryo chondrocytes. Mol Endocrinol 2:1186–1195

Jakowlew SB, Dillard PJ, Sporn MB, Roberts AB (1988c) Nucleotide sequence of chicken transforming growth factor-beta 1 (TGF-β1). Nucleic Acids Res 16:8730

Jakowlew SB, Kondaiah P, Flanders KC, Thompson NL, Dillard PJ, Sporn MB, Roberts AB (1988d) Increased coordinate expression of growth factor mRNA accompanies viral transformation of rodent cells. Oncogene Res 2:135–148

Jennings JC, Mohan S, Linkhart TA, Widstrom R, Baylink DJ (1988) Comparison of the biological activities of TGF-beta 1 and TGF-beta 2: differential activity in endothelial cells. J Cell Physiol 137:167–172

Jetten AM, Shirley JE, Stoner G (1986) Regulation of proliferation and differentiation of respiratory tract epithelial cells by TGF-β. Exp Cell Res 167:539–549

Kehrl JH, Roberts AB, Wakefield LM, Jakowlew SB, Sporn MB, Fauci AS (1986a) Transforming growth factor beta is an important immunomodulatory protein for human B-lymphocytes. J Immunol 137:3855–3860

Kehrl JH, Wakefield LM, Roberts AB, Jakowlew SB, Alvarez-Mon M, Derynck R, Sporn MB, Fauci AS (1986b) Production of transforming growth factor beta by human T lymphocytes and its potential role in the regulation of T cell growth. J Exp Med 163:1037–1050

Kehrl JH, Taylor AS, Delsing GA, Roberts AB, Sporn MB, Fauci AS (1989) Further studies of the role of TGF-β in human B cell function. J Immunol 143:1868–1874

Keller JR, Mantel C, Sing GK, Ellingsworth LR, Ruscetti SK, Ruscetti FW (1988) Transforming growth factor-β1 selectively regulates early murine hematopoietic progenitors and inhibits the growth of IL-3 dependent myeloid leukemia cell lines. J Exp Med 168:737–750

Keller JR, Sing GK, Ellingsworth LR, Ruscetti FW (1989) Transforming growth factor-β: possible roles in the regulation of normal and leukemic hematopoietic cell growth. J Cell Biochem 39:79–84

Keski-Oja J, Lyons RM, Moses HL (1987) Immunodetection and modulation of cellular growth with antibodies against native transforming growth factor-β. Cancer Res 47:6451–6458

Kim S-J, Glick A, Sporn MB, Roberts AB (1989a) Characterization of the promoter region of the human transforming growth factor-β1 gene. J Biol Chem 264:402–408

Kim S-J, Jeang K-T, Glick A, Sporn MB, Roberts AB (1989b) Promoter sequences of the human TGF-β gene responsive to TGF-β1 autoinduction. J Biol Chem 264:7041–7045

Kimchi A, Wang X-F, Weinberg RA, Cheifetz S, Massagué J (1988) Absence of TGF-β receptors and growth inhibitory responses in retinoblastoma cells. Science 240:196–198

Kimelman D, Kirschner M (1987) Synergistic induction of mesoderm by FGF and TGF-β and the identification of an mRNA coding for FGF in the early Xenopus embryo. Cell:51:869–877

Knabbe C, Lippman ME, Wakefield LM, Flanders KC, Kasid A, Derynck R, Dickson RB (1987) Evidence that transforming growth factor-β is a hormonally regulated negative growth factor in human breast cancer cells. Cell 48:417–428

Knecht M, Feng P, Catt K (1987) Bifunctional role of transforming growth factor-β during granulosa cell development. Endocrinology 120:1243–1249

Komm BS, Terpening CM, Benz DJ, Graeme KA, Gallegos A, Korc M, Greene GL, O'Malley BW, Haussler MR (1988) Estrogen binding, receptor mRNA, and biologic response in osteoblast-like osteosarcoma cells. Science 241:81–84

Kondaiah P, Van Obberghen-Schilling E, Ludwig R, Dhar R, Sporn MB, Roberts AB (1988) cDNA cloning of porcine TGF-β: evidence for alternate splicing. J Biol Chem 263:18 313–18 317

Kondaiah P, Sands MJ, Smith JM, Fields A, Roberts AB, Sporn MB, Melton DA (1990) Identification of a novel transforming growth factor-β mRNA in *Xenopus laevis*. J Biol Chem (in press)

Kronke M, Leonard WJ, Depper JM, Arya SK, Wong-Staal F, Gallo RC, Waldmann TA, Greene WC (1984) Cyclosporin A inhibits T cell growth factor gene expression at the level of mRNA transcription. Proc Natl Acad Sci USA 81:5214–5218

Krummel TM, Michna BA, Thomas BL, Sporn MB, Nelson JM, Salzberg AM, Cohen IK, Diegelmann RF (1988) Transforming growth factor-beta (TGF-β) induces fibrosis in a fetal wound model. J Pediatr Surg 23:647–652

Kurokowa M, Lynch K, Podolsky DK (1987) Effects of growth factors on an intestinal epithelial cell line: TGF-β inhibits proliferation and stimulates differentiation. Biochem Biophys Res Commun 142:775–782

Lafyatis R, Remmers EF, Roberts AB, Yocum DE, Sporn MB, Wilder RL (1989) Anchorage-independent growth of synoviocytes from arthritic and normal joints: stimulation by exogenous platelet-derived growth factor and inhibition by transforming growth factor-beta and retinoids. J Clin Invest 83:1267–1276

Laiho M, Saksela O, Andreasen PA, Keski-Oja J (1986) Enhanced production and extracellular deposition of the endothelial-type plasminogen activator inhibitor in cultured human lung fibroblasts by transforming growth factor-β. J Cell Biol 103:2403–2410

Laiho M, Saksela, Keski-Oja J (1987) Transforming growth factor-β induction of type-1 plasminogen activator inhibitor. J Biol Chem 262:17 467–17 474

Lawrence DA, Pircher R, Jullien P (1985) Conversion of a high molecular weight latent beta-TGF from chicken embryo fibroblasts into a low molecular weight active beta-TGF under acidic conditions. Biochem Biophys Res Commun 133:1026–1034

Lawrence WT, Sporn MB, Gorschboth C, Norton JA, Grotendorst GR (1986) The reversal of an Adriamycin induced healing impairment with chemoattractants and growth factors. Ann Surg 203:142–147

Lehnert SA, Akhurst RJ (1988) Embryonic expression pattern of TGF-beta type 1 RNA suggests both paracrine and autocrine mechanisms of action. Development 104:263–273

Leof EB, Proper JA, Getz MJ, Moses HL (1986a) Transforming growth factor type β regulation of actin mRNA. J Cell Physiol 127:83–88

Leof EB, Proper JA, Goustin AS, Shipley GD, DiCorletto PE, Moses HL (1986b) Induction of c-cis mRNA and activity similar to platelet-derived growth factor by transforming growth factor-β: a proposed model for indirect mitogenesis involving autocrine secretion. Proc Natl Acad Sci USA 83:2453–2457

Libby J, Martinez R, Weber MJ (1986) Tyrosine phosphorylation in cells treated with transforming growth factor-β. J Cell Physiol 129:159–166

Like B, Massagué J (1986) The antiproliferative effect of type beta transforming growth factor occurs at a level distal from receptors for growth-activating factors. J Biol Chem 261:13 426–13 429

Lin P, Liu P, Tsao M-S, Grisham JW (1987) Inhibition of proliferation of cultured rat liver epithelial cells at specific cell cycle stages by transforming growth factor-β. Biochem Biophys Res Commun 143:26–30

Lin T, Blaisdell J, Haskell JF (1987) Transforming growth factor-β inhibits Leydig cell steroidogenesis in primary culture. Biochem Biophys Res Commun 146:387–394

Ling N, Ying SY, Ueno N, Shimasaki S, Esch F, Hotta M, Guillemin R (1986) Pituitary FSH is released by a heterodimer of the β-subunits from the two forms of inhibin. Nature 321:779–782

Liu J-M, Davidson JM (1988) The elastogenic effect of recombinant transforming growth factor-beta on porcine aortic smooth muscle cells. Biochem Biophys Res Comm 154:895–901

Lund LR, Riccio A, Andreasen PA, Nielsen LS, Kristensen P, Laiho M, Blasi F, Dano K (1987) Transforming growth factor-β is a strong and fast acting positive regulator of the level of type-1 plasminogen activator inhibitor mRNA in WI-38 human lung fibroblasts. EMBO J 6:1281–1286

Lyons RM, Keski-Oja J, Moses HL (1988) Proteolytic activation of latent transforming growth factor-β from fibroblast-conditioned medium. J Cell Biol 106:1659–1665

Madisen L, Webb NR, Rose TM, Marquardt H, Ikeda T, Twardzik D, Seyedin S, Purchio AF (1988) Transforming growth factor-β2: cDNA cloning and sequence analysis. DNA 7:1–8

Madri JA, Pratt BM, Tucker A (1988) Phenotypic modulation of endothelial cells by transforming growth factor-β depends upon the composition and organization of the extracellular matrix. J Cell Biol 106:1375–1384

Majack RA (1987) Beta-type transforming growth factor specifies organizational behavior in vascular smooth muscle cell cultures. J Cell Biol 105:465–471

Majesky MW, Reidy MA, Twardzik DR, Schwartz SM (1989) Production of type-1 transforming growth factor-β (TGF-β1) during repair of arterial injury. FASEB J 3:A398

Makela TP, Alitalo R, Paulsson Y, Westermark B, Heldin C-H, Alitalo K (1987) Regulation of platelet-derived growth factor gene expression by transforming growth factor β and phorbol ester in human leukemia cell lines. Mol Cell Biol 7:3656–3662

Mansbridge JN, Hanawalt PC (1988) Role of transforming growth factor-beta in the maturation of human epidermal keratinocytes. J Invest Dermatol 90:336–341

Mason AJ, Hayflick JS, Ling N, Esch F, Ueno N, Ying S-Y, Guillemin R, Niall H, Seeburg PH (1985) Complementary DNA sequences of ovarian follicular fluid inhibin show precursor structure and homology with transforming growth factor-β. Nature 318:659–663

Massagué J (1985) Subunit structure of a high-affinity receptor for type β-transforming growth factor. J Biol Chem 260:7059–7066

Massagué J, Like B (1985) Cellular receptors for type beta transforming growth factor. J Biol Chem 260:2636–2645

Massagué J, Kelly B, Mottola C (1985) Stimulation by insulin-like growth factors is required for cellular transformation by type beta transforming growth factor. J Biol Chem 260:4551–4554

Massagué J, Cheifetz S, Endo T, Nadal-Ginard B (1986) Type β transforming growth factor is an inhibitor of myogenic differentiation. Proc Natl Acad Sci USA 83:8206–8210

Masui T, Wakefield LM, Lechner JF, LaVeck MA, Sporn MB, Harris CC (1986) Type β transforming growth factor is the primary differentiation inducing serum factor for normal human bronchial epithelial cells. Proc Natl Acad Sci USA 83:2438–2442

Matrisian LM, Leroy P, Ruhlmann C, Gesnel M-C, Breathnach R (1986) Isolation of the oncogene and epidermal growth factor-induced transin gene: complex control in rat fibroblasts. Mol Cell Biol 6:1679–1686

McCartney-Francis N, Mizel D, Wong H, Wahl L, Wahl S (1988) Transforming growth factor-beta (TGF-β) as an immunoregulatory molecule (Abstract) FASEB J 2:A875

McDonald JA, Quade BJ, Broekelmann TJ, LaChance R, Forsman K, Hasegawa E, Akiyama S (1987) Fibronectin's cell-adhesive domain and an amino-terminal matrix assembly domain participate in its assembly into fibroblast pericellular matrix. J Biol Chem 262:2957–2967

McKeehan WL, Adams PS (1988) Heparin-binding growth factor/prostatropin attenuates inhibition of rat prostate tumor epithelial cell growth by transforming growth factor type beta. In Vitro Cell Dev Biol 24:243–246

McMahon JB, Richards WL, del Campo AA, Song M-K, Thorgiersson SS (1986) Differential effects of transforming growth factor-β on proliferation of normal and malignant rat liver epithelial cells in culture. Cancer Res 46:4665–4671

Mitjavila MT, Vicni G, Villeval JL, Kieffer N, Henri A, Testa U, Breton-Gorius J, Vainchenker W (1988) Human platelet alpha granules contain a nonspecific inhibitor of megakaryocyte colony formation: its relationship to type β transforming growth factor (TGF-β). J Cell Physiol 134:93–100

Miyazono K, Heldin C-H (1989) Interaction between TGF-β1 and carbohydrate structures in its precursor renders TGF-β1 latent. Nature 338:158–160

Miyazono K, Hellman U, Wernstedt C, Heldin C-H (1988) Latent high molecular weight complex of transforming growth factor β1. J Biol Chem 263:6407–6415

Montesano R, Orci L (1985) Tumor-promoting phorbol esters induce angiogenesis in vitro. Cell 42:469–477

Montesano R, Orci L (1988) Transforming growth factor-β stimulates collagen-matrix contraction by fibroblasts: implications for wound healing. Proc Natl Acad Sci USA 85:4894–4897

Morales TI, Roberts AB (1988) Transforming growth factor-β regulates the metabolism of proteoglycans in bovine cartilage organ cultures. J Biol Chem 263:12828–12831

Morera AM, Cochet C, Keramidas M, Chauvin MA, de Peretti E, Benahmed M (1988) Direct regulating effects of transforming growth factor-β on the Leydig cell steroidogenesis in primary cultures. J Steroid Biochem 30:443–448

Moses HL, Branum EL, Proper JA, Robinson RA (1981) Transforming growth factor production by chemically transformed cells. Cancer Res 41:2842–2848

Moses HL, Tucker RF, Leof EB, Coffey RJ, Halper J, Shipley GD (1985) Type-beta transforming growth factor is a growth stimulator and a growth inhibitor. In: Feramisco J, Ozanne B, Stiles C (eds) Cancer Cells, vol 3. Cold Spring Harbor, New York, pp 65–71

Mulder KM, Levine AE, Hernandez X, McKnight MK, Brattain DE, Brattain MG (1988) Modulation of c-myc by transforming growth factor-β in human colon carcinoma cells. Biochem Biophys Res Commun 150:711–716

Mulé JJ, Schwarz SL, Roberts AB, Sporn MB, Rosenberg SA (1988) Transforming growth factor-beta inhibits the in vitro generation of lymphokine-activated killer cells and cytotoxic T cells. Cancer Immunol Immunother 26:95–100

Müller G, Behrens J, Nussbaumer U, Böhlen P, Birchmeier W (1987) Inhibitory action of transforming growth factor-β on endothelial cells. Proc Natl Acad Sci USA 84:5600–5604

Murthy US, Anzano MA, Stadel JM, Grieg R (1988) Coupling of TGF-β induced mitogenesis to G-protein activation in AKR-2B cells. Biochem Biophys Res Commun 152:1228–1235

Mustoe TA, Pierce GF, Thomason A, Gramates P, Sporn MB, Deuel TF (1987) Transforming growth factor type beta induces accelerated healing of incisional wounds in rats. Science 237:1333–1336

Nakamura T, Tomita, Y, Hirai R, Yamaoka K, Kaji K, Ichihara A (1985) Inhibitory effect of transforming growth factor-β on DNA synthesis of adult rat hepatocytes in primary culture. Biochem Biophys Res Commun 133:1042–1060

Niitsu Y, Urushizaki Y, Koshida Y, Terui K, Mahara K, Kohgo Y, Urushizaki I (1988) Expression of TGF-beta gene in adult T cell leukemia. Blood 71:263–266

Noda M, Rodan GA (1987) Type β transforming growth factor (TGF-β) regulation of alkaline phosphatase expression and other phenotype related mRNAs in osteoblastic rat osteosarcoma cells. J Cell Physiol 133:426–437

Noda M, Yoon K, Prince CW, Butler WT, Rodan GA (1988a) Transcriptional regulation of osteopontin production in rat osteosarcoma cells by type β transforming growth factor. J Biol Chem 263:13916–13921

Noda M, Camilliere JJ, Rodan GA (1988b) Transforming growth factor type β promotes bone formation in vivo. J Cell Biol 107:48a

O'Connor-McCourt MD, Wakefield LM (1987) Latent transforming growth factor-β in serum. J Biol Chem 262:14090–14099

Ohta M, Greenberger JS, Anklesaria P, Bassols A, Massagué J (1987) Two forms of transforming growth factor-β distinguished by multipotential hematopoietic progenitor cells. Nature 329:539–541

Olson EN, Sternberg E, Hu JS, Spizz G, Wilcox C (1986) Regulation of myogenic differentiation by type beta transforming growth factor. J Cell Biol 103:1799–1805

Oreffo ROC, Mundy GR, Bonewald LF (1988) Osteoclasts activate latent transforming growth factor-beta and vitamin A treatment increases TGF-β activation. Calcif Tissue Int 42 (Suppl):A15

Overall CM, Wrana JL, Sodek J (1989) Independent regulation of collagenase, 72 kDa-progelatinase, and metalloendoproteinase inhibitor (TIMP) expression in human fibroblasts by transforming growth factor-β. J Biol Chem 264:1860–1869

Owens GK, Geisterfer AAT, Yang YW-H, Komoriya A (1988) Transforming growth factor-β-induced growth inhibition and cellular hypertrophy in cultured vascular smooth muscle cells. J Cell Biol 107:771–780

Padgett RW, St Johnston RD, Gelbart WM (1987) A transcript from a Drosophila pattern gene predicts a protein homologous to the transforming growth factor-beta family. Nature 325:81–84

Pearson CA, Pearson D, Shibahara S, Hofsteenge J, Chiquet-Ehrismann R (1988) Tenascin: cDNA cloning and induction by TGF-β. EMBO J 7:2677–2981

Penttinen RP, Kobayashi S, Bornstein P (1988) Transforming growth factor-β increases mRNA for matrix proteins both in the presence and in the absence of changes in mRNA stability. Proc Natl Acad Sci USA 85:1105–1108

Petkovich PM, Wrana JL, Grigoriadis AE, Heersche JNM, Sodek J (1987) 1,25-Dihydroxyvitamin D3 increases epidermal growth factor receptors and transforming growth factor-β-like activity in a bone-derived cell line. J Biol Chem 262:13424–13428

Pfeilschrifter J, Mundy GR (1987) Modulation of type β transforming growth factor activity in bone cultures by osteotropic hormones. Proc Natl Acad Sci USA 84:2024–2028

Pilatte Y, Bignon J, Lambre CR (1987) Lysosomal and cystosolic sialidases in rabbit alveolar macrophages: demonstration of increased lysosomal activity after in vivo activation with bacillus Calmette-Guerin. Biochim Biophys Acta 923:150–155

Pircher R, Jullien P, Lawrence DA (1986) β-Transforming growth factor is stored in human blood platelets as a latent high molecular weight complex. Biochem Biophys Res Commun 136:30–37

Postlethwaite AE, Keski-Oja J, Moses HL, Kang AH (1987) Stimulation of the chemotactic migration of human fibroblasts by transforming growth factor beta. J Exp Med 165:251–256

Racker E, Resnick RJ, Feldman R (1985) Glycolysis and methylaminoisobutyrate uptake in rat-1 cells transfected with ras or myc oncogenes. Proc Natl Acad Sci USA 82:3535–3538

Raghow R, Postlethwaite AE, Keski-Oja J, Moses HL, Kang AH (1987) Transforming growth factor-β increases steady state levels of type I procollagen and fibronectin messenger RNAs posttranscriptionally in cultured human dermal fibroblasts. J Clin Invest 79:1285–1288

Ranges GE, Figari IS, Espevik T, Palladino MA (1987) Inhibition of cytotoxic T cell development by transforming growth factor β and reversal by recombinant tumor necrosis factor alpha. J Exp Med 166:991–998

Rappolee DA, Brenner CA, Schultz R, Mark D, Werb Z (1988a) Developmental expression of PDGF, TGF-alpha, and TGF-β genes in preimplantation mouse embryos. Science 242:1823–1825

Rappolee DA, Mark D, Banda MJ, Werb Z (1988b) Wound macrophages express TGF-alpha and other growth factors in vivo: analysis of mRNA phenotyping. Science 241:708–712

Redini F, Lafuma C, Pujol J-P, Robert L, Hornebeck W (1988) Effect of cytokines and growth factors on the expression of elastase activity by human synoviocytes, dermal fibroblasts and rabbit articular chondrocytes. Biochem Biophys Res Commun 155:786–793

Reiss M, Sartorelli AC (1987) Regulation of growth and differentiation of human keratinocytes by type β transforming growth factor and epidermal growth factor. Cancer Res 47:6705–6709

Remmers EF, Lafyatis R, Yocum DE, Roberts AB, Sporn MB, Wilder RL (1989) Cytokines and growth regulation of synoviocytes from patients with rheumatoid arthritis and rats with streptococcal cell wall arthritis: antagonistic effects of PDGF and TGF-β's 1 and 2. Growth Factors (in press)

Ristow HJ (1986) BSC-1 growth inhibitor type β transforming growth factor is a strong inhibitor of thymocyte proliferation. Proc Natl Acad Sci USA 83:5531–5534

Rizzino A (1987) Appearance of high affinity receptors for type β transforming growth factor during differentiation of murine embryonal carcinoma cells. Cancer Res 47:4386–4390

Roberts AB, Sporn MB (1984) Cellular biology and biochemistry of the retinoids. In: Sporn MB, Roberts AB, Goodman DS (eds) The retinoids, vol 2. Academic, New York, pp 205–285

Roberts AB, Anzano MA, Lamb LC, Smith JM, Sporn MB (1981) New class of transforming growth factors potentiated by epidermal growth factor. Proc Natl Acad Sci USA 78:5339–5343

Roberts AB, Anzano MA, Meyers CA, Wideman J, Blacher R, Pan Y-C, Stein S, Lehrman SR, Smith JM, Lamb LC, Sporn MB (1983a) Purification and properties of a type beta transforming growth factor from bovine kidney. Biochemistry 22:5692:5698

Roberts AB, Frolik CA, Anzano MA, Sporn MB (1983b) Transforming growth factors from neoplastic and non-neoplastic tissues. Fed Proc 42:2621–2626

Roberts AB, Anzano MA, Wakefield LM, Roche NS, Stern DF, Sporn MB (1985) Type beta transforming growth factor: a bifunctional regulator of cellular growth. Proc Natl Acad Sci USA 82:119–123

Roberts AB, Sporn MB, Assoian RK, Smith JM, Roche NS, Wakefield LM, Heine UI, Liotta LA, Falanga V, Kehrl JH, Fauci AS (1986) Transforming growth factor type-beta: rapid induction of fibrosis and angiogenesis in vivo and stimulation of collagen formation in vitro. Proc Natl Acad Sci USA 83:4167–4171

Roberts AB, Flanders KC, Kondaiah P, Thompson NI, Van Obberghen-Schilling E, Wakefield L, Rossi P, de Crombrugghe B, Heine UL, Sporn MB (1988) Transforming growth factor β: biochemistry and roles in embryogenesis, tissue repair and remodeling, and carcinogenesis. Recent Prog Horm Res 44:157–197

Roberts CJ, Birkenmeier TM, McQuillan JJ, Akiyama SK, Yamada SS, Chen W-T, Yamada KM, McDonald JA (1988) Transforming growth factor-β stimulates the expression of fibronectin and of both subunits of the human fibronectin receptor by cultured human lung fibroblasts. J Biol Chem 263:4586–4592

Robey PG, Young MF, Flanders KC, Roche NS, Kondaiah P, Reddi AH, Termine JD, Sporn MB, Roberts AB (1987) Osteoblasts synthesize and respond to TGF-beta in vitro. J Cell Biol 105:457–463

Rook AH, Kehrl JH, Wakefield LM, Roberts AB, Sporn MB, Burlington DB, Lane HC, Fauci AS (1986) Effects of transforming growth factor β on the functions of natural killer cells: depressed cytolytic activity and blunting of inteferon responsiveness. J Immunol 136:3916–3920

Rosa F, Roberts AB, Danielpour D, Dart LL, Sporn MB, Dawid IB (1988) Mesoderm induction in amphibians: the role of TGF-β2-like factors. Science 239:783–786

Rosen DM, Stempien SA, Thompson AY, Seyedin PR (1988) Transforming growth factor-beta modulates the expression of osteoblast and chondroblast phenotypes in vitro. J Cell Physiol 134:337–346

Rossi P, Karsenty G, Roberts AB, Roche NS, Sporn MB, de Crombrugghe B (1988) A nuclear factor 1 binding site mediates the transcriptional activation of a type I collagen promoter by transforming growth factor-β. Cell 52:405–414

Ruegsegger Veit C, Assoian RK (1988) Identification of transforming growth factor-beta in human ovarian follicular fluid. Endocrinology 122 (Suppl):1227

Ruoslahti E, Pierschbacher MD (1986) Arg-Gly-Asp: a versatile cell recognition signal. Cell 44:517–518

Russell WE, Coffey RJ, Ouellette AJ, Moses HL (1988) Transforming growth factor beta reversibly inhibits the early proliferative response to partial hepatectomy in the rat. Proc Natl Acad Sci USA 85:5126–5130

Saksela O, Moscatelli D, Rifkin DB (1987) The opposing effects of basic fibroblast growth factor and transforming growth factor-beta on the regulation of plasminogen activator activity in capillary endothelial cells. J Cell Biol 105:957–963

Sandberg M, Vuorio T, Hirvonen H, Alitalo K, Vuorio E (1988a) Enhanced expression of TGF-β and c-fos mRNAs in the growth plates of developing human long bones. Development 102:461–470

Sandberg M, Autio-Harmainen H, Vuorio E (1988b) Localization and the expression of types I, III, and IV collagen, TGF-β1 and c-fos genes in developing human calvarial bones. Dev Biol 130:324–334

Segarini PR, Seyedin SM (1988) The high molecular weight receptor to transforming growth factor-β contains glycosaminoglycan chains. J Biol Chem 263: 8366–8730

Segarini PR, Roberts AB, Rosen DM, Seyedin SM (1987) Membrane binding characteristics of two forms of transforming growth factor-beta. J Biol Chem 262:14655–14662

Segarini PR, Rosen DM, Seyedin SM (1989) Binding of TGF-β to cell surface proteins varies with cell type. Mol Endocrinol 3:261–272

Seyedin PR, Segarini PR, Rosen DM, Thompson AY, Bentz H, Graycar J (1987) Cartilage-inducing factor-B is a unique protein structurally and functionally related to transforming growth factor-beta. J Biol Chem 262:1946–1949

Seyedin SM, Thomas TC, Thompson AY, Rosen DM, Piez KA (1985) Purification and characterization of two cartilage-inducing factors from bovine demineralized bone. Proc Natl Acad Sci USA 82:2267–2271

Seyedin SM, Thompson AY, Bentz H, Rosen DM, McPherson JM, Conti A, Siegel NR, Galluppi GR, Piez KA (1986) Cartilage-inducing factor-A. J Biol Chem 261:5693–5695

Sharples K, Plowman GD, Rose TD, Twardzik DR, Purchio AF (1987) Cloning and sequence analysis of simian transforming growth factor-β cDNA. DNA 6:239–244

Shipley GD, Tucker RF, Moses HL (1985) Type β transforming growth factor/growth inhibitor stimulates entry of monolayer cultures of AKR-2B cells into S phase after a prolonged prereplicative interval. Proc Natl Acad Sci USA 82:4147–4151

Shipley GD, Pittelkow MR, Wille JJ, Scott RE, Moses HL (1986) Reversible inhibition of normal human prokeratinocyte proliferation by type β transforming growth factor-growth inhibitor in serum-free medium. Cancer Res 46:2068–2071

Silberstein GB, Daniel CW (1987) Reversible inhibition of mammary gland growth by transforming growth factor-β. Science 237:291–293

Silver IA, Murrills RJ, Etherington DJ (1988) Microelectrode studies on the acid microenvironment beneath adherent macrophages and osteoclasts. Exp Cell Res 175:266–276

Skinner MK, Keski-Oja J, Osteen KG, Moses HL (1987) Ovarian thecal cells produce transforming growth factor-β which can regulate granulosa cell growth. Endocrinology 121:786–792

Slack JMW, Darlington BG, Heath JK, Godsave SF (1987) Mesoderm induction in early *Xenopus* embryos by heparin-binding growth factors. Nature 326:197–200

Smeland EB, Blomhoff HK, Holte H, Ruud E, Beiske K, Funderud S, Godal T, Ohlsson R (1987) Transforming growth factor type β (TGF-β) inhibits G1 to S transition, but not activation of human B lymphocytes. Exp Cell Res 171:213–222

Smiddy WE, Glaser BM, Green R, Connor TB, Roberts AB, Lucas R, Sporn MB (1989) Transforming growth factor beta – a biologic chorioretinal glue. Arch Opthalmol 107:577–580

Spiegel AM (1987) Signal transduction by guanine nucleotide binding proteins. Mol Cell Endocrinol 49:1–16

Sporn MB, Roberts AB (1985) Autocrine growth factors and cancer. Nature 313:745–747

Sporn MB, Roberts AB (1986) Peptide growth factors and inflammation, tissue repair, and cancer. J Clin Invest 78:329–332

Sporn MB, Roberts AB (1988a) Transforming growth factor-beta: new chemical forms and new biological roles. Biofactors 1:89–93

Sporn MB, Roberts AB (1988b) Peptide growth factors are multifunctional. Nature 332:217–219

Sporn MB, Roberts AB, Shull JH, Smith JM, Ward JM, Sodek J (1983) Polypeptide transforming growth factors: isolation from bovine sources and use for wound healing in vivo. Science 219:1329–1331

Sporn MB, Roberts AB, Wakefield LM, de Crombrugghe B (1987) Some recent advances in the chemistry and biology of transforming growth factor-beta. J Cell Biol 105:1039–1045

Sprugel KH, McPherson JM, Clowes AW, Ross R (1987) Effects of growth factors *in vivo*. I. Cell ingrowth into porous subcutaneous chambers. Am J Pathol 129:601–613

Starksen NF, Harsh GR, Gibbs VC, Williams LT (1987) Regulated expression of the platelet-derived growth factor A chain gene in microvascular endothelial cells. J Biol Chem 262:14381–14384

Stern DF, Roberts AB, Roche NS, Sporn MB, Weinberg RA (1986) Differential responsiveness of myc- and ras-transfected cells to growth factors: selective stimulation of myc-transfected cells by EGF. Mol Cell Biol 6:870–877

Strain AJ, Frazer A, Hill DJ, Milner RDG (1987) Transforming growth factor-β inhibits DNA synthesis in hepatocytes isolated from normal and regenerating rat liver. Biochem Biophys Res Commun 145:436–442

Stroobant P, Waterfield MD (1984) Purification and properties of porcine platelet-derived growth factor. EMBO J 12:2963–2967

Tashjian AH, Voelkel EF, Lazzaro M, Singer FR, Roberts AB, Derynck R, Winkler ME, Levine L (1985) Human transforming growth factors alpha and beta stimulate prostaglandin production and bone resorption in cultured mouse calvaria. Proc Natl Acad Sci USA 82:4535–4538

ten Dijke P, Hanson P, Iwata KK, Pieler C, Foulkes JG (1988 a) Identification of a new member of the transforming growth factor-β gene family. Proc Natl Acad Sci USA 85:4715–4719

ten Dijke P, Geurts van Kessel AHM, Foulkes JG, Le Beau MM (1988 b) Transforming growth factor-beta type 3 maps to human chromosome 14, region q23-q24. Oncogene 3:721–724

Thompson NL, Bazoberry F, Speir EH, Casscells W, Ferrans VJ, Flanders KC, Kondaiah P, Geiser AG, Sporn MB (1988) Transforming growth factor beta-1 in acute myocardial infarction in rats. Growth Factors 1:91–99

Thompson NL, Flanders KC, Smith M, Ellingsworth LR, Roberts AB, Sporn MB (1989) Expression of transforming growth factor-β1 in specific cells and tissues of adult and neonatal mice. J Cell Biol 108:661–669

Todaro GJ, Lee DC, Webb NR, Rose TM, Brown JP (1985) Rat type-alpha transforming growth factor: structure and possible function as a membrane receptor. In: Feramisco J, Ozanne B, Stiles C (eds) Cancer cells, vol 3. Cold Spring Harbor, New York, pp 51–58

Tsunawaki S, Sporn M, Ding A, Nathan C (1988) Deactivation of macrophages by transforming growth factor-β. Nature 334:260–262

Tucker RF, Branum EL, Shipley GD, Ryan RJ, Moses HL (1984a) Specific binding to cultured cells of ^{125}I-labeled type β transforming growth factor from human platelets. Proc Natl Acad Sci USA 81:6757–6761

Tucker RF, Shipley GD, Moses HL, Holley RW (1984 b) Growth inhibitor from BSC-1 cells closely related to platelet type beta transforming growth factor. Science 226:705–707

Vale W, Rivier J, Vaughan J, McClintock R, Corrigan A, Woo W, Karr D, Spiess J (1986) Purification and characterization of an FSH releasing protein from porcine ovarian follicular fluid. Nature 321:776–779

Van den Eijnden-van Raaij AJM, Koornneef I, van Zoelen EJJ (1988) A new method for high yield purification of type beta transforming growth factor from human platelets. Biochem Biophys Res Commun 157:16–23

Van Obberghen-Schilling E, Kondaiah P, Ludwig RL, Sporn MB, Baker CC (1987) Complementary deoxyribonucleic acid cloning of bovine transforming growth factor-β1. Mol Endocrinol 1:693–698

Van Obberghen-Schilling E, Roche NS, Flanders KC, Sporn MB, Roberts AB (1988) Transforming growth factor-β1 positively regulates its own expression in normal and transformed cells. J Biol Chem 263:7741–7746

Varga J, Rosenbloom J, Jimenez SA (1987) Transforming growth factor-β. (TGF-β) causes a persistent increase in steady-state amounts of type I and type III collagen and fibronectin mRNAs in normal human dermal fibroblasts. Biochem J 247:597–604

Wahl SM, Hunt DA, Wakefield LM, McCartney-Francis N, Wahl LM, Roberts AB, Sporn MB (1987) Transforming growth-factor beta (TGF-beta) induces monocyte chemotaxis and growth factor production. Proc Natl Acad Sci USA 84:5788–5792

Wahl SM, Hunt DA, Wong HL, Dougherty S, McCartney-Francis N, Wahl LM, Ellingsworth L, Schmidt JA, Hall G, Roberts AB, Sporn MB (1988) Transforming growth factor-β is a potent immunosuppressive agent that inhibits IL-1 dependent lymphocyte proliferation. J Immunol 140:3026–3032

Wakefield LM, Smith DM, Masui T, Harris CC, Sporn MB (1987) Distribution and modulation of the cellular receptor for transforming growth factor-beta. J Cell Biol 105:965–975

Wakefield LM, Smith DM, Flanders KC, Sporn MB (1988) Latent transforming growth factor-β from human platelets. J Biol Chem 263:7646–7654

Wakefield LM, Smith DM, Broz S, Jackson M, Levinson AD, Sporn MB (1989) Recombinant TGF-β1 is synthesized as a two component latent complex that shares some structural features with the native platelet latent TGF-β1 complex. Growth Factors 1:203–218

Wang EA, Rosen V, Cordes P, Hewick RM, Kriz MJ, Luxenberg DP, Sibley BS, Wozney JM (1988) Purification and characterization of other distinct bone-inducing factors. Proc Natl Acad Sci USA 85:9484–9488

Weeks DL, Melton DA (1987) A maternal mRNA localized to the vegetal hemisphere in Xenopus eggs codes for a growth factor related to TGF-β. Cell 51:861–867

Wilcox JN, Derynck R (1988) Developmental expression of transforming growth factors alpha and beta in mouse fetus. Mol Cell Biol 8:3415–3422

Wilke MS, Wille JJ, Pittelkow MR, Scott RE (1988) Biologic mechanisms for the regulation of normal human keratinocyte proliferation and differentiation. Am J Pathol 131:171–181

Wollenberg GK, Semple E, Quinn BA, Hayes MA (1987) Inhibition of proliferation of normal, preneoplastic, and neoplastic rat hepatocytes by transforming growth factor-β. Cancer Res 47:6595–6599

Wozney JM, Rosen V, Celeste AJ, Mitsock LM, Whitters MJ, Kriz RW, Hewick RM, Wang EA (1988) Novel regulators of bone formation: molecular clones and activities. Science 242:1528–1534

Wrana JL, Sodek J, Ber RL, Bellows CG (1986) The effects of platelet-derived transforming growth factor-β on normal human diploid gingival fibroblasts. Eur J Biochem 159:69–76

Wrana JL, Maeno M, Hawrylyshyn B, Yao K-L, Domenicucci C, Sodek J (1988) Differential effects of transforming growth factor-β on the synthesis of extracellular matrix proteins by normal fetal rat calvarial bone cell populations. J Cell Biol 106:915–924

Wrann M, Bodmer S, de Martin R, Siepl C, Hofer-Warbinek R, Frei K, Hofer E, Fontana A (1987) T cell suppressor factor from human glioblastoma cells is a 12.5 KD protein closely related to transforming growth factor-beta. EMBO J 6:1633–1636

Ying S-Y, Becker A, Baird A, Ling N, Ueno N, Esch F, Guillemin R (1986a) Type beta transforming growth factor (TGF-β) is a potent stimulator of the basal secretion of follicle stimulating hormone (FSH) in a pituitary monolayer system. Biochem Biophys Res Commun 135:950–956

Ying S-Y, Becker A, Ling N, Ueno N, Guillemin R (1986b) Inhibin and beta type transforming growth factor (TGF-β) have opposite modulating effects on the follicle stimulating hormone (FSH)-induced aromatase activity of cultured rat granulosa cells. Biochem Biophys Res Commun 136:969–975

Yoshimasa Y, Seino S, Whittaker J, Kakehi T, Kosaki A, Kuzuya H, Imura H, Bell GI, Steiner DF (1988) Insulin-resistant diabetes due to a point mutation that prevents insulin proreceptor processing. Science 240:784–787

CHAPTER 9

Interleukin-1

J. A. SCHMIDT and M. J. TOCCI

A. Introduction

It is now known that there are two classes of IL-1 molecules known as IL-1α and IL-1β, each encoded by a single gene. The IL-1 literature prior to 1984, when the first IL-1 cDNAs were cloned (LOMEDICO et al. 1984; AURON et al. 1984), has been comprehensively reviewed (DINARELLO 1984). This review will therefore largely focus on the literature from 1984 onward.

I. Cells Producing IL-1

Many cells have been described as releasing or containing IL-1 or molecules resembling IL-1. These include epithelial cells [keratinocytes, endothelial cells, mesangial cells, thymic epithelial cells (LUGER et al. 1981; P. T. LE et al. 1987; LOVETT and LARSEN 1988)], connective tissue cells [dermal fibroblasts, chondrocytes (OLLIVIERRE et al. 1986; I. M. LE et al. 1987], cells of neuronal origin [astrocytes, glioma cells (GIULIAN et al. 1986)], and leukocytes [mononuclear phagocytes, granulocytes, T and B lymphocytes, and natural killer (NK) cells (PISTOIA et al. 1986; ACRES et al. 1987a, b; LINDEMANN et al. 1988)]. Transformed cell lines including monocytic leukemia lines P388D1, J774, THP1, and U-937 (KRAKAUER and OPPENHEIM 1983; MATSUSHIMA et al. 1986a), Epstein-Barr virus-(EBV)-transformed human B lymphoblastoid lines (ACRES et al. 1987b), and transformed murine keratinocytes (PAM 212; LUGER et al. 1982) have also been described as producing IL-1 activity. The full panoply of IL-1-producing cells has been previously reviewed (OPPENHEIM and GERY 1982; DURUM et al. 1985).

While many cells appear capable of producing IL-1 (or at least molecules having biological activities similar to IL-1), the cells of the mononuclear phagocyte lineage appear to be the major source of IL-1, for two reasons. First, the mRNA for IL-1β in lipopolysaccharide-activated monocytes comprises approximately 5% of total polyadenylated RNA species (MARCH et al.; AURON et al. 1987; TOCCI et al. 1987) and therefore constitutes a major abundant mRNA species. Second, monocytes appear to have a unique ability to translocate IL-1 molecules across their plasma membrane (see Sect. D). While other cells such as cultured chondrocytes (OLLIVIERRE et al. 1986), keratinocytes (KUPPER et al. 1986), and granulocytes (LINDEMANN et al. 1988) have been shown to contain IL-1 mRNA, very sensitive techniques such as S1 nuclease protection have at times been required. Demonstration of IL-1-like bioactivity has also been difficult and has required the use of glutaraldehyde-fixed cells (see Sect. B. III) to which are added indicator cells responsive to subpicomolar concentrations of IL-1. Demonstration of IL-1 molecules by radioimmunoassay (RIA) or immunoblot techniques in nonmonocytic lines has, with few exceptions, been difficult.

Monocytes progressively lose the ability to produce IL-1 as they mature into macrophages (ELIAS et al.; WEWERS 1987; BERNAUDIN et al. 1988; BUR-CHETT et al. 1988). The major species of IL-1 produced by monocytes is IL-1β at both the mRNA and protein levels (MARCH et al. 1985; GRAY et al. 1986; MALISZEWSKI et al. 1988). Other cells such as cultured human keratinocytes contain equal or higher levels of IL-1α as compared to IL-1β mRNA (KUPPER et al. 1986).

mRNA for IL-1α and IL-1β has been identified by Northern analysis and in situ hybridization in normal human spleen and liver (TOVEY et al. 1988), rheumatoid synovium (DUFF et al. 1988), and the renal cortices of MRL-1pr mice with nephritis (BOSWELL et al. 1988a,b). IL-1β has been found by radioimmunoassay in rheumatoid synovial fluid (SYMONS et al. 1988). Detailed in situ hybridization studies in the C57BL/6 mouse showed IL-1 mRNA throughout the reticuloendothelial system (TAKACS et al. 1988). Polymerase chain reaction (PCR) methodology has also been used to examine tissue for IL-1α transcripts (RAPPOLEE et al. 1988). No increase in plasma IL-1β concentration was observed following intravenous administration of endotoxin (MICHIE et al. 1988). However, a recent report states that IL-1β levels were elevated in the plasma and synovial fluid of patients with rheumatoid arthritis (DUFF et al. 1988). The levels of circulating IL-1 in various diseases should become known as recently developed RIA methodology is introduced into clinical investigation.

II. Stimuli Capable of Releasing IL-1

IL-1 production in the laboratory usually involves the addition of lipo-polysaccharide (LPS), or its active component lipid A (LOPPNOW 1986), to cultured monocytes or macrophages. A minimum 100-fold induction of IL-1 mRNA and protein biosynthesis within 4 h is observed in human monocytes in response to as little as 10 ng/ml of LPS. A large number of other stimuli have been employed including activated T lymphocytes, mitogens, immune complexes, Fc fragments of human immunoglobulin G1 (IgG1; MORGAN et al. 1988), the toxic shock strain of *S. aureus,* muramyl dipeptides, particulates (silica, urate, zymosan, carageenan; DI GIOVINE et al. 1987), ionophores, phorbol diesters and adherence to plastic surfaces (reviewed by DURUM et al. 1985). Care has not always been taken to eliminate endotoxin contamination as contributing to the induction of IL-1 production by these other stimuli. Ultraviolet irradiation (ANSEL et al. 1984; 1988), C5a (OKUSAWA et al. 1987), tumor necrosis factor-α (TNF-α; DINARELLO et al. 1986b), granulo-cyte/macrophage colony stimulating factor (GM-CSF; LINDEMANN et al.

1988), IL-2 (NUMEROF et al. 1988), and IL-1, itself (DINARELLO et al. 1987; GHEZZI and DINARELLO 1988) have also been reported to stimulate IL-1 production. Interferon-γ while a macrophage activator, does not, by itself, stimulate IL-1 production (GERRARD et al. 1987). Evidence has been presented to suggest that certain of these stimuli may differentially regulate IL-1 production and secretion (GERY et al. 1981; see Sect. D). Agents such as prostaglandin E_2 (PGE_2) which cause an increase in cellular cyclic AMP levels appear to block IL-1 production posttranslationally (KNUDSEN et al. 1987). Ongoing study of the regulatory domains of IL-1 genes (Sect. C. II) will hopefully provide a clearer understanding of the mechanisms controlling IL-1 biosynthesis.

III. Control of IL-1 Production In Vivo

Recent data suggest that IL-1 induces the production of glucocorticoid hormones in vivo (BESEDOVSKY et al. 1986). This effect appears not to be due to a direct effect on the adrenocortex but rather is mediated by the release of adrenocorticotropic hormone (ACTH; WOLOSKI et al. 1985). The release of ACTH is due either to a direct effect of IL-1 on the pituitary (BERNTON et al. 1987) or the IL-1 mediated release of corticotropin-releasing factor (CRF; SAPOLSKY et al. 1987; BERKENBOSCH et al. 1987) at the level of the hypothalamus. The results of these studies have been discussed and compared by LUMPKIN (1987). Given that glucocorticoids inhibit IL-1 production (see Sect. C. II), the continued synthesis of IL-1 and certain IL-1 mediated effects (see Sect. G) may be blunted by a negative feedback loop consisting of CRF, ACTH, and hydrocortisol. These data are among the strongest arguing for physiological interplay between the immune and neuroendocrine systems.

B. Molecular Biology of IL-1α and IL-1β

I. Transcripts and Primary Amino Acid Sequences

Complementary DNAs (cDNAs) have been cloned for IL-1α and IL-1β from human (MARCH et al. 1985; AURON et al. 1984), murine (LOMEDICO et al. 1984; GRAY et al. 1986), bovine (MALISZEWSKI et al. 1988) and lapine (FURUTANI et al. 1985; MORI et al. 1988) cells. The mRNAs encoding IL-1α molecules are 2.0–2.3 kilobases (kb) in length while the mRNAs encoding IL-1β molecules are 1.6–1.7 kb in length. All of the mRNAs have large untranslated 3′ domains bearing the TTATTTAT consensus sequence identified by CAPUT et al. (1986). This sequence has been identified in a large number of lymphokine mRNAs and is believed to regulate mRNA half-life (COSMAN 1987).

Each of the translation products encodes a precursor molecule of 269–271 amino acids with a molecular weight of approximately 31 500 (Fig. 1). Hybrid selection and in vitro translation gives precursor proteins of similar molecular weight ($\sim 33\,000$). Similar values are obtained by immunoprecipitation or im-

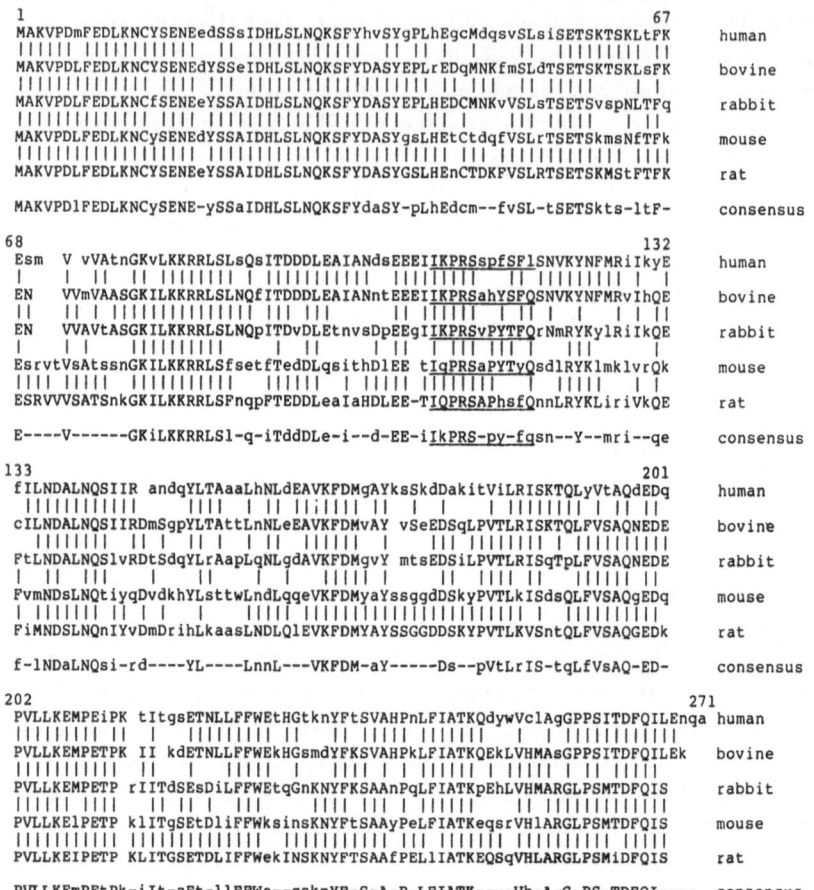

Fig. 1 a–c. Comparisons of IL-1α and IL-1β amino acid sequences form various species. **a** Comparison of IL-1α amino acid sequences from various species. The predicted amino acid sequences were derived from analysis of cloned cDNAs for all the species of IL-1 presented. The degree of IL-1α sequence homology was determined by computer analysis using the INTELLIGENTICS align program. Alignments were determined using the Regions homology algorithm. The sequence numbering is approximate, reflecting differences in the lengths of the proteins from the various species and the introduction of gaps in the sequences to achieve maximum alignment. The sequences encompassing the proteolytic processing sites are *underlined*. The single-letter amino acid code has been used. The consensus sequence shows the degree of conservation among species. *Capital letters* indicate entirely conserved residues; *lower-case letters* designate highly conserved residues or positions where conservative amino acid substitutions occur. **b** Comparison of IL-1β amino acid sequences from various species. The predicted amino acid sequences were derived from analysis of cloned cDNAs from the various species. Homologies were determined as described for **a**. The *underlined* residues encompass the proteolytic processing sites. Sequence numbering is approximate. **c** Comparison of IL-1α and IL-1β amino acid sequences from various species. Alignments were performed as described for **a**. The *consensus* sequence indicates residues which are entirely conserved in all the IL-1 molecules

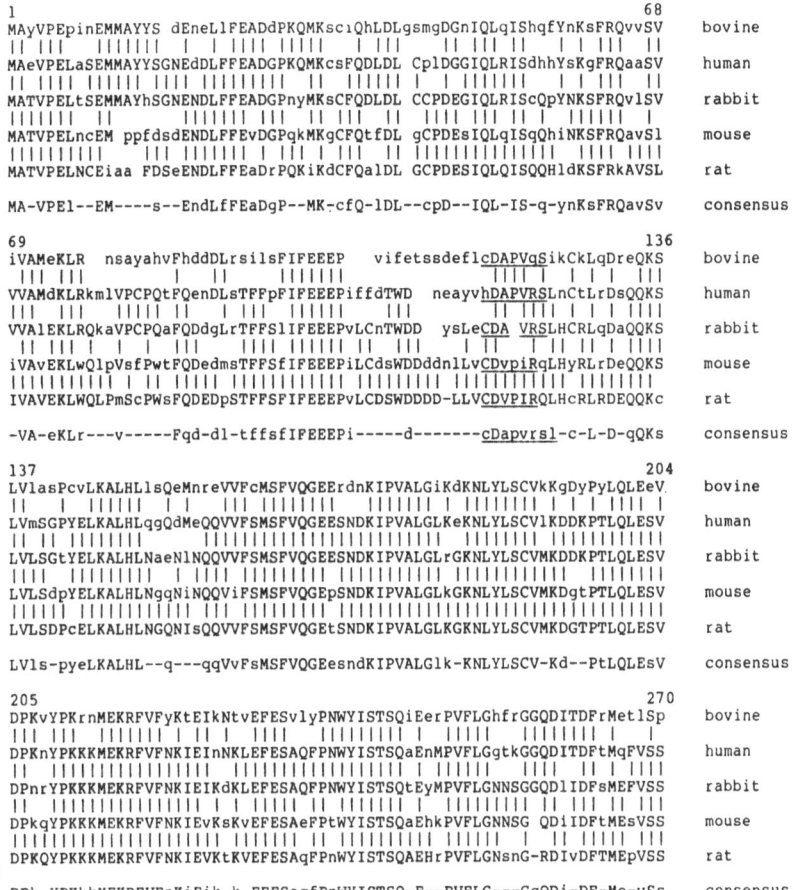

Fig. 1b

munoblot analysis of monocyte lysates (LIMJUCO 1986; GIRI et al. 1985). The identity of the deduced sequences has been confirmed in some cases by amino acid sequence of the native mature proteins purified from monocyte culture fluids (LOMEDICO et al. 1984; CAMERON et al. 1985, 1986; MARCH et al. 1985; MATSUSHIMA et al. 1986 b; GOTO et al. 1988).

There is a high degree (~70%) of conservation of primary amino acid sequence within either the IL-1α or IL-1β family of proteins (Fig. 1 A, B). However, comparison of any IL-1α sequence with any IL-1β sequence reveals a much lower (~25%), albeit significant, degree of homology (Fig. 1 C). This explains why with one exception (KOCK et al. 1986), antisera raised to either IL-1α or IL-1β fail to crossreact with the other species (CAMERON et al. 1986; MASSONE et al. 1988). Nevertheless, IL-1α and IL-1β molecules have similar specific biological activities (RUPP et al. 1986) on most target cells (see Sect. G) and are capable of being bound with similar affinities by all IL-1 receptors thus far identified (see Sect. E). This finding is reminiscent of the relationship

```
          MAKVPDlFEDLKNCYS   ENEdySSaIDHLSLNQKSFYdaSYGsLHEtCtDQfVSLrtSETSKmSnfT      mouse
          |||||| |||||||||   ||| || |||||||||||||  |||  ||  |    ||| ||||||| | |
          MAKVPDmFEDLKNCYS   ENEEdSSsIDHLSLNQKSFYhvSYGPLHEgCMDQsVSLSiSETSKtSkLT      human
          ||||||| ||||||||    ||| |||||||||||||||  ||||  ||   ||  ||| |||| ||||
   α      MAKVPDLFEDLKNCfS   ENEEYSSaIDHLSLNQKSFYDASYEPLHEDCMNKvVSLSTSETSvspnLT      rabbit
          |||||||||||||| |    ||| || ||||||||||||||| |  || |   ||  |||||| |    |
          MAKVPDLFEDLKNCYS   ENEdYSSeIDHLSLNQKSFYDASYEPLrEDqMNKfmSLdTSETSktskLs      bovine
          |||||||||||||||    ||| |||||||||||||||||      |||   ||  |||||||| | |
          MAKVPDLFEDLKNCYS   ENEeYSSAIDHLSLNQKSFYDASYGSLHEnCTDKFVSLRTSETSKMStFT      rat
          || ||                                                                

          MAyVPEpinEMMAYYS   dENELlFEADDPKQMKsciQhLDLgsmgDGnIQLqIShqfYnKsFRQvvSV      bovine
          ||  |||      |      |||  ||| | ||| |  ||   ||       || |||  || |||| || 
   β      MAeVPELaSEMMAYYSGNEdDLFFEADGPKQMKcsFQDLDL CplDGGIQLRISdhhYsKgFRQaaSV      human
          || |||| |||||  |     ||||| ||| ||  |||| |  || | |||||| ||  ||| |||  ||
          MATVPELtSEMMAYhSGNENDLFFEADGPnyMKsCFQDLDL CCPDEGIQLRIScQpYNKSFRQvlSV      rabbit
          ||||||| |||| |  |||  |||||||  |||  |||| |   ||||||||||  |||||||||| ||
          MATVPELncEM ppfdsdENDLFFEvDGPqkMKgCFQtfDL  gCPDEsIQLqISqQhiNKSFRQavSl      mouse
          ||||||||      || |  ||||||  || ||| |||  ||  || ||  |||| || |||||||| |
          MATVPELNCEiaa FDSeENDLFFEaDrPQKiKdCFQalDL  GCPDESIQLQISQQHldKSFRkAVSL      rat
          1                                                                    67
          MA-VP-----------E----------------------------------------------------  consensus
```

```
          FKESrVtvsATssNGKiLKKRRLSfSetfTeDDL  qsithdlEEt IQPRSaPytyqSdlrYklMklvr    mouse
                                                         ─────────
          ||||   |  || ||| |||||||| |   | ||         |   ||||| |  |        |  
          FKESmVV VAT NGKvLKKRRLSLSQsITDDDLEaianDsEEEIIKPRSsPfsFlSNvkYnfMRIIK      human
                                                         ─────────
          | |     ||   || | |||||||| ||| ||       |   ||||| |    | |     ||||
   α      FqE NVVaV  tASGKILKKRRLSLNQpITDvDLEtnvsDpEEgIIKPRSvPYtFQrNmrYkylRIIK      rabbit
                                                         ─────────
          | |  | ||| | ||||||||||||||| | | |||       || ||||| | | |    | |||
          FkE NVVmV  aASGKILKKRRLSLNQfITDdDLEaianntEEeIIKPRSahYsFQsNvkYnfmRvIh      bovine
                                                         ─────────
          | ||      | ||| | ||||||| |   | ||         |    ||||| |  |       || |
          FKESRVVVSATSnkGKILKKRRLSFnqpFTEDDLeaIaHDLEE-TIQPRSAPhsfQnnLRYKLiriVkQE   rat
                                                        ─────────
          |             |            |  |
          iVAMeKLR nsAyahvFhddDLrsilsFIFEEEP  vifEtssdeflcDAPVqSikCKLqDreQKS      bovine
                                                          ─────────
          ||| ||| |     | |||| |     | |||||          ||||| |   |||| || |||
          VVAMdKLRkmlVPCPQtFQenDLsTFFpFIFEEEPiffdTWD  neayvhDAPVRSLnCtLrDsQQKS      human
                                                          ─────────
          | || |||     ||| ||  |  |||  ||||||        | |    ||| ||| || | |||
   β      VVAlEKLRQkaVPCPQaFQDdgLrTFFSlIFEEEPvLCnTWDD  ysLeCDA VRSLHCRLqDaQQKS      rabbit
                                                          ─── ────
          | || ||     ||| | |  ||  ||  ||||||  |   |||     ||  |||  | |  ||||
          iVAvEKLwQlpVsfPwtFQDedmsTFFSfIFEEEPiLCdsWDDddnlLvCDvpiRqLHyRLrDeQQKS      mouse
                                                          ─────────
          | ||       ||  |   ||   ||| | |||||  ||  ||    |||||   ||| || ||||||
          IVAVEKLWQLPmScPWsFQDEDpSTFFSFIFEEEPvLCDSWDDDD-LLVCDVPIRQLHcRLRDEQQKc      rat
                                                          ─────────
          --------------------------------------------------------------------  136
                                                                              consensus
```

```
          qkFvmNDsLNQ tiyqdvdkhYLsttwLndLqqeVKFDMyAYsSggGDDsKyp VtLkISdsQLfVsAQg   mouse
           | | ||||||        |   ||   ||   | |||||  || |    ||  | | || ||| |||
          yEFiLNDALNQ S iiranDQYLtAAaLhNLdeAVKFDMGAYksSkDDaKit ViLRISktQLyVtAQd   human
          | | ||||||| |      ||| | | | || | ||||||  |   || || | || |||| || |||
   α      QEFtLNDALNQ SlvRDtSDQYLrAApLqNLgdAVKFDMG Ymtse EDSiL PVTLRISqTpLFVSAQN   rabbit
          || |||||||   |  | |||| |  | || | ||||||      ||  | | |||||| | ||||||
          QEciLNDALNQ SiiRDmSgpYLtAttLnNLeEAVKFDM VAYvse EDSqL PVTLRISktQLFVSAQN   bovine
          ||  |||||    || |   ||  |  | || |||||||      ||      |||||| | ||||||
          QEFiMNDsLNQnIYvDmDr ihLkaasLNDLQlE VKFDMY AYsSGGGDDsKYPVTLKVSntQLFVSAQG   rat
           |      ||            |        |
          LVlasPcvLKALHLlsQeMnreVVFcMSFVQGEErdnKIPVALGiKdKNLYLSCVkKgDyPyLQLEeV     bovine
          ||    |  ||||||| |  | | || |||||||   |||||| | |||||||||  | |  |||| |
          LVmSGPYELKALHLqgQdMeQQVVFSMSFVQGEESNDKIPVALGKeKNLYLSCV1KDDKPTLQLESV     human
          || |  |||||||| | |  | |||||||||||||||||||| | |||||||||  |||| |||||||
   β      LVLSGtYELKALHLNaeN1NQQVVFSMSFVQGEESNDKIPVALGlrGKNLYLSCVMKDDKPTLQLESV     rabbit
          |||   |||||||| |   | |||||||||||||||||||||   |||||||||| |||| ||||||||
          LVLSdpYELKALHLNgqNiNQQViFSMSFVQGEpSNDKIPVALGLkGKNLYLSCVMKDgtPTLQLESV     mouse
          |||   |||||||| |  | |||| |||||||| ||||||||| | ||||||||||||| |||||||||
          LVLSDPcELKALHLNGQNiSQQVVFSMSFVQGEtSNDKIPVALGLKGKNLYLSCVMKDGTPTLQLESV     rat
          --------L-----------------------------------------------------------  203
                                                                              consensus
```

```
          EDQPVLLKElPEtPKlITGSETdLiFFWksinsKNYFTSaAyPeLFIATKeqsrVhLArGlPSmTDFQIs    mouse
          ||||||||| ||| |||||||| | ||||       ||||| ||| ||||||     ||| | || |||||
          EDQPVLLKEMPEiPKtITGSETnLLFFWEThGtKNYFTSvAhPnLFIATKqdywVcLAgGpPSiTDFQIlenqa human
          || |||||| || ||  | || | ||||||   || ||| || | ||||||      |||| || |||||
   α      EDEPVLLKEMPETPrIITdSEsdiLFFWETqGnKNYFKSaAnPqLFIATKpEhLVHMArGlPSmTDFQ      rabbit
          || |||||| ||| | | | ||  |||||||  ||||| | | | ||||||    |||| || |||||||
          EDEPVLLKEMPETPKII kdEtNlLFFWEkhGsmdYFKSvAhPkLFIATKgEkLVHMAsGpPSiTDFQIS     bovine
          || |||||| |||| ||    |  ||||||   |  |||  | |  ||||||  | |||| || |||||||
          EDkPVLLKEIPETPKLITGSETdLIFFWEKiNSKNYFTSAAfPELlIATKEQSqVHLARGLPSMiDFQIS     rat
              |                |                |
          DPKvYPKrnMEKRFVFyKtEIkNtvEFESvlyPNWYISTSQiEerPVFLGhfrGGQDITDFrMeTlSp     bovine
          |||  |  ||||| || || ||  |  ||||  | |||||||||   ||||||    ||||| | |  | 
          DPKnYPKKKMEKRFVFNKIEInNKLEFESaQFPNWYISTSQaEnMPVFLGgtkGGQDITDFtMqFVSS     human
          ||  ||||||||||||||||||  || |||| ||||||||| ||  ||||||    |||||| | ||||
   β      DPnrYPKKKMEKRFVFNKIEIKdKLEFESaQFPNWYISTSQtEyMPVFLGNNSGGQDlIDFsMEFVSSIlek   rabbit
          ||  ||||||||||||||||| | || |||| |||||||| |  ||||||     || | | | ||||
          DPkqYPKKKMEKRFVFNKIEVKsKvEFESAePtWYISTSQaEhkPVFLGNNSG QDiIDFtMEsVSS       mouse
          ||  ||||||||||||||| | |  |||| | |||||||| ||   |||||   ||  ||| || |||
          DPKQYPKKKMEKRFVFNKIEVKtKVEFESAqFPnWYISTSQAEHrPVFLGNsnG-RDIvDFTMEPVSS       rat
          ----------------------------F--------Y------------------------------  271
                                                                              consensus
```

c

Fig. 1c

between epidermal growth factor and transforming growth factor-α and between the acidic and basic fibroblast growth factors. A low but statistically significant degree of primary sequence homology has been noted between human IL-1β and acidic fibroblast growth factor (GIMENEZ-GALLEGO et al. 1985), but these molecules are not recognized by the same receptors (CHIN et al. 1987).

The primary amino acid sequences of IL-1 molecules are noteworthy in that they lack conserved cysteine residues, asparagine-linked carbohydrate addition sites, and typical hydrophobic leader sequences. IL-1 molecules are similar to the fibroblast growth factors in this latter respect. Chemical modification or replacement of two cysteine residues in human IL-1β had no effect on biological activity (MASUI et al. 1988).

II. Precursor-Product Relationships

Pulse-chase experiments performed in murine macrophages (GIRI et al. 1985) and human monocytes (HAZUDA et al. 1988), as well as immunoblots of human monocytes and their culture fluids (BAYNE et al. 1986; LIMJUCO et al. 1986), have shown that IL-1α and IL-1β are present intracellularly as 31- to 33-kDa precursors and extracellularly as 17.5-kDa mature species. Amino-terminal sequence analysis of mature IL-1α and IL-1β has shown that each is derived from the carboxy-terminal domain of its respective precursor. The amino terminus of mature murine IL-1α was found to be Ser115 (LOMEDICO et al. 1984) while that of mature human IL-1α was found to be Leu119 (CAMERON et al. 1986). The bonds broken to yield these termini, Arg114-Ser115 in the case of murine IL-1α and Phe118-Leu119 in the case of human IL-1α, are consistent with processing by a tryptase and a chymotryptase, respectively. Both of these studies, however, reported the purification of other mature IL-1α species of similar molecular weight whose amino termini were not susceptible to Edman degradation. The amino termini of these other species remain unknown as does the amino terminus of native mature bovine and rabbit IL-1α. The carboxy terminus of human IL-1α has been shown to be Ala271 by mass spectroscopy of a carboxy-terminal peptide (WINGFIELD 1987 b).

The amino terminal of mature native human IL-1β is reported by several groups to be uniquely Ala117 (CAMERON et al. 1985; VAN DAMME et al. 1985; DEWHIRST et al. 1985; MATSUSHIMA et al. 1986 b), while MARCH et al. (1985) report this to be the major amino terminus among several less abundant species. The same amino terminus was observed for rabbit IL-1β (GOTO et al. 1988). These results suggest that the terminus of mature IL-1β molecules results from the cleavage of the Asp116-Ala117 bond, a bond not typically susceptible to cleavage by eukaryotic proteases. Native mature murine and bovine IL-1β have not been purified and therefore their amino termini have not been determined. The carboxy terminus of human IL-1β has been shown to be Ser269 by mass spectroscopy (WINGFIELD et al. 1986).

III. Recombinant IL-1 Molecules and Mutants

The availability of full-length cDNAs for IL-1α and IL-1β precursor molecules has permitted the expression of recombinant IL-1 molecules of various lengths in order to determine the minimum sequence required for maximal biological and receptor binding activity. Expression of residues 115–270 or residues 131–270 of murine IL-1α gave material with specific biological activity equivalent to that of mature native IL-1α (GUBLER et al. 1986; DeCHIARA et al. 1986). However, a more extensive amino-terminal deletion (i.e., residues 144–270) or a carboxy-terminal deletion (i.e., residues 131–257) resulted in at least a 100-fold reduction in biological and receptor binding activity (DeCHIARA et al. 1986). These authors also reported that point mutations made throughout the 131–257 "core" sequence led to further losses of biological activity. In the case of human IL-1α, expression of residues 113–271 or residues 119–271 have resulted in molecules with full biological activity (MARCH et al. 1985; WINGFIELD 1987b; CHIN et al. 1988). Charge heterogeneity exhibited by mature human IL-1α is the result of deamidation (CAMERON et al. 1986; WINGFIELD et al. 1987b). In vitro transcription-translation of the full-length precursor and its mature 17 500 molecular weight form showed that they had equivalent biological and receptor binding activities (KRONHEIM et al. 1986; MOSLEY et al. 1987a, b). This finding has suggested that pre-IL-1α need not be processed in order to be fully active. As in the case of murine IL-1α, however, amino-terminal deletions extending beyond residue 127 or carboxy-terminal deletions terminating before 266 resulted in a steep loss of biological and binding activity (MOSLEY et al. 1987a). Insertion of a five amino acid peptide at position 185 also resulted in loss (30000-fold) of biological activity (ZURAWSKI et al. 1986).

The expression of residues 118–269 of murine IL-1β by HUANG (1988) gave material with a specific biological activity on murine thymocytes that was 10-fold better than reported for recombinant murine IL-1α, residues 115–270, by DeCHIARA et al. (1986). The precursor of murine IL-1β has not been purified or expressed. In the case of human IL-1β, extensive expression work has been done. Unlike the human IL-1α precursor, the human IL-1β precursor has negligible biological or receptor binding activity. This has been shown by testing purified recombinant protein (BLACK et al. 1988; HAZUDA et al. 1989) or the products of in vitro transcription-translation reactions (MOSLEY et al. 1987a, b; JOBLING et al. 1988). Thus the IL-1β precursor must be processed in order to be active (BLACK et al. 1988). Expression and purification of residues 117–269 of human IL-1β gave material with the same specific biological activity and binding activity as native human IL-1β (MARCH et al. 1985; WINGFIELD et al. 1986; TOCCI et al. 1987). Amino-terminal deletions extending beyond residue 120 or carboxy-terminal deletions terminating prior to residue 266 result in a minimum 10-fold reduction in biological and receptor binding activity (MOSLEY et al. 1987a; JOBLING et al. 1988; LILLQUIST et al. 1988). Mutation of Arg120 to Glu120 in human IL-1β reduced binding and biological activity 100-fold while mutation of Ala117-Pro118 to Thr117-Met118 or

Thr117-Pro118 resulted in 7- and 3.8-fold increases in specific biological activity, respectively (HUANG et al. 1987). The des-Ala117 and Met-Ala117 forms of human IL-1β are reported to be 3- to 10-fold less active than the native Ala117 form (WINGFIELD et al. 1987a; YEM et al. 1988). In another report, conservative amino acid substitutions at residues 140, 184, 187, 206, 236–237 had no effect on binding activity. In contrast, three different substitutions at the unique histidine residue at position 146 resulted in a 2- to 100-fold loss in competitive binding activity despite the absence of any conformational change in the mutant proteins as determined by two-dimensional nuclear magnetic resonance (NMR) (MACDONALD et al. 1986).

Another approach to determining the minimal sequence required for full biological activity has been taken by BLACK et al. (1988). Recombinant IL-1β precursor was purified and then cleaved with several proteases. The digests were assayed for biological activity and sequenced following purification by gel electrophoresis. The authors reported that cleavage between residues 75 and 77 by trypsin and between residues 103 and 104 by pancreatic elastase resulted in a 7- to 10-fold increase in specific biological activity as compared to the undigested precursor. Cleavage between residues 111 and 112 by *Staphylococcus aureus* protease resulted in a 300-fold increase whereas cleavage between residues 113 and 114 by chymotrypsin gave a 500-fold increase.

In summary, the above studies suggest that both IL-1α and IL-1β both have core sequences, at least 120 amino acids in length, which are required for detectable biological activity. Contiguous stretches of about 145 residues are required for full activity. Another interpretation of the same results is that the amino- and carboxy-terminal domains are both essential to biological activity and that the function of the intervening residues is to properly orient the termini in space. This latter interpretation is not inconsistent with X-ray crystallography data (see Sect. B. V) or results with antibodies (see Sect. B. IV). However, the earlier proposal that IL-1 activity can be mainly attributed to a single domain made up of a short contiguous stretch of amino acids (DINARELLO et al. 1984a; AURON et al. 1985; ROSENWASSER et al. 1986) now seems unlikely on the basis of more recent studies. Expression of larger IL-1α molecules appears to have no detrimental effect on biological activity whereas extension of the amino terminus of human IL-1β rapidly leads to loss of biological activity (DINARELLO et al. 1986; BLACK et al. 1988). Thus the IL-1β precursor requires processing at or near the Asp116-Ala117 bond in order to obtain full activity. With the exception of a study by PALASZYNSKI (1987; see Sect. B. IV), another consistent finding in these studies is that binding and biological activity track together, i.e., none of the inactive molecules behave as receptor antagonists.

IV. IL-1 Peptides

Peptides of human IL-1α (residues 193–201 and 256–271) and IL-1β (117–128, 197–215, 258–269, and 257–271) have been synthesized for purposes of raising monoclonal antibodies and heterologous antisera (Conlon et al. 1987; Limjuco et al. 1986). While these antisera have proven useful for a variety of applications they did not block biological or binding activity of either ligand. A study by Massone et al. (1988) suggested that monoclonal antibodies raised to mature IL-1β (117–269) recognized regions 133–148 and 251–269 of human IL-1β and blocked biological activity in two different assays. The same study reported that antibodies recognizing residues 218–243 had a moderate inhibitory effect while a single antibody recognizing residues 148–192 had no inhibitory effect. Much higher concentrations of these antibodies were required, however, as compared to affinity-purified antibodies raised in rabbits (Conlon et al. 1987) and large differences (greater than 4-fold) in the inhibitory potency of the active mAbs were not observed. The epitopes recognized by other neutralizing IL-1-specific monoclonal antibodies have not been mapped (Kock et al. 1986; Tanaka et al. 1987; Kasahara et al. 1987; Kenney et al. 1987; Fuhlbrigge et al. 1988; Stya et al. 1988).

Palaszynski (1987) synthesized residues 172–196 and 237–269 of human IL-1β. While these had no intrinsic biological activity, the carboxyteminal peptide was reported to antagonize mature IL-1β in the thymocyte bioassay and bind to murine T lymphocytes with an equilibrium dissociation constant [K_d] of 2 nM. Both peptides were able to block binding of ^{125}I-IL-1β with IC$_{50}$ values of 10^{-6} and $10^{-7}M$, respectively. These findings have not been confirmed. No other peptides have been reported to block the biological or binding activity of IL-1 molecules. Ferreira et al. (1988) reported that a series of human IL-1β peptides (121–134, 140–153, 156–174, 187–204, 190–202, 190–192, 193–195, and 225–243) failed to show agonist or antagonist properties in vitro or in the standardized rabbit pyrogen assay. However, several of these peptides were found to induce hyperalgesia in the rat paw when administered intraperitoneally and peptide 193–195 (KPT), as well as its analog Lys-D-Pro-Thr, given subcutaneously prior to intraperitoneal IL-1β, was found to block IL-1β-evoked hyperalgesia. The mechanism of inhibition remains unclear. Residues 163–171 of human IL-1β (VQGEESNDK) have been reported to exhibit some of the immunostimulatory properties of IL-1β in mice without causing fever or a series of other inflammation-associated changes inducible by the intact IL-1 molecule (Boraschi et al. 1988; Frasca et al. 1988).

V. Structural Studies

The availability of large amounts of recombinant mature human IL-1β has permitted the detailed study of its three-dimensional structure. Hydrodynamic studies have indicated that the protein is nearly spherical (Wingfield et al. 1986) and circular dichroism spectroscopy suggested that the molecule was a member of the all-β class with no α-helical secondary structure (Craig et al. 1987). The circular dichroism spectra of murine and human IL-1β were found to be similar (Huang et al. 1988). Crystals of IL-1β have been obtained by several groups (Gilliland et al. 1987; Carter et al. 1988; Schar et al. 1987) and the structure of the the molecule has been determined at a resolution of 3.0 Å with a crystallographic R-factor of 42.3% (PRIESTLE et al. 1988). PRIESTLE et al. (1988) likened the core of the structure to a distorted tetrahedron whose edges are each formed by two antiparallel β-strands, i.e., each triangular face of the tetrahedron is made up of three pairs of antiparallel β-strands. The IL-1β precursor has been expressed and purified (BLACK et al. 1988; HAZUDA et al. 1989). Studies with proteases and alkylating reagents suggest that the precursor is in an open conformation, in marked contrast to the compact mature IL-1β molecule (HAZUDA et al. 1989). The structure of mature IL-1α has not yet been reported but, like IL-1β, it is found to be a compact structure based on its hydrodynamic properties (WINGFIELD et al. 1987b).

C. Gene Structure of IL-1α and IL-1β

I. Intron-Exon Organization

IL-1α and IL-1β are each the products of single genes. Genomic sequences for human IL-1α (FURUTANI et al. 1986) as well as for murine (TELFORD et al. 1986) and human IL-1β (CLARK et al. 1986; BENSI et al. 1987) have been cloned and sequenced. The intron-exon organization of the genes encoding IL-1α and IL-1β are very similar (see Fig. 2). Both genes consist of seven exons and six introns and the intron-exon boundaries are well conserved (Table 1). In both genes, exon 1 encodes most of the 5' untranslated domain while exons

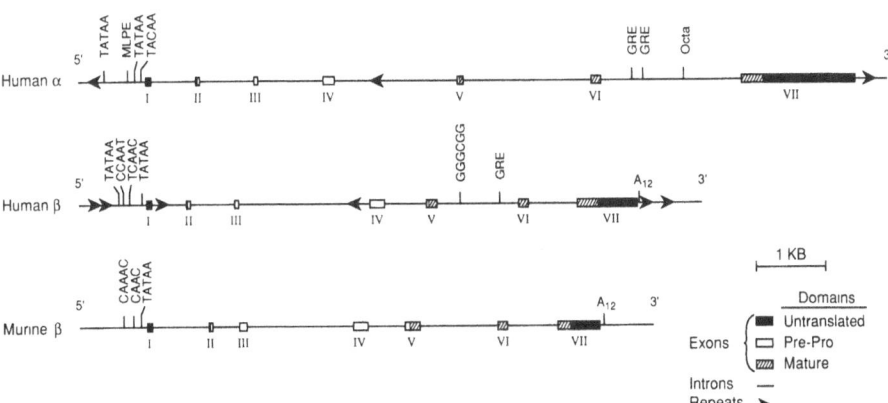

Fig. 2. Organization of IL-1 genomic sequences. The relative size and intron-exon arrangements of the human IL-1α, human IL-1β and murine IL-1β genes are shown. Starting from the transcription intiation sites, the human IL-1α gene is 10.2 kb, the human IL-1β 7.5 kbp, and the murine IL-1β 7.0 kb in length. Exons 1–7 (*I–VII*) correspond to those described in Table 1. The positions of potential TATA and CAAT boxes in the 5' upstream regions of the genes are marked. Other designations are as follows: *GGGCGG,* GC box or SP1 consensus binding site; *Octa,* octamer binding site; *GRE,* glucocorticoid responsive element; *MLPE,* Adenovirus major late promoter element binding site; A_{12}, polydA stretch

Table 1. Exon coding format of IL-1 genomic sequences

Exon	Amino acid residues (no. of residues)		
	Human IL-1α	Human IL-1β	Murine IL-1β
1	5'UT (0)	5'UT (0)	5'UT (0)
2	1– 16 (16)	1– 16 (16)	1– 16 (16)
3	17– 32 (16)	17– 33 (17)	17– 32 (16)
4	33–106 (74)	34–100 (67)	33–100 (68)
5	107–163 (57)	101–155 (55)	101–156 (56)
6	164–205 (42)	156–199 (44)	157–200 (44)
7	206–271 (66)	200–269 (70)	201–270 (70)
	3'UT (0)	3'UT (0)	3'UT (0)
Total amino acids	271	269	270

UT, Untranslated.

2–4 encode the first 106 amino acids of the precursor. Residues 107–271 as well as the untranslated 3' domain are encoded by exons 5–7. The human and murine IL-1β genes show extensive sequence homology in both the exon and intron regions (BENSI et al. 1987). The similarity in organization argues that the IL-1α and IL-1β genes have a common ancestral origin. Indeed, CLARK et al. (1986) have suggested that the IL-1β gene arose by reverse transcription and insertion of an IL-1α-like gene because of the existence of a primordial polyA tail at the 3' end of the human IL-1β gene. No clear association between exon-intron boundaries and functional domains within IL-1 proteins has yet been identified, although the possibility has been raised (AURON et al. 1985).

The genes for murine IL-1α and IL-1β are closely linked on chromosome 2 (D'EUSTACHIO et al. 1987) and the gene for human IL-1β has been localized to the long arm of chromosome 2 at 2q13-2q21 (WEBB et al. 1986).

II. Genomic Regulatory Elements

A TATA box consensus sequence is present at position 31 with respect to the transcription start (CAP) site in the human IL-1β gene and typical CAAT box elements are located at positions −75 and −126. The human IL-1α gene has a poor TATA box consensus sequence (TACAAA) at position −31 and no recognizable CAAT box. This may help to explain the lower IL-1α mRNA levels found in LPS-activated monocytes (see Sect. A. 1). Glucocorticoid regulatory elements are present within intron 6 of the IL-1α gene and intron 5 of the IL-1β gene, perhaps accounting for the effects glucocorticoids have on transcription of the IL-1 genes (see Sect. C. III). Transcriptional elements involved in the LPS activation of the IL-1 genes have not been identified as yet. The 5' upstream regions of both human genes contain sequence homologies to the immunoglobulin core enhancer octamer binding site and E motif (CLARK et al. 1988). Functional analysis of these elements have not yet been performed.

Constructs containing 513 and 1097 bp of upstream sequence from the human IL-1β gene were able to direct CAT (chloramphenicol acetyl transferase) expression in a cell-type-specific manner (CLARK et al. 1988). Electrophoretic band shift assays suggested that cells capable of expressing the IL-1β gene possess nuclear factors capable of binding upstream sequences conserved among IL-1 genes (CLARK et al. 1988). Work is ongoing in several laboratories to identify, using DNA footprinting techniques, those DNA sequences and DNA binding proteins which mediate activation of IL-1 genes by LPS and phorbol esters.

III. Transcriptional Regulation

Stimulation of human monocytes (MATSUSHIMA 1986d; VERMEULEN et al. 1987), THP1 cells (FENTON et al. 1987), and U-937 cells (S. W. LEE et al. 1988) with LPS and other stimuli results in a rapid (2–4 h) and substantial (100-fold) rise in IL-1β mRNA levels. The increase in mRNA levels is due in part to an increased rate of transcription (FENTON et al. 1987; S. W. LEE et al. 1988) and also to an increase in mRNA stability (FENTON et al. 1988). Transcription is blocked by treatment with 100 nM dexamethasone (KNUDSEN 1987; S. W. LEE et al. 1988; LEW et al. 1988). Other effects of dexamethasone on IL-1β mRNA include reduction in half-life (S. W. LEE et al. 1988) and other posttranscriptional effects (KERN et al. 1987; KNUDSEN et al. 1987). Dexamethasone reduces serum levels of IL-1-like activity in vivo (STARUCH and WOOD 1985).

D. Cell Biology of IL-1 Secretion

I. Cellular Physiology

As described in Sect. B, IL-1α and IL-1β are translated as 31.5-kDa precursors which are subsequently processed into 17.5-kDa mature molecules. Processing of the IL-1β precursor is required in order to obtain activity whereas the IL-1α precursor may already be fully active (MOSLEY et al. 1987a, b). In general, the precursors are found intracellularly while the mature species are found in culture supernatants (GIRI et al. 1985; BAYNE et al. 1986; LIMJUCO et al. 1986; AURON et al. 1987; HAZUDA et al. 1988). While low molecular weight species have on occasion been observed in cell lysates (MATSUSHIMA et al. 1986d), this may be due to nonspecific proteolysis during the extraction procedure (AURON et al. 1987; HAZUDA et al. 1988). Similarly, precursor material has been observed extracellularly but chase of radioactivity from extracellular precursor into mature molecules fails to occur (HAZUDA et al. 1988; FUHLBRIDGE et al. 1988). This has led to the suspicion that the precursor material found in culture fluids does not contribute to the formation of mature molecules and may result from cell leakage in culture. Thus the available evidence suggests that maturation of IL-1 precursor molecules occurs synchronously with passage from the interior to the exterior of the cell indicative of a coupled processing-translocation mechanism. The half-life of secretion of human IL-1β from human monocytes was reported to be about three hours whereas that for IL-1α appeared to be significantly longer (HAZUDA et al. 1988).

II. Molecular Biological Studies

A body of information has accumulated suggesting that IL-1 molecules are not secreted via the lumen of the endoplasmic reticulum, Golgi apparatus, and secretory vesicles as classically described by PALADE (1975). Interestingly, IL-1α and IL-1β precursors lack hydrophobic domains typical of signal peptides (Fig. 1). In vitro translation of murine IL-1α mRNA in the presence of dog pancreatic microsomes similarly failed to identify a functional leader sequence (LOMEDICO et al. 1984) such as that concealed in ovalbumin (PALMITER et al. 1978). Light microscopic and electron microscopic techniques have shown that IL-1β is distributed throughout the cytosolic ground substance of LPS-activated monocytes (BAYNE et al. 1986; SINGER et al. 1988) with little or no staining of endoplasmic reticulum lumen, Golgi, or secretory vesicles. COS-7 cells transfected with DNA encoding human IL-1β were unable to secrete IL-1β (MARCH et al. 1985). A Chinese hamster fibroblast cell line stably transfected with DNA encoding human IL-1β (YOUNG et al. 1988) resulted in the accumulation of IL-1β precursor in the cytosol without secretion of IL-1β molecules into the culture medium. Transformation of mouse L cells with cDNAs encoding murine pro-IL-1α or murine pro-IL-1β similarly failed to result in the secretion of IL-1 molecules into the culture medium (FUHLBRIDGE et al. 1988). However, CORBO et al. (1987) claimed that murine L cell transfectants were able to secrete small amounts of mature IL-1β. IL-1 molecules are not glycoproteins although IL-1β has a potential N-linked glycosylation site at Asn123. However, expression of IL-1β in yeast as a fusion protein with the α-mating factor leader sequence resulted in both glycosylation and secretion (BALDARI et al. 1987; ERNST 1988). These data taken together suggest that IL-1 precursor molecules accumulate in the cytosol following translation and that translocation into the culture medium does not involve passage into the ER lumen. Rather, a posttranslational translocation (VERNER and SCHATZ 1988) mechanism appears to be operative in monocytes which involves passage across the plasma membrane of the cell. Whereas posttranslational translocation of molecules from the cytoplasm into mitochondria, peroxisomes, and chloroplasts is well documented (VERNER and SCHATZ 1988), there are no prior examples involving the plasma membrane in eukaryotes (with perhaps the exception of factor A in yeast). Translocation of molecules into these various intracellular compartments is mediated by organelle-specific leader sequences which target the protein to its ultimate destination in the cell by interacting with organellar receptors. We hypothesize that the function of the long amino acid sequences preceding mature IL-1α (i.e., residues 1–112) and IL-1β (residues 1–116) is to target IL-1 molecules to an, as yet, unidentified translocation apparatus on the inner leaflet of the plasma membrane of monocytes.

III. Membrane IL-1

The importance of membrane recognition events in lymphocyte activation and the realization that IL-1 was capable of activating certain populations of lymphocytes, (see Sect. G) led investigators to search for the presence of IL-1 molecules on the surface of monocytes and macrophages (KURT-JONES et al. 1985a, 1986; WEAVER and UNANUE 1986). Surface or membrane IL-1 is generally assayed by briefly fixing IL-1-producing cells with low concentrations of glutaraldehyde, washing, and mixing with T lymphocytes capable of proliferating in response to IL-1. The activity is said to be on the surface because fixation presumably prevents the release of soluble IL-1 from the interior of the cell, though this has recently been called into question (MINNICH-CARRUTH et al. 1989). Membrane IL-1 activity has been found on a large number of activated cells including human monocytes (HURME 1987), B lymphocytes (KURT-JONES et al. 1985b), dendritic cells (NAGELKERKEN et al. 1986), fibroblasts (J. M. LE et al. 1987), and endothelial cells (KURT-JONES et al. 1987a). In support of the membrane localization of IL-1, IL-1 activity has been found in the membrane, crude particulate, and cytosolic fractions of monocytes and macrophages (MATSUSHIMA et al. 1986d; BAKOUCHE et al. 1987; BEUSCHER et al. 1987) and treatment of activated monocytes with trypsin or plasmin was found to release IL-1 activity (MATSUSHIMA et al. 1986d). IL-1 activity has been extracted from membrane preparations with detergent but not concentrated salt or EDTA (KURT-JONES et al. 1985a; BEUSCHER et al. 1987). Membrane IL-1 activity can be blocked with neutralizing antibodies raised to IL-1α (KURT-JONES et al. 1987a; BEUSCHER et al. 1987; FUHLBRIDGE et al. 1988) but not antisera to IL-1β (KURT-JONES et al. 1987a). CONLON et al. (1987) reported that IL-1α but not IL-1β could be identified on the surface of human monocytes using monoclonal antibodies and flow cytometry. Similarly, BAYNE et al. (1986) and SINGER et al. (1988) did not find appreciable levels of IL-1β on the surface of human monocytes using immunofluorescence and electron microscopic techniques. BEUSCHER et al. (1987) and FUHLBRIDGE et al. (1988) reported the presence of [^{35}S]methionine-labeled IL-1α in the plasma membrane fraction of murine peritoneal macrophages and mouse L cells transfected with DNA encoding IL-1α. HAZUDA et al. (1988), however, using the same protocol for membrane isolation as the others, could not detect internally labeled IL-1α or IL-1β in human monocyte membranes and stated that the material found on the crude particulate fraction chased at the same rate as the cytosolic material. HAZUDA et al. (1988) therefore concluded that the material found in the particulate fraction represented cytosolic contamination. BEUSCHER et al. (1987) further reported that IL-1α precursor, but not lysozyme, could be surface labeled on murine peritoneal macrophages and immunoprecipitated. These authors also labeled an approximately 80-kDa protein whose size is similar to IL-1 receptors found on other types of cells (see Sect. E). FUHLBRIGGE et al. (1988) reported that mouse L cells transfected with IL-1α DNA expressed membrane IL-1 activity which could be completely neutralized with a monoclonal antibody directed against IL-1α. Cells transfected with an IL-1β DNA expressed mRNA but no surface activity. These results, taken together, have led a number of investigators to propose that pre-IL-1α is exclusively responsible for surface activity whereas soluble IL-1 activity is due to a mixture of mature IL-1α and IL-1β molecules. A similar hypothesis has recently also been proposed for TNF-α (KRIEGLER et al. 1988). Difficulties with the membrane IL-1 hypothesis include (a) the inability to chase IL-1 molecules through a membrane compartment on their way to the medium (HAZUDA et al. 1988; FUHLBRIDGE et al. 1988) and (b) a biochemical anchoring mechanism for molecules that have no membrane-spanning domain, for example, via linkage to phosphatidylinositol glycans (LOW et al. 1986). However, the recent observation that IL-1α and IL-1β are myristoylated in human monocytes (BURSTEN et al. 1988), that pre-IL-1α is phosphorylated in human monocytes (KOBAYASHI et al. 1988) and that murine IL-1α can be labeled with D[^{14}C]mannose (BRODY and DURUM 1988) may provide mechanisms for membrane association. In any event, the possibility still exists that membrane IL-1 represents the diffusion of IL-1 out of glutaraldehyde-permeabilized cells and that membrane localization represents binding to conventional IL-1 receptors. No contribution to "surface activity" would be expected from IL-1β if IL-1β diffuses out of the fixed cells in the form of an inactive precursor.

E. IL-1 Receptors

I. Binding Studies

The ability of IL-1 to trigger a broad range of biological events at low picomolar concentrations (see Sect. F) has led to the search for IL-1 receptors (IL-1R) on a large number of different cell types. IL-1R have now been documented on leukocytes including T lymphocytes (SHIRAKAWA et al. 1987; LICHTMAN et al. 1988), T-cell lines (DOWER et al. 1985a, b; 1986; KILIAN et al 1986; LOWENTHAL and MACDONALD 1986, 1987; HORUK et al. 1987; BIRD et al. 1987), B-lymphoblastoid lines (MATSUSHIMA et al. 1986c; HORUK et al. (1987), neutrophils (RHYNE et al. 1988), and a large granular lymphocyte line (MATSUSHIMA et al. 1986b); connective tissue cells including lung fibroblasts (CHIN et al. 1987), gingival fibroblasts (QWARNSTROM et al. 1988), 3T3 cells (DOWER et al. 1985a; MIZEL et al. 1987), synovial fibroblasts (BIRD and SAK-LATVAKA 1987; CHIN et al. 1988), and chondrocytes (BIRD and SAKLATVAKA 1986); epithelial cells including keratinocytes (KUPPER et al. 1988; BLANTON et al. 1989) and endothelial cells (THIEME et al. 1987); and in the parenchyma of rat brain (FARRAR et al. 1987a, b). Most cells studied have a few thousand IL-1R per cell with the highest values ($\sim$30 000/cell) having been observed on certain lymphoid lines (LOWENTHAL and MACDONALD 1986; HORUK et al. 1987). Murine thymocytes, while biologically responsive to IL-1, and freshly isolated peripheral blood lymphocytes have very few sites per cell (10–100; DOWER et al. 1985b; SHIRAKAWA et al. 1987). The reported K_d values for IL-1R vary widely (1 pM to 10 nM) and in some reports seem inappropriately high in light of the very low concentrations of IL-1 required for biological activity (DOWER et al. 1985a). The explanation for these discrepancies is not yet clear. The subject of IL-1R has been previously reviewed (DOWER and URDAL 1987).

Despite the low sequence homology between IL-1α and IL-1β (~25%) all IL-1R identified to date are capable of binding both IL-1α and IL-1β with similar affinity. This is in keeping with the observation that IL-1α and IL-1β are equipotent in most in vitro assay systems (RUPP et al. 1986). In general, Scatchard analyses of binding data obtained on most cell types are linear and therefore consistent with a single population of high-affinity sites. However, the identification of high- and low-affinity sites on the same cell has been reported for a murine T-lymphoid line, EL-4 (LOWENTHAL and MACDONALD 1986, 1987) and keratinocytes (KUPPER et al. 1988). Difficulty in detecting high-affinity sites, which are present at only 100–200 sites per cell, may in part explain the seemingly low affinity of IL-1R on cells responsive to low picomolar concentrations of IL-1.

The consistent ability of IL-1α and IL-1β to each completely block the binding of the other has led to the proposal that the receptors for IL-1α and IL-1β are identical (DOWER et al. 1985b, 1986; KILIAN et al. 1986; MATSUSHIMA et al. 1986c; CHIN et al. 1987) and that the IL-1R found on different cell types are similar (DOWER et al. 1985a, 1986). However, HORUK et al. (1987) reported that IL-1R on a human B-lymphoblastoid line (Raji) and a murine T-lymphoid line (EL-4) differed in terms of their molecular weights and their ability to bind mutants of human IL-1β. Similarly, BIRD et al. (1987) reported that IL-1R on murine EL-4 cells and 3T3 cells bound porcine IL-1α with similar affinity but differed 20-fold in terms of their affinity for IL-1β. While these findings need to be extended to diploid cells and membrane preparations of freshly isolated tissues, they suggest that IL-1R subtypes may exist.

Crosslinking experiments have been performed on a number of cells in order to estimate the molecular weight(s) of the IL-1R. IL-1R, whether crosslinked with ^{125}I-IL-1α or ^{125}I-IL-1β, behave as monomers upon reduction and, in general, have exhibited molecular weights of about 80000 (DOWER et al. 1985a, b; BIRD and SAKLATVALA 1986; CHIN et al. 1987; BRON and MACDONALD 1987), although the IL-1R on EBV-transformed human B-lymphoblastoid lines has a molecular weight of about 62000 (MATSUSHIMA et al. 1986c; HORUK et al. 1987). This IL-1R may represent an IL-1R subtype but appears to be abnormal in that it internalizes poorly (HORUK et al. 1987). An additional IL-1R with a molecular weight of about 100000 has been found on murine cells by some authors and has led to the proposal that IL-1R are made up of two noncovalently linked subunits (BIRD et al. 1987). Bands with corrected molecular weights of 115000 (DOWER et al. 1985b) and 200000 (CHIN et al. 1987) have also been noted in single studies.

II. Cell Biology

IL-1R are expressed constitutively on most cells and, in general, their number does not appear to modulate in response to most external stimuli. However, human peripheral blood mononuclear cells have been reported to express 10-fold higher levels of IL-1R in response to glucocorticoids (AKAHOSHI et al. 1988) and and this effect was blocked by actinomycin D or cycloheximide. SHIRAKAWA et al. (1987) similarly reported that stimulation of human T lymphocytes with concanavalin A resulted in a 10-fold increase in IL-1R, albeit to a low level ($\sim$400 sites/cell). Culture of EL-4 cells (MIZEL et al. 1987) or YT cells (MATSUSHIMA et al. 1986b) in the presence of IL-1 resulted in an 80% reduction in the number of IL-1R. IL-1 bound to IL-1R is internalized with a half-life of about 1 h and remains intact intracellularly for several hours (MIZEL et al. 1987; QWARNSTROM et al. 1988). Internalization does not appear to require the accumulation of IL-1R in coated pits (QWARNSTROM et al. 1988). While some of the internalized label is degraded in lysosomes (MATSUSHIMA et al. 1986b), some of the label becomes associated with the cell nucleus (MIZEL et al. 1987; QWARNSTROM et al. 1988). These authors postulate that uptake of IL-1 into the nucleus may be important in stimulating transcription of IL-1-regulated genes. Evidence has been presented suggesting that a portion of IL-1 receptors recycle to the surface of the cell (LOWENTHAL and MACDONALD 1986; BIRD and SAKLATVALA 1987b) but this is controversial (MIZEL et al. 1987).

III. Biochemistry and Molecular Biology

IL-1R have been solubilized in detergent with little or no loss in binding activity (MATSUSHIMA et al. 1986c; PAGANELLI et al. 1987; URDAL et al. 1988). Treatment with N-glycanase has shown that approximately 25% of the 80 000 molecular weight receptor is made up of asparagine-linked carbohydrate (BRON and MACDONALD 1987; URDAL et al. 1988). The solubilized murine receptor is sensitive to chemical modification of arginine quanidinium groups, lysine ε-amino groups, and carboxy groups (URDAL et al. 1988). Sensitivity of the solubilized receptor to reducing agents is controversial (cf. PAGANELLI-PARKER and KILIAN 1988 with URDAL et al. 1988).

Purification of IL-1R from EL-4 cells was achieved by affinity chromatography on IL-1α coupled to Sepharose (URDAL et al. 1988). The same receptor was cloned by expressing cDNA in COS cells which were screened using radioligand and autoradiography (SIMS et al. 1988). Binding of IL-1α and IL-1β to the expressed receptor on COS cells was indistinguishable from that obtained on the parent T-cell line. Whether the expressed IL-1R are functional, however, has yet to be demonstrated. The identity of the cloned receptor was further confirmed by amino-terminal sequence analysis of the purified receptor protein. Northern analysis showed that the IL-1R mRNA in EL-4 cells was at least 5 kb in size and that it was also present in mouse fibroblast lines. The 557 amino acid sequence of the murine IL-1R cloned from EL-4 cells is shown in Fig. 3.

```
-18                    1                                                    41
MENMKVLLGICKMVPLLS-LEIDVCTEYPNQIVLFLSVNEIDIRKCPLTPNKMHGDTIIW
      *                                        *                            101
YKNDSKTPISADRDSRIHQQNEHLWFVPAKVEDSGYYYCIVRNSTYCLKTKVTVTVLEND
                                            *                               161
PGLCYSTQATFPQRLHIAGDGSLVCPYVSYFKDENNELPEVQWYKNCKPLLLDNVSFFGV
                                                                 *          221
KDKLLVRNVAEEHRGDYICRMSYTFRGKQYPVTRVIQFITIDENKRDRPVILSPRNETIE
        *               *                                          *         281
ADPGSMIQLICNVTGQFSDLVYWKWNGSEIETNDPFLAEDYQFVEHPSTKRKYTLITTLN
                                                                            341
ISEVKSQFYRYPFICVVKNTNIFESAHVQLIYPVPDFKNYLIGGFIILTATIVCCVCIYK
                                                                            401
VFKVDIVLWYRDSCSGFLPSKASDGKTYDAYILYPKTLGEGSFSDLDTFVFKLLPEVLEG
                                                                            461
QFGYKLFIYGRDDYVGEDTIEVTNENVKKSRRLIIILVRDMGGFSWLGQSSEEQIAIYNA
                                                                            521
LIQEGIKIVLLELEKIQDYEKMPDSIQFIKQKHGVICWSGDFQERPQSAKTRFYKNLRYQ
                                                                            557
MPAQRRSPLSKHRLLTLDPVRDTKEKLPAATHLPLG.
```

Fig. 3. Amino acid sequence of the murine EL-4 IL-1 receptor predicted from analysis of the cloned cDNA. The sequence between −18 and 1 represents the amino-terminal hydrophobic signal peptide. The mature receptor is 557 amino acids in length. *Asterisks* above each line denote potential N-linked glycosylation sites. The *underlined C residues* connected by *lines* represent potential disulfide bridges as seen in other members of the immunoglobulin supergene family. The *underlined* region represents the membrane-spanning domain

Hydrophobic domains typical of a signal peptide (Met −19 to Ser −1) and a transmembrane domain Asp320-Tyr340 to are present. All seven potential N-linked glycosylation sites lie amino-terminal to the membrane-spanning region. The calculated molecular weight (64 598) is in good agreement with that obtained by sodium dodecylsulfate polyacrylamide gel electrophoresis (SDS-PAGE) of N-glycanase-treated receptor. The presumed cytoplasmic portion, consisting of 217 amino acids, lacks a tyrosine kinase domain or other enzymatic functionality despite the comitogenic properties of IL-1 on certain cells. The extracellular portion contains six cysteines which, in intrachain disulfide linkage, would create three domains. Significant homologies between each of these domains and members of the immunoglobulin superfamily was found (SIMS et al. 1988). Recently, the IL-6 receptor has also been cloned and found to contain a domain homologous to the immunoglobulin superfamily (YAMASAKI et al. 1988).

F. Inhibitors and Antagonists of IL-1

I. Inhibitors

As mentioned in Sect. C. III, glucocorticoids have been shown to interfere with IL-1 production at both the transcriptional and posttranscriptional level.

PGE_2 has been reported to posttranscriptionally interfere with IL-1 production by raising intracellular levels of cyclic AMP (KNUDSEN et al. 1986). Certain inhibitors of cyclooxygenase (CO) and 5-lipoxygenase (5-LO) inhibit the release of IL-1 from monocytes (DINARELLO et al. 1984b; J.C. LEE et al. 1988). However, assay of a number of such compounds shows no correlation between their potency against CO or 5-LO and their ability to interfere with IL-1 production (J.C. LEE et al. 1988). Hence their precise mode of action with respect to inhibition of IL-1 production remains unknown. Inhibitors of protein kinase C and calmodulin-dependent kinase have been reported to block IL-1 production induced by LPS (KOVACS et al. 1988). The human metalloendopeptidase 24.11, also known as enkephalinase, has been reported to degrade and thereby inactivate mature IL-1β (PIERART et al. 1988).

II. Antagonists

A large number of polypeptide bioantagonists of IL-1 have been described in the literature (for review see ROSENSTREICH et al. 1988). Most of these remain poorly characterized and their precise mode of action remains unknown. However, some of these materials have been purified or partially purified and are potentially of interest. Uromodulin, an 85-kDa glycoprotein isolated from the urine of pregnant women, has been reported to inhibit IL-1-mediated biological responses in vitro (MUCHMORE and DECKER 1986, 1987). More recent studies, however, indicate that uromodulin binds both IL-1 and TNF molecules (HESSION et al. 1987) and that uromodulin is identical to the Tamm-Horsfall urinary glycoprotein (PENNICA et al. 1987). Its precise role in vivo remains unclear. Another urinary inhibitor of IL-1 (LIAO et al. 1985; BROWN and ROSENSTREICH 1987) was found only to inhibit certain of the biological effects of IL-1 (KORN et al. 1987). This material has been recently purified and found to have sequence homology with deoxyribonuclease I (ROSENSTREICH et al. 1988). A third urinary inhibitor has been partially purified (BALAVOINE et al. 1986; PRIEUR 1987; SECKINGER et al. 1987 a, b). It is distinct from the others in that it blocks binding of IL-1 to its receptor. It has none of the agonist properties of IL-1 and is not recognized by IL-1 antiserum (SECKINGER et al. 1987 b). Seckinger and colleagues have proposed the novel concept that this material may be a physiological IL-1 receptor antagonist. α-Melanocyte-stimulating hormone has been reported to inhibit some of the biological responses induced by IL-1 in vivo (DAYNES et al. 1987; ROBERTSON et al. 1988).

G. Biochemical Mode of Action

There is as yet no general agreement as to the mode of signal transduction by which IL-1 stimulates cellular responses. As mentioned in Sect. E, knowledge of the primary amino acid sequence of the murine IL-1R did not reveal a catalytic domain. This section will therefore attempt to summarize the published experiments which have addressed the second messenger systems generated or not by IL-1.

I. Ion Fluxes

Studies from a number of laboratories have failed to show an increase in intracellular calcium following stimulation of IL-1-responsive T-lymphoid lines with IL-1 (ABRAHAM et al. 1987; MUKAIDA et al. 1987; DIDIER et al. 1988; FREEDMAN et al. 1988; ROSOFF et al. 1988). Stimulation of human neutrophils likewise had no effect on levels of intracellular calcium (GEORGILIS et al. 1987). However, a progressive rise in intracellular sodium in a pre-B murine cell line was observed following IL-1 exposure (STANTON et al. 1986).

II. Kinases

In addition, IL-1 does not stimulate translocation of protein kinase C or phosphorylation of its substrates (AVISSAR et al. 1985; ABRAHAM et al. 1987; MUKAIDA et al. 1987). Downregulation of protein kinase C activity by chronic exposure to 12-O-tetradecanoylphorbol-13-acetate (TPA) had no effect on the ability of IL-1 to induce IL-6 mRNA in fibroblasts (ZHANG 1988a). A number of laboratories have examined phosphorylation of cell proteins following culture in IL-1. While some laboratories report negative results (AVISSAR et al. 1985; ABRAHAM et al. 1987), others have found evidence for activation of a cyclic nucleotide-dependent kinase. MATSUSHIMA et al. (1987) reported that IL-1 stimulated the serine phosphorylation of a 65-kDa cytosolic protein in dexamethasone-stimulated human monocytes. The labeling was inhibited by nonspecific inhibitors of cyclic nucleotide dependent-protein kinase (H-A) and calmodulin (W-7) but was not sensitive to an inhibitor of protein kinase C (H-7). ZHANG et al. (1988a) found that extracts of IL-1-treated fibroblasts phosphorylated histone protein HII-B and that this was inhibitable by H-8, another nonspecific inhibitor of cyclic nucleotide-dependent protein kinases. However, the induction of IL-6 mRNA by IL-1 was only partially inhibited by H-8. These cells also produced cyclic AMP in response to IL-1 but the authors were not able to demonstrate increases in adenylate cyclase activity in membrane preparations of fibroblasts treated with IL-1. SHIRAKAWA et al. (1988) also reported that IL-1 stimulated an increase in cyclic AMP levels in a human NK cell line (YT) and murine thymocytes. DIDIER et al. (1988), however, reported that cyclic AMP levels in a human T-cell line (Jurkat) did not rise following IL-1 stimulation.

III. Phospholipases

A number of laboratories have examined the ability of IL-1 to activate cellular phospholipases. CHANG et al. (1986) reported that IL-1 induced a 10-fold increase in phospholipase A_2 (PLA$_2$) activity in rabbit chondrocytes. BURCH et al. (1988) reported an 80% increase in PLA$_2$ activity in murine 3T3 cells without any increase in phospholipase (PLC) activity. ABRAHAM et al. (1987) likewise found no increase in PLC activity in a murine T-cell line (LBRM-33 1A5) following exposure to IL-1 as measured by the following parameters: (a) release of inositol phosphates; (b) phosphoinositide hydrolysis; and (c) incorporation of [^{32}P]phosphate into cellular phospholipids. MUKAIDA et al. (1987) and DIDIER et al. (1988) likewise found no PLC response to IL-1 as demonstrated by stable inositol phosphate levels and no increase in diacylglycerol (DAG). Stimulation of murine peritoneal macrophages, however, with IL-1 has been reported to produce a rapid rise in inositol phosphates at the expense of phosphotidylinositol diphosphate (Wijelath et al. 1988). Finally, Rosoff et al. (1988) reported that IL-1 stimulated rapid DAG and phosphorylcholine production from phosphatidylcholine in the absence of phosphoinositide turnover in normal human peripheral blood T cells, a human T-cell line (Jurkat) and a murine T-cell line (EL-4). IL-1R appeared unnecessary for this effect (Rosoff et al. 1988). Clearly, the literature concerning the role of PLC in mediating IL-1 effects is confused, at best, and definitive experiments remain to be done.

IV. Transcriptional Effects

IL-1 stimulates many biological responses at the transcriptional or pretranslational level. Frisch and Ruley (1987) reported that the IL-1-responsive element within the rabbit stromelysin gene was located within a 700-bp 5'-flanking fragment. The DNA binding proteins recognized by these IL-1-sensitive elements have yet to be identified. Transcriptional effects are inferred (cautiously) from IL-1-stimulated increases in mRNA for acute phase proteins (Perlmutter et al. 1986) and IL-6 (Walther 1988; Zhang et al. 1988a). Increased CO biosynthesis by fibroblasts in response to IL-1 (Raz et al. 1988) may also represent regulation at the transcriptional level.

V. Expression of Protooncogenes

Kovacs et al. (1986) and Lin and Vilcek (1987) have reported that IL-1 stimulates a rapid and transient increase in c-*fos* and c-*myc* mRNA levels in human peripheral blood T lymphocytes and human fibroblasts, respectively. Libby et al. (1988) reported that IL-1 induced transient expression of c-*fos* by human smooth muscle cells.

H. Biological Properties

I. Generalizations

It may be useful to make some generalizations before beginning a detailed discussion of the biological activities of IL-1 molecules. First, most of the biological activities to be discussed have been confirmed with pure recombinant mature proteins prepared in *E. coli* by a number of different laboratories (Gubler et al. 1986; March et al. 1985; Dinarello et al. 1986a; Wingfield et al. 1986; Tocci et al. 1987; Meyers et al. 1987). Second, recombinant IL-1α and recombinant IL-1β, though only 25% homologous (see Sect. B), have been found by most workers to be nearly equipotent in most assay systems (for example, see Rupp et al. 1986). However, exceptions have been noted (Lumpkin 1987; Breviario et al. 1988). In some instances lack of apparent potency by either IL-1α or IL-1β may be due to species barriers. For example, human IL-1β was found to be inactive on porcine synoviocytes

(WOOD et al. 1985) and murine IL-1α has low specific activity on human connective tissue cells (GUBLER et al. 1986; HUANG et al. 1988). Third, many of the biological activities of IL-1 may be mediated by hormones and factors released by target cells upon stimulation with IL-1. IL-1 stimulates the release of ACTH (see Sect. A. III), PGE$_2$ (MIZEL et al. 1981), platelet activating factor (VALONE et al. 1988), CSFs (BAGBY et al. 1986, 1988; SIEFF et al. 1987; ZUCALI et al. 1986; ZSEBO et al. 1988), IL-6 (CONTENT et al. 1985; DEWIT et al. 1985; VAN DAMME et al. 1987; SEHGAL et al. 1987; WALTHER et al. 1988) and monocyte-derived neutrophil chemotactic factor (MATSUSHIMA et al. 1988). It is interesting to note in this regard that an increase in vascular permeability stimulated by IL-1 was inhibited in part by administration of platelet activating factor receptor antagonist (RUBIN and ROSENBAUM 1988). Fourth, IL-1 molecules share many of the same biological properties with TNF-α even though these mediators do not bind to the same receptor sites (see, for example, VANDAMME et al. 1987; VALONE et al. 1988; MATSUSHIMA et al. 1988). IL-1 has been reported to downregulate TNF receptors (HOLTMANN and WALLACH 1987). Furthermore IL-1 and TNF-α induce the biosynthesis of an overlapping series of polypeptides in fibroblasts (BERESINI et al. 1988) and when given together may produce either additive or syergistic biological responses (LAST-BARNEY et al. 1988). In vivo, IL-1 and TNF were observed to desensitize to themselves as well as each other (WALLACH et al. 1988).

II. Growth Promoting Activity: IL-1 is a Comitogen

Perhaps the best known biological activity of IL-1 is its ability to stimulate lymphocyte proliferation in culture in conjunction with mitogens such as phytohemagglutinin (GERY et al. 1972). More will be said about this in Sect. H. III. 1. IL-1 also stimulates the proliferation of connective tissue cells (SCHMIDT et al. 1982, 1984; ESTES et al. 1984; POSTLETHWAITE et al. 1984, smooth muscle cells (LIBBY et al. 1988), and mesangial cells (LOVETT et al. 1983, 1986). With the exception of LIBBY et al. (1988), most have found that serum or platelet-derived growth factor must also be present to observe the mitogenic properties of IL-1. IL-1 has also been shown to enhance the proliferative response of immature bone marrow cells to GM-CSF (MOCHIZUKI et al. 1987). In all cases, IL-1 functions as a *co*mitogen requiring the presence of at least one other growth factor or mitogen in order to stimulate growth. The first IL-1R to be cloned lacks a tyrosine kinase domain (see Sect. E. III). The ability of IL-1 to stimulate transient expression of c-*fos* and c-*myc* was discussed in Sect. G. V. These observations, taken together with the work of BITTERMAN et al. (1986), suggest that IL-1 functions to provide competence or "competence-induction" for entry into the cell cycle (PLEDGER et al. 1978). As mentioned in Sect. H. I, IL-1 is able to stimulate the production of other factors which themselves promote growth.

Because several cells are capable of both producing and responding to IL-1 in culture, it has been proposed that IL-1 is an autocrine growth factor. Two studies have suggested that IL-1 functions as an autocrine by blocking growth by the addition of IL-1-specific antibodies to cultured lymphoid lines (Scala et al. 1987; Tartakovsky et al. 1988).

III. Role of IL-1 in Various Organ Systems

1. Immune System

a) B-Lymphocytes

Studies with recombinant IL-1 preparations have confirmed earlier findings (Howard et al. 1983; Falkoff et al. 1983; Lipsky et al. 1983; Booth and Watson 1984) that IL-1 is capable of stimulating B-cell proliferation in culture (Pike and Nossal 1985; Chiplunkar et al. 1986; Jelinek and Lipsky 1987; Freedman et al. 1988). In general, costimulation with antigen or other lymphokines was required. Similarly, recombinant IL-1 preparations have been reported to stimulate B-cell differentiation as measured by antibody production (Pike and Nossal 1985; Chiplunkar et al. 1986; Jelinek and Lipsky 1987; Goud et al. 1988). Once again, as in the case of connective tissue cells, IL-1 by itself has little or no activity in these assays but requires the presence of other stimuli. In no case has IL-1 been shown to be essential for B-cell proliferation or differentiation (Jelinek and Lipsky 1987). For example, Koide and Steinman (1987) showed that freshly isolated B cells proliferated actively in response to LPS without making detectable amounts of IL-1 activity or IL-1α mRNA.

b) T-Lymphocytes

IL-1 is well known to stimulate thymocyte proliferation in conjunction with mitogenic lectins such as phytohemagglutinin (Gery et al. 1972). These early observations, plus the realization that T-cell proliferation in response to soluble antigens required the presence of macrophages or other types of accessory cells, led to the hypothesis that IL-1 was one of the critical macrophage-derived signals required for optimal T-cell proliferation. In support of this hypothesis, IL-1 has been shown to induce the production of both IL-2 (Manger et al. 1985; William et al. 1985; Hagiwara et al. 1987; Mukaida et al. 1987; Hackett et al. 1988) and IL-2R (Kaye et al. 1984; Hagiwara et al. 1987; Hackett et al. 1988; Lubinski et al. 1988) on T cells, T-cell lines, and NK cell lines, suggesting that the growth-promoting effects of IL-1 are indirectly mediated by IL-2.

Recent experiments have cast doubt on the hypothesis that most T cells have an IL-1 requirement. For example, despite an earlier report to the contrary (Nagelkerken et al. 1986), dendritic cells, though active accessory cells for a variety of T lymphocyte-dependent immune responses, seem incapable of producing IL-1 in response to a variety of stimuli (Koide and Steinman 1987; Bhardwaj et al. 1988). Similarly, no IL-1α mRNA could be found in macrophage–T-lymphocyte cocultures proliferating in response to concanavalin A (Koide and Steinman 1987). IL-1 was not able to replace dendritic antigen presenting cells (KOIDE et al. 1987) and, in the absence of antigen presenting cells, had no effect on IL-2 production or mitogen responsiveness (KOIDE et al. 1987). Other workers have not been able to demonstrate the ability of IL-1 to induce IL-2 production or IL-2R expression by unfractionated T cells stimulated with concanavalin A or antibodies to the T-cell antigen receptor (MALEK et al. 1985; GARMAN and RAULET 1987; GARMAN et al. 1987).

The apparent contradictions in the literature with respect to the role of IL-1 in T-cell activation have now been partially resolved by the observation that there are two discrete subsets of helper T cells known as TH1 and TH2 cells. TH1 cells secrete IL-2 whereas TH2 cells secrete IL-4 and each uses its respective lymphokine as its autocrine growth factor (MOSMANN et al. 1986; KURT-JONES et al. 1987b; LICHTMAN et al. 1987). TH1 clones which comprise the overwhelming majority of T cells in the thymus, lymph nodes, and spleen, (POWERS et al. 1988; SIDERAS et al. 1988) are unresponsive to IL-1 (LICHTMAN et al. 1988). This may explain, in part, why unfractionated populations of T cells may have low IL-1 responsiveness (MALEK et al. 1985; KOIDE et al. 1987; GARMAN and RAULET 1987; GARMAN et al. 1987). TH2 cells, in contrast, have an absolute requirement for IL-1 (LICHTMAN et al. 1988; WEAVER et al. 1988). These observations have been corroborated by studies showing that only T helper cells (LOWENTHAL and MACDONALD 1987) and, more specifically, TH2 cells bear IL-1 receptors (GREENBAUM et al. 1988; LICHTMAN et al. 1988). IL-1 appears to provide an independent growth signal to TH2 cells and does not function by stimulating IL-4 production or IL-4 receptor expression (Ho et al. 1987; LICHTMAN et al. 1988). In summary, recent data suggests that IL-1 regulates only a small proportion of T cells, viz., TH2 cells, whose physiological role remains to be defined. Those tumor cell lines which have been described as producing IL-2 and IL-2R in response to IL-1 may not be representative of most normal T cells.

2. Hematopoietic System

Recent studies employing bone marrow cells from 5-fluorouracil-treated mice have demonstrated that IL-1 in conjunction with colony stimulating factors can stimulate the proliferation of precursor cells in vitro as well as speed the recovery of bone marrow regeneration in vivo. MOCHIZUKI et al. (1987) and MOORE and WARREN (1987) presented data that IL-1 is identical to hemopoietin 1. Furthermore, MOORE observed that in vivo administration of IL-1α and G-CSF twice daily accelerated the recovery of blood neutrophil counts by as much as 7 days. Analysis of progenitor cells in colony formation assays showed that erythroid, myeloid, and, to a lesser extent, megakaryocyte progenitors responded dramatically to IL-1 administered alone or in combination with G-CSF. In some analyses, the combination of IL-1α and G-CSF produced additive or synergistic effects. Similarly, WARREN and MOORE (1988) reported that a combination of IL-1 and IL-3 stimulated a 6-fold increase in eosinophil colony forming units from 5-fluorouracil-treated bone-marrow cells.

The ability of IL-1 to protect animals against the lethal effects of radiation may be related to its ability to speed marrow recovery (NETA et al. 1986, 1987, 1988; CASTELLI et al. 1988; MORRISSEY et al. 1988 b), although this appears to be at the expense of thymic hypoplasia (MORRISSEY et al. 1988 a, b). The ability of IL-1 to speed marrow recovery may, in turn, be due to its ability to stimulate the production of colony stimulating factors (BAGBY et al. 1986; ZUCALI et al. 1986; FIBBE et al. 1987; SEGAL et al. 1987; SIEFF et al. 1987; ZSEBO et al. 1988) or the induction of enzymes such as superoxide dismutase which may dissipate oxygen radicals generated by ionizing radiation (WONG and GOEDDEL 1988; MASUDA et al. 1988).

SANTOLI et al. (1987) and ONOZAKI et al. (1988) have presented data showing that IL-1 can interfere with the growth of myeloid leukemia lines in vitro.

3. Host Defense System

Several studies have shown that mice pretreated with IL-1 prior to inoculation with bacteria have higher survival rates (OZAKI et al. 1987; CZUPRYNSKI et al. 1988; VAN DER MEER et al. 1988). In one of these studies, higher survival rate correlated with recovery of fewer infectious organisms (CZUPRYNSKI et al. 1988). In the study by VAN DER MEER (1988), however, the number of organisms cultured from the treated and control groups was similar.

While earlier reports suggested that IL-1 was a chemoattractant for neutrophils (SAUDER et al. 1984) and that IL-1 induced neutrophil degranulation (KLEMPNER et al. 1978), more recent studies with recombinant IL-1 molecules have failed to confirm these activities (SCHRODER et al. 1987; GEORGILIS et al. 1987; SAYERS et al. 1988). Neutrophils are reported to express IL-1R (RHYNE et al. 1988) and produce IL-1 (LINDEMANN et al. 1988; GOTO et al. 1988). While IL-1 is not itself a chemoattractant for neutrophils, injection of IL-1 into skin or peritoneal cavity results in the dramatic accumulation of neutrophils (GRANSTEIN et al. 1985; BECK et al. 1986; CYBULSKY et al. 1988). Thus the stimulation of neutrophil accumulation by IL-1 is likely to be mediated by other chemoattractants induced by IL-1 such as neutrophil activating peptide (SCHRODER et al. 1987; MATSUSHIMA et al. 1988).

4. Cardiovascular–Vascular Endothelial System

IL-1 has a variety of effects on cultured endothelial cells the net result of which is to increase leucocyte-endothelial interactions and to promote a hypercoagulable state. IL-1 treatment of endothelial cells promotes adhesion of neutrophils (BEVILACQUA et al. 1985; BREVARIO et al. 1988), monocytes (BEVILACQUA et al. 1985), lymphocytes of the B and T lineages (CAVENDAR et al. 1986), and tumor cells (DEJANA et al. 1988). This effect appears several hours after initiation of treatment and is inhibited by pretreatment of endothelial cells with cycloheximide. IL-1 has been reported to stimulate the expression of intercellular adhesion molecule-1 (ICAM-1) on the surface of fibroblasts and certain tumor lines (ROTHLEIN et al. 1988) and endothelial leucocyte adhesion molecule (ELAM) on the surface of endothelial cells (POBER et al. 1986). Given that ICAM-1 is regarded as the ligand for one of the major leucocyte integrins, LFA-1, this may be a mechanism by which IL-1 promotes egress of leucocytes from the vascular space (MOSER et al. 1989). The role of platelet activating factor in mediating IL-1 stimulated adhesion is controversial (BREVARIO et al. 1988).

IL-1 stimulates the expression of procoagulant activity on the surface of endothelial cells (BEVILACQUA et al. 1984) and decreases the expression of tissue plasminogen activator while increasing the expression of plasminogen activator inhibitor (PAI-1) (DEJANA et al. 1987; SCHLEEF et al. 1988). All of these effects would appear to promote a hypercoagulable state as seen in clinical conditions such as disseminated intravascular coagulation.

IL-1 stimulates the production of prostacyclin (PGI_2) and PGE_2 by vascular endothelial cells, smooth muscle cells, and dermal fibroblasts (ALBRIGHTSON et al. 1985; DEJANA et al. 1987). Infusion of IL-1 into rabbits resulted in hypotension, decreased peripheral vascular resistance, increased heart rate and cardiac output (OKUSAWA et al. 1988). These effects were all reversed by pretreatment with a CO inhibitor (OKUSAWA et al. 1988).

Endothelial cells and vascular smooth muscle cells have been reported to produce IL-1 (WARNER et al. 1987a, b). Endothelial cells express IL-1 receptors (THIEME et al. 1987). Vascular smooth muscle cells proliferate in response to IL-1 (LIBBY et al. 1988).

5. Pulmonary System

Studies of IL-1 with respect to the pulmonary system have focused mainly on the alveolar macrophage and its ability to produce IL-1 ex vivo. A number of studies have now shown that the alveolar macrophage has a reduced ability to produce IL-1 as compared to freshly isolated blood monocytes (WEWERS et al. 1984, 1987; ELIAS et al. 1985; BERNAUDIN et al. 1988).

The role of IL-1 in lung diseases has yet to be established. Uninvolved lung fragments removed at the time of surgery from patients with tumors release, in culture, material having some of the biological and antigenic properties of IL-1 (BOCHNER et al. 1987). However, early reports that increased amounts of IL-1 are spontaneously released by bronchial lavage cells from patients with sarcoidosis could not be confirmed using highly specific antisera and cDNA probes (WEWERS et al. 1987).

6. Connective Tissue System, Cartilage, and Bone

IL-1 exhibits a number of impressive biological activities on connective tissue cells strongly suggesting that IL-1 plays a role in connective tissue resorption and remodeling. IL-1 stimulates connective tissue cells, including chondrocytes, rheumatoid synovial fibroblasts, and dermal fibroblasts, to make tissue-degrading metalloproteinases including interstitial collagenase and stromelysin (MIZEL et al. 1981; GOWEN et al. 1983, 1984; POSTLETHWAITE et al. 1983; SAKLATVALA et al. 1984, McGUIRE-GOLDRING et al. 1984; MURPHY et al. 1985; McCROSKERY et al. 1985; SCHNYDER et al. 1987; SAUS et al. 1988). Stimulation of these proteases appears to occur at the transcriptional level (McCROSKERY et al. 1985; FRISCH and RULEY 1987; SAUS et al. 1988). IL-1 also stimulates the production of PGE_2 (DAYER et al. 1981) as well as type I collagen and fibronectin by rheumatoid synovial cells (KRANE et al. 1985). In human chondrocytes, IL-1 suppresses expression of types II and IX collagen and stimulates production of types I and III collagen (GOLDRING et al. 1988). Connective tissue cells bear high-affinity receptors for IL-1 (CHIN et al. 1987, 1988; BIRD and SAKLATVALA 1986, 1987; BIRD et al. 1987; MIZEL et al. 1987; QWARNSTROM et al. 1988).

IL-1, in organ culture, stimulates the release of glycosaminoglycans from cartilage explants (SAKLATVALA et al. 1984; KRAKAUER et al. 1985). Injection of IL-1 into the knee joints of rabbits induced cartilage proteoglycan degradation (PETTIPHER et al. 1986) and was found to accelerate the development of collagen-induced arthritis in mice (HOM et al. 1988). JACOBS et al. (1988), however, reported that IL-1 reduced the severity of antigen-induced arthritis in rats. IL-1 has been found in human rheumatoid synovial fluid by RIA (SYMONS et al. 1988).

IL-1 is a potent bone resorptive agent (GOWEN et al. 1983; GOWEN and MUNDY 1986; DEWHIRST et al. 1985; STASHENKO et al. 1987). Sustained infusion of IL-1 in mice caused hypercalcemia and an increase in bone resorptive surface as revealed by histomorphometry (SABATINI et al. 1988).

7. Renal System

Studies examining the role of IL-1 in kidney physiology and pathophysiology have largely focused on mesangial cells cultured from glomeruli. Mesangial cells have been reported to produce IL-1 (LOVETT et al. 1986; LOVETT and LARSEN et al. 1988; WEBER et al. 1987) and to proliferate in response to IL-1 (LOVETT et al. 1983). Increased levels of IL-1 mRNA have been found in cortical extracts of mice with nephritis (BOSWELL 1988 a, b).

8. Endocrine System

The interplay between IL-1 and the adrenocortical axis was described in detail in Sect. A. III. IL-1 has also been noted to prevent loss of corticotropic responsiveness by rat anterior pituitaries to β-adrenergic stimulation in vitro (BOYLE et al. 1988). Furthermore, IL-1 has also been reported to be cytotoxic for isolated rat and human pancreatic β cells (BENDTZEN et al. 1986; MANDRUP-POULSON et al. 1987).

9. Gastrointestinal-Hepatic System

While little data exists concerning the role of IL-1 in regulating the gastrointestinal tract, a large number of studies document that IL-1 regulates secretion of acute phase proteins by the liver. Injection of 100 ng IL-1 into mice leads to a 10-fold or greater increase in serum amyloid A (SAA) and serum amyloid P-component in mice within two hours (RAMADORI et al. 1985; MORTENSON et al. 1988). This is accompanied by an increase in SAA mRNA in liver extracts (RAMADORI et al. 1985). Similarly, administration of IL-1 to rats resulted in a 7-fold increase in mRNA levels for metallothionein in liver (COUSINS and LEINART 1988). Studies employing cultured murine hepatocytes showed an increase in the production of complement factor B and a decrease in the production of albumin in response to IL-1 (RAMADORI et al. 1985). Studies using the human hepatoma lines HepG2 and Hep3B have shown that IL-1 stimulates higher levels of mRNA for metallothionein (KARIN et al. 1985) and the third component of complement (PERLMUTTER et al. 1986; BEUSCHER et al. 1987) while depressing levels of albumin mRNA. More recently, it has become clear that at least one additional cytokine, IL-6, is also important in stimulating acute phase protein production. Studies by BAUMAN et al. (1987) on the human hepatoma line HepG2 have shown that IL-1 and IL-6 regulate different but overlapping groups of acute phase proteins. Further studies will be required to assess the relative roles of IL-1 and IL-6 in the regulation of acute phase protein biosynthesis.

10. Dermatologic System

Freshly isolated keratinocytes, cultured keratinocytes, and keratinocyte lines are capable of expressing IL-1 activity (LUGER et al. 1981; SAUDER et al. 1982) as well as IL-1α and IL-1β mRNA indistinguishable from that found in monocytes (KUPPER et al. 1986; KOIDE and STEINMAN 1987b; ANSEL et al. 1988). However, DEMCZUK (1988) has recently argued that the IL-1-like activity in epidermis is due to molecules other than IL-1 itself. In contrast to monocytes, IL-1α mRNA appears to be the major IL-1 mRNA species found in skin. Expression of IL-1α mRNA in these cells is induced by ultraviolet irradiation (KUPPER et al. 1987; ANSEL et al. 1988). While not certain, keratinocyte-derived IL-1 may function to indirectly recruit neutrophils to the skin in inflammatory skin diseases such as psoriasis (GRANSTEIN et al. 1985).

11. Nervous System

A number of observations support the contention that IL-1 may play a role in the central nervous system. First, cells of CNS origin have been shown to release IL-1-like molecules in culture (FONTANA et al. 1982; GIULIAN et al. 1986). Second, IL-1β has been identified by immunohistochemical techniques in the processes of certain neurons in the human brain (BREDER et al. 1988). Third, IL-1 receptors have been visualized in rat brain slices using ^{125}I-IL-1α (FARRAR et al. 1987a,b). Fourth, IL-1 is a centrally acting endogenous pyrogen (DINARELLO et al. 1986a) and this effect is inhibited by the pituitary hormone α-melanocyte-stimulating hormone (DAYNES et al. 1987; ROBERTSON et al. 1988). Other observations suggesting an interplay between IL-1 and neuropeptides include the ability of neurotensin, substance P, neurokinin A, and neurokinin B to stimulate the production of IL-1-like molecules by macrophages (LEMAIRE 1988; KIMBALL et al. 1988) and the ability of substance P and bradykinin to enhance the ability of IL-1 to stimulate fibroblast proliferation in vitro (KIMBALL and FISHER 1988).

I. Potential Therapeutic Utility of IL-1

The information presented in Sect. H. III suggests that IL-1 may have utility in the treatment of granulocytopenic individuals. This is based on the observation by MOORE and WARREN (1987) that granulocyte counts in mice treated with 5-fluorouracil rebounded more rapidly if the mice were treated with IL-1 on a daily basis. Furthermore, given that this same group of patients is at higher risk for bacterial infection, granulocytopenic patients may also benefit from the reported ability of IL-1 in mice to reduce the incidence of death secondary to bacterial infections (OZAKI et al. 1987; CZUPRYNSKI et al. 1988; VAN DER MEER et al. 1988). The radioprotective effects of IL-1 may allow patients receiving radiation therapy to tolerate larger doses of radiation and thereby attain a higher therapeutic ratio (NETA et al. 1988). The ability of IL-1, in conjunction with other factors, to stimulate fibroblast proliferation suggests that IL-1 may be beneficial in the treatment of wounds. Nevertheless, each of these applications is highly speculative and will require much more detailed study in animal models. Enthusiasm for the potential therapeutic utility of IL-1 must be tempered by the plethora of biological activities attributed to this cytokine, many of which would appear to be detrimental to the host.

J. Conclusions

The next few years of research should greatly expand our understanding of IL-1. The areas of secretion, natural IL-1 inhibitors, transcriptional regulation, receptor cloning and subtypes, and signal transduction will be the most intensively studied. The availability of IL-1-specific monoclonal antibodies and receptor-specific antibodies will identify those processes in which IL-1 plays a central role both in vitro and in vivo. Transgenic animals containing inactivated IL-1 genes or IL-1 genes regulatable in a tissue-specific manner will finally address, in a definitive manner, the roles played by IL-1 in normal physiology and disease.

References

Abraham RT, Ho SN, Barna TJ, McKean DJ (1987) Transmembrane signaling during interleukin 1-dependent T cell activation. Interactions of signal 1- and signal 2-type mediators with the phosphoinositide-dependent signal transduction mechanisms. J Biol Chem 262:2719–2728

Acres RB, Larsen A, Gillis S, Conlon PJ (1987a) Production of IL-1α and IL-1β by clones of EBV transformed, human B cells. Mol Immunol 24:479–485

Acres RB, Larsen A, Conlon PJ (1987b) IL 1 expression in a clone of human T cells. J Immunol 138:2132–2136

Akahoshi T, Oppenheim JJ, Matsushima K (1988) Induction of high-affinity interleukin 1 receptor on human peripheral blood lymphocytes by glucocorticoid hormones. J Exp Med 167:924–936

Albrightson CR, Baenziger NL, Needleman P (1985) Exaggerated human vascular cell prostaglandin biosynthesis mediated by monocytes: role of monokines and interleukin 1. J Immunol 135:1872–1877

Ansel JC, Luger TA, Kock A, Hochstein D, Green I (1984) The effect of in vitro UV irradiation on the production of IL 1 by murine macrophages and P388D1 cells. J Immunol 133:1350–1355

Ansel JC, Luger TA, Lowry D, Perry P, Roop DR, Mountz JD (1988) The expression and modulation of IL-1α in murine keratinocytes. J Immunol 140:2274–2278

Auron PE, Webb AC, Rosenwasser LJ, Mucci SF, Rich A, Wolff SM, Dinarrello CA (1984) Nucleotide sequence of human monocyte interleukin 1 precursor cDNA. Proc Natl Acad Sci 81:7907–7911

Auron PE, Rosenwasser LJ, Matsushima K, Copeland T, Dinarello CA, Oppenheim JJ, Webb AC (1985) Human and murine interleukin-1 possess sequence and structural similarities. J Mol Cell Immunol 2:169–177

Auron PE, Warner SJ, Webb AC, Cannon JG, Bernheim HA, McAdam KJ, Rosenwasser LJ, LoPreste G, Mucci SF, Dinarello CA (1987) Studies on the molecular nature of human interleukin 1. J Immunol 138:1447–1456

Avissar S, Stenzel KH, Novogrodsky A (1985) Selective effects of TPA and IL-1 on protein phosphorylation in murine thymocytes. Cell Immunol 96:462–471

Bagby GC, Dinarello CA, Wallace P, Wagner C, Hefeneider S, McCall E (1986) Interleukin-1 stimulates granulocyte macrophage colony stimulating activity release by vascular endothelial cells. J Clin Invest 78:1316–1323

Bagby GC, Dinarello CA, Neerhout RC, Ridgway D, McCall E (1988) Interleukin 1-dependent paracrine granulopoiesis in chronic granulocytic leukemia of the juvenile type. J Clin Invest 82:1430–1436

Bakouche O, Brown DC, Lachman LB (1987) Subcellular localization of human monocyte interleukin 1: evidence for an inactive precursor molecule and a possible mechanism for IL 1 release. J Immunol 138:4249–4255

Baldari C, Massone A, Macchia G, Telford JL (1987) Differential stability of human interleukin 1 beta fragments expressed in yeast. Protein Eng 1:433–437

Balavoine JF, de Rochemonteix B, Williamson K, Seckinger P, Cruchaud A, Dayer JM (1986) Prostaglandin E₂ and collagenase production by fibroblasts and synovial cells is regulated by urine-derived human interleukin 1 and inhibitor(s). J Clin Invest 78:1120–1124

Bauman H, Richards C, Gauldie J (1987) Interaction among hepatocyte-stimulating factors, interleukin 1, and glucocorticoids for regulation of acute phase plasma proteins in human hepatoma (HepG2) cells. J Immunol 139:4122–4128

Bayne EK, Rupp EA, Limjuco G, Chin J, Schmidt JA (1986) Immunocytochemical detection of interleukin 1 within stimulated human monocytes. J Exp Med 163:1267–1280

Beck G, Habicht GS, Benach JL, Miller F (1986) Interleukin 1: a common endogeneous mediator of inflammation and the local Shwartzman reaction. J Immunol 136:3025–3031

Bendtzen K, Mandrup-Poulsen T, Nerup J, Nielsen JH, Dinarello CA, Svenson M (1986) Cytotoxicity of human pI 7 interleukin-1 for pancreatic islets of Langerhans. Science 232:1545–1547

Bensi G, Raugei G, Palla E, Carinci V, Tornese-Buonamassa D, Melli M (1987) Human interleukin-1 beta gene. Gene 52:95–101

Beresini MH, Lempert MJ, Epstein LB (1988) Overlapping polypeptide induction in human fibroblasts in response to treatment with interferon-α, interferon-γ, interleukin 1α, interleukin 1β, and tumor necrosis factor. J Immunol 140:485–493

Berkenbosch F, Van Oers J, Del Ray A, Tilders F, Besedovsky H (1987) Corticotropin-releasing factor-producing neurons in the rat activated by interleukin 1. Science 238:524–526

Bernaudin JF, Yamauchi K, Wewers MD, Tocci MJ, Ferrans VJ, Crystal RG (1988) Demonstration by in situ hybridization of dissimilar IL-1β gene expression in human alveolar macrophages and blood monocytes in response to lipopolysaccharide. J Immunol 140:3822–3829

Bernton EW, Beach JE, Holaday JW, Smallridge RC, Fein HG (1987) Release of multiple hormones by a direct action of interleukin-1 on pituitary cells. Science 238:519–521

Besedovsky H, Del Rey A, Sorkin E, Dinarello CA (1986) Immunoregulatory feedback between interleukin-1 and glucocorticoid hormones. Science 233:652–654

Beuscher HU, Fallon RJ, Colten HR (1987) Macrophage membrane interleukin 1 regulates the expression of acute phase proteins in human hepatoma hep 3B cells. J Immunol 139:1896–1901

Bevilacqua MP, Pober JS, Majeau GR, Cotran RS, Gimbrone MA (1984) Interleukin 1 (IL-1) induces biosynthesis and cell surface expression of procoagulant activity in human vascular endothelial cells. J Exp Med 160:618–623

Bevilacqua MP, Pober JS, Wheeler ME, Cotran RS, Gimbrone MA (1985) Interleukin 1 acts on cultured human vascular endothelium to increase the adhesion of polymorphonuclear leukocytes, monocytes, and related leukocyte cell lines. J Clin Invest 76:2003–2011

Bhardwaj N, Lau L, Rivelis M, Steinman RM (1988) Interleukin-1 production by monoclear cells from rheumatoid synovial effusions. J Cell Immunol 114:405–423

Bird TA, Saklatvala J (1986) Identification of a common class of high affinity receptors for both types of porcine interleukin-1 on connective tissue cells. Nature 324:263–266

Bird TA, Saklatvala J (1987) Studies on the fate of receptor-bound ^{125}I-Interleukin 1β in porcine synovial fibroblasts. J Immunol 139:92–97

Bird TA, Gearing AJH, Saklatvala J (1987) Murine interleukin-1 receptor: differences in binding properties between fibroblastic and thymoma cells and evidence for a two-chain receptor model. FEBS Lett 225:21–26

Bitterman PB, Wewers MD, Rennard SI, Adelberg S, Crystal RG (1985) Modulation of alveolar macrophage-driven fibroblast proliferation by alternative macrophage mediators. J Clin Invest 77:700–708

Black RA, Kronheim SR, Cantrell M, Deeley MC, March CJ, Prickett KS, Wignall J, Conlon PJ, Cosman D, Hopp TP, Mochizuki DY (1988) Generation of biologically active interleukin-1β by proteolytic cleavage of the inactive precursor. J Biol Chem 263:9437–9442

Blanton RA, Kupper TS, McDougall JK, Dower S (1989) Regulation of interleukin 1 and its receptor in human keratinocytes. Proc Natl Acad Sci USA 86:1273–1277

Bochner BS, Rudledge BK, Schleimer RP (1987) Interleukin 1 production by human lung tissue. Inhibition by anti-inflammatory steroids. J Immunol 139:2303–2307

Booth RJ, Watson JD (1984) Interleukin 1 induces proliferation in two distinct B cell subpopulations responsive to two different murine B cell growth factors. J Immunol 133:1346–1349

Boraschi D, Nencioni L, Villa L, Censini S, Bossu P, Ghiara P, Presentini R, Perin F, Frasca D, Doria G, Forni G, Musso T, Giovarelli M, Ghezzi P, Bertini R, Besedovsky H, Del Rey A, Sipe J, Antoni G, Silvestri S, Tagliabue A (1988) In vivo stimulation and restoration of the immune response by the non inflammatory fragment 163–171 of human interleukin 1β. J Exp Med 168:675–686

Boswell JM, Yui MA, Endres S, Burt DW, Kelley VE (1988a) Novel and enhanced IL-1 gene expression in autoimmune mice with lupus. J Immunol 141:118–124

Boswell JM, Yui MA, Burt DW, Kelley VE (1988b) Increased tumor necrosis factor and IL-1β gene expression in the kidneys of mice with lupus nephritis. J Immunol 141:3050–3054

Boyle M, Yamamoto G, Chen M, Rivier J, Vale W (1988) Interleukin 1 prevents loss of corticotropic responsiveness to β-adrenergic stimulation in vitro. Proc Natl Acad Sci USA 85:5556–5560

Breder CD, Dinarello CA, Saper CB (1988) Interleukin-1 immunoreactive innervation of the human hypothalamus. Science 240:321–324

Breviario F, Bertocchi F, Dejana E, Bussolino F (1988) IL-1-induced adhesion of polymorphonuclear leukocytes to cultured human endothelial cells. J Immunol 141:3391–3397

Brody TB, Durum SK (1988) A plasma membrane anchoring mechanism for IL-1. In: Powanda MC, Oppenheim JJ, Kluger MJ, Dinarello CA (eds) Monokines and other non-lymphocytic cytokines. Liss, New York, pp 101–107 (Progress in leukocyte biology, vol 8)

Bron C, MacDonald HR (1987) Identification of the plasma membrane receptor for interleukin-1 on mouse thymoma cells. FEBS Lett 219:365–368

Brown KM, Rosenstreich DL (1987) Mechanism of action of a human interleukin 1 inhibitor. Cell Immun, 1 105:45

Burch RM, Connor JR, Axelrod J (1988) Interleukin 1 amplifies receptor-mediated activation of phospholipase A$_2$ in 3T3 fibroblasts. Proc Natl Acad Sci USA 85:6306–6309

Burchett SK, Weaver WM, Westall JA, Larsen A, Kronheim S, Wilson CB (1988) Regulation of tumor necrosis factor/cachectin and IL-1 secretion in human monoculear phagocytes. J Immunol 140:3473–3481

Bursten SL, Locksley RM, Ryan JL, Lovett DH (1988) Acylation of monocyte and glomerular mesangial cell proteins. Myristyl acylation of the interleukin 1 precursors. J Clin Invest 82:1479–1488

Cameron P, Limjuco G, Rodkey J, Bennett C, Schmidt JA (1985) Amino acid sequence analysis of human interleukin 1 (IL-1) Evidence for biochemically distinct forms of IL-1. J Exp Med 162:790–801

Cameron PM, Limjuco GA, Chin J, Silberstein L, Schmidt JA (1986) Purification to homogeneity and amino acid sequence analysis of two anionic species of human interleukin 1. J Exp Med 164:237–250

Caput D, Beutler B, Hartog K, Thayer R, Brown-Shimer S, Cerami A (1986) Identification of a common nucleotide sequence in the 3′-untranslated region of mRNA molecules specifying inflammatory mediators. Proc Natl Acad Sci USA 83:1670–1674

Carter DB, Curry KA, Tomich CSC, Yem AW, Deibel MR, Tracey DE, Paslay JW, Carter JB, Theriault NY, Harris PKW, Reardon IM, Jurcher-Neely HA, Heinrikson RL, Clancy LL, Muchmore SW, Watenpaugh KD, Einspahr HM (1988) Crystallization of purified recombinant human interleukin-1β. Proteins 3:121–129

Castelli MP, Black PL, Schneider M, Pennington R, Abe F, Talmadge JE (1988) Protective, restorative, and therapeutic properties of recombinant human IL-1 in rodent models. J Immunol 140:3830–3837

Cavender DE, Haskard DO, Joseph B, Ziff M (1986) Interleukin 1 increases the binding of human B and T lymphocytes to endothelial cell monolayers. J Immunol 136:203–207

Chang J, Gilman SC, Lewis AJ (1986) Interleukin 1 activates phospholipase A$_2$ in rabbit chondrocytes: a possible signal for IL 1 action. J Immunol 136:1283–1287

Chin J, Cameron PM, Rupp E, Schmidt JA (1987) Identification of a high-affinity receptor for native human interleukin 1β and interleukin 1α on normal human lung fibroblasts. J Exp Med 165:70–86

Chin J, Rupp E, Cameron PM, MacNaul KL, Lotke PA, Tocci MJ, Schmidt JA, Bayne EK (1988) Identification of a high-affinity receptor for interleukin 1α and interleukin 1β on cultured human rheumatoid synovial cells. J Clin Invest 82:420–426

Chiplunkar S, Langhorne J, Kaufmann SHE (1986) Stimulation of B cell growth and differentiation by murine recombinant interleukin 1. J Immunol 137:3748–3752

Clark BD, Collins KL, Gandy MS, Webb AC, Auron PE (1986) Genomic sequence for human prointerleukin 1 beta: possible evolution from a reverse transcribed prointerleukin 1 alpha gene. Nucleic Acids Res 14:7897–7914

Clark BD, Fenton MJ, Rey HL, Webb AC, Auron PE (1988) Characterization of cis and transacting elements involved in human proIL-1 beta gene expression. In: Powanda MC, Oppenheim JJ, Kluger MJ, Dinarello CA (eds) Monokines and other non-lymphocytic cytokines. Liss, New York, pp 47–53 (Progress in leukocyte biology, vol 8)

Conlon PJ, Grabstein KH, Alpert A, Prickett KS, Hopp TP, Gillis S (1987) Localization of human mononuclear cell interleukin 1. J Immunol 139:98–102

Content J, De Wit L, Poupart P, Opdenakker G, Van Damme J, Billiau A (1985) Induction of a 26-kDa-protein mRNA in human cells treated with an interleukin-1-related, leukocyte-derived factor. Eur J Biochem 152:253–257

Corbo L, Pizzano R, Scala G, Venuta S (1987) Expression of interleukins in L cells transfected with human DNA. Eur J Biochem 669:674

Cosman D (1987) Control of messenger RNA stability. Immunol Today 8:16–17

Cousins RJ, Leinart AS (1988) Tissue-specific regulation of zinc metabolism and metallothionein genes by interleukin 1. FASEB J 2:2884–2890

Craig S, Schmeissner U, Wingfield P, Pain RH (1987) Conformation, stability, and folding of interleukin 1β. Biochemistry 26:3570–3576

Cybulsky MI, McComb DJ, Movat HZ (1988) Neutrophil leukocyte emigration induced by endotoxin. Mediator roles of interleukin 1 and tumor necrosis factor alpha 1. J Immunol 140:3144–3149

Czuprynski CJ, Brown JF, Yong KM, Cooley AJ, Kurtz RS (1988) Effects of murine recombinant interleukin 1α on the host response to bacterial infection. J Immunol 140:962–968

Dayer JM, Stephenson ML, Schmidt E, Karge W, Krane SM (1981) Purification of a factor from human blood monocyte-macrophages which stimulates the production of collagenase and prostaglandin E_2 by cells cultured from rheumatoid synovial tissues. FEBS Lett 124:253–256

Daynes RA, Robertson BA, Cho BH, Burnham DK, Newton R (1987) α-Melanocyte-stimulating hormone exhibits target cell selectivity in its capacity to affect interleukin 1-inducible responses in vivo and in vitro. J Immunol 139:103–109

DeChiara TM, Yong D, Semionow R, Stern AS, Batula-Bernardo C, Fiedler-Nagy C, Kaffka KL, Kilian PL, Yamazaki S, Mizel SB, Lomedico PT (1986) Structure-function analysis of murine interleukin 1: biologically active polypeptides are at least 127 amino acids long and are derived from the carboxyl terminus of a 270-amino acid precursor. Proc Natl Acad Sci USA 83:8303–8307

Dejana E, Breviario F, Erroi A, Bussolino F, Mussoni L, Gramse M, Pintucci G, Casali B, Dinarello CA, Van Damme J, Mantovani A (1987) Modulation of endothelial cell functions by different molecular species of interleukin 1. Blood 69:695–699

Dejana E, Bertocchi F, Bortolami MC, Regonesi A, Tonta A, Breviario F, Giavazzi R (1988) Interleukin 1 promotes tumor cell adhesion to cultured human endothelial cells. J Clin Invest 82:1466–1470

Demczuk S (1988) IL1/etaf activity and undetectable IL1β mRNA and minimal IL1 α mRNA levels in normal adult heat-separated epidermis. Biochem Biophys Res Commun 156:463–469

D'Eustachio P, Jadidi S, Fuhlbrigge RC, Gray PW, Chaplin DD (1987) Interleukin-1 alpha and beta genes: linkage on chromosome 2 in the mouse. Immunogenetics 26:339–343

Dewhirst FE, Stashenko PP, Mole JE, Tsurumachi T (1985) Purification and partial sequence of human osteoclast-activating factor: identity with interleukin 1β. J Immunol 135:2562–2568

De Wit L, Poupart P, Opdenakker G, Van Damme J, Billiau A (1985) Induction of a 26-kDa-protein mRNA in human cells treated with an interleukin-1-related, leukocyte-derived factor. Eur J Biochem 152:253–257

Didier M, Aussel C, Pellassy C, Fehlmann M (1988) IL-1 signaling for IL-2 production in T cells involves a rise in phosphatidylserine synthesis. J Immunol 141:3078–3080

Di Giovine FS, Malawista SE, Nuki G, Duff GW (1987) Interleukin 1 (IL1) as a mediator of crystal arthritis. Stimulation of T cell and synovial fibroblast mitogenesis by urate crystal-induced IL 1. J Immunol 138:3213–3218

Dinarello CA (1984) Interleukin-1. Rev Infect Dis 6:51–95

Dinarello CA, Clowes GHA, Gordon AH, Saravis CA, Wolff SM (1984a) Cleavage of human interleukin 1: isolation of a peptide fragment from plasma of febrile humans and activated monocytes. J Immunol 133:1332–1338

Dinarello CA, Bishai I, Rosenwasser LJ, Coceani F (1984b) The influence of lipoxygenase inhibitors on the in vitro production of human leukocytic pyrogen and lymphocyte activating factor (IL-1). Int J Immunopharmacol 6:43–50

Dinarello CA, Cannon JG, Mier JW, Bernheim HA, LoPreste G, Lynn DL, Love RN, Webb AC, Auron PE, Reuben RC, Rich A, Wolff S, Putney SD (1986a) Multiple biological activities of human recombinant interleukin 1. J Clin Invest 77:1734–1739

Dinarello CA, Cannon JG, Wolff SM, Bernheim HA, Beutler B, Cerami A, Figari IS, Palladino MA, O'Connor JV (1986b) Tumor necrosis factor (cachectin) is an endogenous pyrogen and induces production of interleukin 1. J Exp Med 163:1433–1450

Dinarello CA, Ikejima T, Warner SJC, Orencole SF, Lonnemann G, Cannon JG, Libby P (1987) Interleukin 1 induces interleukin 1. I Induction of circulating interleukin 1 in rabbits in vivo and in human monoclear cells in vitro. J Immunol 139:1902–1910

Dower SK, Urdal DL (1987) The interleukin-1 receptor. Immunol Today 8:46–51

Dower SK, Call SM, Gillis S, Urdal DL (1985a) Similarity between the interleukin 1 receptors on a murine T-lymphoma cell line and on a murine fibroblast cell line. Proc Natl Acad Sci USA 83:1060–1064

Dower SK, Kronheim SR, March CJ, Conlon PJ, Hopp TP, Gillis S, Urdal DL (1985b) Detection and characterization of high affinity plasma membrane receptors for human interleukin 1. J Exp Med 162:501–515

Dower SK, Kronheim SR, Hopp TP, Cantrell M, Deeley M, Gillis S, Henney CS, Urdal DL (1986) The cell surface receptors for interleukin 1α and interleukin-1β are identical. Nature 324:266–268

Duff GW, Dickens E, Wood N, Manson J, Symons J, Poole S, Di Giovine F (1988) Immunoassay, bioassay and in situ hybridization of monokines in human arthritis. In: Powanda MC, Oppenheim JJ, Kluger MJ, Dinarella CA (eds) Monokines and other non-lymphocytic cytokines. Liss, New York, pp 387–392 (Progress in leukocyte biology, vol 8)

Durum SK, Schmidt JA, Oppenheim JJ (1985) Interleukin-1: an immunological perspective. Annu Rev Immunol 3:263–287

Elias JA, Schreiber AD, Gustilo K, Chien P, Rossmann MD, Lammie PJ, Daniel RP (1985) Differential interleukin 1 elaboration by unfractionated and density fractionated human alveolar macrophages and blood monocytes: relationship to cell maturity. J Immunol 135:3198–3204

Ernst JF (1988) Efficient secretion and processing of heterologous proteins in Saccharomyces cervisiae is mediated solety by the pre-segment of α-factor precursor. DNA 7:355–360

Estes JE, Pledger WJ, Gillespie GY (1984) Macrophage-derived growth factor for fibroblasts and interleukin-1 are distinct entities. J Leukocyte Biol 35:115–129

Falkoff RJM, Muraguchi A, Hong JX, Butler JL, Dinarello CA, Fauci AS (1983) The effects of interleukin 1 on human B cell activation and proliferation. J Immunol 131:801–805

Farrar WL, Kilian P, Hill JM, Ruff MR, Pert CB (1987a) Visualization of cytokine and virus receptors common to the immune and central nervous system. Lymphokine Res 6:29–34

Farrar WL, Kilian PL, Ruff MR, Hill JM, Pert CB (1987b) Visulization and characterization of interleukin 1 receptors in brain. J Immunol 139:459–463

Fenton MJ, Clark BD, Collins KL, Webb AC, Rich A, Auron PE (1987) Transcriptional regulation of the human prointerleukin 1β gene. J Immunol 138:3972–3979

Fenton MJ, Vermeulen MW, Clark BD, Webb AC, Auron PE (1988) Human pro-IL-1β gene expression in monocytic cells is regulated by two distinct pathways. J Immunol 140:2267–2273

Ferreira SH, Lorenzetti BB, Bristow AF, Poole S (1988) Interleukin-1β as a potent hyperalgesic agent antagonized by a tripeptide analogue. Nature 334:698–700

Fibbe WE, VanDamme J, Billiau A, Voogt PJ, Duinkerken N (1987) Purified human interleukin-1 induces GM-CSF release by mononuclear phagocytes. Blood 68:1316–1321

Fontana A, Kristensen F, Dubs R, Gemsa D, Weber E (1982) Production of prostaglandin E and an interleukin-1 like factor by cultured astrocytes and C_6 glioma cells. J Immunol 129:2413–2419

Frasca D, Boraschi D, Baschieri S, Bossu P, Tagliabue A, Adorini L, Doria G (1988) In vivo restoration of T cell functions by human IL-1β or its 163–171 nonapeptide in immunodepressed mice. J Immunol 141:2651–2655

Freedman AS, Freeman G, Whitman J, Segil J, Daley J, Nadler LM (1988) Pre-exposure of human B cells to recombinant IL-1 enhances subsequent proliferation. J Immunol 141:3398–3404

Frisch SM, Ruley HE (1987) Transcription from the stromelysin promoter is induced by interleukin-1 and repressed by dexamethasone. J Biol Chem 262:16300–16304

Fuhlbrigge RC, Sheehan KCF, Schreiber RD, Chaplin DD, Unanue ER (1988) Monoclonal antibodies to murine IL-1α. Production, characterization, and inhibition of membrane-associated IL-1 activity. J Immunol 141:2643–2650

Furutani Y, Notake M, Yamayoshi M, Yamagishi J, Nomura H, Ohue M, Furuta R, Fukui T, Yamada M, Nakamura S (1985) Cloning and characterization of the cDNAs for human and rabbit interleukin-1 precursor. Nucleic Acids Res 13:5869–5882

Furutani Y, Notake M, Fukui T, Ohue M, Normura H, Yamada M, Nakamura S (1986) Complete nucleotide sequence of the gene for human interleukin 1 alpha. Nucleic Acids Res 14:3167–3179

Garman RD, Raulet DH (1987a) Characterization of a novel murine T cell-activating factor. J Immunol 138:1121–1129

Garman RD, Jacobs KA, Clark SC, Raulet HD (1987b) B-cell-stimulatory factor 2 ($\beta_2$infereron) functions as a second signal for interleukin 2 production by mature murine T cells. Proc Natl Acad Sci USA 84:7629–7633

Georgilis K, Schaefer C, Dinarello CA, Klempner MS (1987) Human recombinant interleukin 1β has no effect on intracellular calcium or on functional responses of human neutrophils. J Immunol 138:3403–3407

Gerrard TL, Siegel JP, Dyer DR, Zoon KC (1987) Differential effects of interferon-α and interferon-γ on interleukin 1 secretion by monocytes. J Immunol 138:2535–2540

Gery I, Gershon RK, Waksman B (1972) Potentiation of the T lymphocyte response to mitogens. I. The responding cell. J Exp Med 136:128–142

Gery I, Davis P, Derr J, Krett N, Barranger JA (1981) Relationship between production and release of lymphocyte-activating factor (interleukin 1) by murine macrophages. Effects of various agents. Cell Immunol 64:293–303

Ghezzi P, Dinarello CA (1988) IL-1 induces IL-1. III. Specific inhibition of IL-1 production by IFN-gamma. J Immunol 140:4238–4244

Gilliland GL, Winborne EL, Masui Y, Hirai Y (1987) A preliminary crystallographic study of recombinant human interleukin 1β. J Biol Chem 262:12323–12324

Giulian D, Baker TJ, Shih LCN, Lachman LB (1986) Interleukin 1 of the central nervous system is produced by ameboid microglia. J Exp Med 164:594–604

Gimenez-Gallego G, Rodkey J, Bennett C, Rios-Candelore M, DiSalvo J, Thomas K (1985) Brain-derived acidic fibroblast growth factor: complete amino acid sequence and homologies. Science 230:1385–1388

Giri JG, Lomedico PT, Mizel SB (1985) Studies on the synthesis and secretion of interleukin 1. I. A 33,000 molecular weight precursor for interleukin 1. J Immunol 134:343–349

Goldring MB, Birkhead J, Sandell LJ, Kimura T, Krane SM (1988) Interleukin 1 suppresses expression of cartilage-specific types II and IX collagens and increases types I and III collagens in chondrocytes. J Clin Invest 82:2026–2031

Goto F, Goto K, Ohkawara S, Kitamura M, Mori S, Takahashi H, Sengoku Y, Yoshinaga M (1988) Purification and partial sequence of rabbit polymorphonuclear leukocyte-derived lymphocyte proliferation potentiating factor resembling IL-1β. J Immunol 140:1153–1158

Goud SN, Muthusamy N, Subbarao B (1988) Differential responses of B cells from the spleen and lymph node to TNP-Ficoll. J Immunol 140:2925–2930

Gowen M, Mundy GR (1986) Actions of recombinant interleukin 1, interleukin 2, and interferon-γ on bone resorption in vitro J Immunol 136:2478–2482

Gowen M, Wood DD, Ihrie EJ, McGuire MKB, Russell RGG (1983) An interleukin 1-like factor stimulates bone resorption in vitro. Nature 306:378–380

Gowen M, Wood DD, Ihrie EJ, Meats JE, Russell RG (1984) Stimulation by human interleukin 1 of cartilage breakdown and production of collagenase and proteoglycanase by human chondrocytes but not by human osteoblasts in vitro. Biochim Biophys Acta 797:186–193

Granstein RD, Margolis R, Mizel SB, Sauder DN (1985) In vivo inflammatory activity of epidermal cell-derived thymocyte activating factor and recombinant interleukin 1 in the mouse. J Clin Invest 44:1020

Gray PW, Glaister D, Chen E, Goeddel DV, Pennica D (1986) Two interleukin 1 genes in the mouse: cloning and expression of the cDNA for murine interleukin 1β. J Immunol 137:3644–3648

Greenbaum LA, Horowitz JB, Woods A, Pasqualini, Reich EP, Bottomly K (1988) Autocrine growth of CD4$^+$ T cells. Differential effects of IL-1 on helper and inflammatory T cells. J Immunol 140:1555–1560

Gubler U, Chua AO, Stern AS, Hellmann CP, Vitek MP, Dechiara TM, Benjamin WR, Collier KJ, Dukovich M, Familletti PC, Fiedler-Nagy C, Jenson J, Kaffk K, Kilian PL, Stremlo D, Wittreich BH, Woehle D, Mizel SB, Lomedico PT (1986) Recombinant human interleukin 1α: purification and biological characterization. J Immunol 136:2492–2497

Hackett RJ, Davis LS, Lipsky PE (1988) Comparative effects of tumor necrosis factor-α and IL-1β on mitogen-induced T cell activation. J Immunol 140:2639–2644

Hagiwara H, Huang HJ, Arai N, Herzenberg LA, Arai K, Zlotnik A (1987) Interleukin 1 modulates messenger RNA levels of lymphokines and of other molecules associated with T cell activation in the T cell lymphoma LBRM33-1A5. J Immunol 138:2514–2519

Hazuda DJ, Lee JC, Young PR (1988) The kinetics of interleukin 1 secretion from activated monocytes. Differences between interleukin 1 alpha and interleukin 1 beta. J Biol Chem 263:8473–8479

Hazuda D, Webb RL, Simon P, Young P (1989) Purification and characterization of human recombinant precursor interleukin 1β. J Biol Chem 264:1689–1693

Hession C, Decker JM, Sherblom AP, Kumar S, Yue CC, Mattalliano RJ, Tizard R, Kawashima E, Schmeissner U, Heletky S, Chow EP, Burne CA, Shaw A, Muchmore AV (1987) Uromodulin (Tamm-Horsfall glycoprotein): a renal ligand for lymphokines. Science 237:1479–1484

Ho SN, Abraham RT, Nilson A, Handwerger BS, McKean DJ (1987) Interleukin 1-mediated activation of interleukin 4 (IL 4)-producing T lymphocytes. Proliferation by IL 4-dependent and IL 4-independent mechanisms. J Immunol 139:1532–1540

Holtmann H, Wallach D (1987) Down regulation of the receptors for tumor necrosis factor by interleukin 1 and 4β-phorbol-12 myristate-13-acetate. J Immunol 139:1161–1167

Hom JT, Bendele AM, Carlson DG (1988) In vivo administration with IL-1 accelerates the development of collagen-induced arthritis in mice. J Immunol 141:834–841

Horuk R, Huang JJ, Covington M, Newton RC (1987) A biochemical and kinetic analysis of the interleukin-1 receptor. Evidence for differences in molecular properties of IL-1 receptors. J Biol Chem 262:16275–16278

Howard M, Mizel SB, Lachman L, Ansel J, Johnson B, Paul WE (1983) Role of interleukin 1 in anti-immunoglobulin-induced B cell proliferation. J Exp Med 157:1529–1543

Huang JJ, Newton RC, Horuk R, Matthew JB, Covington M, Pezzella K, Lin YA (1987) Muteins of human interleukin-1 that show enhanced bioactivities. FEBS Lett 223:294–298

Huang JJ, Newton RC, Rutledge SJ, Horuk R, Matthew JB, Covington M, Lin Y (1988) Characterization of Murine IL-1β isolation, expression, and purification. J Immunol 140:3838–3843

Hurme M (1987) Membrane-associated interleukin 1 is required for the activation of T cells in the anti-CD3 antibody-induced T cell response. J Immunol 139:1168–1172

Jacobs C, Young D, Tyler S, Callis G, Gillis S, Conlon PJ (1988) In vivo treatment with IL-1 reduces the severity and duration of antigen-induced arthritis in rats. J Immunol 141:2967–2974

Jelinek DF, Lipsky PE (1987) Enhancement of human B cell proliferation and differentiation by tumor necrosis factor-α and interleukin 1. J Immunol 139:2970–2976

Jobling SA, Auron PE, Gurka G, Webb AC, McDonald B, Rosenwasser LJ, Gehrke L (1988) Biological activity and receptor binding of human prointerleukin-1β and subpeptides. J Biol Chem 31:16372–16378

Karin M, Imbra RJ, Heguy A, Wong G (1985) Interleukin 1 regulates human metallothionein gene expression. Mol Cell Biol 5:2866–2869

Kasahara T, Mukaida N, Shinomiya H, Imai M, Matsushima K, Wakasugi H, Nakano K (1987) Preparation and characterization of polyclonal and monoclonal antibodies against human interleukin 1α (IL 1α). J Immunol 138:1804–1812

Kaye J, Gillis S, Mizel SB, Shevach EM, Malek TR, Dinarello CA, Lachman LB, Janeway CA (1984) Growth of a cloned helper T cell line induced by a monoclonal antibody specific for the antigen receptor: interleukin 1 is required for the expression of receptors for interleukin 2. J Immunol 133:1339–1345

Kenney JS, Masada MP, Eugui EM, Delustro BM, Mulkins MA, Allison AC (1987) Monoclonal antibodies to human recombinant interleukin 1 (IL 1)β: quantitation of IL 1β and inhibition of biological activity. J Immunol 138:4236–4242

Kern JA, Lamb RJ, Reed JC, Daniele RP, Nowell PC (1987) Dexamethasone inhibition of interleukin 1 beta production by human monocytes. Posttranscriptional mechanisms. J Clin Invest 81:237–244

Kilian PL, Kaffka KL, Stern AS, Woehle D, Benjamin WR, Dechiara TM, Gubler U, Farrar JJ, Mizel SB, Lomedico PT (1986) Interleukin 1α and interleukin 1β bind to the same receptor on T cells. J Immunol 136:4509–4514

Kimball ES, Fisher MC (1988) Potentiation of IL-1-induced Balb/3T3 fibroblast proliferation by neuropeptides. J Immunol 141:4203–4208

Kimball ES, Persico FJ, Vaught JL (1988) Substance P, neurokinin A and neurokinin B induce generation of IL-1-like activity in P388D1 cells. J Immunol 141:3564–3569

Klempner MS, Dinarello CA, Gallin JI (1978) Human leukocytic pyrogen induces release of specific granule contents from human neutrophils. J Clin Invest 61:1330–1336

Knudsen PJ, Dinarello CA, Strom TB (1986) Prostaglandins posttranscriptionally inhibit monocyte expression of interleukin 1 activity by increasing intracellular cyclic adenosine monophosphate. J Immunol 137:3189–3194

Knudsen PJ, Dinarello CA, Strom TB (1987) Glucocorticoids inhibit transcriptional and post-transcriptional expression of interleukin 1 in U937 cells. J Immunol 139:4129–4134

Kobayashi Y, Appella E, Yamada M, Copeland TD, Oppenheim JJ, Matsushima K (1988) Phosphorylation of intracellular precursors of human IL-1. J Immunol 140:2279–2287

Kock A, Danner M, Stadler BM, Luger TA (1986) Characterization of a monoclonal antibody directed against the biologically active site of human interleukin 1. J Exp Med 163:463–468

Koide S, Steinman RM (1987) Induction of murine interleukin 1: stimuli and responsive primary cells. Proc Natl Acad Sci USA 84:3802–3806

Koide SL, Inaba K, Steinman RM (1987) Interleukin 1 enhances T-dependent immune responses by amplifying the function of dendritic cells. J Exp Med 165:515–530

Korn JH, Brown KM, Downie E, Liao ZH, Rosenstreich DL (1987) Augmentation of IL 1-induced fibroblast PGE_2 production by a urine-derived IL 1 inhibitor. J Immunol 138:3290–3294

Kovacs EJ, Oppenheim JJ, Young HA (1986) Induction of c-fos and c-myc expression in T lymphocytes after treatment with recombinant interleukin 1-α. J Immunol 137:3649–3651

Kovacs EJ, Radzioch D, Young HA, Varesio L (1988) Differential inhibition of IL-1 and TNF-α mRNA expression by agents which block second messenger pathways in murine macrophages. J Immunol 141:3101–3105

Krakauer T, Oppenheim JJ (1983) Interleukin 1 production by a human acute monocytic leukemia cell line. Cell Immunol 80:223–229

Krakauer T, Oppenheim JJ, Jasin HE (1985) Human interleukin 1 mediates cartilage matrix degradation. Cell Immunol 91:92–95

Krane SM, Dayer JM, Simon LS, Byrne MS (1985) Mononuclear cell-conditioned medium containing mononuclear cell factor (MCF) homologous with interleukin 1, stimulates collagen and fibronectin synthesis by adherent rheumatoid synovial cells: effects of prostaglandin E_1 and indomethacin. Coll Relat Res 5:99–117

Kriegler M, Perez C, DeFay Albert I, Lu SD (1988) A novel form of TNF/cachectin is a cell surface cytotoxic transmembrane protein: ramifications for the complex physiology of TNF. Cell 53:45–53

Kronheim SR, Cantrell MA, Deeley MC, March CJ, Glackin PJ, Anderson DM, Hemenway T, Merriam JE, Cosman D, Hopp TP (1986) Purification and characterization of human interleukin-1 expressed in Escherichia coli. Biotechnology 4:1079–1082

Kupper TS, Ballard DW, Chua AO, McGuire JS, Flood PM, Horowitz MC, Langdon R, Lightfoot L, Gubler U (1986) Human keratinocytes contain mRNA indistinguishable from monocyte interleukin 1α and β mRNA. Keratinocyte epidermal cell-derived thymocyte-activating factor is identical to interleukin 1. J Exp Med 164:2095–2100

Kupper TS, Chua AO, Flood P, McGuire J, Gubler U (1987) Interleukin gene expression in cultured human keratinocytes is augmented by ultraviolet irradiation. J Clin Invest 80:430–436

Kupper TS, Lee F, Birchall N, Clark S, Dower S (1988) Interleukin 1 binds to specific receptors on human keratinocytes and induces granulocyte macrophage colony-stimulating factor mRNA and protein. A potential autocrine role for interleukin 1 in epidermis. J Clin Invest 82:1787–1792

Kurt-Jones EA, Beller DI, Mizel SB, Unanue ER (1985a) Identification of a membrane-associated interleukin 1 in macrophages. Proc Natl Acad Sci USA 82:1204–1208

Kurt-Jones EA, Kiely JM, Unanue ER (1985b) Conditions required for expression of membrane IL 1 on B cells. J Immunol 135:1548–1550

Kurt-Jones EA, Virgin HW IV, Unanue ER (1986) In vivo and in vitro expression of macrophage membrane interleukin 1 in response to soluble and particulate stimuli. J Immunol 137:10–14

Kurt-Jones EA, Fiers W, Pober JS (1987a) Membrane interleukin 1 induction on human endothelial cells and dermal fibroblasts. J Immunol 139:2317–2324

Kurt-Jones EA, Hamberg S, Ohara J, Paul WE, Abbas AK (1987b) Heterogeneity of helper inducer T lymphocytes. I. Lymphokine production and lymphokine responsiveness. J Exp Med 166:1774–1787

Last-Barney K, Homon CA, Faanes RB, Merluzzi VJ (1988) Synergistic and overlapping activities of tumor necrosis factor-α and IL-1. J Immunol 141:527–530

Le JM, Weinstein D, Gubler U, Vilcek J (1987) Induction of membrane-associated interleukin 1 by tumor necrosis factor in human fibroblasts. J Immunol 138:2137–2142

Le PT, Tuck DT, Dinarello CA, Haynes BF, Singer KH (1987) Human thymic epithelial cells produce interleukin 1. J Immunol 138:2520–2526

Lee JC, Griswold DE, Votta B, Hanna N (1988) Inhibition of monocyte IL-1 production by the anti-inflammatory compound. SK and F 86002. Int J Immunopharmacol 10:835–843

Lee SW, Tsou AP, Chan H, Thomas J, Petrie K, Eugui EM, Allison AC (1988) Glucocorticoids selectively inhibit the transcription of the interleukin 1β gene and decrease the stability of interleukin 1β mRNA. Proc Natl Acad Sci USA 85:1204–1208

Lemaire I (1988) Neurotensin enhances IL-1 production by activated alveolar macrophages. J Immunol 140:2983–2988

Lew W, Oppenheim J, Matsushima K (1988) Analysis of the suppression of IL-1α and IL-1β production in human peripheral blood mononuclear adherent cells by a glucocorticoid hormone. J Immunol 140:1895–1902

Liao Z, Haimovitz A, Chen Y, Chan J, Rosenstreich DL (1985) Characterization of a human interleukin 1 inhibitor. J Immunol 134:3882–3886

Libby P, Warner SJC, Friedman GB (1988) Interleukin 1: a mitogen for human vascular smooth muscle cells that induces the release of growth-inhibitory prostanoids. J Clin Invest 81:487–498

Lichtman AH, Kurt-Jones EA, Abbas AK (1987) B-cell stimulatory factor 1 and not interleukin 2 is the autocrine growth factor for some helper T lymphocytes. Proc Natl Acad Sci 84:824–827

Lichtman AH, Chin J, Schmidt JA, Abbas AK (1988) Role of interleukin 1 in the activation of T lymphocytes. Proc Natl Acad Sci USA 85:9699–9703

Lillquist JS, Simon PL, Summers MA, Jonak Z, Young PR (1988) Structure-activity studies of human IL-1β with mature and truncated proteins expressed in Escherichia coli. J Immunol 141:1975–1981

Limjuco G, Galuska S, Chin J, Cameron P, Boger J, Schmidt JA (1986) Antibodies of predetermined specificity to the major charged species of human interleukin 1. Proc Natl Acad Sci USA 83:3972–3976

Lin JX, Vilcek J (1987) Tumor necrosis factor and interleukin-1 cause a rapid and transient stimulation of c-fos and c-myc mRNA levels in human fibroblasts. J Biol Chem 262:11908–11911

Lindemann A, Riedel D, Oster W, Meuer SC, Blohm D, Mertelsmann RH, Herrmann F (1988) Granulocyte/macrophage colony-stimulating factor induces interleukin 1 production by human polymorphonuclear neutrophils. J Immunol 140:837–839

Lipsky PE, Thompson PA, Rosenwasser LJ, Dinarello CA (1983) The role of interleukin 1 in human B cell activation: inhibition of B cell proliferation and the generation of immunoglobulin-secreting cells by an antibody against human leukocyte pyrogen. J Immunol 130:2708–2714

Lomedico PT, Gubler U, Hellmann CP, Dukovich M, Giri JG, Pan YE, Collier K, Semionow R, Chua AO, Mizel SB (1984) Cloning and expression of murine interleukin-1 cDNA in Escherichia coli. Nature 312:458–461

Loppnow H, Brade L, Brade H, Rietschel ET, Kusumoto S, Shiba T, Flad HD (1986) Induction of human interleukin 1 by bacterial and synthetic lipid A. Eur J Immunol 16:1263–1267

Lovett DH, Larsen A (1988) Cell cycle-dependent interleukin 1 gene expression by cultured glomerular mesangial cells. J Clin Invest 82:115–122

Lovett DH, Ryan JL, Sterzel RB (1983) Stimulation of rat mesangial cell proliferation by macrophage interleukin 1. J Immunol 131:2830–2836

Lovett DH, Szamel M, Ryan JL, Sterzel RB, Gemsa D, Resch K (1986) Interleukin 1 and the glomerular mesangium. I. Purification and characterization of a mesangial cell-derived autogrowth factor. J Immunol 136:3700–3705

Low MG, Ferguson MAJ, Futerman AH, Silman I (1986) Covalently attached phosphatidylinositol as a hydrophobic anchor for membrane proteins. Trends Biochem Sci 11:212–215

Lowenthal JW, MacDonald HR (1986) Binding and internalization of interleukin 1 by T cells: direct evidence for high- and low-affinity classes of interleukin 1 receptor. J Exp Med 164:1060–1074

Lowenthal JW, MacDonald HR (1987) Expression of interleukin 1 receptors is restricted to the L3T4$^+$ subset of mature T lymphocytes. J Immunol 138:1–3

Lubinski J, Fong TC, Babbitt JT, Ransone L, Yodoi JJ, Bloom ET (1988) Increased binding of IL-2 and increased IL-2 receptor mRNA synthesis are expressed by an NK-like cell line in response to IL-1. J Immunol 140:1903–1909

Luger TA, Stadler BM, Katz SI, Oppenheim JJ (1981) Epidermal cell (keratinocyte) derived thymocyte activating factor (ETAF). J Immunol 127:1493–1498

Luger TA, Stadler BM, Luger BM, Mathieson BJ, Mage M, Schmidt JA, Oppenheim JJ (1982) Murine epidermal cell-derived thymocyte activating factor resembles murine interleukin 1. J Immunol 125:2147–2152

Lumpkin MD (1987) The regulation of ACTH secretion by IL-1. Science 238:452–454

MacDonald HR, Wingfield P, Schmeissner U, Shaw A, Clore GM, Gronenborn AM (1986) Point mutations of human interleukin-1 with decreased receptor binding affinity. FEBS Lett 209:295–298

Malek TR, Schmidt JA, Shevach EM (1985) The murine IL 2 receptor. III. Cellular requirements for the induction of IL2 receptor expression on T cell subpopulations. J Immunol 134:2405–2413

Maliszewski CR, Baker PE, Schoenborn MA, Davis BS, Cosman D, Gillis S, Cerretti DP (1988) Cloning, sequence and expression of bovine interleukin 1α and interleukin 1β complementary DNAs. Mol Immunol 25:429–437

Mandrup-Poulsen T, Bendtzen K, Dinarello CA, Nerup J (1987) Human tumor necrosis factor potentiates human interleukin 1-mediated rat pancreatic β-cell cytotoxicity. J Immunol 139:4077–4082

Manger B, Weiss A, Weyand C, Goronzy J, Stobo JD (1985) T cell activation: differences in the signals required for IL 2 production by nonactivated and activated T cells. J Immunol 135:3669–3673

March CJ, Mosley B, Larsen A, Cerretti DP, Braedt G, Price V, Gillis S, Henney CS, Kronheim SR, Grabstein K, Conlon PJ, Hopp TP, Cosman D (1985) Cloning, sequence and expression of two distinct human interleukin-1 complementary DNA's. Nature 315:641–647

Massone A, Baldari C, Censini S, Bartalini M, Nucci D, Broaschi D, Telford JL (1988) Mapping of biologically relevant sites on human IL-β using monoclonal antibodies. J Immunol 140:3812–3816

Masuda A, Longo DL, Kobayashi Y, Appella E, Oppenheim JJ, Matsushima K (1988) Induction of mitochondrial manganese superoxide dismutase by interleukin 1. FASEB J 2:3087–3091

Masui Y, Kamogashira T, Hong Y-M, Kikumoto Y, Nakai S, Hirai Y (1988) The role of cysteine residues in human interleukin-1β. In: Marshall GR (ed) Peptides: chemistry and biology. Escom, Leiden, pp 396–398

Matsushima K, Copeland TD, Onozaki K, Oppenheim JJ (1986a) Purifcation and biochemical characteristics of two distinct human interleukins 1 from the myelomonocytic THP-1 cell line. Biochemistry 25:3424–3429

Matsushima K, Yodoi J, Tagaya Y, Oppenheim JJ (1986b) Down-regulation of interleukin 1 (IL 1) receptor expression by IL 1 and fate of internalized ^{125}I-labeled IL 1β in a human large granular lymphocyte cell line. J Immunol 137:3183–3188

Matsushima K, Akahoshi T, Yamada M, Furutani Y, Oppenheim JJ (1986c) Properties of a specific interleukin 1 (IL 1) receptor on human Epstein Barr virus-transformed B lymphocytes: identity of the receptor for IL 1α and IL 1β. J Immunol 136:4496–4502

Matsushima K, Taguchi M, Kovacs EJ, Young HA, and Oppenheim JJ (1986d) Intracellular localization of human monocyte associated interleukin 1 (IL 1) activity and release of biologically active IL 1 from monocytes by trypsin and plasmin. J Immunol 136:2883–2891

Matsushima K, Kobayashi Y, Copeland TD, Akahoshi T, Oppenheim JJ (1987) Phosphorylation of a cytosolic 65-kDa protein induced by interleukin 1 in glucocorticoid pretreated normal human peripheral blood mononuclear leukocytes. J Immunol 139:3367–3374

Matsushima K, Morishita K, Yoshimura T, Lavu S, Kobayashi Y, Lew W, Appella E, Kung HF, Leonard EJ, Oppenheim JJ (1988) Molecular cloning of a human monocyte-derived neutrophil chemotactic factor (MDNCF) and the induction of MDNCF mRNA by interleukin 1 and tumor necrosis factor. J Exp Med 167:1883–1893

McCroskery PA, Arai S, Amento EP, Krane SM (1985) Stimulation of procollagenase synthesis in human rheumatoid synovial fibroblasts by mononuclear cell factor/interleukin-1. FEBS Lett 191:7–12

McGuire-Goldring MB, Meats JE, Wood DD, Ihrie EJ, Ebsworth NM, Russell RGG (1984) In vitro activation of human chondrocytes and synoviocytes by a human interleukin-1-like factor. Arthritis Rheum 27:654–662

Meyers CA, Johanson KO, Miles LM, McDevitt PJ, Simon PL, Webb RL, Chen MJ, Holskin BP, Lillquist JS, Young PR (1987) Purification and characterization of human recombinant interleukin 1β. J Biol Chem 262:11176–11181

Michie HR, Manogue KR, Spriggs DR, Revhaug A, O'Dwyer S, Dinarello CA, Cerami A, Wolff SM, Wilmore DW (1988) Detection of circulating tumor necrosis factor after endotoxin administration. N Engl J Med 318:1481–1486

Minnich-Carruth LL, Suttles J, Mizel SB (1989) Evidence against the existence of a membrane form of murine IL-1α. J Immunol 142:526–530

Mizel SB, Dayer JM, Krane SM, Mergenhagen SE (1981) Stimulation of rheumatoid synovial cell collagenase and prostaglandin production by partially purified lymphocyte-activating factor (interleukin 1). Proc Natl Acad Sci USA 78:2474–2477

Mizel SB, Kilian PL, Lewis JC, Paganelli KA, Chizzonite RA (1987) The interleukin 1 receptor. Dynamics of interleukin 1 binding and internalization in T cells and fibroblasts. J Immunol 138:2906–2912

Mochizuki DY, Eisenman JR, Conlon PJ, Larsen AD, Tushinski RJ (1987) Interleukin 1 regulates hematopoietic activity, a role previously ascribed to hemopoietin 1. Cell Biol 84:5267–5271

Moore MAS, Warren DJ (1987) Synergy of interleukin 1 and granulocyte colony-stimulating factor: in vivo stimulation of stem-cell recovery and hematopoietic regeneration following 5-fluorouracil treatment of mice. Proc Natl Acad Sci USA 84:7134–7138

Morgan EL, Hobbs MV, Noonan DJ, Weigly WO (1988) Induction of IL-1 secretion from human monocytes by Fc region subfragments of human IgG1. J Immunol 140:3014–3020

Mori S, Goto F, Goto K, Ohkawara S, Maeda S, Shimada K, Yoshinaga M (1988) Cloning and sequence analysis of a cDNA for lymphocyte proliferation potentiating factor of rabbit polymorphonuclear leukocytes: identification as rabbit interleukin 1β. Biochem Biophys Res Commun 150:1237–1243

Morrissey PJ, Charrier K, Alpert A, Bressler L (1988a) In vivo administration of IL-1 induces thymic hypoplasia and increased levels of serum corticosterone. J Immunol 141:1456–1463

Morrissey P, Charrier K, Bressler L, Alpert A (1988b) The influence of IL-1 treatment on the reconstitution of the hemopoietic and immune systems after sublethal radiation. J Immunol 140:4204–4210

Mortensen RF, Shapiro J, Lin BF, Douches S, Neta R (1988) Interaction of recombinant IL-1 and recombinant tumor necrosis factor in the induction of mouse acute phase proteins. J Immunol 140:2260–2266

Moser R, Schleiffenbaum B, Groscurth P, Fehr J (1989) Interleukin 1 and tumor necrosis factor stimulate human vascular endothelial cells to promote transendothelial neutrophil passage. J Clin Invest 83:444–455

Mosley B, Dower SK, Gillis S, Cosman D (1987a) Determination of the minimum polypeptide lengths of the functionally active sites of human interleukins 1α and 1β. Proc Natl Acad Sci USA 84:4572–4576

Mosley B, Urdal DL, Prickett KS, Larsen A, Cosman D, Conlon PJ, Gillis S, Dower SK (1987b) The interleukin-1 receptor binds the human interleukin-1α precursor but not the interleukin-1β precursor. J Biol Chem 262:2941–2944

Mosmann TR, Cherwinski H, Bond MW, Geidlin MA, Coffman RL (1986) Two types of murine helper T cell clones. I. Definition according to profiles of lymphokine activities and secreted proteins. J Immunol 136:2348

Muchmore AV, Decker JM (1986) Uromodulin, an immunosuppressive 85 kD glycoprotein isolated from human pregnancy urine, is a high affinity ligand for recombinant IL 1. J Biol Chem 261:13404

Muchmore AV, Decker JM (1987) Evidence that recombinant IL1 α exhibits lectin-like specificity and binds to homogeneous uromodulin via N-linked oligosaccharides. J Immunol 138:2541–2546

Mukaida N, Kasahara T, Yagisawa H, Shioiri-Nakano K, Kawai T (1987) Signal requirement for interleukin 1-dependent interleukin 2 production by a human leukemia-derived HSB.2 subclone. J Immunol 139:3321–3329

Murphy G, Reynolds JJ, Werb Z (1985) Biosynthesis of tissue inhibitor of metalloproteinases by human fibroblasts in culture: stimulation by 12-O-tetradecanoylphorbol 13-acetate and interleukin 1 in parallel with collagenase. J Biol Chem 260:3079–3083

Nagelkerken LM, van Breda Vriesman PJC (1986) Membrane-associated IL 1-like activity on rat dendritic cells. J Immunol 136:2164–2170

Neta R, Douches SD, Oppenheim JJ (1986) Interleukin 1 is a radioprotector. J Immunol 136:2483–2485

Neta R, Sztein MB, Oppenheim JJ, Gillis S, Douches SD (1987) The in vivo effects of interleukin 1. I. Bone marrow cells are induced to cycle after administration of interleukin 1. J Immunol 139:1861–1986

Neta R, Oppenheim JJ, Douches SD (1988) Interdependence of the radioprotective effects of human recombinant interleukin 1α, granulocyte colony-stimulating factor, and murine recombinant granulocyte-macrophage colony-stimulating factor. J Immunol 140:208–211

Numerof RP, Aronson FR, Mier JW (1988) IL-2 stimulates the production of IL-1α and IL-1β by human peripheral blood mononuclear cells. J Immunol 141:4250–4257

Okusawa S, Dinarello CA, Yancey KB, Endres S, Lawley TJ, Frank MM, Burke JF, Gelfand JA (1987) C5a induction of human interleukin 1. Synergistic effect with endotoxin or interferon-γ. J Immunol 139:2635–2640

Okusawa S, Gelfand JA, Ikejima T, Connolly RJ, Dinarello CA (1988) Interleukin 1 induces a shock-like state in rabbits. Synergism with tumor necrosis factor and the effect of cyclooxygenase inhibition. J Clin Invest 81:1162–1172

Ollivierre F, Gubler U, Towle CA, Laurencin C, Treadwell BV (1986) Expression of IL-1 genes in human and bovine chondrocytes: a mechanism for autocrine control of cartilage matrix degradation. Biochem Biophys Commun 141:904–911

Onozaki K, Urawa H, Tamatani T, Iwamura Y, Hashimoto T, Baba T, Suzuki H, Yamada M, Yamamoto S, Oppenheim JJ, Matsushima K (1988) Synergistic interactions of interleukin 1, interferon-β, and tumor necrosis factor in terminally differentiating a mouse myeloid leukemic cell line (ml). Evidence that interferon-beta is an autocrine differentiating factor. J Immunol 140:112–119

Oppenheim JJ, Gery I (1982) Interleukin 1 is more than an interleukin. Immunol Today 3:113–119

Ozaki Y, Ohashi T, Minami A, Nakamura S (1987) Enhanced resistance of mice to bacterial infection induced by recombinant human interleukin-1α. Infect Immun 55:1436–1440

Paganelli KA, Stern AS, Kilian PL (1987) Detergent solubilization of the interleukin 1 receptor. J Immunol 138:2249–2253

Paganelli-Parker K, Kilian PL (1988) Evidence for an essential disulfide bond required for binding activity of the interleukin-1 receptor. In: Powanda MC, Oppenheim JJ, Kluger J, Dinarello CA (eds) Monokines and other non-lymphocytic cytokines. Liss, New York, pp 185–190 (Progress in leukocyte biology, vol 8)

Palade G (1975) Intracellular aspects of the process of protein synthesis. Science 189:347–358

Palaszynski EW (1987) Synthetic C-terminal of IL-1 functions as a binding domain as well as an antagonist for the IL-1 receptor. Biochem Biophys Res Commun 147:204–211

Palmiter RD, Gagnon J, Walsh KA (1978) Ovalbumin: a secreted protein without a transient hydrophobic leader sequence. Proc Natl Acad Sci USA 75:94–98

Pennica D, Kohr WJ, Kuang WJ, Glaister D, Aggarwal BB, Chen EY, Goeddel DV (1987) Identification of human uromodulin as the Tamm-Horsfall urinary glycoprotein. Science 236:83–88

Perlmutter DH, Goldberger G, Dinarello CA, Mizel SB, Colten HR (1986) Regulation of class III major histocompatibility complex gene products by interleukin 1. Science 232:850–852

Pettipher ER, Higgs GA, Henderson B (1986) Interleukin 1 induces leukocytic infiltration and cartilage proteoglycan degradation in the synovial joint. Proc Natl Acad Sci USA 83:8749–8753

Pierart ME, Najdovski T, Appelboom TE, Deschodt-Lanckman MM (1988) Effect of human endopeptidase 24.11 ("enkephalinase") on IL-1-induced thymocyte proliferation activity. J Immunol 140:3808–3811

Pike BL, Nossal GJV (1985) Interleukin 1 can act as a B-cell growth and differentiation factor. Immunology 82:8153–8157

Pistoia V, Cozzolino F, Rubartelli A, Torcia M, Roncella S, Ferrarini M (1986) In vitro production of interleukin 1 by normal and malignant human B lymphocytes. J Immunol 136:1688–1692

Pledger WJ, Stiles CD, Antoniades HN, Scher CD (1978) An ordered sequence of events is required before BALB/C-3T3 cells become committed to DNA synthesis. Proc Natl Acad Sci 75:2839–2843

Pober JS, Bevilacqua MP, Mendrick DL, Lapierre LA, Fiers W, Gimbrone MA (1986) Two distinct monokines, interleukin 1 and tumor necrosis factor, each independently induce biosynthesis and transient expression of the same antigen on the surface of cultured human vascular endothelial cells. J Immunol 136:1680–1687

Postlethwaite AE, Lachman LB, Mainardi CL, Kang AH (1983) Interleukin 1 stimulation of collagenase production by cultured fibroblasts. J Exp Med 157:801–806

Postlethwaite AE, Lachman LB, Kang AH (1984) Induction of fibroblast proliferation by interleukin-1 derived from human monocytic leukemia cells. Arthritis Rheum 27:995–1001

Powers GD, Abbas AK, Miller RA (1988) Frequencies of IL-2 and IL-4-secreting T cells in naive and antigen-stimulated lymphocyte populations. J Immunol 140:3352–3357

Priestle JP, Schar HP, Grutter MG (1988) Crystal structure of the cytokine interleukin-1β. EMBO J 7:339–343

Prieur AM, Griscelli C, Kaufmann MT, Dayer JM (1987) Specific interleukin-1 inhibitor in serum and urine of children with systemic juvenile chronic arthritis. Lancet II:1240–1242

Qwarnstrom EE, Page RC, Gillis S, Dower SK (1988) Binding, internalization, and intracellular localization of interleukin-1β in human diploid fibroblasts. J Biol Chem 263:8261–8269

Ramadori G, Sipe JD, Dinarello CA, Mizel SB, Colten HR (1985) Pretranslational modulation of acute phase hepatic protein synthesis by murine recombinant interleukin 1 (IL-1) and purified human IL-1. J Exp Med 162:930–942

Rappolee DA, Mark D, Banda MJ, Werb Z (1988) Wound macrophages express TGF-α and other growth factors in vivo: analysis by mRNA phenotyping. Science 241:708–712

Raz A, Wyche A, Siegel N, Needleman P (1988) Regulation of fibroblast cyclooxygenase synthesis by interleukin-1. J Biol Chem 263:3022–3028

Rhyne JA, Mizel SB, Taylor RG, Chedid M, McCall CE (1988) Characterization of the human interleukin 1 receptor on human polymorphonuclear leukocytes. Clin Immunol Imunopathol 48:354–361

Robertson B, Dostal K, Daynes R (1988) Neuropeptide regulation of inflammatory and immunologic responses. The capacity of alpha-melanocyte-stimulating hormone to inhibit tumor necrosis factor and IL-1-inducible biologic responses. J Immunol 140:4300–4307

Rosenstreich DL, Tu JH, Kinkade PR, Maurer-Fogy I, Kahn J, Barton RW, Farina PR (1988) A human urine-derived interleukin 1 inhibitor. Homology with deoxyribonuclease 1. J Exp Med 168:1767–1779

Rosenwasser LJ, Webb AC, Clark BD, Irie S, Chang L, Dinarello CA, Gehrke L, Wolff SM, Rich A, Auron PE (1986) Expression of biologically active human interleukin 1 subpeptides by transfected simian COS cells. Proc Natl Acad Sci USA 83:5243–5246

Rosoff PM, Savage N, Dinarello CA (1988) Interleukin-1 stimulates diacylglycerol production in T lymphocytes by a novel mechanism. Cell 54:73–81

Rothlein R, Czajkowski M, O'Neill MM, Marlin SD, Mainolfi E, Merluzzi VJ (1988) Induction of intercellular adhesion molecule 1 on primary and continuous cell lines by pro-inflammatory cytokines. Regulation by pharmacologic agents and neutralizing antibodies. J Immunol 141:1665–1669

Rubin RM, Rosenbaum JT (1988) A platelet-activating factor antagonist inhibits interleukin 1-induced inflammation. Biochem Biophys Res Commun 154:429–436

Rupp EA, Cameron PM, Ranawat CS, Schmidt JA, Bayne EK (1986) Specific bioactivities of monocyte-derived interleukin 1α and interleukin 1β are similar to each other on cultured murine thymocytes and on cultured human connective tissue cells. J Clin Invest 78:836–839

Sabatini M, Boyce B, Aufdemorte T, Bonewald L, Mundy GR (1988) Infusions of recombinant human interleukins 1α and 1β cause hypercalcemia in normal mice. Proc Natl Acad Sci USA 85:5235–5239

Saklatvala J, Pilsworth LMC, Sarsfield SJ, Gavrilovic J, Heath JK (1984) Pig catabolin is a form of interleukin 1. Biochem J 224:461–466

Santoli D, Yang YC, Clark SC, Kreider BL, Caracciolo D, Rovera G (1987) Synergistic and antagonistic effects of recombinant human interleukin (IL) 3, IL-1α, granulocyte and macrophage colony-stimulating factors (G-CSF and M-CSF) on the growth of GM-CSF-dependent leukemic cell lines. J Immunol 139:3348–3354

Sapolsky R, Rivier C, Jamamoto G, Plotsky P, Valew (1987) Interleukin 1 stimulates the secretion of hypothalamic corticotropin-releasing factor. Science 238:522–524

Sauder DN, Carter CS, Katz SI, Oppenheim J (1982) Epidermal cell production of thymocyte activating factor (ETAF). J Invest Dermatol 79:34–39

Sauder DN, Mounessa NL, Katz SI, Dinarello CA, Gallin JI (1984) Chemotactic cytokines: the role of leukocyte pyrogen and epidermal cell thymocyte-activating factor in neutrophil chemotaxis. J Immunol 132:828–832

Saus J, Quinones S, Otani Y, Nagase H, Harris ED, Kurkinen M (1988) The complete primary structure of human matrix metalloproteinase-3. Identity with stromelysin. J Biol Chem 263(14):6742–6745

Sayers TJ, Wiltrout TA, Bull CA, Denn AC III, Pilaro AM, Lokesh B (1988) Effect of cytokines on polymorphonuclear neutrophil infiltration in the mouse. Prostaglandin- and leukotriene-independent induction of infiltration by IL-1 and tumor necrosis factor. J Immunol 141:1670–1677

Scala G, Morrone G, Tamburrini M, Alfinito F, Pastore CI, D'Alessio G, Venuta S (1987) Autocrine growth function of human interleukin 1 molecules on ROHA-9, an EBV-transformed human B cell line. J Immunol 138:2527–2534

Schar HP, Priestle JP, Grutter MG (1987) Crystallization and preliminary X-ray diffraction studies of recombinant human interleukin-1β. J Biol Chem 262:13724–13725

Schleef RR, Bevilacqua MP, Sawdey M, Gimbrone MA, Loskutoff DJ (1988) Cytokine activation of vascular endothelium. Effects of tissue-type plasminogen activator and type 1 plasminogen activator inhibitor. J Biol Chem 263:5797–5803

Schmidt JA, Mizel SB, Cohen D, Green I (1982) Interleukin 1, a potential regulator of fibroblast proliferation. J Immunol 128:2177–2182

Schmidt JA, Oliver CN, Lepe-Zuniga JL, Green I, Gery I (1984) Silica-stimulated monocytes release fibroblast proliferation factors identical to interleukin 1. A potential role for interleukin 1 in the pathogenesis of silicosis. J Clin Invest 73:1462–1472

Schnyder J, Payne T, Dinarello CA (1987) Human monocyte or recombinant interleukin 1's are specific for the secretion of a metalloproteinase from chondrocytes. J Immunol 138:496–503

Schroder JM, Mrowietz U, Morita E, Christophers E (1987) Purification and partial biochemical characterization of a human monocyte-derived, neutrophil-activating peptide that lacks interleukin 1 activity. J Immunol 139:3474–3483

Seckinger P, Williamson K, Balavoine JF, Mach B, Mazzei G, Shaw A, Dayer JM (1987a) A urine inhibibor of interleukin 1 activity affects both interleukin 1α and 1β but not tumor necrosis factor α. J Immunol 139:1541–1545

Seckinger P, Lowenthal JW, Williamson K, Dayer JM, MacDonald HR (1987b) A urine inhibitor of interleukin 1 activity that blocks ligand binding. J Immunol 139:1546–1549

Segal GM, McCall E, Stueve T, Bagby GC (1987) Interleukin-1 stimulates endothelial cells to release multilineage human colony stimulating activity. J Immunol 138:1772–1778

Sehgal PB, Walther Z, Tamm I (1987) Rapid enhancement of β₂-interferon/B-cell differentiation factor BSF-2 gene expression in human fibroblasts by diacylglycerols and the calcium ionophore A23187. Biochemistry 84:3663–3667

Shirakawa F, Tanaka Y, Ota T, Suzuki H, Eto S, Yamashita U (1987) Expression of interleukin 1 receptors on human peripheral T cells. J Immunol 138:4243–4248

Shirakawa F, Yamashita U, Chedid M, Mizel SB (1988) Cyclic AMP-an intracellular second messenger of interleukin 1. Proc Natl Acad Sci USA 85:8201–8205

Sideras P, Funa K, Zalcberg-Quintana I, Xanthopoulos KG, Kisielow P, Palacios R (1988) Analysis of in situ hybridization of cells expressing mRNA for interleukin 4 in the developing thymus and in peripheral lymphocytes from mice. Proc Natl Acad Sci USA 85:218–221

Sieff CA, Tsai S, Faller DV (1987) Interleukin 1 induces cultured human endothelial cell production of granulocyte-macrophage colony-stimulating factor. J Clin Invest 79:48–51

Sims JE, March CJ, Cosman D, Widmer MB, MacDonald HR, McMahan CJ, Grubin CE, Wignall JM, Jackson JL, Call SM, Friend D, Alpert AR, Gillis S, Urdal DL, Dower SK (1988) cDNA expression cloning of the IL-1 receptor, a member of the immunoglobulin superfamily. Science 241:585–589

Singer II, Scott S, Hall GL, Limjuco G, Chin J, Schmidt J (1988) Interleukin 1β is localized in the cytoplasmic ground substance but is largely absent from the Golgi apparatus and plasma membranes of stimulated human monocytes. J Exp Med 167:389–407

Stanton TH, Maynard M, Bomszkty K (1986) Effect of interleukin-1 on intracellular concentration of sodium, calcium, and potassium in 70Z/3 cells. J Biol Chem 261:5699–5701

Staruch MJ, Wood DD (1985) Reduction of serum interleukin-1-like activity after treatment with dexamethasone. J Leukocyte Biol 37:193–207

Stashenko P, Dewhirst FE, Peros WJ, Kent RL, Ago JM (1987) Synergistic interactions between interleukin 1, tumor necrosis factor, and lymphotoxin in bone resorption. J Immunol 138:1464–1468

Stya M, Dower SK, Prickett KS, Gillis S, Conlon PJ (1988) Development and characterization of two neutralizing monoclonal antibodies to human interleukin-1α. J Biol Response Mod 7:162–172

Symons JA, Wood NC, Di Giovine FS, Duff GW (1988) Soluble IL-2 receptor in rheumatoid arthritis. Correlation with disease activity, IL-1 and IL-2 activity. J Immunol 141:2612–2618

Takacs L, Kovacs EJ, Smith MR, Young HA, Durum SK (1988) Detection of IL-1α and IL-1β gene expression by in situ hybridization. Tissue localization of IL-1 mRNA in the normal C57BL/6 mouse. J Immunol 141:3081–3095

Tanaka K, Ishikawa E, Ohmoto Y, Hirai Y (1987) In vitro production of human interleukin 1α and interleukin 1β by peripheral blood mononuclear cells examined by sensitive sandwich enzyme immunoassay. Eur J Immunol 17:1527–1530

Tartakovsky B, Finnegan A, Muegge K, Brody DT, Kovacs EJ, Smith MR, Berzofsky JA, Young HA, Durum SK (1988) IL-1 is an autocrine growth factor in T cell clones. J Immunol 141:3863–3867

Telford JL, Macchia G, Massone A, Carinci V, Palla E, Melli M (1986) Murine interleukin 1β gene: structure and evolution. Nucleic Acids Res 14:9955–9963

Thieme TR, Hefeneider SH, Wagner CR, Burger DR (1987) Recombinant murine and human IL 1α bind to human endothelial cells with an equal affinity, but have an unequal ability to induce endothelial cell adherence of lymphocytes. J Immunol 139:1173–1178

Tocci MJ, Hutchinson NI, Cameron PM, Kirk KE, Norman DJ, Chin J, Rupp EA, Limjuco GA, Bonilla-Argudo VM, Schmidt JA (1987) Expression in Escherichia coli of fully active recombinant human IL 1β: comparison with native human IL 1β. J Immunol 8:1109–1114

Tovey MG, Content J, Gresser I, Gugenheim J, Blanchard B, Guymarho J, Poupart P, Gigou M, Shaw A, Fiers W (1988) Genes for IFN-β-2 (IL-6), tumor necrosis factor, and IL-1 are expressed at high levels in the organs of normal individuals. J Immunol 141:3106–3110

Urdal DL, Call SM, Jackson JL, Dower SK (1988) Affinity purification and chemical analysis of the interleukin-1 receptor. J Biol Chem 263:2870–2877

Valone FH, Epstein LB (1988) Biphasic platelet-activating factor synthesis by human monocytes stimulated with IL-1β, tumor necrosis factor, or IFN-γ. J Immunol 141:3945–3950

Van Damme JM, DeLey M, Opdenakker G, Billiau A, DeSomer P, Van Beeumen J (1985) Homogeneous interferon-inducing 22K factor is related to endogenous pyrogen and interleukin-1. Nature 314:266–268

Van Damme J, Opdenakker G, Simpson RJ, Rubira MR, Cayphas S, Vink A, Billiau A, Van Snick J (1987) Identification of the human 26-kD protein, interferon β₂ (IFN-β₂), as a B cell hybridoma/plasmacytoma growth factor induced by interleukin 1 and tumor necrosis factor. J Exp Med 165:914–919

Van der Meer JWM, Barza M, Wolff SM, Dinarello CA (1988) A low dose of recombinant interleukin 1 protects granulocytopenic mice from lethal gram-negative infection. Immunology 85:1620–1623

Vermeulen MW, David JR, Remold HG (1987) Differential mRNA responses in human macrophages activated by intefereon-γ and muramyl dipeptide. J Immunol 139:7–9

Verner K, Schatz G (1988) Protein translocation across membranes. Science 241:1307–1313

Wallach D, Holtmann H, Engelmann H, Nophar Y (1988) Sensitization and desensitization to lethal effects of tumor necrosis factor and IL-1. J Immunol 140:2994–2999

Walther Z, May LT, Sehgal PB (1988) Transcriptional regulation of the interferon-β_2/B cell differentiation factor BSF-2/hepatocyte-stimulating factor gene in human fibroblasts by other cytokines. J Immunol 140:974–977

Warner SJC, Auger KR, Libby P (1987a) Human interleukin 1 induces interleukin 1 gene expression in human vascular smooth muscle cells. J Exp Med 165:1316–1331

Warner SJC, Auger KR, Libby P (1987b) Interleukin 1 induces interleukin 1. II. Recombinant human interleukin 1 induces interleukin 1 production by adult human vascular endothelial cells. J Immunol 139:1911–1917

Warren DJ, Moore AS (1988) Synergism among interleukin 1, interleukin 3, and interleukin 5 in the production of eosinophils from primitive hemopoietic stem cells. J Immunol 140:94–99

Weaver CT, Unanue ER (1986) T cell induction of membrane IL 1 on macrophages. J Immunol 137:3868–3873

Weaver CT, Hawrylowicz CM, Unanue ER (1988) T helper cell subsets require the expression of distinct costimulatory signals by antigen presenting cells. Proc Natl Acad Sci USA 85:8181–8185

Webb AC, Collins KL, Auron PE, Eddy RL, Nakai H, Byers MG, Haley LL, Henry MW, Shows TB (1986) Interleukin-1 gene (IL1) assigned to long arm of human chromosome 2. Lymphokine Res 5:77–85

Werber HI, Emancipator SN, Tykocinski ML, Sedor JR (1987) The interleukin 1 gene is expressed by rat glomerular mesangial cells and is augmented in immune complex glomerulonephritis. J Immunol 138:3207–3212

Wewers MD, Rennard SI, Hance AJ, Bitterman PB, Crystal RG (1984) Normal human alveolar macrophages obtained by bronchoalveolar lavage have a limited capacity to release interleukin 1. J Clin Invest 74:2208–2218

Wewers MD, Saltini C, Sellers S, Tocci MJ, Bayne EK, Schmidt JA, Crystal RG (1987) Evaluation of alveolar macrophages in normals and individuals with active pulmonary sarcoidosis for the spontaneous expression of the interleukin-1β gene. Cell Immunol 107:479–488

Wijelath ES, Kardasz AM, Drummond R, Watson J (1988) Interleukin-1 induced inositol phospholipid breakdown in murine macrophages: possible mechanism of receptor activation. Biochem Biophys Res Commun 152:392–397

Williams JM, Deloria D, Hansen JA, Dinarello CA, Loertscher R, Shapiro HM, Strom TB (1985) The events of primary T cell activation can be staged by use of sepharose-bound anti-T3 (64.1) monoclonal antibody and purified interleukin 1. J Immunol 135:2249–2255

Wingfield P, Payton M, Tavernier J, Barnes M, Shaw A, Rose K, Simona MG, Demczuk S, Williamson K, Dayer JM (1986) Purification and characterization of human interleukin-1β expressed in recombinant Escherichia coli. Eur J Biochem 160:491–497

Wingfield P, Graber P, Movva NR, Gronenborn AM, MacDonald HR (1987a) N-terminal-methionylated interleukin-1β has reduced receptor-binding affinity. FEBS Lett 215:160–164

Wingfield P, Payton M, Graber P, Rose K, Dayer JM, Shaw AR, Schmiessner U (1987b) Purification and characterization of human interleukin-1α produced in Escherichia coli. Eur J Biochem 165:537–541

Woloski BMRNJ, Smith EM, Meyer WJ III, Fuller GM, Blalock JE (1985) Corticotropin-releasing activity of monokines. Science 230:1035–1037

Wong GHW, Goeddel DV (1988) Induction of manganous superoxide dismutase by tumor necrosis factor: possible protective mechanism. Science 242:941–944

Wood DD, Bayne EK, Goldring MB, Gowen M, Hamerman, Humes JL, Ihrie EJ, Lipsky PE, Staruch MJ (1985) The four biochemically distinct species of human interleukin 1 all exhibit similar biologic activities. J Immunol 134:895–903

Yamasaki K, Taga T, Hirta Y, Yawata H, Kawanishi Y, Seed B, Taniguchi T, Hirano T, Kishimoto T (1988) Cloning and expression of the human interleukin-6 (BSF-21FNβ 2) receptor. Science 241:825–828

Yem AW, Richard KA, Staite ND, Deibel MR (1988) Resolution and biological properties of three N-terminal analogues of recombinant human interleukin-1β. Lymholine Res 7:85–92

Young PR, Hazuda DJ, Simon PL (1988) Human interleukin 1β is not secreted from hamster fibroblasts when expressed constitutively from a transfected cDNA. J Cell Biol 107:447–456

Zhang YH, Lin JX, Yip YK, Vilcek J (1988a) Enhancement of cAMP levels and of protein kinase activity by tumor necrosis factor and interleukin 1 in human fibroblasts: role in the induction of interleukin 6. Proc Natl Acad Sci USA 85:6802–6805

Zhang Y, Lin JX, Vilcek J (1988b) Synthesis of interleukin 6 (interferon-β_2/B cell stimulatory factor 2) in human fibroblasts is triggered by an increase in intracellular cyclic AMP. J Biol Chem 263:6177–6182

Zsebo KM, Yuschenkoff VN, Schiffer S, Chang D, McCall E, Dinarello CA, Brown MA, Altrock B, Bagby GC (1988) Vascular endothelial cells and granulopoiesis. Interleukin-1 stimulates release of G-CSF and GM-CSF. Blood 71:99–103

Zucali JR, Dinarello CA, Oblon DJ, Gross MA, Anderson L, Weiner RS (1986) Interleukin-1 stimulates fibroblasts to produce granulocyte macrophage colony-stimulating activity and prostaglandin E_2. J Clin Invest 77:1857–1863

Zurawski SM, Pope K, Cherwinski H, Zurawski G (1986) Expression in Escherichia coli of synthetic human interleukin-1α genes encoding the processed active protein, mutant proteins, and β-galactosidase fusion proteins. Gene 49:61–68

CHAPTER 10

Interleukin-2

M. Hatakeyama and T. Taniguchi

A. Introduction

Interleukin 2 (IL-2), one of the first lymphokines to be identified, plays a major role in the clonal expansion of T lymphocytes (T cells) by interacting with specific cell surface receptors (IL-2 receptors; Morgan et al. 1976; Smith 1984, 1988; Taniguchi et al. 1986a). Expression of both IL-2 and IL-2 receptor requires T cell activation, triggered primarily by the interaction of the antigen/MHC molecules: T cell receptor complex (CD3/Ti complex). Hence, in effect, clonal proliferation of antigen-specific T cells appears to occur via a process of signal transduction, wherein the specific interaction of antigen/MHC and the T cell receptor complex is converted into a pathway of cellular growth through the operation of the IL-2 system.

In addition to its potent T cell growth-stimulatory activity, multiple biological functions of IL-2 have been described including B cell growth and differentiaton (Waldmann et al. 1984; Jung et al. 1984), generation of lymphokine-activated killer cells (Lotze et al. 1981; Grimm et al. 1982), and augmentation of natural killer cells (Henny et al. 1981). Furthermore, IL-2 also functions as a growth inhibitor in certain neoplastic cells (Hatakeyama et al. 1985; Sugamura et al. 1985). The IL-2 receptor has also been identified on IL-3-dependent bone-marrow cells (Koyasu et al. 1986), macrophages (Malkovsky et al. 1987), epidermal Langerhans cells (Steiner et al. 1986), and oligodendroglial cells (Benveniste and Merrill 1986). These observations suggest that IL-2 delivers various signals to a wide range of cells via its receptor.

To understand the molecular mechanism of IL-2-mediated T cell growth, elucidation of the structure as well as the mechanism of activation of expression of IL-2 and its receptor complex are essential. In this chapter we review recent studies on the structure and expression of IL-2 and its receptor genes.

B. Structure of Human and Mouse IL-2

IL-2 is a 15000-Da glycoprotein produced and secreted by helper-type (CD4$^+$) T cells in response to antigenic or mitogenic stimuli (Morgan et al. 1976). The human IL-2 cDNA was originally isolated from a cDNA library derived from a human T-leukemic line, Jurkat, with the use of the hybridi-

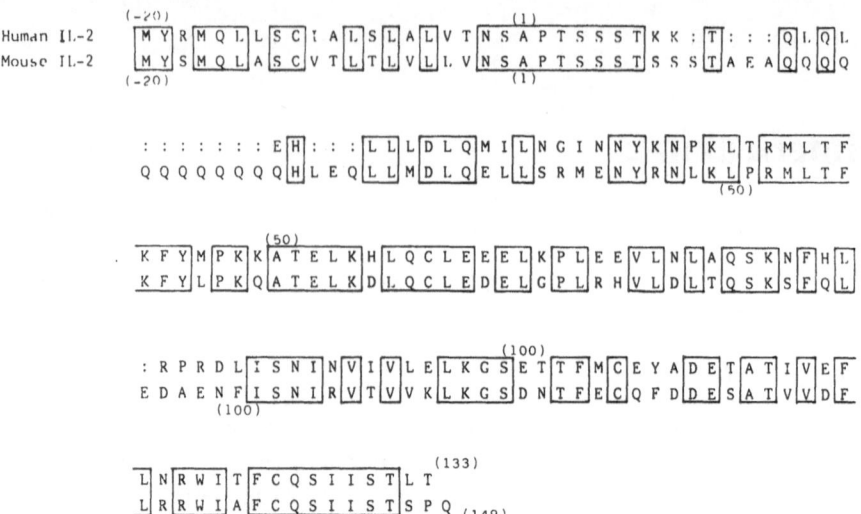

Fig. 1. The primary sequences for human and mouse IL-2. Gaps were introduced to align the sequences for maximal homology. Coincident amino acids are *framed*. For further details, see TANIGUCHI et al. (1983) and KASHIMA et al. (1985)

zation-translation assay (TANIGUCHI et al. 1983). The cDNA obtained was shown to direct the production of biologically active IL-2 when expressed in monkey COS cells. The human cDNA was used to screen for the mouse IL-2 cDNA (KASHIMA et al. 1985). Nucleotide sequence analysis of the human and mouse cDNAs revealed the primary protein structure of the IL-2 molecules from both species. The human IL-2 precursor consists of 153 amino acids and, upon secretion, its signal peptide (20 amino acids) is cleaved to form the mature form, with cysteine residues at positions 58 and 105 forming a disulfide bond essential for its biological activity. The amino acid sequence of mouse IL-2 shows 63% homology with human IL-2 (KASHIMA et al. 1985; YOKATA et al. 1986). It contains a unique stretch of 12 consecutive glutamine repeats which are absent in the human and bovine IL-2 (Fig. 1). Bovine IL-2 has an amino acid homology of 65% with human IL-2 and 50% with murine IL-2 (CERRETTI et al. 1986; REEVES et al. 1986).

Recently, the three-dimentional structure of human IL-2 was elucidated by X-ray crystallography (BRANDHUBER et al. 1987). This study revealed that IL-2 is an α-helical protein without β-sheet structure in the molecule. Although the receptor contact site in the IL-2 molecule is not completely understood at present, studies using IL-2 mutants, IL-2 peptide fragments, or IL-2-specific antibodies provide evidence that the region encompassing the amino-terminal 50–60 residues, containing the first α-helix, is likely to be involved in the binding to its receptor (IL-2Rα, Tac antigen, p55; see below) (ALTMAN et al. 1984; KUO and ROBB 1986; JU et al. 1987).

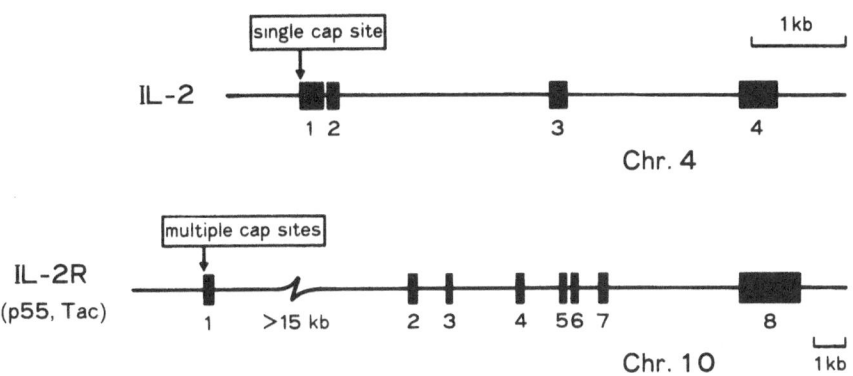

Fig. 2. Organization of the human IL-2 and IL-2 receptor (p55, Tac antigen) genes. The *rectangles* represent exons. For details, see FUJITA et al. (1984), LEONARD et al. (1985), ISHIDA et al. (1985)

C. Structure and Regulation of the IL-2 Gene

IL-2 is produced by a subset of mature T cells, i.e., CD4[+] helper T cells, following stimulation with antigen or mitogen. Hence, the expression of the IL-2 gene is under both developmental (i.e., T cell specific) and environmental control (i.e., dependent on mitogenic signals). It has been documented that production of IL-2 is primarily regulated at the transcriptional level (EFRAT et al. 1982; TANIGUCHI et al. 1983).

The IL-2 gene is present in a single copy in man and mouse, consisting of four exons (Fig. 2; FUJITA et al. 1984; HOLBROOK et al. 1984; FUSE et al. 1984). In man, the gene is located on chromosome 4q (SEIGEL et al. 1984). Structural analysis of the human and mouse IL-2 genes revealed the presence of highly conserved sequences within the 5′ flanking region (FUSE et al. 1984). In the human IL-2 gene, the upstream region spanning from nucleotide residues −129 and −319 from the CAP site contains DNA sequences functioning as an inducible enhancer required for the mitogen-specific IL-2 gene activation in T cells (Fig. 3) (FUJITA et al. 1986; SIEBENLIST et al. 1986; DURAND et al. 1987, 1988). In addition, evidence has been accumulating that the 5′ flanking sequence is functionally subdivided into several pieces and that distinct nuclear factors bind specifically to those subregions (Fig. 3), (TANIGUCHI et al. 1986b; DURAND et al. 1988; NABEL et al. 1988; SHAW et al. 1988). The functional properties of those factors in the T cell-specific activation of the IL-2 gene following antigenic stimulation are yet to be critically examined (see below).

D. The IL-2 Receptor

It is now evident that the functional IL-2 receptor consists of more than one molecule. Multiple components that seem to constitute the high-affinity IL-2 receptor have been identified in both human and mouse T cells (see below).

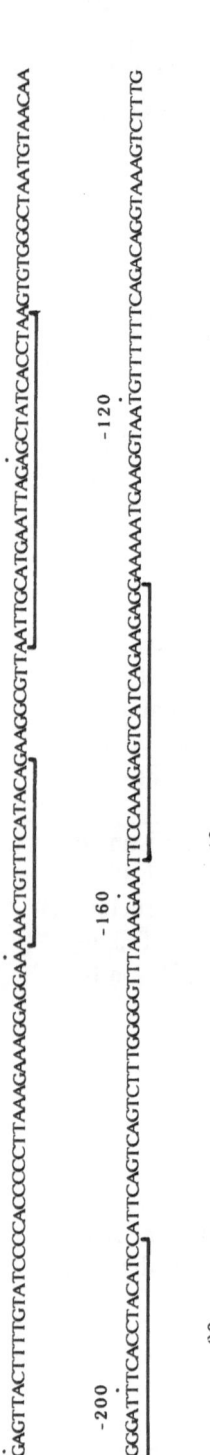

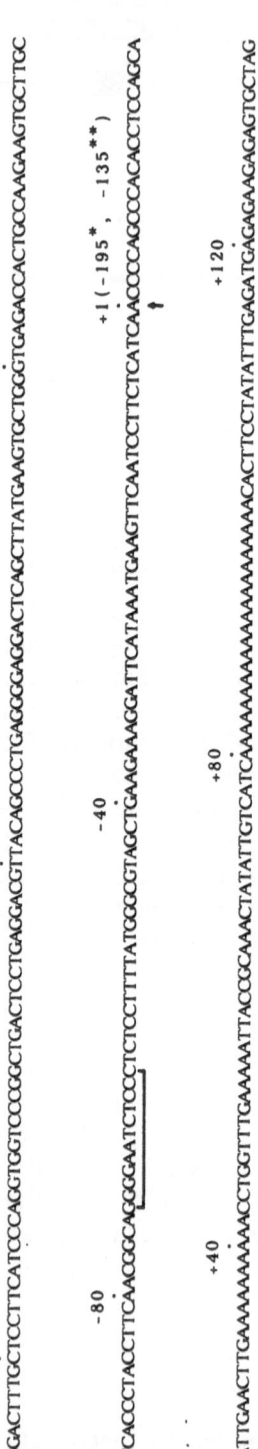

Fig. 3a,b. Functional DNA sequences that regulate IL-2 or IL-2Rα gene expression. **a** 5′-upstream region of the human IL-2 gene (*large arrow*, CAP site). **b** 5′-upstream region of the human IL-2Rα gene. The gene contains multiple CAP sites (*large arrows*, major sites; *small arrows*, minor sites; * Cross et al. 1987; ** Suzuki et al. 1987). In the both genes, additional factor binding sites (regions; *braces*) are yet to be

Since the discovery that a monoclonal antibody, anti-Tac, recognizes a human IL-2 receptor molecule (identified as the Tac antigen), much attention has been focused on the Tac antigen to elucidate the molecular mechanism of IL-2-mediated signal transduction (UCHIYAMA et al. 1981; LEONARD et al. 1982). On the other hand, a series of studies have demonstrated the existence of high- and low-affinity IL-2 binding sites on activated T cells (ROBB et al. 1984) and provided evidence that only the high-affinity IL-2 receptor is capable of transducing the IL-2-mediated extracellular signal(s) (WEISSMAN et al. 1986; FUJII et al. 1986). Molecular cloning of the cDNA encoding the Tac antigen (p55 or IL-2Rα chain; hereafter referred to as IL-2Rα for convenience) revealed that the protein consists of 251 amino acids with a single membrane-spanning region (LEONARD et al. 1984; NIKAIDO et al. 1984; COSMAN et al. 1984). The IL-2Rα shows no significant homology with other growth factor receptors possessing tyrosine kinase activity (i.e., epidermal growth factor receptor, insulin receptor, platelet-derived growth factor receptor, and colony stimulating factor-1 receptor). One of the interesting features of the structure of IL-2Rα is that it contains a cytoplasmic domain consisting of only 13 amino acids. Although IL-2Rα cDNA expression resulted in the generation of a IL-2 binding molecule of 55 kDa in nonlymphoid cells, this receptor was found to manifest only low-affinity (K_d 10–20 nM) ligand binding and could not transduce IL-2-specific signals (GREENE et al. 1985; HATAKEYAMA et al. 1985). In contrast, transfection and expression of the same cDNA in T-lymphoid cells resulted in the expression of functional, high-affinity IL-2 receptors (HATAKEYAMA et al. 1985; KONDO et al. 1986).

These observations support the notion that the biologically functional IL-2 receptor (i.e., the high-affinity IL-2 receptor) requires membrane component(s) other than IL-2Rα, the expression of which to form the "IL-2 receptor complex" is restricted to lymhoid cells. The notion was further substantiated by a series of cell membrane fusion experiments (ROBB 1986). Such a component(s) of IL-2 receptor was subsequently identified both in human and mouse T cells using chemical crosslinking techniques (SHARON et al. 1986; TSUDO et al. 1986; TESHIGAWARA et al. 1987; DUKOVICH et al. 1987; SARAGIVO and MALEK 1987). Surprisingly, the newly identified 70–75-kDa protein (p70–75) also bound IL-2. It is expressed in a variety of lymphoid cells including natural killer lineage cells (DUKOVICH et al. 1987). The second IL-2 receptor (p70–75 or IL-2Rβ chain; hereafter referred to as IL-2Rβ for convenience) has an intermediate affinity for IL-2 when compared with high- and low-affinity IL-2 receptors expressed on activated T cells. Chemical crosslinking experiments with ^{125}I-IL-2 revealed that cells expressing high-affinity IL-2 receptors express both IL-2Rα and IL-2Rβ, while cells expressing only low-affinity IL-2 receptors express only IL-2α. These observations indicate that, the high-affinity IL-2 receptor is composed of IL-2α and IL-2Rβ (Fig. 4). The presence of such a heterodimeric IL-2 receptor complex is further substantiated by equilibrium binding experiments, demonstrating that IL-2 binds to and dissociates very rapidly from the low-affinity IL-2 receptor (i.e., IL-2Rα), whereas the second component of the IL-2 receptor (i.e., IL-2Rβ) as-

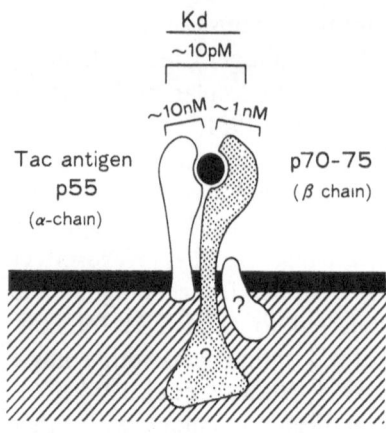

Fig. 4. Current view of the IL-2 receptor complex. Structure of the IL-2Rβ is still obscure at present, but it may contain a large intracytoplasmic domain that is responsible for the signal transduction. It remains to be seen whether any additional components also constitute the "complex". See text for details

sociates and dissociates very slowly with the ligand (Wang and Smith 1987). Thus the high-affinity IL-2 receptor shows kinetic cooperation between the two chains, i.e., it binds IL-2 rapidly and releases it slowly.

Several lines of experimental data suggest that the transmembrane and cytoplasmic domains of IL-2Rα are not essential in IL-2 signal transduction (Hatakeyama et al. 1986; 1987; Kondo et al. 1987). Hence, a likely possibility is that the cytoplasmic domain of IL-2Rβ is responsible, at least in part, for transducing the extracellular IL-2 signal to intracellular signaling pathway(s). It remains to be seen whether additional membrane component(s) (as depicted in Fig. 4) are also involved in the IL-2 receptor signaling. Evidence for the presence of an additional glycoprotein homologous to IL-2Rβ has been presented (Herrmann and Diamantstein 1988).[1]

E. Regulation of IL-2 Receptor Gene Expression

Although IL-2Rα itself is insufficient for IL-2-mediated signal transduction, there is no doubt that it is essential for the formation of a functional IL-2 receptor complex (as described above). The expression of the IL-2Rα gene is tightly regulated. In general, neither IL-2Rα, nor IL-2Rα mRNA is detected in resting T cells (Greene and Leonard 1986; Taniguchi 1988). It is transiently expressed on the cell surface following T cell activation (as well as B cell activation) (Cantrell and Smith 1983). While much of the accumulated evi-

---

[1] During the preparation of this chapter, cDNA encoding the human IL-2Rβ has been isolated and its complete structure determined (Hatakeyama et al. 1989). The predicted mature form of the IL-2Rβ contains 525 amino acid residues of which the N-terminal 214 residues represent a cysteine-rich extracellular region, followed by a single transmembrane region consisting of 25 amino acids. As anticipated, the cytoplasmic region of the IL-2Rβ is 286 amino acids long and it is far larger than that of the IL-2Rα chain (Hatakeyama et al. 1989). Significantly, the primary structure of the cytoplasmic region is well conserved between the human and mouse IL-2Rβs (T. Kono and T. Taniguchi, manuscript in preparation). Unlike many growth factor receptors, the IL-2Rβ lacks a tyrosine kinase domain (Hatakeyama et al. 1989).

dence supports the notion that the genes encoding cytokines but not their receptors are tightly regulated with respect to their expression, the IL-2 system is unique in that expression of both the ligand and receptor genes is under strict control. Hence the regulated expression of the IL-2Rα gene plays a key role in the control of T cell proliferation.

The IL-2Rα subunit is encoded by a single-copy gene on human chromosome 10 which is divided into eight exons spanning more than 25 kilobases (Fig. 2) (LEONARD et al. 1985; ISHIDA et al. 1985). The gene contains multiple CAP sites and regulatory *cis* elements are present within the 5' flanking region that direct mitogen-induced activation of the IL-2Rα gene in T cells (Fig. 3) (MARUYAMA et al. 1987; CROSS et al. 1987; SUZUKI et al. 1987). Although both IL-2 and IL-2Rα are transiently expressed in T cells through similar stimuli, expression of the receptor gene is not as stringently regulated as the IL-2 gene. In normal T cells, for example, the IL-2 receptor gene can be activated in response to 12-*O*-tetradecanoylphorbol-13-acetate (TPA), but the IL-2 gene activation requires stimulation with both calcium ionophore and TPA, suggesting the existence of similar but distinct mechanisms of gene activation for both genes (WEISS et al. 1986). In fact, the 5' regulatory regions that appear to be responsible for the expression of these genes do not show apparent homology with each other, pointing out further that the nuclear transcription factor(s) required for the full activation of IL-2 and the receptor gene are not the same (SHIBUAYA et al. 1989). As described below, recent studies on the transcription factors that function in T cells have revealed interesting features of the gene regulation mechanisms operating in the IL-2 system.

F. Involvement of a Common Regulatory Factor in Control of IL-2 and IL-2Rα Gene Expression

As described above, expression of both the IL-2 and IL-2Rα genes is controlled primarily at the transcriptional level and is induced upon T cell activation. In order to gain insight into the mechanisms of induced IL-2 and IL-2Rα gene expression in T cells, SHIBUYA et al. (1989) examined the possible involvement of a common transcription factor(s) functioning in the expression of the IL-2 and IL-2Rα genes. At least one such factor seems to mediate the induced expression of both genes. Interestingly, the recognition sequences for this factor are significantly diverse in these two genes and are related to those of immunoglobulin (Ig) κ-chain and MHC class I genes (Table 1). Evidence has been provided that this factor indeed binds to the IL-2, IL-2Rα, and Ig sequence elements with different affinities, thereby affecting the magnitude of gene expression (SHIBUYA et al. 1989). In fact, the protein seems to correspond to the 86 kDa protein (NFκB-like factor or HIVEN 86) that binds to IL-2Rα and human immunodeficiency virus long terminal repeat (HIV LTR) sequences (NABEL and BALTIMORE 1987; LEUNG and NABEL 1988; BOHNLEIN et al. 1988). Furthermore, this factor also binds to other cytokine genes, such as

Table 1. Presence of transcription factor (NFκB-like factor, HIVEN 86, TRF-1) recognition motifs within promoter LTR regions of the genes expressed in T cells

Factor binding confirmed		
IL-2	(human)	$^{-204}$GGATTTCACC^{-195}
IL-2R	(human)	$^{-71}$GGAATCTCCC^{-62}
IFN-γ	(human)	$^{-233}$GAATCCCACC^{-224}
IL-6	(human)	$^{-134}$GGATTTTCCC^{-125}
HIV LTR		$^{-89}$GGGACTTTCC^{-80}
		$^{-103}$GGGACTTTCC^{-94}
HTLV-I LTR		$^{+168}$GGAGCCTACC^{+177}
		$^{-154}$GGAAGCCACC^{-145}
Factor binding to be tested		
IL-3	(mouse)	$^{-295}$GAGATTCCAC^{-286}
IL-4	(mouse)	$^{-190}$GGTGTTTCAT^{-181}
GM-CSF	(mouse)	$^{-104}$GAGATTCCAC^{-95}
c-fos	(human)	$^{-278}$GGCCTTTCCC^{-269}

For details, see SHIBUYA et al. (1989).

interleukin-6 (IL-6), interferon-γ (IFN-γ), and human T-cell leukemia virus type I (HTLV-1) LTR sequences (SHIBUYA et al. 1989).

From the data of BOHNLEIN et al. (1988) and SHIBUYA et al. (1989), in Jurkat cells, this factor seems to differ from NFκB as regards its apparent size (BOHNLEIN et al. 1988) and its responsiveness to cycloheximide (SHIBUYA et al. 1989). These reports also indicate that the factor described here functions on a variety of genes and lymphotropic retroviruses. As shown in Table 1, sequence motifs similar to those of IL-2, IL-2R, IL-6, IFN-γ, and HTLV-I genes are also present within the promoter regions of other genes. In this context, it is worth noting that many if not all of the listed genes are also activated by HTLV-I-encoded tax-1 (YOSHIDA and SEIKI 1987; LEUNG and NABEL 1988; BALLARD et al. 1988; BOHNLEIN et al. 1988; SHIBUYA et al. 1989) (see below).

G. Intracellular Signal Transduction in the IL-2 System

IL-2 causes progression to G1 of the cell cycle in activated T cells, associated with a transient expression of cellular protooncogenes such as c-*myc* and c-*myb* (REED et al. 1985; STERN and SMITH 1986). In many growth factor/receptor systems, extracellular interaction of ligand and receptor leads to activation of intracellular second messenger pathways that include the G-protein-coupled cyclic AMP pathway or phosphatidylinositol (PI) turnover, Ca^{2+} mobilization/protein kinase C, and tyrosine kinase. Although activation of protein kinase C and/or G protein by IL-2 stimulation have been suggested (FARRAR and ANDERSON 1985; EVANS et al. 1987), others have failed to identify a second messenger system coupled to the IL-2 receptor. In fact, recent studies provided evidence that protein kinase C is not involved in IL-2-

mediated signaling pathways (LeGRUE 1988; MILLS et al. 1988; VALGE et al. 1988). From this viewpoint, elucidation of the molecular structure of IL-2Rβ may be required before the question of signaling pathways can be answered.

H. Dysregulation of the IL-2 System

The critical role of the IL-2 system in T cell proliferation indicates that abnormal production of IL-2 and/or response to it may lead to clinical disturbances such as autoimmune disease, immune deficiency, and T cell malignancy under certain circumstances. Of these, involvement of the IL-2/IL-2 receptor system has been well documented in a specific type of T cell malignancy. Adult T cell leukemia (ATL) is a fatal hematological disorder that is endemic in southwest Japan and the Caribbean islands. ATL arises from a monoclonal proliferation of peripheral, mature T cells with the CD4$^+$ phenotype (i.e., helper T cells; HATTORI et al. 1981). It may be worth noting that helper T cells are the main, if not the only, source of IL-2. The disease is particularly significant in that a retrovirus, HTLV-I, was identified for the first time as the causative agent for a human T-cell malignancy (WONG-STAAL and GALLO 1985; YOSHIDA and SEIKI 1987). Molecular and biological analysis of HTLV-I revealed that the virus does not contain a typical oncogene derived from cells (SEIKI et al. 1983). On the other hand, the observation that the virus rapidly transforms T cells indicated the existence of a unique transforming mechanism by HTLV-I in vitro that might be distinct from, but partly similar to, its mechanism for leukemogenesis (MIYOSHI et al. 1981). Structural analysis of the HTLV-I proviral genome led to the identification of a unique gene locus called pX between the env gene and 3' LTR sequences (SEIKI et al. 1983). The pX locus encodes three proteins, p21, p27, and p40 (KIYOKAWA et al. 1985). The finding that p40 (hereafter referred to as tax-1) functions as a transcriptional activator for its LTR has led to the hypothesis that the tax protein also acts on host genes, the activation of which is involved in the cellular transformation (SODROSKI et al. 1984; FELBER et al. 1985; FUJISAWA et al. 1985).

The relationship between the IL-2/IL-2 receptor system and ATL became noticeable by the observation that, in almost all cases of ATL cases, the leukemic cells constitutively express relatively large amounts of the IL-2 receptor (IL-2Rα; HATTORI et al. 1981), and that some HTLV-I-transformed T cell lines spontaneously secrete and respond to IL-2 (GOOTENBERG et al. 1981). The idea suggested by the above observations that an IL-2 autocrine mechanism may be involved in the maintenance of ATL was dismissed for a while by the subsequent findings that most of the freshly isolated ATL cells do not produce IL-2 (ARYA et al. 1984). However, it became evident that some leukemic cells freshly isolated from acute, but not chronic, ATL patients produce and respond to IL-2 in an autocrine fashion (ARIMA et al. 1987). This may suggest the essential involvement of the IL-2 system in the acute acceleration (crisis) of ATL.

The above observations indicate that abnormal operation of the IL-2/IL-2 receptor system may be crucial in the development of ATL at a certain stage. In fact, evidence exists that HTLV-I-encoded *tax*-1 *trans*-activates both IL-2 and IL-2Rα genes by acting on the 5' flanking regulatory region of each gene (MARUYAMA et al. 1987; CROSS et al. 1987; SHIBUYA et al. 1989). At present, there is no evidence that *tax*-1 directly binds specific DNA sequences. Accumulating data suggest that it induces and/or activates cellular *trans*-acting factors which otherwise are involved in normal (i.e., mitogen-induced) IL-2/IL-2 receptor gene expression. A candidate of such target proteins is a factor that binds to a critical region located in the 5' flanking sequence of IL-2 and IL-2Rα (see Sect. F).

The mechanism of activation of IL-2 and the IL-2 receptor (p55) by *tax*-1 is rather complex. It has been shown that the *tax*-1-mediated activation of the IL-2Rα gene is stronger than that of the IL-2 gene (INOUE et al. 1986; MARUYAMA et al. 1987; CROSS et al. 1987). Interestingly, transcription of the IL-2 gene is synergistically potentiated by *tax*-1 and physiological T cell activation signals such as T cell receptor (CD3/Ti complex) triggering (MARUYAMA et al. 1987). Taken together, these observations suggest the possibility of dysregulated operation of the IL-2 system in the early stages of ATL leukemogenesis; peripheral helper T cells infected with HTLV-I produce *tax*-1 protein that, by bypassing normal T cell activation pathways, leads to the aberrant expression of IL-2Rα. The infected T cells may proliferate, albeit weakly, without mitogenic stimulation of the cells (provided that *tax*-1 gene is expressed) as a result of IL-2R induction (we must assume in this situation that IL-2Rβ is also expressed in those cells). When a certain mitogenic signal such as T cell receptor triggering occurs, cells produce a high level of IL-2 because of the existence of *tax*-1 (synergistic action as described above). Such a two-step activation of the IL-2 and IL-2 receptor genes may trigger or result in the abnormal proliferation of certain infected T cells (Fig. 5) (MARUYAMA et al. 1987; TANIGUCHI 1988). The aberrant operation of the IL-2 autocrine loop may bring the T cells to a predisposed state for the development of cellular malignancy. In this context, it has been demonstrated that an IL-2-dependent T cell line (CTLL-2) becomes independent of exogenous IL-2 and acquires tumorigenic properties when it is infected by a retrovirus expressing the human IL-2 gene (YAMADA et al. 1987). Furthermore, subsequent analysis of tumor cells grown in nude mice has revealed that cell growth is not inhibited by an antibody against the IL-2 receptor, even though the antibody could block the growth of the original, virus-infected cell transformant (TANIGUCHI 1989). This observation argues for the acquisition of additional growth properties by the infected cells as a result of continuous IL-2 autocrine stimulation. Further studies will be required to analyze the detailed nature of the cells. More recently, the function of the virus LTR and *tax*-1 have been analyzed by stable introduction and expression of a gene construct in which *tax*-1 cDNA is driven by HTLV-I LTR in Jurkat and EL-4 cells. It appears that the function of the LTR is regulated by the cell cycle (effective in G1 and less effective in G0); the expression of *tax*-1 protein as well as of IL-2Rα is also

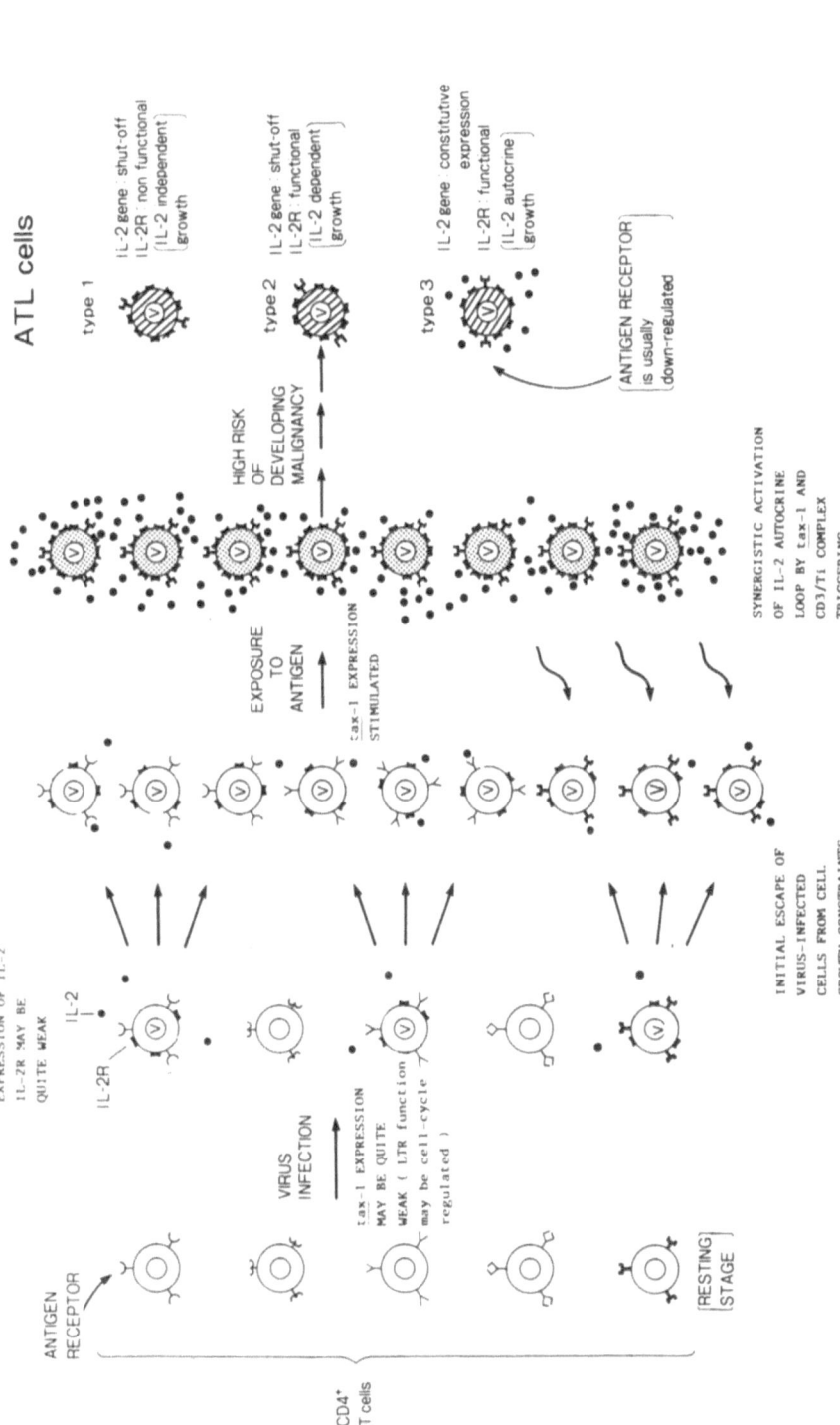

Fig. 5. Possible involvement of an aberrant IL-2 autocrine loop in the early stages of ATL development. Several studies suggest the importance of (a) virus infection of CD4+ T cells, and (b) T cell receptor triggering by antigens (possibly infectious agents or the viral antigens). See text for details

regulated by the cell cycle (Doi et al. 1989). This observation further supports the idea that a mitogenic signal such as T cell receptor trigerring is required for the efficient expression of the LTR-driven *tax*-1 gene.

I. The IL-2 System and Therapeutic Implications

From the accumulated data on the function and regulatory mechanisms operating in the IL-2 system through the process of antigen-specific T cell clonal expansion, it is likely that manipulation of the IL-2 system may allow us to either potentiate or suppress the host immune system, thereby opening possible approaches to immunotherapy.

Antibodies against IL-2 or IL-2 receptors might be useful as potential agents in preventing allograft rejection as well as in suppressing autoimmune diseases (Kirkman et al. 1985; Kelly et al. 1988). Bacterial toxin–IL-2 conjugates that are produced by recombinant DNA techniques would provide another approach to immunosuppression. In this context, it has been reported that IL-2 linked to diphtheria toxin is quite effective in immunosuppression in experimental systems, suggesting the possibility of clinical application of the conjugate molecules (Bacha et al. 1988; Perentesis et al. 1988).

The potential activity of IL-2 in generation of large numbers of tumor-specific lymphocytes has led to a practical approach for adoptive cancer immunotherapy. Potentiation of the host immune responses by IL-2 is under clinical investigation and some of the cancer patients have already benefited quite remarkably (Rosenberg et al. 1986, 1987; Rosenberg 1988). In addition, approaches to immunotherapy may include the use of combinations of cytokines, such as the use of interferons or tumor necrosis factors in conjunction with IL-2 (Rosenberg 1988).

Finally, IL-2 has been used as an adjuvant for stimulating the immune response to vaccines. Two reports have appeared in which the IL-2 gene was inserted with antigen-coding DNA sequences into vaccinia virus. The virally produced IL-2 appears to allow for an effective immune response (Ramshaw et al. 1987; Flexner et al. 1987).

J. Concluding Remarks

The IL-2/IL-2 receptor system is one of the most well-characterized growth factor receptor systems. The availability of molecularly cloned IL-2 and IL-2 receptor (IL-2Rα) genes have made it possible to start unravelling the complicated regulatory mechanisms operating in T cell growth. Furthermore, the system offers important information both for the mechanism of gene expression and intracellular signal transduction. To understand further the mechanism of IL-2 signaling, it has become important to identify and characterize all the component(s) constituting the IL-2 receptor complex. Recently, the cDNA for IL-2Rβ chain has been isolated and characterized. The presence of a large cytoplasmic region lacking a tyrosine kinase domain suggests a novel signal transduction mechanism. Elucidation of the mechanism whereby IL-2 controls growth of T cells may also shed light on the mechanism of cell growth that is mediated by other cytokines. Finally, the regulatory mechanisms of IL-2 and its receptor genes appear to involve complex and multiple protein transcription factors that interact with regulatory DNA sequences. These proteins may function either as activators or repressors, and expression of genes that they control might be determined by the specific combination and/or alignment of these transcription factors. The role of the IL-2/IL-2 receptor complex in ATL is suggestive of the complex biochemical events induced by HTLV-I that lead to oncogenesis.

Acknowledgments. We thank Dr. Edward Barsoumian for valuable comments and Ms. Y. Maeda for excellent assistance. The work is supported in part by a grant-in-aid for special project research, Cancer Bioscience, from the Ministry of Education, Science, and Culture of Japan.

References

Altman A, Cardenas JM, Houghten RA, Dixon FJ, Theofilopoulos AN (1984) Antibodies of predetermined specificity against chemically synthesized peptides of human interleukin 2. Proc Natl Acad Sci USA 81:2176–2180

Arima N, Daitoku Y, Yamamoto Y, Fujimoto K, Ohgaki S, Kojima K, Fukumori J, Matsushita K, Tanaka H, Onoue K (1987) Heterogeneity in response to interleukin 2 and interleukin 2-producing ability of adult T cell leukemic cells. J Immunol 138:3069–3074

Arya SK, Wong-Staal F, Gallo RC (1984) T-cell growth factor gene: lack of expression in human T-cell leukemia-lymphoma virus-infected cells. Science 223: 1086–1087

Bacha P, Williams DP, Waters C, Williams JM, Murphy JR, Storm TB (1988) Interleukin 2 receptor-targeted cytotoxity. J Exp Med 167:612–622

Ballard DW, Bohnlein E, Lowenthal JW, Wano Y, Franza BR, Greene WC (1988) HTLV-I tax induces cellular proteins that activate the NFκB element in the IL-2 receptor α-gene. Science 241:1652–1655

Benveniste EN, Merrill JE (1986) Stimulation of oligodendrogrial proliferation and maturation by interleukin-2. Nature 321:610–613

Bohnlein E, Lowenthal JW, Siekevitz M, Ballard DW, Franza BR, Greene WC (1988) The same inducible nuclear protein regulates mitogen activation of both the interleukin-2 receptor-alpha gene and type 1 HIV. Cell 53:827–836

Brandhuber BJ, Boone T, Kenney WC, McKay DB (1987) Three-dimentional structure of interleukin-2. Science 238:1707–1709

Cantrell DA, Smith KA (1983) Transient expression of interleukin 2 receptors. Consequences for T cell growth. J Exp Med 158:1895–1911

Cerretti DP, McKereghan K, Larsen A, Cantrell MA, Anderson D, Gillis S, Cosman D, Baker PE (1986) Cloning sequence and expression of bovine interleukin 2. Proc Natl Acad Sci USA 83:3223–3227

Cosman D, Ceretti DP, Larsen A, Park L, March C, Dower S, Gillis S, Urdal D (1984) Cloning, sequence and expression of human interleukin-2 receptor. Nature 312:768–771

Cross SL, Feinberg MB, Wolf JB, Holbroch NJ, Wong-Staal F, Leonard WJ (1987) Regulation of the human interleukin-2 receptor α chain promoter: activation of a nonfunctional promoter by the transactivator gene of HTLV-1. Cell 49:47–56

Doi T, Hatakeyama M, Itoh S, Taniguchi T (1989) Transient induction of IL-2 receptor in cultured T cell lines by HTLV-1 LTR-linked *tax*-1 gene. EMBO J 8:1953–1958

Dukovich M, Wano Y, Thuy JB, Katz P, Cullen BR, Kehrl JH, Greene WC (1987) A second human interleukin-2 binding protein that may be a component of high-affinity interleukin-2 receptors. Nature 327:518–522

Durand DB, Bush MR, Margan JG, Weiss A, Crabtree GR (1987) A 275 base pair fragment at the 5′ end of the interleukin 2 gene enhances expression from a heterologous promoter in response to signals from the T cell antigen receptor. J Exp Med 165:395–407

Durand DB, Shaw J-P, Bush MR, Replogle RE, Belagaje R, Crabtree GR (1988) Characterization of antigen receptor, response elements with the interleukin-2 enhancer. Mol Cell Biol 8:1715–1724

Efrat S, Pilo S, Kaempfer R (1982) Kinetics of induction and molecular size of mRNAs encoding human interleukin-2 and γ-interferon. Nature 297:236–239

Evans SW, Beckner SK, Farrar WL (1987) Stimulation of specific GTP binding and hydrolysis activities in lymphocyte membrane by interleukin-2. Nature 325:166–168

Farrar WL, Anderson WB (1985) Interleukin-2 stimulates association of protein kinase C with plasma membrane. Nature 315:233–235

Felber BK, Paskalis H, Kleinman-Ewing C, Wong-Staal F, Pavlakis G (1985) The pX protein of HTLV-1 is a transcriptional activator of its long terminal repeats. Science 229:675–679

Flexner C, Hugin A, Moss B (1987) Prevention of Vaccinia virus infection in immunodeficient mice by vector-directed IL-2 expression. Nature 330:259–262

Fujii M, Sugamura K, Sano K, Nakai M, Sugita K, Hinuma Y (1986) High-affinity receptor-mediated internalization and degradation of interleukin 2 in human T cells. J Exp Med 163:550–562

Fujisawa J, Seiki M, Kiyokawa T, Yoshida M (1985) Functional activation of the long terminal repeat of human T-cell leukemia virus type I by a trans-acting factor. Proc Natl Acad Sci USA 82:2277–2281

Fujita T, Takaoka C, Matsui H, Taniguchi T (1984) Structure of the human interleukin 2 gene. Proc Natl Acad Sci USA 80:7437–7441

Fujita T, Shibuya H, Ohashi K, Yamanishi K, Taniguchi T (1986) Regulation of human interleukin-2 gene: functional DNA sequences in the 5′ flanking region for the gene expression in activated T lymphocytes. Cell 46:401–407

Fuse A, Fujita T, Yasumitsu N, Kashima N, Hasegawa K, Taniguchi T (1984) Organization and structure of the mouse interleukin-2 gene. Nucleic Acids Res 12:9323–9331

Gootenberg JE, Ruscetti FW, Mier JW, Gazdar A, Gallo RC (1981) Human cutaneous T cell lymphoma and leukemia cell lines produce and respond to T cell growth factor. J Exp Med 154:1403–1418

Greene WC, Leonard WC (1986) The human interleukin-2 receptor. Annu Rev Immunol 4:69–96

Greene WC, Robb RJ, Svetlik PB, Rusk CM, Depper JM, Leonard WJ (1985) Stable expression of cDNA encoding the human interleukin-2 receptor in eukaryotic cells. J Exp Med 162:363–368

Grimm EA, Mazumder A, Zhang HZ, Rosenberg SA (1982) Lymphokine-activated killer cell phenomenon. Lysis of natural killer-resistant fresh solid tumor cells by interleukin 2-activated autologous human peripheral blood lymphocytes. J Exp Med 155:1823–1841

Hatakeyama M, Minamoto S, Uchiyama T, Hardy RR, Yamada G, Taniguchi T (1985) Reconstitution of functional receptor for human interleukin-2 in mouse cells. Nature 318:467–470

Hatakeyama M, Minamoto S, Taniguchi T (1986) Intracytoplasmic phosphorylation sites of Tac antigen (p55) are not essential for the conformation, function, and regulation of the human interleukin 2 receptor. Proc Natl Acad Sci USA 83:9650–9654

Hatakeyama M, Doi T, Kono T, Maruyama M, Minamoto S, Mori H, Kobayashi H, Uchiyama T, Taniguchi T (1987) Transmembrane signaling of interleukin 2 receptor: conformation and function of human interleukin 2 receptor (p55)/insulin receptor chimeric molecules. J Exp Med 166:362–375

Hatakeyama M, Tsudo M, Minamoto S, Kono T, Doi T, Miyata T, Miyasaka M, Taniguchi T (1989) Interleukin-2 receptor β chain gene: generation of three receptor forms by cloned human α and β chain cDNA's. Science 244:551–556

Hattori T, Uchiyama T, Toibana T, Takatsuki K, Uchino H (1981) Surface phenotype of Japanese adult T-cell leukemia cells characterized by monoclonal antibodies. Blood 58:645–647

Henny CS, Kuribayashi K, Kern DE, Gillis S (1981) Interleukin-2 augments natural killer cell activity. Nature 291:335–338

Herrmann T, Diamantstein T (1988) The human intermediate-affinity interleukin 2 receptor consists of two distinct, partially homologous glycoproteins. Eur J Immunol 18:1051–1057

Holbrook NJ, Smith KA, Fornace AJ, Comeau CM, Wiskocil RL, Crabtree GR (1984) T-cell growth factor: complete nucleotide sequence and organization of the gene in normal and malignant cells. Proc Natl Acad Sci USA 81:1634–1638

Inoue J, Seiki M, Taniguchi T, Tsuru S, Yoshida M (1986) Induction of interleukin 2 receptor gene expression by p40^x encoded by human T-cell leukemia virus type I. EMBO J 5:2883–2888

Ishida N, Kanamori H, Noma T, Nikaido T, Sabe H, Suzuki N, Shimizu A, Honjo T (1985) Molecular clonig and structure of the human interleukin-2 receptor gene. Nucleic Acids Res 13:7579–7589

Ju G, Collins L, Kaffka KL, Tsien W-H, Chizzonite R, Crowl R, Bhatt R, Kilian PL (1987) Structure-function analysis of human interleukin-2. J Biol Chem 262:5723–5731

Jung LKL, Hara T, Fu SM (1984) Detection and functional studies of p60–65 (Tac antigen) on activated human B cells. J Exp Med 160:1597–1602

Kashima N, Nishi-Takaoka C, Fujita T, Taki S, Yamada G, Hamuro J, Taniguchi T (1985) Unique structure of murine interleukin-2 as deduced from cloned cDNAs. Nature 313:402–404

Kelley VE, Gaulton GN, Hattori M, Ikegami H, Eisenbarth G, Strom TB (1988) Anti-interleukin 2 receptor antibody suppresses murine diabetic insulitis and lupus nephritis. J Immunol 140:59–61

Kirkman RL, Barrett LV, Gaulton GN, Kelly VE, Ythier A, Strom TB (1985) Administration of an anti-interleukin 2 receptor monoclonal antibody prolongs cardiac allograft survival in mice. J Exp Med 162:358–362

Kiyokawa T, Seiki M, Iwashita S, Imagawa K, Shimizu F, Yoshida M (1985) p27^{x-III} and p21^{x-III} proteins encoded by pX sequence of human T-cell leukemia virus type I. Proc Natl Acad Sci USA 82:8359–8363

Kondo S, Shimizu A, Maeda M, Tagaya A, Yodoi J, Honjo T (1986) Expression of functional human interleukin-2 receptor in mouse T cells by cDNA transfection. Nature 320:75–77

Kondo S, Kinoshita M, Shimizu A, Saito Y, Konishi M, Sabe H, Honjo T (1987) Expression and functional characterization of artificial mutants of interleukin-2 receptor. Nature 327:64–67

Koyasu S, Yodoi J, Nikaido T, Tagaya Y, Taniguchi Y, Honjo T, Yahara I (1986) Expression of interleukin 2 receptors on interleukin 3 dependent cell lines. J Immunol 136:984–987

Kuo L-M, Robb RJ (1986) Structure-function relationships for the IL-2 receptor system I. Localization of a receptor binding site on IL-2. J Immunol 137:1538–1543

LeGrue SJ (1988) Does interleukin 2 stimulus-response coupling result in generation of intracellular second messengers? Lymphokine Res 7:187–200

Leonard WJ, Depper JM, Uchiyama T, Smith KA, Waldmann TA, Greene WC (1982) A monoclonal antibody that appears to recognize the receptor for human T cell growth factor; partial characterization of the receptor. Nature 300:267–269

Leonard WJ, Depper JM, Crabtree GR, Rudikoff S, Pumphrey J, Robb RJ, Svetlik PB, Peffer N, Waldmann TA, Greene WC (1984) Molecular cloning and expression of cDNAs for the human interleukin-2 receptor. Nature 311:626–631

Leonard WJ, Depper JM, Kanehisa M, Kronke M, Peffer N, Svetlik PB, Sullivan M, Greene WC (1985) Structure of the human interleukin-2 receptor gene. Science 230:633–639

Leung K, Nabel GJ (1988) HTLV-1 transactivator induces interleukin-2 receptor expression through an NF-kB-like factor. Nature 333:776–778

Lotze MT, Grimm EA, Mazumder A, Strausser SA, Rosenberg SA (1981) Lysis of fresh and cultured autologous tumor by lymphocytes cultured in T-cell growth factor. Cancer Res 41:4420–4425

Malkovsky M, Loveland B, North M, Asherson GL, Gao L, Ward P, Fiers W (1987) Recombinant interleukin-2 directly augments the cytotoxicity of human monocytes. Nature 325:262–264

Maruyama M, Shibuya H, Harada H, Hatakeyama M, Seiki M, Fujita T, Inoue J-I, Yoshida M, Taniguchi T (1987) Evidence for aberrant activation of the interleukin-2 autocrine loop by HTLV-1-encoded p40^x and T3/Ti complex triggering. Cell 48:343–350

Mills GB, Stewart DJ, Mellors A, Gelfand EW (1986) Interleukin 2 does not induce phosphatidylinositol hydrolysis in activated T cells. J Immunol 136:3019–3024

Mills GB, Girard P, Grinstein S, Gelfand EW (1988) Interleukin-2 induces proliferation of T lymphocyte mutants lacking protein kinase C. Cell 55:91–100

Miyoshi I, Kubonishi I, Yoshimoto S, Akagi T, Ohtsuka Y, Shiraishi Y, Nagata K, Hinuma Y (1981) Type C virus particles in a cord T-cell line derived by cocultivating normal human cord leukocytes and human leukemic T cells: Nature 294:770–771

Morgan DA, Ruscetti FW, Gallo RC (1976) Selective in vitro growth of T-lymphocytes from normal human bone marrows. Science 193:1007–1008

Nabel G, Baltimore D (1987) An inducible transcription factor activates expression of human immunodeficiency virus in T cells. Nature 326:711–713

Nabel GJ, Gorka C, Baltimore D (1988) T-cell specific expression of interleukin 2: evidence for a negative regulatory site. Proc Natl Acad Sci USA 85:2934–2938

Nikaido T, Shimizu A, Ishida N, Sabe H, Teshigawara K, Maeda M, Uchiyama T, Yodoi J, Honjo T (1984) Molecular cloning of cDNA encoding human interleukin-2 receptor. Nature 311:631–635

Perentesis JP, Genbauffe FS, Veldman SA, Galeotti GL, Livingston DM, Bodley JM, Murphy J (1988) Expression of diphtheria toxin fragment A and hormone-toxin fusion proteins in toxin-resistant yeast mutants. Proc Natl Acad Sci USA 85:8386–8390

Ramshaw IA, Andrew ME, Phillips SM, Boyle DB, Coupar BEH (1987) Recovery of immunodeficient mice from a vaccinia virus/IL-2 recombinant infection. Nature 329:545–546

Reed JC, Sabath DE, Hoover RG, Prystowsky MB (1985) Recombinant interleukin-2 regulates levels of c-myc mRNA in a cloned murine T lymphocytes. Mol Cell Biol 5:3361–3368

Reeves R, Spies AG, Nissen MS, Buck CD, Weingerg AD, Barr PJ, Magnuson NS, Magnuson JA (1986) Molecular cloning of a functional bovine interleukin 2 cDNA. Proc Natl Acad Sci USA 83:3228–3232

Robb RJ (1986) Conversion of low-affinity interleukin 2 receptors to a high-affinity state following fusion of cell membranes. Proc Natl Acad Sci USA 83:3992–3996

Robb RJ, Greene WC, Rusk CM (1984) Low and high affinity cellular receptors for interleukin-2. Implications for the level of Tac antigen. J Exp Med 154:1455–1474

Rosenberg SA, Spiess P, Lafreniere R (1986) A new approach to the adoptive immunotheraphy of cancer with tumor-infiltrating lymphocytes. Science 233:1318–1321

Rosenberg SA (1988) The development of new immunotherapies for the treatment of cancer using interleukin-2. Ann Surg 208:121–135

Rosenberg SA, Lotze MT, Muul LM, Chang AE, Avis FP, Leitman S, Linehan WM, Robertson CN, Lee RE, Rubin JT, Seipp CA, Simpson CG, White DE (1987) A progress report on the treatment of 157 patients with advanced cancer using lymphokine-activated killer cells and interleukin-2 or high-dose interleukin-2 alone. N Engl J Med 316:889–897

Saragivo H, Malek T (1987) The murine interleukin 2 receptor: irreversible cross-linking of radiolabeled interleukin 2 to high-affinity interleukin 2 receptors reveals a noncovalently associated subunit. J Immunol 139:1918–1926

Seigel LJ, Harper ME, Wong-Staal F, Gallo RC, Nash WG, O'Brien SJ (1984) Gene for T-cell growth factor: location on human chromosome 4q and feline chromosome B1. Science 223:175–178

Seiki M, Hattori S, Hirayama Y, Yoshida M (1983) Human adult T-cell leukemia virus: complete nucleotide sequence of the provirus genome integrated in leukemia cell DNA. Proc Natl Acad Sci USA 80:3618–3622

Sharon M, Klausner RD, Cullen BR, Chizzonite R, Leonard WJ (1986) Novel interleukin-2 receptor subunit detected by cross-linking under high-affinity conditions. Science 234:859–863

Shaw J-P, Utz-PJ, Durand DB, Toole JJ, Emmel EA, Crabtree GR (1988) Identification of a putative regulator of early T cell activation genes. Science 241:202–205

Shibuya H, Yoneyama M, Taniguchi T (1989) Involvement of a common transcription factor in the regulated expression of IL-2 and IL-2 receptor genes. Int Immunol 1:43–49

Siebenlist U, Durand DB, Bressler P, Holbrook NJ, Norris CA, Kamoun M, Kant JA, Crabtree GR (1986) Promoter region of interleukin-2 gene undergoes chromatin structure changes and confers inducibility on chroramphenicol acetyltransferase gene during activation of T cells. Mol Cell Biol 6:3042–3049

Smith KA (1984) Interleukin 2. Annu Rev Immunol 2:319–333

Smith KA (1988) Interleukin 2: inception, impact and implications. Science 240:1169–1176

Sodroski JG, Rosen CA, Haseltine WA (1984) Trans-acting transcriptional activation of the long terminal repeat of human T lymphotropic viruses in infected cells. Science 225:381–385

Steiner G, Tschachler E, Tani M, Malek TR, Shevach EM, Holter W, Knapp W, Wolff K, Stingl G (1986) Interleukin 2 receptors on cultured murine epidermal Langerhans cells. J Immunol 137:155–159

Stern JB, Smith KA (1986) Interleukin-2 induction of T-cell G_1 progression and c-myb expression. Science 233:203–206

Sugamura K, Nakai S, Fujii M, Hinuma Y (1985) Interleukin-2 inhibits in vitro growth of human T cell lines carring retrovirus. J Exp Med 161:1243–1248

Suzuki N, Matsunami N, Kanamori H, Ishida N, Shimizu A, Yaoita Y, Nikaido T, Honjo T (1987) The human IL-2 receptor gene contains a positive regulatory element that functions in cultured cells and cell-free extracts. J Biol Chem 262:5079–5086

Taniguchi T (1988) Regulation of cytokine gene expression. Annu Rev Immunol 6:439–464

Taniguchi T (1989) Structure and regulation of the genes operating in type I IFN and IL-2 systems. In: Yamamura Y (ed) Recent progress in cytokine research. Medical View, Tokyo, pp 75–88

Taniguchi T, Matsui H, Fujita T, Takaoka C, Kashima N, Yoshimoto R, Hamuro J (1983) Structure and expression of a cloned cDNA for human interleukin-2. Nature 302:305–310

Taniguchi T, Matsui H, Fujita T, Hatakeyama M, Kashima N, Fuse A, Hamuro J (1986a) Molecular analysis of the interleukin-2 system. Immunol Rev 92:121–133

Taniguchi T, Fujita T, Hatakeyama M, Mori H, Matsui H, Sato T, Hamuro J, Minamoto S, Yamada G, Shibuya H (1986b) Interleukin-2 and its receptor: structure and functional expression of the genes. Cold Spring Harbor Symp Quant Biol 51:577–586

Teshigawara K, Wang H-M, Kato K, Smith KA (1987) Interleukin 2 high-affinity receptor expression requires two distinct binding proteins. J Exp Med 165:223–238

Tsudo M, Kozak RW, Goldman CK, Waldmann TA (1986) Demonstration of a non-Tac peptide that binds interleukin 2: a potential participant in a multi-chain interleukin 2 receptor complex. Proc Natl Acad Sci USA 83:9694–9698

Uchiyama T, Broder S, Waldmann TA (1981) A monoclonal antibody (anti-Tac) reactive with activated and functionally mature human T cells. I. Production of anti-Tac monoclonal antibody and distribution of Tac(+) cells. J Immunol 126:1393–1387

Valge VE, Wong JGP, Datlof BM, Sinskey AJ, Rao A (1988) Protein kinase C is required for responses to T cell receptor ligands but not to interleukin-2 in T cells. Cell 55:101–112

Waldmann TA, Goldman CK, Robb RJ, Depper JM, Leonard WJ, Sharrow SO, Bongiovanni KF, Korsmeyer SJ, Greene WC (1984) Expression of interleukin 2 receptors on activated human B cells. J Exp Med 160:1450–1466

Wang H-M, Smith KA (1987) The interleukin 2 receptor. Functional consequences of its bimolecular structure. J Exp Med 166:1055–1069

Weiss A, Imboden J, Hardy K, Manger B, Terhost C, Stobo J (1986) The role of the T3/antigen receptor complex in T-cell activation. Annu Rev Immunol 4:593–620

Weissman AM, Harford JB, Svetlik PB, Leonard WJ, Depper JM, Waldmann TA, Greene WC, Klausner RD (1986) Only high-affinity receptors for interleukin 2 mediate internalization of ligand. Proc Natl Acad Sci USA 83:1463–1466

Wong-Staal F, Gallo RC (1985) Human T-lymphotropic retroviruses. Nature 317:395–403

Yamada G, Kitamura Y, Sonoda H, Harada H, Taki S, Mulligan RC, Osawa H, Diamantstein T, Yokoyama S, Taniguchi T (1987) Retroviral expression of the human IL-2 gene in a murine T cell line results in cell growth autonomy and tumorigenicity. EMBO J 6:2705–2709

Yokota T, Arai N, Lee F, Rennick D, Mosmann T, Arai K (1986) Use of a cDNA expression vector for isolation of mouse interleukin 2 cDNA clones: expression of T-cell growth factor activity after transfection of monkey cells. Proc Natl Acad Sci USA 82:68–72

Yoshida M, Seiki M (1987) Recent advances in the molecular biology of HTLV-1: trans-activation of viral and cellular genes. Annu Rev Immunol 5:541–560

CHAPTER 11

Interleukin-3

J. N. IHLE

A. Introduction

Interleukin-3 (IL-3) is one of a large and growing group of growth factors which support the proliferation and differentiation of hematopoietic progenitors as well as cells committed to various myeloid lineages. The term IL-3 was initially introduced to identify a T-cell-derived lymphokine which was capable of inducing the expression of the enzyme 20α-hydroxysteroid dehydrogenase (20αSDH) in cultures of nude mouse splenic lymphocytes (IHLE et al. 1981, 1982a). This assay was developed to identify T-cell factors which might support the proliferation and differentiation of early hematopoietic progenitors capable of committing to T-lineage differentiation (IHLE and WEINSTEIN 1986). With the purification of IL-3 to homogeneity it became evident this lymphokine had a broad spectrum of activities on hematopoietic cells and was equivalent to other biological activities which had been characterized including mast cell growth factor activity, P-cell stimulating factor activity, burst promoting activity, multi-colony stimulating factor, *thy*-1 inducing factor, and WEHI-3 growth factor as well as a number of other activities for which the factors had been less characterized (IHLE et al. 1983; IHLE and WEINSTEIN 1986; IHLE 1986).

Although the initial studies indicated that IL-3 had a variety of biological activities, all appear to stem from the ability of IL-3 to support the proliferation of hematopoietic cells at various stages of differentiation. Indeed, the most convenient assay for IL-3 has been its ability to support the proliferation of a variety of hematopoietic growth factor-dependent cell lines (IHLE et al. 1982b, 1983). IL-3-dependent cell lines were first isolated from long-term bone marrow cultures and more recently from retrovirus-induced myeloid leukemias. The majority of the IL-3-dependent cell lines share in common an absolute requirement for hematopoietic growth factors for the maintenance of viability. In some cases the cells can lose viability at the rate of half the population every 2 h in the absence of IL-3. This remarkable dependency has provided a rapid and sensitive assay for IL-3 as well as providing an excellent system with which to study the mechanism of action of IL-3 on cell growth. These cell lines have also provided insights into the genes which are involved in the differentiation of hematopoietic cells as well as the genes associated with their transformation and altered ability to differentiate.

B. IL-3 Protein Structure

The initial studies on the biology and biochemistry of murine IL-3 utilized protein purified from conditioned media from the murine myelomonocytic cell line WEHI-3 (IHLE et al. 1982a, 1983; CLARK-LEWIS et al. 1984). The purification of IL-3 to homogeneity demanded large volumes of conditioned media and required purifications of up to 10^6-fold. Purified, native, murine IL-3 exists as a monomer with an apparent molecular size of 28-kDa and contains approximately 38% carbohydrate by weight. Some variation in the size of murine IL-3 has been reported which may be due to proteolytic cleavage or differing extents of glycosylation. Studies with native IL-3 as well as recombinant and synthetic IL-3 indicate that glycosylation is not necessary for biological activity in vitro. It has not been determined whether glycosylation affects the biological activity or stability of IL-3 in vivo. Native murine IL-3 has a pI of 4–8. At neutral pH and in low salt IL-3 does not bind to diethylaminoethyl (DEAE) cellulose, a property which has proven quite valuable for purification. Although IL-3 is relatively stable to a wide variety of conditions including pH and temperature, it is quite sensitive to reducing agents and it was this observation which initially suggested the existence of disulfide bonds critical for biological activity.

The specific activity of highly purified, native, murine IL-3 has varied from 2×10^6 to 2×10^8 units/mg protein where a unit is that amount required for half-maximal proliferation of factor-dependent cells or for the induction of 20αSDH. Alternatively specific activities, expressed as the concentration of IL-3 required for half-maximal activity, have been reported in a range from 10^{-12} M to 10^{-14} M (IHLE et al. 1983; CLARK-LEWIS et al. 1984; NICOLA and PETERSON 1986; PARK et al. 1986). The variations are likely due to the difficulty of accurately determining protein concentrations with very small quantities of protein. The specific activities of purified synthetic IL-3 or IL-3 produced by various expression vector systems have been comparable or lower.

The cloning of the IL-3 gene has allowed the use of a variety of expression vector systems as a more convenient starting source for the purification of IL-3. The murine IL-3 gene was idependently cloned by two groups based on its ability to induce the proliferation of IL-3-dependent myeloid cell lines (FUNG et al. 1984) or IL-3-dependent mast cell lines (YOKOTA et al. 1984). The rat IL-3 gene was subsequently cloned based on its homology with the murine gene (COHEN et al. 1986). The cloning of the human gene by nucleic acid homology was not possible because of the considerable divergence of the primate and rodent genes, as described below. The human gene was only cloned after considerable effort, and speculation regarding its existence, from a T-cell expression library based on its ability to induce the proliferation of leukemic cells (YANG et al. 1986).

For biological and biochemical studies, IL-3 has been produced by expression vector systems in mammalian cells (HAPEL et al. 1985; RENNICK et al. 1985), insect cells (MIYAJIMA et al. 1987), yeast cells (F. LEE et al. 1988), and bacteria (KINDLER et al. 1986). Biologically active IL-3 has also been made synthetically (CLARK-LEWIS et al. 1986). In general, the biological activity of IL-3 derived from various sources has been comparable, although there are differences in the type and extent of glycosylation and the extent of proteolytic processing.

The complete amino acid sequence of murine IL-3 has been derived from the sequencing of cDNA clones and is shown in Fig. 1. The precursor protein contains 166 amino acids. The first 26 amino acids encode a typical hydro-

```
Met Val Leu Ala Ser Ser Thr Thr Ser Ile His Thr Met Leu Leu
                        Met  *  Arg Leu Pro Val  *   *
                                            ┌─► Mature Protein
Leu Leu Leu Met Leu Phe His Leu Gly Leu Gln Ala Ser Ile Ser
 *   *  Gln Leu  *  Val Arg Pro  *   *   *   *  Pro Met Thr


Gly Arg Asp Thr His Arg Leu Thr Arg Thr Leu Asn Cys Ser Ser
Gln Thr  -   *  Ser Leu Lys  *  Ser Trp Val  *   *   *  Asn


Ile Val Lys Glu Ile Ile Gly Lys Leu  -   -   -  Pro  -   -
Met Ile Asp  *   *   *  Thr His  *  Lys Gln Pro  *  Leu Pro


     1 ─┬─ 2
 -   -  │Glu Pro Glu Leu Lys Thr Asp Asp Glu Gly Pro Ser Leu
Leu Leu │Asp Phe Asn Asn Leu Asn Gly Glu Asp Gln Asp Ile  *

 2 ─┬─ 3
Arg│Asn Lys Ser Phe Arg Arg Val Asn Leu Ser Lys Phe Val Glu
Met│Glu Asn Asn Leu  *   *  Pro  *   *  Glu Ala  *  Asn Arg


Ser Gln Gly Glu Val Asp Pro Glu Asp Arg Tyr Val Ile Lys Ser
Ala Val Lys Ser Leu  -   -  Gln Asn Ala Ser Ala  *  Glu  *


         3 ─┬─ 4
Asn Leu Gln│Lys Leu Asn Cys Cys Leu Pro Thr Ser Ala Asn Asp
Ile  *  Lys│Asn  *  Leu Pro  *   *   *  Leu Ala Thr Ala Ala


 4 ─┬─ 5
Ser Ala│Leu Pro Gly Val Phe Ile Arg  -   -  Asp Leu Asp Asp
Pro Thr│Arg His Pro Ile His  *  Lys Asp Gly  *  Trp Asn Glu


Phe Arg Lys Lys Leu Arg Phe Tyr Met Val His Leu Asn Asp Leu
 *   *  Arg  *   *  Thr  *   *  Leu Lys Thr  *  Glu Asn Ala


Glu Thr Val Leu Thr Ser Arg Pro Pro Gln Pro Ala Ser Gly Ser
Gln Ala Gln Gln  *  Thr Leu Ser Leu Ala Ile Phe  -   -   -


Val Ser Pro Asn Arg Gly Thr Val Glu Cys
 -   -   -   -   -   -   -   -   -   -
```

Fig. 1. Sequences of murine (*top*) and human (*bottom*) IL-3. The position of the start of the mature murine and human proteins is indicated. The *asterisks* indicate conserved positions and the *dashes* indicate deleted residues. The portions of the protein encoded by the various genomic exons are indicated by the *bars*

phobic leader sequence found in secreted proteins. Protein sequencing of native IL-3 has indicated that the secreted protein begins at either the Ala residue at position 27 (CLARK-LEWIS et al. 1984) or at the Asp residue at position 33 (IHLE et al. 1983). The basis for differences are not known although the later protein may result from secondary proteolytic cleavage. Both derivatives have been shown to have comparable biological activity. In the sequence there exist four consensus sequences for N-linked glycosylation (Asn-X-Thr/Ser). In murine IL-3 it has not been determined which of these sites are glycosylated although in the sequencing of IL-3 there is a dramatic decrease in the amino acid yield at position 42, suggesting that the majority of the molecules are glycosylated at this position. The large amount of serine and threonine also suggests the possibility of O-linked glycosylation.

Using synthetically produced IL-3 a number of points have been made regarding the structure of IL-3 (CLARK-LEWIS et al. 1986). First, the nonglycosylated synthetic protein has nearly full biological activity, suggesting that glycosylation is not required for activity. Secondly, a synthetic derivative containing residues 1–79 of the mature protein retains some biological activity whereas a derivative containing residues 80–140 had no detectable activity, suggesting that the receptor binding domain is in the amino-terminal region of the protein. Lastly, cysteine replacements have shown that the residues at positions 17 and 80 in the mature protein are required for biological activity. Consistent with an important functional role, these residues are conserved in all the IL-3 genes which have been sequenced. It should also be noted that other hematopoietic growth factor genes contain comparably positioned disulfides.

The structures of human, gibbon ape, and rat IL-3 have been deduced by sequencing of cDNA or genomic clones. A comparison of the amino acid sequences of the human and murine genes is shown in Fig. 1 in which the sequences have been separated into the five coding exons of the gene. There is little sequence homology between murine IL-3 and primate IL-3 or with rat IL-3. The amino acid homology between murine and rat IL-3 is 59%. At the nucleic acid level, there is a sequence homology of 76% in the coding region of the gene. The homology between human and murine IL-3 is considerably less. The best amino acid alignment gives an amino acid homology of 29% and if conservative changes are considered the homology is only approximately 38%. At the nucleotide level the overall sequence homology is 45% for the coding region of the gene. Consistent with the lack of homology in the primary sequence, murine IL-3 is not biologically active in assays for rat or human IL-3. Conversely, human and rat IL-3 are not active in biological assays for murine IL-3. In spite of the differences in primary structures the biological activities and apparent function of IL-3 have been highly conserved.

Perhaps the most intriguing question regarding the structure of IL-3 is the basis for the lack of conservation of homology between species. It could be argued that the lack of homology indicates that IL-3 does not play an important role in the regulation of hematopoiesis and therefore has been free to diverge. Surprisingly, however, the function of the protein appears to have been highly conserved. An alternative interpretation is that the concentrations of IL-3 are extremely critical and therefore there are strong selective pressures against too little or too much IL-3. Under these conditions, mutations within the gene which affect the binding properties of the protein and thus determine the biological ranges over which IL-3 is physiologically active could have profound biological effects. Any mutation within the gene must therefore be compensated for by additional mutations within the IL-3 gene or within the receptor for IL-3. Similarly, mutations within the receptor for IL-3 must be compensated for by secondary mutations within the receptor or within IL-3. This "push/pull" mutational response could result in an apparent accelerated rate of divergence and could account for the lack of conservation of the protein among different species.

C. IL-3 Gene Chromosomal Location and Structure

The cellular genes for IL-3 have been cloned from mice (CAMPBELL et al. 1985; TODOKORO et al. 1985; MIYATAKE et al. 1985), rats (COHEN et al. 1986), and humans (YANG et al. 1986). The intron and exon structures for the murine and human genes show considerable similarity, as shown in Fig. 2. Both genes are composed of five relatively small exons. Within both genes there is one relatively large intron which separates the second and third exons. The exons encode 54/53, 14/14, 30/32, 14/14 and 40/51 amino acids for human and murine IL-3, respectively. The location of the exon borders within the sequence of IL-3 are indicated in Fig. 1. Interestingly, the sequences flanking the coding region of the murine and human IL-3 genes show more homology than within the gene. For example the 5' regions, which encompass the putative promoter region of the genes, show 59% homology compared to approximately 45%

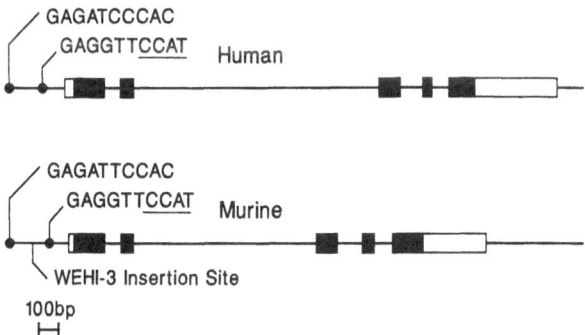

Fig. 2. Structures of the genes for murine (*top*) and human IL-3 (*bottom*). The 5' and 3' noncoding regions are indicated by the *open boxed* areas while the coding regions are indicated by the *solid boxed* areas. The positions of possible regulatory regions are indicated, including a sequence which has been implicated in the regulation of the expression of the gene. The position of the site of integration of an intercisternal A particle in the WEHI-3 cell line is indicated

homology in the coding regions and within the intron sequences. In the 3' non-coding regions of the two genes the homology is approximately 59%. This conservation has suggested that these regions may be important for the regulation of the expression of the gene.

The murine IL-3 gene has been genetically mapped to chromosome 11 by using somatic cell hybrids (IHLE et al. 1987), genetic crosses, and recombinant inbred strains of mice (BUCHBERG et al. 1988). Of particular interest is the observation that three other hematopoietic growth factors also map to murine chromosome 11. Granulocyte-specific colony stimulating factor (G-CSF) maps distal to IL-3 on chromosome 11 (BUCHBERG et al. 1988). Granulocyte/macrophage-specific colony stimulating factor (GM-CSF) also maps to chromosome 11 and is genetically very closely linked to the IL-3 gene. The proximity of the murine IL-3 and GM-CSF genes has been confirmed by pulse-field electrophoresis which has placed the genes within 230 kilobases (kb) (BARLOW et al. 1987). More recently the IL-5 gene has been mapped to murine chromosome 11 (SANDERSON et al. 1988).

The human IL-3 gene has been mapped to chromosome 5 at bands q23–31 by somatic cell hybrid analysis and in situ chromosomal hybridization (LE BEAU et al. 1987). GM-CSF has also been localized to human chromosome 5q and, consistent with the murine data, the human GM-CSF gene is located within 9 kb of the human IL-3 gene and is oriented in the same direction (YANG et al. 1988 a). Human chromosome 5 also contains the locus for CSF-1 (macrophage-specific colony stimulating factor) (PETTENATI et al. 1987), the IL-5 locus (SUTHERLAND et al. 1988), the IL-4 locus (MIYAJIMA et al. 1988), and the c-fms locus (NIENHUIS et al. 1985) which encodes the receptor for CSF-1. The CSF-1 and c-fms loci are distal to the loci for IL-3 and GM-CSF and are not found on murine chromosome 11. Preliminary data indicate that the genes for IL-5 and IL-4 may be very close to one another and within 100 kb of the IL-3 and GM-CSF genes.

The localization of hematopoietic growth factor genes to human chromosome 5q was particularly interesting since nonrandom chromosomal abnormalities have been described which involve this region. In particular, the loss of a part of the long arm of chromosome 5 has been observed in patients with myelodysplastic syndrome or acute nonlymphocytic leukemia secondary to cytotoxic therapy for other diseases (WISNIEWSKI and HIRSCHHORN 1983; TINEGATE et al. 1983; SOKAL et al. 1975; KERKHOFS et al. 1982). Examination of cells from seven patients with 5q— syndrome or patients with acute non-lymphocytic leukemia characterized by a 5q deletion demonstrated that in all cases there was a deletion of the IL-3 gene (LE BEAU et al. 1987; LE BEAU 1987) as well as the GM-CSF gene. In some but not all cases of 5q— syndrome the deletion may also involve the c-fms and CSF-1 genes. The significance of these deletions to the hematopoietic disorders or to development of malignancies is not currently known. In all cases it should be noted that the patients retain one normal chromosome 5 and thus retain one allele for the growth factors. Whether the absence of one allele can cause a significant reduction in the ability of activated T cells to produce IL-3 or GM-CSF is not currently known.

The close physical linkage of the IL-3 and GM-CSF genes has suggested that these genes are evolutionarily related. This is further supported by the general similarity in the structure of the genes and their gene products. However, there is no significant nucleotide or amino acid sequence homology. Nevertheless, it is quite likely that many of the hematopoietic growth factors have evolved from a common growth factor gene through gene duplication and divergence.

D. Regulation of IL-3 Production

The expression of hematopoietic growth factors is highly regulated and reflects the intricate mechanisms involved in the regulation of the daily output of hematopoietic cells. Among the well characterized hematopoietic growth factors, IL-3 is the only factor to be predominantly, if not exclusively, produced by activated T cells in normal cells in mice (IHLE and WEINSTEIN 1986) as well as in man (YANG and CLARK 1988). The T-cell specificity for the production of IL-3 is particularly striking when compared to GM-CSF which, like IL-3, is made by activated T cells, but, unlike IL-3, is also made by stromal fibroblasts and endothelial cells in response to stimulation with IL-1 or tumor necrosis factor (TNF) (YANG et al. 1988 b; BAGBY 1987; M. LEE et al. 1987; KOEFFLER et al. 1987; MUNKER et al. 1986; ZUCALI et al. 1986; BROUDY et al. 1986).

The exclusive production of IL-3 by activated T cells has led to the concept that IL-3 is not involved in normal hematopoietic differentiation but rather may represent a mechanism by which the immune system, during antigenic stimulation, can influence hematopoietic stem cell differentiation. Consistent with this concept, IL-3 is not produced by fetal tissues (AZOULAY et al. 1987) or by bone marrow stromal cells under conditions which support hematopoietic stem cell differentiation (LI et al. 1987; NAPARSTEK et al. 1986; GUALTIERI et al. 1987; BRYANT et al. 1986).

A number of studies have examined the ability of subpopulations of helper T cells to produce various lymphokines. Using cloned T-cell lines, recent studies have demonstrated the existence of distinct classes of murine helper, inducer T-cell populations which vary in the spectrum of lymphokines produced (CHERWINSKI et al. 1987; MOSMANN et al. 1986a). One class is characterized by the lack of IL-2 production and the production of IL-4 whereas a second class produces IL-2 but not IL-4. Both populations have been shown to produce IL-3, indicating that IL-3 production may be a general property of helper, inducer T cells. Studies with primary antigen-reactive T cells have found less of an indication for the coordinate expression of various lymphokines in activated T cells (SANDERSON et al. 1985 b; KELSO and OWENS 1988).

In helper, inducer T-cell lines which produce IL-3, the time course for the production of IL-3 is comparable to other lymphokines (PRYSTOWSKY et al. 1982). Also similar to other lymphokines, IL-3 production requires RNA synthesis, does not require cell proliferation, and is specifically inhibited by glucocorticoids (CULPEPPER and LEE 1985) or cyclosporin A (PALACIOS 1985).

The mechanisms involved in the regulation of IL-3 gene expression in T cells are largely unknown. It was anticipated that the cloning and sequencing of a number of coordinately regulated T-cell-derived lymphokines would allow the identification of common 5′ sequences that might be involved in gene regulation. One potential region, with the consensus sequence of GAGGTTCCA, has been found at approximately 100 and 300 base pairs (bp) 5′ of the transcription start site and is conserved in rodent and primate IL-3 genes (YANG and CLARK 1988). Identical or similar sequences are found at comparable positions in the genes for GM-CSF, IL-2 and interferon-γ which, like IL-3, are produced by activated T cells. However, a similar sequence is also found in the human and murine G-CSF genes which are not produced by activated T cells. Recent studies (SHANNON et al. 1988) have identified a nuclear DNA binding protein which recognizes the sequence GAGATTCCAC. Studies on the regulation of the expression or activity of this transcriptional factor may provide insights into the mechanism of regulation of the lymphokine genes.

One of the notable exceptions to the T-cell production of IL-3 is the constitutive production of the lymphokine by the WEHI-3 myelomonocytic cell line (J. C. LEE et al. 1982). The production of IL-3 by these cells is due to the activation of the expression of the gene by the insertion of an intracisternal A particle genome close to the 5′ region of the gene in an orientation opposite to that of the IL-3 gene (YMER et al. 1985). It has been speculated that this occurred during the evolution of the tumor cell line from a growth factor-dependent stage to an autonomously replicating tumor cell line. More recently (STOCKING et al. 1988) it has been shown that retroviral infection and selection for growth factor independence can often result in the insertional activation of the IL-3 gene and autonomous growth of the cells.

E. Origins and Properties of Hematopoietic Growth Factor-Dependent Cells

The identification, purification, and characterization of IL-3 and its receptor, as well as studies of the mechanisms in signal transduction, have been greatly facilitated by the availability of hematopoietic cell lines which require IL-3 for growth. The characteristics of these cell lines have also provided insights into the properties and growth factor requirements of intermediates in myeloid differentiation. Lastly, IL-3-dependent lines are providing approaches to identifying the genes involved in myeloid differentiation and transformation. Initially GREENBERGER et al. (1979) described the culture conditions which allowed the isolation of IL-3-dependent cells. Using culture conditions which were initially established by DEXTER et al. (1977) it was shown that cell lines could be established which were dependent upon WEHI-3-conditioned media for growth. It was subsequently shown that the relevant factor in the conditioned media was IL-3. The ability to establish such lines was subsequently confirmed by DEXTER et al. (1980).

Primary retrovirus-induced murine leukemias were also shown to be a source of IL-3-dependent cell lines. In a series of studies it was demonstrated that most retrovirally induced myeloid leukemias are dependent on growth factors (IHLE et al. 1984; HOLMES et al. 1985) and continuous factor-dependent cell lines could be readily established with IL-3. The phenotypic properties of the primary tumors and the cell lines indicated that the cells were transformed with regard to their ability to terminally differentiate but were not detectably altered in growth regulation. The genes involved in this type of myeloid transformation are largely unknown; however, rearrangements of the c-*myb* gene and a novel transcriptional regulatory protein of the zinc finger family of proteins have been implicated in transformation as discussed below.

Perhaps the most intriguing growth factor-dependent cell lines described to date have been isolated by PALACIOS et al. (1984, 1987; PALACIOS and STEIN-METZ 1985). These IL-3-dependent cell lines have been suggested to represent pro-T- or pro-B-cell lines. The lines have no rearrangements of T-cell receptor loci or the immunoglobulin loci. However, with differentiation in vivo, rearrangement of T-cell receptor loci occurs in pro-T-cell lines and rearrangement of the immunoglobulin loci occurs in pro-B-cell lines. Comparable cell lines have not been isolated in other laboratories for reasons that are as yet unclear.

Although a variety of murine IL-3-dependent cell lines have been isolated there exist few comparable human cell lines. This is particularly surprising since, like murine myeloid leukemias, human myeloid leukemias are dependent on exogenous growth factors and are particularly responsive to IL-3 (VELLENGA et al. 1987a, b; VELLENGA and GRIFFIN 1987). Recent studies have described the isolation of a few factor-dependent lines (LANGE et al. 1987; VALTIERI et al. 1987; KITAMURA et al. 1988). It is likely that sufficient effort has not been devoted to the isolation of such lines, perhaps because of the lack of sufficient growth factors. With the availability of recombinant sources of growth factors it may be possible to isolate additional human hematopoietic growth factor-dependent cell lines which would be of considerable value for studies dealing with growth regulation and differentiation.

F. Structure of the IL-3 Receptor

A single class of high-affinity receptors for IL-3 has been shown on IL-3-dependent cells (PALASZYNSKI and IHLE 1984; NICOL and METCALF 1985). The number of receptors varies with cell line but generally is in the order of 1000–10000 receptors per cell. Possible exceptions, with increased numbers of receptors, have been noted (SORENSEN et al. 1986). By Scatchard analysis of binding data most studies have determined apparent affinities of 10^{-9}–10^{-11} (PALASZYNSKI and IHLE 1984; PARK et al. 1986). The expression of receptors for IL-3 is highly restricted and in general receptors have only been found on IL-3-dependent cell lines or factor-independent derivatives derived from factor-dependent cells (Palaszynski and Ihle 1984). Interestingly, a number of growth factor-independent myeloid cell lines do not contain detectable receptors for IL-3.

A number of groups have studied the biochemical properties of the IL-3 receptor by using crosslinking reagents (May and Ihle 1986; Park et al. 1986; Nicola and Peterson 1986; Sorensen et al. 1986). In all the studies a major IL-3 crosslinked moiety of approximately 90–95-kDa has been detected, indicating that the major IL-3 binding protein is a 65- to 70-kDa protein. In addition, larger crosslinked species have been variably detected (Park et al. 1986; Nicola and Peterson 1986). One study (Nicola and Peterson 1986) detected a second major crosslinked protein of 60-kDa in addition to a 75-kDa species.

The presence of a 140-kDa membrane-associated protein which is rapidly phosphorylated following IL-3 stimulation (Koyasu et al. 1988; R. Isfort et al. 1988) suggested the possibility that it may be a component of an IL-3 receptor complex containing the 65-kDa protein detected in crosslinking studies. This question has been addressed by determining whether radiolabeled IL-3 becomes associated with a complex which can be isolated with monoclonal antibodies against phosphotyrosine (R.J. Isfort et al. 1988 b). These studies have shown that approximately 25% of the iodinated IL-3 does bind in such a complex. Using glycerol gradients and crosslinking approaches it was further demonstrated that the complex has an apparent molecular mass of 160-kDa and contains only IL-3 and the 140-kDa protein. These studies therefore indicate that the IL-3 receptor may be a 140-kDa membrane tyrosine protein kinase which is rapidly phosphorylated following binding of IL-3 and therefore would appear to have the properties of a number of growth factor receptors. It has been speculated that the major IL-3 binding protein of 65-kDa is a proteolytic degradation product of the 140-kDa protein which retains its membrane localization and ligand binding activity. Other growth factor receptors have been shown to be susceptible to proteolytic degradation and can be cleaved to smaller species which retain ligand binding activity (Leung et al. 1987; Seger et al. 1988).

The further characterization and cloning of the IL-3 receptor could be greatly facilitated by the isolation of monoclonal antibodies against the receptor. Palacios et al. (1986) have described the isolation of two monoclonal antibodies which reacted with IL-3-dependent cell lines and inhibited the growth of the cells in response to IL-3. Both monoclonals detected a cell surface protein of 50–70-kDa. Sugawara et al. (1988) have isolated five monoclonals from spleen cells of nonimmunized, autoimmune mice which selectively bound to IL-3-dependent cells and which supported their growth. The proteins detected by these monoclonal antibodies were not identified. The available data do not definitively establish that any of the monoclonals characterized to date are directed against the IL-3 receptor.

One study has addressed the potential interaction of hematopoietic growth factors through effects on receptor expression (Walker et al. 1985). The studies utilized bone marrow cells and examined the effects of one hematopoietic growth factor on the expression of receptors for other growth factors. The results indicated that there was a hierarchical pattern in the effects. However, the studies are complicated because of the use of mixed cell populations from normal bone marrow. Since a number of the IL-3-dependent cell lines express receptors for other hematopoietic growth factors, it will be important to determine in these cells whether ligand binding to one receptor can alter the expression of a second hematopoietic growth factor receptor.

G. Biochemical Mechanisms in IL-3 Signal Transduction

One of the first studies to address the possible mechanisms by which IL-3 supports the growth of hematopoietic cells noted that an ATP-generating system could partially replace the requirement of factor-dependent cells for IL-3 (WHETTON and DEXTER 1983, 1988). This observation and subsequent studies which demonstrated a rapid decrease in ATP levels following the removal of IL-3 (HASTHORPE et al. 1987; WHETTON and DEXTER 1988) have suggested that one effect of IL-3 is to maintain high ATP concentrations. The observations that IL-3 stimulates lactic acid production and that IL-3-dependent cells are sensitive to 2-deoxyglucose has led to the suggestion (WHETTON and DEXTER 1988) that IL-3 may maintain ATP levels by its effects on glycolysis, perhaps by increasing hexose transport. The biochemical links between the effects on ATP levels, glycolysis, or hexose transport are not currently known.

A potential role for protein kinase C in IL-3 signal transduction was proposed from the observation that IL-3 stimulation resulted in the translocation of protein kinase C from the cytosolic fraction to the membrane fraction, comparable to the response seen when the cells are exposed to phorbol 12-myristate – 13-acetate (PMA; FARRAR et al. 1985). Attempts have been made to reproduce these findings with partial success (WHETTON et al. 1986a) or without success (R.J. ISFORT et al. 1988a). A role for protein kinase C in IL-3 signal transduction is unlikely because of a number of observations. First, neither PMA nor calcium ionophores alone or in combination can replace the requirement for IL-3 for optimal growth or maintenance of viability (WHETTON et al. 1986a; OROSZ et al. 1983). A potential role for protein kinase was also examined by looking for the effects of downregulation of the enzyme by chronic exposure to PMA. Under conditions in which no protein kinase C activity was detectable, the response to IL-3 was unaltered or slightly enhanced (R.J. ISFORT et al. 1988a). Lastly, the pattern of protein phosphorylation has been examined following stimulation of cells with either IL-3 or PMA. While PMA induced the phosphorylation of a number of cellular substrates detectable in two-dimensional gels, stimulation with IL-3 did not result in the phosphorylation of any of these substrates (R. ISFORT et al. 1988, R.J. ISFORT et al. 1988a). Conversely, the tyrosine phosphorylation which is seen following IL-3 stimulation is not induced by PMA (MORLA et al. 1988).

Several studies have also examined the effects of IL-3 on phosphatidylinositol turnover and Ca^{2+} flux with variable results. In growth factor-dependent macrophages neither IL-3 nor CSF-1 was found to induce an increase in intracellular Ca^{2+} or to increase phosphatidylinositol turnover (WHETTON et al. 1986b). In studies of an IL-3-dependent myeloid cell line, stimulation with IL-3 had no effect on phosphatidylinositol turnover (PIERCE et al. 1988). In contrast, one group has observed an effect of IL-3 on intracellular free Ca^{2+} (ROSSIO et al. 1986).

One study (BARTON et al. 1988) has examined the effects of inhibitors of cyclooxygenase and lipoxygenase to evaluate the potential role of arachidonic acid in IL-3 induced proliferation. None of the inhibitors examined inhibited proliferation. Conversely, IL-3 stimulation of cells had no detectable effect on the incorporation of labeled arachidonic acid. These data therefore indicate that arachidonic acid metabolites are not involved in IL-3 signal transduction.

The effects of IL-3 on total cellular protein phosphorylation have been examined. One study reported that IL-3 induced the threonine phosphorylation of a 68-kDa and the phosphorylation of a 20-kDa substrate (S. EVANS et al. 1986). These substrates have not been seen in other studies (R. ISFORT et al. 1988; GARLAND 1988; MORLA et al. 1988). The phosphorylation of a 33-kDa substrate has been detected in IL-3-dependent cells when stimulated with IL-3 but not with IL-4 or tetradecanoylphorbol-acetate (TPA) (GARLAND 1988). The sites of phosphorylation were not determined. The phosphorylation of this substrate has not been detected in other studies (R. ISFORT et al. 1988; MORLA et al. 1988; S. EVANS et al. 1986).

A potential role for tyrosine phosphorylation in IL-3 signal transduction mechanisms was first indicated by studies which demonstrated that Abelson murine leukemia virus (MuLV) could abrogate the requirement for IL-3 of mast cells (PIERCE et al. 1985) or an IL-3-dependent myeloid cell line (COOK et al. 1985). The studies with v-abl have been extended to demonstrate that temperature-sensitive mutants of v-abl confer a temperature-sensitive phenotype for IL-3 dependence (KIPREOS and WANG 1988) (J. N. IHLE, unpublished data). This provides the most direct evidence for a role of tyrosine phosphorylation in IL-3 dependence to date.

In addition to the studies with Abelson MuLV, other tyrosine kinase-containing oncogenes have been shown to abrogate IL-3 dependence. v-erbB has been shown to abrogate IL-3 dependence whereas transfection of cells with a normal epidermal growth factor (EGF) receptor confers the ability to grow in response to EGF (PIERCE et al. 1988). Other activated, oncogenic, growth factor receptors which have been shown to abrogate IL-3-dependence are the v-fms oncogene (WHEELER et al. 1987) and the trk oncogene (R. ISFORT et al. 1988; KIPREOS and WANG 1988).

The effects of the src oncogene on IL-3-dependence have also been examined. WATSON et al. (1987) have shown that cells of the IL-3-dependent lines FD.C1 and 32Dcl-23 become growth factor independent when infected with a v-src-containing murine retroviral construct. These studies further demonstrated that the same construct did not abrogate the requirements of IL-2-dependent derivatives of the IL-3-dependent lines for IL-2. The differences between IL-3- and IL-2-dependent cells suggested the possibility that different signal transduction mechanisms may be involved. However, recent studies (COOK et al. 1987) have demonstrated the *abl* gene can also abrogate IL-2 dependence, suggesting that tyrosine phosphorylation may be involved in IL-2 signal transduction.

In more detailed studies (OVERELL et al. 1987) it was demonstrated that the ability of v-*src* to abrogate IL-3 dependence was not a direct effect but rather a second event was required for complete growth factor abrogation. Whether the "secondary" event can occur independent of the v-*src* oncogene was not determined. In particular, it is possible that in some factor-dependent lines, the "second" event may have occurred and in these cells v-*src* would directly abrogate factor dependence. These possibilities emphasize the concepts that all the factor-dependent lines are transformed to some extent and that the type of transforming events which allowed their establishment may influence their responses to various oncogenes. Another example of the complex interactions that can occur is the observation that in v-*src*-infected cells IL-3 and GM-CSF inhibited *src* protein kinase activity (WATSON et al. 1988) suggesting that *src* may be a substrate for the IL-3 signal transduction mechanism.

In a variety of the above studies it has been demonstrated that, associated with abrogation of growth factor dependence, the cells acquire the ability to grow in nude, athymic mice, indicating that they have become "tumorigenic" and implying that the parental cell lines are not tumorigenic. However, it has been demonstrated (HOLMES et al. 1985) that most murine myeloid leukemias are growth factor dependent, indicating that growth factors are not limiting under conditions in which most leukemias develop. Therefore it is perhaps more appropriate to consider the acquisition of growth factor independence as one possible manifestation of transformation of hematopoietic cells which may not be required for tumorigenicity and clearly is not the only transforming event which can confer tumorigenicity to the cells.

In contrast to tyrosine protein kinase-containing oncogenes, other classes of oncogenes have not been shown to abrogate growth factor dependence. In particular, v-*ras*-infected mast cells retain a requirement for IL-3 for growth (REIN et al. 1985). In addition, v-*mos*-infected IL-3-dependent lines or mast cells retain a requirement for IL-3 for growth (J. N. IHLE, unpublished data). Some of the myeloid tumor cell lines which have an activated c-*myb* require IL-3 for growth (WEINSTEIN et al. 1986) indicating that this nuclear oncogene similarly does not abrogate growth factor dependence.

The effects of *myc*-containing retroviral vectors have been less clear. Initially it was demonstrated that v-*myc*-containing oncogenes abrogated IL-3 and IL-2 dependence (RAPP et al. 1985). Subsequent studies with c-*myc*-containing retroviral vectors (DEAN et al. 1987; CORY et al. 1987; HUME et al. 1988) have found only partial abrogation of IL-3 dependence. The differences are likely due to the levels of expression and possibly to the genetic differences between v-*myc* and c-*myc*.

The above observations with tyrosine protein kinases have promoted a number of studies to look at IL-3-induced tyrosine phosphorylation. Since tyrosine phosphorylation is a minor component of protein phosphorylation,

Table 1. Cellular substrates of tyrosine phosphorylation

Substrate (kDa)	References	Phosphoamino acids	Cellular localization
160	MORLA et al. (1988)	Not determined	Cytoplasmic
150	R. ISFORT et al. (1988), KOYASU et al. (1988)	Tyr, Ser	Membrane
120	R. ISFORT et al. (1988)	Not determined	Not determined
95	MORLA et al. (1988)	Not determined	Cytoplasmic
90	MORLA et al. (1988)	Not determined	Cytoplasmic
85	R. ISFORT et al. (1988)	Not determined	Not determined
70	MORLA et al. (1988), R. ISFORT et al. (1988)	Tyr, Ser, Thr	Cytoplasmic
56–55	MORLA et al. (1988), R. ISFORT et al. (1988)	Tyr, Ser	Cytoplasmic
51	R. ISFORT et al. (1988)	Tyr, Ser	Cytoplasmic
38	R. ISFORT et al. (1988)	Tyr, Ser	Cytoplasmic
28	R. ISFORT et al. (1988)	Not determined	Not determined

immunological reagents have been used to specifically detect phosphotyrosine. This has involved using phosphotyrosine specific antibodies in Western blot analysis (KOYASU et al. 1988; MORLA et al. 1988) or by using monoclonal antibodies against phosphotyrosine to affinity purify phosphotyrosine-containing proteins (R. ISFORT et al. 1988). Using polyvalent rabbit antisera, KOYASU et al. (1988) reported the rapid phosphorylation of a 150-kDa membrane-associated glycoprotein in factor-dependent cells following stimulation with IL-3. The 150-kDa substrate was not detected in cells stimulated with IL-4, GM-CSF, IL-2, or TPA. MORLA et al. (1988) also used a polyvalent antiserum against phosphotyrosine and demonstrated the phosphorylation of a series of cellular substrates indicated in Table 1. Under the conditions of their experiments they did not detect a membrane-associated 150-kDa substrate but did detect a cytoplasmic substrate of 160-kDa which was present in two IL-3-responsive cells but not in two additional lines. In addition, they detected cytoplasmic substrates of 95, 90, 70, and 55-kDa which were found in all the cell lines examined.

By using monoclonal antibodies to phosphotyrosine, R. ISFORT et al. (1988) have detected a number of substrates for tyrosine phosphorylation in cells stimulated with IL-3. As indicated in Table 1, one of the major substrates was a membrane-associated 150-kDa protein which appears to be the same substrate detected by KOYASU et al. (1988). In addition, there were a number of major (70, 56, and 38-kDa) and minor (120, 85, 51, and 28-kDa) cytoplasmic substrates detected. Two of these substrates (70 and 56-kDa appear to be identical to those detected by MORLA et al. (1988).

The demonstration of IL-3-induced tyrosine phosphorylation and abrogation of IL-3 dependence by tyrosine protein kinase-containing oncogenes has suggested the possibility that the oncogenes may abrogate factor dependence by phosphorylating substrates which are normally phosphorylated in response to IL-3 stimulation. This possibility has been examined (R. ISFORT et al. 1988; MORLA et al. 1988) and it does appear that in growth factor-independent cells the substrates which are normally phosphorylated in response to IL-3 are constitutively phosphorylated. However, these studies have been difficult to interpret since the levels of tyrosine phosphorylation are elevated in the transformed cell lines and in most cases the transforming genes phosphorylate a number of additional substrates.

H. Gene Regulation by IL-3

The significance of the IL-3 signal transduction in growth regulation is not known. As indicated above it has been hypothesized that signal transduction may affect biochemical events associated with maintaining ATP levels including hexose transport and glycolysis. In addition, it is clear that IL-3 has an effect on the transcription of cellular genes and that this transcription may also be important for maintenance of viability and proliferation. Studies with serum or platelet-derived growth factor-induced proliferation of fibroblasts have demonstrated that a number of genes are regulated at both the transcriptional and posttranscriptional levels (RITTLING and BASERGA 1987; CHAVRIER et al. 1988; SUKHATME et al. 1987; ZUMSTEIN and STILES 1987; LAU and NATHANS 1985, 1987). The proteins encoded by these genes include potential transcriptional regulatory factors, growth factor-like genes, nuclear protooncogene-related proteins, and cytoskeleton related structural proteins. It is likely that the response of myeloid cells to IL-3 will be equally complex.

One of the first cellular genes to be shown to be regulated by growth factors was the c-*myc* protooncogene (KELLY et al. 1983; GREENBERG and ZIFF 1984). This nuclear oncogene was first characterized as the transforming gene of an avian retrovirus which transformed macrophages in vivo, the avian myeloblastosis virus (BISHOP 1983). Deregulated expression of the c-*myc* gene has been implicated in the transformation of B cells, T cells, and macrophages (CORY 1986). In most IL-3-dependent cells, the levels of c-*myc* transcripts are regulated by IL-3 (CONSCIENCE et al. 1986; DEAN et al. 1987; HAREL-BELLAN and FARRAR 1987). Removal of IL-3 results in a rapid loss of transcripts and stimulation of factor-deprived cells results in the reappearance of transcripts to maximal levels within 30 min, suggesting that some component of the IL-3 signal transduction pathway is involved in regulating the levels of c-*myc* transcripts. A role for tyrosine phosphorylation is suggested by the observation that abrogation of IL-3 dependence by tyrosine protein kinase-containing oncogenes results in constitutively high levels of c-*myc* transcripts. More strikingly, in cells infected with temperature-sensitive mutants of v-*abl* which require IL-3 for growth at the nonpermissive temperature but not at the permissive temperature, IL-3-induced c-*myc* transcription is also temperature sensitive. At the permissive temperature transcription is independent of IL-3 while at the nonpermissive temperature transcription requires IL-3 (J. CLEVELAND, U. RAPP, J. N. IHLE, manuscript in preparation).

The potential significance of IL-3-regulated expression of c-*myc* transcription is indicated by an increased ability to survive in the absence of IL-3 when c-*myc* is constitutively expressed (Dean et al. 1987; Cory et al. 1987; Hume et al. 1988). The role of c-*myc* in maintaining viability is not known; however, a series of observations suggest that part of the effects of c-*myc* gene expression are to control the expression of the ornithine decarboxylase (ODC) gene. Bowlin et al. (1986) initially demonstrated that IL-3 induced rapid increases in ODC and that inhibition of ODC activity inhibited the growth of IL-3-dependent cells. It was subsequently demonstrated that the levels of transcripts for ODC are dependent on IL-3 (Dean et al. 1987). The kinetics of the induction of transcripts were different than the induction of the transcripts for c-*myc* in that ODC transcripts came up later than c-*myc* transcripts. The role of c-*myc* was more definitively demonstrated by the observation that IL-3-dependent cells, infected with retroviral vectors which constitutively express c-*myc*, constitutively expressed transcripts for ODC (Dean et al. 1987). Therefore, it has been proposed that one component of the response to IL-3 involves the induction of transcription of the c-*myc* gene through mechanisms involving tyrosine phosphorylation. c-*myc* in turn regulates the transcription of the ODC gene and results in the ultimate production of a gene product which is required for proliferation.

In addition to c-*myc*, the c-*fos* gene is transiently expressed in response to IL-3 (Conscience et al. 1986; Dean et al. 1987; Harel-Bellan and Farrar 1987). Curiously c-*fos* expression is also transiently induced by serum stimulating the cells or by simply spinning out the cells and resuspending them. The significance of this expression has not been examined.

Although the IL-3 regulation of c-*myc* expression has been implicated in growth regulation it is clear that IL-3 also regulates the expression of genes which are more likely to be involved in differentiation. In particular, most IL-3-dependent cell lines express transcripts derived from one or more of the T-cell γ-receptor loci (Y. Weinstein, J. Cleveland, J. N. Ihle, manuscript in preparation). This transcription is from nonrearranged loci and initiates approximately 200 bp 5′ from the site of recombination with the *V* gene segments. Removal of IL-3 results in the rapid loss of transcripts and addition of IL-3 induces the appearance of transcripts with kinetics which are comparable to those of the induction c-*myc* transcripts. In cells transformed with v-*abl*, expression of transcripts is independent of IL-3, suggesting that tyrosine phosphorylation may be important in signal transduction.

The significance of IL-3 regulation of transcripts from nonreaaranged T-cell receptor loci is not known. However, transcription of nonrearranged immunoglobulin loci has been shown to be required for efficient recombination to occur (Alt et al 1986). More recenctly it has been shown that lipopolysaccharide (LPS) induces the transcription of germline immunoglobulin γ2b. This transcription is specifically inhibited by IL-4 and inhibits switching to this heavy chain isotype (Lutzker et al. 1988). Therefore it is possible that IL-3, through the induction of transcription of T-cell receptor γ loci, could influence the probability of recombination which in turn may be necessary for the commitment to differentiation along the T-cell lineage. In this context it is interesting to note that with the cell lines examined, G-CSF induces proliferation but does not induce transcription of the T-cell receptor γ loci.

IL-3 has also been shown to regulate the levels of transcripts for the 55-kDa (Tac) gene which is thought to encode a subunit of the IL-2 receptor (BIRCHENALL-SPARKS et al. 1986). The significance of this expression is not known, however, since the cells do not have high-affinity receptors for IL-2 and do not proliferate in response to IL-2 at any of the concentrations examined. As above, it might be hypothesized that the regulation of this gene is associated with differentiation rather than growth regulation by IL-3.

From these initial studies it is clear that IL-3 signal transduction will have an important role in the regulation of expression of various genes. It will be important to determine the "link" between ligand binding, tyrosine phosphorylation, and transcriptional activation. The simplest hypothesis might suggest that tyrosine phosphorylation is directly involved in the activation of a transcriptional regulatory protein, but more steps may be involved, such as the tyrosine phosphorylation of a serine protein kinase which then serine phosphorylates a transcriptional regulatory protein. Irrespective, the advances which have been made in the identification, purification, and cloning of transcriptional regulatory proteins will allow direct approaches to study the role of IL-3 signal transduction in gene regulation.

I. Biological Properties of IL-3 In Vitro and In Vivo

From the initial studies it was clear that purified IL-3 had a number of in vitro activities and had been partially purified and studied under a variety of names (IHLE et al. 1983). In in vitro cultures, murine (PRYSTOWSKY et al. 1984; HAPEL et al. 1985; SPIVAK et al. 1985; RENNICK et al. 1985) and human (LEARY et al. 1988; MESSNER et al. 1987) IL-3 have been found to support the proliferation and differentiation of a variety of lineages of cells. Taken together these observations have suggested that IL-3, in contrast to many of the hematopoietic growth factors, can support the proliferation of early hematopoietic stem cells which are capable of becoming committed to differentiation along a variety of lineages.

The hematopoietic lineages are derived from a common pluripotential stem cell which gives rise to cells of the myeloid and lymphoid lineages. This concept has been most elegantly demonstrated by marking progenitor cells in vitro with retroviruses and following the progeny of single cells in vivo after reconstitution of lethally irradiated mice (KELLER et al. 1985; WILLIAMS et al. 1984; JOYNER et al. 1983; LEMISCHKA et al. 1986). Of particular note in these experiments was the lack of evidence for stem cells which were specific for the lymphoid or myeloid lineages. The earliest progenitor cells detectable in in vitro cultures are termed blast cells (NAKAHATA and OGAWA 1982); they have extensive self-renewal capabilities and give rise to differentiated progeny of many of the myeloid lineages. Whether blast cells can also give rise to progenitors for the lymphoid lineages is not known.

The role of various hematopoietic growth factors in the regulation of blast cell proliferation has been examined. Blast cells arise from nondividing bone marrow cells which do not require exogenously added growth factors for maintenance of viability (SUDA et al. 1985, 1983). Once the cells begin to proliferate and differentiate their continued growth is dependent on the addition of IL-3 and to a lesser extent GM-CSF (KOIKE et al. 1987). Paradoxically, G-CSF has also been shown to support multilineage colony formation (SUDA et al. 1987). In the presence of IL-3 the transition from a nondividing to a dividing state is stochastic (NAKAHATA et al. 1982) although recent experiments have demonstrated that both IL-6 (IKEBUCHI et al. 1987) and G-CSF (LEARY et al. 1988) induce cell division of blast cell precursors. Once induced to differentiate in vitro under the conditions normally used, blast cell differentiation results in the production of erythrocytes, megakaryocytes, eosinophils, granulocytes, macrophages, and mast cells.

IL-3 supports the terminal differentiation of several of the myeloid lineages. A number of studies have demonstrated that IL-3 supports the differentiation of megakaryocytes (QUESENBERRY et al. 1985; BURSTEIN 1986) although their differentiation is also supported by GM-CSF as well as factors which may be specific for megakaryocytes (ROBINSON et al. 1987; SPARROW et al. 1987). IL-3 also supports the differentiation of eosinophils (SANDERSON et al. 1985a) although to a lesser extent than that observed with IL-5. Studies of the interaction of IL-3 and IL-5 suggest that IL-3 increases the number of progenitors which become IL-5 responsive (D. J. WARREN and MOORE 1988; YAMAGUCHI et al. 1988).

In contrast to many of the myeloid lineages, cells committed to the erythroid lineage have not been found to terminally differentiate in the presence of IL-3 with the exception of one study (GOODMAN et al. 1985). However, a variety of studies (PRYSTOWSKY et al. 1984; HAPEL et al. 1985; SPIVAK et al. 1985; RENNICK et al. 1985; LEARY et al. 1988; MESSNER et al. 1987) have shown that IL-3 increases the number of progenitors which can become committed to the erythroid lineage and thereby increases the number of differentiated progeny when erythropoietin is present. Consistent with a link between IL-3-responsive stem cells and cells which can proliferate to erythropoietin, IL-3-dependent cell lines have been described which express erythropoietin receptors and which proliferate in response to erythropoietin (TSAO et al. 1988; BRANCH et al. 1987; HARA et al. 1988).

In the initial studies it was demonstrated that purified IL-3 supported the proliferation of fully differentiated mast cells and accounted for much of the activity associated with mast cell growth factor or P-cell stimulating factor (IHLE et al. 1983; RAZIN et al. 1984). More recently, however, it has been demonstrated that there is a second T-cell-derived factor, IL-4, which induces mast cell proliferation (SMITH and RENNICK 1986; RENNICK et al. 1987; F. LEE et al. 1986). It had previously been demonstrated that there exist two types of mast cells termed mucosal mast cells and connective tissue mast cells which can, in part, be distinguished by the type of proteoglycan produced (RAZIN et al. 1982). Recent studies (HAMAGUCHI et al. 1987; NAKAHATA et al. 1986) have indicated that IL-4 specifically induces the proliferation of the connective tissue type of mast cells whereas IL-3 had been shown to support the proliferation of mucosal mast cells. Alternatively, growth factors may influence the phenotype of the cells, as suggested by the observation that in vitro derived IL-3-dependent mucosal mast cells can give rise to the connective tissue type of mast cell in vivo (OTSU et al. 1987).

Although initially identified by its activity on B cells, IL-4 is more active on T cells, early myeloid progenitors, and mast cells (GRABSTEIN et al. 1986; MOSMANN et al. 1986). In our experience the majority of the IL-3-dependent lines proliferate in response to IL-4 although the response is generally less than that observed with IL-3. The ability of individual lines to proliferate to IL-4 does not appear to be correlated to a particular phenotype.

Colony assays and in vitro liquid culture systems in both the presence and absence of serum have indicated that IL-3 can support the terminal differentiation of granulocytes and macrophages. In contrast purified macrophages do not proliferate in response to IL-3 under conditions in which they can respond to CSF-1 (CHEN and CLARK 1986). However, IL-3 increases the expression of CSF-1 receptors, greatly enhances the proliferative capacity of macrophages to respond to suboptimal concentrations of CSF-1, and can cause an increase in the levels of TNF mRNA in monocytes (CANNISTRA et al. 1988). IL-3 does not induce the proliferation of a cloned CSF-1-dependent cell line, BAC.1 (SCHWARZBAUM et al. 1984; D. ASKEW, J. N. IHLE unpublished data). The possible effects of IL-3 on the response to CSF-1 or CSF-1 receptor expression have not been examined nor has it been determined whether BAC.1 cells express receptors for IL-3. Conversely, none of the IL-3-dependent cell lines have been found to be CSF-1 responsive (J. N. IHLE unpublished data), although CSF-1-responsive derivatives can be isolated. The latter observations might suggest that macrophages, like erythroid progenitors, become IL-3 nonresponsive following commitment to the macrophage lineage. The basis for the differentiation of macrophages in colony assays and in liquid cultures may be due, in part, to the endogenous production of CSF-1.

A potential role for IL-3 in the regulation of lymphoid lineage differentiation is much less clear. IL-3 does not induce the proliferation of cells which have committed to either the T- or B-cell lineage including immature thymocytes (PALACIOS and VON BOEHMER 1986). The more important question is whether IL-3-responsive pluripotential stem cells can become committed to the lymphoid lineages. In this case the effects of IL-3 could be comparable to those seen with the erythroid lineage and perhaps other lineages in which IL-3 supports the proliferation of progenitors. Following commitment, however, the cells become completely dependent on lineage-specific growth factors for continued differentiation. For T cells, thymic factors may be required for continued differentiation and B cells may require unique stromal factors for continued differentiation.

The possibility of an effect of IL-3 on early, pro-T- or pro-B-cell differentiation has been most dramatically suggested by a series of studies from the laboratory of PALACIOS (PALACIOS et al. 1984, 1987; PALACIOS and VON BOEHMER 1986; SIDERAS and PALACIOS 1987; PALACIOS and STEINMETZ 1985; PALACIOS and GARLAND 1984). Of particular note has been the isolation of a series of IL-3-dependent cell lines which differentiate along either the T- or B-cell lineages when inoculated in vivo. As IL-3-dependent cells, there is no rearrangement of T-cell receptor loci or immunoglobulin loci. With differentiation in vivo, rearrangement of both loci occurs. Although these experiments are quite provocative, it is important that they be independently reproduced in other laboratories.

A number of other observations have been made which are equally provocative including: (a) the expression of the T-cell receptor γ loci from a nonrearranged configuration and its regulation by IL-3 in factor-dependent cell lines (Y. WEINSTEIN, J. N. IHLE, unpublished data); (b) the expression of the 55-kDa component of the IL-2 receptor on IL-3-dependent cell lines (BIRCHENALL-SPARKS et al. 1986; SIDERAS and PALACIOS 1987); and (c) induction by IL-2 of proliferation of IL-3-dependent mast cell lines (H. S. WARREN et al. 1985) or alternatively isolation of IL-2-depdendent cell clones from IL-3-dependent cells (LEE, GROS et al. 1985; G. S. LE GROS et al. 1987; J. E. LEGROS et al. 1987). Whether these observations support the concept for a role of IL-3 in pro-T- or pro-B-cell differentiation is questionable. Irrespective, continued studies of the growth regulation of pluripotential stem cells and the mechanisms by which they become committed to various lineages are clearly warranted.

The effects of recombinant murine IL-3 in vivo have been examined in a number of studies (LORD et al. 1986; METCALF et al. 1986; KINDLER et al. 1986; KALLAND 1987; KIMOTO et al. 1988). Consistent with the in vitro findings, IL-3 increases hematopoiesis in vivo. The effects most commonly noted were induction of splenomegaly with increases in eosinophils, neutrophils, mast cells, and macrophages. In colony forming assays, increases of 5- to 20-fold in progenitor cells were typically seen. In contrast to the effects on myeloid lineages, the changes detected in the lymphoid lineages were either not detectable or consisted of a suppression. Taken together the studies indicate that in vivo IL-3 is able to increase the number of differentiated progeny of many myeloid lineages.

J. Transformation of IL-3 Lineage Cells

The availability of purified growth factors has allowed the characterization of normal differentiation and growth regulation and thereby has provided the basic information necessary to study transformation of hematopoietic cells. The properties or normal cells would suggest that transformation may involve altering either the growth factor dependence of the cells or their ability to terminally differentiate or both. The distinction between transformation of growth regulation and differentiation is best illustrated by the properties of the IL-3-dependent myeloid cell lines derived from retrovirus-induced leukemias (HOLMES et al. 1985). The phenotypic properties of the cells are similar to those of intermediates in IL-3-supported differentiation of myeloid cells. However, none of the cell lines differentiate spontaneously or in response to a variety of inducers. In contrast, the lines appear unaltered in their requirements for, and response to, hematopoietic growth factors. As noted above, these lines can be "further" transformed to growth factor independence by various oncogenes. The alteration of growth factor requirements might be due to a variety of transforming events. As indicated above, the IL-3 dependence of myeloid leukemia cell lines can be abrogated by retroviruses expressing hematopoietic growth factors in an autocrine mechanism of transformation. Alternatively, a variety of tyrosine protein kinase-containing oncogenes can abrogate growth factor requirements by presumably "short circuiting" the signal transduction pathway. Other mechanisms can involve the activation of a receptor for a factor which is not limiting for cell growth. Alternatively it is conceivable that a growth factor receptor gene could be activated by genomic rearrangements involving a transcriptional regulatory gene which activates receptor gene expression.

In contrast to the studies on growth factor abrogation, relatively few studies have dealt with transformation as it affects differentiation. Since IL-3-dependent myeloid leukemia cell lines have phenotypes comparable to intermediates in the normal differentiation supported by IL-3 in vitro, appropriate transforming viruses should be able to transform normal cells and generate cells with phenotypes comparable to the leukemia cell lines. Studies with v-*ras*, v-*abl*, v-*src*, v-*mos*, v-*raf*, v-*myc*, and c-*myc* retroviral constructs, however, have failed to indicate any effects on the ability of normal cells to differentiate.

ROVERA et al. (1987) have examined the effects of Abelson MuLV on differentiation of myeloid cell lines. Using the 32D-c13 cells it was demonstrated that Abelson MuLV abrogated IL-3 requirements comparable to the other cell lines examined. The 32D-c13, unlike most IL-3-dependent cell lines, differentiate in the presence of G-CSF to granulocytes. Cells exposed to G-CSF were found to be resistant to the transforming effects of Abelson MuLV. In addition, cells transformed by Abelson MuLV could not be induced to differentiate with G-CSF.

An alternative approach to identifying the genes involved in altering differentiation takes advantage of the mechanisms by which replication-competent retroviruses cause transformation. In particular retroviruses have been shown to integrate near or within cellular genes, causing their activation or altered expression. Therefore it is possible to examine the potential role of various protooncogenes by examining cell lines for rearrangements. Alternatively, viral integration sites can be cloned and examined to determine whether viral integrations often occur within a region and therefore might define a potentially important locus. Once common integration sites are defined it is possible to identify the cellular gene whose expression is altered.

Using the above approaches, a number of genes have been identified in studies of murine retrovirus-induced leukemias or leukemia cell lines. Of particular interest are rearrangements of c-*myb*. The *myb* gene was initially identified as the transforming gene of the avian myeloblastosis virus which specifically transforms myeloid cells both in vivo and in vitro. The protoon- cogene c-*myb* is normally expressed in myeloid, T, and B cells and is a nuclear protein of 90-kDa (ROSSON and REDDY 1987). In two myeloid lines (WEINSTEIN et al. 1986, 1987), retroviruses have integrated into the middle of the gene and caused a truncation of the normal transcript of 4 kb to 2 kb. In addition, the protein derived from the rearranged locus is carboxytruncated protein of 45-kDa which retains its nuclear localization. In a series of myeloid cell lines (ROSSON and REDDY 1987), viral integrations occurred in the 5' region of the gene and the major transcripts from the rearranged locus start in the viral 5' long terminal repeats (LTRs) and splice into the c-*myb* gene, result- ing in a 5' deletion (SHEN-ONG et al. 1986). In addition, splicing of the 3' region is altered and results in the introduction of intron sequences in the middle of the gene (ROSSON et al. 1987). The common alteration associated with the ac- tivation of the gene is therefore a 3' truncation or alteration.

The *myb* gene rearrangements have been speculated to alter differentia- tion. The primary effect of v-*myb* on chicken hematopoietic cells in vitro and in vivo is on the ability of the cells to differentiate. Similarly, activation of the *myb* gene in mice is specifically associated with myeloid transformation in which the cells are altered in their ability to terminally differentiate. Although many of the cell lines containing *myb* gene rearrangements are not dependent on growth factors, one of the cell lines, NFS-60, requires IL-3 for growth. Based on these observations and the normal expression of the *myb* gene in hematopoietic cells, it has been proposed that c-*myb* is involved in the regula- tion of gene expression during differentiation and is required for the normal progression of differentiation. Carboxy-truncation of the gene is speculated to interfere with the function of the protein without altering its ability to bind DNA.

The *Evi*-1 gene has also been speculated to affect the differentiation of myeloid cells. The locus for this gene was initially identified as a common site of viral integration in myeloid leukemias in AKXD 23 inbred recombinant mice (MUCENSKI et al. 1988a). In addition, viral insertions in this locus were found in a number of retrovirus-induced, IL-3-dependent myeloid leukemia cell lines. The murine locus was genetically mapped to chromosome 3 (MUCENSKI et al. 1988b) to a region which does not contain any known protooncogenes. Characterization of expression from this locus has shown that viral insertions activate the expression of a novel gene by integrating near 5' noncoding exons.

The structure of the gene, deduced from sequencing of cDNA clones, indi- cates that the gene product is a 120-kDa zinc-finger-containing protein (MORISHITA et al. 1988). The finger family of proteins are characterized by a 28- to 30-amino acid repeating region containing Cys and His residues which coordinate zinc and fold the protein into finger structures (R. M. EVANS and HOLLENBERG 1988). These finger structures are responsible for interacting with DNA. Where the functions are known, finger proteins are involved in transcriptional regulation and many are specifically involved in gene regulation during development. Therefore it is possible that the induced expression of the *Evi*-1 gene affects differentiation by altering the normal transcription that is required for differentiation.

A second common integration site, termed *fim*-3, has been associated with murine myeloid leukemias (BORDEREAUX et al. 1987) and viral integrations in this locus have been found in IL-3-dependent myeloid leukemia cell lines (C. BARTHOLOMEW and J. N. IHLE, unpublished data). This locus maps to murine chromosome 3 and is closely genetically linked to the *Evi*-1 locus such that in backcrosses no recombinants have been detected (A. BUCHBERG, C. BARTHOLOMEW, J. N. IHLE, unpublished data). Viral integrations in the *fim*-3 locus are associated with the activation of the expression of the *Evi*-1 gene, suggesting that the two common integration sites are part of the same gene locus. However, to date no physical linkage has been found between the loci.

Spleen focus forming virus induces an erythroleukemia in which the cells are dependent on erythropoietin for growth and in vitro undergo terminal differentiation normally. When these tumors are transplanted, occasionally continuous tumor cell lines can be isolated which are characterized by the lack of terminal differentiation. These observations have suggested that a second transforming event specifically involves an alteration in the ability of the cells to differentiate. This second event has been shown to be associated with viral insertions into a unique common integration site termed *spi*-1 and the appearance of a new 4-kb transcript from the locus (MOREAU-GACHELIN et al. 1988). The structure of this gene has not been reported.

The above examples indicate the potential value of the approach and suggest that over the next several years it should be possible to identify a number of new transforming genes. By using appropriate cell lines, it should also be possible to focus on a specific phenotype of transformation and in particular to focus on genes that might alter differentiation. Ultimately it will be important to develop retroviral vectors containing potential transforming genes and to specifically determine their effects on normal IL-3-supported differentiation in vitro. From these studies it should also be possible to begin to identify the genes involved in regulation of normal differentiation.

K. Summary

A variety of advances have resulted in the opportunity to study in new and unique ways the mechanisms in the growth and differentiation of myeloid cells. One of the most important advances has been the purification and cloning of the growth factors which are essential for growth and differentiation in vitro as well as in vivo. IL-3 has been particularly useful because of its ability to support the growth of relatively early hematopoietic stem cells as well as support the differentiation of cells committed to a variety of the myeloid lineages. The structure of IL-3, and the structure and location of its gene, are very much like those of a number of the hematopoietic growth factors and suggest that IL-3 is a member of an evolutionarily related family of growth factors.

Studies dealing with the regulation of the expression of hematopoietic growth factors have begun to identify the multiple ways in which hematopoiesis is regulated. Among the hematopoietic growth factors studied to date, IL-3 is unique in its exclusive production by antigen-activated T cells and the lack of any evidence for an essential role in normal hematopoiesis. This has given rise to the concept that there may exist a group of factors which are responsible for "constitutive" hematopoiesis and other factors which are primarily or exclusively used in an "adaptive" response. In particular, IL-3 can be envisioned to be the mechanism by which the immune system can increase the level of hematopoiesis when necessary.

The use of IL-3 to isolate hematopoietic growth factor-dependent cell lines has played an essential role in providing the reagents to better study the mechanisms of hematopoietic growth factor action as well as myeloid lineage differentiation and transformation. Using these cell lines a variety of studies have begun to define the types of cellular changes which occur following ligand binding and which may be involved in signal transduction. A variety of studies have implicated tyrosine phosphorylation as an essential response and indeed tantalizing evidence exists which suggests that the IL-3 receptor may be a member of the tyrosine protein kinase family of genes. In the near future the IL-3 receptor will be cloned to provide more definitive information regarding its function. Like the growth factors themselves, the receptors for hematopoietic growth factors may be expected to have considerable similarity and may also be evolutionarily related.

As with a variety of growth factors, the response of myeloid cells to IL-3 is complex. Indeed, with the growth factor-dependent cell lines, IL-3 is required for the maintenance of viability of the cells and may actually have little to do with the regulation of progression through the cell cycle. With a number of the lines, cells lose viability at the rate of half the population every 2 h in the absence of IL-3. Thus IL-3 may be required for a variety of cellular functions, the absence of any of which could result in cell death. This makes it impossible to identify a single "critical" function which when provided could eliminate the requirement for IL-3.

In addition to any effects that IL-3 has on cellular metabolism, IL-3 is also involved in the regulation of the expression of a variety of genes. Some of the genes, such has the c-*myc* gene, have been implicated in growth while some are more likely to be associated with differentiation. The later genes are particularly interesting and their elucidation will undoubtedly be required to begin to provide insights into the relative roles of hematopoietic growth factors in maintaining the viability of the hematopoietic cells during differentiation versus "directing" differentiation along various pathways or along a particular lineage.

The mechanisms which control the differentiation that occurs when hematopoietic stem cells are cultured in IL-3 are not known. Associated with understanding normal differentiation is the question of the mechanisms involved in the transformation of hematopoietic cells which results in an alteration in the ability to differentiate normally. As discussed above, IL-3 has been essential for the isolation of transformed myeloid cells which can be used to identify the genes which can cause this type of transformation. By identifying these genes it may be possible to gain some insights into the genes which normally control differentiation.

Independent of the importance for basic research, the first major reward from the characterization and cloning of hematopoietic growth factors has been the successes which are occurring in their clinical applications. Perhaps most striking of these have been the results with erythropoietin and with G-CSF. Several companies are currently evaluating the effects of IL-3 in vivo in primates. From the considerable information that exists from both in vitro and murine in vivo studies it is likely that IL-3 will be useful in expanding progenitor populations and may be particularly valuable in combination with other factors which support the differentiation of committed progenitors. Irrespective, the initial successes have fostered the hope that, by using various hematopoietic growth factors or combinations thereof, it may be possible to control hematopoiesis.

References

Alt FW, Blackwell TK, DePinho RA, Reth MG, Yancopoulos GD (1986) Regulation of genome rearrangement events during lymphocyte differentiation. Immunol Rev 89:5–30

Azoulay M, Webb CG, Sachs L (1987) Control of hematopoietic cell growth regulators during mouse fetal development. Mol Cell Biol 7:3361–3364

Bagby GC (1987) Production of multilineage growth factors by hematopoietic stromal cells: an intercellular regulatory network involving mononuclear phagocytes and interleukin-1. Blood Cells 13:147–159

Barlow DP, Bücan M, Lehrach H, Hogan BL, Gough NM (1987) Close genetic and physical linkage between the murine haemopoietic growth factor genes GM-CSF and Multi-CSF (IL3). EMBO J 6:617–623

Barton BE, WoldeMussie E, Wheller L (1988) The role of arachidonic acid metabolism in IL-3-induced proliferation. Immunopharmacol Immunotoxicol 10:35–52

Birchenall-Sparks MC, Farrar WL, Rennick D, Kilian PL, Ruscetti FW (1986) Regulation of expression of the interleukin-2 receptor on hematopoietic cells by interleukin-3. Science 233:455–458

Bishop JM (1983) Cellular oncogenes and retroviruses. Annu Rev Biochem 52:301–354

Bordereaux D, Fichelson S, Sola B, Tambourin PE, Gisselbrecht S (1987) Frequent involvement of the fim-3 region in Friend murine leukemia virus-induced mouse myeloblastic leukemias. J Virol 61:4043–4045

Bowlin TL, McKown BJ, Sunkara PS (1986) Ornithine decarboxylase induction and polyamine biosynthesis are required for the growth of interleukin-2- and interleukin-3-dependent cell lines. Cell Immunol 98:341–350

Branch DR, Turc JM, Guilbert LJ (1987) Identification of an erythropoietin-sensitive cell line. Blood 69:1782–1785

Broudy VC, Kaushansky K, Segal GM, Harlan JM, Adamson JW (1986) Tumor necrosis factor type α stimulates human endothelial cells to produce granulocyte/macrophage colony-stimulating factor. Proc Natl Acad Sci USA 83:7467–7471

Bryant RW, She HS, Ng KJ, Siegel MI (1986) Modulation of the 5-lipoxygenase activity of MC-9 mast cells: activation by hydroperoxides. Prostaglandins 32:615–627

Buchberg AM, Begigian HG, Taylor BA, Brownell E, Ihle JN, Nagata S, Jenkins NA, Copeland NG (1988) Localization of Evi-2 to chromosome 11: linkage to other protooncogene and growth factor loci using interspecific backcross mice. Oncogene Res 2:149–165

Burstein SA (1986) Interleukin 3 promotes maturation of murine megakaryocytes in vitro. Blood Cells 11:469–484

Campbell HD, Ymer S, Fung MC, Young IG (1985) Cloning and nucleotide sequence of the murine interleukin-3 gene. Eur J Biochem 150:297–304

Cannistra SA, Vellenga E, Groshek P, Rambaldi A, Griffin JD (1988) Human granulocyte-monocyte colony-stimulating factor and interleukin 3 stimulate monocyte cytotoxicity through a tumor necrosis factor-dependent mechanism. Blood 71:672–676

Chavrier P, Zerial M, Lemaire P, Almendral J, Bravo R, Charnay P (1988) A gene encoding a protein with zinc fingers is activated during G0/G1 transition in cultured cells. EMBO J 7:29–35

Chen BD-M, Clark CR (1986) Interleukin 3 (IL 3) regulates the in vitro proliferation of both blood monocytes and peritoneal exudate macrophages: synergism between a macrophage lineage-specific colony-stimulating factor (CSF-1) and IL 3. J Immunol 137:563–570

Cherwinski HM, Schumacher JH, Brown KD, Mosmann TR (1987) Two types of mouse helper T cell clone. III. Further differences in lymphokine synthesis between Th1 and Th2 clones revealed by RNA hybridization, functionally monospecific bioassays, and monoclonal antibodies. J Exp Med 166:1229–1244

Clark-Lewis I, Kent SBH, Schrader JW (1984) Purification to apparent homogeneity of a factor stimulating the growth of multiple lineages of hemopoietic cells. J Biol Chem 259:7488–7494

Clark-Lewis I, Aebersold R, Ziltener H, Schrader JW, Hood LE, Kent SB (1986) Automated chemical synthesis of a protein growth factor for hemopoietic cells, interleukin-3. Science 231:134–139

Cohen DR, Hapel AJ, Young IG (1986) Cloning and expression of the rat interleukin-3 gene. Nucleic Acids Res 14:3641–3658

Conscience JF, Verrier B, Martin G (1986) Interleukin-3-dependent expression of the c-myc and c-fos proto-oncogenes in hemopoietic cell lines. EMBO J 5:317–323

Cook WD, Metcalf D, Nicola NA, Burgess AW, Walker F (1985) Malignant transformation of a growth factor-dependent myeloid cell line by Abelson virus without evidence of an autocrine mechanism. Cell 41:677–683

Cook WD, de St Groth BF, Miller JF, MacDonald HR, Gabathular R (1987) Abelson virus transformation of an interleukin 2-dependent antigen-specific T-cell line. Mol Cell Biol 7:2631–2635

Cory S (1986) Activation of cellular oncogenes in hemopoietic cells by chromosome translocation. Adv Cancer Res 47:189–234

Cory, Bernard O, Bowtell D, Schrader S, Schrader JW (1987) Murine c-myc retroviruses alter the growth requirements of myeloid cell lines. Oncogene Res 1:61–76

Culpepper JA, Lee F (1985) Regulation of IL 3 expression by glucocorticoids in cloned murine T lymphocytes. J Immunol 135:3191–3197

Dean J, Cleveland JL, Rapp UR, Ihle JN (1987) Role of myc in the abrogation of IL3 dependence of myeloid FDP-P1 cells. Oncogene Res 1:279–296

Dexter TM, Allen TD, Lajtha LF (1977) Conditions controlling the proliferation of haemopoietic stem cells in vitro. J Cell Physiol 91:335–344

Dexter TM, Garland J, Scott D, Scolnick E, Metcalf D (1980) Growth of factor-dependent hemopoietic precursor cell lines. J Exp Med 152:1036–1047

Evans RM, Hollenberg SM (1988) Zinc fingers: gilt by association. Cell 52:1–3

Evans S, Rennick D, Farrar WL (1986) The multilineage heamopoetic growth factor IL3 and activation of protein kinase C stimulate phosphorylation of common substrates. Blood 68:906–913

Farrar WL, Thomas TP, Anderson WB (1985) Altered cytosol/membrane enzyme redistribution on interleukin-3 activation of protein kinase C. Nature 315:235–237

Fung MC, Hapel AJ, Ymer S, Cohen DR, Johnson RM, Campbell HD, Young IG (1984) Molecular cloning of cDNA for murine interleukin-3. Nature 307:233–237

Garland JM (1988) Rapid phosphorylation of a specific 33-kDa protein (p33) associated with growth stimulated by murine and rat IL3 in different IL3-dependent cell lines, and its constitutive expression in a malignant independent clone. Leukemia 2:94–102

Goodman JW, Hall EA, Miller KL, Shinpock SG (1985) Interleukin 3 promotes erythroid burst formation in serum-free cultures without detectable erythropoietin. Proc Natl Acad Sci USA 82:3291–3295

Grabstein K, Eisenman J, Mochizuki D, Shanebeck K, Conlon P, Hopp T, March C, Gillis S (1986) Purification to homogeneity of B cell stimulating factor. A molecule that stimulates proliferation of multiple lymphokine-dependent cell lines. J Exp Med 163:1405–1414

Greenberg ME, Ziff EB (1984) Stimulation of 3T3 cells induces transcription of the c-fos proto-oncogene. Nature 311:433–438

Greenberger JS, Gans PJ, Davisson PB, Moloney WC (1979) In vitro induction of continuous acute promyelocyte leukemia cell lines by Friend or Abelson murine leukemia viruses. Blood 53:987–1001

Gualtieri RJ, Liang CM, Shadduck RK, Waheed A, Banks J (1987) Identification of the hematopoietic growth factors elaborated by bone marrow stromal cells using antibody neutralization analysis. Exp Hematol 15:883–889

Hamaguchi Y, Kanakura Y, Fujita J. Takeda S, Nakano T, Tarui S, Honjo T, Kitamura Y (1987) Interleukin 4 as an essential factor for in vitro clonal growth of murine connective tissue-type mast cells. J Exp Med 165:268–273

Hapel AJ, Fung MC, Johnson RM, Young IG, Johnson G, Metcalf D (1985) Biologic properties of molecularly cloned and expressed murine interleukin-3. Blood 65:1453–1459

Hara K, Suda T, Suda J, Eguchi M, Ihle JN, Nagata S, Miura Y, Saito M (1988) Bipotential murine hemopoietic cell line (NFS-60) that is responsive to IL-3, GM-CSF, G-CSF, and erythropoietin. Exp Hematol 16:256–261

Harel-Bellan A, Farrar WL (1987) Modulation of proto-oncogene expression by colony stimulating factors. Biochem Biophys Res Commun 148:1001–1008

Hasthorpe S, Carver JA, Rees D, Campbell ID (1987) Metabolic effects of interleukin 3 on 32D c123 cells analyzed by NMR. J Cell Physiol 133:351–357

Holmes KL, Palaszynski E, Fredrickson TN, Morse HC 3d, Ihle JN (1985) Correlation of cell-surface phenotype with the establishment of interleukin 3-dependent cell lines from wild-mouse murine leukemia virus-induced neoplasms. Proc Natl Acad Sci USA 82:6687–6691

Hume CR, Nocka KH, Sorrentino V, Lee JS, Fleissner E (1988) Constitutive c-myc expression enhances the response of murine mast cells to IL-3, but does not eliminate their requirement for growth factors. Oncogene 2:223–226

Ihle JN (1986) Interleukin-3 regulation of the growth and differentiation of hematopoietic lymphoid stem cells. In: Cruse JM, Lewis RE Jr (eds) The year in immunology. Karger, Basel, pp 106–133

Ihle JN, Weinstein Y (1986) Immunological regulation of hematopoietic/lymphoid stem cell differentiation by interleukin 3. Adv Immunol 39:1–50

Ihle JN, Pepersack L, Rebar L (1981) Regulation of T cell differentiation: in vitro induction of 20 α hydroxysteroid dehydrogenase in splenic lymphocytes from athymic mice by a unique lymphokine. J Immunol 126:2184–2189

Ihle JN, Keller J, Henderson L, Klein F, Palaszynski EW (1982a) Procedures for the purification of interleukin-3 to homogeneity. J Immunol 129:2431–2436

Ihle JN, Keller J, Greenberger S, Henderson L, Yetter RA, Morse HC III (1982b) Phenotypic characteristics of cell lines requiring interleukin-3 for growth. J Immunol 129:1377–1383

Ihle JN, Keller J, Oroszlan S, Henderson L, Copeland T, Fitch F, Prystowsky MB, Goldwasser E, Schrader JW, Palaszynski E, Dy M, Lebel B (1983) Biological properties of homogenous interleukin-3. I. Demonstration of WEHI-3 growth factor activity, mast cell growth factor activity, P-cell stimulating factor activity, colony stimulating factor activity and histamine producing cell stimulating factor activity. J Immunol 131:282–287

Ihle JN, Rein A, Mural R (1984) Immunological and virological mechanisms in retrovirus induced murine leukemogenesis. In: Klein G (ed) Advances in viral oncology, vol 4. Raven, New York, pp 95–137

Ihle JN, Silver J, Kozak CA (1987) Genetic mapping of the mouse interleukin 3 gene to chromosome 11. J Immunol 138:3051–3054

Ikebuchi K, Wong GG, Clark SC, Ihle JN, Hirai Y, Ogawa M (1987) Interleukin 6 enhancement of interleukin 3-dependent proliferation of multipotential hemopoietic progenitors. Proc Natl Acad Sci USA 84:9035–9039

Isfort R, Huhn RD, Frackelton AR Jr, Ihle JN (1988) Stimulation of factor-dependent myeloid cell lines with IL-3 induces tyrosine phosphorylation of several cellular substrates. J Biol Chem 263:19 203–19 209

Isfort RJ, Abraham R, May WS, Stevens DA, Frackelton AR Jr, Ihle JN (1988 a) Mechanisms in interleukin-3 dependent growth of factor dependent myeloid leukemia cell lines. In: Ross R, Burgess T, Hunter T (eds) Growth factors and their receptors: genetic control and rational application. Liss, New York (in press)

Isfort RJ, Stevens D, May WS, Ihle JN (1988 b) IL-3 binding to a 140 kd phosphotyrosine containing cell surface protein. Proc Natl Acad Sci USA 85:7982–7986

Joyner A, Keller G, Phillips RA, Bernstein A (1983) Retrovirus transfer of a bacterial gene into mouse haematopoietic progenitor cells. Nature 305:556–558

Kalland T (1987) Physiology of natural killer cells. In vivo regulation of progenitors by interleukin 3. J Immunol 139:3671–3675

Keller G, Paige G, Gilboa E, Wagner EF (1985) Expression of a foreign gene in myeloid and lymphoid cells derived from multipotent haematopoietic precursors. Nature 318:149–154

Kelly K, Cochran B, Stiles CD, Leder P (1983) Cell-specific regulation of the c-myc gene by lymphocyte mitogens and platelet-derived growth factor. Cell 35:603–610

Kelso A, Owens T (1988) Production of two hemopoietic growth factors is differentially regulated in single T lymphocytes activated with an anti-T cell receptor antibody. J Immunol 140:1159–1167

Kerkhofs H, Hagemeijer A, Leeksma CHW, Abels J, Den Ottolander GJ, Somers R, Gerrits WBJ, Langenhuiyen MMA, Von DenBorne AEG, VanHemel JO, Geraedts JPM (1982) The 5q-chromosome abnormality in hematologic disorders: a collaborative study of 34 cases form the Netherlands. Br J Haematol 52:365–381

Kimoto M, Kindler V, Higaki M, Ody C, Izui S, Vassalli P (1988) Recombinant murine IL-3 fails to stimulate T or B lymphopoiesis in vivo, but enhances immune responses to T cell-dependent antigens. J Immunol 140:1889–1894

Kindler V, Thorens B, de Kossodo S, Allet B, Eliason JF, Thatcher D, Farber N, Vassalli P (1986) Stimulation of hematopoiesis in vivo by recombinant bacterial murine interleukin 3. Proc Natl Acad Sci USA 83:1001–1005

Kipreos ET, Wang JYJ (1988) Reversible dependence on growth factor interleukin-3 in myeloid cells expressing temperature sensitive v-abl oncogene. Oncogene Res 2:277–284

Kitamura T, Tange T, Chiba S, Kuwaki T, Mitani K, Urabe A, Takaku F (1989) Establishment and characterization of a unique human cell line that proliferates dependently on GM-CSF, IL-3 or erythropoietin. Blood 73:375–380

Koeffler HP, Gasson J, Ranyard J, Souza L, Shepard M, Munker R (1987) Recombinant human TNF α stimulates production of granulocyte colony-stimulating factor. Blood 70:55–59

Koike K, Ogawa M, Ihle JN, Miyake T, Shimizu T, Miyajima A, Yokota T, Arai K (1987) Recombinant murine granulocyte-macrophage (GM) colony-stimulating factor supports formation of GM and multipotential blast cell colonies in culture: comparison with the effects of interleukin-3. J Cell Physiol 131:458–464

Koyasu SA, Tojo A, Miyajima A, Akiyama T, Kasuga M, Urabe A, Schreurs J, Arai K, Takaku F, Yahara I (1988) Interleukin 3-specific tyrosine phosphorylation of a membrane glycoprotein of M 150 000 in multi-factor-dependent myeloid cell lines. EMBO J 6:3979–3984

Lange B, Valtieri M, Caracciolo D, Mavilio F, Gemperlein I, Griffin C, Emanuel B, Finan J, Nowell P, Rovera G (1987) Growth factor requirements for childhood leukemia: establishment of GM-CSF-dependent cell lines. Blood 70:192–199

Lau LF, Nathans D (1985) Identification of a set of genes expressed during the G0/G1 transition of cultured mouse cells. EMBO J 4:3145–3151

Lau LF, Nathans D (1987) Expression of a set of growth-related immediate early genes in Balb/c 3T3 cells: Coordinate regulation with c-fos and c-myc. Proc Natl Acad Sci USA 84:1182–1186

Leary AG, Yang Y-C, Clark SC, Gasson JC, Golde DW, Ogawa M (1988) Recombinant gibbon interleukin-3 (IL-3) supports formation of human multilineage colonies and blast cell colonies in culture: comparison with recombinant human granulocytic-macrophage colony-stimulating factor (GM-CSF). Blood 71:1759–1763

Le Beau MM (1987) Cytogenetic and molecular analysis of the del(5q) in myeloid disorders: evidence for the involvement of colony-stimulating factor and fms genes. In: Gale RP, Golde DW (eds) Recent advances in leukemia and lymphoma. Liss, New York, pp 71–81

Le Beau MM, Epstein ND, O'Brien SJ, Nienhuis AW, Yang YC, Clark SC, Rowley JD (1987) The interleukin 3 gene is located on human chromosome 5 and is deleted in myeloid leukemias with a deletion of 5q. Proc Natl Acad Sci USA 84: 5913–5917

Lee F, Yokota T, Otsuka T, Meyerson P, Villaret D, Coffman R, Mosmann T, Rennick D, Roehm N, Smith C, Zlotnik A, Arai K (1986) Isolation and characterization of a mouse interleukin cDNA clone that expresses B-cell stimulatory factor 1 activities and T-cell- and mast-cell-stimulating activities. Proc Natl Acad Sci USA 83:2061–2065

Lee F, Abrams J, Arai K et al. (1988) The expression and characterization of recombinant mouse IL-3. In: Schrader JW (ed) Lymphokines 15. Interleukin 3: the panspecific hemopoietin. Academic, New York, pp 163–182

Lee JC, Hapel AJ, Ihle JN (1982) Constitutive production of a unique lymphokine (IL-3) by the WEHI-3 cell line. J Immunol 128:2392–2398

Lee M, Segal GM, Bagby GC (1987) Interleukin-1 induces human bone marrow-derived fibroblasts to produce multilineage hematopoietic growth factors. Exp Hematol 15:983–988

Le Gros GS, Gillis S, Watson JD (1985) Induction of IL2 responsiveness in a murine IL3-dependent cell line. J Immunol 135:4009–4014

Le Gros GS, Shackell P, Le Gros JE, Watson JD (1987) Interleukin 2 regulates the expression of IL2 receptors on Interleukin 3-dependent bone marrow-derived cell lines. J Immunol 138:478–483

Le Gros JE, Jenkins DR, Prestidge RL, Watson JD (1987) Expression of genes in cloned murine cell lines that can be maintained in both interleukin 2- and interleukin 3-dependent growth states. Immunol Cell Biol 65:57–69

Lemischka IR, Raulet DH, Mulligan RC (1986) Developmental potential and dynamic behavior of hematopoietic stem cells. Cell 45:917–927

Leung DW, Spencer SA, Cachianes G, Hammonds RG, Collins C, Henzel WJ, Barnard R, Waters MJ, Wood WI (1987) Growth hormone receptor and serum binding protein: purification, cloning and expression. Nature 330:537–543

Li CL, Culter RL, Johnson GR (1987) Characterization of hemopoietic activities in media conditioned by a murine marrow-derived adherent cell line, B. Ad. Exp Hematol 15:373–381

Lord BI, Molineux G, Testa NG, Kelly M, Spooncer E, Dexter TM (1986) The kinetic response of haemopoietic precursor cells, in vivo, to highly purified, recombinant interleukin-3. Lymphokine Res 5:97–104

Lutzker S, Rothman P, Pollock R, Coffman R, Alt FA (1988) Mitogen- and IL-4-regulated expression of germ-line IG γ2b transcripts: evidence for directed heavy chain class switching. Cell 53:177–184

May WS, Ihle JN (1986) Affinity isolation of the interleukin-3 surface receptor. Biochem Biophys Res Commun 135:870–879

Messner HA, Yamasaki K, Jamal N, Minden MM, Yang YC, Wong GG, Clark SC (1987) Growth of human hemopoietic colonies in response to recombinant gibbon interleukin 3: comparison with human recombinant granulocyte and granulocyte-macrophage colony-stimulating factor. Proc Natl Acad Sci USA 84: 6765–6769

Metcalf D, Begley CG, Johnson GR, Nicola NA, Lopez AF, Williamson DJ (1986) Effects of purified bacterially synthesized murine multi-CSF (IL-3) on hematopoiesis in normal adult mice. Blood 68:46–57

Miyajima A, Schreurs J, Otsu K, Kondo A, Arai K, Maeda S (1987) Use of the silkworm, Bombyx mori, and an insect baculovirus vector for high-level expression and secretion of biologically active mouse interleukin-3. Gene 58:273–281

Miyajima A, Miyatake S, Schreurs J, DeVries J, Arai N, Yokota T, Arai K (1988) Coordinate regulation of immune and inflammatory responses by T cell-derived lymphokines. FASEB J 2:2462–2473

Miyatake S, Yokota T, Lee F, Arai K (1985) Structure of the chromosomal gene for murine interleukin 3. Proc Natl Acad Sci USA 82:316–320

Moreau-Gachelin F, Tavitian A, Tambourin P (1988) Spi-1 is a putative oncogene in virally induced murine erythroleukaemias. Nature 331:277–280

Morishita K, Parker DS, Mucenski ML, Copeland NG, Ihle JN (1988) Retroviral activation of a novel gene encoding a zinc finger protein in IL-3-dependent myeloid leukemia cell lines. Cell 54:831–840

Morla AO, Schreurs J, Miyajima A, Wang JWJ (1988) Hematopoietic growth factors activate the tyrosine phosphorylation of distinct sets of proteins in interleukin-3-dependent murine cell lines. Mol Cell Biol 8:2214–2218

Mosmann T, Cherwinski H, Bond M, Giedlin M, Coffman R (1986a) Two types of murine helper T cell clone. I. Definition according to profiles of lymphokine activities and secreted proteins. J Immunol 136:2438–2457

Mosmann TR, Bond MW, Coffman RL, Ohara J, Paul WE (1986b) T-cell and mast cell lines respond to B-cell stimulatory factor 1. Proc Natl Acad Sci USA 83:5654–5658

Mucenski ML, Taylor BA, Ihle JN, Hartley JW, Morse HC III, Jenkins NA, Copeland NG (1988a) Identification of a common ecotropic viral integration site, Evi-1, in the DNA of AKXD murine myeloid tumors. Mol Cell Biol 8:301–308

Mucenski ML, Taylor BA, Copeland NG, Jenkins NA (1988b) Chromosomal location of Evi-1, A common site of ecotropic viral integration in AKXD murine myeloid tumors. Oncogene Res 2:219–233

Munker R, Gasson J, Ogawa M, Koeffler HP (1986) Recombinant human TNF induces production of granulocyte-monocyte colony-stimulating factor. Nature 323:79–82

Nakahata T, Ogawa M (1982) Identification in culture of a class of hemopoietic colony-forming units with extensive capability of self-renewal and generate multipotential hemopoietic colonies. Proc Natl Acad Sci USA 79:3843–3847

Nakahata T, Gross AJ, Ogawa M (1982) A stochastic model of self-renewal and commitment to differentiation of the primitive hemopoietic stem cells in culture. J Cell Physiol 113:455–458

Nakahata T, Kobayashi T, Ishiguro A, Tsuji K, Naganuma K, Ando O, Yagi Y, Tadokoro K, Akabane T (1986) Extensive proliferation of mature connective-tissue type mast cells in vitro. Nature 324:65–67

Naparstek E, Pierce J, Metcalf D, Shadduck R, Ihle J, Leder A, Sakakeeny MA, Wagner K, Falco J, FitzGerald TJ et al. (1986) Induction of growth alterations in factor-dependent hematopoietic progenitor cell lines by cocultivation with irradiated bone marrow stromal cell lines. Blood 67:1395–1403

Nicola NA, Metcalf D (1985) Binding of iodinated multipotential colony-stimulating factor to normal murine hemopoietic cells. J Cell Physiol 124:313

Nicola NA, Peterson L (1986) Identification of distinct receptors for two hemopoietic growth factors (Granulocyte colony-stimulating factor and multipotential colony-stimulating factor) by chemical cross-linking. J Biol Chem 261:12384–12389

Nienhuis AW, Bunn HF, Turner PH, Gopal TV, Nash WG, O'Brien SJ, Sherr CJ (1985) Expression of the human c-fms proto-oncogene in hematopoietic cells and its deletion in the 5q-syndrome. Cell 42:421–428

Orosz CG, Roopernian DC, Bach FH (1983) Phorbol myristate acetate and in vitro T lymphocyte function. I. PMA may contaminate lymphokine preparations and can interfere with interleukin bioassays. J Immunol 130:1764–1772

Otsu K, Nakano T, Kanakura Y, Asai H, Katz HR, Austen KF, Stevens RL, Galli SJ, Kitamura Y (1987) Phenotypic changes of bone marrow-derived mast cells after intraperitoneal transfer into W/Wv mice that are genetically deficient in mast cells. J Exp Med 165:615–627

Overell RW, Watson JD, Gallis B, Weisser KE, Cosman D, Widmer MB (1987) Nature and specificity of lymphokine independence induced by a selectable retroviral vector expressing v-src. Mol Cell Biol 7:3394–3401

Palacios R (1985) Cyclosporin A inhibits antigen- and lectin-induced but not constitutive production of interleukin 3. Eur J Immunol 15:204–206

Palacios R, Garland J (1984) Distinct mechanisms may account for the growth-promoting activity of interleukin 3 on cells of lymphoid and myeloid origin. Proc Natl Acad Sci USA 81:1208–1211

Palacios R, Steinmetz M (1985) IL-3-dependent mouse clones that express B-220 surface antigen, contain Ig genes in germ-line configuration, and generate B lymphocytes in vivo. Cell 41:727–734

Palacios R, Von Boehmer H (1986) Requirements for growth of immature thymocytes from fetal and adult mice in vitro. Eur J Immunol 16:12–19

Palacios R, Henson G, Steinmetz M, McKearn JP (1984) Interleukin-3 supports growth of mouse pre-B-cell clones in vitro. Nature 309:126–129

Palacios R, Neri T, Brockhaus M (1986) Monoclonal antibodies specific for interleukin 3-sensitive murine cells. J Exp Med 163:369–382

Palacios R, Kiefer M, Brockhaus M, Karjalainen K, Dembic Z, Kisielow P, Von Boehmer H (1987) Molecular, cellular, and functional properties of bone marrow T lymphocyte progenitor clones. J Exp Med 166:12–32

Palaszynski EW, Ihle JN (1984) Evidence for specific receptors for interleukin 3 on lymphokine dependent cell lines established from long-term bone marrow cultures. J Immunol 132:1872–1878

Park LS, Friend D, Gillis S, Urdal DL (1986) Characterization of the cell surface receptor for a multi-lineage colony-stimulating factor (CSF-2α). J Biol Chem 261:205–210

Pettenati MJ, Le Beau MM, Lemons RS, Shima EA, Kawasaki ES, Larson RA, Sherr CJ, Diaz MO, Rowley JD (1987) Assignment of CSF-1 to 5q33.1: evidence for clustering of genes regulating hematopoiesis and for their involvement in the deletion of the long arm of chromosome 5 in myeloid disorders. Proc Natl Acad Sci USA 84:2970–2974

Pierce JH, Di Fiore PP, Aaronson SA, Potter M, Pumphrey J, Scott A, Ihle N (1985) Neoplastic transformation of mast cells by Abelson-MuLV: abrogation of IL-3 dependence by a nonautocrine mechanism. Cell 41:685–693

Pierce JH, Ruggiero M, Fleming TP, Di Fiore PP, Greenberger JS, Varticovski L, Schlessinger J, Rovera G, Aaronson SA (1988) Signal transduction through the EGF receptor transfected in IL-3-dependent hematopoietic cells. Science 239:628–631

Prystowsky MB, Ely JM, Beller DI, Eisenberg L, Goldman J, Goldman M, Goldwasser E, Ihle J, Quintans J, Remold H, Vogel S, Fitch FW (1982) Alloreactive cloned T cell lines. VI. Multiple lymphokine activities secreted by cloned T lymphocytes. J Immunol 129:2337–2344

Prystowsky MB, Otten G, Naujokas MF, Vardiman J, Ihle JN, Goldwasser E, Fitch FW (1984) Multiple hemopoietic lineages are found after stimulation of mouse bone marrow precursor cells with interleukin 3. Am J Pathol 117:171–179

Quesenberry PJ, Ihle JN, McGrath E (1985) The effect of interleukin 3 and GM-CSA-2 on megakaryocyte and myeloid clonal colony formation. Blood 65:214–217

Rapp UR, Cleveland JL, Brightman K, Scott A, Ihle JN (1985) Abrogation of IL-3 and IL-2 dependence by recombinant murine retroviruses expressing v-myc oncogenes. Nature 317:434–438

Razin E, Stevens RL, Akiyama F, Schmid K, Austen KF (1982) Culture from mouse bone marrow of a subclass of mast cells possessing a distinct chondroitin sulfate proteoglycan with glycosaminoglycans rich in N-acetylgalactosamine-4,6-disulfate. J Biol Chem 257:7229–7239

Razin E, Ihle JN, Seldin D, Mencia-Huerta J-M, Katz HR, LeBlance A, Hein A, Caulfield JP, Austen KF, Stevens RL (1984) Interleukin 3: a differentiation and growth factor for the mouse mast cell that contains chondroitin sulfate E proteoglycan. J Immunol 132:1479–1486

Rein A, Keller J, Schultz AM, Holmes KL, Medicus R, Ihle JN (1985) Infection of immune mast cells by Harvey sarcoma virus: immortalization without loss of requirement for interleukin-3. Mol Cell Biol 5:2257–2264

Rennick DM, Lee FD, Yokota T, Arai KI, Cantor H, Nabel GJ (1985) A cloned MCGF cDNA encodes a multilineage hematopoietic growth factor: multiple activities of interleukin 3. J Immunol 134:910–914

Rennick D, Yang G, Muller-Sieburg C, Smith C, Arai N, Takabe Y, Gemmell L (1987) Interleukin 4 (B-cell stimulatory factor 1) can enhance or antagonize the factor-dependent growth of hemopoietic progenitor cells. Proc Natl Acad Sci USA 84:6889–6893

Rittling SR, Baserga R (1987) Regulatory mechanisms in the expression of cell cycle dependent genes. Anticancer Res 7:541–552

Robinson BE, McGrath HE, Quesenberry PJ (1987) Recombinant murine granulocyte macrophage colony-stimulating factor has megakaryocyte colony-stimulating activity and augments megakaryocyte colony stimulation by interleukin 3. J Clin Invest 79:1648–1652

Rossio JL, Ruscetti FW, Farrar WL (1986) Ligand-specific calcium mobilization in IL 2 and IL 3 dependent cell lines. Lymphokine Res 5:163–172

Rosson D, Reddy EP (1987) Mechanism of activation of the myb oncogene in myeloid leukemias. Ann NY Acad Sci 511:219–231

Rosson D, Dugan D, Reddy EP (1987) Aberrant splicing events that are induced by proviral integration: implications for myb oncogene activation. Proc Natl Acad Sci USA 84:3171–3175

Rovera G, Valtieri M, Mavilio F, Reddy EP (1987) Effect of Abelson murine leukemia virus on granulocytic differentiation and interleukin-3 dependence of a murine progenitor cell line. Oncogene 1:29–35

Sanderson CJ, Warren DJ, Strath M (1985a) Identification of a lymphokine that stimulates eosinophil differentiation in vitro. Its relationship to interleukin 3, and functional properties of eosinophils produced in cultures. J Exp Med 162:60–74

Sanderson CJ, Strath M, Warren DJ, O'Garra A, Kirkwood TB (1985b) The production of lymphokines by primary alloreactive T-cell clones: a co-ordinate analysis of 233 clones in seven lymphokine assays. Immunology 56:575–584

Sanderson CJ, Campbell HD, Young IG (1988) Molecular and cellular biology of eosinophil differentiation factor (interleukin-5) and its effects on B cells in man and mouse. Immunol Rec 102:29–50

Schwarzbaum S, Halpern R, Diamond B (1984) The generation of macrophage-like cell lines by transfection with SV40 origin defective DNA. J Immunol 132:1158–1162

Seger R, Yarden Y, Kashles O, Goldblatt D, Schlessinger J, Shaltiel S (1988) The epidermal growth factor receptor as a substrate for a kinase-splitting membranal proteinase. J Biol Chem 263:3496–3500

Shannon MF, Gamble JR, Vadas MA (1988) Nuclear proteins interacting with the promoter region of the human granulocyte/macrophage colony-stimulating factor gene. Proc Natl Acad Sci USA 85:674–678

Shen-Ong GL, Morse HC III, Potter M, Mushinski JF (1986) Two modes of c-myb activation in virus-induced mouse myeloid tumors. Mol Cell Biol 6:380–392 [published erratum appears in Mol Cell Biol 1986:2756]

Sideras P, Palacios R (1987) Bone marrow pro-T and pro-B lymphocyte clones express functional receptors for interleukin (IL) 3 and IL 4/BSF-1 and nonfunctional receptors for IL 2. Eur J Immunol 17:217–221

Smith CA, Rennick DM (1986) Characterization of a murine lymphokine distinct from interleukin 2 and interleukin 3 (IL-3) possessing a T-cell growth factor activity and a mast-cell growth factor activity that synergizes with IL-3. Proc Natl Acad Sci USA 83:1857–1861

Sokal G, Michaux JL, VanDenBergh H, Cordier A, Rodhain J, Ferrant A, Moriau M, Debruyere M, Sonnet J (1975) A new hematologic syndrome with a distinct karyotype: the 5q-chromosome. Blood 45:519–533

Sorensen P, Farber NM, Krystal G (1986) Identification of the interleukin-3 receptor using an iodinatable cleavable, photoreactive cross-linking agent. J Biol Chem 261:9094–9097

Sparrow RL, Swee-Huat O, Williams N (1987) Haemopoietic growth factors stimulating murine megakaryocytopoiesis: interleukin-3 is immunologically distinct from megakaryocyte-potentiator. Leuk Res 11:31–36

Spivak JL, Smith RR, Ihle JN (1985) Interleukin 3 promotes the in vitro proliferation of murine pluripotent hematopoietic stem cells. J Clin Invest 76:1613–1621

Stocking C, Loliger C, Kawai M, Suciu S, Gough N, Ostertag W (1988) Identification of genes involved in growth autonomy of hematopoietic cells by analysis of factor-independent mutants. Cell 53:869–879

Suda T, Suda J, Ogawa M (1983) Proliferative kinetics and differentiation of murine blast cell colonies in culture: evidence for variable G0 periods and constant doubling rates of early pluripotent hemopoietic progenitors. J Cell Physiol 117: 308–318

Suda T, Suda J, Ogawa M, Ihle JN (1985) Permissive role of interleukin 3 (IL-3) in proliferation and differentiation of multipotential hemopoietic progenitors in culture. J Cell Physiol 124:182–190

Suda T, Suda J, Kajigaya S, Nagata S, Asano S, Saito M, Miura Y (1987) Effects of recombinant murine granulocyte colony-stimulating factor on granulocyte-macrophage and blast colony formation. Exp Hematol 15:958–965

Sugawara M, Hattori C, Tezuka E, Tamura S, Ohta Y (1988) Monoclonal autoantibodies with interleukin 3-like activity derived from a MRL/lpr mouse. J Immunol 140:526–530

Sukhatme VP, Kartha S, Toback FG, Taub R, Hoover RG, Tasi-Morris C-H (1987) A novel early growth response gene rapidly induced by fibroblast, epithelial and lymphocyte mitogens. Oncogene Res 1:343–355

Sutherland GR, Baker E, Callen DF, Campbell HD, Young IG, Sanderson CJ, Garson OM, Lopez AF, Vadas MA (1988) Interleukin-5 is at 5q31 and is deleted in the 5q-syndrome. Blood 71:1150–1152

Tinegate H, Gaunt L, Hamilton PJ (1983) The 5q-syndrome: an underdiagnosed form of macrocytic anemia. Br J Haematol 54:103–110

Todokoro K, Yamamoto A, Amanuma H, Ikawa Y (1985) Isolation and characterization of a genomic DDD mouse interleukin-3 gene. Gene 39:103–107

Tsao CJ, Tojo A, Fukamachi H, Kitamura T, Saito T, Urabe A, Takaku F (1988) Expression of the functional erythropoietin receptors on interleukin 3-dependent murine cell lines. J Immunol 140:89–93

Valtieri M, Santoli D, Caracciolo D, Kreider BL, Altmann SW, Tweardy DJ, Gemperlein I, Mavilio F, Lange B, Rovera G (1987) Establishment and characteristics of an undifferentiated human T leukemia cell line which requires GM-CSF for growth. J Immunol 138:4042–4050

Vellenga E, Griffin JD (1987) The biology of acute myeloblastic leukemia. Semin Oncol 14:365–371

Vellenga E, Ostapovicz D, O'Rourke B, Griffin JD (1987a) Effects on recombinant IL-3, GM-CSF, and G-CSF on proliferation of leukemic clonogenic cells in short-term and long-term cultures. Leukemia 1:584–589

Vellenga E, Young DC, Wagner K, Wiper D, Ostapovicz D, Griffin JD (1987b) The effects of GM-CSF and G-CSF in promoting growth of clonogenic cells in acute myeloblastic leukemia. Blood 69:1771–1776

Walker F, Nicola NA, Metcalf D, Burgess AW (1985) Hierarchical down-modulation of hemopoietic growth factor receptors. Cell 43:269–276

Warren DJ, Moore MA (1988) Synergism among interleukin 1, interleukin 3, and interleukin 5 in the production of eosinophils from primitive hemopoietic stem cells. J Immunol 140:94–99

Warren HS, Hargreaves J, Hapel AJ (1985) Some interleukin-3 dependent mast-cell lines also respond to interleukin-2. Lymphokine Res 4:195–204

Watson JD, Le Gros GS, Overell RW, Conlon P, Widmer M, Gillis S (1987) Effect of infection with murine recombinant retroviruses containing the v-src oncogene on interleukin 2- and interleukin 3-dependent growth states. J Immunol 139:123–129

Watson JD, Jenkins DR, Eszes M, Leung E (1988) Effect of granulocyte-macrophage colony-stimulating factor and interleukin 3 on the v-src oncogene. Inhibition of tyrosine kinase activity in the absence of changes in gene expression. J Immunol 140:501–507

Weinstein Y, Ihle JN, Lavu S, Reddy EP (1986) Truncation of the c-myb gene by a retroviral integration in an interleukin-3 dependent myeloid leukemia cell line. Proc Natl Acad Sci USA 83:5010–5014

Weinstein Y, Cleveland JL, Askew DS, Rapp UR, Ihle JN (1987) Insertion and truncation of c-myb by MuLV in a myeloid cell line derived from cultures of normal hematopoietic cells. J Virol 61:2339–2343

Wheeler EF, Askew D, May S, Ihle JN, Sherr CJ (1987) The v-fms oncogene induces factor-independent growth and transformation of the interleukin-3-dependent myeloid cell line FDC-Pl. Mol Cell Biol 7:1673–1680

Whetton AD, Dexter TM (1983) Effect of haemopoetic growth factor on intracellular ATP levels. Nature 303:629–631

Whetton AD, Dexter TM (1988) The mode of action of interleukin 3 in promoting survival, proliferation, and differentiation of hemopoietic progenitor cells. In: Schrader JW (ed) Lymphokines 15 Interleukin 3: The panspecific hemopoietin, Academic Press, Inc., New York, p 355–374

Whetton AD, Heyworth CM, Dexter TM (1986a) Phorbol esters activate protein kinase C and glucose transport and can replace the requirement for growth factor in interleukin-3-dependent multipotent stem cells. J Cell Sci 84:93–104

Whetton AD, Monk PN, Consalvey SD, Downes CP (1986b) The haemopoietic growth factors interleukin 3 and colony stimulating factor-1 stimulate proliferation but do not induce inositol lipid breakdown in murine bone-marrow-derived macrophages. EMBO J 5:3281–3286

Williams DA, Lemischka IR, Nathans DG, Mulligan RC (1984) Introduction of new genetic material into pluripotent haematopoietic stem cells of the mouse. Nature 310:476–480

Wisniewski LP, Hirschhorn K (1983) Acquired partial deletions of the long arm of chromosome 5 in hematologic disorders. Am J Hematol 15:295–310

Yamaguchi Y, Suda T, Suda J, Eguchi M, Miura Y, Harada N, Tominaga A, Takatsu K (1988) Purified interleukin 5 supports the terminal differentiation and proliferation of murine eosinophilic precursors. J Exp Med 167:43–56

Yang YC, Clark SC (1988) Molecular cloning of a primate cDNA and the human gene for interleukin 3. In: Schrader JW (ed) Lymphokines 15. Interleukin 3: the panspecific hemopoietin. Academic, New York, pp 375–391

Yang YC, Ciarletta AB, Temple PA, Chung MP, Kovacic S, Witek-Giannotti JS, Leary AC, Kriz R, Donahue RE, Wong GG, Clark SC (1986) Human IL-3 (multi-CSF): identification by expression cloning of a novel hematopoietic growth factor related to murine IL-3. Cell 47:3–10

Yang YC, Kovacic S, Kriz R, Wolf S, Clark SC, Wellems TE, Nienhuis A, Epstein H (1988a) The human genes for GM-CSF and IL-3 are closely linked in tandem on chromosome 5. Blood 71:958–961

Yang YC, Tsai S, Wong GG, Clark SC (1988b) Interleukin-1 regulation of hematopoietic growth factor production by human stromal fibroblasts. J Cell Physiol 134:292–296

Ymer S, Tucker WQ, Sanderson CJ, Hapel AJ, Campbell HD, Young IG (1985) Constitutive synthesis of interleukin-3 by leukaemia cell line WEHI-3B is due to retroviral insertion near the gene. Nature 317:255–258

Yokota T, Lee F, Rennick D, Hall C, Arai N, Mosmann T, Nabel G, Cantor H, Arai K (1984) Isolation and characterization of a mouse cDNA clone that expresses mast-cell growth-factor activity in monkey cells. Proc Natl Acad Sci USA 81:1070–1074

Zucali JR, Dinarello CA, Oblon DJ, Gross MA, Anderson L, Weiner RS (1986) Interleukin 1 stimulates fibroblasts to produce granulocyte-macrophage colony-stimulating activity and prostaglandin E2. J Clin Invest 77:1857–1863

Zumstein P, Stiles CD (1987) Molecular cloning of gene sequences that are regulated by insulin-like growth factor I. J Biol Chem 262:11252–11260

CHAPTER 12

Interleukin-4

T. YOKOTA, N. ARAI, K.-I. ARAI, and A. ZLOTNIK

A. Introduction

Lymphokines produced by activated helper T cells were originally classified based on the target cells on which they act. Interleukin-2 (IL-2) stimulates predominantly the proliferation of cells belonging to the T-cell lineage, while interleukin-3 (IL-3) and granulocyte-macrophage colony stimulating factor (GM-CSF) stimulate proliferation and differentiation of various hematopoietic progenitor cells. A number of molecules such as interleukin-4 (IL-4), interleukin-5 (IL-5), and interleukin-6 (IL-6) stimulate proliferation and differentiation of cells of the B-lymphocyte lineage. Since many lymphokines are composed of single polypeptide chains, their coding sequences can be isolated by functional expression in appropriate host cells. Based on this expression cloning protocol, many T-cell lymphokine genes have been isolated and their primary structures determined. These studies have revealed that lymphokines are neither lineage specific nor are they confined to a single functional role. Instead, each lymphokine has multiple effects on its target cell(s). The current model describes a regulatory network between lymphoid and hematopoietic cells where various cells produce and react to multifunctional lymphokines (K. ARAI et al. 1986; MIYAJIMA et al. 1988).

IL-4, initially described as a growth factor (HOWARD et al. 1982) for murine B lymphocytes, induces proliferation of B cells when these cells are costimulated with submitogenic doses of immunoglobulin M(IgM)-specific antibodies. Molecular cloning of mouse and human IL-4 cDNAs and characterization of their recombinant products (LEE et al. 1986; NOMA et al. 1986; YOKOTA et al. 1986) revealed that IL-4 is a multifunctional lymphokine which interacts with cells of multiple lineages including T cells, B cells, thymocytes, hematopoietic cells, and fibroblasts. The availability of IL-4 cDNA clones expressible in mammalian cells led to the production of sufficient quantities of purified recombinant proteins to permit detailed studies of its role in lymphocyte and hematopoietic cell regulation. In this review, we describe the molecular biology of the IL-4 gene and the biological properties of its recombinant product.

B. Molecular Cloning of IL-4 cDNA

I. Multiple Biological Activities Produced by Activated Mouse T-Cell Clones

The mouse T-cell clone Cl.Ly1$^+$2$^-$/9 was initially shown to produce several biological activities including stimulation of T-cell and mast-cell proliferation in response to concanavalin A (ConA) (NABEL et al. 1981a–c). Since Cl.Ly1$^+$2$^-$/9 produces no detectable IL-2 mRNA, the T-cell growth factor (TCGF) activity found in the supernatant must be encoded by gene(s) distinct from IL-2 (TCGF II) (LEE et al. 1987). Molecular cloning and expression studies using a Cl.Ly1$^+$2$^-$/9 cDNA library showed that IL-3 accounts for the mast-cell growth factor (MCGF) and multilineage colony-stimulating factor (multi-CSF) activities in Cl.Ly1$^+$2$^-$/9 supernatants (YOKOTA et al. 1984; RENNICK et al. 1985).

However, additional experiments with Cl.Ly1$^+$2$^-$/9-conditioned medium demonstrated the existence of a factor distinct from IL-3 that had MCGF activity and the ability to enhance the MCGF activity of IL-3 (MCGF II) (RENNICK et al. 1985) because even saturating concentrations of recombinant IL-3 do not stimulate a cloned mast cell line to the same extent as conditioned medium from the original T-cell clone (SMITH and RENNICK 1986). Despite multiple biochemical fractionation, the MCGF II activity copurified with a TCGF II activity that is distinct from IL-2 (SMITH and RENNICK 1986). This was supported by the inability of anti-IL-2 antibodies to inhibit the TCGF II activity (MOSMANN et al. 1986a). These results demonstrated that the Cl.Ly1$^+$2$^-$/9 cells produce a factor, distinct from IL-3 and IL-2, with both MCGF and TCGF activities (MCGF II/TCGF II).

Cl.Ly1$^+$2$^-$/9 cells also produce factor(s) that mediated three B-cell-stimulating activities. These include costimulation of anti-IgM-activated B cells (ROEHM et al. 1985), induction of Ia antigen on resting B cells (ROEHM et al. 1984), and enhancement of IgE and IgG1 production by cultures of lipopolysaccharide (LPS)-activated B cells (COFFMAN and CARTY 1986). These studies indicated that Cl.Ly1$^+$2$^-$/9 cells produced B-cell stimulatory factor 1 (BSF-1) (HOWARD et al. 1982; OHARA et al. 1985; NOELLE et al. 1984; ROEHM et al. 1984; VITETTA et al. 1985; COFFMAN et al. 1986).

II. Isolation of IL-4 cDNA Clones

Based on our data suggesting that Cl.Ly1$^+$2$^-$/9 cells produce BSF-1 and a factor with MCGF and TCGF (MCGF II/TCGF II) activities, we used an expression cloning approach to isolate cDNA clones for each of these factors. We have constructed a cDNA library in the pcD expression vector (OKAYAMA and BERG 1983) using mRNA isolated from ConA-induced Cl.Ly1$^+$2$^-$/9 cells. After removing those clones which hybridized with radiolabeled IL-3 and GM-CSF cDNA probes, plasmid pools were transfected into COS7 cells, and cell supernatants were assayed for several bioactivities. Each pool shown

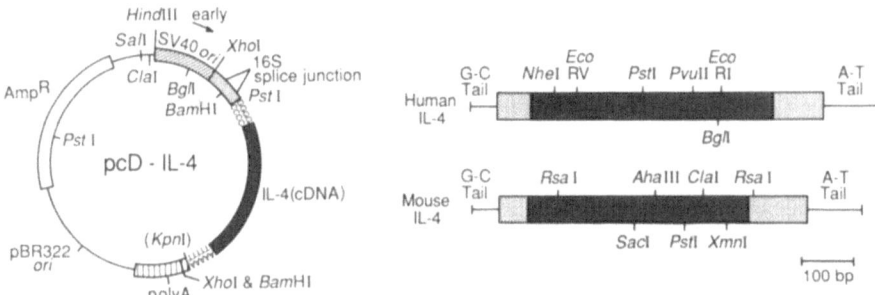

Fig. 1 A, B. Map of human and mouse IL-4 cDNA inserts. **A** pcD IL-4 plasmid. The direction of transcription from the simian virus 40 (SV40) early promoter is indicated by the *arrow*. The structure of the reminder of the plasmid is as described by OKAYAMA and BERG (1983). **B** Restriction endonuclease cleavage maps of the cDNA inserts of the human and mouse IL-4. The long open reading frames are *heavily shaded* and the non-coding regions are *lightly shaded*

to have TCGF and MCGF activities was found also to have Ia-inducing activity on mouse B cells. Thus, there was a perfect correlation between the TCGF, MCGF, and Ia-inducing activities. After several cycles of transfection and assaying transfection supernatants, a single clone was identified and was confirmed to encode a product with MCGF II/TCGF II activity as well as all the activities of BSF-1, including Ia induction and IgG1/IgE enhancement (COFFMAN and CARTY 1986). On the basis of these multiple activities it was proposed that this lymphokine be called interleukin-4 (IL-4) (LEE et al. 1986; NOMA et al. 1986). A restriction endonuclease cleavage map of the IL-4 cDNA is shown in Fig. 1.

The isolation of mouse IL-4 cDNA clone led to the isolation of human IL-4 cDNA clone from activated human T-cell clone cDNA libraries (YOKOTA et al. 1986). The human helper T-cell clone 2F1 produces B-cell growth factor (BCGF), TCGF, GM-CSF and interferon-γ activities after stimulation by either ConA or anti-CD3 antibody and phorbol 12-myristate, 13-acetate (PMA). Using pcD cDNA libraries prepared with mRNA isolated from ConA-stimulated 2F1 cells or from PMA- and ConA-stimulated human peripheral blood lymphocytes, we isolated several cDNA clones using an *Rsa*I fragment of mouse IL-4 cDNA (Fig. 1) as a probe (YOKOTA et al. 1986). Analysis by restriction endonuclease cleavage showed that each of the hybridizing clones was identical in structure and we designated these clones as human IL-4 cDNAs.

III. Structure of IL-4 cDNA Clones and Proteins

The cDNA inserts of mouse and human IL-4 clones are 579 and 614 base pairs (bp) long and contain a single long open reading frame of 140 codons and 153 codons, respectively (Fig. 2). Comparison of the human and mouse IL-4 coding regions deduced from the nucleotide sequences revealed that the

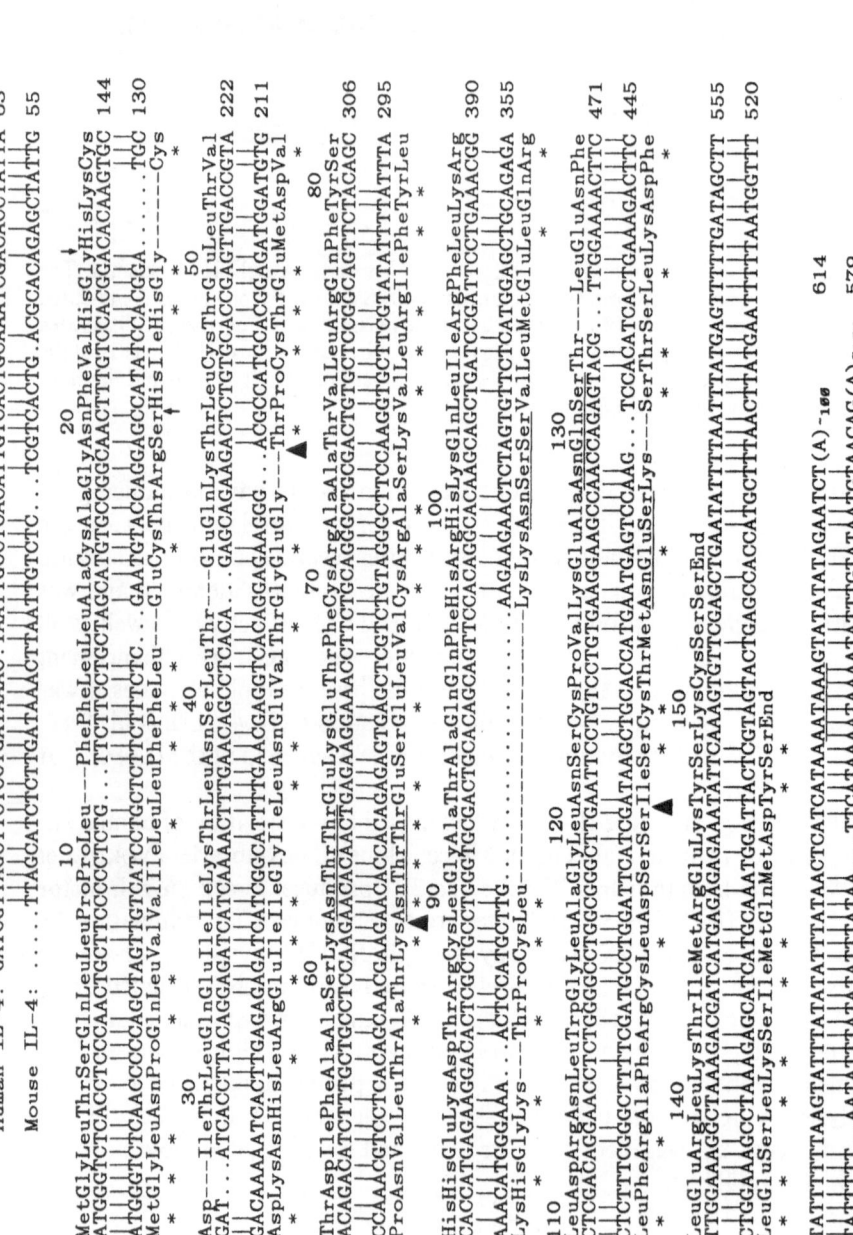

Fig. 2. Nucleotide and predicted amino acid sequences of human and mouse IL-4. *Numbers* above amino acid positions indicate positions of the human IL-4 amino acid sequence. Nucleotide positions are shown on the *right*. Matched nucleotides and amino acids between human and mouse sequences are shown by *vertical bars* and *asterisks*, respectively. Possible N-glycosylation sites are shown by *underlines*. Junctions of exons and introns are indicated by *arrows*. Cleavage sites for processing of signal peptides are indicated by *arrowheads*

regions of the human IL-4 coding sequence from amino acid positions 1 to 90 and from positions 129 to 149 share about 50% homology at the amino acid level with the corresponding regions of the mouse IL-4 coding sequence. These regions and the 5' and 3' untranslated regions of the human and mouse clones share about 70% DNA sequence homology. The regions of human IL-4 covered by amino acid positions 91–128 share very little homology with the corresponding region of mouse IL-4 at either the amino acid or nucleotide sequence levels.

The deduced amino acid sequences of the encoded polypeptides are very hydrophobic at the N termini, characteristic of a secreted protein. Comparison with a proposed consensus sequence for the processing of signal peptides (PERLMAN and HALVORSON 1983; VON HEIJNE 1983) suggests that cleavages of the human and mouse precursor polypeptides would occur following the glycine residue at position 24 and the serine residue at position 20 (Fig. 2), respectively. Therefore, the mature polypeptides of human and mouse IL-4 would be 129 and 120 amino acid residues long, respectively, and both begin with a histidine residue. The predicted N termini of mature mouse IL-4 and human IL-4 have been confirmed by using natural protein purified from T-cell-conditioned medium (GRANSTEIN et al. 1986; OHARA et al. 1987) and recombinant protein expressed in COS7 cells (RAMANATHAN et al. 1988). The deduced molecular weights of mature human and mouse proteins are about 15000 and about 14000, respectively. These predicted human and mouse molecular weights do not take into account potential posttranslational glycosylation of the polypeptides, which are predicted by the presence of two and three potential N-glycosylation sequences (Asn-X-Thr or Asn-X-Ser at positions 62–64 and 129–131 of human protein and at positions 61–63, 91–93, and 117–199 of mouse protein). Biosynthetic labeling of COS7 cells transfected with mouse IL-4 cDNA produced in the cultured supernatants a protein with an apparent molecular weight of 15000 or 19000 (RAMANATHAN et al. 1988) whereas the human IL-4 cDNA clone directs the synthesis of a protein whose molecular weight is 15000, 18000 or 19000. The heterogeneity of mouse and human IL-4 may be due to differences in the degree of glycosylation since each polypeptide has identical N- and C-terminal sequences and endoglycosidase treatment abolished the heterogeneity (LE et al. 1988).

Bacterial expression of mouse and human IL-4 was achieved in either periplasmic membrane or cytoplasm using the expression vectors pIN-III-OmpA2 (GHRAYEB et al. 1984) and pTrpC11 (VAN KIMMENADE et al. 1988), respectively. Bacterially expressed IL-4 was solubilized and activated by in vitro refolding of the protein, including extraction and denaturation with $5M$ guanidine-HCl under carefully controlled redox conditions. Subsequent dilution and slow removal of the denaturant resulted in efficient renaturation of human and mouse IL-4. Gel filtration of the refolded human IL-4 and affinity chromatography of the refolded mouse IL-4 using 11B11 monoclonal antibody (OHARA and PAUL 1985) resulted in an essentially pure and biologically active IL-4 preparation, although bacterially expressed IL-4 has an additional methionine at the N-terminus of the mature IL-4 sequence and is not glycosylated.

Yeast-expressed IL-4 was made by use of the secretion vector pMFα8 (MIYAJIMA et al. 1985, 1986). By constructing fusion genes of mature IL-4 coding sequences with the mating pheromone α-factor (MFα1) gene which contains a MFα1 promoter sequence, hydrophobic signal sequence, and leader sequence, biologically active IL-4 was secreted in the medium, purified, and used for characterization of its properties (SPITS et al. 1987; DEFRANCE et al. 1987b).

Mouse and human IL-4 were also expressed in the hemolymph of silkworms infected in vivo with recombinant virus derived from a baculovirus of *Bombyx mori (Bm)* and in the culture medium of the silkworm-derived *Bm*N cell line infected in vitro (MIYAJIMA et al. 1987). Other investigators also produced recombinant IL-4 in a similar system using *Autographa californica* (W. E. PAUL, personal communication).

IV. Structure of Chromosomal Genes Encoding Mouse and Human IL-4

1. Structure of Mouse and Human IL-4 Genes and Chromosomal Location of the Human IL-4 Gene

The chromosomal DNA segments of the IL-4 gene were isolated based on homology with the IL-4 cDNA sequence. Both mouse and human IL-4 genes are single copies in the haploid genome and are composed of four exons and three introns. The mouse IL-4 gene spans about 6 kilobases (kb) (OTSUKA et al. 1987) and the human IL-4 gene is about 10 kb in size (N. ARAI et al. 1989). Each exon encodes a similar number of amino acid residues in both species (Fig. 3) and the second exon, which is the smallest, is separated from the third exon by the largest intron ($\sim$4000 bp for mouse and 5200 bp for human). By using in situ hybridization (LE BEAU et al. 1988) and by somatic hybrid analysis (N. ARAI et al. 1988), human IL-4 was mapped to the long arm of chromosome 5 at band q23-31, a chromosomal region that is frequently deleted [del(5q)] in patients with myeloid disorders. This cluster of genes includes the IL-3 and GM-CSF (5q23-31), IL-5 (5q23.3-32), CSF-1 (5q33.1), and c-*fms* (5q33.2-33.3, CSF-1 receptor) genes (LE BEAU et al. 1987; HUEBNER

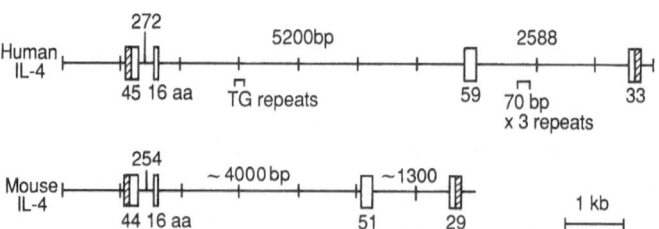

Fig. 3. Schematic representation of human and mouse IL-4 genes. Coding regions of exons are indicated by *open boxes*. Untranslated regions are indicated by *hatched boxes*

et al. 1985; NIENHUIS et al. 1985; SUTHERLAND et al. 1988; LE BEAU et al. 1986a, b; GROFFEN et al. 1983; ROUSSEL et al. 1983). Loss of the whole of chromosome 5 or loss of part of the long arm of chromosome 5 has been observed in malignant cells of patients with acute nonlymphocytic leukemia (ANLL) arising either de novo or secondary to cytotoxic therapy for a previous malignant disease. Mouse GM-CSF and IL-3 genes have been mapped on mouse chromosome 11 (BARLOW et al. 1987). It is tempting to speculate that a family of genes encoding hematopoietic growth factors and their receptors may be clustered within limited region(s) of a chromosome and thereby coordinately regulate hematopoiesis.

2. Nucleotide Sequence of Human IL-4 Gene

We have determined the complete nucleotide sequence of the human IL-4 gene (N. ARAI et al. 1989; Fig. 4). Exon 1 encodes 45 amino acid residues and contains the 5' untranslated region. Exon 2 and exon 3 encode 17 and 59 residues, respectively. Exon 4 encodes 33 residues and contains the 3' untranslated region. Introns 1, 2, and 3 are 272 bp, 5200 bp, and 2588 bp long, respectively. The longest intron, intron 2, contains repeats of TG elements. Although the TG element appears to have enhancer-like activity (HAMADA et al. 1984), we do not have any evidence that suggests a regulatory role of the TG element in the human IL-4 gene. Human IL-4 also contains three tandem repeats (70 bp × 3) in intron 3 but no homology was found with the SV40 enhancer sequence.

3. Comparison of 5' Flanking Sequences of Mouse and Human IL-4 Genes with Other Lymphokine Genes

A conventional TATA-like sequence is located 28 bp upstream of the transcription initiation site of mouse and human IL-4 genes. Extensive nucleotide sequence homology (approximately 85%) is present in the 5' flanking regions, extending more than 500 bp upstream of the TATA box of mouse and human IL-4 genes. In contrast, homology found in the coding region is 70% at the nucleotide level. A relatively high degree of conservation of nucleotide sequences between species in 5' flanking regions has been reported for other lymphokine genes such as GM-CSF (MIYATAKE et al. 1985a; STANLEY et al. 1985), IL-2 (FUJITA et al. 1983; FUSE et al. 1984), and IL-3 (MIYATAKE et al. 1985b; CAMPBELL et al. 1985; COHEN et al. 1986). These regions may be important for the regulated expression during T-cell activation. Although IL-4 is expressed in helper T_H2 cells but not in T_H1 cells and IL-2 is expressed in T_H1 cells but not in T_H2 cells in the mouse system (MOSMANN et al. 1986b; MOSMANN and COFFMAN 1987), IL-4 and IL-2 genes share significant sequence homology in the regions covering more than 200 bp upstream from the TATA box. Similar homology was also found between human IL-4 and human IL-2 genes.

Lymphokine genes are coordinately expressed upon stimulation of T cells by antigen or lectin. It is generally accepted that 5' flanking sequences play a

```
gaattcaataaaaaacaagcaggcgcgtggtggggcactgactaggagggctgatttgtaagtggtaagactgtagctcttttcttaattagctgagatgtgtttaggttccattc                  120
aaaaagtgggcattcctggccaggcagtggtggtcacacctgaatctcagctttgggagactaggtgaggaggatttgagtagcctaggcaacatagt                                    240
gagactcttatctcatcaaaaataaaataaaatagacaggacatggacgatggaggagatcagtaggttggaggagggtgaggg                                                  360
tgcagtgatccctgatcaaacattgcatttcagctgggtgacagagtgagaccctgtctcagaaaaaaaaagtcatcctgaaacctcagaatagactacttgccagggc                          480
ttccttatgggtaaggaccttatgacctgctgggaccccaaactagcctcacctgatacgacctgtccttctcaaaacactaaactggaggaacatgtccccagtgctgggtagg                    600
agagtcgctgttatctgcctctatgcagagaggagcccagatcatcttttccatgacaggacagttccaagatgccacctgtactggagaagccaggtaaaatactttca                        720
agtaaacttttcttgatattactctatcttttcccaggagggactgcattacacaacaaattgcacctgggctctccttctatgcaaagcaaaaagccagcagcccaagctga                      840
taagattaatctaaagagcaaattatggtgataatttgtagttaatttttaaaaggttcatttttccatggtcgtatttcacaggaacatttttacctgttt                               960
gtgaggcattttttctctggaagagagctgctgatggccccaagactcggtgtaacgaaatttccatgtaaactcatttttttaaatcta                                          1080

                                                                                              **
tatatagagatatctttgtcagcattgcatcgttagctctcctgataaactaattgcctcacattgtcactgcaaatcgacacctattaatgggtctcacctcccaactgctctccct            1200
*****                                  Exon 1   M  G  L  T  S  Q  L  L  P  P
ctgtcttcctgctgcatgtgcggcaacttgtccaagtgcgatatacctacgaggatcataaaatttgaacagcctcacgacagaaggtgagaacctatctg                              1320
L  F  F  L  L  A  C  A  G  N  F  V  H  G  H  K  C  D  I  T  L  Q  E  I  I  K  T  L  N  S  L  T  E  E  Q  K

gcaccatctctccagatgtctcggtatgctctcagtatttctaggcatgaaaaacgtaacagctgctagaagagttggaactggtggtggcagtccaggcacacagcgaggctt          1440
ctcctgccactctttttctgagggtttgtagggaagttcccagttgaggggaggtggagctgctcataaggactctctgtccggttggagttaactgtgtcttcttgtctctca          1560

tttctgcctggaccaagactctgtgcaccgagagttgaccgtgaacagcatctttgctgcctccaaggtaagaaagagccgtccacggtctgtttgcaaatggggagatcatcccaaatg          1680
Exon 2   T  L  C  T  E  L  T  V  T  D  I  F  A  A  S  K

tctgaacaagaaacttgtctaatgaaaaacagcgggcccaaataacttaaggtgttagatgtttcaaagaacagagagtctgatcttactcttaagcatgtttggtcttctgg              1800
tttcacttgatttagaagacatgtaatagaaagcttacatgctgtagtcctgactcagatcctggtcaaagaaaagccctctggtttacttacttggcatagtgctggaacgta          1920
ggaggcactcaataaaatgcctgttgaatagagagaattttctggccacatactctcagaaaaccaaatactctcacagaaaccaaatactctcacagaaacagatattgagatgacaggttgagggttgaggagcttcattt          2040
gtctaagagacttcctatggccaacagaaaggatcgcaacgagaccccctcctggttccggttccggcacttcgggtgcctctctggtgctgctgctctcacccc              2160
ggccggccgaactccccagcctcggtcgagagatctcctggcgcagcggcagcatcccgctcgggcacctgcctcatctcgtgctgctcacccc              2280
accctctatctgagtggggagggagatagattgacagctgatagtgcatttctctaactacctgttctaactacctgttctaactacctgttctaacatgactacacgctatcaatagtttgtcatttcgtttgtttgtttcatgga            2400
aacacacgctgagaatgaaagcccccaagccatcaattcacagtgactcatttctctaactacctgtttcatcatgcaaactaggagatgatagccaagtgtatgtggcct              2520
ggtcacactccagcaccagaccatagaacaggccctcaaggccatcctctctcggggcctttagtggactatcttagttatcccggagggagtccgataca              2640
ggctaagctatctcctcagcctccagaatcgtccaccacccgctgacggcgttcgattcgcgtcatcagcagcctcacctcca              2760
ggctcaagctatctccttctccaccagcctccaccggcctctcaagtgggatagcgaccgcaccggggaatggctcttttgtcatttgcaggc          2880
ttgtctgaacttctggctcaagcaatcgtccacctggcctcaaagtgctgggatagtctcacctggcaacagttatatgtgtgtgtgtgtatatgtgt          3000
gtgtgtatatatatgtgtatatatagtgtatatatgtgtatatgtgtgtgtgtgtgtgtatataatctccaagatccatccaaccgatatggctcctactagaagc          3120
caagagtccaccgggttggacctgggtctggaggctgtcgagactgctgagaaggtctaacaaggccagggaggaagggcacctactagaagccaggctggaggaggggtgagggg          3240
ctgagggctggaggtaagactgcctggtgtttagaccccaggctgtggccttcagacatcttcacagctctcgacctcagtttccacatgtgaaga          3360
tatgaaagtgattctgaaggtgatcaaggtattgagatcagctcttgagttagtgcaaagtgtattgagtgatgataaccagcataaccagcagggtgcaggacactgct          3480
gatgattctaagagagggaaccgggtggaaagtgaaagtacatctcggggcaatccatcggggtattcacagaccggcaggacactgct          3600
gccctccgtcctaacctccctttcactcagtcctcactcagtcctcacacacaaacatctctagaataatccccactggctgctgtcactcttacccgtcactcttattgcct            3720
```

```
cccctgaacttcatctcctggagttcacgatctcactcttttcttccctgaagatcagcgactgcttacttacatgttaagatattcagaacagtgaaagtgctatt      3840
ttcaaaacctacaaggtgtatgcagagaaaggtactctttgtgtcccaagaaaaacatctttccaaatccagcctatgattttattctctcggggaacaagaatttagt      3960
atctctaagttgggtagacattctactcttggcagttgctggaagagaggcactggttcaaggtaacacctggtctaggtcctggggttcatatgtcatagtcatcaagggttcatattctacagagagttc      4080
acagcaagtgggagagaagcaagctggctgctgatgaagggttcttgggtggacaagagttggagcgatttctattttaccaaagagagctaaagttcatatattcacagagagttc      4200
cataatgaacctcaaatacctctgttttttgaaggagttcatatacagctagctgacttcacagcaggatggagtaatgatgcagtgcaatcgggctggatttatatgg      4320
cctagtgaggctggtcaagaaccgagttagaactctcacagtcatctcccacagacagaaatctcccagctgggtgttcctgacattcccgggaggcaggcctcttcctggtcact      4440
ccttagcagttcgaactggaggtcagcctgtccagtgcactggcagacacctcagcagctagcagcctaggcagttggagtgcagggtcaggg      4560
caggatggagtggagtccctcgctgcgatacagagcagaaacgttaaggcctttaagctcatttttccatttaatgcagactctt      4680
tcaattctattttatcctgttcctttgaaaaatatcttttagagaggttttttttcctatactatgtggtcatatcgggtcaaaaaaatacgttaaattccaggctcca      4800
agccagcgtttcagaaaaatctcaccaagtttgggtaaaagaagcaaagggctgactttttgttttctgaatctcatgtcccctgcacgatgcatgtgccaccctcca      4920
gacacaggcaccatctgccgcccccatcagccgtgcctccacctgactcgctacaaccagggtctgtttcttggcccccagaccaaagatactgacacactct      5040
tacatttccaactagaataacaggaacgaggatgcactctcagtcagttcattaagtaaatgtctttaacctctgccatggaactacgccccacagggggaagggggaagtctg      5160
tagctgggattcggtggtctgcaggtgctagatttcatacagctcacaaaatgctacacaaataagagtagcacacacaaaagattctaa      5280
gttactcactgccgcttattacagaagacaaaatctgccacgagaggggctgacaaatgacgaccatcgtttcacgcagtctcaatcccatgtctctataacc      5400
accgaaggcttaggaaaatgcttatgtataggtaaaagtaaaagtacaaacagtatcaacagtttttgaaaagtgtgacgatattaccaaaat      5520
attaacgagcaatagtagtacctctggctggtggaggtgatgagttctgattcagaggggctcgggggaggtggtgcaggggttggttctgagatttctgtattctaagagc      5640
ctgaaagctgcttggacacgcgctgctgatagaaaccccctgagcatctgtcttctcgatacaagggacagtcgggggaggctgggggggagtagctttcttttatctgaactgataaagagaggctggt      5760
tcccagtgatgtccatgtctgtccatgagggccacacttgagtacatcaaggccaggtcagactgagaccaatcctctctcggggtctccggacagttgtctc      5880
caagctaagaacctctccaggttgtcatctttctttagtatcaaggaatcaatcccgagacccacaggtgccagaccccatgtggccagacatgtcaatggtccaaaagggcccctt      6000
tgtgtccatcccttctctaagctaaggccccccaggaagagccatggtgaggccagacctctgatcttctctgtgctttgtgctggcaaaagtcttagc      6120
acagagctctgcccaaggctcagatcaatggcaatcatcctgatctctgtgctcctgaagatgtgcactcattattcaacaggatggcag      6240
tcctcactgggccatgtgctgctctctcccatggggaaccaagagcagaacctgactgaagctctatgcgaatagctcatgggtatagagacacggcagca      6360
gattccaggggaaagctggatttttaaaacctctgggaacaagagcagaacctgactgcctttcattagtccccatgtgcttttcataggctcaaatcctccaagaaataacctggagaaaggagcgggag      6480
acgtgggctgtagcgacagatgtccgttcattggggacagtgttgcagaggttgtgcagcgtctctcagtgaatagctcatagcgaatagagtagagacacggcaga      6600
ttctggggaaggctgcgtctgtttgcaattgggagggggttgtgcaattggtgtggcagcagcagctctcttcagtgaatacttgaagacaggtcgtgtagttgagcaaact      6720

cactccatttgtcctcctggaagaagaatcaagaggagagaatctctctccaatgagctggacattgctttcttctctgagaacaaactgagaag      6840
                                                                                                              Exon 3  N  T  T  E  K

gaaaccttcgcaggctgcgacgctgccgtcagttctacacggcaccacatgagaaggacacgttcacgaagcagcagcagttctgatccgattc      6960
E   T   F   C   R   A   A   T   V   L   R   Q   F   Y   S   H   H   E   K   D   T   R   C   L   G   A   T   A   Q   Q   F   H   R   H   K   Q   L   I   R   F

ctgaaacggctcgacaggaactctgggcctggcgggctggttaagctgctgctttggtgagagtcctgctggtccttggtggaacactccttag      7080
L   K   R   L   D   R   N   L   W   G   L   A   G   L

gagctgcagcacccttggtcaaccattcattcactcattcattaagtatttgctgaagttccacaagtgctgggtggttctaggtgctgaggacgtgtcactaaagacagcag      7200
gccgagtccctgttctcatggaatgttctaatgggagagttagaaaacaacatggaaaaatggccagcatgatacggtgcacaaagaaaaacatgaaataaagaacataagagt      7320
catggggagggggtgacttaggtggtgacattatcgagacagttgatttgaggtgagggagctggctggatagtgaggttccaggacagcagatcacaacaggccttta      7440
aggcttaggatggaaatgaactaacttcctgtattttaaagaccatggaggagggacttggcctgagatagggagtcagacagtcacacaaggcctta      7560
gattccaccacgatggagggaacacctgaggttgggcaggacaaagactgtacaattctgattttacgttgattaaaagggtcagtcagtcagtgactggctggctacgtgtggtaaaatgagctgaaa      7680
```

Fig. 4 (continuation see next page)

```
gggggaaagcatagaagcaagatggcctgttgggaggctaccacagtaaaccaggctagagatgatggtggcgtggacagaatgaagcaagatggcctgttgggaggctaccacagtaaa   7800

ccaggctagagatgatggtggcgtggacagaatgaagcaagatggcctgttgggaggctaccacagtaaaccaggctagagatgatggtggcgtggacaaatggagcagttgaggtgaac   7920

agatttgggatatgactaaaaataaaaccagaagatttgctgacagatcggttgtaggggtaagatacaggggaggaaaagatgacctctttgttcctgcccaaacccctctggcgatg   8040
gtcagtactgtttacagagagatgaaagactggcggcaaggcagggctggaggttcagcagaagatcaagagttcaattttgtacatcgtacatgtaaggtggctcttggatagccaagt   8160
gaaggtgttgagaagatggttagaaaagtctggaacttaggggagaggtcagaacttgcaatacaaaaaggagagtccttagatagatactgctgaaaatctgaatgacagaaagggaga   8280
gatcaaaggactgagcctgagatcaacacatggaggtcaggagaggaggatccagccaaggggcctgaggaggagtgaccagtgaggcaggagaacatggagagtgggcggtaccccagg   8400
aagccggtgaggacactcaaggagggagggttgactgtgtcaaatgtactgaaaggacaggtcaggtgaggaccaagaaaggcccctgggtttggctgatggaggccatgggtgaggctg   8520
atgtaaatggagaggcaggaaggaaagcccagctggagtgggctcaccgaggatagggtggcgagaggagacaaagaaggaacagtgagggcagacaactctttgaagatgtttagctat   8640
aaggctgcagagaaactgagcccacagctgcagggtggttatggagtgagggaagctcttttaaggttggggtatacccagcatgttaatgcacctggggaatggtccagtggagcag   8760
gaagaactgaagagagcagaaagaggaagaatcattaggggcagaagtccttgtagcccagagtggatgttatctaatatcgagtggaggaattaattggctttagaggagaacaagga   8880
catgtatcccctctctgggcctatcaccttgtagacaatgggataggtcatgggataggaacttggcacaacacatgttctctcttttaattctctccattatcttatgaagcaggcaag   9000
taggcaaacaattgtcccaactttacaaaagaaactgaagcttttataaattaagtagtacatcctaagcaatacaattaataaatggtagagctgagattcaaactgaagcagtggcct   9120
gggggtagcatctggaatccttcccacctttagggctgctgtgctgcggtgctgctgtttaatggcacagagggccagatgactgaatctctctcagcagtccaggcagtcatgcagaag   9240
gcccagtagagcaccgggcaggtctgagccagcatcttcaagttccaccctgtgagcaagcacttagctgtgacacacttctcgagagactggactcccccccgcgcaacccacccaaaa   9360
gcagataggtaatggtatacagtaaccatttctagaagtgtaagtagtatgcacccaaaataggcaaaacctgctggcctagtgatagagacaactcccagtcaggctagactggaggcc   9480

ttggttttataagtgttcaggtgacaagtgccacagtaggcttgatcaagtagacaggcaggcaagacaaatgcttaccaatgcaagctaatgaaatgtttcttttgcagaattcctgtc   9600
                                                                                                        Exon 4   N  S  C  P
ctgtgaaggaagccaaccagagtacgttggaaaacttcttggaaaggctaaagacgatcatgagagagaaatattcaaagtgttcgagctgaatattttaatttatgagtttttgatagc   9720
  V  K  E  A  N  Q  S  T  L  E  N  F  L  E  R  L  K  T  I  M  R  E  K  Y  S  K  C  S  S  *

tttatttttaagtatttatatatttataactcatcataaaataaagtatatatagaatctaacagcaatggcatttaatgtattggctatgtttacttgacaaatgaaattatggtttg   9840
caacttttagggaaatcaatttagtttaccaagagactataaatgctatggagccaaaac                                                                   9901
```

Fig. 4. Complete nucleotide sequence of the human IL-4 gene and predicted amino acid sequence. *Asterisks*, TATA-like sequence; *thin line*, TG element (positions 2966 to 3077); *thick lines*, 70-bp repeats (7694 to 7902); *single asterisk*, termination codon

Table 1. Common sequence motifs found in the 5' flanking regions of various lymphokine genes

m IL-4		$^{-163}$GGGGTTTCAT–TCCAATTGGTC
h IL-4		$^{-169}$AAGGTTTCAT–TCCTATTGGTC
m IL-3	distal	5'-$^{-265}$GAGATTCCAC–TCAGAGC-3'
	proximal	$^{-84}$GAGGTTCCAT–TCAGATA
m GM-CSF		$^{-76}$GAGATTCCAC–TCAGGTA
h GM-CSF		$^{-73}$GAGATTCCAC–TCAGGTA
m IL-2		$^{-175}$GGGATTTCAC–TCCATTCAGTC
h IL-2		$^{-172}$GGGATTTCAC–TCCATTCAGTC
h IL-5		$^{-75}$TTAGTTTCAC
h IFN-γ		$^{-206}$AGAATCCCAC
h G-CSF		$^{-158}$GAGATTCCAC
	Consensus	5'-RRRRTTYCAY-3'

Location corresponds to the distance from the 5' nucleotide of the TATA box.
m, mouse; h, human.

pivotal role for regulated expression of eukaryotic genes. A consensus sequence (Table 1) between IL-2, IL-3, and GM-CSF (MIYATAKE et al. 1985a; METCALF 1985) is also found in 5' flanking regions of mouse and human IL-4 genes (YOKOTA et al. 1988). However, the significance of these sequences in the expression of the IL-4 gene needs to be determined.

V. Expression of IL-4 mRNA

A careful examination of the lymphokines produced by a panel of activated mouse helper T-cell (T$_H$) clones indicated that there are at least two major subsets of T$_H$ clones, designated T$_H$1 and T$_H$2 (MOSMANN et al. 1986b; MOSMANN and COFFMAN 1987). Both types produce IL-3 and GM-CSF; however, only T$_H$1 produce IL-2 and interferon-γ (IFN-γ) in response to stimulation by antigen or ConA, whereas only T$_H$2 clones produce IL-4 and IL-5 in response to the same stimuli. The mouse T-cell clone Cl.Ly1$^+$2$^-$/9 is a typical T$_H$2 clone. Inducible expression of mouse IL-4 was confirmed by analysis of mRNA isolated from T$_H$2 cells treated or untreated with ConA.

It is not clear whether human helper T-cell clones can be classified into the T$_H$1 and T$_H$2 clones described in the mouse system. So far, almost all CD4$^+$ human T-cell clones isolated from peripheral blood leukocytes of normal individuals are able to produce IL-2, IL-4, and IFN-γ simultaneously in response to ConA, phytohemagglutinin (PHA), or calcium ionophore (A23187) plus PMA (PALIARD et al. 1988). S1 protection analysis showed that the human T-cell clone 2F1, from which we isolated human IL-4 cDNA, not only expressed IL-2, IL-4, and IFN-γ genes but also IL-3 and GM-CSF transcripts upon stimulation by anti-CD3 monoclonal antibody plus PMA. An alloreactive clone, A1, produced IL-4 and IL-5 but no IL-2 or IFN-γ (Jabara et al. 1988). The mechanism underlying these differences in lymphokine secretion patterns is unclear.

C. Biological Activity of IL-4

I. Discovery and Characterization of the B-Cell Activation Properties of IL-4

IL-4 was first discovered as an activity that induced B-lymphocyte proliferation in the presence of anti-μ antibodies. This activity was mediated by a low molecular weight (18 000) molecule that was not IL-2 (Howard et al. 1982), and was therefore called B-cell growth factor (Farrar et al. 1983). It then became apparent that there were other high molecular weight B-cell growth factors (Swain et al. 1983) and to differentiate these molecules the factor that eventually became IL-4 was called B-cell growth factor 1. Shortly thereafter, the ability of this molecule to induce the expression of molecules of the class II major histocompatibility complex (MHC) in resting B cells was described (Roehm et al. 1984; Noelle et al. 1984). Given that this effect was not related to B-cell proliferation, this molecule was renamed B-cell stimulatory factor 1 (BSF-1). The induction of class II MHC molecules in B cells correlated with enhanced antigen presenting ability in these cells. At this time the prevailing view was that IL-2 was the sole T-cell growth factor while BSF-1 was a B-cell growth factor. However, the supernatant from the T-cell clone Cl.Ly1$^+$2$^-$/9 contained a T-cell growth factor activity that was not IL-2 (Smith and Rennick 1986), and at the same time it had mast-cell growth factor activity. Furthermore, this T-cell clone was already known to produce large amounts of BSF-1 (Roehm et al. 1985). Eventually, the molecular cloning of a cDNA encoding a molecule that exhibited two characteristic biological activities of BSF-1 was achieved (Lee et al. 1986; Noma et al. 1986), namely induction of B-cell growth in the presence of anti-μ and induction of class II MHC in rest-

Table 2. Summary of biological activities of recombinant IL-4

T cell	TCGF Thymocyte proliferation
B cell	BCGF I Ia induction IgG1 enhancement IgE enhancement FcεRII induction
Hematopoietic cells	MCGF Macrophage activation Macrophage fusion

ing B cells. In addition, the recombinant material clearly had T-cell growth factor activity as well as mast-cell growth factor activity (LEE et al. 1986). Thus, these observations indicated that BSF-1 had effects on cells of different lineages and is not B-cell specific; the name interleukin-4 was then suggested for this molecule and a committee of the IUIS/WHO supported this nomenclature at the 6th International Immunology Congress (Toronto, Canada). IL-4 should not be confused with another molecule (TRF-1, BCGF II, or eosinophil inducing factor) which is now called interleukin 5 (IL-5) (KINASHI et al. 1986; AZUMA et al. 1986; BOND et al. 1987; COFFMAN et al. 1987; YOKOTA et al. 1987). Multiple biological activities of IL-4 on different target cells including T cells, B cells, thymocytes, and hematopoietic cells are summarized in Table 2.

II. T-Cell Growth Factor Activity of IL-4

1. Activity on Mature T Cells

While the availability of recombinant IL-4 confirmed the T-cell growth factor activity of IL-4, several reports had already provided compelling evidence for this effect. A careful examination of the lymphokines produced by a panel of activated helper T-cell (T_H) clones has revealed that there are at least two major subsets of T_H clones, designated T_H1 and T_H2 (MOSMANN et al. 1986b). MOSMANN and COFFMAN (1987) described two types of T_H cells which produced mutually exclusive T-cell growth factors. Thus, T_H1 cells produce IL-2 and IFN-γ but not IL-4 while T_H2 cells produce IL-4 but not IL-2 or IFN-γ. Both types produce IL-3 and GM-CSF, but only T_H1 produce IL-2 and IFN-γ in response to stimulation by antigen or mitogens (ConA). Using neutralizing monoclonal antibodies against IL-2 and IL-4, these workers demonstrated that some T_H (CD4$^+$8$^-$) clones produced IL-2 while others produced IL-4 by testing ConA-stimulated supernatants of these clones on the indicator cell line HT-2 (WATSON 1979) in the presence or absence of neutralizing antibodies (Fig. 5). The use of these antibodies in this assay provides a monospecific biological assay for assaying IL-2 or IL-4 in the presence of the other lymphokines (MOSMANN et al. 1986a). Another significant difference is that IL-2 induced a significantly higher level of proliferation and stimulation in HT-2 cells than IL-4. SMITH and RENNICK (1986) biochemically purified a T-cell growth factor (IL-4) different from IL-2 actually using a ConA-stimulated supernatant from a T_H2 clone as a source.

IL-4 exhibits T-cell growth factor activity directly on some T-cell clones and lines like HT-2 (MOSMANN et al. 1986b; LEE et al. 1986) or CTLL (SIDERAS and PALACIOS 1987; SEVERINSON et al. 1987). However, in normal T cells, it needs a cofactor for its effect (HU-LI et al. 1987; GRABSTEIN et al. 1987), which can be a mitogen (PHA, ConA) or a phorbol ester like PMA. The latter agent induces no proliferation in normal T cells, while PHA or ConA have to be used in suboptimal amounts to observe the IL-4 growth cofactor effect. Recently the question of whether IL-4 acts directly on T cells was reanalyzed

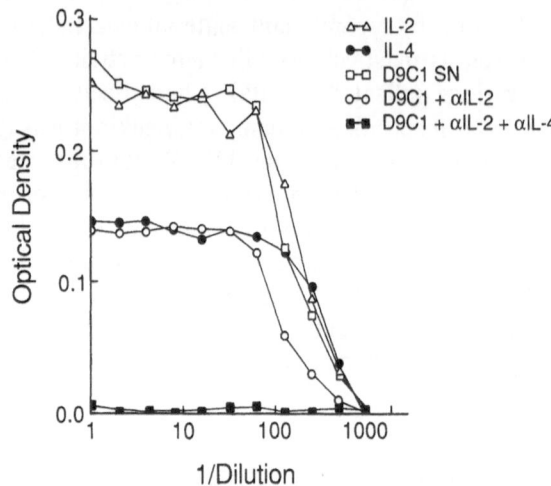

Fig. 5. Monospecific assay for IL-4 using the T-cell clone HT-2. These cells, like most T cells, respond to both IL-2 and IL-4. In this figure, IL-4 (125 U/ml) produces a lower stimulatory signal than IL-2 (100 U/ml) in the MTT (tetrazolium salt) colorimetric assay (MOSMANN et al. 1986a). The supernatant from a T-cell hybridoma (D9C1.12.17) that produces both IL-2 and IL-4 (HAGIWARA et al. 1988) can be tested in this assay. The ConA-stimulated supernatant of D9C1.12.17 (*D9C1 SN*) produces the signal corresponding to IL-2. Addition of anti-IL-2 antibodies (S4B6; MOSMANN et al. 1986a) reveals the IL-4 signal. Further addition of anti-IL-4 monoclonal antibodies (11B11) eliminates all the T-cell growth activity from this supernatant

(BROWN et al. 1988). Using monoclonal antibodies directed against IL-2 or the IL-2 receptor as well as probing for the presence of IL-2 mRNA in the responding cells, these authors concluded that cells growing in IL-4 do not produce detectable IL-2 and, therefore, the effect of IL-4 is probably directly on the responding cells. Most of the above observations were reported for murine cells using mouse IL-4; similar observations have been made in the human system using human IL-4 (SPITS et al. 1987).

2. Activity on Thymocytes

The data which indicated that IL-4 had T-cell growth factor activity prompted several investigators to study its effect on thymocytes. ZLOTNIK et al. (1987b) reported that IL-4, in cooperation with PMA, stimulates the growth of thymocytes. Thymocytes can be subdivided into four populations depending on their expression of CD4 and CD8 (ROTHENBERG and LUGO 1985). The most immature subset are the $CD4^-8^-$ thymocytes and several reports have suggested that IL-2 was not a potent growth factor for these cells (VON BOEHMER et al. 1985). The thymocytes found to proliferate in response to IL-4 and PMA were $CD4^-8^-$, $CD4^+8^-$, and $CD4^-8^+$ (ZLOTNIK et al. 1987b). Interestingly, the $CD4^+8^+$ subset showed minimal or no response to IL-4 and PMA. The $CD4^-8^+$ subset showed significantly higher proliferative response to IL-4 and PMA than the $CD4^+8^-$ and $CD4^-8^-$ ones. The latter observation is interesting since CD4$^-$8$^-$ and CD4$^+$8$^-$ thymocytes produce IL-4 upon stimulation with calcium ionophores and PMA (ZLOTNIK et al. 1987b) while CD4$^-$8$^+$ thymocytes do not (M. FISCHER and A. ZLOTNIK, manuscript in preparation). Day-14 and day-15 fetal murine thymocytes also proliferate to IL-4 and PMA (ZLOTNIK et al. 1987b; PALACIOS et al. 1987). At this time most thymocytes are CD4$^-$8$^-$. Immature (CD4$^-$8$^-$) thymocytes grown in culture with IL-4 and PMA apparently do not differentiate into other phenotypes (LOWENTHAL et al. 1988b) although their expression of IL-4 receptor increases significantly (LOWENTHAL et al. 1988b). One report (PALACIOS et al. 1987) has suggested that fetal day-14 thymocytes differentiate in culture to CD4$^-$8$^+$ thymocytes which become cytotoxic. These observations, which we have been unable to reproduce (GUIDOS et al. 1989), are likely due to the expansion of preexisting CD4$^-$8$^+$ thymocytes present at the beginning of culture. A small population of CD4$^-$8$^+$ thymocytes observed at day 14½–15½ of culture (GUIDO et al. 1989) may represent an intermediate stage of differentiation between CD4$^-$8$^-$ and CD4$^+$8$^+$ thymocytes (PATTERSON and WILLIAMS 1987).

Of particular importance is the fact that IL-4 is a potent growth factor for CD4$^-$8$^-$ thymocytes, both adult and fetal (ZLOTNIK et al. 1987b). A comparison of the growth-promoting activities of IL-2 and IL-4 on these thymocytes indicates that IL-4 is a more potent factor than IL-2. In fact, in day-15 CD4$^-$8$^-$ thymocytes, IL-2 has almost no effect while IL-4 is very potent. Several groups have reported that fetal thymocytes of days 14–15 are able to produce IL-4 upon stimulation with calcium ionophore and PMA or anti-CD3 monoclonal antibodies (RANSOM et al. 1987b; TENTORI et al. 1988; SIDERAS et al. 1988). We have recently obtained evidence measuring the response of single cells to various lymphokines (CHEN et al. 1989) that IL-4 is a better growth factor for CD4$^-$8$^-$ thymocytes than IL-2 while IL-2 is a better growth factor for mature-phenotype, single-positive (CD4$^+$8$^-$ and CD4$^-$8$^+$) thymocytes (Table 3). Taken together, these observations suggest that IL-4 plays a crucial role in early T-cell development.

It appears that not all double-negative thymocytes proliferate to IL-4 and PMA, and only thymocytes that are IL-2 receptor positive proliferated to IL-4 while IL-2 receptor-negative cells apparently did not (TAKEI 1988). This finding is somewhat confusing since it implies that IL-2 receptor-negative cells are the ones responding to IL-2. CD4$^-$8$^-$ thymocytes grown in IL-4 and PMA preferentially express $\gamma\delta$ T-cell receptors, while those grown in IL-2 and PMA preferentially express $\alpha\beta$ T-cell receptors (GAUSE et al. 1988). These reports suggest that different lymphokines may mediate different pathways of T-cell differentiation in the thymus.

Functionally, IL-4 has been reported to be necessary for the induction of cytotoxic T-lymphocyte activity in thymocytes separated using a lobster agglutinin (COLLINS et al. 1988). The monoclonal antibody 11B11 (OHARA and PAUL 1985) inhibited the development of cytotoxic cells in both lobster agglutinin-positive (mature) and -negative (immature) thymocytes. Similar conclusions were reported for cytotoxic T-lymphocyte generation using mature T lymphocytes (WIDMER and GRABSTEIN 1987).

Table 3. Proliferative response of thymocytes to IL-2 or IL-4 in single-cell cultures

Thymocytes	Cultured in[a]	Clones scored	Cloning efficiency[b]	Clone size[b]
Fetal day 15	IL-2	0	0	0
	IL-4	60	38.6 ± 14.8	14.7 ± 12
Adult CD4⁻8⁻	IL-2	16	9.0 ± 6.8	6.0 ± 3.5
	IL-4	23	16.1 ± 3.3	7.6 ± 3.3
Adult CD4⁻8⁺	IL-2	31	60.7 ± 6.6	61.0 ± 52
	IL-4	21	35.7 ± 8.1	37.7 ± 63
Adult CD4⁺8⁻	IL-2	20	39.6 ± 8.8	16.1 ± 8.8
	IL-4	10	12.1 ± 1.9	4.1 ± 1.4
Adult CRT[c]	IL-2	112	61.5 ± 6.8	39.2 ± 39
	IL-4	88	54.0 ± 15.3	19.8 ± 22

[a] All cultures contained ionomycin (0.35 μM) and PMA (10 ng/ml). IL-2 was used at 100 U/ml and IL-4 at 125 U/ml.
[b] Scored on day 4 of culture.
[c] Cortisone-resistant thymocytes used to optimize the cloning conditions.

The effects of IL-4 in the thymus are not restricted to thymocytes and IL-4 induces antigen-presenting ability in cloned thymic macrophages (RANSOM et al. 1987a). These effects will be discussed further in the section on macrophages (Sect. C. VII).

III. Biological Activity on B Cells

As discussed previously, IL-4 was originally discovered and characterized through its costimulatory activity on B cells with anti-μ antibodies (HOWARD et al. 1982). Although a considerable amount of information on the effects of IL-4 on B cells was already available by the time IL-4 was molecularly cloned, the availability of the recombinant material free of endotoxin allowed more definite experiments on B cells. The induction of class II MHC molecule expression in B cells (ROEHM et al. 1984; NOELLE et al. 1984) had indicated that IL-4 was not only a growth factor but a differentiation factor as well. Furthermore, IL-4 induced isotype switching in immunoglobulin-producing B cells from IgM to IgG1 (VITETTA et al. 1985). For these studies it was pivotal to use a monoclonal antibody against IL-4 (OHARA and PAUL 1985) which in turn allowed purification of this molecule to homogeneity (OHARA et al. 1987). The next major step in the characterization of IL-4 as a B-cell differentiation factor was the demonstration that IL-4 induced IgE production by LPS-activated B cells (COFFMAN et al. 1986). Thus, two isotypes are induced in B cells by IL-4: IgG1 and IgE. The latter observation, along with the two types of T_H cells described by MOSMANN et al., suggested that IL-4 was an important molecule mediating allergic responses. Recently, evidence has been presented suggesting that IL-4 directly induces heavy chain switching in B cells, leading to IgE production (LEBMAN and COFFMAN 1988). Furthermore, this effect of IL-4 on IgE production has been demonstrated in vivo (FINKELMAN et al. 1988). IL-4 is also an essential factor for IgE synthesis in human cultures (PENE et al. 1988; DEL PRETE et al. 1988).

Another important effect of IL-4 on B cells has been on the induction of the cell surface receptor for the Fc fragment of IgE (FcεR) in normal human tonsil B cells (KIKUTANI et al. 1986; DEFRANCE et al. 1987a). Similar property of mouse IL-4 was also found in mouse B cells (HUDAK et al. 1987). IL-4 was also found to induce normal B cells to secrete soluble CD23 (nomenclature of the FcεR on B cells; BONNEFOY et al. 1988).

IL-4 is known to act on both resting and activated B cells. For example, resting B cells pretreated with IL-4 for 24 h enter the S phase of the cell cycle quicker than control cells when both groups are challenged with anti-IgM and additional IL-4 (RABIN et al. 1985). Other effects on resting B cells include induction of Ia antigens, as discussed above. In large B cells, IL-4 induces entrance into the cell cycle when used in large amounts (RABIN et al. 1986).

IL-4 production by normal B cells has not been reported, but some transformed B-cell lines can produce IL-4 (O'GARRA et al. 1989).

IV. Relationship of IL-4 Effects with IFN-γ Effects

So far we have not discussed other lymphokines that may inhibit or enhance IL-4 effects. The one lymphokine that it is important to discuss separately is IFN-γ. IFN-γ selectively inhibited the proliferation of B cells to IL-4 and anti-IgM antibodies (MOND et al. 1985) and the ability of IL-4 to induce Ia antigen induction in B cells (MOND et al. 1986). The latter effect was restricted to the IL-4-induced enhancement of Ia expression in B cells and did not affect background Ia expression (IL-4 independent) in B cells (ZLOTNIK et al. 1987a). COFFMAN and CARTY (1986) described that IL-4-mediated induction of

Table 4. IFN-γ inhibits the IL-4-mediated proliferation of CD4$^+$8$^-$ thymocytes

Thymocytes[a]	IFN-γ (U/ml)	CPM (mean ± SD)
Unseparated	–	50 354 ± 1 257
	10	53 435 ± 2 434
	100	47 546 ± 2 449
CD4$^+$8$^-$	–	34 546 ± 1 109
	10	8 423 ± 435
	100	2 331 ± 132
CD4$^-$8$^+$	–	257 467 ± 3 547
	10	234 567 ± 4 235
	100	245 678 ± 3 556

[a] 10^5 thymocytes in 100 μl cultured 3 days in IL-4 125 U/ml and PMA 10 ng/ml.

polyclonal IgE production was inhibited by IFN-γ. Conversely, IL-4 inhibits the IFN-γ-mediated production of B cells producing IgG2α (SNAPPER and PAUL 1987). Similarly, IFN-γ inhibited the IL-4-mediated induction of CD23 (FcgϵR) in B cells (HUDAK et al. 1987). Interestingly, these "IL-4-neutralizing" effects of IFN-γ are not restricted to B cells; it partially inhibits the IL-4-mediated proliferation of CD4$^-$8$^-$ thymocytes (RANSOM et al. 1987b). More recent data (Table 4) indicate that IFN-γ is a powerful inhibitor of the proliferation of CD4$^+$8$^-$ thymocytes but not of CD4$^-$8$^+$ thymocytes when these cells are induced to proliferate with IL-4 and PMA. Among two populations of long-term T-cell clones, T$_H$1 cells produce IFN-γ but not IL-4 while T$_H$2 produce IL-4 but not IFN-γ (MOSMANN et al. 1986a). It appears that the proliferation of T$_H$2 clones to IL-4 is inhibited by IFN-γ (GAJEWSKI and FITCH 1988). Thus, IFN-γ may not only neutralize the biological effects of IL-4 (mostly described so far for B cells) but may in fact inhibit the generation of IL-4-producing cells.

V. Effect of IL-5 on IL-4-Dependent IgE Synthesis

Although mouse IL-5 had little influence on the levels of IgE and IgG1 produced in response to saturating concentrations of IL-4, it could substantially enhance the IgE and IgG1 responses to suboptimal concentrations of IL-4 (COFFMAN et al. 1988). The addition of IL-5 reduced the amount of IL-4 needed for maximum IgE and IgG1 production to levels that are not much above the levels required for other effects. Human IL-5 also enhanced IL-4-induced IgE synthesis 4- to 10-fold at suboptimal IL-4 concentrations (YOKOTA et al. 1987; PENE et al. 1988), and IL-4 induced CD23 expression on normal B cells and the subsequent release of soluble CD23 in the culture supernatant in a dose-dependent way (PENE et al. 1988). The IgE production induced by a combination of IL-4 and IL-5 could also be completely inhibited by IFN-γ and F(ab')$_2$ fragments of anti-CD23 monoclonal antibody 25. Although release of soluble CD23 is required for IL-4 induced IgE production, its regulatory effect seems to be indirect. The precise effects of soluble CD23 on T-cell, B-cell, and monocyte interactions required for IL-4-induced IgE production remain to be determined.

VI. Effects on Mast Cells

The initial reports of the molecular cloning of mouse IL-4 also provided evidence for the direct effects of IL-4 on mast cell survival and proliferation (LEE et al. 1986). Mouse IL-4 enhances mast cell colony formation by bone marrow progenitor cells when used with IL-3 (RENNICK et al. 1987) but does not stimulate mast cell colony formation by itself. However, proliferation of connective tissue mast cells could not be induced by IL-3 alone and requires the presence of both IL-3 and IL-4 (TAKAGI et al. 1989). Interestingly, mast cells have been reported to be able to produce IL-4 (BROWN et al. 1987) and this unexpected observation suggests a T-cell-independent pathway of mast cell activation.

In contrast to the mouse system, neither human IL-3 nor IL-4 stimulates the growth of human mast cells. However, human basophils were detected when either human cord blood or bone marrow cells were cultured with human IL-3 or IL-3 plus IL-4 (SAITO et al. 1988; TADOKORO et al. 1989).

VII. Effects on Macrophages

Effects of recombinant IL-4 on bone marrow macrophages included an increase in Ia antigen expression that resulted in enhanced antigen presenting function (ZLOTNIK et al. 1987a). This effect did not have the magnitude of the Ia increase induced by IFN-γ but was significant nonetheless. Interestingly, IL-4 was found to induce antigen presenting ability in a cloned bone marrow macrophage cell line (14M1.4) but not in the tumor macrophage cell line P388D1, even though the latter line is inducible with IFN-γ (ZLOTNIK et al. 1983). IL-4 also acts as a macrophage activating factor, which induces tumoricidal activity in macrophages and Ia antigen expression in a subset of peritoneal macrophages (CRAWFORD et al. 1987). Furthermore, IL-4 induces antigen-presenting ability in thymic macrophages but not in thymic epithelial cells even though the latter are inducible with IFN-γ (RANSOM et al. 1987a). Thus, it appears that IL-4 induces Ia antigen expression in subsets of macrophages, the significance of which remains to be clarified. In contrast, IFN-γ induces Ia antigen expression in most macrophage populations. This view is supported by a report that IL-4 induces both class I and class II MHC molecule expression in bone marrow macrophages (at both the cell surface and the mRNA level), but these effects could not be demonstrated in thioglycollate-elicited peritoneal macrophages or the myelomonocytic cell line WEHI-3 (STUART et al. 1988). IL-4 also induces Ia antigen expression in human peripheral blood monocytes (TE VELDE et al. 1988).

Another interesting effect of IL-4 is its ability to induce multinucleated cell formation through cell fusion (MCINNES and RENNICK 1988). This effect of IL-4 appears to be what had been reported as "macrophage fusion factor" (GALINDO et al. 1974). These observations suggest that IL-4 may play a role in the development of the granulomatous response.

VIII. Effects on Hematopoietic Progenitor Cells

Both mouse and human IL-4 appear to be unable by themselves to drive the proliferation and differentiation of bone-marrow-derived hematopoietic progenitor cells in soft agar. However, IL-4 strongly influences the development of these colonies when used with other growth factors. For example, mouse IL-4 enhances the development of granulocyte colonies with G-CSF (RENNICK et al. 1987) and of erythrocyte colonies to erythropoietin (PESCHEL et al. 1987). The effects on other cell lineages are more complex. For example, mouse IL-4 inhibits the IL-3-dependent colony formation by granulocyte and macrophage progenitor cells and by multipotential progenitor cells, but it enhances macrophage colony formation in the presence of M-CSF. Likewise, it enhances mast cell colony formation when used with IL-3. Finally, it also inhibits stromal cell-dependent growth of bone marrow-derived pre-B cells (RENNICK et al. 1987). These effects exemplify the complex interactions that IL-4 mediates in hematopoiesis.

IX. Other Effects of IL-4

1. Induction of Lymphokine-Activated Killer Cell Activity

IL-4 has also been examined for its ability to induce lymphokine-activated killer cells (LAK) and different conclusions have been reported on this subject. Murine IL-4 was capable of inducing LAK activity directed against fresh tumor cells (MULE et al. 1987). In contrast, human IL-4 did not induce LAK activity in cell cultures derived from peripheral blood (SPITS et al. 1988). Furthermore, IL-4 added simultaneously with IL-2 inhibited the ability of the latter lymphokine to induce LAK activity. The IL-4 inhibitory effect occurred only when IL-4 was present at the beginning of culture. IL-4 did not inhibit the cytotoxic activity of established natural killer clones. Thus, IL-4 may affect the precursor LAK cells in the human system.

Several explanations are possible for these different conclusions: the source of cells (peripheral blood for the human and spleen for the mouse) were different as well as the target cells used for the assays. Moreover, it is possible that species differences exist in the biological activities of IL-4. These discrepancies suggest caution in extrapolating results in the murine system to the human and vice versa.

2. Effects on Other Cell Lineages

We have already discussed the presence of IL-4 receptors on a variety of cell lineages. R. STRUNK (unpublished) has observed that IL-4 modulates the production of C3 and factor B in human fibroblasts induced by other cytokines (tumor necrosis factor-α, IL-1, etc.). While data on the biological activity of IL-4 on epithelial cells are not available yet, these cells express receptors for IL-4, and we should expect IL-4 to modulate the functions of these cells. For example, IL-4 receptors were described on macrophage lines (OHARA and PAUL 1987) before data on IL-4 effects on macrophages became available (ZLOTNIK et al. 1987 a).

D. The IL-4 Receptor

As described in the previous sections, IL-4 interacts with a broad range of target cells and results in different biological responses in each target cell. Lymphokines are known to exert their actions through binding to specific cell surface receptors. Characterization of the lymphokine-receptor interaction and of the intracellular signal transduction events are important for understanding the mechanism of IL-4 action.

I. The Murine IL-4 Receptor

The presence of specific receptors on several cell lineages was demonstrated using radiolabeled mouse IL-4 (OHARA and PAUL 1987) and Scatchard analysis of the equilibrium binding of radiolabeled IL-4 was used to quantitate the number of receptors on a variety of cells types (PARK et al. 1987a; NAKAJIMA et al. 1987; LOWENTHAL et al. 1988a, b). There appears to be general agreement that the IL-4 receptor is widespread among a variety (both hematopoietic and nonhematopoietic) of cell types and only 5 cells out of 90 tested had undetectable levels of IL-4 receptors (LOWENTHAL et al. 1988a). The receptor itself appears to be a high-affinity receptor with a dissociation constant of 20–80 pM. The number of receptors per cell is low, and ranges from a few hundred to 5600 on the T-cell clone HT-2. This receptor is specific since the binding of ^{125}I-IL-4 is blocked by equimolar amounts of "cold" IL-4 or antibodies against IL-4 (OHARA and PAUL 1985). Other lymphokines (IL-2, IL-3, IFN-γ, and GM-CSF) do not inhibit this binding and neither does human IL-4 (YOKOTA et al. 1988).

Crosslinking studies detect a receptor-ligand complex of approximately 79 kDa, suggesting that the receptor itself is a protein of approximately 65 kDa in the mouse system (OHARA and PAUL 1987; PARK et al. 1987a; LOWENTHAL et al. 1988a). There has been no conclusive evidence for the existence of a low-affinity IL-4 receptor.

II. The Human IL-4 Receptor

In agreement with the murine system, the human IL-4 receptor is expressed in most cells studied regardless of their origin or lineage in relatively small numbers (up to 1200/cell) and with high affinity ($K_d = 70$ pM) (PARK et al. 1987b; CABRILLAT et al. 1987). Again, most other lymphokines fail to block binding of radiolabeled human IL-4 to its receptor, and mouse IL-4 does not block either. The latter finding correlates with the lack of crossreactivity between mouse and human IL-4 in biological assays (MOSMANN et al. 1987). Crosslinking studies demonstrate a 140-kDa cell surface protein (PARK et al. 1987b). More recent crosslinking data suggest that mouse IL-4 receptor is also a 140-kDa protein (M. HOWARD, personal communication).

III. Signaling Events Mediated by IL-4

IL-4, which was initially believed to be a B-cell specific lymphokine (BSF-1), turned out to be the multifunctional lymphokine capable of stimulating T cells, thymocytes, B cells, mast cells, macrophages, and other hematopoietic cells. The biological responses mediated by IL-4 vary depending on the target cells. For example, IL-4 provides a growth-promoting signal to T cells, thymocytes, and mast cells but is inhibitory for macrophage growth. IL-4 also provides growth-promoting as well as differentiation signals to B cells. All these target cells appear to express high-affinity IL-4 receptors. However, the nature of the intracellular signal(s) of IL-4 that induces different biological responses on target cells is unknown.

Several mouse myeloid cell lines that respond to multiple growth factors such as IL-3, GM-CSF, G-CSF, or IL-4 have been described (KOYASU et al. 1987; WEINSTEIN et al. 1986). Using these cells, IL-3 has been shown to induce the phosphorylation of several proteins at tyrosine residues. Under the same experimental conditions, IL-4 also appears to stimulate the protein phosphorylation at tyrosine residues (MORLA et al. 1988). On the other hand, IL-4 induces membrane protein phosphorylation in normal resting B cells (JUSTEMENT et al. 1986). These authors, however, were unable to demonstrate any of a series of common signaling phenomena (Ca^{2+} mobilization, protein kinase C translocation, membrane depolarization, or phosphoinositide metabolism). Thus, it is possible that IL-4 may exert its effects through as yet unknown pathways of cell activation.

E. Concluding Remarks

The data that have appeared within the last 3 years indicate that IL-4 is a major modulator of cell function which probably has important homeostatic effects, not only for the immune system but in other organs as well. However, even with the major advances we have witnessed so far, many important questions remain regarding the consequences of IL-4 action.

I. Dichotomy Between T_H1 and T_H2 Cells

The observation that some long-term T-cell clones produce IL-2 and IFN-γ while others produce IL-4 and IL-5 (MOSMANN et al. 1986a) suggests differential roles for these lymphokines in the regulation of immune responses, with T_H1 more involved in classical delayed hypersensitivity responses and T_H2 mediating more humoral events or allergic (i.e., IgE-mediated) responses. In this respect, the in vivo effects of IFN-γ and IL-4 on IgE production has been especially impressive (FINKELMAN et al. 1988). As predicted from in vitro studies, IL-4 enhanced and IFN-γ inhibited the levels of IgE in mice. However, the patterns of lymphokine production predicted by T_H1 and T_H2 clones have yet to be demonstrated in vivo. In the human system, this distinction between T_H1 and T_H2 has been more difficult to establish (YOKOTA et al. 1988; UMETSU et al. 1988). Furthermore, the existence of other cell types (T_H3?) which would produce "mixed" lymphokine activities between those of T_H1 and T_H2 cannot be ruled out yet. Future experiments should focus on demonstrating in vitro IL-4 effects in vivo, either by the administration of mouse IL-4 or its antibody, or by the use of transgenic animals.

II. Effects of IL-4 on B Cells

The effects of IL-4 on B cells are among the best characterized, given that these were the first effects described for IL-4. However, even in this case we do not have information regarding the possible mechanism of action of IL-4. In a general comment that applies to all cell lineages, it is most likely that IL-4 shows different effects on different subpopulations of B cells that remain to be characterized.

III. Effects on Other Cell Types

One of the most intriguing effects of IL-4 is on macrophages. The information available so far indicates that IL-4 acts on subpopulations of macrophages. The capacity to induce Ia in macrophages suggests, that IL-4 plays a role in the amplification of immune responses in those cases when the predominant antigen presenting cells are macrophages. During the effector phase of the immune response, IL-4 may act as a macrophage activating factor. This aspect of IL-4 function deserves careful study, since the tumoricidal or microbicidal activities of macrophages induced by IFN-γ or IL-4 may determine the outcome of many infections by intracellular parasites or tumors.

The effects of IL-4 on the hematopoietic cells of the bone marrow are puzzling. In this respect, while it may act as a potentiator or inhibitor of the development of several cell lineages, the key element may be its production or availability in situ. Thus, it may be interesting to evaluate the production of IL-4 by bone marrow cells, or by other infiltrating cells.

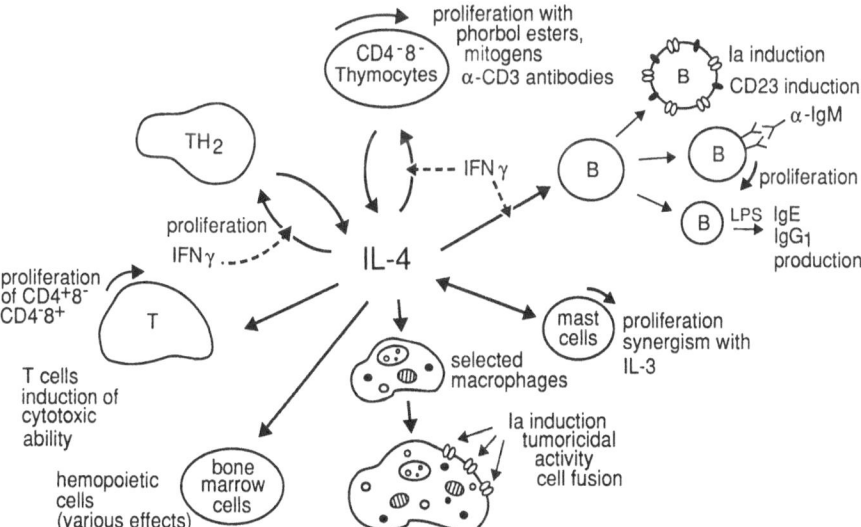

Fig. 6. Summary of the biological activities of IL-4. The *dashed lines* indicate inhibition by IFN-γ

Perhaps one of the most interesting effects of IL-4 is in the developing thymus. In this case, early immature thymocytes have been shown to be able to produce IL-4 and respond to IL-4. These observations strongly indicate a role for IL-4 in the development of the fetal thymus. However, we do not yet know what signal(s) are necessary to activate fetal thymocytes to produce IL-4.

IV. Conclusions

There is no doubt that much progress has been made in our understanding of the role and function of IL-4 in immunity. As shown in Fig. 6, we now know many things about IL-4, and its new name reflects the many actions of this lymphokine. It is both a proliferation and differentiation factor for many cell lineages and acts in concert with many other lymphokines. Thus, while progress has been substantial, the challenge for the future remains to elucidate more networks of lymphokine effects and, more importantly, to correlate them with their role in vivo, with the eventual hope of applying this knowledge to specific clinical problems.

Acknowledgments. The authors would like to thank many colleagues at DNAX and UNICET for communicating their results.

References

Arai K, Yokota T, Miyajima A, Arai N, Lee F (1986) Molecular biology of T-cell-derived lymphokines: a model system for proliferation and differentiation of hemopoietic cells. Bioessays 5:166–171

Arai N, Nomura D, Villaret D, De Waal Malefijt R, Seiki M, Yoshida M, Minoshima S, Fukuyama R, Maekawa M, Kudoh J, Shimizu N, Yokota K, Abe E, Yokota T, Takebe Y, Arai K (1989) Complete nucleotide sequence of the chromosomal gene for human interleukin 4 and its expression. J Immunol 142:274–282

Azuma C, Tanabe T, Konishi M, Kinashi T, Noma T, Matsuda F, Yaoita Y, Takatsu K, Hammarstrom L, Smith CIE, Severinso E, Honjo T (1986) Cloning of cDNA for human T-cell replacing factor (interleukin-5) and comparison with the murine homologue. Nucleic Acids Res 14:9149–9158

Barlow DP, Bucan M, Lehrach H, Hogan BLM, Gough NM (1987) Close genetic and physical linkage between the murine haemopoietic growth factor genes GM-CSF and multi-CSF (IL-3). EMBO J 6:617–623

Bond MW, Schrader B, Mosmann TR, Coffman RL (1987) A mouse T cell product that preferentially enhances IgA production. II. Purification and partial amino acid sequence. J Immunol 139:3691–3696

Bonnefoy JT, DeFrance C, Peronne C, Menetrier F, Rousset J, Pene J, DeVries J, Banchereau J (1988) Human recombinant interleukin 4 induces normal B cells to produce soluble CD23/IgE-binding factor analogous to that spontaneously released by lymphoblastoid B cell lines. Eur J Immunol 18:117–122

Brown M, Pierce J, Watson C, Falco J, Ihle J, Paul W (1987) B cell stimulatory factor 1/interleukin 4 mRNA is expressed by normal and transformed mast cells. Cell 50:809–818

Brown M, Hu-li J, Paul W (1988) IL-4/B cell stimulatory factor 1 stimulates T cell growth by an IL-2-independent mechanism. J Immunol 141:504–511

Cabrillat H, Galizzi J, Djossou OB, Arai N, Yokota T, Arai K, Banchereau J (1987) High affinity binding of human interleukin 4 to cell lines. Biochem Biophys Res Commun 149:995–1001

Campbell HD, Ymer S, Fung M, Young IG (1985) Cloning and nucleotide sequence of the murine interleukin-3 gene. Eur J Biochem 150:297–304

Chen W, Fischer M, Frank G, Zlotnik (1989) Distinct patterns of lymphokine requirement for the proliferation of various subpopulations of activated thymocytes in a single-cell assay. J Immunol 143:1598–1605

Coffman RL, Carty J (1986) A T-cell activity that enhances polyclonal IgE production and its inhibition by interferon-γ. J Immunol 136:949–954

Coffman RL, Ohara J, Bond MW, Carty J, Zlotnik A, Paul WE (1986) B cell stimulatory factor-1 enhances the IgE response of lipopolisaccharide-activated B cells. J Immunol 136:4538–4541

Coffman RL, Schrader B, Carty J, Mosmann TR, Bond MW (1987) A mouse T cell product that preferentially enhances IgA production. I. Biological characterization. J Immunol 139:3685–3690

Coffman RL, Seymour BWP, Lebman DA, Hiraki DD, Christiansen JA, Shrader B, Cherwinski HM, Savelkoul HFJ, Finkelman FD, Bond MW, Mosmann TR (1988) The role of helper T cell products in mouse B cell differentiation and isotype regulation. Immunol Rev 102:5–28

Cohen DR, Hapel AJ, Young IG (1986) Cloning and expression of the rat interleukin-3 gene. Nucleic Acids Res 14:3641–3658

Collins JM, Justement LB, Stedman KE, Zlotnik A, Campbell PA (1988) BSF-1 induces lobster agglutinin 1-separated mouse thymocytes to express CTL activity. J Immunol 141:145–150

Crawford R, Finbloom M, Ohara J, Paul W, Meltzer M (1987) B cell stimulatory factor 1 (interleukin 4) activates macrophages for increased tumoricidal activity and expression of Ia antigens. J Immunol 139:135–141

Defrance T, Aubry J, Rousset F, Vanderbilt B, Bonnefoy Y, Arai N, Takebe Y, Yokota T, Lee F, Arai K, DeVries J, Banchereau J (1987a) Human recombinant interleukin 4 induces Fcε receptors (CD23) on normal B lymphocytes. J Exp Med 165:1459–1467

Defrance T, Vanbervliet B, Aubry J-P, Takebe Y, Arai N, Miyajima A, Yokota T, Lee F, Arai K, de Vries JE, Banchereau J (1987b) B cell growth-promoting activity of recombinant human interleukin 4. J Immunol 139:1135–1141

Del Prete G, Maggi E, Parronchi P, Chetrien I, Tiri A, Macchia D, Ricci M, Banchereau J, de Vries J, Romagnani S (1988) IL-4 is an essential factor the IgE synthesis induced in vitro by human T cell clones and their supernatants. J Immunol 140:4193–4198

Farrar J, Howard M, Fuller-Farrar J, Paul W (1983) Biochemical and physicochemical characterization of mouse B cell growth factor: A lymphokine distinct from interleukin 2. J Immunol 131:1838–1842

Finkelman F, Katona I, Mosmann T, Coffman R (1988) IFNγ regulates the isotypes of Ig-secreted during in vivo humoral immune responses. J Immunol 140:1022–1027

Fujita T, Takaoka C, Matsui H, Taniguchi T (1983) Structure of the human interleukin 2 gene. Proc Natl Acad Sci USA 80:7437–7441

Fuse A, Fujita T, Yasumitsu H, Kashima N, Hasegawa K, Taniguchi T (1984) Organization and structure of the mouse interleukin-2 gene. Nucleic Acids Res 12:9323–9331

Gajewski T, Fitch F (1988) Anti-proliferative effect of IFNγ in immune refulation. I. IFNγ inhibits the proliferation of Th2 but not Th1 murine helper T lymphocyte clones. J Immunol 140:4245–4252

Galindo B, Lazdins J, Castillo R (1974) Fusion of normal rabbit alveolar macrophages induced by supernatant fluids from BCG-sensitized lymph node cells after elicitation by antigen. Infect Immun 9:212–216

Gause W, Takashi T, Mountz J, Finkelman F, Steinberg A (1988) Two independent pathways of T cell activation and differentiation (abstract). FASEB J 2:454

Ghrayeb J, Kimura H, Takahara M, Hsing H, Masui Y, Inoue M (1984) Secretion vectors in E. coli. EMBO J 3:2437–2442

Grabstein K, Eisenman J, Mochizuki D, Shanebeck K, Conlon P, Hopp T, March C, Gillis S (1986) Purification to homogeneity of B cell stimulating factor. J Exp Med 163:1405–1414

Grabstein K, Park L, Morrisey P, Sassenfield H, Price V, Urdal D, Widmer M (1987) Regulation of murine T cell proliferation by B cell stimulatory factor 1. J Immunol 139:1148–1153

Groffen J, Heisterkamp N, Spurr N, Dana S, Wasmuth JJ, Stephenson JR (1983) Chromosomal localization of the human c-fms oncogene. Nucleic Acids Res 11:6331–6339

Guidos C, Ransom J, Fischer M, Weissman I, Zlotnik A (1989) Change in cell surface phenotype and lymphokine production of immature thymocytes after culture with interleukin 4 and phorbol ester. J Autoimmun 2:141–153

Hagiwara H, Yokota T, Luh J, Lee F, Arai K, Arai N, Zlotnik A (1988) Reconstitution of inducible lymphokine production in BW5147-derived T cell hybridomas: evidence that the AKR thymoma BW5147 is able to produce lymphokines. J Immunol 140:1561–1565

Hamada H, Seidman M, Howard BH, Gorman CM (1984) Enhanced gene expression by the poly(dT-dG)·poly(dC-dA) sequence. Mol Cell Biol 4:2622–2630

Howard M, Farrar J, Hilfiker M, Johnson B, Takatsu K, Hamaoka T, Paul WE (1982) Identification of a T-cell derived B cell growth factor distinct from interleukin-2. J Exp Med 155:914–923

Hudak S, Gollnick S, Conrad D, Kehry M (1987) Murine B cell stimulatory factor 1 (interleukin 4) increases expression of the Fc receptor for IgE in B cells. Proc Natl Acad Sci USA 84:4606–4610

Huebner K, Isono M, Croce CM, Golde DW, Kaufman SE, Gasson JC (1985) The human gene encoding GM-CSF is at 5q21-q32, the chromosomal region deleted in the 5q⁻ anomaly. Science 230:1282–1285

Hu-Li J, Shevach E, Mizuguchi J, Ohara J, Mosmann T, Paul W (1987) B cell stimulatory factor 1 (interleukin 4) is a potent costimulant for normal resting T cells. J Exp Med 165:157–172

Jabara H, Ackerman SJ, Vercelli D, Yokota T, Arai K, Abrams J, Dvorak AM, Lavigne MC, Banchereau J, de Vries J, Leung DYM, Geha RS (1988) Induction of interleukin-4-dependent IgE synthesis and eosinophil differentiation by supernatants of a human interleukin-5-dependent helper T cell clone. J Clin Immunol 8:437–446

Justement L, Chen Z, Harris L, Ransom J, Sandoval V, Smith C, Rennick D, Roehm N, Cambier J (1986) Transmembrane signalling by BSF-1 receptors. J Immunol 137:3664–3670

Kikutani H, Inui S, Sato R, Barsumian E, Owaki H, Yamasaki K, Kalaho T, Uchibayashi N, Hardy R, Hirano T, Taunasawa S, Sakiyama S, Suemura M, Kishimoto T (1986) Molecular structure of human lymphocyte receptor for immunoglobulin. Cell 47:657–665

Kinashi T, Harada N, Severinson E, Tanabe T, Sideras P, Konish M, Azuma C, Tominaga A, Bergstedt-Lindquist S, Takahashi E, Matsuda F, Yaoita Y, Takatsu K, Honjo T (1986) Cloning of complementary DNA encoding T-cell replacing factor and identity with B-cell growth factor II. Nature 324:70–73

Koyasu S, Tojo A, Miyajima A, Akiyama T, Kasuga M, Urabe A, Schreurs J, Arai K, Takaku F, Yahara I (1987) Interleukin-3-specific tyrosine phosphorylation of a membrane glycoprotein of Mr 150,000 in multi-factor dependent myeloid cell lines. EMBO J 6:3979–3984

Le HV, Ramanathan L, Labdon JE, Mays-Ichinco CA, Syto R, Arai N, Hoy P, Takebe Y, Nagabhushan TL, Trotta PP (1988) Isolation and characterization of multiple variants of recombinant human interleukin 4 expressed in mammalian cells. J Biol Chem 263:10817–10823

Le Beau MM, Pettenati MJ, Lemons RS, Diaz MO, Westbrook CA, Larson RA, Sherr CJ, Rowley JD (1986a) Assignment of the GM-CSF CSF-1 and FMS genes to human chromosome 5 provides evidence for linkage of a family of genes regulating hematopoiesis and for their involvement in the deletion (5q) in myeloid disorders. Cold Spring Harbor Symp Quant Biol 51:899–909

Le Beau MM, Westbrook CA, Diaz MO, Larson RA, Rowley JD, Gasson JC, Golde DW, Sherr CJ (1986b) Evidence for the involvement of GM-CSF and FMS in the deletion (5q) in myeloid disorders. Science 231:984–987

Le Beau MM, Epstein ND, O'Brien SJ, Nienhuis AW, Yang Y-C, Clark SC, Rowley JD (1987) The interleukin-3 gene is located on human chromosome 5 and is deleted in myeloid leukemias with a deletion of 5q. Proc Natl Acad Sci USA 84:5913–5917

Le Beau MM, Lemons ERS, Espinosa R III, Larson RA, Arai N, Rowley JD (1988) IL-4 and IL-5 map to human chromosome 5 in a region encoding growth factors and receptors and are deleted in myeloid leukemias with a del(5q). Blood 73:647–650

Lebman DA, Coffman RL (1988) Interleukin 4 causes isotype switching to IgE in T cell-stimulated clonal B cell cultures. J Exp Med 168:853–862

Lee F, Yokota T, Otsuka T, Meyerson P, Villaret D, Coffman R, Mosmann T, Rennick D, Roehm N, Smith C, Zlotnik A, Arai K (1986) Isolation and characterization of a mouse interleukin cDNA clone that expresses B-cell stimulatory factor 1 activities and T-cell- and mast-cell-stimulating activities. Proc Natl Acad Sci USA 83:2061–2065

Lee F, Yokota T, Otsuka T, Arai K (1987) Molecular cloning of a mouse T cell lymphokine with T cell, B cell, and mast cell stimulatory activities. In: Goldstein G, Bach J, Wigzell (eds) Immune regulation by characterized polypeptides. Liss, New York, p 397

Lowenthal J, Castle B, Schreurs J, Rennick D, Arai N, Hoy P, Takebe Y, Howard M (1988a) Expression of high affinity receptors for interleukin 4 on hemopoietic and non-hemopoietic cells. J Immunol 140:456–464

Lowenthal JW, Ransom J, Howard M, Zlotnik A (1988b) Upregulation of interleukin-4 receptor expression in immature (L3T4-/LyT2-) thymocytes. J Immunol 140:474–478

McInnes A, Rennick D (1988) Interleukin 4 induces cultured monocytes/macrophages to form giant multinucleated cells. J Exp Med 167:598–611

Metcalf D (1985) The granulocyte-macrophage colony-stimulating factors. Science 229:16–22

Miyajima A, Bond M, Otsu K, Arai K, Arai N (1985) Secretion of mature mouse interleukin-2 by Sacharomyces cerevisiae: use of a general secretion vector containing promoter and leader sequences of the mating pheromone α-factor. Gene 37:155–160

Miyajima A, Otsu K, Schreurs J, Bond M, Abrams J, Arai J, Arai K (1986) Expression of murine and human granulocyte-macrophage colony-stimulating factors in S. cerevisiae: mutagenesis of the potential glycosylation sites. EMBO J 5:1193–1197

Miyajima A, Schreurs J, Otsu K, Kondo A, Arai K, Maeda S (1987) Use of the silkworm, Bombyx mori, and in insect baculovirus vector for high-level expression and secretion of biologically active mouse interleukin-3. Gene 58:273–281

Miyajima A, Miyatake S, Schreurs J, de Vries J, Arai N, Yokota T, Arai K (1988) Coordinate regulation of immune and inflammatory responses by T cell-derived lymphokines. FASEB J 2:2462–2473

Miyatake S, Otsuka T, Yokota T, Lee F, Arai K (1985a) Structure of the chromosomal gene for granulocyte-macrophage colony stimulating factor: comparison of the mouse and human genes. EMBO J 4:2561–2568

Miyatake S, Yokota T, Lee F, Arai K (1985b) Structure of the chromosomal gene for murine interleukin 3. Proc Natl Acad Sci USA 82:316–320

Mond J, Finkelman F, Sarma G, Ohara J, Serrate S (1985) Recombinant interferon γ inhibits the proliferative response stimulated by soluble but not by sepharose-bound anti-immunoglobulin antibody. J Immunol 135:2513–2517

Mond J, Carman J, Sarma G, Ohara J, Finkelman F (1986) Interferon γ suppresses B cell stimulation factor 1 induction of class II MHC determinants on B cells. J Immunol 137:3534–3537

Morla AO, Schreurs J, Miyajima A, Wang JYJ (1988) Hematopoietic growth factors activate the tyrosine phosphorylation of distinct sets of proteins in interleukin-3-dependent murine cell lines. Mol Cell Biol 8:2214–2218

Mosmann TR, Coffman RL (1987) Two types of mouse helper T cell clone. Immunol Today 8:223–227

Mosmann TR, Bond MW, Coffman RL, Ohara J, Paul WE (1986a) T-cell and mast cell lines respond to B-cell stimulatory factor 1. Proc Natl Acad Sci USA 83:5654–5658

Mosmann TR, Cherwinski H, Bond MW, Giedlin MA, Coffman RL (1986b) Two types of murine helper T cell clone. I. Definition according to profiles of lymphokine activities and secreted proteins. J Immunol 126:2348–2357

Mosmann T, Yokota T, Kastelein R, Zurawski S, Arai N, Takebe Y (1987) Species-specificity of T cell stimulating activities of IL-2 and BSF-1 (IL-4): comparison of normal and recombinant mouse and human IL-2 and BSF-1 (IL-4). J Immunol 138:1813–1816

Mule J, Smith C, Rosenberg S (1987) Interleukin 4 (B cell stimulatory factor 1) can mediate the induction of lymphokine activated killer cell activity directed against fresh tumor cells. J Exp Med 166:792–797

Nabel G, Fresno M, Chessman A, Cantor H (1981a) Use of cloned populations of mouse lymphocytes to analyze cellular differentiation. Cell 23:19–28

Nabel G, Galli SJ, Dvorak AM, Dvorak HF, Cantor H (1981b) Inducer T lymphocytes synthesize a factor that stimulates proliferation of cloned mast cells. Nature 291:332–334

Nabel G, Greenberger JS, Sakakeeny MA, Cantor H (1981c) Multiple biologic activities of a cloned inducer T-cell population. Proc Natl Acad Sci USA 78:1157–1161

Nakajima K, Hirano T, Koyama K, Kishimoto T (1987) Detection of receptors for murine B cell stimulatory factor 1 (BSF-1): presence of functional receptors on CBA/N splenic B cells. J Immunol 139:774–779

Nienhuis AW, Bunn HF, Turner PH, Gopal YV, Nash WG, O'Brien SJ, Sherr CJ (1985) Expression of the human c-fms proto-oncogene in hematopoietic cells and its deletion in the 5q⁻ syndrome. Cell 42:421–428

Noelle R, Krammer PH, Ohara J, Uhr JW, Vitetta ES (1984) Increased expression of Ia antigens on resting B cells: a new role for B cell growth factor. Proc Natl Acad Sci USA 81:6149–6153

Noma Y, Sideras P, Naito T, Bergstedt-Lindquist S, Azuma C, Severinson E, Tanabe T, Kinashi T, Matsuda F, Yaoita Y, Honjo T (1986) Molecular cloning of cDNA encoding the murine IgG1 induction factor by a novel strategy using SP6 promoter. Nature 319:640–656

O'Garra A, Barbis D, Harada N, Lee F, Howard M (1989) Constitutive production of lymphokines by cloned murine B-cell lymphomas – CM12 B lymphoma produces interleukin-4. J Mol Cell Immunol 4:149–159

Ohara J, Paul WE (1985) Production of a monoclonal antibody to and molecular characterization of B-cell stimulatory factor-1. Nature 315:333–336

Ohara J, Paul W (1987) Receptors for B cell stimulatory factor 1 expressed on cells of the hemopoietic lineage. Nature 325:537–540

Ohara J, Lahet S, Inman J, Paul WE (1985) Partial purification of murine B cell stimulatory factor-1. J Immunol 135:2518–2524

Ohara J, Coligan J, Zoon K, Maloy W, Paul W (1987) Rapid purification, N-terminal sequencing, and chemical characterization of mouse B cell stimulatory factor 1/interleukin 4. J Immunol 139:1127–1134

Okayama H, Berg P (1983) A cDNA cloning vector that permits expression of cDNA inserts in mammalian cells. Mol Cell Biol 3:280–289

Otsuka T, Villaret D, Yokota T, Takebe Y, Lee F, Arai N, Arai K (1987) Structural analysis of the mouse chromosomal gene encoding interleukin 4 which expresses B cell T cell and mast cell stimulating activities. Nucleic Acids Res 15:333–344

Palacios R, Sideras P, von Boehmer H (1987) Recombinant interleukin 4/BSF-1 promotes growth and differentiation of intrathymic T cell precursors from fetal mice in vitro. EMBO J 6:91–95

Paliard X, de Waal Malefijt R, Yssel H, Blanchard D, Abrams J, de Vries JE, Spits H (1988) Simultaneous production of IL-2, IL-4 and IFN-γ by activated human CD4$^+$ and CD8$^+$ T cell clones. J Immunol 141:849–855

Park L, Friend D, Grabstein K, Urdal D (1987a) Characterization of the high-affinity cell-surface receptor for murine B cell stimulating factor. Proc Natl Acad Sci USA 84:1669–1673

Park L, Friend D, Sassenfeld H, Urdal D (1987b) Characterization of the human B cell stimulatory factor 1 receptor. J Exp Med 166:476–488

Paterson DJ, Williams AF (1987) An intermediate cell in thymocyte differentiation that expresses CD8 but not CD4 antigen. J Exp Med 166:1603–1608

Pene J, Rousset F, Briere F, Chretien I, Bonnefoy JY, Spits H, Yokota T, Arai N, Arai K, Banchereau J, de Vries JE (1988) IgE production by normal human lymphocytes is induced by interleukin 4 and suppressed by interferon γ and α and prostaglandin E$_2$. Proc Natl Acad Sci USA 85:6880–6884

Perlman D, Halvorson HO (1983) A putative signal peptidase recognition site and sequence in eukaryotic and prokaryotic signal peptides. J Mol Biol 167:391–409

Peschel C, Paul W, Ohara J, Green I (1987) Effects of B cell stimulatory factor 1/interleukin 4 on hemopoietic progenitor cells. Blood 70:254–263

Rabin E, Ohara J, Paul W (1985) B cell stimulatory factor 1 (BSF-1) activates resting B cells. Proc Natl Acad Sci USA 82:2935–2939

Rabin E, Mond J, Ohara J, Paul W (1986) B cell stimulatory factor 1 (BSF-1) prepares resting B cells to enter S phase in response to anti-IgM and lipopolisaccharide. J Exp Med 164:517–531

Ramanathan L, Le HV, Labdon JE, Mays-Ichinco CA, Syto R, Arai N, Nagabhushan TL, Trotta PP (1988) Multiple forms of recombinant murine interleukin 4 expressed in COS-7 monkey kidney cells. Biochim Biophys Acta 107:283–288

Ransom J, Fischer M, Mercer L, Zlotnik A (1987a) Lymphokine-mediated induction of antigen-presenting ability in thymic stromal cells. J Immunol 139:2620–2628

Ransom J, Fischer M, Mosmann T, Yokota T, DeLuca D, Schumacher J, Zlotnik A (1987b) Interferon gamma is produced by activated immature thymocytes and inhibits the interleukin-4-induced proliferation of immature thymocytes. J Immunol 139:4102–4108

Rennick DM, Lee FD, Yokota T, Arai K, Cantor H, Nabel G (1985) A cloned MCGF cDNA encodes a multilineage hematopoietic growth factor: Multiple activities of interleukin-3. J Immunol 134:910–914

Rennick D, Yang G, Muller-Sieburg C, Smith C, Arai N, Takebe Y, Gemmell L (1987) Interleukin 4 (B cell stimulatory factor 1) can enhance or antagonize the factor-dependent growth of hemopoietic progenitor cells. Proc Natl Acad Sci USA 84:6889–6893

Roehm NW, Leibson HJ, Zlotnik A, Kappler JW, Marrack P, Cambier JC (1984) Interleukin-induced increase in Ia expression by normal mouse B cells. J Exp Med 160:679–694

Roehm NW, Leibson HJ, Marrack P, Cambier JC, Kappler JE, Rennick DM, Zlotnik A (1985) B lymphocyte activation. In: Sorg C, Schimpl A (eds) Cellular and molecular biology of lymphokines. Academic, New York, p 195

Rothenberg E, Lugo JP (1985) Differentiation and cell division in the mammalian thymus. Dev Biol 112:1–17

Roussel MF, Sherr CJ, Barker PE, Ruddle FH (1983) Molecular cloning of the c-fms locus and its assignment to human chromosome 5. J Virol 48:770–773

Saito H, Hatake K, Dvork AM, Leiferman KM, Donnenberg AD, Arai N, Ishizaka K, Ishizaka T (1988) Selective differentiation and proliferation of hematopoietic cells induced by recombinant human interleukins. Proc Natl Acad Sci USA 85:2288–2292

Severinson E, Naito T, Takumoto H, Fukushima D, Hirano A, Hama K, Honjo T (1987) Interleukin 4 a multifunctional lymphokine activity acting also on T cells. Eur J Immunol 17:67–72

Sideras P, Palacios R (1987) Bone marrow pro-T and pro-B lymphocyte clones express functional receptors for interleukin 3 and IL-4/BSF-1 and nonfunctional receptors for IL-2. Eur J Immunol 17:217–221

Sideras P, Funa K, Zalcberg-Quintana I, Xanthopoulos K, Kisielow P, Palacios R (1988) Analysis by in situ hybridization of cells expressing mRNA for interleukin 4 in the developing thymus and in peripheral lymphocytes from mice. Proc Natl Acad Sci USA 85:218–221

Smith CA, Rennick DM (1986) Characterization of a murine lymphokine distinct from interleukin-2 and interleukin-3 possessing a T cell growth factor activity and a mast cell growth factor that synergizes with IL-3. Proc Natl Acad Sci USA 83:1857–1861

Snapper C, Paul W (1987) Interferon γ and B cell stimulatory factor-1 reciprocally regulate Ig isotype production. Science 236:944–947

Spits H, Yssel H, Takebe Y, Arai N, Yokota T, Lee F, Arai K, Banchereau J, de Vries J (1987) Recombinant IL-4 promotes the growth of human T cells. J Immunol 139:1142–1147

Spits H, Yssel H, Paliard X, Kastelein R, Figdor C, de Vries J (1988) IL-4 inhibits IL-2-mediated induction of human lymphokine-activated killer cells but not the generation of antigen-specific cytotoxic T lymphocytes in mixed leukocyte cultures. J Immunol 141:29–36

Stanley E, Metcalf D, Sobieszczuk P, Gough NM, Dunn AR (1985) The structure and expression of the murine gene encoding granulocyte-macrophage colony stimulating factor: evidence for utilization of alternative promoters. EMBO J 4:2569–2573

Stuart P, Zlotnik A, Woodward J (1988) Induction of class I and class II MHC antigen expression on murine bone marrow-derived macrophages by IL-4 (B cell stimulatory factor 1). J Immunol 140:1542–1547

Sutherland GR, Baker E, Callen DF, Campbell HD, Young IG, Sanderson CJ, Garson OM, Lopez AF, Vadas MA (1988) Interleukin 5 is at 5q31 and is deleted in the 5q⁻ syndrome. Blood 71:1150–1152

Swain S, Howard M, Kappler J, Marrack P, Watson J, Booth M, Wetzel D, Dutton R (1983) Evidence for two distinct classes of murine B cell growth factors with activities in different functional assays. J Exp Med 158:822–835

Tadokoro K, Kohama H, Ohtoshi T, Takafumi S, Suzuki S, Tange T, Yokota T, Miyamoto T (1989) Human recombinant interleukin-4 enhances the in vitro growth of basophils induced by human recombinant interleukin-3. (submitted)

Takagi M, Nakahata T, Kobayashi T, Tsuji K, Koike K, Kojima S, Hirano T, Miyajima A, Arai K, Akabane T (1989) Stimulation of connective tissue-type mast cell proliferation by cross-linking of Fcε receptors. J Exp Med 170:233–244

Takei F (1988) IL-2 and IL-4 stimulate different subpopulations of double-negative thymocytes. J Immunol 141:1114–1119

Tentori L, Pardoll D, Zuniga J, Hu-Li J, Paul W, Bluestone J, Kruisbeek A (1988) Proliferation and production of IL-2 and IL-4 in fetal thymocytes by activation through Thy-1 and CD3. J Immunol 140:1089–1094

te Velde A, Klomp J, Yard B, de Vries J, Figdor C (1988) Modulation of phenotypic and functional properties of human peripheral blood monocytes by IL-4. J Immunol 140:1548–1554

Umetsu D, Jabara H, DeKruyff R, Abbas A, Abrams J, Geha R (1988) Functional heterogeneity among human inducer T cell clones. J Immunol 140:4211–4216

van Kimmenade S, Bond MW, Schmacher JH, Laquoi C, Kasteline RA (1988) Expression, renaturation and purification of recombinant human interleukin 4 from E. coli. Eur J Biochem 173:109–114

Vitetta ES, Ohara J, Myers CD, Layton JE, Krammer PH, Paul WE (1985) Serological biochemical and functional identity of B cell-stimulatory factor 1 and B cell differentiation factor for IgG1. J Exp Med 162:1726–1731

Von Boehmer H, Crisanti A, Kisielow P, Haas W (1985) Absence of growth by most receptor-expressing fetal thymocytes in the presence of interleukin 2. Nature 314:539–540

von Heijne G (1983) Patterns of amino acids near signal-sequence cleavage sites. Eur J Biochem 133:17–21

Watson J (1979) Continuous proliferation of murine antigen-specific helper T lymphocytes in culture. J Exp Med 150:1510–1519

Weinstein Y, Ihle JN, Lavu S, Reddy EP (1986) Truncation of the c-myb gene by a retroviral integration in an interleukin 3-dependent myeloid leukemia cell line. Proc Natl Acad Sci USA 83:5010–5014

Widmer M, Grabstein K (1987) Regulation of cytolytic T-lymphocyte generation by B cell stimulatory factor. Nature 326:795–798

Yokota T, Lee F, Rennick D, Hall C, Arai N, Mosmann T, Nabel G, Cantor H, Arai K (1984) Isolation and characterization of a mouse cDNA clone that expresses mast-cell growth-factor activity in monkey cells. Proc Natl Acad Sci USA 81:1070–1074

Yokota T, Otsuka T, Mosmann T, Banchereau J, Defrance T, Blanchard D, de Vries JE, Lee F, Arai K (1986) Isolation and characterization of a human interleukin cDNA clone homologous to mouse B-cell stimulatory factor 1 that expresses B-cell- and T-cell-stimulatory activities. Proc Natl Acad Sci USA 83:5894–5898

Yokota T, Coffman RL, Hagiwara H, Rennick DM, Takebe Y, Yokota K, Gemmell L, Shrader B, Yang G, Meyerson P, Luh J, Hoy P, Pene J, Briere F, Spits H, Banchereau J, de Vries J, Lee FD, Arai N, Arai K (1987) Isolation and characterization of lymphokine cDNA clones encoding mouse and human IgA-enhancing factor and eosinophil colony-stimulating factor activities: relationship to interleukin 5. Proc Natl Acad Sci USA 84:7388–7392

Yokota T, Arai N, DeVries J, Spits H, Banchereau J, Zlotnik A, Rennick D, Howard M, Takebe Y, Miyatake S, Lee F, Arai K (1988) Molecular biology of interleukin 4 and interleukin 5 genes and biology of their products that stimulate B cells T cells and hemopoietic cells. Immunol Rev 102:137–187

Zlotnik A, Shimonkevitz RP, Gefter ML, Kappler J, Marrack P (1983) Characterization of the Gamma-interferon-mediated induction of antigen-presenting ability in P388D1 cells. J Immunol 131:2814–2820

Zlotnik A, Fischer M, Roehm N, Zipori D (1987a) Evidence for effects of Interleukin 4 (B cell stimulatory factor 1) on macrophages: Enhancement of antigen presenting ability of bone-marrow-derived macrophages. J Immunol 138:4275–4279

Zlotnik A, Ransom J, Frank G, Fischerand M, Howard M (1987b) Interleukin 4 is a growth factor for activated thymocytes: Possible role in T cell ontogeny. Proc Natl Acad Sci USA 84:3856–3860

CHAPTER 13

Interleukin-5

T. Honjo and K. Takatsu

A. Introduction

The immune response to a foreign antigen is regulated by a series of interactions between T lymphocytes (T cells), B lymphocytes (B cells), and macrophages. During this process B cells proliferate and mature into plasma cells which produce antibodies against distinct determinants of the antigen, and the antibodies produced play a key role in the humoral immune response. The B-cell response to an antigen is regulated by a helper T cell responding to, and specific for, the same antigen molecule. Helper T cells recognize antigens in the context of class II major histocompatibility complex (MHC) molecules on accessory cells and/or B cells, and secrete several soluble factors including interleukin-4 (IL-4) and IL-5 which can induce growth and maturation of B cells (reviewed by Howard and Paul 1983; Kishimoto 1985). IL-5 was initially described as a factor that induces terminal differentiation of B cells to immunoglobulin (Ig)-secreting cells, and originally designated as T-cell-replacing factor (TRF; Dutton and Swain 1987; Takatsu 1988; Dutton et al. 1971; Schimpl and Wecker 1972).

The establishment of a TRF-producing T-cell hybrid B151K12 (B151) which does not secrete detectable levels of other lymphokines affecting B cells demonstrated that IL-5 is a lymphokine distinct from IL-1, IL-2, IL-3, IL-4, and interferon-γ (IFN-γ; Takatsu et al. 1980a). IL-5 produced by B151 cells was purified to homogeneity and shown to be an acidic glycoprotein with a molecular mass of 50–60 kDa on gel permeation chromatography (Takatsu et al. 1985). IL-5 was subsequently shown to promote growth of dextran sulfate-stimulated normal B cells or a murine leukemic B-cell line BCL_1, which is known as B-cell growth factor (BCGF) II activity (Harada et al. 1985; Swain 1985).

The recent molecular cloning of complementary DNA (cDNA) encoding murine and human IL-5 has convincingly demonstrated that a single molecule is responsible for both TRF and BCGF II activities (Kinashi et al. 1986; Azuma et al. 1986). Although IL-5 was initially believed to be principally active on B cells, IL-5 has been shown to increase the expression of the IL-2 receptor on antigen-stimulated thymocytes, inducing them into cytotoxic T lymphocytes (CTL) in the presence of IL-2, which is known as killer helper factor activity (Takatsu et al. 1987). IL-5 also induces the terminal differentiation of hematopoietic progenitor cells into eosinophils (Sanderson et al. 1986, 1988).

Studies on IL-5 have provided strong evidence that a single lymphokine exerts a variety of activities on diverse target cells. Together with similar properties of IL-4, these results helped many immunologists escape from earlier beliefs that one lymphokine is required for each activity and that growth factor activity and differentiation factor activity are necessarily different molecular entities.

In this review, we will first briefly summarize the history of the initial discovery of the IL-5 activity that appeared to be involved in B-cell growth and differentiation, and outline the developments that led to the molecular characterization of IL-5. We will also describe the structure and organization of the IL-5 gene. Finally, we will summarize current knowledge about the IL-5 receptor.

B. Historical Background of IL-5

TRF activity was originally assayed by its ability to support the production of IgM against sheep red blood cells (SRBC) by T-cell-depleted normal mouse B cells. Dutton et al. (1971) as well as Schimpl and Wecker (1972) demonstrated that supernatants of mixed lymphocyte cultures or concanavalin A (ConA)-stimulated T cells contain TRF activity. These TRF preparations were found to exert remarkable effects in a number of different assays of T- and B-cell functions (reviewed by Dutton and Swain 1987). Biochemical characterization of TRF took a long time partly because there was a popular (at that time) belief that the different activities should be ascribed to separate biochemical entities and partly because purification of TRF was difficult.

Howard et al. (1982) described the first lymphokine (distinct from IL-2) that had the ability to support B-cell proliferation, and tentatively called it BCGF (later referred to as B-cell stimulatory factor 1, BSF-1). This BCGF synergized with anti-IgM antibodies for inducing DNA synthesis in highly purified resting B cells. Swain and Dutton (1982) reported the second T-cell-derived BCGF (BCGF II) active on murine B cells and distinct from the Howard BCGF. They found that TRF-containing supernatants synergized with IL-2 in in vitro primary synthesis of IgM against sheep erythrocytes and also contained growth promoting activity for BCL_1 and dextran sulfate-stimulated B cells (Swain and Dutton 1982). They called their factor (DL)BCGF. Howard BCGF was shown to have no (DL)BCGF activity, and DL supernatants had no Howard BCGF activity. Such studies suggested the existence of at least two distinct factors affecting B cells, called BCGF I and BCGF II (Swain et al. 1983). Soon after that, Swain (1985) found that their BCGF II could induce Ig secretion from activated B cells and that this activity was copurified with the proliferative activity in a variety of chromatographic separations.

TRF activity had been originally found in the supernatants of $L3T4^+$ T cells from *Mycobacterium tuberculosis*-primed mice (Takatsu et al. 1974, 1980b). This TRF activity was initially assessed by the ability to induce anti-2,4-dinitrophenol (DNP) IgG antibody-secreting cells from T cell-depleted splenic B cells from DNP/keyhole limpet hemocyanin-primed mice. The establishment of a TRF-producing T-cell hybrid B151K12 (B151) by means of fusion between *M. tuberculosis*-primed T cells and murine thymoma BW5147 provided a useful source for purification of TRF (TAKATSU et al. 1980a). Subsequently, in vivo growing murine chronic B-cell leukemia cells (BCL$_1$) were shown to differentiate into IgM-secreting cells upon stimulation with TRF-containing B151 supernatant (PURE et al. 1981).

A homogeneous TRF preparation purified from B151 supernatants appeared to be an acidic glycoprotein which had a smaller molecular mass under reducing conditions (TAKATSU et al. 1985). This purified TRF preparation contained undetectable levels of IL-1, IL-2, IL-3, IL-4, and IFN-γ activities, but did contain the growth-promoting activity of BCL$_1$ cells as well as dextran sulfate-stimulated normal B cells (BCGF II) (HARADA et al. 1985). The BCGF II activity always resided in the same fractions that contained the TRF activity, suggesting that a single molecule is able to induce growth as well as differentiation of B cells.

C. Structure of IL-5 and Its Gene

I. Cloning of IL-5 cDNA

1. SP6 Expression Vector System

Due to limitations in the amount of IL-5 produced by T cells it was difficult to obtain a large amount of the purified IL-5 by conventional biochemical procedures. Since neither structural characterization of TRF nor a mono-

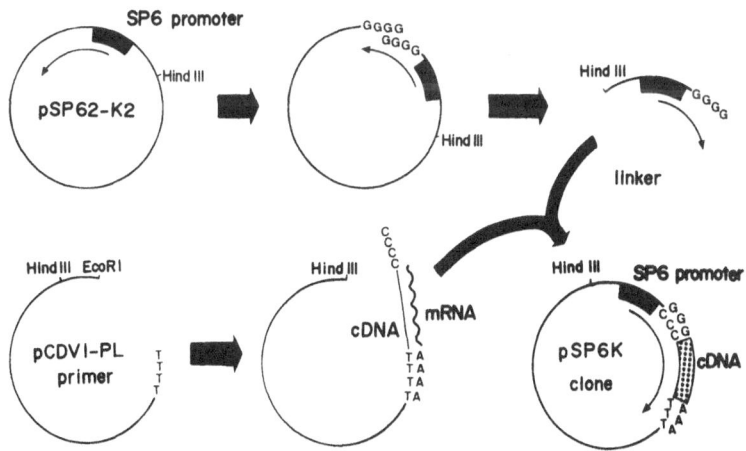

Fig. 1. Strategy for the construction of the SP6 expression vector library. *Filled* and *stippled bars* show the SP6 promoter and double stranded cDNA respectively. *Arrows* indicate the direction of transcription by SP6 polymerase. For further details see Y. NOMA et al. (1986)

clonal antibody against TRF was available at that time, the best strategy was to use an expression vector system that requires only a limited amount of mRNA. We therefore decided to construct a new expression vector containing the SP6 promoter (Y. Noma et al. 1986). The basic strategy of our cDNA library construction is similar to that described by Okayama and Berg (1983) except that the HindIII linker segment containing the SV40 promoter sequence is replaced by another HindIII linker containing the SP6 promoter sequence. To obtain the SP6 promoter HindIII linker, the plasmid pSP62-PL was modified to pSP62-K2. The preparation of the linker and the construction of the cDNA library are shown schematically in Fig. 1. This system was referred to as the pSP6K system. The following differences account for the superiority of the pSP6K system. With the aid of a specific RNA polymerase the pSP6K vector allows the synthesis of up to a few micrograms of mRNA by in vitro transcription. This mRNA is directly injected into *Xenopus* oocytes and the synthesized proteins are tested for biological activity. In this way secretory proteins are synthesized in a system which is mycoplasma-free, free of serum proteins, and highly concentrated. The low proteinase activity in the oocytes makes this system ideal for isolating cDNAs coding for nonsecretory proteins as well. Their activities can be easily analyzed in oocyte homogenates.

The major advantage of the pSP6K system is the high sensitivity that it provides. Using cloned human IL-2 cDNA inserted into the pSP6K vector, 1.6×10^5 U/ml IL-2 activity was obtained in the oocyte supernatant. On the other hand, murine IL-2 cDNA in the $pCDV_1$ vector produced only 6.9×10^3 U/ml IL-2 activity in culture supernatants of COS cells transfected with this clone. In fact, previous cloning with the $pCDV_1$ vector was carried out by pooling 40–50 clones. We could easily detect the activity of TRF by pooling the total cDNA library, consisting of approximately 45000 cDNA clones.

2. Construction of cDNA Libraries

Total RNA was extracted from ConA-activated 2.19 T cells. PolyA$^+$ RNA was obtained by purification with an oligo(dT) cellulose column. PolyA$^+$ RNA was fractionated by sucrose gradient centrifugation. Aliquots of fractionated RNA were microinjected into *Xenopus* oocytes. The oocyte culture supernatants were analyzed for their IgG_1-inducing (IL-4) or TRF (IL-5) activities (Fig. 2). cDNA libraries from polyA + RNA and from the mRNA enriched by the sucrose gradient fractionation were constructed using the pSP6K vector system.

Pools of recombinant plasmid DNA from the SP6 library of 2.19 mRNA were transcribed in vitro into mRNA using SP6 RNA polymerase. mRNA synthesized in vitro was microinjected into *Xenopus* oocytes and the oocytes' culture supernatants were assayed for lymphokine activities. Pools that scored positive in the biological assays were further divided into smaller pools, which were analyzed in the same manner until single cDNA clones capable of directing the synthesis of biologically active lymphokine preparations were ob-

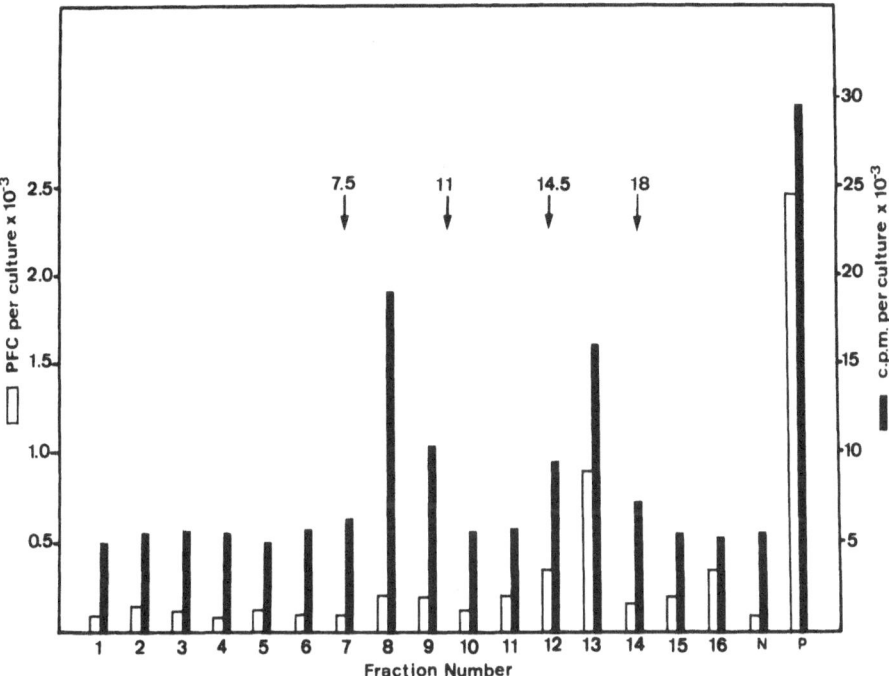

Fig. 2. Effect on BCL$_1$ cells of translational products directed by size-fractionated mRNA of 2.19 T-cell line. PolyA$^+$ RNA was fractionated by centrifugation in a 5%–22% sucrose gradient at 36000 rpm for 15 h. Aliquots from each of the 16 fractions were injected into *Xenopus* oocytes. Incubation media were collected after incubation for 36 h at 20° C and added to the assay medium at a concentration of 1%. *Right-handed scale*, [^{3}H]thymidine incorporation (*filled bars*; cpm per culture $\times 10^{-3}$). *Left-hand scale*, anti-IgM plaque formation response (*open bars*; plaque forming cells (*PFC*) per culture $\times 10^{-3}$ of BCL$_1$ cells). Positive control (*P*) was culture supernatants of B151 hybridoma added to the medium at 50%; negative control (*N*) was culture supernatants from phosphate buffer-injected oocytes. Note that one peak of activity is observed when PFC/culture is analyzed, whereas two peaks of activity (9*S* and 18*S* respectively) are observed when thymidine incorporation is analyzed

tained. pSP6K-mIL4-374 (Y. NOMA et al. 1986) and pSP6K-mTRF (KINASHI et al. 1986) were shown to encode IL-4 and IL-5, respectively. The isolated cDNA clones were subjected to nucleotide sequencing analyses.

3. Structures of the IL-5 cDNA Clone

The nucleotide sequence of the entire IL-5 cDNA was determined by the unidirectional deletion method, using exonucleases III and IV, and the dideoxy method. The IL-5 cDNA codes for a polypeptide chain of 133 residues which contains the N-terminal signal sequence of 20 residues and the secreted core polypeptide with a molecular mass of 12300 kDa (Fig. 3). Three putative N-glycosylation sites as well as three cysteine residues are present in the polypeptide sequence. Homology search between the deduced amino se-

Exon 1

-376 M TAGCCTAACCCTGTTGGAGGTATACATTTGAATACATTTTTTCTCAC----TTTATCAGGAATTGAGTTTAACACATATTAA----AGCAGTGTGGGGCAGGGAGGGGG
-382 H -----TGCCAAGGCTTGGCATTTCCATTTCATTCACTGTCTTCCCACCAGTATTTCA---ATTTCTTTAAGACAGATTAATCTAGCCACAGT----CATAG-------

-285 M GATAAAAAGAAGG--TGCTCAAGAAAAGCCGATCACGCTCCCA-------AGAGTGTGAGCATGGGCGTCTCTAGAGAGATCCGCCATATATG----------CACAAC
-281 H --TAGAACATAGCCGATCTTGAAAAAA-------ACATTCCAATATTTATGTATTTAGCATAAAATTCTGTTTAGTGGTCTACCTTATACTTTGTTTGCACACATC

-187 M TTTTAAGAGAAATT-------CAATAACCAGAATGGAGTGTAAATGTGGATCAAAGTTGTAG-----AAACATTCTTTATGTTATAGAAAATGTCTTTTAAGCAGGGG
-180 H TTTTAAGAGGAAGTTAATTTTCTGATTTTAAGAA----ATGCAAATGTGGGGCAATGAATGATGTATTAACCCAAAGATTCCTTCCGTAATAGAAAATG-TTTTTAAA---GG

-77   M TGGGGGTCAAGATGTTAACTATTATTTAAAGACAAAAAAAAAATGCATTTGTTTGAAGACCCAGGGCACTGGAAACCTGAGTTTCAGGACTGTCGCCTTTATTAGG
-80   H GGGGAAACAGGGATTTT--TATTATTAAAAG---ATAAAAGTAAATTTATTTTTT---AAGATATAAGGCATTGGAAACATTTAGTTTCACGATAT--GCCATTATTAGG

27   M TGTCCTCTATCTGATTGTTAGCAATTATTCATTTCCTCAGAGAGAG--AATAAATGCTTGGGGATTCGCCCTGCTCT-----GCGGCTCTTCCTTGCTGAAGGCCAGCG
27   H CATTCTCTATCTGATTGTTAGAAATTATTCATTTCCTCAAAGACAGAGACAATAAATTGACTGGGGA----CGCAGTCTTGTACTATGCACTTTCTTTGCCAAAGGCAAACG

         M   R   R   M   L   L   H   L   S   V   L   T   L   S   C   V   W   A   T   A   M   E   I   P   M   S   T   V   V
131   M CTGAAGACTTCAGATGCATGAGAAGGATGCTTCTGCACTTGAGTGTTCTGACTCTCAGC-----TGTGTCTGGGCCACTGCCATGGAGATTCCCATGAGCACAGTGGTG
134   H CAGAACGTTTCAGAGCCATG--AGGATGCTTCTGCATTGAGTTTGCTGAGCTGCCTACGTGTATGCCACTGCCATCCCCACAGAAATTCCCACAAGTGCATTGGTG
              M   R   M   L   L   H   L   S   L   A   L   G   A   A   Y   V   Y   A   I   P   T   E   I   P   T   S   A   L   V

         K   E   T   L   Q   L   S   A   H   R   L   L   S   N   E
201   M AAAGAGACCTTGACAGCTGTCCGCCACCAGCTCTGTCTGCTTGACAAGCAATGAGGTAAAGTATAACTTAT
              .   .   .   .
202   H AAAGAGACCTTGGCACTGCTTTCTACTACTCATCGAACTCTGCTGATAGCCAATGAAGGTAATTTT--CTTTAT
         K   E   T   L   A   L   L   S   T   H   R   T   L   L   I   A   N   E

Exon 2

92   M ----ATGACCCTATGATGTTCTTTGCAGACGATGAGGCTTCCTGCTGCCCTACTCATCATAAAAATGTAAGTTATTCTTTACTGCCGTGCTTGCAT
                                     T   M   R   L   P   V   P   T   H   K   N
93   H AAAAATGATTGTAT---TTCCTTTCCTCTCCAGACTCTGAGGATTCCTGCTTCTGTACATAAAAATGTAAGTTAAATTATGATTCAGTAAAATGAT
                                     I   L   R   I   P   V   P   V   H   K   N

exon 3

```
                             H   Q   L   C   I   G   E   I   F   Q   G   L   D   I   L   K   N   Q   T   V   R   G   G   T   V   E   M   L   F
M  TCTTGATAATCTTC---TTTCAGCACCAGCTATGCATTGGAGAAATCTTTCAGGGGCTAGACATACTGAAGAATCAAACTGTCCGTGGGGTACTGTGGAAATGCTATTC      107
       :  :  : :  ::    :::::::  ::: :    ::    ::: :::::: :: :::    : :::    ::: :: :  :::   :: :  :: :::: :::::::   :::::: :
H  -TTTAAAAATTTTCCTCATTTAGCCACCAACTGCACTGAAGAAATCTTTCAGGAATAGGCACACTGGAGAGTCAAACTGTGCAAGGGGTACTGTGGAAAGACTATTC       109
                             H   Q   L   C   T   E   I   F   Q   G   I   G   T   L   E   S   Q   T   V   Q   G   G   T   V   E   R   L   F

                  Q   N   L   S   L   I   K   K   Y   I   D   R   Q   K
M  CAAAACCTGTCATTAATAAAGAAATACATTGACCGCCAAAAAGTAAGTTCCC---                                                         159

H  AAAAACTTGTCCTTAATAAAGAAATACATTGACGGCCAAAAAGTAAGTTACACAC                                                        164
                  K   N   L   S   L   I   K   K   Y   I   D   G   Q   K
```

Exon 4

```
                      E   K   C   G   E   E   R   R   T   R   Q   F   L   D   Y   L   Q   E   F   L   G   V   M   S   T   E
M  ACCTGACAGTC----TGTTCTTTTCACAGAGGAAGTGTGGCGAGGAGACGGAGGCGCAGGCCAGTTCCTGGATTACCTGCAAGAGTTCCTTGGTGTGATGAGTACAGAG      106
       :  :  :::     :::::::::::: :  :::::::  :: :: :: :: ::: ::::: :::: ::::::::: :::::::: ::::  :::: ::: ::  ::  ::
H  ACCT-ATTGTCATTTTTCTTTTTTCACAGAAAAGTGTGGAAGAAGAACGGGAGAGTAAACCAATTCCTAGACTACCTGCAAGAGTTTCTTGGTGTAATGAACACCGAG      109
                      K   K   C   G   E   E   R   R   V   N   Q   F   L   D   Y   L   Q   E   F   L   G   V   M   N   T   E

       W   A   M   E   G   END
M  TGGGCAATGGAAGGCTGAGGCTGACCTGGA-----GCAGCTG--GATTTTGAAAAAGAAAGAGGACATCTCCTTGCAGTGTGAATGAGAGCCAGCCACATGCTGGCCTT     953
       :::  :: :::: ::: ::: :  :::::     ::::: :   ::::::: :   :  :: : : :: ::::  :::: :::  ::::  ::::: ::: :::::::::
H  TGGATAATAGAAAGTTGAGACTAAACTGGTTTGTTGCAGCCAAAGATTTTGGAGGAGAA---GGACATTTACTGCAGTGAAGAATGAGGGCCAAGAAAGAGTCAGGCCTT     216
       W   I   I   E   S   END

M  ACTTCTCCGTGTAACTGAACTTAAGAGAAGCAAAGTAAATACCACAACCTTACTCACCCATGCCAACAGAAAGCATAAAATGTTGGGATGTGTTATTCAGGTATCAGGGTCAC    1063
       :   :::  :  ::   :::::: :::: ::: ::::: ::   : : :  :::      ::   :::::::::::  :   ::   : : :   : : :   :::  :::
H  AATTTCAATATAATTTAACCTTCAGAGGGAAAGTAAATATTTCAGGCATACTGACACTTTGCCA--GAAAGCATAAAAATTCTTAAAATATATTTCAGATATCAGAATCAT     324

M  TGGAGAAGGCCTCCCCCAGTTTACTCCAGGAAAAACAGATGTATGCTTT--ATTTAA-------TTCTGTAAGATGTTCATAT------TATTTATGATTCA            1154
       ::::   : :: :   :::::  ::::::: ::::   :::::::: :    ::::          :::::::::   :: ::::       :::::::::::::
H  TGAAGT-------ATTTTCCTCCAGGCAAAATTGATACTTTTTCTTTATTTATTTCTTATTTAACTTAACCATTCTGTAAAATGTCTGTTAACTTAATAGTATTTATGAAATGGTTA   424

M  GTAAGTT-------AATATTTATTACAACGTATATAATATTCT[AATAAA]GC-AGAAGGGACAACT--CAAATTCATTT                                 1222
       : ::::       :: :::::::::  : ::: :: ::::::  ::::::   :  ::  :::::::   :::: :::
H  AGAATTTGGTAAATTAGTATTTATTTAATGTATGTTGTTC[AATAAA]ACAAAAATAGACAACTGTTCAATTTGCTGC                                   504
```

Fig. 3. Comparison of the nucleotide sequences of the mouse (M) and human IL-5 genes. Gaps were introduced to maximize homology of the two sequences. *Dots* indicate identical nucleotides. *Asterisks* indicate transcription initiation sites determined by S1 mapping analysis. *Boxes* indicate lymphokine consensus, TATA-like, and polyA signal sequences from 5' to 3' in this order. *Horizontal arrows* indicate putative N-glycosylation sites. *Arrowheads* in exon 1 indicate the position of cleavage in precursor polypeptides. *Underlines* in exon 1 indicate the following sequences: Sp1 binding (positions −320 to −316 and positions −302 to −298); purine-rich region (positions −261 to −247 and positions −160 to −143); heat shock consensus (positions −221 to −208); enhancer core-like sequences (positions −114 to −108). The restriction sites used in S1 mapping are also indicated: *XbaI* (positions −313 to −308) and *SacI* (positions 163 to 168) sequences. A *vertical arrow* in the mouse exon 4 indicates the position of a 744-bp insertion which is not shown here (AZUMA et al. 1986). *Arrowheads* in exon 4 indicate the position of polyA attachment

quence of IL-5 and sequences of known proteins including all the lymphokines so far identified failed to show any significant homology except for short sequences of murine IL-3, murine granulocyte-macrophage colony-stimulating factor (GM-CSF) and murine IFN-γ, suggesting a remote phylogenetic relationship between these molecules (KINASHI et al. 1986).

Using the mouse IL-5 cDNA clone as a probe, we isolated the human homolog by screening a cDNA library constructed with polyA$^+$ mRNA extracted from ATL-2 cells (AZUMA et al. 1986). The isolated cDNA clone encodes a polypeptide of 134 residues, containing an N-terminal signal peptide of 19 residues. The nucleotide and amino acid sequence homologies of the coding regions of human and murine IL-5 are 77% and 70%, respectively.

II. Polypeptide Structures of IL-5

1. Murine IL-5

To examine polypeptide structures of murine IL-5, IL-5 mRNA was transiently translated in rabbit reticulocyte lysates. A single polypeptide with a molecular mass of about 14000 kDa was detected. By contrast, the active form of murine IL-5 translated in *Xenopus* oocytes has an apparent molecular mass of 45–50 kDa and migrates to the molecular mass of 25–30 kDa under reducing conditions (Fig. 4). Treatment of recombinant IL-5 with N-glycanase in the presence of 2-mercaptoethanol decreased the molecular mass of monomer IL-5 from 25–30 kDa to 12–14 kDa. Coinjection of tunicamycin and IL-5 mRNA into *Xenopus* oocytes induced the production of 27- to 28-kDa dimer molecules which exert TRF and BCGF II activities, suggesting that an N-linked carbohydrate moiety does not play an essential role in the biological ac-

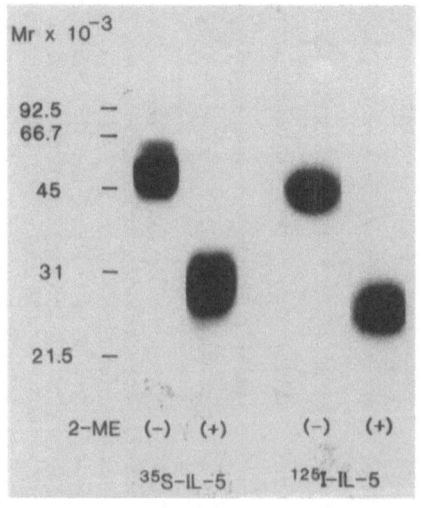

Fig. 4. Autoradiograph of radiolabeled IL-5 analyzed by SDS-PAGE. An aliquot of ^{35}S-methionine-labeled-IL-5 (1×10^5 cpm; *left*) or ^{125}I-labeled IL-5 (1×10^5 cpm; *right*) was boiled with the sample buffer in the presence or absence of 2-mercaptoethanol (*2-ME*) and was applied on polyacrylamide gel with 12.5% acrylamide gels. (Data from MITA et al. 1988)

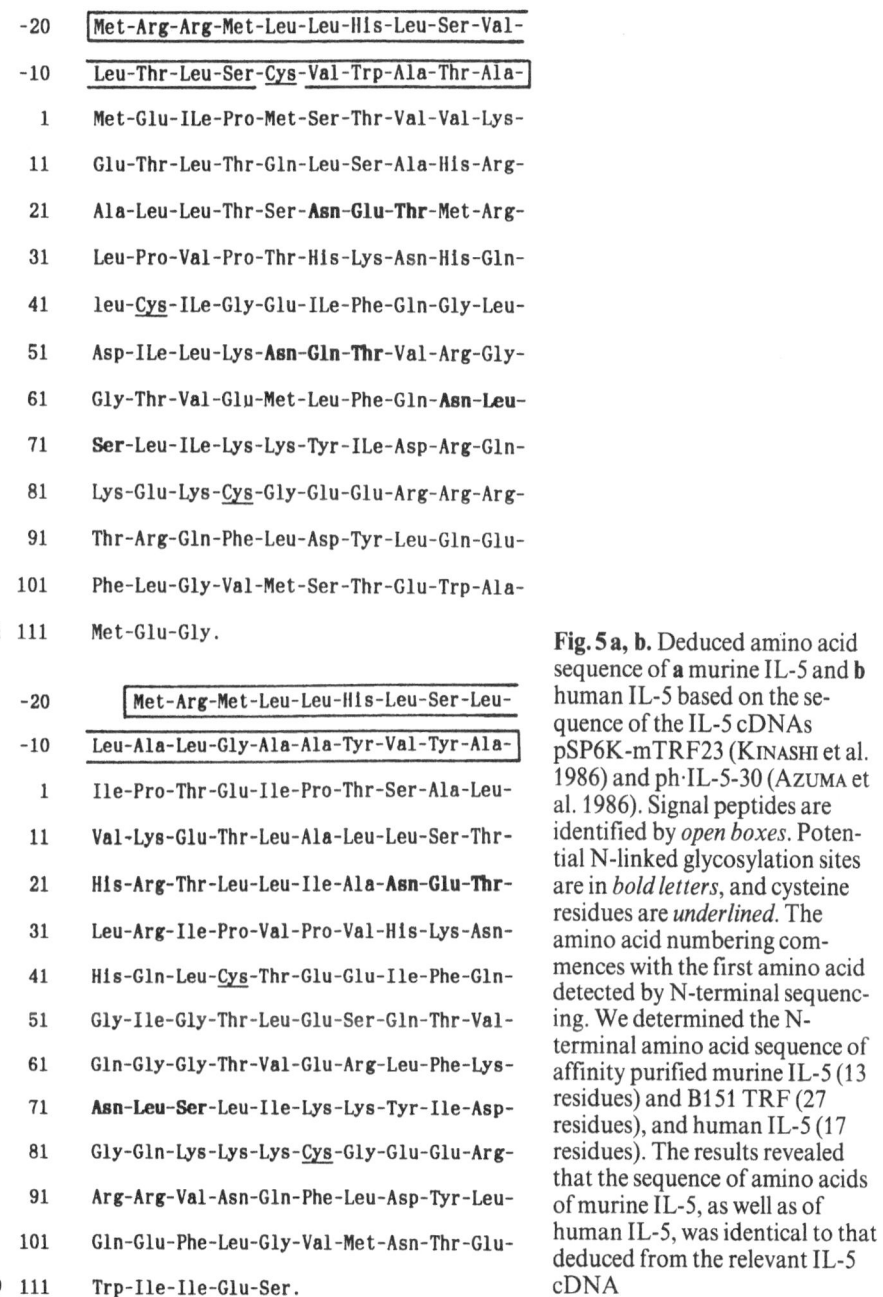

```
-20    │Met-Arg-Arg-Met-Leu-Leu-His-Leu-Ser-Val-
-10    │Leu-Thr-Leu-Ser-Cys-Val-Trp-Ala-Thr-Ala-│
  1     Met-Glu-ILe-Pro-Met-Ser-Thr-Val-Val-Lys-
 11     Glu-Thr-Leu-Thr-Gln-Leu-Ser-Ala-His-Arg-
 21     Ala-Leu-Leu-Thr-Ser-Asn-Glu-Thr-Met-Arg-
 31     Leu-Pro-Val-Pro-Thr-His-Lys-Asn-His-Gln-
 41     leu-Cys-ILe-Gly-Glu-ILe-Phe-Gln-Gly-Leu-
 51     Asp-ILe-Leu-Lys-Asn-Gln-Thr-Val-Arg-Gly-
 61     Gly-Thr-Val-Glu-Met-Leu-Phe-Gln-Asn-Leu-
 71     Ser-Leu-ILe-Lys-Lys-Tyr-ILe-Asp-Arg-Gln-
 81     Lys-Glu-Lys-Cys-Gly-Glu-Glu-Arg-Arg-Arg-
 91     Thr-Arg-Gln-Phe-Leu-Asp-Tyr-Leu-Gln-Glu-
101     Phe-Leu-Gly-Val-Met-Ser-Thr-Glu-Trp-Ala-
```
a 111   Met-Glu-Gly.

```
-20    │Met-Arg-Met-Leu-Leu-His-Leu-Ser-Leu-
-10    │Leu-Ala-Leu-Gly-Ala-Ala-Tyr-Val-Tyr-Ala-│
  1     ILe-Pro-Thr-Glu-Ile-Pro-Thr-Ser-Ala-Leu-
 11     Val-Lys-Glu-Thr-Leu-Ala-Leu-Leu-Ser-Thr-
 21     His-Arg-Thr-Leu-Leu-Ile-Ala-Asn-Glu-Thr-
 31     Leu-Arg-Ile-Pro-Val-Pro-Val-His-Lys-Asn-
 41     His-Gln-Leu-Cys-Thr-Glu-Glu-Ile-Phe-Gln-
 51     Gly-Ile-Gly-Thr-Leu-Glu-Ser-Gln-Thr-Val-
 61     Gln-Gly-Gly-Thr-Val-Glu-Arg-Leu-Phe-Lys-
 71     Asn-Leu-Ser-Leu-Ile-Lys-Lys-Tyr-Ile-Asp-
 81     Gly-Gln-Lys-Lys-Lys-Cys-Gly-Glu-Glu-Arg-
 91     Arg-Arg-Val-Asn-Gln-Phe-Leu-Asp-Tyr-Leu-
101     Gln-Glu-Phe-Leu-Gly-Val-Met-Asn-Thr-Glu-
```
b 111   Trp-Ile-Ile-Glu-Ser.

Fig. 5 a, b. Deduced amino acid sequence of **a** murine IL-5 and **b** human IL-5 based on the sequence of the IL-5 cDNAs pSP6K-mTRF23 (KINASHI et al. 1986) and ph·IL-5-30 (AZUMA et al. 1986). Signal peptides are identified by *open boxes*. Potential N-linked glycosylation sites are in *bold letters*, and cysteine residues are *underlined*. The amino acid numbering commences with the first amino acid detected by N-terminal sequencing. We determined the N-terminal amino acid sequence of affinity purified murine IL-5 (13 residues) and B151 TRF (27 residues), and human IL-5 (17 residues). The results revealed that the sequence of amino acids of murine IL-5, as well as of human IL-5, was identical to that deduced from the relevant IL-5 cDNA

tivity of IL-5 (TAKATSU et al. 1988 a). Taking all the results together, IL-5 is produced first as a 14-kDa monomer, N-glycosylated, and finally dimerized to form the 46-kDa molecule.

Rat anti-mouse IL-5 monoclonal antibodies (TB13 and NC17) blocked the B151-TRF-mediated anti-DNP IgG response of DNP-primed B cells as well as the IgM secretion of BCL_1 cells (HARADA et al. 1987a). Moreover, they also inhibited the proliferation of BCL_1 cells induced by B151 TRF. However, the antibodies did not inhibit IL-1, IL-2, IL-3, and IL-4 activities. TB13 antibody was used for purification of TRF from B151 supernatants. The purified B151 TRF had a molecular mass of 46 kDa estimated by sodium dodecylsulfate polyacrylamide gel electrophoresis (SDS-PAGE) (HARADA et al. 1987b) under nonreducing conditions and migrated to 23–26 kDa under reducing conditions. This observation confirmed our previous observations, and further suggests that B151 TRF comprises a dimer form. Recently, the partial N-terminal amino acid sequence (27 residues) of affinity purified B151 TRF was determined by automated Edman degradation (TAKATSU et al. 1988b). The results revealed that a single N-terminal amino acid (methionine) was found and the sequence of 27 residues is in agreement with the amino acid sequence deduced from the nucleotide sequence of the mouse IL-5 cDNA clone (KINASHI et al. 1986).

We also purified murine IL-5 (using an anti-IL-5 antibody-coupled affinity column) from supernatants of HeLa cells which had been transfected with IL-5 cDNA. The N-terminal 13-residue sequence was identical to that of affinity-purified B151 TRF as well as to the sequence predicted from the IL-5 cDNA. N-terminal methionine in secreted IL-5 is found at position 21 of the amino acid sequence predicted from the cDNA pSP6K-mTRF23 (Fig. 5). Forty-two picograms of purified IL-5 contain one unit of biological activity (MITA et al. 1988).

2. Human IL-5

Recombinant human IL-5 was purified (using an anti-mouse IL-5 monoclonal antibody) from the supernatant of Chinese hamster ovarian (CHO) cells which had been transfected with human IL-5 cDNA, and its N-terminal amino acid sequence was determined. A single amino acid sequence of 19 residues was found that was identical to that of the sequences predicted from the human IL-5 cDNA (MITA et al. 1989a). N-terminal isoleucine in secreted IL-5 was found at position 20 of the amino acid sequence predicted from the cDNA (TSUJIMOTO et al. 1989). Mature human IL-5 in the secreted form would be 115 residues, beginning with the isoleucine residue (Fig. 5).

III. Structure and Organization of the IL-5 Gene

1. Structure of the IL-5 Gene

The chromosomal genes for murine and human IL-5 were isolated using IL-5 cDNAs as probes (TANABE et al. 1987; MIZUTA et al. 1989). Nucleotide sequence analyses of the IL-5 genes as well as their flanking regions show that the IL-5 genes consist of four exons and three introns (Fig. 3). Conserved TATA-like and lymphokine consensus sequences were found at about 30 and 70 base pairs (bp), respectively, upstream of the transcription initiation sites which were determined by S1 mapping analysis and coincided with the 5′ ends of the cloned cDNAs. As the exon-intron organization and the location of the cysteine codons of the IL-5 genes resemble those of the GM-CSF, IL-2, IL-4 genes but are quite distinct from the G-CSF gene, TANABE et al. (1987) suggested that the IL-5 gene might be evolutionarily related to the genes for IL-2, IL-4, and GM-CSF.

2. Chromosomal Location of the IL-5 Gene

TAKAHASHI et al. (1988) mapped the human IL-5 gene on chromosome 5q23.3-31.1 by in situ hybridization. They also mapped the murine IL-4 gene on chromosome 11 using restriction fragment length polymorphism. The GM-CSF and IL-5 genes were mapped to similar location on human chromosome 5 (HUEBNER et al. 1985; LE BEAU et al. 1986, 1987) and shown to cluster within a 230-kilobase (kb) region of murine chromosome 11 (BARLOW et al. 1987). Taking these results together, it is likely that the IL-4 and IL-5 genes are clustered with the GM-CSF and IL-3 genes on human chromosome 5 and murine chromosome 11 and that these genes might be derived by duplications of a common ancestral gene.

3. Expression of IL-5 mRNA in Mouse T Cells

Constitutive expression of 1.7-kb IL-5 mRNA was detected by analysis of mRNA isolated from a TRF-producing B151 hybridoma. IL-5 mRNA expression in B151 was augmented by stimulation with phorbol 12-myristate, 13-acetate (PMA) plus calcium ionophore. Among several cell lines initially analyzed for IL-5 mRNA, EL4 (thymoma) stimulated with PMA and D9 (cloned T cells) treated with ConA were found to produce IL-5 mRNA, with accompanying IL-5 production (TOMINAGA et al. 1988). None of the murine tumor cell lines such as macrophage-monocytic cell line P388D.1, myelomonocytic cell line WEHI-3, B-cell leukemia (BCL$_1$) cells, or myeloma cell line X5563 expressed IL-5 mRNA. Northern blot analysis also revealed that 1.6-kb IL-5 mRNA was found in polyA$^+$ RNA of unstimulated spleen cells from AKR mice and of lymph node cells from MRL *lpr/lpr* mice (YAOITA et al. 1988). Usually spleen cells stimulated with ConA or antigens contained more abundant IL-5 mRNA. IL-5 mRNA expression was not detectable in PMA plus calcium ionophore-stimulated spleen cells, in which remarkable IL-2 mRNA expression was observed.

D. Functional Properties of IL-5

Major functions of IL-5 are summarized in Table 1.

I. Activities on B Cells

1. Roles in B-Cell Response to Antigens

As described previously, IL-5 induces IgM synthesis in BCL$_1$ and in vivo-activated B-cell blasts. IL-5 can induce anti-DNP IgG synthesis by DNP-primed B cells (Table 2). Murine IL-5 stimulates nonprimed B cells and

Table 1. Biological functions of recombinant IL-5

	Target cells	References
1. Induction of differentiation of activated normal B cells and murine chronic B-cell leukemia (BCL₁) cells into Ig (IgM, IgG or IgA) secreting cells	B	Kinashi et al. (1986), Yokota et al. (1987), Karasuyama et al. (1988), R. Matsumoto et al. (1989). Harriman et al. (1988)
2. Induction of increased expression of secretory forms of μ-chain and α-chain	B	Takatsu et al. (1988b), M. Matsumoto et al. (1987)
3. Induction of DNA synthesis in dextran sulfate-stimulated normal B cells and BCL₁ cells (BCGF II)	B	Kinashi et al. (1986), Tominaga et al. (1988)
4. Induction of upregulation of functional IL-2 receptor expression on B cells	B	Harada et al. (1987a), Nakanishi et al. (1988), Loughnan et al. (1987)
5. Augmentation of CTL generation in antigen-stimulated thymocytes in conjunction with IL-2 (killer helper factor)	T	Takatsu et al. (1987)
6. Enhancement of IL-2 receptor expression of T cells	T	Takatsu et al. (1987), T. Noma et al. (1987)
7. Induction of growth and differentiation of eosinophils (eosinophil differentiation factor, eosinophil CSF)	Eo	Yokota et al. (1987), Campbell et al. (1987), Yamaguchi et al. (1988), Lopez et al. (1988)
8. Augmentation of production of O_2^- in and of chemotaxis of eosinophils	Eo	Yamaguchi et al. (1988b), Lopez et al. (1988)

Table 2. Comparison of lymphokines triggering antigen-specific versus polyclonal responses

Recombinant lymphokines (units/ml)					B-cell response (PFC)			
					Antigen-specific		Polyclonal	
IL-5	IL-1	IL-2	IL-4	IFN-γ	Anti-SRBC IgM	Anti-DNP IgG	Resting IgM	LPS-blasts IgM
–	–	–	–	–	9	42	18	22
10	–	–	–	–	26	312	48	246
100	–	–	–	–	51	673	138	591
–	10	–	–	–	58	48	0	35
–	–	50	–	–	28	53	16	39
–	–	–	100	–	6	32	0	28
–	–	–	–	100	0	18	0	12
10	10	–	–	–	182	338	55	262
10	–	50	–	–	168	853	52	348
10	–	–	100	–	25	324	26	228
10	–	–	–	100	10	109	10	89

Data are taken from Takatsu et al. (1988b).

augments anti-SRBC IgM synthesis in the presence of SRBC (Table 2). A synergistic effect between IL-1 or IL-2 and murine IL-5 on anti-SRBC IgM synthesis was also observed (SWAIN 1985; HARADA et al. 1987a; KOYAMA et al. 1988). Polyclonal Ig synthesis was also augmented by IL-5 in B cells stimulated with anti-IgM and IL-4 (HOWARD et al. 1982).

When a single fluorescein-specific B cell was cultured in 10-μl wells in the presence of the T-independent antigen fluorescein-polymerized flagellin, the addition of murine IL-5 markedly increased the frequencies of both proliferating and antibody-secreting clones (ALDERSON et al. 1987). However, murine IL-5 exerted little activity without antigens. IL-4 can also augment the IL-5 activity in this system. Thus in the presence of antigens murine IL-5 can induce a single hapten-specific splenic B cell to proliferate as well as to produce Ig.

It has long been controversial whether a lymphokine can induce resting B cells to mature into Ig-secreting cells without proliferation, i.e., the B-cell maturation factor activity. KARASUYAMA et al. (1988) tested for the B-cell maturation factor activity of various recombinant interleukin preparations using purified resting B cells from BALB/c(nu/nu) mice. Murine IL-5 and IL-2 were found to be active, while IL-1, IL-3, IL-4, and IL-6 were inactive at all concentrations tested. Murine IL-5 was 1000 times more active than IL-2.

It is of great interest whether IL-5 can play some role in B-cell triggering by T cells in a MHC-restricted manner. RASMUSSEN et al. (1988) compared the requirements for B-cell activation in antigen-specific and polyclonal antibody responses by an IL-4-producing helper T cell clone. They found that: (a) the antigen-specific responses involve primarily small B cells, whereas polyclonal responses depend exclusively on large B cells; (b) polyclonal B-cell responses can proceed in the absence of T and B cell interaction, whereas antigen-specific responses require physical interaction of T and B cells; and (c) IL-5 is essential for polyclonal as well as antigen-specific responses because both responses were inhibited by monoclonal anti-IL-5 antibody (TB13). Thus, IL-5 is a critical mediator for Ig secretion by activated B cells, which could be either large B cells freshly isolated from spleen or those initially activated in an IL-4-dependent fashion through the cognate interaction with a helper T-cell clone.

The role of IL-5 in the expression of mRNA for IgM, IgG_1, and IgA was investigated (TAKATSU et al. 1988b; M. MATSUMOTO et al. 1987). BCL_1 or resting B cells cultured with IL-5 expressed increased levels of secreted forms of μ-chain mRNA. B cells cultured with lipopolysaccharide (LPS) plus IL-5 expressed increased levels of the secreted as well as membrane forms of IgM (4-fold). In this system stimulation of secreted forms of γ_1-chain mRNA (2.5-fold) by IL-5 was approximately one-third of that by IL-4. Purified DNP-primed B cells also expressed increased levels of μ-chain mRNA (4-fold), γ_1-chain mRNA (2-fold), and α-chain mRNA (2-fold) upon stimulation with IL-5.

2. Induction of IgA Secretion

Effects of IL-5 on isotype distribution have recently been studied in several laboratories. IL-5 can induce anti-DNP IgG and IgM production by DNP-primed B cells (R. MATSUMOTO et al. 1989; TAKATSU et al. 1988b). IL-5 can also induce anti-DNP IgA synthesis by DNP-primed B cells and polyclonal IgA synthesis by LPS-stimulated splenic B cells as well as Peyer's patch B cells. IL-1, IL-2, IL-3, IL-4, and IFN-γ do not have such activity. In these systems IgA was produced by B-cell populations which had expressed IgA on their surface, suggesting that IL-5 acts as a maturation factor for B cells which have been committed to become IgA-secreting cells rather than as a class switch-inducing factor (BEAGLEY et al. 1988; MATSUMOTO et al. 1989). A selective enhancing effect of murine IL-5 on IgA secretion was also reported from several laboratories (COFFMAN et al. 1987; BOND et al. 1987; MCKENZIE et al. 1987; MURRAY et al. 1987; HARRIMAN et al. 1988).

Human IL-5 added in the form of COS-7 transfection supernatant to peripheral blood B cells enhanced IgA production 3- to 10-fold, but had no effect on IgG and IgM production (YOKOTA et al. 1987). AZUMA et al. (1986) reported that human peripheral blood B cells stimulated with *Staphylococcus aureus* Cowan I and human IL-5 produced significant IgM antibody. CLUTTERBRUCK et al. (1987) reported that human IL-5 barely induced IgM and IgG production in B cells from peripheral blood. More refined experimental systems will clarify these discrepancies.

3. Stimulation of B-Cell Growth

IL-5 together with dextran sulfate stimulates normal B cell proliferation and directly induces DNA synthesis of BCL_1 cells (KINASHI et al. 1986).

We attempted to establish an IL-5-dependent B-cell line. Bone marrow cells were cultured in the Whitlock-Witte culture system, transferred onto a stromal cell line (ST2), and maintained in the presence of IL-5. The ST2 clone was shown to support growth of myeloid and lymphoid precursors (OGAWA et al. 1988). A cell line established (J-87) had a rearranged J_H segment (TOMINAGA et al. 1989) and expressed B-cell lineage markers like B-220 and Ly1, but neither IgM (surface or cytoplasmic) nor Thy1. J-87 proliferates in response to IL-5 or IL-3 only in the presence of ST2 cells. When J-87 was cultured with a low concentration of IL-5 in the absence of ST2 in a limiting dilution condition, the clone T-88 and its subline T88-M were obtained which can be maintained by IL-5 alone in the absence of ST2 (TOMINAGA et al. 1989). The growth of both cell lines is totally IL-5 dependent and none of lymphokines except IL-5 can support their growth. [^{3}H]Thymidine incorporation by T88-M and T-88 reached plateau levels in the presence of 5 pM and 20 pM IL-5, respectively, suggesting that T88-M may have higher numbers of high-affinity IL-5 receptors than T-88. None of J-87, T-88, or T88-M can differentiate into Ig-secreting cells in the presence of IL-5. It appears that certain subsets or stages of early B-cell precursors continue to grow in response to IL-5 without maturation.

4. Induction of IL-2 Receptors

Not only T cells but also B cells can express receptors for IL-2 and respond to IL-2 when activated under appropriate conditions (TSUDO et al. 1984; LOWENTHAL et al. 1985; ZUBLER et al. 1984). DNP-primed B cells cultured in the presence of IL-5 for 4 days expressed increased levels of the IL-2 receptor (L chain) which could react with monoclonal anti-IL-2 receptor antibody (HARADA et al. 1987a). The average numbers of high- and low-affinity sites per cell were 400 and 6500, respectively. Furthermore, IL-2 receptors thus induced are functional because IL-2 can induce those B cells to secrete Ig. IL-5 could induce an increase in the steady-state level of mRNA for the IL-2 receptor L chain by approximately 8-fold (HARADA et al. 1987). Similar findings were made by NAKANISHI et al. (1988) using monoclonal B cells (BCL_1-CL-3) and by LOUGHNAN et al. (1987) using resting B cells.

II. Activities on Other Hematopoietic Cells

1. Eosinophil Colony Stimulating Factor Activity

Eosinophil differentiation factor (EDF) activity was shown to be associated with the BCGF II activity by SANDERSON et al. (1986) using supernatants derived from alloreactive T-cell clones and hybrids. The BCGF II and EDF produced by a T-cell hybrid, NIMP-TH1, copurified in every fractionation procedure employed: both activities are associated with a protein with an approximate molecular mass of 46 kDa and a pI of 5.0 (SANDERSON et al. 1986, 1988).

NIMP-TH1 T cells did not produce IL-1, IL-2, IL-3, IFN-γ, or IL-4. EDF had no effects on purified resting B cells as measured by thymidine uptake, whereas it could induce DNA synthesis as well as Ig secretion by naturally occurring large B cells (O'GARRA et al. 1986). Using single cell culture in liquid, we studied the effects of IL-5 on murine hematopoietic cells at various stages of differentiation (YAMAGUCHI et al. 1988a). The results revealed that murine IL-5 alone acted on untreated bone marrow cells and supported the formation of a small number of colonies, all of which were predominantly eosinophilic. However, it did not support colony formation by spleen cells from 5-fluorouracil-treated mice, in which only primitive stem cells had survived. Eosinophil-containing colonies were formed from these spleen cells in the presence of both IL-5 and G-CSF. In contrast, G-CSF alone did not support any eosinophil colonies. Similar synergy of human IL-5 and G-CSF was shown for eosinophil colony formation from human bone marrow cells (ENOKIHARA et al. 1988). The IL-5 specifically facilitated the terminal differentiation and proliferation of eosinophils. Moreover, IL-5 maintained the viability of mature eosinophils obtained from peritoneal exudate cells of mice infected with parasites, induced the production of superoxide anion in mature eosinophils, and possessed chemotactic activity for eosinophils (YAMAGUCHI et al. 1988a). The synergistic effect of IL-5 and colony-stimulating factors on the expansion of eosinophils is supposed to contribute to the urgent mobilization of eosinophils at the time of helminthic infections and allergic responses.

Campbell et al. (1987) found that recombinant human IL-5 stimulated eosinophil colony formation in semisolid medium and that no other colony type was produced. In contrast, others (Yokota et al. 1987; Enokihara et al. 1988) reported that human IL-5 could induce both eosinophil and mixed eosinophil/basophil colonies from human bone marrow. Human IL-5 has chemotactic activity for eosinophils (Lopez et al. 1988; H. Enokihara et al., unpublished data).

2. Effects on T Cells

IL-5 has killer helper factor activity (Takatsu et al. 1987) which induces generation of CTL from their precursors in peanut agglutinin-binding (PNA$^+$) thymocytes in the presence of stimulator cells and IL-2. The killer helper factor was originally purified from supernatants of antigen-stimulated T cells as well as T-cell hybridoma 2Y4 (Takatsu et al. 1986; Kikuchi et al. 1986), and was shown to be distinct from IL-1, IL-2, IL-3, IL-4, and IFN-γ. Purified killer helper factor from 2Y4 can increase expression of IL-2 receptor L chains on PNA$^+$ thymocytes. IL-5 markedly increased the amount of the IL-2 receptor L-chain mRNA in PNA$^+$ thymocytes which had been stimulated with hapten-modified spleen cells and IL-2. By contrast only a trace amount of 3.5-kb mRNA for the IL-2 receptor L chain was detected in cells stimulated by IL-2 alone (Takatsu et al. 1987). Human IL-5 also induces expression of the IL-2 receptor (Tac antigen) on T cells (T. Noma et al. 1987).

These results, however, do not exclude the possibility that there is another lymphokine which has killer helper factor activity but has no TRF activity. At least four different lymphokines seem to have killer helper factor activity: IL-5, the factor from 2Y4, IL-4, and IL-6 (Takatsu and Kikuchi 1988).

E. Receptors for IL-5

I. Binding Assay by Radiolabeled IL-5

^{35}S-labeled IL-5 was prepared in *Xenopus* oocytes by injection of IL-5 mRNA transcribed from IL-5 cDNA in vitro. ^{35}S-labeled IL-5 was purified from oocyte supernatant using anti-IL-5 monoclonal antibody-coupled beads. IL-5 purified from culture supernatants of HeLa cells which had been transfected with murine IL-5 cDNA was labeled using diiodo-Bolton–Hunter reagent. SDS-PAGE analysis of the ^{35}S-IL-5 and the ^{125}I-IL-5 preparations used in the binding studies is shown in Fig. 4. The specific activities of the ^{35}S-IL-5 and ^{125}I-IL-5 were estimated to be 6.5×10^{15} cpm/mmol and 5.3×10^{14} cpm/mmol, respectively.

Kinetics and specificity of radiolabeled IL-5 binding to BCL$_1$-B20 cells were examined (Mita et al. 1988). As much as 90% of the total ^{125}I-IL-5 bound to BCL$_1$-B20 was inhibited by 100-fold excess unlabeled IL-5. None of the other recombinant lymphokines tested had any measurable inhibitory effect of the binding. Of special note is the finding that IL-3, GM-CSF, and

IFN-γ, which share amino acid sequence homology with IL-5 in short segments at the N-terminal region, did not compete for ^{125}I-IL-5 binding, strongly suggesting that IL-5 binds to different receptors from those for IL-3, GM-CSF, and IFN-γ. The binding of ^{125}I-IL-5 to its receptor was also inhibited by the TB13 monoclonal anti-IL-5 antibody, which is also a potent inhibitor of the biological function of IL-5. Unlabeled, naturally produced, purified B151 IL-5 could inhibit ^{125}I-IL-5 binding to BCL$_1$-B20 cells.

II. Number and Affinity of the IL-5 Receptors

Scatchard plot analysis of the binding data revealed that there were two classes of binding sites with dissociation constants of 66 pM (high affinity) and 12 nM (low affinity) on BCL$_1$-B20 cells (Fig. 6, inset). Average numbers of high- and low-affinity receptor sites were 400 and 7500 binding sites per cell, respectively. The number of high-affinity binding sites for IL-5 on BCL$_1$-B20 cells could be upregulated approximately 3 times by stimulation with LPS for 24 h whereas the number of low-affinity sites remained unchanged. In contrast, the number of high-affinity IL-5 receptors on BCL$_1$-B20 cells decreased to one-third of the control level by culturing them with either IL-5 or IL-2, both of which induce BCL$_1$-B20 cells to secrete IgM. Again, the number of low-affinity binding sites did not change dramatically under the same conditions.

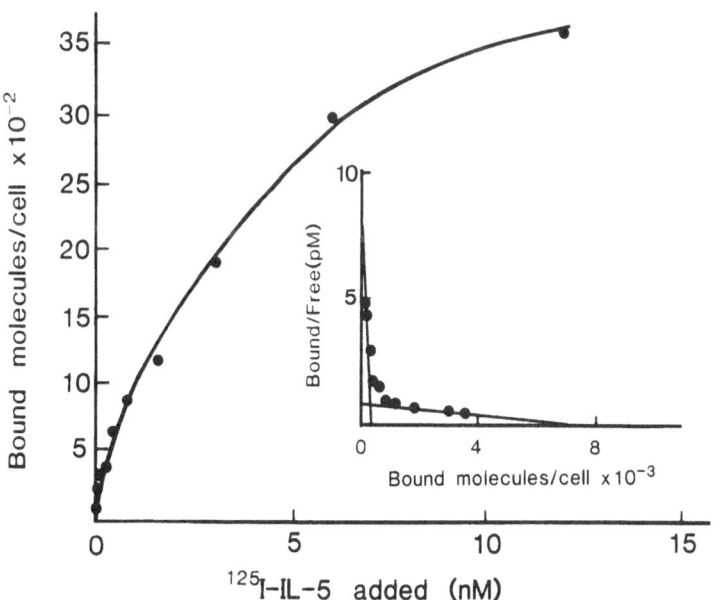

Fig. 6. Scatchard plot analysis of equilibrium binding of ^{125}I-IL-5 to BCL$_1$-B20. BCL$_1$-B20 cells were incubated at 37° C with various amounts of ^{125}I-IL-5 for 10 min. Specific equilibrium binding was determined after subtraction of nonspecific binding and the data were reexpressed as a Scatchard plot (*inset*). (Data from MITA et al. 1988)

    Binding assays for IL-5 on various murine cell lines were carried out (Mita
et al. 1988). Table 3 summarizes IL-5 receptor numbers of various cell types
including primary cells and established cell lines. Among the analyzed B cell
lines, BCL$_1$-B20 and MOPC104E (mouse myeloma) displayed significant
numbers of high-affinity binding sites for IL-5. A mouse thymoma cell line, a
mastocytoma cell line, and a macrophage tumor cell line did not express detectable numbers of IL-5 binding sites. LPS-stimulated normal B cells expressed
detectable numbers of IL-5 binding sites, whereas normal resting B cells, bone
marrow cells, and ConA-stimulated T blasts expressed few, if any. Intriguing-
ly, two early B-cell lines (J-87 and T-88) whose growth is IL-5 dependent expressed substantial numbers of high-affinity binding sites for IL-5. A subline
(T88-M) of T-88 which can respond to 0.5 U IL-5/ml with half-maximal
growth has more high-affinity binding sites for IL-5 than T-88 which requires
more than 5 U IL-5/ml for proliferation. The orders of IL-5 responsiveness
and numbers of high-affinity IL-5 binding sites are in good agreement: T88-

Table 3. Cellular distribution of murine IL-5 receptors

		High-affinity IL-5 receptors	
		Binding sites per cell (mean ± SEM)	K_d (pM) (mean ± SEM)
Primary cells			
Bone marrow cells		<10	n.d.
Splenic B cells		<10	n.d.
LPS-stimulated B blasts		40± 10	217±50
Thymocytes		<10	n.d.
Splenic T cells		<10	n.d.
ConA-stimulated T blasts		<10	n.d.
In vitro cell lines			
Designation	Cell type		
BCL$_1$-B20	B-cell leukemia (in vitro line)	243±100	46±13
BCL$_1$	B-cell leukemia (in vivo line)	120± 52	43±15
MOPC104E	Myeloma (IgM producer)	56± 7	61±10
X5563	Myeloma (γ_{2a} producer)	<10	n.d.
J-87	Stromal cell and IL-5-depen- dent early B-cell line	45± 9	41±17
T-88	IL-5-dependent early B-cell line	76± 11	38±16
T88-M	IL-5-dependent early B-cell line	1400±200	66±20
BAL.17	B-cell lymphoma	<10	n.d.
L10A	B-cell lymphoma	<10	n.d.
MTH	IL-2-dependent chronic T-cell lymphocytic leukemia	<10	n.d.
B151K12	T-cell hybridoma	<10	n.d.
P815	Mastocytoma	<10	n.d.
P388D1	Macrophage tumor	<10	n.d.

Data are taken from Mita et al. (1988) and are slightly modified. n.d., not done.

M, BCL_1-B20, T-88, and MOPC104E. These results are in favor of the hypothesis that the biological effects of IL-5 are mediated through the high-affinity receptor(s) identified by the binding assay.

To further confirm the biological significance of the high-affinity binding of radiolabeled IL-5, various concentrations of ^{35}S-IL-5 were incubated with BCL_1-B20 cells to assay standard TRF activity. The IL-5 concentrations that promoted half-maximal IgM synthesis by BCL_1 cells and that gave the half-maximal high-affinity binding were 2.5 pM and 15 pM, respectively. The rough coincidence between the biological dose-response curve and the high-affinity binding curve of IL-5 suggests that the biological response may be proportional to the extent of IL-5 binding to the high-affinity site. However, the fact that the maximum biological response occurred at the IL-5 concentration that only yields 40%–50% of maximal high-affinity binding does not agree with the report that IL-2 shows similar kinetics for surface binding and biological response (SMITH 1988). IL-5 seems to exert significant biological effects when on average 10–100 molecules of IL-5 are bound to one cell.

III. Affinity Crosslinking of Radiolabeled IL-5 with Its Receptor

Chemical crosslinking of radiolabeled IL-5 to surface receptors was carried out using bivalent lysine-directed crosslinkers, disuccinimidyl suberate (DSS), ethylene glycol *bis*-succinimidyl succinate (EGS), and disuccinimidyl tartarate (DST; MITA et al. 1988, 1989 b). SDS-PAGE analysis of detergent lysates of BCL_1-B20 cells crosslinked with IL-5 revealed a 92.5-kDa band under non-reducing conditions and a 75-kDa band under reducing conditions. The crosslinking of IL-5 to the higher molecular weight species was specific as the addition of unlabeled IL-5 abolished these bands.

When IL-5 was bound to LPS-stimulated BCL_1-B20 cells and crosslinked with DST, two IL-5-crosslinked bands of approximately 92.5 kDa and 160 kDa were observed. The 160-kDa complex was weakened with the use of DSS and was barely visible if EGS was used. In contrast, X5563, which does not have detectable levels of any IL-5 receptors, yielded no visible band under the same conditions. The crosslinked 160-kDa complex was more clearly detected using T88-M cells (MITA et al. 1989 b). As the molecular mass of IL-5 is 46 kDa, it is likely that two proteins of 46.5 kDa and 114 kDa are involved in the high-affinity binding of IL-5. It remains to be seen whether the two proteins are different or oligomeric forms of a common protomer.

F. Summary

Murine IL-5 is an acidic glycoprotein with a molecular mass of 46 kDa and was originally defined as a T-cell replacing factor. Molecular cloning of murine and human IL-5 cDNA revealed that IL-5 cDNA encodes 133 residues for murine IL-5 and 134 residues for human IL-5, with 20 strongly hydrophobic residues for murine IL-5 and 19 residues for human IL-5 at respective N-termini. The mature murine IL-5 molecule is a homodimer; each monomer comprises 113 amino acid residues and is heavily glycosylated. The IL-5 genes consist of four exons and three introns and are located on human chromosome 5 and murine chromosome 11.

Recombinant murine IL-5 can induce IgM, IgG, and IgA synthesis by activated B cells as well as resting B cells, can induce DNA synthesis in activated B cells as well as BCL_1 cells, can support growth of stromal cell-dependent as well as stromal cell-independent early B-cell precursor lines derived from bone marrow precursors, and can induce an increase in the expression of functional IL-2 receptors on resting as well as activated B cells. IL-5 augments CTL generation in antigen-stimulated immature thymocytes in conjunction with IL-2 and induces expression of IL-2 receptors on T cells. It also supports colony formation of eosinophilic precursors in bone marrow, and growth and/or differentiation of mature eosinophils. Furthermore, IL-5 is chemotactic for eosinophils and augments production of superoxide anion by mature eosinophils.

Radiolabeled IL-5 binds within 10 min at 37° C specifically to IL-5-responding cells, such as neoplastic BCL_1 cells and normal early B cell lines (T-88 and T88-M). There are two classes of binding sites with high affinity ($K_d = 66$ pM) and low affinity ($K_d = 12$ nM) for IL-5. Average numbers of high-affinity and low-affinity binding sites are 400 and 7500 per cell, respectively. The number of high-affinity IL-5 binding sites on BCL_1-B20 can be upregulated 3-fold by LPS and downregulated to one-third by IL-5 itself. IL-5 receptors are hardly detectable on normal resting B cells but are expressed on LPS-activated B cells. Treatment of surface-bound radiolabeled IL-5 with bivalent crosslinkers identified membrane polypeptides of approximately 46.5 kDa and 114 kDa to which IL-5 is crosslinked on most IL-5-responding cells under high-affinity conditions.

Acknowledgments. This work was supported by grants from the Ministry of Education, Science, and Culture, from the Special Project for Human Science conducted by the Japanese Agency of Science and Technology, and from Takeda Medical Science Foundation. The authors are grateful to Dr. Toshiyuki Hamaoka for providing various cell lines, and to Drs. Kendall A. Smith and Kimishige Ishizaka for helpful suggestions and discussions during the course of this study. Ms. S. Tachimoto and K. Hirano are acknowledged for their secretarial assistance.

References

Alderson MR, Pike BL, Harada N, Tominaga A, Takatsu K, Nossal GJV (1987) Recombinant T cell replacing factor (interleukin 5) acts with antigen to promote the growth and differentiation of single hapten-specific B lymphocytes. J Immunol 139:2656–2660

Azuma C, Tanabe T, Konishi M, Kinashi T, Noma T, Matsuda F, Yaoita Y, Takatsu K, Hammarström L, Smith CIE, Severinson E, Honjo T (1986) Cloning of cDNA for human T-cell replacing factor (interleukin-5) and comparison with the murine homologue. Nucleic Acids Res 14:9149–9158

Barlow DP, Bucan M, Lehrach H, Hogan BLM, Gough NM (1987) Close genetic and physical linkage between the murine haemopoietic growth factor genes GM-CSF and multi-CSF (IL-3). EMBO J 6:617–623

Beagley KW, Eldridge JH, Kiyono H, Everson MP, Koopman WJ, Honjo T, McGhee JR (1988) Recombinant murine IL-5 induces high rate IgA synthesis in cycling IgA-positive Peyer's patch B cells. J Immunol 144:2035–2042

Bond MW, Shrader B, Mossmann TR, Coffman RL (1987) A mouse T cell product that preferentially enhances IgA production. II. Physiochemical characterization. J Immunol 139:3691–3696

Campbell HD, Tucker WQJ, Hort Y, Martinson ME, Mayo G, Clutterbuck EJ, Sanderson CJ, Young IG (1987) Molecular cloning, nucleotide sequence, and expression of the gene encoding human eosinophil differentiation factor (interleukin 5). Proc Natl Acad Sci USA 84:6629–6633

Clutterbuck E, Shields JG, Gordon G, Smith SH, Boyd A, Callard RE, Campbell HD, Young IG, Sanderson C (1987) Recombinant human interleukin 5 is an eosinophil differentiation factor but has no activity in standard B cell growth factor assay. Eur J Immunol 17:1743–1750

Coffman RL, Shrader B, Carty J, Mossmann TR, Bond MW (1987) A mouse T cell product that preferentially enhances IgA production. II. Biologic characterization. J Immunol 139:3685–3690

Dutton RW, Swain SL (1987) B cell and T cell growth factors. In: Leffert H, Boynton AL (eds) Cell proliferation: recent advances, vol II. Academic, New York, pp 219–242

Dutton RW, Falkoff R, Hirst JA, Hoffman M, Kappler JW, Ketton JR, Lesley JF, Vann D (1971) Is there evidence for a non-antigen specific diffusable chemical mediator from the thymus-derived cell in the initiation of the immune response? Prog Immunol 1:355–368

Enokihara H, Nagashima S, Noma T, Kajitani H, Hamaguchi H, Saito K, Furusawa S, Shishido H, Honjo T (1988) Effect of human recombinant interleukin 5 and G-CSF on eosinophil colony formation. Immunol Lett 18:73–76

Harada N, Kikuchi Y, Tominaga A, Takaki S, Takatsu K (1985) BCGFII activity on activated B cells of a purified murine T cell-replacing factor (TRF) from a T cell hybridoma (B151K12). J Immunol 134:3944–3951

Harada N, Matsumoto M, Koyama N, Shimizu A, Honjo T, Tominaga A, Takatsu K (1987a) T cell replacing factor/interleukin 5 induces not only B-cell growth and differentiation, but also increased expression of interleukin 2 receptor in activated B-cells. Immunol Lett 15:205–215

Harada N, Takahashi T, Matsumoto M, Kinashi T, Ohara J, Kikuchi Y, Koyama N, Severinson E, Yaoita Y, Honjo T, Yamaguchi N, Tominaga A, Takatsu K (1987b) Production of a monoclonal antibody useful in the molecular characterization of murine T-cell-replacing factor B cell growth factor II. Proc Natl Acad Sci USA 84:4581–4585

Harriman GR, Kunimoto DY, Elliott JF, Paektau V, Strober W (1988) The role of IL-5 in IgA B cell differentiation. J Immunol 140:3033–3039

Howard M, Paul WE (1983) Regulation of B cell growth and differentiation by soluble factors. Annu Rev Immunol 1:307–333

Howard M, Farrar J, Hilfiker M, Johnson B, Takatsu K, Hamaoka T, Paul WE (1982) Identification of a T cell-derived B cell growth factor distinct from interleukin 2. J Exp Med 155:914–923

Huebner K, Isobe M, Croce CM, Golde DW, Kaufman SE, Gasson JC (1985) The human gene encoding GM-CSF is at 5q21-q32, the chromosome region deleted in the 5q⁻ anomaly. Science 230:1282–1285

Karasuyama H, Rolink A, Melchers F (1988) Recombinant interleukin 2 or 5, but not 3 or 4, induce maturation of resting mouse B lymphocytes and propagate proliferation of activated B cell blasts. J Exp Med 167:1377–1390

Kikuchi Y, Kato R, Sano Y, Takahashi H, Kanatani T, Takatsu K (1986) Generation of cytotoxic T lymphocytes from thymocyte precursors to trinitrophenyl-modified self antigens. II. Establishment of a T cell hybrid clone constitutively producing killer-helper factor(s) (KHF) and functional analysis of released KHF. J Immunol 136:3553–3560

Kinashi T, Harada N, Severinson E, Tanabe T, Sideras P, Konishi M, Azuma C, Tominaga A, Bergstedt-Lindqvist S, Takahashi M, Matsuda F, Yaoita Y, Takatsu K, Honjo T (1986) Cloning of cDNA for T-cell replacing factor and identity with B-cell growth factor II. Nature 324:70–73

Kishimoto T (1985) Factors affecting B cell growth and differentiation. Annu Rev Immunol 3:133–157

Koyama N, Harada N, Takahashi T, Mita S, Okamura H, Tominaga A, Takatsu K (1988) Role of recombinant interleukin 1 (IL-1) compared to recombinant T-cell-replacing-factor (TRF)/interleukin 5 (IL-5) in B cell differentiation. Immunology 63:277–283

Le Beau MM, Westbrook CA, Diaz MO, Larson RA, Rowley JD, Gasson JC, Golde DW, Sherr CJ (1986) Evidence for the involvement of GM-CSF and FMS in the deletion (5q) in myeloid disorders. Science 231:984–987

Le Beau MM, Epstein ND, O'Brien SJ, Nienhuis AW, Yang YC, Clark SC, Rowley JD (1987) Human Gene Mapping 9:A342

Lopez AF, Sanderson CJ, Gamble JR, Campbell HD, Young IG, Vadas MA (1988) Recombinant human interleukin 5 is a selective activator of human eosinophil function. J Exp Med 167:219–224

Loughnan MS, Takatsu K, Harada N, Nossal GJV (1987) T cell replacing factor (interleukin 5) induces expression of interleukin-2 receptors on murine splenic B cells. Proc Natl Acad Sci USA 84:5399–5403

Lowenthal JW, Zubler RH, Nabholz M, MacDonald HR (1985) Similarities between interleukin-2 receptor number and affinity on activated B and T lymphocytes. Nature 315:669–672

Matsumoto M, Tominaga A, Harada N, Takatsu K (1987) Role of T cell-replacing factor (TRF) in the murine B cell differentiation: induction of increased levels of expression of secreted type IgM mRNA. J Immunol 138:1826–1833

Matsumoto R, Matsumoto M, Mita S, Hitoshi Y, Ando M, Araki S, Yamaguchi N, Tominaga A, Takatsu K (1989) Interleukin 5 induces maturation but not class-switching of surface IgA-positive B cells into IgA-secreting cells. Immunology 66:32–38

McKenzie DT, Filutowicz HI, Swain SL, Dutton RW (1987) Purification and partial sequence analysis of murine B cell growth factor II (interleukin 5). J Immunol 139:2661–2668

Mita S, Hosoya Y, Kubota I, Nishihara T, Honjo T, Takahashi T, Takatsu K (1989a) Rapid methods for purification of human recombinant interleukin-5 (IL-5) using the antimurine IL-5 antibody-coupled immunoaffinity column. J Immunol. Methods in press

Mita S, Harada N, Naomi S, Hitoshi Y, Sakamoto K, Akagi M, Tominaga A, Takatsu K (1988) Receptors for T cell-replacing factor (TRF)/interleukin 5 (IL-5): specificity, quantitation and its implication. J Exp Med 168:863–878

Mita S, Tominaga A, Hitoshi Y, Sakamoto K, Honjo T, Akagi M, Kikuchi Y, Yamaguchi N, Takatsu K (1989b) Characterization of high-affinity receptors for interleukin 5 on interleukin 5-dependent cell lines. Proc Natl Acad Sci USA 86:2311–2315

Mizuta TR, Tanabe T, Nakakubo H, Noma T, Honjo T (1989) Molecular cloning and structure of the mouse interleukin-5 gene. Growth Factors 1:51–57

Murray PD, McKenzie DT, Swain SL, Kagnoff MT (1987) Interleukin 5 and interleukin 4 produced by Peyer's patch T cells selectively enhance immunoglobulin A expression. J Immunol 139:2669–2674

Nakanishi K, Yoshimoto T, Katoh Y, Ono S, Matsui K, Hiroishi K, Noma T, Honjo T, Takatsu K, Higashino K, Hamaoka T (1988) Both B151-TRF1 and interleukin 5 regulate immunoglobulin secretion and IL-2 receptor expression on a cloned B lymphoma line. J Immunol 140:1168–1174

Noma T, Mizuta T, Rosen A, Hirano T, Kishimoto T, Honjo T (1987) Enhancement of the interleukin 2 receptor expression in T cells by multiple B-lymphotropic lymphokines. Immunol Lett 15:249–253

Noma Y, Sideras P, Naito T, Bergstedte-Lindqvist S, Azuma C, Severinson E, Tanabe T, Kinashi T, Matsuda F, Yaoita Y, Honjo T (1986) Cloning of cDNA encoding the murine IgG_1 induction factor by a novel strategy using SP6 promoter. Nature 319:640–646

O'Garra A, Warren DJ, Holman M, Popham AM, Sanderson CJ, Klaus GGB (1986) Interleukin-4 (B cell growth factor-II/eosinophil differentiation factor) is a mitogen and differentiation factor for preactivated murine B lymphocytes. Proc Natl Acad Sci USA 83:5228–5232

Ogawa M, Nishikawa S, Ikuta K, Yamamura F, Naito M, Takahashi K, Nishikawa S (1988) B cell ontogeny in murine embryo studied by a culture system with the monolayer of a stromal cell clone, ST2: B cell progenitor develops first in embryonal body rather than in yolk sac. EMBO J 7:1337–1343

Okayama H, Berg P (1983) A cDNA cloning vector that permits expression of cDNA inserts in mammalian cells. Mol Cell Biol 3:280–289

Pure E, Isakson PC, Takatsu K, Hamaoka T, Swain SL, Dutton RW, Dennert G, Uhr JW, Vitetta ES (1981) Induction of B cell differentiation by T cell factors. I. Stimulation of IgM secretion by products of a T cell hybridoma and a T cell line. J Immunol 127:1953–1958

Rasmussen R, Takatsu K, Harada N, Takahashi T, Bottomly K (1988) T cell-dependent hapten-specific and polyclonal B cell responses require release of interleukin 5. J Immunol 140:705–712

Sanderson CJ, O'Gara A, Warren DJ, Klaus GGB (1986) Eosinophil differentiation factor also has B-cell growth factor activity: proposed name interleukin 4. Proc Natl Acad Sci USA 83:437–440

Sanderson CJ, Gamble JR, Campbell HD, Young IG, Vadas MA (1987) Recombinant human interleukin 5 is a selective activator of human eosinophil function. J Exp Med 167:219–224

Sanderson CJ, Campbell HD, Young IG (1988) Molecular and cellular biology of eosinophil differentiation factor (interleukin 5) and its effect on human and mouse B cells. Immunol Rev 102:51–76

Schimpl A, Wecker E (1972) Replacement of T-cell function by a T-cell product. Nature (New Biol) 237:15–17

Smith KA (1988) Interleukin 2: inception, impact, and implications. Science 240:1169–1176

Swain SL (1985) Role of BCGFII in the differentiation to antibody secretion of normal and tumor B cells. J Immunol 134:3934–3943

Swain SL, Dutton RW (1982) Production of a B cell growth-promoting activity, (DL) BCGF, from a cloned T cell line and its assay on the BCL_1 B cell tumor. J Exp Med 156:1821–1834

Swain SL, Howard M, Kappler J, Marrack P, Watson J, Booth R, Wetzel GD, Dutton RW (1983) Evidence for two distinct classes of murine B cell growth factors with activities in different functional assays. J Exp Med 158:822–835

Takahashi M, Yoshida MC, Satoh H, Hilgers J, Yaoita Y, Honjo T (1988) Chromosomal mapping of the mouse IL-4 and human IL-5 genes. Genomics 4:49–52

Takatsu K (1988) B cell growth and differentiation factors. Proc Soc Exp Biol Med 188:243–258

Takatsu K, Kikuchi Y (1988) Molecular and functional properties of killer-helper factor (KHF) and their roles in the induction of cytotoxic T cell response. Gann Monogr Can Res 34:143–154

Takatsu K, Haba S, Aoki T, Kitagawa M (1974) Enhancing factor on anti-hapten antibody response released from PPDs-stimulated *Tubercle bacilli*-sensitized cells. Immunochemistry 11:107–109

Takatsu K, Tanaka K, Tominaga A, Kumahara Y, Hamoka T (1980a) Antigen-induced T cell-replacing factor (TRF). III. Establishment of T cell hybrid clone continuously producing TRF and functional analysis of released TRF. J Immunol 125:2646–2653

Takatsu K, Tominaga A, Hamaoka T (1980b) Antigen-induced T cell-replacing factor (TRF). I. Functional characterization of a TRF-producing helper T cell subset and genetic studies on TRF production. J Immunol 124:2414–2422

Takatsu K, Harada N, Hara Y, Takahama Y, Yamada G, Dobashi K, Hamaoka T (1985) Purification and physicochemical characterization of murine T cell-replacing factor (TRF). J Immunol 143:382–389

Takatsu K, Kikuchi Y, Kanatani T, Okuno K, Hamaoka T, Tominaga A, Sano Y (1986) Generation of cytotoxic T lymphocytes from thymocyte precursors to tuinitrophenyl-modified self antigens. I. Requirement of both killer-helper factor(s) and interleukin 2 for CTL generation from a subpopulation of thymocytes. J Immunol 136:1161–1170

Takatsu K, Kikuchi Y, Takahashi T, Honjo T, Matsumoto M, Harada N, Yamaguchi N, Tominaga A (1987) Interleukin 5, a T-cell-derived B-cell differentiation factor also induces cytotoxic T lymphocytes. Proc Natl Acad Sci USA 84:4234–4238

Takatsu K, Mita S, Harada N, Matsumoto R, Nishikawa S-I, Tominaga A (1988a) Role of T cell-replacing factor (TRF)/interleukin 5 (IL-5) in the B cell growth and differentiation. In: Klinman N, Howard M, Witte O (eds) B cell development. Liss, New York

Takatsu K, Tominaga A, Harada N, Mita S, Matsumoto M, Kikuchi Y, Takahashi T, Yamaguchi N (1988b) T cell-replacing factor (TRF)/interleukin 5 (IL-5): molecular and functional properties. Immunol Rev 102:107–136

Tanabe T, Konishi M, Mizuta T, Noma T, Honjo T (1987) Molecular cloning and structure of the human interleukin-5 gene. J Biol Chem 262:16 580–16 584

Tominaga A, Matsumoto M, Harada N, Takahashi T, Kikuchi Y, Takatsu K (1988) Molecular properties and regulation of mRNA expression for murine T cell-replacing factor (TRF)/interleukin 5 (IL-5). J Immunol 140:1175–1181

Tominaga A, Nishikawa S-I, Mita S, Ogawa M, Kikuchi Y, Hitoshi Y, Takatsu LK (1989) Establishment of IL-5 dependent early B cell lines by long-term bone marrow cultures. Growth Factors 1:135–146

Tsudo M, Uchiyama T, Uchino H (1984) Expression of Tac antigen on activated normal human B cells. J Exp Med 160:612–617

Tsujimoto M, Adachi H, Kodama S, Tsurnoka N, Yamada Y, Tanaka S, Mita S, Takatsu K (1989) Characterization of recombinant human interleukin 5 produced by Chinese hamster ovary cells. J Biochem (Tokyo) 106:23–28

Yamaguchi Y, Suda T, Suda J, Eguchi M, Miura Y, Harada N, Tominaga A, Takatsu K (1988a) Purified interleukin 5 (IL-5) supports the terminal differentiation and proliferation of murine eosinophilic precursors. J Exp Med 167:43–56

Yamaguchi Y, Hayashi Y, Sugama Y, Miura Y, Kasahara T, Kitamura S, Torisu M, Mita S, Tominaga A, Takatsu K, Suda T (1988b) Highly purified murine interleukin 5 (IL-5) stimulates eosinophil function and prolongs in vitro survival. IL-5 as an eosinophil chemotactic factor. J Exp Med 167:1737–1742

Yaoita Y, Takahashi M, Azuma C, Kanai Y, Honjo T (1988) Biased expression of variable region gene families of the immunoglobulin heavy chain in autoimmune-prone mice. J Biochem 104:337–343

Yokota T, Coffman RL, Hagiwara H, Rennick DM, Takebe Y, Yokota K, Gemmell L, Shrader B, Yang G, Meyerson P, Luh J, Hoy P, Pene J, Briere F, Banchereau J, Vries JD, Lee FD, Arai N, Arai K (1987) Isolation and characterization of lymphokine cDNA clones encoding mouse and human IgA-enhancing factor and eosinophil colony-stimulating factor activities: relationship to interleukin 5. Proc Natl Acad Sci USA 84:7388–7392

Zubler RH, Lowenthal JW, Erard F, Hashimoto N, Devos R, MacDonald HR (1984) Activated B cells express receptors for, and proliferate in response to, pure interleukin 2. J Exp Med 160:1170–1183

CHAPTER 14

Interleukin-6

T. HIRANO and T. KISHIMOTO

A. Introduction

Interleukin-6 (IL-6) is a multifunctional cytokine produced by both lymphoid and nonlymphoid cells (KISHIMOTO and HIRANO 1988 a, b; KISHIMOTO 1988; LE and VILCEK 1989). IL-6 has previously been called by a variety of names, such as B-cell stimulatory factor 2 (BSF-2; HIRANO et al. 1985), interferon-$\beta2$ (IFN-$\beta2$; ZILBERSTEIN et al. 1986), 26-kDa protein (HAEGEMAN et al. 1986), hybridoma/plasmacytoma growth factor (VAN SNICK et al. 1986; NORDAN and POTTER 1986; VAN DAMME et al. 1987 a), and hepatocyte stimulating factor (HSF; ANDUS et al. 1987; GAULDIE et al. 1987) on the basis of its biological activities. Following the molecular cloning of the cDNAs encoding BSF-2, IFN-$\beta2$, and 26-kDa protein (HIRANO et al. 1986; ZILBERSTEIN et al. 1986; HAEGEMAN et al. 1986), it was found that all these molecules were identical and therefore this molecule was called IL-6. It is now known that IL-6 is a polypeptide mediator regulating the immune response, the acute phase reaction, and hematopoiesis. Furthermore, it has been demonstrated that deregulated production of IL-6 is involved in a variety of chronic inflammatory diseases and certain lymphoid malignancies, especially plasmacytoma/myeloma. In fact, it was demonstrated that unregulated expression of the IL-6 gene in B-lineage cells causes development of a massive plasmacytosis histologically indistinguishable from plasmacytoma in transgenic mice with the immunoglobulin heavy chain enhancer–IL-6 gene (SUEMATSAU et al. 1989).

B. Historical Overview

I. B-Cell Stimulatory Factor 2

Upon antigenic stimulation, B cells proliferate and differentiate to antibody forming cells under the control of T cells and macrophages and their effects were found to be replaced by soluble factors (DUTTON et al. 1971; SCHIMPL and WECKER 1972; KISHIMOTO and ISHIZAKA 1973). In the early 1980s it was shown that at least two different kinds of factors, one for growth of activated B cells, B-cell growth factor (BCGF), and the other for antibody induction in B cells, B-cell differentiation factor (BCDF), were involved in the regulation of the B-cell response (YOSHIZAKI et al. 1982). Since then a variety of factors have been reported to be involved in the regulation of proliferation and differentiation of B cells to antibody forming cells (KISHIMOTO 1985). Finally, cDNAs for three factors, IL-4/BSF-1 as an activating factor, IL-5/BCGF II as a growth and differentiation factor, and IL-6/BSF-2 as a differentiation factor, have been cloned and their presence and involvement in B-cell regulation confirmed (KISHIMOTO and HIRANO 1988a). Human BSF-2 was identified as a factor in the culture supernatants of phytohemagglutinin- (PHA-) (MURAGUCHI et al. 1981) or antigen-stimulated (TERANISHI et al. 1982) peripheral mononuclear cells which induced immunoglobulin (Ig) production in Epstein-Barr virus (EBV) transformed B-cell lines and was originally called BCDF. This molecule, BCDF/BSF-2, was found to be separable from other factors such as IL-2 and BCGF (TERANISHI et al. 1982; YOSHIZAKI et al. 1982; HIRANO et al. 1984a). The establishment of a human T-cell hybridoma generating BSF-2 activity confirmed that BSF-2 is a distinct molecule from other cytokines (OKADA et al. 1983). Subsequently it was demonstrated that BSF-2 functions in the late phase of *Staphylococcus aureus* Cowan I (SAC) stimulated normal B cells (HIRANO et al. 1984b; TERANISHI et al. 1984) or leukemic B cells (YOSHIZAKI et al. 1982) to induce Ig production, provided other factors such as IL-2 or BCGF were available. Furthermore, BSF-2 was found to act on B-cell lines at the mRNA level and induce biosynthesis of secretory type Ig (KIKUTANI et al. 1985). BSF-2 was purified to homogeneity (HIRANO et al. 1985) from the culture supernatant of a human T-cell leukemia virus type I (HTLV-I) transformed T-cell line and its partial N-terminal amino acid sequence was determined (HIRANO et al. 1987). Based on these findings, the cDNA encoding human BSF-2 was cloned from a T-cell line (HIRANO et al. 1986).

II. Hybridoma/Plasmacytoma Growth Factor

In 1972, NAMBA and HANAOKA demonstrated that a murine adherent phagocytic cell line produces a growth factor(s) which is required for the MOPC 104E plasmacytoma cell line to grow in vitro (NAMBA and HANAOKA 1972). Growth factors for plasmacytoma were also reported (METCALF 1974; CORBEL and MELCHERS 1984; NORDAN and POTTER 1986). A growth factor(s) for murine hybridomas was found in the supernatant of human endothelial cells (ASTALDI et al. 1980) and human monocytes (AARDEN et al. 1985). A hybridoma growth factor (IL-HP1) derived from a murine helper T-cell clone (VAN SNICK et al. 1986) and a plasmacytoma growth factor derived from murine macrophage cell line, P388D1 (NORDAN et al. 1987), were purified and their partial N-terminal amino acid sequences were determined. The data indicated that both growth factors are identical, although the N-terminal amino acid sequence of the murine hybridoma/plasmacytoma growth factor (HPGF) was found to have no homology with that of human HPGF derived from the osteosarcoma cell line MG-63 (VAN DAMME et al. 1987a) or peripheral blood monocytes (BRAKENHOFF et al. 1987). However, the molecular cloning of murine IL-HP1 demonstrated that murine HPGF has a sequence homology with the human equivalent (VAN SNICK et al. 1988).

III. IFN-β2 and 26-kDa Protein

In 1980, an inducible mRNA species of about 13S coding for a novel human fibroblast-type IFN, named IFN-β2, was reported (WEISSENBACH et al. 1980). It was shown that the isolated cDNA clone for such an induced mRNA was transcribed in vitro into a protein of 23–26-kDa and the RNA selected by hybridization to this cDNA generated IFN with antiviral activity. Since this activity was neutralized by antibodies against IFN-β, the translation product of the 1.3 kilobase (kb) RNA was designated IFN-β2. It was demonstrated that an inducible 13S–14S mRNA species in human fibroblasts can be translated in vitro to a 26-kDa protein which was considered to be identical with IFN-β2; however, the antiviral activity was not detected in the 26-kDa protein (CONTENT et al. 1982). The 26-kDa protein was shown to be induced in

```
                                                -28                        -20
                                                Met AsN Ser Phe Ser Thr Ser Ala Phe Gly Pro Val Ala Phe
GAGAAGCTCTATCTCCCCTCCAGGAGCCCAGCT ATG AAC TCC TTC TCC ACA AGC GCC TTC GGT CCA GTT GCC TTC
              -10                                         50
                                                        ┃  1
Ser Leu Gly Leu Leu Leu Val Leu Pro Ala Ala Phe Pro Ala▼Pro Val Pro Pro Gly Glu Asp Ser Lys
TCC CTG GGG CTG CTC CTG GTG TTG CCT GCT GCC TTC CCT GCC CCA GTA CCC CCA GGA GAA GAT TCC AAA
                        100                          2
10                                            20                          30
Asp Val Ala Ala Pro His Arg GlN Pro Leu Thr Ser Ser Glu Arg Ile Asp Lys GlN Ile Arg Tyr Ile
GAT GTA GCC GCC CCA CAC AGA CAG CCA CTC ACC TCT TCA GAA CGA ATT GAC AAA CAA ATT CGG TAC ATC
6      150                                                        200
                                40                  •••••••••••      50
Leu Asp Gly Ile Ser Ala Leu Arg Lys Glu Thr Cys AsN Lys Ser AsN Met Cys Glu Ser Ser Lys Glu
CTC GAC GGC ATC TCA GCC CTG AGA AAG GAG ACA TGT AAC AAG AGT AAC ATG TGT GAA AGC AGC AAA GAG
                        250                1                          3
Ala Leu Ala Glu AsN AsN Leu AsN Leu Pro Lys Met Ala Glu Lys Asp Gly Cys Phe GlN Ser Gly Phe
GCA CTG GCA GAA AAC AAC CTG AAC CTT CCA AAG ATG GCT GAA AAA GAT GGA TGC TTC CAA TCT GGA TTC
            300                                                       350
80                                            90                          100
AsN Glu Glu Thr Cys Leu Val Lys Ile Ile Thr Gly Leu Leu Glu Phe Glu Val Tyr Leu Glu Tyr Leu
AAT GAG GAG ACT TGC CTG GTG AAA ATC ATC ACT GGT CTT TTG GAG TTT GAG GTA TAC CTA GAG TAC CTC
                            110                  400                120
GlN AsN Arg Phe Glu Ser Ser Glu Glu GlN Ala Arg Ala Val GlN Met Ser Thr Lys Val Leu Ile GlN
CAG AAC AGA TTT GAG AGT AGT GAG GAA CAA GCC AGA GCT GTG CAG ATG AGT ACA AAA GTC CTG ATC CAG
                    450
            130                              140                •••••••••••••
Phe Leu GlN Lys Lys Ala Lys AsN Leu Asp Ala Ile Thr Thr Pro Asp Pro Thr Thr AsN Ala Ser Leu
TTC CTG CAG AAA AAG GCA AAG AAT CTA GAT GCA ATA ACC ACC CCT GAC CCA ACC ACA AAT GCC AGC CTG
            500                         4                            550
150                                           160                          170
Leu Thr Lys Leu GlN Ala GlN AsN GlN Trp Leu GlN Asp Met Thr Thr his Leu Ile Leu Arg Ser Phe
CTG ACG AAG CTG CAG GCA CAG AAC CAG TGG CTG CAG GAC ATG ACA ACT CAT CTC ATT CTG CGC AGC TTT
    8                                         600
                                180
Lys Glu Phe Leu GlN Ser Ser Leu Arg Ala Leu Arg GlN Met
AAG GAG TTC CTG CAG TCC AGC CTG AGG GCT CTT CGG CAA ATG TAGCATGGGCACCTCAGATTGTTGTTGTTAATGGG
                                650                        700
CATTCCTTCTTCTGGTCAGAAACCTGTCCACTGGGCACAGAACTTATGTTGTTGTTCTCTATGGAGAACTAAAAGTATGAGCGTTAGGACACTA
                            750
TTTTAATTATTTTTAATTTATTAATATTTAAATATGTGAAGCTGAGTTAATTTATGTAAGTCATATTTATATTTTAAGAAGTACCACTTGA
    800                        850
AACATTTTATGTATTAGTTTTGAAATAATAATGGAAAGTGGCTATGCAGTTTGAATATCCTTTGTTTCAGAGCCAGATCATTTCTTGGAAA
        900                        950
GTGTAGGCTTACCTCAAATAAATGGCTAACTTATACATATTTTTAAAGAAATATTTATATTGTATTTATATAATGTATAAATGGTTTTTAT
        1000                        1050
ACC┌AATAAA┐TGGCATTTTAAAAAAATTCAGCAAAAAAAAAAAAAAAAAAAAAAAAAAA
        1100
```

Fig. 1. Nucleotide and deduced amino acid sequences of human IL-6 cDNA. *Underlined* amino acids represent the region for which amino acid sequences of purified IL-6 are available (see HIRANO et al. 1987 for the numbering of the peptide fragments shown here as *large numbers* at the left of the underlined sequences). The potential N-glycosylation sites are indicated by *asterisks*. The presumed polyA addition signal sequence is *boxed*. Amino acids are numbered starting at Pro1 of the mature IL-6 protein sequence (HIRANO et al. 1987). (From HIRANO et al. 1986)

fibroblasts upon stimulation with IL-1 (Content et al. 1985; Van Damme et al. 1985). The nucleotide sequences of the cDNAs encoding human IFN-β2 and 26-kDa protein were determined, and showed the identity of these molecules (Zilberstein et al. 1986; Haegeman et al. 1986).

```
                            -24                    -20
                            MET LYS PHE LEU SER ALA ARG ASP PHE HIS PRO VAL
         CACCAAGAACGATAGTCAATTCCAGAAACCGCTATG AAG TTC CTC TCT GCA AGA GAC TTC CAT CCA GTT
                                                                   50
              -10                                   1
         ALA PHE LEU GLY LEU MET LEU VAL THR THR THR ALA PHE PRO THR SER GLN VAL ARG ARG
         GCC TTC TTG GGA CTG ATG CTG GTG ACA ACC ACG GCC TTC CCT ACT TCA CAA GTC CGG AGA
                                            100
              10                    20
         GLY ASP PHE THR GLU ASP THR THR PRO ASN ARG PRO VAL TYR THR THR SER GLN VAL GLY
         GGA GAC TTC ACA GAG GAT ACC ACT CCC AAC AGA CCT GTC TAT ACC ACT TCA CAA GTC GGA
                                    150
              30                            40
         GLY LEU ILE THR HIS VAL LEU TRP GLU ILE VAL GLU MET ARG LYS GLU LEU CYS ASN GLY
         GGC TTA ATT ACA CAT GTT CTC TGG GAA ATC GTG GAA ATG AGA AAA GAG TTG TGC AAT GGC
                            200
              50                                    60
         ASN SER ASP CYS MET ASN ASN ASP ASP ALA LEU ALA GLU ASN ASN LEU LYS LEU PRO GLU
         AAT TCT GAT TGT ATG AAC AAC GAT GAT GCA CTT GCA GAA AAC AAT CTG AAA CTT CCA GAG
         250                                                       300
              70                            80
         ILE GLN ARG ASN ASP GLY CYS TYR GLN THR GLY TYR ASN GLN GLU ILE CYS LEU LEU LYS
         ATA CAA AGA AAT GAT GGA TGC TAC CAA ACT GGA TAT AAT CAG GAA ATT TGC CTA TTG AAA
                                            350
              90                    100
         ILE SER SER GLY LEU LEU GLU TYR HIS SER TYR LEU GLU TYR MET LYS ASN ASN LEU LYS
         ATT TCC TCT GGT CTT CTG GAG TAC CAT AGC TAC CTG GAG TAC ATG AAG AAC AAC TTA AAA
                                    400
              110                           120
         ASP ASN LYS LYS ASP LYS ALA ARG VAL LEU GLN ARG ASP THR GLU THR LEU ILE HIS ILE
         GAT AAC AAG AAA GAC AAA GCC AGA GTC CTT CAG AGA GAT ACA GAA ACT CTA ATT CAT ATC
                            450
              130                           140
         PHE ASN GLN GLU VAL LYS ASP LEU HIS LYS ILE VAL LEU PRO THR PRO ILE SER ASN ALA
         TTC AAC CAA GAG GTA AAA GAT TTA CAT AAA ATA GTC CTT CCT ACC CCA ATT TCC AAT GCT
                                500
              150                           160
         LEU LEU THR ASP LYS LEU GLU SER GLN LYS GLU TRP LEU ARG THR LYS THR ILE GLN PHE
         CTC CTA ACA GAT AAG CTG GAG TCA CAG AAG GAG TGG CTA AGG ACC AAG ACC ATC CAA TTC
         550                                                       600
              170                           180
         ILE LEU LYS SER LEU GLU GLU PHE LEU LYS VAL THR LEU ARG SER THR ARG GLN THR
         ATC TTG AAA TCA CTT GAA GAA TTT CTA AAA GTC ACT TTG AGA TCT ACT CGG CAA ACCTAGT
                                                       650
         GCGTTATGCCTAAGCATATCAGTTTGTGGACATTCCTCACTGTGGTCAGAAAATATATCCTGTTGTCAGGTATCTGACT
                           700
         TATGTTGTTCTCTACGAAGAACTGACAATATGAATGTTGGGACACTATTTTAATTATTTTTAATTTATTGATAATTTAA
         750                           800
         ATAAGTAAACTTTAAGTTAATTTATGATTGATATTTATTATTTTTATGAAGTGTCACTTGAAATGTTATATGTTATAGT
                       850                           900
         TTTGAAATGATAACCTAAAAATCTATTTGATATAAATATTCTGTTACCTAGCCAGATGGTTTCTTGGAATGTATAAGTT
                           950
         TACCTCAATGAATTGCTAATTTAAATATGTTTTTAAAGAAATCTTTGTGATGTATTTTTATAATGTTTAGACTGTCTTC
                1000                                   1050
         AAACAAATAAATTATATTATATTTAAAAAAAA
                1090
```

Fig. 2. Nucleotide and deduced amino acid sequences of murine IL-6 cDNA. Amino acid numbering is based on the N-terminal amino acid sequence of secreted murine IL-6 (HP1) (van Snick et al. 1986). *Underlined* sequences represent regions for which amino acid sequences are available. The presumed polyA addition signal sequence is indicated by a *thick underline*. (From van Snick et al. 1988)

C. Structure of IL-6

Figure 1 shows the nucleotide and deduced amino acid sequences of cloned human IL-6 cDNA (HIRANO et al. 1986). Human IL-6 consists of 212 amino acids including a hydrophobic signal sequence of 28 amino acids. It includes two potential N-glycosylation sites. The N-terminal amino acid residues of T cell line-derived IL-6 and that of osteosarcoma cell line-derived IL-6 (VAN DAMME et al. 1987a) are Pro and Ala, respectively, showing that there is a certain heterogeneity in the cleavage site. The cDNA encoding murine IL-6 was cloned from a T-cell clone (VAN SNICK et al. 1988) and showed that murine IL-6 consists of 211 amino acids with a typical signal sequence of 24 residues (Fig. 2). It includes no N-glycosylation site but several potential O-glycosylation sites. Comparison of the cDNA sequence of murine IL-6 with that of human IL-6 shows a homology of 65% at the DNA level and of 42% at the protein level. Remarkable homology is identified in a 3' untranslated region (>70% homology) which consists predominantly of A and T nucleotides. The AT run has been observed in the 3' untranslated regions of numerous mRNAs for lymphokines, cytokines, and protooncogenes. These sequences are thought to be involved in specific mRNA degradation (SHAW and KAMEN 1986) and important in regulating the stability of the IL-6 mRNA. Furthermore, four cysteine residues are completely conserved and the region (residues 42–102) containing the four cysteine residues showed the highest homology (57%), suggesting that the cysteine-rich middle region of the mature protein may play a critical role in IL-6 activity (Fig. 3). The sequence of IL-6 was compared to other known proteins. Only granulocyte colony-stimulating factor (G-CSF) showed a significant homology with IL-6; the positions of the four cysteine residues of IL-6 match with those of G-CSF (Fig. 3). This suggests a similarity in the tertiary structure of these two molecules and may indicate some functional similarity; IL-6 and G-CSF act not only on hematopoietic stem cells at G_0 of the cell cycle, but also on the murine myeloid leukemic cell line M1, as described in Sect. F. II, although IL-6 and G-CSF act through different receptor molecules expressed on target cells (Sect. E. I).

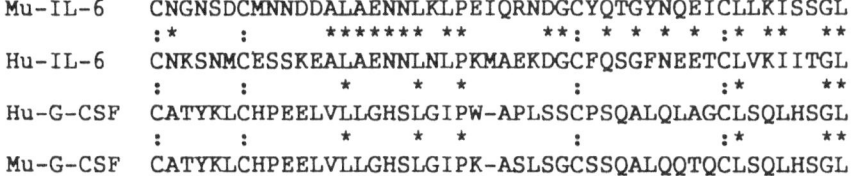

Fig. 3. Conserved cysteine residues among human and murine IL-6 and G-CSF. Homologous amino acid residues are shown by *asterisks*. *Double dots* represent cysteine residues. Sequences of human IL-6, murine IL-6, human G-CSF, and murine G-CSF are taken from HIRANO et al. (1986); VAN SNICK et al. (1988); NAGATA et al. (1986), and TSUCHIYA et al. (1986), respectively

D. Gene Structure and Expression of IL-6

I. Structure of the IL-6 Gene

The chromosomal DNA segments of human (YASUKAWA et al. 1987) and mouse (TANABE et al. 1988) were isolated. The whole nucleotide sequences of the genomic DNAs encoding human and mouse IL-6 are shown in Figs. 4 and 5, respectively. The complete human and mouse IL-6 genes are approximately 5 kb and 7 kb in length, respectively, and both consist of five exons and four introns (Fig. 6). The gene organization of IL-6 shows a distinct similarity with the G-CSF gene. Both genes have the same number of exons

```
                                                                          -1201
                                                              GGATCCTCCTGCAAGAGACACC
                                                                          -1081
ATCCTGAGGGAAGAGGGCTTCTGAACCAGCTTGACCCAATAAGAAATTCTTGGGTGCCGACGCGGACAGAGATTCAGAGCCTAGAGCCGTGCCTGCGTCCGTACTTTCCTTCTAGCTTCT
                                                                          -961
TTTGATTTCAAATCAAGACTTAGAGGGAGAGGGAGCGATAAACACAAACTCTGCAAGATGCCACAAGGTCCTCCTTTGACATCCCCAACAAAGAGGTGAGTAGTAATCTCCCCCCTTTCTG
                                                                          -841
CCCTGAACCAAGTGGGCTTCAGTAATTTCAGGGCTCCAGGAGACTGGGCATGCAGGTGCCGATGAAACAGTGGTGAAGAGACTCAGTGGCAGTGGGGAGAGCACTGGCAGCACAGGCAAA
                                                                          -721
CCTCTGGCACAAGAGCAAAGTCCTACTGGAGATTCCAAGGGTCACTTGGGAGAGGGCAGGCAGCAGCCAACCTCCTCTAAGTGGGCTGAAGCAGGTGAAGAAATGGCAGACAAGCGCGGT
                                                                          -601
GGCAAAAAGGAGTCACACACTCCACCTGGAGACGCCTTGAAGTAACTGCACGAAATTTGAGGGTGGCCAGGCAGTCTACAACAGCCGCTCACAGGGAGAGCCAGAACACAGAAGAACTCA
                                                                          -481
GATGACTGGTAGTATTACCTTCTTCATAATCCAGGCTTGGGGGGCTGCGATGGAGTCAGAGGAAACTCAGTTCAGAACATCTTTGGTTTTTACAAATACAAATTAACTGGAACGCTAAAT
                                                                          -361
TCTAGCCTGTTAATCTGGTCACTGAAAAAAAAATTTTTTTTTTTTCAAAAAACATAGCTTTAGCTTATTTTTTTTCTCTTTGTAAAACTTCGTGCATGACTTCAGCTTTACTCTTTGTCAA
                                                                          -241
GACATGCCAAAGTGCTGAGTCACTAATAAAAGAAAAAAAGAAAGTAAGGAAGAGTGGTTCTGCTTCTTAGCGCTAGCCTCAATGACGACCTAAGCTGCACTTTTCCCCCTAGTTGTGTC
                                                                          -121
TTGCGATGCTAAAGGACGTCACATTGCACAATCTTAATAAGGTTTCCAATCAGCCCCACCCGCTCTGGCCCCACCCTCACCCTCCAACAAGATTTATCAATGTCGGATTTTCCCATGA
                                                                          -1
GTCTCAATATTAGAGTCTCAACCCCCAATAAATATAGGACTTGGACATGTCTGAGGCTCATTCTGCCCTCGAGCCACCGGGAACGAAAGAGAAGCTCTATCTCCCCTCCAGGAGCCCAGCT
1                                                                         120
ATGAACTCCTTCTCCACAAGTAAGTGCAGGAAATCCTTAGCCCTGGAACTGCCAGCCGGTCGAGCCCTGTGTGAGGGAGGGGTGTGTGGCCCAGGGATGCGGGGCGCCAGCAGCAGAGGC
MetAsnSerPheSerThrS
                                                                          240
AGGCTCCCAGCTGTGCTGTCAGTCACCCCTGCGCTCGCTCCCCTCCGGCACAGGCGCCTTCGGTCCAGTTGCCTTCTCCCTGGGGCTGCTCCTGGTGTTGCCTGCTGCCTTCCCTGCCCC
                                             erAlaPheGlyProValAlaPheSerLeuGlyLeuLeuLeuValLeuProAlaAlaPheProAlaPr
                                                                          360
AGTACCCCCAGGAGAAGATTCCAAAGATGTAGCCGCCCCACACAGACAGCCACTCACCTCTTCAGAACGAATTGACAAACAAATTCGGTACATCCTCGACGGCATCTCAGCCCTGAGAAA
oValProProGlyGluAspSerLysAspValAlaAlaProHisArgGlnProLeuThrSerSerGluArgIleAspLysGlnIleArgTyrIleLeuAspGlyIleSerAlaLeuArgLy
                                                                          480
GGAGTGGGAAGGCTTGGCGATGGGGTTGAAGGGCCGGTGCGATGCGTCTCCCCTCCCTGCGTGTGGGGGGGCTGCCTGCATAAGGAGGTCTTTGCTGGGTTCTAGAGCACTGTAGATT
sGlu
                                                                          600
TGAGGCCAACGACCTAGACTGACTTCTGTATTTATCCTTTGCTGGTGTCAGGAGGTTCCTTTCCTTTCTGGAAAATGCAGAATGGGTCTGAAATCCATGCCCACCTTTGGCATGAGCTGA
                                                                          720
GGGTTATTGCTTCTCAGGGCTTCCTTTTCCCTTTCCAAAAAATTAGGTCTGTGAAGCTCCTTTTTGTCCCCGGGCTTTGGAAGGACTAGAAAAGTGCCACCTGAAAGGCATGTTCAGCT
                                                                          840
TCTCAGAGCCAGTTGCAGTACTTTTTGGTTATGTAAACTCAATGGTTAGGATTCCTCAAAGCCATTCCAGCTAAGATTCATACCTCAGAGCCCACCAAAGTGGCAAATCATAAATAGGTTA
                                                                          960
AAGCCATCTCCCCACTTTCAATGCAAGGTATTTTGGTCCTGTTTGGTAGAAAGAAAAGAACACAGGAGGGGAGATTGGGAGCCCAGACTCGAATTCTGGTTCTGCCAAACCAGCCTTGTGA
                                                                          1080
TCTTTGGGTAAATTCCCTACCACCTCTGGACTCCATCAGTAAAATTGGGGGTGGACTAGGTGATCTCATAGATCCTTCCTGCTGGAACATTCTATGGCTTGAATTATATTCTCCTAATTAT
                                                                          1200
TGTCAAAATTGCTGTTATTAAGTATCTACTGTGTGCCAGGCACTTTAAATAAATATTGTGTCTAATCTTCAAAACAAATTTGCAAGGAAGGTTTTTGGAGATAAGCAAACTGAGACTCAG
                                                                          1320
GATTAAGTAACACACCTAAAGTCAAAGGTGAGCTTGGAACTGAACCCAAGTGTGCCCCCACTCCACTGGAATTTGCTTGCCAGGATGCCAATGAGTTGTAGCTTCATTTTTCTTAGAGAC
                                                                          1440
TTTCCTGGCTGTGGTTGAACAATGAAAAGGCCCTCTAGTGGTGTTTGTTTTTAGGGAACTTAGGTGATAACAATTCTGGTATTCTTTCCCAGACATGTAACAAGAGTAACATGTGTGAAAG
                                                              ThrCysAsnLysSerAsnMetCysGluSe
                                                                          1560
CAGCAAAGAGGCACTGGCAGAAAAACAACCTGAACCTTCCAAAGATGGCTGAAAAAGATGGATGCTTCCAATCTGGATTCAATGAGGTACCAACTTGTCGCACTCACTTTTCACTATTCCT
rSerLysGluAlaLeuAlaGluLysAsnAsnLeuAsnLeuProLysMetAlaGluLysAspGlyCysPheGlnSerGlyPheAsnGlu
                                                                          1680
TAGGCCAAAACTTCTCCCTCTTGCATGCAGTCCTGTATACATATAGATCCAGGCAGCAACAAAAAGTGGGTAAATGTAAAGAATGTTATGTAAATTTCATGAGGAGGCCAAGTTCAAGCTT
                                                                          1800
TTTTAAAGGCAGTTTATTCTTGGACAGGTATGGCCAGAGATGGTGCCACTGTGGTGAGATTTTAACAACTGTCAAATGTTTAAAACTCCCACAGGTTTAATTAGTTCATCCTGGGAAAGG
                                                                          1920
TACTCGCAGGGCCTTTTCCCTCTCTGGCTGCCCCTGGCAGGGTCCAGGTCTGCCCTCCCTCCCTGCCCAGCTCATTCTCCACAGTGAGATAACCTGCACTGTCTTCTGATTATTTTATAA
                                                                          2040
AAGGAGGTTCCAGCCCAGCATTAACAAGGGCAAGAGTGCAGGAAGAACATCAAGGGGGACAATCAGAGAAGGATCCCCATTGCCACATTCTAGCATCTGTTGGGCGTTTGGATAAACTAA
                                                                          2160
TTACATGGGGCCTCTGATTGTCCAGTTATTTAAAATGGTGCTGTCCAATGTCCCAAAACATGCTGCCTAAGAGGTACTTGAAGTTCTCTAGAGGAGCAGAGGGAAAAGATGTCGAACTGT
                                                                          2280
```

Fig. 4

```
GGCAATTTTAACTTTTCAAATTGATTCTATCTCCTGGCGATAACCAATTTTCCCACCATCTTTCCTCTTAGGAGACTTGCCTGGTGAAAATCATCACTGGTCTTTTGGAGTTTGAGGTAT
                                                                 GluThrCysLeuValLysIleIleThrGlyLeuLeuGluPheGluValT
                                                                                                              2400
ACCTAGAGTACCTCCAGAACAGATTTGAGAGTAGTGAGGAACAAGCCAGAGCTGTGCAGATGAGTACAAAAGTCCTGATCCAGTTCCTGCAGAAAAAGGTGGGTGTGTCCTCATTCCCTC
yrLeuGluTyrLeuGluAsnArgPheGluSerSerGluGluGlnAlaArgAlaValGlnMetSerThrLysValLeuIleGlnPheLeuGlnLysLys
                                                                                                              2520
AACTTGGTGTGGGGGAAGACAGGCTAAAGACAGTGTCCTGGACAACTCAGGGATGCAATGCCACTTCCAAAAGAGAAGGCTACACGTAAACAAAAGAGTCTGAGAAATAGTTTCTGATTG
                                                                                                              2640
TTATTGTTAAATCTTTTTTTGTTTGTTTGGTTGGTTGGCTCTCTTCTGCAAAGGACATCAATAACTGTATTTTAAACTATATATTAACTGAGGTGGATTTTAACATCAATTTTTAATAGT
                                                                                                              2760
GCAAGAGATTTAAAACCAAAGGCGGGGGGGCGGGCAGAAAAAAGTGCCATCCAACTCCAGCCAGTGATCCACAGAAACAAAGACCAAGGAGCACAAAATGATTTTAAGATTTTAGTCATT
                                                                                                              2880
GCCAAGTGACATTCTTCTCACTGTGGTTGTTTCAATTCTTTTTCCTACCTTTTACCAGAGAGTTAGTTCAGAGAAATGGTCAGAGACTCAAGGGTGGAAAGAGGTACCAAAGGCTTTGGC
                                                                                                              3000
CACCAGTAGCTGGCTATTCAGACAGCAGGGAGTAGACTTGCTGGCTAGCATGTGGAGGAGCCAAAGCTCAATAAGAAGGGGCCTAGAATGAAACCCTTGGTGCTGATCCTGCCTCTGCCA
                                                                                                              3120
TTTCTACTTAAGCCAGGGTTTCTCATATGTTAACATGCTAGGGAATTCCCTGGGCATCTTCTTGTGGTGTGGAGTCTGACTTAGCCAAGCCTCGGGTGGGTTTGAGGGTCAAATTTCACC
                                                                                                              3240
AGGCTTATATCCCTGGTGATGCTGCAGAATTCCAGGACCACACTTGGAGGTTTAAGGCCTTCCACAAGTTACTTATCCCATATGGTGGGTCTATGGAAAGGTGTTTCCCAGTCCTCTTTA
                                                                                                              3360
CACCACCAGATCAGTGGTCTTTCAACAGATCCTAAAGGGATGGTGAGAGGGAAACTGGAGAAAAGTATCAGATTTAGAGGCCATGAAGAACCCATATTAAAATGCCTTTAAGTATGGGCT
                                                                                                              3480
CTTCATTCATATACTAAATATGAACTATGTGCCAGGCATTATTTCATATGCACAGAATACAAACAAATAAGATAGTGATGCTGGTCAGGCTTGGTGGCTCATGCCTGTATTCCCTAAACTT
                                                                                                              3600
TGGGAGCCTAAGGTGAGAACTCCTTGAACTCCTAAGGCCAGGAGTTCAAGACCAGCCTGGATAACATAGCAAGACCCCATCTCTACAAAAAACCAAAACCAAACAAACAAAAATGATAGT
                                                                                                              3720
GGTGCTTCCCTCAGGATGCTTGTGGTCTAATGGGAGACAGAACAGCAAAGGGATGATTAGAAGTTGGTTGCTGTGAGCCAGGCACAGTGCTATATAATCCCAGCGCTATGGGAGGCTGAG
                                                                                                              3840
GTGGGTGGATCATTTAGGCCAGGAGTTTAAGACCAGCCTGGTCAACATGGTAAAACCCCATCTTACTTAAAAATACAAAAAAGTTAGCCAGGCATGGTGGCATACACCTGTAACCCAGCT
                                                                                                              3960
ACTCAGGAGGCTGAGGCACATGAATCACTTGAACCCAGGAGGCAGAGGTTGCTGTGTGCACCACTGCACTCCAGCCTGGGTGACAGAACGAGACCTTGACTCAAAAAAAAAAAAAAGAAGTT
                                                                                                              4080
TGTTGCTATGGAAGGGTCCTACTCAGAGCAGGCACCCCAGTTAATCTCATTCACCCCACATTTCACATTTGAACATCATCCCATAGCCCAGAGCATCCCTCCACTGCAAAGGATTTATTC
                                                                                                              4200
AACATTTAAACAATCCTTTTTTACTTTCATTTTCCTTCAGGCAAAGAATCTAGATGCAATAACCACCCCTGACCCAACCACAAATGCCAGCCTGCTGACGAAGCTGCAGGCACAGAACCAG
                                                 AlaLysAsnLeuAspAlaIleThrThrProAspProThrThrAsnAlaSerLeuLeuThrLysLeuGlnAlaGlnAsnGln
                                                                                                              4320
TGGCTGCAGGACATGACAACTCATCTCATTCTGCGCAGCTTTAAGGAGTTCCTGCAGTCCAGCCTGAGGGCTCTTCGGCAAATGTAGCATGGGCACCTCAGATTGTTGTTGTTAATGGGC
TrpLeuGlnAspMetThrThrHisLeuIleLeuArgSerPheLysGluPheLeuGlnSerSerLeuArgAlaLeuArgGlnMet***
                                                                                                              4440
ATTCCTTCTTCTGGTCAGAAACCTGTCCACTGGGCACAGAACTTATGTTGTTCTCTATGGAGAACTAAAAGTATGAGCGTTAGGACACTATTTTAATTATTTTTAATTTATTAATATTTA
                                                                                                              4560
AATATGTGAAGCTGAGTTAATTTATGTAAGTCATATTTATATTTTAAGAAGTACCACTTGAAACATTTTATGTATTAGTTTTGAAATAATAATGGAAAGTGGCTATGCAGTTTGAATATC
                                                                                                              4680
CTTTGTTTCAGAGCCAGATCATTTCTTGGAAAGTGTAGGCTTACCTCAAATAAATGGCTAACTTATACATATTTTTAAAGAAATATTTATATTGTATTTATATAATGTATAAATGGTTTT
TATACCAATAAATGGCATTTTAAAAAAATTCAGCAACTTTGAGTGTGTCACGTGAAGCTT
```

Fig. 4. The nucleotide sequence of genomic DNA encoding human IL-6. The amino acid sequences of the coding region are also shown. Nucleotides are numbered starting at A of the translation start codon. The TATA-like sequences are *underlined*. Enhancer-like elements found in the IFN-β_1 gene are *boxed*. The sequence homologous to that found in the 5'-flanking region of the IL-2 gene is indicated by a box with a *dashed line*. (From YASUKAWA et al. 1987)

and introns and the sizes of the exons are strikingly similar, although the lengths of the introns differ (YASUKAWA et al. 1987). As described in Sect. C, the primary structure of IL-6 shows a significant sequence homology with G-CSF. Taken together, these findings suggest that the genes for IL-6 and G-CSF might be evolutionarily derived from a common ancestor gene.

The sequence similarity in the coding region of human and mouse IL-6 genes is about 60%, whereas the 3' untranslated region and the first 300 bp sequence of the 5' flanking region are highly conserved (>80%; TANABE et al. 1988). Several sequence blocks with high homology are also found in the introns. The human IL-6 gene contains three transcriptional initiation sites and three TATA-like sequences (YASUKAWA et al. 1987). These TATA boxes are strictly conserved at similar locations relative to the start codon. Furthermore, sequences similar to transcriptional enhancer elements such as the c-*fos* serum responsive element and the consensus sequences for cyclic AMP induction, AP-1 binding, and glucocorticoid receptor binding are identified within the highly conserved 5′ flanking regions both human and mouse genes (YASUKAWA et al. 1987; TANABE et al. 1988; WALTHER et al. 1988) (Fig. 7).

II. Chromosomal Location of the IL-6 Gene and Polymorphism

The gene for human IL-6 was mapped to chromosome 7 (SEHGAL et al. 1986) and has recently been localized to 7p15-21 by in situ hybridization (Y. CHEN et al. 1987). Linkage studies involving the IL-6 gene and 27 other chromosome 7 markers have localized it at 7p21-22. Therefore, the IL-6 gene is most likely to be localized in 7p21 (BOWCOCK et al. 1988).

Fig. 5.

```
TTGTTCCCCATTCTAAGGAGGAATGAAGTATCCACATGTTGGTCTTCCTTCTTCTTGATTTTCTTGTGTTTTGGAAATTGTACCTTGGGTATTCTAAGTTTCTGGGCTAATAATATCCAC 4560
TTATCAGTGAGTGCATATCAAGTGACTTATTTTGTGATTGGGTTACCTCACTAAGGATCATACTCTCCAGATACATCCATTTGACCAAGAATTTCATAAATCCATTGTTTTTAATAGCTG 4680
AATAGTACTCCATTTGTAAATGTACCACATTTTCTGTATCCATTCCTCTGTTGAGGACATCTGGGTTCTTTCCAGCTTCTGGCTATTATGAATAAGGCTGCTATGAACATAGTGGAGCAT 4800
ATGTCCTTATTACCAGTTGGAACATTCTGCGTATATGCCCAGAAGAGGTATTGCTGGATCTTCCGGTAGTACTATGTCTAATTTTTTTGAGGAACAGCCAGACTGATTTCCAGAGTGGTTA 4920
TACAAGCTTGCAATCCCAACAGCAATGGAGGAGTGTTCCTCTTTCTCCACATCCTTACCAGCATCTGCTGTCACCTGAATTTTTGATCTTAGCCATTCTGACTGGTGTGAGGTGGAATCT 5040
CAGGGTTGTTTTGATTTGCATTTCCCTGATGATTAAGGATGTTGAACATTTTTTCAAGTGCTTCTCAGCCATTCAGTATTCCTCAGTTGAGAATTCTTTGTTTAGCTTTGTACACATTTT 5160
TAATGGGGTTATTTGAATTTGAATTTCTGGAGTTCAGCTTCTTGAGCTCTTTGTATATATTGGATATTAGTCCCCTATCAGATTTAGGATTGGTAAAAATCCTTTCCCAATCTGTTGGTG 5280
GCCTTTTTGTCTTTATTGACAGTGTCTTTTCCCTTACAGAAGCTTTGCAATTTTATGAGGTCCATTTGTCGATTCTCGATTTTACTGTACAAGCCATTGCTGTTCTGTTCAGGAATTTTT 5400
CCCCTGTTCCCATATCTACGAGGGTTTTTTTCCCACTTTCTCCTCTATAAATTTCAGTGTCTCTGGTTTTATGTGGAGTTCTTTGATCCACTTAGACTTGAGCTTTGTACAAGGAGATAA 5520
GAATGGATCAATTCGGGCTGGAGAGATGGCTCAGTGGTTAAGAGCACTGACTGCTCTTTCAGAAGTCCTGAGTTCAAATCCCAGCAACATGGTGGTTCATAACCATCTATAATAAGATCT 5640
GATGTCCTTTTCTGGTGTGTCTGAAGACAGCTACAGTGTACTTACATGTAATAAATAAATAAAACTAGTTATTTAAAAAAAAAAAAAAAGAATGGATCAATTCACACTCTTCTACATGAT 5760
AACCACCAGTTGAGCCCGCACCAATTGTTGAAAATGCTCTCTTTTTCCCACTGGATGGTTTTAGCTCCCTTGTCAAAGATCAAGTGACCATAAGTGTGTGGGTTCATTTCTGTGTCTTCA 5880
ATTGTGTTCCATTGACTACCTGTCTGTTGCTGTATCAATACCATGCAGTTTTTTTTTATCACAATTGCTCTGTAGTTTTTTTTTTTTTTTGACAGCACGAATCTTATTCTCAAATTGAATCT 6000
ATTCCTAGAAGAACTGACTTCCTTTTCCATTTACTTATAGGAAATTTGCCTATTGAAAATTTCCTCTGGTCTTCTGGAGTACCATAGCTACCTGGAGTACATGAAGAACAACTTAAAAG 6120
                                                  GluIleCysLeuLeuLysIleSerSerGlyLeuLeuGluTyrHisSerTyrLeuGluTyrMetLysAsnAsnLeuLysA
ATAACAAGAAAGACAAAGCCAGAGTCCTTCAGAGAGATACAGAAACTCTAATTCATATCTTCAACCAAGAGGTGAGTGCTTCCCCATCTCTCATGCAGTGTGGGAAAGAGACACCCGGCA
spAsnLysLysAspLysAlaArgValLeuGlnArgAspThrGluThrLeuIleHisIlePheAsnGlnGlu
CCCTCAGGGTAGCGGCACTTTTTCCAGACAGCTGCTCAGAAGGGAGGAGAGTCTGAACAACAGGCCTTGCTTTGTTTTACTTTGGGGTTTTGTTTGAGGTTCTCTTTTGCAAAGAACATCAA 6360
TACCTGCTTTAAACTGTATTAATAGAATGTTACTAATTGTGTAAGAGGTATGAAAACTATGACAGCCATACATAGTCACCCATTATGAGAGCACAGAGACAAAAGTGACTTTAATATTTA 6480
ATCCTTGGCAAGTGACATTTTTGTAACCAGAGTTCTAATGCAGAGAAGTTTAGCCAAAAGCTAAAATGTCAGGGAAGGAGGGATCAGGGCTTCTGGCTACATTAGCCAGAAGAAGAATGG 6600
TAGATATGAGATAATGACTCAGAGTGTGGGCGAACAAAGCCAGATGCAATAAGAAGGGCCTGGAATGAAACCCTCTTGCTAAGGCTGCTTTTGCCACTTGTAGTTTCTTGCCTTAAACCA 6720
GAGAGTTTTTGCTAAGGCTGCTTTTGCCACTTGTAGTTTCTTGCCTTAAACCAGAGAGTTTCCCAATTTAATGTGCACAGGAACCACTTAGGGTCTTGTTCCAGCAGGGTCTTACTTAGG 6840
AGGTCTAGGAAGGGAACTAAGATTCATTTTTGTAAACAGCTGCTTGTTTATGACCCTGCTATATAGTATTCAGGCTCCTAAGTAGTGGGCAAGCCTTCCAGTTAGTCTTCCCCATCGCAG 6960
TGGGCCCATGGAAGGGTGTTTCCAGACTTCTTCATGCTACCCACACTAAGAGACTCTCAACAGAGTCTGAATGGAAACCACGAAGGAACACATTTGTTTTAGATTCCTCTGTGCCACCTTT 7080
ACTGATGGGAGCTTCTGTTTTCCAGTAGATACAGATGTGTTAGCATGGATGCTTGGATAACAGACAAATAAGATGGTGGTGCTGACCTCTGGACGCTTACTCTCTAGTGGCAGACAGAACA 7200
GTAAGGTTAGAATTCTGTTGCTATTAAAAAACTAATAATTAATCACCTTGAAAAAGAATGGAGTTGTTAGGCATGGGTCTCTCTCGAGTAAGCTTGGAACAAAGCTTCTCCCTGGCTTGG 7320
GTGAGTCAAAGCAGATGGACTTAGCTCGTCTCATTCATTCTAAATTAGAACTTCTTCCCACAGCCCAGAACACGCCACAAGAAAAAAAAAAATGTGCAATATTTAACCAGTCTTTGTTTT 7440
TTCCTCCTTTAGGTAAAAGATTTACATAAAAATAGTCCTTCCTACCCCAATTTCCAATGCTCTCCTAACAGATAAGCTGGAGTCACAGAAGGAGTGGCTAAGGACCAAGACCATCCAATTC 7560
                              ValLysAspLeuHisLysLysIleValLeuProThrProIleSerAsnAlaLeuLeuThrAspLysLeuGluSerGlnLysGluTrpLeuArgThrLysThrIleGlnPhe
ATCTTGAAATCACTTGAAGAATTTCTAAAAGTCACTTTGAGATCTACTCGGCAAACCTAGTGCGTTATGCCCTAAGCATATCAGTTTGTGGACATTCCTCACTGTGGTCAGAAAATATAT 7680
                              IleLeuLysSerLeuGluGluPheLeuLysValThrArgSerThrArgGlnThr***
TCCTGTTGTCAGGTATCTGACTTATGTTGTTCTCTACGAAGAACTGACAATATGAATGTTGGGACACTATTTTAATTATTTTTAATTTATTGATAATTTAAATAAGTAAACTTTAAGTTA 7800
ATTTATGATTGATATTTATTATTTTTTTATGAAGTGTCACTTGAAATGTTATATGTTATAGTTTTGAAATGATAACCTAAAAATCTATTTGATATAAATATTCTGTTACTAGCAGATGGTTC 7920
TTGGAATGTATAAGTTACCTCAATGAATTGCTAATTTAAATATGTTTTTAAAGAAATCTTTGTGATGTATTTTTATAATGTTTAGACTGTCTTCAAACAAATAAATTATATTATATTTAA 8040
AAACCAGTGACTGAAAGACGCATCTCAGCTGGTAAAGTTCTTACCCAACATGAGCAAGGTCCTAAGTTACATCCAAACATCCTCCCCCAAATCAATAATTAAGCACTTTTTATGACATGT 8160
AAAGTTAAATAAGAAGTGAAAGCTGCACATTAGTTAATTTCAGGTCTTGTACATTCTTTTCTGGACTGAGAGTAAGGGATCTAACTAAGCCGCCTTTG 8258
```

Fig. 5. The nucleotide sequence of genomic DNA encoding murine IL-6. The amino acid sequences of the coding region are also shown. The TATA-like and presumed polyA signal sequences are *underlined*. The sequences of alternate purine and pyrimidine are *double or thick underlined*. (From TANABE et al. 1988)

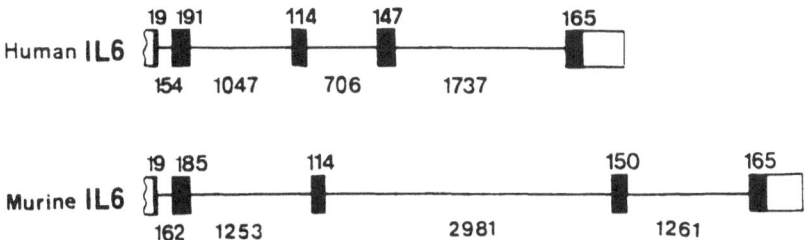

Fig. 6. Comparison of the human and murine IL-6 genes. *Closed* and *open* areas represent the coding and noncoding regions, respectively. (Based on data from YASUKAWA et al. 1987; TANABE et al. 1988)

Human IL6

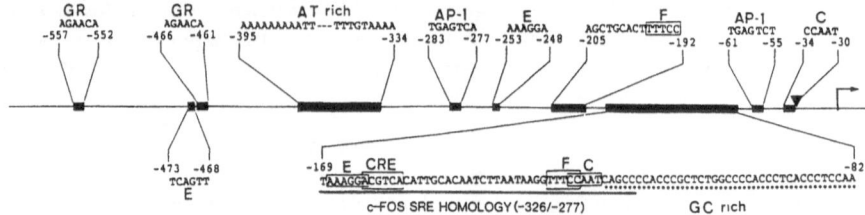

Murine IL6

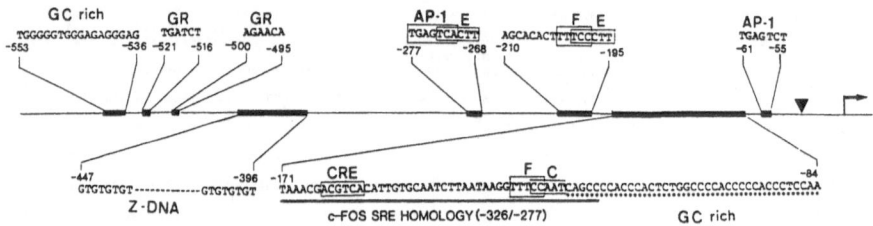

Fig. 7. Schematic summary of the potential regulatory elements in the 5′ flanking region of the human and murine IL-6 genes. *C* represents CCAAT box; *E*, IFN enhancer core sequence; *F*, GGAAA motif; *CRE*, cyclic AMP responsive element; *GR*, glucocorticoid responsive element. (From Tanabe et al. 1988)

Using the enzymes *Msp*I, *Bst*NI and *Bgl*I, three polymorphic systems were reported (Bowcock et al. 1988). The *Msp*I and *Bgl*I polymorphisms are likely to be due to base pair substitutions; the *Bst*NI polymorphism is likely to be due to insertion/deletion of DNA within 0.5 kb of the 3′ end of the fifth exon. The polymorphic *Msp*I and *Bgl*I sites are likely to lie in the vicinity of the fifth exon and in the 5′ flanking region, respectively. The three polymorphisms are separate. *Msp*I and *Bst*NI polymorphisms are observed in Caucasians, pygmies from the Central African Republic and Zaire, Melanesians, and Chinese but at differing frequencies. The *Bgl*I polymorphism is observed in Caucasians and Africans only.

III. Regulation of Expression of the IL-6 Gene

1. IL-6-Producing Cells

It has been found that IL-6 is produced by many different types of lymphoid or nonlymphoid cells such as T cells, B cells, monocytes, fibroblasts, keratinocytes, endothelial cells, and certain tumor cells, either constitutively or in response to various stimuli as summarized in Table 1. The production of IL-6 by normal T cells was demonstrated to be dependent on monocytes and the role of monocytes could be replaced with 12-*O*-tetradecanoylphorbol-13-acetate (TPA) but not with IL-1 (Horii et al. 1988). Monocytes produced IL-6 in the absence of an apparent stimulus and the peak of IL-6 mRNA was

Table 1. Cells producing IL-6

Producer	Stimuli
T cells	PHA plus TPA or PHA plus monocytes
T cell clone	Antigen plus antigen-presenting cells
T cell line (HTLV-I)	Constitutive
B cells	*Staphylococus aureus* Cowan I
Monocytes	LPS, inhibited by dexamethasone
Monocyte cell line P388D1	Constitutive
Fibroblasts	Constitutive, enhanced by IL-1, TNF, PDGF, IFN-β, poly(I)·(C), cycloheximide (CHX), A23187, TPA, prostaglandin E1, forskolin, cholera toxin, dibutyryl cyclic AMP, isobutylmethylxanthine, inhibited by dexamethasone
MG63 osteosarcoma	CHX, poly(I)·(C), IL-1
T24 bladder carcinoma	Constitutive, enhanced by TNF-α, TNF-β, CHX, or poly(I)·(C)
A549 lung carcinoma and 7860 renal carcinoma	TNF-α plus CHX, poly(I)·(C) plus CHX
SK-MG-4 glioblastoma and U373 astrocytoma	IL-1
Cardiac myxoma	Constitutive
Myeloma cells	Constitutive
Mesangial cells	Upon culture
Keratinocytes	Upon culture
Endothelial cells	Upon culture

achieved 5 h following culture, whereas in T cells the peak was at around 48 h after culture initiation, indicating that IL-6 is produced by monocytes and T cells with different kinetics and may exert distinct effects at different phases of the immune response. Helper T cells were shown to be divided into at least two subsets based on differences in the cytokine repertoire (MOSMANN and COFFMAN 1987): T_H1 clones produce IL-2 and IFN-γ, whereas T_H2 clones produce IL-4 and IL-5, for example. It was recently demonstrated that some T_H2 clones produce IL-6 with antigenic stimulation or T-cell mitogens, but none of the T_H1 clones do (HODGKIN et al. 1988). However, this may not apply to human T cells, since a majority of mitogen-stimulated peripheral blood T cells can produce IL-6 (HORII et al. 1988).

2. Stimuli Inducing IL-6 Production

The production of IL-6 is positively or negatively regulated by a variety of stimuli. IL-6 production is induced by T-cell mitogens or antigenic stimulation of T cells or T-cell clones (Horii et al. 1988; Van Snick et al. 1986). Lipopolysaccharide (LPS) enhances IL-6 production in monocytes (Horii et al. 1988; Tosato et al. 1988) and fibroblasts (Helfgott et al. 1987), whereas glucocorticoids inhibit it (Helfgott et al. 1987; Woloski et al. 1985). Various viruses induce IL-6 production in fibroblasts (Sehgal et al. 1988) or in the central nervous system (Frei et al. 1988). Human immunodeficiency virus (HIV) also induces IL-6 production in monocytes (Nakajima et al. 1989). A variety of peptide factors, such as IL-1, tumor necrosis factor (TNF), IFN-β, and platelet-derived growth factor (PDGF) enhance IL-6 production in fibroblasts (Content et al. 1985; Zilberstein et al. 1986; May et al. 1986; Haegeman et al. 1986; Kohase et al. 1987; Van Damme et al. 1987c) and certain tumor cell lines (Content et al. 1985; Wong and Goeddel 1986; Van Damme et al. 1987a).

3. Mechanisms Regulating IL-6 Gene Expression

The mechanism through which IL-6 gene expression is regulated is not known. Phorbol ester compounds known to cause activation of protein kinase C were shown to stimulate IL-6 gene expression in tonsillar lymphocytes (Hirano et al. 1986) and in human fibroblasts (Sehgal et al. 1987), suggesting that the protein kinase C-dependent signal transduction pathway can trigger IL-6 gene expression. IL-6 gene expression appears to be independently regulated through two different biochemical pathways: (1) protein kinase C activation (Sehgal et al. 1987), and (2) a cyclic AMP-dependent pathway (Zhang et al. 1988a, b). The presence of multiple initiation sites and the preferential utilization of a specific initiation site in a variety of cells suggest that different regulatory mechanisms may be responsible for the expression of the IL-6 gene in different tissues (Yasukawa et al. 1987). In fact, sequences similar to several consensus sequences involved in gene activation are present in the 5′ flanking region, as described in Sect. D. I. These consensus elements may be related to IL-6 gene expression mediated by a variety of stimuli.

Utilizing the IL-6 promoter linked to the bacterial chloramphenicol acetyltransferase gene, it was recently reported that the region between -225 and -113 in the IL-6 gene, which contains a DNA sequence similar to the c-fos serum responsive element, appears to contain the major cis-acting regulatory elements responsible for the expression of the IL-6 gene in HeLa cells stimulated with a variety of inducers such as TPA, virus, and IL-1 (Ray et al. 1988). Essentially the same findings were obtained in IL-1-induced IL-6 gene activation in a glioblastoma cell line. Furthermore, the glioblastoma cell line was found to contain several DNA binding proteins. One of these specifically binds to the palindrome sequence ACATTGCACAATCT found in the DNA sequence, similar to the c-fos serum responsive element (S. Akira, unpublished data).

Table 2. Expression of IL-6 receptor on a variety of cells

Cells	Number of receptors/cell
EBV-transformed B-cell lines	200– 3000
Burkitt's lymphoma cell lines	Not detectable
Myeloma cells and cell lines	100–20000
Hepatoma cell lines	2000– 3000
Myeloid leukemia cell lines	2000– 3000
Rat pheochromocytoma cell line (PC12)	∼1200
Resting B cells	Not detectable
Activated B cells	∼ 500
Resting T cells	∼ 300

E. Structure and Expression of the IL-6 Receptor

I. Presence of High- and Low-Affinity Receptors

The specific receptor for IL-6 was found to be expressed on lymphoid as well as nonlymphoid cells in accordance with the multifunctional properties of IL-6 (TAGA et al. 1987; COULIE et al. 1987). Binding of ^{125}I-IL-6 to a B-lymphoblastoid cell line was competitively inhibited by unlabeled IL-6, but not by IL-1, IL-2, IFN-β, IFN-γ, or G-CSF, indicating the presence of a specific receptor for IL-6. The cells expressing IL-6 receptor include EBV-transformed B-cell lines, myeloid cell lines, hepatoma cell lines, myeloma cell lines, resting normal T lymphocytes, and activated normal B lymphocytes (TAGA et al. 1987), as shown in Table 2. Scatchard plot analysis demonstrated the presence of both high- and low-affinity IL-6 receptors on a human myeloma cell line (U266). The number of IL-6 receptors varies in different cells and tissues. The high-affinity receptors were about 10% of the total receptors and the difference in K_d values between high- and low-affinity receptors was approximately 100-fold (YAMASAKI et al. 1988).

II. Structure of the IL-6 Receptor

cDNA encoding human IL-6 receptor capable of expressing both high- and low-affinity binding sites was cloned (YAMASAKI et al. 1988). Figure 8 shows the nucleotide sequence and deduced amino acid sequence of cloned cDNA for the IL-6 receptor. The IL-6 receptor consists of 468 amino acids. A hydropathy plot of the deduced amino acid sequence of the IL-6 receptor shows two major hydrophobic regions, one located between residues 1 and 20 and the other located in the region of residues 359-386. The former is presumably a typical signal peptide and the predicted cleavage site may be between Ala (position 19) and Leu (position 20). The latter is the putative transmembrane domain. There are six potential N-linked glycosylation sites

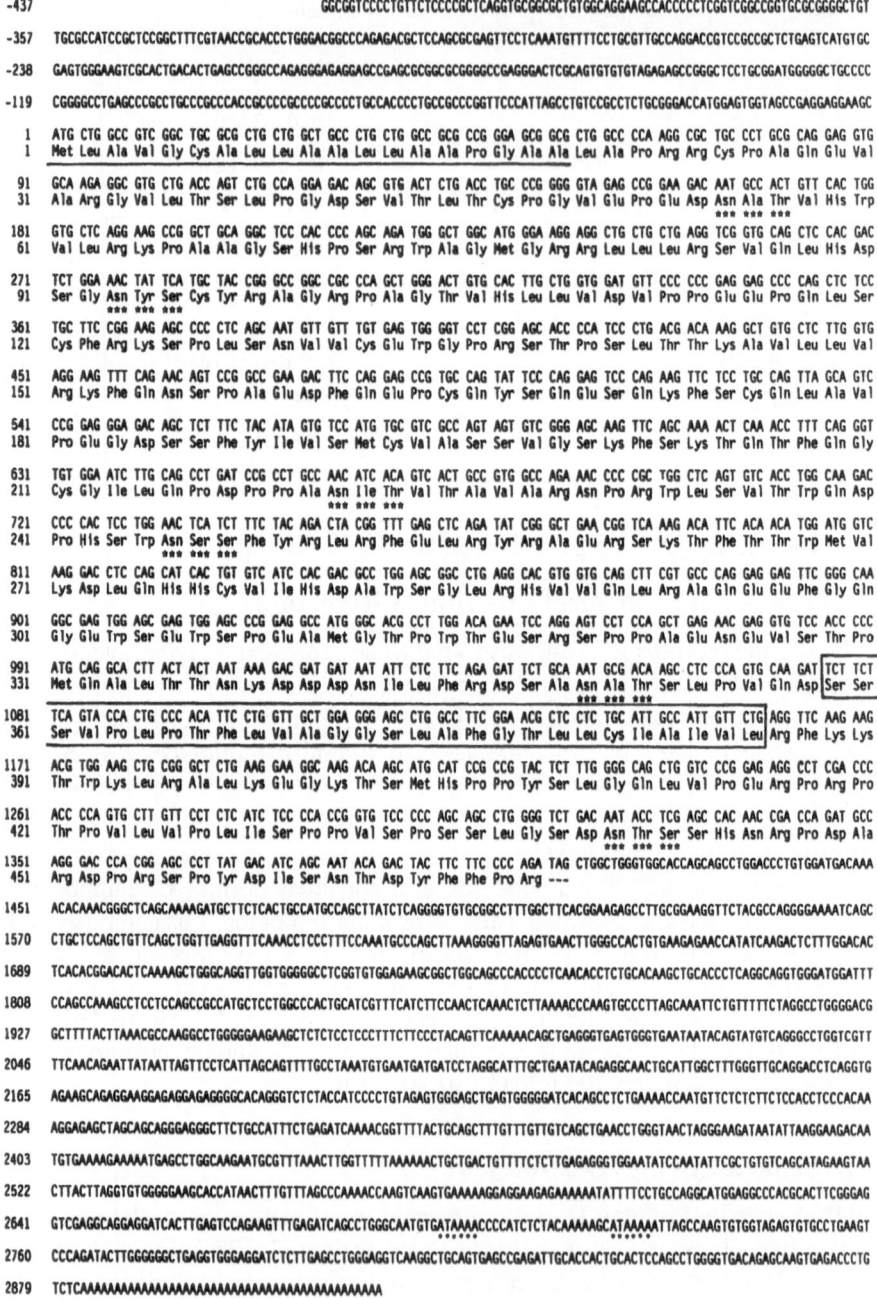

Fig. 8. Nucleotide sequence and deduced amino acid sequence of the human IL-6 receptor cDNA. *Numbers* in the *left margin* of the sequence show positions of nucleotides and amino acids, respectively. *Asterisks*, potential N-glycosylation sites (Asn-X-Ser/Thr); *underlined region*, a presumed signal peptide; *box*, a presumed transmembrane domain; *dots*, a possible polyA addition signal. (From YAMASAKI et al. 1988)

```
β strand          S-------                                                                                    --------S
                  B          C           C'         C''        D       E              F
                  * * *      * *              * *                  3*        * * *         3 3 * *
C2-SET           43        50          60         70            3*      80             90
IL-6 R           VTLTCPGV--EPED-NATVHVLRKPAAGSH-------------------PSEWAGM-----GRRLLRSVQLHDSNYSCYRAG
PDGF R    (III)  IIIRCIV-MGNDVV--NNQWTYPRMK--SGRLV-----------------EPVTDYLFGVPS--RIGSILHFTAELSDSGTYTCNVSV
CSF-1-R (v-fms)  AQIVCSA--SNIDV--NFDVSLRHGDTKLTISQQS---------------DFHDNRYQKV-----LTNLDHVSFQDAGNYSCTATN
Alpha1 B-GP(III) VLTCVA--PLSGV--DFQLRRGE--------------------------KELLVPRSSTSP-DRIFFHLNAVALGDGHYTCRYRL
Fc R    (I)      VTLMCEG-THNPGNS--STQWFHNG------------------------BSIRSQ-----VQASYTFKA-TVNDSGEYRCQMEQ

V-SET
Ig V kappa       ATLSCRASQSI---SNSYLAWYQQKP-SGSPRLLIYGASTRATGIP---ARFSGSGSG----TEFTLTISSLQSEDFAVYYCQQYN
Ig V lambda      VTLTCRSSTGAV-TTSNYANWVQQKP-DHLFTGLIGGTNNRAPGVP---ARFSGSLIG----NKAALTITGAQTEDEAIYFCALWY
Ig V heavy       LSLTCTVSGSTF--SNDYYTWVRQPP-GRGLEWIGYVFYHGTSDDTTPLRSRVTMLVDTS--KNQFSLRLSSVTAADTAVYYCARNL
CD4    (I)       VELTCTASQK---KSIQFHWKNSNQI-KILGNQGSFLTK-GPSK---LNDRADSRRSLWD-QGNFPLIIKNLKIEDSDTYICEVED
Poly Ig R (II)   VTITCPFTYATR--QLKKSFYKVED-----GELVLIIDSSSKEAKDPRYKGRITLQIST-TAKEFTVTIKHVQLNDAGQYVCQSGS
```

Fig. 9. Alignment of IL-6 receptor domain to Ig superfamily protein domains. IL-6 receptor, CSF-1 receptor, and human Ig V kappa chain V-III region sequences were aligned to several proteins of the Ig superfamily by inspection. Alignment of Ig superfamily protein domains is based in the data of WILLIAMS and BARCLAY (1988). *Asterisks*, conserved patterns common to the V, C1, and C2 sets; *§*, common to the V and C2 sets; *#*, common to the C1 and C2 sets (WILLIAMS and BARCLAY 1988). The known locations of β-strands in Ig V domains are marked with *bars* and *capital letters* above the bars. The *numbers* above the alignment represent positions of amino acids of the IL-6 receptor sequence. The position of the putative disulfide bridge within the IL-6 receptor domain is indicated by *S---S*. (From YAMASAKI et al. 1988)

IL-6 Receptor is a member
of Immunoglobulin Superfamily

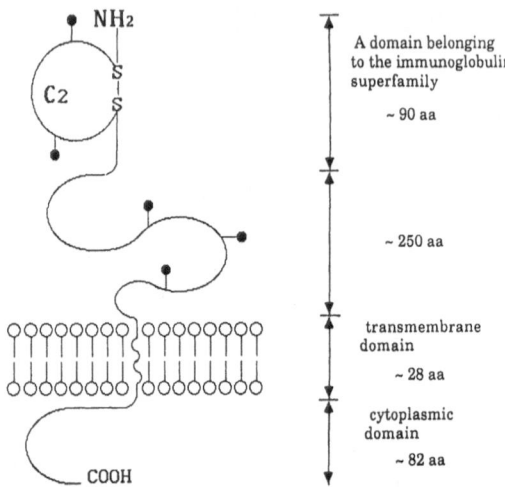

Fig. 10. Schematic model of the IL-6 receptor

(Asn-X-Ser/Thr), as well as 11 cysteine residues. Antibody against a synthetic peptide corresponding to the cytoplasmic domain was prepared and it could precipitate a IL-6 receptor molecule with a molecular mass of 80-kDa, indicating that a mature IL-6 receptor is a glycosylated form of a precursor molecule with a molecular mass of 50-kDa (TAGA et al. 1989).

The sequence of the IL-6 receptor was found to have some homology with several members of the Ig superfamily, including the Ig light chain variable region, the rabbit poly-Ig receptor, the CD4 molecule, and α_{1B}-glycoprotein. Furthermore, inspection of the IL-6 receptor sequence shows that the region between positions ~20 and ~110 fulfills the criteria for the constant 2 (C2) set of the Ig superfamily, as shown in Fig. 9 (WILLIAMS and BARCLAY 1988). A schematic model of the human IL-6 receptor is illustrated in Fig. 10.

The C2 set includes several adhesion molecules, the PDGF receptor, the CSF-1 receptor, the Fcγ receptor, and α_{1B}-glycoprotein (WILLIAMS and BARCLAY 1988). Furthermore, the IL-1 receptor was found to be a member of the Ig superfamily (SIMS et al. 1988). This is of particular interest since receptors for polypeptide growth factors such as PDGF, CSF-1, IL-1, and IL-6 could then be grouped in the Ig superfamily.

It is interesting to note that the IL-6 receptor, unlike certain other growth factors, does not have a tyrosine kinase domain, although IL-6 has been found to be a potent growth factor for myeloma/plasmacytoma cells (see Sects. F.I, H.II). Furthermore, the cytoplasmic domain was found not to be essential for IL-6 binding or for generation of signal transduction. However, it was revealed that there is a possible signal transducer of 130-kDa which associates with the IL-6 receptor after the interaction of IL-6 and its receptor (TAGA et al. 1989).

Table 3. Multiple signals provided by IL-6

Induction of differentiation (or gene expression)
1. Induction of B-cell differentiation or secretory-type Ig production
2. Induction of acute phase protein synthesis
3. Induction of cytotoxic T cell differentiation
4. Induction of IL-2 production and IL-2 receptor expression in T cells
5. Induction of macrophage differentiation
6. Induction of neural cell differentiation
7. Induction of megakaryocyte maturation

Stimulation of cell growth
1. Induction of hybridoma/plasmacytoma/myeloma growth
2. Induction of T-cell growth
3. Enhancement of EBV-transformed B-cell growth
4. Enhancement of hematopoietic stem cell growth
5. Induction of mesangial cell growth

Inhibition of cell growth
1. Inhibition of growth of myeloid leukemic cell lines
2. Inhibition of growth of breast carcinoma cell lines (see L. CHEN et al. 1988)

F. Biological Activities of IL-6

IL-6 has been found to have a wide variety of biological activities not only on lymphoid cells but also on hematopoietic stem cells, hepatocytes, and nerve cells. The signals provided by IL-6 are growth promoting, growth inhibitory, and differentiation inducing activities; it thus shares multifunctional properties with other peptide growth factors (SPORN and ROBERTS 1988). The biological activities of IL-6 are summarized in Table 3.

I. Lymphoid Tissue

1. Effects on B-Lineage Cells

The availability of recombinant IL-6 confirmed the previously demonstrated activities of IL-6 on B cells as BSF-2; IL-6 acts on B cells activated with SAC or pokeweed mitogen (PWM) to induce IgM, IgG, and IgA production (HIRANO et al. 1986, MURAGUCHI et al. 1988). Furthermore, it was demonstrated that IL-6 is absolutely required for antibody production by B cells. In fact, anti-IL-6 antibodies were found to inhibit PWM-induced Ig production in peripheral blood mononuclear cells without any effect on cell proliferation. The antibody was effective even when added on day 4 of an 8-day culture, indicating that IL-6 is one of the essential late-acting factors in PWM-induced Ig production (MURAGUCHI et al. 1988). Human IL-6 was also effective on in vitro as well as in vivo antibody production against sheep red blood cells in mice where the effect of IL-6 was more apparent in the secondary response than the primary response (TAKATSUKI et al. 1988). Mouse IL-6 was also reported to act on mouse B cells activated with anti-Ig or dextran sulfate. In the presence of IL-1, IL-6 induces growth and differentiation not only in B cells activated with anti-Ig but also in dextran sulfate-stimulated and unstimulated B cells (VINK et al. 1988).

As described in Sect. B. II, IL-6 is a growth factor for murine hybridoma/plasmacytoma cells. Furthermore, it was found that IL-6 is an autocrine growth factor for human multiple myeloma cells (KAWANO et al. 1988), as described in more detail in Sect. H. II, and is thought to be involved in the oncogenesis of plasmacytoma/myeloma. IL-6 is practically useful in establishing hybridoma cell lines, as originally reported by ASTALDI et al. (1980). In fact, recombinant IL-6 was found to increase the frequency of development of hybridomas producing monoclonal antibodies and to augment cloning efficiency (MATSUDA et al. 1988). IL-6 also promotes the proliferation of EBV-infected B cells and permits their growth at low cell densities (TOSATO et al. 1988).

2. Effects on T Cells

IL-6 induces IL-2 receptor (Tac antigen) expression in a certain T-cell line (NOMA et al. 1987) and thymocytes (LE et al. 1988) and functions as a second signal for IL-2 production by T cells (GARMAN et al. 1987). IL-6 promotes the growth of human T cells stimulated with PHA (LOTZ et al. 1988; HOUSSIAU et al. 1988) or mouse peripheral T cells of both the L3T4$^+$ and Lyt-2$^+$ subsets (UYTTENHOVE et al. 1988). It also acts on murine thymocytes to induce proliferation (UYTTENHOVE et al. 1988; HELLE et al. 1988; LE et al. 1988). The effects of IL-6 are synergistic with those of IL-1 and TNF, which also induce thymocyte proliferation (LE et al. 1988). IL-6 enhances the proliferative response of thymocytes to IL-4 and TPA (HODGKIN et al. 1988). IL-6 induces not only proliferation but also the differentiation of cytotoxic T lymphocytes (CTL) in the presence of IL-2 from murine as well as human thymocytes and splenic T cells (TAKAI et al. 1988; UYTTENHOVE et al. 1988; OKADA et al. 1988). IL-6 also induces serine esterase, which is required for mediating target cell lysis in the granules of CTL (TAKAI et al. 1988), suggesting a critical role in the differentiation and expression of cytotoxic T-cell function.

II. Hematopoietic Progenitors

Hematopoiesis is regulated by a variety of growth- and differentiation-inducing factors, such as a family of colony stimulating factors (CSFs) or macrophage and granulocyte inducers (MGIs) and interleukins (METCALF 1986; SACHS 1987). The family of these factors represents a hierarchy of growth- and differentiation-inducing proteins for blood cell development: IL-3 induces the proliferation of multipotent hematopoietic progenitors, whereas CSF-1 and G-CSF act on the progenitors committed to the macrophage and granulocyte lineages, respectively; GM-CSF acts on progenitor cells capable of developing into either macrophages or granulocytes; and IL-4 and IL-5 act on mast cell and eosinophil progenitors, respectively (LEE et al. 1986; SANDER-

SON et al. 1986). It was shown that IL-6 and IL-3 act synergistically in support of the proliferation of murine multipotential progenitors (IKEBUCHI et al. 1987). The overall time course of colony formation by spleen cells isolated from mice 4 days after injection of 5-fluorouracil was significantly shortened in cultures containing both IL-3 and IL-6 relative to cultures supported by either of the two factors. At least part of the IL-6 effect results from a decrease in the G_0 period of the individual stem cells. Furthermore, it was demonstrated that IL-6 and IL-3 synergistically increase the number of hematopoietic stem cells in in vitro liquid suspension culture system (OKANO et al. 1989). Moreover, hematopoietic stem cells expanded in vitro by IL-6 and IL-3 were found to exhibit a higher capability to reconstitute the hematopoietic system in vivo than normal bone marrow cells. In fact, in lethally irradiated mice transplanted with untreated bone marrow cells or IL-3-treated cells (2×10^5 cells/mouse), only 10%–20% were alive at day 30. By contrast, transplantation with the same number of bone marrow cells expanded in vitro with both IL-6 and IL-3 induced a marked increase in survival rate (90%). The data indicate that the combination of IL-6 and IL-3 is useful to expand hematopoietic stem cells in vitro and, therefore, could be used in bone marrow transplantation.

Human and mouse myeloid leukemic cell lines such as human histiocytic U937 cells and mouse myeloid M1 cells can be induced to differentiate into macrophages and granulocytes in vitro by several synthetic and natural products (ICHIKAWA 1969; SACHS 1978). Several factors have been identified which can induce differentiation of leukemic cells, such as G-CSF (NICOLA et al. 1983; TOMITA et al. 1986), MGI-2 (LIPTON and SACHS 1981; SACHS 1987), D factor (TOMITA et al. 1984), and leukemia inhibitory factor (LIF; GEARING et al. 1987). G-CSF induces differentiation of the murine myeloid leukemic cell line WEHI-3B D$^+$ and M1 cells and also acts on normal cells to induce proliferation and differentiation. On the other hand, MGI-2, D factor, and LIF, which are capable of inducing the differentiation of M1 cells, do not stimulate the proliferation of normal progenitor cells. LIF was molecularly cloned and found to be a novel factor having no similarity with either G-CSF or IL-6 (GEARING et al. 1987). In addition to these molecules, IL-6 was found to inhibit the growth of human and murine myeloid leukemic cell lines, U937 and M1, resulting in the differentiation of these cells to mature macrophage-like cells (MIYAURA et al. 1988). The untreated control M1 cells were myeloblastic with a large round nucleus; they changed morphologically into macrophage-like cells with abundant cytoplasms upon stimulation with IL-6. IL-6 enhances phagocytosis and expression of $Fc\gamma$ and C3 receptors. These effects of IL-6 are synergistic with those of IL-1. MGI-2 was found to be identical with IL-6 (SHABO et al. 1988). ISHIBASHI et al. (1989a) demonstrated that IL-6 induces maturation of megakaryocytes in vitro. They further showed that administration of IL-6 increased platelet number in mice (ISHIBASHI et al. 1989b). The data indicate that IL-6 functions as a thrombopoietic factor. In accordance with these results, it was found that there was an increase in mature megakaryocytes in bone marrow of IL-6 transgenic mice (SUEMATSU et al. 1989).

III. Acute Phase Responses

The acute phase response is a systemic reaction to inflammation or tissue injury. It is characterized by leukocytosis, fever, increased vascular permeability, alterations in plasma metal and steroid concentrations, and increased levels of acute phase proteins (KUSHNER 1982; KOJI 1985). The biosynthesis of acute phase proteins by hepatocytes is regulated by several factors: IL-1, TNF, and hepatocyte stimulating factor (HSF). IL-1 was originally considered to be the major acute phase regulator, although it would only partially elicit the full acute phase response (KOJI et al. 1984). It was demonstrated that recombinant IL-6 can function as an HSF (ANDUS et al. 1987; GAULDIE et al. 1987) and the activity of crude HSF could be neutralized by anti-IL-6 (ANDUS et al. 1987), indicating that HSF activity is exerted by the IL-6 molecule. IL-6 can induce a variety of acute phase proteins such as fibrinogen, α_1-antichymotrypsin, α_1-acid glycoprotein, and haptoglobin in the human hepatoma cell line HepG2. In addition to those proteins, it induces serum amyloid A, C-reactive protein, and α_1-antitrypsin in human primary hepatocytes (CASTELL et al. 1988 a). The proteins induced in the rat by IL-6 are fibrinogen, cysteine proteinase inhibitor, and α_2-macroglobulin (ANDUS et al. 1987; GAULDIE et al. 1987). However, albumin was negatively regulated by IL-6. In vivo administration of IL-6 in rats induced typical acute phase reactions similar to those induced by turpentine, and IL-6-induced expression of mRNAs for acute phase proteins was more rapid than that induced by turpentine. The results confirmed the in vivo effect of IL-6 in the acute phase reaction (GEIGER et al. 1988). It was also reported that serum levels of IL-6 correlated well with those of C-reactive protein and fever in patients with severe burns (NIJSTEIN et al. 1987), supporting the causal role of IL-6 in the acute phase response.

IV. Neural Cells

IL-1 stimulation of glioblastoma cells or astrocytoma cells was found to induce the expression of IL-6 mRNA (YASUKAWA et al. 1987), suggesting that IL-6 may have certain effects on nerve cells. It is well known that nerve growth factor (NGF) induces a phenotypic shift in chromaffin cells and their neoplastic counterpart, the PC12 cell line, resulting in neural differentiation accompanied by chemical, ultrastructural, and morphological changes. IL-6 was also found to induce the typical differentiation of PC12 cells into neural cells (SATOH et al. 1988): (a) In the presence of IL-6, cell viability was maintained and a change in morphology to neurite-extending cells was observed after several days. (b) IL-6 was found to induce the transient expression of c-*fos* protooncogene and increase in the number of voltage-dependent Na$^+$ channels in PC12 cells. The differentiation induced by IL-6 is similar to that observed with NGF, although IL-6 and NGF utilize completely different receptors on PC12 cells. Moreover, it was found that PC12 cells express about 1200 IL-6 receptors per cell with a K_d of about 1.8×10^{-9} M.

V. Antiviral Activity

IFN-β2 was originally reported as an IFN related to IFN-β (ZILBERSTEIN et al. 1986), as described in Sect. B. III. However, recombinant IL-6 did not show any antiviral activity (POUPART et al. 1987; VAN DAMME et al. 1987b; HIRANO et al. 1988a; REIS et al. 1988). Furthermore, anti-IFN-β did not neutralize the activity of IL-6 in the induction of Ig production in B cells or plasmacytoma growth activity, and anti-IL-6 could not neutralize the antiviral activity of IFN-β (HIRANO et al. 1988; REIS et al. 1988). Moreover, the primary structure and genomic organization of IL-6 have no similarity with those of any IFN (HAEGEMAN et al. 1986; HIRANO et al. 1986; YASUKAWA et al. 1987).

G. Plasma Clearance and Carrier Proteins

The plasma half-life of human IL-6 was determined in rats (CASTELL et al. 1988b). The kinetics of clearance were biphasic; there was a rapid initial disappearance corresponding to a half-life of 3 min, and a second slow one corresponding to a half-life of about 55 min, suggesting the presence of carrier proteins in the plasma which protect IL-6 from degradation, and inhibit its clearance. In fact, binding proteins in plasma were detected (CASTELL et al. 1988; MATSUDA et al. 1989). One of the binding proteins in plasma was identified as α_2-macroglobulin (MATSUDA et al. 1989). α_2-Macroglobulin does not inhibit IL-6 activity or its binding to its homologous receptor. IL-6 bound to α_2-macroglobulin was found to retain its biological activity and was resistant to treatment with proteases, although free IL-6 was easily degraded. These findings indicate that α_2-macroglobulin plays an important role as a carrier protein for IL-6 in serum. Furthermore, it makes IL-6 produced at a local site where inflammation or an immune response is underway available to lymphocytes, hepatocytes, and hematopoietic progenitors. Such a process eventually leads to the induction of coordinated systematic host defense reactions, such as the immune response, the acute phase reaction, and hematopoiesis.

H. IL-6 and Disease

Evidence has accumulated suggesting that unregulated production of IL-6 is involved in a variety of human diseases, such as abnormalities of polyclonal B-cell activation, lymphoid malignancies, and proliferative glomerulonephritis, as summarized in Table 4.

I. Polyclonal B-Cell Activation and Autoimmune Disease

The involvement of IL-6 in human disease was first observed in cardiac myxoma (HIRANO et al. 1987). Cardiac myxoma is a benign intraatrial heart tumor and one-third of the patients show autoimmune symptoms, such as hy-

Table 4. IL-6 and human diseases

Diseases	IL-6 producing cells
1. Polyclonal B-cell abnormalities or autoimmune diseases	
Cardiac myxoma	Myxoma cells
Rheumatoid arthritis	T, B, and synovial cells
Uterine cervical carcinoma	Carcinoma cells
Castleman's disease	B cells
2. IL-6 and malignancies	
Multiple myeloma	Autocrine
Lennert's T-cell lymphoma	Monocytes
3. Others	
Mesangial proliferative glomerulonephritis	Autocrine

pergammaglobulinemia and the presence of various kinds of autoantibodies (SUTTON et al. 1980). Furthermore, these symptoms disappear upon surgical removal of the tumor, suggesting that the myxoma itself or its products are involved in the autoimmune condition of these patients. In fact, the culture supernatants of tumor cells were found to contain high IL-6 activity (HIRANO et al. 1987). Furthermore, IL-6 mRNA could be detected in myxoma cells (HIRANO et al. 1986).

More precise analyses on the relationship between overproduction of IL-6 and polyclonal B-cell activation were performed in patients with Castleman's disease (YOSHIZAKI et al. 1989). This is a chronic disease with benign hyperplastic lymphadenopathy, which is characterized by large lymph follicles with intervening sheets of plasma cells. The patients show fever, hypergammaglobulinemia, and increase in acute phase proteins. Cells in the germinal center of such hyperplastic lymph nodes were found to produce IL-6 constitutively. Dramatic clinical improvement and decrease in serum IL-6 levels were observed following surgical removal of the involved lymph node. The findings showed a correlation between serum IL-6 levels, lymph node hyperplasia, hypergammaglobulinemia, and increased levels of acute phase proteins.

The possible involvement of IL-6 in disease was also demonstrated in rheumatoid arthritis (RA; HIRANO et al. 1988 b). High levels of IL-6 were detected in synovial fluids from the joints of patients with active RA. The synovial fluid cells were found to express IL-6 mRNA. Immunohistochemical analysis demonstrated that T cells as well as B cells infiltrated in the synovial tissues produce IL-6. The data indicate that IL-6 is generated constitutively in RA and its overproduction accounts for some of the local and generalized symptoms of RA, such as production of autoantibody, hypergammaglobulinemia, and elevation of acute phase proteins including serum amyloid A. Furthermore, IL-6 is found to be a growth factor for an EBV-transformed B-lymphoblastoid cell line (TOSATO et al. 1988). This may explain the presence of abnormally elevated numbers of circulating EBV-infected B cells in RA patients.

MRL/lpr mice spontaneously develop autoimmune disease with lymphoid hyperplasia having an infiltration of plasma cells, hypergammaglobulinemia, a high incidence of monoclonal or oligoclonal IgGs, proliferative glomerulonephritis, and arthritis (THEOFILOPOULOS and DIXON 1985). It has been demonstrated that deregulated production of a B-cell stimulatory factor, designated l-BCDF, may be involved in the pathogenesis of autoimmune disease in MRL/lpr, although the molecular nature of l-BCDF is not known (PRUD'HOMME et al. 1984). Recently it was found that serum levels of IL-6 in MRL/lpr mice increase as early as 10 weeks of age. At that time other autoimmune mice such as NZB, $(NZB \times NZW)F_1$, and BXSB do not show any significant increase in the serum level of IL-6, although it is not known whether serum IL-6 levels are increased in these other autoimmune mice when they are older (B. TANG and T. MATSUDA, unpublished data). The data suggest that deregulated gene expression of IL-6 may be involved in polyclonal B-cell activation in MRL/lpr mice. As regards the pathology observed in MRL/lpr mice, the increase in serum IL-6 in MRL/lpr mice is particularly interesting, since deregulated production of IL-6 could be involved in plasmacytoma/myeloma and mesangial proliferative glomerulonephritis, as described in the following sections.

II. Lymphoid Malignancies

1. Plasmacytoma/Myeloma

In 1962, POTTER and BOYCE demonstrated that plasmacytomas are induced in a high percentage of BALB/c mice after intraperitoneal injection of mineral oil or pristane. Furthermore, it was shown that most plasmacytoma cells contain a reciprocal chromosomal translocation that brings the *myc* oncogene near the Ig heavy or light chain gene (OHNO et al. 1979). This translocation apparently results in the deregulation of the *myc* oncogene and in its constitutive expression (MUSHINSKI et al. 1983). The primary plasmacytoma cells can be transplanted to syngeneic hosts but plasmacytoma cells do not grow well in vitro. Furthermore, plasmacytomas arise exclusively in the oil-induced granulomatous tissue which consists of macrophages and neutrophils, suggesting that their growth is dependent on microenvironmental influences provided by the inflammatory cells (CANCRO and POTTER 1976). Moreover, it was shown that peritoneal adherent cells stimulated with pristane produced 50-fold greater amounts of plasmacytoma growth factor, which is now known to be IL-6, as described in Sects. B. II and F. I (NORDAN and POTTER 1986). These facts suggest that IL-6 may be involved in the generation of plasmacytoma and human multiple myeloma as well. In fact, IL-6 was found to be an autocrine growth factor for human multiple myelomas (KAWANO et al. 1988): (a) IL-6 could augment the in vitro proliferation of myeloma cells; (b) myeloma cells constitutively produce IL-6 and express the IL-6 receptor; and

(c) most importantly, anti-IL-6 antibodies inhibit the in vitro proliferation of myeloma cells. The evidence suggests that deregulated expression of the genes encoding IL-6 and/or IL-6 receptor is involved in the oncogenesis of myeloma/plasmacytoma. This was tested utilizing transgenic mice with the human IL-6 gene conjugated to the human Ig heavy chain gene enhancer ($E\mu$; Suematsu et al. 1989). In these transgenic mice, a gradual increase in polyclonal immunoglobulins was detected in the serum. The mice had lymphoma, thymoma, and splenomegaly with a massive plasmacytosis histologically indistinguishable from plasmacytoma. The induction of plasmacytosis was lethal in the transgenic mice. However, the plasma cells were not transplantable to syngeneic mice and were found not to contain apparent c-*myc* gene rearrangement which is observed in almost all pristane-induced plasmacytoma cells. The data conclusively demonstrated that deregulated IL-6 gene expression triggers the polyclonal plasmacytosis. However, additional genetic changes are required for the generation of plasmacytoma.

2. Lennert's T-Cell Lymphoma

IL-6 was shown to be involved in growth of Lennert's T lymphoma cells both in vivo and in vitro (Shimizu et al. 1988). Lennert's lymphoma is a special variant on non-Hodgkin's lymphoma characterized by a massive infiltration of macrophage-derived epithelioid histiocytes. A T-lymphoma cell line established from a patient with Lennert's lymphoma showed macrophage-dependent growth and the effect of macrophages was replaced with macrophage-derived soluble factors. IL-6 supported the in vitro growth of this established T-cell line. Furthermore, anti-IL-6 antibody could completely neutralize the activity of macrophage-derived factor. Considering the massive infiltration of macrophages in lymphoma tissues, this evidence suggests that IL-6 is involved in the in vivo growth of Lennert's lymphoma.

III. Proliferative Glomerulonephritis

Mesangial proliferative glomerulonephritis (PGN) is histologically characterized by proliferation of mesangial cells, suggesting the involvement of a growth factor for mesangial cells in the pathogenesis of the disease (Striker and Striker 1985). Several growth factors, such as PDGF, IL-1, and culture supernatants of macrophages have been suggested as candidates involved in the pathological growth of mesangial cells in glomerulonephritis (Lovett et al. 1983; Melcion et al. 1982; MacCarthy et al. 1985), but none has yet been shown to contribute to the pathogenesis of PGN. It was recently demonstrated that IL-6 is an autocrine growth factor for rat mesangial cells and immunohistochemical analysis shows that mesangial cells from patients with mesangial PGN constitutively produce IL-6 (Horii et al. 1989). Furthermore, IL-6 could be detected in urine samples from patients with PGN, but not from patients with minimal-change nephrotic syndrome or with membranous nephropathy. Moreover, there is a close relationship between the level of urine IL-6 and the progress of PGN. On the other hand, neither the frequency of IL-1-positive samples nor the levels of IL-1 in urine was significantly different in patients and healthy volunteers. All these results suggest that deregulated production of IL-6 in mesangial cells may be responsible for the pathogenesis of PGN. The measurement of IL-6 in patients' urine is useful not only for the differential diagnosis of PGN but also for monitoring the progression of PGN during therapy. This has recently been demonstrated using transgenic mice with human IL-6 gene conjugated to the human $E\mu$ (SUEMATSU et al. 1989). In these transgenic mice, as described in Sect. H. II, plasmacytoma developed, and IL-6 was detected in the serum. Histological analysis demonstrated that these transgenic mice showed a typical PGN. The evidence convincingly demonstrated that unregulated overproduction of IL-6 trigers PGN.

J. Prospects

It has been demonstrated that IL-6 is a multifunctional cytokine regulating the immune response, acute phase reaction, and hematopoiesis, playing a central regulatory role in host defense mechanisms against infections, inflammation, and tissue injuries. IL-6 shows growth-promoting, growth-inhibitory, and differentiation-inducing activities, depending on the target cells. It is produced in various tissues and cells, such as T cells, monocytes, fibroblasts, mesangial cells, endothelial cells, myxoma cells, and various cancer cells, and it has been suggested that the deregulated expression of IL-6 and/or its receptor may actually be involved in the pathogenesis of autoimmune diseases, proliferative glomerulonephritis, and certain lymphoid malignancies, especially plasmacytoma/multiple myelomas. Therefore, future studies on the gene regulation of IL-6 and its receptor should provide critical information on the molecular pathogenesis of those diseases. Furthermore, it will provide new diagnostic methods and treatment approaches to these diseases.

The receptor for IL-6 has been molecularly cloned. Thus studies may become possible on how one cytokine, in this instance IL-6, can provide multiple signals in different tissues and cells. Further studies may reveal a novel mechanism of signal transduction, since the intracytoplasmic portion of the IL-6 receptor does not have a particularly unique structure and IL-6 does not transduce its signal through any known biochemical pathway, such as phosphoinositol turnover, Ca^{2+} influx, or protein phosphorylation.

Acknowledgments. The authors would like to thank their scientific colleagues who aided in the preparation of this review by providing preprints or by allowing us to quote from their unpublished work. We also wish to thank Dr. Edward Barsoumian for his critical review of the manuscript and Ms. M. Harayama and K. Kubota for their secretarial assistance.

References

Aarden LP, Lansdorp, De Groot E (1985) A growth factor for B cell hybridomas produced by human monocytes. Lymphokines 10:175–185

Andus T, Geiger T, Hirano T, Northoff H, Ganter U, Bauer J, Kishimoto T, Heinrich PC (1987) Recombinant human B cell stimulatory factor 2 (BSF-2/IFNβ 2) regulates β-fibrinogen and albumin mRNA levels in Fao-9 cells. FEBS Lett 221:18–22

Astaldi GCB, Janssen MC, Lansdorp PM, Willems C, Zeijlemaker WP, Oosterhof F (1980) Human endothelial culture supernatant (HECS): a growth factor for hybridomas. J Immunol 125:1411–1414

Bowcock AM, Kidd JR, Lathrop M, Danshvar L, May LT, Ray A, Sehgal PB, Kidd KK, Cavallisforza LL (1988) The human "beta-2 interferon/hepatocyte stimulating factor/interleukin-6" gene: DNA polymorphism studies and localization to chromosome 7p21. Genomics 3:8–16

Brakenhoff JPJ, de Groot ER, Evers RF, Pannekoek H, Aarden LA (1987) Molecular cloning and expression of hybridoma growth factor in *Escherichia coli*. J Immunol 139:4116–4121

Cancro M, Potter M (1976) The requirement of an adherent substratum for the growth of developing plasmacytoma cells in vivo. J Exp Med 144:1554–1567

Castell JV, Gomez-Lechon MJ, David M, Hirano T, Kishimoto T, Heinrich PC (1988a) Recombinant human interleukin-6 (IL-6/BSF-2/HSF) regulates the synthesis of acute phase proteins in human hepatocytes. FEBS Lett 232:347–350

Castell JV, Geiger T, Gross V, Andus T, Walter E, Hirano T, Kishimoto T, Heinrich PC (1988b) Plasma clearance, organ distribution, and target cells of interleukin-6/hepatocyte stimulating factor in the rat. Eur J Biochem 177:357–361

Chen L, Mory Y, Zilberstein A, Revel M (1988) Growth inhibition of human breast carcinoma and leukemia/lymphoma cell lines by recombinant IFN-beta2 (BSF-2). Proc Natl Acad Sci USA 85:8037–8041

Chen Y, Ferguson-Smith AC, Newman MS, May LT, Sehgal PB, Ruddle FH (1987) Regional localization of the human B2-interferon gene. Am J Hum Genet 41:A161

Content J, De Wit L, Pierard D, Derynck R, de Clercq E, Fiers W (1982) Secretory proteins induced in human fibroblasts under conditions used for the production of interferon β. Proc Natl Acad Sci USA 79:2768–2772

Content J, De Wit L, Poupart P, Opdenakker G, Van Damme J, Billiau A (1985) Induction of a 26-kDa-protein mRNA in human cells treated with an interleukin-1-related, leukocyte-derived factor. Eur J Biochem 152:253–257

Corbel C, Melchers F (1984) The synergism of accessory cells and of soluble α-factors derived from them in the activation of B cells to proliferation. Immunol Rev 78:51–74

Coulie PG, Vanhecke A, Van Damme J, Cayphas S, Poupart P, De Wit L, Content J (1987) High affinity sites for human 26-kDa protein (interleukin 6, B-cell stimulatory factor-2, human hybridoma plasmacytoma growth factor, interferon-beta 2), different from those of type-1 interferon (alpha, beta) on lymphoblastoid cells. Eur J Immunol 17:1435–1440

Dutton RW, Falkoff R, Hirst JA, Hoffman M, Kappler JW, Kattman JR, Lesley JF, Vann D (1971) Is there evidence for a nonantigen specific diffusable chemical mediator from the thymus cell in the initiation of the immune response? Prog Immunol 1:355–368

Frei K, Leist TP, Meager A, Gallo P, Leppert D, Zinkernagel RM, Fontana A (1988) Production of B cell stimulatory factor-2 and interferon γ in the central nervous system during viral meningitis and encephalitis. J Exp Med 168:449–453

Garman RD, Jacobs KA, Clark SC, Raulet DH (1987) B-cell-stimulatory factor 2 (β 2 interferon) functions as a second signal for interleukin 2 production by mature murine T cells. Proc Natl Acad Sci USA 84:7629–7633

Gauldie J, Richards C, Harnish D, Lansdorp P, Baumann H (1987) Interferon β 2/B-cell stimulatory factor type 2 shares identity with monocyte-derived hepatocyte-stimulating factor and regulates the major acute phase protein response in liver cells. Proc Natl Acad Sci USA 84:7251–7255

Gearing D, Gough NM, King JA, Hilton DJ, Nicola NA, Simpson RJ, Nice EC, Kelso A, Metcalf D (1987) Molecular cloning and expression of cDNA encoding a murine myeloid leukaemia inhibitory factor (LIF). EMBO J 6:3995-4002

Geiger T, Andus T, Klapproth J, Hirano T, Kishimoto T, Heinrich PC (1988) Induction of rat acute-phase proteins by interleukin-6 in vivo. Eur J Immunol 18:717–721

Haegeman G, Content J, Volckaert G, Derynck R, Tavernier J, Fiers W (1986) Structural analysis of the sequence encoding for an inducible 26-kDa protein in human fibroblasts. Eur J Biochem 159:625–632

Helfgott DC, May LT, Sthoeger Z, Tamm I, Sehgal PB (1987) Bacterial lipopolysaccharide (endotoxin) enhances expression and secretion of β 2 interferon by human fibroblasts. J Exp Med 166:1300–1309

Helle M, Brakenhoff JPJ, De Groot ER, Aarden LA (1988) Interleukin 6 is involved in interleukin 1-induced activities. Eur J Immunol 18:957–959

Hirano T, Teranishi T, Onoue K (1984a) Human helper T cell factor(s) III. Characterization of B cell differentiation factor I (BCDFI). J Immunol 132:229–234

Hirano T, Teranishi T, Lin BH, Onoue K (1984b) Human helper T cell factor(s) IV. Demonstration of a human late-acting B cell differentiation factor acting in Staphylococcus aureus Cowan I-stimulated B cells. J Immunol 133:798–802

Hirano T, Taga T, Nakano N, Yasukawa K, Kashiwamura S, Shimizu K, Nakajima K, Pyun KH, Kishimoto T (1985) Purification to homogeneity and characterization of human B cell differentiation factor (BCDF or BSFp-2). Proc Natl Acad Sci USA 82:5490–5494

Hirano T, Yasukawa K, Harada H, Taga T, Watanabe Y, Matsuda T, Kashiwamura S, Nakajima K, Koyama K, Iwamatu A, Tsunasawa S, Sakiyama F, Matsui H, Takahara Y, Taniguchi T, Kishimoto T (1986) Complementary DNA for a novel human interleukin (BSF-2) that induces B lymphocytes to produce immunoglobulin. Nature 324:73–76

Hirano T, Taga T, Yasukawa K, Nakajima K, Nakano N, Takatsuki F, Shimizu M, Murashima A, Tsunasawa S, Sakiyama F, Kishimoto T (1987) Human B cell differentiation factor defined by an anti-peptide antibody and its possible role in autoantibody production. Proc Natl Acad Sci USA 84:228–231

Hirano T, Matsuda T, Hosoi K, Okano A, Matsui H, Kishimoto T (1988a) Absence of antiviral activity in recombinant B cell stimulatory factor 2 (BSF-2). Immunol Lett 17:41–45

Hirano T, Matsuda T, Turner M, Miyasaka N, Buchan G, Tang B, Sato K, Shimizu M, Maini R, Feldman M, Kishimoto T (1988b) Excessive production of interleukin 6/B cell stimulatory factor-2 in rheumatoid arthritis. Eur J Immunol 18:1797–1801

Hodgkin PD, Bond MW, O'Garra A, Frank G, Lee F, Coffman RL, Zlotnik A, Howard M (1988) Identification of IL-6 as a T cell-derived factor that enhances the proliferative response of thymocytes to IL-4 and phorbol myristate acetate. J Immunol 141:151–157

Horii Y, Muraguchi A, Suematu S, Matsuda T, Yoshizaki K, Hirano T, Kishimoto T (1988) Regulation of BSF-2/IL-6 production by human mononuclear cells: macrophage-dependent synthesis of BSF-2/IL-6 by T cells. J Immunol 141:1529–1535

Horii Y, Muraguchi A, Iwano M, Matsuda T, Hirayama T, Yamada H, Fujii Y, Dohi K, Ishikawa H, Ohmoto Y, Yoshizaki K, Hirano T, Kishimoto T (1989) Involvement of interleukin-6 in mesangial proliferative glomerulonephritis. J Immunol 143 (in press)

Houssiau FA, Coulie PG, Olive D, Van Snick J (1988) Synergistic activation of human T cells by interleukin 1 and interleukin 6. Eur J Immunol 18:653–656

Ichikawa Y (1969) Differentiation of a cell line of myeloid leukemia. J Cell Physiol 74:223–234

Ikebuchi K, Wong GG, Clark SC, Ihle JN, Hirai Y, Ogawa M (1987) Interleukin-6 enhancement of interleukin-3-dependent proliferation of multipotential hemopoietic progenitors. Proc Natl Acad Sci USA 84:9035–9039

Ishibashi T, Kimura H, Uchida T, Karyone S, Friese P, Burstein SA (1989a) Human interleukin 6 is a direct promoter of maturation of megakaryocytes in vitro. Proc Natl Acad Sci USA 86:5953–5957

Ishibashi T, Kimura H, Shikama Y, Uchida T, Kariyone S, Hirano T, Kishimoto T, Takatsuki F, Akiyama Y (1989b) Interleukin 6 is a potent thrombopoietic factor in vivo in mice. Blood 74:1241–1244

Kawano M, Hirano T, Matsuda T, Taga T, Horii Y, Iwato K, Asaoku H, Tang B, Tanabe O, Tanaka H, Kuramoto A, Kishimoto T (1988) Autocrine generation and essential requirement of BSF-2/IL-6 for human multiple myelomas. Nature 332:83–85

Kikutani H, Taga T, Akira S, Kishi H, Miki Y, Saiki O, Yamamura Y, Kishimoto T (1985) Effect of B cell differentiation factor (BCDF) on biosynthesis and secretion of immunoglobulin molecules in human B cell lines. J Immunol 134:990–995

Kishimoto T (1985) Factors affecting B cell growth and differentiation. Annu Rev Immunol 3:133–159

Kishimoto T (1988) B cell differentiation – interleukins and receptors for the antibody response. ISI Atlas of Science: Immunology 1:149–154

Kishimoto T, Hirano T (1988a) Molecular regulation of B lymphocyte response. Annu Rev Immunol 6:485–512

Kishimoto T, Hirano T (1988b) A new interleukin with pleiotropic activities. Bioessays 9:11–15

Kishimoto T, Ishizaka K (1973) Regulation of antibody responses in vitro. VII. Enhancing soluble factors for IgG and IgE antibody response. J Immunol 111:1194–1205

Kohase M, May LT, Tamm I, Vilcek J, Sehgal PB (1987) A cytokine network in human diploid fibroblasts: interactions of β-interferons, tumor necrosis factor, platelet-derived growth factor, and interleukin-1. Mol Cell Biol 7:273–280

Koji A (1985) Definition and classification of acute-phase proteins. In: Gordon AH, Koji A (eds) The acute phase response to injury and infection. Elsevier, Amsterdam, p 139 (Research monographs in cell and tissue physiology, vol 10)

Koji A, Gauldie J, Regoeczi E, Sauder DN, Sweeney GD (1984) The acute-phase response of cultured rat hepatocytes. System characterization and the effect of human cytokines. Biochem J 224:505–514

Kushner I (1982) The phenomenon of the acute phase response. Ann NY Acad Sci 389:39–48

Le J, Vilcek J (1989) Interleukin 6. In: Barr PJ (ed) Growth factors and lymphokines: production and clinical application. Dekker, New York (in press)

Le J, Fredrickson G, Reis LFL, Diamantsein T, Hirano T, Kishimoto T, Vilcek J (1988) Interleukin 2-dependent and interleukin 2-independent pathways of regulation of thymocyte function by interleukin 6. Proc Natl Acad Sci USA 85:8643–8647

Lee F, Yokota T, Otsuka T, Meyerson P, Villaret D, Coffman R, Mosmann T, Rennick D, Roehm N, Smith C, Zlotnik A, Arai L (1986) Isolation and characterization of a mouse interleukin cDNA clone that expresses BSF-1 activities and T cell and mast cell stimulatory activities. Proc Natl Acad Sci USA 83: 2061–2065

Lipton JH, Sachs L (1981) Characterization of macrophage- and granulocyte-inducing proteins for normal and leukemic myeloid cells produced by the Krebs ascites tumor. Biochim Biophys Acta 673:552–569

Lotz M, Jirik F, Kabouridis R, Tsoukas C, Hirano T, Kishimoto T, Carson DA (1988) BSF-2/IL-6 is a costimulant for human thymocytes and T lymphocytes. J Exp Med 167:1253–1258

Lovett DH, Ryan JL, Sterzel RB (1983) Stimulation of rat mesangial cell proliferation by macrophage interleukin 1. J Immunol 131:2830–2836

MacCarthy EP, Hsu A, Ooi BS (1985) Modulation of mouse mesangial cell proliferation by macrophage products. Immunology 56:695–699

Matsuda T, Hirano T, Kishimoto T (1988) Establishment of a IL-6/BSF-2-dependent cell line and preparation of anti-IL-6/BSF-2 monoclonal antibodies. Eur J Immunol 18:951–956

Matsuda T, Hirano T, Nagasawa S, Kishimoto T (1989) Identification of α2-macroglobulin as a carrier protein for interleukin 6. J Immunol 142:148–152

May LT, Helfgott DC, Sehgal PB (1986) Anti-β-interferon antibodies inhibit the increased expression of HLA-B7 mRNA in tumor necrosis factor-treated human fibroblasts: structural studies of the β_2-interferon involved. Proc Natl Acad Sci USA 83:8957–8961

Melcion C, Lachman L, Killen PD, Morel-Maroger L, Stricker GE (1982) Mesangial cells, effect of monocyte products on proliferation and matrix synthesis. Transplant Proc 14:559–564

Metcalf D (1974) The serum factor stimulating colony formation in vitro by murine plasmacytoma cells: response to antigens and mineral oil. J Immunol 113:235–243

Metcalf D (1986) The molecular biology and functions of the granulocyte-macrophage colony-stimulating factors. Blood 67:257–267

Miyaura C, Onozaki K, Akiyama Y, Taniyama T, Hirano T, Kishimoto T, Suda T (1988) Recombinant human interleukin 6 (B-cell stimulatory factor 2) is a potent inducer of differentiation of mouse myeloid leukemia cells (M1). FEBS Lett 234:17–21

Mosmann TR, Coffman RL (1987) Two types of helper T-cell clone: implications for immune regulation. Immunol Today 8:223–227

Muraguchi A, Kishimoto T, Miki Y, Kuratani T, Kaieda T, Yoshizaki K, Yamamura Y (1981) T cell-replacing factor (TRF)-induced IgG secretion in human B blastoid cell line and demonstration of acceptors for TRF. J Immunol 127:412–416

Muraguchi A, Hirano T, Tang B, Matsuda T, Horii Y, Nakajima K, Kishimoto T (1988) The essential role of B cell stimulatory factor 2 (BSF-2/IL-6) for the terminal differentiation of B cells. J Exp Med 167:332–344

Mushinski JF, Bauer SR, Potter M, Reddy EP (1983) Increased expression of myc-related oncogene mRNA characterizes most BALB/c plasmacytomas induced by pristane and Abelson murine leukemia virus. Proc Natl Acad Sci USA 80:1073–1077

Nagata S, Tsuchiya M, Asano S, Kaziro Y, Yamazaki T, Yamamoto O, Hirata Y, Kubota N, Oheda M, Nomura H, Ono M (1986) Molecular cloning and expression of cDNA for human granulocyte colony-stimulating factor. Nature 319:415–418

Nakajima K, Martinez-Maza O, Hirano T, Nishanian P, Salazar-Gonzalez JF, Fahey JL, Kishimoto T (1989) Induction of interleukin 6(BSF-2/IFN-β 2) production by the human immunodeficiency virus (HIV). J Immunol 142:144–147

Namba Y, Hanaoka M (1972) Immunocytology of cultured IgM-forming cells of mouse. I. Requirement of phagocytic cell factor and its role in antibody formation. J Immunol 109:1193–1200

Nicola NA, Metcalf D, Matsumoto M, Johnson GR (1983) Purification of a factor inducing differentiation in murine myelomonocytic leukemia cells. Identification as granulocyte colony-stimulating factor. J Biol Chem 258:9017–9023

Nijstein MWN, De Groot ER, Ten Duis HJ, Klasen HJ, Hack CE, Aarden LA (1987) Serum levels of interleukin-6 and acute phase responses. Lancet ii: 921

Noma T, Mizuta T, Rosen A, Hirano T, Kishimoto T, Honjo T (1987) Enhancement of the interleukin 2 receptor expression on T cells by multiple B-lymphotropic lymphokines. Immunol Lett 15:249–253

Nordan RP, Potter M (1986) A macrophage-derived factor required by plasmacytomas for survival and proliferation in vitro. Science 233:566–569

Nordan RP, Pumphrey JG, Rudikoff S (1987) Purification and NH_2-terminal sequence of a plasmacytoma growth factor derived from the murine macrophage cell line P388D1. J Immunol 139:813–817

Ohno S, Babonits M, Wiener F, Spira J, Klein G, Potter M (1979) Non-random chromosome changes involving the Ig gene-carrying chromosomes 12 and 6 in pristane-induced mouse plasmacytomas. Cell 18:1001–1007

Okada M, Sakaguchi N, Yoshimura N, Hara H, Shimizu K, Yoshida N, Yoshizaki K, Kishimoto S, Yamamura Y, Kishimoto T (1983) B cell growth factor (BCGF) and B cell differentiation factor from human T hybridomas: two distinct kinds of BCGFs and their synergism in B cell proliferation. J Exp Med 157:583–590

Okada M, Kitahara M, Kishimoto S, Matsuda T, Hirano T, Kishimoto T (1988) BSF-2/IL-6 functions as killer helper factor in the in vitro induction of cytotoxic T cells. J Immunol 141:1543–1549

Okano A, Suzuki C, Takatsuki F, Akiyama Y, Kake K, Ozawa K, Hirano T, Kishimoto T, Nakahata T, Asano S (1989) In vitro expansion of murine pluripotent hemopoietic stem cell population in response to interleukin 3 and interleukin 6: application to bone marrow transplantation. Transplantation 48:495–498

Potter M, Boyce C (1962) Induction of plasma cell neoplasms in strain Balb/c mice with mineral oil and mineral oil adjuvants. Nature 193:1086–1087

Poupart P, Vandenabeele P, Cayphas S, Snick JV, Haegeman G, Kruys V, Fiers W, Content J (1987) B cell growth modulating and differentiating activity of recombinant human 26-kd protein (BSF-2, HuIFN-β 2, HPGF). EMBO J 6:1219–1224

Prud'homme GJ, Fieser TM, Dixon FJ, Theofilopoulos AN (1984) B cell-tropic interleukins in murine systemic lupus erythematosus (SLE). Immunol Rev 78:159–183

Ray A, Tatter SB, May LT, Sehgal PB (1988) Activation of the human "β 2-interferon/hepatocyte-stimulating factor/interleukin 6" promoter by cytokines, viruses, and second messenger agonists. Proc Natl Acad Sci USA 85:6701–6705

Reis LL, Le J, Hirano T, Kishimoto T, Vilcek J (1988) Antiviral action of tumor necrosis factor in human fibroblasts is not mediated by B cell stimulatory factor 2/IFN-β 2, and is inhibited by specific antibodies to IFN-β. J Immunol 140:1566–1570

Sachs L (1978) Control of normal cell differentiation and the phenotypic reversion of malignancy in myeloid leukemia. Nature 274:535–539

Sachs L (1987) The molecular control of blood cell development. Science 238:1374–1379

Sanderson CL, O'Garra A, Warren DJ, Klaus GGB (1986) Eosinophil differentiation factor also has B-cell growth factor activity: proposed name interleukin 4. Proc Natl Acad Sci USA 83:437–440

Satoh T, Nakamura S, Taga T, Matsuda T, Hirano T, Kishimoto T, Kaziro Y (1988) Induction of neural differentiation in PC12 cells by B cell stimulatory factor 2/interleukin 6. Mol Cell Biol 8:3546–3549

Schimpl A, Wecker E (1972) Replacement of T cell function by a T cell product. Nature 237:15–17

Sehgal PB, Zilberstein A, Ruggieri R-M, May LT, Ferguson-Smith A, Slate DL, Revel M, Ruddle F (1986) Human chromasome 7 carries the β 2 interferon gene. Proc Natl Acad Sci USA 83:5219–5222

Sehgal PB, Walther Z, Tamm I (1987) Rapid enhancement of β 2-interferon/B-cell differentiation factor BSF-2 gene expression in human fibroblasts by diacylglycerols and the calcium ionophore A23187. Proc Natl Acad Sci USA 84:3633–3667

Sehgal PB, Helfgott DC, Santhanam U, Tatter SB, Clarick RH, Ghrayeb J, May LT (1988) Regulation of the acute phase and immune responses in viral disease. J Exp Med 167:1951–1956

Shabo Y, Lotem J, Rubinstein M, Revel M, Clark SC, Wolf SF, Kamen R, Sachs L (1988) The myeloid blood cell differentiation-inducing protein MGI-2A is interleukin 6. Blood 72:2070–2073

Shaw G, Kamen R (1986) A conserved AU sequence from the 3' untranslated region of GM-CSF mRNA mediates selective degradation. Cell 46:659–667

Shimizu S, Hirano T, Yoshioka K, Sugai S, Matsuda T, Taga T, Kishimoto T, Konda S (1988) Interleukin 6 (B cell stimulatory factor 2)-dependent growth of a Lennert's lymphoma-derived T cell line (KT-3). Blood 72:1826–1828

Sims JE, March CJ, Cosman D, Widmer MB, MacDonald HR, McMahan CJ, Grubin CE, Wignall JM, Jackson JL, Call SM, Friend D, Alpert AR, Gillis S, Urdal DL, Dower SK (1988) cDNA expression cloning of the IL-1 receptor, a member of the immunoglobulin superfamily. Science 241:585–589

Sporn MB, Roberts AB (1988) Peptide growth factors are multifunctional. Nature 332:217–219

Striker GE, Striker LJ (1985) Biology of disease: glomerular cell culture. Lab Invest 53:122–131

Suematsu S, Matsuda T, Aozasa K, Akira S, Nakao N, Ohno S, Miyazaki J, Jamamura K, Hirano T, Kishimoto T (1989) IgG1 plasmacytosis in interleukin 6 transgenic mice. Proc Natl Acad Sci USA 86:7547–7551

Sutton MGSJ, Mercier L, Giuliani ER, Lie JT (1980) Atrial myxomas: a review of clinical experience in 40 patients. Mayo Clin Proc 55:371–376

Taga T, Kawanishi K, Hardy RR, Hirano T, Kishimoto T (1987) Receptors for B cell stimulatory factor 2 (BSF-2): quantitation, specificity, distribution and regulation of the expression. J Exp Med 166:967–981

Taga T, Hibi M, Hirata Y, Yamasaki K, Yasakawa K, Matsuda T, Hirano T, Kishimoto T (1989) Interleukin-6 (IL-6) triggers the association of its receptor (IL-6-R) with a possible signal transducer, gp 130. Cell 58:573–581

Takai Y, Wong GG, Clark SC, Burakoff SJ, Herrmann SH (1988) B cell stimulatory factor-2 is involved in the differentiation of cytotoxic T lymphocytes. J Immunol 140:508–512

Takatsuki F, Okano A, Suzuki C, Chieda R, Takahara Y, Hirano T, Kishimoto T, Hamuro J, Akiyama Y (1988) Human recombinant interleukin 6/B cell stimulatory factor 2 (IL-6/BSF-2) augments murine antigen-specific antibody responses in vitro and in vivo. J Immunol 141:3072–3077

Tanabe O, Akira S, Kamiya T, Wong GG, Hirano T, Kishimoto T (1988) Genomic structure of the murine IL-6 gene: high degree conservation of potential regulatory sequences between mouse and human. J Immunol 141:3875–3881

Teranishi T, Hirano T, Arima N, Onoue K (1982) Human helper T cell factor(s) (ThF). II. Induction of IgG production in B lymphoblastoid cell lines and identification of T cell-replacing factor-(TRF) like factor(s). J Immunol 128:1903–1908

Teranishi T, Hirano T, Lin BH, Onoue K (1984) Demonstration of the involvement of interleukin 2 in the differentiation of Staphylococcus aureus Cowan I-stimulated B cells. J Immunol 133:3062–3067

Theofilopoulos AN, Dixon FJ (1985) Murine models of systemic lupus erythematosus. Adv Immunol 37:269–390

Tomita M, Yamamoto-Yamaguchi Y, Hozumi M (1984) Purification of a factor inducing differentiation of mouse myeloid leukemic M1 cells from conditioned medium of mouse fibroblast L929 cells. J Biol Chem 259:10978–10982

Tomita M, Yamamoto-Yamaguchi Y, Hozumi M, Okabe T, Takaku F (1986) Induction by recombinant human granulocyte colony-stimulating factor of differentiation of mouse myeloid leukemic M1 cells. FEBS Lett 207:271–275

Tosato G, Seamon KB, Goldman ND, Sehgal PB, May LT, Washington GC, Jones KD, Pike SE (1988) Monocyte-derived human B-cell growth factor identified as interferon-β 2 (BSF-2, IL-6). Science 239:502–504

Tsuchiya M, Shigetaka A, Kaziro Y, Nagata S (1986) Isolation and characterization of the cDNA for murine granulocyte colony-stimulating factor. Proc Natl Acad Sci USA 84:7633–7637

Uyttenhove C, Coulie PG, Van Snick J (1988) T cell growth and differentiation induced by interleukin-Hp1/IL-6, the murine hybridoma/plasmacytoma growth factor. J Exp Med 167:1417–1427

Van Damme J, De Ley M, Opdenakker G, Billiau A, De Somer P (1985) Homogeneous interferon-inducing 22K factor is related to endogenous pyrogen and interleukin-1. Nature 314:266–268

Van Damme J, Opdenakker G, Simpson RJ, Rubira MR, Cayphas S, Vink A, Billiau A, Snick JV (1987a) Identification of the human 26-kD protein, interferon β 2 (IFNβ 2), as a B cell hybridoma/plasmacytoma growth factor induced by interleukin 1 and tumor necrosis factor. J Exp Med 165:914–919

Van Damme J, De Ley M, Van Snick J, Dinarello CA, Billiau A (1987b) The role of interferon-β 1 and the 26-kDa protein (interferon-β 2) as mediators of the antiviral effect of interleukin 1 and tumor necrosis factor. J Immunol 139:1867–1872

Van Damme J, Cayphas S, Opdenakker G, Billiau A, Van Snick J (1987c) Interleukin 1 and poly(rI)·poly(rC) induce production of a hybridoma growth factor by human fibroblasts. Eur J Immunol 17:1–7

Van Snick J, Cayphas S, Vink A, Uyttenhove C, Coulie PG, Rubira MR, Simpson RJ (1986) Purification and NH$_2$-terminal amino acid sequence of a T-cell-derived lymphokine with growth factor activity for B-cell hybridomas. Proc Natl Acad Sci USA 83:9679–9683

Van Snick J, Vink A, Cayphas S, Uyttenhove C (1987) Interleukin-HP1, a T cell-derived hybridoma growth factor that supports the in vitro growth of murine plasmacytomas. J Exp Med 165:641–649

Van Snick J, Cayphas S, Szikora J-P, Renauld J-C, Van Roost E, Boon T, Simpson RJ (1988) cDNA cloning of murine interleukin-HP1: homology with human interleukin 6. Eur J Immunol 18:193–197

Vink A, Coulie PG, Wauters P, Nordan RP, Van Snick J (1988) B cell growth and differentiation activity of interleukin-HP1 and related murine plasmacytoma growth factors. Synergy with interleukin 1. Eur J Immunol 18:607–612

Walther Z, May LT, Sehgal PB (1988) Transcriptional regulation of the interferon-β 2/B cell differentiation factor BSF-2/hepatocyte-stimulating factor gene in human fibroblasts by other cytokines. J Immunol 140:974–977

Weissenbach J, Chernajovsky Y, Zeevi M, Shulman L, Soreq H, Nir U, Wallach D, Perricaudet M, Tiollais P, Revel M (1980) Two interferon mRNA in human fibroblasts: in vitro translation and *Escherichia coli* cloning studies. Proc Natl Acad Sci USA 77:7152–7156

Williams AF, Barclay AN (1988) The immunoglobulin superfamily – domains for cell surface recognition. Annu Rev Immunol 6:381–405

Woloski BM, Smith EM, Meyer WJ, Fuller GM (1985) Corticotropin-releasing activity of monokines. Science 230:1035–1037

Wong GHW, Goeddel DV (1986) Tumor necrosis factors α and β inhibit virus replication and synergize with interferons. Nature 323:819–822

Yamasaki K, Taga T, Hirata Y, Yawata H, Kawanishi Y, Seed B, Taniguchi T, Hirano T, Kishimoto T (1988) Cloning and expression of the human interleukin-6 (BSF-2/IFNβ) receptor. Science 241:825–828

Yasukawa K, Hirano T, Watanabe Y, Muratani K, Matsuda T, Kishimoto T (1987) Structure and expression of human B cell stimulatory factor 2 (BSF-2/IL-6) gene. EMBO J 6:2939–2945

Yoshizaki K, Nakagawa T, Kaieda T, Muraguchi A, Yamamura Y, Kishimoto T (1982) Induction of proliferation and Ig production in human B leukemic cells by anti-immunoglobulins and T cell factors. J Immunol 128:1296–1301

Yoshizaki K, Matsuda T, Nishimoto N, Kuritani T, Taeho L, Aozasa K, Nakahata T, Kawai H, Tagoh H, Komori T, Kishimoto S, Hirano T, Kishimoto T (1989) Pathological significance of interleukin 6 (IL-6/BSF-2) in Castleman's disease. Blood 74:1360–1367

Zhang Y, Lin J-X, Vilcek J (1988a) Synthesis of interleukin 6 (interferon-β 2/B cell stimulatory factor 2) in human fibroblasts is triggered by an increase in intracellular cyclic AMP. J Biol Chem 263:6177–6182

Zhang Y, Lin J-X, Yip YK, Vilcek J (1988b) Enhancement of cAMP levels and of protein kinase activity by tumor necrosis factor and interleukin 1 in human fibroblasts: role in the induction of interleukin 6. Proc Natl Acad Sci USA 85:6802–6805

Zilberstein A, Ruggieri R, Korn JH, Revel M (1986) Structure and expression of cDNA and genes for human interferon-beta-2, a distinct species inducible by growth-stimulatory cytokines. EMBO J 5:2529–2537

CHAPTER 15

Colony-Stimulating Factor 1
(Macrophage Colony-Stimulating-Factor)

C. J. SHERR and E. R. STANLEY

A. Colony-Stimulating Factor 1, a Mononuclear Phagocyte Growth Factor

The study of mononuclear phagocytes has been greatly aided by the purification, molecular cloning, and vector-mediated production of biologically active colony-stimulating factors that regulate the proliferation and support the viability of monocytes, macrophages, and their bone marrow progenitors. The term "colony stimulating factor" (CSF) was first applied to a group of impure growth factors obtained from tissues and biological fluids and from culture media conditioned by certain cell lines. These compounds were initially recognized through their ability to stimulate the formation of colonies composed of differentiated myeloid cells from single bone marrow-derived precursor cells plated in semisolid medium (PLUZNIK and SACHS 1965; BRADLEY and METCALF 1966; ICHIKAWA et al. 1966). Colonies did not develop in the absence of such factors, and their patterns of differentiation were found to be governed by the source of CSF, suggesting that multiple growth factors were responsible for the types of colonies obtained (ICHIKAWA et al. 1966; METCALF et al. 1967, 1974; METCALF 1969).

Among the known colony-stimulating activities, one form of CSF found in serum (ROBINSON et al. 1967), urine (STANLEY et al. 1970), certain tissues (SHERIDAN and STANLEY 1971), and in conditioned media from some established cell lines (BRADLEY and SUMNER 1968; WORTON et al. 1969) was found to stimulate the development of only macrophage colonies and was designated CSF-1 (STANLEY), M-CSF (METCALF), or MGI-1M (SACHS) (reviewed in METCALF 1985; SACHS 1987). This hemopoietin, discovered independently in another guise, was also termed "macrophage growth factor" (MGF), because it stimulated adherent peritoneal macrophages to divide (VIROLAINEN and DEFENDI 1967; MAUEL and DEFENDI 1971 a, b; CIFONE and DEFENDI 1974). Copurification of CSF-1 and MGF activities from L cell-conditioned media and comparison of their biological, physicochemical, and antigenic properties showed them to be identical to each other (STANLEY et al. 1976). Modification of the purification scheme yielded highly purified CSF-1 preparations that consisted predominantly of a 70-kilodalton (kDa) glycoprotein composed of two similarly charged polypeptide chains of 35 kDa assembled through disulfide bonds (STANLEY and HEARD 1977; DAS and STANLEY 1982). The purified hormone was able to stimulate the proliferation of macrophages derived from peritoneal and pleural exudates, alveoli, lymph nodes, thymus, spleen, and liver, as well as circulating monocytes (Stanley et al. 1978). Purified CSF-1 was later shown to regulate pleiotypic functional responses of macrophages and their hematopoietic progenitors and to support their viability. CSF-1 is distinguished by its lineage specificity and physicochemical properties from granulocyte-macrophage colony stimulating factor (GM-CSF) and from multi-CSF (interleukin-3, or IL-3), both of which stimulate more primitive myeloid elements, including mononuclear phagocyte progenitors (reviewed in Metcalf 1985; Clark and Kamen 1987; Sachs 1987).

B. Assay and Purification of CSF-1

I. Bioassay and Purification

The original and still commonly used bioassay for CSF-1 activity involves the stimulation of macrophage colony formation by single murine bone marrow progenitors plated in semisolid culture medium (Metcalf 1970). A biological unit of CSF-1 corresponds to the amount of growth factor necessary to induce a single macrophage colony and represents 0.44 fmol of purified murine CSF-1 (Stanley and Guilbert 1981; Stanley 1985). Proliferation of colonies in bioassays is detectable at CSF-1 concentrations as low as 1 pM (ca. 90 pg/ml), whereas maximal stimulation is observed at 250 pM. The colony assay is more sensitive than the original MGF assay system (Stanley et al. 1976), which depends upon [^{3}H]thymidine incorporation by stimulated peritoneal macrophages, as determined by autoradiography, and the percentage of total cells labeled, recorded as a stimulation index (Virolainen and Defendi 1967). In neither case are these bioassays absolutely specific for CSF-1, because GM-CSF and IL-3 also stimulate the proliferation of cells of the mononuclear phagocyte lineage.

Using the assay for colony formation by murine bone marrow progenitors, CSF-1 was first purified from mouse L cell-conditioned medium by sequential chromatographic and molecular sieving procedures (reviewed in Stanley and Guilbert 1981). The medium was concentrated by vacuum rotary evaporation, chromatographed on diethylaminoethyl (DEAE) cellulose, filtered on Sephadex G-200, adsorbed and eluted from concanavalin A–Sepharose, and fractionated by gradient gel electrophoresis, resulting in an approximately 1100-fold purification with 25% yield (Stanley and Heard 1977). Batch purification steps involving DEAE-cellulose or calcium phosphate gels have been found to be more convenient initial steps, particularly if large volumes of starting material are used. The latter procedures are rapid and result in equally high yields. Zonal sedimentation (Waheed and Shadduck 1979) or the use of calcium phosphate chromatography (Stanley and Guilbert 1981) can also be substituted for gradient gel electrophoresis in the final purification step. More recently, methods involving the use of immunoaffinity chromatography with immobilized antibodies to CSF-1 have been employed to purify the murine growth factor (WAHEED and SHADDUCK 1982; STANLEY 1985). With these protocols, 20 l murine L cell-conditioned medium yields about 400 µg purified CSF-1. Human urinary CSF-1 has been purified by similar methods, yielding preparations of equivalent specific activity (DAS et al. 1981; STANLEY 1985).

II. Radioimmunoassay and Radioreceptor Assay

Antisera to human or murine CSF-1 raised in rabbits (STANLEY et al. 1970, 1976; SHADDUCK and METCALF 1975) can be titered by neutralization of CSF-1 bioactivity or by their ability to combine with ^{125}I-labeled CSF-1. Methods for radioiodination of the purified growth factor with retention of its biological activity were critical in developing radioimmunoassay (RIA) procedures for its rapid detection (STANLEY 1979; SHADDUCK and WAHEED 1979). Because the RIA detects only the biologically active growth factor, there is concordance between the activities of CSF-1 determined both by bioassay and immunoassay; the latter can be most conveniently used to determine the concentration of CSF-1 throughout its purification or in biological fluids and tissue extracts.

Iodinated CSF-1 can also be used to quantitate the growth factor by a competition radioreceptor assay (RRA) in which solutions containing unknown amounts of ligand compete for binding of subsaturating quantities of ^{125}I-labeled CSF-1 to its receptor on a macrophage cell line. At 4° C, the high-affinity interaction between radioiodinated CSF-1 and its cell surface receptor on peritoneal macrophages is highly specific, saturable, and essentially irreversible (GUILBERT and STANLEY 1980), and ligand-receptor complexes are neither rapidly internalized nor degraded (see Sect. G. IV). The competitive RRA is therefore performed at 4° C, and cells are pretreated with competitor for 1 h prior to incubation with ^{125}I-labeled CSF-1 in order to maximize the sensitivity of the assay. Like the RIA, the RRA is standardized by generating a competition curve using known concentrations of purified CSF-1. Under optimal conditions, both assays are equally sensitive (DAS et al. 1980). Because CSF-1 can be iodinated to specific activities as high as 3×10^8 dpm per microgram with retention of antigenic and biological activity (STANLEY 1985), the assays can detect picogram quantities of growth factor and are slightly more sensitive than bioassays based on bone marrow cell colony formation. Detailed experimental procedures for radioiodination and for both RIA and RRA are summarized in detail elsewhere (STANLEY and GUILBERT 1981; STANLEY 1985).

C. Molecular Cloning
and Predicted Primary Structure of CSF-1

I. Initial Cloning Strategy

The amino-terminal 12 amino acids of human CSF-1 and the amino-terminal 42 amino acids of mouse CSF-1 were determined for the growth factors purified from human urine and from the conditioned medium of mouse L cells. Degenerate oligonucleotide probes predicted from the N-terminal polypeptide sequences were used to isolate partial human CSF-1 genomic clones. The nucleotide sequences of the latter were determined, and an exact 32-base oligonucleotide complementary to exon coding sequences was then employed to isolate human CSF-1 clones from a MIA PaCa-2 cDNA library. One clone 1.6 kilobases (kb) in length was isolated that encoded CSF-1 activity in a transient expression system (KAWASAKI et al. 1985). Nucleotide sequence analysis predicted that this cDNA (designated pcCSF-17) encoded a leader sequence of 32 amino acids followed by an additional 224 amino acids that included the predicted amino terminus of the purified polypeptide as well as the 20 internal residues determined by protein sequencing (Fig. 1). This form of human CSF-1 minus its signal peptide is referred to here as CSF-1^{224}.

When the 1.6-kb CSF-1 cDNA was used as a probe to analyze mRNA from MIA PaCa cells, longer polyadenylated RNA species ranging in size to about 4 kb were detected. Southern blotting analysis revealed the presence of only a single CSF-1 gene, suggesting that the mRNAs of different lengths were produced by alternative splicing (KAWASAKI et al. 1985; see Sect. E. II).

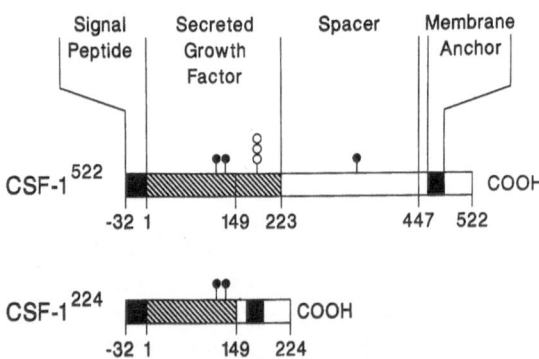

Fig. 1. Human CSF-1 precursors encoded by 4-kb and 1.6-kb mRNAs. The diagrams are based on the amino acid sequences of CSF-1^{522} and CSF-1^{224} shown in Fig. 2. Amino acids defining junctions of different domains are *numbered* with the amino-terminus of the secreted growth factor defined as residue 1. Hydrophobic segments corresponding to the signal peptide (residues -32 to -1) and the membrane-spanning anchor are shown as *black bars,* and residues within the secreted growth factor are indicated by *hatching.* Sites for addition of N-linked oligosaccharides (*filled circles*) and O-linked sugars (*open circles*) are also shown. The 298 amino acid segment from residues 149 to 447 is missing in CSF-1^{224}

Analysis of mRNA from different cellular sources demonstrated that the most ubiquitous form of CSF-1 mRNA is approximately 4 kb in length. With probes based on the sequence of pcCSF-17, a human cDNA corresponding to the 4-kb mRNA was molecularly cloned (WONG et al. 1987).

II. Primary Structure of Human and Murine CSF-1

The 4-kb CSF-1 cDNA contained a translational reading frame of 1662 nucleotides, encoding a predicted 554 amino acid polypeptide as well as 2200 nucleotides of 3' untranslated sequences. As shown in Fig.1, the predicted polypeptide included a 32 amino acid signal peptide followed by 522 additional residues. The coding region of this cDNA (Fig. 2) differed from that of pcCSF-17 by the presence of 894 additional base pairs that specify amino acids 150–447 (indicated by the vertical bars in Fig. 2). The absence of this 298 amino acid coding block, together with additional variations within their 3' untranslated sequences, accounts for the differences in length of the 1.6-kb and 4-kb CSF-1 human cDNA clones. Since the 1.6-kb pcCSF-17 cDNA encodes a biologically active form of CSF-1, amino acids 150–447 are not required for biological activity.

Probes from pcCSF-17 were used to isolate CSF-1 cDNAs from murine L929 fibroblasts (DeLAMARTER et al. 1987; LADNER et al. 1988). The longest murine clone was also about 4 kb in length. In addition, LADNER and coworkers (1988) isolated a biologically active 2.3-kb cDNA that differed from the longer murine clone only in its 3' untranslated sequences. Unlike the 1.6-kb human clone, the shorter murine cDNA encodes a precursor of 552 amino acids (including its signal peptide), which is indistinguishable from that encoded by the 4-kb cDNA. As shown in Fig. 2, the murine clones, like the 4-kb human cDNA, encode a segment of 295 amino acids analogous to the 298 amino acids that are absent from human CSF-1^{224}. Although this region of the murine precursors is three amino acids shorter than that encoded by the human 4-kb CSF-1 cDNA, a glutamic acid residue at position 512 in the predicted murine polypeptide is missing from its human counterpart. Thus, murine CSF-1 is only two amino acids shorter than the human polypeptide when the sequences are maximally aligned (Fig. 2).

The overall sequence similarity between human and murine CSF-1 is 69.5%. The highest degree of similarity occurs within amino-terminal residues 1–149 (80.5%), and the lowest (64%) is found within amino acids 150–447, which are dispensable for biological activity. The latter region is composed of 16% proline residues, whereas the former contains only three prolines. Twelve conserved cysteines are clustered within the amino-terminal 226 residues, including two within the signal peptide and 10 interspersed between amino acids 7 and 226. These cysteine residues are important not only in forming interchain disulfide bonds, but also in precursor dimerization (see Sect. D). The predicted amino acid sequences of human CSF-1^{522} and murine CSF-1^{520} include several canonical sites (Asn-X-Ser/Thr, X being any amino acid except

```
-32   Human  CSF-1⁵²²                    Met Thr Ala Pro Gly Ala Ala Gly Arg Cys Pro Pro
      Murine CSF-1⁵²⁰                    --- --- --- Arg --- --- --- --- --- --- --- Ser

-20   Thr Thr Trp Leu Gly Ser Leu Leu Leu Leu Val Cys Leu Leu Ala Ser Arg Ser Ile Thr
      Ser --- --- --- --- --- Arg --- --- --- --- --- --- --- Met --- --- --- --- Ala

  1   Glu Glu Val Ser Glu Tyr Cys Ser His Met Ile Gly Ser Gly His Leu Gln Ser Leu Gln
      Lys --- --- --- --- His --- --- --- --- --- --- Asn --- --- --- Lys Val --- ---

 21   Arg Leu Ile Asp Ser Gln Met Glu Thr Ser Cys Gln Ile Thr Phe Glu Phe Val Asp Gln
      Gln --- --- --- --- --- --- --- --- --- --- --- Ala --- --- --- --- --- --- ---

 41   Glu Gln Leu Lys Asp Pro Val Cys Tyr Leu Lys Lys Ala Phe Leu Leu Val Gln Asp Ile
      --- --- --- Asp --- --- --- --- --- --- --- --- --- Phe --- --- --- --- --- ---

 61   Met Glu Asp Thr Met Arg Phe Arg Asp Asn Thr Pro Asn Ala Ile Ala Ile Val Gln Leu
      Ile Asp Glu --- --- --- --- Lys --- --- --- --- --- Asn --- Thr Glu Arg ---

 81   Gln Glu Leu Ser Leu Arg Leu Lys Ser Cys Phe Thr Lys Asp Tyr Glu Glu His Asp Lys
      --- --- --- --- Asn Asn --- Asn --- --- --- --- --- --- --- --- --- Gln Asn ---

101   Ala Cys Val Arg Thr Phe Tyr Glu Thr Pro Leu Gln Leu Leu Glu Lys Val Lys Asn Val
      --- --- --- --- --- --- His --- --- --- --- --- --- --- --- --- Ile --- --- Phe

121   Phe Asn Glu Thr Lys Asn Leu Leu Asp Lys Asp Trp Asn Ile Phe Ser Lys Asn Cys Asn
      --- --- --- --- --- --- --- --- Glu --- --- --- --- --- --- --- Thr --- --- ---

141   Asn Ser Phe Ala Glu Cys Ser Ser Gln Asp Val Val Thr Lys Pro Asp Cys Asn Cys Leu
      --- --- --- --- Lys --- --- --- Arg --- --- --- --- --- --- --- --- --- --- ---

161   Tyr Pro Lys Ala Ile Pro Ser Ser Asp Pro Ala Ser Val Ser Pro His Gln Pro Leu Ala
      --- --- --- --- Thr --- --- --- --- --- --- --- Ala --- --- --- --- --- Pro ---

181   Pro Ser Met Ala Pro Val Ala Gly Leu Thr Trp Glu Asp Ser Glu Gly Thr Glu Gly Ser
      --- --- --- --- --- Leu --- --- --- Ala --- Asp --- --- Gln Arg --- --- --- ---

201   Ser Leu Leu Pro Gly Glu Gln Pro Leu His Thr Val Asp Pro Gly Ser Ala Lys Gln Arg
      --- --- --- --- Ser --- Leu --- --- Arg Ile Glu --- --- --- --- --- --- --- ---

221   Pro Pro Arg Ser Thr Cys Gln Ser Phe Glu Pro Pro Glu Thr Pro Val Val Lys Asp Ser
      --- --- --- --- --- --- --- Thr Leu --- Ser Thr --- Gln --- Asn His Gly --- Arg

241   Thr Ile Gly Gly Ser Pro Gln Pro Arg Pro Ser Val Gly Ala Phe Asn Pro Gly Met Glu
240   Leu Thr Glu Asp --- ( ) Gln Pro His --- --- Ala --- Gly Pro Val --- --- Val ---

261   Asp Ile Leu Asp Ser Ala Met Gly Thr Asn Trp Val Pro Glu Glu Ala Ser Gly Glu Ala
260   --- --- --- Glu --- Ser Leu --- --- --- --- --- Leu --- --- --- --- --- --- ---

281   Ser Glu Ile Pro Val Pro Gln Gly Thr Glu Leu Ser Pro Ser Arg Pro Gly Gly Gly Ser
280   --- --- Gly Phe Leu Thr --- Glu Ala Lys Phe --- --- --- Thr --- Val --- --- ---

301   Met Gln Thr Glu Pro Ala Arg Pro Ser Asn Phe Leu Ser Ala Ser Ser Pro Leu Pro Ala
300   Ile --- Ala --- Thr Asp --- --- ( ) Arg Ala --- --- --- --- ( ) --- Phe --- Lys
```

Fig. 2

```
321  Ser Ala Lys Gly Gln Gln Pro Ala Asp Val Thr Gly Thr Ala Leu Pro Arg Val Gly Pro
319  --- Thr Glu Asp --- Lys --- Val --- Ile Thr Asp Arg Pro --- Thr Glu --- Asn ---


341  Val Arg Pro Thr Gly Gln Asp Trp Asn His Thr Pro Gln Lys Thr Asp His Pro Ser Ala
339  Met --- --- Ile --- --- Thr Gln --- Asn --- --- Glu --- --- --- Gly Thr --- Thr


361  Leu Leu Arg Asp Pro Pro Glu Pro Gly Ser Pro Arg Ile Ser Ser Leu Arg Pro Gln Gly
359  --- Arg Glu --- His Gln --- --- --- --- --- His --- Ala Thr Pro Asn --- --- Arg


381  Leu Ser Asn Pro Ser Thr Leu Ser Ala Gln Pro Gln Leu Ser Arg Ser His Ser Ser Gly
379  Val --- --- ( ) Ala --- Pro Val --- --- Leu Leu --- Pro Lys --- --- --- Trp ---


401  Ser Val Leu Pro Leu Gly Glu Leu Glu Gly Arg Arg Ser Thr Arg Asp Arg Arg Ser Pro
398  Ile --- --- --- --- --- --- --- --- --- Lys --- --- --- --- --- --- --- --- ---


421  Ala Glu Pro Glu Gly Gly Pro Ala Ser Glu Gly Ala Ala Arg Pro Leu Pro Arg Phe Asn
418  --- --- Leu --- --- --- Ser --- --- --- --- --- --- --- --- --- Val Ala --- --- ---


441  Ser Val Pro Leu Thr Asp Thr│Gly His Glu Arg Gln Ser Glu Gly Ser Ser Ser Pro Gln
438  --- Ile --- --- --- --- --─│--- --- Val Glu --- His --- --- --- --- Asp --- ---


461  Leu Gln Glu Ser Val Phe His Leu Leu Val Pro Ser Val Ile Leu Val Leu Leu Ala Val
458  Ile Pro --- --- --- --- --- --- --- --- --- Gly Ile --- --- --- --- --- Thr ---


481  Gly Gly Leu Leu Phe Tyr Arg Trp Arg Arg Arg Ser His Gln Glu Pro Gln Arg Ala Asp
478  --- --- --- --- --- --- Lys --- Lys Trp --- --- --- Arg Asp --- --- Thr Leu ---


501  Ser Pro Leu Glu Gln Pro Glu Gly Ser Pro Leu Thr Gln Asp ( ) Asp Arg Gln Val Glu
498  --- Ser Val Gly Arg --- --- Asp --- Ser --- --- --- --- Glu Asp Arg Gln --- ---


520  Leu Pro Val END
518  --- --- --- END
```

Fig. 2. Predicted amino acid sequences of human CSF-1[522] and mouse CSF-1[520]. Residues missing from human CSF-1[224] are bracketed by *vertical bars*. Sites for addition of N-linked oligosaccharides are indicated by *underlining* and the hydrophobic membrane-spanning region is designated by the *overline*. In order to maximally align the sequences, four gaps were introduced, as shown by open *parentheses*. (Data taken from KAWASAKI et al. 1985; WONG et al. 1987; DELAMARTER et al. 1987)

Pro) for the addition of asparagine-linked oligosaccharide chains (shown by filled circles in Fig. 1 and underlined in Fig. 2). Two conserved sites are found within residues 1–149, and an additional site is located in the spacer segment. A site (Asn-Pro-Ser) at residues 383–385, noted by WONG et al. (1987) in human CSF-1[522], is unlikely to serve as a carbohydrate acceptor and has not been conserved in murine CSF-1[520].

An internal hydrophobic sequence of 23 amino acids is located at residues 464–486 of human CSF-1[522] (overlined in Fig. 2) and in the corresponding region of murine CSF-1[520]. This hydrophobic segment is retained in human CSF-1[224], because the missing 298 amino acid element terminates at amino acid 447 (Fig. 1). The hydrophobic region is directly followed by the amino acids Arg-Trp-Arg-Arg-Arg in human CSF-1 and by Lys-Trp-Lys-Trp-Arg in murine CSF-1, reminiscent of integral transmembrane proteins that contain a membrane-spanning hydrophobic domain followed by positively charged residues (SABATINI et al. 1982). Based on these consideration, KAWASAKI et al. (1985) first suggested that the product of the 1.6-kb human CSF-1 precursor might be synthesized as a membrane-bound polypeptide that is released from the membrane by proteolysis. Indeed, such processing events occur for all known CSF-1 precursors studied to date (RETTENMIER et al. 1987; HEARD et al. 1987; RETTENMIER and ROUSSEL 1988; MANOS 1988; also see Sect. D).

D. Mechanisms of CSF-1 Posttranslational Processing and Secretion

I. Processing of CSF-1^{224}

Based on a consideration of their predicted primary structures, CSF-1 precursors should be directed to membrane-bound polyribosomes through their amino-terminal signal peptides and translocated across the membrane of the endoplasmic reticulum (ER) during their synthesis. Translation of the internal hydrophobic segment would immobilize the nascent polypeptide in the membrane with its amino-terminal portion in the ER cisterna and its carboxy-terminal tail in the cytoplasm. All cysteine residues and all sites for addition of N-linked oligosaccharide chains would then be localized within the ER.

RETTENMIER and coworkers (1987), using retroviral expression vectors to produce high levels of human CSF-1 in mouse NIH 3T3 cells, experimentally confirmed that CSF-1^{224} is synthesized as a membrane-bound precursor. The polypeptide undergoes cotranslational glycosylation and is rapidly assembled into disulfide-linked homodimers of 64 kDa containing mannose-rich, asparagine-(N)-linked oligosaccharide chains. Modification of N-linked carbohydrate from high-mannose forms to complex oligosaccharide chains during intracellular transport yields a 68-kDa cell surface dimer, a portion of which is released from the plasma membrane by proteolysis (Table 1). The soluble extracellular growth factor is only 44 kDa in mass and is composed of two 22-kDa polypeptide chains each estimated to contain two N-linked oligosaccharides. After treatment with endoglycosidases specific for N-linked oligosaccharide chains, the apparent molecular mass of each subunit is reduced to 19 kDa, whereas no evidence for O-linked sugars was obtained after digestion with O-glycosidases. Cleavage of the 68-kDa homodimer from the cell surface is very inefficient, and the majority of molecules remain membrane-bound and cell-associated. When the 1.6-kb pcCSF-17 cDNA was truncated 5' to the predicted membrane-spanning segment by insertion of a termination residue after codon 158 (corresponding to Ser456 in CSF-1^{522}), a soluble, biologically active 44-kDa glycosylated homodimer was efficiently secreted (HEARD et al. 1987). These experiments indicated that CSF-1^{224} is im-

Table 1. Properties of two forms of human CSF-1

Poly-peptide[a]	Molecular mass of mature glycosylated precursor (kDa)	Oligosaccharide chains		Molecular mass of membrane-bound homodimer (kDa)		Molecular mass of soluble homodimeric growth factor (kDa)
		N-linked[b]	O-linked[c]	Endoplasmic reticulum	Plasma membrane	
CSF-1[522]	70	Yes	Yes	ca. 140	Not detected	86[e]
CSF-1[224]	34	Yes	No	64[d]	68	44[f]

[a] The primary translation products are 32 amino acids longer due to the presence of amino-terminal signal peptides.
[b] Oligosaccharides attached to asparagine at canonical sites (Asn-X-Ser/Thr).
[c] Oligosaccharides attached to Ser or Thr.
[d] The immature glycosylated subunit in the endoplasmic reticulum contains mannose-rich oligosaccharide chains and has an apparent molecular mass of 32 kDa.
[e] The 86-kDa homodimer is composed of two identical 43-kDa subunits, each 223 amino acids in length.
[f] The 44-kDa homodimer is composed of two identical 22-kDa subunits, each ca. 158 amino acids in length.

mobilized through its predicted hydrophobic anchor and is released by proteolytic cleavage at a site just amino-terminal to its membrane-spanning segment (Fig. 1). Residues 1–158 of the small CSF-1 precursor are therefore sufficient for biological activity.

II. Processing of CSF-1[522]

Analogous studies with the 4-kb human CSF-1 cDNA revealed differences in its processing (RETTENMIER and ROUSSEL 1988; MANOS 1988). Like CSF-1[224], CSF-1[522] is synthesized as an integral transmembrane glycoprotein of 70 kDa and undergoes rapid disulfide-mediated assembly into homodimers. However, the homodimer acquires both N-linked and O-linked sugars during its transport and is cleaved from the membrane within the intracellular secretory compartment before its release from cells. Thus, plasma membrane-bound forms are not detected, and the proteolyzed homodimer is efficiently secreted into the medium as an 86-kDa glycoprotein composed of two 43-kDa subunits (Table 1). When treated with endoglycosidases specific for both N- and O-linked sugars, the apparent molecular mass of each polypeptide subunit was reduced to 26 kDa. Because the molecular mass of the endoglycosidase-treated subunit derived from CSF-1[224] was only 19 kDa (see Sect. D. I), cleavage of CSF-1[522] must occur carboxy-terminal to residue 158.

WONG and coworkers (1987) also detected homodimers of 70–90 kDa containing both N- and O-linked carbohydrate chains in the medium of CHO cells engineered to produce the human CSF-1[522] precursor. The soluble secreted molecules displayed physical properties similar to those of natural human CSF-1 (WU et al. 1979; MOTOYOSHI et al. 1982; DAS and STANLEY 1982; CSEJTEY and BOOSMAN 1986). Sequencing of all of the major tryptic peptides of the purified human urinary growth factor revealed that its carboxy – terminus extended beyond residue 189 (WONG et al. 1987). Sequencing analysis of the carboxy terminus of the secreted molecule indicated that it terminates at residue 223 (CLARK and KAMEN 1987). It seems likely that residues 158–223 include the sites of O-linked oligosaccharide addition, since the product of CSF-1^{224} lacks O-linked chains (Fig. 1). Moreover, because the product of CSF-1^{522} can bind to specific proteoglycan(s) after its release from cells (RETTENMIER and ROUSSEL 1988), sequences required for intermolecular interactions between the growth factor and cell matrix components also appear to reside in its unique carboxy-terminal tail. Both IL-3 and GM-CSF can adsorb to heparan sulfate, and the bound growth factors are biologically active in stimulating hematopoiesis (R. ROBERTS et al. 1988). Similar interactions between CSF-1 and proteoglycans may also be important in modulating its biologic activity in vivo.

Because the proteolytic cleavage site of CSF-1^{522} resides within a sequence that is not present in CSF-1^{224}, the smaller precursor remains relatively resistant to proteolysis and can only be inefficiently released from the plasma membrane through the action of extracellular proteases that cleave the molecule at an alternative site (see Sect. D. I). Assuming that an mRNA corresponding to the 1.6-kb transcript encoding CSF-1^{224} is induced in certain cell types under physiologic conditions (RALPH et al. 1986), both secreted and membrane-bound forms of the growth factor might play different functional roles in interacting with CSF-1 receptors. Hence, possibilities exist for endocrine, paracrine, and cell-to-cell interactions between CSF-1 and its receptor.

E. Chromosomal Localization and Organization of the Human CSF-1 Gene

I. Linkage of CSF-1 and Its Receptor on Human Chromosome 5

Probes derived from pcCSF-17 cDNA were used to map the human CSF-1 gene to the long arm of chromosome 5 at band 5q33.1 (PETTENATI et al. 1987). The c-*fms* protooncogene which encodes the CSF-1 receptor (SHERR et al. 1985; see Sect. G) maps to an adjacent region of the chromosome at 5q33.2-34 (ROUSSEL et al. 1983; GROFFEN et al. 1983; LE BEAU et al. 1986 b). Other linked genes include those for GM-CSF (HUEBNER et al. 1985) and multi-CSF (LE BEAU et al. 1987) at 5q21–32; IL-5 at 5q31 (SUTHERLAND et al. 1988); endothelial cell growth factor (acidic fibroblast growth factor) at 5q31.3–33.2 (JAYE et al. 1986); the monocyte differentiation antigen CD14 at 5q23–31 (GOYERT et al. 1988); the *Egr*-1 product, a putative transcriptional regulatory protein, at 5q23–31 (SUKHATME et al. 1988); the β_2-adrenergic receptor at 5q31–q32 (KOBILKA et al. 1987); and the B-type receptor for platelet-derived growth factor (PDGF) at q31–32 (YARDEN et al. 1986). The multi-CSF and GM-CSF genes map only 9 kb apart at the centromeric end of the gene cluster (YANG et al. 1988), whereas the genes encoding the structurally related CSF-1 and PDGF receptors (see Sect. G. II) have now been shown to map in a tandem array at the telomeric end and are much closer to one another than previously suspected (M. ROBERTS et al. 1988). Interstitial deletions of the long arm of human chromosome 5 have been frequently observed in the bone marrow cells of patients with myelodysplasias, refractory anemia ("5q⁻ syndrome"), and therapy-related acute myeloid leukemia (reviewed in LE BEAU et al. 1986a), and may etiologically involve one or more of these genes.

II. Exon–Intron Structure and Alternative Splicing

The human CSF-1 gene consists of 10 exons distributed over 21 kb of DNA (Fig. 3A; KAWASAKI and LADNER 1989). Exons 1–8 specify the CSF-1 polypeptide, whereas exons 9 and 10 encode 3′ nontranslated sequences. Most of the coding exons are small, but exon 6 is alternatively spliced and is either 131 or 1025 base pairs in length.

Within 400 base pairs of the mRNA initiation site in exon 1 are putative transcriptional regulatory motifs. There is a potential Goldberg-Hogness box (TTAAA) at position -26 to -21 and a CATAAA element at position -54 to -49 (KAWASAKI and LADNER 1989). Residues -126 to -80 contain an extended series of alternating T and G residues, upstream of which are two GGCGGG boxes at positions -129 to -124 and -147 to -142. Finally, the sequences AGGAAAG at position -317 to -311 and GGGAAAG at -377 to -371 are reminiscent of a consensus enhancer core motif, TGGAAAG

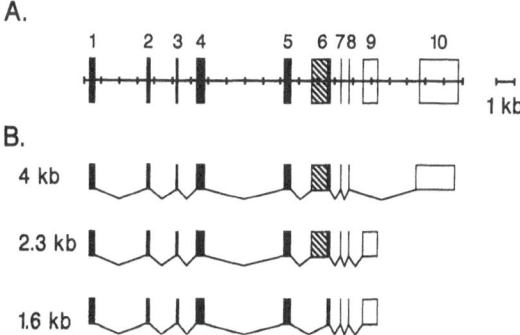

Fig. 3A, B. Organization of the human CSF-1 locus and origin of alternatively spliced mRNAs. **A** The CSF-1 gene is 21 kb in length and is composed of 10 exons (*bars*). Coding exons 1–8 are indicated by *filled bars* and noncoding exons 9 and 10 by *open bars*. Exon 6 (*partially hatched*) is alternatively spliced. **B** Splicing of three alternative CSF-1 mRNAs. The differences in the mRNAs are due to alternative splicing of coding exon 6 and of noncoding exons 9 and 10

(LAIMINS et al. 1983). The functional importance of these sequences in regulating the transcription of CSF-1 mRNA has not been determined.

CSF-1 mRNA molecules of different lengths arise through alternative splicing (Fig. 3 B). All introns begin with a GT dinucleotide, end with an AG dinucleotide, and contain pyrimidine-rich sequences immediately 5' to the acceptor boundaries, consistent with splice consensus signals (KAWASAKI and LADNER 1989). The different mRNA molecules can be accounted for by alternative use of 3' noncoding regions and by the 894 base pair element absent from the original 1.6-kb pcCSF-17 clone. The alternative utilization of exon 9 (679 base pairs) or exon 10 (2 kb) accounts for the different 3' untranslated regions (LADNER et al. 1987) and results in transcripts that are either 4 kb or 2.3 kb in length. The 4-kb transcript is generally the most abundant mRNA detected in different tissues (WONG et al. 1987). The loss of the 894 base pair element in pcCSF-17 stems from the use of an alternative splice acceptor site within exon 6 that results in an in-frame deletion. The latter mRNA also has the shorter of the two 3' untranslated sequences, resulting in a clone only 1.6 kb in length (Fig. 3 B). AU-rich sequences in the 3' untranslated regions of certain transcripts can contribute to mRNA lability (SHAW and KAMEN 1986; CAPUT et al. 1986). These are found in the 4-kb mRNA at three positions and are absent from the 2.3-kb mRNA, but their potential effect on mRNA turnover has not yet been determined. The turnover of 4-kb CSF-1 transcripts is markedly decreased in the presence of cycloheximide, an inhibitor of protein synthesis (HORIGUCHI et al. 1988), suggesting that labile RNases may be responsible for their relative instability.

F. CSF-1 Gene Expression

CSF-1 is produced during fetal development by extraembryonic tissues and may be responsible for the early appearance of macrophage precursors in the fetus (AZOULAY et al. 1987). Remarkably, the levels of uterine CSF-1 rise more than 1000-fold during pregnancy and are highest at parturition (BARTOCCI et al. 1986). Northern blots revealed the presence of the 2.3-kb murine CSF-1 mRNA in uterine tissues, and in situ hybridization localized its synthesis within the secretory epithelial cells of the endometrium (POLLARD et al. 1987). CSF-1 produced by the uterus may bind to receptors expressed on placental trophoblasts (MÜLLER et al. 1983; RETTENMIER et al. 1986) and may therefore play a role in placental development that differs from its known function in hematopoiesis.

CSF-1 transcripts are ubiquitous in adult tissues and have been detected in human liver and placenta (WONG et al. 1987), in murine and feline liver, lung, brain, and spleen (RETTENMIER et al. 1985a; RAJAVASHISTH et al. 1987), and in stromal cells from bone marrow (LANOTTE et al. 1982). Because many established fibroblast cell lines synthesize CSF-1 (STANLEY and HEARD 1977), mesenchymal cells are likely to be responsible for the synthesis of the growth factor in many organs. In addition, the CSF-1 gene may be upregulated in response to PDGF (ROLLINS et al. 1988). CSF-1 enters the circulation and is normally present at concentrations of several hundred units per milliliter of serum; the concentration of circulating CSF-1 is regulated by CSF-1 receptors on liver and splenic macrophages that specifically bind the growth factor, which is then endocytosed and destroyed intracellularly (BARTOCCI et al. 1987).

Human endothelial cells, which normally do not transcribe CSF-1 mRNA, increase transcription of the gene after stimulation with bacterial lipopolysaccharide (LPS), tumor necrosis factor-α (TNF-α), or IL-1 (SEELENTAG et al. 1987). Similarly, circulating blood monocytes do not normally express CSF-1 mRNA, but adherence (HASKILL et al. 1987) or addition of phorbol esters such as phorbol 12-myristate, 13-acetate (PMA; HORIGUCHI et al. 1986; RAMBALDI et al. 1987) induce CSF-1 expression by increasing the transcription of the 4-kb mRNA species (HORIGUCHI et al. 1988). Synthesis of the growth factor is also induced by GM-CSF, interferon-γ, or TNF-α (HORIGUCHI et al. 1987; RAMBALDI et al. 1987; OSTER et al. 1987). CSF-1 is therefore likely to be released at sites of vascular injury and inflammation, and its actions complement the activities of other cytokines that modulate the inflammatory response (see Sect. I, below). Since mononuclear phagocytes express CSF-1 receptors, induction of growth factor synthesis in these cells may stimulate their proliferation through either autocrine or paracrine mechanisms.

G. The CSF-1 Receptor (c-*fms* Protooncogene Product)

I. Tissue-Specific Expression

The actions of CSF-1 are mediated by its interaction with a single class of high-affinity receptors expressed primarily on mononuclear phagocytes (GUILBERT and STANLEY 1980). During maturation of cells of this lineage, the number of CSF-1 receptors (CSF-1R) per cell increases and is highest on mature circulating monocytes and tissue macrophages (GUILBERT and STANLEY 1986). However, in spite of their higher receptor number, more mature mononuclear phagocytes have a lower proliferative potential in response to CSF-1 (VAN DER ZEIJST et al. 1978). Unlike receptors for GM-CSF and multi-CSF (IL-3), which are routinely detected in a range of less than 1000 binding sites per cell, the number of CSF-1R on mature monocytes and bone marrow-derived macrophages is generally between 1×10^4 and 5×10^4 per cell (GUILBERT and STANLEY 1986). Autoradiographic methods detect binding sites for CSF-1 on macrophages, regardless of their tissue of origin (BYRNE et al. 1981). However, a role for the receptor in other cell types has not as yet been rigorously excluded, and the expression of CSF-1R in placental trophoblasts probably connotes another role for the growth factor (see Sect. F, above).

```
001   MGPRALLVLL  MATAWHAQGV  PVIQPSGPEL  VVEPGTTVTL  RCVGNGSVEW    feline  v-fms
      ---GV--L--  V-----G--I  ---E--V---  --K--A----  ----------    human   c-fms
      MELGPP----  L--V--G--A  ---E------  -----E----  ---S------    mouse   c-fms


051   DGPISPHWNL  DLDPPSSILT  TNNATFQNTG  TYHCTEPGNP  RGGNATIHLY
      ---P----T-  YS-GS----S  ----------  --R-----D-  L--S-A----
      ------I-T-  -PES-G-T--  -S----K---  --R---LED-  MA-ST-----


101   VKDPARPWKV  LAQEVTVLEG  GDALLPCLLT  DPALEAGVSL  VRVRGRPVLR
      --------N-  -----V-F-D  ----------  --V-------  -------LM-
      -----HS-NL  -------V--  QE-V----I-  ----KDS---  M-EG--Q---


151   QTNYSFSPWH  GFTIHKAKFI  ENHVYQCSAR  VDGRTVTSMG  IWLKVQKDIS
      ----------  -----R----  QSQD-----L  MG--K-M-IS  -R-----V-P
      K-V-F----R  -SI-R---VL  DSNT-V-KTM  -N--ES--T-  -----NRVHP


201   GPATLTLEPA  ELVRIQGEAA  QIVCSASNID  VNFDVSLRHG  DTKLTISQQS
      --PA---V--  -----R----  -------SV-  -----F-Q-N  N---A-P---
      E-PQIK---S  K----R----  ------T-AE  -G-N-I-KR-  ----E-PLN-


251   DFHDNRYQKV  LTLNLDHVSF  QDAGNYSCTA  TNAWGNHSAS  MVFRVVESAY
      ---N------  -----Q--D-  -H------V-  S-VQ-K--T-  -F--------
      --Q-Y--K--  RA-S-NA-D-  ----I---V-  S-DV-TRT-T  -N-Q------


301   SNLTSEQSLL  QEVTVGEKVD  LQVKVEAYPG  LESFNWTYLG  PFSDYQD   K
      L--S---N-I  -------GLN  -K-M------  -QG-------  ----W-PEP-
      L---------  ---S--DSLI  -T-HAD---S  IQHY------  --FED-R   -


349   LDFVTIKDTY  RYTSTLSLPR  LKRSESGRYS  FLARNAGGQN  ALTFELTLRY
351   -ANA-T----  -H-F------  --P--A----  -----P--WR  ----------
349   -E-I-QRAI-  ---FK-F-N-  V-A--A-Q-F  LM-Q-KA-W-  N---------


399   PPEVRVTMTL  INGSDTLLCE  ASGYPQPSVT  WVQCRSHTDR  CDESAGLVLE
401   ----S-IW-F  ----G----A  -------N--  -L--SG----  ---AQV-QVW
399   ----S--WMP  V----V-F-D  V---------  -ME--G----  ---AQA-H-W


449   DSHS EVLSQ  VPFYEVIVHS  LLAIGTLEHN  RTYECRAFNS  VGNSSQTFWP
451   -DPYP-----  E--HK-T-Q-  --TVE-----  Q------H--  --SG-WA-I-
449   NDTHP-----  K--DK--IQ-  Q-P--P-K--  M--F-KTH--  ------Y-RA


498   ISIGAHTPLP  AELLFTPVLL  TCMSIMALLL  LLLLLLLYKY  KQKPKYQVRW
501   --A----HP-  --F-----VV  A---------  ----------  ----------
499   V-L-QSKQ--  D-S-----VV  A---V-S--V  ----------  ----------


548   KIIESYEGNS  TTFIDPTQLP  YNEKWEFPRN  NLQFGKTLGT  GAFGKVVEAT
551   ----------  ----------  ----------  ---------A  ----------
549   ----R-----  Y---------  ----------  ---------A  ----------


598   AFGLGKEDAV  LKVAVKMLKS  TAHADEKEAL  MSELKIMSHL  GQHENIVNLL
601   ----------  ----------  ----------  ----------  ----------
599   ----------  ----------  ----------  ----------  ----------
```

Fig. 4

```
648  GACTHGGPVL  VITEYCCYGD  LLNFLRRQAE  AMPGPSLSVG  QDPEAGAGYK
651  ----------  ----------  -------K--  --L-----P-  ----G-VD--
649  ----------  -Y--------  H------K--  --H-----P-  --S-GDSS--


698  NIHLEKKYVR  RDSGFSSQGV  DTYVEMRPVS  TSSSNDSFSE  EDLGKEDGRP
701  ----------  ----------  ----------  ---  ------  Q--D------
699  ----------  ----------  ----------  ----  ---FK  Q--D--HS--


748  LELRDLLHFS  SQVAQGMAFL  ASKNCIHRDV  AARNVLLTSG  RVAKIGDFGL
750  ----------  ----------  ----------  --------N-  H---------
748  ---W------  ----------  ----------  ----------  H---------


798  ARDIMNDSNY  IVKGNARLPV  KWMAPESIFD  CVYTVQSDVW  SYGILLWE1F
800  ----------  ----------  ----------  ----------  ----------
798  ----------  V-----  ---  ----------  --I------  ----------


848  SLGLNPYPGI  LVNSKFYKLV  KDGYQMAQPA  FAPKNIYSIM  QACWALEPTR
850  ----------  ----------  ----------  ----------  ---------H
847  ----------  H--N------  ---------V  ----------  -S--D-----


898  RPTFQQICSL  LQKQAQEDRR  VPNYTNLPSS  SSSRLLRPWR  GPPL***
900  ---------F  --E-------  ERD-------  -R-GGSGSSS  SELEEESSSE
897  --------F-  --E--RLE--  DQD-A-----  GG-SGSDSGG  GSSGGSSSEP


950  HLTCCEQGDI  AQPLLQPNNY  QFC*** human
947  EEESSSEHLA  CCEPGDIAQP  LLQPNNYQFC  *** mouse
```

Fig. 4. Predicted amino acid sequences of human and murine CSF-1R and of the v-*fms* oncogene product. The single hydrophobic membrane-spanning region (*overlined*) bisects the amino-terminal extracellular and carboxy-terminal intracellular domains. (Data taken from HAMPE et al. 1984; COUSSENS et al. 1986; ROTHWELL and ROHRSCHNEIDER 1987)

CSF-1R is one of a family of growth factor receptors with an intrinsic tyrosine-specific protein kinase activity and is encoded by one of the known protooncogenes, c-*fms* (SHERR et al. 1985). A carboxy-terminally truncated and mutated cDNA copy of the c-*fms* gene was transduced as the v-*fms* oncogene in two strains of feline sarcoma viruses (FeSV) and is responsible for cell transformation and tumor formation induced by these viruses (reviewed in RETTENMIER and SHERR 1988; SHERR 1988; SHERR et al. 1988). Both the v-*fms*- and c-*fms*-coded polypeptides are synthesized as integral transmembrane glycoproteins oriented with their amino-terminal ligand-binding domains in the ER cisternae and their carboxy-terminal tyrosine kinase domains in the cytoplasm. The polypeptides are cotranslationally glycosylated, acquiring N-linked oligosaccharide chains in their amino-terminal domains; modification of sugar chains during transport of the glycoproteins to the cell surface results in an apparent increase in their molecular mass (ANDERSON et al. 1984; ROUS-SEL et al. 1984; RETTENMIER et al. 1985b, 1986).

II. Primary Structure of CSF-1R and the v-*fms* Oncogene Product

Biologically active c-*fms* cDNAs, obtained from human placental (COUSSENS et al. 1986) and murine pre-B cell libraries (ROTHWELL and ROHRSCHNEIDER 1987), show sequence similarity to each other and to the feline v-*fms* oncogene product (HAMPE et al. 1984). As shown in Fig. 4, human CSF-1R consists of a 512 amino acid extracellular ligand-binding domain (including the 19 amino acid signal peptide), a single 25 amino acid membrane-spanning segment, and a 435 amino acid tyrosine kinase domain. Although the v-*fms* gene product retains the complete extracellular ligand-binding domain of CSF-1R (COUSSENS et al. 1986; WHEELER et al. 1986b; SACCA et al. 1986), the distal carboxy-terminal amino acids of human CSF-1R are replaced by completely unrelated residues in the v-*fms*-coded glycoprotein. In addition, the feline v-*fms* and c-*fms* gene products differ by 9 scattered point mutations (WOOLFORD et al. 1988).

Among the known tyrosine-specific protein kinases, CSF-1R is most closely related to the B- and A-type PDGF receptors (YARDEN et al. 1986; MATSUI et al. 1989; CLAESSON-WELCH et al. 1989) and to the product of the c-*kit* protooncogene (BESMER et al. 1986; YARDEN et al. 1987), a putative receptor for an unidentified ligand. Both the CSF-1R and B-type PDGF-R genes map in a closely linked tandem array on human chromosome 5 (M. ROBERTS et al. 1988; also see Sect. E. I), whereas c-*kit* and the A-type PDGF-R gene map to human chromosome 4 (YARDEN et al. 1987; MATSUI et al. 1989). CSF-1R, PDGF receptors, and the c-*kit* product share a distinctive pattern of cysteine spacing within their extracellular domains that includes sequences characteristic of the immunoglobulin gene superfamily (A. F. WILLIAMS and BARCLAY 1988). In addition, their enzymatic domains are interrupted by hydrophilic "spacers" of 74–104 amino acids that are unique to each receptor, suggesting that they may function in substrate recognition (YARDEN et al. 1987). Analogous sequences are not found in the kinase domains of other receptors or oncogene products that have intrinsic tyrosine kinase activities. In human CSF-1R (Fig. 4), the spacer sequences, represented by residues 679–751, interrupt blocks of close sequence homology with both the PDGF receptor and the c-*kit* product (maximal throughout amino acids 563–678 and 752–906). Residues 563–678 include a consensus sequence for ATP binding (Gly-X-Gly-X-X-Gly, residues 589–594) as well as the ATP binding site itself (Lys 616; DOWNING et al. 1989). The carboxy-terminal tail of human CSF-1R (amino acids 907–972) shows no homology to other tyrosine kinases and acts to negatively regulate receptor activity (see Sect. G. III).

III. Transforming Potential of Human CSF-1R

Transduction of the human (ROUSSEL et al. 1987) or mouse (ROTHWELL and ROHRSCHNEIDER 1987) c-*fms* genes into mouse fibroblasts renders the cells responsive to CSF-1, thereby providing formal genetic proof that c-*fms* encodes CSF-1R. When retroviral vectors expressing the human c-*fms* and CSF-1 cDNAs were introduced together into mouse NIH 3T3 cells, the cells were transformed through an autocrine mechanism and were tumorigenic in nude mice (ROUSSEL et al. 1987; HEARD et al. 1987). Moreover, an "activating mutation" involving a substitution of serine for leucine at position 301 in the extracellular domain of human CSF-1R enables it to transform mouse NIH 3T3 cells in the absence of ligand (ROUSSEL et al. 1988a, b). A similar mutation at codon 301 together with a second amino acid substitution at codon 374 renders the feline c-*fms* gene transforming in rat-1 fibroblasts (WOOLFORD et al. 1988). Although the activating mutation does not perturb the high-affinity CSF-1 binding site, it induces a change in the receptor that mimics an effect of ligand binding and results in constitutive phosphorylation of CSF-1R on tyrosine in the absence of CSF-1. The potency of the activating mutation is further increased by truncation of the CSF-1R carboxy-terminal tail (as in the v-*fms* gene product) or, alternatively, by a second point mutation that changes tyrosine 969 to a phenylalanine residue (BROWNING et al. 1986; ROUSSEL et al. 1987). However, the latter mutation is in itself insufficient to activate human CSF-1R to oncogenic status. Taken together, these results suggest that the receptor can be divided into four functional regions: (1) an extracellular ligand binding pocket, (2) a region in the extracellular domain (that includes Leu301) which is important in transducing the ligand-induced signal across the membrane, (3) the intracellular tyrosine kinase domain, and (4) the carboxy-terminal tail which acts to negatively regulate kinase activity through an as yet unknown mechanism.

IV. Binding Affinity of CSF-1R for Ligand

At 4° C, macrophages display a single class of saturable high-affinity binding sites for CSF-1 (GUILBERT and STANLEY 1980; STANLEY and GUILBERT 1981). Chemical crosslinking of ^{125}I-labeled CSF-1 to its binding sites revealed that the receptor was a unique cell surface polypeptide (MORGAN and STANLEY 1984) with an apparent molecular weight in agreement with that later predicted from translation of cDNA clones (COUSSENS et al. 1986; ROTHWELL and ROHRSCHNEIDER 1987; WOOLFORD et al. 1988). Ligand-receptor complexes could be immunoprecipitated with antisera raised to the v-*fms* gene product, providing the first evidence that c-*fms* encoded CSF-1R (SHERR et al. 1985; SACCA et al. 1986). Although the number of ^{125}I-labeled CSF-1 molecules bound after removal of free ligand remains stable for up to 72 h at 4° C, the number of unoccupied sites does not remain constant but gradually decreases (STANLEY and GUILBERT 1981). A typical analysis of ligand-receptor interactions requires determinations of the equilibrium concentrations of bound and free ligand at different ligand concentrations (SCATCHARD 1949) and rests on the assumption that the concentration of binding sites remains constant. At CSF-1 concentrations in the picomolar range, however, the rate of ^{125}I-labeled CSF-1 binding is sufficiently slow to allow unoccupied binding sites to disappear before all of the free ligand can be irreversibly bound. The result is an apparent equilibrium with persistence of free radiolabeled CSF-1 even when the total number of ligand molecules is lower than the total number of binding sites. Thus, the dissociation constants determined by equilibrium binding at 4° C are overestimated $K_{d,app} = 2-5 \times 10^{-11}$ M), whereas direct kinetic analysis argues for a single class of binding sites with significantly higher binding affinities ($K_d \simeq 10^{-13}$ M). Interestingly, there is a marked temperature dependence of the dissociation constant ($K_d \simeq 4 \times 10^{-10}$ M at 37° C) (GUILBERT and STANLEY 1986).

V. Receptor Downregulation and Turnover

Binding of CSF-1 activates the receptor kinase and leads to autophosphorylation of the receptor on tyrosine, phosphorylation of heterologous substrates, and a subsequent loss of CSF-1 binding sites from the cell surface due to internalization and degradation of receptor-ligand complexes (downregulation). Ligand-induced downmodulation requires receptor tyrosine kinase activity (DOWNING et al. 1989). The macrophage-mediated destruction of CSF-1 can be reversed with inhibitors of energy metabolism or of lysosomal fusion, indicating that CSF-1 is degraded intracellularly (STANLEY and GUILBERT 1981). At 37° C in the absence of extracellular ligand, the cell surface form of CSF-1R turns over with a half-life of 2–3 h, but exposure of cells to saturating concentrations of CSF-1 leads to the complete degradation of immunoprecipitable receptors within 15 min (WHEELER et al. 1986a). The rapid disappearance of prebound cell surface ^{125}I-labeled CSF-1 following a temperature shift from 2° C to 37° C results from two competitive first-order reactions involving internalization and ligand dissociation. Internalization occurs at a 6-fold faster rate than dissociation (GUILBERT and STANLEY 1986). Downregulated receptors can be initially replaced from a cryptic intracellular pool and later by de novo synthesis. Since macrophages rapidly degrade extracellular ligand at 37° C, CSF-1 must be added in sufficient amounts to compensate for the loss of the growth factor. Interpretation of dose–response data should therefore include a consideration of the ligand degradation rate (STANLEY and GUILBERT 1981; TUSHINSKI et al. 1982).

CSF-1R can also be degraded in response to the phorbol ester PMA, whereas the v-*fms* gene product is relatively refractory to PMA treatment (CHEN et al. 1983; GUILBERT et al. 1983; WHEELER et al. 1986a; ROUSSEL et al. 1988a). This suggests that agents which activate protein kinase C might "transmodulate" unoccupied CSF-1R molecules. The diminution in CSF-1 binding sites observed in response to treatment of macrophage progenitors with other cytokines (WALKER et al. 1985) or of peritoneal macrophages with bacterial LPS (GUILBERT and STANLEY 1984) might reflect this mechanism. The underlying processes governing ligand- and PMA-induced receptor turnover differ from one another because their effects can be dissociated by use of certain c-*fms*/v-*fms* receptor chimeras (ROUSSEL et al. 1988a). Moreover, when protein kinase C is itself downmodulated by chronic PMA treatment, the turnover of newly synthesized CSF-1R molecules appearing at the cell surface remains sensitive to ligand. Recent evidence indicates that protein kinase C does not phosphorylate CSF-1R in response to PMA treatment but, rather, activates a protease that cleaves the receptor near its transmembrane segment, thereby releasing the intact ligand-binding domain from the cell (DOWNING et al. 1989).

H. Receptor Tyrosine Kinase Activity and Signal Transduction

In macrophage membrane preparations incubated in vitro with $[\gamma^{32}P]ATP$, manganese, and purified CSF-1 (SHERR et al. 1985), or after CSF-1R solubilization and purification (YEUNG et al. 1987), the stimulated CSF-1 receptor is phosphorylated on tyrosine residues. Exposure of intact macrophages to CSF-1 at 37° C similarly activates the receptor kinase in vivo and leads to receptor phosphorylation and to the phosphorylation of heterologous substrates on tyrosine within 1 min of ligand addition (DOWNING et al. 1988; SENGUPTA et al. 1988). This process initiates a series of intracellular events that can culminate in mitogenesis. The physiologic targets for the receptor kinase remain unknown, although CSF-1 induces the phosphorylation of a series of cellular substrates (JUBINSKY et al. 1988; DOWNING et al. 1988). If studied at low temperature, differences in the order of appearance of the phosphorylated proteins can be observed (SENGUPTA et al. 1988). One of these putative substrates was recently shown to be a serine/threonine kinase encoded by the c-*raf* protooncogene (MORRISON et al. 1988). This suggests that transmission of signals from the cell surface to the nucleus could potentially involve the intermediary participation of other protein kinases that modulate the action of transcriptional regulatory factors through serine/threonine phosphorylation.

In the presence of CSF-1, responsive mononuclear phagocytes have a finite capacity for cell division that becomes more limited as the cells mature (VAN DER ZEIJST et al. 1978). The growth factor is required through the entire G1 phase of the cell cycle in order for bone marrow-derived macrophages to enter S phase, but CSF-1 is not required for progression through S, G2, and M (TUSHINSKI and STANLEY 1985). Complete removal of CSF-1 from culture medium containing 15% fetal calf serum induces these cells to enter a quiescent state and ultimately to die within several days. However, readdition of the growth factor enables the cells to move synchronously into S phase with a lag time of 12 h and to continue to proliferate exponentially. Cells cultured in submitogenic doses of the growth factor can be maintained in a quiescent state for many weeks and can be stimulated to enter S phase in about 10 h when the concentration of CSF-1 is increased.

CSF-1 induces pleiotropic effects and initiates both immediate and temporally delayed responses during G1. Among the most rapid are alterations in membrane structure, including ruffling, formation of filopodia, vesiculation, and an increase in pinocytic vacuoles (TUSHINSKI et al. 1982). These are accompanied by a rise in intracellular pH through an effect on the amiloride-sensitive Na^+, H^+ antiporter with compensatory stimulation of Na^+, K^+-ATPase activity (VAIRO and HAMILTON 1988). CSF-1 increases the transcription of several "immediate-early" genes, including the protooncogenes c-*fos* and c-*myc*, within minutes to a few hours after stimulation (BRAVO et al. 1987; ORLOFSKY and STANLEY 1987). The growth factor governs cell metabolism by increasing the uptake of 2-deoxyglucose through a carrier-facilitated D-glucose transport system (HAMILTON et al. 1988a) and increasing the rate of protein synthesis within 1 h of addition and subsequently decreasing the rate of protein degradation (TUSHINSKI and STANLEY 1983). LPS and concanavalin A can also induce intracellular alkalinization, Na^+, K^+-ATPase activity, and 2-deoxyglucose transport in bone marrow-derived macrophages without stimulating mitogenesis (VAIRO and HAMILTON 1988; HAMILTON et al. 1988b); in fact, LPS appears to suppress the DNA synthetic response (HAMILTON et al. 1988b). These elements of the immediate-early response are therefore not sufficient to trigger mitogenesis. Similarly, the rapid inductive effects of CSF-1 on transcription do not make cells competent to progress through G1, since the requirement for CSF-1 persists even after these events are completed. While some of the metabolic effects of the growth factor are necessary for proliferation, others may be crucial for cell survival. If the dependence of macrophages on CSF-1 during the G1 phase of their cell cycle depends solely on signals initiated through the receptor kinase, different physiologic substrates for CSF-1R might be progressively elaborated as the cells prepare to replicate their DNA and divide.

J. Physiologic Effects and Agents That Modulate CSF-1 Activity

I. Direct and Synergistic Effects

CSF-1 is part of a cytokine network that regulates not only hematopoiesis but also the inflammatory response. Although CSF-1 does not activate macrophages (C. F. NATHAN et al. 1984), it augments the production of other macrophage cytokines induced during host defense against infection. CSF-1 stimulates the release of IL-1 (MOORE et al. 1980), granulocyte colony stimulating factor (G-CSF, MOTOYOSHI et al. 1982; METCALF and NICOLA 1985), interferon (MOORE et al. 1984a; M. K. WARREN and RALPH 1986), and TNF (M. K. WARREN and RALPH 1986), and potentiates the release of plasminogen activator (LIN and GORDON 1979; HAMILTON et al. 1980), thromboplastin (LYBERG et al. 1987), and biocidal oxygen metabolites (WING et al. 1985). It also protects macrophages from the lytic effects of vesicular stomatitis infection (LEE and WARREN 1987), stimulates killing of *Candida albicans* (KARBASSI et al. 1987), and promotes tumor cell cytolysis (WING et al. 1982; RALPH and NAKOINZ 1987). Apparent effects of CSF-1 on cells of other myeloid lineages in vivo may therefore be secondary to the induction of other growth factors produced by the responding macrophages.

    CSF-1 can also synergize with other growth factors that affect macrophage differentiation and proliferation both in vitro and in vivo (BARTELMEZ and STANLEY 1985; CHEN and CLARK 1986; STANLEY et al. 1986; IKEBUCHI et al. 1987; MOCHIZUKI et al. 1987; D. WARREN and MOORE 1987; BROXMEYER et al. 1987; HAMILTON et al. 1988b; CHEN et al. 1988). In the presence of hemopoietins that act on early myeloid progenitors, including IL-1, IL-3, IL-6, and GM-CSF, CSF-1 will stimulate primitive and presumably multipotent cells to form macrophage colonies. In agar cultures of mouse bone marrow cells, the maximum plating efficiency for the primitive precursors is achieved with a combination of IL-1, IL-3, and CSF-1. In its action on committed precursors, human CSF-1 stimulates the formation of both human and mouse macrophage colonies, whereas mouse CSF-1 is species-specific. Paradoxically, the human growth factor is more potent in stimulating the proliferation of mouse bone marrow progenitors than human cells (WAHEED and SHADDUCK 1982; MOTOYOSHI et al. 1982; DAS and STANLEY 1982). This may reflect poorly defined requirements for the assay of human bone marrow precursors or, alternatively, the necessity for synergism with other hemopoietins in this system (RALPH et al. 1986; CHEN et al. 1988).

II. CSF-1 Antagonists

Although the appearance of macrophages and lymphocytes at sites of tissue injury is the *sine qua non* of the chronic inflammatory response, not all of the potent effectors released by macrophages potentiate the biological activity of the participating cells. For example, prostaglandins of the E series (PGE_1 and PGE_2), which exert their physiologic effects by activating adenyl cyclase, inhibit macrophage colony formation by bone marrow-derived cells (KURLAND and BOCKMAN 1978; KURLAND et al. 1978 a, b, 1979; N. WILLIAMS 1979; HAMILTON 1983; MOORE et al. 1984 b; HUME and GORDON 1984). KURLAND and coworkers (1978 a) showed that release of PGE by adherent human monocyte feeder layers inhibited bone marrow cell colony formation (CFU-GM) in semisolid medium. The inhibitory effects could be reversed by treatment of the monocytes by indomethacin, an inhibitor of cyclooxygenase and, hence, PGE synthesis. They further demonstrated that murine macrophages cultured in mouse WEHI-3 conditioned medium (now known to contain IL-3) were induced to release a dialyzable inhibitor (presumed to be PGE) of CFU-GM, and concluded that CSF-induced (IL-3 induced?) PGE production by monocytes could act as a negative feedback mechanism to limit excessive mononuclear phagocyte proliferation. MOORE and coworkers (1984 a, b) subsequently found that bone marrow cultures exposed to purified CSF-1 also produced interferon which, in turn, inhibited macrophage colony formation. PGE, interferon, and dexamethasone are now known to antagonize the CSF-1-induced proliferative response of peritoneal and bone marrow-derived macrophages (HAMILTON 1983; HUME and GORDON 1984; MOORE et al. 1984 a, b). In vivo effects of administration of PGE_2 on CFU-GM may be multiphasic, based on recent evidence that marrow and spleen cells from PGE_2-treated mice can suppress myelopoiesis after transfer to cyclophosphamide-treated recipient animals (PELUS and GENTILE 1988).

PGE_2 similarly inhibits DNA synthesis by SV40-immortalized murine macrophages (C. O. ROCK and S. JACKOWSKI, personal communication) that are strictly dependent on CSF-1 for their proliferation and survival in vitro (MORGAN et al. 1987). The inhibitory effect is apparent even when the cells are grown in serum-free medium containing CSF-1 as the only exogenously added growth factor. As expected, cyclic AMP levels were found to be increased in response to PGE_2 treatment of these cells, and other pharmacological agents (e.g., dibutyryl cyclic AMP, 3-isobutyl-1-methylxanthine, forskolin) that increase intracellular cyclic AMP also inhibit cell growth. Thus, signals generated by CSF-1 through the receptor tyrosine kinase appear to act antagonistically to agents that directly induce cyclic AMP, and vice versa. Some investigators have suggested that CSF-1 has the capacity to elicit macrophage PGE_2 production (KURLAND et al. 1978a, b, 1979). However, SV40-immortalized macrophages starved of CSF-1 synthesized and secreted high levels of PGE_2, whereas PGE_2 synthesis was immediately reduced after fully mitogenic doses of CSF-1 were added to the cultures (C. O. ROCK and S. JACKOWSKI, personal communication). One possibility, then, is that CSF-1 induces some of its effects by negatively regulating PGE_2 release and cyclic AMP synthesis. This is consistent with recent reports that pertussis toxin, which catalyzes ADP-ribosylation of a guanine nucleotide binding protein (G_i) that negatively regulates adenyl cyclase, inhibits the growth of CSF-1-responsive cells (HE et al. 1988).

K. Clinical Implications

The production of recombinant CSFs has permitted their in vivo testing, first in mice and then in primates and humans. Erythropoietin, GM-CSF, and G-CSF have been in clinical trials for several years, and their applications have been generally directed toward bone marrow failure syndromes, including those induced by radiation and chemotherapy (reviewed in SIEFF 1987; D. G. NATHAN and SIEFF 1989). Such trials have already shown that erythropoietin and G-CSF have clinical efficacy without major toxicity, even when administered intravenously at high physiologic doses. The development of clinical protocols for the use of CSF-1 are just under way (GARNICK and O'REILLY 1989). In view of the lineage-specific action of CSF-1 on monocytes, macrophages, and their progenitors, one would imagine that future applications might stress the role of this cytokine as a modulator of the inflammatory response rather than as a hemopoietin. In particular, specific CSF-1 antagonists might prove useful as anti-inflammatory or immunosuppressive drugs, because they should affect the functional integrity of macrophages themselves and might limit the ability of macrophages to release interleukins that activate lymphocytes and granulocytes.

On another front, the fact that the CSF-1 receptor is encoded by the c-*fms* protooncogene suggests that malignancies affecting cells of the mononuclear phagocyte lineage (i.e., a subset of acute myeloid leukemias, AML) might be etiologically related to abnormalities in ligand-receptor interactions. In mice, inappropriate activation of the CSF-1 gene and consequent autocrine regulation by CSF-1 has been shown to be a secondary event in the development of mononuclear phagocyte tumors in vivo (BAUMBACH et al. 1987). Northern analysis and in situ hybridization studies have indicated that blast cells from a subset of human AML cases constitutively express both CSF-1 and CSF-1R

(WAKAMIYA et al. 1987; RAMBALDI et al. 1988), suggesting that an autocrine mechanism may similarly be important in inducing their unregulated growth. As might be expected, CSF-1R is expressed on blast cells from about 30% of patients with AML, and receptor expression generally correlates with evidence of monocytic differentiation. In certain cases, however, CSF-1R is found to be expressed in AML blasts of both children and adults that lack morphologic, immunophenotypic, or histochemical properties of monocytic cells, and are either poorly differentiated or granulocytic in character (ASHMUN et al. 1989). In mice, retroviral insertions 5' to the c-*fms* protooncogene can activate high levels of receptor expression in immature myeloid cells and etiologically contribute to leukemia (GISSELBRECHT et al. 1987), but in humans, no evidence of c-*fms* rearrangement was observed, even in receptor-positive cases that lacked evidence of monocytic differentiation (ASHMUN et al. 1989). Studies of the as yet uncharacterized c-*fms* promoter/enhancer will be important in determining whether the gene is appropriately regulated in the latter AML cases. Finally, the recent demonstration that the human c-*fms* gene can be activated to transform cells by point mutations in its extracellular domain (ROUSSEL et al. 1988 b; WOOLFORD et al. 1988) raises the possibility that similar mutations affecting receptor function might contribute to AML. The tools are now at hand to explore these different possibilities at a molecular level.

The study of CSFs, beginning as "factorology" less than a quarter century ago, has been brought to fruition in an era when cell biological, biochemical, and molecular biological techniques could be effectively combined to study them. Further advances in our understanding of growth factor–receptor biology should now bear directly on the processes that govern signal transduction and should ultimately yield novel approaches to the treatment of proliferative and inflammatory disorders.

Acknowledgments. This work was supported by NIH grants CA26504, CA32551 (to E.R.S.), and CA47064 (to C.J.S.), and by the Howard Hughes Medical Institute and the American Lebanese Syrian Associated Charities (ALSAC) of St. Jude Children's Research Hospital (C.J.S.). Dr. E. Richard Stanley is the Belfer Chairman of Developmental Biology at Albert Einstein College of Medicine. Dr. Charles J. Sherr is the Herrick Foundation Chairman of Tumor Cell Biology at St. Jude Children's Research Hospital and an Investigator of the Howard Hughes Medical Institute.

References

Anderson SJ, Gonda MA, Rettenmier CW, Sherr CJ (1984) Subcellular localization of glycoproteins encoded by the viral oncogene v-*fms*. J Virol 51:730–741

Ashmun RA, Look AT, Roberts WM, Roussel MF, Seremetis S, Ohtsuka M, Sherr CJ (1989) Monoclonal antibodies to the human CSF-1 receptor (c-*fms* proto-oncogene product) detect epitopes on normal mononuclear phagocytes and on human myeloid leukemic blast cells. Blood 73:827–837

Azoulay M, Webb CG, Sachs L (1987) Control of hematopoietic cell growth regulators during mouse fetal development. Mol Cell Biol 7:3361–3364

Bartelmez SH, Stanley ER (1985) Synergism between hemopoietic growth factors (HGFs) detected by their effects on cells bearing receptors for a lineage specific HGF: assay of hemopoietin-1. J Cell Physiol 122:370–378

Bartocci A, Pollard JW, Stanley ER (1986) Regulation of colony-stimulating factor 1 during pregnancy. J Exp Med 164:956–961

Bartocci A, Mastrogiannis DS, Migliorati G, Stockert RJ, Wolkoff AW, Stanley ER (1987) Macrophages specifically regulate the concentration of their own growth factor in the circulation. Proc Natl Acad Sci USA 84:6179–6183

Baumbach WR, Stanley ER, Cole MD (1987) Induction of clonal monocyte-macrophage tumors in vivo by a mouse c-*myc* retrovirus: rearrangement of the CSF-1 gene as a secondary transforming event. Mol Cell Biol 7:664–671

Besmer P, Murphy JE, George PC, Qiu F, Bergold PJ, Lederman L, Snyder HW Jr, Brodeur D, Zuckerman EE, Hardy WD (1986) A new acute transforming feline retrovirus and relationship of its oncogene v-*kit* with the protein kinase gene family. Nature 320:415–421

Bradley TR, Metcalf D (1966) The growth of mouse bone marrow cells in vitro. Aust J Exp Biol Med Sci 44:287–299

Bradley TR, Sumner MA (1968) Stimulation of mouse bone marrow colony growth in vitro by conditioned medium. Aust J Exp Biol Med Sci 46:607–618

Bravo R, Neuberg M, Burckhardt J, Almendral J, Wallich R, Müller R (1987) Involvement of common and cell type-specific pathways in c-*fos* gene control: stable induction by cAMP in macrophages. Cell 48:251–260

Browning PJ, Bunn HF, Cline A, Shuman M, Nienhuis AW (1986) "Replacement" of COOH-terminal trunction of v-*fms* with c-*fms* sequences markedly reduces transformation potential. Proc Natl Acad Sci USA 83:7800–7804

Broxmeyer HE, Williams DE, Cooper S, Shadduck RK, Gillis S, Waheed A, Urdal DL, Bicknell DC (1987) Comparative effects in vivo of recombinant murine interleukin-3, natural murine colony stimulating factor-1, and recombinant murine granulocyte-macrophage colony-stimulating factor on myelopoiesis in mice. J Clin Invest 79:721–730

Byrne PV, Guilbert LJ, Stanley ER (1981) Distribution of cells bearing receptors for a colony-stimulating factor (CSF-1) in murine tissues. J Cell Biol 91:848–853

Caput D, Beutler B, Hartog K, Thayer R, Brown-Shimer S, Cerami A (1986) Identification of a common nucleotide sequence in the 3′-untranslated region of mRNA molecules specifying inflammatory mediators. Proc Natl Acad Sci USA 83:1670–1674

Chen BD, Clark CR (1986) Interleukin-3 (IL-3) regulates the in vitro proliferation of both blood monocytes and peritoneal exudate macrophages. Synergism between a macrophage lineage-specific colony stimulating factor (CSF-1) and IL-3. J Immunol 137:563–570

Chen BD-M, Lin HS, Hsu S (1983) Tumor-promoting phorbol esters inhibit the binding of colony stimulating factor (CSF-1) to murine peritoneal exudate macrophages. J Cell Physiol 116:207–212

Chen BD-M, Clark CR, Chou T (1988) Granulocyte-macrophage colony stimulating factor stimulates monocyte and tissue macrophage proliferation and enhances their responsiveness to macrophage colony stimulating factor. Blood 71:997–1002

Cifone M, Defendi V (1974) Cyclic expression of a growth conditioning factor (MGF) on the cell surface. Nature 252:151–153

Claesson-Welsh L, Eriksson A, Moren A, Severinsson L, Ek B, Ostman A, Betsholtz C, Heldin C-H (1989) Cloning and expression of the human A-type platelet-derived growth factor (PDGF) receptor establishes structural similarity to the B-type PDGF receptor. Proc Natl Acad Sci USA 86:4917–4921

Clark SC, Kamen R (1987) The human hematopoietic colony-stimulating factors. Science 236:1229–1237

Coussens L, Van Beveren C, Smith D, Chen E, Mitchell RL, Isacke CM, Verma IM, Ullrich A (1986) Structural alteration of viral homologue of receptor protooncogene *fms* at carboxyl terminus. Nature 320:277–280

Csejtey J, Boosman A (1986) Purification of human macrophage colony stimulating factor (CSF-1) from medium conditioned by pancreatic carcinoma cells. Biochem Biophys Res Commun 138:238–245

Das SK, Stanley ER (1982) Structure-function studies of a colony stimulating factor (CSF-1). J Biol Chem 257:13679–13684

Das SK, Stanley ER, Guilbert LJ, Forman LW (1980) Discrimination of a colony stimulating factor subclass by a specific receptor on a macrophage cell line. J Cell Physiol 104:359–366

Das SK, Stanley ER, Guilbert LJ, Forman LW (1981) Human colony-stimulating factor (CSF-1) radioimmunoassay: resolution of three subclasses of human colony-stimulating factors. Blood 58:630–641

DeLamarter JF, Hession C, Semon D, Gough NM, Rothenbuhler R, Mermod J-J (1987) Nucleotide sequence of a cDNA encoding murine CSF-1 (macrophage-CSF). Nucleic Acids Res 15:2389–2390

Downing JR, Rettenmier CW, Sherr CJ (1988) Ligand-induced tyrosine kinase activity of the colony stimulating factor-1 receptor in a murine macrophage cell line. Mol Cell Biol 8:1795–1799

Downing JR, Roussel MF, Sherr CJ (1989) Ligand and protein kinase C downmodulate the colony-stimulating factor-1 receptor by independent mechanisms. Mol Cell Biol 9:2890–2896

Garnick MB, O'Reilly RJ (1989) Clinical promise of new hematopoietic growth factors: M-CSF, IL-3, IL-6. Hematology/Oncology Clinics NA 3:495–509

Gisselbrecht S, Fichelson S, Sola B, Bordereaux D, Hampe A, André C, Galibert F, Tambourin P (1987) Frequent c-*fms* activation by proviral insertion in mouse myeloblastic leukaemias. Nature 329:259–261

Goyert SM, Ferrero E, Rettig WJ, Yenamandra AK, Obata F, Le Beau MM (1988) The CD14 monocyte differentiation antigen maps to a region encoding growth factors and receptors. Science 239:497–500

Groffen J, Heisterkamp N, Spurr N, Dana S, Wasmuth JJ, Stephenson JR (1983) Chromosomal localization of the human c-*fms* oncogene. Nucleic Acids Res 11: 6331–6399

Guilbert LJ, Stanley ER (1980) Specific interaction of murine colony-stimulating factor with mononuclear phagocytic cells. J Cell Biol 85:153–159

Guilbert LJ, Stanley ER (1984) Modulation of receptors for the colony-stimulating factor CSF-1 by bacterial lipopolysaccharide and CSF-1. J Immunol Methods 73:17–28

Guilbert LJ, Stanley ER (1986) The interaction of ^{125}I-colony-stimulating factor-1 with bone marrow-derived macrophages. J Biol Chem 261:4024–4032

Guilbert LJ, Nelson DJ, Hamilton JA, Williams N (1983) The nature of 12-O-tetra decanoylphorbol-13-acetate (TPA) stimulated hemopoiesis. Colony stimulating factor (CSF) requirement for colony formation and the effect of TPA on ^{125}I-CSF-1 binding. J Cell Physiol 115:276–282

Hamilton JA (1983) Glucocorticoids and prostaglandins inhibit the induction of macrophage DNA synthesis by macrophage growth factor and phorbol ester. J Cell Physiol 115:67–74

Hamilton JA, Stanley ER, Burgess AW, Shadduck RK (1980) Stimulation of macrophage plasminogen activator activity by colony-stimulating factors. J Cell Physiol 103:435–445

Hamilton JA, Vairo G, Lingelbach SR (1988a) Activation and proliferation signals in murine macrophages: stimulation of glucose uptake by hemopoietic growth factors and other agents. J Cell Physiol 134:405–412

Hamilton JA, Vairo G, Nicola NA, Burgess A, Metcalf D, Lingelbach S (1988b) Activation and proliferation signals in murine macrophages. Synergistic interactions between the hemopoietic growth factors and with phorbol ester for DNA synthesis. Blood 71:1574–1580

Hampe A, Gobet M, Sherr CJ, Galibert F (1984) The nucleotide sequence of the feline retroviral oncogene v-*fms* shows unexpected homology with oncogenes encoding tyrosine-specific protein kinases. Proc Natl Acad Sci USA 81:85–89

Haskill S, Warren MK, Becker S, Ladner MB, Johnson C, Eierman D, Ralph P, Mark DF (1987) Adherence induces CSF-1 gene expression in monocytes. J Leukocyte Biol 42:359

C.J. SHERR and E. R. STANLEY

He Y, Hewlett E, Temeles D, Quesenberry P (1988) Inhibition of interleukin-3 and colony-stimulating factor 1-stimulated marrow cell proliferation by pertussis toxin. Blood 71:1187–1195

Heard JM, Roussel MF, Rettenmier CW, Sherr CJ (1987) Synthesis, post-translational processing, and autocrine transforming activity of a carboxylterminal truncated form of colony stimulating factor-1. Oncogene Res 1:423–440

Horiguchi J, Warren MK, Ralph P, Kufe D (1986) Expression of the macrophage specific colony stimulating factor (CSF-1) during human monocytic differentiation. Biochem Biophys Res Commun 141:924–930

Horiguchi J, Warren MK, Kufe D (1987) Expression of the macrophage-specific colony-stimulating factor in human monocytes treated with granulocyte-macrophage colony-stimulating factor. Blood 69:1259–1261

Horiguchi J, Sariban E, Kufe D (1988) Transcriptional and post-transcriptional regulation of CSF-1 gene expression in human monocytes. Mol Cell Biol 8:3951–3954

Huebner K, Isobe M, Croce CM, Golde DW, Kaufman SE, Gasson JC (1985) The human gene encoding GM-CSF is at 5q21–q32, the chromosome region deleted in the 5q⁻ anomaly. Science 230:1282–1285

Hume D, Gordon S (1984) The correlation between plasminogen activator activity and thymidine incorporation in mouse bone marrow-derived macrophages. Exp Cell Res 150:347–355

Ichikawa Y, Pluznik DH, Sachs L (1966) In vitro control of the development of macrophage and granulocyte colonies. Proc Natl Acad Sci USA 56:488–495

Ikebuchi K, Wong GG, Clark SC, Ihle JN, Hirai Y, Ogawa M (1987) Interleukin-6 enhancement of interleukin-3 dependent proliferation of multipotential hematopoietic progenitors. Proc Natl Acad Sci USA 84:9035–9039

Jaye M, Howk R, Burgess W, Ricca GA, Chiu IM, Ravera MW, O'Brien SJ, Modi WS, Maciag T, Drohan WN (1986) Human endothelial cell growth factor: cloning, nucleotide sequence, and chromosome localization. Science 233:541–545

Jubinsky PT, Yeung Y-G, Sacca R, Li W, Stanley ER (1988) Colony stimulating factor-1 stimulated macrophage membrane protein phosphorylation. In: Kudlow JE, Maclennan DH, Bernstein A, Gottlieb AI (eds) Biology of growth factors: molecular biology, oncogenes, signal transduction and clinical applications. Plenum, New York, pp 75–90

Karbassi A, Becker JM, Foster JS, Moore RN (1987) Enhanced killing of *Candida albicans* by murine macrophages treated with macrophage colony stimulating factor: evidence for augmented expression of mannose receptors. J Immunol 139:417–421

Kawasaki ES, Ladner MB (1989) Molecular biology of macrophage colony-stimulating factor (M-CSF). In: Dexter D, Garland JM, Testa NG (eds) Cellular and molecular biology of colony stimulating factors. Dekker, New York (in press)

Kawasaki ES, Ladner MB, Wang AM, Van Arsdell J, Warren MK, Coyne MY, Schweickart VL, Lee MT, Wilson KJ, Boosman A, Stanley ER, Ralph P, Mark DF (1985) Molecular cloning of a complementary DNA encoding human macrophage-specific colony stimulating factor (CSF-1). Science 230:291–296

Kobilka BK, Dixon RAF, Frielle T, Dohlman HG, Bolanowski MA, Sigal IS, Yang-Feng TL, Francke U, Caron MG, Lefkowitz R (1987) cDNA for the human β2-adrenergic receptor: a protein with multiple membrane-spanning domains and encoded by a gene whose chromosomal location is shared with that of the receptor for platelet-derived growth factor. Proc Natl Acad Sci USA 84:46–50

Kurland JI, Bockman R (1978) Prostaglandin E production by human blood monocytes and mouse peritoneal macrophages. J Exp Med 147:952–957

Kurland JI, Bockman R, Broxmeyer HE, Moore MAS (1978a) Limitation of excessive myelopoiesis by the intrinsic modulation of macrophage derived prostaglandin E. Science 199:552–555

Kurland JI, Broxmeyer HE, Pelus LM, Bockman RS, Moore MAS (1978b) Role for monocyte-macrophage-derived colony stimulating factor and prostaglandin E in the positive and negative feedback control of myeloid stem cell proliferation. Blood 52:388–407

Kurland JI, Pelus LM, Ralph P, Bockman RS, Moore MAS (1979) Induction of prostaglandin E synthesis in normal and neoplastic macrophages: role for colony stimulating factor(s) distinct from effects on myeloid progenitor cell proliferation. Proc Natl Acad Sci USA 76:2326–2330

Ladner MB, Martin GA, Noble JA, Nikoloff DM, Tal R, Kawasaki ES, White TJ (1987) Human CSF-1: gene structure and alternative splicing of mRNA precursors. EMBO J 6:2693–2698

Ladner MB, Martin GA, Noble JA, Wittman VP, Shadle PJ, Warrens MK, McGrogan M, Stanley ER (1988) cDNA cloning and expression of murine CSF-1 from L929 cells. Proc Natl Acad Sci USA 85:6706–6710

Laimins LA, Kessel M, Rosenthal N, Khoury G (1983) Viral and cellular enhancer elements. In: Gluzman Y, Shenk T (eds) Communication in molecular biology: enhancers and eukaryotic gene expression. Cold Spring Harbor Laboratory, New York, pp 28–87

Lanotte M, Metcalf D, Dexter TM (1982) Production of monocyte/macrophage colony stimulating factor by preadipocyte cell lines derived from murine bone marrow stroma. J Cell Physiol 112:123–127

Le Beau MM, Pettenati MJ, Lemons RS, Diaz MO, Westbrook CA, Larson RA, Sherr CJ, Rowley JD (1986a) Assignment of the GM-CSF, CSF-1, and *fms* genes to human chromosome 5 provides evidence for linkage of a family of genes regulating hematopoiesis and for their involvement in the deletion (5q) in myeloid disorders. Cold Spring Harbor Symp Quant Biol 51:899–909

Le Beau MM, Westbrook CA, Diaz MO, Larson RA, Rowley JD, Gasson JC, Golde DW, Sherr CJ (1986b) Evidence for the involvement of *GM-CSF* and *fms* in the deletion (5q) in myeloid disorders. Science 231:984–987

Le Beau MM, Epstein ND, O'Brien SJ, Nienhuis AW, Yang Y-C, Clark SC, Rowley JD (1987) The interleukin 3 gene is located on human chromosome 5 and is deleted in myeloid leukemias with a deletion of 5q. Proc Natl Acad Sci USA 84:5913–5917

Lee MT, Warren MK (1987) CSF-1-induced resistance to viral infection in murine macrophages. J Immunol 138:3019–3022

Lin H-S, Gordon S (1979) Secretion of plasminogen activator by bone marrow-derived mononuclear phagocytes and its enhancement by colony-stimulating factor. J Exp Med 150:231–245

Lyberg T, Stanley ER, Prydz H (1987) Colony stimulating factor-1 induces thromboplastin activity in murine macrophages and human monocytes. J Cell Physiol 132:367–370

Manos MM (1988) Expression and processing of a recombinant human macrophage colony-stimulating factor in mouse cells. Mol Cell Biol 8:5035–5039

Matsui T, Heideran M, Miki T, Popescu N, La Rochelle W, Kraus M, Pierce J, Aaronson SA (1989) Isolation of a novel receptor cDNA establishes the existence of two PDGF receptor genes. Science 243:800–804

Mauel J, Defendi V (1971a) Regulation of DNA synthesis in mouse macrophages. I. Sources action and purification of the macrophage growth factor (MGF). Exp Cell Res 65:33–42

Mauel J, Defendi V (1971b) Regulation of DNA synthesis in mouse macrophages. II. Studies on mechanisms of action of the macrophage growth factor. Exp Cell Res 65:377–385

Metcalf D (1969) Studies on colony formation in vitro by mouse bone marrow cells. I. Continuous cluster formation and relation of clusters to colonies. J Cell Physiol 74:323–332

Metcalf D (1970) Studies on colony formation in vitro by mouse bone marrow cells. II. Action of colony stimulating factor. J Cell Physiol 76:89–99

Metcalf D (1985) The granulocyte-macrophage colony stimulating factors. Science 229:16–22

Metcalf D, Nicola A (1985) Synthesis by mouse peritoneal cells of G-CSF, the differentiation inducer for myeloid leukemia cells: stimulation by endotoxin, M-CSF and multi-CSF. Leuk Res 9:35–50

Metcalf D, Bradley TR, Robinson W (1967) Analysis of colonies developing in vitro from mouse bone marrow cells stimulated by kidney feeder layers or leukemic serum. J Cell Physiol 69:93–108

Metcalf D, Parker J, Chester HM, Kincade PW (1974) Formation of eosinophilic-like granulocytic colonies by mouse bone marrow cells in vitro. J Cell Physiol 84:275–289

Mochizuki DY, Eisenman JR, Conlon PJ, Larsen AD, Tushinski RJ (1987) Interleukin 1 regulates hematopoietic activity, a role previously ascribed to hemopoietin 1. Proc Natl Acad Sci USA 84:5267–5271

Moore RN, Oppenheim JJ, Farrar JJ, Carter CS Jr, Waheed A, Shadduck RK (1980) Production of lymphocyte-activating factor (interleukin 1) by macrophages activated with colony-stimulating factors. J Immunol 125:1302–1305

Moore RN, Larsen HS, Horohov DW, Rouse BT (1984a) Endogenous regulation of macrophage proliferative expansion by colony stimulating factor-induced interferon. Science 223:178–180

Moore RN, Pitruzzello FJ, Larsen HS, Rouse BT (1984b) Feedback regulation of colony stimulating factor (CSF-1)-induced macrophage proliferation by endogenous E prostaglandins and interferon-α/β. J Immunol 133:541–543

Morgan CJ, Stanley ER (1984) Chemical crosslinking of the mononuclear phagocyte specific growth factor CSF-1 to its receptor at the cell surface. Biochem Biophys Res Commun 119:35–41

Morgan CJ, Pollard W, Stanley ER (1987) Isolation and characterization of a cloned growth factor dependent macrophage cell line, BAC12F5. J Cell Physiol 130:420–427

Morrison DK, Kaplan DR, Rapp U, Roberts TM (1988) Signal transduction from membrane to cytoplasm: growth factors and membrane-bound oncogene products increase *raf*-1 phosphorylation and associated protein kinase activity. Proc Natl Acad Sci USA 85:8855–8859

Motoyoshi K, Suda T, Kusumoto K, Takaku F, Miura Y (1982) Granulocyte-macrophage colony stimulating and binding activities of purified human urinary colony stimulating factor to murine and human bone marrow cells. Blood 60:1378–1386

Müller R, Slamon DJ, Adamson ED, Tremblay JM, Müller D, Cline MJ, Verma IM (1983) Transcription of c-*onc* genes c-*ras*[ki] and c-*fms* during mouse development. Mol Cell Biol 3:1062–1069

Nathan CF, Prendergast TJ, Wiebe ME, Stanley ER, Platzer E, Remold HG, Welte K, Rubin BY, Murray HW (1984) Activation of human macrophages. Comparison of other cytokines with interferon-γ. J Exp Med 160:600–605

Nathan DG, Sieff CA (1989) The clinical applications of hematopoietic growth factors. In: Ross R, Burgess T, Hunter T (eds) Proceedings of the UCLA symposia conference on growth factors and their receptors. New York (in press)

Orlofsky A, Stanley ER (1987) CSF-1-induced gene expression in macrophages: dissociation from the mitogenic response. EMBO J 6:2947–2952

Oster W, Lindemann A, Horn S, Mertelsmann R, Herrmann F (1987) Tumor necrosis factor (TNF)-alpha but not TNF-beta induces secretion of colony stimulating factor for macrophages (CSF-1) by human monocytes. Blood 70:1700–1703

Pelus LM, Gentile PS (1988) In vivo modulation of myelopoiesis by prostaglandin E_2. III. Induction of suppressor cells in marrow and spleen capable of mediating inhibition of CFU-GM proliferation. Blood 71:1633–1640

Pettenati MJ, Le Beau MM, Lemons RS, Shima EA, Kawasaki ES, Larson RA, Sherr CJ, Diaz MO, Rowley JD (1987) Assignment of *CSF-1* to 5q331: evidence for clustering of genes regulating hematopoiesis and for their involvement in the deletion of the long arm of chromosome 5 in myeloid disorders. Proc Natl Acad Sci USA 84:2970–2974

Pluznik DH, Sachs L (1965) The cloning of normal "mast" cells in tissue culture. J Cell Comp Physiol 66:319–324

Pollard JW, Bartocci A, Arceci R, Orlofsky A, Ladner MB, Stanley ER (1987) Apparent role of the macrophage growth factor CSF-1 in placental development. Nature 330:484–486

Rajavashisth TB, Eng R, Shadduck RK, Waheed A, Ben-Avram CM, Shively JE, Lusis AJ (1987) Cloning and tissue-specific expression of mouse macrophage colony-stimulating factor mRNA. Proc Natl Acad Sci USA 84:1157–1161

Ralph P, Nakoinz I (1987) Stimulation of macrophage tumoricidal activity by the growth and differentiation factor CSF-1. Cell Immunol 105:270–279

Ralph P, Warren MK, Lee MT, Csejtey J, Weaver JF, Broxmeyer HE, Williams DE, Stanley ER, Kawasaki ES (1986) Inducible production of human macrophage growth factor, CSF-1. Blood 68:633–639

Rambaldi A, Young DC, Griffin JD (1987) Expression of the M-CSF (CSF-1) gene by human monocytes. Blood 69:1409–1413

Rambaldi A, Wakamiya N, Vellenga E, Horiguchi J, Warren MK, Kufe D, Griffin JD (1988) Expression of the macrophage colony stimulating factor and c-*fms* genes in human acute myeloblastic leukemia cells. J Clin Invest 81:1030–1035

Rettenmier CW, Sherr CJ (1988) The *fms* oncogene. In: Reddy EP, Skalka AM, Curren T (eds) The oncogene handbook. Elsevier, Amsterdam, pp 73–100

Rettenmier CW, Roussel MF (1988) Differential processing of colony-stimulating factor-1 precursors encoded by two human cDNAs. Mol Cell Biol 8:5026–5034

Rettenmier CW, Chen JH, Roussel MF, Sherr CJ (1985a) The product of the c-*fms* proto-oncogene: a glycoprotein with associated tyrosine kinase activity. Science 228:320–322

Rettenmier CW, Roussel MF, Quinn CO, Kitchingman GR, Look AT, Sherr CJ (1985b) Transmembrane orientation of glycoproteins encoded by the v-*fms* oncogene. Cell 40:971–981

Rettenmier CW, Sacca R, Furman WL, Roussel MF, Holt JT, Nienhuis AW, Stanley ER, Sherr CJ (1986) Expression of the human c-*fms* proto-oncogene product (colony-stimulating factor-1 receptor) on peripheral blood mononuclear cells and choriocarcinoma cell lines. J Clin Invest 77:1740–1746

Rettenmier CW, Roussel MF, Ashmun RA, Ralph P, Price K, Sherr CJ (1987) Synthesis of membrane-bound colony-stimulating factor-1 (CSF-1) and downmodulation of CSF-1 receptors in NIH 3T3 cells transformed by cotransfection of the human CSF-1 and c-*fms* (CSF-1 receptor) genes. Mol Cell Biol 7:2378–2387

Roberts M, Look AT, Roussel MF, Sherr CJ (1988) Tandem linkage of human CSF-1 receptor (c-*fms*) and PDGF receptor genes. Cell 55:655–661

Roberts R, Gallagher J, Spencer E, Allen TD, Bloomfield F, Dexter TM (1988) Heparan sulphate bound growth factors: a mechanism for stromal cell mediated haemopoiesis. Nature 332:376–378

Robinson W, Metcalf D, Bradley TR (1967) Stimulation by normal and leukemic mouse sera of colony formation in vitro by mouse bone marrow cells. J Cell Physiol 69:83–92

Rollins BJ, Morrison ED, Stiles CD (1988) Cloning and expression of JE, a gene inducible by platelet-derived growth factor and whose product has cytokine-like properties. Proc Natl Acad Sci USA 85:3738–3842

Rothwell VM, Rohrschneider LR (1987) Murine c-*fms* cDNA; cloning, sequence analysis, and retroviral expression. Oncogene Res 1:311–324

Roussel MF, Sherr CJ, Barker PE, Ruddle FH (1983) Molecular cloning of the c-*fms* locus and its assignment to human chromosome 5. J Virol 48:770–773

Roussel MF, Rettenmier CW, Look AT, Sherr CJ (1984) Cell surface expression of v-*fms*-coded glycoproteins is required is required for transformation. Mol Cell Biol 4:1999–2009

Roussel MF, Dull TJ, Rettenmier CW, Ralph P, Ullrich A, Sherr CJ (1987) Transforming potential of the c-*fms* proto-oncogene (CSF-1 receptor). Nature 325:549–552

Roussel MF, Downing JR, Ashmun RA, Rettenmier CW, Sherr CJ (1988a) CSF-1 mediated regulation of a chimeric c-*fms*/v-*fms* receptor containing the v-*fms*-coded tyrosine kinase domain. Proc Natl Acad Sci USA 85:5903–5907

Roussel MF, Downing JR, Rettenmier CW, Sherr CJ (1988b) A point mutation in the extracellular domain of the human CSF-1 receptor (c-*fms* proto-oncogene product) activates its transforming potential. Cell 55:979–988

Sabatini DD, Kreibich G, Morimoto T, Adesnik M (1982) Mechanisms for the incorporation of protein in membranes and organelles. J Cell Biol 92:1–22

Sacca R, Stanley ER, Sherr CJ, Rettenmier CW (1986) Specific binding of the mononuclear phagocyte colony-stimulating factor CSF-1 to the product of the v-*fms* oncogene. Proc Natl Acad Sci USA 83:3331–3335

Sachs L (1987) The molecular control of blood cell development. Science 238:1374–1379

Scatchard G (1949) The attractions of proteins for small molecules and ions. Ann NY Acad Sci 51:660–672

Seelentag WK, Mermod JJ, Montesano R, Vassalli P (1987) Additive effects of interleukin 1 and tumor necrosis factor-alpha on the accumulation of the three granulocyte and macrophage colony-stimulating factor mRNAs in human endothelial cells. EMBO J 6:2261–2265

Sengupta A, Liu W-K, Yeung YG, Yeung DCY, Frackelton AR, Stanley ER (1988) Identification and subcellular localization of proteins that are rapidly phosphorylated in tyrosine in response to colony stimulating factor 1. Proc Natl Acad Sci USA 85:8062–8066

Shadduck RK, Metcalf D (1975) Preparation and neutralization characteristics of an anti-CSF antibody. J Cell Physiol 86:247–252

Shadduck RK, Waheed A (1979) Development of a radioimmunoassay for colony stimulating factor. Blood Cells 5:421–431

Shaw G, Kamen R (1986) A conserved AU sequence from the 3' untranslated region of GM-CSF mRNA mediates selective degradation. Cell 46:659–667

Sheridan JW, Stanley ER (1971) Tissue sources of bone marrow colony stimulating factor. J Cell Physiol 78:451–459

Sherr CJ (1988) The *fms* oncogene. BBA reviews on cancer. Elsevier, Amsterdam 948:225–243

Sherr CJ, Rettenmier CW, Sacca R, Roussel MF, Look AT, Stanley ER (1985) The c-*fms* proto-oncogene product is related to the receptor for the mononuclear phagocyte growth factor CSF-1. Cell 41:665–676

Sherr CJ, Rettenmier CW, Roussel MF (1988) Macrophage colony stimulating factor CSF-1 and its proto-oncogene encoded receptor. Cold Spring Harbor Symp Quant Biol 53:521–530

Sieff CA (1987) Hematopoietic growth factors. J Clin Invest 79:1549–1557

Stanley ER (1979) Colony-stimulating factor (CSF) radioimmunoassay: detection of a CSF subclass stimulating macrophage production. Proc Natl Acad Sci USA 76:2969–2973 [See also erratum ibid (1979) 76:5411]

Stanley ER (1985) The macrophage colony stimulating factor CSF-1. In: Colowick SP, Kaplan NO (eds) Immunochemical techniques: effectors and mediators of lymphoid cells. Harcourt Brace Jovanovich Academic, New York, pp 564–587 (Methods in Enzymology, vol 116)

Stanley ER, Guilbert LJ (1981) Methods for the purification, assay, characterization and target cell binding of a colony stimulating factor (CSF-1). J Immunol Methods 42:253–284

Stanley ER, Heard PM (1977) Factors regulating macrophage production and growth. Purification and some properties of the colony stimulating factor from medium conditioned by mouse L cells. J Biol Chem 252:4305–4312

Stanley ER, McNeill TA, Chan SH (1970) Antibody production to the factor in human urine stimulating colony formation in vitro by bone marrow cells. Br J Haematol 18:585–590

Stanley ER, Cifone M, Heard PM, Defendi V (1976) Factors regulating macrophage production and growth: identity of colony-stimulating factor and macrophage growth factor. J Exp Med 143:631–647

Stanley ER, Chen DM, Lin H-S (1978) Induction of macrophage production and proliferation by a purified colony stimulating factor. Nature 274:168–170

Stanley ER, Bartocci A, Patinkin D, Rosendaal M, Bradley TR (1986) Regulation of very primitive multipotent hematopoietic cells by hemopoietin-1. Cell 45: 667–674

Sukhatme VP, Cao X, Chang LC, Tsai-Morris C-H, Stamenkovich D, Ferreira PCP, Cohen DR, Edwards SA, Shows TB, Curran T, Le Beau MM, Adamson ED (1988) A zinc finger-encoding gene coregulated with c-*fos* during growth and differentiation and after cellular depolarization. Cell 53:37–43

Sutherland GR, Baker E, Callen DF, Campbell HD, Young IG, Sanderson CJ, Garson OM, Lopez AF, Vadas MA (1988) Interleukin-5 is at 5q31 and is deleted in the 5q⁻ syndrome. Blood 71:1150–1152

Tushinski RJ, Stanley ER (1983) The regulation of macrophage protein turnover by a colony stimulating factor (CSF-1). J Cell Physiol 116:67–75

Tushinski RJ, Stanley ER (1985) The regulation of mononuclear phagocyte entry into S phase by the colony stimulating factor CSF-1. J Cell Physiol 122:221–228

Tushinski RJ, Oliver IT, Guilbert LJ, Tynan PW, Warner JR, Stanley ER (1982) Survival of mononuclear phagocytes depends on a lineage-specific growth factor that the differentiated cells selectively destroy. Cell 28:71–81

Vairo G, Hamilton JA (1988) Activation and proliferation signals in murine macrophages: stimulation of Na⁺K⁺-ATPase activity by hematopoietic growth factors and other agents. J Cell Physiol 134:13–24

Van der Zeijst BAM, Stewart CC, Schlesinger S (1978) Proliferative capacity of mouse peritoneal macrophages in vitro. J Exp Med 147:1253–1266

Virolainen M, Defendi V (1967) Dependence of macrophage growth in vitro upon interaction with other cell types. Wistar Inst Symp Monogr 7:67–85

Waheed A, Shadduck RK (1979) Purification and properties of L cell-derived colony stimulating factor. J Lab Clin Med 94:180–194

Waheed A, Shadduck RK (1982) Purification of colony stimulating factor by affinity chromatography. Blood 60:238–244

Wakamiya N, Horiguchi J, Kufe D (1987) Detection of c-*fms* and CSF-1 RNA by in situ hybridization. Leukemia 1:518–520

Walker F, Nicola NA, Metcalf D, Burgess A (1985) Hierarchical down-modulation of hemopoietic growth factor receptors. Cell 43:269–276

Warren D, Moore MAS (1987) Synergy of interleukin 1 and granulocyte colony-stimulating factor: in vivo stimulation of stem cell recovery and hematopoietic regeneration following 5-fluorouracil treatment of mice. Proc Natl Acad Sci USA 84:7134–7138

Warren MK, Ralph P (1986) Macrophage growth factor CSF-1 stimulates human monocyte production of interferon tumor necrosis factor and colony stimulating activity. J Immunol 137:2281–2285

Wheeler EF, Rettenmier CW, Look AT, Sherr CJ (1986a) The v-*fms* oncogene induces factor independence and tumorigenicity in CSF-1 dependent macrophage cell line. Nature 324:377–380

Wheeler EF, Roussel MF, Hampe A, Walker MH, Fried VA, Look AT, Rettenmier CW, Sherr CJ (1986b) The amino-terminal domain of the v-*fms* oncogene product includes a functional signal peptide that directs synthesis of a transforming glycoprotein in the absence of feline leukemia virus *gag* sequences. J Virol 59:224–233

Williams AF, Barclay AN (1988) The immunoglobulin superfamily – domains for cell surface recognition. Annu Rev Immunol 6:381–405

Williams N (1979) Preferential inhibition of murine macrophage colony formation by prostaglandin E. Blood 53:1089–1094

Wing EJ, Waheed A, Shadduck RK, Nagle LS, Stephenson K (1982) Effect of colony stimulating factor on murine macrophages. Induction of anti-tumor activity. J Clin Invest 69:270–276

Wing EJ, Ampel NM, Waheed A, Shadduck RK (1985) Macrophage colony-stimulating factor (M-CSF) enhances the capacity of murine macrophages to secrete oxygen reduction products. J Immunol 135:2052–2056

Wong GG, Temple PA, Leary AC, Witek-Giannotti JS, Yang Y-C, Ciarletta AB, Chung M, Murtha P, Kriz R, Kaufman RJ, Ferenz CR, Sibley BS, Turner KJ, Hewick RM, Clark SC, Yanai N, Yokota H, Yamada M, Saito M, Motoyoshi K, Takaku F (1987) Human CSF-1: molecular cloning and expression of 4 kb cDNA encoding the human urinary protein. Science 235:1504–1508

Woolford J, McAuliffe A, Rohrschneider LR (1988) Activation of the feline c-*fms* proto-oncogene: Multiple alterations are required to generate a fully transformed phenotype. Cell 55:965–977

Worton RG, McCulloch EW, Till JE (1969) Physical separation of hemopoietic stem cells from cells forming colonies in culture. J Cell Physiol 74:171–182

Wu MC, Cini JK, Yunis AA (1979) Purification of a colony-stimulating factor from cultured pancreatic carcinoma cells. J Biol Chem 254:6226–6228

Yang Y-C, Kovacic S, Kriz R, Wolf S, Clark SC, Wellems TE, Nienhuis A, Epstein N (1988) The human genes for GM-CSF and IL-3 are closely linked in tandem on chromosome 5. Blood 71:958–961

Yarden Y, Escobedo JA, Kuang WJ, Yang-Fang TL, Daniel TO, Tremble PM, Chen EY, Ando ME, Harkins RN, Francke U, Fried VA, Ullrich A, Williams LT (1986) Structure of the receptor for platelet-derived growth factor helps define a family of closely related growth factor receptors. Nature 323:226–232

Yarden Y, Kuang WJ, Yang-Feng T, Coussens L, Munemitsu TJ, Dull TJ, Chen E, Schlessinger J, Francke U, Ullrich A (1987) Human proto-oncogene c-*kit*: a new cell surface receptor tyrosine kinase for an unidentified ligand. EMBO J 6:3341–3351

Yeung YG, Jubinsky PT, Sengupta A, Yeung DCY, Stanley ER (1987) Purification of the colony-stimulating factor 1 receptor and demonstration of its tyrosine kinase activity. Proc Natl Acad Sci USA 84:1268–1271

CHAPTER 16

Granulocyte Colony-Stimulating Factor

S. NAGATA

A. Introduction

Granulocyte colony-stimulating factor (G-CSF) is one of the colony-stimulating factors (CSF) which stimulate colony formation from bone marrow cells. Of the four well-known CSFs, G-CSF is the one that specifically regulates proliferation, differentiation, survival, and activation of cells of the restricted neutrophilic granulocyte lineage (METCALF 1985, 1986, 1987; SACHS 1987). Mouse G-CSF is also called MGI-1G (macrophage and granulocyte inducer type 1, granulocyte) while human G-CSF has been designated pluripotent CSF (pluripoetin) or granulocyte-macrophage colony-stimulating factor β (GM-CSFβ). Recombinant DNA technology has been used to elucidate the molecular and genetic nature of G-CSF (CLARK and KAMEN 1987) and in vitro and in vivo functions of G-CSF have been extensively studied using recombinant G-CSF. Recent clinical application of G-CSF has shown that this hormone is valuable in the treatment of patients suffering from neutropenia. In this article, current information on G-CSF gene structure and function will be reviewed.

B. Purification of Natural G-CSF

I. Purification of Mouse G-CSF

G-CSF was initially described on the basis of its ability to induce terminal differentiation in a murine myelomonocytic leukemia cell line (WEHI-3B), and was designated granulocyte-macrophage differentiation factor (GM-DF; BURGESS and METCALF 1980). This activity was detected in sera of mice and in media conditioned by organs from mice previously injected with an endotoxin (NICOLA and METCALF 1981). Since medium conditioned by the lungs exhibits the highest specific activity of GM-DF, this material was used to purify the factor which induced the differentiation of WEHI-3B cells (NICOLA et al. 1983; NICOLA 1985). Using three successive ordinary column chromatographies (salting out, Phenyl-Sepharose, and Bio-Gel P-60) followed by two high-performance liquid chromatography (HPLC) columns (Phenyl-silica and TSK gel filtration), the factor was purified about 500000-fold with a yield of 30%. Since the purified factor could be used to stimulate the formation of granulocytic colonies exclusively from bone marrow cells, the factor was redesignated G-CSF. The purified G-CSF has a relative molecular mass (M_r) of 24000–25000 and is believed to contain sugar moieties including sialic acids, since endoglycosidase reduces the M_r of the molecule, and its isoelectric point shifts significantly after treatment with neuraminidase. The molecule is relatively stable to extreme pH levels (pH 2–10), temperature (50% loss of activity at 70° C for 30 min), and with strong denaturing agents (6 M guanidine hydrochloride, 8 M urea, 0.1% sodium dodecylsulfate, SDS).

II. Purification of Human G-CSF

During purification of GM-CSFs in human placental conditioned medium, GM-CSFs were separated into two fractions, GM-CSFα and GM-CSFβ (NICOLA et al. 1979). Partially purified GM-CSFβ was found to stimulate colonies consisting mainly of neutrophils from bone marrow cells, and compete with mouse G-CSF for the mouse G-CSF receptor (NICOLA et al. 1985). GM-CSFβ was therefore considered as the human counterpart of mouse G-CSF. ASANO et al. (1980) found granulocytosis in some patients suffering from malignant cancer. Since nude mice bearing transplanted human carcinoma cells also exhibit granulocytosis, it was concluded that the carcinoma cells themselves are responsible for granulocytosis, and suggested that the carcinoma cells produce a CSF which acts on mouse cells. One such human carcinoma cell was established as a cell line (designated CHU-2) and a large amount of colony-stimulating activity was detected in medium conditioned by CHU-2 cells (NOMURA et al. 1986). One of the CSFs produced by the CHU-2 cells was purified to a homogeneous state using ordinary column chromatography and HPLC. Since the purified CSF stimulated exclusively granulocytic colony formation from human bone marrow cells as well as mouse bone marrow cells, it was designated human G-CSF. Human G-CSF is a glycoprotein of M_r 19 000 with an isoelectric point of 6.1.

Human bladder carcinoma 5637 cells were found to independently produce the factor which stimulates colony formation of not only neutrophilic granulocytes but erythroid cells, megakaryocytes, and macrophages as well (WELTE et al. 1985). The purified factor (M_r 18 000) was found to induce differentiation of human promyelocytic leukemia cell line HL-60 and murine myelomonocytic leukemia cell line WEHI-3B D$^+$, and was designated pluripotent CSF (pluripoetin) because of its effects on these multipotential progenitor cells; however, it was shown later that (STRIFE et al. 1987; WELTE et al. 1987a) when bone marrow cells depleted of mature myeloid and lymphoid cells were used as the target cells, the factor supported exclusively colony formation of neutrophilic granulocytes. This result indicates that one of the factors produced by 5637 bladder carcinoma cells is G-CSF and the pluripotent activity of this factor is an indirect effect, mediated by accessory cells.

C. Primary Structure of G-CSF

I. Isolation and Characterization of Human G-CSF cDNA

The partial amino acid sequence of G-CSF purified from human CHU-2 or 5637 bladder carcinoma was determined, and using an oligonucleotide probe,

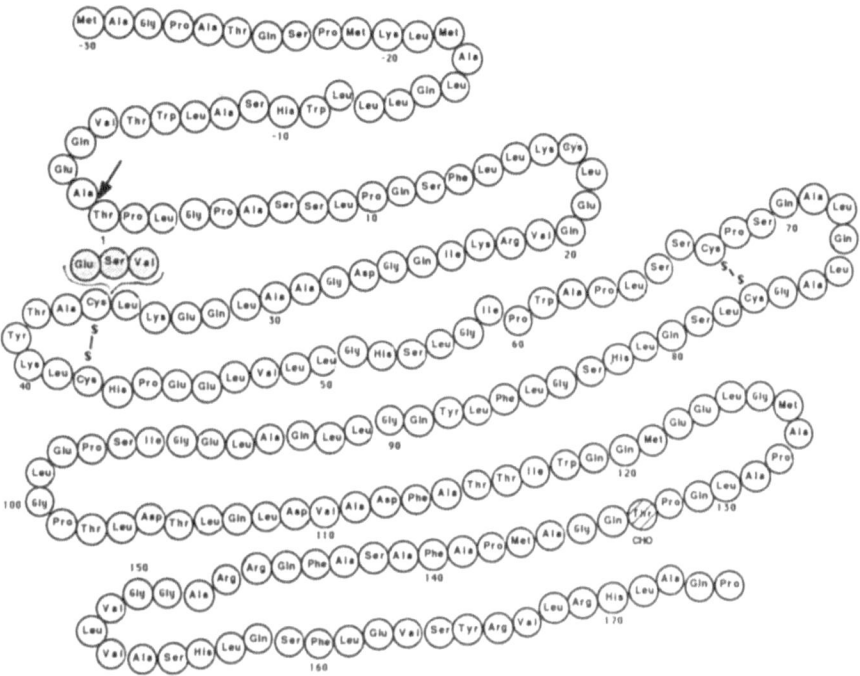

Fig. 1. The primary structure of human G-CSF. The *numbering* of the amino acid sequence is for the version of human G-CSF with three amino acids deleted and starts from the N terminus of the mature protein. The minor molecule of human G-CSF has an insertion of three amino acids (Val-Ser-Glu) at the 35th amino acid. The glycosylation site (Thr133) of human G-CSF is indicated by *CHO*

human G-CSF cDNAs were isolated independently from cDNA libraries constructed with mRNA from CHU-2 or 5637 cells (NAGATA et al. 1986a, b; SOUZA et al. 1986). Nucleotide sequence analysis of the cDNAs revealed the primary amino acid sequence of human G-CSF shown in Fig. 1. There are two different human G-CSF mRNAs coding for 207 or 204 amino acids, of which 30 amino acids at the N terminus are the signal sequence for the secretion of G-CSF. The two human G-CSF molecules are identical except that three amino acids are deleted (or inserted) around the 35th amino acid residue from the N-terminus of the mature G-CSF protein. The mature human G-CSF protein consisting of 174 amino acids has an M_r of 18671 and is at least 20 times more active than the one consisting of 177 amino acids (M_r 18987; NAGATA et al. 1986b). These two different G-CSFs are coded by two different mRNAs produced by alternative splicing from the single precursor RNA. S1 mapping analysis and Northern hybridization analysis of the mRNAs from human CHU-2 (NAGATA et al. 1986b) and 5637 bladder carcinoma cells (ZSEBO et al. 1986) have indicated that more than 80% of the G-CSF molecules produced in these cell lines are the proteins which have the three amino acid deletion. It is not yet known whether two different G-CSF molecules are actually expressed in the primary cellular sources of G-CSF in the human body.

The amino acid sequence of human G-CSF contains no sequence for N-glycosylation (Asn-X-Ser/Thr) and a recent analysis of human recombinant G-CSF produced in mammalian cells has indicated that human G-CSF is O-glycosylated at Thr residue 133 (in the three amino acid deleted version; OHEDA et al. 1988). The sugar moiety attached to the recombinant human G-CSF molecule produced in chinese hamster ovary cells is N-acetyl-neuraminic acid α (2-6) [galactose β (1-3)] N-acetylgalactosamine (OHEDA et al. 1988), which is identical to that found in the natural human G-CSF protein from 5637 bladder carcinoma cells (SOUZA et al. 1986).

The primary structure of human G-CSF has no apparent homology with those of GM-CSF, interleukin-3 (IL-3), and macrophage CSF (M-CSF) which, like G-CSF, stimulates the growth of neutrophilic granulocytes and/or macrophages. Furthermore, it has little homology with other cytokines and lymphokines such as interferon (IFN) and tumor necrosis factor (TNF); however, human G-CSF exhibits significant homology with IL-6, which stimulates the proliferation and differentiation of B cells (KISHIMOTO and HIRANO, this volume). As shown in Fig. 2, the region from amino acid residues 20 to 93 of human G-CSF is similar (44.6% homology considering conservative amino acid replacements as homologous) to IL-6 (HIRANO et al. 1986). Especially noteworthy is the fact that the positions of four cysteine residues are identical between the two. Since Cys36 and Cys42, and Cys64 and Cys74 of human G-CSF are connected by disulfide bonds (Fig. 1), IL-6 may also have two disulfide bonds. G-CSF and IL-6 may also have similar tertiary structures. Human G-CSF, however, has no effect on B cells, and G-CSF does not compete with IL-6 for the IL-6 receptor (TAGA et al. 1987).

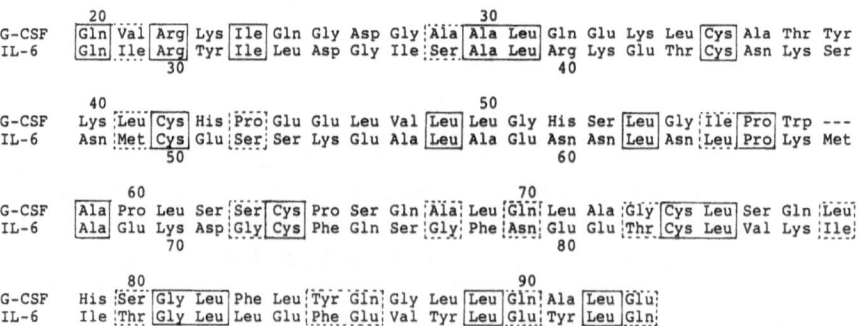

Fig. 2. Comparison of the amino acid sequences of human G-CSF and human IL-6. Identical amino acids are enclosed within *solid lines* while amino acid residues regarded as favored substitutions are enclosed within *dotted lines*. Favored amino acid substitutions are defined as pairs of residues belonging to one of the following groups: Ser, Thr, Pro, Ala, and Gly; Asn, Asp, Gln, and Glu; His, Arg, and Lys; Met, Ile, Leu and Val

II. Isolation and Characterization of Murine G-CSF cDNA

In a murine system, it was also found that some carcinoma cells constitutively produce proteins having CSF activity (SAKAI et al. 1984; JOHNSON et al. 1985). Northern hybridization analysis of mRNA from one of these cell lines, mouse fibrosarcoma NFSA, showed a band hybridizing with human G-CSF cDNA. The mouse cDNA homologous to human G-CSF cDNA was isolated from a cDNA library constructed with mRNA from mouse NFSA fibrosarcoma cells (TSUCHIYA et al. 1986). The amino acid sequence predicted from the nucleotide sequence of the cloned mouse G-CSF cDNA agrees with the partial amino acid sequence determined with the purified natural mouse G-CSF (SIMPSON et al. 1987), and suggests that the N-terminal amino acid of mature G-CSF starts from the valine residue indicated in Fig. 3. Mouse G-CSF consists of 178 amino acids in the mature protein and 30 amino acids in the signal sequence. The M_r of mature G-CSF is 19061, and the amino acid sequence of mouse G-CSF is 72.6% homologous with that of human G-CSF. This value is much larger than the values observed between human and mouse GM-CSF (58%) and IL-3 (29%), and corresponds well with the fact that GM-CSF and IL-3, but not G-CSF, are species specific in mice and humans. Unlike human G-CSF, there is a single type of mouse G-CSF which corresponds to the version of human G-CSF with the three amino acid deletion. The four cysteine residues which are connected by the disulfide bonds in human G-CSF also exist in the mouse G-CSF molecule (Fig. 3). The amino acid sequence of mouse G-CSF contains no sequence for N-glycosylation, which suggests that the sugar moiety of mouse G-CSF is attached via O-glycosylation.

G-CSF mRNA has a relatively long 3' noncoding sequence (about 800 bases), and approximately 70 nucleotides downstream from the termination

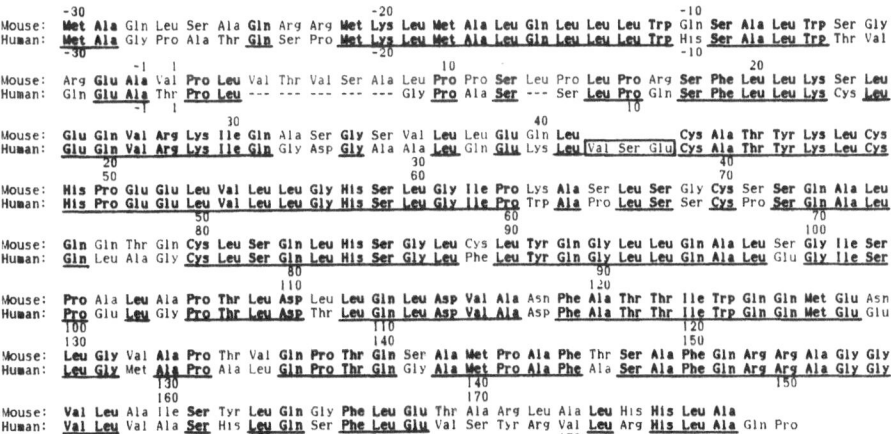

Fig. 3. Comparison of mouse and human G-CSF amino acid sequences. The two sequences are aligned to achieve the maximum homology between them, and identical amino acids are *underlined*. The human G-CSF amino acid sequence is the one which shows the three amino acids inserted. The three amino acids Val-Ser-Glu at residue numbers 36–38 (*boxed*) cannot be found on most human G-CSF molecules

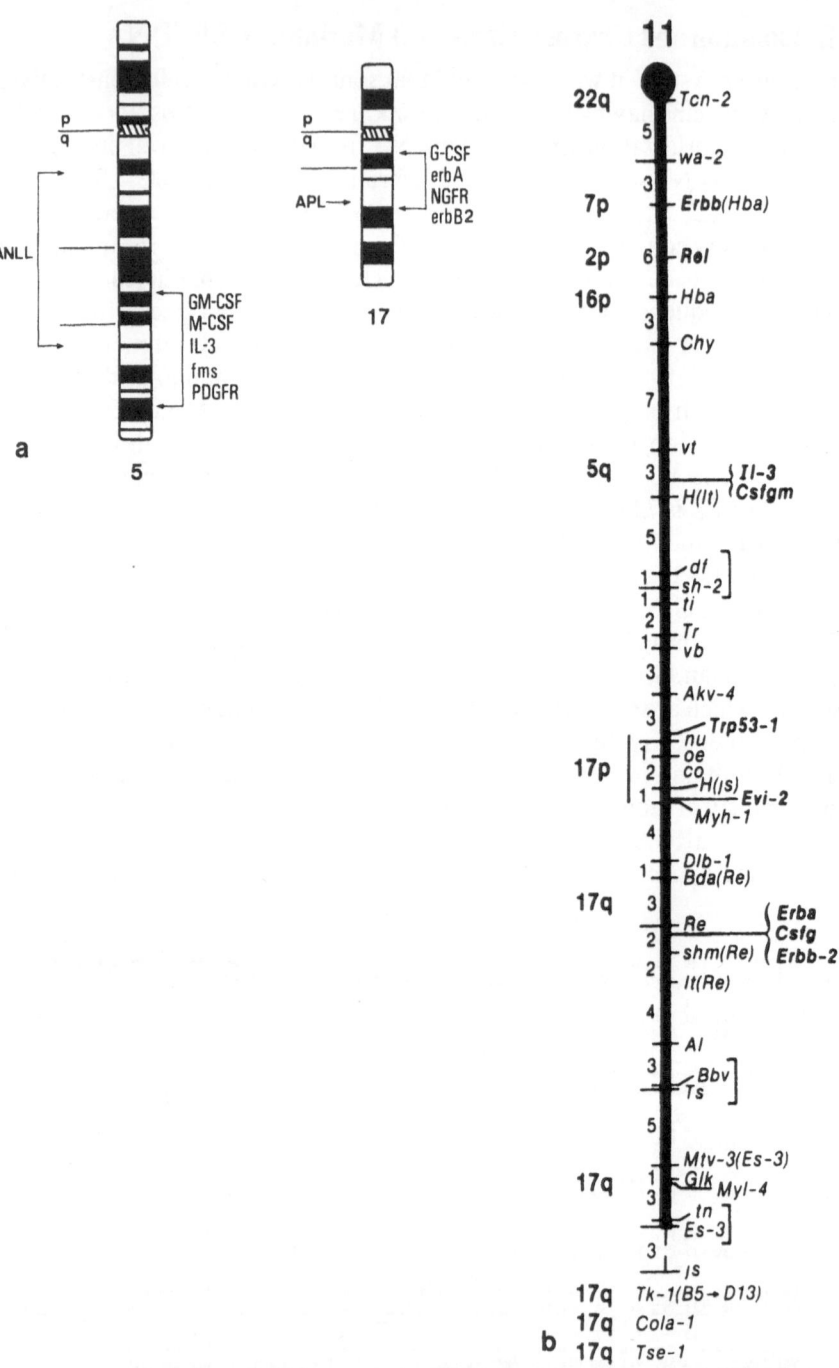

Fig. 4a, b. Localization of CSF genes on chromosomes. **a** Localization of CSF genes on human chromosomes. The breakpoint of translocation t(15; 17) in acute promyelocytic leukemia cells and the deleted region in acute nonlymphocytic leukemia cells are indicated by *APL* and *ANLL*, respectively. **b** Localization of CSF genes on mouse chromosomes. *Bold numbering* to the *left* of the chromosome indicates the location on the human chromosomes of the designated loci

codon are highly homologous between human and mouse G-CSF mRNAs. In this region, the ATTTA motif is repeated five times. Since this sequence is usually found on the 3′ noncoding region of mRNAs which are transiently expressed (SHAW and KAMEN 1986), G-CSF mRNA may also be posttranscriptionally regulated by selective degradation.

D. Chromosomal Gene Structure for G-CSF

I. Location of the G-CSF Gene on Human and Mouse Chromosomes

Southern hybridization analysis of human genomic DNA with human G-CSF cDNA has indicated that there is only one chromosomal gene for G-CSF per haploid genome. The single human G-CSF gene was found to be located on chromosome 17 (KANDA et al. 1987) by Southern hybridization analysis of DNAs from human-mouse somatic cell hybrids and flow-sorted human chromosomes using human G-CSF cDNA as a probe. Furthermore, in situ hybridization of human metaphase chromosomes with the human G-CSF cDNA probe indicated that the gene encoding human G-CSF is on the q21–q22 region of chromosome 17 (KANDA et al. 1987; SIMMERS et al. 1987; see Fig. 4a). In this region, there are genes for nerve growth factor receptor, c-*erb* B2, and c-*erb* A. On the other hand, the genes for GM-CSF, M-CSF, and IL-3 are on human chromosome 5. The q21–q22 region of human chromosome 17 is known to be involved in the translocation of t(15; 17) (q23; 21) in human acute promyelocytic leukemia. Southern hybridization analysis and in situ hybridization of metaphase chromosomes from acute promyelocytic leukemia cells indicated that the G-CSF gene is proximal to the breakpoint in the translocation observed in acute promyelocytic leukemia (SIMMERS et al. 1987; M. TSUCHIYA et al., unpublished results).

There is also only one G-CSF chromosomal gene per haploid genome in the murine system. The location of the G-CSF gene on murine chromosomes was determined using interspecific backcross analysis (BUCHBERG et al. 1988). As shown in Fig. 4b, the gene for G-CSF *(Csfg)* is assigned to the distal part of mouse chromosome 11 while the genes for GM-CSF and IL-3 are located on the proximal part of the same chromosome. Comparison of the genes on mouse chromosome 11 with those on human chromosomes suggests that several translocations and inversions have occurred during evolution (A. M. BUCHBERG et al., unpublished results).

II. Structure of the G-CSF Chromosomal Gene

1. Structure of the Human G-CSF Chromosomal Gene

The chromosomal gene for human G-CSF was isolated from a human gene library and characterized by nucleotide sequence analysis (NAGATA et al. 1986b). The human G-CSF gene consists of about 2500 nucleotides and is

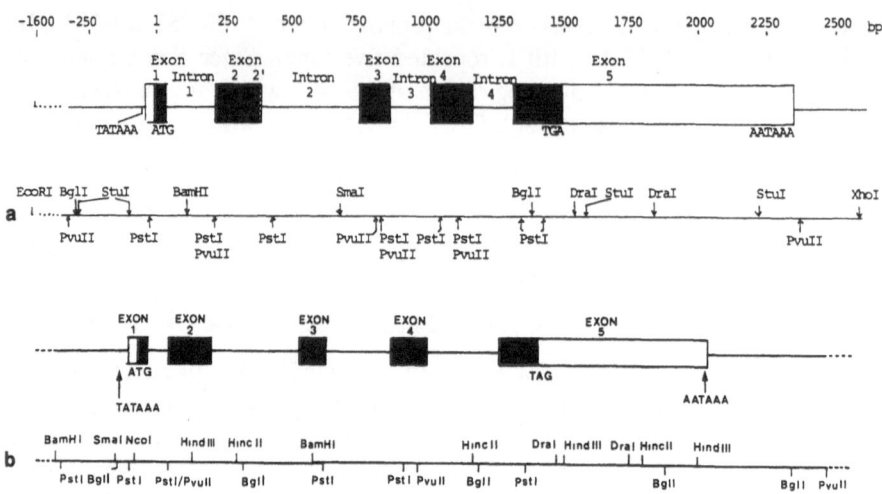

Fig. 5a, b. The chromosomal genes for G-CSF. The organization of **a** human and **b** mouse G-CSF chromosomal genes is shown. The *filled boxes* indicate the coding regions while the *open boxes* indicate the noncoding regions. The *lines* between them represent introns. The major recognition sites for restriction enzymes are given under the genes

split by four introns (Fig. 5). As shown in Fig. 6, at the 5′ terminus of intron 2, two donor sequences for splicing are arranged in tandem, 9 base pairs (bp) apart. The mRNA coding 207 amino acid G-CSF utilizes the second donor sequence while the mRNA coding 204 amino acid G-CSF is generated using the first splice donor sequence. The 3′ splice acceptor site is used to produce both mRNAs. Although the two splice donor sequences in intron 2 correspond equally well with the consensus sequence for splicing, the splicing of G-CSF precursor RNA occurs more frequently at the first splice donor sequence in CHU-2, 5637, and MIA PaCa-2 cells constitutively producing G-CSF (NAGATA et al. 1986b; ZSEBO et al. 1986; DEVLIN et al. 1987).

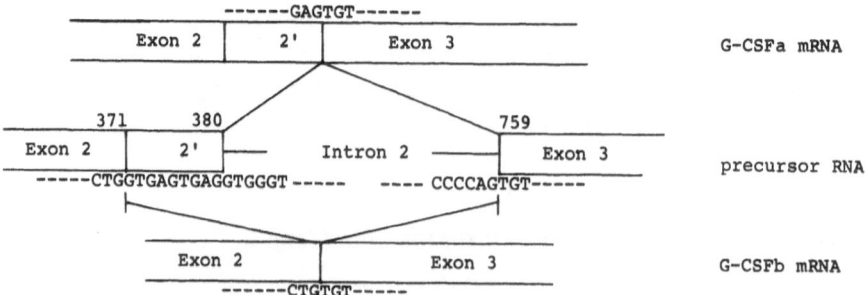

Fig. 6. Schematic representation of alternative splicing of human G-CSF precursor RNA. The structure of the splice junction in intron 2 of the human G-CSF chromosomal gene is shown. Exons and introns are represented by *open bars* and *lines*, respectively

Corresponding to the significant homology of the primary structure between G-CSF and IL-6 (Fig. 2), the gene organization of human G-CSF and IL-6, especially the size of each exon, is very similar (YASUKAWA et al. 1987). These data suggest that G-CSF and IL-6 genes might have evolved from a common ancestor gene.

2. Structure of the Mouse G-CSF Chromosomal Gene

The chromosomal gene structure of mouse G-CSF was characterized using DNA fragments isolated from a mouse gene library (TSUCHIYA et al. 1987a). As shown in Fig. 5b, the organization of the mouse G-CSF gene is very similar to that of the human G-CSF gene, that is, the mouse gene consists of 5 exons spanning about 2.5 kilobases (kb), and the positions and sizes of the introns in the mouse G-CSF gene are similar to those in the human G-CSF gene. However, little homology was found in the sequences of the corresponding introns of human and mouse G-CSF genes except for the intron–exon junction portion which is necessary for RNA splicing. Furthermore, in contrast with the human G-CSF gene, the 5′ donor sequence of intron 2 is not repeated in the mouse G-CSF gene and a single type of mRNA corresponding to the 9 nucleotide shorter version of human mRNA is produced from the mouse G-CSF precursor RNA.

3. Promoter of the G-CSF Chromosomal Gene

The 5′ flanking sequences of the human G-CSF chromosomal gene are very similar to those of the mouse G-CSF gene. As shown in Fig. 7, the 5′ flanking region extending approximately 310 nucleotides upstream of the ATG initiation codon shows about 80% homology between human and mouse G-CSF

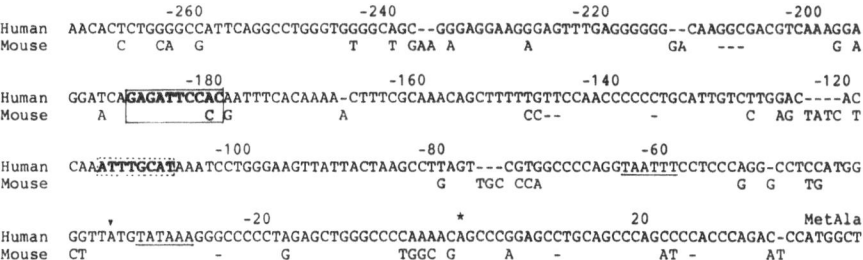

Fig. 7. The promoter sequence of human and mouse G-CSF chromosomal genes. The 5′ flanking regions of human and mouse G-CSF genes are compared. In aligning the sequences, gaps (–) were introduced to illustrate the maximum homology. The number 1 of the nucleotide sequence is at the major transcription initiation site (asterisk). The minor transcription initiation site is marked by a *triangle*. TATA-like sequences are *underlined*. The consensus decanucleotide present on the promoters of not only G-CSF but also of GM-CSF and IL-3 genes is indicated by a *solid box* while the OTF binding site is shown by a *dotted box*

genes, whereas the coding region shows 69% homology at the nucleotide sequence level (Tsuchiya et al. 1986, 1987a). Since the upstream sequences beyond this point have little homology, it is likely that the 5′ flanking sequence of about 300 nucleotides is responsible for the regulated expression of the G-CSF gene.

The transcription of mRNA starts at two positions, one at −35 and the other at −71 nucleotides upstream of the ATG initiation codon. About 25 nucleotides upstream from each transcription initiation site (CAP site), TATA-like sequences are present on human and mouse G-CSF genes. S1 mapping analysis of mouse G-CSF mRNA revealed that more than 90% of the mRNA starts from the first transcription initiation site located at −35 (Tsuchiya et al. 1987a).

On the 300 nucleotide-long promoter region of the G-CSF gene, there are several elements which seem to be important for the promoter activity. As shown in Fig. 7, the decanucleotide (GAGRTTCCAC) found in the promoter regions of the GM-CSF and IL-3 genes (Stanley et al. 1985; Miyatake et al. 1985) is present at −225 to −215 on the human and mouse G-CSF genes. The conservation of this decanucleotide sequence among G-CSF, GM-CSF, and IL-3 genes is fairly significant although the relative position of the decanucleotide with respect to the TATA box is different among those genes (Fig. 8). Furthermore, this sequence is significantly homologous to the NF-κB element (RGGGRA/CTYYCC) which is present on the promoters of immunoglobulin kappa chain, human immunodeficiency virus, β_2-microglobulin, class I major histocompatibility complex (MHC), and IL-2 receptor genes (Leung and Nabel 1988). Since this decanucleotide sequence on the G-CSF gene is essential for its promoter activity (Nishizawa et al., unpublished results), it will be interesting to characterize the nuclear factor which binds to it. In addition to the decanucleotide sequence mentioned above, the promoter of the G-CSF gene contains the OTF binding site (ATTTGCAT) at around 110 bp upstream of the transcription initiation site (Fig. 7). OTF binding sites were found upstream of the TATA box on the promoter of several mammalian genes including immunoglobulin and histone

```
hG·CSF      AGAGATTCCACA ····( 145)····TATAAA

mG·CSF      AGAGATTCCcCG ····(151)····TATAAA

hGM·CSF     GGAGATTCCACA····(61)···· TATTTA

mGM·CSF     GGAGATTCCACA····(67)···· TATAAG

mIL·3 (1)   GGAGGTTCCAtG····(74)····.
                                        TATAAG
      (2)   tGAGATTCCACt····(258)····'

rIL·3 (1)   GGAGGTTCCAtG····(79)····.
                                        TATAAG
      (2)   AGAGGTTCCAgt····(262)····'

consensus   RₗGAGRₗTTCCACRₗ              TATA box
sequence
```

Fig. 8. Consensus sequences in the 5′ flanking region of hemopoietic growth factor genes. Abbreviations: *h*, human; *m*, mouse; *r*, rat

genes, and were shown to be essential for their promoter activity (FLETCHER et al. 1987; SCHEIDEREIT et al. 1987). Whether or not the OTF binding site on the G-CSF promoter is involved in the expression of the G-CSF gene remains to be studied. Further analysis of the G-CSF promoter may reveal additional sequences which are required for the expression of the G-CSF gene.

E. Expression of the G-CSF Gene

I. Inducible Expression of the G-CSF Gene

Granulopoiesis occurs during an inflammatory process. G-CSF activity can be detected in the serum of mice previously injected with endotoxins or infected with bacteria (BURGESS and METCALF 1980; CHEERS et al. 1988). Peritoneal macrophages from humans and mice can be induced in vitro to produce G-CSF in response to bacterial endotoxins such as *E. coli* lipopolysaccharide (METCALF and NICOLA 1985; VELLENGA et al. 1988). Furthermore, recombinant human TNF-α and recombinant human interferon-γ (IFN-γ) can act cooperatively to induce monocytes to produce G-CSF (LU et al. 1988). In addition to macrophages and monocytes, fibroblasts and endothelial cells can be stimulated to produce G-CSF using IL-1 or TNF-α (KOEFFLER et al. 1987; SEELENTAG et al. 1987; BROUDY et al. 1987; KAUSHANSKY et al. 1988). The stimulating activities of TNF-α and IL-1 on endothelial cells are additive (SEELENTAG et al. 1987). TNF-β (lymphotoxin) is at least 1000 times less potent than TNF-α (KOEFFLER et al. 1987). The TNF-α-stimulated induction of expression of the G-CSF gene is transcriptionally and posttranscriptionally regulated and is insensitive to cycloheximide (SEELENTAG et al. 1987; KOEFFLER et al. 1988). The induction of CSFs by IL-1 and TNF-α was also demonstrated in vivo by administering recombinant IL-1 and TNF-α to mice (VOGEL et al. 1987). These results suggest that in the inflammatory process, endotoxins stimulate macrophages to produce not only G-CSF but several monokines as well, including IL-1 and TNF-α. IL-1 and TNF-α then induce the release of G-CSF from fibroblasts and endothelial cells. The G-CSF thus accumulated in the serum seems to be responsible for the granulocytosis that accompanies inflammation.

In addition to the granulocytosis observed during inflammation, neutrophilic granulocytes must be produced daily in bone marrow to replace short-lived circulating neutrophils in the blood. Little is known, however, about the mechanisms regulating hematopoiesis within the bone marrow. Recently, stroma cell lines were established from human and mouse bone marrow cells (RENNICK et al. 1987; FIBBE et al. 1988). These stroma cells are capable of supporting the proliferation and differentiation of progenitor cells for neutrophilic granulocytes, and low levels of G-CSF and GM-CSF activity are detected in the supernatants of these cultures. The production of G-CSF and GM-CSF by stroma cells can be increased 70- to 200-fold by treating cells with lipopolysaccharide or IL-1 (RENNICK et al. 1987; FIBBE et al. 1988). The

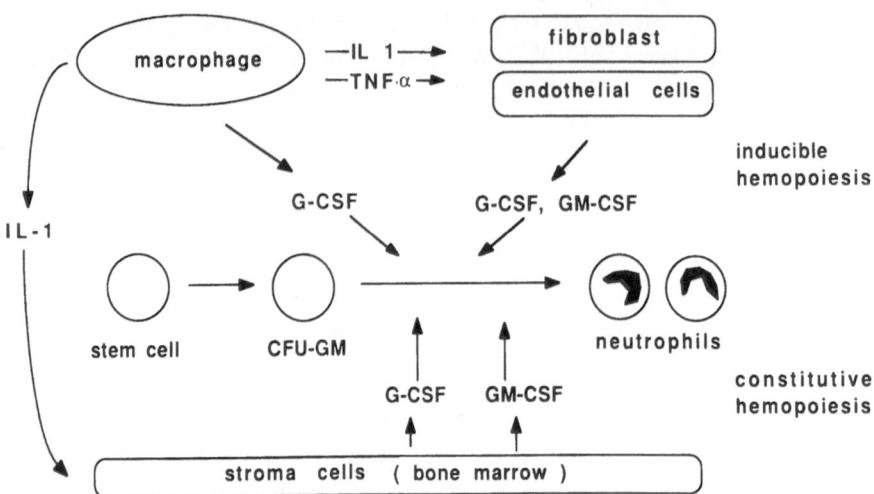

Fig. 9. Inducible granulopoiesis during the inflammatory process and constitutive granulopoiesis in bone marrow

G-CSF and GM-CSF produced by the stroma cells are structurally similar to those produced by macrophages or T cells. These data suggest that G-CSF and GM-CSF are involved in normal granulopoiesis in bone marrow. Figure 9 schematically illustrates inducible granulopoiesis during the inflammatory process and constitutive granulopoiesis in bone marrow. Further studies using macrophages, endothelial cells, and stroma cells may reveal the network of lymphokines and monokines which are involved in proliferation and differentiation of blood cells.

II. Constitutive Expression of the G-CSF Gene in Carcinoma Cells

It is well known that malignant tumors are often accompanied by marked leukocytosis without apparent bacterial infection (HUGHES and HIGBY 1952). When some of these tumor cells are transplanted into nude mice or syngeneic mice, they cause remarkable granulocytosis (ASANO et al. 1980; JOHNSON et al. 1985; EGAMI et al. 1986). These tumor cells were established as cell lines and were characterized for CSF production (NOMURA et al. 1986; EGAMI et al. 1986; SAKAI et al. 1984). Subsequently, several other solid tumor cell lines have also been found to produce CSF (WELTE et al. 1985; GABRILOVE et al. 1985; LILLY et al. 1987; DEVLIN et al. 1987; TWEARDY et al. 1987). Table 1 lists cell lines which constitutively produce CSFs. One of the CSFs produced in the human tumor cell lines shown in Table 1 seems to be G-CSF since it can be used to stimulate specific colony formation of neutrophilic granulocytes from mouse bone marrow cells and can induce terminal differentiation of mouse myeloid leukemia WEHI-3B D$^+$ or 32D clone 3 cells. In some cases, expression of G-CSF mRNA was confirmed by Northern hybridization or cloning of the cDNA. mRNA for GM-CSF was also detected in some tumor cells

Table 1. Carcinoma cell lines producing CSFs

	Origin	G-CSF activity	G-CSF mRNA (cDNA)	GM-CSF activity or mRNA	References
Human cell lines					
CHU-2	Oral cavity (squamous)	Yes	Yes	Yes	NOMURA et al. (1986)
5637	Bladder	Yes	Yes	Yes	WELTE et al. (1985)
LD-1	Melanoma	Yes	Yes	ND	LILLY et al. (1987)
MIA PaCa-2	Pancreas	Yes	Yes	ND	DEVLIN et al. (1987)
U87 MG	Glioblastoma multiforme	Yes	ND	ND	TWEARDY et al. (1987)
SW756	Cervix (squamous)	Yes	ND	ND	TWEARDY et al. (1987)
BRO	Melanoma	Yes	ND	ND	TWEARDY et al. (1987)
GBK-1	Gall bladder	Yes	ND	ND	EGAMI et al. (1986)
SK-HEP-1	Hepatoma (epithelial)	Yes	ND	Yes	GABRILOVE et al. (1985)
Mouse cell lines					
NFSA	Fibrosarcoma	Yes	Yes	ND	SAKAI et al. (1984)

ND, no data.

(Table 1), suggesting that both G-CSF and GM-CSF genes are active in these cells. The mechanism of constitutive expression of the G-CSF gene in these tumor cells is not yet known; however, it was suggested recently that the constitutive expression of the G-CSF gene in human squamous carcinoma CHU-2 cells is not due to rearrangement of the G-CSF gene but to constitutive expression (or activation) of the nuclear factors which work on the promoter region of the G-CSF gene (M. TSUCHIYA et al., unpublished results).

In addition to the solid tumor cells mentioned above, some acute myeloblastic leukemia (AML) cells produce G-CSF constitutively. For example, YOUNG et al. (1988) found the G-CSF transcript in six out of 22 AML cells, especially cells at the M1 stage (according to the French-American-British classification). In most cases (not all), the G-CSF secreted by AML cells is responsible for their autonomous growth and it has been suggested that normal hematopoietic progenitor cells themselves may express CSF at a certain stage of differentiation (YOUNG et al. 1988). In two out of 18 cases with AML cells, CHENG et al. (1988) found rearrangement of the G-CSF gene while YOUNG et al. (1988) did not observe rearrangement in all 22 AML cells examined. Further investigation is required to better understand the mechanism of constitutive expression of the G-CSF gene in AML cells.

F. Production of Recombinant G-CSF

Using their cDNAs, human and mouse recombinant G-CSFs were produced in *E. coli* (Souza et al. 1986; Shimamura et al. 1987; Devlin et al. 1988) and mammalian cells (Tsuchiya et al. 1987b). Since *E. coli* and mammalian cells carrying G-CSF cDNAs produce G-CSF at an efficiency of 20–50 mg/l culture, recombinant G-CSF was easily purified by gel filtration and hydrophobic column chromatography at a yield of 20%–50% (Tsuchiya et al. 1987b; Devlin et al. 1988). This efficiency may be compared with the value obtained during purification of G-CSF from natural sources, that is, 5 µg from 800 mice (Nicola et al. 1983). The human G-CSF produced in *E. coli* does not contain a sugar moiety and has a slightly lower M_r than natural G-CSF. On the other hand, human G-CSF produced in mammalian cells has the same M_r as natural G-CSF (Tsuchiya et al. 1986), and the sugar moiety attached to recombinant human G-CSF is identical to that of natural G-CSF (Oheda et al. 1988). Recombinant mouse G-CSF produced in mammalian cells has the same M_r (25 000) as the natural protein, which is much larger than the value (19 062) predicted from its amino acid sequence, suggesting that mouse G-CSF is heavily glycosylated (Fukunaga et al., unpublished results). Although nonglycosylated human G-CSF produced in *E. coli* is as active as natural G-CSF or recombinant G-CSF produced by mammalian cells, the stability of the G-CSF molecule may differ between the glycosylated and nonglycosylated forms.

G. Function of G-CSF

I. In Vitro Function of G-CSF

G-CSF stimulates the formation of colonies which consist mainly of neutrophils in semisolid cultures of bone marrow cells. Other types of colonies are occasionally observed in G-CSF-supported cultures but these seem to be indirect effects mediated by accessory cells (Strife et al. 1987; Welte et al. 1987a). G-CSF is not species specific in humans and mice; that is, human G-CSF works on mouse bone marrow cells as well as human bone marrow cells, and vice versa (Nicola et al. 1985; Tsuchiya et al. 1987b). The concentration of G-CSF required to stimulate the half-maximal response is 10^{-11}–10^{-12} M, and the sizes and numbers of colonies reach a maximum after 7 days of incubation. This time course contrasts with that observed with GM-CSF-supported cultures where the maximum number of colonies occurs at day 14. A low concentration of TNF-α or TNF-β inhibits colony formation stimulated by G-CSF but not by GM-CSF (Barber et al. 1987a, b). These results suggest that either G-CSF and GM-CSF act on different granulocyte progenitor cells or that TNF-α and TNF-β selectively influence the biochemical reactions caused by G-CSF.

G-CSF affects mature cells as well as progenitor cells of neutrophils. Like GM-CSF, G-CSF can enhance the survival of mature neutrophils (Begley et al. 1986), and can activate neutrophils to kill tumor cells as shown in antibody-dependent cellular cytotoxic assays (LOPEZ et al. 1983; PLATZER et al. 1985). Preincubation of normal mature neutrophils or mature neutrophils from patients with myelodysplastic syndromes with G-CSF causes a significant increase in O_2^- production in response to the bacterial chemotactic peptide *N*-formyl-methionyl-leucyl-phenylalanine (YUO et al. 1987; KITAGAWA et al. 1987). Furthermore, G-CSF can induce synthesis of alkaline phosphatase in neutrophilic granulocytes from patients with myelodysplastic syndromes or chronic myelogeneous patients (YUO et al. 1987; CHIKKAPPA et al. 1988). For this induction of alkaline phosphatase in neutrophils by G-CSF, de novo synthesis of protein and RNA, but not DNA, is required. Since O_2^- production and alkaline phosphatase are essential for the neutrophils to function, these results suggest that G-CSF is actually involved in the final activation of mature neutrophils.

It is known that several mouse and human myeloid leukemia cell lines respond to G-CSF. For example, G-CSF effectively stimulates mouse myelomonocytic leukemia WEHI-3B D^+ cells to proliferate and differentiate into monocytes in semisolid agar culture (NICOLA et al. 1983). This activity is unique to G-CSF and can be used to distinguish G-CSF from other CSFs such as GM-CSF, M-CSF, and IL-3. The other mouse myeloid leukemia cells, 32D clone 3 cells, can be induced to differentiate into neutrophilic granulocytes using G-CSF in liquid culture, while IL-3 supports the proliferation of this cell line (VALTIERI et al. 1987; ROVERA et al. 1987). In accord with the normal differentiation of neutrophilic granulocytes, several enzymes (for example, lactoferrin, myeloperoxidase, and chloroacetate esterase) specific for neutrophilic granulocytes are transiently expressed during the differentiation process of WEHI-3B D^+ and 32D clone 3 cells (ROVERA et al. 1987; MORISHITA et al. 1987b) (Fig. 10). Furthermore, the expression of c-*myc* and c-*myb* genes is suppressed during the later stages of monocytic differentiation of WEHI-3B D^+ while the expression of c-*fos* increases markedly during this process (GONDA and METCALF 1984). Mouse M1 cells can also be induced to differentiate using G-CSF although a relatively high concentration of G-CSF is needed for this (TOMIDA et al. 1986; HILTON et al. 1988).

G-CSF stimulates the proliferation of one of the IL-3-dependent mouse myeloid leukemia cells called NFS-60 without inducing differentiation (WEINSTEIN et al. 1986). In NFS-60 cells, the c-*myb* and c-*evi*-1 loci are rearranged by the integration of Cas-Br-M-MLV virus, and recently it was shown that the *evi*-1 gene codes for a zinc finger protein (MORISHITA et al. 1988). Further studies using several mouse myeloid leukemia cell lines such as WEHI-3B D^+, 32D clone 3, and NFS-60, together with cloned genes of c-*myb, evi*-1, and myeloperoxidase (MORISHITA et al. 1987a, b), may reveal the mechanism of differentiation of myeloid cells at the molecular level.

Several groups have studied the effect of G-CSF on primary human myeloid leukemia cells. Some acute myeloid leukemia cells (BEGLEY et al. 1987; PEBUSQUE et al. 1988a), especially cells which have a translocation involving chromosome 17, t(15; 17), can be induced to proliferate in response to

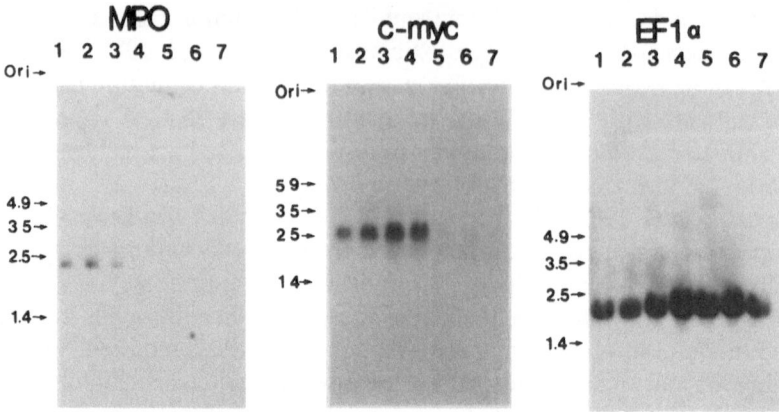

Fig. 10. Regulation of the expression of myeloperoxidase and c-*myc* in WEHI-3B D⁺ cells by G-CSF. WEHI-3B D⁺ cells were incubated in the presence of human G-CSF for various lengths of time. By using about 20 µg RNA, Northern hybridization was carried out with myeloperoxidase (*MPO*) cDNA, c-*myc*, and elongation factor-1α (*EF1*α) cDNA

G-CSF (PEBUSQUE et al. 1988 b). Except for one report (SOUZA et al. 1986), most human acute myeloid leukemia cells did not differentiate into mature cells with G-CSF (BEGLEY et al. 1987; PEBUSQUE et al. 1988 a, b).

II. Receptor for G-CSF

The receptor for G-CSF can be found on myeloid leukemia cells such as WEHI-3B D⁺, NFS-60, and HL-60 which respond to G-CSF, but not on cells which do not respond to G-CSF (NICOLA and METCALF 1984). In bone marrow or spleen cells, the receptors are essentially restricted to cells of neutrophilic granulocyte lineage. No receptors were found on the cell surface of erythroid, lymphoid, and eosinophilic cells (NICOLA and METCALF 1985). Mature neutrophils also have receptors for G-CSF, which corresponds well with the fact that G-CSF is able to activate mature cells. The number of receptors is around 300–1000 per cell, and the G-CSF molecule binds to its receptor with a dissociation constant (K_d) of about 100 pM (NICOLA and METCALF 1985), which is much higher than the concentration (~10 pM) required for the half-maximal biological response. These results may indicate that the biological responses caused by G-CSF occur at low levels of receptor occupancy. The G-CSF receptor on bone marrow cells is downregulated not only by G-CSF, but also by IL-3 and GM-CSF (WALKER et al. 1985). Chemical crosslinking of ¹²⁵I-labeled murine G-CSF with the receptor on mouse WEHI-3B D⁺ cells has suggested that the mouse G-CSF receptor is a single subunit protein of M_r 150000 (NICOLA and PETERSON 1986). Purification and molecular cloning of the receptor will be valuable in elucidating the molecular mechanism associated with the binding of G-CSF to the receptor.

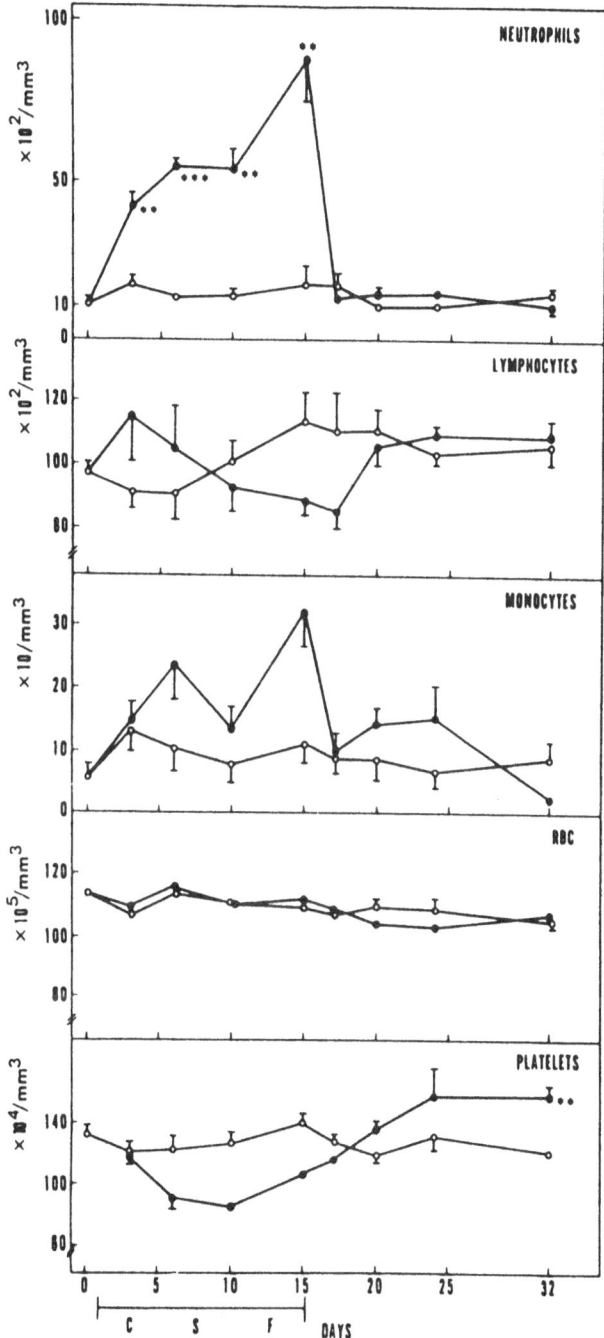

Fig. 11. In vivo effects of human G-CSF in mice. Mice were injected daily subcutaneously with G-CSF (*filled circles*) or saline (*open circles*) for 15 days (day 1 to day 15). At the indicated time, mice were killed and their blood examined

III. In Vivo Function of G-CSF

As was previously mentioned, granulocytosis has been observed in nude mice bearing human carcinoma cells producing G-CSF (ASANO et al. 1980). In addition, remarkable granulopoiesis was also observed in mice, hamsters, and monkeys when G-CSF was administered (TSUCHIYA et al. 1987b; COHEN et al. 1987; WELTE et al. 1987a,b; TAMURA et al. 1987). Figure 11 shows typical results in a mouse system (TAMURA et al. 1987). The in vivo effect of G-CSF is specific for neutrophilic granulocytes, and no changes were observed in other blood cells such as erythrocytes, monocytes, or lymphocytes. The injected G-CSF was cleared in a biexponential manner with an estimated distribution half-life of 0.5 h and an elimination half-life of 3.8 h (COHEN et al. 1987). The effect of G-CSF on granulopoiesis can be controlled by the amount of G-CSF injected into animals. The number of neutrophils in the blood returns to normal levels when the administration of G-CSF ceases. When G-CSF was injected into animals with neutropenia induced by treatment with chemical drugs such as 5-fluorouracil or cyclophosphamide, G-CSF markedly accelerated the recovery of granulopoiesis (COHEN et al. 1987; TAMURA et al. 1987; WELTE et al. 1987a,b). Furthermore, the administration of G-CSF to mice pretreated with a cytotoxic drug greatly increased the resistance of the mice against infections caused by *E. coli, Staphylococcus aureus,* and *Candida albicans* (MATSUMOTO et al. 1987).

IV. Clinical Application of G-CSF

Encouraged by results obtained with the animal model system mentioned above, clinical application of G-CSF is currently under way. In a phase I/II study of G-CSF, patients with advanced malignancy receiving cytotoxic chemotherapy were treated with G-CSF (BRONCHUD et al. 1987; GABRILOVE et al. 1988; MORSTYN et al. 1988). G-CSF administration following chemotherapy remarkably reduced the period of neutropenia caused by chemotherapy. In another example, G-CSF was administered to patients who had received bone marrow transplants followed by irradiation or cytotoxic agents such as cyclophosphamide (KODO et al. 1988). Recovery of neutrophilic granulocytes in these patients after bone marrow transplantation was greatly accelerated using G-CSF. In both cases, G-CSF had almost no adverse side effects; at most, there may have been slight bone pain which was well tolerated by patients. Since delayed recovery of neutrophilic granulocytes after chemotherapy or bone marrow transplantation often causes severe bacterial and fungal infection, these results suggest that the administration of G-CSF may prove to be beneficial in chemotherapy and bone marrow transplantation therapy.

Acknowledgments. I thank Drs. Y. Kaziro and C. Weissmann for encouragement and discussion. The work in our laboratory was carried out in a collaboration with Dr. S. Asano, M. Tsuchiya, and K. Morishita, and supported in part by Grants-in-Aid from the Ministry of Education, Science and Culture of Japan, and Special Coordination Funds of the Science and Technology Agency of the Japanese Government. I also thank Ms. M. Ikeda for her secretarial assistance.

References

Asano S, Sato N, Mori M, Ohsawa N, Kosaka N, Ueyama Y (1980) Detection and assessment of human granulocyte-macrophage colony-stimulating factor (GM-CSF) producing tumors by heterotransplantation into nude mice. Br J Cancer 41:689–694

Barber KE, Crosier PS, Gillis S, Watson JD (1987a) Human granulocyte-macrophage progenitors and their sensitivity to cytotoxins: analysis by limiting dilution. Blood 70:1773–1776

Barber KE, Crosier PS, Watson JD (1987b) The differential inhibition of hemopoietic growth factor activity by cytotoxins and interferon-γ. J Immunol 139:1108–1112

Begley CG, Lopez AF, Nicola NA, Warren DJ, Vadas MA, Sanderson CJ, Metcalf D (1986) Purified colony-stimulating factors enhance the survival of human neutrophils and eosinophils in vitro: a rapid and sensitive microassay for colony-stimulating factors. Blood 68:162–166

Begley CG, Metcalf D, Nicola NA (1987) Primary human myeloid leukemia cells: comparative responsiveness to proliferative stimulation by GM-CSF or G-CSF and membrane expression of CSF receptors. Leukemia 1:1–8

Bronchud MH, Scarffe JH, Thatcher N, Growther D, Souza LM, Alton NK, Testa NG, Dexter TM (1987) Phase I/II study of recombinant human granulocyte colony-stimulating factor in patients receiving intensive chemotherapy for small cell lung cancer. Br J Cancer 56:809–813

Broudy VC, Kaushansky K, Harlan JM, Adamson JW (1987) Interleukin 1 stimulates human endothelial cells to produce granulocyte-macrophage colony-stimulating factor and granulocyte colony-stimulating factor. J Immunol 139:464–468

Buchberg AM, Bedigian HG, Taylor BA, Brownell E, Ihle JN, Nagata S, Jenkins NA, Copeland NG (1988) Localization of Evi-2 to chromosome 11: linkage to other proto-oncogene and growth factor loci using interspecific backcross mice. Oncogene Res 2:149–166

Burgess AW, Metcalf D (1980) Characterization of a serum-factor stimulating the differentiation of myelomonocytic leukemic cells. Int J Cancer 26:647–654

Cheers C, Haigh AM, Kelso A, Metcalf D, Stanley ER, Young AM (1988) Production of colony-stimulating factors (CSFs) during infection: separate determination of macrophage-, granulocyte-, granulocyte-macrophage-, and multi-CSFs. Infect Immun 56:247–251

Cheng GYM, Kelleher CA, Miyauchi J, Wang C, Wong G, Clark SC, McCulloch EA, Minden MD (1988) Structure and expression of genes of GM-CSF and G-CSF in blast cells from patients with acute myeloblastic leukemia. Blood 71:204–208

Chikkappa G, Wang GJ, Santella D, Pasquale D (1988) Granulocyte colony-stimulating factor (G-CSF) induces synthesis of alkaline phosphatase in neutrophilic granulocytes of chronic myelogeneous leukemia patients. Leuk Res 12:491–498

Clark SC, Kamen R (1987) Human hematopoietic colony-stimulating factors. Science 236:1229–1237

Cohen AM, Zsebo KM, Inoue H, Hines D, Boone TC, Chazin VR, Tsai L, Ritch T, Souza LM (1987) In vivo stimulation of granulopoiesis by recombinant human granulocyte colony-stimulating factor. Proc Natl Acad Sci USA 84:2484–2488

Devlin JJ, Devlin PE, Myambo K, Lilly MB, Rado TA, Warren MK (1987) Expression of granulocyte colony-stimulating factor by human cell lines. J Leuk Biol 41:302–306

Devlin PE, Drummond RJ, Toy P, Mark DF, Watt KWK, Devlin JJ (1988) Alteration of amino-terminal codons of human granulocyte-colony-stimulating factor increases expression levels and allows efficient processing by methionine aminopeptidase in Escherichia coli. Gene 65:13–22

Egami H, Sakamoto K, Yoshimura R, Kikuchi H, Akagi M (1986) Establishment of a cell line of gallbladder carcinoma (GBK-1) producing human colony stimulating factor. Jpn J Cancer Res 77:168–176

Fibbe WE, Damme J, Billiau A, Goselink HM, Voogt PJ, Eeden G, Ralph P, Altrock BW, Falkenburg JHF (1988) Interleukin 1 induces human marrow stromal cells in long-term culture to produce granulocyte colony-stimulating factor and macrophage colony-stimulating factor. Blood 71:430–435

Fletcher C, Heintz N, Roeder RG (1987) Purification and characterization of OTF-1, a transcription factor regulating cell cycle expression of a human histone H2b gene. Cell 51:773–781

Gabrilove JL, Welte K, Lu L, Castro-Malaspina H, Moore MAS (1985) Constitutive production of leukemia differentiation, colony-stimulating, erythroid burst-promoting, and pluripoietic factors by a human hepatoma cell line: characterization of the leukemia differentiation factor. Blood 66:407–415

Gabrilove JL, Jakubowski A, Scher H, Sternberg C, Wong G, Grous J, Yagoda A, Fain K, Moore MAS, Clarkson B, Oettgen HF, Alton K, Welte K, Souza L (1988) Effect of granulocyte colony-stimulating factor on neutropenia and associated morbidity due to chemotherapy for transitional-cell carcinoma of the urothelium. N Engl J Med 318:1414–1422

Gonda T, Metcalf D (1984) Expression of *myb, myc* and *fos* protooncogenes during the differentiation of a murine myeloid leukaemia. Nature 310:249–251

Griffin LD, Rambaldi A, Vellenga E, Young DC, Ostapovicz D, Cannistra SA (1987) Secretion of interleukin-1 by acute myeloblastic leukemia cells in vitro induces endothelial cells to secrete colony stimulating factors. Blood 70:1218–1221

Hilton DJ, Nicola NA, Gough NM, Metcalf D (1988) Resolution and purification of three distinct factors produced by krebs ascites cells which have differentiation-inducing activity on murine myeloid leukemic cell lines. J Biol Chem 19:9238–9243

Hirano T, Yasukawa K, Harada H, Taga T, Watanabe Y, Matsuda S, Kashiwamura S-I, Nakajima K, Koyama K, Iwamatsu A, Tsunasawa S, Sakiyama F, Matsui H, Takahara Y, Taniguchi T, Kishimoto T (1986) Complementary DNA for a novel human interleukin (BSF-2) that induces B lymphocytes to produce immunoglobin. Nature 324:73–76

Hughes WF, Higby CS (1952) Marked leukocytosis resulting from carcinomatosis. Ann Intern Med 37:1085–1088

Johnson GR, Whitehead R, Nicola NA (1985) Effects of a murine mammary tumor on in vivo and in vitro hemopoiesis. Int J Cell Cloning 3:91–105

Kanda N, Fukushige S-I, Murotsu T, Yoshida MC, Tsuchiya M, Asano S, Kaziro Y, Nagata S (1987) Human gene coding for granulocyte colony-stimulating factor is assigned to the q21–q22 region of chromosome 17. Somatic Cell Mol Genet 13:679–684

Kaushansky K, Lin N, Adamson JW (1988) Interleukin 1 stimulates fibroblasts to synthesize granulocyte-macrophage and granulocyte colony-stimulating factors. J Clin Invest 81:92–97

Kitagawa S, Yuo A, Souza LM, Saito M, Miura Y, Takaku F (1987) Recombinant human granulocyte colony-stimulating factor enhances superoxide release in human granulocytes stimulated by the chemotactic peptide. Biochem Biophys Res Commun 144:1143–1146

Kodo H, Tajika K, Takahashi S, Ozawa K, Asano S, Takaku F (1988) Acceleration of neutrophilic granulocyte recovery after bone-marrow transplantation by administration of recombinant human granulocyte colony-stimulating factor. Lancet 2(8601):38

Koeffler HP, Gasson J, Ranyard J, Souza L, Shepard M, Munker R (1987) Recombinant human TNFα stimulates production of granulocyte colony-stimulating factor. Blood 70:55–59

Koeffler HP, Gasson J, Tobler A (1988) Transcriptional and posttranscriptional modulation of myeloid colony-stimulating factor expression by tumor necrosis factor and other agents. Mol Cell Biol 8:3432–3438

Leung K, Nabel GJ (1988) HTLV-1 transactivator induces interleukin-2 receptor expression through an NF-κB-like factor. Nature 333:776–778

Lilly MB, Devlin PE, Devlin JJ, Rado TA (1987) Production of granulocyte colony-stimulating factor by a human melanoma cell line. Exp Hematol 15:966–971

Lopez AF, Nicola NA, Burgess AW, Metcalf D, Battye FL, Sewell WA, Vadas M (1983) Activation of granulocyte cytotoxic function by purified mouse colony-stimulating factors. J Immunol 131:2983–2988

Lu L, Walker D, Graham CD, Waheed A, Shadduck RK, Broxmeyer HE (1988) Enhancement of release from MHC class II antigen-positive monocytes of hematopoietic colony stimulating factors CSF-1 and G-CSF by recombinant human tumor necrosis factor-alpha: synergism with recombinant human interferon-gamma. Blood 72:34–41

Matsumoto M, Matsubara S, Matsuno T, Tamura M, Hattori K, Nomura H, Ono M, Yokota T (1987) Protective effect of human granulocyte colony-stimulating factor on microbial infection in neutropenic mice. Infect Immun 55:2715–2720

Metcalf D (1985) The granulocyte-macrophage colony-stimulating factors. Science 229:16–22

Metcalf D (1986) The molecular biology and functions of the granulocyte-macrophage colony-stimulating factors. Blood 67:257–267

Metcalf D (1987) The molecular control of normal and leukaemic granulocytes and macrophages. Proc R Soc Lond 230:389–423

Metcalf D, Nicola NA (1985) Synthesis by mouse peritoneal cells of G-CSF, the differentiation inducer for myeloid leukemia cells: stimulation by endotoxin, M-CSF and multi-CSF. Leuk Res 1:35–50

Miyatake S, Otsuka T, Yokota T, Lee F, Arai K (1985) Structure of the chromosomal gene for granulocyte-macrophage colony stimulating factor: Comparison of the mouse and human genes. EMBO J 4:2561–2568

Morishita K, Kubota N, Asano S, Kaziro Y, Nagata S (1987a) Molecular cloning and characterization of cDNA for human myeloperoxidase. J Biol Chem 262:3844–3851

Morishita K, Tsuchiya M, Asano S, Kaziro Y, Nagata S (1987b) Chromosomal gene structure of human myeloperoxidase and regulation of its expression by granulocyte-colony stimulating factor. J Biol Chem 262:15 208–15 213

Morishita K, Parker DS, Mucenski ML, Jenkins NA, Copeland NG, Ihle JN (1988) Retroviral activation of a novel gene encoding a zinc finger protein in IL-3-dependent myeloid leukemia cell lines. Cell 54:831–840

Morstyn G, Campbell L, Souza LM, Alton NK, Keech J, Green M, Sheridan W, Metcalf D, Fox R (1988) Effect of granulocyte colony stimulating factor on neutropenia induced by cytotoxic chemotherapy. Lancet 1 (8587):667–672

Nagata S, Tsuchiya M, Asano S, Kaziro Y, Yamazaki T, Yamamoto O, Hirata Y, Kubota N, Oheda M, Nomura H, Ono M (1986a) Molecular cloning and expression of cDNA for human granulocyte colony stimulating factor. Nature 319:415–418

Nagata S, Tsuchiya M, Asano S, Yamamoto O, Hirata Y, Kubota N, Oheda M, Nomura H, Yamazaki T (1986b) The chromosomal gene structure and two mRNAs for human granulocyte colony-stimulating factor. EMBO J 5:575–581

Nicola NA (1985) Granulocyte colony stimulating factor. Methods Enzymol 116:600–619

Nicola NA, Metcalf D (1981) Biochemical properties of differentiation factors for murine myelomonocytic leukemic cells in organ conditioned media – separation from colony-stimulating factors. J Cell Physiol 109:253–264

Nicola NA, Metcalf D (1984) Binding of the differentiation-inducer, granulocyte-colony-stimulating factor, to responsive but not unresponsive leukemic cell lines. Proc Natl Acad Sci USA 81:3765–3769

Nicola NA, Metcalf D (1985) Binding of ^{125}I-labeled granulocyte colony-stimulating factor to normal murine hemopoietic cells. J Cell Physiol 124:313–321

Nicola NA, Peterson L (1986) Identification of distinct receptors for two hemopoietic growth factors (granulocyte colony-stimulating factor and multipotential colony-stimulating factor) by chemical cross-linking. J Biol Chem 261:12 384–12 389

Nicola NA, Metcalf D, Johnson GR, Burgess AW (1979) Separation of functionally distinct human granulocyte-macrophage colony-stimulating factors. Blood 54:614–627

Nicola NA, Metcalf D, Matsumoto M, Johnson GR (1983) Purification of a factor inducing differentiation in murine myelomonocytic leukemia cells: identification as granulocyte colony-stimulating factor (G-CSF). J Biol Chem 258:9017–9023

Nicola NA, Begley CG, Metcalf D (1985) Identification of the human analogue of a regulator that induces differentiation in murine leukaemic cells. Nature 314:625–628

Nomura H, Imazeki I, Oheda M, Kubota N, Tamura M, Ono M, Ueyama Y, Asano S (1986) Purification and characterization of human granulocyte colony-stimulating factor (G-CSF). EMBO J 5:871–876

Oheda M, Hase S, Ono M, Ikenaka T (1988) Structure of the sugar chains of recombinant human granulocyte-colony-stimulating factor produced by chinese hamster ovary cells. J Biochem 103:544–546

Pebusque M-J, Lopez M, Torres H, Carotti A, Guilbert L (1988a) Growth response of human myeloid leukemia cells to colony-stimulating factors. Exp Hematol 16:360–366

Pebusque M-J, Lafage M, Lopez M, Mannoni P (1988b) Preferential response of acute myeloid leukemias with translocation involving chromosome 17 to human recombinant granulocyte colony-stimulating factor. Blood 72:257–265

Platzer E, Welte K, Gabrilove JL, Lu L, Harris P, Mertelsmann R, Moore MAS (1985) Biological activities of a human pluripotent hematopoietic colony stimulating factor on normal and leukemic cells. J Exp Med 162:1788–1801

Rennick D, Yang G, Gemmell L, Lee F (1987) Control of hemopoiesis by a bone marrow stromal cell clone: lipopolysaccharide- and interleukin-1-inducible production of colony stimulating factors. Blood 69:682–691

Rovera G, Valtieri M, Mavilio F, Reddy EP (1987) Effect of abelson murine leukemia virus on granulocytic differentiation and interleukin-3 dependence of a murine progenitor cell line. Oncogene 1:29–35

Sachs L (1987) The molecular control of blood cell development. Science 238:1374–1379

Sakai N, Shikita M, Tsuneoka K, Bessho M, Hirashima K (1984) Macrophage colony-stimulating factor and granulocyte colony-stimulating factor separated from fibrosarcoma tissue in mice. Jpn J Cancer Res 75:355–361

Scheidereit C, Heguy A, Roeder RG (1987) Identification and purification of a human lymphoid-specific octamer-binding protein (OTF-2) that activates transcription of an immunoglobulin promoter in vitro. Cell 51:783–793

Seelentag WK, Mermod J-J, Montesano R, Vassalli P (1987) Additive effects of interleukin 1 and tumor necrosis factor-α on the accumulation of the three granulocyte and macrophage colony-stimulating factor mRNAs in human endothelial cells. EMBO J 6:2261–2265

Shaw G, Kamen R (1986) A conserved AU sequence from the 3′ untranslated region of GM-CSF mRNA mediates selective mRNA degradation. Cell 46:659–667

Shimamura M, Kobayashi Y, Yuo A, Urabe A, Okabe T, Komatsu Y, Itoh S, Takaku F (1987) Effect of human recombinant granulocyte colony-stimulating factor on hematopoietic injury in mice induced by 5-fluorouracil. Blood 69:353–355

Simmers RN, Webber LM, Shannon MF, Garson OM, Wong G, Vadas MA, Sutherland GR (1987) Localization of the G-CSF gene on chromosome 17 proximal to the breakpoint in the t(15; 17) in acute promyelocytic leukemia. Blood 70:330–332

Simpson RJ, Nice EC, Nicola NA (1987) Structural studies on the murine granulocyte colony-stimulating factor. Biol Chem Hoppe Seyler 368:1327–1331

Souza LM, Boone TC, Gabrilove J, Lai PH, Zsebo KM, Murdock DC, Chazin VR, Bruszewski J, Lu H, Chen KK, Barendt J, Platzer E, Moore MAS, Mertelsmann R, Welte K (1986) Recombinant human granulocyte colony-stimulating factor: effects on normal and leukemic myeloid cells. Science 232:61–65

Stanley E, Metcalf D, Sobieszczuk P, Gough NM, Dunn AR (1985) The structure and expression of the murine gene encoding granulocyte-macrophage colony stimulating factor; evidence for utilization of alternative promoters. EMBO J 4:2569–2573

Strife A, Lambek C, Wisniewski D, Gulati S, Gasson JC, Golde DW, Welte K, Gabrilove JL, Clarkson B (1987) Activities of four purified growth factors on highly enriched human hematopoietic progenitor cells. Blood 69:1508–1523

Taga T, Kawanishi Y, Hardy RR, Hirano T, Kishimoto T (1987) Receptors for B cell stimulatory factor 2: Quantitation, specificity, distribution, and regulation of their expression. J Exp Med 166:967–981

Tamura M, Hattori K, Nomura H, Oheda M, Kubota N, Imazeki I, Ono M, Ueyama Y, Nagata S, Shirafuji N, Asano S (1987) Induction of neutrophilic granulocytosis in mice by administration of purified human native granulocyte colony-stimulating factor (G-CSF). Biochem Biophys Res Commun 142:454–460

Tomida M, Yamamoto-Yamaguchi Y, Hozumi M, Okabe T, Takaku F (1986) Induction by recombinant human granulocyte colony-stimulating factor of differentiation of mouse myeloid leukemic M1 cells. FEBS Lett 207:271–275

Tsuchiya M, Asano S, Kaziro Y, Nagata S (1986) Isolation and characterization of the cDNA for murine granulocyte colony-stimulating factor. Proc Natl Acad Sci USA 83:7633–7637

Tsuchiya M, Kaziro Y, Nagata S (1987a) The chromosomal gene structure for murine granulocyte colony-stimulating factor. Eur J Biochem 165:7–12

Tsuchiya M, Nomura H, Asano S, Kaziro Y, Nagata S (1987b) Characterization of recombinant human granulocyte colony-stimulating factor produced in mouse cells. EMBO J 6:611–616

Tweardy DJ, Caracciolo D, Valtieri M, Rovera G (1987) Tumor-derived growth factors that support proliferation and differentiation of normal and leukemic hemopoietic cells. Ann NY Acad Sci 511:30–38

Valtieri M, Tweardy DJ, Carraci D, Johnson K, Mavilio F, Altman S, Santoli D, Rovela G (1987) Cytokine-dependent granulocyte differentiation: regulation of proliferative and differentiative responses in a murine progenitor cell line. J Immunol 138:3829–3835

Vellenga E, Rambaldi A, Ernst TJ, Ostapovicz D, Griffin JD (1988) Independent regulation of M-CSF and G-CSF gene expression in human monocytes. Blood 71:1529–1532

Vogel SN, Douches SD, Kaufman EN, Neta R (1987) Induction of colony stimulating factor in vivo by recombinant interleukin 1α and recombinant tumor necrosis factor α. J Immunol 138:2143–2148

Walker F, Nicola NA, Metcalf D, Burgess AW (1985) Hierarchical down-modulation of hemopoietic growth factor receptors. Cell 43:269–276

Weinstein Y, Ihle JN, Lavu S, Reddy EP (1986) Truncation of the c-myb gene by a retroviral integration in an interleukin 3-dependent myeloid leukemia cell line. Proc Natl Acad Sci USA 83:5010–5014

Welte K, Platzer E, Lu L, Gabrilove JL, Levi E, Mertelsmann R, Moore MAS (1985) Purification and biochemical characterization of human pluripotent hematopoietic colony-stimulating factor. Proc Natl Acad Sci USA 82:1526–1530

Welte K, Bonilla MA, Gabrilove JL, Gillio AP, Potter GK, Moore MAS, O'Reilly RJ, Boone TC, Souza LM (1987a) Recombinant human granulocyte-colony stimulating factor: In vitro and in vivo effects on myelopoiesis. Blood Cells 13:17–30

Welte K, Bonilla MA, Gillio AP, Boone TC, Potter GK, Gabrilove JL, Moore MAS, O'Reilly RJ, Souza LM (1987b) Recombinant human granulocyte colony-stimulating factor: Effects on hematopoiesis in normal and cyclophosphamide-treated primates. J Exp Med 165:941–948

Yasukawa K, Hirano T, Watanabe Y, Muratani K, Matsuda T, Nakai S, Kishimoto T (1987) Structure and expression of human B cell stimulatory factor-2 (BSF-2/IL-6) gene. EMBO J 6:2939–2945

Young DC, Demetri GD, Ernst TJ, Cannistra SA, Griffin JD (1988) In vitro expression of colony-stimulating factor genes by human acute myeloblastic leukemia cells. Exp Hematol 16:378–382

Yuo A, Kitagawa S, Okabe T, Urabe A, Komatsu Y, Itoh S, Takaku F (1987) Recombinant human granulocyte colony-stimulating factor repairs the abnormalities of neutrophils in patients with myelodysplastic syndromes and chronic myelogenous leukemia. Blood 70:404–411

Zsebo KM, Cohen AM, Murdock DC, Boone TC, Inoue H, Chazin VR, Hines D, Souza LM (1986) Recombinant human granulocyte colony stimulating factor: Molecular and biological characterization. Immunobiology 172:175–184

Granulocyte-Macrophage Colony-Stimulating Factor

A. W. Burgess

A. Introduction

The control of blood cell production is modulated directly or indirectly by more than ten different hematopoietic growth factors (METCALF 1984). Many names have been assigned to each of the hematopoietic growth factors: although a specific set of names is used throughout this article, the alternative names can be found in other reviews (NICOLA and VADAS 1984; METCALF 1985; ARAI et al. 1986; MORSTYN and BURGESS 1988). Six of these growth factors appear to be capable of stimulating the proliferation of hematopoietic progenitor cells – multi-colony-stimulating factor (multi-CSF, also called interleukin-3), granulocyte-macrophage colony-stimulating factor (GM-CSF), granulocyte colony-stimulating factor (G-CSF), macrophage colony-stimulating factor (M-CSF, also called CSF-1), interleukin-5 (also called eosinophil differentiation factor), and erythropoietin. The existence of these different growth factors had been apparent for almost 20 years, but in the last 5 years all of these molecules have been purified, cloned, and expressed. We know that there is no structural relationship between the regulators, but there is a considerable overlap in their biological activities. Multi-CSF, GM-CSF, and interleukin-5 all stimulate the production of eosinophils (METCALF et al. 1986; LOPEZ et al. 1986); multi-CSF (IHLE et al. 1982; BURGESS et al. 1980), GM-CSF (BURGESS et al. 1977a, 1986), and M-CSF (STANLEY and GUILBERT 1981) stimulate the production of macrophages; and multi-CSF, GM-CSF, and G-CSF lead to the production and stimulation of neutrophils. The redundancy in these functional activities occurs at the level of cell surface receptor interactions, so it is necessary to consider some of the target cell receptor biochemistry for several of the hematopoietic growth factors as well as considering the properties of a single molecule such as GM-CSF.

Whilst this review will consider the biology and biochemistry of one particular hematopoietic growth factor – GM-CSF – the reader may need to extend many of these observations to include the effects of the other hematopoietic growth factors and response modifiers such as interleukin-1, interleukin-6, or tumor necrosis factor. This is particularly important when considering the action of GM-CSF on mature cells – eosinophils, neutrophils, or macrophages. Many of the actions of GM-CSF on mature cells appear to prime the cells to other biological response modifiers, e.g., antibodies, chemotactic agents, or bacterial cell wall products.

Although GM-CSF was discovered by its effects in vitro, the availability of considerable amounts of recombinant GM-CSF has enabled scientists and clinicians to confirm that GM-CSF acts as a hematopoietic growth factor in vivo. Indeed the beneficial effects of GM-CSF hold promise for accelerating bone marrow recovery, amplifying the antibody-dependent tumoricidal effects of monocytes, and increasing the effectiveness of neutrophils to protect against bacterial infections.

Much of this information has appeared in a rush and there has been little time to investigate the detailed molecular properties of GM-CSF. However, we have a detailed knowledge of the GM-CSF gene and the expression of this gene in vitro. It has been possible to produce retroviruses encoding GM-CSF and these constructs have been used to induce constitutive production of GM-CSF by myeloid cells (LANG et al. 1985) and even several strains of transgenic mice (LANG et al. 1987). In both sets of experiments the overproduction of GM-CSF has indicated likely roles for GM-CSF – the first in leukemogenesis and the second in macrophage-mediated tissue destruction. Our present knowledge of the physiology of GM-CSF is still hampered by its extraordinary potency and its low abundance. Hopefully, antibodies or molecular probes will soon be available to investigate the production of GM-CSF in the steady state.

B. Discovery and Nomenclature

The initial experiments which detected hematopoietic growth factors (BRADLEY and METCALF 1966; PLUZNICK and SACHS 1965) could not distinguish between the properties of a single growth factor or mixtures of molecules. However, it was clear that there were hematopoietic growth factors with distinctly different biological specificities. The first sources of CSF were generated in vitro from cell suspensions, organ cultures, cell lines, or extracts of pregnant mouse uterus. The conditioned medium from one fibroblast cell line (mouse L cells) (STANLEY et al. 1975) stimulated the production of macrophages – a similar specificity to the pregnant mouse uterus extract (BRADLEY et al 1971). The initial search for a convenient source of a hematopoietic growth factor which stimulated murine bone marrow cells in vitro was only partially successful: neither human urine extracts nor mouse L-cell-conditioned medium produced a hematopoietic growth factor to stimulate a significant proportion of the granulocyte progenitor cells. Eventually, it was discovered that when murine tissues were cultured for 12–48 h significant levels of growth factor were released into the medium (SHERIDAN and METCALF 1973). Mouse lung conditioned medium stimulated the formation of both granulocyte and macrophage colonies and one of the active components was purified and termed granulocyte-macrophage colony-stimulating factor (GM-CSF) (BURGESS et al. 1977a). Lymphocytes activated by a mitogen also produce hematopoietic growth factors capable of stimulating the production of both granulocytes and macrophages. The initial steps in the purification of

GM-CSF disguised the fact that the conditioned medium also contained another hematopoietic growth factor: granulocyte colony-stimulating factor (G-CSF). Murine GM-CSF was the first hematopoietic growth factor to be purified to homogeneity (BURGESS et al. 1977a, 1986) and its biological specificity has been studied in considerable detail (METCALF et al. 1986). Using the N-terminal amino acid sequence of murine GM-CSF (SPARROW et al. 1985) to predict oligonucleotide probes for screening cDNA libraries, it was possible to construct a full-length cDNA clone for murine GM-CSF (GOUGH et al. 1984). These clones defined the complete amino acid sequence for murine GM-CSF and by expressing this clone in COS cells it was possible to prove that this molecule was responsible for the biological activities ascribed to GM-CSF.

Murine GM-CSF does not stimulate human bone marrow progenitor cells. It proved to be a difficult process to locate an appropriate source of GM-CSF to stimulate human hematopoietic progenitor cells. Although monkey lungs produced some GM-CSF, there was always insufficient source material to attempt the complete purification of this GM-CSF. Human placental conditioned medium (BURGESS et al. 1977b) was a convenient source of both GM-CSF and G-CSF (NICOLA et al. 1979b), but whilst this was a useful source of these for many biological assays, homogeneous preparations of GM-CSF from human placental conditioned medium proved difficult to obtain. A human T-lymphoid cell line produced sufficient amounts of a human GM-CSF to allow its purification (GASSON et al. 1984). Molecular clones corresponding to human GM-CSF have been obtained (WONG et al. 1985a) and again expression of these clones in COS cells or bacteria produced a recombinant form of GM-CSF which had all the characteristics of human GM-CSF. Human GM-CSF does not stimulate murine bone marrow progenitor cells.

C. Biology

GM-CSF stimulates all cells in the granulocyte, macrophage, and eosinophil lineages. The earliest progenitor cells stimulated by murine or human GM-CSF are multipotential and capable of producing granulocytes, macrophages, megakaryocytes, erythroid cells, and eosinophils. Even at low concentrations (50 pM), recombinant human GM-CSF stimulates colonies from purified multipotential progenitor cells and in the presence of erythroprotein blast-forming units are also stimulated by GM-CSF (SIEFF et al. 1985). It is important to note that GM-CSF is required for the survival of hematopoietic progenitor cells (BURGESS et al. 1982), i.e., in the presence of sufficient quantities of GM-CSF (> 20 pM) the normal progenitor cells proliferate and differentiate and in the absence of GM-CSF the progenitor cells die (METCALF and MERCHAV 1982). GM-CSF also stimulates the final divisions of the most mature progenitor cells in these lineages. Interestingly, GM-CSF is required for all of the divisions of the early progenitor cells; the more mature progenitor cells only require GM-CSF for part of the last cell cycle. The differential responsiveness of immature and mature progenitor cells is illustrated by delivering GM-CSF continuously or as a 24-h pulse (BEGLEY et al. 1988). Immature blast cells and promyelocytes (the most mature progenitor cells still capable of proliferation) can be prepared from human bone marrow by density separation and cell sorting. The light density cells are separated on a sorter to select the cells with high zero-angle scatter and low right-angle scatter and an antibody, WEMG11, distinguishes between the blasts and promyelocytes. When either of these populations of cells is cultured in the absence of GM-CSF, there is no proliferation. In the presence of GM-CSF, almost 10% of the blast cells form clones in 7 days and 20% of the promyelocytes proliferate within 5 days. An initial 24-h pulse is not sufficient to induce blast cell proliferation, but when the promyelocytes are stimulated for 24 h the cloning efficiency is still 20% at 5 days. The more mature cells are able to traverse two or three rounds of proliferation after a GM-CSF pulse.

The proliferative actions of GM-CSF are not limited to granulocyte/macrophage, granulocyte, macrophage, or erythroid progenitor cells. The initial divisions of the erythroid and megakaryocyte progenitor cells are stimulated by GM-CSF (KANNOURAKIS and JOHNSON 1988), but mature cells in these lineages are only produced if either erythropoietin or a megakaryocyte stimulating factor are also included in the cultures (METCALF et al. 1980; MIGLIACCIO et al. 1987). In vitro GM-CSF alone only produces granulocytes, macrophages, and eosinophils. GM-CSF is not simply a proliferative stimulus for hematopoietic progenitor cells. GM-CSF can extend the lifespan and activate mature granulocytes (DI PERSIO et al. 1988 b), macrophages (CHEN et al. 1988), and eosinophils (BEGLEY et al. 1986; STANLEY and BURGESS 1983). Indeed, the activation of mature cells appears to be an important function of GM-CSF. In older patients the neutrophil counts are in the normal range, but the response to formylmethionylleucylproline (FMLP, a chemotactic peptide) is deficient – only 30% of the neutrophils depolarize in vitro. When these neutrophils are primed with GM-CSF 80% depolarize in response to FMLP (FLETCHER and GASSON 1988). The effects on neutrophils and eosinophils are similar: in the presence of chemotactic peptide GM-CSF primes superoxide production, phagocytosis, and lysozyme secretion (WEISBART et al. 1987; LINDEMANN et al. 1988). These effects would increase the effectiveness of each neutrophil. However, GM-CSF is also likely to cause the accumulation of neutrophils at sites of infection or blood vessel damage (WEISBART et al. 1985; WANG et al. 1987), thus accentuating the protective effects of the activated neutrophils. Finally, it appears that GM-CSF-treated neutrophils secrete interleukin-1, which is likely to recruit other cells capable of fighting infections (LINDEMANN et al. 1988).

Antibody-dependent cell-mediated cytotoxicity by neutrophils (LOPEZ et al. 1983; VADAS et al. 1983) and macrophages is stimulated by GM-CSF (GRABSTEIN et al. 1986). GM-CSF induces macrophages to secrete a number of biological response modifiers including tumor necrosis factor (CANNISTRA et al. 1987). Indeed it appears that the release of tumor necrosis factor is responsible for the tumoricidal activity of activated macrophages. When medium conditioned by macrophages is added to an antibody-dependent cytotoxicity assay only 20% of the target cells are killed; however, if the macrophages are treated with either GM-CSF or interleukin-3 the conditioned medium increases the killing to 60% (CANNISTRA et al. 1988). If antibodies which neutralize tumor necrosis factor are added to the assay the antibody-dependent cytotoxicity is eliminated.

It has been known for some time that macrophage killing is accentuated by interferon-γ, but this effect is dependent on the presence of low concentrations (0.1 ng/ml) of lipopolysaccharide (GRABSTEIN et al. 1986). No such accessory signals are required for the activation of macrophage tumoricidal activity. In vitro GM-CSF stimulates macrophages to kill intracellular parasites such as *Leishmania tropica* (HANDMAN and BURGESS 1979; WEISER et al. 1987) and *Trypanosoma cruzi* (REED et al. 1987). It is difficult to use the in vitro results to predict the effects of GM-CSF in vivo. Although there is killing of parasites with high concentrations of GM-CSF, similar concentrations of GM-CSF in vivo are likely to cause many other effects (e.g., immobilization and secretion of tumor necrosis factor) which may alter the physiological outcome. If GM-CSF causes more macrophages to accumulate at the site of a parasite invasion (e.g., *L. tropica*), the increased number of host cells needs to be balanced against the increased rate of killing.

GM-CSF was detected originally by its proliferative action on hematopoietic progenitor cells (PLUZNICK and SACHS 1965; BRADLEY and METCALF 1966). For almost two decades the only quantitative assay for GM-CSF required the use of bone marrow cells suspended in soft agar. The number, type, and size of colonies is proportional to the concentration of GM-CSF in the culture. At low concentrations (< 10 pM) GM-CSF stimulates preferentially the proliferation of macrophage progenitor cells. As the concentration of GM-CSF is increased the number of cells in these macrophage colonies increases and extra colonies appear. At concentrations of around 200 pM many of these extra colonies contain both macrophages and neutrophils, but there are also pure neutrophil colonies. Although the frequency of eosinophil progenitors is quite low, GM-CSF stimulates these colonies with a similar potency to most of the neutrophil precursors. At very high concentrations (> 10 nM) GM-CSF induces another class of neutrophil progenitors. These progenitors form large multicentric colonies, but it is not clear why such a high concentration of GM-CSF is required to stimulate these progenitors (METCALF 1984).

The bone assays were sufficient before the existence of the other hematopoietic growth factors had been established, but considerable care is required in purifying the GM-CSF before these cultures can be used quantitatively. Many unfractionated tissue extracts, conditioned media from cell lines, or sera from antigen-treated animals contain mixtures of GM-CSF, M-CSF, and/or G-CSF (BURGESS and METCALF 1980; JOHNSON et al. 1985). Whilst mixtures of G-CSF and M-CSF stimulate a similar spectrum of colonies to GM-CSF in the bone marrow assay, there may be no GM-CSF in the sample!

Indeed, human serum samples suspected of containing GM-CSF are particularly difficult to assay using bone marrow cells (FOSTER et al. 1968; MYERS and ROBINSON 1975; SHADDUCK and NAGABHUSHAN 1971). The colony types stimulated by human G-CSF, multi-CSF (interleukin-3), and GM-CSF are quite similar. Contaminating substances such as endotoxin in sera, urine, or tissue samples induce the production of G-CSF by human cells (GALELLI et al. 1985) which interfere with attempts to quantitate the CSFs using this bioassay.

Several murine cell lines (e.g., FDCP-1) have been developed which are dependent for their survival on interleukin-3 or GM-CSF (DEXTER et al. 1980). These cells can be used to quantitate murine GM-CSF in samples which are free of interleukin-3. When interleukin-3 is present this can be quantitated separately using a cell line which responds only to interleukin-3 (GREENBERGER et al. 1983) and the GM-CSF determined as the extra activity on FDCP-1 cells. These assays are capable of detecting murine GM-CSF in the range of 50 pM to 1 nM. The human cell lines dependent on human GM-CSF are not readily available (MORSTYN and BURGESS 1988). Although radioreceptor assays can be used to quantitate both murine GM-CSF (WALKER and BURGESS 1985) and human GM-CSF (GASSON et al. 1986), these assays require considerable amounts of radiolabeled GM-CSF and are effective only when the concentration of GM-CSF is greater than 200 pM. Whilst the GM-CSF receptor only binds the one molecule, i.e., GM-CSF, the GM-CSF receptor can be downregulated on bone marrow cells by multi-CSF (interleukin-3) and to a lesser extent by M-CSF (WALKER et al. 1985). This transmodulation is rapid at 37° C and even at room temperature considerable care is required to avoid false positives in a GM-CSF radioreceptor assay.

Monoclonal antibodies are now available which recognize human GM-CSF and a sensitive immunosorbent assay (ELISA) has been developed (CEBON et al. 1988). The GM-CSF ELISA can be used directly to quantitate GM-CSF levels in tissue extracts or serum down to 20 pM. The sensitivity and lack of crossreactivity with other growth factors or biological response modifiers make this monoclonal assay the preferred method for quantitating GM-CSF.

For many years it appeared that the effects of GM-CSF were restricted to hematopoietic cells; however, evidence is accumulating that some epithelial cells and bone cells also respond to GM-CSF. The precursors for osteoclasts are derived from the hematopoietic stem cells (ASH et al. 1980), so it is not all that surprising that GM-CSF stimulates the number of osteoclast-like cells formed in long-term bone marrow cultures (MACDONALD et al. 1986). GM-CSF synergizes with 1,25-dihydroxyvitamin D_3 to further increase the number of multinucleated osteoclasts.

D. Biochemistry

Murine and human GM-CSF have been purified to homogeneity (BURGESS et al. 1977a; SPARROW et al. 1985; GASSON et al. 1984), the mRNAs correspond-

```
                    10                    20                   30      33
Humanᵃ   A  P  A  R  S  P  S  P  S  T  Q  P  W  E  H  V  N  A  I  Q  E  A  R  R  L  L  N  L  S  R  D  T  A
Gibbonᵇ  A  P  S  R  S  P  S  P  S  R  Q  P  W  E  H  V  N  A  I  Q  E  A  R  R  L  L  N  L  S  R  D  T  A
Mouseᶜ   A  P  T  R  S  P  I  T  V  T  R  P  W  K  H  V  E  A  I  K  E  A  -  -  -  L  N  L  L  D  D  M  P

         34             40                   50                   60            66
Human    A  E  M  N  E  T  V  E  V  I  S  E  M  F  D  L  Q  E  P  T  C  L  Q  T  R  L  E  L  Y  K  Q  G  L
Gibbon   A  E  I  N  E  T  V  E  V  V  S  E  M  F  D  L  Q  E  P  T  C  L  Q  T  R  L  E  L  Y  K  Q  G  L
Mouse    V  T  L  N  E  E  V  E  V  V  S  N  E  F  S  F  K  K  L  T  C  V  Q  T  R  L  K  I  F  E  Q  G  L

         67       70                   80                    90                   99
Human    R  G  S  L  T  K  L  K  G  P  L  T  M  M  M  A  S  H  Y  K  Q  H  C  P  P  P  T  P  E  T  S  C  A  T  Q
Gibbon   R  G  S  L  T  K  L  K  G  P  L  T  M  M  M  A  S  H  Y  K  Q  H  C  P  P  P  T  P  E  T  S  C  A  T  Q
Mouse    R  G  N  F  T  K  L  K  G  A  L  N  M  T  A  S  Y  Y  Q  T  Y  C  P  P  T  P  E  T  D  C  E  T  Q

         100                  110                  120              127
Human    T  I  T  F  E  S  F  K  E  N  L  K  D  F  L  L  V  I  P  F  D  C  W  E  P  V  Q  E
Gibbon   I  I  T  F  E  S  F  K  E  N  L  K  D  F  L  L  V  T  P  F  D  C  W  E  P  V  Q  G
Mouse    V  T  T  Y  A  D  F  I  D  S  L  K  T  F  L  T  D  I  P  F  E  C  K  K  P  V  Q  K
```

Fig. 1. Amino acid sequence homology between GM-CSFs from three different mammals: [a]Wong et al. (1985a); [b]Wong et al. (1985b); [c]Gough (1985)

ing to both have been molecularly cloned (Gough et al. 1984, 1985; Wong et al. 1985a; Lee et al. 1985), and both have been expressed as biologically active molecules in *E. coli* (DeLamarter et al. 1985; Libby et al. 1987; Burgess et al. 1987). The natural GM-CSFs and recombinant GM-CSFs have identical activities in the bone marrow factor-dependent cell line, neutrophil activation, and ELISA assays. The full amino acid sequences for murine (Gough et al. 1985), gibbon (Wong et al. 1985b), and human GM-CSF (Wong et al. 1985a; Lee et al. 1985) have been determined by extrapolation from the sequences of the corresponding mRNAs (see Fig.1). Although there are few regions of complete identity, the disulphide bonded cysteines are conserved (Shrimser et al. 1987) and there are two regions where five consecutive residues are identical in all these species. The most conserved stretch is a proline-rich sequence between cysteines 88 and 96. Interestingly, this region is part of the epitope for a monoclonal antibody which is capable of neutralizing the activity of GM-CSF. The leader sequences for both mouse and human GM-CSF suggest the possibility of a transmembrane form of the molecule (Gough et al. 1985), but as yet only secreted and cytoplasmic forms of GM-CSF have been detected. Several monoclonal anti-GM-CSF antibodies have been prepared and characterized based on Western blot analysis of T lymphocytes synthesizing GM-CSF. Although the secreted forms have molecular weights between 14000 and 28000, high-mannose forms of pro-GM-CSF were detected in the cytoplasm. The GM-CSF secreted into mouse lung conditioned medium is proteolytically cleaved to remove both the signal peptide and the next six amino acids (Sparrow et al. 1985). The GM-CSF secreted by the mouse lymphoid cell lines LB-3 is processed without further cleavage and thus the amino acid sequence starts at Ala 30 of the precursor sequence (Burgess and Nice 1985). Several modifications at the N terminus have been detected, but no effect on the biological activity is apparent. Only limited information is available concerning the conformation of GM-CSF or the amino acid residues which are critical for its function. Predictive algorithms suggest that more than 50% of the residues of GM-CSF are bundled into α helices (PARRY et al. 1988). Circular dichroism analysis supports this prediction (47% α helix), but also indicates the presence of a β sheet structure (WINGFIELD et al. 1988). Although the first six N-terminal residues are not required for the activity of murine GM-CSF, there is some evidence to suggest that the amino acid residues in human GM-CSF between 14 and 25 are important for its activity (CLARK-LEWIS et al. 1988). When residues 122-127 are omitted from the synthesis of human GM-CSF there is no signifiant diminution of its biological activity. The minimal fragment for full biological activity of human GM-CSF would appear to be 14-122. If GM-CSF is reduced and carboxymethylated it loses all biological activity – both disulfide bonds need to be intact to permit receptor binding. Several monoclonal antibodies have been produced which neutralize the activity of human GM-CSF. The epitope for one of these antibodies recognizes the amino acids associated with disulfide bonds from residues 88 to residue 121 (NICE et al., in preparation). Presumably this region of the molecule is involved in receptor recognition.

Since native GM-CSF, bacterially synthesized GM-CSF, and recombinant GM-CSF secreted from yeast all have identical biological activities in vitro and in vivo, the carbohydrate moieties cannot be important for receptor binding, signal processing, or pharmacological clearance. Bacterially derived GM-CSF is cleared from the serum of mice with an initial half-life of 6–10 min (METCALF et al. 1987 a). Human GM-CSF is cleared from patients sera in two phases: a rapid 8- to 10-min phase and a longer phase (approximately 90 min; LIESCHKE et al. 1989). Most organs and tissues produce GM-CSF. A superficial analysis of CSFs from different sources shows forms with molecular weights ranging from 23 000 to more than 200 000. However, under dissociating conditions, after removal of sialic acid, a single derivative of 23 000 daltons is found (NICOLA et al. 1979 a).

Although GM-CSF appears to be made by a range of cell types (monocytes, endothelial cells, T lymphocytes, and fibroblasts) no GM-CSF appears to be stored in these cells. There are several T-lymphoblastoid cell lines which can be induced to express GM-CSF (KELSO and GOUGH 1987). Induction of GM-CSF secretion by T-lymphocytes can be affected by antigen, antibodies against the T-cell receptor, lectins such as phytohemagglutinin (LU et al. 1988), or interleukin-1 (HERMANN et al. 1988). In all cases these stimuli appear to act by causing the release of intracellular Ca^{2+} and the subsequent activation of protein kinase C (KOZUMBO et al. 1987). The induction of *transcription* of GM-CSF mRNA in T lymphocytes by lectins does *not* require protein synthesis.

Other cells can also be stimulated to produce GM-CSF: Mitogen-stimulated B lymphocytes (MERCHAV et al. 1987), macrophages stimulated with bacterial lypopolysaccharide (THORENS et al. 1987), endothelium (MALONE et al. 1988; SEELENTAG et al. 1987; SIEFF et al. 1987), and even fibroblasts (KOURY and PRAGNELL 1982; KOURY et al. 1983). Interactions between these cell populations also induce the production of GM-CSF – a process which may be particularly important during tissue remodeling or inflammation.

GM-CSF is encoded by a single gene in mouse (GOUGH et al. 1984) and human (MIYATAKE et al. 1985), and even when low-stringency hybridization conditions are used, only a single gene is detected by Southern analysis. The gene is on chromosome 11 in the mouse (GOUGH et al. 1984) while the human GM-CSF gene is located between bands G21 and G31 on chromosome 5. Interestingly, this region of chromosome 5 also encodes the M-CSF, interleukin-3, interleukin-4, interleukin-5, and the endothelial cell growth factor genes. GM-CSF and interleukin-3 are linked closely in both the human and mouse genomes (BARLOW et al. 1987; YANG et al. 1988). Despite this close linkage these genes are not always regulated coordinately.

A single mRNA is produced from the four exons of the GM-CSF gene. Two slightly different transcriptional start sites have been proposed (MIYATAKE et al. 1985; STANLEY et al. 1985): 32 or 35 bases before the first methionine residue of the murine GM-CSF prohormone. At the 3′ end of the mRNA there are over 300 untranslated bases – this region of the mRNA contains a $(U_N A)$ motif which ensures a rapid turnover (SHAW and KAMEN 1986). The human mRNA also has 5′ and 3′ extensions (MIYATAKE et al. 1985; CHAN et al. 1986). The most conserved nucleotide sequence between the murine and human GM-CSF mRNAs occurs in the 5′ untranslated region proposed to control the breakdown of the mRNA (Fig. 2).

The sequences important for the control of GM-CSF expression have been difficult to define. Initial reports indicated that two heptameric sequences 5′ of the murine GM-CSF coding sequence appeared to be involved (SHANNON et al. 1988). These two regions interact with nuclear protein extracts in a gel retardation assay (SHANNON et al. 1988). However, DNAase I footprinting has implicated another controlling region for the human GM-CSF gene (NIMER et al. 1988). This last region is only 30–40 nucleotides from the start of

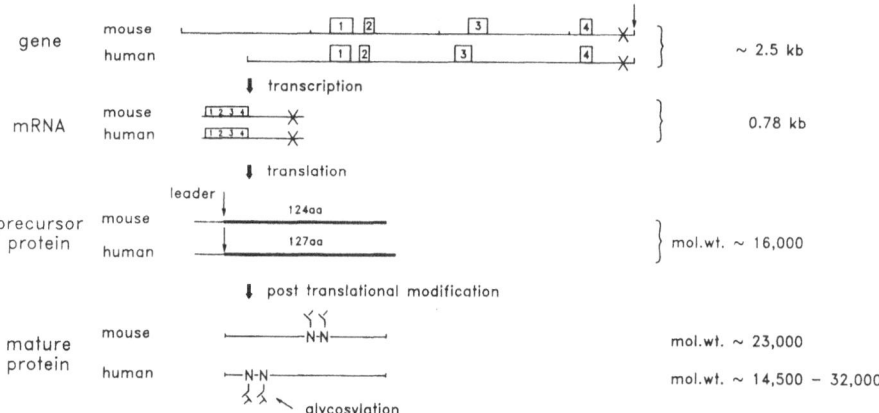

Fig. 2. The mouse and human GM-CSF genes both consist of four exons (*boxes*) and a long 3′ untranslated region containing a sequence (*X*) which controls the breakdown of the mRNA (SHAW and KAMEN 1986). The leader sequence is removed during the secretion process producing different glycosylated forms of GM-CSF

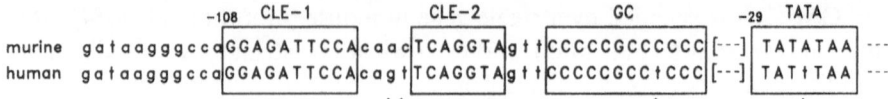

Fig. 3. The promoter regions of the human and murine GM-CSF genes contain a number of controlling elements including the two conserved lymphokine elements (*CLE*), the GC-rich box which contains an "SP1"-like element, and the TATA initiation sequence (MIYATAKE et al. 1988)

the coding region and actually encompasses part of the TATA box (Fig. 3). More recently, there has been a careful study of the action of the $p40^x$ protein (which is encoded by the human T-cell leukemia virus type I) on the GM-CSF gene (MIYATAKE et al. 1988). The two conserved lymphokine elements (CLE-1 and CLE-2), as well as a GC-rich box, all appeared to be important sites for signals which activate the GM-CSF gene.

GM-CSF acts on its target cells via cell surface receptors (WALKER and BURGESS 1985). Although there is some confusion about the molecular properties of this receptor, it is generally agreed that most cells have less than 5000 receptors per cell. Radioiodinated murine GM-CSF binds to cells in bone marrow, spleen, and the peritoneum (WALKER and BURGESS 1985) as well as myeloid cell lines (WALKER and BURGESS 1985; PARK et al. 1986a). Depending on the quality of the ^{125}I-labeled GM-CSF, a small number of high-affinity receptors ($K_d \sim 20$ pM) and about 10 times as many low-affinity receptors ($K_d \sim 1$ nM) can be detected. Using disuccimidyl suberimidate to crosslink the radiolabeled GM-CSF to its receptor on *bone marrow cells*, a single species with a molecular weight of about 50 000 was observed (WALKER and BURGESS 1985; PARK et al. 1986a). However, on some cell lines the murine GM-CSF receptor appears to have a molecular weight of 130 000 (PARK et al. 1986a). It has been suggested that the smaller molecular weight receptor detected on bone marrow cells is a degradation product, but more experiments are required to unravel this discrepancy. Human ^{125}I-labeled GM-CSF binds specifically to normal myeloid cells, myeloid cell lines, and myeloid leukemias (GASSON et al. 1986; PARK et al. 1986b; KELLEHER et al. 1988). Neutrophils, eosinophils, and monocytes all bind ^{125}I-labeled GM-CSF, but erythroid cells, lymphocytes, and mast cells do not appear to have receptors. Again, there appear to be both high- (K_d 20 pM) and low-affinity ($K_d \sim 1$ nM) receptors (N. A. NICOLA, personal communication), but considerable care is required in preparing the ^{125}I-labeled GM-CSF if the high-affinity receptors are to be detected. The human GM-CSF receptor on the myeloid leukemic cell line HL-60 has an apparent molecular weight of 84 000 (DI PERSIO et al. 1988a). Although the only known ligand for the GM-CSF receptor is GM-CSF, the display of both the murine and human GM-CSF receptors can be modulated by other HGFs. When bone marrow cells are treated with interleukin-3 at 37° C, there is a rapid transmodulation of the high-affinity GM-CSF receptors (WALKER et al. 1985; NICOLA 1987). Although the functional significance of this transmodulation is not known, there is a good correlation between the biological responses of different hematopoietic growth factors and their transmodulatory effects (NICOLA 1987).

E. Pharmacological Studies

Bacterially synthesized murine GM-CSF has been administered to mice both i.p. and i.v. The short half-life of GM-CSF in vivo (<5 min) requires that several i.p. injections of GM-CSF be administered to modulate hematopoiesis (METCALF et al. 1987a). At daily doses of 6-200 ng GM-CSF per mouse there was a significant increase of monocytes, neutrophils, and eosinophils in the peritoneal cavity. The number of circulating and peritoneal lymphocytes decreased in the mice receiving GM-CSF. The macrophages which accumulated in the peritoneum as a result of the GM-CSF injections were more active phagocytes than peritoneal macrophages from untreated mice (METCALF et al. 1987a; MORRISEY et al. 1988). Careful analysis of the liver and lung revealed elevated numbers of neutrophils and eosinophils in the GM-CSF-treated mice. Although the number of monocytes in the peritoneum can be as much as 100 times the number found in the peritoneum of normal mice no mitotic figures were observed in the monocyte population. Thus, the monocytes appear to be synthesized in the bone marrow or spleen and accumulate in the peritoneum. It is possible that the i.p. injection of GM-CSF immobilizes trafficking neutrophils, eosinophils, and monocytes as well as increasing the tissue half-life of these mature cells.

Continuous infusion of human GM-CSF into monkeys elevates the levels of white blood cells (DONAHUE et al. 1987; MAYERS et al. 1987; KRUMWIEH et al. 1988). Although some of the GM-CSF-treated monkeys also developed a reticulocytosis (DONAHUE et al. 1987), this was not a general finding in other studies (KRUMWIEH et al. 1988). Antibodies against human GM-CSF developed during the course of injection, but these antibodies did not interfere with the hematopoietic responses. After priming monkeys with recombinant human interleukin-3 for 5 days, the responses to recombinant human GM-CSF were considerably greater (KRUMWIEH et al. 1988). More than twice the number of white blood cells were produced compared to GM-CSF alone. The increase was mainly in neutrophils and eosinophils. There was no apparent effect on platelet numbers in the monkeys treated with either interleukin-3 or GM-CSF alone. However, in the monkeys treated sequentially with interleukin-3 and then GM-CSF platelet levels were elevated 3-fold (KRUMWIEH et al. 1988).

Synergistic effects of interleukin-3 and GM-CSF on hematopoiesis in mice have also been reported (BROXMEYER et al. 1987). In these experiments, hematopoiesis was suppressed by pretreatment with lactoferrin and the administration of a mixture of interleukin-3 and GM-CSF increased the number of hematopoietic progenitor cells. GM-CSF is not a radioprotective agent by itself but in combination with interleukin-1 or tumor necrosis factor, GM-

CSF accelerates hematopoietic recovery to such an extent that it confers increased protection against radiation (Neta et al. 1988). Furthermore, GM-CSF improves bone marrow engraftment after irradiation of mice (Blazar et al. 1988) and reduces the period of neutropenia in irradiated and bone marrow-transplanted monkeys (Monroy et al. 1987; Neinhuis et al. 1987).

Chronic exposure of mice to GM-CSF has been achieved by creating mice bearing a GM-CSF transgene (Lang et al. 1987) or by transplanting mice with bone marrow previously infected by a virus encoding GM-CSF (Johnson et al. 1989). This transgene was driven by a Moloney long terminal repeat (LTR) promoter and marked with φ_x DNA to facilitate the detection of mRNA expressed from the transgene. GM-CSF transgenic mice have opaque eyes. During the first few days of neonatal life the eye is normally filled with hyalocytes (a "macrophage-like" cell) which phagocytose the structural network, leaving the vitreous. Eight days after birth these hyalocytes usually die, leaving a clear, almost acellular, vitreous humor. In the GM-CSF transgenic mice the hyalocytes persist, destroying the retinal layers, lining the lens, and filling the vitreous humor, thus causing a distinct opacity in the eye (Lang et al. 1987; Cuthbertson and Lang 1989).

The GM-CSF transgenic mice appear to develop normally for the first 4–5 weeks postnatally, but after that time many of the mice develop tremors and some begin to die. Although the peripheral blood white cell counts were close to normal, autopsy revealed large accumulations of macrophages in the peritoneum and infiltration of the muscles by activated macrophages. Using the φ_x probe it was possible to demonstrate the presence of the transgenic GM-CSF mRNA in these activated macrophages. Interestingly, there was no detectable GM-CSF mRNA from the transgene in the bone marrow of the GM-CSF transgenic mice (even the bone marrow monocytes were devoid of GM-CSF mRNA). It appears that the expression of the transgene only occurs in tissue macrophages. The peritoneal macrophages in the older GM-CSF transgenic mice were active phagocytes and in contrast to macrophages from normal mice produced significant levels of mRNA for tumor necrosis factor (Lang 1989). The increased phagocytosis and the cytotoxicity of tumor necrosis factor from the macrophages might be expected to account for most of the tissue destruction and consequently the pathology observed in the GM-CSF transgenic mice.

Using the retrovirus which encodes GM-CSF it is possible to induce constitutive production of GM-CSF in a range of cell types (Lang et al. 1985). Indeed, it is possible to infect murine bone marrow or fetal liver with the GM-CSF virus to induce endogenous production of the growth factor. When the infected bone marrow producing GM-CSF is transplanted into syngeneic, irradiated recipients, the mice develop high levels of GM-CSF in their sera. There is a large increase in peripheral blood neutrophils, eosinophils, and monocytes, as well as an accumulation of macrophages in the peritoneum, liver, lungs, and other organs. These macrophages appear to be activated phagocytically and cause tissue destruction in the same way as the transgenic macrophages.

F. GM-CSF and Leukemia

A feature of acute myeloid leukemia is the block to the normal maturation of the blast cells (BENNETT et al. 1976). This block may be due to a disturbance in the production of autocrine growth factors and/or to the biochemical pathways associated with the induction of maturation. At present the evidence for the existence of functional autocrine inducers of normal maturation is rather circumstantial, so it is not surprising that the molecular nature of the lesions inhibiting the differentiation responses are still unclear. In some cases, normal differentiation and maturation can be achieved by leukemic cells, but for most leukemias chemotherapy appears to eradicate the leukemic clone (FEARON et al. 1986).

Bone marrow cells have been transformed to leukemic cells in a number of ways, for example, split-dose X-irradiation or retroviral transformation. When murine bone marrow or fetal liver cells are infected with the GM-CSF virus, some autocrine-driven proliferation and maturation occurs, but there is no evidence of either immortalization or transformation (JOHNSON et al. 1989). Immortalized myeloid cell lines have been formed from long-term cultures of murine bone marrow (GREENBERGER et al. 1983; DEXTER et al. 1980). In particular, the factor-dependent cell lines 32D (GREENBERGER et al. 1983) and FDCP-1 (DEXTER et al. 1980) have been derived. Neither of these cell lines is leukemogenic when injected into syngeneic mice, but in vitro in the presence of interleukin-3 or, in the case of FDCP-1 cells, in the presence of either interleukin-3 or GM-CSF, these cells grow immortally. These factor-dependent cell lines are blocked along the maturation pathway, but no growth has been detected in vivo. Factor-dependent cells can be converted to a factor-independent phenotype (which are almost invariably tumorigenic) by spontaneous transformation (SCHRADER and CRAPPER 1983), the Abelson leukemia virus (COOK et al. 1985; PIERCE et al. 1985), or the retrovirus encoding and expressing GM-CSF (LANG et al. 1985). It has been determined that many of the spontaneously transformed FDCP-1 cells have mutations leading to the autocrine production of GM-CSF (SCHRADER and CRAPPER 1983). However, the v-*abl* tyrosine kinase appears to short-circuit the requirement for activation of the GM-CSF receptor.

Although these factor-independent murine leukemia models are interesting in their own right, they are not ideal models for the study of leukemogenesis. At the time of diagnosis most myeloid leukemias require exogenous GM-CSF to proliferate (MOORE et al. 1973), i.e., they are not autonomous (MIYAUCHI et al. 1988). In some myeloid leukemias, the blast cells will not proliferate in vitro even in the presence of an exogenous source of the appropriate GM-CSF.

It *is* possible to convert *non*-leukemic, factor-dependent murine cells to leukemic, factor-dependent murine cells by infecting the cells with a retrovirus which encodes and expresses the polyoma middle-T antigen (METCALF et al. 1987b). Some of the middle-T transformants are clearly factor-independent leukemias, some appear to be "paracrine" leukemias (the cells capable of feeding each other), and others whilst being leukemic are still factor dependent in vitro. This latter class of polyoma middle-T-transformed factor-dependent cells appears to resemble the phenotype of many human myeloid leukemias.

Why does the middle-T antigen allow these cells to grown in vivo? Perhaps there is low-level production of GM-CSF (compare with the effects of *src*-related on chicken myeloid cells; ADKINS et al. 1984) and that is sufficient growth factor to synergize with another factor(s) in vivo. Perhaps in vivo there are other growth factors for early hematopoietic cells and the middle-T antigen induces the factor-dependent cells into a responsive state. It should be noted that many leukemic cells in blast crisis proliferate in vivo, but even with GM-CSF some of these cells will not grow in vitro, suggesting that there are still some hematopoietic growth factors to be identified.

Several murine leukemic cell lines can be induced to differentiate in vitro (METCALF 1980; ICHIKAWA 1969). Although G-CSF, interleukin-6, and leukemia inhibitory factor are considered to be the most potent inducers of differentiation, GM-CSF also induces differentiation of several such cell lines (METCALF 1980; LOTEM and SACHS 1988). Indeed, in vivo GM-CSF and interleukin-3 appear to be the most effective differentiation inducers for one particular murine leukemia (LOTEM and SACHS 1988).

Several studies have been published on the frequency of GM-CSF gene translocations and mRNA expression by leukemic cells (OSTER et al. 1988; YOUNG et al. 1987; HUEBNER et al. 1985; CHENG et al. 1988). Some studies indicated that more than 40% of myeloid leukemias had abnormally arranged GM-CSF genes or overexpression of GM-CSF mRNA. However, it is still not clear whether autocrine production of GM-CSF is an important feature of human myeloid leukemia, as almost all primary leukemias require exogenous GM-CSF for growth in vitro.

G. Clinical Studies with GM-CSF

Sufficient human GM-CSF has been produced from chinese hamster ovary cells, yeast, and *E. coli* for phase I/II clinical trials (VADHAN-RAJ et al. 1987; MERTELSMANN and COSMAN 1988; LIESCHKE et al. 1989). All three forms of GM-CSF are biologically active and the presence or absence of the carbohydrate moiety appears to make remarkably little difference to its action or pharmacokinetics (CEBON et al. 1989). A wide variety of patients have been used in these initial trials: acquired immunodeficiency syndrome (GROOPMAN et al. 1987), myelodysplastic syndrome (VADHAN-RAJ et al. 1987; LIESCHKE et al. 1989; SOCINSKI et al. 1988; DUHRSEN et al. 1989), and bone marrow transplantation patients (BRANDT et al. 1988). In all of these patients GM-CSF causes an increase in the number of white blood cells. At low doses of GM-CSF the rise is due usually to increased numbers of neutrophils, but at higher doses the monocytes and eosinophils also increase (LIESCHKE et al. 1989). The optimal route of administration has not been defined as yet. However, pharmacokinetic studies revealed that after a subcutaneous single dose of GM-CSF (10 µg/kg) concentrations of greater than 1 ng/ml were sustained in the serum for more than 12 h (CEBON et al. 1989). Subcutaneous administration of GM-CSF has been associated with local rashes in some patients.

While there has been one report that the administration of GM-CSF helped to resolve neutropenia for some patients with myelodysplasia (VADHAN-RAJ et al. 1987), in other reports GM-CSF treatment appears to cause a worsening of the leukemic state (MERTELSMANN and COSMAN 1987). At doses of GM-CSF above 15 µg/kg, significant toxicities (e.g., bone pain, fever, edema, rashes, and serositides including pericarditis) (LIESCHKE et al. 1989).

Since GM-CSF can activate macrophages, it has been suggested that GM-CSF could induce the killing of tumor cells by promoting antibody-dependent cytotoxicity (VADAS et al. 1983). Although systemic administration of GM-CSF may not be effective as an anticancer treatment, intraperitoneal infusion of GM-CSF together with monoclonal antibodies may be able to stimulate an antibody-mediated cytotoxic effect on ovarian cancers. GM-CSF accelerates the recovery of bone marrow cellularity in transplant patients (BRANDT et al. 1988) and patients on cytotoxic therapy and there are some indications that GM-CSF might protect patients with acquired immunodeficiency syndrome (GROOPMAN et al. 1987). The cellular specificity of the hematopoietic growth factors makes them potentially powerful modulators of blood cell production. Hopefully, the GM-CSF protection can be combined with new or even current therapies to help cancer patients. If dose schedules for drugs such as adriamycin can be increased significantly by using GM-CSF to rescue bone marrow function, there is a real possibility of the development of improved treatments for diseases such as breast cancer.

Twenty years ago, GM-CSF was simply a curious diversion for experimental hemotologists; today it is being used in research clinics for the treatment of cancer patients. An excellent collection of small papers summarizing the present status of the chemistry, biochemistry, biology, and clinical areas of the CSFs has been published (SEILER et al. 1988). The progress in our understanding of this molecule is a remarkable tribute to the industry, creativity, and cooperation of medical scientists.

References

Adkins B, Leutz A, Graf T (1984) Autocrine growth induced by src-related oncogenes in transformed chicken myeloid cells. Cell 39:439–445

Arai K, Yokota T, Miyajima A, Arai N, Lee F (1986) Molecule biology of T-cell derived lymphokines: a model system for proliferation and differentiation of hemopoietic cells. Bioessays 5:166–171

Ash P, Loutit JF, Townsend KMS (1980) Osteoclasts derived from hemopoietic stem cells. Nature 283:669–670

Barlow DP, Bucan M, Lehrach H, Hogan BLM, Gough NM (1987) Close genetic and physical linkage between the murine hemopoietic growth factor genes GM-CSF and multi-CSF (IL-3). EMBO J 6:617–624

Begley CG, Lopez AF, Nicola NA, Warren DJ, Vadas MA, Sanderson CJ, Metcalf D (1986) Purified colony-stimulating factors enhance the survival of human neutrophils and eosinophils in vitro: a rapid and sensitive microassay for colony-stimulating factors. Blood 68:162–166

Begley CG, Nicola NA, Metcalf D (1988) Proliferation of normal human promyelocytes and myelocytes after a single pulse stimulation by purified GM-CSF or G-CSF. Blood 71:640–645

Bennett JM, Catovsky D, Daniel MT, Flandrin G, Galton DAG, Gralnick HR, Sultan C (1976) Proposals for the classification of the acute leukemias: French-American-British (FAB) Cooperative Group. Br J Haematol 33:451–458

Blazar BR, Widmer MB, Soderling CC, Urdal DL, Gillis S, Robison LL, Vallera DA (1988) Augmentation of donor bone marrow engraftment in histoincompatible murine recipients by granulocyte/macrophage colony-stimulating factor. Blood 71:320–328

Bradley TR, Metcalf D (1966) The growth of mouse bone marrow cells in vitro. Aust J Exp Biol Med Sci 44:287–300

Bradley TR, Stanley ER, Sumner MA (1971) Factors from mouse tissues stimulating colony growth of mouse bone marrow cells in vitro. Aust J Exp Biol Med Sci 49:595–603

Brandt SJ, Peters WP, Atwater SK, Kurtzberg J, Borowitz MJ, Jones RB, Shpall EJ, Bast RC Jr, Gilbert CJ, Oette DH (1988) Effect of recombinant human granulocyte-macrophage colony-stimulating factor on hematopoietic reconstitution after high-dose chemotherapy and autologous bone marrow transplantation. N Engl J Med 318:869–876

Broxmeyer HE, Williams DE, Hangoc G, Cooper S, Gillis S, Shadduck RK, Bicknell DC (1987) Synergistic myelopoietic actions in vivo after adminsitration to mice of combinations of purified natural murine colony-stimulating factor-1, recombinant murine interleukin-3, and recombinant murine granulocyte/macrophage colony-stimulating factor. Proc Natl Acad Sci USA 84:3871–3875

Burgess AW, Metcalf D (1980) Characterization of a serum factor stimulating the differentiation of myelomonocytic leukemic cells. Int J Cancer 26:647–654

Burgess AW, Nice EC (1985) Murine granulocyte-macrophage colony-stimulating factor. Methods Enzymol 116:588–600

Burgess AW, Camakaris J, Metcalf D (1977a) Purification and properties of colony stimulating factor from mouse lung-conditioned medium. J Biol Chem 252:1978–2003

Burgess AW, Wilson EMA, Metcalf D (1977b) Stimulation by human placental conditioned medium of hemopoietic colony formation by human bone marrow cells. Blood 49:573–583

Burgess AW, Metcalf D, Russell SHM, Nicola NA (1980) Granulocyte-macrophage, megakaryocyte, eosinophil and erythroid colony-stimulating factors produced by mouse spleen cells. Biochem J 185:301–314

Burgess AW, Nicola NA, Johnson GR, Nice EC (1982) Colony-forming cell proliferation: a rapid and sensitive assay system for murine granulocyte and macrophage colony-stimulating factors. Blood 60:1219–1223

Burgess AW, Metcalf D, Sparrow LG, Simpson RJ, Nice EC (1986) Granulocyte/macrophage colony-stimulating factor from mouse lung conditioned medium. Biochem J 235:805–814

Burgess AW, Begley CG, Johnson GR, Lopez AF, Williamson DJ, Mermod JJ, Simpson RJ, Schmitz A, DeLamarter JF (1987) Purification and properties of bacterially synthesized human granulocyte-macrophage colony stimulating factor. Blood 69:43–51

Cannistra SA, Ramboldi A, Spriggs DR, Herrmann F, Kufe D, Griffin JD (1987) Human granulocyte-macrophage colony-stimulating factor induces expression of the tumor necrosis factor gene by the U937 cell line and by normal human monocytes. J Clin Invest 99:1720–1728

Cannistra SA, Vellenga E, Groshek P, Rambaldi A, Griffin JD (1988) Human granulocyte-monocyte colony-stimulating factor and interleukin 3 stimulate monocyte cytotoxicity through a tumor necrosis factor-dependent mechanism. Blood 71:672–676

Cebon J, Dempsey P, Fox R, Kannourakis G, Bonnem E, Burgess AW, Morstyn G (1988) Pharmacokinetics of human granulocyte-macrophage colony stimulating factor (hGM-CSF) using a sensitive immunoassay. Blood 72:1340–1347

Chan JY, Slamon DJ, Nimer SD, Golde DW, Gasson JC (1986) Regulation of expression of human granulocyte/macrophage colony-stimulating factor. Proc Natl Acad Sci USA 83:8669–8673

Chen BD-M, Mueller M, Chon T-H (1988) Role of granulocyte/macrophage colony-stimulating factor in the regulation of murine alveolar macrophage proliferation and differentiation. J Immunol 141:139–144

Cheng GYM, Kelleher CA, Miyauchi J, Wang C, Wong G, Clark SC, McCulloch EA, Minden MD (1988) Structure and expression of genes of GM-CSF and G-CSF in blast cells from patients with acute myeloblastic leukemia. Blood 71:204–208

Clark-Lewis I, Lopez AF, To LB, Vadas MA, Schrader JW, Hood LE, Kent SBH (1988) Structure-function studies of human granulocyte-macrophage colony-stimulating factor. J Immunol 141:881–889

Cook WD, Metcalf D, Nicola NA, Burgess AW, Walker F (1985) Malignant transformation of a growth factor-dependent myeloid cell line by Abelson virus without evidence of an autocrine mechanism. Cell 41:677–683

Cuthbertson RA, Lang RA (1989) Developmental ocular disease in GM-CSF transgenic mice is mediated by autostimulated macrophages. Dev Biol 134:119–129

DeLamarter JF, Mermod JJ, Liang CM, Eliason JF, Thatcher DR (1985) Recombinant murine GM-CSF from E. coli has biological activity and is neutralized by a specific anti-serum. EMBO J 4:2575–2581

Dexter TM, Garland J, Scott D, Scolnick E, Metcalf D (1980) Growth of factor-dependent hemopoietic precursor cell lines. J Exp Med 152:1036–1047

Di Persio J, Billing P, Kaufman S, Eghtesady P, Williams RE, Gasson JC (1988a) Characterization of the human granulocyte-macrophage colony-stimulating factor receptor. J Biol Chem 263:1834–1841

Di Persio JF, Billing P, Williams R, Gasson JC (1988b) Human granulocyte-macrophage colony-stimulating factor and other cytokines prime human neutrophils for enhanced arachidonic acid release and leukotriene B4 synthesis. J Immunol 140:4315–4322

Donahue RE, Wang EA, Stone DK, Kamen R, Wong GG, Sehgal PK, Nathan DG, Clark SC (1986) Stimulation of haematopoiesis in primates by continuous infusion of recombinant human GM-CSF. Nature 321:827–875

Duhrsen U, Villeval J-L, Boyd J, Kannourakis G, Morstyn G, Metcalf D (1988) Effects of recombinant human granulocyte-colony stimulating factor on hematopoietic progenitor cells in cancer patients. Blood 72:2074–2081

Fearon ER, Burke PJ, Schiffer CA, Zehnbauer BA, Vogelstein B (1986) Differentiation of leukemia cells to polymorphonuclear leukocytes in patients with acute non-lymphocytic leukemia. N Engl J Med 315:15–24

Fletcher MP, Gasson JC (1988) Enhancement of neutrophil function by granulocyte-macrophage colony-stimulating factor involves recruitment of a less responsive subpopulation. Blood 71:652–658

Foster R, Metcalf D, Robinson WA, Bradley TR (1968) Bone marrow colony stimulating activity in human sera. Br J Haematol 15:147–159

Galelli A, Dosne A, Morin A, Dubor F, Chedid L (1985) Stimulation of human endothelial cells by synthetic muramyl peptides: production of colony-stimulating activity (CSA). Exp Hematol 13:1157–1163

Gasson JC, Weisbart RH, Kaufman SE, et al. (1984) Purified human granulocyte-macrophage colony-stimulating factor: direct action on neutrophils. Science 226:1339–1342

Gasson JC, Kaufman SE, Weisbart RH, Tomonaga M, Golde DW (1986) High affinity binding of granulocyte-macrophage colony-stimulating factor to normal and leukemic human myeloid cells. Proc Natl Acad Sci USA 83:669–673

Gough NM, Gough J, Metcalf D, Kelso A, Grail D, Nicola NA, Burgess AW, Dunn AR (1984) Molecular cloning of cDNA encoding a murine haematopoietic growth regulator, granulocyte-macrophage colony stimulating factor. Nature 309:763–767

Gough NM, Metcalf D, Gough J, Grail D, Dunn AR (1985) Structure and expression of the mRNA for murine granulocyte-macrophage colony stimulating factor. EMBO J 4:645–653

Grabstein KH, Urdal DL, Tushinski RJ, Mochizuki DY, Price VL, Cantrell MA, Gillis S, Conlon PJ (1986) Induction of macrophage tumoricidal activity by granulocyte-macrophage colony-stimulating factor. Science 232:506–508

Greenberger JS, Sakakeeny MA, Humphries RK, Eaves CJ, Ecker RJ (1983) Demonstration of permanent factor-dependent multi-potential (erythroid/neutrophil/basophil) hematopoietic progenitor cell lines. Proc Natl Acad Sci USA 80:2931–2935

Groopman JE, Mitsuyasu RT, DeLeo MJ, Oette DH, Golde D (1987) Effect of recombinant human granulocyte-macrophage colony-stimulating factor on myelopoiesis in the acquired immunodeficiency syndrome. N Engl J Med 317:593–598

Handman E, Burgess AW (1979) Stimulation by granulocyte-macrophage colony stimulating factor of *Leishmania tropica* killing by macrophages. J Immunol 122:1134–1137

Hermann F, Oster W, Meuer SC, Lindermann A, Mertelsmann RH (1988) Interleukin 1 stimulates T lymphocytes to produce granulocyte-monocyte colony stimulating factor. J Clin Invest 81:1415–1418

Huebner K, Isobe M, Crocec M, Golde DW, Kaufman SE, Gasson JC (1985) The human gene encoding GM-CSF is at 5q29-q29, the chromosome region deleted in the 5q⁻ anomaly. Science 230:1282–1285

Ichikawa Y (1969) Differentiation of a cell line of myeloid leukemia. J Cell Physiol 74:223–234

Ihle JN, Keller J, Henderson L, Klein F, Palaszynski E (1982) Procedures for the purification of interleukin-3 to homogeneity. J Immunol 129:2431–2436

Johnson GR, Whitehead RH, Nicola NA (1985) Effects of a murine mammary tumor on in vivo and in vitro hemopoiesis. Int J Cell Cloning 3:91–105

Johnson GR, Gonda TJ, Metcalf D, Hariharan IK, Cory S (1989) A lethal myeloproliferative syndrome in mice transplanted with bone marrow cells infected with a retrovirus expressing granulocyte-macrophage colony-stimulating factor. EMBO J 8:441–448

Kannourakis G, Johnson GR (1988) Fractionation of subsets of BFu-e from normal human bone marrow: responsiveness to erythropoietin, human placental conditioned medium or granulocyte-macrophage colony-stimulating factor. Blood 71:758–765

Kelleher CA, Wong GG, Clark SC, Schendel PF, Minden MD, McCulloch EA (1988) Binding of iodinated recombinant human GM-CSF to the blast cells of acute myeloblastic leukemia. Leukemia 2:211–215

Kelso A, Gough N (1987) Expression of hemopoietic growth factor genes in murine T lymphocytes. In: Webb DR, Goeddel DV (eds) The lymphokines, vol 3. Academic, New York, pp 209–238

Koury MJ, Pragnell IB (1982) Retroviruses induce granulocyte-macrophage colony stimulating activity in fibroblasts. Nature 299:638–640

Koury MJ, Balmain A, Pragnell IB (1983) Induction of granulocyte-macrophage colony-stimulating activity in mouse skin by inflammatory agents and tumor promoters. EMBO J 2:1877–1882

Kozumbo WJ, Harris DT, Gromkowski S, Cerottini J-C, Cerutti PA (1987) Molecular mechanisms involved in T cell activation. II. The phosphatidylinositol signal-transducing mechanism mediates antigen-induced lymphokine production but not interleukin 2-induced proliferation in cloned cytoxic T lymphocytes. J Immunol 138:606–612

Krumwieh D, Weinmann E, Seiler FR (1988) Human recombinant derived IL-3 and GM-CSF in hematopoiesis of normal cynomolgous monkeys. Behring Inst Res Commun 83:250–257

Lang RA, Metcalf D, Gough NM, Dunn AR, Gonda TJ (1985) Expression of a haemopoietic growth factor cDNA in a factor dependent cell line results in autonomous growth and tumorigenicity. Cell 43:531–542

Lang RA, Metcalf D, Cuthbertson RA, Lyons I, Stanley E, Kelso A, Kannourakis G, Williamson DJ, Klintworth GK, Gonda TJ, Dunn AR (1987) Transgenic mice expressing a hemopoietic growth factor gene (GM-CSF) develop accumulations of macrophages, blindness and a fatal syndrome of tissue damage. Cell 51:675–686

Lang RA (1989) Studies on the aberrant expression of hemopoietic growth factors in vitro and in vivo. PhD thesis, University of Melbourne

Lee F, Yokota T, Otsuka T, Gemmell L, Larson N, Luh J, Arai K, Rennick D (1985) Isolation of a cDNA for a humans granulocyte-macrophage colony stimulating factor by functional expression in mammalian cells. Proc Natl Acad Sci USA 82:4360–4364

Libby RT, Braedt G, Kronheim SR, March CJ, Urdal DL, Chiaverotti TA, Tushinski RJ, et al. (1987) Expression and purification of native human granulocyte-macrophage colony-stimulating factor from an *Escherichia coli* secretion vector. DNA 6:221–229

Lieschke GJ, Maher D, Cebon J, O'Connor M, Green M, Sheridan W, Boyd A, Rallings M, Bonnem E, Metcalf D, Burgess AW, McGrath K, Fox RM, Morstyn G (1989) Effects of subcutaneously administered bacterially-synthesized recombinant human granulocyte-macrophage colony-stimulating factor in patients with advanced malignancy. Ann Intern Med 110:357–364

Lindemann A, Riedel D, Oster W, Meuer SC, Blohn D, Mertelsmann RM, Hermann F (1988) Granulocyte/macrophage colony-stimulating factor induces interleukin 1 production by human polymorphonuclear neutrophils. J Immunol 140:837–839

Lopez AF, Nicola NA, Burgess AW, Metcalf D, Battye FL, Sewell WA, Vadas MA (1983) Activation of granulocyte cytotoxic function by purified mouse colony-stimulating factors. J Immunol 131:2983–2988

Lopez AF, Begley CG, Williamson DJ, Warren DJ, Vadas MA, Sanderson CJ (1986) Murine eosinophil differentiation factor. J Exp Med 163:1085–1099

Lotem J, Sachs L (1988) In vivo control of differentiation of myeloid leukemic cells by recombinant granulocyte-macrophage colony-stimulating factor and interleukin-3. Blood 71:375–382

Lu L, Srour EF, Warren DJ, Walker D, Graham CD, Walker EB, Jansen J, Broxmyer HE (1988) Enhancement of release of granulocyte and granulocyte-macrophage colony-stimulating factors from phytohemagglutinin-stimulated sorted subsets of human T-lymphocytes by recombinant human tumor necrosis factor α. J Immunol 141:201–207

MacDonald BR, Mundy GR, Clark S, Wang EA, Kuehl TJ, Stanley ER, Roodman GD (1986) Effects of human recombinant CSF-GM and highly purified CSF-1 on the formation of multi-nucleated cells with osteoclast characteristics in long-term bone marrow cultures. J Bone Mineral Res 1:227–233

Malone DG, Pierce JH, Falko JP, Metcalfe DD (1988) Production of granulocyte-macrophage colony-stimulating factor by primary cultures of unstimulated rat microvascular endothelial cells. Blood 71:684–689

Mayers P, Lam C, Obenbhaus H, Lichl L, Bessemer J (1987) Recombinant human GM-CSF induces leukocytosis and activates peripheral blood polymorphonuclear neutrophils in non-human primates. Blood 70:206–213

Merchav S, Naylor A, Tatarsky I (1987) Human stem cell colony-stimulating activity (CFU-GEMM) in medium conditioned by leukemic B-lymphocytes. Exp Hematol 15:115–118

Mertelsmann R, Cosman D (1988) Colony-stimulating factors in vivo and in vitro. Immunol Today 9:97–98

Metcalf D (1980) Clonal extinction of myelomonocytic leukemic cells by serum from mice injected with endotoxin. Int J Cancer 25:225–233

Metcalf D (1984) The hemopoietic colony stimulating factors. Elsevier, Amsterdam

Metcalf D (1985) The granulocyte macrophage colony stimulating factors. Science 229:16–22

Metcalf D, Merchav S (1982) Effects of GM-CSF deprivation on precursors of granulocytes and macrophages. J Cell Physiol 112:411–418

Metcalf D, Johnson GR, Burgess AW (1980) Direct stimulation by purified GM-CSF of the proliferation of multipotential and erythroid precursor cells. Blood 55:138–147

Metcalf D, Burgess AW, Johnson GR (1986) In vitro actions on hemopoietic cells of recombinant murine GM-CSF purified after production in *E. coli*: comparison with purified native GM-CSF. J Cell Physiol 128:421–431

Metcalf D, Begley CG, Williamson DJ, Nice EC, DeLamarter J, Mermod JJ, Thatcher D, Schmidt A (1987a) Hemopoietic responses in mice injected with purified recombinant murine GM-CSF. Exp Hematol 15:1–9

Metcalf D, Roberts TM, Cherington V, Dunn AR (1987b) The in vitro behaviour of hemopoietic cells transformed by polyoma middle T antigen parallels that of primary human myeloid leukemic cells. EMBO J 6:3703–3709

Migliaccio AR, Bruno M, Migliaccio G (1987) Evidence for direct action of biosynthetic (recombinant) GM-CSF on erythroid progenitors in serum-free culture. Blood 70:1867–1871

Miyatake S, Otsuka T, Yokota T, Lee F, Arai K (1985) Structure of the chromosomal gene for granulocyte-macrophage colony stimulating factors: comparison of the mouse and human genes. EMBO J 4:2561–2568

Miyatake S, Seiki M, Yoshida M, Arai K-I (1988) T-cell activation signals and human T-cell leukemia virus type I-encoded p40^x protein activate the mouse granulocyte-macrophage colony-stimulating factor gene through a common DNA element. Mol Cell Biol 8:5581–5587

Miyauchi J, Kelleher CA, Wong GG, Yang Y-C, Clark SC, Minkin S, Minden MD, McCulloch EA (1988) The effects of combinations of the recombinant growth factors GM-CSF, G-CSF, IL-3 and CSF-1 on leukemic blast cells in suspension culture. Leukemia 2:382–387

Monroy RL, Skelly RR, MacVittie TJ, Davis TA, Sauber JJ, Clark SC, Donahue RE (1987) The effect of recombinant GM-CSF on the recovery of monkeys transplanted with autologous bone marrow. Blood 70:1696–1699

Moore MAS, Williams N, Metcalf D (1973) In vitro colony formation by normal and leukemic human hematopoietic cells: characterization of the colony-forming cells. JNCI 50:603–623

Morrisey PJ, Bressler L, Charrier K, Alpert A (1988) Response of resident murine peritoneal macrophages to in vivo administration of granulocyte-macrophage colony-stimulating factor. J Immunol 140:1910–1915

Morstyn G, Burgess AW (1988) Hemopoietic growth factors: review. Cancer Res 48:5624–5637

Myers AM, Robinson WA (1975) Colony stimulating factor levels in human serum and urine following chemotherapy. Proc Soc Exp Biol Med 148:694–700

Neinhuis AW, Donahue RE, Karlsson S, Clark SC, Agricola B, Antinoff N, Pierce JE, Turner P, Anderson WF, Nathan DG (1987) Recombinant human granulocyte-macrophage colony-stimulating factor (GM-CSF) shortens the period of neutropenia after autologous bone marrow transplantation in a primate model. J Clin Invest 80:573–577

Neta R, Oppenheim JJ, Douches SD (1988) Interdependence of the radioprotective effects of human recombinant interleukin alpha, tumor necrosis factor alpha, granulocyte colony-stimulating factor, and murine recombinant granulocyte-macrophage colony-stimulating factor. J Immunol 140:108–111

Nicola NA (1987) Why do hemopoietic growth factor receptors interact with each other? Immunol Today 8:134–140

Nicola NA, Vadas M (1984) Hemopoietic colony-stimulating factors. Immunol Today 5:76–80

Nicola NA, Burgess AW, Metcalf D (1979a) Similar molecular properties of granulocyte-macrophage colony stimulating factors produced by different mouse organs in vitro and in vivo. J Biol Chem 254:5290–5299

Nicola NA, Metcalf D, Johnson GR, Burgess AW (1979b) Separation of functionally distinct human granulocyte-macrophage colony-stimulating factors. Blood 54:614–627

Nimer SD, Morita EA, Martis MJ, Washsman W, Gasson J (1988) Characterization of the human granulocyte-macrophage colony-stimulating factor promoter region by genetic analysis: correlation with DNase I footprinting. Mol Cell Biol 8:1979–1984

Oster W, Lindemann A, Ganser A (1988) Constitutive expression of hematopoietic growth factor genes by acute myeloblastic leukemia cells. Behring Inst Mitt 83:68–69

Park LS, Friend D, Gillis S, Urdal DL (1986a) Characterization of the cell surface receptor for granulocyte-macrophage colony-stimulating factor. J Biol Chem 261:4177–4183

Park LS, Friend D, Gillis S, Urdal DL (1986b) Characterization of the cell surface receptor for human granulocyte-macrophage colony-stimulating factor. J Exp Med 164:251–262

Parry DAD, Minasian E, Leach SJ (1988) Conformational homologies among cytokines: Interleukins and colony stimulating factors. J Mol Recogn 1:107–110

Pierce JH, di Fiore PP, Aaronson SA, Potter P, Pumphrey J, Scott A, Ihle JN (1985) Neoplastic transformation of mast cells by Abelson-MuLV: abrogation of IL-3 dependence by a nonautocrine mechanism. Cell 41:685–693

Pluznick DH, Sachs L (1965) The cloning of normal mast cells in tissue culture. J Cell Comp Physiol 66:319–324

Reed SG, Nathan CF, Pihl OL, Rodricks P, Shanebeck K, Conlon PJ, Grabstein KH (1987) Recombinant granulocyte-macrophage colony-stimulating factor activates macrophages to inhibit *Trypanosoma cruzi* and release hydrogen peroxide – comparison with interferon gamma. J Exp Med 166:1734–1746

Schrader JW, Crapper RM (1983) Autogenous production of a hemopoietic growth factor, persisting-cell-stimulating factor, as a mechanism for transformation of bone marrow-derived cells. Proc Natl Acad Sci USA 30:6892–6896

Seelentag WK, Mermod J-J, Montesano R, Vassalli P (1987) Additive effects of interleukin 1 and tumour recrosis factor-α on the accumulation of the three granulocyte and macrophage colony-stimulating factor mRNA's in human endothelial cells. EMBO J 6:2261–2265

Seiler FR, Henney CS, Krumwieh D, Schultz G (eds) (1988) Colony stimulating factors. Behring Inst Mitt 83

Shadduck RK, Nagabhushan NG (1971) Granulocyte colony stimulating factor. I. Response to acute granulocytopenia. Blood 38:559–568

Shannon MF, Gamble JR, Vadas MA (1988) Nuclear proteins interacting with the promoter region of the human granulocyte-macrophage colony-stimulating factor gene. Proc Natl Acad Sci USA 85:674–678

Shaw G, Kamen R (1986) A conserved AU sequence from the 3' untranslated region of GM-CSF mRNA mediates selective mRNA degradation. Cell 46:659–667

Sheridan JW, Metcalf D (1973) A low molecular weight factor in lung-conditioned medium stimulating granulocyte and monocyte colony formation in vitro. J Cell Physiol 81:11–23

Shrimser JL, Rose K, Simona MG, Wingfield P (1987) Characterization of human and mouse granulocyte-macrophage colony-stimulating factors derived from *Escherichia coli*. Biochem J 247:195–199

Sieff CA, Emerson SG, Donahue RE, Nathan DG, Wang EA, Wong GG, Clark SC (1985) Human recombinant granulocyte-macrophage colony-stimulating factor: a multilineage hematopoietin. Science 230:1171–1173

Sieff CA, Tsai S, Faller DV (1987) Interleukin-1 induces cultured human endothelial cell production of granulocyte-macrophage colony stimulating factor. J Clin Invest 79:48–51

Socinski MA, Cannistra SA, Elias A, Antman KH, Schnipper L, Griffin JD (1988) Granulocyte-macrophage colony stimulating factor expands the circulating haemopoietic progenitor cell compartment in man. Lancet i:1194–1198

Sparrow LG, Metcalf D, Hunkapiller MW, Hood LE, Burgess AW (1985) Purification and partial amino acid sequence of asialo murine granulocyte-macrophage colony stimulating factor. Proc Natl Acad Sci USA 82:292–296

Stanley ER, Guilbert LJ (1981) Methods for the purification assay, characterization and target cell binding of a colony stimulating factor (CSF-1). J Immunol Methods 42:253–284

Stanley ER, Hansen G, Woodstock J, Metcalf D (1975) Colony stimulating factor and the regulation of granulopoiesis and macrophage production. Fed Proc 34:2272–2278

Stanley E, Metcalf D, Sobieszczuk P, Gough NM, Dunn AR (1985) The structure and expression of the murine gene encoding granulocyte-macrophage colony-stimulating factor: evidence for utilization of alternative promoters. EMBO J 4:2569–2573

Stanley IJ, Burgess AW (1983) Granulocyte-macrophage colony stimulating factor stimulates the synthesis of membrane and nuclear proteins in murine neutrophils. J Cell Biochem 23:241–258

Thorens B, Mermod J-J, Vassalli P (1987) Phagocytosis and inflammatory stimuli induce GM-CSF mRNA in macrophages through post-transcriptional regulation. Cell 48:671–679

Vadas MA, Nicola NA, Metcalf D (1983) Activation of antibody-dependent cell-mediated cytotoxicity of human neutrophils and eosinophils by separate colony stimulating factors. J Immunol 130:795–799

Vadhan-Raj S, Keating M, LeMaistre A, Hittelman WN, McCredie K, Trujillo JM, Broxmeyer HE, Henney C, Gutterman JU (1987) Effects of recombinant human granulocyte-macrophage colony-stimulating factor in patients with myelodysplastic syndromes. N Engl J Med 317:1545–1552

Walker F, Burgess AW (1985) Specific binding of radioiodinated granulocyte-macrophage colony stimulating factor to hemopoietic cells. EMBO J 4:933–939

Walker F, Nicola NA, Metcalf D, Burgess AW (1985) Hierarchical downmodulation of hemopoietic growth factor receptors. Cell 43:269–276

Wang JM, Colella S, Allavena P, Mantovani A (1987) Chemotactic activity of human recombinant granulocyte-macrophage colony-stimulating factor. Immunology 60:439–444

Weisbart RM, Golde DW, Clark SC, Wong GG, Gasson JC (1985) Human granulocyte-macrophage colony-stimulating factor is a neutrophil activator. Nature 314:361–363

Weisbart RH, Kwan L, Golde DW, Gasson JC (1987) Human GM-CSF primes neutrophils for enhanced oxidative metabolism in response to the major physiological chemoattractants. Blood 69:18–21

Weiser WY, van Niel A, Clark SC, David JR, Remold H (1987) Recombinant human granulocyte/macrophage colony-stimulating factor activates intracellular killing of *Leishmania donovani* by human monocyte-derived macrophages. J Exp Med 166:1436–1446

Wingfield P, Graber P, Monnen P, Craig S, Pain RH (1988) The conformation and stability of recombinant-derived granulocyte-macrophage colony stimulating factors. Eur J Biochem 173:65–72

Wong GG, Witek JS, Temple PA, Wilkens KM, Leary AC, Luxenberg DP, Jones SS, Brown EL, Kay RM, Orr EC, Shoemaker C, Golde DW, Kaufman RJ, Hewick RM, Wang EA, Clark SC (1985a) Human GM-CSF: molecular cloning of the complementary DNA and purification of the natural and recombinant proteins. Science 228:810–815

Wong GG, Witek JS, Temple PA, Wilkens KM, Leary AG, Luxenberg DP, Jones SS, Brown EL, Kay RM, Orr EC, Shoemaker C, Golde DW, Kaufmann RJ, Hewick RM, Clark SC, Wang EA (1985b) Molecular cloning of human and gibbon T-cell-derived GM-CSF cDNAs and purification of the natural and recombinant human proteins. In: Feramisco J, Ozanne B, Stiles C (eds) Cancer cells. Cold Spring Harbor, New York, p 235

Yang Y-C, Kovac S, Kriz R, Wolf S, Clark SC, Wellems TE, Nienhuis A, Epstein N (1988) The human genes for GM-CSF and IL-3 are closely linked in tandem on chromosome 5. Blood 71:958–961

Young DC, Wagner K, Griffin JD (1987) Constitutive expression of the granulocyte-macrophage colony-stimulating factor gene in acute myeloblastic leukemia. J Clin Invest 79:100–106

CHAPTER 18

Erythropoietin:
The Primary Regulator of Red Cell Formation

E. Goldwasser, N. Beru, and D. Smith

A. Introduction

The role of a humoral factor in a feedback system that regulates the rate of red blood cell formation was suggested in 1906 (Carnot 1906). The reported experiments indicated that rabbits responded to blood loss by forming some substance, found in the circulation, that signaled the blood forming system to accelerate red cell formation. Verification of this concept was delayed for more than 30 years (Erslev et. al. 1953; Krumdieck 1943). This factor, named erythropoietin (epo), has been the subject of many reviews and symposium volumes (Anagnostou 1985; Gordon et al. 1971; Graber and Krantz 1978; Jelkmann 1986; Krantz and Jacobson 1970; Miyake et al. 1977; Nakao et al. 1975; Orlic and Lobue 1984; Rich 1987; Sherwood 1984; Spivak 1986) over the past two or three decades and its role in the regulation of erythropoiesis well documented. The present review will, therefore, be selective rather than comprehensive and will concern, for the most part, advances made in the more recent past as research and accumulation of information have accelerated, largely due to the use of the techniques of molecular biology in this field.

B. Chemistry

Our understanding of the structure of human epo followed from the first purification of it from urine of patients with aplastic anemia. This preparation, homogeneous by sodium dodecylsulfate (SDS) gel electrophoresis, consisted of two forms, separable by chromatography on hydroxylapatite, with essentially the same potency (70000–120000 units/mg protein; Miyake et al. 1977). We later found that these two forms of epo had somewhat different carbohydrate compositions (Dordal et al. 1985). It is possible that one form was derived from the other either during the excretion in the urine, i.e., some deglycosylation in the kidney or bladder, or by loss of carbohydrate during the purification process. In any case, it would appear that the form with the lower carbohydrate content could represent a trivial artifact.

The role of the carbohydrate in the biological activity of epo is still not completely settled but it appears that it may be required for in vivo stability rather than for ability to bind to specific receptors on hematopoietic precursor cells. The small amount of data available suggest that deglycosylated epo, whether of urinary (u-epo) or recombinant (r-epo) origin, has significant biological activity when assayed in vitro but none when assayed in vivo (DORDAL et al. 1985; EGRIE et al. 1986). Enzymically deglycosylated u-epo tends to aggregate to a multimeric form without biological activity, but the small amount of monomeric deglycosylated u-epo remaining has enhanced activity on bone marrow cells in culture (DORDAL et al. 1985). This finding agrees with that reported some years ago (GOLDWASSER et al. 1974) showing that removal of terminal sialic acid residues resulted in loss of in vivo activity but increased activity in bone marrow cell cultures. It would appear that interaction with receptors may be facilitated by removal of sialic acid and/or most of the oligosaccharides. In a recent paper WOJCHOWSKI et al. (1987) showed that r-epo (human) expressed in insect cells by use of a baculovirus vector has a significantly lower molecular weight than that expressed in mammalian cells, probably because of less carbohydrate. This epo preparation has essentially full activity in an in vitro assay (WOJCHOWSKI et al. 1987) and when treated with N-glycanase lost at least 80% of its activity. No explanation of the difference in the two sets of experiments is yet at hand, but it appears likely that aggregation could have caused the loss of activity.

Purified human u-epo was shown to have two internal disulfide bonds which are required for appropriate folding of denatured u-epo to be renatured to the biologically active conformation (WANG et al. 1985). It was also used to determine the primary structure of human u-epo by direct amino acid sequencing (LAI et al. 1986). The sequence so determined was the same as that predicted from the nucleotide sequence of the coding regions of the human epo gene (JACOBS et al. 1985; LIN et al. 1985). Figure 1 shows the amino acid sequences of epo from three species. These are the only ones available at the time of writing. The mouse and monkey sequences are inferred from the nucleotide sequences. For human r-epo there are three potential sites for N-glycosylation (asparagines 24, 38, and 83). Results of selective N-glycanase hydrolysis of r-epo (EGRIE et al. 1986) show that all three asparagines are glycosylated. Serine 126 is a potential site of O-glycosylation. Similar enzymic hydrolysis with O-glycanase (after sialidase treatment) resulted in a decrease in molecular size and verified the presence of an O-linked oligosaccharide. Automated amino acid sequencing showed no residue detected at position 126 (LAI et al. 1986), presumably because the amino acid had the sugar attached. This permitted the assignment of the O-linked carbohydrate to the serine at position 126.

In a recent paper RECNY et al. (1987) found that both r-epo and u-epo lack the C-terminal arginine predicted from the DNA sequence and found earlier by direct amino acid sequencing in a different preparation of u-epo (LAI et al. 1986). They suggest that des-Arg 166 epo may be the naturally circulating form of the human hormone, although it has not been determined what the structure of circulating plasma epo is.

Comparisons of the amino acid sequences of human, monkey, and mouse epos show a high degree of identity (MCDONALD et al. 1986). Human and monkey epos are 92% identical with respect to amino acids. The correspond-

```
-27
MET GLY VAL HIS GLU CYS PRO ALA TRP LEU TRP LEU LEU LEU SER LEU LEU
--- --- --- --- --- --- --- --- --- --- --- --- --- --- --- --- VAL
--- --- --- PRO --- ARG --- THR     --- LEU --- --- --- --- --- ---
                                            +1
SER LEU PRO LEU GLY LEU PRO VAL LEU GLY ALA PRO PRO ARG LEU ILE CYS
--- --- --- --- --- --- --- -°- PRO --- --- --- --- --- --- --- ---
LEU ILE --- --- --- --- --- --- --- CYS --- --- --- --- --- --- ---
         10                                        20              *
ASP SER ARG VAL LEU GLU ARG TRY LEU LEU GLU ALA LYS GLU ALA GLU ASN
--- --- --- --- --- --- --- --- --- --- --- --- --- --- --- --- ---
--- --- --- --- --- --- --- --- ILE --- --- --- --- --- --- --- ---
                         30                           *     40
ILE THR THR GLY CYS ALA GLU HIS CYS SER LEU ASN GLU ASN ILE THR VAL
VAL --- MET --- --- SER --- SER --- --- --- ASN --- --- --- --- ---
VAL --- MET --- --- --- --- GLY PRO ARG --- SER --- --- --- --- ---
                                      50
PRO ASP THR LYS VAL ASN PHE TYR ALA TRP LYS ARG MET GLU VAL GLY GLN
--- --- --- --- --- --- --- --- --- --- --- --- --- --- --- --- ---
--- --- --- --- --- --- --- --- --- --- --- --- --- --- --- GLU GLU
    60                                           70
GLN ALA VAL GLU VAL TRP GLN GLY LEU ALA LEU LEU SER GLU ALA VAL LEU
--- --- --- --- --- --- --- --- --- --- --- --- --- --- --- --- ---
--- --- ILE --- --- --- --- --- --- SER --- --- --- --- --- ILE ---
                 80              *                            90
ARG GLY GLN ALA LEU LEU VAL ASN SER SER GLN PRO TRP GLU PRO LEU GLN
--- --- --- --- VAL --- ALA --- --- --- --- --- PHE --- --- --- ---
GLN ALA --- --- --- --- ALA --- --- --- --- --- PRO --- THR --- ---
                                 100
LEU HIS VAL ASP LYS ALA VAL SER GLY LEU ARG SER LEU THR THR LEU LEU
--- --- MET --- --- --- ILE --- --- --- --- --- ILE --- --- --- ---
--- --- ILE--- ---- --- ILE --- --- --- --- --- --- --- SER --- ---
110                                      120                     **
ARG ALA LEU GLY ALA GLN LYS GLU ALA ILE SER PRO PRO ASP ALA ALA SER
--- --- --- --- --- ---     --- --- --- --- LEU --- --- --- --- ---
--- VAL --- --- --- --- --- LEU MET --- --- --- --- THR THR PRO
         130                                        140
ALA ALA PRO LEU ARG THR ILE THR ALA ASP THR PHE ARG LYS LEU PHE ARG
--- --- --- --- --- --- --- --- --- --- --- --- CYS --- --- --- ---
PRO --- --- --- --- --- LEU --- VAL --- --- --- CYS --- --- --- ---
                         150                                     160
VAL TYR SER ASN PHE LEU ARG GLY LYS LEU LYS LEU TYR THR GLY GLU ALA
--- --- --- --- --- --- --- --- --- --- --- --- --- --- --- --- ---
--- --- ALA --- --- --- --- --- --- --- --- --- --- --- --- --- VAL
                 166
CYS ARG THR GLY ASP ARG
--- --- ARG --- --- ---
--- --- ARG --- --- ---
```

Fig. 1. Amino acid sequences of human (*top*), monkey (*middle*), and mouse (*bottom*) erythropoietins. Sequences of the leader peptide as well as the mature proteins are shown. Numbering is from the N terminus of the human protein. Differences from the human sequence are indicated. The monkey epo +1 amino acid as determined by N-terminal analysis of the protein expressed in Chinese hamster ovary cells is indicated by a *small open circle*. *Single asterisks* indicate the sites of potential N-linked glycosylation. The O-glycosylation site is indicated by *double asterisks*

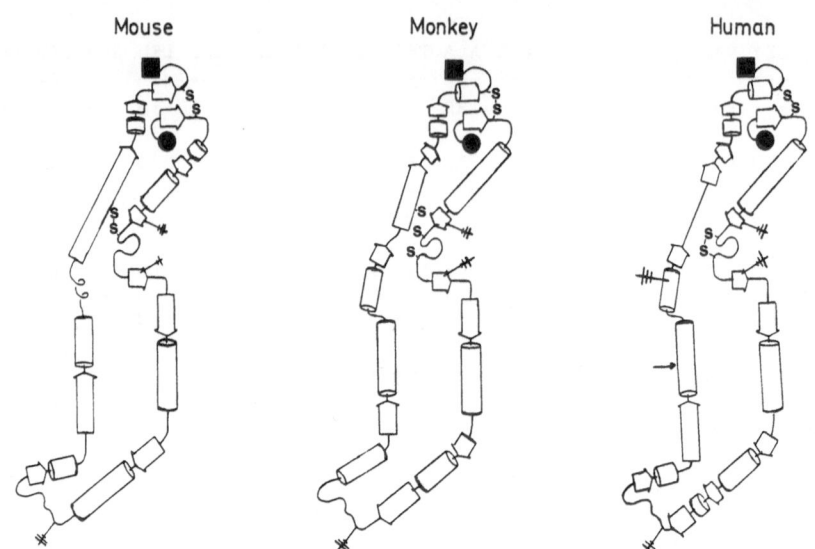

Fig. 2. Predicted secondary structures of three erythropoietins as determined by the method of CHOU and FASMAN (1978). Each N terminus is shown as a *solid circle* and each C terminus as a *solid square*

ing figures for the human/mouse and monkey/mouse comparisons are 80% and 82%, respectively. Some possibly significant differences exist, however: in the human and monkey epo molecules, the serine at position 126 is a potential site of O-glycosylation, verified in human; in the mouse, this serine is replaced by a proline. Also, in the monkey, the amino acid corresponding to the lysine at position 116 in the human epo is missing. A brief evolutionary analysis of the nucleotide sequences of the three cloned epos was presented by McDONALD et al. (1986).

Prediction of secondary structure by the CHOU and FASMAN (1978) method indicates substantial conservation of structures, as might be expected from the conservation of sequence (Fig. 2). A newer method of secondary structure prediction (K. BERNDT, E. T. KAISER and F. KEZDY, unpublished) applied to human epo suggests a four-helix model with the hydrophobic surfaces of the amphiphilic helixes facing each other in close approximation (Fig. 3).

Recombinant human epo expressed in Chinese hamster ovary cells and purified from the culture medium was compared to u-epo by circular dichroism, UV absorbance, and fluorescence spectroscopy with no differences found between the two preparations (DAVIS et al. 1987). In the same paper, the authors used the sedimentation equilibrium method to find a molecular weight of 30 400. With the peptide molecular weight (from the amino acid sequence including the C-terminal arginine) of 18 398 the carbohydrate content was determined to be 39%, which is significantly higher than the previous estimate of 31% for u-epo which was based on a holomolecular weight of

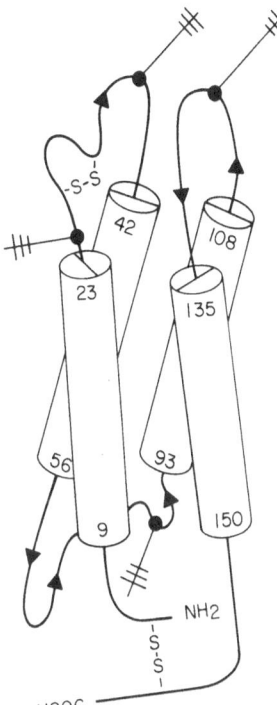

Fig. 3. Predicted secondary structure (K. BERNDT et al., unpublished) of human erythropoietin

34000 determined by SDS gel electrophoresis (DORDAL et al. 1985). A corrected value for carbohydrate content based on a molecular weight of 30400 would be 35% for u-epo, which is in fair agreement with the 39% calculated for r-epo (DAVIS et al. 1987).

The structures of the individual oligosaccharide chains, both O-linked and N-linked, of r-epo were determined by fast atom bombardment–mass spectrometry and methylation analysis (H. SASAKI et al. 1987) and by exoglycosidase digestion, Smith degradation, and methylation analysis (TAKEUCHI et al. 1988). The major component of the O-linked oligosaccharide on Ser 126 is *N*-acetyl neuraminic acid 2→3 galactose β 1→3 (*N*-acetyl neuraminic acid 2→6) *N*-acetyl galactosamine–serine. The N-linked oligosaccharides consist of the following (expressed as percentage of total carbohydrate): biantennary 1.4%, triantennary 10%, triantennary with one *N*-acetyl lactosaminyl repeat 3.5%, tetraantennary 31.8%, tetraantennary with one such repeat 32.1%, with two repeats 16.5%, or with three repeats 4.7%. The tetraantennary chains for the most part are di- or trisialated with α 2→3 linked sialic acid. An example of the structure of the predominant species (31.8%) of tetraantennary N-linked oligosaccharide with one sialic acid per penultimate galactose residue is seen in Fig. 4. The same authors find that there is only a slight difference in sialic acid content between u-epo and r-epo, with no other difference in carbohydrate found.

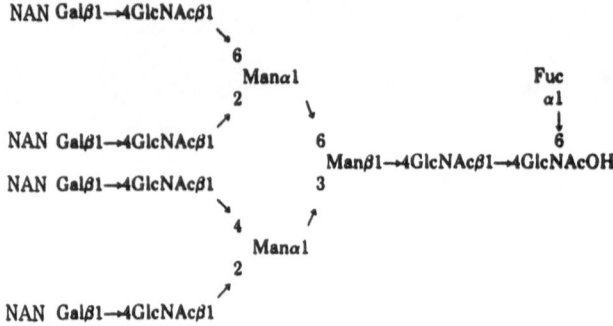

Fig. 4. An example of the structure (H. SASAKI et al. 1987) of the predominant species of tetraantennary N-linked oligosaccharide with one sialic acid per penultimate galactose residue

No definitive study of the active site (or sites) of epo has been published although older evidence had indicated that tryptophan (LOWY and KEIGHLY 1968), tyrosine, and α- or ε-amino groups (GOLDWASSER 1981) may be required for biological activity. In addition, the disulfide bonds, or the conformation imposed on epo by the disulfide bridging, are important for activity (WANG et al. 1985).

Analysis of the amino acid sequence of human epo by the HOPP and WOODS (1981) method shows an average hydrophilicity value of 1.83 for the region 18–23 (Glu-Ala-Lys-Glu-Ala-Glu) representing the most probable antigenic site (Fig. 5). Studies using antisera raised against synthetic peptides indicate that an antibody made to the N-terminal 26 residues does not impair biological function (SYTKOWSKI and DONAHUE 1987). Of five other antibodies made against small synthetic peptides, only those directed against residues 99–118 and 111-129 had any inhibitory effect on epo action. These regions of the epo protein do not have either a tryptophan or a tyrosine although there is one lysine at residue 116. The current state of knowledge does not permit an unequivocal interpretation of these seemingly inconsistent observations; it is

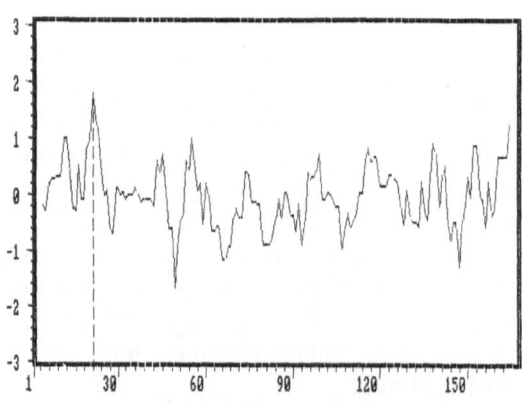

Fig. 5. Predicted antigenic site of human erythropoietin (HOPP and WOODS 1981) done by computer using software from Intelligenetics, Inc., Mountain View, CA, USA. *Ordinate* is amino acid number; *abscissa* is hydrophilicity value

clear that much more data are needed before the structure responsible for the biological activity of epo can be identified.

Some important information has been derived from experiments with site-directed mutagenesis experiments using the ratio of activity by an in vitro assay (GOLDWASSER et al. 1975) to activity by radioimmunoassay (SHERWOOD and GOLDWASSER 1979) as the indicator. The substitution of Leu at residue 2 for Pro has no effect on activity, nor did Phe at 49 for Tyr, or Glu at 52 for Lys. The replacement of Cys 33 with Pro (as it is in the mouse; McDONALD et al. 1986) resulted in decreased biological activity (LIN 1987). This suggests that the loop formed by the 29–33 disulfide bond is not exactly replaced by putting a turn in the peptide chain with a proline. It might also suggest that pure mouse epo would have a lower potency than human epo.

C. Molecular Biology

I. Cloning and Analysis of the Erythropoietin Gene

The human epo gene was cloned recently (JACOBS et al. 1985; LIN et al. 1985) using a mixture of oligonucleotide probes based on the sequences of selected tryptic peptides of pure human u-epo. Subsequently the old world monkey (*Macaca fascicularis*) epo gene was cloned from a cDNA library using the same approach (LIN et al. 1986) as was the mouse epo gene from a genomic library using the monkey epo cDNA probe (McDONALD et al. 1986) and a human epo cDNA probe cloned from a fetal liver cDNA library (SHOEMAKER and MISTOCK 1986). These papers show that there is a single copy of the epo gene per haploid genome and therefore that the renal and hepatic genes are identical.

Comparison of the human and mouse epo genes shows that the overall organization of the gene is conserved (Fig. 6), i.e., the number of exons and introns is the same and exon length as well as sequence is well conserved (McDONALD et al. 1986). Intron length and sequence, on the other hand, are not conserved (beyond the first 10–20 nucleotides flanking the exon boundaries), with the exception of the first intron which exhibits moderate sequence conservation. The 5' and 3' untranslated sequences also show substantial conservation (SHOEMAKER and MISTOCK 1986). The mouse and the monkey coding sequences are 79% and 94% identical to the human coding sequence, respectively (LIN et al. 1986; SHOEMAKER and MISTOCK 1986).

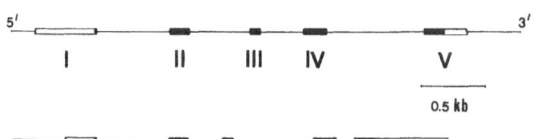

Fig. 6. Organization of the mouse (*top*) and human (*bottom*) epo genes. *Open boxes* indicate exons; *filled boxes* indicate the coding regions. The positions of the exact transcript ends are uncertain

A 600 base pair (bp) region upstream of the protein initiation codon of the human epo gene did not contain any promotor-like sequences such as TATA or CCAAT boxes (LIN et al. 1985; LIN 1987). This appears to be true of the promoter region of mouse epo gene as well, although there are sequences with some homology to these elements (MCDONALD et al. 1986). Similarly, in both the human and monkey epo genes, the normal polyadenylation sequence AAUAAA is not present (LIN 1987). The site of polyadenylation of the mouse transcript has not been determined (MCDONALD et al. 1986; SHOEMAKER and MISTOCK 1986). In the mouse the start site of transcription has been determined by S1 nuclease mapping to about 230 nucletides upstream of the initiator codon (SHOEMAKER and MISTOCK 1986). In all three species, the 5' un-translated region of the transcripts appears to be longer than most mRNAs studied and translation starts at the second AUG in the mRNA instead of the first as is the case in most mRNAs (KOZAK 1984).

II. Chromosomal Localization of the Erythropoietin Gene and Its Restriction Fragment Length Polymorphism

Using a human epo cDNA probe and Southern blot analysis on DNA isolated from a number of human–Chinese hamster (LAW et al. 1986) and human–mouse somatic cell hybrids (WATKINS et al. 1986) containing reduced numbers and different combinations of human chromosomes, the epo gene has been localized to chromosome 7. Further, Southern analysis of DNA from a cell hybrid containing a translocated derivative of chromosome 7 has localized the epo gene to 7pter-q22 (WATKINS et al. 1986). Localization of the epo gene to the region q11-q22 of chromosome 7 was also accomplished by in situ hybridization to metaphase chromosomes (LAW et al. 1986). In the mouse, the epo gene has been localized to chromosome 5 (C. LACOMBE, M. G. MATTEI, D. SIMON, J. L. GUENET, and P. TAMBOURIN, personal communication).

When the cDNA probe was used to screen samples of genomic DNA from a number of individuals, restriction fragment length polymorphism (RFLP) was observed in those digested with the restriction enzymes HindIII (LAW et al. 1986; WATKINS et al. 1986) as well as HinfI (WATKINS et al. 1986). Each enzyme revealed two alleles and family studies have shown that these RFLPs display segregation patterns consistent with Mendelian inheritance, thus making them useful genetic markers for this region of the human genome.

III. Expression of the Erythropoietin Gene

The availability of specific molecular probes for epo has finally made it possible to confirm the finding made in 1957 (JACOBSON et al. 1957) that the kidney is the major source of epo in the adult animal. Using mouse genomic probes for Northern blots, epo mRNA was shown to be present in the kidneys

of rats 10 h after bleeding. No epo message was detected in the kidney of un-
bled rats nor in the salivary glands, thymuses, spleens, and livers from bled or
control rats (BERU et al. 1986). The epo mRNA comigrated with the 18S
ribosomal RNA on denaturing gels, indicating that it is about 1.85 kilobases
in size. When cobalt chloride was used to stimulate epo production (BERU et
al. 1986), epo mRNA was always detected in the kidney of treated animals
and, to a much smaller extent, sporadically in the liver. Again, no epo message
was detected in the salivary gland or spleen with or without cobalt chloride
(BERU et al. 1986). Similar results have been obtained in response to anemia
induced by bleeding (BONDURANT and KOURY 1986) as well as in response to
hypoxia induced by exposure of the experimental animals to a hypobaric
chamber (NISHIDA et al. 1986).

In time-course studies using male Long-Evans rats, epo mRNA first appears in the kidney some time between 3 and 6 h after cobalt injection. Serum
epo levels at 3 h are the same as in unstimulated animals, but rise 7-fold at 6 h
after injection (BERU et al. 1986), consistent with the mRNA findings. In
female BDF1 mice, a similar time-course study showed that the earliest time
epo mRNA could be detected in the kidney was at 10 h after cobalt injection
and this also paralleled serum epo levels. That no increase in epo levels is seen
prior to the appearance of mRNA in the kidney may mean that epo gene expression is regulated at the transcriptional level or related to the stability of
the epo message.

The availability of cloned mouse epo gene probes has also been instrumen-
tal in the search for the cell type that synthesizes epo. Two papers report that
in situ hybridization studies on kidney sections from mice made profoundly
anemic indicate that peritubular cells of the cortex and, to a lesser extent, the
outer medulla are the sites of epo production (KOURY et al. 1987; LACOMBE et
al. 1987a). These cells are nonglomerular and nontubular and were tentatively
identified as capillary endothelial cells. On the other hand, in situ hybridization studies done in our laboratory suggest that the cells of the proximal
tubule are the source of epo (D. SMITH and E. GOLDWASSER, unpublished).
Clearly more work is required to definitely determine the cell type within the
kidney responsible for epo secretion. Similar studies on fetal liver should also
be available in the near future.

IV. Studies of Erythropoietin Gene Expression in Transformed Cells

Two mouse cell lines which secrete epo constitutively into the culture medium
have been studied in the hope that understanding the mechanism of constitutive expression may lead to an understanding of the normal regulation of epo
gene expression. These cell lines, designated IW32 and NN10, are
erythroleukemic cell lines isolated from spleens of mice infected with two different biologically cloned viruses, the former with a virus derived from a
helper-independent ecotropic Friend murine leukemia virus (FMuLV) and the
latter with a virus derived from a complete Friend virus (anemia strain; CHOP-

PIN et al. 1984, 1985; TAMBOURIN et al. 1983). Both cell lines constitutively secrete epo and contain epo mRNA of normal size (BERU et al. 1986; LACOMBE et al. 1987b). Southern blot analysis has shown that the IW32 cell line has a normal as well as a rearranged and amplified epo gene (MCDONALD et al. 1987). The amplification may be related to the polyploidy of these transformed cells. The NN10 cell line, on the other hand, contains only the normal epo gene (BERU et al. 1989b; LACOMBE et al. 1987b).

The restriction maps of both the normal and rearranged epo loci of IW32 cells which have been studied in more detail are depicted in Fig. 7. The rearranged locus contains an apparently normal epo gene and downstream region but an altered upstream region which contains PstI, BamHI, and XbaI sites not present upstream of the normal epo gene. Moreover, it does not contain a different PstI site that is present upstream of the normal gene. The entire upstream regions of both loci shown in Fig. 7 have now been sequenced and the rearrangement breakpoint localized to 530 bp upstream of the XbaI site found immediately 5' to the gene (BERU et al. 1989b).

Studies of chromatin structure by DNase I hypersensitivity methods (GOLDWASSER et al. 1987; MCDONALD et al. 1987) have shown that the normal epo locus is relatively resistant to DNase I, while the rearranged epo locus is hypersensitive. The DNase I hypersensitive site maps as shown in Fig. 7. Thus it would appear that IW32 cells express the rearranged but not the normal gene since nuclease hypersensitivity is generally associated with actively transcribing genes.

There was no integration of the FMuLV provirus within the region 4.5 kb upstream and 9.5 kb downstream of the epo gene at the rearranged locus and thus the epo gene was not brought under the influence of viral long terminal repeat regulatory elements. Similarly, no transduction of the epo gene into the

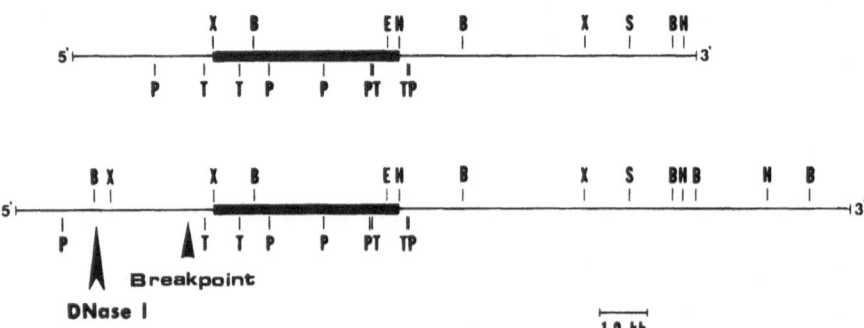

Fig. 7. Restriction maps of two clones containing the normal (*top*) and the rearranged (*bottom*) epo gene loci of IW32 cells. Fine-structure PstI and BstXI analysis was limited to the immediate region of the epo gene which is contained within the *thick line*. The rearrangement breakpoint deduced from sequencing (BERU et al. 1989a) as well as the DNase I hypersensitive site (MCDONALD et al. 1987) are indicated. X, Xba I; B, BamHI; E, EcoRI; H, HindIII; S, Sal I; P, Pst I; T, Bst XI

retroviral genome has been found. Thus these two possible explanations had to be discounted as mechanisms by which the epo gene at the rearranged locus was activated (McDONALD et al. 1987). Sequence analysis of the upstream region of the rearranged locus encompassing the rearrangement breakpoint indicates that the epo gene at this locus may have been activated because the rearrangement event has brought a novel transcriptionally active gene into close proximity and opposite transcriptional orientation to the epo gene (BERU et al. 1989 a) and activation could be due to bidirectional transcription. The mechanism by which the epo gene in NN10 cells has been activated toward constitutive expression is still under study.

D. Specific Erythropoietin Receptors

The first indication that receptors for epo were expressed on erythropoietic precursor cells was indirect; showing that the effect of impure epo on transcription by unfractionated rat marrow cells in vitro was abolished by prior trypsin treatment of the cells (CHANG and GOLDWASSER 1973). Confirmation of the existence of specific receptors had to wait until pure epo and a homogeneous epo-responsive cell population were available. Pure u-epo labeled with ^{3}H in modified sialic acid residues (KRANTZ and GOLDWASSER 1984) was used with purified cells from the spleens of mice infected with the anemic variant of the Friend erythroleukemic virus (FVA cells; KOURY et al. 1984). The use of tritium at this time was dictated by the evidence that iodinated u-epo was devoid of biological activity (GOLDWASSER 1981) whereas the ^{3}H-labeled material was fully active (KRANTZ and GOLDWASSER 1984). Because of the low specific radioactivity of the epo, high cell concentrations were used in order to obtain reasonable count rates and the results may have been distorted by the high cell number. Once it was found that specific binding of ^{125}I-epo to FVA cells occurred, even though there might be uncertainty about biological activity, the question could be reexamined. The later experiments (SAWYER et al. 1987 a) suggested that FVA cells have two classes of receptors with K_d values of 0.09 nM and 0.6 nM, with about 300 of the former and 500–700 of the latter per cell. These FVA cells show maximal response to epo with respect to hemoglobin synthesis when about 180 receptors are occupied, indicating that it is the higher affinity class that is biologically effective.

Binding of epo to cell surface receptors is followed by internalization of the ligand-receptor complex and the intracellular, iodinated epo is then broken down and released from the cells into the medium (SAWYER et al. 1987 a). Whether the internalization step is necessary for the biological action of epo, or whether the binding of epo to the external domain of the receptor is sufficient for biological action on appropriate target cells, still remains to be determined.

Another binding study (MUFSON and GESNER 1987) used r-epo biolabeled with ^{35}S by Chinese hamster ovary cells containing an amplified human epo gene. The cells used to bind epo in these experiments were enriched from the spleens of mice made anemic with phenylhydrazine. The K_d found for the 100–200 receptors on these cells was found to be 0.75 nM, although the data at the lower levels of bound epo (< 0.02 fmol/10^6 cells) may have been too sparse to permit detection of any second class of higher affinity receptors. The authors also showed that asialo-r-epo was a better competitor for receptors than native epo with these cells. Our own recent data (Y.-J. DONG and E. GOLDWASSER, unpublished) indicate that with FVA cells the K_d value for asialo-r-epo is significantly lower than that of native epo. This would confirm the observation cited earlier (GOLDWASSER et al. 1974) that asialo-epo has greater intrinsic activity than native epo when assayed in vitro. In addition, MUFSON and GESNER (1987) found that ^{125}I-labeled r-epo was not significantly internalized whereas the ^{35}S-labeled r-epo was almost completely internalized under the same conditions. The difference between these findings and those reported by SAWYER et al. (1987a) using FVA cells may be due to different cells or to the degree of iodination (I atoms/molecule), which was not specified by MUFSON and GESNER (1987) although they reported that the iodinated r-epo they used had less than 10% of the original biological activity. SAWYER et al. (1987a) showed that at 0.6 I/epo molecule they found 95% of the original biological activity but at 4 I/epo molecule binding activity was decreased by 90%.

SAKAGUCHI et al. (1987) reported that the cell line DA-1 (species was not indicated), which is IL-3 dependent, can be converted to epo dependence by growth in the presence of r-epo. These cells [termed DA-1(cl.14)] then express about 130 epo receptors per cell with a K_d of 0.5 nM. Similarily, BOUSSIOS and BERTLES (1987) reported that hamster yolk sac erythroid cells have a mean of about 700 epo receptors per cell with a K_d of 0.2 nM. Cells infected with the nonanemic, polycythemic, variant of Friend virus also have been reported to bind epo with a K_d of 0.5 nM and 300–600/cell (MAYEUX et al. 1987a). The same authors showed that rat fetal liver cells have about 490 receptors with a K_d of 0.17 nM. Similarly, TODOKORO et al. (1987) showed that an erythroleukemic cell line responsive to epo has about 470 receptors per cell with a K_d of 0.15 nM. In still another study of epo receptors (R. SASAKI et al. 1987), murine erathroleukemic cells were reported to change both the K_d and the number of receptors when induced with dimethylsulfoxide.

Table 1. The molecular weight of epo receptors reported in the literature

Size of receptors (kDa)	References
63, 94, 119	TODOKORO et al. (1987)
120, 140	R. SASAKI et al. (1987)
78, 94	MAYEUX et al. (1987b)
85, 100	SAWYER et al. (1987b)
95, 110	TOJO et al. (1987)
90, 105	DONG and GOLDWASSER (unpublished)

A number of publications have reported on the results of crosslinking studies to determine the apparent molecular weight of epo receptors. In general two labeled components were found with the sizes indicated in Table 1. Despite an unusually large number of publications (for this hormone) it still is clear that much work needs to be done before epo receptors are well characterized. The common thread in all of these studies is that whether cells are responsive to epo or not, they express a small number of surface receptors.

E. Mode of Action

The mechanism of action of epo on bone marrow target cells has been studied extensively by a number of methods, yet the complex course of biochemical events leading to the expression of the erythroid phenotype by these cells has still not been delineated. The actions of epo may be viewed in terms of its effects upon pluripotent stem cells towards lineage commitment, differentiation, and proliferation, as well as on its effect on committed erythroid cells. There is some evidence that epo acts to commit pluripotent cells to the erythroid phenotype (GOLDWASSER 1984a, b). Supporting this concept is evidence that epo can compete in vitro with nonerythroid colony-stimulating factors in committing precursor cells towards their respective lineages (VAN ZANT and GOLDWASSER 1979). Nonetheless, the mechanism of hematopoietic commitment, erythroid or otherwise, is unknown, and until this mechanism is fully understood, the role of epo in erythroid commitment will remain controversial. In contrast, there is some information about the events of epo-induced erythroid differentiation and proliferation; processes where its regulatory action is well established and which appear to proceed in an orderly and predictable manner. This discussion will, therefore, address what is known of the role that epo plays in normal erythroid cell differentiation and proliferation.

In earlier studies, crude epo preparations containing endotoxins and/or colony-stimulating factors, which have been shown to influence the effects of epo, made interpretations uncertain. In addition, difficulties arose because of the use of assays in which the cellular components were heterogeneous with respect to their responsiveness to epo (e.g., bone marrow cells) or capable of secreting epo (e.g., fetal liver cells). These problems have been reviewed elsewhere (VAN ZANT and GOLDWASSER 1984). More recent studies of the events following interaction of cells with epo have used mouse spleen cells, either virally transformed or erythropoietic by virtue of the induction of anemia, as a source of relatively homogeneous precursor cells for study. One price of the improved specificity that these models provide is the caution required in extrapolating events from virally transformed cells and spleen cells to the phenomena characteristic of normal adult bone marrow.

The earliest known event following the exposure of target cells to epo is an accumulation of calcium ions in the cytosol. This was demonstrated in studies of FVA cells within 1 min of addition of a high concentration of epo (2 units/ml) at $4°$ C (SAWYER and KRANTZ 1984). In a recent paper MILLER et al.

(1988) showed that individual erythroid precursor cells increased their intracellular calcium concentration as a result of exposure to epo. This increase was due to mobilization of intracellular calcium rather than to calcium entry. The effect of calcium in erythroid differentiation of CFU-E supports the potentially important role of calcium. MISTI and SPIVAK (1979) showed that colony formation increased when cultured with carboxylic ionophores, which increase cytosolic calcium, and decreased in the presence of EGTA chelation of calcium, an effect that may reflect an action of calcium on receptor binding of epo (MISTI and SPIVAK 1979). These results are not unchallenged; BONNANOU-TZEDAKI et al. (1987) found no evidence of calcium flux with enriched bone marrow erythroblast fractions at physiologic temperatures and epo concentrations. Nonetheless, BONNANOU-TZEDAKI et al. (1987) found a calcium-associated effect when intracellular calcium levels were manipulated by pharmacologic means. An epo response measured by enhanced DNA synthesis (a somewhat late effect) was blunted by calcium chelation with EGTA, by calcium channel blockade with verapamil, and by the intracellular calcium antagonist TMB-8 (BONNANOU-TZEDAKI et al. 1987). Artificially induced changes in cytosolic calcium levels within cells not exposed to epo do not induce erythroid differentiation (BONNANOU-TZEDAKI et al. 1987; MISTI and SPIVAK 1979). This suggests that the role of calcium is probably accessory to the differentiation and proliferation process. It is unknown whether alterations in intracellular calcium in erythroid precursor cells might involve a calmodulin- or protein kinase C-mediated process (BONNANOU-TZEDAKI et al. 1987).

The possible role of sodium ions in erythroid proliferation and differentiation is even more uncertain. Exposure of CFU-E to pharmacologic doses of Na^+/K^+ ATPase inhibitors have had variable and conflicting effects on colony formation depending upon the concentrations of the inhibitors used (GALLICHIO et al. 1982; SPIVAK et al. 1980).

There has been no general agreement in the literature as to the role of cyclic nucleotides in mediating the action of epo. The most persuasive evidence, however, discussed in detail by SPIVAK (1986), argues against cyclic AMP as a second messenger in this process. Nor has a role for cyclic GMP been found (GRABER et al. 1979).

Although the earliest events in precursor cells remain speculative, later occurrence leading to the erythroid phenotype have been characterized to some extent. Using posthypoxic, erythropoietic spleen cells, the timing of some of the developmental processes following in vivo exposure to exogenous epo has been studied (reviewed by SPIVAK 1986). Increased RNA polymerase II activity occurs within 0.5–2.0 h, followed by increases in activity of RNA polymerase I and synthesis of nonhistone proteins (3–12 h). There was a second increase in RNA polymerase II and DNA polymerase-α activities and in histone synthesis between 12 and 24 h after epo injection. Maximal DNA synthesis occurred after about 48 h.

The synthesis of hemoglobin, is, as expected, influenced by the action of epo. A difference in the level of globin mRNA between marrow cells in short-term culture from suppressed or stimulated rats was seen within 2 h of the addition of epo (SAHR and GOLDWASSER 1983). With FVA cells, in vitro, the earliest β-globin mRNA was seen 6 h after exposure to epo (BONDURANT et al. 1985). NIJHOF et al. (1987) showed that globin mRNA transcription in purified CFU-E can be induced by epo in less than 1 h. This suggests that globin gene expression may be one of the earliest, if not primary, effects of epo. The effect of epo on globin gene transcription probably precedes the effect of epo upon heme synthesis (see below). Expression of both α- and β-globin genes appears to be coordinately regulated, but the mechanism of epo action on the expression of these genes is still unknown.

Rats with an excess of red cells have marrow cells very greatly suppressed with respect to red cell formation. Such marrow cells in vitro have sufficient quantities of all of the enzymes needed to provide heme for the initial days of hemoglobin synthesis although they synthesize almost no hemoglobin (BERU and GOLDWASSER 1985). Added epo stimulates the appearance of porphobilinogen deaminase without any discernible effects upon the other enzymes of the heme synthetic pathway (BERU and GOLDWASSER 1985). The synthesis of this apparent rate-limiting enzyme occurred between 20 and 40 h after the addition of epo.

SAWYER and KRANTZ (1986) have shown that FVA cells, after additon of epo, have an increased number of transferrin binding sites by 6 h, and within 24 h the number is doubled. Study of suppressed rat marrow cells in culture showed that epo had a stimulatory effect on synthesis of several components of the red cell membrane and on some components that were only transiently expressed during differentiation (TONG and GOLDWASSER 1981). KOURY et al. (1986), studying the effects of epo on the synthesis of other proteins on the mature red blood cell using FVA cells, showed de novo synthesis of bands 3 (12 h) and 4.1 (48 h) as a result of epo stimulation. Spectrin, constitutively produced prior to hormone exposure, accumulated sharply in these cells 12 h after the addition of epo.

F. Clinical Aspects

CARNOT and DEFLANDRE's early experiments (CARNOT 1906) led them to describe the first human clinical trial of what later was known as epo and they published their results the same year (CARNOT and DEFLANDRE 1906). Their observations supported the notion of a circulating erythropoietic factor that was effective in correcting clinical anemia. This work, in retrospect, appears seriously flawed. An intravenous injection of serum from anemic rabbits led to an increase in red blood count from a pretreatment 5.5 million cells/mm^3 to an impossible 12 million cells/mm^3 cells *after only 3 days*. It is unclear why they obtained such rapid responses. Nevertheless, the hope that a soluble blood growth factor could be used to treat anemia has intrigued physicians for decades. Today, 80 years after the observations of CARNOT and DEFLANDRE, epo is undergoing clinical trials, and appears to work as successfully as had been hoped in the correction of certain anemias (but at more physiologically plausible rates).

In spite of this important advancement, our understanding of the regulation of epo secretion is still vastly incomplete. The current model for the regulation of hematopoiesis involves a feedback system which regulates the rate of red blood cell production to the need for oxygen in peripheral tissues (general discussion by ERSLEV 1983). The need for oxygen is determined by an "oxygen sensor" located in the kidney, possibly within the cells capable of synthesizing epo. The epo synthesizing cells, determined in two published reports using the technique of in situ hybridization, appear in the renal cortex and are located within the interstitial spaces next to renal tubules (KOURY et al. 1987; LACOMBE et al. 1987a). These results are somewhat at variance with data showing that transformed renal cells of tubular origin can secrete epo in culture. Tubular cells, by virtue of their high oxygen utilization, might be better candidates than interstitial cells for oxygen sensing (ERSLEV et al. 1985). The mechanism of oxygen sensing is unknown. After an oxygen deficit is sensed, epo is secreted, circulates to the bone marrow, and stimulates erythroid cells to proliferate and provide a long-term compensation for the perceived oxygen deficit.

Conceivably, perturbations in this feedback model due to changes in environment or health would be reflected in changes in epo production. Assuming relatively uniform clearance of this hormone, production changes would be reflected in serum epo levels. Unfortunately, clinical correlations with serum epo levels have been problematic. Many early studies have been flawed by unreliable or insensitive assays for epo (this has been discussed by KOEFFLER and GOLDWASSER (1981). Additionally, single-point-in-time measurements of epo may be misleading in a dynamic system where there are other acute means of compensation for changes in oxygen delivery (e.g., changes in 2,3-diphosphoglycerate levels, cardiac output, and renal blood flow). Lastly, epo levels must be interpreted relative to the degree of anemia. The normal erythropoietic regulatory axis is characterized by an inverse log-linear relationship between serum epo level and hematocrit (ERSLEV et al. 1987). Epo-hematocrit values not on this log-linear curve indicate an "inappropriate" epo response, lending support to the notion that epo excess or deficiency contributes to the hemopathy. Data on the curve indicates an "appropriate" physiologic degree of epo secretion in the face of the hematologic disorder.

The problems mentioned above notwithstanding, a review of the literature indicates that serum epo levels can be useful in classifying patients with respect to their clinical anemia or polycythemia. The clinical anemias are listed in Table 2. Epo-deficiency anemias occur in patients with inadequate secretion of epo to support erythropoiesis at normal levels. While epo hypercatabolism or elimination might be a cause for this type of anemia, this process has never been described. The anemia of end-stage renal disease (SHERWOOD and GOLDWASSER 1979), and possibly the anemia of prematurity (SHANNON et al. 1987), are the only anemias where epo deficiency is felt to be the major contributor to the cause of the anemia. The epo response to the anemia associated with

Table 2. Clinical anemias

Erythropoietin-deficient anemias[a]
Anemia of end-stage renal disease
Anemia of prematurity
Anemia of hypothyroidism
Anemia of malnutrition
?Sickle cell anemia

Erythropoietin-sufficient anemias
Aplastic anemia
Iron deficiency anemia
Megaloblastic anemia
Pure red cell aplasia
Thalassemias
Myelodysplastic syndromes

---

[a] Anemias that are felt wholly or in part to result from a relative deficiency of erythropoietin.

rheumatoid arthritis has also been studied. This disorder is a chronic inflammatory disease, with a relatively well-defined and homogeneous pathogenesis. As such, the epo response of this anemia might be reflective of the large group of disorders collectively referred to as anemia of chronic inflammatory disease. Rheumatoid arthritis patients have been reported to have both relatively low (BAER et al. 1987) and normal (BIRGEGARD et al. 1987; COTES et al. 1980) serum epo levels. These patients have been reported to respond to pharmacologic doses of recombinant human epo (R. T. MEANS et al., personal communication). While low epo levels correctly characterize physiologic causes of some anemias, it is not clear that this is a pathologic deficiency. The low epo levels in patients with the anemia of hypothyroidism may be appropriate for the lower metabolism and oxygen demands in this condition; likewise, epo levels found associated with the anemia of prematurity and protein malnutrition may not be maladaptive. High serum epo levels in non-epo-deficient anemias reflect a correct response for the erythropoietic regulatory axis, usually to a condition of bone marrow failure. This is seen in aplastic anemia, nutritional anemias (iron, vitamin B_{12}, and folate), pure red cell aplasia, and the thalassemias. The serum epo levels found in patients with the anemia associated with sickle cell disease have been variously reported as being low (SHERWOOD et al. 1986) and normal (ERSLEV et al. 1987), possibly reflecting variable subclinical renal impairment in these patients.

The clinical polycythemias (Table 3) can be viewed on the basis of relatively high or relatively normal epo levels for the degree of polycythemia. Invariably, absolute serum epo levels are normal or low (< 30 milliunits/ml) in conditions of autonomous production of red blood cells such as polycythemia vera (KOEFFLER and GOLDWASSER 1981), suggesting a normal physiologic response of the oxygen sensing mechanism. In contrast, epo-excess

Table 3. Clinical polycythemias

Low-erythropoietin polycythemias
  Polycythemia vera
  Primary erythrocytosis (some)

Erythropoietin-excessive polycythemias
Physiologic
  Chronic obstructive pulmonary disease
  Cyanotic heart disease
  Sleep apnea
  High-affinity hemoglobinopathy
  Smoking
  Localized renal hypoxia

Nonphysiologic
  Ectopic epo secretion by tumors
  Primary erythrocytosis (some)

Other causes
  Androgen therapy
  Cushing's disease
  Bartter's syndrome
  Postrenal transplantation

polycythemias are caused by a number of mechanisms, some of which are physiologically appropriate, others not. Patients with polycythemia and physiologically increased production of epo frequently have a cardiopulmonary disorder (chronic obstructive pulmonary disease, cyanotic heart disease, sleep apnea, or Pickwickian syndrome), a hemoglobinopathy (abnormally high affinity hemoglobin or increased carboxyhemoglobin levels secondary to smoking), or a renal disorder causing localized hypoxia (renovascular disease, renal cysts, and rarely hydronephrosis, glomerulonephritis, pyeloncephritis, and essential hypertension). Ecotopic, nonphysiologic production of epo has been seen in a number of neoplasms. Most commonly involved tumors are hypernephromas and hepatomas, although the syndrome of ectopic epo secretion has also been reported with cerebellar hemangioblastomas, Wilm's tumor, leiomyoma, pheochromocytoma, and androgen-producing tumors. Serum epo concentrations in patients with familial erythrocytosis have been varied (KULKARIN et al. 1985). Some families have relatively high epo levels; others, with relatively normal levels, are felt to be unusually sensitive to epo (KULKARIN et al. 1985). Additional conditions associated with clinical polycythemia are listed on Table 3. In the evaluation of polycythemia of unclear etiology, multiple epo level determinations may be necessary to verify an abnormally elevated epo level (COTES et al. 1986). Moreover, patients with chronic obstructive pulmonary disease, in contrast to normal individuals, may have diurnal variations in serum epo titers (MILLER et al. 1981).

From the time it was first shown that epo was made in the kidney (JACOBSON et al. 1957), the inference had been that the anemia of chronic renal disease might be alleviated by administration of exogenous epo. This was supported by the findings that patients with this anemia had low epo titers (SHERWOOD and GOLDWASSER 1979). The desirability of treatment of the anemia with epo was heightened by the realization that if transfusions could be replaced by epo administration, serious problems such as iron overload, sensitization to red cell antigens, hepatitis, and now AIDS, could be eliminated.

In three articles (ESCHBACH et al. 1987; SCHAEFER et al. 1988; WINEARLS et al. 1986) the data are quite convincing that the anemia of chronic renal disease in patients on hemodialysis can be corrected and the requirement for transfusion eliminated. In these studies, recombinant human epo was given to patients requiring hemodialysis who were moderately or severely anemic with hematocrits <25% (ESCHBACH et al. 1987; WINEARLS et al. 1986) or <30% (SCHAEFER et al. 1988). Care was taken to exclude patients with other causes of their anemias, such as hemolysis, blood loss, nutritional anemias, or aluminium toxicity. Epo was administered in each study as an intravenous bolus following dialysis 3 days per week. Doses varied between studies, but low doses of epo (ranging from 3 to 24 units/kg body weight) were given initially and escalated to higher doses contingent on hematologic response. The anemia of all patients was effectively corrected in a dose-dependent manner, particularly when doses greater than 15 units/kg were given. At the higher dose range, hematocrits could be increased by as much as 10 percentage points within 3 weeks (ESCHBACH et al. 1987). This response was attenuated by iron deficiency in one patient, but was restored upon the administration of iron supplementation (ESCHBACH et al. 1987). In the absence of a placebo control arm to these studies, it is difficult to differentiate the possibly deleterious effects of large doses of epo and the complications that hemodialysis patients experience routinely. Collectively, the most common complications seen in 27 patients with end-stage renal disease were worsening hypertension (four patients), clotting in arteriovenous fistulas (two patients), hypertensive encephalopathy (one patient), and hyperkalemia (two patients). Long-term observations and a placebo-controlled trial are needed to determine whether these effects could be the consequences of increasing hematocrit in these patients.

Judgment of the clinical usefulness of epo in other disorders must await the results of future trials. There are ongoing trials for the anemias associated with malignancies, chemotherapy, and rheumatoid arthritis. Another possible therapeutic use of epo derives from the observations made by AL-KHATTI et al. (1987) showing that circulating fetal hemoglobin can be markedly increased in adult baboons by administration of exogenous epo. This finding might be the eventual basis of a therapy for patients with sickle cell anemia where fetal hemoglobin has been known to diminish the occurrence of sickling episodes.

References

Al-Khatti A, Veith R, Papayannopoulou T, Fritsch E, Goldwasser E, Stamatoyannopoulos G (1987) Stimulation of fetal hemoglobin synthesis by erythropoietin in baboons. N Engl J Med 317:415–420

Anagnostou A (1985) Erythropoietin: a hematopoietic hormone produced by the kidney. Semin Nephrol 5:104–114

Baer AN, Dessypris EN, Goldwasser E, Krantz SB (1987) Blunted erythropoietin response to anaemia in rheumatoid arthritis. Br J Haematol 66:559–564

Beru N, Goldwasser E (1985) The regulation of heme biosynthesis during erythropoietin induced erythroid differentiation. J Biol Chem 260:9251–9257

Beru N, McDonald J, Lacombe C, Goldwasser E (1986) Expression of the erythropoietin gene. Mol Cell Biol 6:2571–2575

Beru N, McDonald J, Goldwasser E (1989a) Erythropoietin gene activation by translocation to near a novel active gene. DNA 8:253:259

Beru N, McDonald J, Goldwasser E (1989b) Studies of the constitutive expression of the mouse erythropoietin gene. Ann NY Acad Sci 554:29–35

Birgegard G, Hallgren R, Caro J (1987) Serum erythropoietin in rheumatoid arthritis and other inflammatory arthridities. Br J Haematol 65:479–483

Bondurant MC, Koury MJ (1986) Anemia induces accumulation of erythropoietin mRNA in the kidney and liver. Mol Cell Biol 6:2731–2733

Bondurant MC, Lind RN, Koury MJ, Ferguson ME (1985) Control of globin gene transcription by erythropoietin in erythroblasts from Friend virus-infected mice. Mol Cell Biol 5:675–683

Bonnanou-Tzedaki SA, Sohi MK, Arnstein H (1987) The role of cAMP and calcium in the stimulation of proliferation of immature erythroblasts by erythropoietin. Exp Cell Res 170:276–289

Boussios T, Bertles JF (1987) Receptors specific for erythropoietin on yolk-sac erythroid cells. In: Stammatoyannopoulos G, Nienhuis AW (eds) Developmental control of globin gene expression. Liss, New York, pp 35–41

Carnot P (1906) Sur le mécanisme d'hyperglobulie provoquée par le sérum d'animaux en rénovation sanguine. C R Acad Sci (Paris) 111:344–346

Carnot P, Deflandre C (1906) Sur l'activité hemopoiètique du sérum au cours de la régénération du sang. C R Acad Sci (Paris) 143:348–387

Chang CS, Goldwasser E (1973) On the mechanism of erythropoietin-induced differentiation. XII. A cytoplasmic protein mediating induced nuclear RNA synthesis. Dev Biol 34:246–254

Choppin JC, Lacombe C, Casadevall N, Muller O, Tambourin P, Varet B (1984) Characterization of erythropoietin produced by IW32 murine erythroleukemia cells. Blood 64:341–347

Choppin JC, Casadevall N, Lacombe C, Wendling F, Goldwasser E, Berger R, Tambourin P, Varet B (1985) Production of erythropoietin by cloned malignant murine erythroid cells. Exp Hematol 13:610–615

Chou PY, Fasman GD (1978) Empirical predictions of protein conformation. Annu Rev Biochem 47:251–276

Cotes PM, Brozovic B, Mansell M, Samson DM (1980) Radioimmunoassay of erythropoietin in human serum: validation and application of an assay system. Exp Hematol [Suppl 8] 8:292

Cotes PM, Dore CJ, Liu Yin JA, Lewis SM, Messinezy M, Pearson TC, Reid C (1986) Determination of serum immunoreactive erythropoietin in the investigation of erythrocytosis. N Engl J Med 315:283–287

Davis JM, Arakawa T, Strickland TW, Yphantis DA (1987) Characterization of recombinant human erythropoietin produced in Chinese hamster ovary cells. Biochem 26:2633–2638

Dordal MS, Wang FF, Goldwasser E (1985) The role of carbohydrates in erythropoietin action. Endocrinology 116:2293–2299

Egrie J, Strickland TW, Lane J, Aoki K, Cohen AM, Smalling R, Trail G, Lin FK, Browne JK, Hines DK (1986) Characterization and biological effects of recombinant human erythropoietin. Immobiology 172:213–224

Erslev AJ (1983) Production of erythrocytes. In: Williams WJ, Beutler E, Erslev AJ, Lichtman MA (eds) Hematology. McGraw-Hill, New York, pp 370–373

Erslev AJ, Lavietes PH, van Wagenen G (1953) Erythropoietic stimulation induced by "anemic" serum. Proc Soc Exp Biol Med 83:548–550

Erslev AJ, Caro J, Besarab A (1985) Why the kidney? Nephron 41:213–216

Erslev AJ, Wilson J, Caro J (1987) Erythropoietin titers in anemic, nonuremic patients. J Lab Clin Med 109:429–433

Eschbach JW, Egrie JC, Downing MR, Browne JK, Adamson JW (1987) Correction of the anemia of end-stage renal disease with recombinant human erythropoietin: results of a combined phase I and II clinical trial. N Engl J Med 316:73–77

Gallichio VS, Chen MG, Murphy MJ (1982) Modulation of murine in vitro erythroid and granulocyte colony formation by ouabain, digoxin, and theophylline. Exp Hematol 10:682–688

Goldwasser E (1981) Erythropoietin and red cell differentiation. In: Cunnigham D, Goldwasser E, Watson J, Fox CF (eds) Control of cellular division and development, part A. Liss, New York, pp 487–494

Goldwasser E (1984a) The action of erythropoietin as the inducer of erythroid differentiation. In: Cronkite EP, Daniak N, Palek J, McCaffrey RP, Quesenberry P (eds) Hematopoietic stem cell physiology. Liss, New York, pp 77–84

Goldwasser E (1984b) Erythropoietin and its mode of action. Blood Cells 10:147–162

Goldwasser E, Kung CK-H, Eliason JF (1974) On the mechanism of erythropoietin induced differentiation. XII. The role of sialic acid in erythropoietin action. J Biol Chem 249:4202–4206

Goldwasser E, Eliason JF, Sikkema D (1975) An assay for erythropoietin in vitro at the mulliunit level. Endocrinology 97:315–323

Goldwasser E, McDonald J, Beru N (1987) The molecular biology of erythropoietin and the expression of its gene. In: Rich IN (ed) Molecular and cellular aspects of erythropoietin and erythropoiesis. Springer, Berlin Heidelberg New York, pp 849–858

Gordon AS, Condorelli M, Peschle C (eds) (1971) Regulation of erythropoiesis. Ponte, Milan

Graber SE, Krantz SB (1978) Erythropoietin and the control of red cell production. Annu Rev Med 29:51–66

Graber SE, Bomboy SE, Salmon WD, Krantz SB (1979) Evidence that endotoxin is the cyclic 3′:5′-cGMP-promoting factor in erythropoietin preparations. J Lab Clin Med 93:25–31

Hopp TP, Woods KR (1981) Prediction of protein antigenic determinants from amino acid sequences. Proc Natl Acad Sci USA 78:3824–3828

Jacobs K, Shoemaker C, Rudersdorf R, Neill SD, Kaufman RJ, Mufson A, Seehra AJ, Jones SS, Hewick R, Fritsch EF, Kawakita M, Shimizu T, Miyake T (1985) Isolation and characterization of genomic and cDNA clones of human erythropoietin. Nature 313:806–810

Jacobson LO, Goldwasser E, Fried W, Plzak LF (1957) Role of the kidney in erythropoiesis. Nature 179:633–634

Jelkmann W (1986) Renal erythropoietin: properties and production. Rev Physiol Biochem Pharmacol 104:140–215

Koeffler HP, Goldwasser E (1981) Erythropoietin radioimmunoassay in evaluating patients with polycythemia. Ann Intern Med 94:44–47

Koury M, Sawyer ST, Bondurant MC (1984) Splenic erythroblasts in anemia-inducing Friend disease: a source of cells for studies of erythropoietin-mediated differentiation. J Cell Physiol 121:526–532

Koury MJ, Bondurant MC, Mueller TJ (1986) The role of erythropoietin in the production of principal erythrocyte proteins other than hemoglobin during terminal differentiation. J Cell Physiol 126:259–265

Koury ST, Bondurant MC, Koury MJ (1987) Localization of erythropoietin synthesizing cells in murine kidneys by in situ hybridization. Blood 71:524–527

Kozak M (1984) Compilation and analysis of sequences upstream from the translational start site in eukaryotic mRNA's. Nucleic Acids Res 12:857–872

Krantz SB, Goldwasser E (1984) Specific binding of erythropoietin to spleen cells infected with the anemia strain of Friend virus. Proc Natl Acad Sci USA 81:7574–7578

Krantz SB, Jacobson LO (1970) Erythropoietin and the regulation of erythropoiesis. University of Chicago Press, Chicago

Krumdieck N (1943) Erythropoietic substance in the serum of anemic animals. Proc Soc Exp Biol Med 54:14–17

Kulkarin V, Richey K, Howard D, Daniak N (1985) Heterogeneity of erythropoietin-dependent erythrocytosis: Case report and synopsis of primary erythrocytosis syndromes. Br J Haematol 66:751–758

Lacombe C, da Silva JL, Bruneval P, Fournier J-G, Wendling F, Casadevall N, Camilleri J-P, Bariety J, Varet B, Tambourin P (1987a) Peritubular cells are the site of erythropoietin synthesis in the murine hypoxic kidney. J Clin Invest 81:620–623

Lacombe C, Casadevall N, Choppin J, Muller O, Goldwasser E, Varet B, Tambourin P (1987b) Erythropoietin production and erythropoietin receptors on murine erythroleukemia cell lines. In: Rich IN (ed) Molecular and cellular aspects of erythropoietin and erythropoiesis. Springer, Berlin Heidelberg New York, pp 61–72

Lai PH, Everett R, Wang FF, Arakawa T, Goldwasser E (1986) Structural characterization of human erythropoietin. J Biol Chem 261:3116–3121

Law ML, Cai G-Y, Lin F-K, Wei Q, Huang S-Z, Hartz JH, Morse H, Lin C-H, Jones C, Kao F-T (1986) Chromosomal assignment of the human erythropoietin gene and its DNA polymorphism. Proc Natl Acad Sci USA 83:6920–6924

Lin FK (1987) The molecular biology of erythropoietin. In: IN Rich (ed) Molecular and cellular aspects of erythropoietin and erythropoiesis. Springer, Berlin Heidelberg New York, pp 23–36

Lin FK, Suggs S, Lin CH, Browne J, Smalling R, Egrie J, Chen K, Fox G, Martin F, Stabinsky Z, Badrawi S, Lai PH, Goldwasser E (1985) Cloning and expression of the human erythropoietin gene. Proc Natl Acad Sci USA 82:7580–7584

Lin FK, Lin C-H, Lai P-H, Browne JK, Egrie JC, Smalling R, Fox GM, Chen KK, Castro M, Suggs S (1986) Monkey erythropoietin gene: cloning, expression and comparison with the human erythropoietin gene. Gene 44:201–209

Lowy P, Keighly G (1968) Inactivation of erythropoietin by Koshland's tryptophan reagent and by membrane filtration. Biochim Biophys Acta 160:413–419

Mayeux P, Billat C, Jacquot R (1987a) Murine erythroleukaemia cells (Friend cells) possess high-affinity binding sites for erythropoietin. FEBS Lett 211:229–233

Mayeux P, Billat C, Jacquot R (1987b) The erythropoietin receptor of rat erythroid progenitor cells. J Biol Chem 262:13985–13990

McDonald J, Lin FK, Goldwasser E (1986) Cloning, sequencing and evolutionary analysis of the mouse erythropoietin gene. Mol Cell Biol 6:842–848

McDonald J, Beru N,. Goldwasser E (1987) Rearrangement and expression of erythropoietin genes in transformed mouse cells. Mol Cell Biol 7:365–370

Miller BA, Scaduto RC, Tillotson DL, Botti JJ, Cheung JY (1988) Erythropoietin stimulates a rise in intracellular free calcium concentration in single early human erythroid precursors. J Clin Invest 82:309–315

Miller ME, Garcia JF, Cohen RA, Cronkite EP, Moccia G, Acevedo J (1981) Diurnal levels of immunoreactive erythropoietin in normal subjects and subjects with chronic lung disease. Br J Haematol 49:189–200

Misti J, Spivak JL (1979) Erythropoiesis in vitro. Role of calcium. J Clin Invest 64:1573–1579

Miyake T, Kung CK-H, Goldwasser E (1977) Purification of human erythropoietin. J Biol Chem 252:5558–5564

Mufson RA, Gesner TG (1987) Binding and internalization of recombinant erythropoietin in murine erythroid precursor cells. Blood 69:1485–1490

Nakao K, Fisher JW, Takaku F (1975) Erythropoiesis. University of Tokyo Press, Tokyo

Nijhof W, Wierenga P, Sahr K, Beru N, Goldwasser E (1987) Induction of globin mRNA transcription in differentiating precursor cells. Exp Hematol 15:779–784

Nishida J, Hirai H, Kubota M, Lin F-K, Okabe T, Urabe A, Takaku F (1986) Detection of erythropoietin message in hypoxic mouse kidney. Jpn J Exp Med 56:321–323

Orlic D, LoBue J (eds) (1984) Controls of hematopoiesis. Blood Cells 10

Recny MA, Scoble HA, Kim Y (1987) Structural characterization of natural human urinary and recombinant DNA-derived erythropoietin. J Biol Chem 262:17156–17163

Rich IN (ed) (1987) Molecular and cellular aspects of erythropoietin and erythropoiesis. Springer, Berlin Heidelberg New York

Sahr K, Goldwasser E (1983) The effects of erythropoietin on the biosynthesis of translatable globin mRNA. In: Goldwasser E (ed) Regulation of hemoglobin biosynthesis. Elsevier, New York, pp 153–161

Sakaguchi M, Koishihara Y, Tsuda H, Fujimoto K, Shibuya K, Kawakita M, Takatsuki K (1987) The expression of functional erythropoietin receptors on an interleukin-3 dependent cell line. Biochem Biophys Res Commun 146:7–12

Sasaki H, Bothner B, Dell A, Fukuda M (1987) Carbohydrate structure of erythropoietin expressed in Chinese hamster ovary cells by a human erythropoietin cDNA. J Biol Chem 262:12059–12076

Sasaki R, Yanagawa S, Hitomi K, Chiba H (1987) Characterization of erythropoietin receptors of murine erythroid cells. Eur J Biochem 168:43–48

Sawyer ST, Krantz SB (1984) Erythropoietin stimulates $^{45}Ca^{2+}$ uptake in Friend virus-infected erythroid cells. J Biol Chem 259:2769–2774

Sawyer ST, Krantz SB (1986) Transferrin receptor number, synthesis and endocytosis during erythropoietin-induced maturation of Friend virus-infected erythroid cells. J Biol Chem 261:9187–9195

Sawyer ST, Krantz SB, Goldwasser E (1987a) Binding and receptor-mediated endocytosis of erythropoietin in Friend virus-infected erythroid cells. J Biol Chem 262:5554–5562

Sawyer ST, Krantz SB, Luna J (1987b) Identification of the receptor for erythropoietin by cross-linking to Friend virus-infected erythroid cells. Proc Natl Acad Sci USA 84:3690–3694

Schaefer RM, Kurner B, Zech M, Krahn R, Heidland A (1988) Therapy of renal anemia with recombinant human erythropoietin. Dtsch Med Wochenschr 113(4):125–129

Shannon KM, Naylor GS, Torkildon JC, Clemons GK, Schaffner V, Goldman SL, Lewis K, Bryant P, Phibbs R (1987) Circulating erythroid progenitors in the anemia of prematurity. N Engl J Med 317:728–733

Sherwood JB (1984) The chemistry and physiology of erythropoietin. Vitam Horm 41:161–211

Sherwood JB, Goldwasser E (1979) A radioimmunoassay for erythropoietin. Blood 53:935–946

Sherwood JB, Goldwasser E, Chilcote R, Carmichael LP, Nagel RL (1986) Sickle cell anemia patients have low erythropoietin levels for their degree of anemia. Blood 67:46–49

Shoemaker CB, Mistock LD (1986) Murine erythropoietin gene: Cloning, expression, and human gene homology. Mol Cell Biol 6:849–858

Spivak JL (1986) The mechanism of action of erythropoietin. Int J Cell Cloning 4:139–166

Spivak JL, Misti J, Stuart R, Sharkis SJ (1980) Suppression and potentiation of mouse hematopoietic progenitor cell proliferation by ouabain. Blood 56:315–317

Sytkowski AJ, Donahue KA (1987) Immunochemical studies of human erythropoietin using site-specific anti-peptide antibodies. J Biol Chem 262:1161–1165

Takeuchi M, Takasaki S, Miyazaki H, Kato T, Hoshi S, Kochibe N, Kobata A (1988) Comparative study of the asparagine-linked sugar chains of human erythropoietin purified from urine and the culture medium of recombinant Chinese hamster ovary cells. J Biol Chem 263:3657–3663

Tambourin P, Casadevall N, Choppin J, Lacombe C, Heard JM, Fichelson S, Wendling F, Varet B (1983) Production of erythropoietin-like activity by a murine erythroleukemia cell line. Proc Natl Acad Sci USA 80:6269–6273

Todokoro K, Kanazawa S, Amanuma H, Ikawa Y (1987) Specific binding of erythropoietin to its receptor on responsive mouse erythroleukemia cells. Proc Natl Acad Sci USA 84:4126–4130

Tojo A, Fukamachi H, Kasuga M, Urabe A, Takaku F (1987) Identification of erythropoietin receptors on fetal liver erythroid cells. Biochem Biophys Res Commun 148:443–448

Tong BD, Goldwasser E (1981) The formation of erythrocyte membrane proteins during erythropoietin-induced differentiation. J Biol Chem 256:12666–12672

Van Zant G, Goldwasser E (1979) Competition between erythropoietin and colony-stimulating factor for target cells in mouse marrow. Blood 53:946–965

Van Zant G, Goldwasser E (1984) Erythropoietin and its target cells. In: Guroff G (ed) Growth and maturation factors. Wiley, New York, pp 3–36

Wang FF, Kung CK-H, Goldwasser E (1985) Some chemical properties of human erythropoietin. Endocrinology 116:2286–2292

Watkins PC, Eddy R, Hoffman N, Stanislovitis P, Beck AK, Galli J, Vellucci V, Gusella JF, Shows TB (1986) Regional assignment of the erythropoietin gene to human chromosome region 7pter→q22. Cell Genet 42:214–218

Winearls CG, Oliver DO, Pippard MJ, Reid C, Downing MR, Cotes PM (1986) Effect of human erythropoietin derived from recombinant DNA on the anaemia of patients maintained by chronic haemodialysis. Lancet 2:1175–1177

Wojchowski DM, Orkin SH, Sytkowski AJ (1987) Active human erythropoietin expressed in insect cells using a baculovirus vector: a role for N-linked oligosaccharide. Biochim Biophys Acta 910:224–232

Appendix A. Alternate Names for Growth Factors

Activins (A, AB, B)
    FSH releasing protein (FRP)

Colony-stimulating factor 1 (CSF-1)
    Macrophage colony-stimulating factor (M-CSF)

Epidermal growth factor (EGF)
    Urogastrone

Erythropoietin (eop; EP)
    Hemopoietine
    Erythrocyte stimulating factor (ESF)

Fibroblast growth factors (FGF)
    Acidic fibroblast growth factor (aFGF)
    Basic fibroblast growth factor (bFGF)
    Brain-derived growth factor (BNDF; BDGF)
    Heparin binding growth factor (HBGF)
    Endothelial cell growth factor (ECGF)
    Retina-derived growth factor (RDGF)
    Eye-derived growth factor (EDGF)
    Kidney angiogenic factor (KAF)
    Adrenal growth factor (AGF)
    Corpus luteum angiogenic factor (CLAF)
    Ovarian growth factor (OGF)
    Placental angiogenic factor (PAG)
    Hepatocyte growth factor (HGF)
    Myogenic growth factor (MGF)
    Cartilage-derived growth factor (CDGF)
    Bone growth factor (BGF)
    Seminiferous growth factor (SGF)
    Prostatropin (PGF)
    Tumor-derived growth factor (TDGF)
    Hepatoma-derived growth factor (HDGF)
    Melanoma-derived growth factor (MDGF)
    Mammary tumor-derived growth factor (MTGF)

Gastrin-releasing peptide (GRP)
    Mammalian bombesin

Granulocyte colony-stimulating factor (G-CSF)
    Macrophage/granulocyte inducer type 1, granulocyte (MG-1G)
    Pluripotent colony-stimulating factor (pluripoetin)
    Granulocyte-macrophage colony-stimulating factor-β (GM-CSFβ)

Granulocyte-macrophage colony-stimulating factor (GM-CSF)
    Macrophage-granulocyte inducer
    Colony-stimulating factor 2 (CSF-2)

Inhibin

Insulin-like growth factor I (IGF-I)
    Somatomedin A
    Somatomedin C
    Basic somatomedin

Insulin-like growth factor II (IGF-II)
    Multiplication stimulating activity (MSA)

Interferon-α (IFN-α)
    Leukocyte (Le) interferon
    Type I interferon

Interferon-β (IFN-β)
    Fibroblast (F) interferon
    Type I interferon

Interferon-γ (IFN-γ)
    Immune interferon
    T interferon
    Type II interferon

Interleukin-1 (IL-1)
    Interleukin-1 α
    Interleukin-1 β
    Lymphocyte activating factor (LAF)

Interleukin-2 (IL-2)
    T-cell growth factor (TCGF)

Interleukin-3 (IL-3)
    Mast cell growth factor
    P-cell stimulating factor
    Multi-colony-stimulating factor
    Burst promoting activity
    WEHI-3 hematopoietic growth factor
    Thy-1 inducing factor
    Histamine cell-producing stimulating factor
    20α-dehydrogenase-inducing factor

Interleukin-4 (IL-4)
    B-cell stimulatory factor 1 (BSF-1)
    T-cell growth factor (TCGF)

B-cell growth factor I (BCGF-I)
Mast-cell growth factor (MCGF)

Interleukin-5 (IL-5)
T-cell replacing factor (TRF-1)
B-cell growth factor II (BCGF II)
B-cell differentiation factor μ (BCDFμ)
Eosinophil differentiation factor (EDF)
IgA enhancing factor (IgA-EF)
B-cell maturation factor (BMF)
B-cell growth and differentiation factor (BGDF)

Interleukin-6 (IL-6)
B-cell stimulatory factor 2 (BSF-2)
Interferon-β2
26-kDa protein
Hepatocyte stimulating factor
Hybridoma/plasmacytoma growth factor
Interleukin-HP1
Macrophage granulocyte inducer type 2

Lymphotoxin
Tumor necrosis factor-β (TNF-β)

Müllerian inhibiting substance (MIS)
Anti-Müllerian hormone (AMH)
Müllerian inhibiting factor (MIF)

Nerve growth factor (NGF)

Platelet-derived endothelial cell growth factor (PD-ECGF)

Platelet-derived growth factor (PDGF-AA, PDGF-BB, PDGF-AB)

Transforming growth factor-α (TGF-α)
Sarcoma growth factor (SGF)

Transforming growth factor-β (TGF-β)
Cartilage inducing factor-A (CIF-A)
Cartilage inducing factor-B (CIF-B)
BSC-1 growth inhibitor (BSC-1 GI)
Differentiation inhibitor (DI)
Polyergin

Tumor necrosis factor (TNF)
Cachectin
Tumor necrosis factor-α (TNF-α)

Appendix B. Chromosomal Locations of Growth Factors/Growth Factor Receptors

Human

chromosome 1   Nerve growth factor (p22.1)
Transforming growth factor-$\beta2$ (q41)

chromosome 2   Inhibin-α (distal portion of long arm)
Inhibin-β_B (near centromere on short arm)
Interleukin-1β (2q13–2q21)
Transforming growth factor-α (p11–p13)

chromosome 4   Basic fibroblast growth factor
Epidermal growth factor (q25–q27)

chromosome 5   Acidic fibroblast growth factor
Colony-stimulating factor 1 (q33.1)
Colony-stimulating factor-1 receptor (q33.2–33.3)
Granulocyte-macrophage colony-stimulating factor (q23–31)
Interleukin-3 (q23–q31)
Interleukin-4 (q23–31)
Interleukin-5 (q23.3–32)
Platelet-derived growth factor receptor (β-subunit)

chromosome 6   Interferon-γ receptor

chromosome 7   Epidermal growth factor receptor (p14–p12)
Erythropoietin (q11–q22)
Inhibin-β_A
Interleukin-6 (p21)
Platelet-derived growth factor A chain

chromosome 9   Interferon-α
Interferon-β

chromosome 11   Insulin-like growth factor II

chromosome 12   Insulin-like growth factor I
Interferon-γ

chromosome 14   Transforming growth factor-$\beta3$ (q24)

chromosome 17   Granulocyte colony-stimulating factor (q21–q22)
Nerve growth factor receptor (q12–q22)

chromosome 18   Gastrin-releasing peptide (q21)

chromosome 19   Müllerian inhibiting substance
                 Transforming growth factor-β1 (q13)

chromosome 21   Interferon-α/β receptor

chromosome 22   Platelet-derived growth factor B chain

Mouse

chromosome 1   Transforming growth factor-β2

chromosome 2   Interleukin-1α
                 Interleukin-1β

chromosome 3   Epidermal growth factor
                 Nerve growth factor

chromosome 4   Interferon-α
                 Interferon-β

chromosome 5   Erythropoietin
                 Interleukin-6

chromosome 7   Transforming growth factor-β1

chromosome 10   Interferon-γ
                 Interferon-γ receptor

chromosome 11   Epidermal growth factor receptor
                 Granulocyte colony-stimulating factor
                 Granulocyte-macrophage colony-stimulating factor
                 Interleukin-3
                 Interleukin-5
                 Nerve growth factor receptor

chromosome 12   Transforming growth factor-β3

chromosome 16   Interferon-α/β receptor

chromosome 18   Colony stimulating growth factor 1 receptor

Subject Index

Handbook of Experimental Pharmacology

Eds. G.V.R.Born, P.Cuatrecasas, H.Herken, A.Schwartz

Volume 94: **C.S.Cooper** (Ed.)

Chemical Carcinogenesis and Mutogenesis

Part I: 1989. Approx. 600 pp. 86 figs. Hardcover in preparation ISBN 3-540-51182-2

Part II: 1989. Approx. 515 pp. 16 figs. 17 tabs. Hardcover in preparation ISBN 3-540-51183-0

Volume 93: **D.Ganten** (Ed.)

Pharmacology of Antihypertensive Therapeutics

1989. in preparation ISBN 3-540-50427-3

Volume 92: **P.Cuatrecasas, S.J.Jacobs** (Eds.)

Insulin

1991. Hardcover in preparation
ISBN 3-540-50319-6

Volume 91: **L.E.Bryan** (Ed.)

Microbial Resistance to Drugs

1989. XVII, 451 pp. 39 figs. Hardcover
DM 460,- ISBN 3-540-50318-8

Volume 90: **U.Trendelenburg, N.Weiner** (Eds.)

Catecholamines

Part I: 1988. XX, 571 pp. 43 figs. Hardcover
DM 450,- ISBN 3-540-18904-1

Part II: 1989. XXII, 488 pp. 34 figs. Hardcover
DM 450,- ISBN 3-540-19117-8

Volume 89: **E.M.Vaughan Williams** (Ed.)

Antiarrhythmic Drugs

Coeditor: T.J.Campbell

1989. XXVII, 650 pp. 113 figs. Hardcover
DM 480,- ISBN 3-540-19239-5

Volume 88: **D.B.Calne** (Ed.)

Drugs for the Treatment of Parkinson's Disease

1989. XXIV, 599 pp. 62 figs. Hardcover
DM 480,- ISBN 3-540-50041-3

Volume 87:

Pharmacology of the Skin

Part I: M.W.Greaves, S.Shuster (Eds.)

Pharmacology of Skin Systems. Autocoids in Normal and Inflamed Skin

1989. XXIX, 510 pp. 78 figs. Hardcover DM 480,-
ISBN 3-540-19403-7

Part II: M.W.Greaves, S.Shuster (Eds.)

Methods, Absorption, Metabolism and Toxicity Drugs and Diseases

1989. XXXVII, 587 pp. 55 figs. Hardcover
DM 580,- ISBN 3-540-50277-7

Volume 86: **V.P.Whittaker** (Ed.)

The Cholinergic Synapse

1988. XXV, 762 pp. 98 figs. Hardcover DM 680,-
ISBN 3-540-18613-1

Volume 85: **M.A.Bray, J.Morley** (Eds.)

The Pharmacology of Lymphocytes

1988. XXII, 626 pp. 69 figs. Hardcover DM 680,-
ISBN 3-540-18609-3

Volume 84: **K.Bartmann** (Ed.)

Antituberculosis Drugs

1988. XXIV, 566 pp. 68 figs. Hardcover DM 590,-
ISBN 3-540-18139-3

Springer-Verlag Berlin
Heidelberg New York London
Paris Tokyo Hong Kong

Springer

Progress in Clinical Biochemistry and Medicine

Eds.: E. Beaulieu, D. T. Forman, M. Ingelman-Sundberg, L. Jaenicke, J. A. Kellen, Y. Nagai, G. F. Springer, L. Träger, L. Will-Shahab, J. L. Wittliff

Volume 10

Ruthenium and Other Non-Platinum Metal Complexes in Cancer Chemotherapy

With contributions by numerous experts

1989. X, 226 pp. 99 figs. 47 tabs.
Hardcover ISBN 3-540-51146-6

Volume 9

Calcitonins – Physiological and Pharmacological Aspects

With contributions by M. Azria, J. Engel, S. Grünwald, P. Hilgard, U. Niemeyer, M. Peukert, G. P. Pfeifer, J. Pohl, H. Sindermann

1989. VII, 106 pp. 21 figs. 12 tabs.
Hardcover ISBN 3-540-51097-4

Volume 8
F. Salvatore, A. Roda, L. Sacchetti (Eds.)

Clinical Biochemistry in Hepatobiliary Diseases

Proceedings of the International Satellite Symposium, Bologna, Italy, 1988

With contributions by numerous experts

1989. IX, 196 pp. 72 figs. 23 tabs.
Hardcover ISBN 3-540-50705-1

Volume 7

Drug Concentration Monitoring · Microbial Alpha-Glucosidase Inhibitors: Chemistry, Biochemistry and Therapeutic Potential – Plasminogen Activators: Molecular Properties, Biological Cell Function and Clinical Application

With contributions by M. B. Bottorff, W. E. Evans, I. Hillebrand, B. Junge, L. Müller, W. Puls, D. D. Schmidt, E. Truscheit, H. Will

1988. VII, 148 pp. 56 figs. 19 tabs.
Hardcover ISBN 3-540-19002-3

Volume 6
R. D. Smith, P. S. Wolf, J. R. Regan, S. R. Jolly

The Emergence of Drugs Which Block Calcium Entry

1988. VIII, 154 pp. 20 figs. Hardcover
ISBN 3-540-18620-4

Distribution rights for all volumes for the socialist countries: Akademie-Verlag, Berlin

Springer-Verlag Berlin
Heidelberg New York London
Paris Tokyo Hong Kong

Springer